Mass Transfer Operations

Theory and Applications

Second Editon

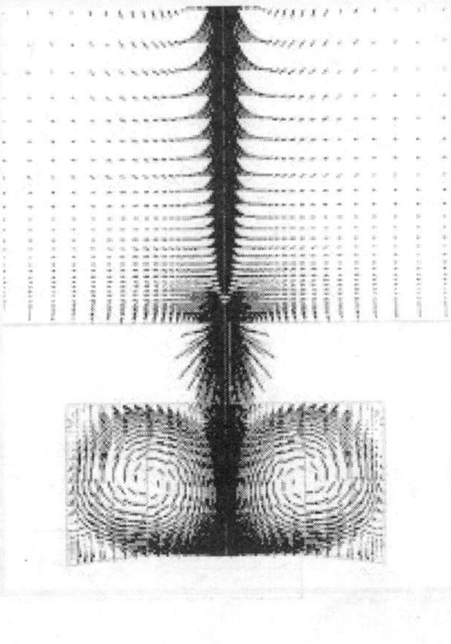

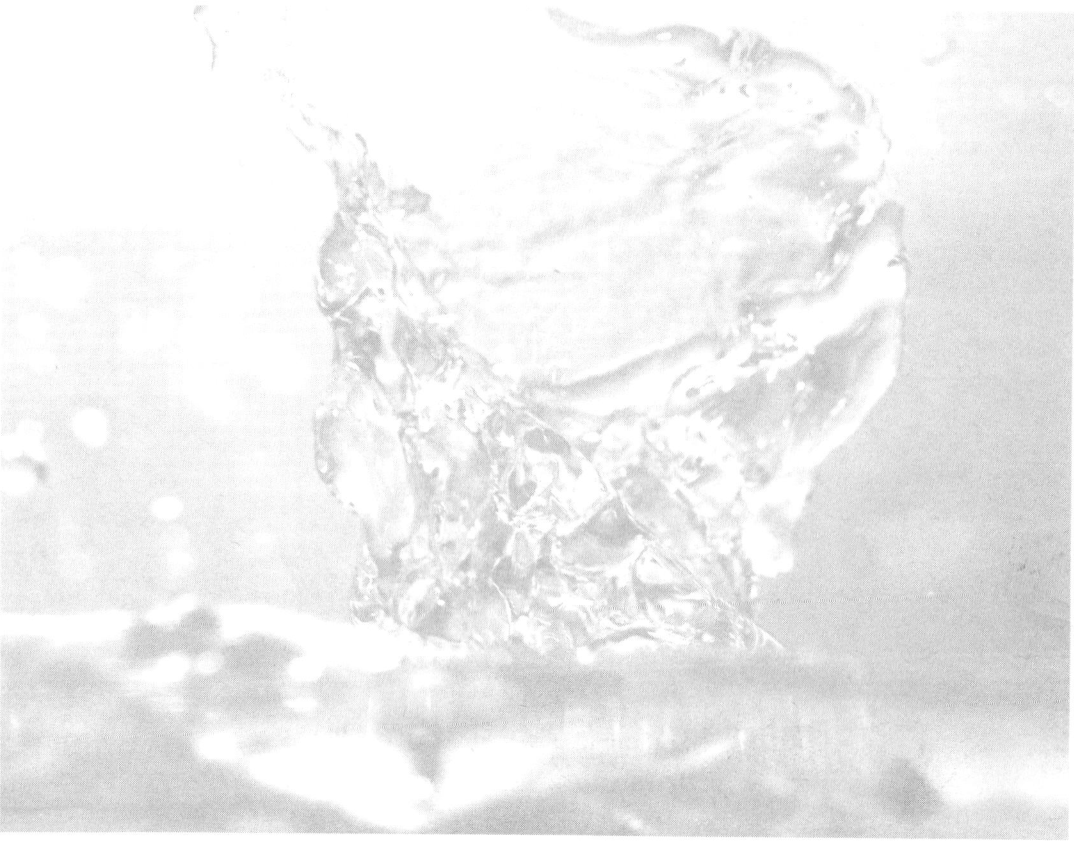

Mass Transfer Operations

Theory and Applications

Second Editon

KV Narayanan ME, PhD
Former Professor and Head
Department of Chemical Engineering
Former Principal, Government Engineering College
Thrissur, Kerala

B Lakshmikutty MTech, PhD
Professor
Department of Chemical Engineering
Government Engineering College
Thrissur, Kerala

CBSPD

CBS Publishers & Distributors Pvt Ltd

New Delhi • Bengaluru • Chennai • Kochi • Kolkata • Lucknow • Mumbai
Gujarat • Hyderabad • Jharkhand • Nagpur • Patna • Pune • Uttarakhand

Disclaimer

Science and technology are constantly changing fields. New research and experience broaden the scope of information and knowledge. The authors have tried their best in giving information available to them while preparing the material for this book. Although, all efforts have been made to ensure optimum accuracy of the material, yet it is quite possible some errors might have been left uncorrected. The publisher, the printer and the authors will not be held responsible for any inadvertent errors, omissions or inaccuracies.

Mass Transfer Operations

Theory and Applications

Second Editon

ISBN: 978-93-5466-609-4

Second Edition: 2024
Reprint 2026
First Edition: 2014
Reprint 2017, 2020

Published by **Satish Kumar Jain** and produced by **Varun Jain** for

CBS Publishers & Distributors Pvt Ltd
4819/XI Prahlad Street, 24 Ansari Road, Daryaganj, New Delhi 110 002, India
Ph: 011-23289259, 23266838 Website: www.cbspd.com
e-mail: delhi@cbspd.com

Corporate Office: 204 FIE, Industrial Area, Patparganj, Delhi 110 092
Ph: 011-4934 4934 Fax: 011-4934 4935 e-mail: publishing@cbspd.com; publicity@cbspd.com

Branches

- **Bengaluru:** Seema House 2975, 17th Cross, K.R. Road, Banasankari 2nd Stage, Bengaluru 560 070, Karnataka, India
  Ph: +91-80-26771678/79 Fax: +91-80-26771680 e-mail: bangalore@cbspd.com
- **Chennai:** 18/8B, Subbarayan Street, Shenoy Nagar, Chennai 600 030, Tamil Nadu, India
  Ph: +91-44-42032115, 26681266 e-mail: chennai@cbspd.com
- **Kochi:** 42/1325, 1326, Power House Road, opposite KSEB, Power House, Ernakulam 682 018, Kochi, Kerala, India
  Ph: +91-484-4059061-65 Fax: +91-484-4059065 e-mail: kochi@cbspd.com
- **Kolkata:** 147, Hind Ceramics Compound, 1st Floor, Nilgunj Road, Belghoria, Kolkata-700 056, West Bengal, India
  Ph: +91-33-25633055-56 e-mail: kolkata@cbspd.com
- **Lucknow:** Basement, Khushnuma Complex, 7-Meerabai Marg (behind Jawahar Bhawan), Lucknow 226 001, UP, India
  Ph: +91-522-4000032 e-mail: tiwari.lucknow@cbspd.com
- **Mumbai:** PWD Shed. Gala No. 25/26, Ramchandra Bhatt Marg, Next to JJ Hospital Gate No. 2, Opp. Union Bank of India Noorbaug Mumbai-400 009, Maharashtra, India
  Ph: +91-22-66661880/89 e-mail: mumbai@cbspd.com

Representatives

- **Gujarat** 0-9879558667 • **Hyderabad** 0-9885175004 • **Jharkhand** 0-9811541605 • **Nagpur** 0-8692091830
- **Patna** 0-9334159340 • **Pune** 0-9664372571 • **Uttarakhand** 0-9716462459

Printed at Glorious Printers, Delhi, India

Preface to the Second Editon

Separation of mixtures into individual constituents is one of the most challenging tasks in chemical process industries. These separations are achieved mainly by mass transfer operations. Mass transfer can be treated as processes involving transfer of material through another on a molecular scale. Process industries employ mass transfer operations for purification, isolation, recovery, refining, concentration, or enrichment of the materials at various stages of their processing. There is scarcely any chemical process industry that does not require some sort of raw material pretreatment or product purification steps in the course of the manufacturing process; separation operations are required for both these steps. These operations also play a significant role in the treatment of effluents to meet the environmental regulations. No wonder that in chemical process industries the investment on mass transfer equipment far exceeds the investment on the reaction vessels and related equipment in which the actual transformation of the raw materials into products is carried out. Industrial processes consisting entirely of separation processes are also not uncommon.

Chemical engineering curriculum gives due importance to mass transfer operations both at the undergraduate and graduate levels. In most of the Indian universities, the subject is taught in two or three semesters at the undergraduate levels. During several years of experience of teaching chemical engineering, we felt that the books on mass transfer which present the subject material in a manner that can sustain the interest of the average student without sacrificing the theoretical and analytical rigour demanded by the subject are a few. The present work which explains the theory, principles and applications of mass transfer techniques in detail is an attempt to address this inadequacy in a student-friendly way. The preliminary draft of the book has its genesis in the lecture notes prepared and revised several times by the first author during the time he taught the subject to the undergraduate chemical engineering students of Government Engineering College, Thrissur, Kerala. The wide acceptance of these lectures among the students had been the stimulus for writing this book.

Mass Transfer Operations: Theory and Applications is intended as a textbook for undergraduate courses in chemical engineering, petrochemical engineering, biotechnology and courses in other related areas such as agriculture and food engineering, fire and safety engineering, environmental engineering, etc. The book begins with an introductory chapter on mass transfer that gives an overall view of the materials covered in the book. Chapters 2 and 3 deal with the principles of molecular diffusion and theory of convective and interphase mass transport. Individual chapters are devoted from Chapters 4 to 11 for detailed discussion of the theory and application of the different separation techniques such as gas absorption, distillation, humidification and dehumidification, extraction, leaching, drying, crystallisation and adsorption. State-of-the-art developments and trends in the design and construction of separation equipment and principles of their operation are covered in detail. Membrane separations offer tremendous potential as the technology of future separations. Chapter 12 is devoted to an exhaustive coverage of the principles and applications of membranes and membrane separations.

The book provides a balanced mix of fundamentals and applications. Numerical problems are provided in each chapter as solved examples and the book contains about 157 such examples. They help the readers to reinforce and strengthen their foundation in the theory and enhance their understanding of the principles discussed. They also give them an idea on the range and scope of the practical design and analysis problems that they may have to face in their professional life. With a view to enhance their ability to interpret and analyse new situations successfully, a large number of exercise problems with answers are provided at the end of each chapter. More than 660 such practice problems are included in the book with the simple logic that practice makes one perfect. Hundreds of objective multiple choice questions provided in the book will act as stimulant to the thought process in the students and enable them to assess their foundation in the subject. For those preparing for competitive examinations, they provide an exhaustive tool for a review of their foundation in the subject of mass transfer.

We are thankful to CBS Publishers & Distributors. We would like to put on record the sincere efforts of Mr YN Arjuna (Senior Vice President Publishing, Editorial and Publicity), and his team comprising Ms Ritu Chawla (GM Production), for bringing out the book in the present form.

KV Narayanan
B Lakshmikutty

Contents

Chapter 1

Introduction

1.1 MASS TRANSFER

Chemical process industries are concerned with the transformation of raw materials into useful products. The products include a wide variety of materials such as petroleum products and petrochemicals, inorganic and organic chemicals, paints and pigments, fertilizers and pesticides, pulp and paper, rubber and plastics, soaps and detergents, synthetic fibres, drugs and pharmaceuticals, minerals, metals and semiconductors. These products are either directly used in our day-to-day life, or they serve as raw materials or intermediates for further processing. The materials undergo several processing steps which include physical operations and chemical conversions, before the final products of the desired quality and specifications are obtained. The raw materials reaching a processing unit may require some pre-processing treatments. The products leaving the reactors may sometimes be contaminated by the undesired by-products that have to be removed, or they should be treated to recover valuable side products. The unconverted reactants and the impurities brought into the system by the feed materials have to be separated from the products. These operations, whether they are the pre-processing of raw materials, recovery of the by-products or the final purification of the desired products, involve certain separation operations.

The objectives of the separation operations are purification, isolation, recovery, refining, concentration, or enrichment of the materials at various stages of their processing. The separation is usually achieved by exploiting the differences in one or other of the properties of the substances such as the size, density, volatility, solubility, diffusivity, electrical charge, etc. The separation processes in the chemical process industries fall into one of the following broad categories.

- Mechanical separations
- Diffusional separations

Mechanical separations are generally used for separating one phase from another. For example, filtration is used to separate a solid phase from a liquid phase. They employ the differences in the size, density, velocity, etc. to effect separation. Filtration, screening, sedimentation, magnetic separation, centrifuging, etc. are examples of mechanical separations. Diffusional separations also known as *mass transfer operations*, on the other hand, are used for separating the components of a homogeneous solution. The differences in the properties such as the volatility, solubility, diffusivity, etc. of the constituent species, are utilised in achieving the separation. The concentration difference existing in a homogeneous phase can cause movement of material from one region to another. This transfer of material from the region of higher concentration to the region of lower concentration is termed diffusion which is the basic mechanism of mass transfer. However, this transfer is incapable to achieve separation of the given substance into the products with distinct properties. Since mixing of substances to form a solution is a highly irreversible process, the separation of a homogeneous solution into its components will not occur spontaneously. For separation to be effective, the material is to be transferred to a second phase. The second phase necessary to effect the separation is either generated from the original solution by supplying energy or it is introduced into the system as an external agent. Certain components of the original solution get transferred to the new phase, thus effecting separation. Thus the mass transfer operations generally involve the diffusion process within a homogeneous phase followed by the interphase transport. Majority of the separations in chemical process industries are carried out by mass transfer; a variety of mass transfer operations such as distillation, absorption, extraction, adsorption, crystallisation, etc. are employed for this purpose. Mass transfer operations play such an important role in chemical processing that in a large scale chemical process industry, the capital investment on the mass transfer equipment generally far exceeds the investment on the reaction vessels and the related equipment.

Mass transfer operations are employed for separation of the components of a solution. The solution so separated may be a solid, gas or a liquid. In the majority of separations, the transfer of mass occurs between the two insoluble phases which are in *direct contact*. The common mass transfer operations falling under this category are distillation, liquid–liquid extraction, gas absorption, leaching, drying, adsorption, humidification and dehumidification, crystallisation, ion-exchange, etc. If the phases are allowed to be in contact for a sufficiently long time, they eventually reach the equilibrium state, a state in which there is no further net transfer of material. In the *membrane separation operations*, the phases are prevented from coming into contact and kept separated from each other by a semipermeable membrane; the membrane acts as a barrier to the hydrodynamic flow of the phases, but permits through it, the migration of certain components of the solution in preference to the others. Microfiltration, ultrafiltration, reverse osmosis, electrodialysis, pervaporation and gas permeation are examples of membrane separations. Since the membranes achieve the separation by size exclusion as well as by diffusion, their role as separation techniques lie somewhere between the mechanical separations and the conventional mass transfer operations.

In mass transfer, material is transferred on a molecular scale, and this transfer is quite different from the bulk transport of material from one location to another. Moving the material by means of say, a pump or a blower, involves the bulk transfer of material facilitated by the pressure difference created externally. This bulk transport of mass is distinct from the mass transfer, which involves the relative movement of the components due to a concentration difference. For example, inhalation of air by the lungs is bulk transport where as transfer of oxygen from the air to the blood by diffusion is mass transfer. When ammonia is absorbed from a mixture of ammonia and air using water as the solvent, ammonia moves relative to the air from the gas phase to the liquid phase. When oil is extracted from the oil cake using a solvent such as hexane, there is relative movement of components between the solid and the liquid phases. The transfer of ammonia is due to the concentration difference for the ammonia existing between the gas and the liquid phases and the transfer of oil from the oil cake is due to the concentration difference between the solid and the liquid phases. In these examples, the transfer of ammonia is by the mass transfer operation called *gas absorption* and the transfer of oil is by the operation called *leaching*.

We know that the heat transfer between two planes comes to a stop when the temperatures at the planes become identical. Therefore, we can treat the temperature difference to be the true driving force in heat transfer. We can see that the driving forces in the gas absorption and leaching are the concentration differences between the two insoluble phases. The concentration difference between the two immiscible phases which are in direct contact is responsible for the transfer of materials in the operations such as absorption, distillation, humidification, extraction, leaching, drying, adsorption and crystallisation which are discussed in Chapters 4 through 11 of this book. However, the concentration difference cannot be treated as the true driving force for mass transfer between phases; if it were so, the concentration in both phases would have been the same when the equilibrium is established and the transfer of material no longer occurs. This does not happen in mass transfer. For example, in the absorption of ammonia from an ammonia-air mixture using water as the solvent, the concentration of ammonia in the gas and liquid phases will be different even after the phases were allowed to be in contact for an infinitely long time for the equilibrium to be attained.

1.2 THEORETICAL FOUNDATION

Equilibrium

Thermodynamics requires that when different phases of a solution coexist in equilibrium, the *chemical potentials* of the components in the different phases should be equal. Since the chemical potentials in different phases are related to the concentrations differently, the concentrations of the components in the phases need not be equal when this criterion of equilibrium is satisfied. Concentration difference is thus a measure of the driving force for mass transfer between the phases, but not the true driving force. The true driving force is determined by the difference in the chemical potential of the species in the different phases. Thus thermodynamics, particularly the phase equilibrium thermodynamics plays a very important role in determining the driving force for mass transfer. The discussion of the mass transfer operations in the following chapters will begin with a detailed analysis of the equilibrium characteristics of the system under consideration.

While thermodynamic equilibrium determines the theoretical extent of a given mass transfer operation, the actual rate of mass transfer depends upon other additional factors. The rate is influenced by the flow pattern within the system and the diffusivities of the constituent species.

Molecular Diffusion and Interphase Transport

Chapter 2 discusses the theory of molecular diffusion which is the basic mechanism of mass transfer in a homogeneous phase. Molecular diffusion is the movement of molecules from the region of higher concentration to the region of lower concentration when a concentration gradient exists in a homogeneous solution; the ultimate condition attained by the system is a uniformity of concentration throughout. Molecular diffusion in mass transfer is in several ways analogous to conduction in heat transfer and the molecular transport of momentum in momentum transfer. The Fick's law of diffusion predicts a direct relationship between the rate of mass transport and the concentration gradient, just as the Fourier's law of conduction gives the direct proportionality between the rate of heat conduction and the temperature gradient and Newton's law of viscosity predicts the relationship between the rate of momentum transfer and the velocity gradient. Chapter 2 also discusses the influence of temperature and pressure on mass diffusivity which can be treated as the transport coefficient of mass. The various techniques available for the estimation of mass diffusivity in the gaseous and the liquid phases are also reviewed in this chapter. Chapter 3 is concerned with the convective transport and the interphase mass transfer. Molecular diffusion is a very slow process; even in a homogeneous phase originally consisting of regions of non-uniform compositions, it will take a very long time for attaining the uniformity of concentration, if the transfer were due to the molecular diffusion alone. We cannot wait for such an infinitely long time for the completion of the process; neither can we have infinitely large vessels for carrying out the transfer process. In order to carry out the process in a reasonable time and in equipment of manageable sizes, the practicable solution is to increase the rate of transfer by introducing the convective transport in the system by external means. In all the separation equipment used in industries, the mass transfer takes place predominantly by the convective transport. The mass transfer coefficient serves as an index for measuring the rate of convective transport. The concept of mass transfer coefficient, the different types of mass transfer coefficients in use and their interrelationship are explained in Chapter 3. The chapter also discusses how the concept of mass transfer coefficient is different from that of the heat transfer coefficient, the theories of mass transfer and the methods of estimating the coefficient for different geometrical systems.

1.3 EQUIPMENT AND METHODS OF OPERATION

Direct contact separation methods always involve bringing into contact two immiscible phases that are initially not in equilibrium. Mass gets transferred from one phase to the other until equilibrium is established between these two phases. The phases in contact can be a gas, a liquid or a solid. Thus we have gas–liquid contact operations, liquid–liquid contact operations, solid–liquid contact operations and gas–solid contact operations. The basic principles of interphase transport, the two-resistance theory, the different modes of contacting the phases, salient features of *stage contact* and *continuous contact* are also discussed in Chapter 3 under interphase transport.

Based on the mode of operation, the mass transfer processes can be classified as *batch process*, *semi-batch process* or *continuous process*. In batch operation, definite quantities of the materials are introduced into the equipment for intimate mixing, the phases are allowed to be in contact for a sufficiently long time and the product streams are withdrawn when the desired extent of separation is achieved. Batch operations are *unsteady-state operations* in the sense that the concentration of the phases at a given point inside the equipment change with the time. In contrast, in continuous operation, the feed streams are introduced continuously and products are also withdrawn continuously from the equipment. Since the concentrations at a given point in the vessel remain constant and do not change with the time, the continuous operations are also *steady-state operations*. In the semi-batch operation, one of the phases is allowed to flow continuously through the equipment which holds a definite quantity or batch of the other phase. *Drying* a wet solid by passing hot air continuously across a batch of the wet feed taken on a tray is a semi-batch operation.

In the continuous operation, the changes in the concentration of the phases can be achieved either continuously or intermittently in stages. The *plate column* is an example of a stage contact equipment for gas–liquid and liquid–liquid contact and the *packed column* is a continuous contact equipment. In the plate columns, the tall vertical vessel is separated into different stages by means of the horizontal plates which allow the intimate contact of the gas and liquid streams or two immiscible liquid streams. The streams are in contact only on the plates and after mass transfer, they are separated and led to the next stage as distinct phases. The streams thus undergo concentration changes intermittently in discrete quantities only over the plates. In packed columns, on the other hand, streams are in contact at all points inside the vessel and they undergo concentration change continuously. A wide variety of stage contact and continuous contact equipment are in operation and they are discussed in detail when specific operations are considered in Chapters 4 through 11. Table 1.1 gives a list of different separation operations, their major industrial applications and important equipment for carrying out these operations.

Gas–liquid Contact Operations

In *gas absorption*, components of a gas mixture are separated by absorbing in liquid solvents. The absorption operation makes use of the difference in the solubility of the components in the liquid. The components with greater solubilities get dissolved in the solvent in preference to the others, thereby achieving the separation. Gas absorption is a typical gas–liquid contact operation which can be carried out as a multistage continuous operation in plate columns or continuous contact operation in packed columns. Chapter 4 discusses the gas absorption in detail with an elaborate account of the gas–liquid contact equipment which are also applicable for the other gas–liquid contact operations. *Stripping or*

Table 1.1: Mass transfer operations and their applications

Operation	Phases in Contact	Major Equipment	Applications
Gas Absorption	Gas–liquid	Plate columns, packed columns, venturi scrubber, spray towers, bubble columns, centrifugal absorbers	Separation of gases, removal of pollutants and impurities from gases, recovery of valuable chemicals.
Stripping	Gas–liquid	Plate columns, packed columns, spray towers, bubble columns, centrifugal contactors	Regeneration of absorption solvent, recovery of volatile components from oils
Distillation (Fractionation)	Vapour–liquid	Plate columns, packed columns	Petroleum refining, separation of alcohol water solution into components, liquefied air into oxygen and nitrogen, etc.
Humidification/ dehumidification	Gas–liquid	Spray chambers, packed columns, cooling towers	Liquid cooling, gas cooling, air conditioning, solvent recovery, gas drying, water desalination
Extraction	Liquid–liquid	Mixer-settler assemblies, spray, packed and plate columns, rotary-impeller columns, pulse columns, rotating-disk contactors, centrifugal contactors	Separation of close boiling liquids, azeotropes, liquids of poor relative volatility, high boiling mixtures, heat sensitive materials, pharmaceuticals, removal of colour.
Leaching	Solid–liquid	Bollmann extractor, diffuser, extraction battery, Kennedy extractor, rotocell, Pachuca tank, Bonnoto extractor, Hildebrandt extractor	Metallurgical industries, food, pharmaceutical and biological industries, oil extraction from seeds, extraction of sugar from sugar beets, tannin extraction, etc.
Drying	Solid–gas	Tray, truck, agitated pan, rotating batch, vacuum, freeze, turbo, rotary, tunnel, flash, conveyor, fluidized bed and spray dryer	Drying of paper, salt, chemicals, timber, pharmaceutical products, fertilizers, soap, detergents, milk, sugar, etc.
Adsorption	Solid–gas	Fixed bed, moving bed (hypersorber), combined fluidised bed-moving bed adsorber.	Dehumidification of gases, odour removal from industrial gases, pollution control, solvent recovery, separation of nitrogen from air, separation of hydrogen from synthetic gases, control of automobile emissions
Adsorption	Solid–liquid	Agitated vessels, moving bed, (hypersorber), fixed bed, simulated moving bed adsorber	Removal of moisture from gasoline, organic solvents etc., removal of sulphur compounds from organic solutions, removal of organic components from drinking water, decolourisation of petroleum products, sugar solutions, syrups, vegetable oils, etc. removal of objectionable taste and odour, process effluent treatment, recovery of vitamins.
Crystallisation	Solid–liquid	Agitated batch, Swenson-Walker, Oslo evaporative crystalliser, forced circulation evaporator, draft tube-baffle crystallisers	Pharmaceutical and fine chemicals industries, agrochemical industries, cosmetics industries, food industries, etc. desalination of sea water.

desorption is the opposite of absorption; while in absorption, the mass transfer is from the gas to the liquid, in stripping, the transfer is from the liquid to the gas. Ammonia can be absorbed from the ammonia air mixture using water as the solvent; the aqueous ammonia solution can be stripped off the ammonia by a stripping agent such as a hot gas or superheated steam.

Distillation is the most widely used method of separating components of a liquid mixture; its acceptability being attributed to the fact that it is a direct separation technique that does not require an external agent for effecting separation. What is achieved by introducing a solvent or a separating agent in other operations, is achieved in distillation, by the application of heat. When a liquid mixture is boiled, the resulting vapour and liquid will have different compositions, and these compositions are different from the composition of the original feed as well. This difference in the composition of the vapour and liquid phases depends on the volatilities of the constituents of the mixture. In *fractionation*, the vapour produced is continuously enriched in the more volatile component by contacting with a portion of the liquid, called *reflux*, which is obtained by the condensation of the vapour in an external condenser. Similarly, the liquid produced is stripped of the more volatile component by bringing it in contact with the vapours produced by boiling the liquid in an external reboiler. Ultimately, the fractionation columns yield a distillate product containing mostly the more volatile components, and a bottom product containing mostly the less volatile or the high-boiling components. Distillation is a vapour-liquid contact operation where both the vapour and the liquid are produced from the same feed stream. A thorough knowledge of the vapour-liquid equilibrium (VLE) is required for the design and analysis of the distillation processes and equipment. A detailed discussion of VLE is highly warranted and is included in Chapter 5 which contains an exhaustive coverage of distillation theory, design and applications.

Humidification and *dehumidification* operations are also gas–liquid contact operations, the gas is air and the liquid is pure water, in majority of practical applications. In humidification, the gas gets humidified by receiving the vapours of the liquid with which it is contacted. Since the liquid phase consists of a single component, the mass transfer is restricted to the gas phase; however, the rate of transfer is influenced not only by the mass transfer, but by the heat transfer as well. The concentration difference in the gas phase alone cannot represent the driving force in such a process which is governed by the combined heat and mass transfer mechanisms. Enthalpy difference between the bulk gas and the interface between the gas and the liquid serves as the driving force. In dehumidification, the gas phase from which the vapours are to be removed, is contacted with the liquid phase of the vapour; the vapours get condensed and transferred to the liquid. In humidification, the mass transfer is from the liquid to the gas, whereas in dehumidification, the transfer is in the reverse direction. Chapter 6 covers these operations in detail preceded by a detailed study of the psychrometry, an essential prerequisite for a better understanding of these operations.

Liquid–Liquid Contact Operation

If materials distribute between two immiscible liquid phases in different proportions, liquid solutions consisting of these materials can be separated by *extraction*. Chapter 7 deals with liquid–liquid extraction in detail. Extraction is a liquid–liquid contact operation and its importance as an industrial separation is second only to distillation. As an example of the liquid extraction, we can consider the separation of an aqueous solution of acetic acid by contacting with a solvent such as ethyl acetate with which water is only partially miscible. The preferential distribution of acetic acid between the aqueous and acetate phases results in the separation of the original mixture. In the majority of extraction operations, three components are present in both the phases, and therefore, the ternary liquid–liquid equilibrium forms the thermodynamic basis for the design and analysis of extraction operations. Extraction operations are carried out in equipment similar in construction to the gas–liquid contact equipment. Thus plate columns and packed columns are common in extraction as well, though some specialized equipment are also in use.

Transfer from or to Solids

Operations such as leaching, adsorption, crystallisation, ion exchange, drying, etc. fall under this category. *Leaching* is a solid–liquid contact operation in which the transfer of solutes occurs from the solid phase to the liquid phase, as in the recovery of the oil from the oil-cakes using an organic solvent, or in the extraction of gold from gold ores using the cyanide solution. Since solids are available in different sizes and shapes, the equipment handling solids such as the leaching equipment are in use in a wide variety of designs, which are discussed in detail in Chapter 8. This is true for drying operations discussed in Chapter 9 also. *Drying* is the removal of moisture from the solids by evaporation. In many industries, drying operation is one of the last processes employed just before packaging and dispatch of the products. When moisture is evaporated into a gas stream that is in contact with the solid, the drying becomes a gas–solid contact operation. In general, the drying process is influenced by the simultaneous transfer of heat and mass.

Adsorption involves transfer of solutes onto the surfaces of the solids known as adsorbents from a liquid phase or a gas phase and therefore, can be considered as solid–fluid contact operations. The equilibria in the adsorption operation are popularly known as the *adsorption isotherms*, for representing which theoretical as well as empirical models are available. Adsorption isotherm, adsorption systems, different methods of adsorption operation and equipment are discussed in Chapter 10. *Crystallisation* from solutions is a mass transfer operation in which the solute is transferred from the solution onto solids. *Nucleation and crystal growth* are two distinct processes which govern the rate of crystallisation. The nucleation denotes the birth of crystals and the growth results due to mass transfer. Crystallisation is an important separation technique since it yields relatively very pure products with minimum energy consumption. Chapter 11 deals with crystallisation from solutions.

1.4 MEMBRANE SEPARATIONS

Membrane separations which were considered as unconventional separation techniques, have gained wide acceptability in recent years. Considered to be very unreliable, unselective, slow and expensive operation in the past, membrane separation techniques have now undergone significant transformation. They find wide applications now in industries such as chemical, petrochemical, pharmaceutical, paper, textile, food and beverage, environmental and electronic industries, apart from their application in the medical field as in artificial kidney for haemodialysis and blood-oxygenators. This dramatic development is due to the concerted efforts of physical chemists, material scientists and chemical engineers in developing and adapting membrane materials with high *selectivity and permeability*, in the fabrication methods for high-flux asymmetric or composite membranes, and in the construction of membrane modules capable of accommodating large membrane surface into small volume.

Based on the materials, the membranes can be broadly classified as organic and inorganic membranes. Organic membranes include membranes made from natural or synthetic polymers and the inorganic membranes include ceramic, metal, carbon and zeolite membranes. Membrane separations offer several advantages over the conventional separation techniques. They are ideally suited for the separation of the heat-sensitive materials, as the separation is generally carried out at moderate temperatures. Equipment is modular, compact, require less capital investment, and is less energy intensive than the usual separation operations. The separation is purely physical and requires no foreign substances as separating agents.

The major membrane separation operations are microfiltration, ultrafiltration, nanofiltration, reverse osmosis, dialysis, electrodialysis, pervaporation and gas permeation. These operations involve transfer between two phases, liquid or gas, the direct contact of the phases being prevented by separating them by semipermeable barrier called membrane. Separation of the constituents is achieved by taking advantage of their differences in the rates of movement through the membranes, which in turn depend on the molecular size, shape, electric charge and/or the sorption characteristics of the materials. As such, different driving forces are used to facilitate transport of substances through the membrane. Examples are the applications of high pressure, the maintenance of a concentration gradient on both sides of the membrane and the introduction of an electric potential. The microfiltration, ultrafiltration, nanofiltration, reverse osmosis, pervaporation and the gas permeation are pressure–driven operations, dialysis and osmosis are concentration–driven operations and electrodialysis is an operation involving electric potential gradient across the membrane.

Microfiltration, ultrafiltration, nanofiltration and reverse osmosis are generally referred to as the membrane filtration processes since all rely on the size exclusion principle for separation. The size range of particles retained by the membrane filtration techniques are shown in Fig. 1.1. Table 1.2 gives the classification of membrane separation techniques based on the size range of particles retained. The different driving forces used for effecting separation are also listed in the table.

The membrane separations are discussed in Chapter 12 with emphasis on the basic principles behind the separations, the membranes used, their characteristics and the major industrial applications.

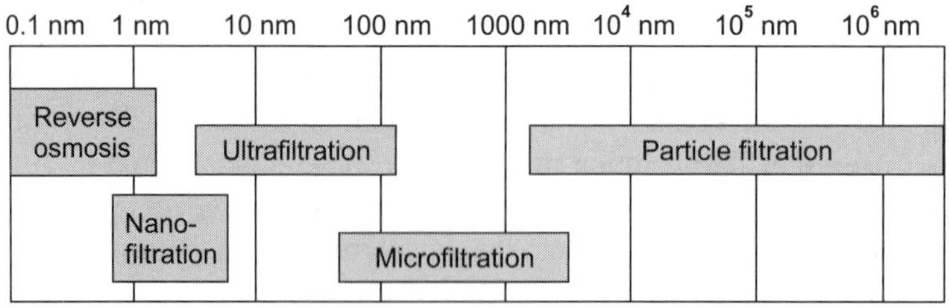

Fig. 1.1: Range of particle sizes separated by membrane filtration techniques

Table 1.2: Classification of membrane processes based on size range of particles separated and driving force of separation

Process	Driving force	Size range of particles retained
Microfiltration	Pressure gradient	> 50 nm
Ultrafiltration	Pressure gradient	> 3 nm
Nanofiltration	Pressure gradient	> 1 nm
Reverse Osmosis	Pressure gradient	> 0.1 nm
Electrodialysis	Electric field gradient	> 5 nm
Dialysis	Concentration gradient	> 5 nm
Pervaporation	Pressure gradient	> 1 nm
Gas permeation	Pressure gradient	> 1 nm

Chapter 2

Molecular Diffusion

Molecular diffusion is the primary mechanism of mass transfer, just as heat conduction is the basic mode of heat transfer. Diffusion involves random movement of individual molecules because of thermal agitation. When concentration gradient exists in a homogeneous gas, liquid or solid mixture, the molecules move from the region of higher concentration to the region of lower concentration till equilibrium is established; the state of equilibrium is characterised by uniformity of concentration throughout. For illustration, consider what happens to a drop of soluble red dye placed in a beaker containing water. We can visualize molecular diffusion from the spread of the red colour to the water surrounding the dye. Ultimately a dynamic equilibrium is achieved at which the concentration of the dye is uniform throughout the solution. Similarly, the molecular diffusion makes the coffee uniformly sweet when a lump of sugar is added to a cup of coffee. The pleasant fragrance of the perfume kept somewhere reaches our nostrils by molecular diffusion.

Molecular diffusion is not restricted to homogeneous solutions. Mass transfer operations usually involve transfer of materials between phases. In interphase mass transfer, molecular diffusion occurs first from one phase to the interface and then from the interface to the second phase. In the absorption operation in which ammonia is absorbed from a mixture of ammonia and air into water, ammonia diffuses from the bulk of the gas to the gas–liquid interface and then from the interface to the bulk of liquid. In the leaching operation for the recovery of soya bean oil from seeds using hexane as the solvent, diffusion of oil takes place through the solid to the solid–solvent interface, followed by diffusion of oil through the liquid from the interface.

In interphase mass transfer, the rate of mass transfer is governed by the rate of diffusion; however, the extent of transfer is governed by phase equilibrium because no mass transfer can occur when the phases are in thermodynamic equilibrium. The criterion of phase equilibrium requires that the chemical potential of the components be the same in all phases under equilibrium condition. This criterion will be satisfied even when the concentration of the components in different phases are not equal, because concentrations are related to chemical potentials in different ways for different phases. It is the equality of chemical potential and not the uniformity of concentration, which is the criterion of phase equilibrium. The true driving force for mass transfer between phases, therefore, is the difference in the chemical potential of the components in different phases. However, for individual phases, the concentration will be uniform throughout the phase under equilibrium. The difference in the concentration can be treated as the true driving force for diffusional processes in individual phases.

2.1 DEFINITIONS OF DIFFUSION TERMS

Concentrations

The concentration of different constituents of a mixture can be expressed as mass (or moles) of the components per unit volume of the mixture. The mass concentration of species i is denoted by ρ_i and it represents the mass in kilograms of the component i in one cubic meter of the solution. The total mass concentration ρ is the mass density of the solution.

$$\rho = \sum_1^n \rho_i \qquad \text{...(2.1)}$$

The molar concentration C_i of component i is the number of moles of component i per unit volume of the solution and in SI system has units kmol/m^3. The total molar concentration of the mixture is represented by C and it is related to C_i as

$$C = \sum_1^n C_i \qquad \text{...(2.2)}$$

It may be noted that the molar concentration is related to the mass concentration as follows:

$$C_i = \frac{\rho_i}{M_i} \qquad \text{...(2.3)}$$

M_i is the molecular weight of component i. It is frequently more convenient to express the concentrations as mass or mole fractions. The mass fraction ω_i is the ratio of the mass of species i to the total mass of the mixture. In terms of mass concentrations, the mass fraction can be defined as

$$\omega_i = \frac{\rho_i}{\rho} \qquad \text{...(2.4)}$$

The mole fraction x_i is the ratio of the moles of component i to the total moles of the mixture. In terms of molar concentrations

$$x_i = \frac{C_i}{C} \qquad \text{...(2.5)}$$

Sum of all mole fractions (or mass fractions) in a mixture is unity.

$$\sum_1^n x_i = 1 \quad \text{or} \quad \sum_1^n \omega_i = 1$$

In a mixture of n components, $n - 1$ mole fractions need be fixed independently. The mole fraction of the remaining constituent is one minus the sum of the mole fractions of the $n - 1$ constituents. In a binary mixture, if x is the mole fraction of component **A**, the mole fraction of component **B** is $1 - x$ and therefore it is not necessary to retain the subscripts representing the constituents in the mole fraction terms.

In dealing with gas mixtures, the assumption of ideal gas behaviour permits us to relate the molar concentrations to partial pressures as

$$C_i = \frac{n_i}{V} = \frac{\bar{p}_i}{RT} \qquad \text{...(2.6)}$$

The overall molar concentration of the solution is

$$C = \sum_1^n C_i = \sum_1^n \frac{\bar{p}_i}{RT} = \frac{P}{RT} \qquad \text{...(2.7)}$$

$\bar{p}_i$ is the partial pressure of gas i and P is the total pressure.

Velocity

Diffusion involves the movement of individual species through a fluid by the random, individual movement of molecules. In multicomponent solutions, several components will be diffusing in the same direction. Even in binary mixtures, both components have to diffuse if uniformity in concentration is to be achieved. As a result, several velocities are used to describe the movements of individual substances and the total mixture. The velocity, or to be more precise, the macroscopic velocity of a component is obtained by summing up all the instantaneous velocities of the component, resolving it in the direction perpendicular to the interface and by dividing by the number of molecules of the component. For component **A**, the macroscopic velocity is denoted by u_A. This is the velocity of species **A** with reference to stationary coordinate axes or u_A is the velocity apparent to an observer at rest. Since the components may possess different mobilities, the mixture velocity is obtained by averaging the velocities of all the constituents.

Mass average velocity, *u*. This is the mass fraction weighted average of the velocities of the components constituting the mixture.

$$u = \sum_{1}^{n} \omega_i u_i \qquad \text{...(2.8)}$$

where u_i denotes the absolute velocities of species i and ω_i is the mass fraction of species i.

Molar average velocity, *U*. This is the mole fraction weighted average of the velocities of the components constituting the mixture.

$$U = \sum_{1}^{n} x_i u_i \qquad \text{...(2.9)}$$

where x_i is the mole fraction of species i.

Velocity of a particular component relative to the average velocity of the mixture is termed diffusion velocity of that component. The diffusion velocity of species i relative to the mass average velocity of the mixture (u_i') is

$$u_i' = u_i - u \qquad \text{...(2.10)}$$

and the diffusion velocity relative to the molar average velocity of the mixture (U_i') is

$$U_i' = u_i - U \qquad \text{...(2.11)}$$

Flux

The rate at which a material is transferred by molecular diffusion is directly proportional to the area normal to the direction of transfer. In dealing with diffusional processes, it is customary to express the rate as mass or molar flux which can be defined as the amount of the particular species measured in mass or molar units that pass a unit area normal to the direction of transfer in unit time. Flux is a vector quantity and has units kg/m^2 h or kmol/m^2 h. In terms of concentration of species and velocity of diffusion, flux can be written as

$$\text{flux} = \text{concentration} \times \text{velocity} \qquad \text{...(2.12)}$$

Flux can be defined in different ways. The one referred to the molar average velocity of the fluid (J_A) and another referred to the stationary coordinate axis (N_A) are important in studying mass transfer. Consider a binary mixture made up of components **A** and **B**. Suppose that the mixture is moving by bulk flow in the same direction in which species **A** is diffusing due to concentration difference. This bulk flow may arise due to an externally introduced pressure difference or due to the molecular diffusion of different species in the mixture. The rate of transfer of species **A** is calculated as the sum of two contributions: molecular diffusion contribution and bulk flow contribution. The molecular diffusion contribution is directly related to the concentration gradient for species **A** and the bulk flow contribution is the fraction of species **A** in the mixture that is transferred by convective transport.

With reference to a fixed location in space, component **A** is transferred across a unit area in the direction of transfer at a rate N_A. N_A is the flux of species **A** with reference to a stationary observer or with reference to stationary coordinate axes.

$$N_A = C_A u_A \qquad \text{...(2.13)}$$

where u_A is the velocity of component **A** in the mixture. In contrast, the flux J_A is the rate of transfer of **A** across a reference plane moving at a velocity equal to the molar average velocity of the mixture. Denoting the diffusion velocity of species **A** by U'_A,

$$J_A = C_A U'_A \qquad \text{...(2.14)}$$

where $U'_A = u_A - U$ and U is the molar average velocity of the mixture. The flux of **A** sensed by an observer will depend upon whether he is stationary or whether he is moving with the mixture. Consider for example, another case where there is no bulk flow or convective transport of the entire phase. The system as a whole is stagnant. Species **A** diffuses from the region of higher concentration to that of lower concentration. The flux of **A** in this case is entirely due to molecular diffusion. It is clear that if the molar average velocity of the mixture were zero, or if the mixture were stagnant, both J_A and N_A are identical. In general,

$$N_A = J_A + x_A N \qquad \text{...(2.15)}$$

where N is the total rate of transfer of the fluid by bulk flow and x_A is the mole fraction of **A** in the mixture. The flux J_A is a measure of the diffusion capabilities or is characteristic of the nature of constituents whereas the flux N_A is more important in the design of equipments as it reflects the actual rate at which material is transferred.

2.2 FICK'S LAW OF DIFFUSION

The following generalisations can be made regarding molecular diffusion:

1. The rate of molecular diffusion is directly proportional to the surface normal to the direction of transfer.
2. The rate can be expressed as a flux for the component diffusing.
3. Diffusion occurs in the direction of decreasing concentration and is proportional to the negative concentration gradient.

These observations are mathematically expressed by the Fick's law of diffusion, which states that the molar flux of diffusion in the z-direction of component **A** in solution with **B** is directly proportional to the concentration gradient.

$$J_A = -D_{AB} \frac{dC_A}{dz} \qquad \text{...(2.16)}$$

Here J_A is the molar flux of component **A** in the z-direction relative to the molar average velocity of the fluid, in kmol/m^2 h or kmol/m^2 s, C_A is the molar concentration of **A** in kmol/m^3 and $\frac{dC_A}{dz}$ is the concentration gradient in the z-direction. The proportionality constant D_{AB} is the *mass diffusivity* or *diffusion coefficient* of component **A** diffusing through component **B** and has units m^2/h or m^2/s. Since $C_A = x_A C$, where x_A is the mole fraction of species **A** and C is the overall molar concentration, Fick's law can be written in terms of mole fraction as

$$J_A = -CD_{AB} \frac{dx_A}{dz} \qquad \text{...(2.17)}$$

The negative sign in Eqs. (2.16) and (2.17) indicate that the diffusion occurs in the direction of decreasing concentration or the negative concentration gradient and ensures that the flux is positive in the direction of diffusion.

The mass diffusivity (D_{AB}) is the transport coefficient for mass just as kinematic viscosity (ν) is the transport coefficient for momentum and thermal diffusivity (α) is the transport coefficient for heat. These three transport coefficients are analogous and all have identical dimensions. They are also termed as molecular diffusivity of momentum (ν), molecular diffusivity of heat (α) and molecular diffusivity of mass (D_{AB}). In the SI system, all have units m²/s. The basic equations for momentum flux, heat flux and mass flux are written in terms of the respective transport coefficients in analogous fashion as given below:

$$\tau_{zx} = -\nu \frac{d(\rho u_x)}{dz}$$ Newton's law of viscosity

$$q = -\alpha \frac{d(\rho C_p T)}{dz}$$ Fourier's law of heat conduction

$$J_A = -D_{AB} \frac{dC_A}{dz}$$ Fick's law of diffusion.

Binary Mass Diffusivities

The diffusion coefficient D_{AB} is a characteristic of the substance **A** as well as the substance **B,** the other constituent of the mixture. The value of diffusion coefficient is strongly influenced by the physical state of the solution in which diffusion is occurring, i.e. whether diffusion is in the gas, liquid or solid solution. Diffusivities are influenced by the environmental conditions such as temperature, pressure and concentration of the constituents. The diffusion coefficients are generally higher for gases than for liquids. For solids, the coefficients are much smaller. The range of diffusivity values are approximately as follows:

Solids : 10^{-14} to 10^{-10} m²/s
Liquids : 10^{-10} to 10^{-9} m²/s
Gases : 5×10^{-6} to 10^{-5} m²/s

Diffusivities of gases at low density are generally independent of concentration, whereas the liquid and the solid diffusivities are strongly concentration dependent. Mass diffusivity of gases increases with temperature and varies inversely with pressure. For liquids and solids diffusivity increase with temperature, but are insensitive to pressure. The diffusivities determined experimentally provide the most reliable values for practical calculations. In the absence of experimental data or when experimental measurements are difficult to carry out, semi-theoretical expressions or published correlations may be used. Some such correlations are reviewed under Section 2.7.

Fick's Law in Terms of Stationary Coordinates

In terms of the velocity and concentrations, the flux J_A is given by Eq. (2.14)

$$J_A = C_A U'_A = C_A(u_A - U)$$

Rewriting the above equation, we get

$$C_A u_A = J_A + C_A U \quad \text{...(2.18)}$$

By Eq. (2.9), the molar average velocity U is

$$U = x_A u_A + x_B u_B = \frac{C_A}{C} u_A + \frac{C_B}{C} u_B \quad \text{...(2.19)}$$

Substituting this in Eq. (2.18), we get

$$C_A u_A = J_A + \frac{C_A}{C}(C_A u_A + C_B u_B) \quad \text{...(2.20)}$$

By Eq. (2.13)

$$C_A u_A = N_A \text{ and } C_B u_B = N_B \quad \text{...(2.21)}$$

Noting that $N_A + N_B = N$, the net transfer of mass with reference to a stationary observer as given by Eq. (2.20) can be simplified as

$$N_A = J_A + \frac{C_A}{C}(N_A + N_B)$$

$$= J_A + x_A N \qquad \text{...(2.22)}$$

Equation (2.22) reveals that the diffusion flux N_A with reference to a stationary coordinate axes can be treated as made up of two terms: J_A, the flux due to molecular diffusion and $x_A N$, the flux of **A** resulting from the bulk motion of the fluid. Thus, the Fick's law with reference to a fixed location in space can be written as

$$N_A = -D_{AB}\frac{dC_A}{dz} + x_A N = -CD_{AB}\frac{dx_A}{dz} + x_A N \qquad \text{...(2.23)}$$

Relationship between D_{AB} and D_{BA}

A concentration gradient for component **A** in a binary solution automatically implies that concentration gradient exists for component **B** in the opposite direction. The Fick's law can be written for component **B** as

$$N_B = -D_{BA}\frac{dC_B}{dz} + x_B N \qquad \text{...(2.24)}$$

Adding Eqs. (2.23) and (2.24) and noting that $N_A + N_B = N$, we get

$$N_A + N_B = -D_{AB}\frac{dC_A}{dz} + x_A N - D_{BA}\frac{dC_B}{dz} + x_B N \qquad \text{...(2.25)}$$

or

$$N = -D_{AB}\frac{dC_A}{dz} - D_{BA}\frac{dC_B}{dz} + (x_A + x_B)N \qquad \text{...(2.26)}$$

Since $x_A + x_B = 1$, Eq. (2.26) simplifies as

$$-D_{AB}\frac{dC_A}{dz} - D_{BA}\frac{dC_B}{dz} = 0 \qquad \text{...(2.27)}$$

i.e.,

$$-D_{AB}\frac{dC_A}{dz} = D_{BA}\frac{dC_B}{dz} \qquad \text{...(2.28)}$$

$$-D_{AB}\frac{dx_A}{dz} = D_{BA}\frac{dx_B}{dz} \qquad \text{...(2.29)}$$

This means that in a binary mixture

$$J_A = -J_B \qquad \text{...(2.30)}$$

Since the sum of the mole fractions is unity, we have

$$\frac{dx_A}{dz} = -\frac{dx_B}{dz} \qquad \text{...(2.31)}$$

Substituting this in Eq. (2.29), we have

$$D_{AB} = D_{BA} \qquad \text{...(2.32)}$$

Equation (2.32) means that for a binary mixture of **A** and **B**, the diffusivity coefficient D_{AB} of **A** diffusing into **B** is the same as D_{BA} of **B** diffusing into **A.**

Eddy Diffusion

Molecular diffusion is the mechanism by which mass transfer occurs in stagnant fluids and fluids in laminar motion. However, it is a very slow process and we may have to wait for a very long time for achieving equilibrium or uniformity of concentration by this process. In contrast, if turbulence is provided in the fluid by means of an agitator, the uniformity in concentration can be achieved in a very short time. Here, the transfer of solute is largely due to eddy diffusion. Just as uniformity of temperature is achieved in flow system by convective heat transfer, uniformity of concentration is achieved by eddy diffusion created by mechanical stirring and convective movement of the fluid. In eddy diffusion, large macroscopic aggregates of fluid called eddies carry material from one location to another. The eddies promote rapid mixing of the fluid and constitute a highly effective method for levelling off the concentration difference within the fluid stream. Consequently, the rate of eddy diffusion is several hundred times greater than the rate of molecular diffusion. It may be noted that even in turbulent transport, the basic mechanism is still molecular diffusion because within each eddy, uniformity of concentration is achieved by molecular transport.

2.3 STEADY-STATE MOLECULAR DIFFUSION IN GASES

Most of the industrial operations involve steady-state operation, i.e. continuous flow through an apparatus with time-invariant conditions of temperature, pressure, concentrations, and flow rates at any given point. Diffusion under such conditions implies the continuous supply and removal of the diffusing material. Two cases of steady state diffusion are of immense practical importance in mass transfer studies. They are equimolar counter diffusion and diffusion through stagnant gases.

Equimolar Counter Diffusion

In equimolar counter diffusion, the number of moles of **A** moving in one direction is equal to the number of moles of **B** moving in the opposite direction. With reference to a stationary observer, there is no net molar flow in either direction or, the net motion of the entire phase is zero. Mathematically, we have the condition

$$N_A = -N_B \text{ or } N = N_A + N_B = 0$$

Equimolar counter diffusion prevails in diffusive mixing of two gases and in rectification. Consider the distillation of a binary mixture of **A** and **B** using plate column. The vapour rising upwards and the liquid flowing downward are brought into intimate contact on the plates. As a result, the vapour gets partially condensed and the liquid gets partially vaporised. If the molar latent heat of vaporisation of **A** and **B** are approximately equal, for each mole of **B** condensed, one mole of **A** gets vaporised. The less volatile component **B** diffuses from the vapour to liquid and the more volatile component diffuses from the liquid to the vapour phase. Thus, under ideal condition, rectification involves equimolar counter diffusion.

Since $N_A = -N_B$, Fick's law of diffusion given by Eq. (2.23) simplifies to

$$N_A = -CD_{AB}\frac{dy_A}{dz} \qquad \text{...(2.33)}$$

In Eq. (2.33) symbol y_A is used to represent mole fraction of **A** in the gas phase instead of x_A. Eq. (2.33) can be rewritten as

$$N_A dz = -CD_{AB} dy_A \qquad \text{...(2.34)}$$

Assume that at two planes at distances z_1 and z_2 from the origin, the concentrations of **A** are y_{A1} and y_{A2}, respectively. Equation (2.34) can be integrated as

$$\int_{z_1}^{z_2} N_A dz = -\int_{y_{A1}}^{y_{A2}} CD_{AB} dy_A \qquad \text{...(2.35)}$$

For steady-state diffusion, N_A is constant along the length of the diffusion path. The diffusivity D_{AB} of gases is independent of concentration and the overall concentration C is constant so long as temperature and pressure are kept constant. Integration under these conditions gives

$$N_A Z = -CD_{AB}(y_{A2} - y_{A1})$$
$$= CD_{AB}(y_{A1} - y_{A2}) \quad \text{...(2.36)}$$

In Eq. (2.36), $Z = z_2 - z_1$, is the length of diffusion path. On rearranging Eq. (2.36), we get

$$N_A = \frac{CD_{AB}}{Z}(y_{A1} - y_{A2}) \quad \text{...(2.37)}$$

If the ideal gas law is applicable, we can express mole fractions and concentrations in terms of partial pressure and total pressure as

$$y_A = \frac{\bar{p}_A}{P} \qquad C = \frac{P}{RT} \quad \text{...(2.38)}$$

Substituting Eq. (2.38) into Eq. (2.37), we get

$$N_A = \frac{D_{AB}}{RTZ}(\bar{p}_{A1} - \bar{p}_{A2}) \quad \text{...(2.39)}$$

Equations (2.37) or (2.39) can be used to calculate the mass flux under equimolar counter diffusion.

Concentration profile: Under steady-state conditions, the flux N_A is constant along the diffusion path, so that

$$\frac{dN_A}{dz} = 0 \quad \text{...(2.40)}$$

Substituting Eq. (2.33) into Eq. (2.40), we get

$$\frac{d^2 y_A}{dz^2} = 0 \quad \text{...(2.41)}$$

The solution for Eq. (2.41) is

$$y_A = C_1 z + C_2 \quad \text{...(2.42)}$$

where C_1 and C_2 are constants. These are evaluated using the boundary conditions

$$y_A = y_{A1} \text{ at } z = z_1$$
$$y_A = y_{A2} \text{ at } z = z_2$$

We can see that

$$C_1 = \frac{y_{A1} - y_{A2}}{z_1 - z_2} \text{ and } C_2 = \frac{y_{A2} z_1 - y_{A1} z_2}{z_1 - z_2} \quad \text{...(2.43)}$$

Substituting Eq. (2.43) into Eq. (2.42) and after rearrangement, we get the concentration distribution as

$$\frac{y_A - y_{A1}}{y_{A1} - y_{A2}} = \frac{z - z_1}{z_1 - z_2} \quad \text{...(2.44)}$$

where y_A is the mole fraction of **A** at a distance z along the diffusion path. Equation (2.44) can be equivalently written in terms of partial pressure of **A** as

$$\frac{\bar{p}_A - \bar{p}_{A1}}{\bar{p}_{A1} - \bar{p}_{A2}} = \frac{z - z_1}{z_1 - z_2} \quad \text{...(2.45)}$$

The partial pressure varies linearly with distance along the diffusion path. The concentration distribution under equimolar counter diffusion is shown graphically in Fig. 2.1.

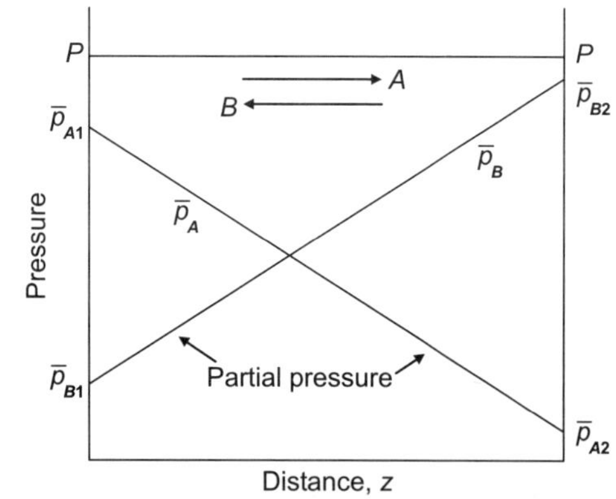

Fig. 2.1: Concentration distribution in equimolar counter diffusion

Example 2.1: Alcohol (**A**) vapour is diffusing through a layer of water (**B**) vapour under equimolar counter diffusion at 308 K and 101.3 kPa. The concentration of alcohol at two planes 0.3 mm apart are 60% (mole) and 10% (mole), respectively. The diffusivity of alcohol in water vapour is 1.8×10^{-5} m²/s, calculate the rate of diffusion of alcohol and water vapour in kg/h m².

***Solution*:** The molar flux under equimolar counter diffusion is calculated using Eq. (2.37)

$$N_A = \frac{CD_{AB}}{Z}(y_{A1} - y_{A2})$$

The molar concentration C is calculated using Eq. (2.38)

$$C = \frac{P}{RT} = \frac{101.3}{8.314 \times 308} = 0.0396 \frac{\text{kmol}}{\text{m}^3}$$

The mole fraction of alcohol at the two points are $y_{A1} = 0.60$ and $y_{A2} = 0.10$.

Diffusivity of alcohol in water vapour, $D_{AB} = 1.8 \times 10^{-5}$ m²/s.

Length of diffusion path, $Z = 0.3$ mm $= 0.3 \times 10^{-3}$ m

Substituting the values in Eq. (2.37), we get the molar flux of diffusion of alcohol

$$N_A = \frac{0.0396 \times 1.8 \times 10^{-5}}{0.3 \times 10^{-3}}(0.6 - 0.1) = 1.188 \times 10^{-3} \frac{\text{kmol}}{\text{m}^2 s}$$

Since $N_A = -N_B$,

$$\text{the molar flux of diffusion of water} = N_B = -N_A = -1.188 \times 10^{-3} \frac{\text{kmol}}{\text{m}^2 s}$$

By multiplying N_A by the molecular weight of alcohol (MW = 46) and N_B by molecular weight of water (MW = 18), the mass flux in kg/m²s are obtained which are converted into kg/m²h by multiplying by 3600.

$$\text{Rate of diffusion of alcohol} = 1.188 \times 10^{-3} \times 46 \times 3600 = 196.73 \frac{\text{kg}}{\text{m}^2\text{h}}$$ **(Answer)**

$$\text{Rate of diffusion of water} = -1.188 \times 10^{-3} \times 18 \times 3600 = -76.98 \frac{\text{kg}}{\text{m}^2\text{h}}$$ **(Answer)**

Example 2.2: A tube, 0.25 m long and having a diameter 1.5 mm connects two tanks one of which is filled with nitrogen gas and the other with hydrogen. The pressure is maintained constant at 101.3 kPa and the temperature at 298 K. Nitrogen and hydrogen diffuse in the opposite directions. The mole fraction of nitrogen in the tanks were 0.7 and 0.25 respectively, at one particular instant during diffusion. The mass diffusivity of nitrogen-hydrogen mixture is 7.84×10^{-5} m²/s. Determine the following:

(a) The rate of diffusion of nitrogen in kmol/h and kg/h.

(b) The velocity of species with reference to stationary coordinates as function of concentration.

***Solution*:** (a) For equimolar counterdiffusion, the molar flux of diffusion is given by Eq. (2.37)

$$N_A = \frac{CD_{AB}}{Z}(y_{A1} - y_{A2})$$

For the particular instant during diffusion, the mole fraction of nitrogen in the two tanks are:

$$y_{A1} = 0.7, y_{A2} = 0.25$$

The molar concentration of the mixture is calculated assuming ideal gas behaviour:

$$C = \frac{P}{RT} = \frac{101.3}{8.314 \times 298} = 0.0409 \frac{\text{kmol}}{\text{m}^3}$$

The mass diffusivity $D_{AB} = 7.84 \times 10^{-5}$ m²/s.

Substituting these values in Eq. (2.37),

$$N_A = \frac{0.0409 \times 7.84 \times 10^{-5}}{0.25}(0.7 - 0.25) = 5.772 \times 10^{-6} \frac{\text{kmol}}{\text{m}^2\text{s}}$$

The rate of diffusion of nitrogen in kmol/s is obtained by multiplying N_A by the area of cross section of the tube.

Area of cross section of the tube, $A = \frac{\pi}{4}D^2 = \frac{\pi}{4}(1.5 \times 10^{-3})^2 = 1.7672 \times 10^{-6}\,\text{m}^2$

Therefore, rate of diffusion $= N_A \times A = 5.772 \times 10^{-6} \times 1.7672 \times 10^{-6} = 1.02 \times 10^{-11} \frac{\text{kmol}}{\text{s}}$

$$= 1.02 \times 10^{-11} \times 3600 = 3.672 \times 10^{-8} \frac{\text{kmol}}{\text{h}}$$ **(Answer)**

Since the molecular weight of nitrogen is 28, the rate of diffusion on mass basis is

$$= 3.672 \times 10^{-8} \times 28 = 1.028 \times 10^{-6} \frac{\text{kg}}{\text{h}}$$ **(Answer)**

(b) The molar flux N_A is related to the velocity of species **A** with reference to a stationary observer as

$$N_A = C_A u_A \qquad \ldots(2.13)$$

where u_A is the velocity of component **A** in the mixture.

Substituting $C_A = y_A C$ in Eq. (2.13), we get

$$u_A = \frac{N_A}{y_A C} = \frac{5.772 \times 10^{-6}}{0.0409 y_A} = \frac{1.41 \times 10^{-4}}{y_A}\,\text{m/s}$$ **(Answer)**

The flux of component **B** is equal and opposite to that of **A**. Therefore, $N_B = -5.772 \times 10^{-6}$ kmol/m²s.

The species velocity for B is

$$u_B = \frac{N_B}{y_B C} = -\frac{5.772 \times 10^{-6}}{0.0409 y_B} = -\frac{1.41 \times 10^{-4}}{y_B}\,\text{m/s}$$ **(Answer)**

y_A and y_B are the mole fractions of **A** and **B** at a given point along the diffusion path.

Diffusion of A through Stagnant B

In mass transfer operations such as gas absorption and humidification, one component in a gaseous mixture will be transported relative to fixed plane, such as a liquid interface, and the other component will undergo no net movement. Consider the absorption of a soluble component **A** from a mixture of gases **A** and **B** into a liquid solvent as shown in Fig. 2.2. A typical example is the absorption of ammonia (**A**) from its mixture with air (**B**) into water. The solute gas **A** diffuses to the gas-liquid interface and gets dissolved in the liquid. The insoluble gas **B** undergoes no net movement with respect to the interface. But since a concentration gradient exists for the diffusion of **B** as well, there will be a bulk flow of the gas towards the interface in which **B** is transported in such quantities as to counterbalance its diffusional flux away from the interface. With reference to a stationary observer, **A** alone is diffusing and **B** appears to be stagnant, i.e. flux $N_B = 0$.

This is the mechanism of mass transfer in the evaporation from a free surface. The vapour moves away from the surface, and the air has no net movement. The diffusion cells used for the determination of diffusivity of gases are also based on this mechanism. Figure 2.3 shows schematically an Arnold Diffusion cell.

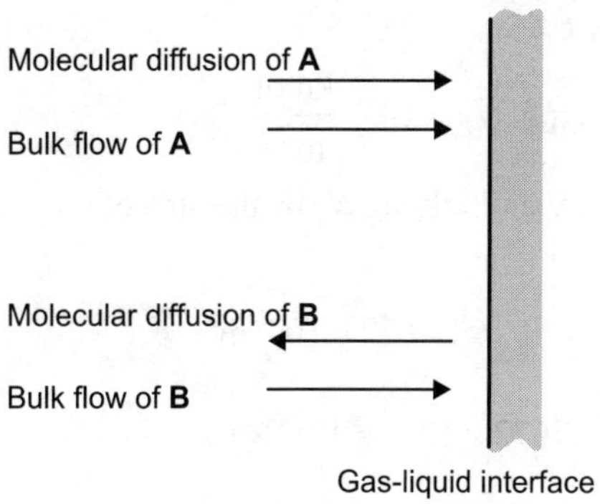

Fig. 2.2: Diffusion of A through stagnant B

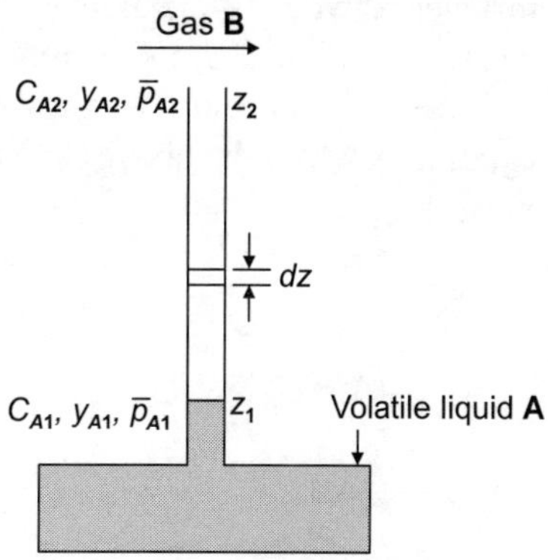

Fig. 2.3: Arnold diffusion cell

The cell consists of a narrow tube which is partially filled with the volatile liquid **A** maintained at a constant temperature T and pressure P. The diffusivity of the vapour **A** in gas **B** is to be measured. The gas **B** is allowed to flow across the open end of the tube. It is assumed that the liquid level in the tube is maintained at a fixed position and the gas **B** has negligible solubility in liquid **A.**

Vapours of **A** diffuses from the liquid surface into the gas phase with flux

$$J_A = -CD_{AB}\frac{dy_A}{dz} \qquad \text{...(2.46)}$$

Substance **B** is also diffusing and it is in the direction opposite to that of **A.** With reference to the molar average velocity of the fluid, the flux of **B** is

$$J_B = -CD_{AB}\frac{dy_B}{dz} \qquad \text{...(2.47)}$$

Since **B** is insoluble in the liquid, the diffusion of **B** towards the interface tends to increase the total concentration at the interface. This provides the necessary driving force for bulk flow of the entire mixture away from the interface. The amount of **B** carried away in this manner in the bulk flow is equal to the moles of **B** diffusing towards the interface as given by Eq. (2.47). Though component **B** is diffusing with reference to the average velocity of the fluid, **B** will appear to be stagnant with reference to a stationary observer. Species **B** is diffusing through the stream as a whole, but the stream itself is moving in the opposite direction, so that relative to the interface, **B** is at rest, i.e., $N_B = 0$ or $N = N_A + N_B = N_A$

Let $N_B = 0$ or $N = N_A$ in Eq. (2.23).

$$N_A = -CD_{AB}\frac{dy_A}{dz} + y_A N_A \qquad \text{...(2.48)}$$

Equation (2.48) clearly indicates that component **A** is transported away from the interface by molecular diffusion as well as by bulk flow. Rearrange Eq. (2.48) into the following forms:

$$N_A(1 - y_A) = -CD_{AB}\frac{dy_A}{dz} \qquad \text{...(2.49)}$$

$$N_A dz = -\frac{CD_{AB}}{1 - y_A}dy_A \qquad \text{...(2.50)}$$

It can be easily shown that under the steady-state condition, since there is neither accumulation nor depletion of material and material is neither generated nor consumed by reaction or by other means anywhere along the diffusion path, the flux N_A is constant throughout the gas phase. The diffusion path is from level z_1 where mole fraction of the diffusing component is y_{A1} and that of non-diffusing component is y_{B1} to level z_2 where mole fraction of **A** is y_{A2} and that of **B** is y_{B2}. It may be noted that the length of diffusion path is $Z = z_2 - z_1$.

Integrating between the above limits, Eq. (2.50) becomes

$$N_A \int_{z_1}^{z_2} dz = -CD_{AB} \int_{y_{A1}}^{y_{A2}} \frac{dy_A}{1-y_A} \qquad \text{...(2.51)}$$

Or

$$N_A = \frac{CD_{AB}}{Z} \ln \frac{1-y_{A2}}{1-y_{A1}} \qquad \text{...(2.52)}$$

Since $y_A + y_B = 1$,

$$N_A = \frac{CD_{AB}}{Z} \ln \frac{y_{B2}}{y_{B1}} \qquad \text{...(2.53)}$$

Equation (2.53) gives the molar flux of component **A** when it is diffusing through stagnant **B**. The flux given by Eq. (2.53) can be expressed in terms of the mole fraction difference driving force as given by Eq. (2.37) for the equimolar counter diffusion. Let us denote the logarithmic mean of y_{B2} and y_{B1} by y_{BM}. Then,

$$y_{BM} = \frac{y_{B2} - y_{B1}}{\ln(y_{B2}/y_{B1})} = \frac{y_{A1} - y_{A2}}{\ln(y_{B2}/y_{B1})} \qquad \text{...(2.54)}$$

In writing the second equality in Eq. (2.54) note that we have used the relation $y_B = 1 - y_A$. Introducing Eq. (2.54) into Eq. (2.53), we get

$$N_A = \frac{CD_{AB}}{Zy_{BM}}(y_{A1} - y_{A2}) \qquad \text{...(2.55)}$$

If the gas phase can be assumed to be an ideal gas, we have

$$C = \frac{P}{RT};\ y_A = \frac{\bar{p}_A}{P};\ y_B = \frac{\bar{p}_B}{P};\ y_{BM} = \frac{\bar{p}_{BM}}{P} \qquad \text{...(2.56)}$$

Substituting Eq. (2.56) into Eqs. (2.53) and (2.55) we get the following relations:

$$N_A = \frac{D_{AB}P}{RTZ} \ln \frac{\bar{p}_{B2}}{\bar{p}_{B1}} \qquad \text{...(2.57)}$$

$$N_A = \frac{D_{AB}P}{RTZ\,\bar{p}_{BM}}(\bar{p}_{A1} - \bar{p}_{A2}) \qquad \text{...(2.58)}$$

In Eq. (2.58), $\bar{p}_{BM}$ is the logarithmic average of the partial pressures of the non-diffusing gas **B** at the two sections.

$$\bar{p}_{BM} = \frac{\bar{p}_{B2} - \bar{p}_{B1}}{\ln(\bar{p}_{B2}/\bar{p}_{B1})} = \frac{(P - \bar{p}_{A2}) - (P - \bar{p}_{A1})}{\ln(\bar{p}_{B2}/\bar{p}_{B1})} = \frac{\bar{p}_{A1} - \bar{p}_{A2}}{\ln(\bar{p}_{B2}/\bar{p}_{B1})} \qquad \text{...(2.59)}$$

Concentration profile: Since steady-state conditions are applicable N_A is constant along the diffusion path and the following relation is applicable.

$$\frac{dN_A}{dz} = 0 \qquad \text{...(2.40)}$$

From Eq. (2.50), we have

$$N_A = -\frac{CD_{AB}}{1-y_A}\frac{dy_A}{dz} \qquad \text{...(2.60)}$$

Substituting Eq. (2.60) into Eq. (2.40), and noting that C and D_{AB} are constant, we get

$$\frac{d}{dz}\left(\frac{1}{1-y_A}\frac{dy_A}{dz}\right)=0 \qquad \ldots(2.61)$$

Equation (2.61) is a second order differential equation, the solution of which is

$$-\ln(1-y_A)=C_1 z+C_2 \qquad \ldots(2.62)$$

where C_1 and C_2 are constants. The constants are evaluated using the boundary conditions

$$y_A = y_{A1} \text{ at } z = z_1$$
$$y_A = y_{A2} \text{ at } z = z_2$$

$$C_1=\frac{1}{z_1-z_2}\ln\frac{1-y_{A2}}{1-y_{A1}} \text{ and } C_2=\frac{1}{z_1-z_2}\ln\frac{(1-y_{A2})^{z_2}}{(1-y_{A1})^{z_1}} \qquad \ldots(2.63)$$

Substituting Eq. (2.63) into Eq. (2.62) and simplifying, we get

$$\frac{1-y_A}{1-y_{A1}}=\left(\frac{1-y_{A2}}{1-y_{A1}}\right)^{\frac{z-z_1}{z_2-z_1}} \qquad \ldots(2.64)$$

Since $y_B = 1 - y_A$, Eq. (2.64) may be written as

$$\frac{y_B}{y_{B1}}=\left(\frac{y_{B2}}{y_{B1}}\right)^{\frac{z-z_1}{z_2-z_1}} \qquad \ldots(2.65)$$

In terms of partial pressures, Eq. (2.65) can be written as

$$\frac{P-\bar{p}_A}{P-\bar{p}_{A1}}=\left(\frac{P-\bar{p}_{A2}}{P-\bar{p}_{A1}}\right)^{\frac{z-z_1}{z_2-z_1}} \qquad \ldots(2.66)$$

Equations (2.65) and (2.66) indicate a logarithmic concentration profile for both components as depicted in Fig. 2.4.

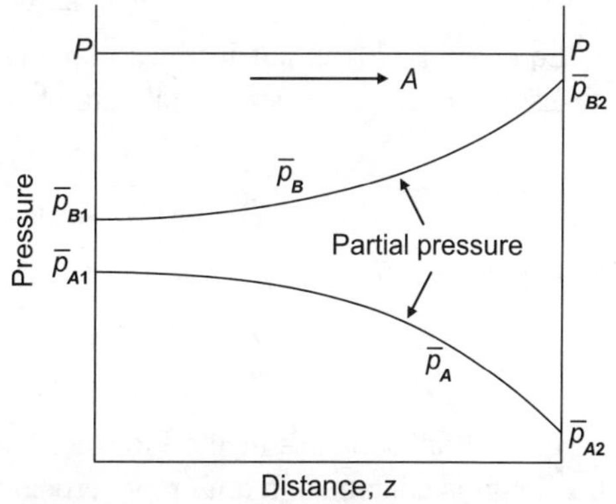

Fig. 2.4: Concentration profile for diffusion of **A** through stagnant **B**

Pseudo-Steady-State Diffusion of A through Stagnant B

We have derived Eq. (2.53) on the assumption that the liquid level is maintained and the length of the diffusion path is constant. If the liquid level is allowed to fall with time, a pseudo-steady-state approximation may be used for the calculation. Initially (i.e. at time $t = t_0$), the level of liquid was at a distance z_1 from the top of the tube. (Level 'p' in Fig. 2.5). In a time interval of t s, the level has fallen by a very small depth due to the evaporation and the subsequent diffusion of the vapours (Level 'q' in the figure).

If the fall in level during the interval $t - t_0$ is only a negligible fraction of the total length of the diffusion path, and $t - t_0$ is a relatively long period of time, the molar flux of component **A** may be evaluated by Eq. (2.53)

$$N_A=\frac{CD_{AB}}{z}\ln\frac{y_{B2}}{y_{B1}} \qquad \ldots(2.53)$$

where $z = z_1 - z_2$, the length of diffusion path at time t.

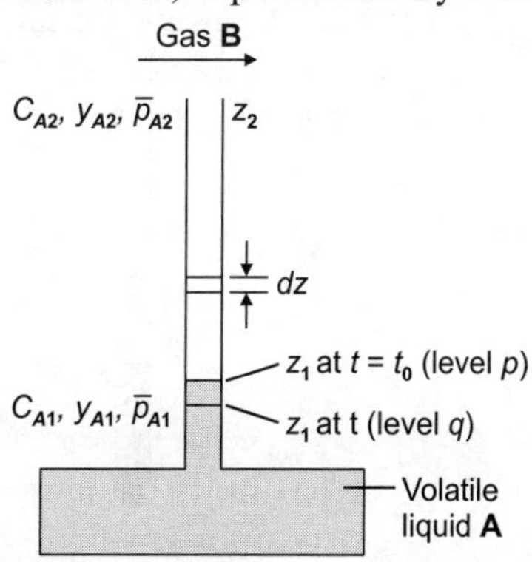

Fig. 2.5: Pseudo-steady-state diffusion

If the liquid level falls by a height dz in time dt seconds, the molar flux can be evaluated from the following equation as well.

$$N_A = \frac{\rho_{A,L}}{M_A}\frac{dz}{dt} \qquad \text{...(2.67)}$$

Here, $\rho_{A,L}$ is the density of the liquid **A** and M_A is its molecular weight. Combining Eqs. (2.53) and (2.67) we get

$$\frac{\rho_{A,L}}{M_A}\frac{dz}{dt} = \frac{CD_{AB}}{z}\ln\frac{y_{B2}}{y_{B1}} \qquad \text{...(2.68)}$$

Equation (2.68) can be rearranged as

$$dt = \frac{\rho_{A,L}}{M_A CD_{AB}\ln(y_{B2}/y_{B1})} zdz \qquad \text{...(2.69)}$$

Eq. (2.69) is integrated from $t = 0$ to $t = t$, and $z = z_0$ to $z = z_t$. The result is

$$t = \frac{\rho_{A,L}}{M_A CD_{AB}\ln(y_{B2}/y_{B1})}\frac{z_t^2 - z_0^2}{2} \qquad \text{...(2.70)}$$

Equation (2.70) is put in the following form which will be useful for the determination of the diffusivity from the experimental data obtained using diffusion cells such as the Arnold cell.

$$D_{AB} = \frac{\rho_{A,L}}{M_A Ct\ln(y_{B2}/y_{B1})}\frac{(z_t^2 - z_0^2)}{2} \qquad \text{...(2.71)}$$

Replacing $\ln(y_{B2}/y_{B1})$ in favour of y_{BM}, Eq. (2.71) becomes

$$D_{AB} = \frac{\rho_{A,L}y_{BM}(z_t^2 - z_0^2)}{2M_A Ct(y_{A1} - y_{A2})} \qquad \text{...(2.72)}$$

In the Winkelmann's method for measuring the diffusivity of the vapours of a volatile liquid in air, the liquid is contained in a narrow vertical tube maintained at constant temperature and an air stream is passed over the top of the tube (Fig. 2.6a). The fall in level of liquid is noted against time. The experimental result is used to evaluate the diffusivity by rearranging Eq. (2.72).

$$(z_t^2 - z_0^2) = \frac{2M_A CtD_{AB}(y_{A1} - y_{A2})}{\rho_{A,L}y_{BM}} \qquad \text{...(2.73)}$$

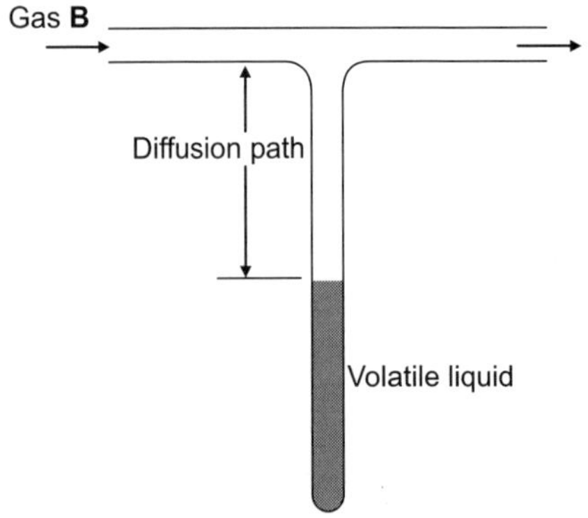

Fig. 2.6: (a) Winkelmann's method for determination of diffusivity

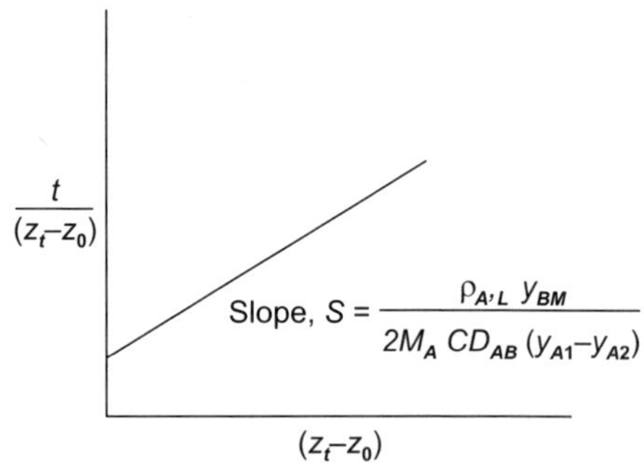

Fig. 2.6: (b) Plot of $\frac{t}{(z_t - z_0)}$ versus $(z_t - z_0)$ for Winkelmann's method

$$(z_t - z_0)(z_t - z_0 + 2z_0) = \frac{2M_A C\, tD_{AB}(y_{A1} - y_{A2})}{\rho_{A,L} y_{BM}} \qquad \ldots(2.74)$$

Or

$$\frac{t}{(z_t - z_0)} = \frac{\rho_{A,L} y_{BM}}{2M_A CD_{AB}(y_{A1} - y_{A2})}(z_t - z_0) + \frac{\rho_{A,L} y_{BM} z_0}{M_A CD_{AB}(y_{A1} - y_{A2})} \qquad \ldots(2.75)$$

A graph is plotted with $\frac{t}{(z_t - z_0)}$ on the *y*-axis and $(z_t - z_0)$ on the *x*-axis as in Fig. 2.6b.

The slope of the straight line is *S*, where

$$S = \frac{\rho_{A,L} y_{BM}}{2M_A CD_{AB}(y_{A1} - y_{A2})} \qquad \ldots(2.76)$$

from which the diffusivity is calculated.

$$D_{AB} = \frac{\rho_{A,L} y_{BM}}{2M_A CS(y_{A1} - y_{A2})} \qquad \ldots(2.77)$$

Example 2.3: Ethanol (**A**) is diffusing through air (**B**) under steady-state condition, with air non-diffusing, at temperature 313 K and pressure 101.3 kPa. The partial pressure of ethanol at two planes 5 mm apart are 8 kPa and 4 kPa. The diffusivity of ethanol in air is 1.45×10^{-5} m²/s. Calculate:

(a) The rate of diffusion of ethanol per unit area of the planes

(b) The mole fraction of ethanol as a function of the distance along the direction of diffusion. Use the result to evaluate the mole fraction at the centre of the diffusion path.

***Solution*:** (a) Equations (2.53), (2.55), (2.57) or (2.58) may be used to evaluate the rate of diffusion per unit area. Consider Eq. (2.57):

$$N_A = \frac{D_{AB}P}{RTZ} \ln \frac{\bar{p}_{B2}}{\bar{p}_{B1}} \qquad \ldots(2.57)$$

Here, $D_{AB} = 1.45 \times 10^{-5}$ m²/s; $P = 101.3$ kPa; $T = 313$ K; $Z = 5 \times 10^{-3}$ m

The partial pressure of ethanol at point 1 is $\bar{p}_{A1} = 8$ kPa

The partial pressure of ethanol at point 2 is $\bar{p}_{A2} = 4$ kPa

Therefore, the partial pressure of air at points 1 and 2 are:

$$\bar{p}_{B1} = P - \bar{p}_{A1} = 101.3 - 8 = 93.3 \text{ kPa}$$

$$\bar{p}_{B2} = P - \bar{p}_{A2} = 101.3 - 4 = 97.3 \text{ kPa}$$

Substitute the above values in Eq. (2.57):

$$N_A = \frac{1.45 \times 10^{-5} \times 101.3}{8.314 \times 313 \times 5 \times 10^{-3}} \ln \frac{97.3}{93.3} = 4.739 \times 10^{-6} \frac{\text{kmol}}{\text{m}^2\text{s}} \qquad \textbf{(Answer)}$$

(b) Equation (2.64) gives the concentration profile along the diffusion path.

$$\frac{1 - y_A}{1 - y_{A1}} = \left(\frac{1 - y_{A2}}{1 - y_{A1}}\right)^{\frac{z - z_1}{z_2 - z_1}} \qquad \ldots(2.64)$$

Here,

$$y_{A1} = \frac{\bar{p}_{A1}}{P} = \frac{8}{101.3} = 0.07897 \text{ at } z = z_1 = 0$$

$$y_{A2} = \frac{\bar{p}_{A2}}{P} = \frac{4}{101.3} = 0.03949 \text{ at } z = z_2 = 5$$

The mole fraction of ethanol y_A at a distance z from point 2 is calculated using Eq. (2.64):

$$\frac{1 - y_A}{1 - 0.07897} = \left(\frac{1 - 0.03949}{1 - 0.07897}\right)^{\frac{z}{5}} = \left(\frac{0.9605}{0.9210}\right)^{z/5}$$

$$\frac{1 - y_A}{0.9210} = (1.0429)^{z/5}$$

which can be rearranged as

$$y_A = 1 - 0.9210 \times (1.0429)^{z/5}$$ **(Answer)**

Substituting $z = 2.5$ mm, we get the mole fraction of ethanol at the centre of the diffusion path.

$$y_A = 0.0595$$ **(Answer)**

Example 2.4: Water at 293 K and 101.3 kPa is contained in a narrow vertical tube. The level of liquid is maintained 150 mm below the open end, across which air with a humidity of 0.002 kg water per kg dry air is passed at 293 K and 101.3 kPa. If the diffusivity of water vapour in air is 2.5×10^{-5} m²/s, and the vapour pressure of water is 2.34 kPa, calculate the rate of evaporation of water. What will be the rate of evaporation if dry air is passed over the tube?

***Solution*:** Let us use Eq. (2.53) for calculating the rate of diffusion of water through stagnant air layer.

$$N_A = \frac{CD_{AB}}{Z} \ln \frac{y_{B2}}{y_{B1}} \quad \ldots(2.53)$$

$$C = \frac{P}{RT} = \frac{101.3}{8.314 \times 293} = 0.041585 \frac{\text{kmol}}{\text{m}^3}$$

$$D_{AB} = 2.5 \times 10^{-5} \text{ m}^2/\text{s}$$

$$Z = 150 \times 10^{-3} \text{ m}$$

The gas layer in contact with the liquid is assumed to be saturated with water vapour so that, the partial pressure of water vapour is its vapour pressure. $\bar{p}_{A1}$ = vapour pressure at 293 K = 2.34 kPa. Therefore,

$$y_{A1} = \frac{\bar{p}_{A1}}{P} = \frac{2.34}{101.3} = 0.0231$$

Since air contains 0.002 kg water vapour per kg dry air, the mole fraction of water vapour in the air is

$$y_{A2} = \frac{0.002/18}{0.002/18 + 1/29} = 0.0032$$

Therefore, $y_{B1} = 0.9769$ and $y_{B2} = 0.9968$

Substituting values in Eq. (2.53) gives

$$N_A = \frac{4.1585 \times 10^{-2} \times 2.5 \times 10^{-5}}{150 \times 10^{-3}} \ln \frac{0.9968}{0.9769}$$

$$N_A = 1.3978 \times 10^{-7} \text{ kmol/m}^2 \text{ s}$$ **(Answer)**

If the air used is dry, $y_{A2} = 0$, $y_{B2} = 1$.

$$N_A = \frac{4.1585 \times 10^{-2} \times 2.5 \times 10^{-5}}{150 \times 10^{-3}} \ln \frac{1}{0.9769} = 1.6198 \times 10^{-7} \text{ kmol/m}^2 \text{ s}$$

This means there would be an increase of about 16% in the rate of evaporation if the air used were dry. **(Answer)**

Example 2.5: The diffusivity of benzene in air is measured by Stefan's method. A vertical glass tube is filled with benzene to a depth of 20 mm from the top open end. It took 190 hours for the level to fall to 80 mm from the top. The temperature and pressure are maintained at 298 K and 100 kPa. Density of liquid benzene is 875 kg/m^3 and the vapour pressure of benzene is 12.6 kPa at 298 K. Calculate the diffusivity of benzene in air at 298 K.

***Solution*:** A pseudo-steady-state approximation for diffusion of **A** through stagnant gas **B** gives

$$D_{AB} = \frac{\rho_{A,L}(z_t^2 - z_0^2)}{2M_A Ct \ln(y_{B2}/y_{B1})} \quad \text{...(2.71)}$$

which may be used for calculating diffusivity.

Here, ρ_{AL} = Density of liquid benzene = 875 kg/m^3

M_A = Molecular weight of benzene = 78 (kg/kmol)

$$C = \frac{P}{RT} = \frac{100}{8.314 \times 298} = 0.04036 \text{ kmol/m}^3$$

$$t = 190 \times 3600 \text{ s}$$

$$z_o = 20 \text{ mm} = 0.02 \text{ m}; \quad z_t = 80 \text{ mm} = 0.08 \text{ m}$$

The gas layer at the surface of the liquid is saturated with benzene vapour so that the partial pressure of benzene there is equal to its vapour pressure. Therefore, $\bar{p}_{A1}$ = 12.6 kPa.

$$y_{A1} = \frac{12.6}{100} = 0.126$$

and the air is assumed free of benzene so that $y_{A2} = 0$ or $y_{B2} = 1$.

Use these values in Eq. (2.71):

$$D_{AB} = \frac{\rho_{A,L}(z_1^2 - z_0^2)}{2M_A C\, t \ln(y_{B2}/y_{B1})} = \frac{875(0.08^2 - 0.02^2)}{2 \times 78 \times 0.04036 \times 190 \times 3600 \times \ln(1/0.874)}$$

$$= 9.052 \times 10^{-6} \text{ m}^2/\text{s}$$ **(Answer)**

Example 2.6: Diffusivity of chloroform ($CHCl_3$) in air at 298 K and 100 kPa is measured by the Winkelmann's method. The fall in level is reported against time as given below:

Time, min	0	24	56	190	550	1065	1820	2775	3810	5040	6480
Level, mm	0	2	4	10	20	30	40	50	60	70	80

The concentration of chloroform at the surface of the tube may be assumed zero. The density of chloroform liquid is 1485 kg/m^3 and the vapour pressure of chloroform at 298 K is 26 kPa. Find the diffusivity of chloroform in air.

***Solution*:** We have shown that the pseudo-steady-state diffusion equation can be rearranged to the following form:

$$\frac{t}{(z_t - z_0)} = \frac{\rho_{A,L} y_{BM}}{2M_A CD_{AB}(y_{A1} - y_{A2})}(z_t - z_0) + \frac{\rho_{A,L} y_{BM} z_0}{M_A CD_{AB}(y_{A1} - y_{A2})} \quad \text{...(2.75)}$$

Equation (2.75) indicates that when a graph is plotted with $\frac{t}{(z_t - z_0)}$ on the y-axis and $(z_t - z_0)$ on the x-axis, a straight line will be obtained with slope $S = \frac{\rho_{A,L} y_{BM}}{2 M_A C D_{AB} (y_{A1} - y_{A2})}$

From the experimental data $\frac{t}{(z_t - z_0)}$ and $(z_t - z_0)$ are calculated and tabulated as given below:

$(z_t - z_0)$, mm	2	4	10	20	30	40	50	60	70	80
$\frac{t}{(z_t - z_0)}$, min/mm	12	14	19	27.5	35.5	45.5	55.5	63.5	72	81

A graph is plotted with $\frac{t}{(z_t - z_0)}$ on the y-axis and $(z_t - z_0)$ on the x-axis.

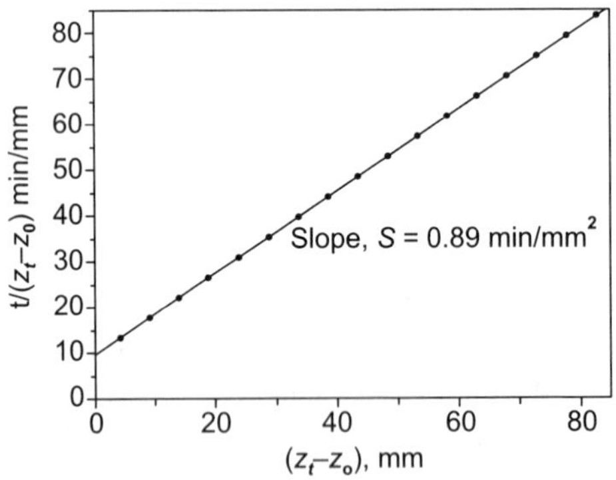

Fig. 2.7: Determination of slope in Winkelmann's method (Example 2.6)

The slope S is measured. $S = 0.89$ min/(mm)^2 (from the graph).

$$S = \frac{\rho_{A,L} y_{BM}}{2 M_A C D_{AB} (y_{A1} - y_{A2})} = 0.89 \times \frac{60}{(10^{-3})^2} = 5.34 \times 10^7 \frac{\text{s}}{\text{m}^2}$$

$$C = \frac{P}{RT} = \frac{100}{8.314 \times 298} = 0.0404 \frac{\text{kmol}}{\text{m}^3}$$

$$y_{A1} = \frac{\bar{p}_{A1}}{P} = \frac{26}{100} = 0.26 \quad y_{B1} = 1 - 0.26 = 0.74$$

$$y_{A2} = 0, \; y_{B2} = 1.0$$

$$y_{BM} = \frac{y_{B2} - y_{B1}}{\ln(y_{B2}/y_{B1})} = \frac{1 - 0.74}{\ln(1/0.74)} = 0.8635$$

Now slope S can be written as

$$\frac{1485 \times 0.8635}{2 \times 119.35 \times 0.0404 \times D_{AB}(0.26 - 0)} = 5.34 \times 10^7 \frac{\text{s}}{\text{m}^2}$$

On solving this we get, $D_{AB} = 9.58 \times 10^{-6}$ m^2/s **(Answer)**

Example 2.7: An open beaker is filled with ethyl alcohol at 293 K to within 6 mm from the top. A gentle breeze of dry air at 293 K and 100 kPa is blown across the mouth of the beaker. The vapour pressure of ethanol at 293 K is 5.73 kPa, and the diffusivity of ethanol in air is 1.33×10^{-5} m^2/s. Determine:

(a) Initial rate of evaporation in kmol/m^2 s.
(b) Initial concentration profile and mole fraction of ethanol at a point 3 mm above the liquid surface.
(c) What percent of the mass transfer flux is due to molecular diffusion?
(d) The diffusion velocities and the species velocities of ethanol and air relative to stationary coordinates at a point 3 mm above the liquid surface.
(e) The time in hours for the level in the beaker to fall by 10 mm from the initial level, if the density of the liquid is 789 kg/m^3.

***Solution*:** (a) The rate of evaporation for diffusion of one component through the other non-diffusing component may be calculated by Eq. (2.55)

$$N_A = \frac{CD_{AB}}{Zy_{BM}}(y_{A1} - y_{A2}) \quad \text{...(2.55)}$$

$$C = \frac{P}{RT} = \frac{100}{8.314 \times 293} = 0.04105 \frac{\text{kmol}}{\text{m}^3}$$

$$D_{AB} = 1.33 \times 10^{-5} \text{ m}^2/\text{s}$$

$Z = 6$ mm $= 6 \times 10^{-3}$ m (initial length of diffusion path)

$$y_{A1} = \frac{\bar{p}_{A1}}{P} = \frac{5.73}{100} = 0.0573 \quad y_{B1} = 1 - 0.0573 = 0.9427$$

$$y_{A2} = 0, \; y_{B2} = 1.0$$

$$y_{BM} = \frac{y_{B2} - y_{B1}}{\ln(y_{B2}/y_{B1})} = \frac{1 - 0.9427}{\ln(1/0.9427)} = 0.9711$$

Now Eq. (2.55) gives

$$N_A = \frac{0.04105 \times 1.33 \times 10^{-5} \times (0.0573 - 0)}{6 \times 10^{-3} \times 0.9711} = 5.3691 \times 10^{-6} \frac{\text{kmol}}{\text{m}^2 s} \quad \textbf{(Answer)}$$

(b) The concentration profile is

$$\frac{1 - y_A}{1 - y_{A1}} = \left(\frac{1 - y_{A2}}{1 - y_{A1}}\right)^{z - z_1/z_2 - z_1} \quad \text{...(2.64)}$$

Knowing that $y_A = y_{A1}$ at $z = z_1 = 0$ and $y_A = y_{A2}$ at $z = z_2 = 6$

$$\frac{1 - y_A}{1 - 0.0573} = \left(\frac{1}{1 - 0.0573}\right)^{z/6} = \left(\frac{1}{0.9427}\right)^{z/6}$$

That is,
$$\ln\frac{1 - y_A}{0.9427} = \frac{z}{6}\ln\left(\frac{1}{0.9427}\right) = 9.8345 \times 10^{-3} z$$

Rearranging the above equation, we get the concentration profile as

$$y_A = 1 - 0.9427\exp(9.8345 \times 10^{-3} z) \quad \textbf{(Answer)}$$

To get the mole fraction at a height 3 mm above the liquid surface, let $z = 3$ mm in the above equation. The result is

$$y_A = 0.0291 \quad \textbf{(Answer)}$$

(c) The molar flux of ethyl alcohol (**A**) can be written as

$$N_A = J_A + y_A(N_A + N_B)$$

where the first term on the right side is the contribution due to molecular diffusion and the second term that due to bulk flow. Since $N_B = 0$,

$$N_A = J_A + y_A N_A \quad \text{or} \quad J_A = (1 - y_A)N_A$$

Thus at a height 3 m above the liquid surface

$$J_A = (1 - 0.0291)N_A = 5.2130 \times 10^{-6}\ \text{kmol/m}^2\text{s}.$$

Therefore, the flux due to molecular diffusion alone is $\dfrac{5.2130}{5.3694} \times 100 = 97.1\%$ of the flux in the presence of bulk flow. **(Answer)**

(d) The flux, velocity and concentration are related as

$$N_A = u_A C_A$$

$$J_A = U'_A C_A$$

where u_A and U'_A are the species velocity and diffusion velocity, respectively for component **A**, and C_A is the concentration of **A** at the given point. It is already shown that at a height 3 mm from the liquid surface, $y_A = 0.0291$ so that $C_A = y_A C = 0.0291 \times 0.04105 = 1.19456 \times 10^{-3}$ kmol/m³. We can calculate N_A and J_A at $z = 3$ mm by

$$N_A = \frac{CD_{AB}}{Zy_{BM}}(y_{A1} - y_{A2})$$

$$J_A = \frac{CD_{AB}}{Z}(y_{A1} - y_{A2})$$

Here $y_{A1} = 0.0291$, $y_{B1} = 0.9709$, $y_{A2} = 0$, $y_{B2} = 1.0$, $y_{BM} = 0.9854$

Therefore,

$$N_A = 5.369 \times 10^{-6}\ \text{kmol/m}^2\text{s},\ J_A = 5.213 \times 10^{-6}\ \text{kmol/ m}^2\text{s}$$

$$u_A = \frac{N_A}{C_A} = \frac{5.369 \times 10^{-6}}{1.19456 \times 10^{-3}} = 4.5 \times 10^{-3}\ \frac{\text{m}}{\text{s}}$$ **(Answer)**

$$U'_A = \frac{J_A}{C_A} = \frac{5.213 \times 10^{-6}}{1.19456 \times 10^{-3}} = 4.36 \times 10^{-3}\ \frac{\text{m}}{\text{s}}$$ **(Answer)**

For component **B**, $N_B = 0$; Therefore, $u_B = 0$ **(Answer)**

$$J_B = -J_A = -5.213 \times 10^{-6}\ \frac{\text{kmol}}{\text{m}^2\text{s}}$$

$$U'_B = \frac{J_B}{C_B} = \frac{-5.213 \times 10^{-6}}{0.9709 \times 0.04105} = -1.308 \times 10^{-4}\ \frac{\text{m}}{\text{s}}$$ **(Answer)**

(e) Equation (2.70) gives the time t required for the level to fall from the initial value z_0 to the final value z_t as

$$t = \frac{\rho_{A,L}(z_t^2 - z_0^2)}{2M_A CD_{AB}\ln(y_{B2}/y_{B1})}$$

$$t = \frac{789(0.016^2 - 0.006^2)}{2 \times 46 \times 0.04105 \times 1.33 \times 10^{-5}\ln(1/0.9427)} = 5.85641 \times 10^4\ \text{s} = 16.27\ \text{h}$$

It takes 16.27 h for the liquid level to drop to 16 mm from the initial level 6 mm. **(Answer)**

2.4 A GENERAL EQUATION FOR STEADY-STATE DIFFUSION

Starting from the Fick's law of diffusion [Eq. (2.23)], a general equation can be developed for molar flux of a component **A** under steady-state diffusion.

$$N_A = -D_{AB}\frac{dC_A}{dz} + y_A(N_A + N_B) \qquad ...(2.23)$$

$$= -D_{AB}\frac{dC_A}{dz} + \frac{C_A}{C}(N_A + N_B) \qquad ...(2.78)$$

Equation (2.78) is rearranged to

$$CN_A - C_A(N_A + N_B) = -CD_{AB}\frac{dC_A}{dz} \qquad ...(2.79)$$

$$\frac{dz}{CD_{AB}} = \frac{-dC_A}{CN_A - C_A(N_A + N_B)} \qquad ...(2.80)$$

Noting C to be constant, the equation is integrated between limits z_1 to z_2 and C_{A1} to C_{A2}.

$$\frac{Z}{CD_{AB}} = \frac{1}{N_A + N_B}\ln\frac{CN_A - C_{A2}(N_A + N_B)}{CN_A - C_{A1}(N_A + N_B)} \qquad ...(2.81)$$

Here $Z = z_2 - z_1$. Multiply both sides by N_A and rearrange the resulting equation. The result is

$$N_A = \frac{CD_{AB}}{Z}\frac{N_A}{N_A + N_B}\ln\frac{CN_A - C_{A2}(N_A + N_B)}{CN_A - C_{A1}(N_A + N_B)} \qquad ...(2.82)$$

Or

$$N_A = \frac{CD_{AB}}{Z}\frac{N_A}{N_A + N_B}\ln\frac{\dfrac{N_A}{(N_A + N_B)} - \dfrac{C_{A2}}{C}}{\dfrac{N_A}{(N_A + N_B)} - \dfrac{C_{A1}}{C}} \qquad ...(2.83)$$

Noting that $y_A = \frac{C_A}{C}$, Eq. (2.83) becomes

$$N_A = \frac{CD_{AB}}{Z}\frac{N_A}{N_A + N_B}\ln\frac{\dfrac{N_A}{(N_A + N_B)} - y_{A2}}{\dfrac{N_A}{(N_A + N_B)} - y_{A1}} \qquad ...(2.84)$$

Equation (2.84) can be used for calculating molar flux of diffusion of **A** under steady-state condition for all cases if the relationship between N_A and N_B are known. For equimolar counter diffusion, this relation is $N_A = -N_B$ and for diffusion of **A** though stagnant **B**, $N_B = 0$. Now consider the rate of diffusion of oxygen to the surface of a spherical particle of coal where the following reaction occurs.

$$2C + O_2 = 2CO$$

For each mole of oxygen diffusing to the coal surface, two moles of carbon monoxide will diffuse back so that $N_B = -2\,N_A$. Eq. (2.84) simplifies as

$$N_A = -\frac{CD_{AB}}{Z}\ln\frac{1 + y_{A2}}{1 + y_{A1}} \qquad ...(2.85)$$

2.5 DIFFUSION IN MULTICOMPONENT MIXTURES

Since the theoretical analysis of diffusion in multicomponent mixtures leads to very complex equations, the rate of diffusion in such systems is conveniently estimated by means of equations analogous to those derived for the diffusion in binary systems, replacing the binary diffusivity by an 'effective diffusivity'. For example, for diffusion of **A** through a multicomponent system made up of components **A**, **B**, **C**, **D**, etc. where all components other than **A** are stagnant, we can estimate the molar flux of **A** by

$$N_A = \frac{CD_{A,m}}{Z} \ln \frac{1 - y_{A2}}{1 - y_{A1}} \qquad \text{...(2.86)}$$

which is analogous to Eq. (2.52) for diffusion of **A** through stagnant **B**. $D_{A,m}$ in Eq. (2.86) is the effective diffusivity of **A** in the multicomponent mixture. Whereas the binary diffusion coefficients in ideal gas systems are properties of the gas pair, the effective diffusion coefficient $D_{A,m}$ of species **A** in a multicomponent mixture depends not only on the nature and concentrations, but also on the fluxes of the other components present. $D_{A,m}$ can be related to fluxes of the different species present in the mixture and the binary diffusivities as follows:

$$D_{A,m} = \frac{N_A - y_A \sum_{i=A}^{n} N_i}{\sum_{i=A}^{n} \frac{y_i N_A - y_A N_i}{D_{Ai}}} \qquad \text{...(2.87)}$$

where D_{Ai} are the binary diffusivities. The effective diffusivity may vary from one end of the diffusion path to the other, but a linear variation with distance can be assumed for practical calculations.

When **A** is diffusing in a mixture where all other components are non-diffusing, we have $N_i = 0$ for $i \neq$ **A**. Equation (2.87) can be simplified for this case as

$$D_{A,m} = \frac{1 - y_A}{\sum_{i=B}^{n} \frac{y_i}{D_{Ai}}} \qquad \text{...(2.88)}$$

where y_i is the mole fraction of species i. The mole fraction of i on **A**-free basis is $y_i' = (y_i)/(1 - y_i)$. Equation (2.88) can be rewritten in terms of y_i'.

$$D_{A,m} = \frac{1}{\sum_{i=B}^{n} \frac{y_i'}{D_{Ai}}} \qquad \text{...(2.89)}$$

Example 2.8: Oxygen is diffusing through a non-diffusing gas mixture of methane and hydrogen that are present in the mole ratio 2:1 at 100 kPa and 273 K. At this condition, the diffusivity of oxygen in methane is 1.86×10^{-5} m²/s and diffusivity of oxygen in hydrogen is 6.99×10^{-5} m²/s. Calculate the rate of diffusion of oxygen per unit area across two planes 2 mm apart, if the partial pressures of oxygen at these planes are 20 kPa and 10 kPa, respectively.

Solution: The effective diffusivity for oxygen in the gas mixture is evaluated by Eq. (2.89).

$$D_{A,m} = \frac{1}{\sum_{i=B}^{n} \frac{y_i'}{D_{Ai}}} \qquad \text{...(2.89)}$$

Equation (2.89) can be written as

$$D_{A,m} = \frac{1}{\frac{y_B'}{D_{AB}} + \frac{y_C'}{D_{AC}}}$$

Here, oxygen (**A**) is diffusing through a mixture of methane (**B**) and hydrogen (**C**). y'_B is the mole fraction of methane on oxygen-free basis = 2/3 and y'_C is the mole fraction of hydrogen on oxygen-free basis = 1/3. $D_{AB} = 1.86 \times 10^{-5}$ m²/s and $D_{AC} = 6.99 \times 10^{-5}$ m²/s.

$$D_{A,m} = \frac{1}{\dfrac{2/3}{1.86\times10^{-5}} + \dfrac{1/3}{6.99\times10^{-5}}} = 2.4624\times10^{-5}\ \frac{\text{m}^2}{\text{s}}$$

$$C = \frac{P}{RT} = \frac{100}{8.314\times273} = 0.04406\ \frac{\text{kmol}}{\text{m}^3}$$

$$Z = 2\ \text{mm} = 2\times10^{-3}\ \text{m}$$

$$y_{A1} = 0.2,\ (1 - y_{A1}) = 0.8$$

$$y_{A2} = 0.1,\ (1 - y_{A2}) = 0.9$$

These values are substituted in Eq. (2.86):

$$N_A = \frac{CD_{A,m}}{Z}\ln\frac{1-y_{A2}}{1-y_{A1}} = \frac{0.04406\times2.4624\times10^{-5}}{2\times10^{-3}}\ln\frac{0.9}{0.8} = 6.3893\times10^{-5}\ \frac{\text{kmol}}{\text{m}^2\text{s}}$$ **(Answer)**

2.6 MOLECULAR DIFFUSION IN LIQUIDS

Molecular diffusion in a liquid is considerably slower than diffusion in gases. Molecules in a liquid are so closely packed that the molecules of the diffusing solute **A** will collide with molecules of **B** more frequently than in gases and hence diffusion in liquids becomes slower than that in gases. The order of magnitudes of the liquid diffusivities are smaller by 10^{-5} times than the diffusivities of gases. However, the diffusion rates in the liquids are only about 10^{-2} times smaller than that in the gases, because the diffusion rate is dependent on the product of the molar density and diffusivity for a given concentration gradient and the molar density of a liquid is 10^3 times that of a gas. Because of the closer spacing of molecules the attractive forces between molecules also have profound influence in liquid phase diffusion.

Since the kinetic theory of liquids is not yet fully developed, diffusion rates in liquids are estimated using equations analogous to those used for gases. In the case of binary gases, the assumption of constant concentration and diffusivity introduce little error; it is not so for liquids. The diffusivity as well as densities in liquids vary appreciably with concentration. Since this relationship is not accurately known, average values are used for concentration and diffusivity.

For equimolar counter diffusion in liquids, we can modify Eq. (2.37) as

$$N_A = \frac{D_{AB}}{Z}(C_{A1} - C_{A2}) = \frac{C_{av}D_{AB}}{Z}(x_{A1} - x_{A2}) \qquad ...(2.90)$$

The symbol x is used to represent mole fraction in liquids. C_{av} represents the average of the concentrations at two planes at distance Z apart.

$$C_{av} = \frac{(\rho_1/M_1) + (\rho_2/M_2)}{2} \qquad ...(2.91)$$

In Eq. (2.91) ρ_1 and ρ_2 are the densities at planes 1 and 2, respectively, where the average molecular weights are M_1 and M_2, respectively. For diffusion of **A** through stagnant **B** in the liquid phase, we can write analogous to Eq. (2.55),

$$N_A = \frac{C_{av}D_{AB}}{Zx_{BM}}(x_{A1} - x_{A2}) \qquad ...(2.92)$$

where
$$x_{BM} = \frac{(x_{B2} - x_{B1})}{\ln(x_{B2}/x_{B1})} \quad \ldots(2.93)$$

For dilute solutions, x_{BM} is close to 1 and C is almost constant. Equation (2.92) becomes

$$N_A = \frac{CD_{AB}}{Z}(x_{A1} - x_{A2}) \quad \ldots(2.94)$$

Example 2.9: Calculate the rate of diffusion of sodium chloride at 298 K through a stationary film of water 1 mm thick where the concentrations are 20 (weight)% and 8 (weight)%, respectively on either side of the film. The diffusivity of sodium chloride in water is 1.35×10^{-9} m²/s. Densities of 20% and 8% NaCl solutions are 1145.3 kg/m³ and 1054.1 kg/m³, respectively.

***Solution*:** Equation (2.92) can be used for calculating rate of diffusion of sodium chloride (**A**) across the film of non-diffusing water (**B**).

$$N_A = \frac{C_{av} D_{AB}}{Z x_{B,M}}(x_{A1} - x_{A2}) \quad \ldots(2.92)$$

The average concentration of the solution is evaluated as the average of concentrations at point 1 and point 2.

$$C_{av} = \frac{(\rho_1/M_1) + (\rho_2/M_2)}{2}$$

ρ_1 is the density of the solution at point 1 where the average molecular weight is M_1 and ρ_2 is density of the solution at point 2 where average molecular weight of the solution is M_2.

$$\rho_1 = 1145.3\frac{\text{kg}}{\text{m}^3} \text{ and } \rho_2 = 1054.1\frac{\text{kg}}{\text{m}^3}$$

M_1 is the average molecular weight of the solution of composition 20% NaCl (MW = 58.45) and 80% water (MW = 18).

$$M_1 = \frac{100}{\dfrac{20}{58.45} + \dfrac{80}{18}} = 20.89\frac{\text{kg}}{\text{kmol}}$$

M_2 is the average molecular weight of the solution containing 8% NaCl and 92% water.

$$M_2 = \frac{100}{\dfrac{8}{58.45} + \dfrac{92}{18}} = 19.06\frac{\text{kg}}{\text{kmol}}$$

Now,
$$C_{av} = \frac{(\rho_1/M_1) + (\rho_2/M_2)}{2} = \frac{(1145.3/20.89) + (1054.1/19.06)}{2} = 55.065\frac{\text{kmol}}{\text{m}^3}$$

x_{A1} and x_{A2} are the mole fractions of NaCl at point 1 and point 2, respectively.

$$x_{A1} = \frac{20/58.45}{(20/58.45 + 80/18)} = 0.0715$$

$$x_{A2} = \frac{8/58.45}{(8/58.45 + 92/18)} = 0.0261$$

$$x_{B1} = (1 - 0.0715) = 0.9285$$

$$x_{B2} = (1 - 0.0261) = 0.9739$$

$$x_{BM} = \frac{(0.9739 - 0.9285)}{\ln(0.9739/0.9285)} = 0.951$$

Equation (2.92) now gives

$$N_A = \frac{55.065 \times 1.35 \times 10^{-9}}{1 \times 10^{-3} \times 0.951}(0.0715 - 0.0261) = 3.5488 \times 10^{-6} \frac{\text{kmol}}{\text{m}^2\text{s}} \qquad \textbf{(Answer)}$$

2.7 PREDICTION OF BINARY DIFFUSIVITIES

Diffusivity of Gases

Semi-theoretical correlations were derived for predicting diffusivity of gases using kinetic theory of gases. Utilizing the Lennard-Jones potential to account for the forces of attraction and repulsion between molecules, the diffusion coefficient of non polar, non reacting binary gases is given by the Hirschfelder-Bird-Spotz equation (Bird et al, 1960):

$$D_{AB} = \frac{2.6628 \times 10^{-4} T^{3/2}}{P M_{AB}^{1/2} \sigma_{AB}^2 \Omega_D} \qquad \ldots(2.95)$$

where D_{AB} is in m^2/s, T is the absolute temperature in K, P is the absolute pressure in N/m^2, σ_{AB} is the Lennard-Jones parameter representing molecular separation at collision in nm, and Ω_D is a dimensionless function of temperature T and energy of molecular interaction ε_{AB} between **A** and **B**. M_{AB} is evaluated from M_A and M_B which are the the molecular weights of **A** and **B,** respectively as

$$M_{AB} = 2\left(\frac{1}{M_A} + \frac{1}{M_B}\right)^{-1} \qquad \ldots(2.96)$$

The Lennard-Jones parameters for the binary pair, σ_{AB} and ε_{AB} can be estimated by combining the corresponding values of the pure species.

$$\sigma_{AB} = \frac{1}{2}(\sigma_A + \sigma_B) \qquad \ldots(2.97)$$

$$\varepsilon_{AB} = (\varepsilon_A \times \varepsilon_B)^{1/2} \qquad \ldots(2.98)$$

If experimental data are not available, σ and ε values can be estimated empirically using the following relations

$$\sigma = 1.18 V^{1/3} \qquad \ldots(2.99)$$

$$\frac{\varepsilon}{\kappa} = 1.21 T_b \qquad \ldots(2.100)$$

where V is the molar volume in m^3/kmol at T_b, the normal boiling point in K and σ in nm. The Lennard-Jones potential for a few substances are listed in Table 2.1. Ω_D values are listed as function of $\kappa T/\varepsilon_{AB}$ where κ is Boltzman constant, in Table 2.2. For rigid spheres, Ω_D is unity at all temperatures.

Table 2.1: Lennard-Jones potential*

Material	σ, nm	ε/κ, K	Material	σ, nm	ε/κ, K
Acetylene, C_2H_2	0.4221	185	Hydrogen chloride, HCl	0.3305	360
Air	0.3617	97	Hydrogen iodide, HI	0.4123	324
Argon, Ar	0.3418	124	Iodine, I_2	0.4982	550
Benzene, C_6H_6	0.5270	440	Krypton, Kr	0.3498	225
Bromine, Br_2	0.4268	520	Methane, CH_4	0.3822	137
i-Butane, *i*-C_4H_{10}	0.5341	313	Methanol, CH_3OH	0.3585	507
n-Butane, *n*-C_4H_{10}	0.4997	410	Methyl chloride, CH_3Cl	0.3375	855
Carbon dioxide, CO_2	0.3996	190	Methylene chloride, CH_2Cl_2	0.4759	406

(Contd.)

Ttable 2.1 contd.

Material	σ, nm	ε/κ, K	Material	σ, nm	ε/κ, K
Carbon disulfide, CS_2	0.4438	488	Mercuric iodide, HgI_2	0.5625	691
Carbon monoxide, CO	0.3590	110	Mercury, Hg	0.2898	851
Carbon tetrachloride, CCl_4	0.5881	327	Neon, Ne	0.2789	35.7
Carbonyl sulphide, COS	0.413	335	Nitric oxide, NO	0.3470	119
Chlorine, Cl_2	0.4115	357	Nitrogen, N_2	0.3681	91.5
Chloroform, $CHCl_3$	0.5430	327	Nitrous oxide, N_2O	0.3879	220
Cyclohexane, C_6H_{12}	0.6093	324	*n*-Nonane, C_9H_{20}	0.8448	240
Ethane, C_2H_6	0.4418	230	*n*-Octane, C_8H_{18}	0.7451	320
Ethanol, C_2H_5OH	0.4455	391	Oxygen, O_2	0.3433	113
Ethylene, C_2H_4	0.4232	205	*n*-Pentane, *n*-C_5H_{12}	0.5769	345
Fluorine, F_2	0.3653	112	Propane, C_3H_8	0.5061	254
Helium, He	0.2576	10.2	Sulphur dioxide, SO_2	0.4290	252
n-Heptane, *n*-C_7H_{16}	0.8880	282	Water, H_2O	0.2649	356
n-Hexane, *n*-C_6H_{14}	0.5909	413	Xenon, Xe	0.4055	229
Hydrogen, H_2	0.2915	38			

* Adapted from Bird R.B., Stewart W.E., Lightfoot E.N., Transport Phenomena, John Wiley, New York, 1960. The values are computed from viscosity data and are applicable to temperature above 100 K

Table 2.2: Functions for prediction of transport properties of gases at low pressures*

$\kappa T/\varepsilon_{AB}$	Ω_D	$\kappa T/\varepsilon_{AB}$	Ω_D	$\kappa T/\varepsilon_{AB}$	Ω_D
0.30	2.662	1.65	1.153	4.00	0.8836
0.35	2.476	1.70	1.140	4.10	0.8788
0.40	2.318	1.75	1.128	4.20	0.8740
0.45	2.184	1.80	1.116	4.30	0.8694
0.50	2.066	1.85	1.105	4.40	0.8652
0.55	1.966	1.90	1.094	4.50	0.8610
0.60	1.877	1.95	1.084	4.60	0.8568
0.65	1.798	2.00	1.075	4.70	0.8530
0.70	1.729	2.10	1.057	4.80	0.8492
0.75	1.667	2.20	1.041	4.90	0.8456
0.80	1.612	2.30	1.026	5.0	0.8422
0.85	1.562	2.40	1.012	6.0	0.8124
0.90	1.517	2.50	0.9996	7.0	0.7896
0.95	1.476	2.60	0.9878	8.0	0.7712
1.00	1.439	2.70	0.9770	9.0	0.7556
1.05	1.406	2.80	0.9672	10.0	0.7424
1.10	1.375	2.90	0.9576	29.0	0.6640
1.15	1.346	3.00	0.9490	30.0	0.6232
1.20	1.320	3.10	0.9406	40.0	0.5960
1.25	1.296	3.20	0.9328	50.0	0.5756
1.30	1.273	3.30	0.9256	60.0	0.5596
1.35	1.253	3.40	0.9186	70.0	0.5464
1.40	1.233	3.50	0.9120	80.0	0.5352
1.45	1.215	3.60	0.9058	90.0	0.5256
1.50	1.198	3.70	0.8998	100.0	0.5170
1.55	1.182	3.80	0.8942		
1.60	1.167	3.90	0.8888		

* Adapted from Bird R.B., Stewart W.E., Lightfoot E.N., Transport Phenomena, John Wiley, New York, 1960.

Equation (2.38) is made more general and reliable by Wilke-Lee modification (Reid et al, 1987). The modified equation is

$$D_{AB} = \frac{(3.03 - 0.98/M_{AB}^{1/2})(10^{-4})T^{3/2}}{PM_{AB}^{1/2}\,\sigma_{AB}^2\,\Omega_D} \qquad \text{...(2.101)}$$

where D_{AB} is in m^2/s, P is in N/m^2 and σ in nm.

The Chapman-Enskog model describing diffusion coefficient in binary gas mixtures at low to moderate pressures has been derived based on the kinetic theory. The following empirical modification proposed by Fuller, Schettler and Giddings (1966) for the Chapman-Enskog model uses empirical constants derived from experimental data. The equation is found to be more accurate and easy to use for estimating binary gas diffusivities.

$$D_{AB} = \frac{1.43(10^{-2})T^{1.75}}{PM_{AB}^{1/2}\left[(\Sigma_v)_A^{1/3} + (\Sigma_v)_B^{1/3}\right]} \qquad \text{...(2.102)}$$

where D_{AB} is in m^2/s, P is in N/m^2. Σ_V is found for each component by summing the atomic diffusion volumes. Atomic diffusion volumes determined by a regression analysis of experimental data are given in Table 2.3. The experimental data were compared with the results obtained by Eq. (2.102), over a very wide range of temperatures and the average deviation was found to be only 5.4% (Reid et al, 1987). The binary diffusivities of some typical gas pairs are given in Table 2.4.

Effect of Temperature, Pressure and Composition on Diffusivity of Gases

At low pressures where the gases can be assumed ideal, Eq. (2.101) indicates that

$$D_{AB} \propto \frac{T^{3/2}}{\Omega_D} \qquad \text{...(2.103)}$$

Taking the variation of Ω_D with temperature also into account, the above equation indicates that in general D_{AB} varies as $T^{1.5}$ to T^2. This result is in agreement with the empirical method suggested by Fuller in which $D_{AB} \propto T^{1.75}$. For rigid spheres, since Ω_D is unity at all temperatures, $D_{AB} \propto T^{1.5}$. D_{AB} varies inversely with pressure for pressures up to about 1500 kPa. At low pressures, the mass diffusivity is independent of composition. At high pressures, when the deviation of the gas from the ideal behaviour is appreciable, small, but finite changes in D_{AB} with composition were observed.

Example 2.10: Estimate the diffusivity of CO_2 in air at 317.2 K and 101.3 kPa by the following methods:

(a) Hirschfelder equation
(b) Wilke-Lee equation
(c) Fuller, et al. equation

Compare the result with the experimental value of 1.77×10^{-5} m^2/s.

Solution: (a) Hirschfelder equation:

$$D_{AB} = \frac{2.6628 \times 10^{-4}\,T^{3/2}}{PM_{AB}^{1/2}\,\sigma_{AB}^2\,\Omega_D}$$

Component-**A** = CO_2; Component-**B** = Air

Here, T = 317.2 K, P = 101.3 kPa, M_A = 44, M_B = 29

By Eq. (2.96)

$$M_{AB} = 2\left(\frac{1}{M_A} + \frac{1}{M_B}\right)^{-1} = 2\left(\frac{1}{44} + \frac{1}{29}\right)^{-1} = 34.9589$$

The Lennard-Jones parameters are obtained from Table 2.1.

Table 2.3: Atomic and structural diffusion volumes

Atomic and structural diffusion volume increments			
C	15.9	F	14.7
H	2.31	Cl	21.0
O	6.11	Br	21.9
N	4.54	I	29.8
Aromatic ring	–18.3	S	22.9
Heterocyclic ring	–18.3		
Diffusion volumes of simple molecules			
He	2.67	CO	18.0
Ne	5.98	CO_2	26.7
Ar	16.2	N_2O	35.9
Kr	24.5	NH_3	20.7
Xe	32.7	H_2O	13.1
H_2	6.12	SF_6	71.3
D_2	6.84	Cl_2	38.4
N_2	18.5	Br_2	69.0
O_2	16.3	SO_2	41.8
Air	19.7		

Table 2.4: Binary gas diffusivities*

Gas pair, A-B	T, K	$D_{AB} \times 10^5$, m²/s	Gas pair, A-B	T, K	$D_{AB} \times 10^5$, m²/s
Air-Acetone	293	0.92	CO_2–water	307.2	1.98
Air-Ammonia	273	1.98	CO–nitrogen	373	3.18
	298	2.26	Helium-benzene	423	6.10
Air-CO_2	317.2	1.77	Helium-methane	298	6.75
Air-Chloroform	298	0.95	Helium-methanol	423	10.32
Air-ethanol	293	1.33	Helium-water	307.1	9.02
	313	1.45	Hydrogen-ammonia	298	7.83
Air-Benzene	298	0.905		533	21.49
Air-helium	317.2	7.65	Hydrogen-cyclohexane	288.6	3.19
Air-n-hexane	328	0.93	Hydrogen-methane	288	6.94
Air-Toluene	332	9.2	Hydrogen-nitrogen	298	7.84
Air-water	293	2.5	Nitrogen-Ammonia	298	2.33
	302	2.7		358	3.32
	313	2.88	Nitrogen-benzene	311.3	1.02
Argon-ammonia	333	2.53	Nitrogen-cyclohexane	288.6	0.731
Argon-hydrogen	242.2	5.62	Nitrogen-SO_2	263	1.04
	806	48.6	Nitrogen-water	352.1	2.56
Argon-methane	298	2.02	Oxygen-benzene	311.3	1.01
Benzene-Toluene	362	0.59	Oxygen-CCl_4	296	0.749
CO_2 -nitrogen	298	1.67	Oxygen-cyclohexane	288.6	0.746
CO_2 -oxygen	293.2	1.53	Oxygen-water	352.3	3.52

* Adapted from Seader J.D., Henley E.J., Separation Process Principles, 2nd Ed., John Wiley, (2006).

For CO_2, $\sigma_A = 0.3996$ nm; $\varepsilon_A/\kappa = 190$ K

For air, $\sigma_B = 0.3617$ nm; $\varepsilon_B/\kappa = 97$ K

$$\sigma_{AB} = \frac{1}{2}(\sigma_A + \sigma_B) = \frac{1}{2}(0.3996 + 0.3617) = 0.3807\text{nm}$$

$$\frac{\varepsilon_{AB}}{\kappa}=\left(\frac{\varepsilon_A}{\kappa}\times\frac{\varepsilon_B}{\kappa}\right)^{1/2}=(190\times 97)^{1/2}=135.76\text{ K}$$

$$\frac{\varepsilon_{AB}}{\kappa T}=\frac{135.76}{317.2}=0.428$$

or
$$\frac{\kappa T}{\varepsilon_{AB}}=2.3365$$

Corresponding to $\frac{\kappa T}{\varepsilon_{AB}}=2.3365$, Ω_D value is read from Table 2.2. Ω_D = 1.021. (This is obtained by linear interpolation between $\frac{\kappa T}{\varepsilon_{AB}}=2.3$ and $\frac{\kappa T}{\varepsilon_{AB}}=2.4$ for which Ω_D values are 1.026 and 1.012.)

Substituting the values in the Hirschfelder-Bird-Spotz equation:

$$D_{AB}=\frac{2.6628\times 10^{-4}T^{3/2}}{PM_{AB}^{1/2}\sigma_{AB}^2\Omega_D}=\frac{2.6628\times 10^{-4}\times 317.2^{3/2}}{1.013\times 10^5\times 34.9589^{1/2}\times(0.38065)^2\times 1.021}$$

$$=1.6978\times 10^{-5}\ \frac{\text{m}^2}{\text{s}}$$ **(Answer)**

(b) Wilke-Lee equation

$$D_{AB}=\frac{(3.03-0.98/M_{AB}^{1/2})(10^{-4})T^{3/2}}{PM_{AB}^{1/2}\ \sigma_{AB}^2\Omega_D}$$

$$=\frac{(3.03-0.98/34.9589^{1/2})\times(10^{-4})\times 317.2^{3/2}}{1.013\times 10^5\times 34.9589^{1/2}\times(0.38065)^2\times 1.021}=1.8257\times 10^{-5}\ \text{m}^2/\text{s}$$ **(Answer)**

(c) Fuller et al. equation

$$D_{AB}=\frac{1.43(10^{-2})T^{1.75}}{PM_{AB}^{1/2}\left[(\Sigma_v)_A^{1/3}+(\Sigma_v)_B^{1/3}\right]^2}$$

The diffusion volumes for CO_2 and air are obtained from Table 2.3.
$(\Sigma_v)_{CO_2}=26.7$ and $(\Sigma_v)_{\text{Air}}=19.7$.

$$D_{AB}=\frac{1.43(10^{-2})\times 317.2^{1.75}}{1.013\times 10^5\times 34.9589^{1/2}\left[(26.7)^{1/3}+(19.7)^{1/3}\right]^2}=1.7584\times 10^{-5}\ \frac{\text{m}^2}{\text{s}}$$ **(Answer)**

The values calculated by the different equations differ from the reported value as given below:

$$\text{Hirschfelder equation}=\frac{1.70-1.77}{1.77}\times 100=-4\%$$

$$\text{Wilke-Lee equation}=\frac{1.83-1.77}{1.77}\times 100=3.4\%$$

$$\text{Fuller, et al. equation}=\frac{1.76-1.77}{1.77}\times 100=-0.6\%$$

Example 2.11: Molar volume of liquid ethanol at normal boiling point of 351.4 K = 0.0592 m³/kmol. The Lennard-Jones parameters for air are σ = 0.3617 nm; ε/κ = 97 K.

(a) Estimate the Lennard-Jones parameters for ethanol.
(b) Estimate the diffusivity of ethanol vapour through air at 293 K and 101.3 kPa using the Wilke-Lee equation.
(c) Estimate the diffusivity of ethanol vapour through air at 293 K and 101.3 kPa using the Fuller, et al. equation.

Compare the result with the literature value for diffusivity of ethanol vapour of 1.33 × 10^{-5} m^2/s.

Solution: (a) The Lennard-Jones parameters are given by Eqs. (2.99) and (2.100). The normal boiling point T_b = 351.4 K and the molar volume of ethanol at the normal boiling point V = 0.0592 m^3/kmol.

$$\sigma_A = 1.18V^{1/3} = 1.18 \times (0.0592)^{1/3} = 0.4599$$

$$\frac{\varepsilon_A}{\kappa} = 1.21T_b = 1.21 \times 351.4 = 425.19 \text{ K}$$

(b) For air, the parameters are: $\sigma_B = 0.3617\,\text{nm};\ \dfrac{\varepsilon_B}{\kappa} = 97$ K

$$\sigma_{AB} = \frac{1}{2}(\sigma_A + \sigma_B) = \frac{1}{2}(0.4599 + 0.3617) = 0.4108\,\text{nm}$$

$$\frac{\varepsilon_{AB}}{\kappa} = \left(\frac{\varepsilon_A}{\kappa} \times \frac{\varepsilon_B}{\kappa}\right)^{1/2} = (425.19 \times 97)^{1/2} = 203.085 \text{ K}$$

$$\frac{\varepsilon_{AB}}{\kappa T} = \frac{203.085}{293} = 0.6931 \text{ or } \frac{\kappa T}{\varepsilon_{AB}} = 1.4428$$

Corresponding to $\dfrac{\kappa T}{\varepsilon_{AB}} = 1.4428$, Ω_D value is read from Table (2.2). Ω_D = 1.2176

$$M_A = 46;\ M_B = 29;\ M_{AB} = 2\left(\frac{1}{M_A} + \frac{1}{M_B}\right)^{-1} = 2\left(\frac{1}{46} + \frac{1}{29}\right)^{-1} = 35.573$$

The Wilke-Lee equation gives

$$D_{AB} = \frac{(3.03 - 0.98/M_{AB}^{1/2})(10^{-4})T^{3/2}}{PM_{AB}^{1/2}\sigma_{AB}^2\Omega_D}$$

$$= \frac{(3.03 - 0.98/35.57^{1/2}) \times (10^{-4}) \times 293^{3/2}}{1.013 \times 10^5 \times 35.57^{1/2} \times (0.4108)^2 \times 1.2176} = 1.1566 \times 10^{-5}\,\frac{\text{m}^2}{\text{s}}$$

Diffusivity of ethanol-air system at 293 K = 1.157 × 10^{-5} m^2/s **(Answer)**

(c) The diffusion volume of air is $(\Sigma_v)_{\text{Air}} = 19.7$

The diffusion volume of ethanol is computed as the sum of atomic volumes. The atomic volumes are obtained from Table 2.3

$$(\Sigma_v)_{C_2H_5OH} = 2 \times (15.9) + 6 \times (2.31) + 6.11 = 51.77$$

$$D_{AB} = \frac{1.43(10^{-2})T^{1.75}}{PM_{AB}^{1/2}\left[(\Sigma_v)_A^{1/3} + (\Sigma_v)_B^{1/3}\right]^2}$$

$$D_{AB} = \frac{1.43(10^{-2}) \times 293^{1.75}}{1.013 \times 10^5 \times 35.57^{1/2}\left[(19.7)^{1/3} + (51.77)^{1/3}\right]^2} = 1.189 \times 10^{-5}\,\frac{\text{m}^2}{\text{s}}$$ **(Answer)**

We can see that the diffusivity predicted by the Wilke-Lee equation is 13.5% less than the literature value whereas the diffusivity given by the Fuller equation varies by 10.5%

Diffusivity of Liquids

Since the molecules get densely packed in liquids and are strongly affected by force fields of neighbouring molecules the diffusivity in liquids are smaller than that in gases. Liquid diffusivities are orders of magnitude 10^{-5} times the diffusivities of gases. Table 2.5 lists binary liquid diffusivities at infinite dilution for some typical systems. Since there is no rigorous model for representing liquid state, diffusion coefficients in liquid state are difficult to estimate. However, for diffusion of large, spherical molecules of very dilute solute (**A**) diffusing through a stationary solvent (**B**) of small molecules, based on the hydrodynamic theory, the Stoke's–Einstein equation predicts the diffusivity as given below:

$$D_{AB} = \frac{\kappa T}{6\pi\mu_B r_A} \qquad \text{...(2.104)}$$

where r_A is the radius of solute molecules, μ_B is the viscosity of the solvent and T is the absolute temperature. This equation is found to represent the diffusivity of colloidal particles or large round molecules diffusing through a solvent which behaves as a continuum relative to the diffusing species.

Table 2.5: Experimental binary liquid diffusivities for solutes **A**, at infinite dilution in solvents **B**[*]

Solute, A	Solvent, B	Temperature, K	Diffusivity, D_{AB}, $m^2/s \times 10^9$
Acetic acid	Water	293	1.19
Acetone		298	1.28
Aniline		293	0.92
Carbon dioxide		298	2.00
Ethanol		288	1.00
Methanol		288	1.26
Sodium chloride		291	1.26
Allyl alcohol	Ethanol	293	0.98
Benzene		298	1.81
Oxygen		303	2.64
Pyridine		293	1.10
Water		298	1.24
Acetic acid	Benzene	298	2.09
Cyclohexane		298	2.09
Ethanol		288	2.25
n-Heptane		298	2.10
Toluene		298	1.85
Carbon tetrachloride	*n*-Hexane	298	3.70
Methyl ethyl ketone		303	3.74
Propane		298	4.87
Toluene		298	4.21
Acetic acid	Acetone	288	2.92
		313	4.04
Formic acid		298	3.77
Nitrobenzene		293	2.94
Water		298	4.56

* (Reid R.C., Prausnitz J.M., Poling B.E., *The Properties of Gases and Liquids, 4th Ed., McGraw Hill, New York (1987)*).

Although Eq. (2.104) has limited success in predicting liquid diffusivities, it has served as a basic model for the development of more reliable empirical correlations for binary liquid mixtures where both solute and solvent molecules are of approximately the same size. One such correlation which gives reasonably good prediction for small solute molecules is the Wilke-Chang equation (Wilke and Chang, 1955):

$$D_{AB} = \frac{1.173\times10^{-16}(\phi_B M_B)^{1/2}T}{\mu_B v_A^{0.6}} \qquad \text{...(2.105)}$$

Here D_{AB} is in m²/s, μ_B is the viscosity of the solution in kg/m s, v_A is the molal volume of the solute at the normal boiling point in m³/kmol, and Φ_B is an association parameter for solvent **B**. Recommended values for association factors are given below:

Solvent	Φ_B
Water	2.6
Methanol	1.9
Ethanol	1.5
Unassociated solvents (such as hydrocarbons)	1.0

The method is not applicable when the solute is water. The molecular volume v_A can be estimated by summing the atomic volumes. Atomic volumes are listed in Table 2.6.

Table 2.6: Atomic volume contributions of elements and molecular volumes of common light gases at the normal boiling point*

Atomic volume, (m³/kmol) × 10³ **Element**		**Atomic volume, (m³/kmol) × 10³** **Ring**	
C	14.8	Three-membered, as in ethylene oxide	– 6
H	3.7	Four-membered	– 8.5
O (except as below)	7.4	Five-membered	–11.5
Doubly bonded as carbonyl	7.4	Six-membered	–15
Coupled to two other elements:		Naphthalene ring	–30
As in aldehydes, ketones	7.4	Anthracene ring	–47.5
As in methyl esters	9.1	**Molecular volume, (m³/kmol) × 10³**	
As in methyl ethers	9.9	**Compound**	
As in ethyl esters	9.9	Air	29.9
As in ethyl ethers	9.9	O_2	25.6
As in higher esters	11.0	N_2	31.2
As in higher ethers	11.0	Br_2	53.2
As in acids (-OH)	12.0	Cl_2	48.4
Joined to S, P, N	8.3	CO	30.7
N		CO_2	34.0
Doubly bonded	15.6	H_2	14.3
In primary amines	10.5	H_2O	18.8
In secondary amines	12.0	H_2S	32.9
Br	27.0	NH_3	25.8
Cl in RCHClR	24.6	NO	23.6
Cl in RCl (terminal)	21.6	N_2O	36.4
F	8.7	SO_2	44.8
I	37.0		
S	25.6		
P	27.0		

* Seader J.D., Henley E.J., Separation Process Principles, 2nd Ed., John Wiley, (2006)

Using regression analysis, Hayduk and Minhas proposed for infinite dilution diffusion coefficients, several correlations depending on the type of solute-solvent system. These give better agreement than the Wilke-Chang equation with experimental values for non electrolyte solutions. For nonaqueous (nonelectrolyte) solutions they proposed the following equation:

$$D_{AB} = \frac{1.0973 \times 10^{-15} T^{1.29} (\mathbf{P}_B^{0.5} / \mathbf{P}_A^{0.42})}{\mu_B^{0.92} v_B^{0.23}} \quad \text{...(2.106)}$$

Here D_{AB} is the diffusivity at infinite dilution in m²/s, T is the absolute temperature in K, μ_B is the viscosity of the solvent in kg/m s, and v_B is the molar volume in m³/kmol. **P** is known as the parachor which is defined as

$$\mathbf{P} = v\sigma^{1/4} \quad \text{...(2.107)}$$

Given the liquid molar volume in m³/kmol and the surface tension in N/m, the units of parachor is $\left(\frac{\text{m}^3}{\text{kmol}}\right)\left(\frac{\text{N}}{\text{m}}\right)^{1/4}$. At ambient conditions **P** is normally treated as constant. Table 2.7 lists the Parachor values for some typical compounds.

Table 2.7: (a) Parachors for representative compounds* (Parachor in (m³/kmol)(N/m)$^{(1/4)}$ × 10^4)

Compound	Parachor	Compound	Parachor	Compound	Parachor
Acetic acid	233.3	Chlorobenzene	434.8	Methyl amine	170.5
Acetone	287.2	Diphenyl	675.7	Methyl formate	246.5
Acetonitrile	217.0	Ethane	197.0	Naphthalene	555.7
Acetylene	157.6	Ethylene	176.9	n-Octane	622.9
Aniline	416.8	Ethyl butyrate	524.8	1-Pentene	388.0
Benzene	365.1	Ethyl ether	376.5	1-Pentyne	368.1
Benzonitrile	458.8	Ethyl mercaptan	289.7	Phenol	393.5
n-Butyric acid	371.8	Formic acid	166.6	*n*-Propanol	294.1
Carbon disulfide	255.4	Isobutyl benzene	649.8	Toluene	436.6
Cyclohexane	425.5	Methanol	157.9	Triethyl amine	529.6

* Meissner, Chem. Eng. Prog.. **45,** 149-153 (1949).

There are certain restrictions in the use of the Hayduk and Minhas correlations:

1. The correlation is not valid for viscous solvents (μ_B above about 20 to 30 centipoise).
2. If the solute is an organic acid and the solvent is not water, methanol, or butanol, the acid should be considered a dimer with twice the values of $\mathbf{P}_A$ and v_A.
3. For non polar solutes diffusing into monohydroxy alcohols, the values of v_B and $\mathbf{P}_B$ should be multiplied by a factor of 8 μ_B where μ_B is the solvent viscosity in centipoises.

It has been reported that the Hayduk and Minhas correlation gave relatively small errors in predicting the infinite dilution diffusivity values.

Effect of Viscosity, Temperature and Composition on Liquid Diffusivity

Most of the correlations for liquid phase diffusion coefficients assume that the diffusivity in dilute systems varies inversely with the viscosity of the solvent. This is in agreement with the Stoke's-Einstein equation. However, it has been observed that over wide viscosity ranges, the diffusivity does not decrease in proportion to the increase in solvent viscosity; rather, $D_{AB} \alpha \mu_B^n$, where n varies, usually from – 0.5 to –1.

The correlations presented for liquid systems apply to the diffusion of the solute present in the solution in very dilute proportions. Unlike gases, in which diffusion is almost composition independent, the effect of composition on liquid diffusivity is complex, sometimes showing strong positive or negative deviation from linearity.

Table 2.7: (b) Structural contributions for estimating the parachor*

Carbon-hydrogen	Contribution	R-[-CO-]-R′ (ketone)	Contribution
C	9.0	R+ R′ = 2	51.3
H	15.5	R+ R′ = 3	49.0
CH_3	55.5	R+ R′ = 4	47.5
CH_2 in $-(CH_2)n$		R + R′ = 5	46.3
$n < 12$	40.0	R + R′ = 6	45.3
$n > 12$	40.3	R + R′ = 7	44.1
		-CHO	66
Alkyl groups			
1-Methylethyl	133.3	O (not noted above)	20
1-Methylpropyl	171.9	N (not noted above)	17.5
1-Methylbutyl	211.7	S	49.1
2-Methylpropyl	173.3	P	40.5
1-Ethylpropyl	209.5	F	26.1
1,1-Dimethylethyl	170.4	Cl	55.2
1,2-Dimethylpropyl	207.5	Br	68.0
1,2-Dimethylpropyl	207.9	I	90.3
1,1,2-Trimethylpropyl	243.5	Ethylenic bonds:	
C_6H_5	189.6	Terminal	19.1
		2,3-position	17.7
Special groups:		3,4-position	16.3
-COO-	63.8		
-COOH-	73.8	Triple bond	40.6
-OH	29.8		
$-NH_2$	42.5	**Ring closure:**	
-O-	20.0	Three-membered	12.
$-NO_2$	74	Four-membered	6.0
$-NO_3$ (nitrate)	93	Five-membered	3.0
$-CO(NH_2)$	91.7	Six-membered	0.8

* Seader J.D., Henley E.J., Separation Process Principles, 2nd Ed., John Wiley, (2006)

The effect of temperature on liquid diffusivity is accounted for in the Stokes-Einstein and Wilke Chang Equations by the relation, $\frac{D_{AB}\mu_B}{T}$ = constant. Since μ_B decreases exponentially with temperature, D_{AB} is expected to increase exponentially with temperature. For a dilute solution of water in ethanol, the diffusivity of water increases by a factor of almost 20 when the absolute temperature is increased by 50%. However, over wide temperature ranges, it is usually preferred to assume that

$$D_{AB} = A \exp\frac{-B}{T} \qquad \ldots(2.108)$$

Example 2.12: Determine the diffusivity of acetic acid in a dilute aqueous solution at 293 K using Wilke-Chang equation and compare with the experimental value of 1.19×10^{-9} m²/s. The viscosity of water at 293 K is 1.01 cP.

Solution: The Wilke-Chang equation for diffusivity of liquids is

$$D_{AB} = \frac{1.173\times10^{-16}(\phi_B M_B)^{1/2} T}{\mu_B v_A^{0.6}}$$

For water, the association parameter $\Phi_B = 2.6$, M_B, molecular weight of water = 18, T = 293 K, $\mu_B = 1.01 \times 10^{-3}$ kg/m s. v_A is the molal volume of acetic acid at the normal boiling point = 2 (atomic volume of carbon) + 4 (atomic volume of hydrogen) + 2 (atomic volume of oxygen) = $2 \times (14.8 \times 10^{-3}) + 4 \times (3.7 \times 10^{-3}) + 2 \times (7.4 \times 10^{-3}) = 59.2 \times 10^{-3}$ m³/kmol

$$D_{AB} = \frac{1.173 \times 10^{-16} (2.6 \times 18)^{1/2} \times 293}{1.01 \times 10^{-3} \times 0.0592^{0.6}} = 1.2693 \times 10^{-9} \frac{\text{m}^2}{\text{s}} \quad \textbf{(Answer)}$$

The estimated value is greater than the experimental value by $\dfrac{1.27 - 1.19}{1.19} \times 100 = 6.7\%$

Example 2.13: Estimate the diffusivity of acetic acid (**A**) in benzene (**B**) at 298 K and infinite dilution, using the Hayduk-Minhas correlation. Viscosity of benzene is 0.6 cP and the molar volume of benzene is 0.096 m³/kmol.

***Solution*:** The Hayduk-Minhas correlation for diffusivity is

$$D_{AB} = \frac{1.0973 \times 10^{-15} T^{1.29} (\mathbf{P}_B^{0.5} / \mathbf{P}_A^{0.42})}{\mu_B^{0.92} v_B^{0.23}} \quad \text{...(2.106)}$$

T = 298 K, $\mu_B = 0.6 \times 10^{-3}$ kg/m s, v_B = 0.096 m³/kmol. The parachor values are taken from Table 2.7a.

$$\mathbf{P}_A = 233.3 \times 10^{-4} \left(\frac{\text{m}^3}{\text{kmol}}\right)\left(\frac{\text{N}}{\text{m}}\right)^{1/4} \quad \mathbf{P}_B = 365.1 \times 10^{-4} \left(\frac{\text{m}^3}{\text{kmol}}\right)\left(\frac{\text{N}}{\text{m}}\right)^{1/4}$$

The Parachor value of organic acid should be multiplied by 2 for use in Eq. (2.106). Therefore, $\mathbf{P}_A = 466.6 \times 10^{-4} \left(\dfrac{\text{m}^3}{\text{kmol}}\right)\left(\dfrac{\text{N}}{\text{m}}\right)^{1/4}$. With these values Eq. (2.106) gives

$$D_{AB} = \frac{1.0973 \times 10^{-15}\, 298^{1.29} \left[(365.1 \times 10^{-4})^{0.5} / (466.6 \times 10^{-4})^{0.42}\right]}{(0.6 \times 10^{-3})^{0.92} \times 0.096^{0.23}} = 1.864 \times 10^{-9} \frac{\text{m}^2}{\text{s}}$$

The diffusivity of acetic acid in benzene at infinite dilution is $D_{AB} = 1.864 \times 10^{-9} \dfrac{\text{m}^2}{\text{s}}$ **(Answer)**

2.8 DIFFUSIVITY IN SOLIDS

The mechanism of diffusion in solids is very complex and depends on the diffusing atom, molecule or ion, the nature of the solid structure, whether it is porous or non porous, crystalline or amorphous, and the type of solid material, whether it is metallic, ceramic, polymeric, biological or cellular. In spite of all these complexities, Fick's law can be used to describe the diffusion if an experimentally measured diffusivity is available. If the diffusing species is a gas, its solubility in the solid must also be known.

Porous Solids

Diffusion of gaseous and liquid solutes through porous solids is important in operations such as membrane separation, leaching, adsorption and drying. Another important area where diffusion through pores is important is in the analysis and design of reactors using porous solid catalysts. Following three mechanisms are proposed for diffusion through porous solids:

1. **Fick's diffusion:** Ordinary molecular diffusion through pores which present tortuous paths hinder the movement of large molecules when their diameter is more than 10% of the pore diameter. When the mean free path for diffusion is small compared to pore diameter, molecule-molecule collisions predominate and molecule-wall collisions are few. Ordinary molecular

diffusion holds good under this situation. If the diffusion in solid is considered to be due to the diffusion occurring only in the fluid in the pores, it can be represented using Fick's law with an effective diffusivity, D_{eff}. The effective diffusivity may be related to the ordinary diffusion coefficient as

$$D_{eff} = \frac{D_{AB}\varepsilon}{\tau} \quad \text{...(2.109)}$$

where ε is the porosity of the solid and τ is the tortuosity of the pore path. The tortuosity is defined as the ratio of the pore length to the length of the pore if it were straight in the direction of diffusion. The effective diffusivity, D_{eff} is based on the total cross sectional area of the porous solid rather than the cross sectional area of the pores and on the straight path rather than the tortuous path through the pores. The effective diffusivity is either determined experimentally or estimated using Eq. (2.109) in which the experimentally determined ε and τ are used along with the ordinary molecular diffusivity values estimated using suitable correlations.

2. **Knudsen diffusion**: It involves collision of diffusing gaseous molecules with the pore walls when the pore diameter and pressure are such that the molecular mean free path is large compared to the pore diameter. Knudsen diffusion occurs when the size of the pores is of the order of the mean free path of the diffusing molecules. When the mean free path is large in comparison to the pore diameter, the molecule-wall collisions become important. A relationship which describes Knudsen diffusion and the diffusion coefficient has been developed using the kinetic theory of gases.

$$N_A = \frac{D_{K,eff}}{z}(C_{A1} - C_{A2}) \quad \text{...(2.110)}$$

where z is the length of the diffusion path. For diffusion of gases, the Knudsen diffusivity is independent of pressure and is calculated from the relation

$$D_{K,eff} = 97.0\bar{r}\left(\frac{T}{M_A}\right)^{1/2} \quad \text{...(2.111)}$$

Here M_A is the molecular weight of **A**, $\bar{r}$ is the average pore radius in m, and the diffusivity is in m^2/s.

3. **Surface diffusion**: This type of diffusion involves the jumping of molecules, adsorbed on the pore walls, from one adsorption site to another based on a surface concentration driving force.

Polymers

Thin, dense, nonporous polymer membranes are widely used to separate gas and liquid mixtures. The mechanism of diffusion is strongly dependent on the type of polymer, whether it be crystalline or amorphous, and if the latter, whether glassy or rubbery. Commercial crystalline polymers are about 20% amorphous. It is mainly through the amorphous region that diffusion occurs. The transport of gaseous species through polymer membrane is usually characterised by the *solution diffusion mechanism*.

Gas species: Fick's law in the following integrated form can be applied to compute the mass transfer flux for diffusion of gaseous species.

$$N_i = \frac{P_{Mi}}{(z_2 - z_1)}(\bar{p}_{i1} - \bar{p}_{i2}) \quad \text{...(2.112)}$$

where P_{Mi} is called the *permeability* of species i, defined as the product of diffusivity and solubility and $\bar{p}_i$ is the partial pressure of the gas species at the polymer surface.

$$P_{Mi} = D_i H_i \qquad \ldots(2.113)$$

where D_i is the diffusivity and H_i is the Henry's law constant for species i that represents its solubility.

Liquid species: For diffusion of liquid species, the mass flux is given by

$$N_A = \frac{P_{Mi}}{z_2 - z_1}(c_{i1} - c_{i2}) \qquad \ldots(2.114)$$

The liquid permeability P_{Mi} is evaluated as the product of diffusivity and partition coefficient.

$$P_{Mi} = K_i D_i$$

where K_i the equilibrium partition coefficient, is equal to the ratio of the concentration in the polymer to the concentration c_i in the liquid adjacent to the polymer surface. Diffusion and permeability in crystalline polymers depend on the degree of crystallinity. Polymers that are 100% crystalline permit little or no diffusion of gases and liquids.

Cellular Solids

Cellular solids such as wood, cork, sponge, and coral exist in nature. Synthetic cellular structures include honeycombs, and foams made from polymers, metals, ceramics and glass. Wood is the most widely used cellular solid and it consists of lignin, cellulose, hemicellulose and minor amounts of organic chemicals and elements. Green wood contains up to about 25% moisture in the cell walls and cavities. Movement of liquids and gases through wood and wood products takes time during drying and treatment with preservatives, fire retardants, and other chemicals. This movement takes place by capillary, pressure permeability, and diffusion. Nevertheless, wood is not highly permeable because, the cell voids are largely discreet and lacks direct interconnections. Instead, communication among cells is through circular openings spanned by thin membranes with submicrometer-sized pores, called pits, and to a smaller extent, across the cell walls.

Diffusion of water in wood is complex. Moisture content determines degree of swelling or shrinkage. Water is held in the wood in different ways. It may be physically adsorbed on cell walls in monomolecular layers, condensed in pre-existing or transient cell capillaries, or absorbed in cell walls to form a solid solution. Because of the practical importance of lumber drying rates, most diffusion coefficients are measured under drying conditions in the radial direction across the fibres.

Crystalline Solids

The mechanism of diffusion in crystalline solids depend on the crystalline structure and nature of the solute. The following are the mechanism of diffusion in crystalline solids:

1. ***Interstitial mechanism*:** Interstitial sites are places between the atoms of a crystal lattice. Small diffusing solute atoms may pass from an interstitial site to the next when the matrix atoms of the crystal lattice move apart temporarily to provide the necessary space.

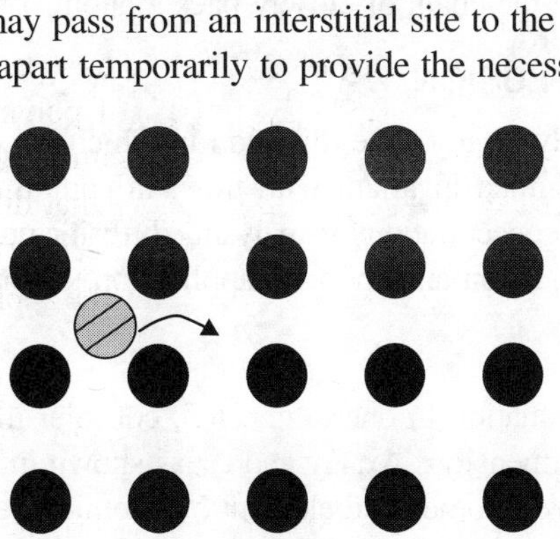

Fig. 2.8: (a) Diffusion within crystals: Interstitial mechanism

2. ***Vacancy mechanism*:** If there are unoccupied or vacant sites in the lattice sites, an atom in the adjacent site may jump into such a vacancy. The diffusing atom continues to move through the crystal by a series of exchanges with vacancies which appear adjacent to it from time to time. This mechanism is prominent in face-centred cubic metals and alloys. This mechanism is used to explain diffusion in body centred cubic material, as well as ionic compounds and oxides.

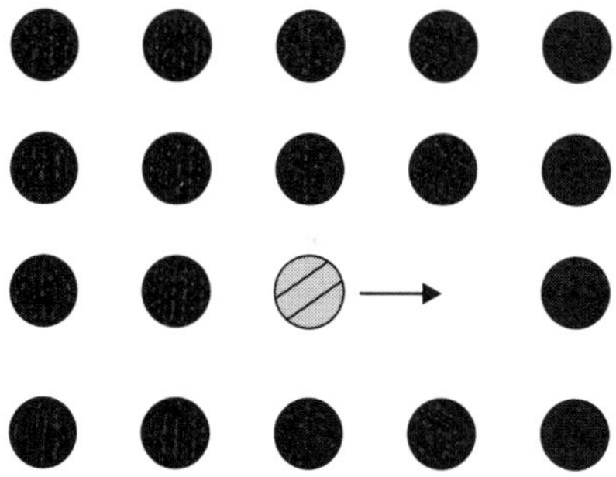

Fig. 2.8: (b) Diffusion within crystals: vacancy mechanism

3. ***Interstitialcy mechanism*:** A large atom occupying an interstitial site pushes one of its lattice neighbours into an interstitial position and moves into the vacancy thus produced.

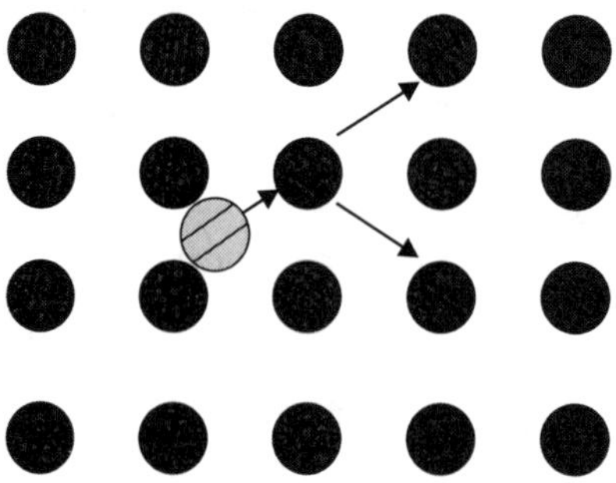

Fig. 2.8: (c) Diffusion within crystals: Interstitialcy mechanism

4. ***Diffusion along grain boundaries***: Diffusion occurs along grain boundaries or crystal interfaces and dislocations.
5. ***Crowd-ion mechanism***: A diffusion flux is produced by the displacement of atoms from their equilibrium positions in the chain of closely packed atoms by an extra atom in the chain.

2.9 UNSTEADY-STATE DIFFUSION

So far, we have considered only steady-state diffusion in which the concentration at any given point along the diffusion path is assumed invariant with time. In certain cases we may have to take into account the time dependency of concentration for analysing diffusion problems. Therefore it is worthwhile to derive the basic diffusion equation and show its application to some typical practical cases.

Basic Diffusion Equation

To derive the basic diffusion equation in one dimension, consider material **A** entering and leaving a cubical fluid element having dimensions Δx, Δy and Δz as shown in Fig. 2.9. Assuming that material is not generated or consumed within the fluid element by chemical reaction, the conservation of mass can be written as

$$\text{rate of input of } \mathbf{A} - \text{rate of output of } \mathbf{A} = \text{rate of accumulation of } \mathbf{A} \qquad \ldots(2.115)$$

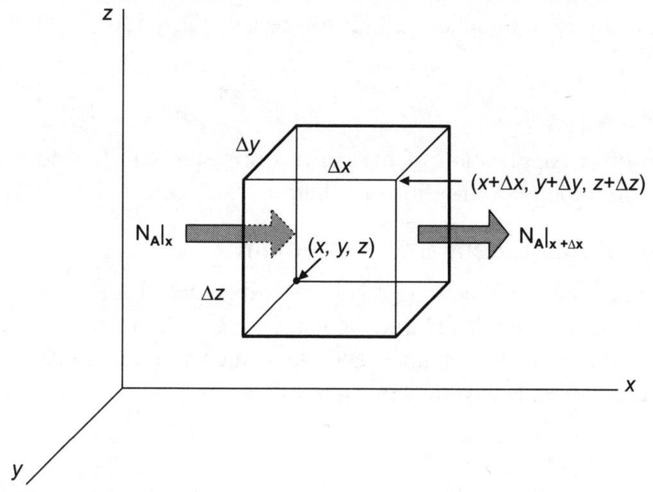

Fig. 2.9: Analysis of unsteady-state diffusion in one dimension

Consider the diffusion of **A** through the fluid volume element in the x- direction only. Assume that the total molar concentration C and the diffusivity D_{AB} are constant. Moles of **A** entering the face at a distance x from the origin in unit time is

$$\text{rate of input of } \mathbf{A} = N_A\big|_x \Delta y \Delta z \qquad \text{...(2.116)}$$

Material **A** leaves the volume element through the face at $x = x + \Delta x$. The rate at which **A** leaves the element is

$$\text{rate of output of } \mathbf{A} = N_A\big|_{x+\Delta x} \Delta y \Delta z \qquad \text{...(2.117)}$$

Rate of accumulation of **A** is the product of volume of the fluid element and the rate of change of concentration.

$$\text{Rate of accumulation of } \mathbf{A} = (\Delta x \Delta y \Delta z)\frac{\partial C_A}{\partial t} \qquad \text{...(2.118)}$$

Substituting Eqs.(2.116), (2.117) and (2.118) into Eq. (2.115) and rearranging, we get

$$\left(N_A\big|_{x+\Delta x} - N_A\big|_x\right)\Delta y \Delta z = -(\Delta x \Delta y \Delta z)\frac{\partial C_A}{\partial t}$$

$$\frac{N_A\big|_{x+\Delta x} - N_A\big|_x}{\Delta x} = -\frac{\partial C_A}{\partial t} \qquad \text{...(2.119)}$$

Taking limits as Δx approaches zero, Eq. (2.119) becomes

$$\frac{\partial N_A}{\partial x} = -\frac{\partial C_A}{\partial t} \qquad \text{...(2.120)}$$

For equimolar counter diffusion and for one-way diffusion in dilute solutions, Fick's law of diffusion in the x-direction is

$$N_A = -D_{AB}\frac{\partial C_A}{\partial x} \qquad \text{...(2.121)}$$

Substituting Eq. (2.121) into Eq. (2.120) and assuming D_{AB} to be constant, we get

$$\frac{\partial C_A}{\partial t} = D_{AB}\frac{\partial^2 C_A}{\partial x^2} \qquad \text{...(2.122)}$$

Equation (2.122) is referred to as Fick's second law of diffusion. This equation can be solved for a number of boundary conditions. Equation (2.122) can be generalized for the three dimensional case as

$$\frac{\partial C_A}{\partial t} = D_{AB}\left(\frac{\partial^2 C_A}{\partial x^2} + \frac{\partial^2 C_A}{\partial y^2} + \frac{\partial^2 C_A}{\partial z^2}\right) \qquad \ldots(2.123)$$

As an illustration of the application of the general diffusion equation to specific problems, the unsteady-state diffusion through semi-infinite medium is discussed in the following sections.

Transient Diffusion through Semi-infinite Medium

Consider a semi-infinite medium which extends in the z-direction from $z = 0$ to $z = \infty$. Let the initial concentration in the medium be uniform throughout at $C_A = C_{A0}$. At time $t = 0$, the surface of the semi-infinite medium is instantaneously brought to and maintained at a concentration $C_A = C_{As}$. Diffusion occurs in the z-direction, which is governed by the Fick's second law, Eq. (2.123), which simplifies to:

$$\frac{\partial C_A}{\partial t} = D_{AB}\frac{\partial^2 C_A}{\partial z^2} \qquad \ldots(2.124)$$

Figure 2.10 depicts the concentration-time history as a function of distance from the face maintained at uniform concentration C_{As}.

The initial and boundary conditions are:

$C_A = C_{A0}$ at $t = 0$, for all z

$C_A = C_{As}$ at $z = 0$, for all t

$C_A = C_{A0}$ as $z \to \infty$, for all t

Equation (2.124) is amenable to analytical solution. Either the Laplace transform method or the method of combination of variables is applicable. The solution is

$$\frac{C_A - C_{A0}}{C_{As} - C_{A0}} = 1 - \text{erf}\left(\frac{z}{2\sqrt{D_{AB}t}}\right) \qquad \ldots(2.125)$$

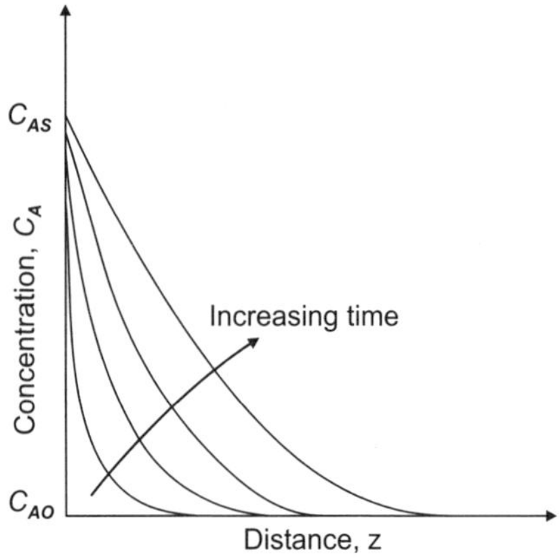

Fig. 2.10: Transient diffusion in semi-infinite solid

The error function is listed in most spreadsheet programs as well as in handbooks. Equation (2.125) is used to compute the concentration in the semi-infinite medium as a function of time and distance from the surface, assuming no bulk flow. It is applicable to diffusion in solids, stagnant liquids and gases when the concentration of the diffusing solute is small. For example, such problems as the absorption of oxygen from air in the aeration of a lake or the solid-phase diffusion process involved in the case hardening of mild steel in the carburizing atmosphere, etc. can be analyzed using Eq. (2.125).

REVIEW QUESTIONS

1. State and explain Fick's law of molecular diffusion.
2. What are the momentum transfer and heat transfer analogues of Fick's law?
3. Distinguish between J_A and N_A and state how these are expressed by Fick's law.
4. Bring out the similarity between heat, mass and momentum transport.
5. Distinguish between equimolar counter diffusion and diffusion through stagnant gas.
6. Starting from the Fick's law of diffusion, derive an equation to calculate the rate of mass transfer for diffusion of **A** through stagnant **B**.

7. Explain mass diffusivity. What is the effect of temperature and pressure on the mass diffusivity of gases?
8. Distinguish between molecular diffusion and eddy diffusion.
9. Show that for molecular diffusion in a binary gas mixture, $D_{AB} = D_{BA}$
10. Describe the effect of temperature and pressure on the binary mass diffusivity in gases.
11. Derive a general relationship for calculating steady-state flux for molecular diffusion in binary mixtures and deduce the equation for diffusion of **A** through stagnant **B**.
12. Develop expressions for the concentration profile for diffusion of **A** through stagnant **B** and equimolar counter diffusion.
13. How is effective diffusivity in a multicomponent gaseous mixture defined? Derive mean diffusivity of component **A** in a multi component mixture when all components other than **A** are stagnant.
14. Derive the necessary relation for experimental determination of diffusivity by Winkelmann's method.
15. In what way the equations for calculating steady-state flux for molecular diffusion in liquids different from those for gases?
16. Discuss briefly the effects of viscosity, temperature and composition on liquid diffusivity.
17. What are the different factors that make the analysis of diffusion in solids complex?
18. Explain the mechanism of diffusion through porous solids.
19. Explain Knudsen diffusion.
20. Explain the different mechanism of diffusion in crystalline solids.
21. Explain with reference to diffusion through porous solids the terms, Fick's diffusion, Knudsen diffusion and Surface diffusion.
22. Define partition coefficient and liquid permeability with reference to diffusion in polymeric solids.
23. State Fick's second law of diffusion. Where do you find its application?

MULTIPLE CHOICE QUESTIONS

2.1 Which of the following is the true driving force for mass transfer?

A. Concentration difference
B. Difference in chemical potential
C. Difference in temperature and pressure
D. Difference in enthalpies of phases.

2.2 The units of diffusion coefficient is

A. m^2/s B. m/s
C. $kmol/m^2$ s D. kg/m^2s

2.3 In a binary mixture which one of the following is true?

A. $D_{AB} = D_{BA}$ B. $x_A D_{AB} = x_B D_{BA}$
C. $D_{BA} = D_{AB} + y_A(N_A + N_B)$ D. $D_{AB} = 2D_{BA}$

2.4 The molar concentration of species *i* in a mixture of ideal gases (ρ is density, *p* is partial pressure and *n* is number of moles, *M* is molecular weight, *V* is volume of the mixture and *T* is the temperature) is

A. $C_i = \frac{\rho_i}{M_i}$ B. $C_i = \frac{n_i}{V}$

C. $C_i = \frac{p_i}{RT}$ D. All the above.

2.5 The units of molar flux in SI system is

A. $kmol/m^2s$ B. $kmol/m^2$
C. kmol/s D. m^2/s

2.6 Molar flux of a component can be written as
A. flux = diffusivity × concentration difference
B. flux = concentration × velocity
C. flux = diffusivity × velocity
D. none of the above

2.7 In molecular diffusion, according to Fick's law, the ratio of mass flux to concentration gradient is
A. the diffusion coefficient
B. the molar average velocity
C. constant
D. the eddy diffusivity

2.8 The flux J_A and N_A will be identical
A. if **A** alone diffuses and **B** is stagnant
B. if molar average velocity of the mixture is zero
C. if flux is measured relative to stationary coordinates
D. for all the above cases

2.9 Which one of the following statements is true regarding fluxes J_A and N_A?
A. J_A is measured with reference to stationary observer
B. N_A is a characteristic of the diffusing species alone and is not influenced by the convective transport of the fluid
C. N_A should always be greater than J_A
D. J_A is the flux of diffusion relative to an observer moving with the fluid stream.

2.10 Which of the following is not true regarding molecular diffusion?
A. The rate is directly proportional to the surface normal to the direction of transfer.
B. The rate can be expressed as a flux for the component diffusing.
C. Flux is negative since concentration gradient is negative.
D. Diffusion occurs in the direction of decreasing concentration and is proportional to the negative concentration gradient.

2.11 Fick's first law of diffusion is
A. $J_A = -D_{AB}\dfrac{dC_A}{dz}$
B. $N_A = -D_{AB}\dfrac{dC_A}{dz}$
C. $J_A = -CD_{AB}\dfrac{dC_A}{dz}$
D. $J_A = CD_{AB}\dfrac{dx_A}{dz}$

2.12 Fick's second law of diffusion is
A. $\dfrac{\partial C_A}{\partial t} = D_{AB}\left(\dfrac{\partial C_A}{\partial x} + \dfrac{\partial C_A}{\partial y} + \dfrac{\partial C_A}{\partial z}\right)$
B. $\dfrac{\partial C_A}{\partial t} = D_{AB}\left(\dfrac{\partial^2 C_A}{\partial x^2} + \dfrac{\partial^2 C_A}{\partial y^2} + \dfrac{\partial^2 C_A}{\partial z^2}\right)$
C. $\dfrac{\partial^2 C_A}{\partial t^2} = D_{AB}\left(\dfrac{\partial^2 C_A}{\partial x^2} + \dfrac{\partial^2 C_A}{\partial y^2} + \dfrac{\partial^2 C_A}{\partial z^2}\right)$
D. $\dfrac{\partial C_A}{\partial t} = -D_{AB}\left(\dfrac{\partial^2 C_A}{\partial x^2} + \dfrac{\partial^2 C_A}{\partial y^2} + \dfrac{\partial^2 C_A}{\partial z^2}\right)$

2.13 Diffusivity of gases
A. increases with increase of P and T
B. increases with increase of P but decreases with increase of T
C. decreases with increase of P and T
D. decreases with increase of P but increases with increase of T

2.14 Diffusivity of a binary gas mixture is 2.83×10^{-5} m²/s at 300 K. What will be the approximate value for the diffusivity at 600 K?
A. 1.43×10^{-5} m²/s
B. 5.66×10^{-5} m²/s
C. 8×10^{-5} m²/s
D. 1.68×10^{-5} m²/s.

2.15 Which of the following represents a probable value for the diffusion coefficient in m^2/s of acetic acid in liquid benzene?

A. 2.09×10^{-4} B. 2.09×10^{-5}

C. 2.09×10^{-9} D. 2.09×10^{-12}

2.16 If the diffusivity of ammonia in nitrogen is 2.2×10^{-5} m^2/s and the diffusivity of ammonia in hydrogen is 5.0×10^{-5} m^2/s, the mean diffusivity of ammonia diffusing in a stagnant mixture containing nitrogen and hydrogen in the mole ratio 1:4 is

A. 3.6×10^{-5} m^2/s B. 4.0×10^{-5} m^2/s

C. 2.9×10^{-5} m^2/s D. 2.2×10^{-5} m^2/s

2.17 The diffusivities of gases with rigid spherical molecules vary with temperature as

A. T^2 B. $T^{1.5}$

C. T D. $T^{1.75}$

2.18 Diffusion coefficient in a binary gas mixture at low pressure varies

A. directly with P B. directly with P^2

C. inversely with P D. independent of P

2.19 Molecular diffusivity of a liquid

A. increases with temperature B. decreases with temperature

C. may increase or decrease with temperature

D. is independent of temperature

2.20 Given that C is the molar concentration of the mixture, C_A is the concentration of **A**, u_A is the velocity of **A** and U is the molar average velocity of the mixture, ρ is the density and M is the molecular weight of the mixture, the molar flux of the total mixture (N) is equal to

A. Cu_A B. C_AU

C. $U\frac{\rho}{M}$ D. AUC

2.21 For steady-state diffusion of **A** through stagnant **B**, $N_A/(N_A + N_B)$ is

A. 0 B. 1

C. ∞ D. 1/2

2.22 For steady-state equimolar counter diffusion, $N_A/(N_A + N_B)$ is

A. 0 B. 1

C. ∞ D. 1/2

2.23 A practical situation where equimolar counter diffusion is closely approached is in

A. gas absorption B. liquid extraction

C. rectification D. dialysis

2.24 Component **A** is diffusing in a mixture of **A** and **B**. The flux referred to stationary coordinates is equal to flux due to molecular diffusion if

A. mass transfer is accompanied by chemical reaction

B. component **B** is non-diffusing

C. equimolar counter diffusion prevails

D. molecular mean free path is high

2.25 In the absorption of ammonia from a mixture of ammonia and air in water, the mechanism of diffusion is

A. diffusion of **A** through stagnant **B** B. equimolar counter diffusion

C. unsteady-state diffusion D. none of the above

2.26 In equimolar counter diffusion, the concentration variation along the diffusion path is
A. logarithmic B. exponential
C. linear D. invariant

2.27 In diffusion of species **A** through stagnant **B**, the rate of transfer of **A** is
A. entirely due to molecular diffusion
B. due to molecular diffusion supplemented by bulk transport
C. independent of concentration gradient
D. characterised by $N = 0$

2.28 Which one of the following statements is true with reference to diffusion in gases?
A. Diffusivity is directly proportional to temperature
B. Diffusivity increases linearly with pressure
C. Diffusivity is independent of composition at low pressures
D. Diffusivity is independent of composition at high pressures.

2.29 The ratio of diffusion coefficients in a gas to that in a liquid is of the order of
A. 10^5 B. 10^{-5}
C. 10^{-2} D. 10^2

2.30 For molecular diffusion in an ideal gas mixture, the partial pressure distribution under steady-state condition is non-linear in the case of
A. equimolar counter diffusion B. unimolal unidirectional diffusion
C. both the above D. none of the above

2.31 The concentration distribution for diffusion of **A** through stagnant **B** under steady-state is
A. linear B. parabolic
C. logarithmic D. hyperbolic

2.32 The diffusivity in liquids is very small in comparison with gases. However, the rate of diffusion in liquids is not that small as expected. Why?
A. Because in liquids the mean free path is small.
B. Because molar density in liquids is high
C. Because concentration difference in liquids is high.
D. Because liquids are incompressible in comparison to gases.

2.33 The diffusivity of solute **A** in a liquid solvent **B** depends on the viscosity of liquid **B**. This dependency is given as $D_{AB} \propto \mu_B^n$ where, the value of n is approximately
A. $0 \le n \le 2$ B. $1 \le n \le 2$
C. $-1 \le n \le -0.5$ D. $-2 \le n \le -1$

2.34 In the equation $N_A = \dfrac{CD_{AB}}{Z\, y_{B,M}}(y_{A1} - y_{A2})$, the term $y_{B,M}$ is a correction term to account for
A. the variation of concentration along the diffusion path
B. the bulk flow contribution to rate of diffusion
C. the variation of diffusivity with mole fraction of **B**
D. none of the above.

2.35 Methane (**A**) diffuses to the catalyst surface where it is cracked to hydrogen (**B**) which diffuses back according to the reaction $CH_4 \rightarrow C + 2H_2$. The diffusion flux of methane is given by

A. $N_A = -\dfrac{CD_{AB}}{Z}\ln\dfrac{1+y_{A2}}{1+y_{A1}}$ B. $N_A = \dfrac{CD_{AB}}{Z}\ln\dfrac{1+y_{A2}}{1+y_{A1}}$

C. $N_A = \dfrac{CD_{AB}}{Z\, y_{B,M}}(y_{A1} - y_{A2})$ D. $N_A = -\dfrac{CD_{AB}}{Z}\ln\dfrac{1-y_{A2}}{1-y_{A1}}$

2.36 Oxygen is diffusing through a mixture of oxygen and carbon dioxide at 101.3 kPa and 293 K with carbon dioxide non-diffusing. Concentration of oxygen at two planes 2 mm apart are 20% and 10%, respectively. The diffusivity of oxygen in carbon dioxide is 1.53×10^{-5} m^2/s . The molar flux of oxygen in kmol/m^2 s is

A. 3.2×10^{-5} B. 3.7×10^{-5}
C. 3.5×10^{-3} D. 4.2×10^{-3}

2.37 The following figure depicts steady state one dimensional diffusion of water vapour from the surface of water taken in a conical flask at room temperature. Derive the governing equation for determining the concentration profile of water vapour in the gas medium. Neglect the change of level of water due to condensation. The temperature of the gas and the liquid media are identical and constant.

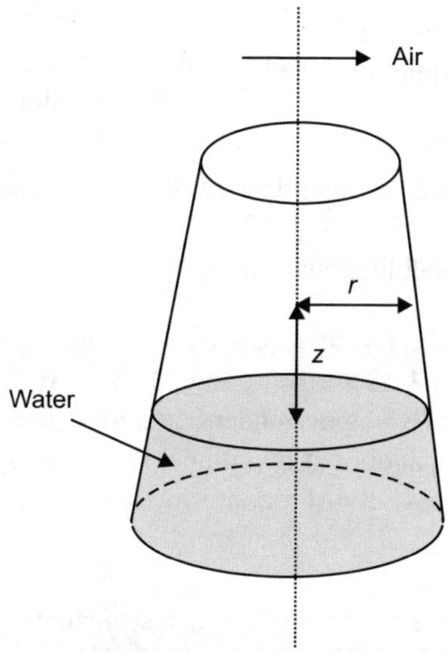

A. $\frac{d}{dz}\left[r^2 \frac{CD_{WA}}{1-x_W}\frac{dx_w}{dz}\right]=0$

B. $\frac{d}{dz}\left[\frac{CD_{WA}}{1-x_W}\frac{dx_w}{dz}\right]=0$

C. $\frac{d}{dz}\left[r\frac{CD_{WA}}{1-x_W}\frac{dx_w}{dz}\right]=0$

D. $\frac{d}{dz}\left[rCD_{WA}\frac{dx_w}{dz}\right]=0$

2.38 A pure drug is administered as a sphere and as a cube. The amount of drug is the same in the two tablets. Assuming that the shape and size do not influence the mass transfer, the ratio of rate of dissolution in water at $t = 0$ for the cubic to spherical tablet is

A. 0.54 B. 1.04
C. 1.24 D. 1.94

2.39 Ammonia (component 1) is evaporating from a partially filled bottle into surrounding air (component 2). The liquid level in the bottle and the concentration of ammonia at the top of the bottle are maintained constant. N_i is the molar flux relative to a fixed location in space and J_i is the molar flux with respect to the average molar velocity of the constituent species in the gas phase. Assume that the air in the bottle is stagnant. Which one of the following is correct?

A. N_1 = constant, $N_2 = 0$, $J_1 + J_2 = 0$ B. $N_1 + N_2 = 0$, $J_1 + J_2 = 0$
C. $N_1 + N_2 = 0$, J_1 = constant, $J_2 = 0$ D. N_1 = constant, $N_2 = 0$, J_1 = constant, $J_2 = 0$

EXERCISES

Steady-state molecular diffusion in gases

2.1 An open tube, 1 mm in diameter and 200 mm long, has pure hydrogen blowing across one end and pure nitrogen blowing across the other. The temperature is 298 K and pressure 101.3 kPa.

Diffusivity of nitrogen-hydrogen mixture is 7.84×10^{-5} m^2/s. Assuming equimolar counter diffusion, what will be the rate of transfer of hydrogen into the nitrogen stream in kmol/s. Also plot the mole fraction of hydrogen against distance from the end of the tube past which nitrogen is blown. (**Answer:** 1.2588×10^{-11} kmol/s)

2.2 A mixture of benzene and toluene is being separated in a rectifying column. At certain point in the column, the vapour contains 80% (mole) benzene and concentration of benzene on the vapour side of the interface is 91% (mole). The temperature at this point is 362 K and pressure is 101.3 kPa. The diffusion in the vapour phase is confined to a stagnant film of thickness 2.5 mm, calculate the rate at which benzene and toluene are transferred across unit area of the interface in kg/s. The diffusivity of the mixture of benzene and toluene at 362 K is 5.9×10^{-6} m^2/s. Assume equimolar counter diffusion.

(**Answer:** 6.816×10^{-4} kg/s benzene, 8.041×10^{-4} kg/s toluene)

2.3 Two bulbs are connected by a tube, 2 mm in diameter and 200 mm in length. Initially the bulb at end 1 contains Argon and the bulb at end 2 contains xenon. The pressure and temperature are maintained at constant at 378 K and 101.3 kPa. At a certain time after allowing diffusion to occur between the two bulbs, the argon content of the gas at end 1 and at end 2 are 75% (mole) and 20% (mole). If the binary diffusion coefficient is 1.8×10^{-5} m^2/s. Calculate:

(a) The rates and direction of mass transfer of argon and xenon in kmol/s

(b) The species velocities relative to stationary coordinates in m/s.

(c) The molar average velocity of the mixture.

(**Answer:** (a) 1.596×10^{-6} kmol/m^2 s in the positive z-direction for argon and negative z-direction for xenon (b) $4.95 \times 10^{-5}/y_A$ and $-4.95 \times 10^{-5}/y_B$)

2.4 Alcohol (A) vapour is diffusing through a layer of water (B) vapour under equimolar counter diffusion at 308 K and 101.3 kPa. The concentration of alcohol at two planes 0.3 mm apart are 80% (mole) and 10% (mole), respectively. The diffusivity of alcohol in water is 1.8×10^{-5} m^2/s, calculate:

(a) The rate of diffusion of alcohol and water vapour in kg/h

(b) The rate of diffusion of alcohol in kg/h if the diffusion is occurring through a stagnant water vapour layer. (**Answer:** (a) 275.4 kg/h, 107.8 kg/h (b) 591.8 kg/h)

2.5 Ammonia is diffusing through a uniform tube 0.1 m long containing nitrogen gas maintained at 298 K and 100 kPa pressure. At the one end of the tube, the partial pressures of ammonia is 10 kPa and at the other end 5 kPa, respectively. The diffusivity of ammonia in nitrogen is 2.33 $\times 10^{-5}$ m^2/s. Calculate the molar flux of ammonia. (**Answer:** 4.702×10^{-7} kmol/m^2 s)

Diffusion of A through stagnant gas B

2.6 For determination of diffusion coefficient of n-heptane in air, a column of diameter 2 mm is filled with the liquid at 294 K and 101.3 kPa. If the diffusivity is 7.1×10^{-6} m^2/s, what rate of decrease of the liquid level can be expected when the meniscus is 10 mm from the top of the tube? The vapour pressure and density of *n*-heptane are 5.06 kPa and 660 kg/m^3, respectively. (**Answer:** 0.816 mm/h)

2.7 An open beaker is filled with benzene at 298 K to within 8 mm of the top. The temperature and pressure are maintained constant at 298K and 100 kPa, respectively. The evaporated benzene diffuses to the top and is carried away in the air. The vapour pressure of benzene at 298 K is 13.3 kPa, and the diffusivity of benzene in air is 9.05×10^{-6} m^2/s Determine:

(a) Rate of evaporation in kmol/m^{2}s.

(b) Initial concentration profile and mole fraction of benzene at a point 4 mm above the liquid surface.

(c) What percent of the mass–transfer flux is due to molecular diffusion?

(d) The diffusion velocities and the species velocities of benzene and air relative to stationary coordinates at a point 4 mm above the liquid surface.

(**Answer:** (a) 6.5160×10^{-6} kmol/m^2 s (b) $y_A = 1 - 0.867 \exp(0.01784z)$, 0.069 (c) 93.2% (d) 2.34×10^{-3} m/s, 2.18×10^{-3} m/s for benzene, 0, -1.61×10^{-4} m/s for air)

2.8 At a certain point in the absorption column in which ammonia is absorbed from its mixture with air at 100 kPa and 298 K, the partial pressure of ammonia in the gas phase was 7 kPa and the ammonia concentration at the gas-liquid interface was negligible. The diffusion is assumed to be confined to a gas film 1 mm thick. Determine the rate of transfer of ammonia at that point in the column, if the diffusivity of ammonia in air is 2.26×10^{-5} m^2/s. If the pressure is doubled, how would the rate of transfer be affected? (**Answer:** 6.62×10^{-5} kmol/m^2 s, not affected)

2.9 Liquid acetone is contained in a capillary tube at 292 K and 102 kPa. The initial level of acetone was at 20 mm from the top of the tube. The vapour pressure of acetone at 292 K is 22.7 kPa and density of the liquid acetone is 790 kg/m^3. Determine the fall in level of acetone after 2 hours if the diffusivity of acetone in air at 292 K is 9.0×10^{-6} m^2/s. (**Answer:** 2.4 mm).

2.10 The diffusivity of water vapour in air is determined experimentally in a capillary tube maintained at a pressure of 101.3 kPa and temperature of 333 K. Air circulated is free of water vapour. If it took 39 h for the liquid to fall from the initial level 125 mm from the end open to atmosphere to 130 mm, calculate the diffusivity of water vapour in air. Vapour pressure of water at 333 K is 19.9 kPa. (**Answer:** 3.152×10^{-5} m^2/s)

2.11 An open circular tank 7 m in diameter contains benzene at 295 K which is exposed to atmosphere. The diffusion of benzene can be assumed to be occurring through a stagnant air film 5 mm thick in contact with the liquid surface. Concentration of benzene at the outer surface of the film is assumed to be zero. The vapour pressure of benzene at 295 K is 11 kPa and the density of liquid benzene is 880 kg/m^3. The diffusivity of benzene in air at 295 K is 9.05×10^{-6} m^2/s. What is the loss of benzene from the tank in cubic meters per day? (**Answer:** 2.53 m^3/day)

2.12 Carbon dioxide is diffusing through nitrogen at 273 K and 101.3 kPa across two planes 3 m away where the concentration of carbon dioxide are 20% (mole) and 2% (mole), respectively. Diffusivity of carbon dioxide in nitrogen is 1.44 x 10^{-5} m^2/s. The gas phase as a whole is stationary.

(a) What is the molar flux of carbon dioxide?

(b) What is the net mass flux in kg/m^2 h?

(c) At what speed in m/s an observer has to move so that the net mass flux relative to him would be zero?

(d) At what speed, would he have to move so that relative to him nitrogen is stationary?

(e) What would be the molal flux of carbon dioxide relative to the observer under conditions in (d)?

(**Answer:** (a) 1.388×10^{-4} kmol/m^2 h (b) 2.21×10^{-3} kg/m^2 h (c) 4.65×10^{-7} m/s (d) 9.71×10^{-7} m/s (e) 4.33×10^{-8} kmol/m^2 s.)

2.13 The diffusivity of toluene in air is measured by Stefan's method. A vertical glass tube is filled with toluene to a depth of 20 mm from the top open end. It took 280.5 hours for the level to fall to 80 mm from the top. The temperature and pressure are maintained at 310 K and 100 kPa. Density of liquid toluene is 850 kg/m^3 and the vapour pressure of toluene is 7.64 kPa at 310 K. Calculate the diffusivity of toluene in air at 310 K. (**Answer:** 8.9×10^{-6} m^2/s)

2.14 Chloropicrin ($CCl_3\ NO_2$) is contained in a tube of cross sectional area 2.5×10^{-4} m^2 to a depth 100 mm from the top at a temperature 300K and pr[illegible] [illegible]3 kPa. The vapour pressure of chloropicrin is 3.17 kPa and the diffusivity of chloropicrin in air is 8.8×10^{-4} m^2/s. Calculate the rate of evaporation of chloropicrin in kg/s. (**Answer:** 4.672×10^{-7} kg/s)

2.15 The diffusivity of carbon tetrachloride in oxygen is determined by evaporating it into a stream of oxygen at 273 K and 100 kPa. The cross sectional area of the tube is 8.2×10^{-5} m^2 and the liquid level in the tube is maintained at 200 mm from the top of the tube. The vapour pressure of carbon tetrachloride at 273 K is 4.4 kPa. It has been observed that 2.18×10^{-8} m^3 of tetrachloride is evaporated in 10 hours after attaining steady state. Determine the diffusivity of carbon tetrachloride in air at 273 K. Density of the liquid is 1540 kg/m^3.

(**Answer:** 7.46×10^{-6} m^2/s)

2.16 Air at 298 K and 100 kPa with a dew-point temperature of 278 K flows past the open end of a vertical tube filled with liquid water to a level 12.5 mm from the top. The diffusivity of water in air is 2.56×10^{-5} m^2/s.

(a) How long it will take for the liquid level in the tube to drop to 25 mm from the top.

(b) Make a plot of liquid level in the tube as function of time.

Vapour pressure of water at 278 K is 0.87 kPa and Vapour pressure at 298 K is 3.17 kPa.

(**Answer:** (a) 149.12 h (b) $Z_t^2 = 1.5625 \times 10^{-4} + 3.1435 \times 10^{-6} t$)

2.17 Water in an open dish exposed to dry air at 298 K and 100 kPa is found to vaporise at a constant rate of 0.4 kg/h m^2. Assuming the water surface to be at 298 K, calculate the effective gas film thickness. Diffusivity of water vapour in air is 2.56×10^{-5} m^2/s and vapour pressure of water at 298 K is 3.17 kPa. (**Answer:** 5.39 mm)

2.18 A narrow tube was filled with acetone to 11 mm from the top and maintained at a temperature 293 K and pressure 100 kPa in a gentle current of air. If the level of liquid has fallen to 20.5 mm in 5 hours, calculate the diffusivity of acetone in air. The vapour pressure of acetone in air at 293 K is 24 kPa and density of liquid acetone is 790 kg/m^3. (**Answer:** 1.005×10^{-5} m^2/s)

2.19 Ammonia is diffusing at a constant rate through a stagnant layer of air film 1 mm thick, through a semipermeable membrane, and then through a stagnant layer of hydrogen 1 mm thick. The concentration of ammonia at the outer boundary of air layer is 0.75% (volume) and that at the outer boundary of hydrogen layer is 0.001% (volume). The temperature is 298 K and pressure is 100 kPa. The resistance of the membrane to diffusion is negligible. Diffusivity of ammonia in air is 2.26×10^{-5} m^2/s and that of ammonia in hydrogen is 7.83×10^{-5} m^2/s. Calculate:

(a) Rate of diffusion of ammonia in kmol/m^{2}s

(b) Partial pressure of ammonia at the membrane.

(**Answer:** (a) 5.302×10^{-6} kmol/m^2 s (b) 0.1692 kPa)

2.20 A 5 m diameter cylindrical tank was filled with acetone to a level of 25 mm from the top and is exposed to ambient air at 20°C and 101.3 kPa pressure. Vapour pressure of acetone at 20°C is 25 kPa and density of acetone liquid is 790 kg/m^3. The diffusivity of acetone at 20°C is 9.2×10^{-6} m^2/s. Calculate:

(a) The fall in level of acetone in five hours.

(b) The loss in rupees, if the cost of acetone is Rs. 25/L.

(**Answer:** (a) 5.2 mm (b) Rs. 2547.6)

2.21 Chloroform vapour is diffusing from the surface of the liquid contained in a capillary in to ambient air at 298 K and 101.3 kPa. The diffusivity of chloroform in air at this condition is 9.5×10^{-6} m^2/s. The vapour pressure of chloroform at 298 K is 26 kPa. If the length of the diffusion path is 70 mm from the surface of the liquid, estimate the rate of steady state diffusion in kmol/(m^{2}s). (**Answer:** 1.6467×10^{-6} kmol/m^{2}s)

2.22 Oxygen is diffusing through carbon dioxide under steady-state condition, with CO_2 non-diffusing at temperature 293 K and pressure 101.3 kPa. The partial pressure of oxygen at two planes

3 mm apart are 15 kPa and 7.5 kPa. The diffusivity of oxygen in CO_2 is 1.53×10^{-5} m^2/s. Calculate the rate of diffusion per unit area of the planes. (**Answer:** 1.768×10^{-5} $kmol/m^2s$)

2.23 A 1 mm thick layer of water is being evaporated to the surrounding quiescent air, both air and water being at 302 K and 100 kPa. Air is having a humidity 0.002 kg water per kilogram dry air. The evaporation is assumed to take place by molecular diffusion through a gas film 6 mm thick. The mean diffusivity of water vapour in air is 2.7×10^{-5} m^2/s. The vapour pressure of water at 302 K is 4 kPa. Determine the time required for complete evaporation. (**Answer**: 2.29 h)

2.24 A 2-mm thick layer of water is being evaporated at 298 K and 100 kPa to the surrounding quiescent air at the same temperature and pressure and having a humidity 0.002 kg water per kilogram dry air. The evaporation is assumed to take place by molecular diffusion through a gas film 5 mm thick. The mean diffusivity of water vapour in air is 2.56×10^{-5} m^2/s. The vapour pressure of water at 298 K is 3.2 kPa. Determine the time required for evaporation of half the quantity of water. (**Answer:** 2.55 h).

2.25 Water is evaporating from an open bowl, 0.3 m in diameter, at 350 K and the vapour are removed as they are formed by air flowing across the bowl. If the resistance to mass transfer is equivalent to that of a 1 mm layer for conditions of molecular diffusion, what will be the rate of cooling due to evaporation? The water can be considered as well mixed and the water equivalent of the system is equal to 10 kg. The diffusivity of water vapour in air may be taken as 2.0×10^{-5} m^2/s. The vapour pressure of water at 350 K is 42 kPa and the latent heat of vaporisation of water = 2316 kJ/kg. (**Answer:** 1.57 K/min)

2.26 Diffusivity of chloroform at 298 K and 100 kPa is measured by an Arnold cell. Chloroform surface was initially at a height of 50 mm from the top and after 10 hours the liquid surface dropped by 6.3 mm. The concentration of chloroform at the surface of the tube is negligible and at the liquid surface, air is saturated with chloroform vapours. If the liquid density is 1485 kg/m^3 and the vapour pressure at 298 K is 26 kPa, determine the diffusivity of chloroform in air. (**Answer:** 9.52×10^{-6} m^2/s)

2.27 In measuring the diffusivity of carbon tetra chloride in air at 321 K and 101.3 kPa, the fall in level of liquid in the tube is measured against time. The results are given below:

Time, min	0	40	125	364	690	1120	1650	2280	3010	3840	4770	5800
Level, mm	0	4	10	20	30	40	50	60	70	80	90	100

The concentration of carbon tetrachloride at the surface of the tube may be assumed zero. The density of carbon tetrachloride liquid is 1540 kg/m^3 and the vapour pressure of carbon tetrachloride at 321 K is 37.60 kPa. (**Answer:** 9.47×10^{-6} m^2/s)

Diffusion in multicomponent mixtures

2.28 In an ammonia-air mixture, the concentrations of ammonia at two points 100 mm apart are 20 percent (mol) and 5 percent (mol) respectively. The system is maintained at a total pressure of 101.3 kPa and a temperature of 298 K. Evaluate the following:

(a) The rate of diffusion assuming that ammonia alone diffuses.

(b) The rate of diffusion assuming equimolar counter diffusion.

(c) The concentration of ammonia in mol percent at a distance 50 mm away from point 1 in the direction of mass transfer, for conditions in part (a) and (b).

(d) If air is replaced by a mixture of nitrogen and hydrogen in the ratio 1:3, other conditions remaining the same, what would be the effective diffusivity and rate of diffusion, if ammonia alone is diffusing.

At 298 K the diffusivity of ammonia in air is 2.26×10^{-5} m²/s, the diffusivity of ammonia in nitrogen is 2.33×10^{-5} m²/s and the diffusivity of ammonia in hydrogen is 7.83×10^{-5} m²/s.

(**Answer:** (a) 1.585×10^{-6} kmol/m² s (b) 1.3865×10^{-6} kmol/m²s (c) 0.128, 0.125 (d) 3.461×10^{-6} kmol/m²s)

2.29 Ammonia is diffusing through a stagnant gas mixture containing one-third nitrogen and two-thirds hydrogen by volume. The total pressure is 200 kPa and the temperature is 327 K. Calculate the rate of diffusion of ammonia through a film of gas 1 mm thick, if the concentration of ammonia at the film surface are 10% and 5%, respectively. At 200 kPa pressure and 298 K the diffusivity of ammonia in nitrogen is 1.17×10^{-5} m²/s and diffusivity of ammonia in hydrogen is 3.92×10^{-5} m²/s. (**Answer**: 1.005×10^{-4} kmol/m² s)

2.30 Methane (**A**) is diffusing through non diffusing mixture of argon (**B**) and helium (**C**) at 298 K and 100 kPa. The partial pressures of methane, argon and helium are 40 kPa, 40 kPa and 20 kPa, respectively, at plane 1 and 10 kPa, 60 kPa and 30 kPa, respectively at plane 2. If the planes are 5 mm apart and the diffusivities are $D_{AB} = 2.02 \times 10^{-5}$ m²/s, and $D_{AC} = 6.75 \times 10^{-5}$ m²/s, calculate the molar flux of methane. (**Answer:** 8.6261×10^{-5} kmol/m²s)

2.31 Oxygen is allowed to diffuse through a gas mixture containing one-fifth hydrogen and four-fifth methane by volume at a pressure of 100 kPa and temperature of 293 K. Diffusivity of oxygen in hydrogen is 7.1×10^{-5} m²/s and diffusivity of oxygen in methane is 1.88×10^{-5} m²/s. What would be the rate of diffusion of oxygen through a gas film 3 mm thick when the concentration of oxygen at the surfaces are 12% (volume) and 7% (volume), respectively.

(**Answer:** 1.6667×10^{-5} kmol/m²s)

Steady-state molecular diffusion in liquids

2.32 Calculate the rate of diffusion of acetic acid across a film of water at 293 K. The film is 1 mm thick and the water is non diffusing. The concentration of acetic acid across the film are 15% (weight) and 5%(weight) respectively. The diffusivity of acetic acid in water is 1.19×10^{-9} m²/s. Density of 15% solution is 1020 kg/m³ and the density of 5% solution is 1006 kg/m³. (**Answer:** 2.241×10^{-6} kmol/m²s)

2.33 Calculate the rate of diffusion of sugar in kg/m² s through a film of coffee 1 mm thick when the concentrations are 15% (weight) and 5% (weight) respectively, on either side of film. Assume that the diffusivity of sugar through the coffee under the given conditions to be 7×10^{-10} m²/s and the density of 10% solution to be 1013 kg/m³.Assume the properties of coffee to be the same as that of water. Molecular weight of sugar = 342.

(**Answer:** 7.855×10^{-5} kg/m²s)

2.34 An ethanol – water solution in the form of a stagnant film 2 mm thick at 293 K is in contact at one surface with an organic solvent in which ethanol is soluble and water is insoluble. At point 1 the concentration of ethanol is 20% (weight) and the solution density is 968.6 kg/m³. At point 2 the concentration of ethanol is 10% (weight) and the density is 981.9 kg/m³. The diffusivity of ethanol in water is 1.0×10^{-9} m²/s. Calculate the steady-state flux.

(**Answer:** 1.249×10^{-6} kmol/m²s)

2.35 The solute HCl is diffusing through a thin film of water 2 mm thick at 293 K. The concentration of HCl at one boundary of the film is 14% (weight) and at the other boundary is 6%(weight). If the density of the 14% solution is 1067.5 kg/m³ and the density of the 6% solution is 1027.9 kg/m³. The diffusivity of HCl in water is 2.5×10^{-9} m²/s. Calculate the molar flux of HCl in water in kmol/m² s. (**Answer:** 3.193×10^{-6} kmol/m²s)

2.36 Sodium chloride at 298 K is diffusing through a stagnant film of water 1.5 mm thick when the concentrations are 25% (weight) and 15% (weight) on the opposite sides of the film. The rate of diffusion of NaCl is 2.21×10^{-6} kmol/m^{2}s. The density of 25% and 15% solutions are 1186 kg/m^3 and 1107 kg/m^3, respectively. Determine the diffusivity of NaCl through water.

(**Answer:** 1.349×10^{-9} m^2/s)

2.37 It is desired to extract methanol from a methanol-water solution. Assume that water is insoluble in the extracting solution, that all mass transfer takes place by diffusion of methanol through a thin film of thickness 0.005 m and that the concentration of methanol on the water side is 15% (weight) and on the other side 6% (weight). The diffusivity may be assumed to be 1.6×10^{-9} m^2/s. The densities of the pure liquids are 990 kg/m^3 for water and 800 kg/m^3 for methanol. Estimate the molar flux of methanol. (**Answer:** 9.2×10^{-7} kmol/m^{2}s).

2.38 A film of ammonia-water solution at 278 K and 4 mm thick is in contact at one surface with an organic liquid. The concentration of ammonia in the organic phase is held constant and is such that the equilibrium concentration of ammonia in the water at this interface is 2% (weight) ammonia and the concentration of ammonia in water at the other end of the film 4 mm away is 10 % (weight). The densities of the aqueous ammonia solution are 991.7 kg/m^3 and 961.7 kg/m^3 respectively, for 2% and 10% solution. Water and the organic are insoluble in each other. The diffusion coefficient of ammonia in water is 1.24×10^{-9} m^2/s. Calculate the molar flux.

(**Answer:** 1.516×10^{-6} kmol/m^{2}s)

Prediction of binary diffusivities

2.39 Estimate the diffusivity of CO_2 in air at 293 K and 101.325 kPa by the following methods:
(a) Hirschfelder equation
(b) Wilke-Lee equation
(c) Fuller, et al. equation
Compare the result with the experimental value of 1.55×10^{-5} m^2/s.

(**Answer:** (a) 1.468×10^{-5} m^2/s (b) 1.579×10^{-5} m^2/s (c) 1.53×10^{-5} m^2/s)

2.40 Estimate the diffusivity of argon in methane at 298 K and 101.3 kPa using
(a) Hirschfelder equation
(b) Wilke-Lee equation
(c) Fuller, et al. equation

(**Answer:** (a) 2.1×10^{-5} m^2/s (b) 2.23×10^{-5} m^2/s (c) 2.12×10^{-5} m^2/s)

2.41 The diffusivity of chloroform in air at 298K and 101.3 kPa is reported as 9.5 x 10^{-6} m^2/s. Evaluate the diffusivity by the following methods and compare it with experimental value.
(a) Hirschfelder equation
(b) Wilke-Lee equation
(c) Fuller, et al. equation

(**Answer:** (a) 8.43×10^{-6} m^2/s (b) 9.14×10^{-6} m^2/s (c) 8.93×10^{-6} m^2/s)

2.42 Estimate the diffusion coefficient for benzene - oxygen mixture at 311 K and 202.6 kPa using the method of
(a) Wilke-Lee
(b) Fuller, et al.
Compare the result with the experimental value of 5.05×10^{-6} m^2/s

(**Answer:** (a) 4.955×10^{-6} m^2/s (b) 4.88×10^{-6} m^2/s)

2.43 Estimate the diffusivity of nitrogen (**A**)-benzene (**B**) at 311.3 K and 300 kPa using the method of Fuller, et al. Compare with the experimental value of 3.4×10^{-6} m^2/s

(**Answer:** 3.36×10^{-6} m^2/s)

2.44 *n*-Butanol (**A**) is diffusing through air (**B**) at 101.3 kPa and 273 K. Using the Fuller et al. method, estimate the diffusivity D_{AB}. (**Answer:** 7.674×10^{-6} m^2/s)

2.45 Calculate the diffusivity of ammonia in nitrogen at 358 K and 200 kPa using Hirschfelder equation and Fuller, et al. equation and compare with the experimental value of 1.66×10^{-5} m²/s. For ammonia, $\sigma_A = 0.29$ nm, $\frac{\varepsilon_A}{\kappa} = 558.3\,\text{K}$.

(**Answer:** 1.56×10^{-5} m²/s Hirschfelder, 1.58×10^{-6} m²/s Fuller et al).

2.46 Use the Wilke-Chang equation to estimate the diffusivity of aniline (**A**) in aqueous solution at 293 K. Compare the result with the experimental value of 9.2×10^{-10} m²/s. The viscosity of water at 293 K is 1.01 cP and the molar volume of liquid aniline at its normal boiling point of 457.6 K is 0.107 m³/kmol. (**Answer:** 8.9×10^{-10} m²/s)

2.47 Estimate using Wilke-Chang equation, the diffusivity of ethanol in a dilute solution with water at 298 K given that the atomic volumes of carbon, hydrogen, and oxygen are respectively, 14.8×10^{-3}, 3.7×10^{-3}, 7.4×10^{-3} m³/kmol and viscosity of a solution containing 50 mol alcohol per m³ of water is 1.45 cP. (**Answer:** 8.97×10^{-10} m²/s)

2.48 Calculate the diffusivity of ethanol in dilute solution in water at 298 K. Viscosity of the solution is 1.45 cP. (**Answer:** 8.99×10^{-10} m²/s).

2.49 Calculate the diffusivity of mannitol ($C_6H_{14}O_6$) in dilute solution in water at 293 K. Viscosity of the solution is 1.00 cP. (**Answer:** 6.47×10^{-10} m²/s)

2.50 Estimate the diffusion coefficient of acetone in water at 298 K and 101.3 kPa using the Wilke-Chang equation. Compare the result with the experimental value of 1.28×10^{-9} m²/s. The viscosity of water is 0.89 cP. (**Answer:** 1.277×10^{-9} m²/s)

2.51 Determine the diffusivity of acetic acid in acetone at infinite dilution at 313 K and compare the result with the experimental value of 4.04×10^{-9} m²/s. The viscosity of acetone is 0.27 cP and the molar volume of acetone is 77.5×10^{-3} m³/kmol. (**Answer:** 3.857×10^{-9} m²/s)

2.52 Determine the diffusivity of formic acid in benzene at infinite dilution at 298 K. The viscosity of benzene 0.6 cP and the molar volume of benzene is 96×10^{-3} m³/kmol.

(**Answer:** 2.147×10^{-9} m²/s)

Chapter 3

Convective Transport and Interphase Mass Transfer

Mass transfer operations involve the transfer of material between two immiscible phases. The phases in contact may be a gas and a liquid as in gas absorption, distillation, humidification and dehumidification operations, two immiscible liquids, as in liquid–liquid extraction, a gas and a solid as in drying, a liquid and a solid as in leaching, etc. The mechanisms of mass transfer in stagnant fluids and laminar flow systems are well understood and can be described by the molecular diffusion models discussed in Chapter 2. The molecular diffusion models are, however, inadequate to explain the mass transfer phenomena in fluids in turbulent flow past solid surfaces and transfer of mass across phase boundaries when two immiscible phases are in contact. The mass fluxes in these situations are represented in terms of mass transfer coefficients.

The rate of transfer and the conditions under which equilibrium is established between the phases are two important aspects in the analysis of mass transfer operations and in the design and sizing of mass transfer equipment. In the following chapters the individual mass transfer operations are considered where the equilibrium and the diffusional phenomena characteristic to each operation are discussed in detail. In this chapter we define mass transfer coefficients and introduce methods for evaluation of the coefficients for different geometric configurations and flow systems. Also certain basic features of interphase mass transfer common to all these operations are discussed with examples.

3.1 MASS TRANSFER COEFFICIENTS

Consider turbulent flow of water through a circular pipe the inside of which is coated with a fused salt that is only slightly soluble in water. The solute gets transferred from the wall to the water stream. In the region immediately adjacent to the surface, the flow is laminar and in this region, mass transfer is due to molecular diffusion. Outside this laminar sub-layer and occupying the major portion of the fluid cross section, is the turbulent core in which diffusion of solute is primarily due to eddy diffusion. Large aggregates of fluid, called eddies move rapidly from one position to the other carrying with them the dissolved salt. Though molecular diffusion also is present in the turbulent core, its contribution to the total mass transfer is negligible compared to the transfer by eddy diffusion. In this region, the rate of transfer of solute is proportional to eddy diffusivity, E_D and the concentration gradient in the radial direction $\frac{dC}{dr}$. In the region very close to the pipe wall, the turbulence approaches zero and the rate of transfer by eddy diffusion is negligible in comparison to the rate of transfer by molecular diffusion, which in turn is proportional to the mass diffusivity D_{AB} and the concentration gradient $\frac{dC}{dr}$. Transfer of solute from the pipe wall to the fluid in turbulent flow is thus partly due to molecular diffusion and partly due to eddy diffusion; for both, the transfer rates are proportional to the concentration gradient. Thus it is possible to write the mass flux in turbulent transport analogous to Fick's law of molecular diffusion:

$$J_A = -\left(D_{AB} + E_D\right)\frac{dC_A}{dz} \qquad \text{...(3.1)}$$

While molecular diffusivity D_{AB} dominates in the region near the wall (the diffusional sub layer), the eddy diffusivity E_D is prominent in the turbulent core. At a given temperature and pressure, the molecular diffusivity D_{AB} is constant. However, the eddy diffusivity E_D depends not only on pressure, temperature and nature of the constituents, but is a function of position and flow patterns in the fluid. Though Eq. (3.1) appears to be theoretically sound, it is not of much practical use for calculation of mass transfer rates, as it is not easy to describe the flow patterns mathematically. Also, the relative importance of eddy diffusion and molecular diffusion in the overall process is not known. For these reasons it is customary to describe the mass transfer in fluids in turbulent flow in terms of mass transfer coefficients. The mass transfer coefficient k is defined as

$$k = \frac{\text{mass flux}}{\text{concentration difference}} \qquad \text{...(3.2)}$$

Generally, the mass transfer coefficient k is used to represent mass flux for diffusion of one component when the other component is stagnant and the mass transfer coefficient k' is used to represent mass flux due to equimolar counter diffusion. Thus we have for diffusion of **A** through stagnant **B** (one-way diffusion),

$$\text{mass or molar flux} = k \times \text{concentration difference} \qquad \text{...(3.3)}$$

and for equimolar counter diffusion

$$\text{mass or molar flux} = k' \times \text{concentration difference} \qquad \text{...(3.4)}$$

This definition of mass transfer coefficient is analogous to Newton's law of cooling which defines the heat transfer coefficient h as:

$$h = \frac{\text{heat flux}}{\text{temperature difference}} = \frac{q}{\Delta T} \qquad \text{...(3.5)}$$

where q is the heat flux and ΔT is the temperature difference. In heat transfer, the driving force for transfer is the temperature difference ΔT. The driving forces in mass transfer can be expressed in a number of ways. These include the difference in the molar concentration (ΔC), the difference in the partial pressures of the diffusing component in the case of gaseous systems ($\Delta \bar{p}$), and the difference in the mole fractions of the diffusing component (Δy or Δx). Depending upon the concentration difference driving force used for indicating the mass flux, the mass transfer coefficients can be represented by different symbols, having different units. Thus k_c represents the mass transfer coefficient in gaseous systems with molar concentration difference driving force.

$$k_C = \frac{N_A}{\Delta C_A} = \frac{N_A}{C_{Ai} - C_{Ab}} \qquad \text{...(3.6)}$$

Here, C_{Ai} is the molar concentration of **A** in the fluid-solid interface and C_{Ab} is the concentration of **A** in the bulk of the fluid. Substituting the units of N_A (kmol/m^2 s) and C (kmol/m^3), the units for k_c can be determined:

$$\frac{N_A\left(\frac{\text{kmol}}{\text{m}^2\text{s}}\right)}{\Delta C_A\left(\frac{\text{kmol}}{\text{m}^3}\right)} = k_c\left(\frac{\text{m}}{\text{s}}\right) \quad ...(3.7)$$

The mass transfer coefficients k_y and k_G are defined in a similar manner choosing the difference in mole fractions Δy or the difference in partial pressures $\Delta\bar{p}$, respectively, for the driving force for mass transfer. Different types of mass transfer coefficients, their units and their interrelationships are given in Table 3.1. The table also lists k_x and k_L, the liquid phase counterparts of mass transfer coefficients k_y and k_c. The relationship among mass transfer coefficients k_y, k_G and k_c are obtained as given below:

$$N_A = k_y(y_{Ai} - y_{Ab}) \quad ...(3.8)$$

$$N_A = k_c(C_{Ai} - C_{Ab}) = k_c C(y_{Ai} - y_{Ab}) \quad ...(3.9)$$

$$N_A = k_G(\bar{p}_{Ai} - \bar{p}_{Ab}) = k_G P(y_{Ai} - y_{Ab}) \quad ...(3.10)$$

Comparison of Eqs. (3.8), (3.9) and (3.10) reveals that

$$k_y = k_c C = k_G P \quad ...(3.11)$$

Table 3.1: Definition of mass transfer coefficients and their interrelationships

Driving force	Coefficient	Rate equation: Equimolar counter diffusion	Rate equation: Diffusion in stagnant gas	Units of coefficient in SI
		Gases		
Mole fraction difference	k'_y or k_y	$N_A = k'_y \Delta y_A$	$N_A = k_y \Delta y_A$	kmol/[m² s (mole fraction)]
Partial pressure difference	k'_G or k_G	$N_A = k'_G \Delta\bar{p}_A$	$N_A = k_G \Delta\bar{p}_A$	$\frac{\text{kmol}}{\text{m}^2\text{s(Pa)}}$
Molar concentration difference	k'_c or k_c	$N_A = k'_c \Delta C_A$	$N_A = k_c \Delta C_A$	$\frac{\text{m}}{\text{s}}$
	$k_y = k_c C = k_G P = \left(k_c \frac{P}{RT}\right)$		$k'_y = k'_c C = k'_G P = k'_c \left(P/RT\right)$	
		Liquids		
Mole fraction difference	k'_x or k_x	$N_A = k'_x \Delta x_A$	$N_A = k_x \Delta x_A$	$\frac{\text{kmol}}{\text{m}^2\text{s}}$
Molar concentration difference	k'_L or k_L	$N_A = k'_L \Delta C_A$	$N_A = k_L \Delta C_A$	$\frac{\text{m}}{\text{s}}$
	$k_x = k_L C = \left(k_L \frac{\rho}{M}\right)$		$k'_x = k'_L C = \left(k'_L \frac{\rho}{M}\right)$	

For equimolar counter diffusion, equations similar to Eqs. (3.6) through (3.11) can be written with k replaced by k'. These are given below:

$$N_A = k'_y(y_{Ai} - y_{Ab}) \quad ...(3.12)$$

$$N_A = k'_c(C_{Ai} - C_{Ab}) = k'_c C(y_{Ai} - y_{Ab}) \quad ...(3.13)$$

$$N_A = k'_G(\bar{p}_{Ai} - \bar{p}_{Ab}) = k'_G P(y_{Ai} - y_{Ab}) \quad ...(3.14)$$

Comparison of Eqs. (3.12), (3.13) and (3.14) reveal that

$$k'_y = k'_c C = k'_G P \quad \text{...(3.15)}$$

Analogous to the film coefficient of heat transfer developed for a fluid in turbulent flow past a solid surface, the film coefficient of mass transfer also can be defined. Let us first consider the heat transfer from the solid surface to the fluid (Fig. 3.1). Let the temperature difference between the solid surface and the bulk fluid be ΔT. The fall in temperature is very steep in the stagnant fluid layer (the viscous sublayer) close to the solid surface. The temperature levels off rapidly beyond the viscous layer due to convective currents and the mixing motion of the eddies.

If the entire temperature gradient is assumed to be confined in a thin stagnant fluid layer of thickness δ within which the mechanism of heat transfer is assumed to be conduction alone, the heat flux can be estimated using Fourier's law of heat conduction:

$$q = k\frac{\Delta T}{\delta} \quad \text{...(3.16)}$$

where q is the heat flux, k is the thermal conductivity of the fluid, and δ is the thickness of the fictitious film. Since the film is immeasurably thin, δ is not available for using Eq. (3.16) for estimating the convective heat flux. The thermal conductivity and film thickness are combined in the single quantity h known as the film coefficient of heat transfer.

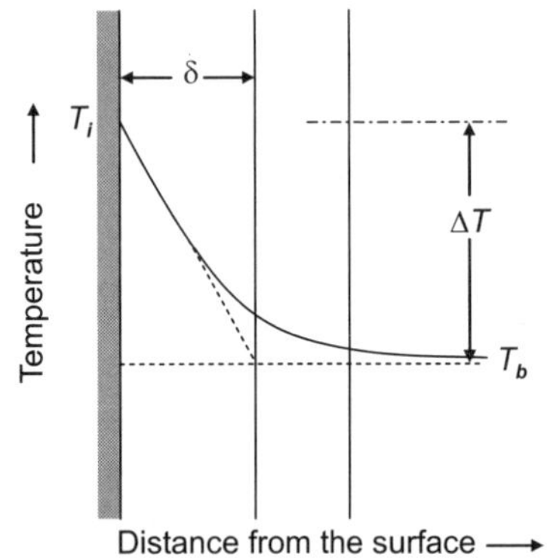

Fig. 3.1: Film concept of heat transfer coefficient

$$h = \frac{k}{\delta} \quad \text{...(3.17)}$$

so that Eq. (3.16) becomes

$$q = h\Delta T \quad \text{...(3.18)}$$

Equations (3.17) and (3.18) show how the film concept of heat transfer coefficient is developed using the Fourier's law of heat conduction. Now consider the transfer of solute from the surface of the solid to a fluid moving in turbulent flow past the surface (Fig. 3.2). The concentration of the solute decreases as the distance from the wall increases. In the film theory of mass transfer, the entire concentration change from C_{Ai} to C_{Ab} is assumed to be confined within an effective film of thickness Z. It is also assumed that within this fictitious film, the change in concentration with distance is linear as in steady-state equimolar counter diffusion. If the concept of film coefficient is extended to represent the convective transport of mass, the film coefficient of mass transfer by analogy with Eq. (3.17) would be $\dfrac{CD_{AB}}{Z}$.

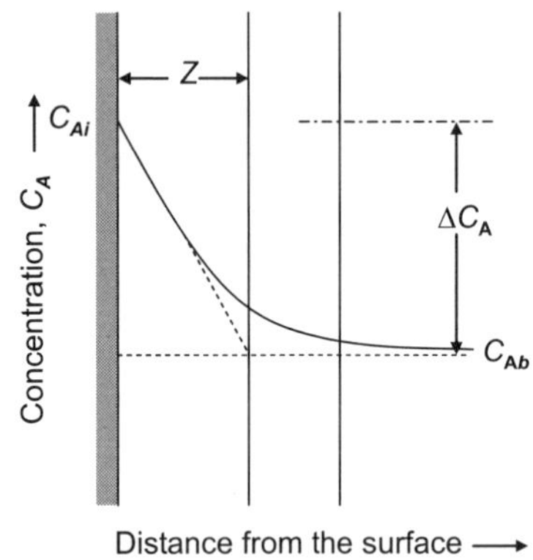

Fig. 3.2: Film concept of mass transfer coefficient

For equimolar counter diffusion we have Eq. (2.37) to represent molar flux.

$$N_A = \frac{CD_{AB}}{Z}(y_{Ai} - y_{Ab}) \quad \text{...(2.37)}$$

We have defined mass transfer coefficient k' by Eq. (3.12) as

$$N_A = k'_y(y_{Ai} - y_{Ab}) \quad \text{...(3.12)}$$

Comparison Eq. (2.37) with Eq. (3.12) reveals that $k'_y = \frac{CD_{AB}}{Z}$, or the mass transfer coefficient k'_y also represents the film coefficient of mass transfer. However, for diffusion of **A** through stagnant **B**, the flux is given by Eq. (2.55) as

$$N_A = \frac{CD_{AB}}{Zy_{B,M}}(y_{Ai} - y_{Ab}) \qquad ...(2.55)$$

and the flux in terms of mass transfer coefficient k_y is

$$N_A = k_y(y_{Ai} - y_{Ab}) \qquad ...(3.8)$$

which imply that

$$k_y y_{B,M} = \frac{CD_{AB}}{Z} = k'_y \qquad ...(3.19)$$

k_y will represent the film coefficient for mass transfer only when $y_{B,M}$ is unity, that is, when mass transfer is taking place in very dilute solutions. $y_{B,M}$ as we have seen is a correction term to account for the bulk flow contribution to molar flux, N_A. Therefore, if the bulk flow is absent, we have

$$k_y = k'_y = \frac{CD_{AB}}{Z} \qquad ...(3.20)$$

The mass transfer coefficients k_y, k_G and k_c are related as given by Eq. (3.11)

$$k_y = k_C C = k_G P \qquad ...(3.11)$$

Using Eq. (3.19) in the above result, we can derive the following relationships as well:

$$k_y y_{B,M} = k_C C\, y_{B,M} = k_G P\, y_{B,M} = \frac{CD_{AB}}{Z} = k'_y \qquad ...(3.21A)$$

or,

$$k_y y_{B,M} = k_C C_{B,M} = k_G \bar{p}_{B,M} = \frac{CD_{AB}}{Z} = k'_y \qquad ...(3.21B)$$

While the heat transfer coefficient determined as the ratio of heat flux to the temperature gradient also represents the film coefficient of heat transfer, the mass transfer coefficient measured experimentally as the ratio of mass flux to the concentration difference ($k_y = N_A/\Delta y$) will represent the film coefficient of mass transfer only for cases involving low mass transfer rates or cases where bulk flow contribution is negligible. The k values determined as the ratio of mass flux to concentration difference, should be multiplied by the bulk flow correction term $y_{B,M}$ or its equivalent before applying to mass transfer under conditions different from those prevailing in the experimental evaluation.

Example 3.1: Show that the mass transfer coefficients k_x and k_L for liquid phases are related as $k_x = \left(k_L \frac{\rho}{M}\right)$

***Solution*:** Corresponding to Eqs. (3.8) and (3.9), we have for diffusion of **A** through stagnant liquid **B**

$$N_A = k_x(x_{Ai} - x_{Ab}) \qquad ...(3.22)$$

$$N_A = k_L(C_{Ai} - C_{Ab}) = k_L C(x_{Ai} - x_{Ab}) \qquad ...(3.23)$$

Comparing Eqs. (3.22) and (3.23), we see that

$$k_x = k_L C \qquad ...(3.24)$$

The molar concentration is related to the density and molecular weight of the solution:

$$C = \frac{\rho}{M} \qquad ...(3.25)$$

Substituting Eq. (3.24) into Eq. (3.25), we get

$$k_x = k_L\left(\frac{\rho}{M}\right) \qquad ...(3.26)$$

Example 3.2: Prove that the mass transfer coefficients in equimolar counterdiffusion in gases, the coefficients k'_G and k'_C, are related as $k'_G = \frac{k'_C}{RT}$.

***Solution*:** From Eq. (3.15) we see that

$$k'_C C = k'_G P \qquad ...(3.27)$$

Since $C = \frac{P}{RT}$, Eq. (3.27) becomes

$$k'_C \frac{P}{RT} = k'_G P \qquad ...(3.28)$$

or

$$\frac{k'_C}{RT} = k'_G \qquad ...(3.29)$$

Example 3.3: Show that the mass transfer coefficients k'_x, k_x and k_L are related as

$$k'_x = k_x x_{B,M} = k_L C x_{B,M}$$

***Solution*:** For diffusion of **A** through stagnant **B** in the liquid phase, the molar flux is given by Eq. (2.92)

$$N_A = \frac{CD_{AB}}{Zx_{B,M}}(x_{A1} - x_{A2}) \qquad ...(3.30)$$

which may be written equivalently as

$$N_A = \frac{D_{AB}}{Zx_{B,M}}(C_{A1} - C_{A2}) \qquad ...(3.31)$$

In terms of the mass transfer coefficient k_x and k_L, the molar flux can be written as

$$N_A = k_x(x_{A1} - x_{A2}) \qquad ...(3.32)$$

$$N_A = k_L(C_{A1} - C_{A2}) \qquad ...(3.33)$$

Comparing Eq. (3.30) with Eq. (3.32) and Eq. (3.31) with Eq. (3.33), we get

$$k_x x_{B,M} = k_L C x_{B,M} = \frac{CD_{AB}}{Z} \qquad ...(3.34)$$

The film coefficient for liquid systems can be represented by

$$k'_x = \frac{CD_{AB}}{Z} \qquad ...(3.35)$$

Substituting Eq. (3.35) into Eq. (3.34) we get

$$k'_x = k_x x_{B,M} = k_L C x_{B,M} \qquad ...(3.36)$$

Example 3.4: Dry air is passed across a flat surface at 300 K and 100 kPa. The surface is covered with a thin film of an organic liquid which gets evaporated into the air stream. The mass transfer coefficient k'_y is estimated to be 1.2×10^{-5} kmol/m² s. Assume one-way diffusion. Determine:

(a) The rate of vaporisation of the liquid if the vapour pressure is 10 kPa at 300 K.

(b) The mass transfer coefficient k_y.

***Solution*:** Temperature $T = 300$ K, Pressure $P = 100$ kPa

$$k'_y = 1.2 \times 10^{-5}\ \text{kmol/m}^2\text{s}$$

(a) Mass transfer occurs from the surface of the liquid to the air. The air layer in contact with the liquid is assumed to be saturated with the vapours of the liquid so that the partial pressure of the vapour at the surface is equal to the vapour pressure, or $\bar{p}_{A1} = P_A^s = 10$ kPa. Since air passed across the surface is free of the vapours of the liquid, $\bar{p}_{A2} = 0$

The mole fraction of the vapours of the liquid are:

$$y_{A1} = \frac{\bar{p}_{A1}}{P} = \frac{10}{100} = 0.1 \text{ and } y_{A2} = 0$$

The mole fraction of air are:

$$y_{B1} = 1 - y_{A1} = 1 - 0.1 = 0.9$$
$$y_{B2} = 1 - y_{A2} = 1 - 0 = 1.0$$

$$y_{B,M} = \frac{y_{B2} - y_{B1}}{\ln \dfrac{y_{B2}}{y_{B1}}} = \frac{1.00 - 0.90}{\ln \dfrac{1.00}{0.90}} = 0.9491$$

For diffusion of **A** through stagnant **B** (one way diffusion), the rate of mass transfer is given by

$$N_A = \frac{CD_{AB}}{Zy_{B,M}}(y_{A1} - y_{A2}) = \frac{k'_y}{y_{B,M}}(y_{A1} - y_{A2})$$

The rate of vaporisation of the liquid is

$$N_A = \frac{k'_y}{y_{B,M}}(y_{A1} - y_{A2}) = \frac{1.2 \times 10^{-5}}{0.9491} \times (0.1 - 0) = 1.264 \times 10^{-6} \frac{\text{kmol}}{\text{m}^2\text{s}} \qquad \textbf{(Answer)}$$

(b) Mass transfer coefficients k_y and k'_y are related as $k_y y_{B,M} = k'_y$

$$k_y = \frac{k'_y}{y_{B,M}} = \frac{1.2 \times 10^{-5}}{0.9491} = 1.264 \times 10^{-5} \frac{\text{kmol}}{\text{m}^2\text{s}} \qquad \textbf{(Answer)}$$

Example 3.5: Air is blown countercurrent to water which is flowing down as a thin film along a solid surface at 101.3 kPa and 273 K. The air is dry, and $k_y = 30$ kmol/m^2 h. Diffusivity of water vapour in air is 2.25×10^{-5} m^2/s. The vapour pressure of water at 273 K is 610 Pa. Assuming one-way diffusion, calculate:

(a) The molar flux N_A.
(b) The values of k_c and k_G.
(c) The film coefficient k'_y and the thickness of the film.

***Solution*:** Mass transfer coefficient $k_y = 30$ kmol/m^{2}h. Mass diffusivity of water (**A**) in air (**B**) is $D_{AB} = 2.25 \times 10^{-5}$ m^2/s

$$T = 273 \text{ K}, P = 101.3 \text{ kPa}$$

The air layer in contact with water is assumed saturated with water vapour so that the partial pressure of water there is equal to the vapour pressure of water at 273 K. $\bar{p}_{A1} = P_A^S = 610$ Pa. Therefore, the mole fraction of water at the air-water interface is

$$y_{A1} = \frac{P_A^S}{P} = \frac{610}{101.3 \times 10^3} = 6.02 \times 10^{-3}$$

The mole fraction of air at the interface $= y_{B1} = 1 - 6.02 \times 10^{-3} = 0.9940$

Since the air is dry, mole fraction of water in the bulk air is 0. That is $y_{A2} = 0$. Mole fraction of air in the bulk, $y_{B2} = 1 - y_{A2} = 1.0$

(a) $N_A = k_y(y_{A1} - y_{A2}) = 30(6.02 \times 10^{-3} - 0) = 0.1806\,\text{kmol/m}^2\text{h}$ **(Answer)**

(b) Equation (3.11) gives the relationship between k_c and k_y.

$$k_y = k_c C \text{ or } k_c = \frac{k_y}{C}$$

Assuming that the gas phase behaves as an ideal gas, the molar concentration C is calculated as

$$C = \frac{P}{RT} = \frac{101.3}{8.314 \times 273} = 4.46 \times 10^{-2}\ \text{kmol/m}^3$$

$$k_c = \frac{k_y}{C} = \frac{30}{4.46 \times 10^{-2}} = 672.645 \frac{\text{m}}{\text{h}}$$ **(Answer)**

Equation (3.11) also gives

$$k_y = k_G P \text{ or } k_G = \frac{k_y}{P} = \frac{30}{101.3 \times 10^3} = 2.96 \times 10^{-4} \frac{\text{kmol}}{\text{m}^2\text{h Pa}}$$ **(Answer)**

(c) Equation (3.20) relates k_y and k_y'

$$k_y y_{B,M} = k_y'$$

where y_{BM} is the logarithmic mean of y_{B1} and y_{B2}.

$$y_{B,M} = \frac{y_{B2} - y_{B1}}{\ln \dfrac{y_{B2}}{y_{B1}}} = \frac{1 - 0.9940}{\ln \dfrac{1}{0.9940}} = 0.9970$$

$$k_y y_{B,M} = k_y' = 30 \times 0.9970 = 29.91 \frac{\text{kmol}}{\text{m}^2\text{h}}$$ **(Answer)**

The mass transfer coefficient k_y' also represents the film coefficient of mass transfer.

$$k_y' = \frac{CD_{AB}}{Z}$$

$$29.91/3600 = \frac{4.46 \times 10^{-2} \times 2.25 \times 10^{-5}}{Z}$$

where Z is the thickness of the film through which diffusion is occurring.

$$Z = \frac{4.46 \times 10^{-2} \times 2.25 \times 10^{-5}}{29.91/3600} = 1.208 \times 10^{-4}\,\text{m} = 0.12\ \text{mm}$$ **(Answer)**

3.2 THEORIES OF MASS TRANSFER

In most of the practical applications of mass transfer, material is transferred across a phase boundary. The rate of mass transfer between the two phases depends on the physical properties of the phases, the concentration difference, the interfacial area and the degree of turbulence. A number of theories have been proposed to explain the mechanism of mass transfer at the phase boundary. Some of them are reviewed in the following sections.

Film Theory

The film concept of mass transfer was already discussed briefly in the section on mass transfer coefficient. The concentration profile in turbulent flow when the solute is dissolving from the surface and is diffusing to the main stream is shown in Fig. 3.2. The concentration of the solute decreases as the distance from the wall increases. Three regions can be visualized in the flowing stream. In the stagnant layer of fluid immediately adjacent to the solid surface, most of the mass transfer occurs by molecular diffusion; this layer is known as diffusional sub-layer. No eddies are present in this layer. A large concentration drop occurs across the viscous sub-layer as is clear from the figure. Major portion of the fluid cross section is the turbulent core in which the mass transfer is mainly due to eddy transport. The contribution of molecular diffusion to mass transport is insignificant compared to that of eddy transport. In between the diffusional sub-layer and the turbulent core, is the buffer zone, where the mass transfer by both molecular diffusion and eddy diffusion are significant.

It can be seen that the concentration profile from C_{Ai} adjacent to the solid surface to C_{Ab}, the average bulk composition is very steep in the diffusional sub-layer and thereafter the concentration levels off rapidly. The exact shape of the concentration-distance relation curve is dependent on the relative magnitudes of eddy diffusivity and molecular diffusivity at that point. The film concept of mass transfer proposed by Nernst (1904) assumes that the principal resistance to overall transfer process lies in the region near the wall. In the film theory of mass transfer, the entire concentration change from C_{Ai} to C_{Ab} is assumed to be confined within an effective film of thickness Z. It is a fictitious length offering the same resistance to molecular diffusion as that encountered in the combined processes of molecular diffusion and eddy diffusion. It is also assumed that within this film, the change in concentration with distance is linear as in steady-state equimolar counter diffusion. We have seen that if the concept of film coefficient of heat transfer is extended to represent the convective transport of mass, the film coefficient of mass transfer should be $\frac{CD_{AB}}{Z}$.

In the treatment of interphase mass transfer, it is assumed that a film of stagnant fluid similar to that existing near a solid boundary is present at the fluid-fluid interface within which, the mechanism of mass transfer is molecular diffusion. Then the quantity $\frac{CD_{AB}}{Z}$ will represent film coefficient of mass transfer which is related to k_y or k_y' as given by Eq. (3.19)

$$k_y y_{B,M} = \frac{CD_{AB}}{Z} = k_y' \qquad \text{...(3.19)}$$

The film theory is simple to comprehend and easy to apply. It suggests that the mass transfer coefficient is directly proportional to the diffusion coefficient D_{AB}. This linear dependency of the mass transfer coefficient on the diffusivity is the main limitation of the film theory. The experimental data, predicts a dependency of $(D_{AB})^n$ where n ranges from 0.5 to 0.75. Using the Chilton-Colburn analogy, discussed later in this chapter, it is seen that $k_c \propto D_{AB}^{2/3}$ which is in better agreement with experimental data. These observations suggest that the film thickness Z should be a function of D_{AB}. Even though the film model cannot be used effectively for prediction and correlation, it is useful because of its mathematical simplicity for predicting and analysing the effects of such complicating factors as the simultaneous chemical reaction at the interface and the influence of high solute concentration and fluxes. Also, it is a useful model for membranes as they satisfy the film assumptions, and is reasonably good as a first approximation for highly turbulent fluids near solid surfaces, where the existence of a thin stagnant layer is a reality.

Penetration Theory

The penetration theory was proposed by Higbie (1935). Instead of the stagnant film at the interface as proposed in the film theory, Higbie suggested that eddies move from the bulk of the fluid stream to the interface at definite intervals, they stay at the interface for a short but fixed period of time. During this fixed time of exposure, the solute diffuses into the eddy by molecular diffusion, after which the eddies move away from the interface to get mixed with the bulk stream. Thus according to this theory, a fluid element whose initial concentration is the same as the concentration of the bulk stream (C_{Ab}), when exposed to the second fluid phase, equilibrium is established instantaneously at the interface. A process of unsteady-state molecular diffusion occurs for a fixed time of exposure, after which the fluid element is replaced by fresh one and moves to and gets mixed with the bulk stream. Thus the eddies are alternately static and moving.

In Higbie's experiments, a slug-like bubble of carbon dioxide was allowed to rise through a column of water. It was assumed that the length of time each element of the surface in the liquid is exposed to the gas was equal to the time taken by the gas bubble to pass it. Absorption took place during this period by unsteady state molecular diffusion. The depth of the liquid element is assumed to be infinite in comparison with the depth of penetration of the solute. The concentration-time-distance history in the liquid which was exposed to diffusion of a soluble gas can be plotted as shown in Fig. 3.3. In the figure, the percentage saturation of the liquid is plotted against the distance for different exposure times on arbitrary units.

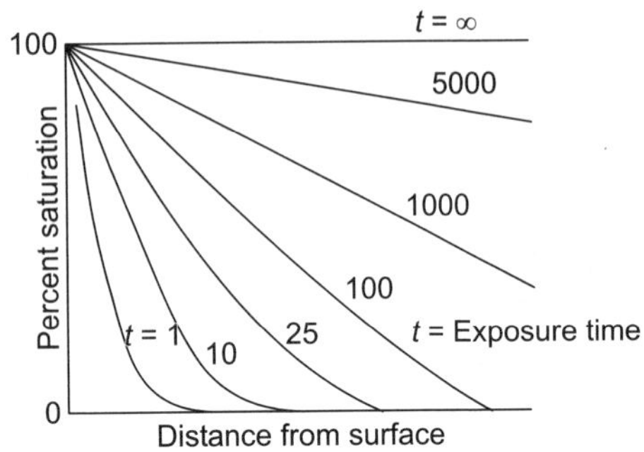

Fig. 3.3: Solute penetration as function of exposure time

The concentration profile is such that at any time, the effective depth of liquid which contains an appreciable concentration of solute can be specified, and hence the theory is known as penetration theory. The rate of diffusion is governed by Fick's second law (*see* Eq. (2.122):

$$\frac{\partial C_A}{\partial t} = D_{AB}\frac{\partial^2 C_A}{\partial z^2} \quad \text{...(3.37)}$$

The following boundary conditions are applicable:

$$C_A = C_{Ab} \text{ at } t = 0, \text{ for all } z$$
$$C_A = C_{Ai} \text{ at } z = 0, \text{ for } t > 0$$
$$C_A = C_{Ab} \text{ as } z \to \infty, \text{ for all } t$$

The solution of Eq. (3.37) can be written as

$$\frac{C_A - C_{Ab}}{C_{Ai} - C_{Ab}} = 1 - \text{erf}\left(\frac{z}{2\sqrt{D_{AB}t}}\right) \quad \text{...(3.38)}$$

where t is the exposure time for the fluid element. For bubbles, the contact time of the liquid surrounding the bubble is assumed to be the ratio of bubble diameter to bubble rise velocity. For liquid spray, the contact time is the time taken by the droplets to fall through the gas. The depth of penetration is defined as the distance at which the concentration change is one percent of the final value. The molar flux under this condition is

$$N_A = 2\sqrt{\frac{D_{AB}}{\pi t}}(C_{Ai} - C_{Ab}) \quad \text{...(3.39)}$$

Compare this with the following expression for mass flux written in terms of mass transfer coefficient:

$$N_A = k_L(C_{Ai} - C_{Ab}) \quad \text{...(3.40)}$$

Thus, according to the penetration theory,

$$k_L = 2\sqrt{\frac{D_{AB}}{\pi t}} \quad \text{...(3.41)}$$

which predicts that the mass transfer coefficient is proportional to the square root of the diffusion coefficient. If the dependency of k_L to D_{AB} is written as $k_L \propto D_{AB}^n$, the penetration theory sets the value of n to be 0.5, which is generally the lower limit of the observed range of values of n. The penetration theory is most useful when mass transfer involves bubbles or droplets and for flow over random packing.

Surface Renewal Theory

Penetration theory with constant exposure time fails to represent the mass transfer occurring in devices such as stirred tanks, the bubble and spray columns where bubbles and droplets of varying sizes are present. Danckwerts (1951) suggested an improvement of the penetration theory by assuming a residence time distribution in the place of the constant exposure time. He suggested that liquid surface elements will not be exposed to the gas for a constant duration; instead, a random distribution of ages is more probable for the fluid elements at the interface. That is, the interface will be made up of surface elements of different exposure-time histories. The probability of any element of surface getting destroyed and mixed with the bulk of the fluid is independent of the age of the element. Since the rate of solute penetration depends on the exposure time, an average rate should be determined by summing up the individual values. Indicating the fractional rate of replacement of elements by s, the surface-age distribution function is shown to be

$$\phi = se^{-st} \qquad \text{...(3.42)}$$

Here ϕ represents the probability that any element of area is exposed for time t before being replaced by fresh fluid from the bulk. The rate of absorption is

$$N_A = \sqrt{D_{AB}s}(C_{Ai} - C_{Ab}) \qquad \text{...(3.43)}$$

which relates k_L and D_{AB} as

$$k_L = \sqrt{D_{AB}s} \qquad \text{...(3.44)}$$

For the mass transfer coefficient, the surface renewal theory also predicts the same dependency on molecular diffusivity as the penetration theory. The numerical value of s is difficult to estimate, although we know that it will increase as the fluid becomes more turbulent. In packed column, s will be of the same order of magnitude as the ratio of the velocity of the liquid flowing over the packing to the length of packing.

Film-Penetration Theory

The film-penetration theory proposed by Toor and Marchello (1958) combines certain features of the film theory and the penetration theory. It predicts the dependency of the mass transfer coefficient k_L on the diffusivity as $k_L \propto D_{AB}^n$, with n varying from 0.5 to 1.0. The theory assumes that the entire resistance to mass transfer resides in a laminar film of fixed thickness as in film theory, but the mass transfer is regarded as an unsteady state process as in penetration theory. Eddies move between the bulk fluid and this film. Fresh surface is formed at intervals by fluid which is brought from the bulk stream to the interface by the eddies. The material diffuses into the film and gets immediately mixed with the bulk of the fluid. For short time of exposure, no material will reach the far side of the film as postulated in the penetration theory. For sufficiently longer periods of exposure, the conditions are identical to those expected in the film theory.

The rate of diffusion is still governed by Fick's second law,

$$\frac{\partial C_A}{\partial t} = D_{AB}\frac{\partial^2 C_A}{\partial z^2} \qquad \text{...(3.37)}$$

The following boundary conditions are now applicable:

$$C_A = C_{Ab} \text{ at } t = 0, \text{ for all } z$$
$$C_A = C_{Ai} \text{ at } z = 0, \text{ for } t > 0$$
$$C_A = C_{Ab} \text{ as } z \to Z, \text{ for all } t$$

Using Danckwerts surface-age distribution function [Eq. (3.42)], an expression is obtained which relates k_c to Z, D_{AB} and s. This reduces to Eq. (3.41) for the penetration model when $\frac{D_{AB}}{Z^2 s}$ is small, and to the film model when $\frac{D_{AB}}{Z^2 s}$ is large.

Example 3.6: A pure gas is being absorbed in a liquid solvent using a gas-liquid contact equipment. The average mass flux over the time of exposure is found to be 8.0 × 10^{-3} kmol/m^2 h. The solubility of the gas in the liquid is 4.0 × 10^{-2} kmol/m^3. The diffusivity of the gas in the liquid is 2.0 × 10^{-9} m^2/s. If the penetration theory is found to be applicable, determine the time of exposure.

***Solution*:** The liquid layer in contact with the gas (**A**) can be assumed to be saturated with the gas with concentration of **A** same as the saturation value, 4.0 × 10^{-2} kmol/m^3. The concentration of **A** in the bulk liquid is 0. That is,

$$C_{A1} = 4.0 \times 10^{-2} \frac{\text{kmol}}{\text{m}^3} \text{ and } C_{A2} = 0.$$

$$N_A = 8.0 \times 10^{-3} \text{ kmol/m}^2\text{h}$$

$$N_A = k_L(C_{A1} - C_{A2})$$

$$k_L = \frac{N_A}{C_{A1} - C_{A2}} = \frac{8 \times 10^{-3}}{3600(0.04 - 0)} = 5.56 \times 10^{-5} \frac{\text{m}}{\text{s}}$$

According to the penetration theory,

$$k_L = 2\sqrt{\frac{D_{AB}}{\pi t}} \quad \text{...(3.41)}$$

Rearranging this equation,

$$t = \frac{D_{AB}}{\pi}\left(\frac{2}{k_L}\right)^2 = \frac{2 \times 10^{-9}}{\pi} \times \left(\frac{2}{5.56 \times 10^{-5}}\right)^2 = 0.82 \text{ s}$$

Time of exposure = 0.82 s **(Answer)**

Example 3.7: A gas **A** is being absorbed in water in a countercurrent packed tower. The mole fraction of **A** at the gas-liquid interface and in the bulk liquid are, respectively, 2.5 × 10^{-3} and 2.0 × 10^{-4}. The average flux over the time of exposure is 3.0 × 10^{-2} kmol/m^{2}h and the diffusivity of the gas **A** in water is 1.7 × 10^{-9} m^2/s. If the mechanism of mass transfer is represented by the surface renewal theory, determine:

(a) Fractional rate of surface renewal
(b) Average residence time
(c) Surface-age distribution function.

***Solution*:** The mass flux for absorption of gas **A** into water is expressed in terms of the mass transfer coefficient k_L as

$$N_A = k_L(C_{A1} - C_{A2}) = k_L\left(\frac{\rho}{M}\right)(x_{A1} - x_{A2}) \quad \text{...(A)}$$

Assuming that the density and molecular weight of the solution are the same as those of pure water,

$$\rho = 1000\frac{\text{kg}}{\text{m}^3} \text{ and } M = 18 \text{ kg/kmol.}$$

The mole fraction of gas **A** at the interface and bulk liquid are $x_{A1} = 2.5 \times 10^{-3}$ and $x_{A2} = 2 \times 10^{-4}$ Substituting these values in Eq. A, we get,

$$k_L = \frac{N_A}{\left(\frac{\rho}{M}\right)(x_{A1} - x_{A2})} = \frac{3\times10^{-2}}{3600\left(\frac{1000}{18}\right)(2.5\times10^{-3} - 2\times10^{-4})}$$

$$= 6.52\times10^{-5}\ \frac{\text{m}}{\text{s}}$$

(a) According to surface renewal theory

$$k_L = \sqrt{D_{AB}s} \qquad \ldots(3.44)$$

which gives the fractional rate of surface renewal as

$$s = \frac{(k_L)^2}{D_{AB}} = \frac{(6.52\times10^{-5})^2}{1.7\times10^{-9}} = 2.5\ \text{s}^{-1} \qquad \textbf{(Answer)}$$

(b) The average residence time is the reciprocal of the fractional rate of surface renewal

$$t = \frac{1}{s} = \frac{1}{2.5} = 0.4\ \text{s} \qquad \textbf{(Answer)}$$

(c) Surface-age distribution function is expressed as

$$\phi(t) = s e^{-st} \qquad \ldots(3.42)$$

$$\phi = 2.5e^{-2.5t} \qquad \textbf{(Answer)}$$

Example 3.8: The rate at which solid dissolves from a flat plate of solid benzoic acid into water flowing past the solid at 300 K and atmospheric pressure is found to be 1.8×10^{-6} kmol/m^2 s. The solubility of benzoic acid in water is 2.95×10^{-2} kmol/m^3. The diffusivity of benzoic acid in water is 1.245×10^{-9} m^2/s. Assuming dilute solutions, calculate:

(a) The mass transfer coefficient k_L
(b) The film thickness using the film theory
(c) Contact time for the penetration theory
(d) The fractional rate of surface renewal for surface renewal theory

***Solution*:** C_{A1}, concentration of benzoic acid at the solid-liquid interface = solubility in water = 2.95×10^{-2} kmol/m^3

C_{A2}, concentration of benzoic acid in the bulk liquid = 0
D_{AB} = diffusivity of benzoic acid in water = 1.245×10^{-9} m^2/s
N_A = rate of dissolution of benzoic acid = 1.8×10^{-6} kmol/m^{2}s

(a) The mass transfer coefficient k_L is given by

$$N_A = k_L(C_{A1} - C_{A2})$$

so that
$$k_L = \frac{N_A}{C_{A1} - C_{A2}} = \frac{1.8\times10^{-6}}{2.95\times10^{-2}} = 6.1\times10^{-5}\ \frac{\text{m}}{\text{s}} \qquad \textbf{(Answer)}$$

(b) Assuming that the solution is dilute, the bulk flow contribution is neglected. Therefore,

$$k_L = k_L' = 6.1 \times 10^{-5} \text{ m/s}$$

From Eqs. (3.24) and (3.35) we see that

$$k_x' = k_L' C = \frac{CD_{AB}}{Z} \quad \text{Or,} \quad k_L' = \frac{D_{AB}}{Z}$$

where Z is the thickness of the film.

$$Z = \frac{D_{AB}}{k_L'} = \frac{1.245 \times 10^{-9}}{6.1 \times 10^{-5}} = 2.04 \times 10^{-5} \text{m}$$

$Z = 2.04 \times 10^{-2}$ mm **(Answer)**

(c) According to the penetration theory,

$$t = \frac{D_{AB}}{\pi}\left(\frac{2}{k_L}\right)^2 = \frac{1.245 \times 10^{-9}}{\pi} \times \left(\frac{2}{6.1 \times 10^{-5}}\right)^2 = 0.426 \text{ s}$$

Contact time = 0.426 s **(Answer)**

(d) The fractional rate of surface renewal is given by

$$s = \frac{k_L^2}{D_{AB}} = \frac{(6.1 \times 10^{-5})^2}{1.245 \times 10^{-9}} = 2.99 \text{ s}^{-1}$$ **(Answer)**

3.3 DIMENSIONAL ANALYSIS

Dimensional analysis is a mathematically simple, but powerful tool for explaining physical processes that are not amenable to a formal theoretical development. This technique is intermediate between a formal mathematical development and a complete empirical study. It is based on the fact that if a theoretical equation exists among the variables to represent a physical process, that equation must be dimensionally homogeneous and can be written equivalently in terms of a lesser number of dimensionless groups involving the variables. For instance, the physical process of transfer of mass from the walls of a tube to a fluid moving in turbulent flow is a function of six variables. By dimensional analysis it has been shown that these variables can be grouped into three dimensionless groups and the process can be described in terms of these groups instead of the original six variables.

The importance of dimensional analysis will be apparent when we attempt to correlate experimental data. To develop a correlation for the influence of different variables on the mass transfer coefficient, we may have to perform separate experiments by varying the tube diameter, the liquid density, the velocity of flow, etc. one at a time, while keeping others constant. With dimensional analysis, however, the experimental investigation is considerably simplified as the results of one series of tests can be applied to a variety of other problems. The major limitation of dimensional analysis is that the results obtained by this method are incomplete unless supplemented by experiments. Though its contribution towards understanding the intricacies of the physical process is very little, the dimensional analysis offers considerable simplification in the interpretation and correlation of experimental data.

Dimensionless Groups in Mass Transfer

Dimensionless parameters generally used for correlating convective heat transfer data are the Nusselt number, the Prandtl number, the Reynolds number, etc. Some of these groups along with some newly defined dimensionless groups such as the Schmidt number, the Sherwood number, the Lewis number, etc. are useful in the correlation of convective mass transfer data.

Schmidt number

The molecular diffusivity of momentum, also known as the kinematic viscosity, is represented by v and is equal to the ratio of the absolute viscosity (μ) to the density (ρ) of the fluid. It has the same dimension L^2/T (or units, m^2/s) as the molecular diffusivity of mass D_{AB}. The Schmidt number (Sc) is defined as the ratio of the molecular diffusivity of momentum to the molecular diffusivity of mass.

$$Sc = \frac{\text{momentum diffusivity}}{\text{mass diffusivity}} = \frac{\nu}{D_{AB}} = \frac{\mu}{\rho D_{AB}} \quad \text{...(3.45)}$$

The role of Schmidt number in convective mass transfer is analogous to that of Prandtl number (Pr) in convective heat transfer. The Prandtl number is defined as the ratio of the momentum diffusivity to thermal diffusivity (α).

$$Pr = \frac{\text{momentum diffusivity}}{\text{thermal diffusivity}} = \frac{\nu}{\alpha} = \frac{C_p\mu}{k} \quad \text{...(3.46)}$$

Schmidt number can be treated as a fluid property, as the quantities involved in its definition are all flow properties. At moderate pressures, the values of Schmidt number of a gas in air range from 0.5 to 2.0. For ideal gases, the Schmidt number is independent of pressure. This is because the viscosity of the gas is independent of pressure and the effects of pressure on density and diffusivity mutually cancel. Temperature has only very small effect on Schmidt number because viscosity μ and ρD_{AB} both change with temperature as $T^{0.7–0.8}$. For liquids, the Schmidt number values are much higher than those of gases. For self diffusion in water at 298 K, Schmidt number is 297; it is typically of the order of 10^3 to 10^4 for more viscous liquids and even for water with slowly diffusing solutes. For diffusion of solutes at infinite dilution, D_{AB} is of the order of 10^{-9} m²/s, μ is approximately 10^{-3} kg/m s, and density is approximately 10^3 kg/m³, which means Schmidt number is approximately 10^3. The Schmidt number decreases with increasing temperature because the diffusivity of liquids increases and viscosity decreases with temperature.

Lewis number

Lewis number is defined as the ratio of molecular diffusivity of heat (thermal diffusivity) to the molecular diffusivity of mass (mass diffusivity).

$$Le = \frac{\text{thermal diffusivity}}{\text{mass diffusivity}} = \frac{\alpha}{D_{AB}} = \frac{k}{C_p\rho D_{AB}} \quad \text{...(3.47)}$$

Lewis number finds application in the study of processes involving simultaneous transfer of mass and heat such as humidification and drying operations. Lewis number can also be treated as the ratio of Schmidt number to Prandtl number.

$$Le = \frac{k}{C_p\rho D_{AB}} = \frac{(\mu/\rho D_{AB})}{(C_p\mu/k)} = \frac{Sc}{Pr} \quad \text{...(3.48)}$$

Just as Schmidt number, the Lewis number can also be treated as a property of the diffusing system. For air-water vapour system, Le is approximately unity.

Sherwood number

Sherwood number is a dimensionless mass transfer coefficient and is defined as

$$Sh = \frac{k_c' L}{D_{AB}} \quad \text{...(3.49)}$$

where k_c' is the film mass transfer coefficient $k_c' = \dfrac{D_{AB}}{Z}$. For dilute solutions and low rate of mass transfer, $k_c' = k_c$, and Sherwood number is also equal to

$$Sh = \frac{k_c L}{D_{AB}} \qquad \ldots(3.50)$$

If bulk flow contribution is appreciable, and the assumption of dilute solution is not applicable, $k_c' = k_c y_{B,M}$. Under this condition,

$$Sh = \frac{k_c y_{B,M} L}{D_{AB}} \qquad \ldots(3.51)$$

In the above equations, L is the significant linear dimension for the given flow system. For flow through tubes, L is the diameter of the tube D, and for flow past plane surfaces, L is the length from the leading edge.

It can be shown that Sherwood number is the ratio of the concentration gradient in the fluid immediately in contact with the surface to a reference concentration gradient. Consider the transfer of solute from the surface of the solid to the fluid flowing along the surface (*see* Fig. 3.4).

Let C_{Ai} be the concentration of the diffusing component at the surface and C_{Ab} is the undisturbed concentration in the bulk fluid. The mass transfer flux is given by

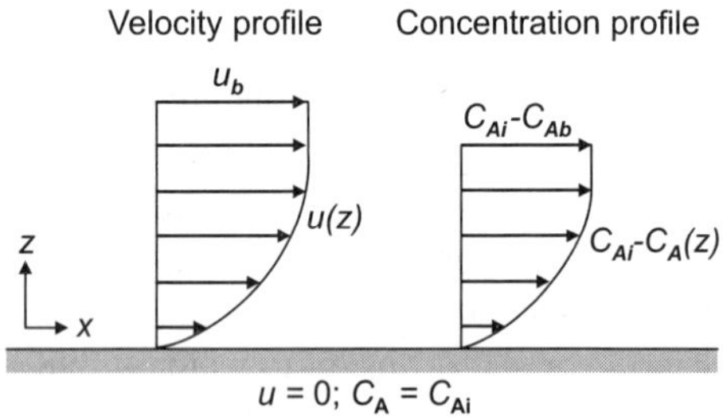

Fig. 3.4: Velocity and concentration profile in fluid over solid surface

$$N_A = k_c (C_{Ai} - C_{Ab}) \qquad \ldots(3.52)$$

The rate of mass transfer at the surface can be described by Fick's law of molecular diffusion

$$N_A = -D_{AB} \left.\frac{dC_A}{dz}\right|_{z=0} \qquad \ldots(3.53)$$

Since C_{Ai} is constant, Eq. (3.53) can be written as

$$N_A = D_{AB} \left.\frac{d(C_{Ai} - C_A)}{dz}\right|_{z=0} \qquad \ldots(3.54)$$

Since both Eqs. (3.52) and (3.54), represent the molar flux of component **A** leaving the surface and entering the fluid, we can write

$$k_c (C_{Ai} - C_{Ab}) = D_{AB} \left.\frac{d(C_{Ai} - C_A)}{dz}\right|_{z=0} \qquad \ldots(3.55)$$

Rearrange Eq. (3.55) as follows:

$$\frac{k_c}{D_{AB}} = \frac{\left.\dfrac{d(C_{Ai} - C_A)}{dz}\right|_{z=0}}{(C_{Ai} - C_{Ab})} \qquad \ldots(3.56)$$

Multiplying both sides by L, the significant dimension, we get

$$Sh = \frac{k_c L}{D_{AB}} = \frac{\left.\dfrac{d(C_{Ai} - C_A)}{dz}\right|_{z=0}}{\dfrac{(C_{Ai} - C_{Ab})}{L}} \qquad \ldots(3.57)$$

The right-hand-side of Eq. (3.57) is in fact, the ratio of concentration gradient at the surface to an overall concentration gradient. It also means that the Sherwood number is the ratio of the resistance to molecular transport to the resistance to convective transport. It is analogous to the Nusselt number for heat transfer, which is defined as $Nu = \frac{hL}{k}$, where h is the heat transfer coefficient and k is the thermal conductivity of the fluid.

Reynolds number

The forces acting on a unit mass of fluid element under a given flow situation are acceleration or inertial force $\left(\frac{u^2}{L}\right)$, the gravity force ($g$), the pressure force $\left(\frac{P}{\rho L}\right)$, and the viscous force $\left(\frac{\nu u}{L^2}\right)$ where u is the velocity of flow of the fluid. All these have dimensions $\left(\frac{L^2}{T}\right)$ so that the ratio between any two will form a dimensionless group. The Reynolds number is defined as the ratio of the inertial forces to viscous forces acting on a fluid element.

$$Re = \frac{\text{inertial forces}}{\text{viscous forces}} = \frac{Lu}{\nu} = \frac{Lu\rho}{\mu}$$

The Reynolds number is the most important dimensionless group used extensively in the study of momentum, heat and mass transport.

Dimensional Analysis in Convective Mass transfer

Consider the transfer of mass from the walls of a circular tube to the fluid flowing through it in forced convection. The transfer is due to a concentration difference $C_{Ai} - C_A$ from the tube surface to the fluid. The variables influencing the transport process, their symbol and dimensions are listed below:

Variable	Symbol	Dimensions
Tube diameter	D	L
Fluid density	ρ	ML^{-3}
Fluid viscosity	μ	$ML^{-1}T^{-1}$
Fluid diffusivity	D_{AB}	L^2T^{-1}
Fluid velocity	u	LT^{-1}
Mass transfer coefficient	k_c	LT^{-1}

Setting aside the mass transfer coefficient k_c, which is the quantity of primary interest in this process, the other variables can be classified into three groups: variables descriptive of the system geometry (D), the fluid properties (ρ, μ, D_{AB}) and the dynamics of flow (u).

According to the *Buckingham-π theorem*, the functional relationship between n variables required to describe a system, can be equivalently written in terms of n-m dimensionless groups, where m is the number of primary dimensions involved. In the present problem, there are six variables and three primary dimensions (M, L, T). Therefore, the system can be represented in terms of three dimensionless groups, say, π_1, π_2 and π_3, which are known as π groups.

$$\phi(\pi_1, \pi_2, \pi_3) = 0 \qquad ...(3.58)$$

Or
$$\pi_1 = f(\pi_2, \pi_3) \qquad ...(3.59)$$

Three of the variables will appear in all the dimensionless groups and these are known as the core variables. Of the three remaining variables, one each will be present in different groups. The variable whose effect we want to isolate and study is excluded from the core group. Also the core group variables should satisfy the following requirements:

1. These variables will contain all the primary dimensions.
2. They should not form a dimensionless group among themselves.
3. No two should have identical dimensions.

With these restrictions in mind, we may choose D_{AB}, ρ and D as the core variables so that the π-groups can be formed as

$$\pi_1 = D_{AB}^{a_1} \rho^{b_1} D^{c_1} k_c \qquad ...(3.60)$$

$$\pi_2 = D_{AB}^{a_2} \rho^{b_2} D^{c_3} u \qquad ...(3.61)$$

$$\pi_3 = D_{AB}^{a_3} \rho^{b_3} D^{c_3} \mu \qquad ...(3.62)$$

where a_i, b_i and c_i are constants. Writing π_1 [Eq. (3.60)] in terms of its dimensions,

$$(MLT)^0 = (L^2T^{-1})^{a_1} (ML^{-3})^{b_1} L^{c_1} (LT^{-1}) \qquad ...(3.63)$$

Equating the exponents of the fundamental dimensions on both sides of the above equation, we have
exponent of $M = 0 = b_1$
exponent of $L = 0 = 2a_1 - 3b_1 + c_1 + 1$
exponent of $T = 0 = -a_1 - 1$
The solution of the above yields $a_1 = -1$, $b_1 = 0$, $c_1 = 1$ so that

$$\pi_1 = D_{AB}^{-1} \rho^0 D^1 k_c = \frac{k_c D}{D_{AB}}$$, which is the Sherwood number, *Sh*.

Similarly, it can be shown that

$$\pi_2 = \frac{Du}{D_{AB}}$$

$$\pi_3 = \frac{\mu}{\rho D_{AB}}$$, the Schmidt number, *Sc*

Dividing π_2 by π_3, we get,

$$\frac{\pi_2}{\pi_3} = \frac{Du\rho}{\mu}$$, the Reynold's Number, *Re*

Equation (3.59) can be written as $\pi_1 = f\left(\frac{\pi_2}{\pi_3}, \pi_3\right)$, so that we have

$$Sh = f(Re, Sc) \qquad ...(3.64)$$

The dimensional analysis on its own cannot provide a theoretically sound and practically useful mathematical equation for estimating the mass transfer coefficient. Experiments are needed to develop correlations that are useful for practical calculations. Experimental evidences indicate the following functional relationship between different variables:

$$Sh = a\, Re^b Sc^c \qquad ...(3.65)$$

where *a*, *b* and *c* are empirical constants, with values dependent on the geometry of the flow system and the flow regimes.

3.4 BOUNDARY LAYER THEORY

In the film and penetration models, the mass transfer through an interface between two immiscible fluids was considered. The boundary layer model is applicable when the mass transfer occurs between the solid surface and the flowing fluid, such as a solid subliming into a gas stream or a solid dissolving into a liquid stream.

Consider a fluid with a concentration of solute C_b approaching a flat plate with a free stream velocity u_b. The velocity of the fluid is zero at the interface between the solid and the fluid and it increases with distance in the z-direction. The velocity change from zero at the wall to the main stream value u_b is confined to a thin layer of fluid known as the *hydrodynamic boundary layer* or *velocity boundary layer*. The boundary layer separates the region of constant velocity u_b and the region of lower velocity as shown by the dashed line in Fig. 3.5. The thickness of the boundary layer δ is arbitrarily taken as the distance in the z-direction where the velocity reaches 99% of the free-stream velocity.

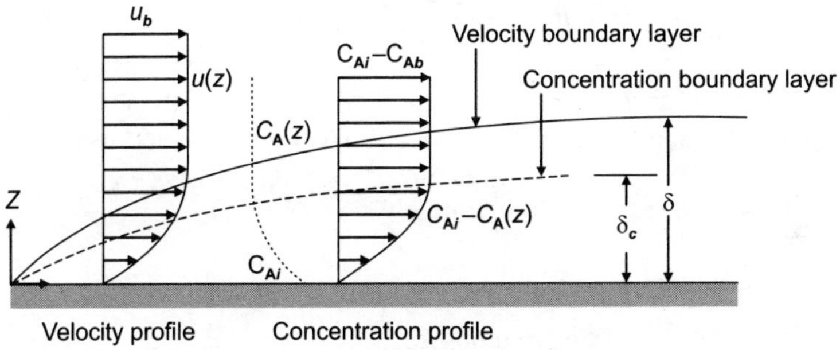

Fig. 3.5: Laminar flow boundary layers on flat surface

The thickness of the boundary layer increases with distance x from the leading edge of the plate. For small values of x, the flow is laminar in the boundary layer. At larger values of x, the flow becomes turbulent. The distance from the leading edge where the flow in the boundary layer changes from laminar to turbulent is the transition length x_t. The transition Reynolds number is generally taken to be about 5×10^5 so that

$$\frac{x_t u_b \rho}{\mu} = 5 \times 10^5$$

Even in the region in which the boundary layer is turbulent, a very thin film of fluid known as laminar sub-layer or viscous sub-layer exists, within which the velocity gradient is very large.

Similar to the hydrodynamic boundary layer, a concentration boundary layer also develops within which the concentration changes from C_i at the wall to C_b at the main stream. The thickness of the concentration boundary layer δ_c can be defined as the distance in the z-direction at which the concentration of solute in the fluid reaches 99% of the free stream concentration. Considering the region where only laminar boundary layer exists and assuming that the velocity of the fluid is not disturbed by diffusion in the z-direction, the thicknesses of the hydrodynamic boundary layer and the concentration boundary layer are roughly related to the Schmidt number by the following expression:

$$\frac{\delta}{\delta_c} = Sc^{1/3} \qquad \text{...(3.66)}$$

The thickness of the hydrodynamic boundary layer increases with distance along the leading edge as

$$\frac{\delta}{x} = \frac{4.96}{Re_x^{1/2}} \qquad \text{...(3.67)}$$

An expression for mass transfer coefficient can be derived by solving simultaneously, the Navier-Stokes equation in the x-direction, the equation of energy, the equation of continuity for the diffusing component **A** and the equation of continuity for the total material. The resulting equation for the local mass transfer coefficient for laminar flow at a distance x from the leading edge of the plate, $k_{c,x}$ is given by

$$Sh_x = 0.332 Re_x^{1/2} Sc^{1/3} \qquad \text{...(3.68)}$$

where Sh_x is a local Sherwood number and Re_x is a local Reynolds number, both correspond to a location at a distance x from the leading edge and Sc is the Schmidt number. Thus, for low rates of mass transfer,

$$Sh_x = \frac{k_{cx} x}{D_{AB}} \qquad \text{...(3.69)}$$

$$Re_x = \frac{u_b \rho x}{\mu} \qquad \text{...(3.70)}$$

The average Sherwood number, Sh_{av}, over a length L where the flow in the boundary layer is laminar, can be obtained by substituting Eqs. (3.69) and (3.70) into Eq. (3.68) and using the relation

$$k_{c,av} = \frac{1}{L}\int_0^L k_{c,x}\, dx \qquad \text{...(3.71)}$$

The result can be rearranged as

$$Sh_{av} = 0.664 Re_L^{1/2} Sc^{1/3} \qquad \text{...(3.72)}$$

where

$$Sh_{av} = \frac{k_{c,av} L}{D_{AB}} \qquad \text{...(3.73)}$$

$$Re_L = \frac{u_b \rho L}{\mu} \qquad \text{...(3.74)}$$

Thus we see that the average mass transfer coefficient over a length L of a flat surface is twice the local mass transfer coefficient evaluated at a distance L from the leading edge. Equation (3.72) also reveals that k_c varies as $D_{AB}^{2/3}$ in the region where flow in the boundary layer is laminar.

When the flow in the boundary layer is turbulent, it is not easy to develop exact expressions for mass transfer coefficient. Boundary layer theory, in such situations serves merely as a guide for developing empirical correlations. Analogy between heat, mass and momentum transport are relied upon in situations where flow in the boundary layer is turbulent, to develop expressions for mass transfer coefficient from those available for heat transfer for similar geometry. An empirical correlation suitable for the estimation of the local mass transfer coefficient for turbulent boundary layer is

$$Sh_x = 0.0292 Re_x^{0.8} Sc^{1/3} \qquad \text{...(3.75)}$$

Since the flow in the boundary layer is laminar up to the transition length x_t, and turbulent beyond the transition length, the average mass transfer coefficient can be evaluated utilizing Eq. (3.68) and (3.75) in

$$k_{c,av} = \frac{1}{L}\left(\int_0^{x_t} k_{c,l}\, dx + \int_{x_t}^{L} k_{c,t}\, dx\right) \qquad \text{...(3.76)}$$

where $k_{c,l}$ is the local mass transfer coefficient in laminar flow and $k_{c,t}$ is the local mass transfer coefficient in turbulent flow.

$$k_{c,av} = \frac{1}{L}\left(\int_0^{x_t} 0.332\frac{D_{AB}}{x}Re_x^{1/2}Sc^{1/3}dx + \int_{x_t}^{L} 0.0292\frac{D_{AB}}{x}Re_x^{0.8}Sc^{1/3}dx\right) \quad \text{...(3.77)}$$

Rohsenow and Choi (1961) give the average mass transfer coefficient for a transition Reynolds number equal to 3.2×10^5 as

$$\frac{k_{c,av}L}{D_{AB}} = 0.037\,Sc^{1/3}(Re_L^{0.8} - 15{,}500) \quad \text{...(3.78)}$$

Example 3.9: Air at 373 K and 101.3 kPa is flowing along a 1-m long flat plate made of naphthalene. Naphthalene sublimes into the air stream under this condition. It is observed that at the trailing edge of the plate, the Reynolds number is the transition Reynolds number, 5×10^5. The diffusivity of naphthalene vapour in air is 9.4×10^{-6} m²/s, and the vapour pressure of naphthalene in air at 373 K is 1.33 kPa. Viscosity of air is 2.15×10^{-2} cP and its density is 0.95 kg/m³. Assuming dilute solutions, determine:

(a) The velocity of air.
(b) The thickness of the velocity and concentration boundary layers at the trailing edge.
(c) The average mass transfer coefficient $k_{c,av}$.
(d) The rate of mass transfer of naphthalene into air for 1 m width of the plate.

***Solution*:** $T = 373$ K, $P = 101.3$ kPa

Diffusivity of naphthalene in air, $D_{AB} = 9.4 \times 10^{-6}$ m²/s

Vapour pressure of naphthalene, $P_A^S = 1.33$ kPa

Viscosity of air, $\mu = 2.15 \times 10^{-2}$ cP $= 2.15 \times 10^{-5}$ kg/m s

Density of air, $\rho = 0.95$ kg/m³

(a) At the trailing edge of the plate, $x = L = 1$ m

Since the Reynolds number at the trailing edge of the plate is equal to the transition Reynolds number, $Re_{x=L} = 5\times10^5$

$$Re_{x=L} = \frac{u_b \rho x}{\mu} = 5\times10^5$$

That is,

$$\frac{u_b \times 0.95\times1}{2.15\times10^{-5}} = 5\times10^5, \text{ where } u_b \text{ is the velocity of air.}$$

This equation is solved for u_b.

$u_b = 11.316$ m/s **(Answer)**

(b) Thickness of the velocity boundary layer is related to the distance from the leading edge, x and the Reynolds number at x by Eq. (3.67) as

$$\frac{\delta}{x} = \frac{4.96}{Re_x^{1/2}}$$

At the trailing edge,

$$\frac{\delta}{1} = \frac{4.96}{(5\times10^5)^{1/2}}$$

so that

$$\delta = \frac{4.96}{(5\times10^5)^{1/2}}\times1\times10^3 = 7.01 \text{ mm}$$ **(Answer)**

The thicknesses of velocity and concentration boundary layers are related as

$$\frac{\delta}{\delta_c} = Sc^{1/3} \quad \text{...(3.66)}$$

The Schmidt number is calculated using $Sc = \dfrac{\mu}{\rho D_{AB}}$

$$Sc = \frac{\mu}{\rho D_{AB}} = \frac{2.15 \times 10^{-5}}{0.95 \times 9.4 \times 10^{-6}} = 2.4076$$

Equation (3.66) now gives

$$\delta_c = \frac{\delta}{Sc^{1/3}} = \frac{7.01}{(2.4076)^{1/3}} = 5.23 \text{ mm} \quad \textbf{(Answer)}$$

(c) Since the Reynolds number at $x = L$ is equal to the transition Reynolds number, the flow in the boundary layer is laminar over the entire length of the surface. For laminar flow conditions, the average mass transfer coefficient $k_{c,av}$ is given by

$$Sh_{av} = 0.664 Re_L^{1/2} Sc^{1/3} \quad \text{...(3.72)}$$

Substituting *Re* and *Sc* in Eq. (3.72),

$$Sh_{av} = 0.664(5 \times 10^5)^{1/2}(2.4076)^{1/3} = 629.286$$

Assuming low rates of mass transfer,

$$Sh_{av} = \frac{k_{c,av} L}{D_{AB}} \quad \text{...(3.73)}$$

Therefore,

$$k_{c,av} = \frac{Sh_{av} D_{AB}}{L} = \frac{629.286 \times 9.4 \times 10^{-6}}{1} = 5.915 \times 10^{-3} \frac{\text{m}}{\text{s}} \quad \textbf{(Answer)}$$

(d) The rate of mass transfer is

$$\text{Rate} = N_A A = k_{c,av} A(C_{A1} - C_{A2})$$

Since the width and length of the surface are each 1 m, $A = 1 \text{ m}^2$.

Vapour pressure of naphthalene in air at 373 K is 1.33 kPa. The air at the gas-solid interface is saturated with the naphthalene vapours, the partial pressure $\bar{p}_A = 1.33$ kPa. Therefore,

$$C_{A1} = \frac{\bar{p}_A}{RT} = \frac{1.33}{8.314 \times 373}$$

$$= 4.289 \times 10^{-4} \frac{\text{kmol}}{\text{m}^3}$$

Air passed across the plate is assumed free of naphthalene, so that $C_{A2} = 0$. Therefore,

$$\text{Rate} = 5.915 \times 10^{-3} \times 1 \times (4.289 \times 10^{-4} - 0)$$

$$= 2.5369 \times 10^{-6} \frac{\text{kmol}}{\text{s}} \quad \textbf{(Answer)}$$

Example 3.10: Nitrogen gas at 290 K and 101.3 kPa is passed across a plastic sheet during its manufacture to vaporise and remove excess styrene. The sheets are 0.75 m long. Nitrogen flows with a velocity 10 m/s. The diffusivity of styrene vapour in nitrogen at this condition is 7×10^{-6} m²/s, and its vapour pressure is 670 Pa. Determine the rate of vaporisation from the sheet given that the density and viscosity of nitrogen are respectively, 1.14 kg/m^3 and 1.786 × 10^{-5} kg/m s.

***Solution*:** The length of the sheet L = 0.75 m.

Velocity of nitrogen gas, u_b = 10 m/s

Density of nitrogen, ρ = 1.14 kg/m^3

Viscosity of nitrogen, μ = 1.786 × 10^{-5} kg/m s

Reynolds number evaluated at $x = L$ is

$$Re_{x=L} = \frac{u_b \rho x}{\mu} = \frac{10 \times 1.14 \times 0.75}{1.786 \times 10^{-5}}$$

$$= 4.78723 \times 10^5$$

Since the Reynolds number is below the transition Reynolds number of 5 × 10^5, the flow is laminar and Eq. (3.72) can be used for the evaluation of mass transfer coefficient.

$$Sh_{av} = 0.664 Re_L^{1/2} Sc^{1/3} \qquad \text{...(3.72)}$$

The Schmidt number is given by

$$Sc = \frac{\mu}{\rho D_{AB}} = \frac{1.786 \times 10^{-5}}{1.14 \times 7 \times 10^{-6}} = 2.238$$

Substituting Re and Sc in Eq. (3.72), we get

$$Sh_{av} = 0.664 Re_L^{1/2} Sc^{1/3}$$

$$Sh_{av} = 0.664(4.78723 \times 10^5)^{1/2}(2.238)^{1/3} = 600.94$$

For low rate of mass transfer,

$$Sh_{av} = \frac{k_{c,av} L}{D_{AB}} \qquad \text{...(3.73)}$$

$$\frac{k_{c,av} 0.75}{7 \times 10^{-6}} = 600.94$$

which gives $k_{c,av}$ = 5.609 × 10^{-3} m/s

The rate of mass transfer from unit area is denoted by N_A and it is related to the mass transfer coefficient as

$$N_A = k_{c,av}(C_{A1} - C_{A2})$$

At the solid-gas interface, the nitrogen layer is saturated with styrene vapours and the partial pressure of styrene is equal to the vapour pressure, 670 Pa. Or, $\bar{p}_{A1} = P_A^S = 670$ Pa

The concentration of styrene at the gas-solid interface is

$$C_{A1} = \frac{\bar{p}_{A1}}{RT} = \frac{670 \times 10^{-3}}{8.314 \times 290} = 2.7789 \times 10^{-4} \frac{\text{kmol}}{\text{m}^3}$$

Since the nitrogen passed over the sheet contains no styrene, C_{A2} = 0.

Therefore rate of evaporation is

$$N_A = k_{c,av}(C_{A1} - C_{A2}) = 5.609 \times 10^{-3} \times 2.7789 \times 10^{-4}$$

$$= 1.559 \times 10^{-6} \text{ kmol/m}^2\text{s}$$

(Answer)

3.5 ANALOGY BETWEEN HEAT, MASS AND MOMENTUM TRANSFER

The analogy expressions provide relationships between mass transfer coefficient, heat transfer coefficient and friction factor for flow. These equations help us to deduce considerable amount of information on mass transfer coefficient from the information available on heat transfer coefficient for flow systems of same geometry. Fluid friction and heat transfer experiments are relatively easier to perform than the mass transfer experiments. The results of experimental investigation on fluid friction and heat transfer can be utilized for improving our understanding of mass transfer phenomena and for predicting behaviour of systems for which only limited quantitative data on mass transfer coefficient are available.

The following conditions should be satisfied for developing the analogy expressions:

(a) The physical properties are constant.
(b) No energy or mass is produced within the system by means of chemical reaction.
(c) Energy is neither absorbed nor emitted by the system.
(d) The rate of mass transfer is low so that the velocity profile is not affected by mass transfer.
(e) There is no viscous dissipation.

Eddy Diffusivity of Momentum, Heat and Mass

The cross section of a fluid stream in turbulent flow consists of three zones: the laminar sub layer, the buffer zone and the turbulent core. The first one existing immediately next to the wall, is the laminar sublayer or viscous layer where momentum transfer is entirely due to viscosity. Eddies are not present here. The second is a buffer zone in which the combined viscous and turbulent momentum transfer occurs. The bulk of the fluid cross section is the turbulent core in which the momentum transfer is due to the Reynolds stresses caused by the mixing motion of the eddies. Within the viscous sub layer, heat transfer is mainly due to conduction and mass transfer is due to molecular diffusion. The basic equations – Newton's law of viscosity, Fourier's law of conduction and Fick's law of diffusion – are applicable in the viscous sublayer:

$$\tau_{zx} = -\nu \frac{d(\rho u_x)}{dz} \qquad \text{Newton's law of viscosity} \qquad \ldots(3.79)$$

$$q = -\alpha \frac{d(\rho C_p T)}{dz} \qquad \text{Fourier's law of conduction} \qquad \ldots(3.80)$$

$$J_A = -D_{AB} \frac{dC_A}{dz} \qquad \text{Fick's law of diffusion} \qquad \ldots(3.81)$$

The total shear stress in the turbulent core can be written as the sum of the laminar shearing stress and turbulent shearing stress:

$$\tau = -(\nu + \varepsilon_M)\frac{d}{dz}(\rho u_x) \qquad \ldots(3.82)$$

where ν is the kinematic viscosity of the fluid and ε_M is the eddy diffusivity of momentum or the turbulent exchange coefficient of momentum.

The transfer of heat in to a fluid in turbulent flow can be represented in a similar fashion. The total heat flux at any given point is the sum of heat flux by conduction and turbulence.

$$q = -(\alpha + \varepsilon_H)\frac{d}{dz}(\rho C_p T) \qquad \ldots(3.83)$$

The heat flux by conduction is proportional to α, the thermal diffusivity, and the heat flux by turbulence is proportional to ε_H, the eddy diffusivity of heat or turbulent exchange coefficient for temperature.

In a turbulent stream, the moving eddies transport mass from one location to another just as they transport momentum and heat energy. The total flux of component **A** is sum of the contributions by molecular diffusion and eddy diffusion.

$$J_A = -(D_{AB} + \varepsilon_D)\frac{dC_A}{dz} \quad \text{...(3.84)}$$

where ε_D is known as the eddy diffusivity of mass.

The kinematic viscosity (ν), the thermal diffusivity (α), and the mass diffusivity (D_{AB}) are true fluid properties and are constant for a particular fluid at fixed conditions of temperature and pressure. However, the eddy diffusivity of momentum (ε_M), eddy diffusivity of heat (ε_H) and eddy diffusivity of mass (ε_D) depend on the fluid velocity and geometry of the flow system and are sensitive to the location in the fluid with respect to the solid boundaries and intensity of turbulence. The contributions due to the molecular diffusivities dominate in the region near the wall, while that due to the eddy diffusivities dominate in the turbulent core.

Linear Distribution of Flux

The distribution of shear stress in a stream of fluid through a circular tube is linear with radius for both laminar and turbulent flow. This can be mathematically expressed as

$$\frac{\tau}{\tau_w} = \frac{r}{r_w} \quad \text{...(3.85)}$$

where r is the radial distance from the centre of the tube and r_w is the radius of the tube. The distribution of heat flux and mass flux also can be approximated by equations analogous to Eq. (3.85) so that we have

$$\frac{\tau}{\tau_w} = \frac{q}{q_w} = \frac{N_A}{N_{Aw}} = \frac{r}{r_w} \quad \text{...(3.86)}$$

where q_w, $N_{A,w}$ are the heat flux and mass flux, respectively, at the interface between the solid surface and the fluid. We can easily deduce from the above result the following relation

$$\frac{\tau}{q} = \frac{\tau_w}{q_w} \quad \text{...(3.87)}$$

$$\frac{\tau}{N_A} = \frac{\tau_w}{N_{Aw}}$$

Reynolds Analogy

Osborne Reynolds (1874) was the first to recognise the analogy between the convective transfer of heat and momentum and his expressions are readily extended to mass transfer process. The Reynolds model ignores the existence of laminar sublayer and the buffer zone and assumes that the entire fluid cross section is occupied by the turbulent core.

Dividing Eq. (3.82) by Eq. (3.83) we get

$$\frac{\tau}{q} = \frac{\nu + \varepsilon_M}{C_p(\alpha + \varepsilon_H)}\frac{du_x}{dT} \quad \text{...(3.88)}$$

In a similar manner dividing Eq. (3.82) by (3.84), we get

$$\frac{\tau}{N_A} = \frac{\rho(\nu + \varepsilon_M)}{(D_{AB} + \varepsilon_D)}\frac{du_x}{dC_A} \quad \text{...(3.89)}$$

Utilizing Eq. (3.87), the left-hand-side of the Eqs. (3.88) and (3.89) can be replaced in terms of the fluxes at the surface.

$$\frac{\tau_w}{q_w} = \frac{\nu + \varepsilon_M}{C_p(\alpha + \varepsilon_H)} \frac{du_x}{dT} \quad \ldots(3.90)$$

$$\frac{\tau_w}{N_{A_w}} = \frac{\rho(\nu + \varepsilon_M)}{(D_{AB} + \varepsilon_D)} \frac{du_x}{dC_A} \quad \ldots(3.91)$$

The basic assumption of Reynolds analogy is that the ratio between two molecular diffusivities is equal to the ratio between the corresponding eddy diffusivities with numerical value equal to unity. Thus

$$\frac{\nu}{\alpha} = \frac{\varepsilon_M}{\varepsilon_H} = 1$$

$$\frac{\nu}{D_{AB}} = \frac{\varepsilon_M}{\varepsilon_D} = 1 \quad \ldots(3.92)$$

By definition, $\frac{\nu}{\alpha}$ is the Prandtl number and $\frac{\nu}{D_{AB}}$ is the Schmidt number. Therefore, Reynolds analogy will be applicable only for fluids for which $Sc = Pr = 1$. With these simplifying assumptions, Eqs. (3.90) and (3.91) reduce to

$$\frac{\tau_w}{q_w} = \frac{1}{C_p} \frac{du_x}{dT} \quad \ldots(3.93)$$

$$\frac{\tau_w}{N_{A_w}} = \rho \frac{du_x}{dC_A} \quad \ldots(3.94)$$

The shear stress at the wall is related to the fluid friction factor. The Fanning friction factor f is the ratio of the shear stress at the wall to the product of density and velocity head. Or,

$$\tau_w = \frac{f \rho u_b^2}{2} \quad \ldots(3.95)$$

where u_b is the bulk average velocity of the fluid stream. The heat flux at the surface is related to the heat transfer coefficient (h) and the temperature difference between the surface and the fluid.

$$q_w = h(T_i - T_b) \quad \ldots(3.96)$$

The mass flux at the interface is related to the mass transfer coefficient (k_c) and the concentration difference between the solid surface and the fluid.

$$N_{Aw} = k_c(C_{Ai} - C_{Ab}) \quad \ldots(3.97)$$

Substituting Eqs. (3.95) and (3.97) into Eq. (3.94) and rearranging, we get

$$\frac{k_c}{u_b^2} du_x = \frac{f}{2} \frac{dC_A}{C_{Ai} - C_{Ab}} \quad \ldots(3.98)$$

On integrating the above equation from the solid surface where $u_x = 0$ and $C_A = C_{Ai}$, to a location in the fluid where the velocity is the bulk average velocity u_b and concentration is C_{Ab},

$$\frac{k_c}{u_b} = \frac{f}{2} \quad \ldots(3.99)$$

The ratio $\frac{k_c}{u_b}$ is known as the Stanton number for mass transfer, denoted by St_D.

Now Eq. (3.99) can be written as:

$$St_D = \frac{k_c}{u_b} = \frac{f}{2} \qquad \text{...(3.100)}$$

In a similar manner, by substituting Eqs. (3.96) and (3.95) into Eq. (3.93) and rearranging, we have

$$\frac{h}{C_p \rho u_b^2} du_x = \frac{f}{2} \frac{dT}{T_i - T_b} \qquad \text{...(3.101)}$$

If the equation is integrated from the solid surface where $u_x = 0$ and $T = T_i$, to a location in the fluid where the velocity is the bulk average velocity u_b and temperature is T_b,

$$\frac{h}{C_p \rho u_b} = \frac{f}{2} \qquad \text{...(3.102)}$$

The ratio $\frac{h}{C_p \rho u_b} = \frac{h}{C_p G}$ where G is the mass velocity of the fluid, is called the Stanton number for heat transfer and is denoted by St_H. Thus,

$$St_H = \frac{h}{C_p \rho u_b} = \frac{f}{2} \qquad \text{...(3.103)}$$

Combining Eqs. (3.100) and (3.103) we can write the Reynolds analogy between heat, mass and momentum transfer as

$$St_H = St_D = \frac{f}{2} \qquad \text{...(3.104)}$$

Stanton Number: The Stanton number can be shown to be the ratio of Sherwood number to the product of Schmidt number and Reynolds number.

$$\frac{Sh}{ReSc} = \frac{\dfrac{k_c D}{D_{AB}}}{\dfrac{Du\rho}{\mu} \times \dfrac{\mu}{\rho D_{AB}}} = \frac{k_c}{u} = St_D \qquad \text{...(3.105)}$$

The mass transfer coefficient k_c corresponds to low rates of mass transfer.

It can be shown that St_H is actually the ratio of Nusselt number for heat transfer to the product of Reynolds number and Prandtl number.

$$\frac{Nu}{RePr} = \frac{\dfrac{hD}{k}}{\dfrac{Du\rho}{\mu} \times \dfrac{C_p \mu}{k}} = \frac{h}{C_p \rho u} = \frac{h}{C_p G} = St_H \qquad \text{...(3.106)}$$

The main limitation of Reynolds analogy is due to the inherent simplifying assumptions used in its derivation, viz., ignoring the existence of the laminar sublayer. It is found to be a good approximation only for fluids with Schmidt number close to unity. Schmidt number will be unity if the mass and momentum diffusivities are equal, which is approximately true in the case of gas phase transfer processes. The Reynolds analogy also requires that drag forces should be wholly viscous in nature and no form drag is present, which will be true for flow in closed conduits and for exterior flow without any boundary layer separation.

Prandtl Analogy

Taylor and Prandtl pointed out the major limitations of Reynolds analogy and suggested a modification by taking into account the laminar sublayer also for developing the relationship between heat and momentum transfer. Thus the fluid cross section consisted of two regions, first, immediately close to the wall a laminar sublayer in which the shear stress is purely viscous, and the second, the turbulent core for which the Reynolds analysis is applicable. The resulting analogy between heat, mass and momentum transfer can be written as:

$$St_D = \frac{\frac{f}{2}}{1+\frac{u_L}{u_b}(Sc-1)} \quad \text{...(3.107)}$$

$$St_H = \frac{\frac{f}{2}}{1+\frac{u_L}{u_b}(Pr-1)} \quad \text{...(3.108)}$$

where u_L is the velocity at the outer boundary of the laminar sub layer and u_b is the average bulk velocity of the fluid stream. For $Pr = Sc = 1$, the Prandtl analogy reduces to the Reynolds analogy.

The ratio u_L/u_b depends on the geometry of the flow system. For flow through smooth tubes, the Prandtl mixing length theory proposes

$$\frac{u_L}{u_b} = 5\sqrt{\frac{f}{2}} \quad \text{...(3.109)}$$

so that

$$St_D = \frac{\frac{f}{2}}{1+5\sqrt{\frac{f}{2}}(Sc-1)} \quad \text{...(3.110)}$$

$$St_H = \frac{\frac{f}{2}}{1+5\sqrt{\frac{f}{2}}(Pr-1)} \quad \text{...(3.111)}$$

Though Eqs. (3.110) and (3.111) provide definite improvement over the Reynolds analogy, they still fail to represent the mass transfer coefficient at high Schmidt numbers.

Chilton-Colburn Analogy

The experimental data on heat transfer coefficient, mass transfer coefficient and friction factor were correlated empirically by Chilton and Colburn (1934) to obtain the analogy equations of great practical utility. The Chilton-Colburn analogy is expressed in terms of Colburn *J*-factors as

$$J_H = J_D = \frac{f}{2} \quad \text{...(3.112)}$$

where J_H is the Colburn *J*-factor for heat and J_D is the Colburn *J*-factor for mass. The *J*-factor for mass transfer is defined as

$$J_D = St_D Sc^{2/3} \quad \text{...(3.113)}$$

where St_D is the Stanton number for mass transfer. The Colburn *J*-factor for heat is defined as

$$J_H = St_H Pr^{2/3} \quad \text{...(3.114)}$$

where St_H is the Stanton number for heat transfer. The *J*-factors are uniquely determined by the geometric configuration and the Reynolds numbers. The friction factor for flow is related to the Reynolds number as

$$f = aRe^b \quad \text{...(3.115)}$$

where *a* and *b* are empirical constants with values dependent on the geometry and the Reynolds number. Several such correlations have been established using the experimental data and a few of them are listed in the Table 3.2. These correlations are reasonably accurate for Schmidt numbers and Prandtl numbers in the range 0.5 to 10. The correlations provided by the Chilton-Colburn analogy for laminar flow over flat surfaces are found to be in exact agreement with the equations developed by boundary layer analysis.

In drawing analogies, it has been assumed that the rate of diffusion is directly proportional to the concentration difference, which implies that there is no flow in the direction in which diffusion is occurring. Strictly speaking this is true only for equimolar counter diffusion; therefore the mass transfer coefficient appearing in the above equations should have been k_c'. Thus Stanton number for mass transfer is $St_D = \dfrac{k_c'}{u_b}$ and the Sherwood number is $Sh = \dfrac{k_c' D}{D_{AB}}$. In writing the dimensionless numbers as $St_D = \dfrac{k_c}{u_b}$ and $Sh = \dfrac{k_c D}{D_{AB}}$ we have assumed that the bulk flow contribution to the rate of diffusion is negligible or the diffusing material is present in very dilute proportions. However, in absorption and desorption process there is a net flow in the direction of diffusion, since the component **A** is diffusing in a direction in which **B** has no component velocity. Attempts have been made to correct the transfer coefficients to allow for significant net mass transfer. This has been done through the bulk flow correction term $y_{B,M}$. Thus the Sherwood number and the Stanton number for mass transfer are written as

$$St_D = \frac{k_c y_{B,M}}{u_b} \quad \text{...(3.116)}$$

$$Sh = \frac{k_c y_{B,M} D}{D_{AB}} \quad \text{...(3.117)}$$

Evidently, $y_{B,M}$ approaches unity as the concentration of diffusing species becomes very low, or when the solution is very dilute.

Other Analogies

Von Kármán and Martinelli modified the Taylor-Prandtl analogy by taking into account the buffer zone in addition to the viscous sub layer and the turbulent core. Deissler gave a modification to the Reynolds analogy by gradually reducing the eddy diffusivity as the wall surface is approached. These modifications are discussed in detail by Knudsen and Katz (1958). The modification suggested by Friend and Metzner (1958) gives better accuracy up to Prandtl numbers and Schmidt numbers as high as 3000. The mass transfer part of the Friend and Metzner correlation is

$$St_D = \frac{\frac{f}{2}}{1.2 + 11.8\sqrt{\frac{f}{2}}(Sc - 1)Sc^{-1/3}} \quad \text{...(3.118)}$$

Churchill and Zajic (2002) using a time- averaged conservation of energy, derived a turbulent-flow form of Fourier's law of conduction and in a similar manner, a turbulent-flow form of Fick's law of diffusion. The Churchill and Zajic equation for convective mass transfer in straight, circular pipe for Schmidt numbers greater than 1 can be written as

$$Sh = \frac{1}{\left(\frac{Pr_t}{Pr}\right)\frac{1}{Sh_t} + \left[1 - \left(\frac{Sc_t}{Sc}\right)^{2/3}\right]\frac{1}{Sh_\infty}} \quad \text{...(3.119)}$$

where

$$Sc_t = \text{turbulent Schmidt number} = 0.85 + \frac{0.015}{Sc} \quad \text{...(3.120)}$$

$$Sh_t = \text{Sherwood number for } (Sc = Sc_t) = \frac{Re\left(\frac{f}{2}\right)}{1 + 145\left(\frac{2}{f}\right)^{-5/4}} \quad \text{...(3.121)}$$

$$Sh_\infty = \text{Sherwood number for } (Sc = \infty) = 0.07343\left(\frac{Sc}{Sc_t}\right)^{1/3} Re\left(\frac{f}{2}\right)^{1/2} \quad \text{...(3.122)}$$

The Churchill–Zajic equation is a theoretical equation free of empiricism to a large extent and is recommended for estimating mass transfer coefficient for flow through tubes at high Schmidt numbers.

Example 3.11: Using Reynolds analogy, determine the mass transfer coefficient k_C for absorption of a soluble component from a mixture of gases if the heat transfer coefficient for the system is 100 W/m^2 K. The specific heat of the gas is 2.0 kJ/kg K and density is 1.2 kg/m^3. The concentration of the soluble component is sufficiently low so that the effect of bulk flow on mass transfer is negligible.

***Solution*:** If Reynolds analogy is applicable, by Eq. (3.104)

$$St_H = St_D$$

where St_H is the Stanton number for heat transfer and St_D is the Stanton number for mass transfer.

$$St_H = \frac{h}{C_p G} = \frac{h}{C_p \rho u_b}$$

$$St_D = \frac{k_c}{u_b}$$

Substituting in Eq. (3.104),

$$\frac{h}{C_p \rho u_b} = \frac{k_c}{u_b}$$

which gives the mass transfer coefficient as

$$k_c = \frac{h}{C_p \rho} = \frac{100}{2.0 \times 10^3 \times 1.2} = 4.17 \times 10^{-2} \, \frac{\text{m}}{\text{s}} \qquad \textbf{(Answer)}$$

Example 3.12: An air stream at approximately atmospheric temperature and pressure and containing low concentration of CS_2 vapour is flowing at 40 m/s through a circular tube of 50 mm diameter. The inside of the tubes is covered with a thin film of liquid and both heat and mass transfer are taking place between the gas stream and the liquid film. The film heat transfer coefficient is found to be 100 W/m^2 K. Specific heat of air = 1.0 kJ/kg K, viscosity of air = 2×10^{-2} cP, diffusivity of CS_2 in air = 1.1×10^{-5} m^2/s, thermal conductivity of air = 2.4×10^{-2} W/m K. Density of air = 1.04 kg/m^3. Assuming Colburn analogy, determine the mass transfer coefficient k_y.

Solution: Diameter of tube, $D = 50 \times 10^{-3}$ m; velocity of air, $u = 40$ m/s; specific heat of air; $C_P =$ 1.0 kJ/kg K; viscosity of air, $\mu = 2 \times 10^{-5}$ kg/m s; density of air = 1.04 kg/m^3; diffusivity of CS_2 in air, $D_{AB} = 1.1 \times 10^{-5}$ m^2/s; thermal conductivity, $k = 2.4 \times 10^{-2}$ W/m K; heat transfer coefficient, $h =$ 100 W/m^2 K.

According to Colburn analogy,

$$J_H = J_D \qquad \text{...(3.112)}$$

The Colburn J-factor for heat transfer, J_H is evaluated as

$$J_H = St_H Pr^{2/3} \qquad \text{...(3.114)}$$

The Stanton number for heat transfer is calculated as:

$$St_H = \frac{h}{C_p G} = \frac{h}{C_p \rho u_b} = \frac{100}{1 \times 10^3 \times 1.04 \times 40} = 2.404 \times 10^{-3}$$

The Prandtl number, Pr is

$$Pr = \frac{C_p \mu}{k} = \frac{1 \times 10^3 \times 2 \times 10^{-5}}{2.4 \times 10^{-2}} = 0.833$$

Substituting the values in Eq. (3.114), we get the J-factor for heat transfer.

$$J_H = 2.404 \times 10^{-3} \times (0.833)^{2/3} = 2.13 \times 10^{-3}$$

By Colburn analogy J_D and J_H are equal,

$$J_D = J_H = 2.13 \times 10^{-3}$$

Since

$$J_D = St_D Sc^{2/3},$$

$$\frac{k_c}{u_b} Sc^{2/3} = 2.13 \times 10^{-3} \qquad \text{...(A)}$$

Schmidt number Sc is now evaluated.

$$Sc = \frac{\mu}{\rho D_{AB}} = \frac{2 \times 10^{-5}}{1.04 \times 1.1 \times 10^{-5}} = 1.75$$

Substituting this in Eq. (A), k_c is evaluated.

$$k_c = \frac{2.13 \times 10^{-3} \times 40}{(1.75)^{2/3}} = 5.87 \times 10^{-2} \, \frac{\text{m}}{\text{s}}$$

$$k_y = k_c C = k_c \left(\frac{\rho}{M}\right)$$

where ρ is the density of the solution and M is the average molecular weight of the solution.

Assuming the properties of air, we get

$$k_y = k_c \left(\frac{\rho}{M}\right) = 5.87 \times 10^{-2} \times \left(\frac{1.04}{29}\right) = 2.11 \times 10^{-3} \text{ kmol/m}^2 \text{ s}$$

Mass transfer coefficient, $k_y = 2.11 \times 10^{-3}$ kmol/m²s **(Answer)**

Example 3.13: (a) At 298 K and 101.3 kPa, the rate of evaporation of water from the outside of cylinders into air stream flowing parallel to the axis of the cylinders was found to obey

$$\frac{wL}{P_A^S - \bar{p}_A} = 3.769 \times 10^{-5} (u_b L)^{0.8}$$

where P_A^S is the vapour pressure of water at the surface temperature, kPa, $\bar{p}_A$ is the partial pressure of water in air stream in kPa, w is the water evaporated in kg/m² s, u_b is the velocity of air in m/s, L is the length of cylinder, m. At the given conditions, viscosity and density of air are, respectively 1.8464×10^{-5} kg/m s and 1.1769 kg/m³ and the diffusivity of water vapour in air is 2.55×10^{-5} m²/s. Transform the equation into the form $J_D = f(Re)$, where Re is the Reynolds number based on the cylinder length.

(b) Carbon dioxide is passed along a naphthalene cylinder 0.1 m in diameter and 1 m long at a velocity of 6 m/s at 101.3 kPa and 373 K. The vapour pressure of naphthalene at these conditions is 1.33 kPa and the diffusivity of naphthalene in CO_2 at 373 K is 8.2×10^{-6} m²/s. The density and viscosity of CO_2 are, respectively, 1.45 kg/m³ and 1.82×10^{-5} kg/m s. Calculate the rate of sublimation of naphthalene in kg/h.

***Solution*:** The Colburn J-factor for mass transfer, J_D is defined as

$$J_D = St_D Sc^{2/3} \qquad \text{...(3.113)}$$

For the conditions when the bulk flow contribution to the rate of diffusion is negligible or the diffusing material is present in very dilute proportions, $St_D = \frac{k_c}{u_b}$ or

$$J_D = \frac{k_c}{u_b} \times Sc^{2/3} \qquad \text{(A)}$$

The flux of vaporisation can be written in terms of mass transfer coefficient and the concentration difference driving force.

$$N_A = k_G (P_A^S - \bar{p}_A) \qquad \text{(B)}$$

Where k_G is the mass transfer coefficient and P_A^S is the partial pressure of water at the air-water interface which is equal to the vapour pressure of water and $\bar{p}_A$ is the partial pressure of water in the air stream. $N_A = \frac{w}{18}$, where w is the kg of water evaporated in unit time per unit area.

$$N_A = \frac{w}{18} = k_G (P_A^S - \bar{p}_A) \qquad \text{(C)}$$

Since $\frac{k_c}{RT} = k_G$ [*See* Eq. (3.29)], Eq. (C) becomes

$$N_A = \frac{w}{18} = \frac{k_c}{RT} (P_A^S - \bar{p}_A) \qquad \text{(D)}$$

Equation (D) may be rearranged as

$$\frac{w}{P_A^S - \bar{p}_A} = 18 \times \frac{k_c}{RT} \qquad \text{Or} \qquad \frac{wL}{P_A^S - \bar{p}_A} = 18 \times \frac{k_c L}{RT} \tag{E}$$

It is given that

$$\frac{wL}{P_A^S - \bar{p}_A} = 3.769 \times 10^{-5} (u_b L)^{0.8} \tag{F}$$

Comparing (E) and (F), we have

$$18 \times \frac{k_c L}{RT} = 3.769 \times 10^{-5} (u_b L)^{0.8} \tag{G}$$

Equation (G) can be rearranged after multiplying both sides by $\dfrac{Sc^{2/3}}{u_b}$.

$$\frac{k_c}{u_b} Sc^{2/3} = 3.769 \times 10^{-5} \times \frac{RT}{18} (u_b L)^{-0.2} Sc^{2/3} \tag{H}$$

Comparing with Eq. (A) we can see that

$$J_D = 3.769 \times 10^{-5} \times \frac{RT}{18} (u_b L)^{-0.2} \mathrm{Sc}^{2/3}$$

$$J_D = 3.769 \times 10^{-5} \times \frac{RT}{18} \left(\frac{u_b \rho L}{\mu} \right)^{-0.2} \left(\frac{\mu}{\rho} \right)^{-0.2} Sc^{2/3}$$

$$J_D = 3.769 \times 10^{-5} \times \frac{RT}{18} \left(\frac{\mu}{\rho} \right)^{-0.2} Sc^{2/3} Re^{-0.2} \tag{J}$$

At the given conditions, viscosity and density of air are, respectively 1.8464 × 10^{-5} kg/m s and 1.1769 kg/m^3 and the diffusivity of water vapour in air is 2.55 × 10^{-5} m^2/s. The Schmidt number is

$$Sc = \frac{\mu}{\rho D_{AB}} = \frac{1.8464 \times 10^{-5}}{1.1769 \times 2.55 \times 10^{-5}} = 0.615$$

Substituting the values in Eq. (J), we get

$$J_D = 3.769 \times 10^{-5} \times \frac{8.314 \times 298}{18} \left(\frac{1.8464 \times 10^{-5}}{1.1769} \right)^{-0.2} (0.615)^{2/3} Re^{-0.2}$$

$$= 3.4286 \times 10^{-2} (Re)^{-0.2} \tag{K}$$

The relationship between J_D and the Reynolds number is

$$J_D = 3.4286 \times 10^{-2} (Re)^{-0.2}$$ **(Answer)**

(b) Diameter of the cylinder, D = 0.1 m; length of cylinder, L = 1 m; velocity of gas, u_b = 6 m/s; vapour pressure at the gas-solid interface, P_A^S = 1.33 kPa; diffusivity of naphthalene in CO_2 = 8.2 × 10^{-6} m^2/s; density of CO_2 gas, ρ = 1.45 kg/m^3; viscosity of CO_2 gas, μ = 1.82 × 10^{-5} kg/m s.

The mass transfer coefficient is given by Eq. (K).

$$J_D = 3.4286 \times 10^{-2} (Re)^{-0.2}$$

The Reynolds number, $Re = \dfrac{L u_b \rho}{\mu} = \dfrac{1.0 \times 6 \times 1.45}{1.82 \times 10^{-5}} = 4.78 \times 10^4$

This is substituted in Eq. (K) to get J_D

$$J_D = 3.4286 \times 10^{-2}(4.78 \times 10^4)^{-0.2} = 3.974 \times 10^{-3}$$

That is,

$$St_D\, Sc^{2/3} = 3.974 \times 10^{-3} \tag{L}$$

The Schmidt number can now be evaluated.

$$Sc = \frac{\mu}{\rho D_{AB}} = \frac{1.82 \times 10^{-5}}{1.45 \times 8.2 \times 10^{-6}} = 1.5307$$

Substituting in (L), we get the Stanton number.

$$St_D = 3.974 \times 10^{-3} \times (1.5307)^{-2/3} = 2.992 \times 10^{-3}$$

Since $St_D = k_c/u_b$, the mass transfer coefficient is now evaluated.

$$k_c = 2.992 \times 10^{-3} \times u_b = 2.992 \times 10^{-3} \times 6 = 1.795 \times 10^{-2}\ \frac{\text{m}}{\text{s}}$$

Therefore, by Eq. (3.29),

$$k_G = \frac{k_c}{RT} = \frac{1.795 \times 10^{-2}}{8.314 \times 373} = 7.426 \times 10^{-6}\ \text{kmol/m}^2\ \text{s kPa}$$

The flux of vaporisation is given by

$$N_A = k_G(P_A^S - \bar{p}_A) = 7.426 \times 10^{-6}(1.33 - 0) = 9.877 \times 10^{-6}\ \text{kmol/m}^2\text{s}$$

The rate of vaporisation is obtained by multiplying the flux by area of mass transfer.

$$\text{Rate of vaporisation} = N_A A = 9.877 \times 10^{-6} \pi D L$$

$$= 9.877 \times 10^{-6} \times \pi \times 0.1 \times 1 = 3.103 \times 10^{-6}\ \text{kmol/s}$$

$$= 1.43\ \text{kg/h}$$

Rate of vaporisation = 1.43 kg/h **(Answer)**

3.6 WETTED-WALL COLUMN

Wetted-wall towers are one of the simplest pieces of equipment used for the determination of mass transfer coefficients. In wetted-wall columns, the area of contact between the phases is known accurately and the boundary layer separation does not occur. The wetted wall column is illustrated schematically in Fig. 3.6. It consists of a wetted-wall section made of glass about 50 mm diameter and 1.5 m long. Liquid is introduced at the top circular section of the tower which serves as an overflow weir and distributes the liquid as an even film along the walls. The liquid flows down the surface of the tube by gravity in a thin film and is removed from the slightly flared lower end through the seal. Gas is admitted at the bottom and flows through the inside of the tube countercurrent to the liquid film. Calming sections are provided at the ends to prevent turbulence.

The rate of evaporation of the liquid into the gas stream is measured by noting the flow rate and the inlet and exit compositions of the gas stream. Knowing the interfacial area of mass transfer, which is roughly the same as the inside surface area of the tube, the mass transfer coefficient for the gas phase can be evaluated. The studies using wetted-wall column are of great importance because of the direct utility of the data in many practical problems of engineering design. Also, the results obtained from such columns throw light on the mechanism of mass transfer.

Gilliland and Sherwood (1934) studied the vaporisation of different liquids into air and provided the following correlation for the gas phase mass transfer coefficient.

$$Sh_{av} = 0.023 Re^{0.83} Sc^{0.44} \quad ...(3.123)$$

where the Sherwood number is

$$Sh_{av} = \frac{k_c y_{B,M} D}{D_{AB}} \quad ...(3.124)$$

where D is the inner diameter of the pipe and $y_{B,M} = \frac{\bar{p}_{B,M}}{P}$ is the logarithmic mean composition of the non-diffusing gas **B**. The Gilliland-Sherwood equation is found to be applicable over the range $2000 < Re < 35000$ and $0.6 < Sc < 2.5$.

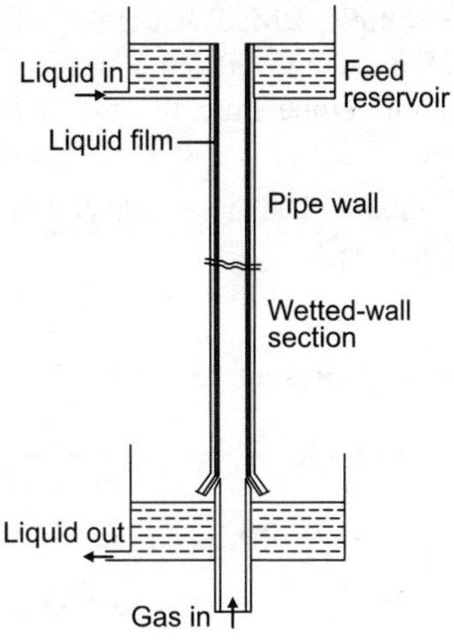

Fig. 3.6: Schematic representation of a wetted-wall column

Linton and Sherwood (1950) carried out investigations by dissolution of sparingly soluble solids such as benzoic acid into water which led to the following modification of the Eq. (3.123) which is applicable to Schmidt numbers as high as 3000.

$$Sh_{av} = 0.023 Re^{0.83} Sc^{1/3} \quad ...(3.125)$$

3.7 MASS TRANSFER CORRELATIONS

Some representative correlations for relatively simple geometric configurations with applicable range of *Sc* and *Re* are listed in Table 3.2.

Table 3.2: Mass transfer correlations for selected geometric configurations

System characteristics	Correlation	Applicability	
		Sc	*Re*
Flow inside straight circular tube	$J_D = 0.023 Re^{-0.17}$	0.6 – 3000	4000 – 60000
Flow normal to circular tube	$J_D = 0.600 Re^{-0.487}$	0.6 – 3000	50 – 50000
Unconfined flow over flat plate	$J_D = 0.664 Re^{-0.5}$		< 50000
	$J_D = 0.037 Re^{-0.2}$		$5 \times 10^5 - 5 \times 10^8$
Confined gas flow over flat plate	$J_D = 0.11 Re^{-0.29}$		2600 – 22000
Flow past single spheres	$J_D = 0.37 Re^{-0.4}$		$20 - 10^5$
Flow through packed bed of spherical particles	$J_D = \frac{2.06}{\varepsilon} Re^{-0.575}$	0.6	90 – 4000
	$J_D = \frac{20.4}{\varepsilon} Re^{-0.815}$	0.6	5000 – 10300
	$J_D = \frac{1.09}{\varepsilon} Re^{-0.667}$	168 – 70600	0.0016 – 55
	$J_D = \frac{0.25}{\varepsilon} Re^{-0.31}$	168 – 70600	5 – 1500
For fluidized bed of spheres	$J_D = \frac{0.4548}{\varepsilon} Re^{-0.4069}$		10 – 4000

(**Note:** *The Reynolds number for use in the above equations is* $Re = xu\rho/\mu$, *where x is the tube diameter for flow inside and outside tubes, length of flat surface measured from the leading edge for unconfined flow over flat plates, equivalent diameter for flow space for flow over confined plates, and particle diameter for flow over spherical particle or flow through packed and fluidised beds.*)

Example 3.14: What is the effective thickness of gas film for the evaporation of water into air in a 25-mm diameter wetted wall column at a Reynolds number of 25,000 and temperature 302 K given that at this temperature, the diffusivity of water vapour in air is 2.7×10^{-5} m²/s, density of air is 1.17 kg/m³ and viscosity is 1.85×10^{-5} kg/ms. The pressure is 100 kPa.

***Solution*:** The mass transfer data for wetted-wall column are well correlated by the Gilliland-Sherwood equation:

$$Sh_{av} = 0.023 Re^{0.83} Sc^{0.44}$$

Reynolds number, $Re = 25{,}000$

$$\text{Schmidt number, } Sc = \frac{\mu}{\rho D_{AB}} = \frac{1.85 \times 10^{-5}}{1.17 \times 2.7 \times 10^{-5}} = 0.586$$

Average Sherwood number is evaluated using Gilliland equation.

$$Sh_{av} = 0.023(25000)^{0.83}(0.586)^{0.44} = 81.263$$

$$Sh_{av} = \frac{k'_{c,av} D}{D_{AB}} = 81.263$$

The average mass transfer coefficient $k'_{c,av}$ is determined from the Sherwood number.

$$k'_{c,av} = \frac{81.263 \times D_{AB}}{D} = \frac{81.263 \times 2.7 \times 10^{-5}}{25 \times 10^{-3}}$$

$$= 8.7764 \times 10^{-2} \frac{\text{m}}{\text{s}}$$

The film coefficient of mass transfer, is related to the film thickness Z.

$$k'_c = \frac{D_{AB}}{Z} = 8.7764 \times 10^{-2}$$

$$Z = \frac{D_{AB}}{8.7764 \times 10^{-2}} \times 10^3 = 0.308 \text{ mm}$$ **(Answer)**

Example 3.15: Ammonia is absorbed from air by a dilute sulphuric acid solution in a 25-mm diameter wetted-wall tower at 298 K and 100 kPa. Air flow rate is 2×10^{-3} kg/s, and average partial pressure of ammonia is 4.0 kPa. Diffusivity of ammonia in air at 298 K is 2.26×10^{-5} m²/s. Density and viscosity of air at 298K and 100 kPa are 1.177 kg/m³ and 1.85×10^{-5} kg/m s, respectively. Partial pressure of ammonia at the interface is negligible. Estimate the mass transfer coefficient k_y and effective film thickness.

***Solution*:** Temperature, $T = 298$ K; pressure, $P = 100$ kPa; diameter, $D = 25 \times 10^{-3}$ m.

The mass flow rate of air, $\dot{m} = 2 \times 10^{-3}$ kg/s; partial pressure of ammonia in the bulk, $\bar{p}_{A1} = 4.0$ kPa; partial pressure of ammonia at the interface, $\bar{p}_{A2} = 0$; density of air, $\rho = 1.177$ kg/m³; viscosity of air, $\mu = 1.85 \times 10^{-5}$ kg/ms

Gilliland-Sherwood equation is used for evaluating the mass transfer coefficient

$$Sh_{av} = 0.023 Re^{0.83} Sc^{0.44}$$

Velocity of air is first determined for evaluating the Reynolds number:

$$u = \frac{\dot{m}}{(\pi/4) \times D^2 \times \rho} = \frac{2 \times 10^{-3}}{(\pi/4) \times (25 \times 10^{-3})^2 \times 1.177} = 3.46 \frac{\text{m}}{\text{s}}$$

Reynolds number, $$Re = \frac{Du\rho}{\mu} = \frac{25\times10^{-3}\times3.46\times1.177}{1.85\times10^{-5}} = 5.503\times10^3$$

Schmidt number, $$Sc = \frac{\mu}{\rho D_{AB}} = \frac{1.85\times10^{-5}}{1.177\times2.26\times10^{-5}} = 0.695$$

Sherwood number, $$Sh_{av} = 0.023Re^{0.83}Sc^{0.44}$$

$$= 0.023(5.503\times10^3)^{0.83}(0.695)^{0.44} = 24.94$$

$$Sh_{av} = \frac{k'_{c,av}D}{D_{AB}} = 24.94$$

$$k'_{c,av} = \frac{24.94\times D_{AB}}{D} = \frac{24.94\times2.26\times10^{-5}}{25\times10^{-3}}$$

$$= 2.255\times10^{-2}\,\frac{\text{m}}{\text{s}}$$

$$k'_y = k'_cC = k'_c\frac{P}{RT}$$

$$= 2.255\times10^{-2}\times\frac{100}{8.314\times298} = 9.102\times10^{-4}\,\frac{\text{kmol}}{\text{m}^2\text{s}}$$

$$k_y = \frac{k'_y}{y_{B,M}} \tag{A}$$

where $$y_{BM} = \frac{y_{B2}-y_{B1}}{\ln\dfrac{y_{B2}}{y_{B1}}}$$

Mole fraction of ammonia in the air stream, $y_{A1} = \dfrac{P_A^S}{P} = \dfrac{4}{100} = 0.04$

Mole fraction of ammonia at the interface, $y_{A2} = 0$

$$y_{B1} = 1 - y_{A1} = 0.96 \text{ and } y_{B2} = 1 - y_{A2} = 1.0$$

$$y_{BM} = \frac{y_{B2}-y_{B1}}{\ln\dfrac{y_{B2}}{y_{B1}}} = \frac{1-0.96}{\ln\dfrac{1}{0.96}} = 0.9799$$

Substituting in (A),

$$k_y = \frac{k'_y}{y_{B,M}} = \frac{9.102\times10^{-4}}{0.9799} = 9.289\times10^{-4}\text{ kmol/m}^2\text{s}$$ **(Answer)**

Effective film thickness Z is given by

$$Z = \frac{D_{AB}}{k'_c} = \frac{2.26\times10^{-5}\times10^3}{2.255\times10^{-2}} = 1\text{ mm}$$ **(Answer)**

Example 3.16: Air at 320 K and 200 kPa is passed through a tower packed with naphthalene spheres 12 mm in diameter at a rate of 2 m/s, based on empty cross section of the tower. The vapour pressure of naphthalene is 74 Pa. If the porosity of the bed is 40 percent, determine the rate of evaporation of naphthalene in kg/min from 1 m^3 of bed. Diffusivity of naphthalene in air at 320 K and 200 kPa is 3.274×10^{-6} m^2/s, density of air is 2.2064 kg/m^3, viscosity is 1.939×10^{-5} kg/m s. Use $J_D = \dfrac{2.06}{\varepsilon} Re^{-0.575}$.

Solution: Temperature, $T = 320$ K; pressure, $P = 200$ kPa; velocity of air, $u = 2$ m/s; vapour pressure of naphthalene, $P_A^S = 74 \times 10^{-3}$ kPa; porosity of bed, $\varepsilon = 0.40$; diffusivity of naphthalene in air, $D_{AB} = 3.274 \times 10^{-6}$ m^2/s; density of air, $\rho = 2.2064$ kg/m^3; viscosity of air, $\mu = 1.939 \times 10^{-5}$ kg/ms.

$$\text{Reynolds number, } Re = \frac{D_p u \rho}{\mu} = \frac{12 \times 10^{-3} \times 2 \times 2.2064}{1.939 \times 10^{-5}} = 2.731 \times 10^3$$

$$\text{Schmidt number, } Sc = \frac{\mu}{\rho D_{AB}} = \frac{1.939 \times 10^{-5}}{2.2064 \times 3.274 \times 10^{-6}} = 2.6842$$

For flow of gases through packed columns in the Reynolds number range, 90 – 4000, the mass transfer coefficient is correlated as

$$J_D = \frac{2.06}{\varepsilon} Re^{-0.575}$$

Substitute the Reynolds number and porosity values,

$$J_D = \frac{2.06}{\varepsilon} Re^{-0.575} = \frac{2.06}{0.4} \times (2.731 \times 10^3)^{-0.575} = 5.444 \times 10^{-2}$$

But,
$$J_D = St_D Sc^{2/3}$$

$$St_D = J_D \times Sc^{-2/3} = 5.444 \times 10^{-2} \times (2.6842)^{-2/3} = 2.8187 \times 10^{-2}$$

Since
$$St_D = k_c'/u_b,$$

$$k_c' = St_D u = 2.8187 \times 10^{-2} \times 2 = 5.6374 \times 10^{-2} \text{ m/s}$$

Assuming dilute solutions,

$$k_c = k_c' = 5.6374 \times 10^{-2} \text{ m/s}$$

The rate of evaporation is obtained as

$$N_A = k_c (C_{A1} - C_{A2}) = k_c \times C(y_{A1} - y_{A2}) = k_c \left(\frac{P}{RT}\right)(y_{A1} - y_{A2})$$

$$= 5.6374 \times 10^{-2} \times \left(\frac{200}{8.314 \times 320}\right)(y_{A1} - y_{A2})$$

$$y_{A1} = \frac{P_A^S}{P} = \frac{74 \times 10^{-3}}{200} = 3.7 \times 10^{-4}$$

$$y_{A2} = 0$$

$$N_A = 5.6374 \times 10^{-2} \times \left(\frac{200}{8.314 \times 320}\right)(3.7 \times 10^{-4} - 0) = 1.568 \times 10^{-6}\ \text{kmol/m}^2\text{s}$$

Area for mass transfer, surface area of particles,

$$a = \frac{6(1-\varepsilon)}{D_P} = \frac{6 \times 0.6}{12 \times 10^{-3}} = 3 \times 10^2\ \text{m}^2$$

Rate of vaporisation is $N_A aM$ where M is the molecular weight of naphthalene (= 128).

$$\text{Rate} = 1.568 \times 10^{-6} \times 3 \times 10^2 \times 128 \times 60 = 3.61 \frac{\text{kg}}{\text{min}}$$ **(Answer)**

3.8 PHASE EQUILIBRIUM

In interphase mass transfer, the material is transferred between two immiscible phases. Two important aspects in the analysis of all mass transfer operation are the rate of transfer and the equilibrium between the phases. Rate of mass transfer depends on the resistance to transfer and the driving force. The driving force for mass transfer is measured by the concentration difference between phases. As it is the departure from equilibrium that provides the necessary concentration gradient for mass transfer, the phase equilibrium and the conditions governing it are important considerations in interphase mass transfer. It is important to recognize that at equilibrium, the composition of the contacting phases need not be equal; instead, the chemical potentials of the constituents in all the phases should be identical. It is the equality of the chemical potential, and not concentration which, causes the net transfer of solute between phases to stop.

To get an insight into the significance of phase equilibrium in interphase mass transfer, consider the equilibrium between gas and liquid phases that govern the gas absorption operations. Let us consider a specific example: the dissolution of ammonia from a mixture of ammonia and air in water. Let a mixture of ammonia and air in which the partial pressure of ammonia is $\bar{p}_A$, and the mole fraction is y_A, is passed continuously past pure water contained in a well agitated open vessel. Ammonia gets transferred from the gas to the liquid and as a result, the concentration of ammonia in the liquid increases. The transfer of ammonia from the gas to the liquid continues till the chemical potential of ammonia in the liquid is the same as that in the gas phase. The composition of the liquid will not change with further contact of the gas. This is the state of equilibrium between the gas phase and the liquid phase and at this state, the liquid is saturated with the gas. Let the mole fraction in the liquid under this condition is x_A and that in the gas phase is y_A. x_A and y_A are the equilibrium compositions for the given conditions of temperature and pressure. These compositions are independent of the amount of water and gas contacted. As an example, for a partial pressure of ammonia in the gas phase equal to, say, 26.6 kPa at atmospheric pressure and 303 K ($y_A = 0.263$), the mole fraction of ammonia in the liquid phase will be 0.145 under equilibrium. If ammonia-air mixture containing 26.3% (volume) ammonia is brought into contact with water containing 14.5% (mol) ammonia in a closed container, no change in concentration occurs in either phase, as the phases are already in equilibrium. Instead, if the mole fraction of ammonia is greater than 0.263 in the gas phase and the liquid phase contained 0.145 mole fraction ammonia, ammonia will be transferred from the gas to the liquid till a new set of equilibrium compositions are established. On the other hand, if the gas mole fraction were less than 0.263, ammonia will be released from the liquid and transferred to the gas phase till new equilibrium conditions are established. Three sets of equilibrium values for ammonia-water system are given below for the sake of illustration:

x_A	0.09	0.145	0.250
y_A	0.131	0.263	0.657

Equilibrium Distribution Diagram

The relationship between concentrations of the phases in equilibrium can be expressed mathematically as

$$y = f(x) \quad ...(3.126)$$

where y is the concentration of one of the phases (say, G-phase) and x is the concentration of the other phase (say, L-phase) which is in equilibrium with the former. In gas-absorption, x and y are concentration of the liquid and gas phases in equilibrium. This relationship is conveniently represented graphically by means of the equilibrium distribution diagram or simply, the equilibrium diagram. For gas absorption equilibrium, the distribution diagram is as shown in Fig. 3.7.

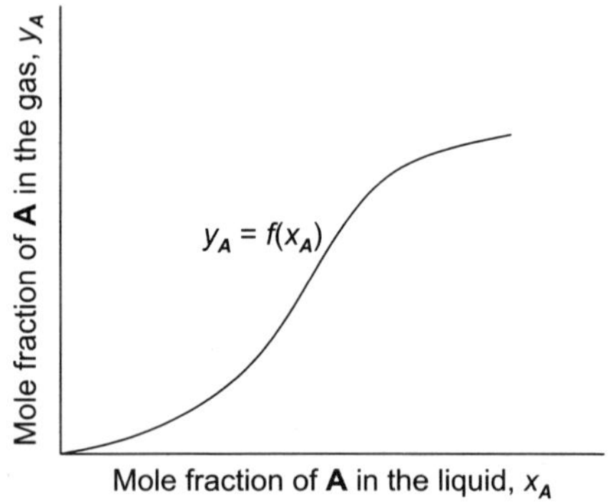

Fig. 3.7: Equilibrium distribution of solute between gas and liquid

In Fig. 3.7, the mole fraction of solute in the gas phase (y_A) is taken as the ordinate and mole fraction in the liquid (x_A) is taken as the abscissa. The concentrations of the phases can be represented in a number of ways, and, as a consequence, distribution diagrams with different co-ordinates are common. For example, the absorption equilibrium curve can be plotted with partial pressure of solute ($\bar{p}_A$) instead of y_A, as ordinate and molar concentration (C_A, kmol/m^3) as abscissa instead of x_A.

For distillation operations, the equilibrium distribution curve relates mole fraction of the more volatile component (MVC) in the liquid phase to the mole fraction of the more volatile component in the vapour phase as shown in Fig. 3.8.

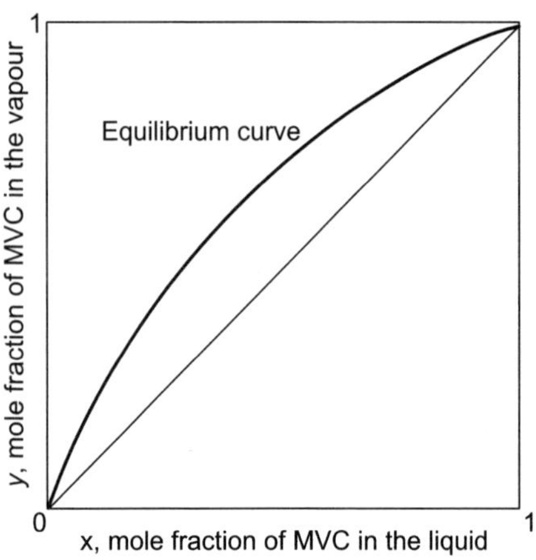

Fig. 3.8: Vapour–liquid equilibrium curve

In extraction operation, the solute is distributed between two immiscible liquid phases – a solvent rich phase and a solvent-lean phase. This equilibrium can be represented graphically in a number of ways. One of the methods is to take the weight fraction of solute in the solvent-lean phase on the x-axis and weight fraction in the solvent rich phase on the y-axis as shown in Fig. 3.9.

For adsorption operations in which a solute is removed from a gas stream by adsorbing on an adsorbent such as activated carbon, the equilibrium can be represented as a plot of concentration on the solid phase versus concentration in the gas phase. Thus we plot the mass of solute adsorbed on unit mass of adsorbent on the x-axis versus partial pressure of solute in the vapour phase on the y-axis as shown in Fig. 3.10.

The equilibrium relationship must be established for the analysis of mass transfer operations and design of separation equipment; reliable experimental results are the most dependable sources for equilibrium data. Reid, et al. (1987) give an excellent review of sources of data and prediction methods for vapour–liquid and liquid–liquid equilibria required for distillation, absorption, and extraction operations.

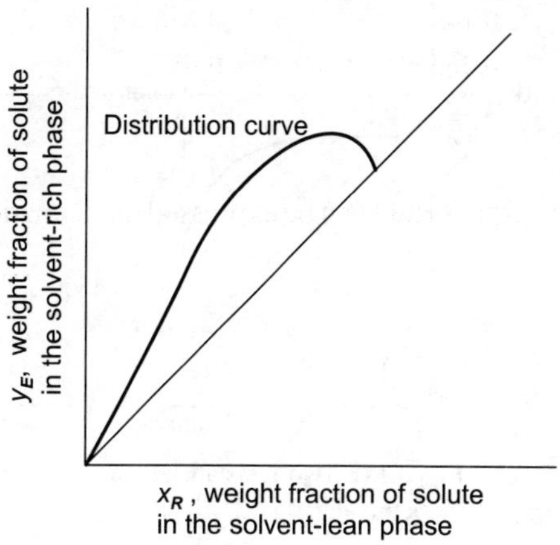

Fig. 3.9: Distribution of solute between two immiscible liquid phases

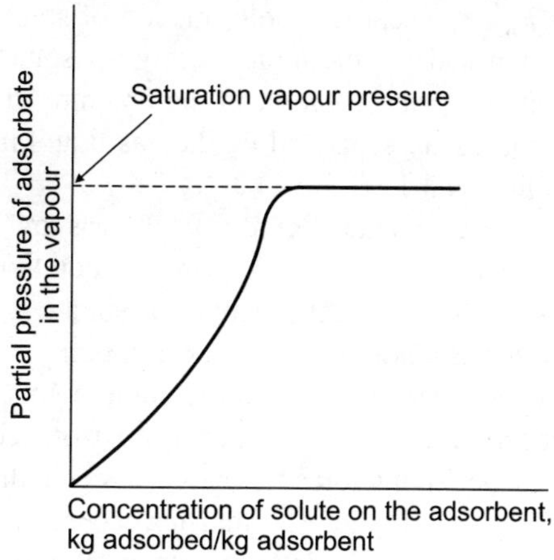

Fig. 3.10: Adsorption isotherm

There are certain simple situations where equilibrium relationship can be readily established from theoretical considerations. When the liquid is an ideal solution and the gas phase behaves as an ideal gas, the equilibrium vapour and liquid compositions are related as

$$y = x\frac{P^S}{P} \qquad \text{...(3.127)}$$

where y and x are respectively, the mole fractions of the vapour and liquid-phases in equilibrium and P^S and P are the vapour pressure and total pressure of the system. Equation (3.127) is obtained by combining Raoult's law applicable for ideal liquid solution and Dalton's law applicable for ideal vapour phase.

Very dilute solutions also exhibit a form of ideal behaviour represented by the Henry's law.

$$y = x\frac{H}{P} \qquad \text{...(3.128)}$$

where H is known as Henry's law constant. Ideal solution behaviour and application of Raoult's law and Henry's law will be discussed in detail in Chapters 4 and 5. For a detailed discussion on phase equilibria and for a review of equations applicable for real solutions, books on Chemical Engineering Thermodynamics should be consulted. However, the following basic concepts are common to all systems involving distribution of components between phases.

1. For a given temperature and pressure, there exists a set of equilibrium compositions which is represented graphically by a distribution diagram.
2. For a system in equilibrium, there is no net diffusion of components between the phases.
3. For a system not in equilibrium, diffusion will occur between the phases leading to a state of equilibrium where the chemical potentials for any given component are equal in all phases.

3.9 THE TWO-RESISTANCE THEORY

Consider a wetted-wall column used for absorption of a solute **A** from a gas flowing upwards to a liquid flowing down as a thin film along the walls. As a specific example, we may consider the absorption of ammonia from its mixture with air in water. The gas enters the bottom of the column and moves up in contact with the liquid which is admitted at the top. The solute concentration of the gas decreases steadily as it moves up because of transfer into the liquid, and the solute content in the liquid increases as it moves down. However, the composition of the gas and liquid phases at any point in the column are constant under steady-state operation.

Let y_{AG} represent the mole fraction of solute **A** in the gas stream and x_{AL} the mole fraction of solute **A** in the liquid at any given location in the column. The gas and liquid phases are separated by the gas-liquid interface as shown in Fig. 3.11.

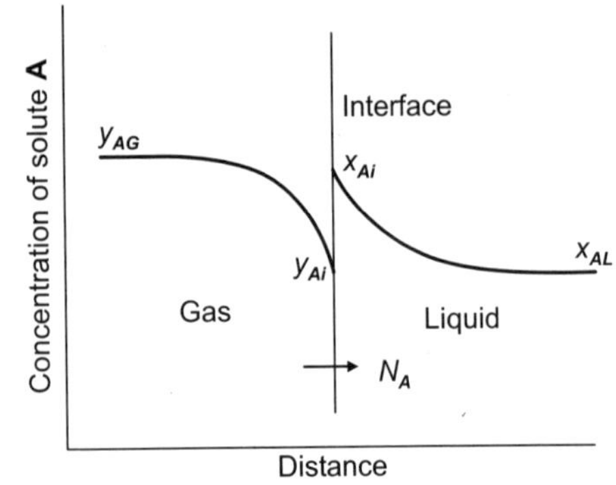

Fig. 3.11: Two-resistance theory for interphase mass transfer

The solute gets transferred from the gas to the liquid as the concentrations y_{AG} and x_{AL} are not equilibrium compositions. The chemical potential of solute **A** in the gas stream corresponding to its concentration y_{AG} is greater than the chemical potential in the liquid corresponding to its concentration x_{AL}. This departure from equilibrium provides the driving force for mass transfer in the direction from the gas to liquid. The transfer occurs first from the bulk of the gas to the interface, then from the interface to the bulk of the liquid. On the gas side, the driving force for this transfer is $y_{AG} - y_{Ai}$, where y_{Ai} is the mole fraction of component **A** at the gas side of the interface, and on the liquid side the driving force is $x_{Ai} - x_{AL}$, where x_{Ai} is the mole fraction of **A** at the liquid side of the interface.

The two fluids are in motion relative to the interface between them, and two films are visualized through which solute must diffuse in series. The two-resistance theory (Whitman, 1923) is used to explain interphase mass transfer. The theory assumes that

1. The rate of mass transfer between the two phases is controlled by the rate of diffusion through the phases on each side of the interface.
2. There is no resistance to the transfer of the diffusing component across the interface.

Since no resistance to mass transfer exists at the interface, the concentrations at the interface, x_{Ai} and y_{Ai} are equilibrium compositions. It means that the chemical potentials corresponding to these concentrations for the respective phases are equal. Equilibrium at the interface does not require x_{Ai} to be equal to y_{Ai}. Equilibrium may well be attained, if x_{Ai} is greater than or less than y_{Ai}. The transfer is from gas to liquid even when x_{Ai} is greater than y_{Ai} as shown in Fig. 3.11.

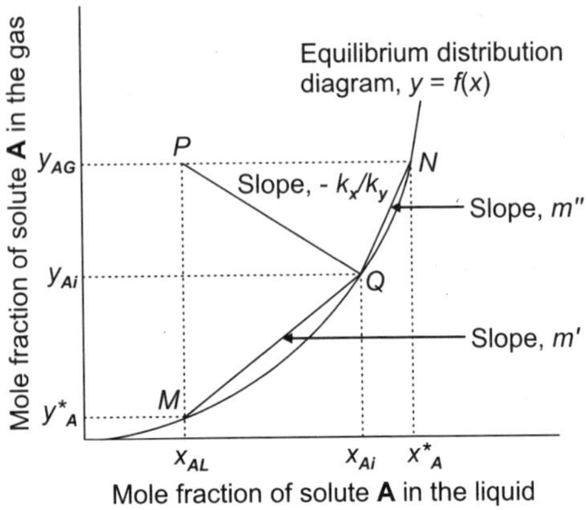

Fig. 3.12: Concentration driving force in interphase mass transfer

The bulk average composition of the gas and liquid phases can be represented graphically on the equilibrium distribution diagram as shown in Fig. 3.12. Since x_{AL} and y_{AG} are not equilibrium compositions, the point P whose coordinates represent these bulk compositions will be located away from the equilibrium distribution curve for the system. Point P will be above the equilibrium curve, if the transfer is from gas to liquid, and below the equilibrium curve, if the transfer is in the reverse direction. Since x_{Ai} and y_{Ai} are equilibrium compositions the point Q representing the interface compositions will lie on the equilibrium curve.

Let the local mass transfer coefficients for the gas and liquid phases be, respectively, k_y and k_x. Assume that the steady-state conditions prevail so that the rate at which the material **A** reaches the interface from the gas side equals the rate at which it is removed from the interface to the bulk liquid and there is no accumulation or depletion of material at the interface. The flux of **A** in terms of the mass transfer coefficients and the concentration difference can then be written as

$$N_A = k_y(y_{AG} - y_{Ai}) = k_x(x_{Ai} - x_{AL}) \qquad \ldots(3.129)$$

Equation (3.129) can be rearranged as

$$\frac{y_{AG} - y_{Ai}}{x_{AL} - x_{Ai}} = -\frac{k_x}{k_y} \qquad \text{...(3.130)}$$

The terms on the left side of Eq. (3.130) is the slope of straight line joining point P (x_{AL}, y_{AG}) and point Q (x_{Ai}, y_{Ai}). Therefore, the concentrations at the interface at a given location in the equipment where the bulk compositions are known can be determined graphically using Eq. (3.130). On the diagram locate point P, the coordinates of which are given by the bulk compositions, and from P draw a straight line with slope $-\frac{k_x}{k_y}$, which intersects the equilibrium curve at Q. The coordinates of Q are the interface compositions. Alternatively, Eq. (3.130) can be solved analytically for interface compositions if the equilibrium is represented by an equation of the form $y = f(x)$.

Limitations: The assumption of equilibrium at the interface may not be valid when gradients of interfacial tension are established during the mass transfer. These gradients will give rise to interfacial turbulence and to increase in mass transfer coefficients, a phenomenon known as *Marangoni effect*. The Marangoni effect may be reduced by adding surfactants, which tend to concentrate at the surface, at times introducing an interfacial resistance to mass transfer. The additional resistance introduced by the presence of the contaminants at the surface spoils the validity of the two-resistance theory. The heat transfer resulting from the unequal heats of solution in the two phases may lead to an increase or decrease of temperature at the interface, which in turn, may give rise to equilibrium compositions at the interface different from the values expected relative to the bulk phase temperatures. The occurrence of chemical reactions at the interface, that is not encountered when the individual coefficients are measured is another factor which may offset the validity of the two-resistance theory.

Overall Mass Transfer Coefficient

We have seen how the mass flux is determined knowing the mass transfer coefficient for individual phases, the bulk phase composition and the interfacial compositions. The bulk phase compositions can be determined easily by sampling and analysis. However, it is very difficult to measure the interfacial compositions experimentally. Therefore, it will be convenient if the mass flux is expressed in terms of the overall driving force, rather than the concentration difference confined to the individual phases. This is made possible by defining the overall mass transfer coefficient as

rate of mass transfer = overall coefficient × overall driving force

In gas absorption in which a gas stream of composition y_{AG} is contacted with a liquid stream of composition x_{AL}, one may think that the overall driving force is $y_{AG} - x_{AL}$. If it is so, the driving force would have vanished when equilibrium is established between the phases. But, we know that y_{AG} need not be equal to x_{AL}, even if the phases were in equilibrium and $y_{AG} - x_{AL}$ need not be zero. (*See* Example 3.17). Criteria of phase equilibrium requires only that the chemical potentials be equal in both phases under equilibrium. Since chemical potentials are differently related to compositions depending upon the physical states of the fluid, the phases should be compared on the same basis for the sake of determining the true overall driving force for mass transfer in terms of concentrations of the contacting phases. For example, the overall driving force can be the difference between composition of the gas phase, y_{AG} and the gas phase equilibrium equivalent of the given liquid composition, y_A^*. The latter is the concentration of a fictitious gas phase which is equivalent to the given liquid or it is as good a measure of x_{AL} as x_{AL} itself and has the advantage that it is on the same basis as y_{AG}. The equilibrium equivalent y_A^* can be determined corresponding to bulk liquid composition x_{AL} from the equilibrium relationship Eq. (3.126) or the distribution diagram for the system. Thus y_A^* is the y-coordinate of point M in Fig. 3.12.

The rate of interphase mass transfer can now be measured as

$$N_A = K_y (y_{AG} - y_A^*) \qquad \text{...(3.131)}$$

where K_y is the overall gas phase mass transfer coefficient. Similarly we can have an overall liquid-phase mass transfer coefficient K_x, using which, the rate of mass transfer would be

$$N_A = K_x (x_A^* - x_{AL}) \qquad \text{...(3.132)}$$

Here, x_A^* is the liquid-phase equilibrium equivalent of the gas phase concentration y_{AG}. It is the concentration of a fictitious liquid phase that is in equilibrium with a gas phase of concentration y_{AG}. Thus x_A^* is the x-co-ordinate of point N in Fig. 3.12.

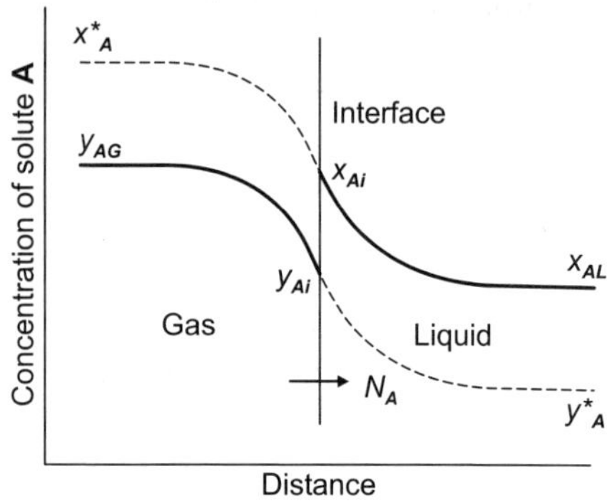

Fig. 3.13: Overall driving force in interphase transfer

From Fig. 3.12 we can see that

$$y_{AG} - y_A^* = (y_{AG} - y_{Ai}) + (y_{Ai} - y_A^*) \qquad \text{...(3.133)}$$

$$= (y_{AG} - y_{Ai}) + m'(x_{Ai} - x_{AL})$$

m' is the slope of the chord MQ or it is the slope of the equilibrium curve if the equilibrium relationship could be represented by a straight line between M and Q.

Let us now derive the relationship between individual mass transfer coefficients and the overall coefficient. The molar flux N_A can be written as

$$N_A = k_y (y_{AG} - y_{Ai})$$

$$= k_x (x_{Ai} - x_{AL}) \qquad \text{...(3.134)}$$

$$= K_y (y_{AG} - y_A^*)$$

Equation (3.134) gives the following:

$$y_{AG} - y_{Ai} = \frac{N_A}{k_y}$$

$$x_{Ai} - x_{AL} = \frac{N_A}{k_x} \qquad \text{...(3.135)}$$

$$y_{AG} - y_A^* = \frac{N_A}{K_y}$$

Substituting Eq. (3.135) into Eq. (3.133), we get

$$\frac{N_A}{K_y} = \frac{N_A}{k_y} + m' \frac{N_A}{k_x}$$

Or

$$\frac{1}{K_y} = \frac{1}{k_y} + m' \frac{1}{k_x} \qquad \text{...(3.136)}$$

In a similar manner, for the liquid phase,

$$x_A^* - x_{AL} = (x_A^* - x_{Ai}) + (x_{Ai} - x_{AL})$$

$$= \frac{y_{AG} - y_{Ai}}{m''} + (x_{Ai} - x_{AL}) \qquad \text{...(3.137)}$$

Substituting Eq. (3.135) into Eq. (3.137), we get

$$\frac{N_A}{K_x} = \frac{1}{m''}\frac{N_A}{k_y} + \frac{N_A}{k_x}$$

Or,
$$\frac{1}{K_x} = \frac{1}{m''}\frac{1}{k_y} + \frac{1}{k_x} \quad \ldots(3.138)$$

Equation (3.136) and (3.138) reveal that the individual transfer coefficients and overall coefficients are related in such a way that the overall resistance to mass transfer is the sum of the resistances in the individual phases. Also,

$$\frac{\text{Resistance in gas phase}}{\text{Total resistance in both phases}} = \frac{1/k_y}{1/K_y}$$

$$\frac{\text{Resistance in liquid phase}}{\text{Total resistance in both phases}} = \frac{1/k_x}{1/K_x}$$

Expressions relating the overall mass transfer coefficient to individual coefficients defined with reference to different concentration terms have similar forms as Equations (3.136) and (3.138). Thus it is possible to have the following relationship between individual and overall coefficients:

$$\frac{1}{K_G} = \frac{1}{k_G} + m'\frac{1}{k_L} \quad \ldots(3.139)$$

$$\frac{1}{K_L} = \frac{1}{m''}\frac{1}{k_G} + \frac{1}{k_L} \quad \ldots(3.140)$$

Here, m' and m'' are the slope of the chords when the partial pressure $\bar{p}_A$ is plotted on *y*-axis against the molar concentration C_A on *x*-axis. If the equilibrium relationship can be expressed by Henry's law, $\bar{p}_A = H_A C_A$, then m' and m'' in Eqs. (3.139) and (3.140) can be replaced by H_A.

In the strict sense, the above relationships – Eqs. (3.136), (3.138, (3.139) and (3.140) – are valid only for straight equilibrium lines.These are not valid if the equilibrium line is curved or *m* is not constant. Even if the local coefficients remain constant, the overall coefficient may vary if *m* varies from one end of the column to the other. For real solutions *m* is constant only for dilute systems. Even if the equilibrium lines are not straight, we may have to depend on the use of overall coefficients, because it is impossible to measure interfacial compositions for use with individual phase coefficients.

Gas-Phase and Liquid-Phase Controlled Systems

It is possible to arrive at some qualitative conclusions on the influence of the slope of the equilibrium curve on the overall mass transfer coefficient. Consider Eq. (3.136) and assume approximately equal numerical values for k_x and k_y. If the solute in the gas phase is readily soluble in the liquid, a small concentration in the gas phase will provide a very large concentration in the liquid phase. The equilibrium curve will be very flat and m' will be very small as shown in Fig. 3.14.

In Eq. (3.136), m'/k_x becomes negligibly small and the major resistance to mass transfer lies in the gas phase, and we say that the mass transfer is gas-phase controlled.

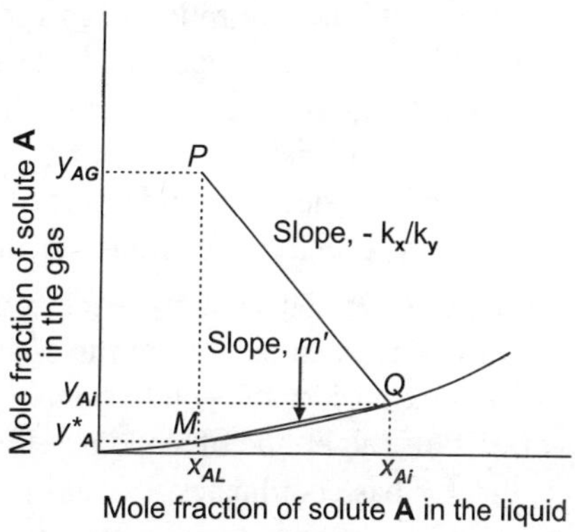

Fig. 3.14: Equilibrium curve for the gas-phase controlled system

For the gas-phase controlled situations,

$$(y_{AG} - y_A^*) \approx (y_{AG} - y_{Ai}) \text{ and } \frac{1}{K_y} = \frac{1}{k_y}$$

and no significant change in K_y will be observed by increasing the liquid phase coefficient k_x. For such systems, only by increasing the gas phase mass transfer coefficient, or by decreasing the gas phase resistance to mass transfer, any improvement in the overall coefficient can be brought about.

Now consider the solute **A** to be only sparingly soluble in the liquid, as for example, dissolution of CO_2 in water. Even with very high solute concentration in the gas phase, the equilibrium liquid side concentration will be small. The equilibrium curve is very steep with large m'' values as shown in Fig. 3.15.

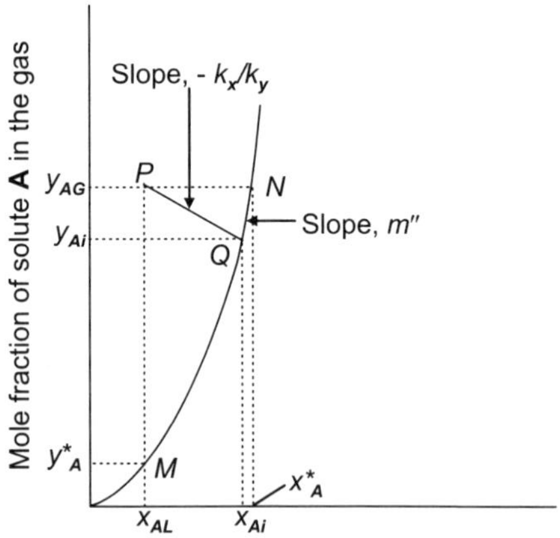

Fig. 3.15: Equilibrium curve for the liquid-phase controlled system

The mass transfer process shown in Fig. 3.15 is liquid-phase controlled. The first term on the right-hand-side of Eq. (3.138) is negligible, the major resistance to mass transfer resides with the liquid and $\frac{1}{K_x} = \frac{1}{k_x}$ and $x_A^* - x_{AL} \approx (x_{Ai} - x_{AL})$. In such situations, the rate of mass transfer may be improved by improving the liquid side coefficient. A large influence of temperature on the overall coefficient is a definite indication that the system is liquid-phase controlled, because, it is the liquid side coefficient that is more sensitive to temperature change, than the gas side coefficient (Treybal, 1981).

It is also possible that the major resistance to mass transfer may change from gas-phase controlled at one end of the apparatus to the liquid-phase controlled at the other end as illustrated in Fig. 3.16. For example, at the bottom end of the absorption column where the gas and liquid phases are rich in solute, the liquid-side resistance dominate (point p'), whereas, at the top end of the absorber, the gas and liquid phases are lean in solute (point p'') and the absorption is gas-phase controlled.

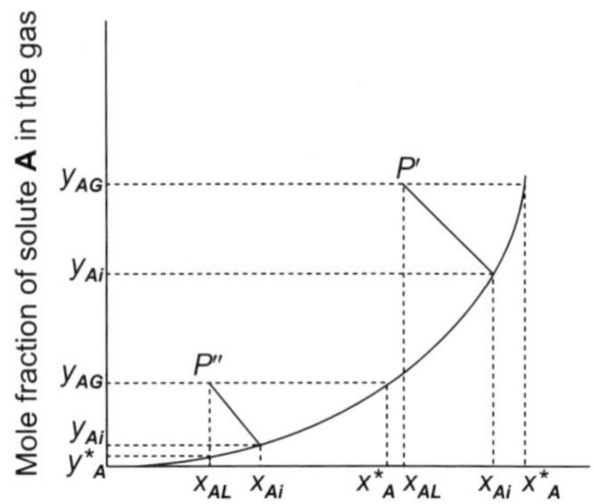

Fig. 3.16: Liquid side and gas side resistances dominate

Finally, it may be noted that to decide which phase is the controlling one, the magnitude of m (the equilibrium characteristics) is not the only criteria to be followed; the relative magnitudes of k_x and k_y are also important. Our primary concern in this section was only the cases where k_x and k_y are almost equal.

Example 3.17: Show that $y_{AG} - x_{AL}$ is a function of m, the slope of the equilibrium curve, and need not be zero even when equilibrium is established between the gas and liquid phases.

***Solution*:** Consider the equilibrium curve as shown in Fig. 3.17. Let us assume that the equilibrium curve is a straight line of slope m.

Point P has coordinates, x_{AL} and y_{AG}. Point Q has coordinates x_{Ai} and y_{Ai}.

$$\begin{aligned} y_{AG} - x_{AL} &= (y_{AG} - y_{Ai}) + (y_{Ai} - x_{Ai}) + (x_{Ai} - x_{AL}) \\ &= (y_{AG} - y_{Ai}) + (m-1)x_{Ai} + (x_{Ai} - x_{AL}) \end{aligned} \quad \text{(A)}$$

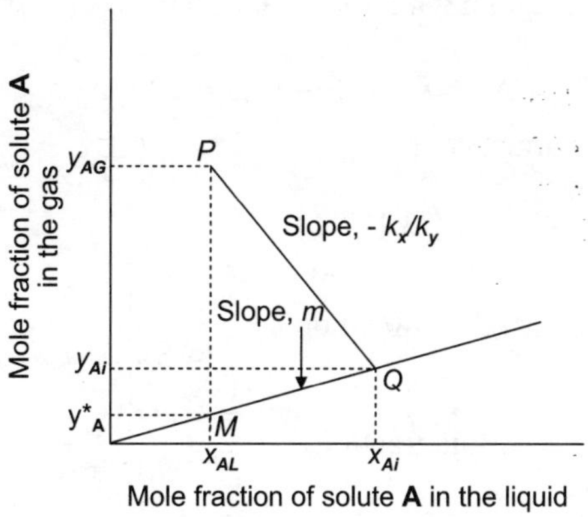

Fig. 3.17: Equilibrium curve for Example 3.17

As equilibrium is approached, concentration becomes uniform within a homogeneous phase and therefore, y_{AG} approaches y_{Ai} and x_{AL} approaches x_{Ai}, so that Eq. (A) becomes

$$y_{AG} - x_{AL} = (m-1)x_{Ai} \qquad \text{...(B)}$$

Equation (B) reveals that even under equilibrium, the concentration difference driving force will not become zero.

Example 3.18: From a mixture of gases made up of constituents **A** and **B** the solute **A** is being absorbed into water in a wetted wall column. The mass transfer coefficient in the gas phase was determined to be 34.67 kmol/m^2 h (bar) and that in the liquid phase was found to be 40 kmol/m^2 h (mole fraction). At one point in the column, laboratory analysis of the gas sample showed 20% (mole) **A** in the gas phase and the liquid sample analysed 5% mole. The pressure in the column is 1 bar and the temperature is 293 K. Assume that **B** is insoluble in water and the bulk flow effect is negligible. The equilibrium data is as follows:

Mole fraction of **A** in liquid, x_A	0.05	0.10	0.15	0.20	0.25	0.30
Partial pressure of **A** in gas, $\overline{p}_A$, bar	0.0565	0.1035	0.1771	0.2916	0.4680	0.7330

(a) Estimate the concentration of **A** at the interface and the rate of absorption.

(b) Determine the overall mass transfer coefficient K_G and K_x and identify the controlling resistances.

***Solution*:** The equilibrium diagram is plotted as shown in Fig. 3.18a with mole fraction of **A** (x_A) on the x-axis and the partial pressure of **A** in the gas phase ($\overline{p}_A$) on the y-axis.

The mole fraction of **A** in the gas phase is 0.2 and in the liquid is 0.05.

$$y_{AG} = 0.2; \; x_{AL} = 0.05$$

Since we have chosen the partial pressure in the gas $\overline{p}_A$ for representing gas phase concentration and mole fraction of **A**, x_A, for representing the liquid concentration, the bulk concentrations will be represented by point P with coordinates x_{AL} and $\overline{p}_{AG}$.

$$x_{AL} = 0.05; \; \overline{p}_{AG} = y_{AG} \times P = 0.2 \times 1 = 0.2 \text{ bar}$$

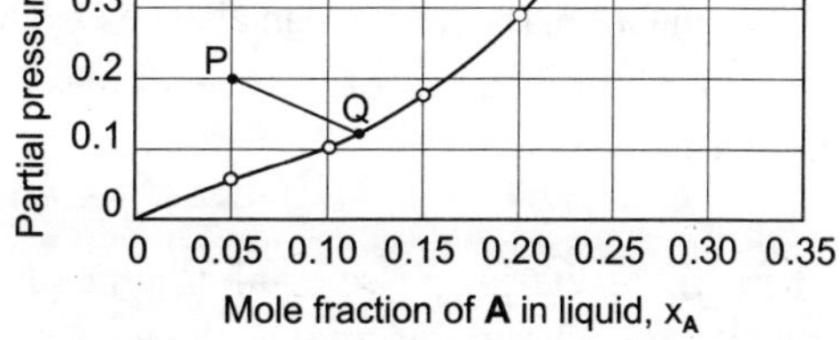

Fig. 3.18: (a) Equilibrium curve for Example 3.18

(a) Point P is marked on the diagram with coordinates (0.05, 0.2). Corresponding to Eq. (3.129) we can write

$$N_A = k_G(\bar{p}_{AG} - \bar{p}_{Ai}) = k_x(x_{Ai} - x_{AL}) \qquad \text{...(3.129 A)}$$

Equation (3.129 A) can be rearranged as

$$\frac{\bar{p}_{AG} - \bar{p}_{Ai}}{x_{AL} - x_{Ai}} = -\frac{k_x}{k_G} \qquad \text{...(3.130 A)}$$

Here, $k_G = 34.67$ kmol/(m^2 h bar), $k_x = 40$ kmol/(m^{2}h mole fraction).

$$-\frac{k_x}{k_G} = \frac{-40}{34.67} = -1.154 \text{ bar/(mole fraction)}$$

A straight line with slope $-\dfrac{k_x}{k_G} = -1.154$ is drawn from point P, which intersect the equilibrium curve at point Q, the coordinates of which give the interface compositions x_{Ai} and $\bar{p}_{Ai}$. The construction is shown on an enlarged scale in Fig. 3.18b.

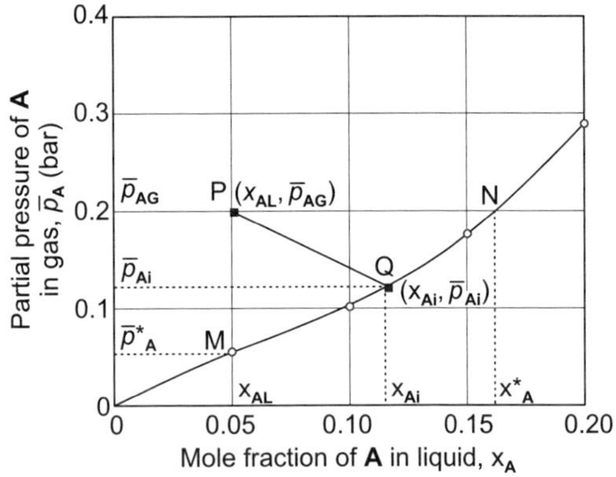

Fig. 3.18: (b) Equilibrium curve on an enlarged scale for Example 3.18

From the graph it is seen that

$$\bar{p}_{Ai} = 0.125 \text{ bar}, \; x_{Ai} = 0.115 \qquad \textbf{(Answer)}$$

Since $\bar{p}_{Ai} = y_{Ai}P$ and $P = 1$ bar, we can write

$$y_{Ai} = 0.125, \; x_{Ai} = 0.115 \qquad \textbf{(Answer)}$$

The rate of mass transfer is calculated using Eq. (3.129 A).

$$N_A = k_G(\bar{p}_{AG} - \bar{p}_{Ai})$$

$$= 34.67 \times (0.2 - 0.125) = 2.6 \frac{\text{kmol}}{\text{m}^2\text{h}}$$

Also

$$N_A = k_x(x_{Ai} - x_{AL})$$

$$= 40 \times (0.115 - 0.05) = 2.6 \frac{\text{kmol}}{\text{m}^2\text{h}} \qquad \textbf{(Answer)}$$

(b) We have the bulk compositions

$$x_{AL} = 0.05; \; \bar{p}_{AG} = 0.2 \text{ bar}$$

and the interface compositions $x_{Ai} = 0.115$ and $\bar{p}_{Ai} = 0.125$ bar. The individual phase coefficients are $k_G = 34.67$ kmol/(m^{2}h bar), $k_x = 40$ kmol/(m^{2}h mole fraction). The rate of mass transfer was calculated using Eq. (3.129 A).

$$N_A = k_G(\bar{p}_{AG} - \bar{p}_{Ai}) = k_x(x_{Ai} - x_{AL}) = 2.6 \frac{\text{kmol}}{\text{m}^2\text{h}}$$

The rate of mass transfer can also be calculated using Eq. (3.131) which can be transformed into the following form in terms of the overall mass transfer coefficient K_G and the partial pressure difference driving force.

$$N_A = K_G(\bar{p}_{AG} - \bar{p}^*_A) \qquad \text{...(3.131 A)}$$

where $\bar{p}^*_A$ is the gas phase equilibrium equivalent of the bulk liquid composition x_{AL}. $\bar{p}^*_A$ is obtained from the equilibrium diagram as the y-coordinate of point M for which the x-coordinate is $x_{AL} = 0.05$. From the Fig. 3.18b $\bar{p}^*_A = 0.0565$ bar. Substituting the values in Eq. (3.131 A), we get

$$2.6 = K_G(\bar{p}_{AG} - \bar{p}_A^*) = K_G(0.20 - 0.0565)$$

which gives K_G = 18.12 kmol/(m^2 h bar) **(Answer)**

Similarly we can calculate the overall liquid-phase mass transfer coefficient K_x. For this, the rate of mass transfer would be

$$N_A = K_x(x_A^* - x_{AL}) \quad \text{...(3.132)}$$

x_A^* is the liquid-phase equilibrium equivalent of the bulk gas phase composition. Here x_A^* is the x-coordinate of point N in the Fig. 3.18b for which the y-coordinate is $\bar{p}_{AG}$ = 0.2 bar. x_A^* = 0.161. Substituting the values in Eq. (3.132) we get $2.6 = K_x(0.161 - 0.05)$, which gives

$$K_x = 23.42 \text{ kmol/(m}^2\text{h mole fraction)}$$ **(Answer)**

The ratio of the gas phase resistance to the overall resistance is

$$\frac{1/k_G}{1/K_G} = \frac{1/34.67}{1/18.12} = 0.522.$$

It means that 52.2% of the total resistance to mass transfer lies in the gas phase and 41.8% lies in the liquid phase. **(Answer)**

Example 3.19: Chlorine water for pulp bleaching is being prepared by absorption of chlorine in water in packed tower operating at 293 K and 1.0 × 10^5 Pa pressure. At one point in the tower, the chlorine pressure in the gas is 5 × 10^4 Pa and the concentration in the liquid is 1.12 kg/m^3. Data on the solubility of chlorine in water at 293 K are as follows:

Chlorine pressure in Pa × 10^{-4}	0.674	1.21	2.63	4.96	9.77
Solubility in kmol Cl_2/m^3 × 10^2	1.58	2.25	3.73	5.94	10.21(a)

(a) If 75% of the resistance to mass transfer lies in the liquid phase, what are the interfacial compositions?

(b) If the overall liquid phase mass transfer coefficient is 2 × 10^{-5} m/s, determine the individual coefficients k_L and k_G and the overall gas phase mass transfer coefficient K_G.

***Solution*:** The bulk conditions are

$$\bar{p}_{AG} = 5 \times 10^4 \text{ Pa and}$$

$$C_{AL} = 1.12 \frac{\text{kg}}{\text{m}^3} = \frac{1.12}{70.9} = 1.58 \times 10^{-2} \frac{\text{kmol}}{\text{m}^3}$$

The equilibrium curve is plotted with the solubility of chlorine in water, C_A, on the x-axis and the partial pressure of chlorine in the gas, $\bar{p}_A$, on the y-axis. (Fig. 3.19)

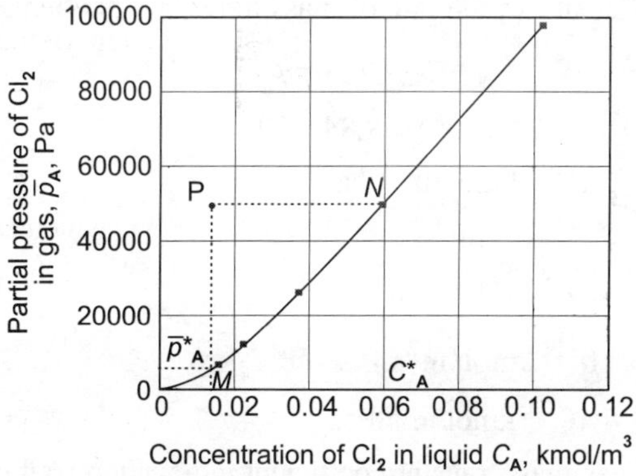

Fig. 3.19: Equilibrium curve for Example 3.19

The equilibrium equivalent of the bulk liquid ($\bar{p}_A^*$) is read from the diagram corresponding to C_{AL}.

$$\bar{p}_A^* = 6.74 \times 10^3 \text{ Pa}$$

The equilibrium equivalent of the bulk gas (C_A^*) is read from the diagram corresponding to $\bar{p}_{AG}$.

$$C_A^* = 6 \times 10^{-2} \text{ kmol/m}^3.$$

(a) Since 75% of the total resistance lies in the liquid,

$$1/k_L = 0.75 \times 1/K_L$$

or

$$k_L = \frac{K_L}{0.75} \quad \text{(A)}$$

The rate of mass transfer is given by

$$N_A = K_L(C_A^* - C_{AL}) = k_L(C_{Ai} - C_{AL}) \quad \text{(B)}$$

Substituting the values, we get

$$K_L(6 \times 10^{-2} - 1.58 \times 10^{-2}) = \frac{K_L}{0.75} \times (C_{Ai} - 1.58 \times 10^{-2})$$

Solving the above equation, we get

$$C_{Ai} = 4.895 \times 10^{-2} \text{ kmol/m}^3$$

$\bar{p}_{Ai}$ is the composition of the gas in equilibrium with the liquid of composition C_{Ai}. This is obtained from the equilibrium diagram. $\bar{p}_{Ai} = 3.7 \times 10^4$ Pa.

The interface compositions are

$$C_{Ai} = 4.90 \times 10^{-2} \text{ kmol/m}^3, \ \bar{p}_{Ai} = 3.7 \times 10^4 \text{ Pa.}$$ **(Answer)**

(b) From Eq. (A),

$$k_L = \frac{K_L}{0.75} = \frac{2 \times 10^{-5}}{0.75} = 2.67 \times 10^{-5} \text{ m/s}$$

The rate of mass transfer is

$$N_A = K_L(C_A^* - C_{AL}) = 2 \times 10^{-5}(6 \times 10^{-2} - 1.58 \times 10^{-2}) = 8.84 \times 10^{-7} \text{ kmol/m}^2\text{s}$$

In terms of individual coefficient k_G, the rate of mass transfer is written as

$$N_A = k_G(\bar{p}_{AG} - \bar{p}_{Ai})$$

That is $\quad k_G(5 \times 10^4 - 3.7 \times 10^4) = 8.84 \times 10^{-7}$

which gives $\quad k_G = 6.8 \times 10^{-11}$ kmol/m²s Pa. **(Answer)**

In terms of the overall coefficient K_G, the rate of mass transfer is written as

$$N_A = K_G(\bar{p}_{AG} - \bar{p}_A^*)$$

That is $\quad K_G(5 \times 10^4 - 6.74 \times 10^3) = 8.84 \times 10^{-7}$

which gives $\quad K_G = 2.04 \times 10^{-11}$ kmol/m² s Pa. **(Answer)**

Thus

$k_L = 2.67 \times 10^{-5}$ m/s **(Answer)**

$k_G = 6.8 \times 10^{-11}$ kmol/m²s Pa. **(Answer)**

$K_G = 2.04 \times 10^{-11}$ kmol/m²s Pa. **(Answer)**

Example 3.20: The overall gas mass transfer coefficient in a wetted wall column used for absorption of ammonia from an air ammonia mixture was found to be 2.5×10^{-9} kmol/(m² s Pa). The rate of mass transfer at a point where the gas phase concentration was 3% ammonia by volume, was measured to be 7.4×10^{-6} kmol/m^2 s. The temperature was 293 K and pressure was 101.3 kPa. At this temperature, the equilibrium partial pressure is given by $\overline{p}_A = 2\times10^3 C_A$ where $\overline{p}_A$ is in Pa and C_A is in kmol/m^3.

Calculate:

(a) The concentration of ammonia in the liquid, kmol/m^3.

(b) The interfacial concentrations if 80% of the total diffusion resistance lies in the gas phase.

(c) The mass transfer coefficients k_G, k_L, K_L.

***Solution*:** The overall mass transfer coefficient $K_G = 2.5 \times 10^{-9}$ kmol/(m^{2}s Pa).

The mole fraction of ammonia in the bulk is 0.03. Therfore,

$$\overline{p}_{AG} = y_{AG} \times P$$

$$= 0.03\times101.3\times10^3 = 3.039\times10^3 \text{ Pa}$$

The equilibrium equivalent of $\overline{p}_{AG}$ is C_A^*. Or $\overline{p}_{AG}$ and C_A^* satisfy the equilibrium relationship

$$\overline{p}_A = 2\times10^3 C_A$$

Therefore,

$$C_A^* = \frac{\overline{p}_{AG}}{2\times10^3} = \frac{3.039\times10^3}{2\times10^3} = 1.5195\frac{\text{kmol}}{\text{m}^3}$$

The rate of mass transfer $N_A = 7.4 \times 10^{-6}$ kmol/m^{2}s.

(a) $N_A = K_G(\overline{p}_{AG} - \overline{p}_A^*)$ where $\overline{p}_A^*$ is the equilibrium equivalent of C_{AL}.

$$7.4\times10^{-6} = 2.5\times10^{-9}(3.039\times10^3 - \overline{p}_A^*)$$

Solving this, we get $\overline{p}_A^* = 79$ Pa.

Since C_{AL} is the composition of the liquid in equilibrium with this gas phase, C_{AL} and $\overline{p}_A^*$ should satisfy the equilibrium relationship. Therefore,

$$\overline{p}_A^* = 2\times10^3 C_{AL}$$

$$79 = 2\times10^3 C_{AL}$$

Or $$C_{AL} = 3.95 \times 10^{-2} \text{ kmol/m}^3$$ **(Answer)**

(b) If the gas phase accounts for 80% of the total resistance,

$$\frac{1}{k_G} = 0.8\times\frac{1}{K_G}$$

$$k_G = \frac{K_G}{0.8} \quad \text{(A)}$$

Rate of mass transfer is given as

$$N_A = k_G(\overline{p}_{AG} - \overline{p}_{Ai}) = \frac{K_G}{0.8}(\overline{p}_{AG} - \overline{p}_{Ai})$$

$$K_G(3.039\times10^3 - 79) = \frac{K_G}{0.8}(3.039\times10^3 - \overline{p}_{Ai})$$

That is,

$$\overline{p}_{Ai} = 671 \text{ Pa}$$

C_{Ai} and $\overline{p}_{Ai}$ should obey the equilibrium relationship $\overline{p}_A = 2\times10^3 C_A$

Therefore,

$$C_{Ai} = \frac{\bar{p}_{Ai}}{2\times10^3} = \frac{671}{2\times10^3} = 0.336\frac{\text{kmol}}{\text{m}^3}$$

Thus

$$\bar{p}_{Ai} = 671 \text{ Pa and } C_{Ai} = 0.336\frac{\text{kmol}}{\text{m}^3} \qquad \textbf{(Answer)}$$

(c) From Eq. (A),

$$k_G = \frac{K_G}{0.8} = \frac{2.5\times10^{-9}}{0.8} = 3.125\times10^{-9}\frac{\text{kmol}}{\text{m}^2\text{s Pa}} \qquad \textbf{(Answer)}$$

$$k_L = \frac{N_A}{C_{Ai} - C_{AL}} = \frac{7.4\times10^{-6}}{0.336 - 3.95\times10^{-2}}$$

$$= 2.5\times10^{-5}\frac{\text{m}}{\text{s}} \qquad \textbf{(Answer)}$$

Similarly,

$$K_L = \frac{N_A}{C_A^* - C_{AL}} = \frac{7.4\times10^{-6}}{1.5195 - 3.95\times10^{-2}}$$

$$K_L = 5.0\times10^{-6}\frac{\text{m}}{\text{s}} \qquad \textbf{(Answer)}$$

Overall Mass Transfer Coefficient—General Case

The mass transfer coefficients involved in the equations developed above refer to situations in which equimolar counter diffusion or diffusion through stagnant, but dilute solutions are involved. To be applicable to other cases, these equations require modification. For instance, if we are going to apply the mass transfer coefficients evaluated using correlations applicable for equimolar counter diffusion or dilute systems, the mass transfer coefficient should be treated as k_y' and k_x', so that

$$N_A = \frac{k_y'}{(1-y_A)_{iM}}(y_{AG} - y_{Ai}) = \frac{k_x'}{(1-x_A)_{iM}}(x_{Ai} - x_{AL})$$
$$= \frac{K_y'}{(1-y_A)_{iM}}(y_{AG} - y_A^*) = \frac{K_x'}{(1-x_A)_{iM}}(x_A^* - x_{AL}) \qquad \ldots(3.141)$$

Rearranging Eq. (3.141) we can get

$$(y_{AG} - y_{Ai}) = \frac{N_A}{k_y'/(1-y_A)_{iM}}$$

and

$$(x_{Ai} - x_{AL}) = \frac{N_A}{k_x'/(1-x_A)_{iM}} \qquad \ldots(3.142)$$

Or

$$\frac{y_{AG} - y_{Ai}}{x_{Ai} - x_{AL}} = \frac{k_x'/(1-x_A)_{iM}}{k_y'/(1-y_A)_{iM}} \qquad \ldots(3.143)$$

The left-side of Eq. (3.143) is the slope line PQ which is the line joining (x_{AL}, y_{AG}) and (x_{Ai}, y_{Ai}) in Fig. 3.20. A trial-and-error procedure is involved in using Eq. (3.143) for evaluating interfacial compositions. As a first approximation, $\frac{(1-y_A)_{iM}}{(1-x_A)_{iM}}$ is assumed unity and a straight line is drawn with slope $-\frac{k'_x}{k'_y}$ to intersect the equilibrium curve at Q the co-ordinates of which give new interfacial compositions x_{Ai} and y_{Ai}. These are substituted on the right-hand-side of Eq. (3.143) to get the slope of line PQ for the second trial. The coordinates of the new point of intersection give the new set of interfacial concentrations. This procedure is repeated till there is no appreciable change in the values of x_{Ai} and y_{Ai} from the previous trial.

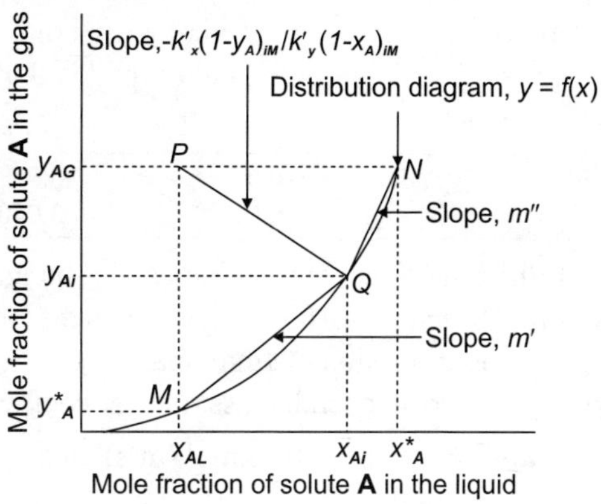

Fig. 3.20: Overall mass transfer coefficient – general case

From Fig. 3.20 it is clear that

$$y_{AG} - y^*_A = (y_{AG} - y_{Ai}) + (y_{Ai} - y^*_A)$$

$$= (y_{AG} - y_{Ai}) + m'(x_{Ai} - x_{AL}) \quad \ldots(3.144)$$

Using Eq. (3.141) we can write Eq. (3.144) as

$$\frac{N_A}{K'_y/(1-y_A)_{*M}} = \frac{N_A}{k'_y/(1-y_A)_{iM}} + m'\frac{N_A}{k'_x/(1-x_A)_{iM}} \quad \ldots(3.145)$$

which can be rearranged to the following forms:

$$\frac{1}{K'_y/(1-y_A)_{*M}} = \frac{1}{k'_y/(1-y_A)_{iM}} + \frac{m'}{k'_x/(1-x_A)_{iM}} \quad \ldots(3.146)$$

$$\frac{1}{K'_x} = \frac{1}{k'_y}\frac{(1-y_A)_{iM}}{(1-y_A)_{*M}} + \frac{m'}{k'_x}\frac{(1-x_A)_{iM}}{(1-y_A)_{*M}} \quad \ldots(3.147)$$

Equations (3.146) and (3.147) simplify to

$$\frac{1}{K'_x} = \frac{1}{k'_y} + \frac{m'}{k'_x} \quad \ldots(3.148)$$

when we are dealing with dilute solutions. This is same as Eq. (3.136). The liquid-phase counter part of Eqs. (3.146) and (3.147) can be developed easily. The results are:

$$\frac{1}{K'_x/(1-x_A)_{*M}} = \frac{1}{m''k'_y/(1-y_A)_{iM}} + \frac{1}{k'_x/(1-x_A)_{iM}} \quad \ldots(3.149)$$

$$\frac{1}{K'_x} = \frac{1}{m''k'_y}\frac{(1-y_A)_{iM}}{(1-x_A)_{*M}} + \frac{1}{k'_x}\frac{(1-x_A)_{iM}}{(1-x_A)_{*M}} \quad \ldots(3.150)$$

[**Note:** The mass transfer coefficient k'_x and k'_y are equivalent to the film coefficient of mass transfer defined as CD_{AB}/Z as pointed out earlier. The film coefficient of mass transfer is denoted by F by some authors. Thus the individual coefficients k'_x and k'_y and the overall coefficients are K'_x and K'_y become F_L, F_G and F_{OL} and F_{OG}, respectively.]

Example 3.21: A wetted wall column is used for absorption of ammonia from an ammonia-air mixture using water as the solvent at 300 K and 101.3 kPa pressure. The local mass transfer coefficients evaluated from the correlations obtained with dilute solutions are $k'_x = 1.6 \times 10^{-3}$ kmol/(m^2 s mole fraction) and $k'_y = 1.5 \times 10^{-3}$ kmol/(m²s mole fraction). The mole fraction of ammonia in the bulk gas phase is 0.8 and that in the bulk liquid is 0.05. The equilibrium distribution data for NH_3 at 300 K is given below.

x_A	0.0	0.05	0.10	0.25	0.30
y_A	0.0	0.0707	0.1347	0.591	0.920

If the bulk flow contribution cannot be ignored, calculate:

(a) The interfacial compositions and the local mass flux for absorption of ammonia.

(b) The local overall mass transfer coefficient K'_y

Solution: $k'_x = 1.6 \times 10^{-3}$ kmol/(m²s), and $k'_y = 1.5 \times 10^{-3}$ kmol/(m²s)

$$y_{AG} = 0.8, x_{AL} = 0.05$$

(a) The equilibrium curve is drawn and the point P representing bulk concentrations is marked.

By Eq. (3.143),

$$\frac{y_{AG} - y_{Ai}}{x_{Ai} - x_{AL}} = \frac{k'_x(1-y_A)_{iM}}{k'_y(1-x_A)_{iM}}$$

It means that the straight line drawn with slope $-\frac{k'_x(1-y_A)_{iM}}{k'_y(1-x_A)_{iM}}$ will intersect the equilibrium curve at point Q with coordinates x_{Ai} and y_{Ai}. Here,

$(1-y_A)_{iM}$ = logarithmic mean of $(1-y_{AG})$ and $(1-y_{Ai})$

$(1-x_A)_{iM}$ = logarithmic mean of $(1-x_{AL})$ and $(1-x_{Ai})$

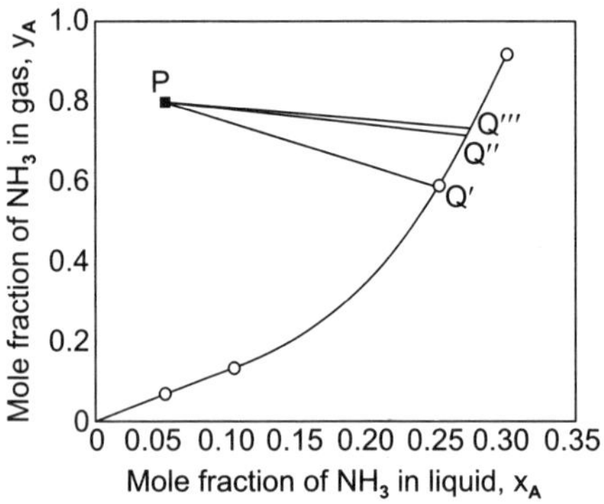

Fig. 3.21: Equilibrium curve for ammonia-water system (Example 3.21)

A trial-and-error method is needed to use Eq. (3.143), because the slope of the line PQ involves the interfacial compositions which are unknown. For the first trial $(1-y_A)_{iM}$ and $(1-x_A)_{iM}$ are assumed to be unity and the line PQ' is drawn with slope $-\frac{k'_x}{k'_y} = -1.067$. The coordinates of point of intersection are $x_{Ai} = 0.25$, $y_{Ai} = 0.59$. These values are used for evaluating $(1-y_A)_{iM}$ and $(1-x_A)_{iM}$ for the second trial.

$$(1-y_A)_{iM} = \frac{(1-y_{Ai})-(1-y_{AG})}{\ln\frac{1-y_{Ai}}{1-y_{AG}}} = \frac{(1-0.59)-(1-0.80)}{\ln\frac{1-0.59}{1-0.8}} = 0.2925$$

$$(1-x_A)_{iM} = \frac{(1-x_{Ai})-(1-x_{AL})}{\ln\frac{1-x_{Ai}}{1-x_{AL}}} = \frac{(1-0.25)-(1-0.05)}{\ln\frac{1-0.25}{1-0.05}} = 0.8461$$

For the second trial the slope of line PQ'' is

$$-\frac{k'_x(1-y_A)_{iM}}{k'_y(1-x_A)_{iM}} = -\frac{1.6 \times 0.2925}{1.5 \times 0.8461} = -0.3688$$

The coordinates of new point of intersection Q'' are $x_{Ai} = 0.27$, $y_{Ai} = 0.717$. The calculations are repeated with new $(1-y_A)_{iM}$ and $(1-x_A)_{iM}$ and new slope $-\frac{k'_x(1-y_A)_{iM}}{k'_y(1-x_A)_{iM}}$ till there is no appreciable change in the values of x_{Ai} and y_{Ai} from the previous trial. The desired result is obtained in the third attempt and the line PQ''' gives the interface compositions as:

$$x_{Ai} = 0.276 \text{ and } y_{Ai} = 0.732 \qquad \textbf{(Answer)}$$

Using Eq. (3.141),

$$N_A = \frac{k'_y}{(1-y_A)_{iM}}(y_{AG} - y_{Ai}) = 1.5\times10^{-3}\frac{0.8-0.732}{(1-y_A)_{iM}}$$

where

$$(1-y_A)_{iM} = \frac{(1-0.732)-(1-0.80)}{\ln\dfrac{1-0.732}{1-0.8}} = 0.2323$$

$$N_A = \frac{1.5\times10^{-3}(0.8-0.732)}{0.2323} = 4.39\times10^{-4}\ \text{kmol/m}^2\text{s} \qquad \textbf{(Answer)}$$

(b) The overall mass transfer coefficient K'_y can be evaluated using Eq. (3.147)

$$\frac{1}{K'_y} = \frac{1}{k'_y}\frac{(1-y_A)_{iM}}{(1-y_A)_{*M}} + \frac{m'}{k'_x}\frac{(1-x_A)_{iM}}{(1-y_A)_{*M}}$$

The slope of the equilibrium curve is evaluated as

$$m' = \frac{y_{Ai}-y_A^*}{x_{Ai}-x_{AL}} = \frac{0.732-0.0707}{0.276-0.05} = 2.926$$

$$(1-y_A)_{*M} = \frac{(1-y_A^*)-(1-y_{AG})}{\ln\dfrac{1-y_A^*}{1-y_{AG}}} = \frac{(1-0.0707)-(1-0.80)}{\ln\dfrac{1-0.0707}{1-0.8}} = 0.4748$$

The above values are substituted in Eq. (3.147):

$$\frac{1}{K'_y} = \frac{1}{1.5\times10^{-3}}\frac{0.2323}{0.4748} + \frac{2.926}{1.6\times10^{-3}}\frac{0.8319}{0.4748} = 3.5303\times10^3\ \text{m}^2\text{s/kmol}$$

$$K'_y = 2.8326\times10^{-4}\ \text{kmol/m}^2\text{s} \qquad \textbf{(Answer)}$$

3.10 METHOD OF CONTACTING OF THE PHASES

Mass transfer operations are generally carried out either as stage contact operation or as continuous contact operation. In stage contact operations, the two phases are brought into contact in devices which we call stages. The streams undergo change in composition in discrete steps.

A plate column or a tray tower (Fig. 3.22) used for gas absorption or distillation operation is an example of a multistage unit in which each plate or tray acts as a stage. On the other hand, in continuous contact operation, the fluids are in contact throughout the equipment and concentration changes are effected in a continuous fashion. A packed column (Fig. 3.23) in which two insoluble fluids flow in opposite directions in contact with each other is a typical continuous differential contact unit.

Stagewise Contact Operations

A stage is any mechanical unit or combination of such units in which the following operations are performed:

(1) Two immiscible phases are brought into intimate contact.
(2) Mass transfer occur and the diffusing components get redistributed between the phases.
(3) The phases are allowed to separate mechanically.

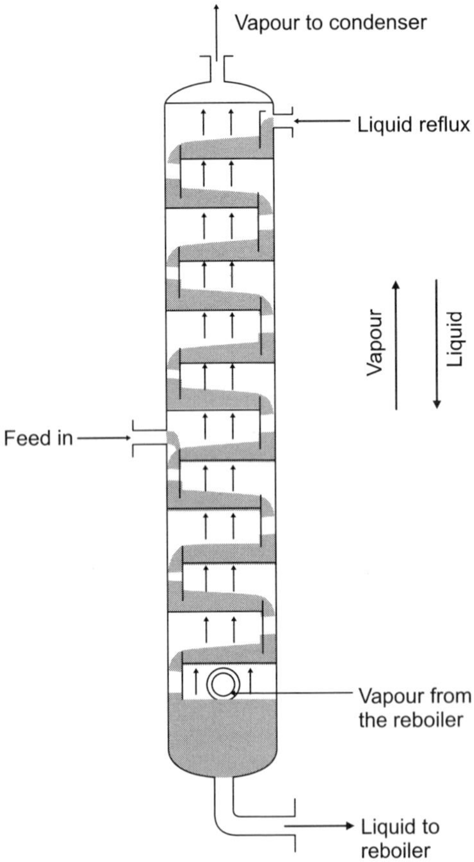

Fig. 3.22: A plate column for distillation

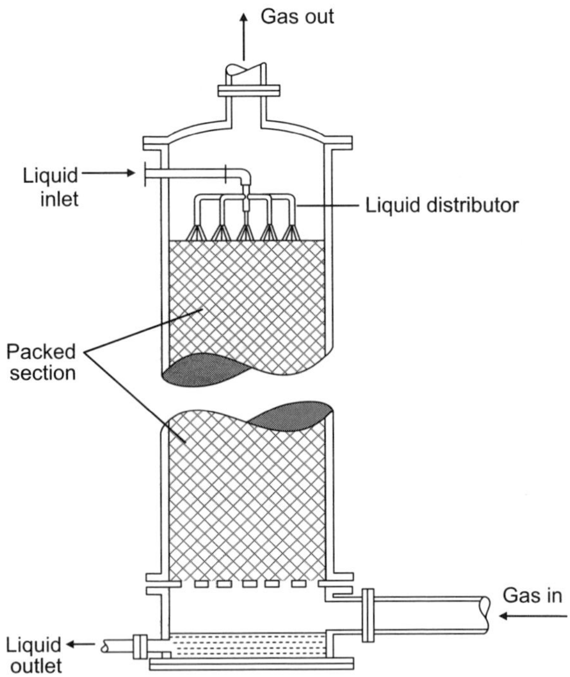

Fig. 3.23: Packed bed absorption column

A single stage separation operation is represented schemati-cally as in Fig. 3.24. The two immiscible phases represented as *G*-phase and *L*-phase are brought into contact in a mechanical device represented by the circle in Fig. 3.24. The solute concentration of the phases are y_o and x_o. Since the phases are not in equilibrium, mass transfer occurs between them, and after the desired concentration change is attained, the phases are separated. The *G*-phase and *L*-phase leaving the unit have solute mole fractions y_1 and x_1, respectively.

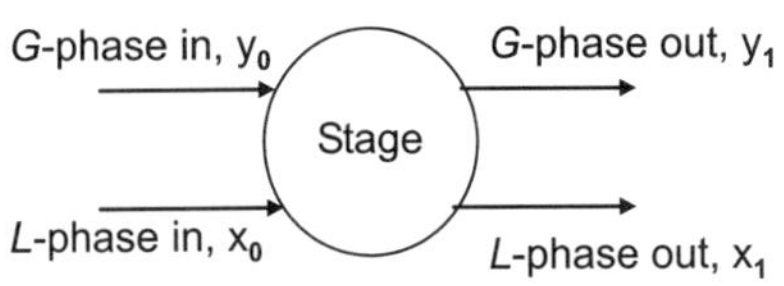

Fig. 3.24: Single-stage separation

An *ideal stage* is a device in which the phases are allowed to remain in contact for a period long enough to establish thermodynamic equilibrium and then the phases leave with equilibrium compositions. An ideal stage is also known as an *equilibrium stage* or a *theoretical stage*. In an actual operation, we will not be able to provide such a long time of contact and therefore, the concentration change in an actual stage will be less than that achieved with an ideal stage. In the Fig. 3.25, the composition of the phases entering the stage are represented by coordinates of point M (x_0, y_0) and the effluents from the stage are represented by the coordinates of point N (x_1, y_1). For ideal stage, the effluents could be represented by point N' (x_e, y_e) lying on the equilibrium curve.

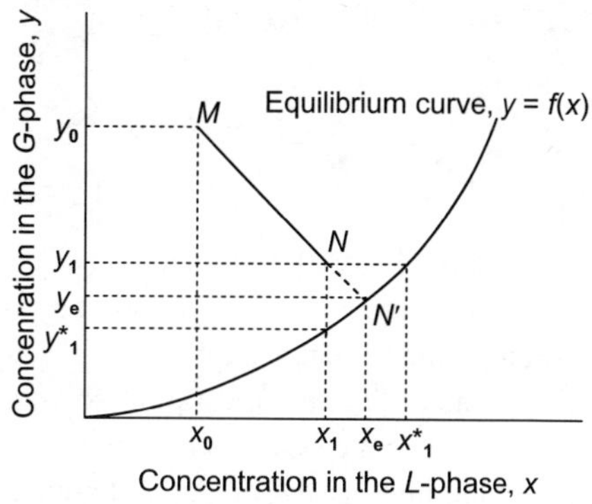

Fig. 3.25: Single-stage operation on equilibrium coordinates

The extent of mass transfer can be improved over and above that which is possible with a single stage by multiple contact in which several stages are interconnected. In multistage operation, several stages are interconnected with the streams flowing from one stage to the other in either countercurrent or cross current fashion. The cocurrent operation, where both streams are flowing in the same direction from one stage to the other as shown in Fig. 3.26, can never bring about a concentration change in the streams greater than that possible with a single equilibrium stage.

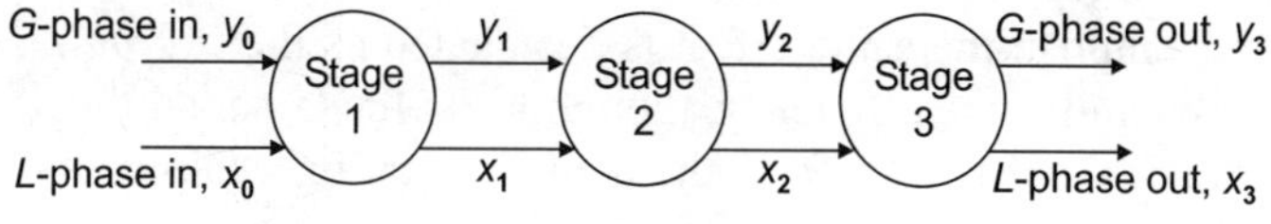

Fig. 3.26: Cocurrent multistage operation

Thus, a continuous cocurrent operation is, in effect, a single-stage operation. In order to achieve separation to an extent greater than that is possible with a single stage, either a cross current or a counter current operation should be used.

As an example of crosscurrent operation, let us consider the leaching of oil from oil seeds using an organic liquid solvent (Fig. 3.27).

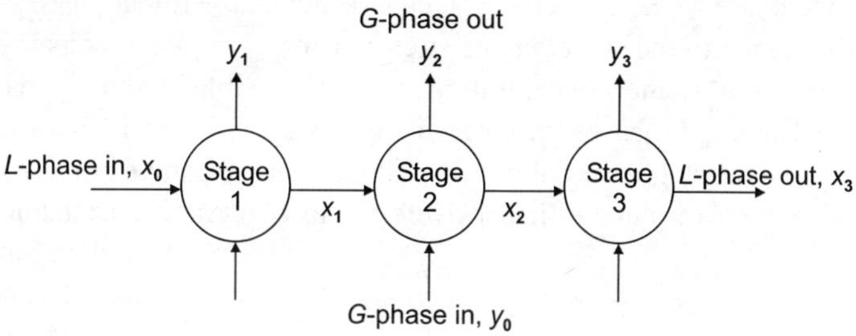

Fig. 3.27: Three-stage crosscurrent operation

The solids (L-phase in the figure) leaving the first stage are admitted to the second stage and from the second stage, are taken to the third and so on. The solids thus move from one stage to the other. In each stage, the solids are contacted with fresh batches of solvent (G-phase in the figure). In this fashion, the recovery of solute from the feed can be considerably improved. Cross-flow cascade is used in separation operation such as leaching, extraction, drying and adsorption.

The most frequently used mode of operation involves counterflow of contacting phases. A countercurrent multistage cascade is shown in Fig. 3.28.

The stages are numbered from left to right in Fig. 3.28. Figure shows the L-phase (the feed) admitted to the first stage and the G-phase (the separating agent) to the last stage. They flow in the opposite

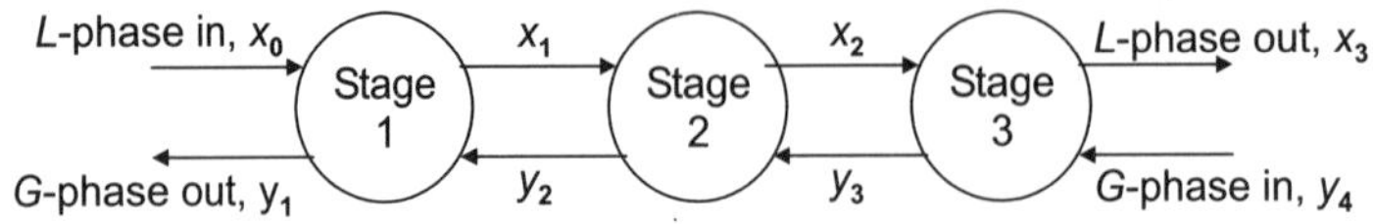

Fig. 3.28: Multistage countercurrent cascade

directions. Thus, the feed to the second stage will be the *L*-phase leaving the first stage and the separating agent for the second stage will be the *G*-phase leaving the third stage. The last traces of solute, which are normally very difficult to separate, are removed in the last stage where the *L*-phase is in contact with the most pure solvent; hence separation is very effective in countercurrent cascades. Countercurrent cascade gives high recovery and high purity of products. For a given ratio of flow rates of contacting phases and percent recovery of solute, countercurrent multistage operation requires lesser number of stages than other arrangements. With the same number of stages, greater separation can be achieved with counterflow cascades than with the cross-flow arrangement.

Stage efficiency

The effectiveness of a multistage unit is measured by the overall stage efficiency. The ratio of the number of theoretical stages required to bring about a certain separation to the number of actual stages required to accomplish the same effect is designated as the *overall stage efficiency* of the multistage unit. The overall stage efficiency, though lacks fundamental theoretical basis and is to be determined on the basis of results from the actual operating equipment, is easy to apply and hence is widely used.

The *stage efficiency* is defined in order to compare the performance of an actual stage with a theoretical stage. The stage efficiency measures the fractional approach to equilibrium that is achieved in an actual stage. The ratio of line lengths *MN/MN′* in Fig. 3.25 in a way reflects stage efficiency, as it is an indication of the ratio of the actual amount of solute transfer to the amount of solute that would be transferred if equilibrium could be attained between the contacting phases. The Murphree stage efficiency is defined in a slightly different fashion. It is defined as the ratio of the actual concentration change effected in one phase to the concentration change attainable if that phase leaves in equilibrium with the actual concentration of the other leaving stream. Thus $y_0 - y_1$ is the actual concentration change for the *G*-phase and $y_0 - y_1^*$ is the concentration change possible if the *G*-phase were leaving in equilibrium with the leaving *L*-phase. Referring to Fig. 3.25, y_1^* is the composition of the fictitious *G*-phase that would leave the stage in equilibrium with the average composition of the *L*-phase actually leaving the stage. Thus the Murphree efficiency referred to *G*-phase is defined as

$$E_{MG} = \frac{y_0 - y_1}{y_0 - y_1^*} \qquad \ldots(3.151)$$

In a similar fashion we can define the Murphree stage efficiency referred to the *L*-phase as

$$E_{ML} = \frac{x_1 - x_0}{x_1^* - x_0} \qquad \ldots(3.152)$$

Here, x_1^* is the composition of the fictitious *L*-phase that would leave the stage in equilibrium with the average composition of the *G*-phase actually leaving the stage.

Example 3.22: Show that if the equilibrium curve is a straight line with slope *m*,

$$E_{MG} = \frac{E_{ML}}{E_{ML}(1-S)+S} = \frac{E_{ML}}{E_{ML}\left(1-\frac{1}{A}\right)+\frac{1}{A}}$$

The absorption factor A and the stripping factor S are defined as $A = L/mG = 1/S$.

Solution: Let G be the molar flow rate of the G-phase and L that of the L-phase. Assume that the concentration of the solute in these phases are small so that the flow rates remain constant before and after separation. The moles of solute transferred from the G-phase is $G\ (y_0 - y_1)$ and the moles of solute received by the L-phase is $L\ (x_1 - x_0)$. The solute balance for the single-stage operation is

$$G\ (y_0 - y_1) = L\ (x_1 - x_0) \qquad \text{...(3.153)}$$

The slope of the material balance line MN (the operating line for single stage operation) is obtained by rearranging Eq. (3.154).

$$\frac{y_0 - y_1}{x_0 - x_1} = -\frac{L}{G} \qquad \text{...(3.154)}$$

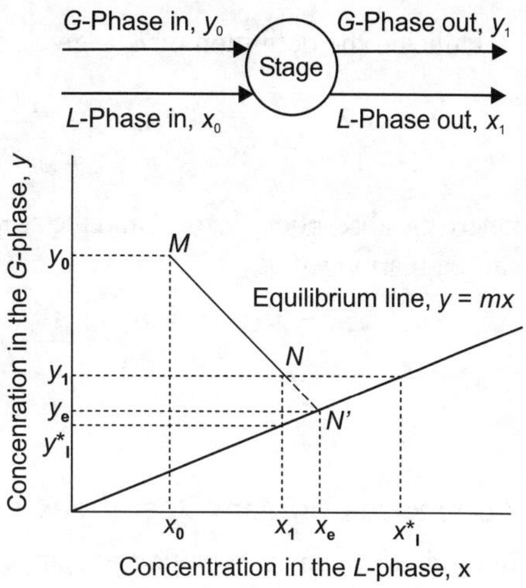

Fig. 3.29: Equilibrium relationship for Example 3.22

The Murphree stage efficiency referred to the G-phase is given by Eq. (3.151).

$$E_{MG} = \frac{y_0 - y_1}{y_0 - y_1^*} \qquad \text{...(3.151)}$$

From the figure, it can be seen that

$$y_0 - y_1^* = (y_0 - y_1) + (y_1 - y_1^*) \qquad \text{...(3.155)}$$

Substituting Eqs. (3.154) and (3.155) into Eq. (3.151), we get

$$E_{MG} = \frac{y_0 - y_1}{y_0 - y_1^*} = \frac{(L/G)(x_1 - x_0)}{(L/G)(x_1 - x_0) + (y_1 - y_1^*)} \qquad \text{...(3.156)}$$

From Fig. 3.29, it is evident that the slope of the equilibrium line is

$$m = \frac{y_1 - y_1^*}{x_1^* - x_1} \qquad \text{...(3.157)}$$

Substituting Eq. (3.157) into Eq. (3.156), we get

$$E_{MG} = \frac{(L/G)(x_1 - x_0)}{(L/G)(x_1 - x_0) + m(x_1^* - x_1)} \qquad \text{...(3.158)}$$

Divide the numerator and denominator of the above equation by $(L/G)(x_1^* - x_0)$. The result is

$$E_{MG} = \frac{\dfrac{(x_1 - x_0)}{(x_1^* - x_0)}}{\dfrac{(x_1 - x_0)}{(x_1^* - x_0)} + \dfrac{mG}{L}\dfrac{(x_1^* - x_1)}{(x_1^* - x_0)}} \qquad \text{...(3.159)}$$

$$E_{MG} = \frac{\dfrac{(x_1 - x_0)}{(x_1^* - x_0)}}{\dfrac{(x_1 - x_0)}{(x_1^* - x_0)} + \dfrac{mG}{L}\dfrac{(x_1^* - x_0) - (x_1 - x_0)}{(x_1^* - x_0)}} \qquad \text{...(3.160)}$$

Utilising the definition of E_{ML} given by Eq. (3.152), Eq. (3.160) simplifies as

$$E_{MG} = \frac{E_{ML}}{E_{ML} + \frac{1}{A}(1 - E_{ML})} = \frac{E_{ML}}{E_{ML} + S(1 - E_{ML})} \quad \text{...(3.161)}$$

where the absorption factor A and the stripping factor S are related as $A = L/mG = 1/S$. Equation (3.161) can be rearranged as

$$E_{MG} = \frac{E_{ML}}{E_{ML}\left(1 - \frac{1}{A}\right) + \frac{1}{A}} = \frac{E_{ML}}{E_{ML}(1 - S) + S} \quad \text{...(3.162)}$$

Continuous Contact Operations

In continuous contact operation, the phases are in contact over the entire effective length of the equipment, in contrast to stage contact equipment in which the fluids are contacted only intermittently over stages. There is no repeated separation and recontacting as is common with stage contact equipment. The continuous contact units are also termed differential contact equipment, as the concentration of the streams change continuously. The extent of mass transfer depends strongly upon the rate of mass transfer and time of contact. Equilibrium between phases serves to determine the driving force and hence influence the rate of mass transfer.

From the above discussion it is clear that the mass transfer process can be considered either in terms of the number of stages it represents or in terms of the mass transfer rates. Stage contact equipment are modelled based on the concept of equilibrium stages and continuous contact units are modelled based on the concept of diffusional rate processes. The computation of the number of equilibrium stages requires only the phase equilibrium and material balance relationships. The number of equilibrium stages and appropriate stage efficiency will enable one to determine the size of the column. For example, the number of stages in a multistage unit can be determined as

$$N = \frac{N_P}{\eta}$$

where N is the number of actual plates, N_P is the number of theoretical stages and η is the overall stage efficiency of the column. The height of the effective mass transfer section of the column is determined by the number of stages N.

Differential changes in composition occur in continuous contact units, and the unit will be described in terms of mass transfer coefficients or equivalent terms. The earlier method of design of continuous contact units used the *height equivalent of a theoretical plate* (HETP). For a packed column, HETP is the height of a packed section that will bring about the same change in composition as that achieved in a theoretical plate. Then the height of a packed column is given as

$$Z = (\text{HETP}) \times N_P \quad \text{...(3.163)}$$

where Z the effective height of packed column and N_P is the number of theoretical stages required to achieve the desired separation if it is carried out in a plate column. By treating the separation in a packed column at par with that in a plate column, the HETP concepts ignores the fundamental difference between the continuous contact and stage contact. Also HETP is to be determined experimentally and in the case of packed columns, depends on the nature, type and size of packings, as well as the flow rates and concentration of different streams. Since an enormous amount of experimental data are required to cover these different conditions and they are difficult to collect, the HETP method is now abandoned in favour of the *transfer unit method*. One transfer unit brings about a concentration change in one of the contacting phases, the extent of which is approximately equal to the average driving force existing between the phases in contact at that point. The number of such transfer units (NTU) required for the given separation multiplied by the height of a transfer unit (HTU) gives the effective height of the continuous contact equipment. Thus,

$$Z = (\text{HTU}) \times (\text{NTU}) \quad \ldots(3.164)$$

In the case of packed columns, the height of a transfer unit depends on the nature and type of packing materials, the flow rates of the streams and the mass transfer coefficient. The number of transfer units depends on the average driving force for mass transfer and the total concentration change to be effected in the phases.

REVIEW QUESTIONS

1. Prove that the mass transfer coefficients k_y and k_c are related as $k_y = k_c \dfrac{P}{RT}$
2. Prove that the mass transfer coefficients k_y' and k_c are related as $k_y' = k_c \dfrac{\bar{p}_{B,M}}{RT}$
3. What are the general principles underlying the film, penetration and surface renewal theories for mass transfer across phase boundaries?
4. Give the basic differential equation and the boundary conditions for describing the film, penetration, surface renewal and film-penetration theories of mass transfer coefficient
5. Show that according to the penetration theory, the instantaneous rate of mass transfer per unit area N_A at some time after the commencement of transfer is given by
$$(N_A)_t = \Delta C \sqrt{\frac{D_{AB}}{\pi t}}$$
6. Obtain expression for the average rate of transfer on the basis of Higbie and Danckwerts assumptions.
7. Distinguish between k_y and k_y'. Explain the circumstances both k_y and k_y' will represent the film coefficient $\dfrac{CD_{AB}}{Z}$?
8. The heat transfer coefficient h can be defined as $h = \dfrac{q}{\Delta T} = \dfrac{k}{\delta}$ where q is the heat flux, ΔT, the temperature difference, k is the thermal conductivity and δ is the film thickness. Is it possible to define mass transfer coefficient k in a similar fashion as mass transfer coefficient $= \dfrac{N_A}{\Delta C} = \dfrac{D_{AB}}{Z}$, where Z is the film thickness?
9. Show that the average mass transfer coefficient over a given length L is twice the local mass transfer coefficient evaluated at L when the flow in the boundary layer is laminar.
10. What are the general conditions to be satisfied by the system for the analogy equations to be valid?
11. Derive Reynolds analogy between heat, mass and momentum transfer
12. What are the assumptions inherent in the derivation of Reynolds analogy?
13. Explain the Taylor-Prandtl analogy and how it is an improvement over Reynolds analogy.
14. State and explain Colburn analogy between heat, mass and momentum transfer.
15. Explain the advantages of wetted-wall columns for mass transfer studies.
16. Explain how in gas-liquid contact operations, the gas phase mass transfer coefficient is determined using wetted-wall columns.

17. What is the criterion of equilibrium between phases?
18. What are the basic features of the two-resistance theory?
19. How do you determine the interfacial compositions according to the two-resistance theory?
20. What are the major limitations of the two-resistance theory?
21. Derive the relationship between the individual phase coefficients and the overall coefficient. Explain why this relationship is fully correct only for cases where the equilibrium curve is straight.
22. Discuss gas phase and liquid-phase controlled systems in mass transfer.
23. If the overall coefficient is very sensitive to temperature changes, what conclusion do you derive regarding the controlling resistance?
24. What do you mean by "stage" in stage contact operations?
25. Define ideal stage and stage efficiency. What do you mean by the overall stage efficiency?
26. What are the essential differences between the mass transfer models used for the analysis of stage contact and continuous contact operations?
27. What is the advantage of multistage operation over single stage operation?
28. Explain the concept of height equivalent of theoretical plate (HETP). How is it useful in the design of separation equipment? Why is the HETP method abandoned as a design method?
29. What do you understand by the transfer unit? Distinguish between height and number of transfer units.

MULTIPLE CHOICE QUESTIONS

3.1 When two phases attain equilibrium due to mass transfer between them,

A. concentration of all components will be equal throughout.
B. mole fraction of different components will be equal in both phases.
C. chemical potential of any constituent will be the same in both phases.
D. partial pressures in both phases will be identical for any constituent.

3.2 The true driving force for mass transfer between phases is

A. difference in concentration between phases
B. difference in temperature and pressure
C. difference in temperature, pressure and concentration
D. difference in chemical potential of the components

3.3 Pure aniline is evaporating through a stagnant film of 1mm thickness at 300 K and total pressure of 100 kPa. The vapour pressure of aniline at 300 K is 0.1 kPa. The total molar concentration under these conditions is 40.1mol/m^3. The diffusivity of aniline in air is 0.74×10^{-5} m^2/s. The numerical value of mass transfer coefficient is 7.4×10^{-3}. Its units are

A. m/s
B. cm/s
C. mol/(m^2 sPa)
D. kmol/(m^{2}sPa)

3.4 The Reynolds Number of the liquid was increased 100 fold for a laminar falling film used for gas liquid contacting. Assuming penetration theory is applicable, the fold increase in the mass transfer coefficient (k_c) for the same system is

A. 100
B. 10
C. 5
D. 1

3.5 The surface renewal frequency in Danckwerts model of mass transfer is given by (k_L mass transfer coefficient, m/s)

A. $(k_L^2 D_{AB})^{0.5}$
B. $k_L^2 D_{AB}$
C. k_L^2 / D_{AB}
D. k_L / D_{AB}^2

3.6 Penetration theory states that the mass transfer coefficient (where D_{AB} is diffusivity and t is time) is equal to

A. $(D_{AB}\,t)^{0.5}$ B. $(D_{AB}/t)^{0.5}$

C. $(4D_{AB}/\pi\,t)^{0.5}$ D. $(4D_{AB}/t)^{0.5}$

3.7 The mass transfer coefficient for a solid sphere of radius 'a' dissolving in a large volume of quiescent liquid, in which D_{AB} is the diffusivity of the solute, is

A. D_{AB}/a B. $D_{AB}/2a$

C. Proportional to $D_{AB}^{0.5}$ D. dependent on the Reynolds number

3.8 Mass transfer coefficient k, according to penetration theory varies with mass diffusivity as

A. $D_{AB}^{0.5}$ B. D_{AB}

C. $1/D_{AB}$ D. $D_{AB}^{1.5}$

3.9 According to film theory, for equimolar counter diffusion, the mass transfer coefficient is

A. $\dfrac{CD_{AB}}{Z}$ B. $\dfrac{CD_{AB}}{Z\,y_{BM}}$

C. $\propto D_{AB}^{0.5}$ D. None of the above.

3.10 At 750 K and 1 atm, the approximate value of the Schmidt number for air is

A. 0.01 B. 0.1

C. 1 D. 10

3.11 If the Prandtl number is greater than the Schmidt number

A. the thermal boundary layer lies inside the concentration boundary layer

B. the concentration boundary layer lies inside the thermal boundary layer

C. the thermal and concentration boundary layers are of equal thickness

D. the hydrodynamic boundary layer is thicker than the thermal and concentration boundary layers

3.12 Lewis number =1 signifies

A. $Pr = Sc$ B. $Pr = Re$

C. $Sc = Re$ D. $Nu = Sh$

3.13 The Reynolds analogy for heat and mass transfer is applicable when

A. Schmidt number is unity B. Schmidt number = Prandtl number

C. Prandtl number is unity D. all the above are true

3.14 For flow over a flat plate, the thicknesses of hydrodynamic (δ) and concentration (δ_c) boundary layer for laminar flow are related as

A. $\delta = \delta_c$ B. $\delta = \delta_c\ \mathrm{Sc}$

C. $\delta = \delta_c\ \mathrm{Sc}^{(1/3)}$ D. $\delta = \delta_c\ \mathrm{Sc}^{(-1/3)}$

3.15 Sherwood number for flow in a pipe can be expressed as the ratio of the concentration gradient at .... to the overall concentration gradient

A. the interface between the laminar sub-layer and the buffer zone

B. the interface between the buffer zone and turbulent core

C. the wall

D. the entrance to the pipe

3.16 Experiments were conducted to determine the flux of a species **A** in a stagnant medium across a gas–liquid interface. The overall mass transfer coefficient based on the liquid side for dilute systems for the above was estimated to be 4×10^{-3} kmol/m²s. The equilibrium data for the system is given as $y = 2x$. The flux across the interface (in kmol/m^{2}s) for bulk concentration of **A** in gas phase and liquid phase as $y = 0.4$ and $x = 0.01$, respectively is

A. 5.6×10^{-4} B. 8.5×10^{-4}
C. 5.6×10^{-3} D. 8.5×10^{-3}

3.17 The dimensionless group in mass transfer that is equivalent to the Prandtl number in heat transfer is

A. Nusselt number B. Sherwood number
C. Schmidt number D. Stanton number

3.18 The Reynolds analogy for momentum, heat and mass transfer is best applicable for

A. gases in turbulent flow B. gases in laminar flow
C. liquids in turbulent flow D. liquids and gases in laminar flow

3.19 For mass transfer from a flat surfaces into the fluid flowing past the surface, which of the following is not true regarding a laminar boundary layer?

A. Local Sherwood number is twice the average Sherwood number
B. The mass transfer coefficient varies as $D_{AB}^{2/3}$
C. The Sherwood number varies as $D_{AB}^{-1/3}$
D. The hydrodynamic boundary and the concentration boundary layer will have same thickness if Schmidt number is unity.

3.20 For transfer of mass from the surface into the flowing fluid, the coefficient decreases with distance from the leading edge because,

A. the concentration gradient at the surface decreases with distance
B. the diffusivity decreases with distance
C. the driving force for mass transfer decreases
D. all of the above

3.21 Schmidt number is the ratio of

A. momentum diffusivity to mass diffusivity
B. mass diffusivity to thermal diffusivity
C. momentum diffusivity to thermal diffusivity
D. mass diffusivity to momentum diffusivity

3.22 The Chilton-Colburn analogy states that

A. $St_D Sc^{1/3} = \frac{f}{2}$ B. $St_D Sc^{1/3} = f$

C. $St_D Sc^{2/3} = \frac{f}{2}$ D. $Sh Sc^{2/3} = \frac{f}{2}$

(St_D = Stanton number, Sh = Sherwood number, Sc = Schmidt number)

3.23 Which of the following is the definition of Lewis number?

A. $\frac{k}{C_p \rho D_{AB}}$ B. $\frac{Sc}{Pr}$

C. Ratio of thermal diffusivity to mass diffusivity

D. All the above

3.24 What is the dimensionless group in heat transfer that corresponds to Schmidt number in mass transfer?

A. Nusselt number B. Prandtl number
C. Fourier number D. Sherwood number

3.25 Stanton number for mass transfer is defined as

A. $\dfrac{k_c D}{D_{AB}}$ B. $\dfrac{Sh}{RePr}$

C. $\dfrac{Sh}{PrSc}$ D. $\dfrac{Sh}{ReSc}$

3.26 One of the following statements is wrong regarding the Reynolds analogy. Which one?

A. Fluid cross section is assumed to be consisting of only turbulent core
B. Applicable for fluids for which $Sc = Pr$
C. Applicable only for $Sc = Pr = 1$
D. Velocity, concentration and thermal boundary layers have same thickness

3.27 For flow of water at 5 m/s over a flat surface at atmospheric temperature and pressure, roughly up to what distance from the leading edge, the flow in the boundary layer can be expected to remain laminar?

A. 10 m B. 1 m
C. 0.1 m D. 0.01 m

3.28 If Chilton-Colburn analogy is applicable to the system, what would be the ratio of the heat transfer coefficient to the mass transfer coefficient k_c?

A. $C_p \rho Le^{-1/3}$ B. $\dfrac{Sc}{RePr}$

C. $C_p \rho Le^{2/3}$ D. $C_p \rho Le^{1/3}$

3.29 Simultaneous heat and mass transfer is occurring in afluid flowing over a flat plate. The concentration boundary layer will coincide with the thermal boundary layer, when

A. $Sc = Nu$ B. $Sh = Nu$
C. $Sh = Pr$ D. $Sc = Pr$

3.30 In an interphase mass transfer process, the lesser the solubility of a given solute in a liquid, the higher are the chances for the transfer process to be

A. liquid phase resistance controlled B. gas phase resistance controlled
C. chemical reaction controlled D. driven by a nonlinear driving force

3.31 In interphase heat transfer process, the equilibrium state corresponds to equality of temperatures in the two phases, while the condition for equilibrium in interphase mass transfer process is

A. equality of concentrations B. equality of chemical potentials
C. equality of activity coefficients D. equality of mass transfer coefficients

3.32 In two-resistance theory of interphase mass transfer, which one is not true?

A. Interface is in equilibrium
B. Interface offers no resistance to mass transfer
C. The interfacial concentrations are equal as the interface offers no resistance to mass transfer
D. Mass transfer rate is controlled by resistances on the two sides of the interface.

3.33 Marangoni effect

A. deals with criterion of phase equilibria.
B. deals with interfacial turbulence
C. liquid–liquid mass transfer
D. absorption with chemical reaction.

3.34 In the expression $N_A = K_y(y_{AG} - y_A^*)$, the concentration y_A^* is a measure of

A. the interfacial composition on the gas side
B. bulk concentration of the gas phase

C. bulk concentration of the liquid phase

D. equilibrium equivalent of the given bulk gas concentration.

3.35 For gas phase controlled mass transfer, which one is not true?

A. The gas is readily soluble in liquid

B. The gas is sparingly soluble in liquid

C. The equilibrium curve is flat

D. Overall resistance to mass transfer is equal to the resistance on the gas side

3.36 If the overall coefficient is very sensitive to temperature changes,

A. the mass transfer is likely to be liquid phase controlled

B. mass transfer with chemical reaction occurs

C. the resistances on the two sides of the interface are approximately equal

D. the mass transfer is accompanied by evolution of heat.

3.37 Which one of the following statements is not true regarding an ideal stage?

A. The effluents from the stage are in equilibrium

B. The Murphree stage efficiency is 100%

C. The input streams are having equilibrium compositions

D. Contacting of phases and separation of the effluents occur in an ideal stage.

3.38 Murphree stage efficiency in a gas-liquid contact operation is 100%, the actual concentration change in the gas phase is the same as

A. the concentration change in the liquid phase.

B. the concentration change in the gas stream if it leaves in equilibrium with the effluent liquid

C. the concentration change in the gas stream if it leaves in equilibrium with the liquid entering

D. the concentration change possible for the gas phase if an infinite time of contact is provided.

3.39 Which of the following statements is true regarding transfer unit of a packed column?

A. Number of transfer unit depends on the type and size of packing

B. Height of transfer unit is unity if the average driving force available is equal to the concentration change

C. Height of a transfer unit is the same as the height equivalent of a theoretical plate (HETP)

D. The number of transfer units is determined by the concentration change and driving force.

3.40 For improving the rate of absorption of a sparingly soluble gas in liquid,

A. the gas-side coefficient should be increased

B. the gas-side coefficient should be decreased

C. liquid-side coefficient should be increased

D. liquid-side coefficient should be decreased

3.41 The height equivalent of theoretical plate (HETP) of a packed column varies with

A. type and size of packing
B. flow rate of each fluid
C. concentration of each fluid
D. all of the above

3.42 Overall stage efficiency is the ratio of the

A. number of ideal trays required to the number of real trays required

B. number of real trays required to the number of ideal trays required

C. number of ideal trays to the number of transfer units

D. actual concentration change to the maximum concentration change possible.

EXERCISES

Mass Transfer Coefficients

3.1 A film of liquid covering a flat surface is being evaporated into dry air passed across the surface at 298 K and 200 kPa. The air in contact with the surface is saturated with the vapours of liquid which exerts a vapour pressure of 20 kPa. The mass transfer coefficient k_y' is found to be 7.0×10^{-5} kmol/m^{2}s. Assume one-way diffusion. Calculate:

(a) The rate of vaporisation, N_A

(b) The mass transfer coefficients k_G' and k_c'

(c) The mass transfer coefficients k_G and k_y

(**Answer:** (a) 7.375×10^{-6} kmol /(m^{2}s) (b) 3.5×10^{-10} kmol/m^{2}s Pa, 8.67×10^{-4}m/s (c) 3.688×10^{-10} kmol/m^{2}s Pa, 7.375×10^{-5} kmol/m^{2}s)

3.2 Pure liquid **A** is vaporising into pure gas **B** flowing across the liquid surface at 298 K and 2 bar. The vapour pressure of **A** is 20 kPa at 298 K. **B** is insoluble in **A**. The film coefficient k_y' has been estimated to be 6.5×10^{-5} kmol/m^{2}s. Calculate:

(a) the mass transfer coefficient k_y and k_G

(b) The rate of vaporisation.

(**Answer:** (a) 6.85×10^{-5} kmol /m^2 s, 3.43×10^{-10} kmol /m^{2}s Pa (b) 6.85×10^{-6} kmol/m^{2}s)

3.3 A gas **A** is being absorbed from air into water in a countercurrent packed absorption tower. At certain point in the tower, the mole fraction of **A** at the gas-liquid interface was 2.5×10^{-3} and that in the bulk liquid was 2.0×10^{-4}. The mass flux at this point was 3.0×10^{-2} kmol/m^2 h. If the diffusivity of **A** in water is 1.7×10^{-9} m^2/s, determine the mass transfer coefficient k_L and the film thickness, neglecting bulk flow effect. (**Answer:** 6.52×10^{-5} m /s, 2.61×10^{-2} mm)

3.4 Small spheres of solid benzoic acid are dissolved in water in an agitated tank. If the Sherwood number is nearly constant at a value of 4.0, show how the time for complete dissolution varies with initial size of particles. How much time would be required for 100 μm particles to dissolve completely in pure water at 298 K? Solubility of benzoic acid in water is 0.43 g/100g water and diffusivity of benzoic acid in water is 1.21×10^{-9} m^2/s. Density of benzoic acid is 1266 kg/m^3.

(**Answer:** 2.55 min)

3.5 Hydrogen sulphide is absorbed from its mixture with air in a wetted wall column using water flowing as a thin film at 303 K and 150 kPa. The gas phase mass transfer coefficient assuming no bulk flow is estimated to be 9.6×10^{-4} m/s. At certain point, the mole fraction of H_2S in the liquid side of the gas-liquid interface is 2×10^{-5} and the partial pressure of H_2S in the gas is 5 kPa. The Henry's law $\bar{p}_A = H_A x_A$ where $\bar{p}_A$ is the partial pressure of H_2S in the gas, kPa and H_A = Henry's law constant = 609 kPa/mole fraction. Calculate the rate of absorption of H_2S.

(**Answer:** 1.92×10^{-6} kmol/m^2 s)

3.6 A sugary substance **A** is added to a pot of milk initially containing no **A**. The contents are stirred vigorously by a spoon so that the concentration of **A**, C_A, is uniform everywhere. The mass transfer coefficient for the transfer of **A** into the liquid is 10^{-4} m/s. Solid **A** is added in great excess compared to the saturation capacity of milk to dissolve **A**. Assume that the solid-liquid interfacial area stays constant throughout the dissolution process and is given by a = 0.1 m^2. Volume of the vessel is 10^{-3} m^3.

(a) Derive an expression for C_A versus time t.

(b) Find time taken for $\frac{C_A}{C_A^*} = 0.95$ (**Answer:** (a) $t = 100 \ln \frac{1}{1 - C_A/C_A^*}$ (b) 5 min)

Theories of mass transfer

3.7 Carbon dioxide is stripped from water by air in a wetted-wall column. At certain location, where the pressure is 100 kPa, and the temperature is 300 K, the mass transfer flux of CO_2 is 8.5 kmol/m^2 h. The partial pressure of CO_2 are 80 kPa at the interface and 1 kPa in the bulk gas. The diffusivity of CO_2 in air at these conditions is 1.63×10^{-5} m^2/s. Using film theory, calculate k_c for the gas phase and the film thickness. (**Answer:** 7.45×10^{-2} m/s, 0.44 mm)

3.8 In an experimental study, carbon dioxide is absorbed in water using a wetted wall column. Assuming that the penetration theory represents the rate of absorption, determine the time at which the concentration of carbon dioxide reaches one per cent of the saturation value at a depth of 0.1 mm given that the diffusivity of carbon dioxide in water is 1.5×10^{-9} m^2/s.

(**Answer:** 0.5 s)

3.9 A gas **A** is being absorbed in water in a countercurrent packed tower. If the penetration theory provides a realistic model for the mass transfer, determine the contact time. The mole fraction of **A** at the gas-liquid interface and in the bulk liquid are, respectively, 2.5×10^{-3} and 2.0×10^{-4}. The average flux over the time of exposure is 3.0×10^{-2} kmol/m^2 h and the diffusivity of the gas **A** in water is 1.7×10^{-9} m^2/s. (**Answer:** 0.51 s)

3.10 Carbon dioxide is recovered from its mixture with air by absorption in water in a packed column. At a certain location in the column, where the partial pressure of CO_2 at the interface is 10^3 kPa and the concentration in the bulk liquid is negligible, the absorption rate is 8.3×10^{-2} kmol/m^2 h. The diffusivity of CO_2 in water is 2.0×10^{-9} m^2/s. Henry's law for CO_2 is $\bar{p} = Hx$, where H, the Henry's law constant, is 6.2×10^4 kPa. Assume that the density of solution is the same as the density of water at atmospheric pressure and neglecting bulk-flow contribution, calculate:

(a) The liquid-phase mass transfer coefficient and the film thickness.

(b) Contact time for penetration theory

(c) Average eddy residence time and the probability distribution for the surface renewal theory

(**Answer:** (a) 1.43×10^{-3} kmol/m^{2}s, 0.078 m
(b) 3.84 s (c) 3.02 s, $\varphi(t) = 0.3313e^{-0.3313t}$)

3.11 A pure gas is bubbling through a column of liquid at a sufficiently slow rate that the bubbles are well separated from each other. The bubbles are spherical, are 3 mm in diameter, and rise with constant velocity of 0.24 m/s. At any given time, ten bubbles can be seen in the liquid, the liquid volume being 3.14×10^{-4} m^3. The liquid contains no dissolved gas to start with and may be assumed to be well mixed at all times. The mass transfer coefficient for this situation is given by Higbie's Penetration theory as

$$k_L = (4D/\pi t)^{0.5}$$

where

D = diffusivity of the gas in the liquid, 2×10^{-9} m^2/s and

t = time of contact between liquid elements and gas, s

Determine the time it takes for the liquid to attain 50% of the saturation concentration.

(**Answer:** 28.43 min)

3.12 Using surface renewal model, calculate the following:

(a) At what age of a surface element is the mass transfer rate equal to the mean value for the whole surface for a surface renewal rate s of $0.01 s^{-1}$?

(b) For what proportion of the total mass transfer is surface of an age exceeding 10 s responsible?

(**Answer:** (a) 31.8 s (b) 65.5% of total surface)

3.13 For the diffusion of carbon dioxide at atmospheric pressure and temperature of 293 K, at what time the concentration of solute 1 mm below the surface reach 1 percent of the value at the surface? At that time, what will the mass transfer rate be:

(a) At the free surface?
(b) At a depth of 1 mm?
Diffusivity of $CO_2 = 1.5 \times 10^{-9}$ m²/s and Henry's law constant for CO_2 in water is 1.44×10^5 kPa/mole fraction.

(**Answer:** 50.3 s (a) $1.204 \times 10^{-7}\ \frac{\text{kmol}}{\text{m}^2\text{s}}$ (b) $4.38 \times 10^{-9}\ \frac{\text{kmol}}{\text{m}^2\text{s}}$)

3.14 The gas phase mass transfer coefficient for the absorption of ammonia into water from a mixture of composition NH_3: 20%, N_2: 73%, H_2: 7% is found experimentally to be 0.03 m/s. What would be the coefficient for a mixture of composition NH_3: 5%, N_2: 60%, H_2: 35%? Assume that the solubility of nitrogen and hydrogen in water to be negligible. Temperature and pressure remain the same in both cases. Assume film model with constant film thickness in both cases. Diffusivity of NH_3 in $N_2 = 2.3 \times 10^{-5}$ m²/s and diffusivity of NH_3 in $H_2 = 7.8 \times 10^{-5}$ m²/s.

(**Answer:** 0.038 m/s)

Boundary layer theory

3.15 Ethyl alcohol is being evaporated from a 1-m long flat surface into an air stream flowing across the surface at a velocity 2 m/s. Air is at 289 K and 101.3 kPa. Vapour pressure of alcohol is 4.05 kPa. Kinematic viscosity of air at 289 K is 1.48×10^{-5} m²/s and the mass diffusivity of alcohol in air is 1.26×10^{-5} m²/s. Determine the amount of alcohol evaporating from one square meter surface of the bench in one minute. (**Answer:** 1.51×10^{-2} kg/m² min)

3.16 Water is taken in a pan 4 m long to a depth of 25 mm and is exposed to a wind blowing at a rate of 1500 m/h at 293 K and 101.3 kPa. Under these conditions, the vapour pressure of water is 2.34 kPa, the mass diffusivity of water vapour in air is 2.5×10^{-5} m²/s, and the kinematic viscosity of air is 1.6×10^{-5} m²/s. Find the time required to evaporate all of the water. (**Answer:** 348 h)

3.17 Water at 300K is flowing with a velocity 0.5 m/s across a flat surface made of solid benzoic acid 0.6 m long in the direction of flow. The solubility of benzoic acid in water is 3.0×10^{-2} kmol/m³, the diffusivity of benzoic acid in water is 1.245×10^{-9} m²/s. Viscosity of water at 300 K is 0.855 cP and density is 996 kg/m³. Estimate the mass transfer coefficient k_L and the flux N_A.

(**Answer:** 7.20×10^{-6} m/s, 2.16×10^{-7} kmol/m² s)

3.18 Ethyl alcohol is being evaporated from a 1-m long flat surface into an air stream flowing across the surface at a velocity 10 m/s. Air is at 289 K and 101.3 kPa. Vapour pressure of alcohol is 4.05 kPa. Kinematic viscosity of air at 289 K is 1.48×10^{-5} m²/s and the mass diffusivity of alcohol in air is 1.26×10^{-5} m²/s. Determine the amount of alcohol evaporating from one square meter surface of the bench in one minute. Do the calculations
(a) assuming a transition Reynolds number of 5×10^5 and use Eq. (3.77)
(b) assuming a transition Reynolds number of 3.2×10^5 and use Eq. (3.77)
(c) using the correlation of Rohsenow and Choi.

(**Answer:** (a) 0.052 kg/m² min (b) 0.071 kg/m² min (c) 0.07 kg/m² min)

3.19 Air at 373 K and 101.3 kPa is flowing with a velocity of 5 m/s along a flat plate made of naphthalene. The plate is 3 m long in the direction of flow. Naphthalene sublimes into the air stream under this condition. The diffusivity of naphthalene vapour in air is 9.4×10^{-6} m²/s, and the vapour pressure of naphthalene in air at 373 K is 1.33 kPa. Viscosity of air is 0.0215 cP and its density is 0.95 kg/m³. Calculate:
(a) The transition length
(b) The thickness of the hydrodynamic and concentration boundary layer at the transition point
(c) The average mass transfer coefficient $k_{c,av}$ for the laminar boundary layer

(d) The rate of mass transfer of naphthalene into air for 1 m width of the plate for the conditions given in part (c)

(**Answer:** (a) 2.26 m (b)15.87 mm, 11.84 mm (c) 2.614×10^{-3} m/s (d) 2.54×10^{-6} kmol/s)

Analogy between heat, mass and momentum transfer

3.20 Water is flowing across a flat plate of solid benzoic acid (C_6H_5COOH) with a velocity of 0.06 m/s at 300K. The plate is 0.25 m long in the direction of flow. The average heat transfer coefficient is correlated by the relation

$$J_H = 0.664\,\text{Re}^{-0.5}$$

Estimate the average mass transfer coefficient k_L and the flux N_A if Colburn analogy is valid for the system.

The solubility of benzoic acid in water at 300 K is 3.6 kg/m^3 and the diffusivity is 1.245×10^{-9} m^2/s. Density and viscosity of water at 300 K are 996 kg/m^3 and 0.87 cP, respectively.

(**Answer:** 3.85×10^{-6} m/s, 1.14×10^{-7} kmol/m^{2}s)

3.21 Wet-bulb-thermometer is a simple device used to determine humidity of air. It is an ordinary mercury-in-glass thermometer with the bulb covered with a wet wick. When dry air was passed across a wet-bulb-thermometer, the thermometer indicated 293 K. At 293 K, the vapour pressure of water is 2.34 kPa, latent heat of vaporisation of water is 2452 kJ/kg, density of air is 1.21 kg/m^3, specific heat of air is 1.005 kJ/kg K, Prandtl number is 0.72 and Schmidt number is 0.61. Determine the temperature of dry air. (**Answer:** 332 K)

3.22 A pan containing water is placed in a position where it is exposed to a 16 km/h wind. The pan contains a uniform depth of 12.5 mm water and has a length of 2.5 m in the wind direction. The water in the pan is at a constant temperature of 288 K. The moisture content of the air stream is 0.005 kg water per kg dry air. Find the time required to evaporate all the water. The heat transfer coefficient for the system is given by

$$J_H = 0.664\,\text{Re}^{-0.5} \text{ for Re} < 80000$$
$$J_H = 0.036\,\text{Re}^{-0.2} \text{ for Re} > 5 \times 10^5$$

The kinematic viscosity of air is 1.5×10^{-5} m^2/s and diffusivity of water vapour in air is 2.44×10^{-5} m^2/s. Vapour pressure of water at 288 K is 1.7 kPa. (**Answer:** 35.16 h)

3.23 The mass transfer coefficient for sublimation of solid naphthalene to air at 303 K and 100 kPa was correlated as a function of superficial mass velocity of air as

$$k_G = 9.0 \times 10^{-10}\, G^{0.56}$$

With G representing superficial mass velocity in kg/m^{2}s and k_G is mass transfer coefficient kmol/m^{2}s Pa. Estimate the mass transfer coefficient for the evaporation of water into hydrogen gas for the same geometrical arrangements when hydrogen flows at a superficial velocity of 15 m/s at 303 K, 200 kPa pressure.

Diffusivity of naphthalene in air at 303 K, 100 kPa = 8.33×10^{-6} m^2/s, diffusivity of water vapour in hydrogen at 303 k and 200 kPa = 5.19×10^{-5} m^2/s. Kinematic viscosity of air is 1.62×10^{-5} m^2/s, kinematic viscosity of hydrogen is 1.13×10^{-4} m^2/s, viscosity of air = 1.83×10^{-5} kg/m s, density of hydrogen at 200 kPa, 303 K = 0.1587 kg/m^3.

(**Answer:** 9.74×10^{-9} kmol/m^{2}s Pa)

3.24 The mass transfer coefficient for dissolution of benzoic acid into water is measured experimentally by flowing water through a tube made of cinnamic acid. The tube diameter is 50 mm and the Reynolds number for flow is 40,000 and Schmidt number is 700. The Stanton number was found to be 5.57×10^{-5}. Compare the result with the value calculated using the following analogies:

(a) Reynolds analogy

(b) Colburn analogy
(c) Friend-Metzner analogy
(d) Churchill-Zajic equation

(**Answer:** (a) 2.805×10^{-3} (b) 3.558×10^{-5} (c) 5.566×10^{-5} (d) 5.17×10^{-5})

3.25 Under laminar flow conditions, the mass transfer coefficient and heat transfer coefficient were measured to be 1.28×10^{-5} m/s and 2.82 kW/m^2 K, respectively for a particular system. The molecular diffusivity = 1.5×10^{-9} m^2/s, viscosity = 1 cP, density = 1000 kg/m^3, thermal conductivity = 0.48 W/m K, specific heat = 4.2 kJ/kg K. Calculate the ratio of the velocity in the fluid where the laminar sub-layer terminates to the stream velocity. (**Answer:** 0.2)

Mass transfer correlations

3.26 Calculate the mass transfer coefficient and rate of mass transfer in kg/min from a 25 mm diameter sphere of naphthalene to air at 315 K and atmospheric pressure flowing at a velocity of 0.30 m/s. The diffusivity of naphthalene in air is 6.92×10^{-6} m^2/s and the vapour pressure of solid naphthalene is 74 Pa. Viscosity of air is 1.93×10^{-5} kg/m s and density is 1.1 kg/m^3.
For gases flowing past solid spheres, for $1 < \text{Re} < 48{,}000$, the mass transfer coefficient is given by $Sh = 2 + 0.552\,\text{Re}^{0.53}\,Sc^{1/3}$

(**Answer:** $5.72 \times 10^{-3}\,\dfrac{\text{m}}{\text{s}}$, 2.44×10^{-6} kg/min)

3.27 Water flows down the inside wall of a 1-m long wetted-wall column while air flows upward through the core. The column diameter is 25 mm and dry air enters at a rate of 7 kg/m^2 s at 310K. Water temperature is maintained constant at 294 K at which the vapour pressure of water is 2.5 kPa. The pressure is atmospheric and the mass transfer coefficient may be assumed constant. Calculate the average partial pressure of air leaving the tower.
At 310 K, the viscosity of air is 1.8928×10^{-5} kg/m s, density of air is 1.14 kg/m^3 and diffusivity of water vapour in air is 2.75×10^{-5} m^2/s. (**Answer:** 1.608 kPa)

3.28 What is the effective thickness of gas film for the evaporation of ethanol into air in a 25-mm diameter wetted wall column at a Reynolds number of 25,000 and temperature 302 K given that at this temperature, the diffusivity of ethanol vapour in air is 1.45×10^{-5} m^2/s, density of air is 1.17 kg/m^3 and viscosity is 1.85×10^{-5} kg/ms. The pressure is 100 kPa.

(**Answer:** 0.234 mm)

3.29 A 50-mm diameter wetted-wall column is being used to strip CO_2 from an aqueous solution by air stream flowing at 0.8 m/s. At one point in the column, the concentration of CO_2 in the air stream is 1% (mole). At he same point in the column, the concentration of CO_2 in the water is 0.5 % (mole). Determine the average gas-mass transfer coefficient k_y and the mass flux at the point in the column. The column is operated at 1000 kPa and 298 K. The mass diffusivity of CO_2 in air at 273 K and 1000 kPa is 1.6×10^{-6} m^2/s and viscosity of air at 298 K is 0.0185 cP. At 298 K, the Henry's law constant for CO_2 in water is 1.64×10^5 kPa/mole fraction CO_2 in solution. Density of air is 11.77 kg/m^3. (**Answer:** 2.8×10^{-3} kmol/m^2 s, 2.267×10^{-3} kmol/m^2 s)

3.30 Liquid **A** is being vaporised into air flowing at a rate of 2×10^{-3} kg/s in a wetted-wall column at 70 kPa. The observed rate of evaporation is 1.2×10^{-5} kmol/s. The partial pressure of a in the vapour is 10 kPa and the vapour pressure of **A** is 20 kPa. The mass diffusivity of **A** in air is 2.5×10^{-5} m^2/s.
The column is to be used for evaporation of liquid **B** at the same temperature as in the previous case, but at a pressure of 100 kPa. The air flow rate is 1.7×10^{-3} kg/s. The partial pressure of **B** in the gas is 4 kPa and the vapour pressure is 7.5 kPa. The binary diffusivity of **B** in air is 7.2×10^{-6} m^2/s. Determine the rate of vaporisation of **B**, kmol/s. (**Answer:** 1.39×10^{-6} kmol/s)

3.31 Pure water at a velocity of 0.1 m/s is flowing at 300 K past a flat surface of solid benzoic acid of length 0.4 m. The solubility of benzoic acid in water is 2.95×10^{-2} kmol/m^3. The diffusivity of benzoic acid in water is 1.245×10^{-9} m^2/s. Viscosity of the solution is $\mu = 8.71 \times 10^{-4} \frac{\text{kg}}{\text{m s}}$ and density is $\rho = 996 \frac{\text{kg}}{\text{m}^3}$. Assume dilute solutions. Calculate:

(a) Mass transfer coefficient k_L (b) The film thickness using the film theory

(c) Contact time for the penetration theory

(d) The fractional rate of surface renewal for surface renewal theory

(**Answer:** (a) 3.929×10^{-6} m/s (b) 0.32 mm (c) 102.7 s (d) 0.0124 s^{-1})

3.32 Pure water at a rate of 5.5×10^{-7} m^3/s is admitted to a column packed with 6.5 mm diameter benzoic acid spheres. The total surface area of the spheres in the bed is 1.2×10^{-2} m^2 and the bed porosity is 0.44. The column diameter is 65 mm. The solubility of benzoic acid in water is 4.3 kg/m^3 and the diffusivity is 1.25×10^{-9} m^2/s. The viscosity and density of water are 0.87 cP and 997 kg/m^3, respectively. Assume dilute solutions and calculate:

(a) The mass transfer coefficient k_c, given that the correlation $J_D = \frac{1.09}{\varepsilon} \text{Re}^{-0.667}$ is applicable

(b) The concentration of benzoic acid in the exit water in kmol/m^3.

(**Answer:** (a) 4.53×10^{-6} m/s (b) 3.32×10^{-3} kmol/m^3)

3.33 Estimate the mass transfer coefficient in a stream of air at 320 K and 200 kPa flowing past the following objects made of solid naphthalene at 1.5 m/s. The diffusivity of naphthalene in air is 5.16×10^{-6} m^2/s at 273 K and 100 kPa.

(a) Air flow parallel to a flat plate, 150 mm long

(b) Air flow past a sphere of diameter 15 mm

The following correlations are applicable:

$$J_D = 0.664\,\text{Re}^{-0.5} \text{ (flat plate)}$$
$$J_D = 0.037\,\text{Re}^{-0.4} \text{ (single sphere)}$$

(**Answer:** (a) 2.87×10^{-3} m/s (b) 1.035×10^{-2} m/s)

3.34 Air at 101.3 kPa and 373 K flows at a Reynolds number of 50000 normal to a long, circular, 25 mm diameter cylinder made of naphthalene. Vapour pressure of naphthalene is 1.33 kPa, viscosity of air is 0.0215 cP, density of air is 0.95 kg/m^3 and diffusivity of naphthalene in air is 9.4×10^{-6} m^2/s. Calculate the sublimation flux, kmol/s m^2. Use $J_D = 0.600\,\text{Re}^{-0.487}$

(**Answer:** 3.36×10^{-5} kmol/m^2 s)

3.35 Ammonia is absorbed from an air stream flowing at 5m/s by the wet surface of an object placed in the flowing stream. No data on mass transfer coefficient is available. However, heat transfer tests with the same object under similar flow conditions indicate a heat transfer coefficient value of 5.2×10^{-2} kW/m^2 K. The heat capacity and density of air are 1.005 kJ/kg K and 1.1 kg/m^3. Estimate the value of k_c for the absorption process. For air, Pr = 0.74, Sc = 0.61.

(**Answer:** 5.35×10^{-2} m/s)

3.36 Estimate the mass transfer coefficient k_G for evaporation of water into a turbulent air stream flowing at 3m/s at 100 kPa and 311 K for the following cases:

(a) A horizontal flat plate 0.5 m long

(b) A wetted-wall tube 25 mm inside diameter

(c) A 25-mm diameter cylinder placed normal to the air stream

(d) A single sphere 25 mm in diameter

Assume driving force to be 10 kPa, so that $\bar{p}_{BM}$ may be taken as 95 kPa. For air, viscosity is 1.8 × 10^{-5} kg/m s and density is 1.14 kg/m^3; diffusivity of water vapour in air is 2.75 × 10^{-5} m^2/s. The following correlations may be used:

$J_D = 0.664\,\mathrm{Re}^{-0.5}$ (flat plates) $\quad J_D = 0.023\,\mathrm{Re}^{-0.17}\,Sc^{0.11}$ (wetted-wall tube)

$J_D = 0.600\,\mathrm{Re}^{-0.487}$ (cylinder) $\quad J_D = 0.37\,\mathrm{Re}^{-0.4}$ (single sphere)

(**Answer:** (a) 3.81 × 10^{-6} kmol/m^2 s kPa (b) 9.07 × 10^{-7} kmol/m^2 s kPa (c) 1.72 × 10^{-5} kmol/m^2 s kPa (d) 2.21 × 10^{-5} kmol/m^2 s kPa)

3.37 The mass flux from a 50 mm diameter naphthalene ball, placed in stagnant air at 313 K and atmospheric pressure is 3.72 × 10^{-6} kmol/m^2 s. Assume the vapour pressure of naphthalene to be 74 kPa at 313 K and negligible bulk concentration of naphthalene in air. If air starts blowing across the surface of naphthalene ball at 3 m/s, what will the mass transfer rate, all other conditions remaining the same?

For spheres: $\mathrm{Sh} = 2.0 + 0.6\,(\mathrm{Re})^{0.5}\,(Sc)^{0.33}$

The viscosity and density of air are 1.8 × 10^{-5} kg/m s and 1.123 kg/m^3

(**Answer:** 8.44 × 10^{-3} kmol/m^2 s)

Interphase mass transfer

3.38 Water is being vaporised into air in a wetted wall column of 15 mm diameter and 0.85 m long at 101.3 kPa and 298 K. Air is admitted to the column at 298 K at a rate 7.5 × 10^{-4} m^3/s and with a partial pressure of water of 836 Pa. The partial pressure of water in the exit air is 2680 kPa. The vapour pressure of water at 298 K is 3.166 kPa. Calculate:

(a) Rate of mass transfer of water into air

(b) The overall mass transfer coefficient K_G.

(**Answer:** (a) 5.75 × 10^{-7} kmol/s (b) K_G = 1.23 × 10^{-5} kmol/m^2 s kPa)

3.39 In the absorption of a solute from a gas mixture into a liquid solvent, if the liquid film resistance is 5 times that of the gas film, by how much would the rate of absorption change if the liquid film coefficient could be doubled without changing other parameters? What would be the effect of doubling the gas film coefficient? (**Answer:** 71.4%, 9.1%)

3.40 In an experimental study of the absorption of ammonia by water in a wetted-wall column, the value of overall mass transfer coefficient K_G was found to be 1.7 kmol/m^2 h bar. At one point in the column, the gas contained 10% mole NH_3 and the liquid phase concentration was 0.05 kmol/m^3 of solution. The temperature was 293 K and the total pressure was 1 bar. Eighty five percent of the total resistance to mass transfer was found to be in the gas phase. If Henry's law constant at 293 K is 9.4 × 10^{-3} bar/kmol ammonia per m^3 of solution, calculate:

(a) The individual film coefficients k_G, and k_L (b) The interfacial compositions $\bar{p}_{Ai}$, C_{Ai}.

(**Answers:** (a) k_G = 2 kmol/m^2 h bar, k_L = 0.1065 m/ h, (b) $\bar{p}_{Ai}$ = 0.0154 bar, C_{Ai} =1.6382 kmol/m^3)

3.41 Ammonia is being absorbed into water from an ammonia-air mixture using a wetted-wall column. The overall gas mass transfer coefficient K_G = 6.0 × 10^{-2} kmol/m^2 h kPa. The tower was operated at 290 K and 101.3 kPa. At the top of the tower, the gas leaving the column contained 1% by volume NH_3, and the entering liquid which this gas contacted was pure water. For dilute solutions of ammonia in water at 290 K, the equilibrium partial pressure is given by $\bar{p}_A = 1.6C_A$ where $\bar{p}_A$ is in kPa and C_A is the concentration of ammonia in water in kmol/m^3 of solution. If the gas phase offered 80% of the resistance to mass transfer, calculate (a) k_G (b) $\bar{p}_{Ai}$ and C_{Ai} (c) k_L (d) K_L

(**Answer:** (a) 7.5 × 10^{-2} kmol/(m^2 h kPa) (b) $\bar{p}_{Ai}$ = 0.2026 kPa, C_{Ai} = 0.1266 kmol/m^3 (c) 0.48 m/h (d) 9.6 × 10^{-2} m/h)

3.42 In an experimental study of absorption of ammonia by water, the overall mass transfer coefficient K_G was found to be 1.6×10^{-8} kmol/(m^2 s Pa). At one point in the column the gas contained 8 mol% ammonia and the liquid composition was 6.5×10^{-2} kmol NH_3/m^3 solution. The temperature was 293 K and pressure was 100 kPa. 80% of the total resistance was in the gas phase. If Henry's law constant is 9.3×10^2 Pa/(kmol/m^3), calculate the individual film coefficients and interfacial concentrations.

(**Answer:** $k_G = 2 \times 10^{-8}$ kmol/(m^2 s Pa), $k_L = 7.45 \times 10^{-5}$ m/s, $\overline{p}_{Ai} = 1650$ Pa and $C_{Ai} = 1.77$ kmol/m^3)

3.43 In a wetted wall column, where ammonia was stripped from an ammonia water solution into air stream, the overall liquid mass transfer coefficient K_L was 1.5×10^{-2} m/h. At a point in the column, the concentration of ammonia in the liquid was 5 kmol/m^3 and the partial pressure of ammonia in the gas stream was 6 kPa. For dilute solution of ammonia in water, the equilibrium relation is $\overline{p}_A = 1.56C_A$ where $\overline{p}_A$ is in kPa and C_A is in kmol/m^3. If the gas phase offered 75% of the total resistance, calculate (a) individual mass transfer coefficients (b) interfacial compositions.

(**Answer:** (a) $k_G = 1.28 \times 10^{-2}$ kmol/(m^{2}h kPa), $k_L = 6 \times 10^{-2}$ m/h, $\overline{p}_{Ai} = 7.35$ kPa and $C_{Ai} = 4.71$ kmol/m^3)

3.44 In an experimental study of the absorption of NH_3 from a mixture with N_2 in a wetted wall column, the value of overall mass transfer coefficient K_G was found to be 2.75×10^{-8} kmol/m^2 s bar. At certain point in the column the gas contained 10% ammonia by volume while the liquid phase concentration was 6.5×10^{-2} kmol/m^3 solution. If 85% of the resistance to mass transfer is in the gas phase and Henry's law coefficient for this system under operating conditions is 9.35×10^{-3} bar/(kmol ammonia/ m^3 of solution). Calculate

(a) The interfacial composition (b) The individual mass transfer coefficients.

The temperature and pressure of operation are 293 K and 1 bar respectively.

(**Answer:** (a) $\overline{p}_{Ai} = 1.55 \times 10^{-2}$ bar and $C_{Ai} = 1.66$ kmol/m^3
(b) $k_L = 1.7 \times 10^{-9}$ m/s, $k_G = 3.24 \times 10^{-8}$ kmol/m^{2}s bar)

3.45 Solute A is being transferred from the liquid to the gas phase. The equilibrium relation is given by

$$y_A = 0.75x_A$$

At certain point in the apparatus, the concentration of A in the gas and liquid are 45% (mole) and 90% (mole), respectively. The gas film coefficient is 10 kmol/hm^2 (mole fraction). Seventy percent of the total resistance to mass transfer lie with the gas phase. Evaluate:

(a) the molar flux of **A**
(b) the interfacial compositions x_{Ai} and y_{Ai}
(c) The overall mass transfer coefficient, K_y

(**Answer:** (a) 1.575 kmol/m^2 h (b) $x_{Ai} = 0.81$ and $y_{Ai} = 0.608$
(c) $K_y = 7$ kmol/m^2 h)

3.46 A liquid mixture of benzene and toluene is being separated in a packed distillation column. At a point in the column, the liquid phase contained 40 mole percent benzene and the gas phase contained 50 mole percent benzene. The mass transfer rate of benzene is 0.25 kmol/h m^2. The relative volatility of benzene-toluene system is 2.54 and is assumed constant. If 60 percent of the resistance to mass transfer lies with the gas phase, determine the interfacial compositions.

(**Answer:** $x_{Ai} = 0.349$ and $y_{Ai} = 0.577$)

3.47 Air is dried by absorbing the water vapour into a stream of H_2SO_4 in a packed column. Air enters the column with a relative humidity of 50% and leaves with a relative humidity of 10%. The acid enters at the top of the column with a concentration 70% (weight) H_2SO_4 and leaves with a concentration of 55% H_2SO_4. Experimental evidence indicates that $k_G = 10$ kmol/m² h bar and $k_L = 0.021$ m/h and they may be assumed to remain constant. The temperature is assumed to remain constant at 298 K. Determine for the conditions at the bottom of the column,

(a) Interfacial composition $\bar{p}_{Ai}$ and C_{Ai}
(b) Rate of mass transfer, kmol/m² h
(c) Percent total diffusional resistance that lie in the gas phase
(d) Overall mass transfer coefficient K_G and K_L

Vapour pressure of water at 298 K is 3.17 kPa. The equilibrium data at 298 K for the H_2SO_4 – water system are:

% RH	0.8	2.3	5.2	9.8	17.2	26.24	36.8	46.8	56.8
% weight H_2SO_4	80.0	75.0	70.0	65.0	60.0	55.0	50.0	45.0	40.0

Density of sulphuric acid solutions:

%weight H_2SO_4	80.0	75.0	70.0	65.0	60.0	55.0	50.0	45.0	40.0
Density at 298 K kg/m³	1722	1664	1606	1549	1494	1441	1391	1344	1299

(**Answer:** (a) $\bar{p}_{Ai} = 1.12 \times 10^{-2}$ bar and $C_{Ai} = 38$ kmol/m³ (b) $N_A = 4.65 \times 10^{-2}$ kmol/m² h (c) 61.6% (d) $K_G = 6.16$ kmol/m² h bar, $K_L = 8.35 \times 10^{-3}$ m/h)

3.48 Air is dried by absorbing the water vapour into a stream of H_2SO_4 in a packed column. Air enters the column with a relative humidity of 50% and leaves with a relative humidity of 10%. The acid enters at the top of the column with a concentration 70% (weight) H_2SO_4 and leaves with a concentration of 55% H_2SO_4. Experimental evidence indicates that $k_G = 10$ kmol/m² h bar and $k_L = 0.021$ m/h and they may be assumed to remain constant. The temperature is assumed to remain constant at 298 K. Determine for the conditions at the top of the column,

(a) Interfacial composition $\bar{p}_{Ai}$ and C_{Ai}
(b) Rate of mass transfer, kmol/m² h
(c) Percent total diffusional resistance that lie in the gas phase
(d) Overall mass transfer coefficient K_G and K_L

Vapour pressure of water at 298 K is 3.17 kPa. The equilibrium data at 298 K for the H_2SO_4 – water system are:

% RH	0.8	2.3	5.2	9.8	17.2	26.24	36.8	46.8	56.8
% weight H_2SO_4	80.0	75.0	70.0	65.0	60.0	55.0	50.0	45.0	40.0

Density of sulphuric acid solutions:

%weight H_2SO_4	80.0	75.0	70.0	65.0	60.0	55.0	50.0	45.0	40.0
Density at 298 K kg/m³	1722	1664	1606	1549	1494	1441	1391	1344	1299

(**Answer:** (a) $\bar{p}_{Ai} = 1.85 \times 10^{-3}$ bar and $C_{Ai} = 27.4$ kmol/m³ (b) $N_A = 1.32 \times 10^{-2}$ kmol/m² h (c) 86.8% (d) $K_G = 8.68$ kmol/m² h bar, $K_L = 3.74 \times 10^{-3}$ m/h)

3.49 Sulphur dioxide is absorbed in water using a column. At one point in the column the gas contained 10% SO_2 by volume and was in contact with liquid containing 0.4% weight SO_2. The temperature was 323 K and the total pressure 100 kPa. The overall transfer coefficient based on gas concentration was 7.36×10^{-10} kmol/m²s (N/m²). Of the total diffusion resistance, 45% lay in the gas phase and 55% in the liquid. Equilibrium data at 323 K are:

Kg SO_2/100 kg water	0.2	0.3	0.5	0.7
p_{SO_2}, Pa	3865	6131	11063	15861

Assume that the density of the solution is 990 kg/m^3. Calculate:

(a) Overall coefficient K_L

(b) The individual coefficients k_G and k_L

(c) Interfacial compositions $\bar{p}_{Ai}$ and $C_{,i}$

(**Answer:** (a) $K_L = 1.3 \times 10^{-4}$ m/s (b) $k_G = 1.636 \times 10^{-9}$ kmol/(m^{2}s Pa), $k_L = 2.08 \times 10^{-4}$ m/h (c) $\bar{p}_{Ai} = 9.366 \times 10^3$ Pa , $C_{,i} = 6.7 \times 10^{-2}$ kmol/m^3)

3.50 A wetted wall absorption tower of 25 mm ID is fed with water as the wall liquid and ammonia – air mixture as the central core gas. At a particular level in the tower, the ammonia concentration in the bulk gas is 0.8 mole fraction, and that in the bulk liquid is 0.05 mole fraction. The temperature is 300 K and the pressure is 1 bar. The local mass transfer coefficient in the liquid k_x is 1.40×10^{-4} kmol/m^2 s (mole fraction) and the local Sherwood number for the gas is 4. Diffusivity of ammonia in air is 2.3×10^{-5} m^2/s. Equilibrium relationship for ammonia-air-water is as follows.

x_A	0.0	0.05	0.10	0.25	0.30
y_A	0.0	0.06	0.135	0.591	0.92

Calculate:

(a) The interfacial compositions

(b) The local mass transfer flux

(c) The local overall mass transfer coefficient K_G

(**Answer:** (a) $x_{Ai} = 0.254$, $y_{Ai} = 0.607$ (b) 2.847×10^{-5} kmol/m^2 s (c) $K_G = 3.847 \times 10^{-5}$ kmol/m^2 s bar)

3.51 The solute **A** is being absorbed from a gas mixture of **A** and **B** in a wetted wall tower with the liquid flowing as a film downward along the wall. The mole fraction of **A** in the bulk gas and the bulk liquid are 0.38 and 0.1, respectively. The equilibrium data are:

x	0	0.05	0.10	0.15	0.2	0.25	0.3	0.35
y	0	0.022	0.05	0.087	0.131	0.187	0.267	0.385

The solute **A** diffuses through stagnant **B** in the gas phase and then through a non-diffusing liquid. The film coefficients are $k_y = 1.5 \times 10^{-3}$ kmol/m^2 s and $k_x = 2.0 \times 10^{-3}$ kmol/m^2 s. Assuming bulk flow effect is negligible, calculate the interface concentrations and flux.

(**Answer:** $x_{Ai} = 0.25$, $y_{Ai} = 0.18$, $N_A = 3.0 \times 10^{-4}$ koml/m^2 s)

3.52 Sulphur dioxide is absorbed into water in a packed column at 323K and 101.3 kPa. The equilibrium solubility of SO_2 in water is given by

$$y = 30x + 6750x^2$$

where x and y are mole fractions of SO_2 in the liquid and the gas, respectively. The liquid phase and gas phase mass transfer coefficients are $k_x = 10$ kmol/m^{2}h and $k_y = 8$ kmol/m^2 h. At a certain point in the apparatus the bulk concentrations are, $y_{AG} = 0.085$ and $x_{AL} = 0.001$. Calculate:

(a) The interfacial mole fractions

(b) The overall mass transfer coefficient K_x

(c) The mass flux in kmol/m^2 h

(**Answer:** (a) $x_{Ai} = 1.94 \times 10^{-3}$ and $y_{Ai} = 8.38 \times 10^{-2}$ (b) $K_x = 9.79$ kmol/m^2 h (c) $N_A = 9.2 \times 10^{-3}$ kmol/m^2 h)

3.53 In the desorption of component **A** from an aqueous solution into an air stream, at a particular point in the tower, the partial pressure of **A** in the gas phase is 1.33 kPa and the concentration in the liquid is 4 kmol/m^3. The overall mass transfer coefficient K_G was equal to 2.65×10^{-3} kmol/(m^2 h kPa). Sixty percent of the total resistance to mass transfer lie in the gas film and the remaining in the liquid film. Henry's law constant was found to be 760 kPa/ (kmol/m^3 of solution). Determine:

(a) The gas-film coefficient, k_G (b) The liquid film coefficient, k_L

(c) The mass flux N_A (d) The interfacial concentrations

(**Answer:** (a) 4.417 × 10^{-3} kmol/m^2 h kPa (b) 5.035 m/h (c) 8.05 kmol/h (d) $\overline{p}_{Ai}$ = 1.822 × 10^3 kPa and C_{Ai} = 2.397 kmol/m^3)

3.54 The solute is being absorbed from a gas mixture of **A** and **B** in a wetted-wall tower with the liquid flowing as a film downward along the wall. At a certain point in the tower, the bulk gas concentration y_{AG} = 0.38 and bulk liquid concentration x_{AL} = 0.10. The tower is operated at 298 K and 101.3 kPa. Equilibrium data is as follows:

x_A	0	0.05	0.10	0.15	0.20	0.25	0.30	0.35
y_A	0	0.022	0.05	0.087	0.131	0.187	0.267	0.385

The solute **A** diffuses through stagnant **B** in the gas phase and then through a non diffusing liquid.

The film mass transfer coefficients predicted using correlations for dilute solutions are, k'_x = 2 × 10^{-3} kmol/m^2 s and k'_y = 1.5 × 10^{-3} kmol/m^2 s. Calculate the interfacial compositions x_{Ai} and y_{Ai} and the flux N_A.

(**Answer:** x_{Ai} = 0.256, y_{Ai} = 0.198, N_A = 3.86 × 10^{-4} koml/m^2 s)

3.55 The solute is being absorbed from a gas mixture of **A** and **B** in a wetted-wall tower with the liquid flowing as a film downward along the wall. At a certain point in the tower, the bulk gas concentration y_{AG} = 0.38 and bulk liquid concentration x_{AL} = 0.10. The tower is operated at 298 K and 101.3 kPa. Equilibrium data is as follows:

x_A	0	0.05	0.10	0.15	0.20	0.25	0.30	0.35
y_A	0	0.022	0.05	0.087	0.131	0.187	0.267	0.385

The solute **A** diffuses through stagnant **B** in the gas phase and then through a non diffusing liquid. The film mass transfer coefficients predicted using correlations for dilute solutions are, k'_x = 2 × 10^{-3} kmol/m^{2}s and k'_y = 1.5 × 10^{-3} kmol/m^{2}s. Calculate the overall mass transfer coefficient K'_y, the flux, and the percent resistance in the gas and liquid films.

(**Answer:** K'_y = 8.99 × 10^{-4} kmol/m^2 s, N_A = 3.84 × 10^{-4} kmol/m^2 s, 54.8% in gas phase)

3.56 An SO_2–air mixture is being scrubbed with water in a packed tower operating at 293 K and 101.3 kPa. Solute free water enters the top of the tower at a constant rate of 500 kg/h and the liquid leaving contains 0.6 kg SO_2 per 100 kg SO_2 free water. The partial pressure of SO_2 in the spent gas leaving the top is 3.07 kPa. The mole ratio of water to air is 25. The individual liquid and gas mass transfer coefficients are found to be constant with the values k_L = 0.4 m/h and k_G = 9.0 × 10^{-3} kmol/m^2 h kPa. Assume that the density of solution is approximately the same as that of water. The solubility of SO_2 in water at 293 K are as follows:

kg SO_2/100kg H_2O	0.02	0.05	0.10	0.15	0.20	0.30	0.50	0.70	1.0
Partial pressure of SO_2, kPa	0.067	0.16	0.43	0.77	1.13	1.88	3.47	5.20	7.86

Determine:

(a) The percent of SO_2 in the entering gas that is absorbed in the tower

(b) The liquid concentration at the interface at a point in the column where the bulk liquid concentration is 0.001 kmol SO_2 per kmol water.

(**Answer:** (a) 57.4% (b) 0.0016 kmol SO_2 per kmol water)

3.57 The solute is being absorbed from a gas mixture of **A** and **B** in a wetted-wall tower with the liquid flowing as a film downward along the wall. At a certain point in the tower, the bulk gas concentration $y_{AG} = 0.25$ and bulk liquid concentration $x_{AL} = 0.05$. The tower is operated at 298 K and 101.3 kPa. Equilibrium data is as follows:

x_A	0	0.05	0.10	0.15	0.20	0.25	0.30	0.35
y_A	0	0.022	0.05	0.087	0.131	0.187	0.267	0.385

The solute **A** diffuses through stagnant **B** in the gas phase and then through a non diffusing liquid.

The film mass transfer coefficients predicted using correlations for dilute solutions are, $k'_x = 2 \times 10^{-3}$ kmol/m^2 s and $k'_y = 1.5 \times 10^{-3}$ kmol/m^{2}s. Calculate:

(a) The interfacial mole fractions and the flux N_A

(b) Overall mass transfer coefficient K'_y and K_y

(c) Overall mass transfer coefficient K'_x and K_x

(**Answer:** (a) $x_{Ai} = 0.17$, $y_{Ai} = 0.1035$, $N_A = 2.68 \times 10^{-4}$ koml/m^{2}s
(b) $K'_y = 1.01 \times 10^{-3}$ koml/m^{2}s and $K_y = 1.1824 \times 10^{-3}$ koml/m^{2}s
(c) $K'_x = 9.2 \times 10^{-4}$ koml/m^{2}s and $K_x = 1.117 \times 10^{-3}$ koml/m^{2}s)

Chapter 4

Gas Absorption

Gas absorption is an important gas-liquid contact operation in which the soluble components of a gas mixture are transferred into a liquid solvent. Absorption operations are also known as gas scrubbing or gas washing. It is used as an industrial separation operation for removing contaminants, pollutants and impurities from gases and for recovery of valuable chemicals. Ammonia is absorbed from its mixture with air producing ammonia-free air and acetone can be recovered from acetone-air mixture producing acetone-free air. In both cases, water is used as the solvent and the separation is possible because ammonia and acetone are soluble in water and air is relatively insoluble. Solvents such as alkaline solutions, hydrocarbon oils, ethanolamines, etc. can also be used as the solvent for absorption. For example, sulphur dioxide present in the flue gases can be removed by absorption in alkaline solutions. Naturally occurring hydrocarbon gases can be freed off objectionable hydrogen sulphide by absorbing in alkaline solutions. In the hydrogenation of edible oils, hydrogen gas is first absorbed in oils. Benzene and toluene vapours can be recovered from by-product coke oven gases by absorbing in hydrocarbon oils. In its simplest form, absorption involves transfer of a solute (**A**) through a stagnant non-diffusing gas (**B**) into a liquid (**C**).

Stripping or desorption is the reverse process of gas absorption. In absorption, the transfer of solute is from the gas to the liquid, whereas in stripping, the transfer is from the liquid to the gas. An example is the stripping of oils by contacting with superheated steam. Volatile components in the oil get selectively transferred to steam. In absorption, the liquid solvent is the separating agent added to bring about separation whereas in stripping the separation is achieved by introducing a gas stream. For regeneration and recycling of solvent used for absorption, industries frequently combine stripping operations with absorption as shown in Fig. 4.1. Thus, the hydrocarbon oil used for absorption of benzene and toluene vapours in coke oven gases can be regenerated by contacting the solution leaving the absorption towers with steam. The benzene and toluene vapours absorbed in the oil are transferred to the gas stream and the regenerated oil is recycled to the absorption unit.

Absorption can be classified into two broad categories: physical absorption and chemical absorption. In the former no significant chemical reaction occurs between the solute and the solvent and the process is purely physical. Absorption using water or hydrocarbon oil as solvents generally fall into this category. Absorption of solutes such as ammonia, acetone, ethanol and sulphur dioxide in water, benzene and toluene vapours in hydrocarbon oil are examples of physical absorption. In the second category, chemical reaction occurs between the solute being absorbed and the solvent. Irreversible chemical reaction occurs when acidic gases such as carbon dioxide, hydrochloric acid, hydrogen sulphide, etc. are absorbed in alkaline solution. When carbon dioxide or hydrogen sulphide is absorbed using monoethanolamine and diethanolamine or when oxides of nitrogen or chlorine are absorbed in water, the reaction occurring in the liquid phase is reversible. Reaction between the solute and the solvent is found to increase the rate of absorption and the capacity and selectivity of the solvent for absorption.

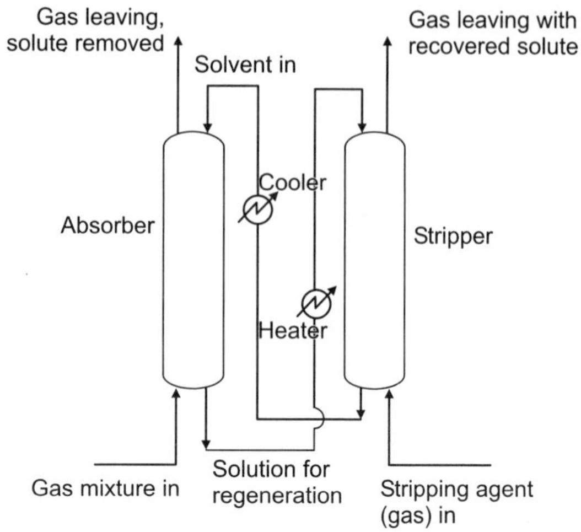

Fig. 4.1: Absorption and stripping operation

Absorption and stripping are gas-liquid contact operations. Another important gas-liquid separation operation is distillation, and it will be worthwhile here to compare distillation with absorption and stripping. Distillation operation is discussed in detail in Chapter 5. In distillation operation, the gas and liquid phases are generated by the partial vaporisation and condensation of the feed mixture, while the liquid phase in gas absorption and the gas phase in stripping are external agents introduced into the system to effect separation. Distillation is carried out at the boiling point of the mixture, whereas the absorption is a low-temperature operation and occurs well below the boiling point. Ideally, equimolar counter diffusion occurs in binary distillation operation because molecules diffuse in both directions. In absorption, gas molecules are diffusing into the liquid and the movement in the opposite direction is negligible; the diffusion mechanism is the diffusion of the solute through stagnant gas.

4.1 ABSORPTION EQUIPMENT

The absorption of a gas in a liquid is generally conducted in packed columns, plate columns, centrifugal absorbers, spray columns and bubble columns. Packed columns and plate columns constitute the majority of industrial absorbers. In plate columns, the gas phase is dispersed in a continuous stream of liquid whereas in packed columns the gas phase is continuous and the liquid stream is dispersed to provide large interfacial areas.

Plate Columns

Plate columns, also known as tray columns, are multistage contact equipment commonly used for gas-liquid contact operations such as gas absorption and distillation. A plate column is shown in Fig. 4.2a. These are vertical cylindrical towers with diameter ranging from 300 mm to more than 9 m for typical industrial applications.

The column is divided into a number of compartments by means of metallic plates or trays. The liquid and gas streams are contacted in stepwise fashion on the trays. On each tray, the fluids come into contact, interphase diffusion occurs and finally the phases get separated from each other. Since these operations constitute the three basic requirements of a device commonly designated in mass transfer terminology as a stage, each tray can be treated as a stage in a countercurrent multistage cascade. The column can be as tall as to accommodate a large number of plates, at times exceeding one hundred. Plate spacing may range typically from 150 mm to 2 m. The liquid admitted at the top of the column flows downward by gravity. The liquid flows across the plate, over an outlet weir, into a downcomer which takes the liquid to the plate below by gravity as shown in Fig. 4.2b.

The downcomer usually occupies 10 to 15 per cent of the column cross section. In plates of medium sizes, the downcomers are located alternatively on opposite sides of successive plates, so that the liquid

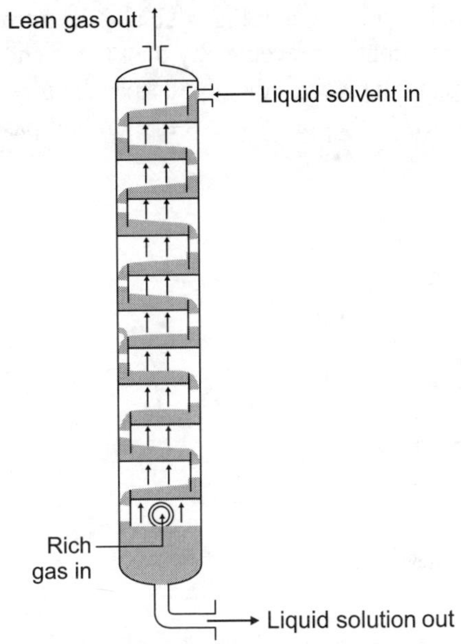

Fig. 4.2: (a) Plate column for gas absorption

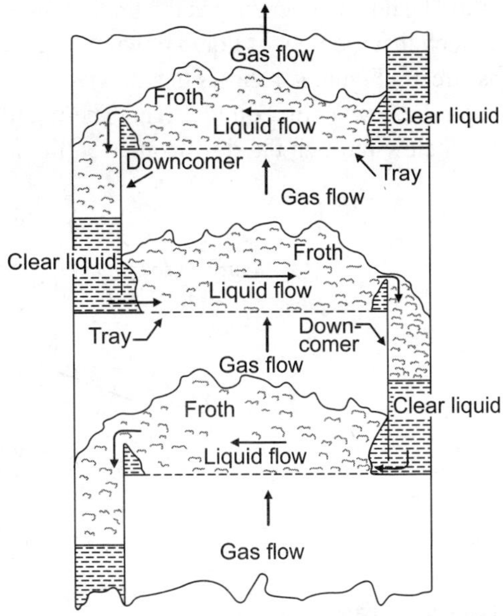

Fig. 4.2: (b) Flow pattern in sieve plate column

traverses the full width of each plate (Fig. 4.3a). In larger plates, the downcomers are placed alternately at the centre and sides of successive plates, so that the liquid need traverse only half the diameter of the plates (Fig. 4.3c). Radial flow type downcomers (Fig. 4.3b) are also used in large-diameter towers.

On each plate, the liquid is contacted with the gas stream which flows upwards. Several designs have been proposed for the plates to facilitate uniform distribution of the gas, intimate contact between the gas and the liquid and easy disengagement of gas and liquid streams after mass transfer. The sieve plates, bubble-cap plates and valve trays are the most common. In sieve trays, for example, the gas issues through tiny perforations on the trays providing sufficient agitation in the pool of liquid over the plate so that intimate contact between gas and liquid is established. Since the gas and liquid reaching a plate are not in equilibrium, mass transfer occurs between the phases when these phases are brought into contact on the plates. If the time of contact between the gas and the liquid is sufficient and the mixing occurring on the plate is perfect, the gas stream leaving the plate and the liquid flowing down after disengagement will be in equilibrium and the plate can be treated as an *ideal or theoretical plate*. In the following sections the important types of trays and their features are briefly discussed.

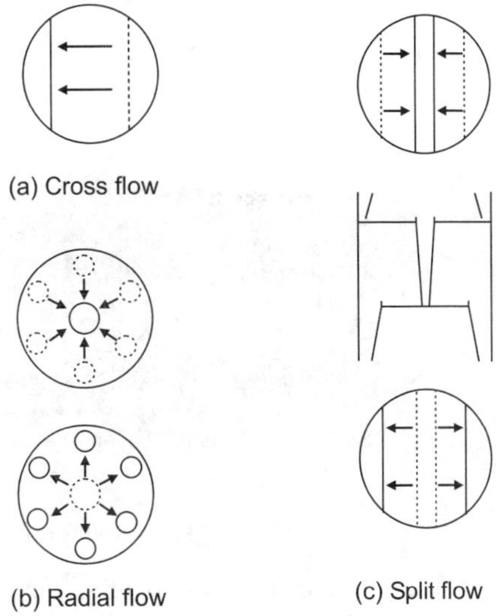

Fig. 4.3: Downcomer types

Sieve tray

These are metallic plates provided with tiny perforations through which gas flows upward. Diameter of the perforations ranges from 3 mm to 12 mm. The area of the perforations generally varies between 5 to 15% of the tray area. An overflow weir helps to maintain sufficient depth for the pool of liquid over the plate. The downcomer carries the liquid to the plate below as shown in Fig. 4.2b and the rate of gas flow is such that the flow of liquid down through the perforations is prevented by the kinetic energy of the gas stream. Figure 4.4 shows the top view of a sieve plate. It receives the liquid from the tray above it through the downcomer on the right. The liquid flows across the tray towards the left and passes to the plate below through the downcomer on the left.

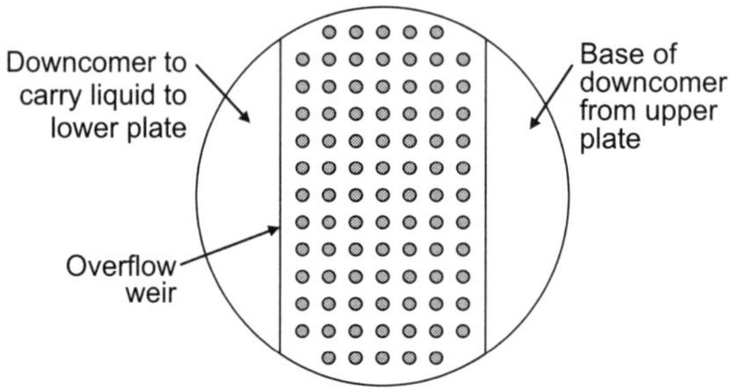

Fig. 4.4: Top view of a sieve tray

Bubble-cap tray

In bubble-cap trays (Fig. 4.5), the caps are provided on risers or chimneys of diameter ranging from 50 to 80 mm, through which the gas flow up. The caps are generally 75 mm to 150 mm in diameter and they allow the gas to change the direction of flow by making the gas to turn around and flow down through the space between the cap and the riser.

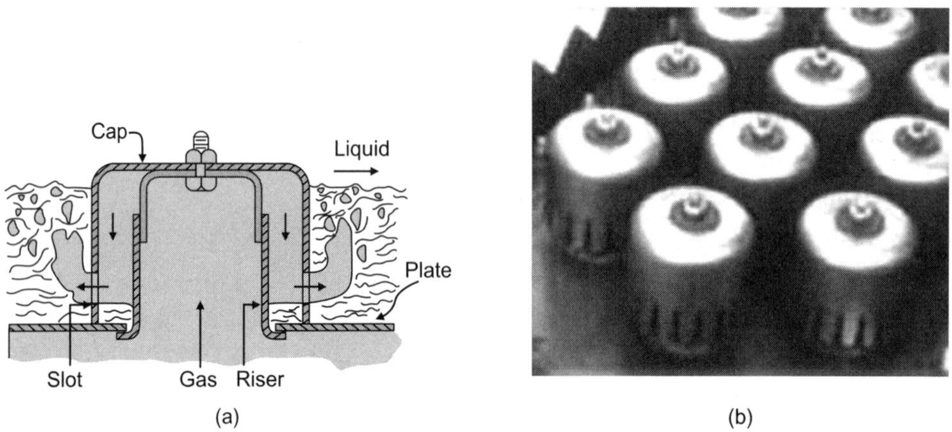

Fig. 4.5: (a) Bubble-cap tray (b) Close-up view of bubble caps

The liquid on the plate is of sufficient depth as to nearly submerge the caps. The edges of the caps are serrated or perforated with slots to break up the gas stream into bubbles. The gas emerges as bubbles through these narrow slots into the pool of liquid to form a froth. Bubble-cap trays were in use for quite a long time before these were abandoned in favour of sieve plates or valve trays.

Valve tray

Valve tray (Fig. 4.6) is a cross between the sieve tray and the bubble cap tray and now finds wide industrial applications. Each of the openings on the tray, the diameter of which ranges from 25 mm to 50 mm is provided with a lift valve cover which prevents flow of liquid down through the openings and permits flow of gas upwards at almost constant linear velocity through the liquid pool. The position of the valve cap riser is determined by the momentum of the gas flowing upwards. With no vapour flow, the valve sits on the tray covering the hole. At high gas velocities, the riser is fully open and at low velocities, the riser is lowered and hence partly closed. Since the flow area is controlled by the position of the riser, the velocity of gas issuing through the valves remain constant and is independent of the amount of vapour flowing through the tower. The problem of *weeping* encountered in sieve trays at low gas rates can be greatly reduced in valve trays. Also the gas is directed by the caps almost horizontally, thereby providing thorough mixing in the liquid than that achieved in sieve trays. Valve trays can be operated at a greater range of flow rates than the sieve trays.

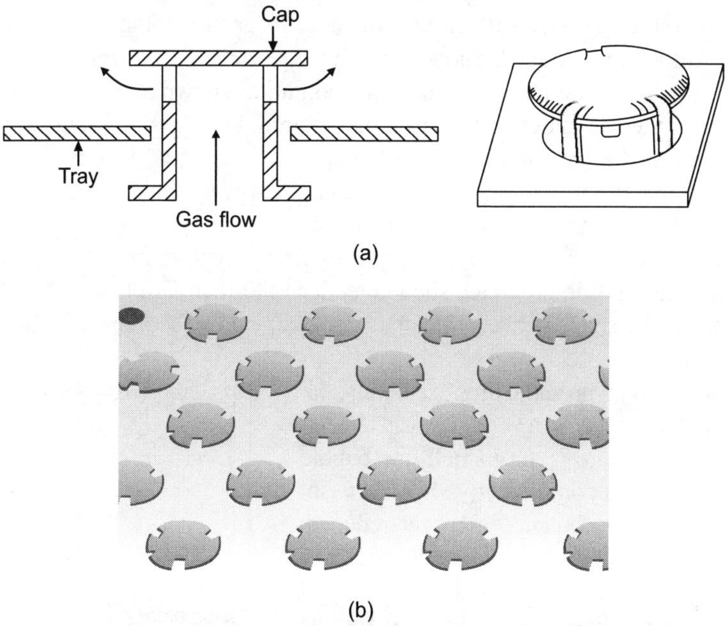

Fig. 4.6: (a) Valve cap (b) Arrangement of valves on the tray

These different types of trays are compared in Table 4.1.

Table 4.1: Comparison of tray types

	Sieve trays	Valve trays	Bubble-cap trays
Relative cost	1.0	1.2	2.0
Pressure drop	Lowest	Intermediate	Highest
Efficiency	Lowest	Highest	Highest
Vapour capacity	Highest	Highest	Lowest
Turndown ratio	2	4	5

Factors Affecting Column Performance

The primary function of a plate is to provide perfect contact between gas and liquid. An *ideal or theoretical plate* is the one in which the effluent liquid and gas streams are in equilibrium. The plate efficiency denotes the performance of an actual plate in comparison with an ideal plate. For an actual plate to approach ideal behaviour, it should have high tray efficiency. A high tray efficiency will result if (1) the liquid and gas were allowed to be in contact for sufficiently long time (2) the interfacial surface between the liquid and gas phases is large and (3) the intensity of turbulence is high. By providing a deep pool of liquid over the plate and a very high gas velocity, the above requirements can be achieved to a large extent. Deep pool of liquid over the plate guarantees long contact between the gas and the liquid, as the gas bubbles travel a greater distance through the liquid before disengagement. A high gas velocity through the liquid ensures thorough dispersion of the gas in the liquid and formation of smaller bubbles with high interfacial area per unit of gas volume.

However, a high gas rate and deep pool of liquid introduce certain complications in the operation of the column such as large pressure drop for the gas stream. The higher the pressure drop, the greater is the spacing to be maintained between plates in order to provide sufficient hydraulic head necessary to facilitate liquid down-flow from plate to plate. The level of liquid in the downcomer assumes an elevated position to assist flow of liquid from the upper plate (the low pressure region) to the plate below (the high pressure region). With further increase in gas or liquid rate, the liquid level in the downcomer rises further, forcing the liquid to enter the upper tray. In the worst scenario, the liquid will fill the entire space between the trays, resulting in a condition known as *flooding*. This may cause the liquid to exit through the gas outlet at the top of the column. High gas pressure drop also necessitates high power requirement for gas circulation. *Entrainment* of the liquid is another problem associated with high gas rate. Small droplets of liquids will be carried away by the gas stream into the plate above, thereby reducing the concentration change brought about by mass transfer; consequently, the plate efficiency also gets reduced. *Priming* is an exaggerated form of entrainment that is experienced at high gas velocities when gas and liquid which show a tendency to foam are handled in the column. When priming occurs, the entire space between the plates is occupied by the foam and the liquid gets carried away to the plate above.

To eliminate unduly high pressure drop, generally, the column is not to be operated with a very high gas rate. At the same time, if the gas rate is very small, the bubble size may become large and consequently, the interfacial area and the intensity of turbulence may get reduced. Also, the liquid will flow relatively unagitated and may pass over the tray without being contacted by the gas. All these affect the mass transfer process adversely and the tray efficiency therefore, will be low. In sieve trays, too low a gas rate may also result in conditions known as *weeping* in which the liquid drain through the perforations rather than flow across the plate into the downcomer and *dumping* in which the entire liquid flow down through perforations and none reaches the downcomer. On the other hand, if the liquid rate is low, the gas rising up through the perforations may push the liquid away without allowing proper contact between the gas and the liquid, a condition known as *coning*.

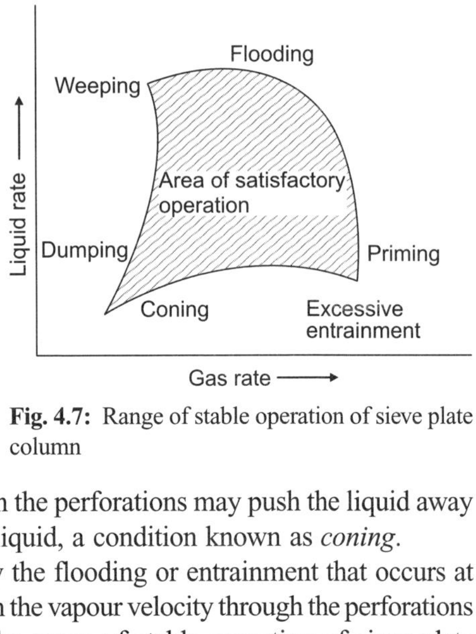

Fig. 4.7: Range of stable operation of sieve plate column

The stable operating range of a sieve plate is limited by the flooding or entrainment that occurs at the high vapour velocities and by weeping and dumping when the vapour velocity through the perforations falls below a certain minimum value. Figure 4.7 shows the range of stable operation of sieve plate columns. The ratio of maximum gas velocity to the minimum is known as the *turndown ratio* and is a measure of the flexibility in column operation. As evident from Table 4.1 sieve plates operate in a narrow range of vapour velocities in comparison with other plates.

Packed Columns

Packed columns are vertical columns (Fig. 4.8) packed with materials that provide large interfacial area for mass transfer on the one hand and allow the liquid to flow as a thin film on the other. These are used for continuous contact of gas and liquid in both cocurrent and countercurrent flow. The liquid which is distributed over the packing by means of liquid distributors trickles down through the bed.

The gas flows upward through the open channels not occupied by the packing or the liquid. In the packed columns, there are no identifiable stages to provide equilibrium between gas and liquid. The gas and liquid are in continuous contact and undergo change in concentrations continuously as against plate columns in which the transfer is carried out in stages.

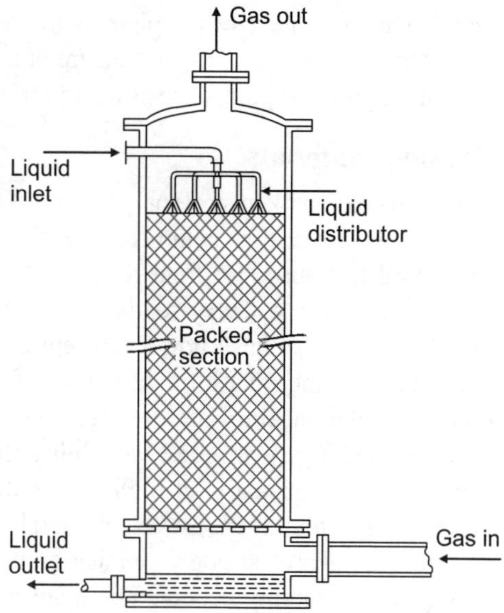

Fig. 4.8: Packed bed absorption column

Liquid distributors

For proper distribution of liquid, different types of liquid distributors are in use, two of which are shown in Fig. 4.9. Figure 4.9a shows a system of troughs and weirs on which the liquid admitted is subdivided into a number of separate streams and Fig. 4.9b shows perforated ring type distributor. Horizontal trays fitted with downcomer tubes for liquid flow, spray nozzles, etc. are other types of distributors in common use. For handling liquids containing suspended solids, spray nozzles are not recommended because they may get clogged by the solid particles. Perforated ring type distributors are used when gas rate is relatively high and the liquid rate is low. Even open ended pipes can be used as liquid distributors with the liquid directed towards certain targets laid on the packing.

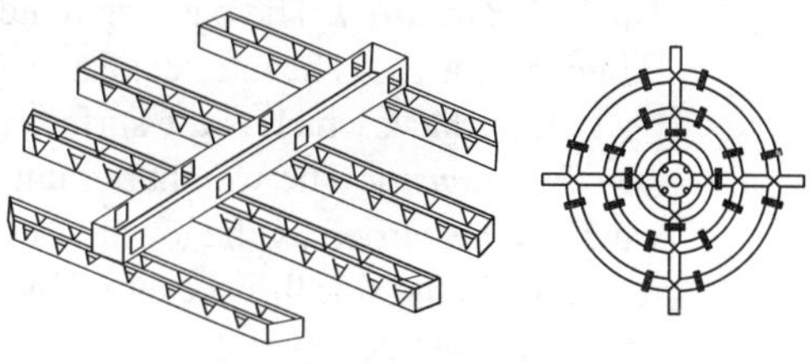

(a) (b)

Fig. 4.9: Liquid distributors (a) Weir-trough distributor (b) Ring type distributor

Liquid redistributors

The liquid even after proper distribution at the top of the column, shows a tendency to flow towards the tower wall. The liquid tends to flow along localized paths leaving much of the packing surface dry or covered with stagnant film. This effect is known as *channelling*. In towers of moderate size, channelling can be minimized by keeping the diameter of the tower at least 8 times the packing diameter. In large towers, liquid redistribution is necessary to eliminate channelling completely. Therefore, in towers with depth of packing more than about 6 m, it is customary to divide the packing into several sections each provided with liquid redistributors which receive the liquid falling from the upper section and distribute evenly on the packed section below.

Packing supports

The column is provided with a gas inlet and a distributing space below the packing support. The packing is supported at the bottom of the tower by metal screen or grid, below which the gas is admitted. The screen type of packing support is generally corrugated to give strength, with large open area so that flooding does not occur at the support. A grid made up of relatively widely spaced bars on which a few layers of large Raschig or partition rings are stacked constitute a simple form of support. The gas-injection plate packing supports are designed to provide separate passageways for gas and liquid. This is achieved by a plate as the one shown in Fig. 4.9c. The gas inlets to the bed are at a higher level than the liquid exit points as shown in the figure.

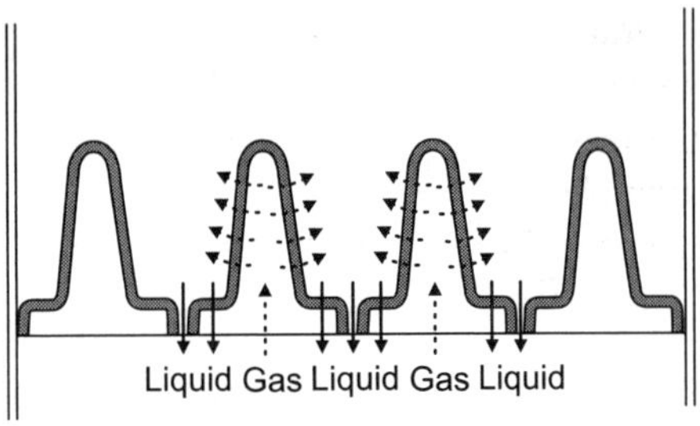

Fig. 4.9: (c) Gas-injection plate packing support

Packing materials

Lumps of coke, broken stone or quartz were the packings used in old generation columns. These materials give very small free space for fluid flow and have high bulk density that lead to heavy loads on the tower and its foundations. These have now been replaced by special fabricated packings that offer 70 to 90 % free space and very large interfacial area for gas-liquid mass transfer. Metal packings are preferred because of their superior strength and remarkable wettability. Ceramic packings with good wettability though poor in strength, are favoured for use in corrosive environments particularly at high temperatures, where plastics cannot be used. These are useful for handling most liquids except alkalies and hydrofluoric acid. Carbon rings are useful for all applications that do not involve strongly oxidising atmosphere. Plastic packings are inexpensive and relatively strong, but wettability is poor. These are not recommended for use with certain organic solvents and with oxygen-bearing gases at moderately high temperatures, as they may get deteriorated under these conditions. The desirable properties of tower packings are:

1. *Large surface area*: The packing materials should provide large interfacial surface between the liquid and the gas. The interfacial area for mass transfer will normally be smaller than the specific packing surface, which is the surface area of packing per unit volume of packed surface.
2. *Low bulk density*: The weight per unit volume of tower packing should be low so that the total weight to be carried by the tower is small. Also, in the case of random packings, this will ensure reduction in the side thrust exerted by the materials on the walls and therefore, saving in the cost of tower construction.
3. *Large free cross section*: The packing material should provide large free area for flow of gas so that the frictional loss and thereby the power requirement for gas circulation are reduced. A small free cross section for flow leads to high velocity for a given throughput of the gas and make the tower more prone to flooding and loading.
4. *Good wetting characteristics*: The packing should have good wetting properties.
5. *Small liquid hold up*: With the exception of absorption involving slow chemical reaction or absorption of a sparingly soluble gas, generally it is desirable that the packing material provide low liquid hold up in the tower.
6. *Chemical inertness*: Packing materials should not react chemically with the materials being handled in the column.
7. *Good structural strength*: The packing materials should have reasonable structural strength to facilitate easy handling and installation.
8. *Resistance to corrosion*: The material should be corrosion resistant.
9. *Low cost*: The materials should not be expensive.

The tower packings can be broadly classified into two groups: random packing and regular packings (Fig. 4.10).

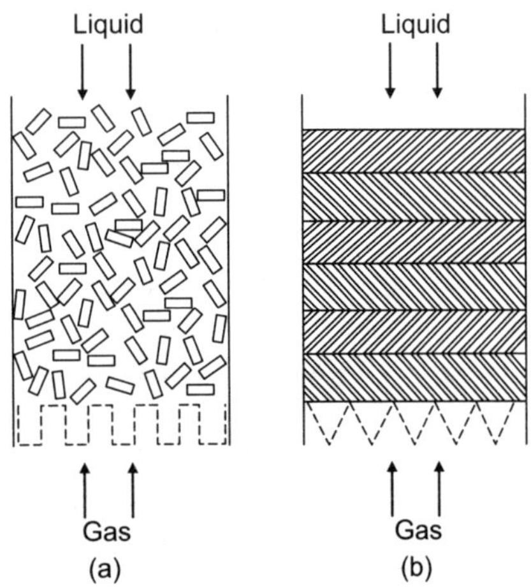

Fig. 4.10: (a) Random packing (b) Regular packing

Random Packings. Figure 4.11 illustrates random packings that are in common use. Random packings are usually dumped into the tower. The advantage of random packing is that by dumping into the tower, the mass becomes homogeneous and the tendency for channelling of the gas and liquid streams is minimized. To avoid breakage during packing, usually the empty tower is first filled with water before the packing pieces are dropped into it.

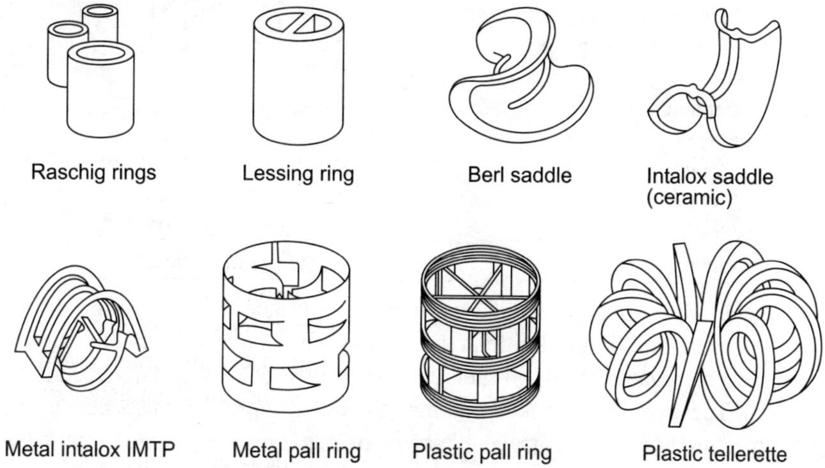

Fig. 4.11: Random packings

The most popular random packings are the Raschig rings, which are hollow cylinders with height equal to the diameter with size ranging from 6 mm to 100 mm. Raschig rings are made of glass, carbon, ceramic materials, or metal. Berl and Intalox saddles are packing materials shaped in such a way that they adopt a random arrangement on dumping with no lines of contact between adjacent pieces. These packings though expensive, are more efficient than rings and cause a lower pressure drop. They are made of chemical stoneware or plastic and are available in sizes ranging from 6 to 75 mm. Lessing rings and partition rings are also used as random packing in many applications.

Raschig rings and Berl saddles are now replaced by specialised metal and plastic rings that provide more surface area, higher flow capacity and lower pressure drop. Pall rings, metal Bialecki rings, ceramic Intalox saddles, etc. belong to such specialized packings. Random packings of lattice-work design known as "high-efficiency" packings have been introduced recently, which offer still higher interfacial area per unit volume and lower pressure drop per unit height than the above types of materials. These include Cascade Mini-rings (metal, plastic and ceramic), Nutter rings (metal, plastic), Hiflow rings (metal, plastic and ceramic), Levapak (metal), tri-packs (metal). Plastic and ceramic Super Intalox saddles are also available. Metal Intalox saddles-IMPT (Intalox Metal Tower Packing) have a very open structure. The most recent innovations in random packings – Fleximax, Raschig super-rings – promote turbulence and even wetting of the packing, both contributing to maintain lower pressure drop and high mass transfer efficiency. Table 4.2 lists the properties of common random packings.

Regular packings. Regular packings, though slightly more expensive than random packings, offer the advantage that they allow greater flow rates and low pressure drop for gases. Regular packings are either stacked packing or structured packings. Large diameter Raschig rings can be stacked in towers and serve as regular stacked packings. When it is desired to have large free space for gas or liquid flow as in cooling towers, wood grids or hurdles are used as packing. In comparison with random packings, the chances for channelling of liquid is less in stacked packing since the flow channels are vertical. Flooding velocities for regular or stacked packings are generally considerably greater than those for random packing.

Table 4.2: Properties of random packings of nominal size 25 mm

Type of packing	Material	Bulk density	Bed porosity	Specific surface, m^2/m^3
Raschig rings	Ceramic	721	0.68	190
Raschig rings	Metal (wall thickness, 0.8 mm)	620	0.92	207
Pall rings	Metal (mild steel)	480	0.94	210
Pall rings	Plastic (polypropylene)	88	0.90	207
Berl saddles	Ceramic	721	0.68	260
Intalox saddles	Ceramic	670	0.73	256
Super Intalox	Plastic	96	0.90	207
Intalox	Metal	224	0.967	135F
Hy-Pak	Metal	304	0.96	177
Tellerettes	Plastic		0.93	190
VSP rings	Metal		0.975	200

The structured regular packings provide large interfacial surface at low pressure drops. These may be broadly classified as knitted or non-knitted type, both arranged in a segmented form or spiral form. In the spiral form, the corrugated sheets of metal gauze, knitted or woven wire screens are rolled as a fabric into cylinders. These are then stacked one over the other to provide the necessary bed height. In the rigid type of structured packing, corrugated sheets of metal or plastics are arranged to form open channels (Fig. 4.12). Most common ones are made of sheets of perforated corrugated metal with adjacent sheets arranged so that liquid spreads over their surface while vapour flows through the channels formed by the corrugations. The porosity of corru-gated structured packings ranges from 0.93 to 0.97 and specific surface area from 200 to 250 m^2/m^3. The Sulzer BX packing, fabricated from metal gauze provides a specific surface area of 500 m^2/m^3 with a porosity of 0.90. Though the capacity and efficiency of the structured packings are high, the capital investment is very high when compared with other packings.

Fig. 4.12: Structured packing

Pressure drop in packed columns

Since the cost of gas circulation is a major operating cost, due consideration should be given to the pressure drop in packed columns when we make a choice of packing materials for specific applications. Usually the pressure drop is less if the packings are of large sizes, but the height of packed section will be greater, as the interfacial surface is correspondingly reduced. For different types of same-size packings, the pressure drop is less for the one which gives larger void fraction.

Figure 4.13 illustrates the pressure drop through the column as a function of liquid and gas rates. The pressure drop increases as square of the velocity of gas (or as $G^{1.8}$, to be more accurate) at low gas rates irrespective of the liquid rate and the curve of ΔP vs G is linear on logarithmic coordinates as shown in Fig. 4.13. However, at high gas rates, the curves bend up-wards slightly and the pressure drop rises more quickly almost as $G^{2.5}$. For the region MN in the figure, the liquid flow interferes with the gas flow and the liquid holdup in the bed steadily increases. As the free space available for flow of gas steadily diminishes, the pressure drop also increases in this region and the curve begins to deviate from the linear behaviour. This condition represented by the region MN is known as *loading*. In this condition, a layer of liquid may appear above the packing or the liquid may fill the tower transforming the system into gas dispersed – liquid continuous one and slugs of foam may rise rapidly through the packing. At the loading point, the straight line representing the variation of ΔP with flow rates passes into a curve as the curve along the line *MM* in the figure.

For the gas rate corresponding to *flooding*, as indicated by the line *NN*, there is very steep increase in ΔP for a given liquid rate. The curve of ΔP versus gas flow rate becomes vertical above the flooding point. Above this gas rate, the liquid cannot flow downward, and is pushed up by the effluent gas. *The flooding point can be defined as the gas velocity at which a liquid layer forms at the top of the packing, and the liquid hold up in the tower increases abruptly.* Since the tower cannot be operated in a flooded condition, the gas and liquid rates should be below those corresponding to flooding.

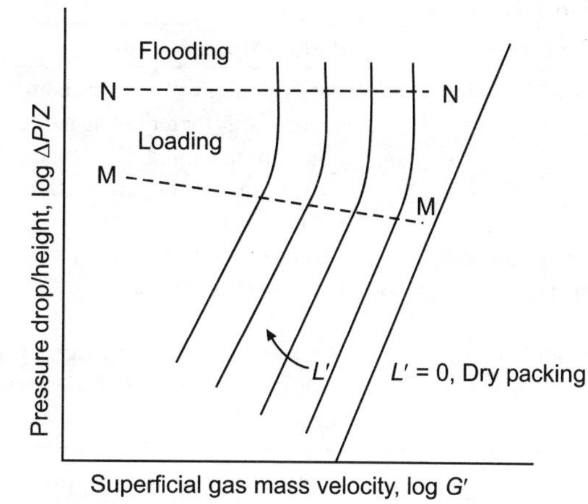

Fig. 4.13: Pressure drop through packed columns

If the liquid flow rate is high, the loading point (*M*) and the flooding point (*N*) are reached at a lower gas rate than those reached at a low liquid rate through the column. Since a reasonable liquid holdup promotes interfacial contact, the column is best operated in the region *MN*. The optimum economic gas velocity should be arrived at by an economic balance between the cost of power and the fixed charges on the equipment and it is found to be about one half of the flooding velocity.

The flooding characteristics of a packing are important in the design of columns, as they determine the allowable gas and liquid rates which in turn determine the minimum diameter of the column for a given gas or liquid rate. In some cases, excessive entrainment may occur at gas rates well below flooding rates and the column capacity is limited by entrainment.

Packed Columns versus Plate Columns

The relative merits and demerits of plate towers and packed towers are summarised in Table 4.3.

Table 4.3: Comparison of plate and packed column performance

Feature	Plate column	Packed column
Pressure drop	Gas pressure drop is high.	Gas pressure drop is low for columns designed for similar duty.
Liquid holdup	High	Low holding time, suitable for handling liquids that deteriorate at high temperatures.
Corrosive fluids	Costly materials of construction such as corrosion resistant alloy steel, glass, etc. are required for the column.	Simple and cheap construction.
Flooding	Occurs at relatively high liquid rates.	Occurs at low liquid rates.
Liquid/Gas ratio	More suitable for handling small L/G values.	Preferred for high L/G values.
Cleaning	Easily cleaned of deposited solids by providing manholes.	Get clogged by accumulated sediments and require costly cleaning and refilling.
Side streams	Easily removed.	Removal is relatively difficult.
Channelling	Not encountered.	Occurs due to the uneven distribution of liquid and gas over the tower cross section.
Foaming	Foaming may occur because gas is bubbled through the liquid.	Not affected.

(Contd.)

Table 4.3 contd.

Feature	Plate column	Packed column
Liquid cooling	Easily done as plates can be provided with cooling coils for removing the heat of absorption from the liquid. Cooling can be done by passing the liquid leaving through external coolers also.	Liquid cooling is relatively difficult.
Temperature change	Damage to the tower due to temperature change is not serious.	Can damage the tower as the fragile packing such as ceramic and graphite tend to be crushed under temperature changes.
Floor loading	The total weight of the tower is generally less than that of the packed tower designed for the same duty.	With the exception of plastic packed towers, the total weight is generally high.

The relative merits of the plate tower and packed tower for a given purpose are properly determined only by comparison of the cost figures resulting from the detailed design for each type.

Agitated Vessels

Agitated vessels generate swarms of bubbles which provide enormous surface per unit volume of gas-liquid dispersion. As a result, large values of liquid-side volumetric mass transfer coefficient are achieved. They are usually operated batch-wise and are limited to the equivalent of one theoretical plate because of the thorough mixing of the liquid. Agitated vessels find wide application in chemical industry for batch oxidation, chlorination, hydrogenation and aerobic fermentation. Figure 4.14 illustrates an agitated vessel for gas-liquid contacting. It is equipped with a motor driven stirrer and a gas sparger. The liquid depth is approximately the same as the tank diameter. The impeller may be a flat paddle, marine propeller, or a turbine. The sparger may be a perforated pipe ring or a single tube delivering the gas under the impeller near the tip. The impeller is generally located at one-third of the distance from the bottom to the liquid level. Vertical baffles are attached to the walls to increase turbulence and minimize swirling and vortex formation at the liquid surface.

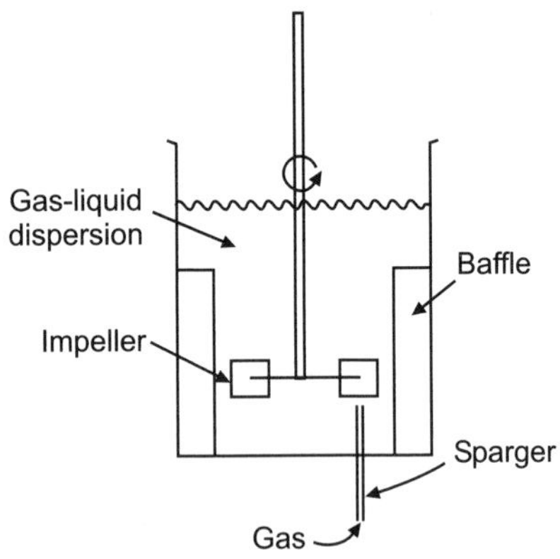

Fig. 4.14: Agitated vessel for gas-liquid contact

The gas is admitted at a very low superficial velocity, generally between 0.003 m/s and 0.08 m/s. The gas is dispersed as small bubbles by the impeller, the bubble size being determined by the impeller tip speed. At moderate to high impeller tip speeds (2.5 to 15 m/s), the bubble size is independent of the design of the sparger. Mean bubble size varies between 1 mm and 5 mm in aqueous solutions. The bubble size increases with viscosity of the liquid. Gas holdup increases with agitation and gas flow and is independent of the vessel diameter. However, at low agitation the gas holdup increases with tip speed and power input per unit volume. The higher the gas holdup, the larger will be the volumetric mass transfer coefficient k_La as the holdup directly affects the interfacial area a.

Bubble Columns

Bubble columns, or sparged vessels are columns constructed with a large ratio of height to diameter. They provide large interfacial surface per unit volume of the gas-liquid dispersion by generating small bubbles. For small diameter vessels, the sparger located at the bottom may simply be an open tube through which the gas is issued. For vessels with diameter exceeding about 0.3 m, a better gas distribution is ensured by means of several orifices through which gas is issued. These are widely employed as air-agitated fermenter. The liquid is agitated by the sparged gas and not by any external device. The bubble columns can be operated with no liquid flow or with either cocurrent or counter current flow of gas and liquid (Fig. 4.15).

The gas-pressure drop is high in tall columns and as a result, the cost of compression of the gas stream represents the major operating cost in such columns. The main advantage of the bubble column is that it is less expensive than other gas-liquid contact equipment. The bubble swarm provide enormous surface for gas-liquid contacting. There is rapid liquid back-mixing and the liquid concentration is uniform from top to bottom even in columns with a height to diameter ratio of 12 or more. Since it is difficult to prevent back-mixing, the column can never be equivalent to more than one or two equilibrium stages. The gas pressure drop is relatively high and the gas throughput is very small. The bubble column is recommended for the absorption of solutes with very low solubility and for the absorption of solute involved in very slow chemical reactions in the liquid phase.

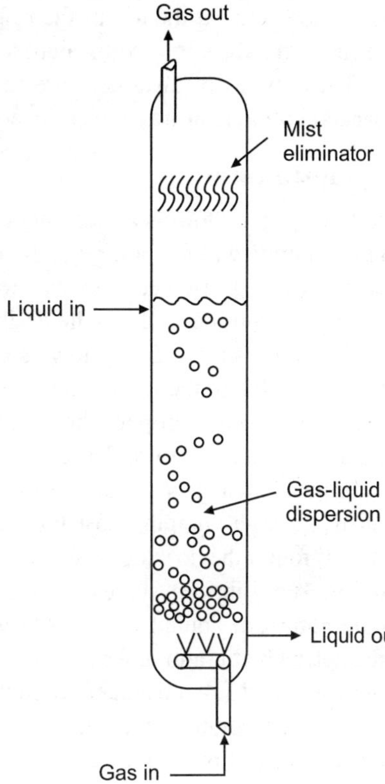

Fig. 4.15: Bubble column for gas-liquid contact

The Centrifugal Absorber

The principle of a centrifugal absorber is illustrated in Fig. 4.16. A set of concentric rings attached to a rotating shaft intermeshes with another set of stationary concentric rings. The liquid that is admitted at the centre of the plate moves up though the first ring and splashes over to the baffle and then falls into the trough between the rings. It then moves up through the second ring and in a similar way passes from ring to ring through the unit. The gas is admitted at the top if cocurrent flow is desired and at the bottom if the countercurrent flow is desired. The gas is admitted at the top if

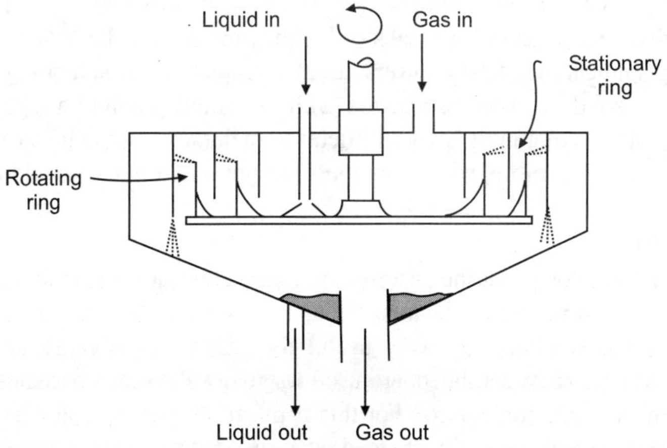

Fig. 4.16: Centrifugal absorber

cocurrent flow is desired and at the bottom if the countercurrent flow is desired. Most of the mass transfer between the gas and liquid is found to occur at the point where the gas comes into contact with the liquid spray at the top of the rings. Very high mass transfer rate is achieved even with small rates of rotation. Absorption equivalent to several equilibrium stages can be attained in small units.

These types of absorbers are recommended when the head room required for accommodating packed columns or tray towers is not available or if the residence time is to be very short.

Spray Towers

Industrial spray towers are usually empty towers equipped with a number of nozzles (Fig. 4.17). The gas is admitted at the bottom and the liquid is sprayed at the top. The nozzles disperse the liquid as a coarse spray distributed over the cross section of the tower. The size of the liquid drops depends on the design of the nozzles and on the pressure of the liquid supply. Spray towers are advantageous when the solute is highly soluble in the liquid phase. The main advantage of these devices is the low pressure drop for the gas phase. The major drawback is that the liquid droplets tend to coalesce as they move down and consequently the interfacial area gets considerably reduced. Another limitation is the relatively high pumping cost for the liquid owing to the large pressure drop through the nozzles. Also, the unit is equivalent to merely than one or two equilibrium stages in terms of separation. The chances for entrainment of liquid by the gas leaving the column are high. Unless the diameter length ratio is very small, the gas will be thoroughly mixed by the spray and full advantage of counterflow cannot be attained.

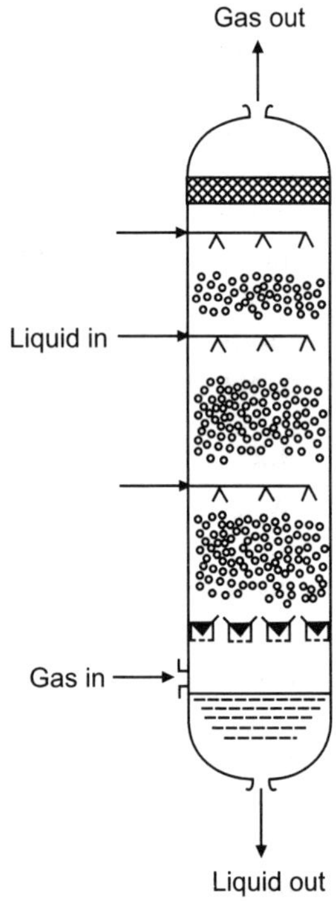

Fig. 4.17: Spray tower

For short columns, the efficiency of spray chamber is found to be very close to that of packed tower. For height of the column greater than, say 1.2 m, the efficiency drops rapidly due to the coalescence of liquid droplets. Although there is considerable turbulence in the gas phase, the circulation of liquids within the drops is very poor and consequently the resistance of the liquid film is generally high. Spray towers are therefore useful when the absorption is gas-phase controlled as for example, absorption of ammonia in water. Spray towers are advantageous in some situations where packed columns are unsuitable. For example, if the gas stream undergoes cooling from a very high temperature to a very small value, the packings may disintegrate due to thermal shock. Spray towers may be employed for such applications. Spray towers are also employed for scrubbing dust-laden gases and gases which produce solid deposits on reaction with the liquid. Though a high interfacial surface is available at the top of the column, it is very difficult to maintain it at the lower portion of the column because of coalescence. Therefore, spray chambers are not much in use as industrial absorbers.

Venturi Scrubbers

Venturi scrubbers are employed for the absorption of gases by liquids containing suspended particles. These are similar in construction to ejectors. Because they have relatively open design compared to other scrubbers, venturi scrubbers are very useful for absorption of gases and for the removal of particulate matter. Thus venturi scrubbers are used for removal of dust particles from gases and find wide applications in air pollution control. For this application, packed columns and spray towers are unsuitable as the gas passages get easily clogged by suspended particles. An example is the absorption of sulphur dioxide from furnace gases in slurries of limestone, lime or magnesia, where a chemical reaction occurs between the solute and the suspended particle. In short, venturi scrubbers are particularly useful (i) when scaling will be a problem, (ii) when the gas stream contains a high concentration of particles and the particles have a tendency to plug the openings and (iii) when the gaseous contaminant is very soluble or chemically reactive with the liquid.

Venturi scrubber consists of three sections: a converging section, a throat section and a diverging section (Fig. 4.18). The inlet gas stream enters the converging section and, as the area decreases, the gas velocity increases. Liquid is introduced either at the throat or at the entrance to the converging section. Figure shows a unit employing cocurrent flow of gas and liquid in which the gas is drawn into the throat of a venturi by a stream of liquid sprayed into the convergent duct section. The absorption occurs mainly in the throat section as the inlet gas stream mixes with the fog of tiny liquid droplets. These types of scrubbers can handle inlet gas flows as large as 90000 m^3/h.

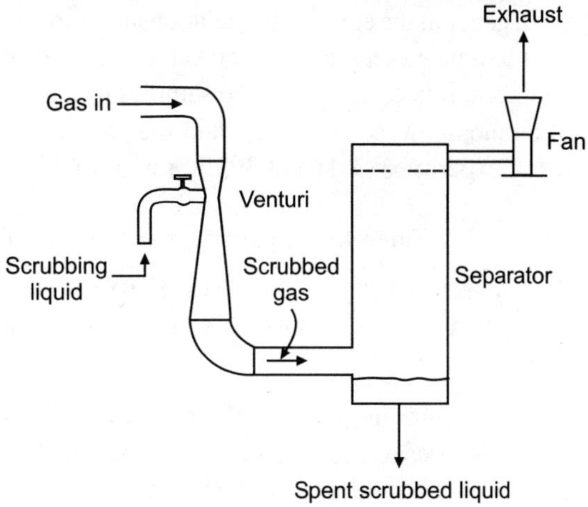

Fig. 4.18: Venturi scrubber system

4.2 ABSORPTION EQUILIBRIUM: SOLUBILITY OF GASES

When a gas mixture is brought in contact with a liquid and if the two phases are not in equilibrium, transfer of solute occurs between the gas and liquid. The rate at which this transfer occurs depends on the driving force for mass transfer which is determined by the extent or the degree of departure from equilibrium between the phases. Therefore, for the design and operation of absorption systems, it is essential to have a good knowledge of the equilibrium characteristics of gas-liquid systems.

The equilibrium between phases is governed by the phase rule which states that $F = C - \pi + 2$, where C is the number of components constituting the system, π is the number of phases, and F is the number of degrees of freedom. By the number of degrees of freedom, we mean the number of variables that need be specified for defining the conditions of the system uniquely. The properties that are generally used to determine the conditions are the temperature, pressure and the composition of the phases. Consider that a solute gas is being absorbed from its mixture with an insoluble gas into a liquid solvent. As an example, consider the absorption of ammonia into water from a mixture of ammonia and air. Here, there are three components and two phases. Therefore, by the phase rule, the number of degrees of freedom is 3. If the absorption is carried out at a specified temperature and pressure, only one variable can be set arbitrarily. That is, if we specify the concentration of the gas phase, the composition of the liquid in equilibrium with the gas phase is automatically determined. It means that at constant temperature and pressure, there exists a definite relationship between the concentration of the solute in the gas phase (y_A) and the concentration in the liquid phase, which is in equilibrium with the gas (x_A). This relationship is conveniently represented graphically by means of the equilibrium curve. In the case of absorption equilibrium, the equilibrium curve is nothing but the solubility curve for gases in liquid. If a gaseous solute (say, ammonia) from a mixture with an insoluble gas (say, air) is allowed to come in equilibrium with a non-volatile liquid (say, water) at constant temperature and pressure, the resulting concentration of solute gas in the liquid (x_A) is the gas solubility when the gas phase concentration of the solute is y_A. The gas phase concentration can also be represented by the partial pressure of the solute in the gas $\bar{p}_A$.

In general, the equilibrium relationship is to be established experimentally. Consider a certain amount of ammonia gas, air and water taken in a closed container and shaken well at constant temperature until equilibrium is established. The samples of gas and liquid are withdrawn to measure the mole fraction of ammonia in the liquid (x_A) and the partial pressure of ammonia in the gas ($\bar{p}_A$). Table 4.4 gives a set of experimental data at 303 K for ammonia-air-water system.

Table 4.4: Equilibrium distribution of ammonia between air and water at 303 K

x_A	0.0126	0.0208	0.0309	0.0405	0.0503	0.0737	0.096	0.137	0.175	0.210	0.241	0.297
$\bar{p}_A$, kPa	1.53	2.57	3.95	5.34	6.8	10.6	14.66	23.86	34.66	46.92	60.5	95.8

The equilibrium curve (solubility curve) for ammonia-air-water system shown in Fig. 4.19 is the graphical presentation of the experimental data given in Table 4.4. The solubility of the soluble components of the gas mixture in the liquid is not appreciably influenced by the presence of insoluble components such as air. Therefore, the same curve will represent the solubility of pure ammonia in water at 303 K, provided that the *y*-axis represents the total pressure of ammonia gas in contact with the saturated solution of ammonia in water.

The solubility of components of a gas mixture is not affected by the total pressure, for pressures up to about 5 bar. However, temperature influences the solubility of gases in liquids considerably. Increase in temperature results in the decrease of the equilibrium solubility. For a given partial pressure of ammonia in the gas, the saturation concentration of ammonia represented by its mole fraction in the liquid is higher at lower temperature than that at a higher temperature.

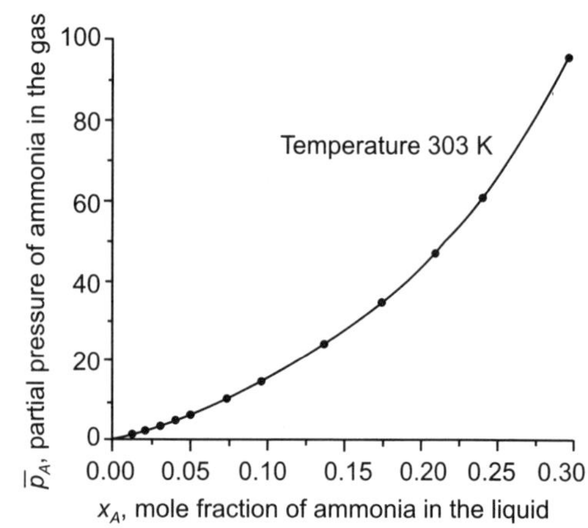

Fig. 4.19: Equilibrium solubility of ammonia in water at 303 K

Thus the equilibrium curve for absorption of ammonia in water at 293 K will lie below that at 303 K as shown in Fig. 4.20. Alternatively, the partial pressure of gas in equilibrium with a specified liquid composition is high at higher temperature. For example, the partial pressure exerted by ammonia in the gas phase in equilibrium with a saturated solution of ammonia in water containing 24.1 mole % ammonia is 60.5 kPa at 303 K whereas it is 39.7 kPa at 293 K.

The effect of temperature on the equilibrium solubility of gases in liquids can be explained by thermodynamic principles that govern phase equilibrium. Generally, heat is evolved when gas dissolves in a liquid. If the temperature of a system in equilibrium is increased, those changes will occur in the system that results in the absorption of heat. Therefore, by increasing the temperature, the solubility of the gas in the liquid decreases. For this reason, the absorption towers are generally operated at low temperatures. Frequently it would be necessary to provide cooling coils in the tower to carry away the

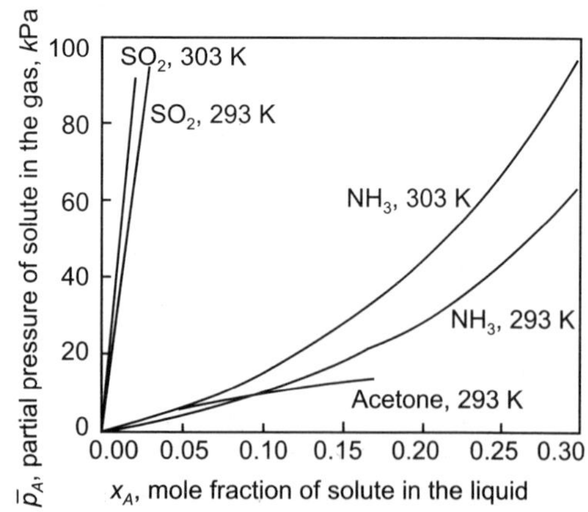

Fig. 4.20: Equilibrium solubility of gases in water

heat of absorption and prevent increase of temperature in the column. It may be noted that at pressures above 1 bar, the solubility of certain low molecular weight gases such as hydrogen, oxygen, nitrogen, methane, etc. in water increase with temperature above 373 K.

Solubility curves of different gases in water are shown in Fig. 4.20. Table 4.5 gives mole fractions of ammonia and sulphur dioxide in water at 293 K, that will develop equilibrium partial pressures of 10, 20, 40 and 60 kPa in the gas phase. For a given partial pressure of the solute gas, the equilibrium concentration in the liquid of a readily soluble gas is higher than that of a sparingly soluble gas as given in Table 4.5. Alternatively, if the partial pressure of a gas at a given liquid concentration is high, the gas is relatively insoluble in the liquid, whereas if the partial pressure is low, the gas solubility in the liquid is very high. That is, for gases which are highly soluble in water, the curves will be flat and lie closer to the x-axis representing the liquid concentration and for those sparingly soluble, the curves will be steep and will lie closer to the y-axis representing the gas concentration as shown in Fig. 4.20.

Table 4.5: Solubility of ammonia and sulphur dioxide in water at 293 K

	10 kPa	20 kPa	40 kPa	60 kPa
x_A (SO_2)	0.0035	0.0066	0.0123	0.018
x_A (NH_3)	0.10	0.161	0.240	0.291

Since, the departure from the equilibrium provides the driving force for absorption, the analysis of absorption operation cannot be carried out without the data on the gas-liquid equilibrium. For example, assume that a mixture of SO_2 and air in which the partial pressure of SO_2 is 40 kPa is brought into contact with pure water. The gas is passed as a continuous stream in contact with the liquid. SO_2 gets absorbed in water till the mole fraction of SO_2 in the liquid becomes equal to the equilibrium concentration $x = 0.0123$. If instead of pure water, the same gas were brought into contact with an aqueous solution of SO_2 with a mole fraction 0.0123, no absorption occurs because, the liquid and gas phases are having equilibrium compositions. The driving force in this case is zero because the phases are in equilibrium. If the partial pressure of SO_2 in the gas phase in contact with the liquid containing 0.0123 mole fraction SO_2 is less than 40 kPa, then the driving force for mass transfer is in the opposite direction and as a consequence, SO_2 is transferred from the liquid to the gas phase and the liquid gets stripped of SO_2.

The equilibrium data is generally obtained experimentally. However, for ideal solutions, it is possible to express the equilibrium relationships as simple mathematical relationships derived from the vapour pressures of the pure constituents known as Raoult's law. A similar relationship exists for even non-ideal solutions when the solute is present in very small proportions, which is known as Henry's law. Raoult's law and Henry's law are used to represent the equilibrium in the case of ideal solutions and dilute non-ideal solutions, respectively.

Ideal Solutions and Raoult's Law

For a solute gas which forms an ideal solution with a liquid solvent, the equilibrium relationship for the gas-liquid system can be established using the vapour pressure data. Ideal solutions are distinguished by the following characteristics:

(i) The intermolecular forces existing in the pure state of the components and in the solution should be of the same order of magnitude. This condition requires that the constituents of ideal solutions are chemically similar substances.

(ii) There is no volume change on mixing the components constituting the ideal solution. Or, the volume of the solution varies linearly with composition.

(iii) There is no change in enthalpy or temperature during mixing of the components to form an ideal solution. No heat is absorbed or evolved during the formation of an ideal solution from its constituents; the only heat effect is the heat of condensation that results when the gas dissolves in the liquid.

(iv) The constituents obey Raoult's law which states that the partial pressure exerted by the components in the gas phase is directly proportional to the concentration of the component in the liquid. That is,

$$\bar{p}_A = x_A P_A^S \quad \text{...(4.1)}$$

where x_A is the mole fraction of the component, $\bar{p}_A$ is the partial pressure and P_A^S is the vapour pressure of the pure substance.

Only substances which are similar in size, structure and chemical nature can be expected to form ideal solutions. Substances such as optical isomers of organic compounds meet these requirements. However, for practical purposes, solutions formed by adjacent members of the homologous series of organic compounds can be treated as ideal solutions. For example, mixtures of benzene and toluene, n-heptane and n-hexane, propane and butane, methanol and ethanol, etc. can be treated as ideal solutions. It is clear that for ideal solutions, the equilibrium solubility expressed as mole fraction (x_A) is independent of the properties of the solvent. When more than one component of a multicomponent mixture are soluble in the liquid, the solubility of a component is not influenced by the other constituents if ideal solution is formed on dissolution. The concept of ideal solutions is therefore of great advantage when dealing with multicomponent systems. For example, the absorption of paraffinic hydrocarbons such as propane and butane in hydrocarbon oil takes place independently and the equilibrium can be represented accurately by assuming ideal solution behaviour.

Dilute Solutions: Henry's Law

Real solutions deviate significantly from the ideal behaviour expressed by the Raoult's law. The actual partial pressure of ammonia over aqueous solutions at 293 K and the partial pressure predicted using Raoult's law at 293 K are plotted in Fig. 4.21. The difference between these two is a clear indication of the marked deviation of ammonia-water system from ideal behaviour. It is also obvious from Fig. 4.21 that when the concentration of ammonia in the solution tends to zero, the variation of partial pressure with mole fraction is linear so that we can write

$$\bar{p}_A = x_A K_A \quad \text{...(4.2)}$$

The above relation is the mathematical statement of Henry's law which is found to be applicable to almost all real solutions in the dilute region. It states that the partial pressure of the solute in the gas phase in equilibrium with the liquid is directly proportional to the concentration in the liquid and the proportionality constant K_A is known as the Henry's law constant.

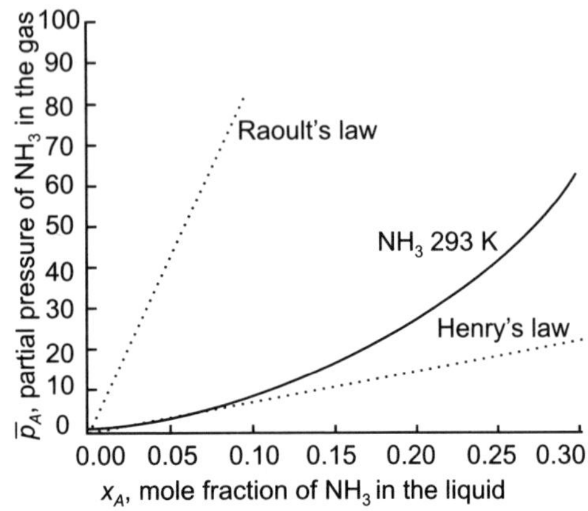

Fig. 4.21: Partial pressure of NH_3 predicted by Raoult's law and Henry's law

Assuming the gas phase to be ideal, the partial pressure can be expressed as the product of the mole fraction in the gas phase and the total pressure so that Eq. (4.2) becomes

$$y_A P = x_A K_A \quad \text{...(4.3)}$$

or

$$y_A = x_A \frac{K_A}{P} \quad \text{...(4.4)}$$

$$y_A = m x_A \quad \text{...(4.5)}$$

where m is a constant. As concentrations increase, the actual partial pressure of the gas deviates markedly from that predicted by the Henry's law. Essentially all liquids will obey Henry's law as solute mole fraction tends to zero, but many will deviate from the law above 0.01-0.02 mole fraction. And almost all liquids deviate above 0.1 mole fraction. But in some exceptional cases, Henry's law is found to be obeyed quite well up to $x_i = 0.5$. The deviation from Henry's law at higher concentrations may be due to the chemical interaction of the solute gas with the solvent liquid, or due to electrolyte dissociation as is the case with ammonia in water.

Since the solubility of the gas in a liquid is generally very low, the mole fraction of a gas in a saturated liquid solution is very small. The solute gas obeys Henry's law; therefore its partial pressure in the gas would be directly proportional to its mole fraction in the liquid, the proportionality constant being the Henry's law constant. In other words, the mole fraction or the solubility of the gas in the liquid is proportional to the partial pressure of the gas over the liquid. The molar solubility of the gas in that case is the inverse of the Henry's law constant multiplied by the partial pressure of the gas above the solution.

$$x_A = \frac{\overline{p}_A}{K_A} \qquad \ldots(4.6)$$

The Henry's law constant is temperature dependent and it generally increases with increasing temperature. This is in line with our earlier observation that the solubilities of gases generally decrease with increasing temperature. Henry's law constant is found to be independent of total pressure up to a total pressure of about 500 kPa. Table 4.6 gives the Henry's law constant values for gases dissolved in water.

Table 4.6: Henry's law constants for gases in aqueous solutions*
(Henry's law : $y = mx$; values given are $m \times 10^{-4}$)

T, K	N_2	H_2	CO	O_2	CH_4	C_2H_6	CO_2	H_2S
273.2	5.3588	5.8653	3.5658	2.5832	2.2691	1.2764	0.07375	0.02715
283.2	6.7668	6.4427	4.4775	3.3125	3.0086	1.9146	0.10535	0.03718
293.2	8.1445	6.9188	5.4297	4.0621	3.8089	2.6642	0.14385	0.04893
303.2	9.3601	7.3848	6.2806	4.8118	4.5484	3.4645	0.18842	0.06169
313.2	10.5352	7.6076	7.0505	5.4196	5.2676	4.2850	0.23603	0.07547

* [National Research Council, International Critical tables, Vol III, McGraw Hill, NY (1929)]

Example 4.1: The Henry's law constant for oxygen in water at 298 K is 4.437×10^4 bar. Estimate the concentration of oxygen in aqueous solution in equilibrium with a gas phase in which the partial pressure of oxygen is 0.30 bar at 298 K.

***Solution*:** Equation (4.2) gives the equilibrium relationship between the gas and liquid when Henry's law is obeyed.

$$\overline{p}_A = x_A K_A \qquad \ldots(4.2)$$

Here, a gas phase with partial pressure of oxygen 0.30 bar is in equilibrium with an aqueous solution. The Henry's law constant is 4.437×10^4 bar. Therefore, the concentration of the aqueous phase in equilibrium with the gas is x_A, such that

$$x_A = \frac{\overline{p}_A}{K_A}$$

Substituting $\overline{p}_A = 0.30$ bar and $K_A = 4.437 \times 10^4$ bar, we get the mole fraction of oxygen in the liquid at equilibrium.

$$x_A = \frac{\bar{p}_A}{K_A} = \frac{0.30}{4.437 \times 10^4} = 6.76 \times 10^{-6}$$

Since the solution is very dilute, the molecular weight of the solution may be assumed to be the same as that of water. Thus the equilibrium solubility of oxygen may be expressed as 6.76×10^{-6} moles of oxygen per mole of solution. This is equivalent to $6.76 \times 10^{-6} \times 32$ kg oxygen per 18 kg solution or the equilibrium solubility is 1.2×10^{-5} kg oxygen per kg of solution or 12 parts per million parts (ppm).

(Answer)

Example 4.2: *n*-Heptane and toluene form ideal solutions. At 350 K, the vapour pressure of toluene is 34.8 kPa and that of *n*-heptane is 51.4 kPa. Determine the composition of the liquid and vapour in equilibrium at 350 K and 45 kPa.

***Solution*:** Ideal solutions obey Raoult's law. Raoult's law states that

$$\bar{p}_A = x_A P_A^S \qquad \text{...(4.1)}$$

where x_A is the mole fraction of the component **A**, $\bar{p}_A$ is the partial pressure and P_A^S is the vapour pressure of component **A** (*n*-Heptane). For component **B** (toluene), the Raoult's law gives

$$\bar{p}_B = x_B P_B^S = (1 - x_A) P_B^S \qquad \text{...(4.7)}$$

The total pressure is the sum of partial pressures $\bar{p}_A$ and $\bar{p}_B$. Adding the above two equations, we get

$$\bar{p}_A + \bar{p}_B = P = x_A P_A^S + (1 - x_A) P_B^S$$

Rearranging this equation, we get

$$x_A = \frac{P - P_B^S}{P_A^S - P_B^S} \qquad \text{...(4.8)}$$

Dalton's law is applicable to the vapour phase on the assumption that the vapour behaves as an ideal gas. It states that

$$\bar{p}_A = y_A P$$

Combining this with Eq. (4.1), we get

$$y_A = \frac{\bar{p}_A}{P} = \frac{x_A P_A^S}{P} \qquad \text{...(4.9)}$$

We have P_A^S = 51.4 kPa, P_B^S = 34.8 kPa, P = 45 kPa. Substituting in Eq. (4.8),

$$x_A = \frac{P - P_B^S}{P_A^S - P_B^S} = \frac{45 - 34.8}{51.4 - 34.8} = 0.614$$

Substituting $x_A = 0.614$ in Eq. (4.9), we get

$$y_A = \frac{x_A P_A^S}{P} = \frac{0.614 \times 51.4}{45} = 0.702$$

At equilibrium the liquid contains 61.4 % *n*-Heptane (mole fraction = 0.614) and the vapour contains 70.2% *n*-Heptane (mole fraction 0.702).

(Answer)

Example 4.3: Ammonia-air mixtures of varying concentration of ammonia are brought in contact with pure water and at equilibrium the gas and liquid phase compositions were measured at constant temperature. The partial pressure exerted by ammonia in the gas phase versus composition of aqueous solution in equilibrium at a pressure of 101.3 kPa are tabulated below.

$\bar{p}_{NH_3}$, kPa	1.6	2.4	4.3	6.7	9.3	22.1	30.3
kg NH_3/100 kg water	2	3	5	7.5	10	20	25

Represent the equilibrium graphically as

(a) partial pressure $\bar{p}_{NH_3}$ versus liquid mole fraction x_{NH_3}

(b) mole fraction in the gas phase, y_{NH_3} versus liquid mole fraction, x_{NH_3}

(c) Y_{NH_3} (moles of ammonia in the gas phase per mole of air) versus X_{NH_3} (moles of ammonia in the liquid per mole of water).

***Solution*:** (a) The second row of the equilibrium data gives concentration of the liquid in kg of ammonia per 100 kg water, which can be readily converted to mole fraction, x_{NH_3}. For example,

$$\frac{2 \text{ kg NH}_3}{100 \text{ kg water}} = \frac{2/17 \text{ kmol NH}_3}{100/18 \text{ kmol water}} = \frac{2/17 \text{ kmol NH}_3}{(100/18 + 2/17) \text{ kmol solution}}$$

$$x_{NH_3} = \frac{2/17}{(100/18 + 2/17)} = 0.0207$$

The calculation is repeated for other values. The results are summarized in the table below.

x_{NH_3}	0.0207	0.0308	0.0503	0.0736	0.0957	0.1748	0.2093
$\bar{p}_{NH_3}$, kPa	1.6	2.4	4.3	6.7	9.3	22.1	30.3

The equilibrium curve is plotted in Fig. 4.22a with $\bar{p}_{NH_3}$ as the ordinate and x_{NH_3} as the abscissa.

(b) Partial pressure can be converted to mole fraction y_{NH_3} as $y_{NH_3} = \dfrac{\bar{p}_{NH_3}}{P}$

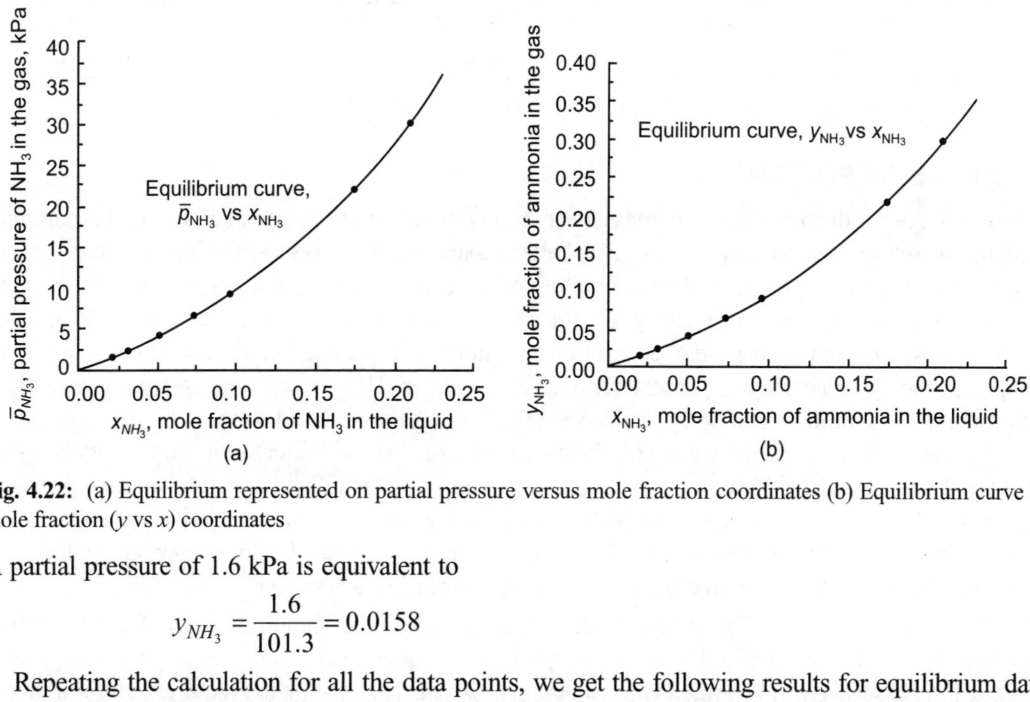

Fig. 4.22: (a) Equilibrium represented on partial pressure versus mole fraction coordinates (b) Equilibrium curve on mole fraction (*y* vs *x*) coordinates

A partial pressure of 1.6 kPa is equivalent to

$$y_{NH_3} = \frac{1.6}{101.3} = 0.0158$$

Repeating the calculation for all the data points, we get the following results for equilibrium data

x_{NH_3}	0.0207	0.0308	0.0503	0.0736	0.0957	0.1748	0.2093
y_{NH_3}	0.0158	0.0237	0.0424	0.0661	0.0918	0.2182	0.2991

The equilibrium curve on mole fraction coordinates is plotted as shown in Fig. 4.22b

(a) x_{NH_3} represents moles of ammonia per mole of aqueous solution and $(1 - x_{NH_3})$ represents the moles of water per mole of solution. X_{NH_3} which denotes moles of ammonia per mole of water in the solution, can be written as

$$X_{NH_3} = \frac{x_{NH_3}}{1 - x_{NH_3}}$$

Similarly y_{NH_3} represents moles of ammonia per mole of ammonia-air mixture and Y_{NH_3} denotes the moles of ammonia per mole of air in the mixture,

$$Y_{NH_3} = \frac{y_{NH_3}}{1 - y_{NH_3}}$$

For example, consider the data set (x_{NH_3} = 0.2182, y_{NH_3} = 0.1748). The corresponding values for X_{NH_3} and Y_{NH_3} are calculated as

$$X_{NH_3} = \frac{0.2182}{1 - 0.2182} = 0.2791 \text{ and}$$

$$Y_{NH_3} = \frac{0.1748}{1 - 0.1748} = 0.2118$$

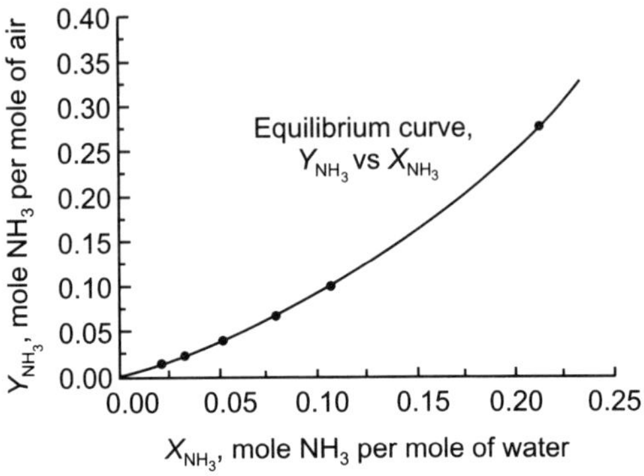

Fig. 4.22: (c) Equilibrium curve on mole ratio (*Y* vs *X*) coordinates

The calculations are repeated for other points in the table and the following results are obtained:

X_{NH_3}	0.0212	0.0318	0.0529	0.0794	0.1059	0.2118	0.2647
Y_{NH_3}	0.0161	0.0243	0.0443	0.0708	0.1011	0.2790	0.4268

The equilibrium curve on *Y* vs *X* is plotted as shown in Fig. 4.22c.

4.3 CHOICE OF SOLVENT

In absorption, or for that matter, in any mass transfer operation that makes use of an external separating agent, the search for a good solvent that gives better separation is always worthwhile. If the principal purpose of absorption operation is the removal of certain constituents from a gas mixture, the solvent that serves the purpose most effectively should be employed. However, such a choice will not be possible in cases where the purpose of absorption operation is the production of a specific solution. For example, in the manufacture of hydrochloric acid by absorbing HCl in water, the solvent is specified by the nature of the product and no choice is possible. The commonly used solvents for absorption are water, hydrocarbon oils and aqueous solutions of acids and bases. Water is the most widely used solvent as it is easily available and is least expensive. However, it is ineffective for absorption of certain gases. For example, for the absorption of benzene and toluene vapours from by-product coke oven gas, water is not effective and hydrocarbon oil is preferred as the solvent. In the following section, we discuss the factors to be considered for the choice of solvent for absorption.

(1) *Solubility of the gas*: The solubility of the gas in the liquid should be high. The higher the solubility, the larger is the driving force for absorption. Consequently, the rate of absorption is high and the size of the absorber will be small. Good solubility also means less solvent requirement for a given absorption. For ideal solutions, though the solubility expressed as mole fraction in the saturated solution is independent of the solvent, the solvent with the low molecular weight is preferred, if a choice is possible among various solvents which form ideal solution with the solute gas. The amount of solvent required in kilograms will be less for solvents having low molecular weight.

(2) *Chemical reactivity*: The solubility of the solute gas in the liquid is generally high, if the solvent reacts chemically with the solute. However, if the solvent is to be regenerated for reuse, the chemical reaction should be reversible. Alkaline solutions are very effective for absorption of hydrogen sulphide gas. Here, regeneration of the solvent by stripping off the absorbed gas will be difficult as an irreversible chemical reaction occurs between the sulphide and the alkali. On the other hand, if the absorption is carried out using ethanolamine solution, the solvent can be easily regenerated in a stripping operation.

(3) *Volatility of the solvent*: The loss of solvent by vaporisation can be reduced if the volatility of the solvent is low. Also, the recovery of the absorbent from the solution is easier if the volatility is low. Therefore, the vapour pressure of the solvent should be as low as possible. This requirement is sometimes overlooked in favour of the superior solubility of the gas in certain volatile solvents. The solvent loss in such cases can be reduced by carrying out the absorption in two sections of the tower as shown in Fig. 4.23, using the high volatile solvent (HVS) in the primary section of the absorption column and subsequently recovering the solvent vapours in the gas leaving this section in a less volatile solvent (LVS) in the secondary portion of the column.

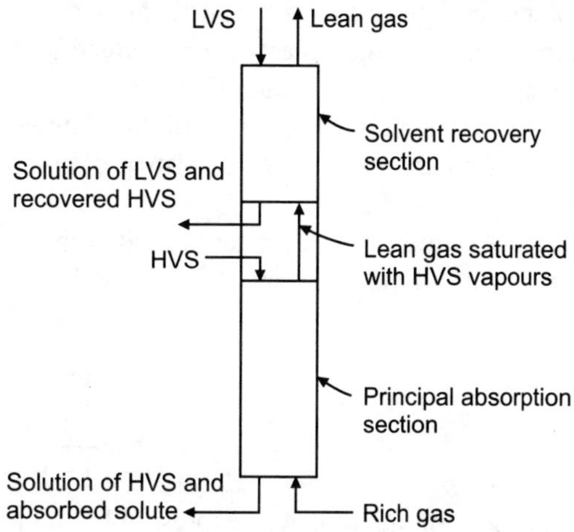

Fig. 4.23: Absorber with a recovery section for volatile solvent

Thus an aqueous solution of sodium phenolate (HVS) can be used for absorption of hydrogen sulphide and the vapours of phenol leaving with the sulphide-free gas can be recovered using water (LVS). Hydrocarbons are at times absorbed using a volatile solvent oil and the solvent vapours subsequently recovered using non-volatile oil.

(4) *Corrosiveness*: The solvent should not be corrosive to common materials of construction of the absorber and other solvent handling units.

(5) *Viscosity*: The solvent viscosity should be low so that the rate of absorption and rate of heat transfer are high. The low viscosity also leads to low pressure drop so that the cost of pumping and circulation of the solvent are low.

(6) *Toxicity*: The solvent should not be toxic.

(7) *Flammability*: Safety considerations demand that the solvent should be nonflammable.

(8) *Freezing point*: The freezing point of the solvent should be as low as possible so that it can be stored and handled in cold weather conditions without solidification.

(9) *Chemical stability*: A good solvent should resist decomposition or deterioration that may result with continuous use. The absorbent life is thus increased and the make-up solvent requirement is reduced.

(10) *Cost*: The solvent should be available in plenty at low cost.

4.4 MATERIAL BALANCE

The ratio of the flow rates of the gas to the liquid, the terminal concentrations of the gas and liquid phases, and the driving force for absorption are factors that determine the size of absorption columns. Material balance helps us to determine the material requirements for a given separation and allows us to follow the variation of driving force with position in the column. Driving force at any location in an absorption column is dependent on the difference between the concentrations of solute in the gas and the liquid. These concentrations vary from one end of the column to the other and this change is continuous for packed columns and is stepwise for plate columns. The change in concentration results in change in the driving force throughout the column, and this change can be estimated by material balance.

Concentration. The gas phase and liquid phase compositions can be expressed in different ways. The relationship between the different concentration terms are discussed here. The concentration of soluble component or diffusing component (designated as component **A**) in the gas is commonly expressed by its mole fraction represented by the symbol y. It is the number of moles of **A** present in one mole of the gas mixture. Gas-phase concentration can also be expressed by the mole ratio Y. It is defined as the number of moles of diffusing component per mole of non-diffusing components in the gas. For binary gas mixture consisting of components **A** and **B**, the mole ratio Y is the ratio of moles of **A** to moles of **B**.

Since Y moles of **A** are associated with one mole of other components in the mixture, the mole fraction y and the mole ratio Y are related as

$$y = \frac{Y}{1+Y} \qquad \text{...(4.10)}$$

$$Y = \frac{y}{1-y} \qquad \text{...(4.11)}$$

If the concentration of **A** is expressed by the partial pressure $\bar{p}_A$ in the gas, then,

$$y = \frac{\bar{p}_A}{P} \qquad \text{...(4.12)}$$

where P is the total pressure. In terms of partial pressures, the mole ratio Y is the ratio of partial pressure of **A** to the partial pressure of other constituents.

$$Y = \frac{\bar{p}_A}{P-\bar{p}_A} \qquad \text{...(4.13)}$$

Let G be the total number of moles in the gas mixture and G_S be the moles of non-diffusing gases in the mixture, then the moles of **A** in the mixture is Gy in terms of mole fraction or G_SY in terms of mole ratio. Thus

$$Gy = G_SY \qquad \text{...(4.14)}$$

The moles of non-diffusing components is

$$G_S = G(1-y) \qquad \text{...(4.15)}$$

$$G_S = \frac{G}{(1+Y)} \qquad \text{...(4.16)}$$

Composition of liquids also can be expressed in a similar way. The mole fraction of solute in the liquid is denoted by x and the mole ratio by X. These are related as

$$x = \frac{X}{1+X} \qquad \text{...(4.17}$$

$$X = \frac{x}{1-x} \qquad \text{...(4.18)}$$

Let L be the moles of liquid in which L_S is the moles of non-diffusing component (solvent). Then Lx and L_SX represent the moles of solute **A** in the liquid. We have the following relationships analogous to Eqs. (4.14), (4.15) and (4.16).

$$Lx = L_S X \quad \text{...(4.19)}$$

$$L_S = L(1-x) \quad \text{...(4.20)}$$

$$L_S = \frac{L}{(1+X)} \quad \text{...(4.21)}$$

Operating Line for Countercurrent Absorbers

Consider a countercurrent absorber as shown in Fig. 4.24. Assume that the operation is carried out at constant pressure, constant temperature and under continuous steady-state flow conditions.

The molar flow rates (kmol/h) of gas and liquid are indicated by the symbols G and L, respectively. In the analysis of continuous contact towers such as packed columns, G and L will be used to represent molar flow rates per unit area of tower cross section, kmol/m^{2}h. It is assumed that only one component (the solute **A**) is transferred and the other components in the gas are non-diffusing. The mole fraction of solute in the gas is represented by y and the mole ratio by Y. Similarly, x and X represent respectively, the mole fraction and mole ratio of solute in the liquid. The flow rates and compositions at the bottom of the column are represented by the suffix 1 and those at the top by suffix 2. The total molar flow rate of gas G is not constant; it decreases from G_1 at the bottom to G_2 at the top, because solute gets transferred to liquid as the gas flows up. The total molar flow of liquid increases from L_2 at the top to L_1 at the bottom, because the liquid receives the solute transferred from the gas as it flows down the column. However the molar flow rates of insoluble non-diffusing gases, G_S kmol/h remain unchanged throughout the column. Similarly, L_S represents the molar flow rate of non-volatile solute-free liquid which is also constant throughout the column.

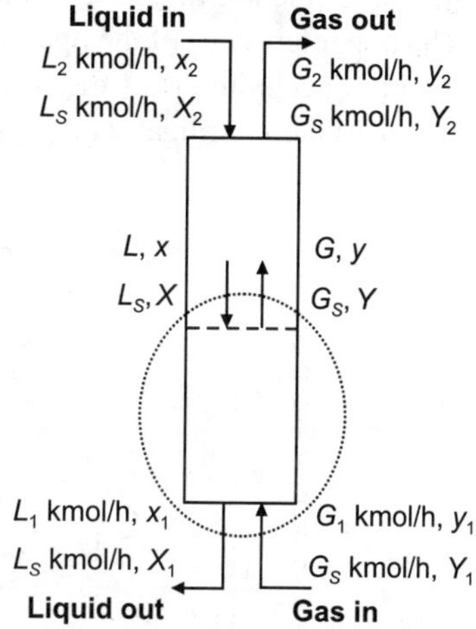

Fig. 4.24: Material balance in absorber/stripper

The solute balance for the column can be written as:
Solute entering the column = Solute leaving the column.

$$G_1y_1 + L_2x_2 = G_2y_2 + L_1x_1 \quad \text{...(4.22)}$$

Equation (4.22) can be rearranged as

$$G_1y_1 - L_1x_1 = G_2y_2 - L_2x_2 \quad \text{...(4.23)}$$

Consider any arbitrary section in the column where the molar flow rates of gas and liquid are G and L, respectively and the mole fraction of solute in the liquid and gas are x and y, respectively. Now consider the material balance from the bottom of the column to this intermediate point.

$$G_1y_1 + Lx = Gy + L_1x_1 \quad \text{...(4.24)}$$

The above equation may be rearranged as

$$y = \frac{L}{G}x + \frac{G_1y_1 - L_1x_1}{G} \quad \text{...(4.25)}$$

Substituting Eq. (4.23) into the above, we get

$$y = \frac{L}{G}x + \frac{G_2y_2 - L_2x_2}{G} \quad \text{...(4.26)}$$

For absorbers using pure solute-free solvent, $x_2 = 0$, and Eq. (4.26) simplifies to

$$y = \frac{L}{G}x + \frac{G_2}{G}y_2 \quad \text{...(4.27)}$$

Equations (4.25), (4.26) and (4.27) are the material balance equations for countercurrent absorbers. The material balance equation can be plotted on equilibrium coordinates and the resulting curve is known as the *operating line*. The slope of the curve is equal to the liquid to gas ratio in the column. Since this ratio is not constant, the operating line is not straight when plotted on mole fraction coordinates. The curve *AB* in Fig. 4.25 represents the operating line for absorber. Operating line is the graphical representation of the material balance and the coordinates of any point on the operating line represent the mole fraction of solute in the liquid and gas crossing each other at the given point in the column. The upper terminus of the operating line (point *B*) has coordinates (x_1, y_1), which represents the gas and liquid concentrations at the lower end of the absorber and the lower terminus of the operating line (point *A*) has coordinates (x_2, y_2), which represents the concentrations at the upper end of the absorber.

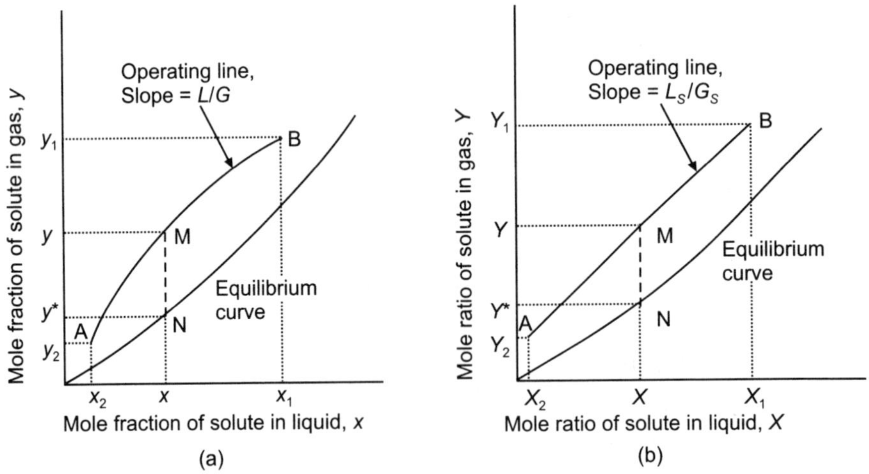

Fig. 4.25: Operating line for absorber (a) on mole fraction coordinates (b) on mole ratio coordinates

Consider point *M* on the operating line and the point *N* on the equilibrium curve, both points having the liquid composition *x* as the abscissa. The *y*-coordinate of point *M* lying on the operating line is *y*, the mole fraction of solute in the gas phase which is in contact with the liquid of concentration *x* and that of point *N* on the equilibrium curve is y^*, the concentration of a gas phase which is in equilibrium with the liquid. In absorption, the solute concentration in the gas phase is greater than the equilibrium value and therefore, *y* is greater than y^* for any given *x* and the solute is transferred from the gas to liquid. For absorption operations, therefore, the operating line lies above the equilibrium curve. The vertical distance *MN* is $y - y^*$, the driving force at the point under consideration.

The operating line can be constructed in mole ratio coordinates which will be more convenient for many graphical computations. The material balance can be taken in terms of the mole ratios and solute free gas and liquid flow rates. Thus solute entering the absorber is $G_S Y_1 + L_s X_2$ and the solute leaving is $G_S Y_2 + L_S X_1$. Therefore, the solute balance for the total absorber is

$$G_S Y_1 + L_S X_2 = G_S Y_2 + L_S X_1 \quad \text{...(4.28)}$$

For the lower portion of the absorber, shown by the envelope in Fig. 4.24, the solute balance is

$$G_S Y_1 + L_S X = G_S Y + L_S X_1 \quad \text{...(4.29)}$$

Equation (4.29) can be rearranged as

$$Y = \frac{L_S}{G_S}X + \left(Y_1 - \frac{L_S}{G_S}X_1\right) \quad \text{...(4.30)}$$

which is the equation for the operating line in mole ratio coordinates. Utilizing the result of Eq. (4.28), the operating line equation, Eq. (4.30) can be modified as

$$Y = \frac{L_S}{G_S}X + \left(Y_2 - \frac{L_S}{G_S}X_2\right) \quad \text{...(4.31)}$$

And for the special case of absorption utilizing pure solvent, $X_2 = 0$, and the above equation simplifies to

$$Y = \frac{L_S}{G_S}X + Y_2 \quad \text{...(4.32)}$$

The L_S and G_S are the molar flow rates of non-diffusing components and therefore remain constant. The slope of the operating line L_S/G_S is constant throughout the column and the operating line on mole ratio coordinates is a straight line as shown by the line *AB* in Fig. 4.25b where the equilibrium curve is also plotted on mole ratios. Thus coordinates of point *A* are (X_1, Y_1) and those of point *B* are (X_2, Y_2). The operating line can be easily drawn knowing the concentrations of the gas and liquid at the two ends of the absorber, or by knowing the concentrations at one end and the slope L_S/G_S. The construction of operating line on mole fraction coordinates is not that easy, because it is not a straight line. It is necessary to know the coordinates of several points on the operating line, which means that the concentrations at several locations in the absorber should be calculated by material balance. However, for dilute solutions,

$$X \approx x, \quad Y \approx y, \quad \frac{L_S}{G_S} \approx \frac{L}{G}$$

and for all practical purposes involving dilute solutions, the operating line plotted on mole fraction coordinates can also be treated as straight line.

Operating Line for Strippers

The operating line for stripping columns can be constructed in the same way as for absorbers. The liquid solution to be stripped off is admitted at the top of the absorber and the stripping agent is the gas phase which is admitted at the lower end of the absorber. Gas and liquid flow countercurrent to each other and wherever they are in contact, the solute gets transferred from the liquid to gas. The gas leaving at the top of the column is richer in solute and the liquid leaving at the bottom is almost free of solute. Adopting the same symbols as in absorbers to represent the gas and liquid flow rates and their concentrations, we have

$$G_1y_1 + Lx = Gy + L_1x_1 \quad \text{...(4.33)}$$

which may be rearranged as

$$y = \frac{L}{G}x + \frac{G_1y_1 - L_1x_1}{G} \quad \text{...(4.34)}$$

If the stripping gas is admitted free of component **A**, $y_1 = 0$ and the operating line becomes

$$y = \frac{L}{G}x - \frac{L_1}{G}x_1 \quad \text{...(4.35)}$$

Equation (4.34), when plotted on *x*-*y* coordinates gives the operating line of the stripper. It is not a straight line as the slope *L*/*G* is not constant. The operating line for the stripper as shown in the Fig. 4.26 will lie below the equilibrium curve as the transfer is from the liquid to the gas.

Alternatively, the material balance can be written as,

$$G_SY_1 + L_SX = G_SY + L_SX_1 \quad \text{...(4.36)}$$

which can be rearranged as

$$Y = \frac{L_S}{G_S}X + \left(Y_1 - \frac{L_S}{G_S}X_1\right) \quad \text{...(4.37)}$$

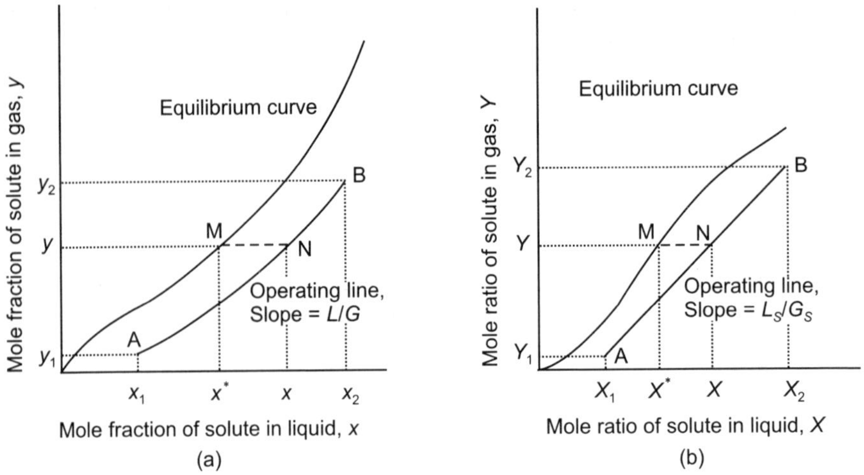

Fig. 4.26: Operating line for stripper (a) on mole fraction coordinates (b) on mole ratio coordinates

Equation (4.37) is the equation for a straight line of slope L_S/G_S and the line passes through the points representing the terminal concentrations A (X_1, Y_1) and B (X_2, Y_2). The lower terminus of the operating line represents conditions at the lower end of the stripper and the upper terminus represents the conditions at the upper end. Line MN gives the concentration difference driving force x–x^* (or X–X^*) for the stripper at a given point, where x (or X) is the concentration of the liquid phase brought into contact with the gas of composition y (or Y) and x^* (or X^*) is the concentration of the liquid phase equilibrium equivalent of the gas phase. At all points in the column, the concentration of solute in the liquid is greater than the equilibrium value, $x > x^*$, the solute transfer occurs from liquid to gas, and the operating line should lie below the equilibrium curve.

Example 4.4: A countercurrent stripper is used for regenerating wash oil (Molecular weight = 260) which is used as solvent for absorbing benzene from coal gas. The wash oil-benzene solution containing 10.0 mol% benzene is admitted in a stripper at a rate 1800 kg/h. Superheated steam at 400 K and 100.0 kPa is the stripping agent. The steam is supplied at a rate 45 kg/h. The regenerated solvent should contain not more than 0.5 mol% benzene. The oil-benzene solution may be assumed to behave as ideal solution. The vapour pressure of benzene at 400 K is 350 kPa.

(a) Calculate the concentration of benzene in the exit gas stream expressed as mole %

(b) Plot the equilibrium curve and the operating line on (i) mole fraction coordinates (ii) mole ratio coordinates.

***Solution*:** Figure 4.27 shows the flow rates and compositions of various streams entering and leaving the stripper.

The gas is admitted at a rate 45 kg/h so that

$$G_1 = \frac{45}{18} = 2.5 \text{ kmol/h}$$

Since the super heated steam is free of benzene, $y_1 = 0$. Therefore the benzene-free flow rate of gas G_S is the same as G_1.

$$G_S = G_1(1 - y_1) = G_1 = 2.5 \text{ kmol/h}$$

Wash-oil benzene solution enters the stripper at 1800 kg/h and it contains 10% benzene. The average molecular weight of the solution is

$$MW = 0.10 \times 78 + 0.90 \times 260 = 241.8$$

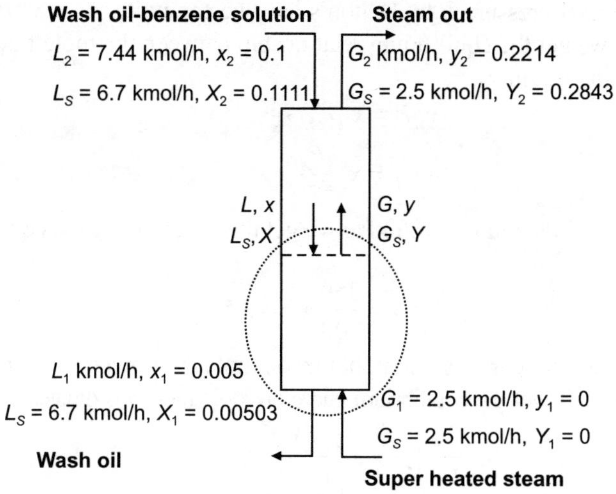

Fig. 4.27: Schematic representation of stripper in Example 4.4

Therefore,

$$L_2 = \frac{1800}{241.8} = 7.44 \text{ kmol/h}$$

The mole fraction of benzene in the solution entering the stripper, $x_2 = 0.1$. Therefore, the solute free liquid rate L_S is

$$L_S = L_2(1 - x_2) = 7.44 \times 0.90 = 6.7 \text{ kmol/h}$$

Let the mole fraction of benzene in the exit liquid from the stripper be 0.005. That is $x_1 = 0.005$.

The concentration of benzene can be expressed in terms of mole ratios which are readily calculated from the mole fractions. Thus

$$Y_1 = \frac{y_1}{1-y_1} = 0 \qquad X_1 = \frac{x_1}{1-x_1} = \frac{0.005}{1-0.005} = 0.00503 \quad X_2 = \frac{x_2}{1-x_2} = \frac{0.10}{1-0.10} = 0.1111$$

(a) The concentration of benzene in the exit gas is calculated by taking a solute balance on the stripper.

$$G_S(Y_2 - Y_1) = L_S(X_2 - X_1) \tag{A}$$

The mole ratio Y_2 is the number of moles of benzene per mole of steam in the exit gas. Rearranging Eq. (A), we get

$$Y_2 = Y_1 + \frac{L_S}{G_S}(X_2 - X_1) \tag{B}$$

$Y_2 = 0 + \dfrac{6.7}{2.5}(0.1111 - 0.00503) = 0.2843$ mol benzene per mole of steam. The mole fraction in the exit gas is

$$y_2 = \frac{Y_2}{1+Y_2} = \frac{0.2843}{1+0.2843} = 0.2214$$

which means that the exit gas contains 22.14% benzene. **(Answer)**

(b) To plot the equilibrium curve on mole fraction coordinates, consider the equilibrium relationship. Since Raoult's law is obeyed, we have $\bar{p}_A = x_A P_A^s$ where $\bar{p}_A$ is the partial pressure of benzene in the vapour and P_A^s is the vapour pressure of benzene at 400 K. Since the partial pressure in the vapour phase is related to the total pressure P by Dalton's law, we can write $\bar{p}_A = y_A P$. Combining Raoult's law and Dalton's law, we get the equilibrium relationship between the mole fraction of benzene in the vapour y_A and that in the liquid x_A.

$$y_A P = x_A P_A^s$$

or

$$y_A = x_A \frac{P_A^s}{P}$$

Substituting $P_A^s = 350$ kPa and $P = 100$ kPa, we get the equation for the equilibrium curve on mole fraction coordinates:

$$y_A = \frac{350}{100} x_A = 3.5 x_A \tag{C}$$

Obviously, the equilibrium curve is a straight line on mole fraction coordinates. This is plotted as the upper curve in Fig. 4.28a. To get the equilibrium curve on mole ratio coordinates, Eq. (C) can be written as

$$\frac{Y_A}{1+Y_A} = 3.5 \frac{X_A}{1+X_A} \tag{D}$$

Assume different values for X and calculate the corresponding values for Y using Eq. (D). For example when $X = 0.05$,

$$\frac{Y_A}{1+Y_A} = 3.5 \frac{0.05}{1+0.05} \quad \text{and therefore, } Y_A = 0.2$$

The results of these calculations are summarized in the following table.

X_A	0	0.02	0.04	0.06	0.08	0.10	0.12
Y_A	0	0.0737	0.1556	0.2470	0.3501	0.4667	0.60

The plot of Y_A versus X_A gives the equilibrium curve on mole ratio coordinates, the upper curve in Fig. 4.28b.

Now consider the material balance from the bottom of the column to the intermediate section in Fig. 4.27 where gas and liquid compositions are Y and X, respectively.

$$G_S(Y - Y_1) = L_S(X - X_1) \tag{E}$$

Since $Y_1 = 0$, and $X_1 = 0.00503$, Eq. (E) gives

$$Y = Y_1 + \frac{L_S}{G_S}(X - X_1) \quad \text{or} \quad Y = \frac{L_S}{G_S} X - \frac{L_S}{G_S} X_1 = \frac{6.7}{2.5} X - \frac{6.7}{2.5}(0.00503)$$

That is

$$Y = 2.68X - 0.01348 \tag{F}$$

Equation (F) is the operating line on mole ratio coordinates which is a straight line of slope 2.68. This is plotted as the lower line (AB) in Fig. 4.28b. The operating line passes through the terminal concentrations (0.00503,0) and (0.1111,0.2843).To get the operating line on mole fraction coordinates, Eq. (F) is written in terms of mole fractions x and y.

$$\frac{y}{1-y} = 2.68 \frac{x}{1-x} - 0.01348 \tag{G}$$

Equation (G) reveals that the operating line on mole fraction coordinates is not a straight line. To get the points on the operating line, assume values for x (from 0.005 to 0.10) and calculate y using Eq. (G). The results are given in the table below:

x_A	0.005	0.01	0.02	0.04	0.06	0.08	0.10
y_A	0	0.013	0.040	0.089	0.136	0.180	0.221

The operating line on mole fraction coordinates is plotted as the lower curve (AB) in Fig. 4.28a using the x and y values in this table. It passes through point A (0.005,0) and point B (0.1, 0.221)

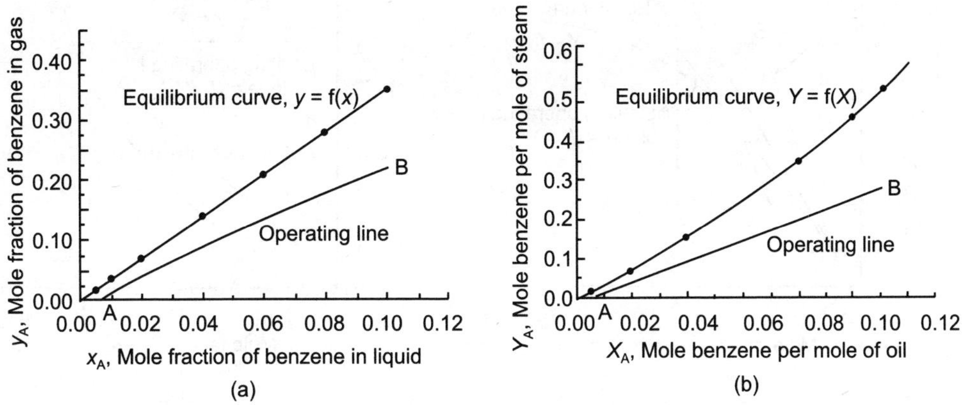

Fig. 4.28: (a) Equilibrium and operating line on mole fraction coordinates for the stripper in Example 4.4. (b) Equilibrium and operating line on mole ratio coordinates for the stripper in Example 4.4.

Limiting Liquid-to-Gas Ratio

Absorber

In the design of absorbers, the gas flow rate, gas inlet and exit compositions and the quality of the solvent used for absorption are generally specified. That means G_1, y_1, y_2 and x_2 (or the corresponding solute-free quantities G_S, Y_1, Y_2, and X_2) are generally specified by process requirements. For example, if an absorber is to be designed to treat 100 kmol/h of ammonia-air mixture to reduce the ammonia content from 10% (mole) to 0.5% (mole) using pure water as absorbent, we have $G_1 = 100$, $y_1 = 0.10$, $y_2 = 0.005$, and $x_2 = 0$. It follows that $G_S = 90$, $Y_1 = 0.111$, $Y_2 = 0.00503$, $X_2 = 0$. We have to decide the quantity of solvent (L_S) to be used for absorption before proceeding further with the design. The solute concentration of the exit liquid (X_1) is dependent on the amount of solvent used. It is obvious that the solution leaving the absorber gets diluted by increasing the amount of solvent used. Since the slope of the operating line is equal to the ratio L_S/G_S, by changing the solvent quantity, the slope also changes and the operating line move either away from the equilibrium curve or towards the equilibrium curve depending upon whether the solvent quantity is increased or decreased.

The equilibrium curve and the operating line on mole ratio coordinates are plotted in Fig. 4.29. As mentioned earlier, the lower terminus of the operating line represents the conditions at the upper end of the absorber. Since X_2 and Y_2 are known, the lower end of the operating line A is located as shown in the figure. The operating line ends at point B for which the coordinates are the liquid and gas compositions at the lower end of the absorber. The composition of the gas admitted is Y_1 and therefore point B lies on the horizontal line at Y_1. The exact location of B depends on the slope of the line AB (L_S/G_S) which depends on the amount of solvent used for absorption since G_S is fixed by process requirement. For a lower quantity of solvent, the slope decreases, the operating line moves toward the equilibrium curve as shown by AB' in the figure. The driving force which is measured by the vertical distance between the operating line and the equilibrium curve decreases as these curves come closer. Thus the driving force is smaller when the operating line is AB' than that when the operating line is AB. Consequently, the gas and liquid should be in contact for a longer time in continuous contact unit or should be brought into contact on larger number of plates in stage contact equipment both implying that a taller absorber would be required for carrying out the specified task. However, the cost of the solvent,

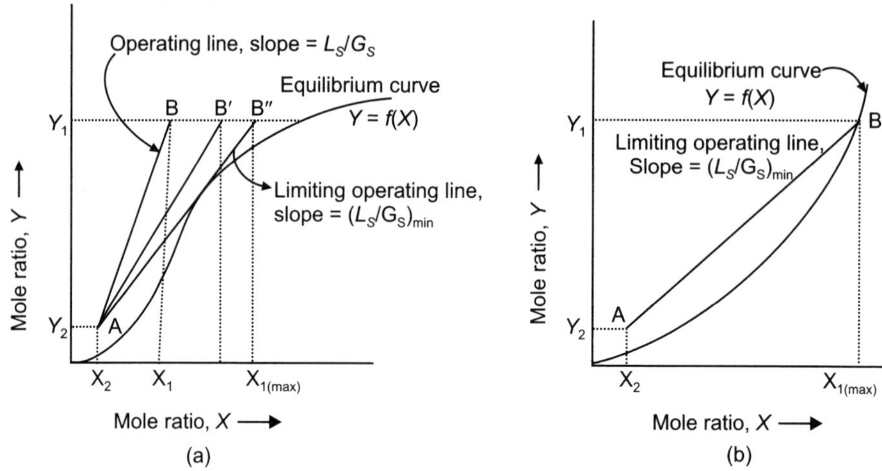

Fig. 4.29: Limiting operating line and minimum L/G ratio for absorbers

the cost of the solvent circulation and regeneration are reduced by using lesser quantity of solvent. On the other hand by using a large quantity of solvent, the driving force is increased and a shorter column would be sufficient. The x-coordinate of B gives the liquid exit composition X_1 which is smaller for large L_S/G_S values. The solution leaving the column is very dilute and regeneration of solvent by stripping will be expensive as huge quantities of very dilute solution are to be treated.

Thermodynamics sets a lower limit to the amount of solvent that can be used in the column. As shown in Fig. 4.29a, for the limiting liquid rate, the operating line AB'' becomes tangential to the equilibrium curve and the driving force vanishes at the point of contact. An infinitely tall tower will be required to bring out the necessary separation. The operating line AB'' is the limiting operating line and its slope gives minimum (L_S/G_S) ratio for the column. The composition of the exit liquid corresponding to the minimum quantity of solvent used is X_1 (max). The actual operating solvent rate L_S used in the column should be greater than the minimum L_S given by the slope of the limiting operating line. The minimum slope $(L_S/G_S)_{\text{min}}$ is computed as

$$\left(\frac{L_S}{G_S}\right)_{\text{min}} = \frac{Y_1 - Y_2}{X_1(\text{max}) - X_2} \quad \ldots(4.38)$$

from which minimum solvent rate can be determined. The limiting operating line need not always be tangential to the equilibrium curve. If the equilibrium curve is concave upwards as shown in Fig. 4.29b, the limiting operating line is the one that passes through the intersection of the equilibrium curve and the horizontal at $Y = Y_1$. The exit liquid composition, X_1(max) is the concentration of liquid that is in equilibrium with the gas of composition Y_1 and the minimum solvent requirement can be computed by Eq. (4.38). In this case, X_1(max) is the abscissa of the point on the equilibrium curve whose ordinate is Y_1 and the right-hand side of Eq. (4.38) can be determined even without constructing the limiting operating line.

The actual solvent flow rate chosen generally lies between 1.1 to 2 times the minimum solvent rate computed as above.

Stripper

In the case of strippers, the liquid rate L_s, its concentration at the inlet X_2 and that at the exit X_1 and the purity of the stripping agent Y_1 are the quantities that are specified by the process. The rate at which the stripping gas is to be admitted is to be decided by the designer. The height of the packed column or the number of plates in the plate column depends on this choice.

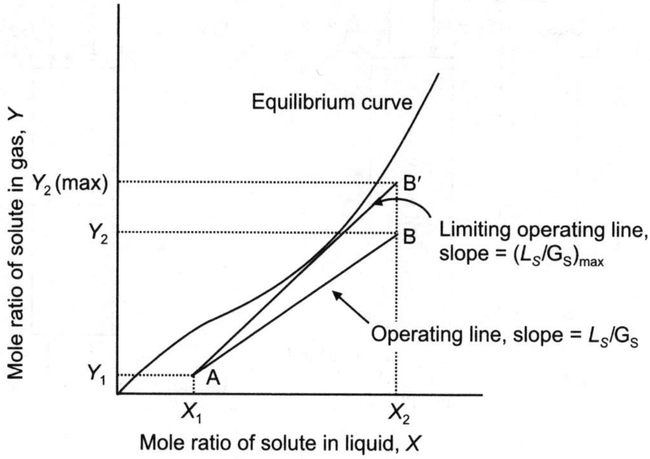

Fig. 4.30: Minimum gas flow for strippers

The lower terminus of the operating line A is fixed by the process specifications (Fig. 4.30). The operating line AB can be drawn if the gas rate G_S is decided. Slope of the line is L_S/G_S. The upper terminus B lies on the vertical erected at X_2, the liquid exit concentration which is also specified by the process. The ordinate of B is the exit gas concentration. To determine the minimum gas rate permissible, the slope of the limiting operating line which is drawn from point A as tangent to the equilibrium curve (shown by line AB') is determined. For minimum gas rate, the L_S/G_S is maximum and the slope of the limiting operating line is

$$\left(\frac{L_S}{G_S}\right)_{\max} = \frac{Y_2\,(\max) - Y_1}{X_2 - X_1} \qquad \ldots(4.39)$$

in which Y_2 (max) is the ordinate of B'.

If the equilibrium curve is concave downwards, the limiting operating line passes through the intersection of the equilibrium curve and the vertical line at X_2. In this case, the exit gas concentration Y_2 (max) corresponding to minimum G_S is the concentration of the gas in equilibrium with liquid concentration X_2. As Y_2(max) is the ordinate of the point on the equilibrium curve whose abscissa is X_2, the right-hand side of Eq. (4.39) can be determined even without drawing the limiting operating line.

Operating Line for Cocurrent Absorbers

Cocurrent absorbers are not common for industrial applications. When the absorber size happens to be unusually large for construction as a single tower, it will be divided into two towers with the first one operated countercurrently and the second cocurrently as shown in Fig. 4.31a. This is done to save on the cost of very large diameter pipe required to carry the gas from the top of the first tower to the bottom of the second if the latter is to be operated countercurrently. Cocurrent units can also be used for the absorption of pure gas in a liquid solvent and for absorption in which rapid irreversible chemical reaction occurs between the solute and the solvent.

A solute balance for the entire column (Fig. 4.31b) gives

$$G_S\,(Y_1 - Y_2) = L_S\,(X_2 - X_1) \qquad \ldots(4.40)$$

or

$$\frac{Y_1 - Y_2}{X_1 - X_2} = -\frac{L_S}{G_S} \qquad \ldots(4.41)$$

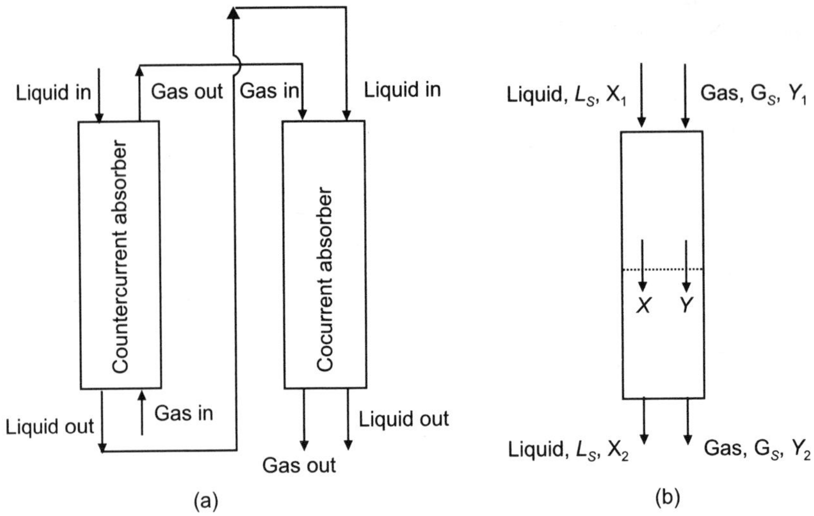

Fig. 4.31: (a) Cocurrent absorber as part of a cocurrent-countercurrent sections of tall absorber (b) Material balance on cocurrent absorber

This is the equation for the operating line for cocurrent absorber. It passes through the points *A* (X_1,Y_1) and *B* (X_2,Y_2) which represent the terminal conditions. Its slope is negative of the liquid to gas ratio for the absorber. By restricting the solute balance from top of the column to an intermediate section within the column where the liquid and gas concentrations are *X* and *Y*, we get

$$G_S(Y_1 - Y) = L_S(X - X_1) \quad \ldots(4.42)$$

$$\frac{Y_1 - Y}{X_1 - X} = -\frac{L_S}{G_S} \quad \ldots(4.43)$$

Equation (4.43) means that all points with the coordinates *X* and *Y*, the concentration of the liquid and gas at any given location in the absorber, lie on the operating line *AB*. The exit streams will approach equilibrium if the streams are allowed to be in contact for a sufficiently long time. An infinitely tall tower can produce the exit liquid and gas in equilibrium as shown by the point *C* in Fig. 4.32. For lower liquid rates, the operating line shifts towards the right, and the liquid exit composition increases. It is not possible to reduce the liquid rate beyond that given by the slope of the line *AD* in Fig. 4.32. If the liquid rate is reduced further, it is not possible to bring down the exit gas composition to Y_2 even if an infinitely tall tower is provided.

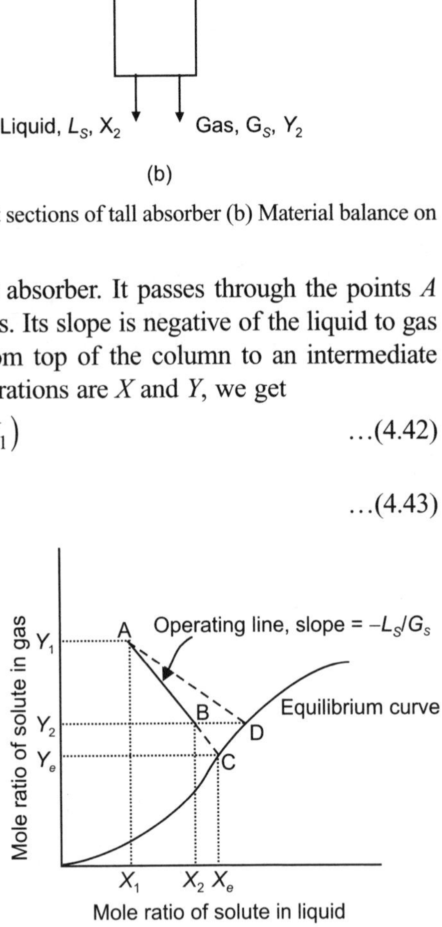

Fig. 4.32: Operating line for cocurrent absorber

Example 4.5: The gas leaving an alcohol fermenter consists of 10% CO_2 and 90% air. 80% of the CO_2 in the gas is to be recovered by absorbing in a water solution of triethanolamine containing 0.01 mole CO_2 per mole of CO_2 free amine solution. Absorption is carried out at 298 K. Amine is supplied 40% in excess of the minimum. The equilibrium data for 298 K is: (X = mole CO_2/mole amine; Y = mole CO_2/mole air).

X	0.01	0.02	0.03	0.04	0.05	0.06	0.07	0.08	0.09	0.10	0.11
Y	0.003	0.008	0.015	0.023	0.032	0.043	0.055	0.068	0.083	0.099	0.12

Determine for treating 100 kmol/h feed gas:

(a) The minimum solvent required, kmol/h

(b) The solvent required, kmol/h

(c) Mole percent of CO_2 in the exit liquid

Solution: The absorber is shown in the Fig. 4.33.

The gas enters at 100 kmol/h with a concentration of CO_2 = 0.10 mole fraction. Therefore,

$$G_1 = 100 \text{ kmol/h}; \; y_1 = 0.10$$

The flow rate of solute-free gas, G_S and the mole ratio (moles of CO_2 per mole of air) Y_1 can now be evaluated.

$$G_S = G_1(1-y_1) = 100 \times 0.90 = 90 \text{ kmol/h};$$

$$Y_1 = \frac{y_1}{(1-y_1)} = \frac{0.10}{0.90} = 0.1111$$

Since 80% of CO_2 is absorbed in amine solution, the exit gas contains 20% of the CO_2 entering the column. Therefore,

$$Y_2 = 0.20\,Y_1 = 0.20 \times 0.1111 = 0.02222.$$

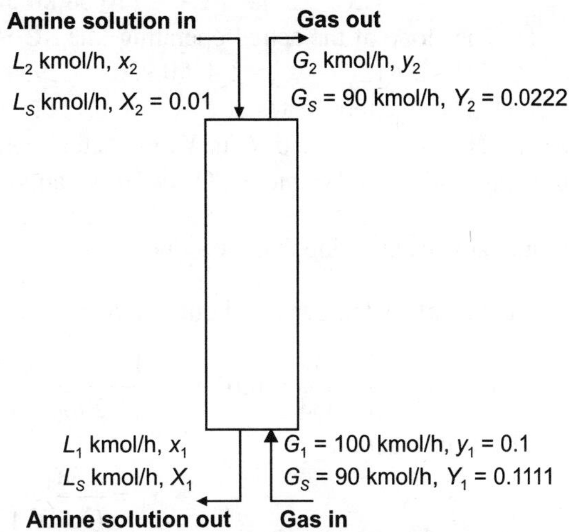

Fig. 4.33: Absorber in Example 4.5

Amine solution entering the column contains 0.01 mole CO_2 per mole of CO_2–free amine solution,

$$X_2 = 0.01$$

(a) The equilibrium curve is drawn as shown in Fig. 4.34 by taking mole ratio Y_A on the y-axis and X_A on the x-axis. Point A represents the lower terminus of the operating line with coordinates (X_2 = 0.01 and Y_2 = 0.0222). The location of the upper terminus is dependent on the quantity of solvent used for

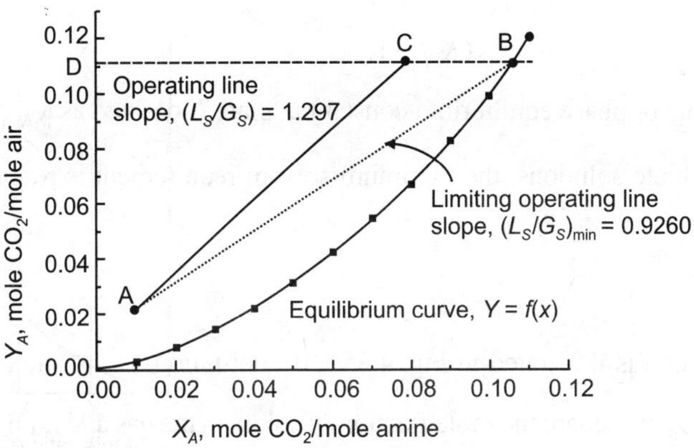

Fig. 4.34: Limiting and actual operating lines for Example 4.5

absorption. However, it should lie on the horizontal line with $Y = Y_1$ = 0.1111 (shown by the dotted line DB in figure). Corresponding to the minimum quantity, the operating line touches the equilibrium curve at point B. Line AB is the limiting operating line whose slope is $(L_S/G_S)_{min}$. From the graph, the coordinates of point B can be determined and the slope $(L_S/G_S)_{min}$ calculated. The coordinates of B are (X_1(max), Y_1). Y_1 = 0.1111, X_1(max) = 0.106 (from graph)

$$\left(\frac{L_S}{G_S}\right)_{min} = \frac{Y_1 - Y_2}{X_1(\max) - X_2} = \frac{0.1111 - 0.0222}{0.106 - 0.01} = 0.926$$

Since G_S = 90 kmol/h, L_S (min) = 0.926 × 90 = 83.34 kmol/h
The minimum solvent requirement is 83.34 kmol/h. **(Answer)**

(b) The actual solvent requirement is 40 % in excess of the minimum. Therefore,

L_S = 83.34 × 1.4 = 116.68 kmol/h **(Answer)**

(c) The slope of the actual operating line AC is L_S/G_S, which is 40% in excess of $(L_S/G_S)_{min}$. $(L_S/G_S) = 1.4\,(L_S/G_S)_{min} = 1.4 \times 0.926 = 1.2964$. The operating line is drawn from point A with slope 1.2964. The line terminates at $Y = Y_1 = 0.1111$, the upper terminus of the operating line C is located easily. The x-coordinate of C is X_1, the actual composition of the exit liquid from the absorber. From the graph, $X_1 = 0.0786$ mole CO_2 per mole amine.

[Alternatively, knowing that the slope $\left(\dfrac{L_S}{G_S}\right) = \dfrac{Y_1 - Y_2}{X_1 - X_2} = 1.2964$, and all values except X_1 are known, X_1 can be calculated even without drawing the operating line AC.

$$X_1 = X_2 + \frac{Y_1 - Y_2}{1.2964} = 0.01 + \frac{0.1111 - 0.0222}{1.2964} = 0.0786 \text{ mole } CO_2 \text{ per mole amine]}$$

The mole fraction

$$x_1 = \frac{X_1}{(1+X_1)} = \frac{0.0786}{1.0786} = 0.0729$$

That is, the mole percent of CO_2 in the exit liquid is 7.29% **(Answer)**

Example 4.6: The concentration of solute entering a plate column is reduced from mole ratio Y_{N+1} to Y_1 using a liquid solvent which enters the column containing solute mole ratio X_0 and leaves the column with a mole ratio X_N. The equilibrium curve is concave upwards throughout.

(a) Show that the minimum solvent requirement can be computed as

$$\left(L_S\right)_{min} = \frac{G_S(Y_{N+1} - Y_1)}{\dfrac{Y_{N+1}}{K_N + Y_{N+1}(K_N - 1)} - X_0}$$

where K_N is the K-value or phase equilibrium constant at plate N defined as $K_N = \dfrac{y_N}{x_N}$.

(b) Show that for dilute solutions, the minimum solvent requirement is roughly

$$\left(L_S\right)_{min} = \frac{G_S(y_{N+1} - y_1)}{\dfrac{y_{N+1}}{K_N} - x_0}$$

***Solution*:** A plate column is illustrated in Fig. 4.35a. By definition, the K-value on the Nth plate,

$K_N = \dfrac{y_N}{x_N}$, where y_N and x_N are the mole fraction of solute in the gas and liquid leaving the Nth plate.

Since the plates are ideal, y_N and x_N are equilibrium concentrations.

$$K_N = \frac{y_N}{x_N} = \frac{Y_N/(1+Y_N)}{X_N/(1+X_N)} \qquad \text{...(4.44)}$$

For minimum liquid rate, the operating line passes through the point on the equilibrium curve for which the ordinate is Y_{N+1} as shown in Fig. 4.35b. Therefore X_N and Y_{N+1} are equilibrium concentrations, or

$$K_N = \frac{y_{N+1}}{x_N} = \frac{Y_{N+1}/(1+Y_{N+1})}{X_N/(1+X_N)} \qquad \text{...(4.45)}$$

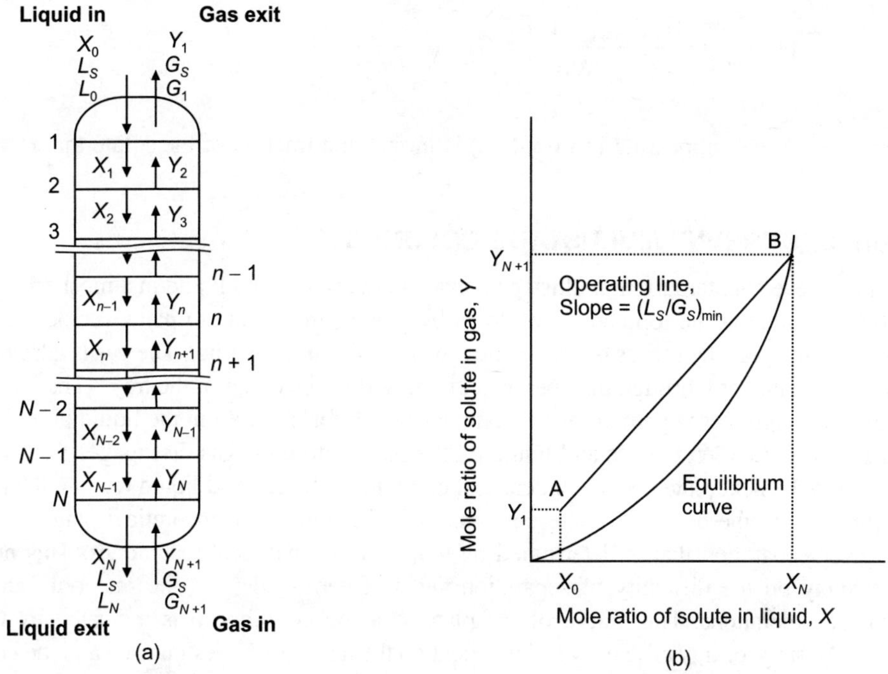

Fig. 4.35: Multistage column and limiting operating line for Example 4.6

Rearranging Eq. (4.45), we get

$$X_N = \frac{Y_{N+1}}{K_N + Y_{N+1}(K_N - 1)} \qquad \ldots(4.46)$$

The slope of the limiting operating line is

$$\left(\frac{L_S}{G_S}\right)_{min} = \frac{Y_{N+1} - Y_1}{X_N - X_0} \qquad \ldots(4.47)$$

Substituting X_N as given by Eq. (4.46) in Eq. (4.47), we get

$$\left(\frac{L_S}{G_S}\right)_{min} = \frac{Y_{N+1} - Y_1}{Y_{N+1}/\left[K_N + Y_{N+1}(K_N - 1)\right] - X_0} \qquad \ldots(4.48)$$

or

$$(L_S)_{min} = \frac{G_S(Y_{N+1} - Y_1)}{Y_{N+1}/\left[K_N + Y_{N+1}(K_N - 1)\right] - X_0} \qquad \ldots(4.49) \qquad \textbf{(Answer)}$$

(b) For dilute solutions, $Y \approx y$ and $X \approx x$, the above equation becomes

$$(L_S)_{min} = \frac{G_S(y_{N+1} - y_1)}{y_{N+1}/\left[K_N + y_{N+1}(K_N - 1)\right] - x_0} \qquad \ldots(4.50)$$

Further, since y is very small, $K_N + y_{N+1}(K_N - 1) \approx K_N$ so that

$$(L_S)_{min} = \frac{G_S(y_{N+1} - y_1)}{y_{N+1}/K_N - x_0} \qquad \ldots(4.51) \qquad \textbf{(Answer)}$$

If the solvent used for absorption is free of solute, $x_0 = 0$ so that

$$(L_S)_{\min} = \frac{G_S(y_{N+1} - y_1)}{y_{N+1}/K_N} = G_S K_N \frac{(y_{N+1} - y_1)}{y_{N+1}} \qquad \text{...(4.52)}$$

The term $\dfrac{(y_{N+1} - y_1)}{y_{N+1}}$ appearing in Eq. (4.52) is in fact, the fraction of the solute that is absorbed.

4.5 COUNTERCURRENT MULTISTAGE COLUMNS

Plate towers achieve countercurrent multistage contact between gas and liquid in a most effective way. The liquid is admitted at the top and flows down by gravity from plate to plate. A pool of liquid is maintained over the plate by means of an overflow weir, as discussed earlier. The gas is admitted at the bottom and passes upward through the openings of one sort or the other on the trays. The gas emerges from the plate through these openings as tiny bubbles which form a froth in the liquid pool. Sufficient contact is achieved between the gas and liquid on the plate before the gas disengages from the liquid and passes on to the next plate above. Under ideal conditions, the gas and liquid leaving the plate will be in equilibrium. In the design of plate columns, it is the number of theoretical plates required to achieve a given separation that is determined by graphical or analytical techniques. This number is dependent solely on the difficulty of separation and is determined from the material balance and equilibrium considerations. The failure of the plate to attain equilibrium is indicated by the stage efficiencies. The stage or tray efficiency is dependent on the mechanical design used and the conditions of operation. Knowing the number of theoretical plates and the stage efficiency, the number of actual plates can be determined. The height of the column is dependent on the number of plates and the spacing between the trays. The diameter of the column is dependent on the absolute flow rates of gas and liquid through the column.

Number of Plates by Graphical Construction

Consider a plate column shown schematically in Fig. 4.36. The plates are numbered treating the top plate as plate 1 and the bottommost plate as plate N, where N is the number of theoretical plates in the column. Plate n is any general plate in the column. The concentration of gas and liquid are y and x, respectively and the flow rates are G and L, respectively. The suffix appearing on these symbols indicate the plate from which the streams are originated. Thus the gas and liquid leaving the plate n are G_n and L_n and the mole fraction of solute in the gas and liquid are, respectively y_n and x_n and the mole ratios Y_n and X_n. Following this convention, the mole ratio of solute in the liquid and gas entering the plate n are Y_{n+1} and X_{n-1}, respectively. Similarly the mole ratio of solute in the liquid and gas entering the column are X_0 and Y_{N+1}, respectively and the concentration of liquid and gas leaving the column are X_N and Y_1, respectively. The solute-free gas flow rate is constant at G_S and the solute-free liquid rate at L_S throughout the column.

A solute balance for the absorber gives

$$G_S(Y_{N+1} - Y_1) = L_S(X_N - X_0) \qquad \text{...(4.53)}$$

which can be written as

$$\frac{Y_{N+1} - Y_1}{X_N - X_0} = \frac{L_S}{G_S} \qquad \text{...(4.54)}$$

This is the equation for the operating line on mole ratio terms. The line passes through the points A and B representing the terminal concentrations of the absorber and its slope is L_S/G_S. The upper

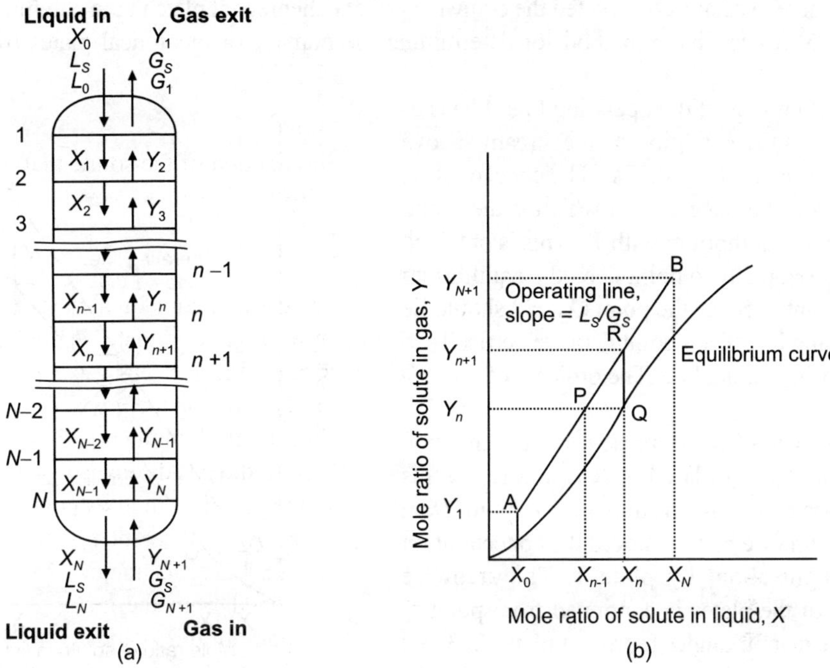

Fig. 4.36: Multistage absorption column

terminus of the operating line has coordinates (X_N, Y_{N+1}) and the lower terminus has coordinates (X_0, Y_1). The operating line and the equilibrium curve are plotted in the Fig. 4.36b in terms of mole ratio of solute in the liquid and the gas.

Now consider solute balance from the top of the column up to and including plate *n*.

$$G_S(Y_{n+1} - Y_1) = L_S(X_n - X_0) \quad \text{...(4.55)}$$

which on rearrangement gives

$$\frac{Y_{n+1} - Y_1}{X_n - X_0} = \frac{L_s}{G_s} \quad \text{...(4.56)}$$

Equation (4.56) means that the slope of the straight line joining point (X_1, Y_0) and (X_n, Y_{n+1}) is equal to L_S/G_S. This line and the operating line *AB* have one point in common and have same slope which implies that the point (X_n, Y_{n+1}) lies on the operating line *AB* since X_n and Y_{n+1} are the concentrations of liquid and gas crossing each other between plates *n* and *n* +1. Since *n* represents any plate in the column, all points, the coordinates of which represent concentrations of liquid and gas streams crossing each other between any two adjacent plates in the column such as (X_1, Y_2), (X_2, Y_3), (X_{n-1}, Y_n), etc. lie on the operating line. In short, we can say that the operating line passes through all points having coordinates of the type (X_n, Y_{n+1}) whereas the equilibrium curve passes through points having coordinates of the type (X_n, Y_n) because X_n and Y_n are the concen-trations of streams leaving the ideal plate *n*. Referring to Fig. 4.36b, let *P* be the point on the operating line with coordinates X_{n-1} and Y_n the concentration of the liquid and gas between two adjacent plates *n*-1 and *n*. Move horizontally from point *P* to point *Q* on the equilibrium curve. The ordinate of *Q* is Y_n, its abscissa should be X_n as it lies on the equilibrium curve. Move upward vertically from point *Q* to point *R* on the operating line. The *x*-coordinate of *R* is X_n and therefore the *y*-coordinate should be Y_{n+1}. Thus point *P* represents the composition of the liquid and gas streams above plate *n* and the point *R* represents those compositions below plate *n*. That is, by constructing a triangle such as Δ*PQR* between the operating line and the equilibrium curve, we have constructed the equivalent of one theoretical plate in the column. This is the basis for the McCabe-Thiele method for determining the number of theoretical plates by graphical construction.

The lower terminus of the operating line *A* has coordinate (X_0, Y_1) the condition of the streams above the top plate or plate 1 (Fig. 4.37a). The concentration of the liquid leaving plate 1 is X_1 which is the liquid concentration in equilibrium with Y_1. This is obtained by moving horizontally to point *C* on the equilibrium curve from point *A*. Since the point (X_1, Y_2) should lie on the operating line, from point *C* move vertically to point *D* on the operating line. The ordinate of point *D* should be Y_2.

D has coordinates which represent the concentration of gas and liquid below plate 1. ΔACD now represents plate 1 because by moving from point *A* to point *D* on the operating line we have achieved the concentration change brought about by plate 1. The stepwise construction of the plates is continued from point *D* and the subsequent triangles represent plates 2, 3 and so on. The triangle whose corner coincides with the upper terminus of the operating line *B*, represents the bottom plate *N*. The number of theoretical plates required to effect the desired separation is equal to the number of triangles constructed.

If the vertex of the last triangle falls beyond the upper terminus of the operating line *B*, as shown in the Fig. 4.37b, the number of theoretical plates would include the fraction of the last triangle that falls below the operating line. The following formula can be used for calculating the fraction of the theoretical plate that is to be included. Referring to the figure, fraction of the theoretical plate = *MO*/*MN*, where the vertical line at *B* divides the base *MN* of the last but incomplete triangle at *O*.

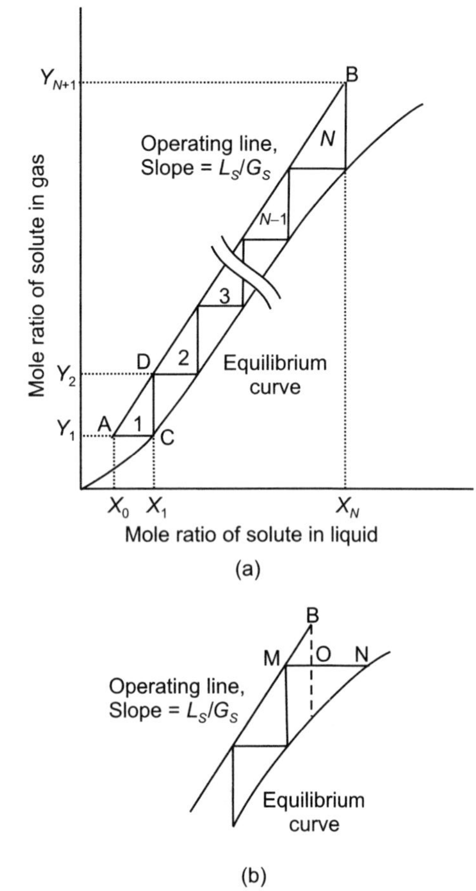

Fig. 4.37: (a) Graphical construction for the number of theoretical plates of absorber (b) Estimating the fraction of theoretical stage

The number of theoretical plates required to effect a given separation is dependent on the position of the operating line on the diagram relative to the equilibrium curve. The closer the operating line to the equilibrium curve, the larger the number of theoretical plates. Corresponding to the minimum L_S/G_S ratio, the operating line coincides with the equilibrium curve, and an infinite number of plates would be required to effect the desired separation. As the operating line moves away from the equilibrium curve corresponding to larger values of liquid to gas ratio, the number of plates decreases.

Number of Stages of Stripper

The McCabe Thiele method can be used to find the number of plates of a stripper. Here, the operating line lies below the equilibrium curve. The construction is as shown in Fig. 4.38. We carry out the stepwise construction for theoretical plates starting from point *B*, the upper terminus of the operating line and move towards point *A*, the lower terminus.

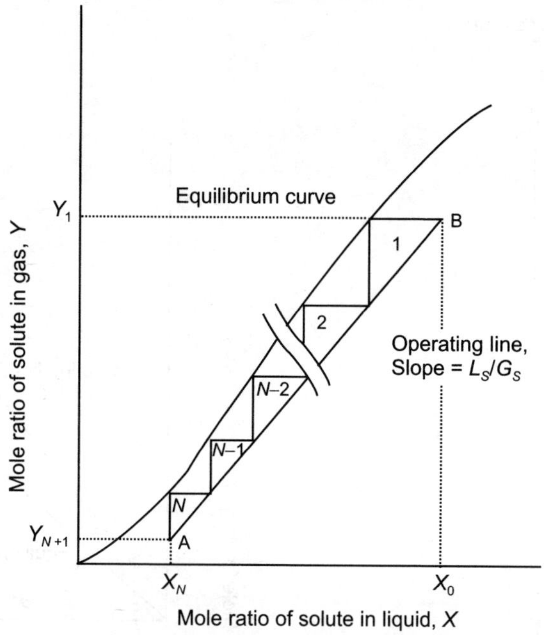

Fig. 4.38: Graphical construction for the number of theoretical plates of stripper

Analytical Method: The Kremser Equation

Consider a plate column as shown in Fig. 4.39. When the operating line and the equilibrium curves are straight lines, the number of theoretical plates can be determined by the algebraic method. For dilute solutions, the mole fraction of solute is very small so that mole fraction and mole ratio become almost the same. Also the flow of liquid and gas undergo no significant change during absorption.

$$X \approx x \quad \text{and } Y \approx y$$

$$L_N = L_0 = L \approx L_s$$

$$G_{N+1} = G_1 = G \approx G_S$$

As a result, a straight operating line is obtained even on mole fraction coordinates. For dilute solutions, the equilibrium can be generally represented by Henry's law so that the equilibrium curve will be a straight line on mole fraction coordinates (Fig. 4.40).

The Kremser equation is derived assuming that the operating line and equilibrium curve are straight lines. Consider the nth plate from the top of the countercurrent absorption tower. Since the plate is ideal, the concentration of the gas and liquid leaving this plate are equilibrium concentrations related by the linear relationship $y = mx$. Therefore, we have

$$y_n = mx_n \qquad \text{...(4.57)}$$

Here m is the slope of the equilibrium line which passes through the origin. (The analysis is also applicable for the equilibrium curve which is straight only in the region of interest. That is, from (x_0, y_1) to (x_N, y_{N+1}), in which case, the equilibrium relationship is represented by $y_n = mx_n + c$)
A solute balance for plate n is

$$L(x_n - x_{n-1}) = G(y_{n+1} - y_n) \qquad \text{...(4.58)}$$

Both x_n and x_{n-1} in Eq. (4.58) can be replaced in terms of y_n and y_{n-1} using the equilibrium relationship $y_n = mx_n$ and $y_{n-1} = mx_{n-1}$.

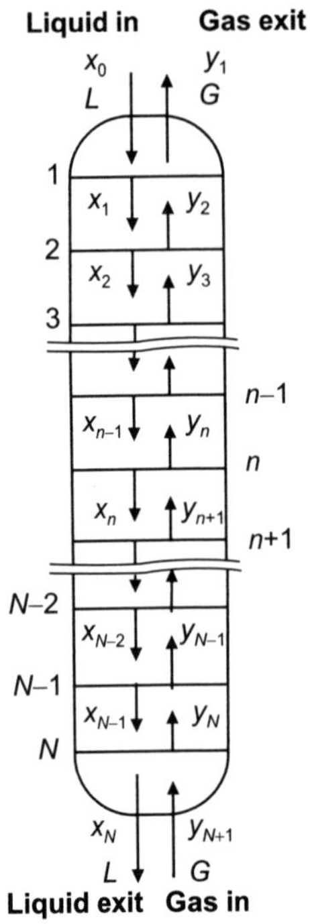

Fig. 4.39: Analytical method to find the number of stages of multistage absorption column

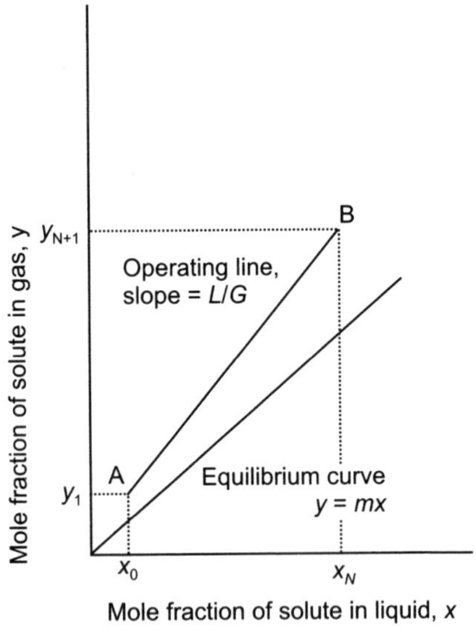

Fig. 4.40: Operating line and equilibrium line when analytical method is applicable

$$\frac{L}{m}(y_n - y_{n-1}) = G(y_{n+1} - y_n) \qquad \ldots(4.59)$$

which on rearranging gives

$$y_n + \frac{L}{mG}(y_n) = y_{n+1} + \frac{L}{mG}(y_{n-1}) \qquad \ldots(4.60)$$

$$y_n(1+A) = y_{n+1} + A(y_{n-1}) \qquad \ldots(4.61)$$

Equation (4.61) can be written as

$$y_n = \frac{y_{n+1} + A(y_{n-1})}{(1+A)} \qquad \ldots(4.62)$$

where $A = L/mG$ is known as the absorption factor for the column. Put $n = 1$ in Eq. (4.62).

$$y_1 = \frac{y_2 + A\,y_0}{(1+A)} \qquad \ldots(4.63)$$

Here y_0 is the concentration of a fictitious gas stream that is in equilibrium with the entering liquid. $y_0 = m\,x_0$.

For $n = 2$, Eq. (4.62) gives

$$y_2 = \frac{y_3 + A\,y_1}{(1+A)} \qquad \ldots(4.64)$$

Replace y_1 in Eq. (4.64) using Eq.(4.63), we get

$$y_2 = \frac{(1+A)y_3 + A^2\,y_0}{(1+A+A^2)} \qquad \ldots(4.65)$$

Now let $n = 3$ in Eq. (4.62) and use Eq. (4.65) for replacing y_2. The result is

$$y_3 = \frac{(1 + A + A^2)y_4 + A^3 y_0}{(1 + A + A^2 + A^3)} \quad ...(4.66)$$

In a similar manner for n^{th} plate, we can write

$$y_n = \frac{(1 + A + A^2 + ... + A^{n-1})y_{n+1} + A^n y_0}{(1 + A + A^2 + ... + A^n)} \quad ...(4.67)$$

The sum of the n terms of the geometric series starting with 1 and common ratio A is given by the following formula

$$(1 + A + A^2 + ... + A^{n-1}) = \frac{A^n - 1}{A - 1} \quad ...(4.68)$$

Using this result, Eq. (4.67) can be simplified as

$$y_n = \frac{(A^n - 1)y_{n+1} + A^n (A - 1) y_0}{(A^{n+1} - 1)} \quad ...(4.69)$$

Consider a solute balance from top of the column to the plate n:

$$L(x_n - x_0) = G(y_{n+1} - y_1) \quad ...(4.70)$$

Using the equilibrium relationship $y_n = mx_n$ and $y_0 = mx_0$, we have

$$\frac{L}{m}(y_n - y_0) = G(y_{n+1} - y_1) \quad ...(4.71)$$

Or

$$y_n = \frac{(y_{n+1} - y_1)}{A} + y_0 \quad ...(4.72)$$

Comparing Eqs. (4.69) and (4.72) we see that

$$\frac{(A^n - 1)y_{n+1} + A^n (A - 1) y_0}{(A^{n+1} - 1)} = \frac{(y_{n+1} - y_1)}{A} + y_0 \quad ...(4.73)$$

Subtract y_0 from both sides of Eq. (4.73) and rearrange. The result is

$$\frac{(A^n - 1)(y_{n+1} - y_0)}{(A^{n+1} - 1)} = \frac{(y_{n+1} - y_1)}{A} \quad ...(4.74)$$

or

$$\frac{(y_{n+1} - y_1)}{(y_{n+1} - y_0)} = \frac{(A^{n+1} - A)}{(A^{n+1} - 1)} \quad ...(4.75)$$

Let $n = N$ and we get

$$\frac{(y_{N+1} - y_1)}{(y_{N+1} - y_0)} = \frac{(A^{N+1} - A)}{(A^{N+1} - 1)} \quad ...(4.76)$$

Equations (4.75) and (4.76) are known as the Kremser-Brown-Souders equation or simply the Kremser equation. On the numerator of the left-hand-side of Eq. (4.76) we have $(y_{N+1} - y_1)$ which is the actual concentration change the gas stream is subjected to and on the denominator, we have $(y_{N+1} - y_0)$ which is the maximum concentration change possible for the gas stream that could be achieved if the gas were leaving in equilibrium with the liquid entering the column. Kremser equation can be used for calculating the concentration of gas stream y_n for given n or for determining the number of plates required when the terminal compositions are known.

Equation (4.76) can be rearranged to get the following alternative forms for Kremser equation:

$$\frac{(y_1 - y_0)}{(y_{N+1} - y_0)} = \frac{(y_1 - mx_0)}{(y_{N+1} - mx_0)} = \frac{(A-1)}{(A^{N+1} - 1)} \quad \text{...(4.77)}$$

$$N = \frac{\log\left[\dfrac{(A-1)}{A}\left(\dfrac{y_{N+1} - mx_0}{y_1 - mx_0}\right) + \dfrac{1}{A}\right]}{\log A} \quad \text{...(4.78)}$$

When the operating line and the equilibrium curve are straight parallel lines, $L/G = m$ and $A = 1$. Then the Kremser Eqs. (4.76) and (4.78) take the following simple forms:

$$\frac{(y_{N+1} - y_1)}{(y_{N+1} - y_0)} = \frac{N}{N+1} \quad \text{...(4.79)}$$

$$N = \frac{y_{N+1} - y_1}{y_1 - mx_0} \quad \text{...(4.80)}$$

The Kremser equations can be derived in terms of the liquid-phase compositions as well.

$$\frac{(x_0 - x_N)}{(x_0 - y_{N+1}/m)} = \frac{\left(\dfrac{1}{A}\right)^{N+1} - \dfrac{1}{A}}{\left(\dfrac{1}{A}\right)^{N+1} - 1} \quad \text{...(4.81)}$$

$$N = \frac{\log\left[\left(\dfrac{x_0 - y_{N+1}/m}{x_N - y_{N+1}/m}\right)(1 - A) + A\right]}{\log\left(\dfrac{1}{A}\right)} \quad \text{...(4.82)}$$

Equations (4.81) and (4.82) are preferred for use in the design of strippers with $1/A$ replaced by S, the stripping factor. The stripping factor is

$$S = \frac{1}{A} = \frac{mG}{L}$$

Kremser equation can be represented graphically as shown in Fig. 4.41. Equation (4.79) can be modified as

$$\frac{(y_1 - y_0)}{(y_{N+1} - y_0)} = \frac{(A-1)}{(A^{N+1} - 1)} \quad \text{...(4.83)}$$

The left-hand-side of Eq. (4.83) is the ordinate of the graph. Such diagrams are convenient for determining the number of plates for any chosen value of absorption factor. The same graph can be used for calculation of theoretical plates of strippers with the parameter A replaced by $1/S$ and the ordinate by $\dfrac{x_N - y_{N+1}/m}{x_0 - y_{N+1}/m}$.

As long as the operating line and equilibrium curve are straight lines, the absorption factor A is constant. However, for practical application of the Kremser equation, the variation in A can be accommodated by taking the average A for use in the equations to be the geometric mean of absorption factors at the top and bottom of the column. Kremser equation can be written in terms of mole ratios as well, provided the equilibrium curve and the operating lines are straight lines in mole ratio coordinates. On mole ratio coordinates the Kremser equation is

$$N = \frac{\log\left[\frac{(A-1)}{A}\left(\frac{Y_{N+1} - mX_0}{Y_1 - mX_0}\right) + \frac{1}{A}\right]}{\log A} \quad \ldots(4.84)$$

Absorption Factor

The factor A appearing in the Kremser equation is the absorption factor and the factor S, which is the reciprocal of the absorption factor is called the stripping factor. The absorption factor can be viewed as the ratio of the slope of the operating line to that of the equilibrium line.

$$A = \frac{L/G}{m} = \frac{L}{mG}$$

If the operating line and the equilibrium curve are parallel, the absorption factor will be unity. If the absorption factor is less than unity, the operating line may converge with the equilibrium curve at the bottom of the column. This happens because the slope of the operating line (L/G) is less than the slope of the equilibrium curve, m. The lower end of the operating line can never come closer to the equilibrium curve, which means that the gas leaving the top of the column can never come in equilibrium with the solvent admitted. As a result, for values of A less than unity, as evident from the Fig. 4.41, the extent of absorption is limited because very large number of plates are required to achieve a good recovery, and even for infinite number of plates the recovery is not complete. From Eq. (4.83) we have for small values of A,

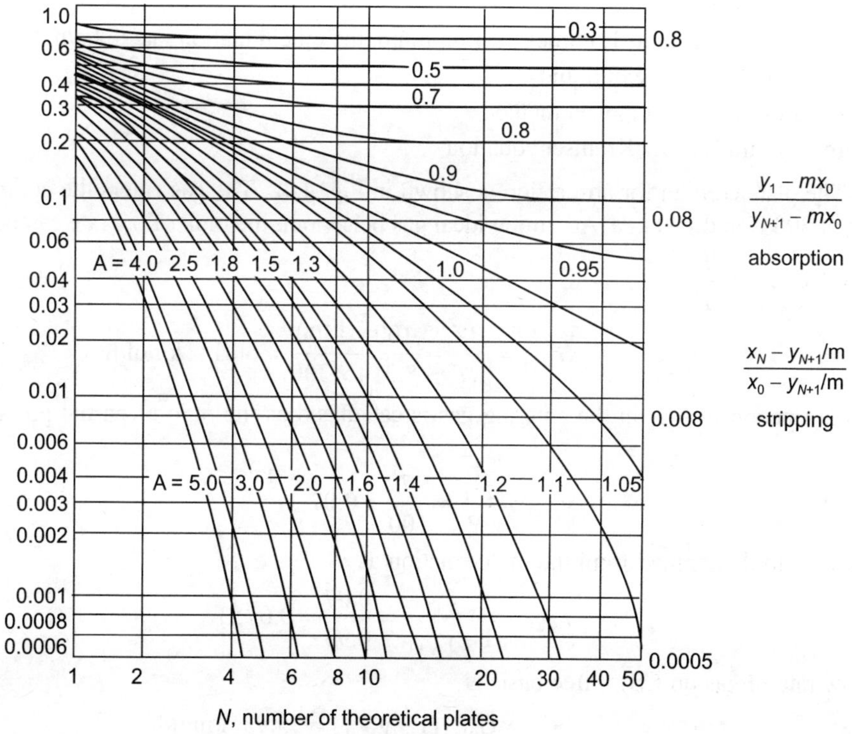

Fig. 4.41: Graphical presentation for Kremser equation for calculating number of theoretical plates

$$\frac{(y_1 - y_0)}{(y_{N+1} - y_0)} = 1 - A, \quad \text{as } N \to \infty \qquad \text{...(4.85)}$$

The ordinate of the graph reaches an asymptotic value for a large number of plates, thus severely limiting the degree of absorption. For example, for $A = 0.3$, seventy percent of the solute remains in the gas leaving the column even if the entering liquid is free of solute.

If absorption factor is greater than unity, the upper end of the operating line moves away from the equilibrium curve and the number of plates decreases for a given absorption. It is clear from the Fig. 4.41 that for values of A greater than unity, a high degree of absorption results either by increasing the number of plates or by increasing the absorption factor. A large value of A means a large solvent throughput and therefore a high cost of pumping and circulation of solvent. Also since the solution gets diluted, the recovery of solute from the liquid leaving the absorber and regeneration of solvent for further use will be difficult. We have to treat large quantities of very dilute solution for separation and therefore, the cost of solvent regeneration is high. At the same time, as pointed out earlier, the number of trays and the cost of equipment decrease. In arriving at an optimum absorption factor for use in the column, we may have to consider these opposing cost tendencies. It is suggested that the most economical absorption factor lies in the range 1.25 - 2.0, around 1.4 being the ideal one in most cases.

Example 4.7: Carbon disulphide is to be recovered from a nitrogen-CS_2 mixture by scrubbing with an absorbent oil (Molecular weight 180) in a plate column. The gas enters the column at a rate of 1500 m^3/h at 300 K and 100 kPa and the partial pressure of CS_2 in the entering gas is 7 kPa. The CS_2 content is to be reduced to 0.5%. The solvent oil entering the column is free of CS_2. The CS_2-oil solutions may be assumed to obey Raoult's law. The vapour pressure of CS_2 at 300 K is 51.6 kPa. Assuming isothermal operation, determine:

(a) The minimum L_S/G_S ratio

(b) For a liquid-to-gas ratio 1.5 times the minimum, the oil circulation rate in kg/h

(c) Mole % of CS_2 in the exit liquid

(d) Number of trays by graphical method

(e) Number of trays using Kremser equation

***Solution*:** The plate column for absorption is shown in Fig. 4.42. The gas enters the column at a rate 1500 m^3/h at 300 K and 100 kPa. Assuming ideal gas behaviour, the molar flow rate can be calculated using the relation $n = \dfrac{PV}{RT}$.

Thus
$$G_{N+1} = \frac{PV}{RT} = \frac{100 \times 1500}{8.314 \times 300} = 60.14 \text{ kmol/h}$$

The mole fraction of CS_2 in the entering gas is calculated as the ratio of partial pressure to total pressure.

$$y_{N+1} = \frac{\bar{p}_A}{P} = \frac{7}{100} = 0.07$$

The mole ratio determined from the mole fraction as

$$Y_{N+1} = \frac{y_{N+1}}{1 - y_{N+1}} = \frac{0.07}{0.93} = 0.0753$$

The flow rate of gas on CS_2 – free basis is

$$G_S = G_{N+1}(1 - y_{N+1}) = 55.93 \text{ kmol/h}$$

The CS_2 content in the gas leaving the column is 0.5%. Therefore,

$$y_1 = 0.005 \text{ and } Y_1 = \frac{0.005}{0.995} = 0.00503$$

The solute free gas flow rate G_S is constant throughout the column. The total flow rate of the gas at the top of the column is G_1 which may be calculated as

$$G_1 = G_S(1 + Y_1) = 55.93 \times (1 + 0.00503) = 56.21 \text{ kmol/h}$$

The solvent enters free of CS_2 at the top.

$$x_0 = 0, X_0 = 0 \text{ and } L_S = L_0$$

The known compositions and flow rates are indicated in the Fig. 4.42.

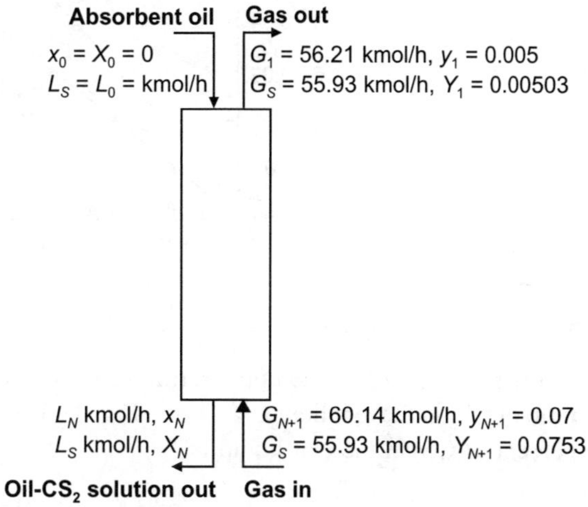

Fig. 4.42: Absorption column in Example 4.7

The equilibrium relationship is given by Raoult's law.

$$y = \frac{P_A^S}{P} x = \frac{51.6}{100} x = 0.516x$$

(a) To find the minimum L_S/G_S value, the equilibrium curve is plotted on mole ratio co-ordinates. The equilibrium relationship in mole ratio coordinates is

$$\frac{Y}{1+Y} = 0.516 \frac{X}{1+X}$$

To plot the equilibrium curve, coordinates of various points on the curve are to be obtained. Assume various X values and obtain the corresponding Y values using the above equation. For example, when $X = 0.06$,

$$\frac{Y}{1+Y} = 0.516 \frac{0.06}{1+0.06} = 0.0292, \text{ so that } Y = 0.03$$

The results of these calculations are given below:

X	0	0.02	0.04	0.06	0.08	0.10	0.12	0.14	0.16
Y	0	0.0102	0.0202	0.03	0.0397	0.0492	0.0585	0.0677	0.0766

The equilibrium curve is plotted and the point A (0, 0.00503), the lower terminus of the operating line is marked as shown in Fig. 4.43. The limiting operating line AB is drawn tangential to the equilibrium curve.

The slope of line AB is the minimum L_S/G_S ratio permissible for the column. From Fig. 4.43, the x-coordinate of point B, X_N(max) = 0.157. Substituting this in Eq. (4.38)

$$\left(\frac{L_S}{G_S}\right)_{\min} = \frac{Y_{N+1} - Y_1}{X_N(\max) - X_0} = \frac{0.0753 - 0.00503}{0.157 - 0} = 0.4476 \qquad \textbf{(Answer)}$$

(b) For a liquid rate 1.5 times the minimum,

$$\left(\frac{L_S}{G_S}\right) = 1.5\left(\frac{L_S}{G_S}\right)_{\min} = 1.5 \times 0.4476 = 0.6714$$

Since G_S = 55.93 kmol/h, L_S = 0.6714 × 55.93 = 37.55 kmol/h.

Since the molecular weight of oil is 180, the oil circulation rate = 37.55 × 180 = 6759 kg/h

(Answer)

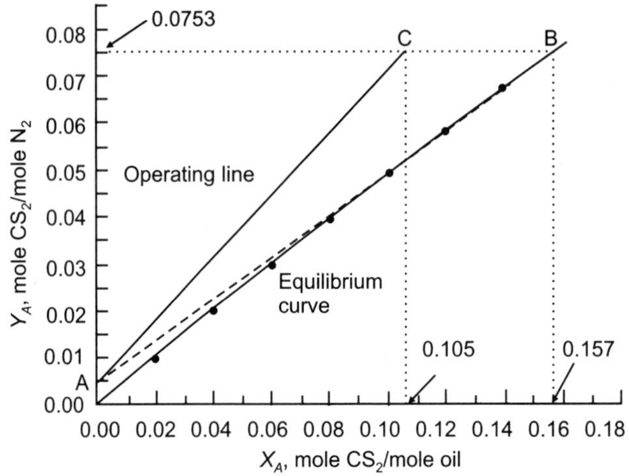

Fig. 4.43: Operating line and Equilibrium curve for the absorber in Example 4.7

(c) x-coordinate of the upper terminus of the operating line (point C) in the figure gives the mole ratio of CS_2 in the exit liquid, X_N. From Fig. 4.43, $X_N = 0.105$ mole CS_2 per mole oil. Therefore the mole percent of CS_2 in the exit liquid is

$$(x_N) \times 100 = \frac{X_N}{1 + X_N} \times 100 = \frac{0.105}{1.105} \times 100 = 9.5\% \qquad \textbf{(Answer)}$$

(d) The number of theoretical stages are determined by graphical construction as shown in the Fig. 4.44. A horizontal line is drawn from point A, the lower terminus of the operating line which represents the conditions above plate 1. The intersection of this line with the equilibrium curve is X and a vertical line is constructed at X, which meets the operating line at Y. The triangle ΔAXY represent the first plate in the column. The construction is continued till the upper terminus of the operating line point B is reached. The integral number of plates is found to be 5.

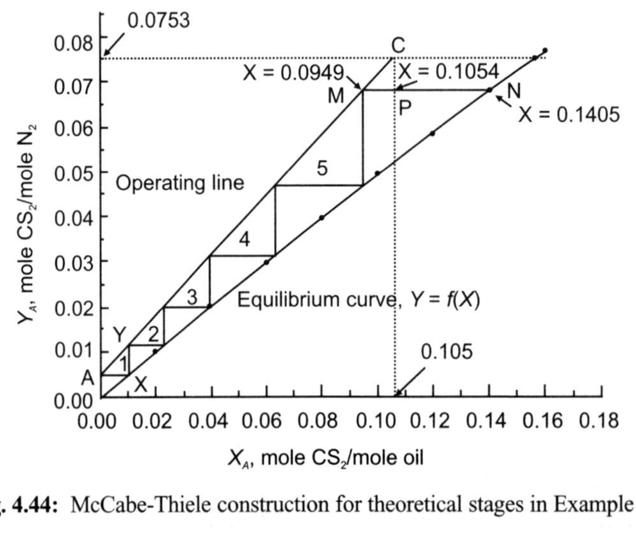

Fig. 4.44: McCabe-Thiele construction for theoretical stages in Example 4.7

The fraction of the theoretical plate is determined as the ratio of the line length PM to the line length MN. By noting down the x-coordinates of M, N and P, the fractional stage can be evaluated as

$$\text{Fractional stage} = \frac{\text{line } MP}{\text{line } MN} = \frac{(0.1054 - 0.0949)}{(0.1405 - 0.0949)} = 0.23$$

The number of theoretical plates = 5 + 0.23 = 5.23 **(Answer)**

(d) The number of theoretical plates can be determined by Kremser equation (Eq. 4.86) assuming that the operating line and equilibrium curve are straight lines on mole fraction coordinates.

$$N = \frac{\log\left[\frac{(A-1)}{A}\left(\frac{y_{N+1} - mx_0}{y_1 - mx_0}\right) + \frac{1}{A}\right]}{\log A} \qquad \ldots(4.86)$$

The absorption factor A to be used here is taken to be the geometric mean of the absorption factors at the top and bottom of the column $A = (A_1 A_N)^{0.5}$. Where

$$A_1 = \frac{L_1}{mG_1} \approx \frac{L_0}{mG_1} \text{ and } A_N = \frac{L_N}{mG_N} \approx \frac{L_N}{mG_{N+1}}$$

$L_0 = L_S = 37.55$ kmol/h; $G_1 = 56.21$ kmol/h; $m = 0.516$

$$A_1 = \frac{L_0}{mG_1} = \frac{37.55}{0.516 \times 56.21} = 1.2946$$

$L_N = L_S(1 + X_N) = 37.55(1 + 0.105) = 41.49$ kmol/h; $G_{N+1} = 60.14$ kmol/h

$$A_N = \frac{L_N}{mG_{N+1}} = \frac{41.49}{0.516 \times 60.14} = 1.3370$$

$$A = (A_1 A_N)^{0.5} = (1.2946 \times 1.3370)^{0.5} = 1.3156$$

Substitute the values in Eq. (4.86)

$$N = \frac{\log\left[\frac{(1.3156-1)}{1.3156}\left(\frac{0.07-0}{0.005-0}\right) + \frac{1}{1.3156}\right]}{\log 1.3156} = 5.16$$

The number of theoretical plates by Kremser equation = 5.16. **(Answer)**

It is very close to the number obtained by the graphical method because the assumption of straight operating line does not introduce serious error as the solutions involved are very dilute.

Example 4.8: The substance **A** (Molecular weight = 17) is stripped off from an aqueous solution containing 17.2% by weight **A** in a plate column using air. If 90% of **A** is to be removed, calculate for 1000 kg/h of feed solution,

(a) Minimum gas rate, kmol/h

(b) For a gas rate 1.5 times the minimum, the number of ideal stages required.

Equilibrium data (x, y are mole fractions of **A**)

x	0.0196	0.0476	0.0909	0.1304	0.1667	0.20	0.2308
y	0.0303	0.0663	0.1189	0.1825	0.2674	0.3711	0.4792

***Solution*:** The concentration and flow rates specified are indicated in the schematic diagram of the stripper (Fig. 4.45).

Inlet concentration of the liquid = 17.2% (weight) **A**.

The composition on mole basis is $\frac{17.2/17}{17.2/17 + 82.8/18} = 18.03\%$ **A**

Therefore, $x_0 = 0.1803 . X_0 = \dfrac{0.1803}{1-0.1803} = 0.22$

When 90% **A** is recovered, the exit liquid will contain only 10% of the **A** entering the column.

Therefore, $X_N = 0.10 \times 0.22 = 0.022$

The solution feed rate = 1000 kg/h. The average molecular weight of a solution containing 0.1803 mole **A** and (1 – 0.1803) mole water per mole of solution is

$$MW = 0.1803 \times 17 + (1 - 0.1803) \times 18 = 17.8197$$

The molar flow rate of the solution is

$$L_0 = \frac{1000}{17.8197} = 56.12\,\text{kmol/h}$$

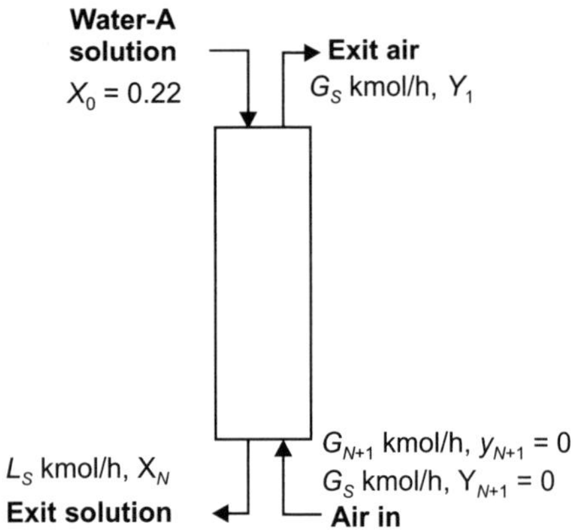

Fig. 4.45: Stripper in Example 4.8

Therefore, solute-free solution feed rate,

$$L_S = 56.12 \times (1 - 0.1803) = 46.0\,\text{kmol/h}$$

Since the air entering the column is free of solute, $Y_{N+1} = 0$.

The equilibrium data is converted into mole ratios using the following relations:

$$X = \frac{x}{1-x} \text{ and } Y = \frac{y}{1-y}$$

The following results are obtained:

X	0.02	0.05	0.10	0.15	0.20	0.25	0.30
Y	0.031	0.071	0.135	0.223	0.365	0.590	0.920

(a) The equilibrium curve is plotted as shown in Fig. 4.46. The lower end of the operating line, point A has coordinates (X_N, Y_{N+1}). Point A (0.022,0) is marked on the x-axis. The upper end of the operating line has coordinates (X_0,Y_1). Since X_0 is specified to be 0.22, the operating line terminates on the vertical line at $X = 0.22$. The slope of the operating line is L_s/G_s and depending on the quantity G_s employed operating line upper terminus C can lie anywhere on the vertical at $X = 0.22$. For minimum gas rate, the slope L_S/G_S is maximum and the limiting operating line is tangential to the equilibrium curve as shown by line AB. The slope of the limiting operating line is calculated using the coordinates of B.

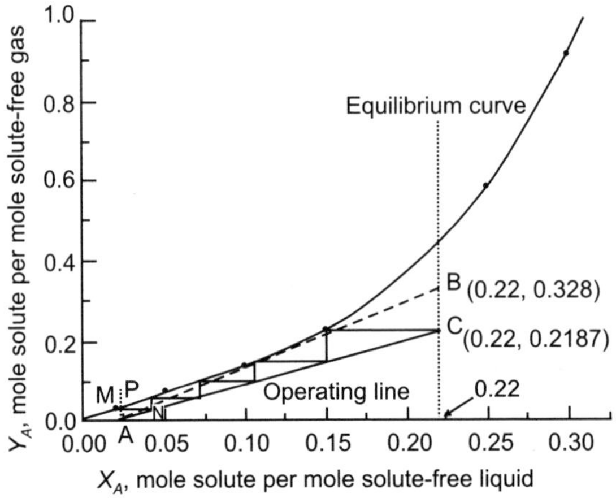

Fig. 4.46: The graphical construction for the number of theoretical plates in Example 4.8

$$\left(\frac{L_S}{G_S}\right)_{max} = \frac{0.328-0}{0.220-0.022} = 1.6566$$

The minimum gas rate is calculated as

$$G_S(\text{min}) = \left(\frac{L_s}{1.6566}\right) = \frac{46.0}{1.6566} = 27.77 \text{ kmol/h}$$

(b) When the gas rate employed is 1.5 times the minimum G_S, the slope of the operating line is

$$\left(\frac{L_S}{G_S}\right) = \frac{(Ls/Gs)_{max}}{1.5} = 1.1044$$

The operating line *AC* is drawn with the desired slope as shown in Fig. 4.46. McCabe-Thiele stepwise construction is carried out beginning with point *C* and terminating at point *A*. The number of stages from the figure is 4 + fractional stage. The fraction is evaluated as the ratio of the line length *PN* to the line length *MN* in Fig. 4.46. The fraction is found to be 0.76. Therefore number of ideal plates required = 4.76 **(Answer)**

Example 4.9: The gas leaving an alcohol fermenter consists of 10% CO_2 and 90% air. 80% of the CO_2 in the gas is to be recovered by absorbing in a water solution of triethanolamine containing 0.02 mole CO_2 per mole of amine solution. Absorption is carried out in a 4-plate column at 298 K. The equilibrium data for 298 K is: (X = mole CO_2/mole amine; Y = mole CO_2/mole air)

X	0.01	0.02	0.03	0.04	0.05	0.06	0.07	0.08	0.09	0.10	0.11
Y	0.003	0.008	0.015	0.023	0.032	0.043	0.055	0.068	0.083	0.099	0.12

Determine for 100 kmol/h feed gas:

(a) The amine solution required

(b) Mole percent of CO_2 in the exit liquid

***Solution*:** An absorption column having 4 ideal plates is to treat a gas to absorb CO_2 using an amine solution. $G_{N+1} = 100$ kmol/h

The mole fraction of CO_2 in the entering gas is

$$y_{N+1} = 0.10$$

The mole ratio determined from the mole fraction as

$$Y_{N+1} = \frac{y_{N+1}}{1-y_{N+1}} = \frac{0.10}{0.90} = 0.1111$$

The flow rate of gas on CO_2-free basis is

$$G_S = G_{N+1}(1-y_{N+1}) = 90 \text{ kmol/h}$$

20% of CO_2 in the entering gas leaves the column unabsorbed. Therefore,

$$Y_1 = 0.2 \times 0.1111 = 0.02222$$

The solvent admitted at the top contains 0.02 mole CO_2 per mole of amine. $X_0 = 0.02$

The compositions and flow rates are indicated in Fig. 4.47.

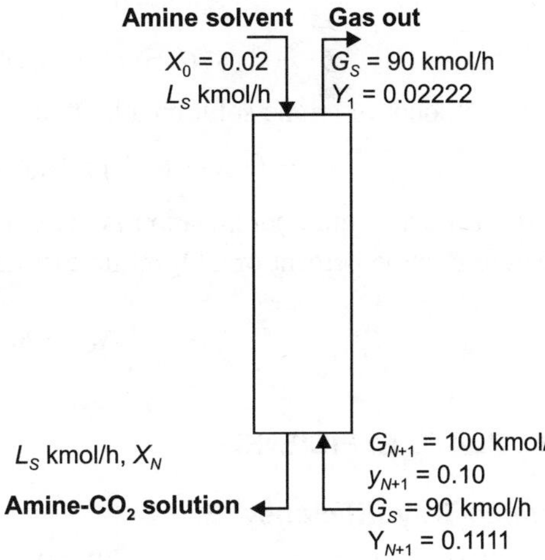

Fig. 4.47: Absorber for Example 4.9

Equilibrium curve on mole ratios is plotted as shown in Fig. 4.48. The lower end of the operating line is at point *A* whose coordinates are given by the composition of the gas and liquid above plate 1.

Point A is thus located with coordinates ($X_0 = 0.02$, $Y_1 = 0.0222$). The upper terminus of the operating line, point B should lie on the horizontal at $Y_{N+1} = 0.1111$. The operating line AB is to be drawn such that exactly 4 numbers of ideal plates can be constructed between the operating line and the equilibrium curve. The slope of the operating line gives L_S/G_S. Thus L_S/G_S is to be determined by trial and error. In Fig. 4.48 is shown the operating line AB which gives exactly 4 numbers of triangles when McCabe-Thiele stepwise construction for theoretical plates is carried out. The coordinates of point B are noted. These are $X_N = 0.0968$, $Y_{N+1} = 0.1111$.

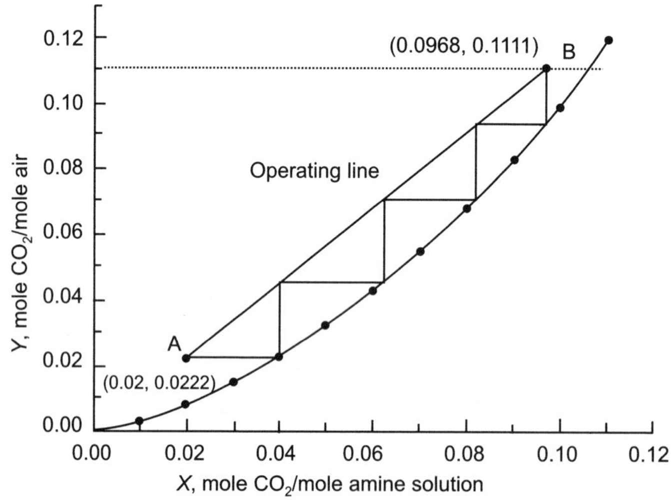

Fig. 4.48: Graphical construction for operating line in Example 4.9

(a) The slope of the operating line is calculated as

$$\frac{L_S}{G_S} = \frac{0.1111 - 0.0222}{0.0968 - 0.02} = 1.1576$$

$$L_S = 1.1576 \times 90 = 104.18 \text{ kmol/h}$$

The amount of amine solution admitted is L_0 kmol/h.

$$L_0 = L_s \times (1 + X_0) = 106.26 \text{ kmol/h}$$ **(Answer)**

(b) The exit liquid composition is given by $X_N = 0.0968$ kmol CO_2 per mole of amine solution. Therefore, mole percent of CO_2 in the exit liquid is

$$x_N \times 100 = \frac{X_N}{1 + X_N} \times 100 = 8.8\%$$ **(Answer)**

4.6 PLATE EFFICIENCY

Overall Tray Efficiency, E_o

Overall tray efficiency, or the overall column efficiency is defined as the ratio of the number of theoretical plates calculated to achieve a given separation to the number of actual plates required for the same separation. Thus,

$$E_o = \frac{\text{Number of theoretical plates}}{\text{Number of actual plates}} \quad \ldots(4.87)$$

The overall stage efficiency depends on the geometry and design of the trays, rates and path of flow of the liquid and the gas streams and the composition and properties of these streams. The overall stage efficiency is found to depend strongly on the liquid viscosity and the solubility of the diffusing substance in the liquid absorbent. The most common method for estimating the overall stage efficiency is by comparing with performance data from actual operating columns. For absorbers and strippers, E_o is found to be very low, less than about 50%. A high stage efficiency is achieved, if the liquid viscosity is low and the gas solubility is high.

An empirical relationship is proposed by Drickamer and Bradford (1943) for estimating the overall stage efficiency of columns handling petroleum hydrocarbons in terms of the average viscosity of the feed.

$$E_o = 0.17 - 0.616 \log_{10} \sum \left[x_f \frac{\mu_L}{\mu_W} \right] \quad \ldots(4.88)$$

where x_f is the mole fraction of solute in the feed, μ_L is the viscosity at the mean temperature, μ_w is the viscosity of water at 293 K (approximately, 10^{-3} kg/m s). O'Connel (1946) obtained a more general correlation by using a parameter that included the liquid density and the Henry's law constant in addition to the liquid viscosity. Later the Henry's law constant in the correlation were replaced by K-values by Edmister (1949) and Lockhart and Leggett (1958). An O'Connell type plot is given in Fig. 4.49 for estimating the overall plate efficiency in absorption columns.

The following empirical equation is found to represent the data presented in Fig. 4.49 reasonably well.

$$\log E_o = -0.8237 - 0.9525 \log\left(\frac{K_i M_L \mu_L}{\rho_L}\right) - 0.0896 \left[\log\left(\frac{K_i M_L \mu_L}{\rho_L}\right)\right]^2 \quad \ldots(4.89)$$

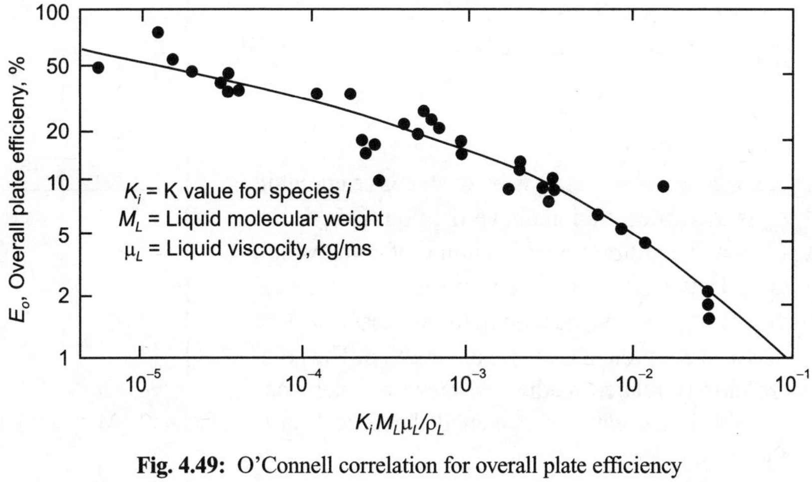

Fig. 4.49: O'Connell correlation for overall plate efficiency

where K_i = K value for species i, M_L = liquid molecular weight, μ_L = liquid viscosity, kg/m s, and ρ_L = density of liquid.

Murphree Plate Efficiency

Among the semi-theoretical models that have been proposed to predict the overall stage efficiency, the Murphree plate efficiency is relatively simple and is widely used. The plate efficiency is defined for each component in the mixture and for each plate. These efficiency values are utilised for determining the conditions for each tray and averaged for estimating the overall plate efficiency. Consider the n th plate in the tower as shown in Fig. 4.50.

The Murphree plate efficiency referred to the gas phase is the ratio of the actual change in the gas phase concentration to the change which would have resulted if the gas phase actually reached a state of equilibrium with the liquid. The Murphree efficiency of the nth plate referred to the gas phase is defined as

$$E_{MG} = \frac{y_{n+1} - y_n}{y_{n+1} - y_n^*} \quad \ldots(4.90)$$

In the above equation, y_n^* is the concentration of the gas-phase leaving if it were in equilibrium with the actual exit liquid concentration x_n. The Murphree tray efficiency referred to the liquid phase, E_{ML} can be defined in a similar manner:

$$E_{ML} = \frac{x_n - x_{n-1}}{x_n^* - x_{n-1}} \quad \ldots(4.91)$$

where x_n^* is the concentration of the liquid if it were leaving in equilibrium with the exit gas of composition y_n.

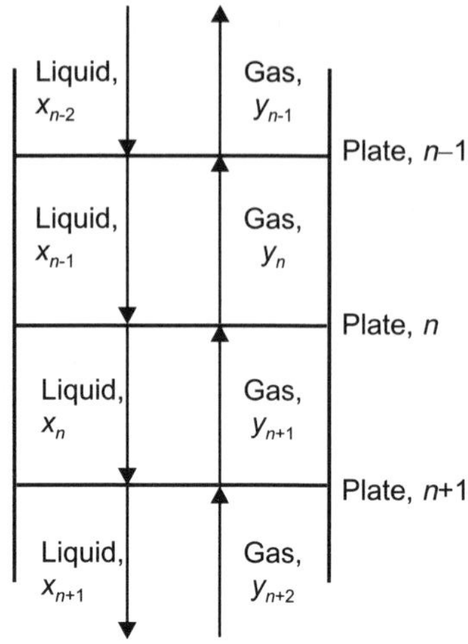

Fig. 4.50: Concentration changes on the nth plate in the absorption tower

Murphree Point Efficiency

Murphree tray efficiency represents the true physical picture only if the liquid on each tray is well mixed. The Murphree point efficiency is defined in order to take care of the incomplete mixing of liquid on the trays. It is assumed that liquid composition varies in the horizontal flow direction starting the downcomer, but is uniform in the vertical direction. Consider Fig. 4.51.

The Murphree point efficiency is defined as

$$E_{OG} = \frac{y_{p,n+1} - y_{p,n}}{y_{p,n+1} - y_{p,n}^*} \quad \ldots(4.92)$$

where $y_{p,n+1}$ is the concentration of the gas below the given point in the liquid and $y_{p,n}$ is the concentration above the point. $y_{p,n}^*$ is the gas phase composition in equilibrium with a liquid of composition $x_{p,n}$. Since the liquid is assumed completely mixed in the vertical direction, $x_{p,n}$ is constant at a given location in the vertical direction and therefore $y_{p,n}^*$ is also constant in the vertical direction. The point efficiency is more closely related to the resistance to interphase diffusion and therefore it has got a more sound theoretical basis than the plate efficiency.

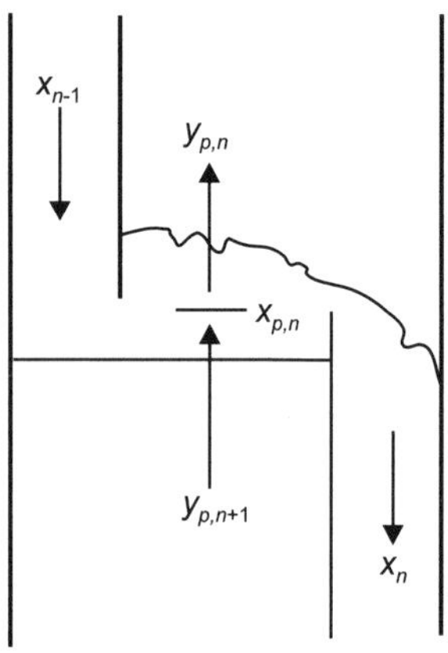

Fig. 4.51: Murphree point efficiency

Relation between Column Efficiency and Murphree Efficiency

In large diameter columns, the distance traversed by the liquid across the plate from the inlet weir to the overflow weir is so long that the concentration of solute in the liquid near the inlet weir and that in the liquid near the overflow weir are considerably different. Consequently, even the vapour entering the plate with a uniform concentration will be subjected to absorption of the solute to different extents. The gas crossing the liquid near the inlet weir will undergo greater concentration change than the gas near the liquid exit from the plate. The average concentration of the exit gas, y_n in Eq. (4.90) may exceed the concentration $y_{p,n}^*$ which is the gas phase equivalent of the liquid leaving the plate. Therefore, the

Murphree plate efficiency may exceed 100%. However, the point efficiency, by definition, cannot exceed 100% and is always less than the plate efficiency.

The Murphree point efficiency being a fundamental quantity closer to the actual mass transfer process than other types of efficiencies, it serves as a basis for deriving semi-theoretical models to estimate the overall column efficiency. The following assumptions are made regarding the flow behaviour of vapour and liquid:

1. The liquid is assumed well mixed in the vertical direction. The liquid concentration is constant at x_p from the tray deck to the top of the froth.
2. The vapour reaching a plate n is assumed to be of uniform concentration y_{n+1} across the plate cross section
3. The vapour passing upward through the liquid along the vertical direction is in plug flow. There is no mixing of the vapour while in contact with the liquid.

Let G kmol/m^2 s of total gas with mole fraction of solute y_{n+1} reach plate n on the tower and leaves with mole fraction y_n. Since the solute absorbed is so small in comparison with total flow rate, the molar flow rate of the gas G can be assumed constant over the plate. Consider a differential fluid element, unit area in cross section and of height dh_L in the vertical direction along the liquid pool. Let the local concentration of solute in the gas reaching the plate in this direction is $y_{p,n+1}$ and that in the gas leaving the liquid pool is $y_{p,n}$. The solute concentration of the gas at the differential section is y_p. The concentration in the liquid in the vertical direction is assumed constant at x_p, the gas-phase equilibrium equivalent of which is $y^*_{p,n}$. Let a (m^2/m^3) be the interfacial area of mass transfer per unit volume between gas and liquid. The gas undergoes a differential change in concentration dy while it crosses the differential section. The rate of absorption is

$$-Gdy = K_y(adh_L)(y_p - y^*_p) \quad \ldots(4.93)$$

where K_y is the overall mass transfer coefficient. Rearranging this equation, we get

$$-\frac{dy}{y_p - y^*_p} = \frac{K_y a}{G} dh_L \quad \ldots(4.94)$$

Integration of the above equation leads to

$$-\int_{y_{p,n+1}}^{y_{p,n}} \frac{dy}{y_p - y^*_p} = \frac{K_y a}{G} \int_0^{h_L} dh_L \quad \ldots(4.95)$$

$$\ln \frac{y_{p,n+1} - y^*_p}{y_{p,n} - y^*_p} = \frac{K_y a h_L}{G} \quad \ldots(4.96)$$

The left-hand-side of Eq. (4.96) can be treated as the number of overall gas transfer unit denoted by N_{tOG}. The concept of transfer unit and now it is utilized to determine the height of continuous contact units are discussed in detail in the coming sections. Equation (4.96) can be written as

$$\frac{y_{p,n+1} - y^*_p}{y_{p,n} - y^*_p} = e^{N_{toG}} \quad \ldots(4.97)$$

$$\frac{y_{p,n} - y^*_p}{y_{p,n+1} - y^*_p} = e^{-N_{toG}} \quad \ldots(4.98)$$

We can rearrange Eq. (4.98) as

$$\frac{y_{p,n+1} - y_{p,n}}{y_{p,n+1} - y^*_p} = 1 - e^{-N_{toG}} \quad \ldots(4.99)$$

Utilizing the definition of Murphree point efficiency given by Eq. (4.92), the above equation becomes

$$E_{OG} = 1 - e^{-N_{toG}} \quad ...(4.100)$$

Equation (4.100) is the relationship between the Murphree point efficiency and the number of transfer units. The relationship between the Murphree tray efficiency and the point efficiency can be derived by integrating the local E_{OG} values over the entire tray. For the limiting cases in which the liquid is completely back mixed so that the liquid composition is constant at x_n everywhere on the plate,

$$E_{MG} = E_{OG} \quad ...(4.101)$$

For plug flow of liquid across the tray with no longitudinal diffusion, it was shown that (Lewis, 1936)

$$E_{MG} = \frac{L}{mG}\left(e^{\frac{mG}{L}E_{OG}} - 1\right) \quad ...(4.102)$$

where L and G are the liquid and gas molar flow rates and m is the slope of the equilibrium line.

The plate efficiency is considerably reduced when the strong liquid from the lower plate is carried away by entrainment to the upper plate where it gets mixed with weak liquid. The apparent Murphree vapour efficiency corrected for entrainment E_{MGE} is lower than the efficiency in the absence of entrainment. The following simple, but approximate relation can be derived showing the effect of the amount of entrainment E and the L/G ratio on the apparent vapour efficiency.

$$E_{MGE} = \frac{E_{MG}}{1 + E_{MG}\dfrac{\varepsilon G}{L}} \quad ...(4.103)$$

where ε is the moles of liquid entrained per mole of dry vapour.

If E_{MG} is the same for all plates in a column and if the operating and equilibrium lines are straight, the relationship between the overall column efficiency and Murphree plate efficiency is given as

$$E_O = \frac{\ln\left\{1 + E_{MG}\left[\left(\dfrac{mG}{L}\right) - 1\right]\right\}}{\ln\left(\dfrac{mG}{L}\right)} \quad ...(4.104)$$

Example 4.10: Air is used for stripping ammonia from water solution in a plate column operated at 303 K and 101.3 kPa. Air enters a plate free of ammonia at a molar rate twice that of liquid. The liquid entering contains 0.2 mole% ammonia. If the Murphree gas efficiency of the plate is 80% and the equilibrium relationship is $\bar{p}_{NH_3}(\text{kPa}) = 129 x_{NH_3}$ at 303 K, find the exit water composition.

***Solution*:** Consider the nth plate in the stripping column (Fig. 4.52).

The Murphree efficiency of the plate is given by Eq. (4.90) which can be written as

$$E_{MG} = \frac{y_n - y_{n+1}}{y_n^* - y_{n+1}}$$

The liquid entering contains 0.2 mole% ammonia. $x_{n-1} = 0.002$; Air enters a plate free of ammonia. $y_{n+1} = 0$.

The equilibrium relationship is given as $\bar{p}_{NH_3} = 129 x_{NH_3}$ or

$$y^* = \frac{129}{101.3}x = 1.2734\, x.$$

y_n^* is the concentration of the gas leaving the plate in equilibrium with the liquid leaving. Therefore,

$$y_n^* = 1.2734\, x_n$$

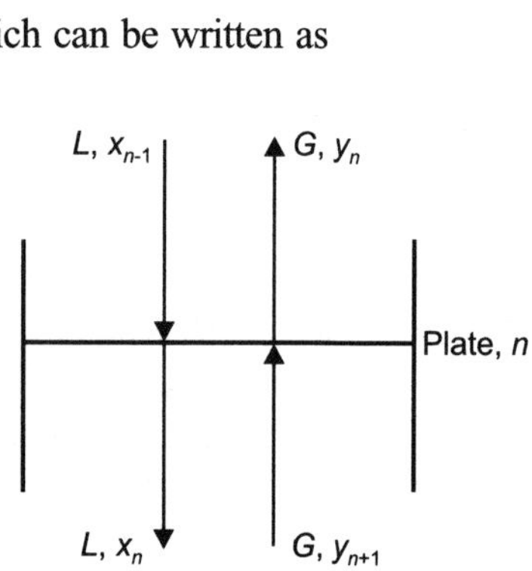

Fig. 4.52: The ideal plate in Example 4.10

In Eq. (4.90), $E_{MG} = 0.80$

$$E_{MG} = 0.80 = \frac{y_n - 0}{1.2734\,x_n - 0}$$

which can be simplified as

$$y_n = 1.0187x_n \tag{A}$$

The ammonia balance on the plate gives

$$G(y_n - y_{n+1}) = L(x_{n-1} - x_n) \tag{B}$$

In writing this equation, it is assumed that the changes in the gas and liquid flow rates over the plate are negligible. Also it is given that $G = 2L$. Equation (B) can be simplified as

$$2L(y_n - 0) = L(0.002 - x_n)$$

$$y_n = 0.001 - 0.5x_n \tag{C}$$

Solving Eq. (A) and (C) simultaneously,

$$1.5187x_n = 0.001$$

Or the composition of the exit liquid, $x_n = 0.00066$. The exit liquid contains 0.066% ammonia.

(Answer)

Example 4.11: Gas leaving an alcohol fermenter consists of 98% CO_2 and 2% ethanol. 97% alcohol is absorbed in pure water in a plate column operated at 303 K and 110 kPa. The gas is admitted at a rate 180 kmol/h and water rate is 1.3 times the minimum. Equilibrium can be represented by $y = 0.57x$. The overall stage efficiency is 30%. Estimate the average vapour efficiency.

***Solution*:** Equation (4.104) relates the Murphree plate efficiency to the overall stage efficiency

$$E_O = \frac{\ln\left\{1 + E_{MG}\left[\left(\frac{mG}{L}\right) - 1\right]\right\}}{\ln\left(\frac{mG}{L}\right)} \qquad \ldots(4.104)$$

The overall stage efficiency $E_0 = 0.30$.

The mole fraction of ethanol in the entering gas is 0.02.

$$y_{N+1} = 0.02 \text{ and } Y_{N+1} = \frac{0.02}{0.98} = 0.0204 \approx 0.02$$

The gas leaving the column contains 3% of the ethanol in the entering gas. Therefore,

$$Y_1 = 0.03 \times 0.02 = 0.0006 = y_1$$

Since the solutions are very dilute, it is assumed that the flow rates of gas and liquid remain constant throughout. $G_{N+1} = G_1 = G_S = 180$ kmol/h

The solutions being very dilute, the mole ratio and mole fraction are identical and the limiting operating line can be drawn on the mole fraction coordinates as shown in Fig. 4.53. The slope of the limiting operating line can be approximated as

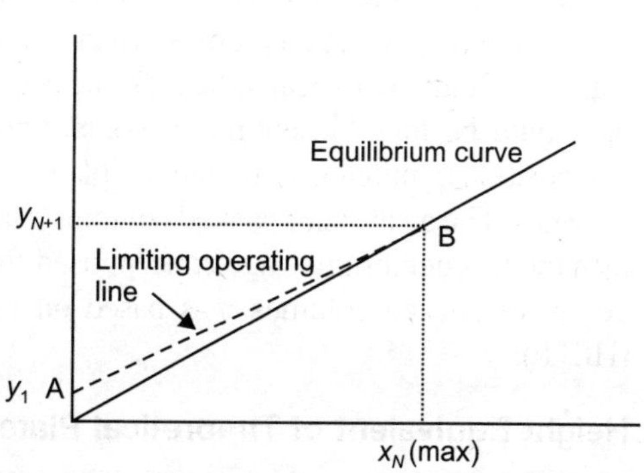

Fig. 4.53: The minimum L/G ratio in Example 4.11

$$\left(\frac{L}{G}\right)_{\min} = \frac{y_{N+1} - y_1}{x_N(\max) - x_0}$$

$x_N(\max)$ is the liquid composition in equilibrium with a gas of composition y_{N+1}. Since the equilibrium relation is $y = 0.57\ x$,

$$x_N(\max) = y_{N+1} / 0.57 = 0.0351$$

$$\left(\frac{L}{G}\right)_{\min} = \frac{y_{N+1} - y_1}{x_N(\max) - x_0} = \frac{0.02 - 0.0006}{0.0351 - 0} = 0.5527$$

The actual L/G ratio is

$$\left(\frac{L}{G}\right) = 0.5527 \times 1.3 = 0.7185$$

Since $G = 180$ kmol/h, $L = 0.7185 \times 180 = 129.33$ kmol/h.

$$\left(\frac{L}{mG}\right) = \frac{0.7185}{0.57} = 1.2605$$

Substituting in Eq. (4.104)

$$E_0 = 0.30 = \frac{\ln\left\{1 + E_{MG}\left[\left(\frac{mG}{L}\right) - 1\right]\right\}}{\ln\left(\frac{mG}{L}\right)} = \frac{\ln\left\{1 + E_{MG}\left[\frac{1}{1.2605} - 1\right]\right\}}{\ln\left(\frac{1}{1.2605}\right)}$$

$$0.30 = \frac{\ln\left\{1 + E_{MG}\left[\frac{1}{1.2605} - 1\right]\right\}}{\ln\left(\frac{1}{1.2605}\right)} = -\frac{\ln(1 - 0.2067 E_{MG})}{0.2315}$$

Solving the above equation, we get $E_{MG} = 0.3246$, that is, the Murphree vapour efficiency is 32.46%.

(Answer)

4.7 ABSORPTION IN PACKED COLUMNS

Packed columns are the most commonly used continuous contact equipment for gas absorption. They are preferred over plate columns when (1) the column diameter is small (less than say, 0.6 m), (2) corrosive fluids are to be handled, (3) the pressure drop should be kept minimum and (4) the liquid hold up should be low. In continuous contact equipment, the gas and liquid phases undergo change in composition continuously unlike the plate towers where the concentration changes are brought about in stages. The analysis of packed columns is best carried out by the rate concept of mass transfer, rather than by the equilibrium stage concept used for sizing the plate columns. However, the older method of design of packed columns was based on the concept of the height equivalent of theoretical plate (HETP).

Height Equivalent of Theoretical Plate (HETP)

The height equivalent of a theoretical plate (HETP) is the height of the packed section that would bring about separation to the same extent as that achieved in an ideal or theoretical plate. For example, the change in the gas phase concentration in a packed section of height equal to the HETP would be identical to the concentration change that the gas phase undergoes on an ideal plate. Thus the height of a packed column can be estimated as the product of the height equivalent of a theoretical plate and the number of theoretical plates required to bring about the desired concentration change. If it requires N theoretical plates for absorption using a plate column, the same absorption can be carried out in a packed column of height Z, where

$$Z = (HETP) \times N$$

Though the concept of theoretical plate is easy to apply, it does not have a theoretical basis as it ignores the fundamental difference between the stage contact and continuous contact mechanisms. HETP is a very complex function of many variables affecting the absorption in packed columns and it is difficult to have appropriate values for use in a given application. It depends on the concentration, is strongly influenced by the type and size of packing, and varies with the flow rates of gas and liquid phases. Therefore a choice of an appropriate HETP value is possible only if a very large amount of experimental data is at hand. In practice, HETP values are estimated using Eq. (4.105) by back calculation using experimental data on laboratory size columns.

$$HETP = \frac{Z}{N} \qquad \ldots(4.105)$$

The HETP method is now largely abandoned in favour of the method based on the concept of transfer units as the latter have got more sound theoretical basis, being related directly to mass transfer coefficients.

The Concept of Transfer Unit

The height of a packed column to effect a desired separation depends on the rate of absorption which in turn is determined by the rate of transfer of mass through the liquid and gas phases. The concept of transfer unit is used to estimate the height of a packed column. The height of packed column is estimated as the product of the number of transfer units and the height of a transfer unit.

$$Z = HTU \times NTU \qquad \ldots(4.106)$$

where HTU is the height and NTU is the number of transfer units. The height of the transfer unit is different from the height equivalent of a theoretical plate. One transfer unit can bring about a concentration change in one of the phases to an extent that is approximately equal to the average driving force existing at that section. The height of a transfer unit is therefore approximately equal to the height of the packed section that will bring about a concentration change in any of the phases equal to the average driving force existing there. HTU depends on the properties and characteristics of the fluids, size and type of packing, the flow rates and concentration of the phases, etc. It is a measure of the separation effectiveness of a particular packing for the materials being processed. The number of transfer unit on the other hand is a measure of the difficulty of separation and it is equal to the number of times the average driving force for mass transfer divides into the total concentration change of gas or liquid phases. For example, if the number of transfer unit based on the gas side is 4, the gas phase undergoes a concentration change which is 4 times the average driving force in the column.

Height of Packed Column

Consider unit area of cross section of a packed column of height Z m (Fig. 4.54a). The liquid and gas flow rates are L and G kmol/m^2 h and the mole fraction of solute **A** in the liquid and gas phases are represented by x and y, respectively. The flow rates and concentrations are indicated with suffix 1 to represent the conditions at the bottom of the column and the conditions at the top with suffix 2. Let 'a' be the interfacial surface per unit volume of packing (m^2/m^3). It is assumed that the absorption is isothermal and no chemical reaction is involved.

Consider a packed section of thickness dz at a distance z from the top of the column. The mole fraction of solute in the liquid and gas crossing each other at this section are x and y, respectively. The solute content in the gas phase changes by $d(Gy)$ as the gas passes this section. The mass flux is obtained by dividing this quantity by the interfacial area of mass transfer. For the packed section of unit cross section and height dz, the area of mass transfer is $a\,dz$. Therefore, the mass flux is

$$N_A = \frac{d(Gy)}{adz} \qquad \text{...(4.107)}$$

Since both G and y change with distance z, the differential $d(Gy)$ is evaluated by replacing G in terms of G_S, where G_S is the flow rate of non-diffusing gas, kmol/m^2 h and is constant.

$$G = \frac{G_S}{1-y} \qquad \text{...(4.108)}$$

Using Eq. (4.108), Eq. (4.107) can be written as

$$N_A = \frac{G_S dy}{(1-y)^2 a\,dz} \qquad \text{...(4.109)}$$

Noting that $G_S = G(1-y)$, Eq. (4.109) becomes

$$N_A = \frac{Gdy}{(1-y)a\,dz} \qquad \text{...(4.110)}$$

The mass flux can be expressed in terms of mass transfer coefficients as well. In terms of gas-phase coefficient k_y and concentration difference in the gas phase (Fig. 4.54b), the rate equation is

$$N_A = k_y(y - y_i) \qquad \text{...(4.111)}$$

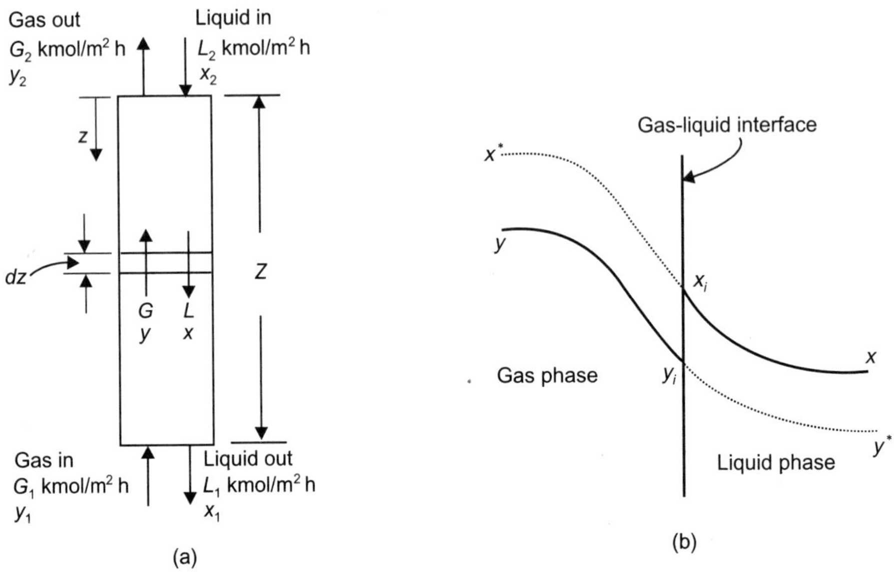

Fig. 4.54: (a) Absorption in packed column (b) Mechanism of interphase transfer in absorption

Comparing Eqs. (4.110) and (4.111), we get

$$k_y(y - y_i) = \frac{Gdy}{(1-y)a\,dz} \qquad \text{...(4.112)}$$

Rearranging Eq. (4.112), the height of the packed section is

$$dz = \frac{Gdy}{k_y a(1-y)(y-y_i)} \quad \text{...(4.113)}$$

Instead of individual mass transfer coefficient k_y, if mass flux were written in terms of the overall mass transfer coefficient K_y, corresponding to Eq. (4.111) we have Eq. (4.114).

$$N_A = K_y(y-y^*) \quad \text{...(4.114)}$$

where y^* is the equilibrium equivalent of the bulk liquid phase concentration x. Comparing Eq. (4.110) with Eq. (4.114) and rearranging, we get

$$dz = \frac{Gdy}{K_y a(1-y)(y-y^*)} \quad \text{...(4.115)}$$

Equation (4.113) forms the basis for the most general and rigorous method for the determination of height of the packed column and Eq. (4.115) forms the basis for the simplified techniques for calculation of the height.

Simplified techniques

(a) ***Using the overall transfer unit:*** Consider Eq. (4.115) and multiply the numerator and denominator by the drift factor $(1-y)_{*M}$, which is the logarithmic mean of $(1-y)$ and $(1-y^*)$.

$$dz = \frac{G(1-y)_{*M}\,dy}{K_y a(1-y)_{*M}(1-y)(y-y^*)} \quad \text{...(4.116)}$$

Integrating Eq. (4.116) we get

$$\int_0^Z dz = \int_{y_2}^{y_1} \frac{G(1-y)_{*M}\,dy}{K_y a(1-y)_{*M}(1-y)(y-y^*)} \quad \text{...(4.117)}$$

In absorption operation where diffusion of **A** occurs through stagnant non-diffusing gas, the quantity $K_y a(1-y)_{*M}$ is the volumetric mass transfer coefficient corrected for bulk flow which is concentration independent, but, vary with mass velocity G. G decreases as gas flows from bottom to top due to absorption. The mass transfer coefficient is related to gas velocity and therefore, $K_y a(1-y)_{*M}$ also decreases from bottom to the top. Therefore, $\dfrac{G}{K_y a(1-y)_{*M}}$ can be treated as a constant and it can be taken outside the integral in Eq. (4.117). Thus

$$Z = \frac{G}{K_y a(1-y)_{*M}} \int_{y_2}^{y_1} \frac{(1-y)_{*M}\,dy}{(1-y)(y-y^*)} \quad \text{...(4.118)}$$

The term $\dfrac{G}{K_y a(1-y)_{*M}}$ has the dimensions of length. This quantity is referred to as the height of an overall gas transfer unit (*HTU*) and is denoted as H_{tOG}. H_{tOG} is assumed constant for the column and if any variation is there, it is taken into account by using the arithmetic mean of values at the inlet and exit to the column. The integral $\displaystyle\int_{y_2}^{y_1} \frac{(1-y)_{*M}\,dy}{(1-y)(y-y^*)}$ is referred to as the number of overall gas transfer units (*NTU*) and is denoted by N_{tOG}.

Thus Eq. (4.118) can be written as

$$Z = HTU \times NTU = H_{tOG} \times N_{tOG} \quad \text{...(4.119)}$$

Thus the height of packed column is determined as the product of the height and number of transfer units. Noting that for dilute systems, the drift factor is approximately unity, the height of overall gas transfer unit is evaluated by

$$H_{tOG} = \frac{G}{K_y a(1-y)_{*M}} \approx \frac{G}{K_y a} \qquad \ldots(4.120)$$

The number of transfer units is

$$N_{tOG} = \int_{y_2}^{y_1} \frac{(1-y)_{*M}\, dy}{(1-y)(y-y^*)} \qquad \ldots(4.121)$$

Equation (4.121) can be simplified by taking $(1-y)_{*M}$ to be the arithmetic mean of $(1-y)$ and $(1-y^*)$, instead of the logarithmic mean.

$$N_{tOG} = \int_{y_2}^{y_1} \frac{[(1-y)+(1-y^*)]dy}{2(1-y)(y-y^*)} = \frac{1}{2}\int_{y_2}^{y_1} \frac{dy}{y-y^*} + \frac{1}{2}\int_{y_2}^{y_1} \frac{(1-y^*)dy}{(1-y)(y-y^*)}$$

$$= \frac{1}{2}\int_{y_2}^{y_1} \frac{dy}{y-y^*} + \frac{1}{2}\int_{y_2}^{y_1} \frac{dy}{(1-y)} + \frac{1}{2}\int_{y_2}^{y_1} \frac{dy}{(y-y^*)} \qquad \ldots(4.122)$$

$$= \int_{y_2}^{y_1} \frac{dy}{y-y^*} + \frac{1}{2}\int_{y_2}^{y_1} \frac{dy}{(1-y)} \qquad \ldots(4.123)$$

$$= \int_{y_2}^{y_1} \frac{dy}{y-y^*} + \frac{1}{2}\ln\frac{1-y_2}{1-y_1} \qquad \ldots(4.124)$$

The second term of Eq. (4.124) is calculated analytically from the terminal concentrations of the gas phase and the first term is determined by graphical integration.

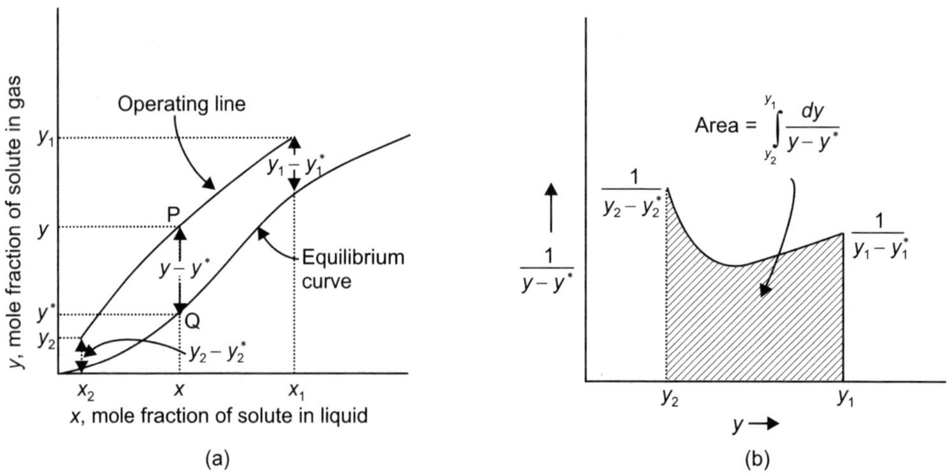

Fig. 4.55: Estimation of the height of the packing by the area method

Figure 4.55a shows the equilibrium curve and operating lines plotted on mole fraction coordinates. Let P be a point on the operating line with coordinates (x, y) where x and y are concentrations of gas and liquid crossing each other at any location in the packed column. Point Q on the equilibrium curve vertically below P has coordinates (x, y^*) where y^* is the concentration of a fictitious gas phase in equilibrium with the liquid. Therefore, the distance PQ is the driving force for mass transfer $y - y^*$. The driving force varies from point to point in the column with $y_1 - y_1^*$ at the bottom of the column and $y_2 - y_2^*$ at the top. Choose different values of y on the operating line ranging from y_2 to y_1 and measure the corresponding values of $y - y^*$ as the vertical distance between the equilibrium curve and the operating line. A graph such as the one shown in Fig. 4.55b results when the reciprocal of $y - y^*$ is plotted on the y-axis against y on the x-axis. The area under the curve between the limits y_2 and y_1 is the integral in Eq. (4.124).

For dilute gases, contribution of the second term of Eq. (4.124) becomes insignificant and can be neglected. Therefore,

$$N_{tOG} \approx \int_{y_2}^{y_1} \frac{dy}{y - y^*} \qquad \text{...(4.125)}$$

Equation (4.124) has the mole ratio counterpart as given below:

$$N_{tOG} = \int_{Y_2}^{Y_1} \frac{dY}{Y - Y^*} + \frac{1}{2} \ln \frac{1 + Y_2}{1 + Y_1} \qquad \text{...(4.126)}$$

Since the operating line on X-Y coordinates is a straight line, Eq.(4.125) is convenient for calculation of N_{tOG} by graphical integration.

The physical significance of transfer unit is obvious from Eq. (4.125). If the concentration difference driving force in the column is almost constant and can be represented by $(y - y^*)_{\text{mean}}$, then Eq. (4.125) becomes

$$N_{tOG} \approx \frac{y_1 - y_2}{(y - y^*)_{\text{mean}}} \qquad \text{...(4.127)}$$

The number of overall gas transfer units is the number of times the average driving force divides into the total concentration change of the gas phase. One transfer unit can achieve a concentration change equal to the average driving force in the column. If very high purity products are required, the number of transfer units will be very large. The height of a transfer unit, on the other hand, as evident from Eq. (4.119), is a measure of the separation effectiveness of the packing materials. For high values of the interfacial area of mass transfer a, and high rates of mass transfer, the height of transfer unit will be small. The height of transfer unit can be viewed as a mass transfer coefficient with dimension of length and its value generally lies in the range 0.15 to 1.5 m.

The method based on the overall coefficient is suitable when the equilibrium distribution curve is straight and ratio of individual phase mass transfer coefficients is constant. If the equilibrium curve is not straight, the overall coefficient is not constant and cannot be used in the design unless the concentration changes encountered in the actual equipment and the values used in the experimental investigation of the coefficient are approximately the same.

(b) *Simplification assuming dilute solutions and linear equilibrium*: We have pointed out that for dilute solutions, the number of overall transfer units can be evaluated as

$$N_{tOG} = \int_{y_2}^{y_1} \frac{dy}{y - y^*} \qquad \text{...(4.128)}$$

For dilute solutions, the operating line is straight on mole fraction coordinates so that

$$y = m_1 x + c_1 \qquad \text{...(4.129)}$$

where m_1 is the slope of the operating line and it is equal to the ratio of the molar flow rates of the liquid and the gas.

$$m_1 = L/G = \frac{y_1 - y_2}{x_1 - x_2} \qquad \text{...(4.130)}$$

If the equilibrium relationship is linear in the region of interest, we have

$$y^* = m_2 x + c_2 \qquad \text{...(4.131)}$$

The driving force for mass transfer $y - y^*$ at any location in the column is determined by subtracting Eq. (4.131) from Eq. (4.129).

$$y - y^* = (m_1 - m_2)x + (c_1 - c_2) \qquad \text{...(4.132)}$$

Denoting the constant $(m_1 - m_2)$ by m and $(c_1 - c_2)$ by c, we have

$$y - y^* = mx + c \qquad \text{...(4.133)}$$

Equation (4.133) means that the driving force for absorption varies linearly with concentration x, and it can be represented by a straight line with slope m where

$$m = \frac{(y_1 - y_1^*) - (y_2 - y_2^*)}{x_1 - x_2} \qquad \text{...(4.134)}$$

Using Eq. (4.133) to replace $y - y^*$ and Eq. (4.129) to replace dy in Eq. (4.128), we get

$$N_{tOG} = \int_{y_2}^{y_1} \frac{dy}{y - y^*} = \int_{x_2}^{x_1} \frac{m_1 dx}{mx + c} = \frac{m_1}{m} \ln \frac{mx_1 + c}{mx_2 + c} \qquad \text{...(4.135)}$$

Noting that $y_1 - y_1^* = mx_1 + c$, $y_2 - y_2^* = mx_2 + c$ and using Eq. (4.130) and Eq. (4.134), we get

$$N_{tOG} = \frac{y_1 - y_2}{(y_1 - y_1^*) - (y_2 - y_2^*)} \ln \frac{y_1 - y_1^*}{y_2 - y_2^*} \qquad \text{...(4.136)}$$

Eq. (4.136) can be written as

$$N_{tOG} = \frac{y_1 - y_2}{(y - y^*)_M} \qquad \text{...(4.137)}$$

where $(y - y^*)_M$ denotes the logarithmic mean of the concentration difference driving forces at the top and bottom of the column.

$$(y - y^*)_M = \frac{(y_1 - y_1^*) - (y_2 - y_2^*)}{\ln \dfrac{y_1 - y_1^*}{y_2 - y_2^*}} \qquad \text{...(4.138)}$$

The height of the transfer unit is given by Eq. (4.113).

$$H_{tOG} = \frac{G}{K_y a} \qquad \text{...(4.113)}$$

The height of the packed column is

$$Z = H_{tOG} N_{tOG} = \frac{G}{K_y a} \frac{(y_1 - y_2)}{(y - y^*)_M} \qquad \text{...(4.139)}$$

Equation (4.139) can be rearranged as

$$G(y_1 - y_2) = K_y (aZ)(y - y^*)_M \qquad \text{...(4.140)}$$

Equation (4.140) means that the rate of absorption (kmol **A** absorbed in unit time) is calculated as the product of the overall mass transfer coefficient, the area of mass transfer and the average concentration difference. This in fact is analogous to the heat transfer rate equation which relates the rate of heat transfer as the product of the overall heat transfer coefficient, the area of heat transfer and the logarithmic mean temperature difference.

$$Q = U(A)(\Delta T)_{LM} \qquad \text{...(4.141)}$$

(c) *Simplification assuming dilute solutions and Henry's law*: The operating line is a straight line with slope L/G.

$$y = \frac{L}{G}x + c \qquad \text{...(4.142)}$$

Since $y = y_2$ when $x = x_2$, we have

$$c = y_2 - \frac{L}{G}x_2 \qquad \text{...(4.143)}$$

The equation for the operating line - Eq. (4.142) - can now be modified as

$$y = \frac{L}{G}x + \left(y_2 - \frac{L}{G}x_2\right) \qquad \text{...(4.144)}$$

Since Henry's law is applicable, the equilibrium relationship is a modified form of Eq. (4.131) with $c_2 = 0$.

$$y^* = mx \qquad \text{...(4.145)}$$

The driving force $y - y^*$ is

$$y - y^* = y - mx \qquad \text{...(4.146)}$$

Replacing x in Eq. (4.146) using Eq. (4.144), we get

$$y - y^* = y - \frac{mG}{L}\left(y - y_2 + \frac{L}{G}x_2\right) = y\left(1 - \frac{1}{A}\right) + \left(\frac{y_2}{A} - mx_2\right) \qquad \text{...(4.147)}$$

Here $A = L/mG$ is the absorption factor. Now the number of transfer units is

$$N_{tOG} = \int_{y_2}^{y_1} \frac{dy}{y - y^*} = \int_{y_2}^{y_1} \frac{dy}{y\left(1 - \frac{1}{A}\right) + \left(\frac{y_2}{A} - mx_2\right)}$$

$$N_{tOG} = \frac{A}{A-1} \ln\left[\frac{y_1\left(1 - \frac{1}{A}\right) + \left(\frac{y_2}{A} - mx_2\right)}{y_2 - mx_2}\right] \qquad \text{...(4.148)}$$

Equation (4.148) may be rearranged as

$$N_{tOG} = \frac{A}{A-1} \ln\left[\frac{y_1 - mx_2}{y_2 - mx_2}\left(1 - \frac{1}{A}\right) + \frac{1}{A}\right] \qquad \text{...(4.149)}$$

Equation (4.148) can also be put in another convenient form as given below:

$$\frac{y_1 - mx_2}{y_2 - mx_2} = \frac{e^{N_{tOG}\left(1 - \frac{1}{A}\right)} - \left(\frac{1}{A}\right)}{1 - \frac{1}{A}} \qquad \text{...(4.150)}$$

The above equations can be represented graphically as shown in Fig. 4.56 which can be used for a quick estimate of the number of transfer unit required for a given separation. The column height is strongly dependent on the liquid flow rate as shown by the marked variation of N_{tOG} with the absorption factor A for a specified separation. Colburn (1939) suggests that the optimum value of A for absorption lies between 1.25 and 2.

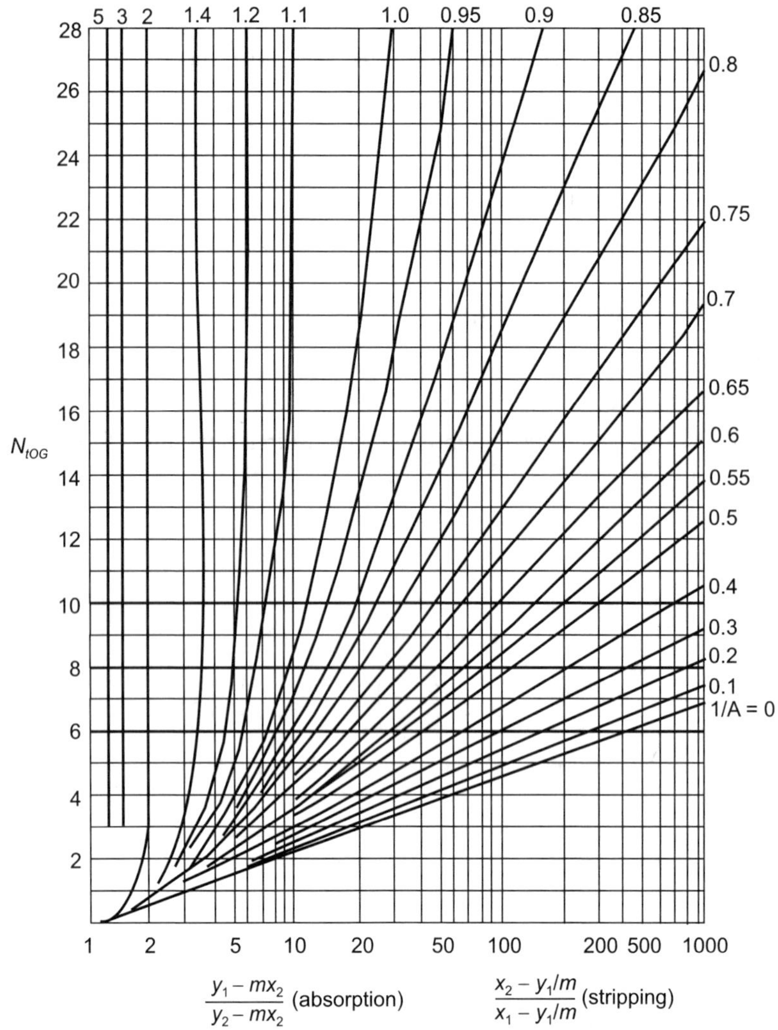

Fig. 4.56: Number of transfer units for absorbers (or strippers) with constant absorption factor (or stripping factor)

(d) *Baker's graphical method*: Baker method assumes that the equilibrium curve and the operating line are straight lines over the length of one transfer unit and the driving force for absorption can be approximated as the arithmetic average of the driving forces. Consider the operating line and equilibrium curve plotted as shown in Fig. 4.57. The construction line MN is drawn vertically halfway between the equilibrium curve and the operating line. Now from point A on the operating line let the horizontal line AD is drawn such that $AC = CD$, where C is the intersection of line AD with line MN. The line MN divides the vertical distance such as FE in two equal parts, FC and CE. A vertical line is drawn from point D to intersect the operating line at G. The length FE represents the average driving force existing between points A and G. Assuming the operating line to be linear between A and G, it can be shown that the triangle ADG will represent an overall gas transfer unit. This section of the absorption column brings about a concentration change $y_G - y_D$ (represented by the line length GD) in the gas phase. Since

$$\frac{AC}{AD} = \frac{1}{2} = \frac{FC}{GD}$$

we get

$$GD = 2FC$$

Since $FC = CE$, $2FC = FE$ and we have $GD = FE$. That is, the concentration change in the gas phase (GD) is equal to the average driving force for absorption (EF) between points A and G. Therefore the section of the column represented by the triangle ADG represents one transfer unit. In a similar manner, the second transfer unit represented by the triangle GHK in the figure is constructed by drawing the line GH such that $GO = OH$ and by constructing the vertical line HK. The steps GHK and KLB in Fig. 4.57 represent the second and third transfer units. The number of such steps that can be constructed between the termini of the operating line is the number of overall gas transfer units required for a specified separation.

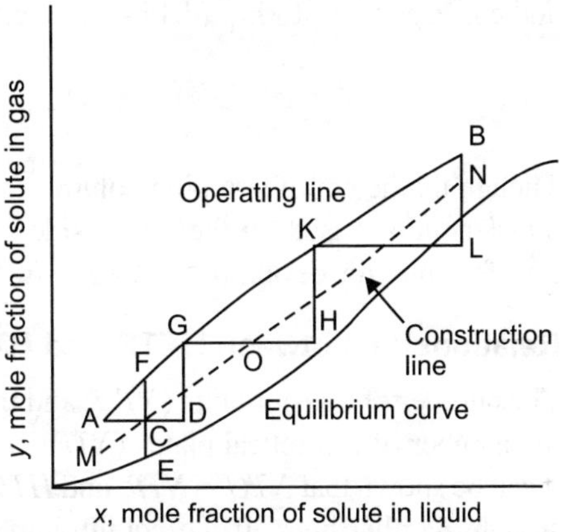

Fig. 4.57: Baker's graphical method for the number of transfer units

***Height of the Packed Column in terms of the Liquid Phase Transfer Unit*:** In the previous paragraphs, the transfer units were discussed and analysed with reference to the gas phase. The liquid phase counterparts can also be easily derived. The liquid-side based equations are preferred when the major resistance to mass transfer lies within the liquid phase. The height of the column can be determined as the product of the height and number of overall liquid transfer units.

$$Z = H_{tOL} \times N_{tOL} \qquad \text{...(4.151)}$$

where the height of overall liquid transfer unit

$$H_{tOL} = \frac{L}{K_x a(1-x)_{*M}} \approx \frac{L}{K_x a} \qquad \text{...(4.152)}$$

The number of transfer units are

$$N_{tOL} = \int_{x_2}^{x_1} \frac{(1-x)_{*M}\,dx}{(1-x)(x^*-x)} \qquad \text{...(4.153)}$$

$$= \int_{x_2}^{x_1} \frac{dx}{x^*-x} + \frac{1}{2}\ln\frac{1-x_1}{1-x_2} \qquad \text{...(4.154)}$$

For dilute solutions with Henry's law representing the equilibrium,

$$N_{tOL} = \frac{1}{1-A}\ln\left[\frac{x_2 - y_1/m}{x_1 - y_1/m}(1-A) + A\right] \qquad \text{...(4.155)}$$

The number of liquid transfer units can be determined by the Baker's method by drawing the construction line MN halfway horizontally between the equilibrium curve and the operating line so that the line MN will bisect the vertical leg of the steps.

***Height and Number of Transfer Units for Strippers*:** The height of transfer unit can be estimated by the equations already developed for absorbers. Since the direction of mass transfer is from the liquid to the gas, the equations for the number of transfer units gets slightly modified. For example, corresponding to Eq. (4.124) we can write

$$N_{tOG} = \int_{y_1}^{y_2} \frac{dy}{y^* - y} + \frac{1}{2} \ln \frac{1 - y_1}{1 - y_2} \qquad \ldots(4.124\text{ A})$$

and corresponding to Eq. (4.154) we can have

$$N_{tOL} = \int_{x_2}^{x_1} \frac{dx}{x - x^*} + \frac{1}{2} \ln \frac{1 - x_1}{1 - x_2} \qquad \ldots(4.154\text{ A})$$

Though the height of stripping column can be determined as the product of the number and height of transfer units, based on the liquid side or the gas side, it is customary to use the liquid-side-based transfer units for design of strippers.

Relationship between HETP and HTU

The number of transfer units (*NTU*) and the height of transfer unit (*HTU*) should not be confused with the number of theoretical plates (*NTP*) and height equivalent of a theoretical plate (*HETP*). However, it can be shown that $NTU = NTP$, and $HTU = HETP$ when the operating line and equilibrium curve are straight parallel lines. If the equilibrium curve and operating lines are straight but not parallel, the following relationships between *HTU* and *HETP* and between *NTU* and *NTP* can be derived.

$$HETP = H_{tOG} \frac{\ln(1/A)}{1/A - 1} \qquad \ldots(4.156)$$

$$N_{toG} = N \frac{\ln(1/A)}{1/A - 1} \qquad \ldots(4.157)$$

Equation (4.157) predicts that $N_{toG} > N$ for $A > 1$ and $N_{toG} < N$ for $A < 1$.

Relationship between the Overall and Individual Transfer Units

We have seen in the previous chapter (Refer Eq. (3.136)) that the overall mass transfer coefficient and individual mass transfer coefficients are related by

$$\frac{1}{K_y a} = \frac{1}{k_y a} + \frac{m}{k_x a} \qquad \ldots(4.158)$$

The height of overall transfer unit was defined by Eq. (4.120).

$$H_{tOG} = \frac{G}{K_y a} \qquad \ldots(4.120)$$

In a similar manner, we can define individual transfer units based on the gas side or the liquid side. Let H_{tG} denote the height of gas transfer unit and H_{tL} represent the height of transfer unit based on the liquid side. Then,

$$H_{tG} = \frac{G}{k_y a} \qquad \ldots(4.159)$$

$$H_{tL} = \frac{L}{k_x a} \qquad \ldots(4.160)$$

where G and L are the molar flow of gas and liquid, respectively, per unit area of tower cross section for the gas and liquid phases, respectively and $k_y a$ and $k_x a$ are the volumetric mass transfer coefficients for the gas and liquid phases, respectively. Equations (4.120), (4.159) and (4.160) lead to the following expressions for mass transfer coefficients.

$$\frac{1}{K_y a} = \frac{H_{tOG}}{G} \quad \text{...(4.161a)}$$

$$\frac{1}{k_y a} = \frac{H_{tG}}{G} \quad \text{...(4.161b)}$$

$$\frac{1}{k_x a} = \frac{H_{tL}}{L} \quad \text{...(4.161c)}$$

Substituting Eqs. (4.161a), (4.161b) and (4.161c) into Eq. (4.158), we get

$$H_{tOG} = H_{tG} + \frac{mG}{L} H_{tL} \quad \text{...(4.162)}$$

For the transfer unit based on the liquid side, we can show that

$$H_{tOL} = H_{tL} + \frac{L}{mG} H_{tG} \quad \text{...(4.163)}$$

In the design of absorbers based on overall transfer units, the height of an overall gas transfer unit is estimated either by using the overall coefficient given by Eq. (4.120), or by using the relationship between heights of individual and overall transfer units, Eq. (4.163).

Example 4.12: Show that for dilute solutions, the number of transfer units based on the gas phase for cocurrent gas absorption in a packed tower is given by

$$N_{tOG} = \frac{A}{A-1} \ln \frac{y_1 - mx_1}{y_2 - mx_2}$$

where gas and liquid enter the top and subscripts 1 and 2 represent top and bottom conditions, respectively.

***Solution*:** Consider cocurrent absorber of unit cross section and packed height Z m as shown in Fig. 4.58. The flow rates and concentrations are indicated with suffix 1 to represent the conditions at the top of the column and with suffix 2 at the bottom. Let 'a' be the interfacial surface per unit volume of packing (m²/m³). It is assumed that the absorption is isothermal and no chemical reaction is involved.

Consider a packed section of thickness dz at a distance z from the bottom of the column. The mole fraction of the solute in the liquid and gas crossing each other at this section are x and y, respectively. The mass flux at this section is given by Eq. (4.164).

$$N_A = \frac{G\,dy}{(1-y)\,a\,dz} \quad \text{...(4.164)}$$

The mass flux can be expressed in terms of overall mass transfer coefficient K_y and the overall driving force for mass transfer as given by the following equation.

$$N_A = K_y(y - y^*) \quad \text{...(4.165)}$$

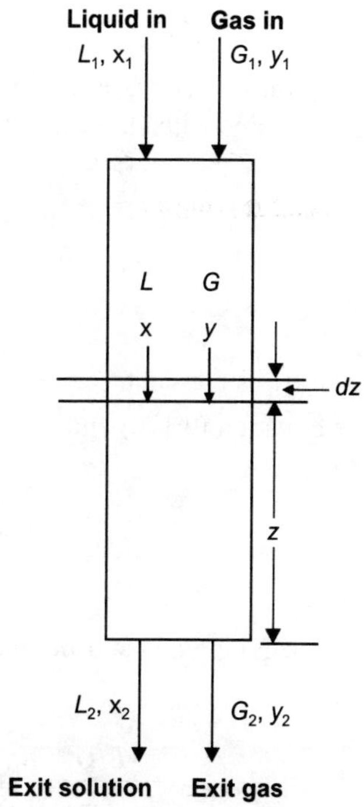

Fig. 4.58: Cocurrent absorption (Example 4.12)

where y^* is the equilibrium equivalent of the bulk liquid phase con-centration x. Comparing Eq. (4.164) with Eq. (4.165) and rearranging, we get

$$dz = \frac{Gdy}{K_y a(1-y)(y-y^*)} \qquad \text{...(4.166)}$$

Assuming that the solutions are very dilute, $(1-y) \approx 1$, so that Eq. (4.166) gives

$$Z = \frac{G}{K_y a}\int_{y_2}^{y_1} \frac{dy}{(y-y^*)} = H_{tOG} \times N_{tOG} \qquad \text{...(4.167)}$$

in which

$$H_{tOG} = \frac{G}{K_y a}$$

$$N_{tOG} = \int_{y_2}^{y_1} \frac{dy}{(y-y^*)} \qquad \text{...(4.168)}$$

For cocurrent operation we have already shown that the operating line is a straight line of slope $-L_S/G_S$ on mole ratio basis. For dilute solutions, the mole ratio and mole fractions are almost identical and the operating line will be a straight line with slope $-L/G$ on mole fraction coordinates, where $L = L_S$ = constant and $G = G_S$ = constant.

$$y = -\frac{L}{G}x + c \qquad \text{...(4.169)}$$

and

$$dy = -\frac{L}{G}dx \qquad \text{...(4.170)}$$

Because Henry's law is applicable for dilute solutions, the equilibrium curve is assumed to be a straight line with slope m.

$$y^* = mx \qquad \text{...(4.171)}$$

The driving force $y - y^*$ is obtained by subtracting Eq. (4.171) from Eq. (4.169).

$$y - y^* = -\left(\frac{L}{G} + m\right)x + c \qquad \text{...(4.172)}$$

where c is constant.

Equation (4.170) and Eq. (4.172) are substituted into Eq. (4.168) and we get

$$N_{tOG} = \int_{y_2}^{y_1} \frac{dy}{(y-y^*)} = -\frac{L}{G}\int_{x_2}^{x_1} \frac{dx}{-\left(\frac{L}{G}+m\right)x + c} \qquad \text{...(4.173)}$$

Integrating and substituting the limits, we get

$$N_{tOG} = \frac{L/G}{(L/G+m))}\ln\frac{-\left(\frac{L}{G}+m\right)x_1 + c}{-\left(\frac{L}{G}+m\right)x_2 + c} = \frac{L/mG}{(L/mG+1)}\ln\frac{\left(-\frac{L}{G}x_1 + c\right) - mx_1}{\left(-\frac{L}{G}x_2 + c\right) - mx_2} \qquad \text{...(4.174)}$$

Since the absorption factor $A = L/mG$ and by Eq. (4.169) $y_1 = -\frac{L}{G}x_1 + c$ and $y_2 = -\frac{L}{G}x_2 + c$, Eq. (4.174) simplifies as

$$N_{tOG} = \frac{A}{(A+1)} \ln \frac{y_1 - mx_1}{y_2 - mx_2} \qquad \ldots(4.175) \qquad \textbf{(Answer)}$$

Example 4.13: An inert gas contaminated with hydrogen sulphide is to be scrubbed with an aqueous solution of triethanolamine in a countercurrent packed column operated at 300 K and 101.3 kPa to reduce H_2S content from the initial concentration of 0.03 mole H_2S per mole of inert gas to one-hundredth of this value. Solvent enters the tower free of H_2S and leaves with a concentration of 0.013 mole H_2S per mole solvent. The flow of inert gas is 50.0 kmol/m² h. The equilibrium is described by the linear relation $y = 2\,x$ where x and y are mole fractions of H_2S. The overall mass transfer coefficient $K_y a = 125$ kmol/m³ h (mole fraction). Determine:

(a) The rate at which solvent is introduced, kmol/m² h and

(b) The height of the absorber by assuming dilute solutions.

***Solution*:** The flow rate of gas is $G_s = 50$ kmol/m² h and the inlet concentration $Y_1 = 0.03$ kmol H_2S/kmol inert gas. Exit concentration is one hundredth of this value. $Y_2 = 0.0003$

$$G_1 = G_s(1+Y_1) = 50.0 \times (1+0.03) = 51.5 \text{ kmol/m}^2\text{h}$$

$$G_2 = G_s(1+Y_2) = 50.0 \times (1+0.0003) = 50.015 \text{ kmol/m}^2\text{h}$$

Mole fraction of H_2S in the inlet and exit gas are

$$y_1 = \frac{Y_1}{1+Y_1} = \frac{0.03}{1.03} = 0.0291, \; y_2 = \frac{Y_2}{1+Y_2} = \frac{0.0003}{1.0003} = 0.0003$$

Solvent enters free of H_2S. Therefore, $X_2 = 0;\; x_2 = 0$

Solution leaving contains 0.013 mole H_2S per mole solvent. $X_1 = 0.013,\; x_1 = \frac{0.013}{1.013} = 0.0128$

(a) The solvent flow rate can be estimated by a H_2S balance for the tower:

$$G_s(Y_1 - Y_2) = L_s(X_1 - X_2)$$

$$50.0(0.03 - 0.0003) = L_s(0.013 - 0) \qquad \text{(A)}$$

Solving Eq. (A), we get $L_s = 114.23 \text{ kmol/m}^2\text{h}$

The total liquid flow rates are

$$L_2 = L_s(1+X_2) = 114.23 \times (1+0) = 114.23 \text{ kmol/m}^2\text{h} \qquad \textbf{(Answer)}$$

$$L_1 = L_s(1+X_1) = 114.23 \times (1+0.013) = 115.71 \text{ kmol/m}^2\text{h}$$

The flow rates and composition of the streams are represented on Fig. 4.59.

(b) Assuming dilute solutions, the height of overall gas transfer unit is estimated using Eq. (4.119)

$$H_{tOG} = \frac{G}{K_y a(1-y)_{*M}} \approx \frac{G}{K_y a}$$

The average gas flow rate

$$G = \frac{G_1 + G_2}{2} = \frac{51.5 + 50.015}{2} = 50.758 \text{ kmol/m}^2\text{h}$$

The volumetric mass transfer coefficient is $K_y a = 125 \text{ kmol/m}^3\text{h}$ (mole fraction). Substituting in Eq. (4.119), we get

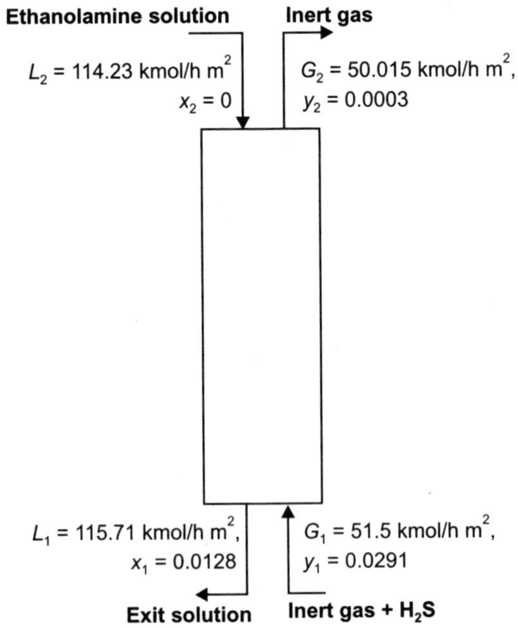

Fig. 4.59: Absorption tower for Example 4.13

$$H_{tOG} = \frac{G}{K_y a} = \frac{50.758}{125} = 0.406 \text{ m}.$$

Since the solutions are dilute and the equilibrium relationship is linear, let us use the logarithmic mean driving force method for evaluation of the number of transfer units.

Driving force at the bottom of the column is $y_1 - y_1^*$ where y_1^* is the equilibrium equivalent of the liquid of composition x_1. Since the equilibrium relationship is $y^* = 2x$,

$$y_1^* = 2x_1 = 2 \times 0.0128 = 0.0256$$

$$y_1 - y_1^* = 0.0291 - 0.0256 = 0.0035$$

Driving force at the top of the column is $y_2 - y_2^*$, where y_2^* is the equilibrium equivalent of the liquid of composition x_2. Since $x_2 = 0$, $y_2^* = 0$.

$$y_2 - y_2^* = 0.0003 - 0 = 0.0003$$

The logarithmic mean driving force $(y - y^*)_M$ can now be evaluated.

$$(y - y^*)_M = \frac{(y_1 - y_1^*) - (y_2 - y_2^*)}{\ln \dfrac{y_1 - y_1^*}{y_2 - y_2^*}} = \frac{0.0035 - 0.0003}{\ln \dfrac{0.0035}{0.0003}} = 0.0013$$

The number of transfer unit by Eq. (4.137) is

$$N_{tOG} = \frac{y_1 - y_2}{(y - y^*)_M} = \frac{0.0291 - 0.0003}{0.0013} = 22.15$$

The height of the packed section is $Z = H_{tOG} \times N_{tOG} = 0.406 \times 22.15 = 9 \text{ m}$ **(Answer)**

Example 4.14: Acetone-air mixture containing 5% (mole) acetone is scrubbed with pure water in a countercurrent packed column of diameter 1.5 m in order to remove 95% of the acetone. The gas enters at a rate 3000 kg/h and the liquid rate is 50% in excess of the minimum. The overall mass transfer coefficient K_Ga is 1.1 kmol/m^3 h kPa. Equilibrium data at the operating conditions are (x, y are mole fractions of acetone)

x	0.01	0.02	0.03	0.04
y	0.02	0.0382	0.0552	0.0697

Determine:

(a) The liquid rate in kg/h
(b) The mole percent of acetone in the exit liquid
(c) The height of the absorber by graphical integration.

***Solution*:** The absorber is represented by Fig. 4.60.

The acetone-air mixture contains 5% (mole) acetone. $y_1 = 0.05$

$$Y_1 = \frac{y_1}{1-y_1} = \frac{0.05}{1-0.05} = 0.0526$$

The average molecular weight of the entering acetone ($MW = 58$) – air ($MW = 29$) mixture is

$$MW = 0.05 \times 58 + 0.95 \times 29 = 30.45$$

The molar flow rate of the gas is $\dfrac{3000}{30.45} = 98.52\ \text{kmol/h}$.

Since the determination of the height of transfer unit requires the molar flow rate to be expressed per unit area of tower cross section,

$$G_1 = \frac{\text{Molar flow rate}}{\text{area of cross section}} = \frac{98.52}{\frac{\pi}{4} \times D^2} = \frac{98.52}{\frac{\pi}{4} \times 1.5^2} = 55.75\ \text{kmol/hm}^2$$

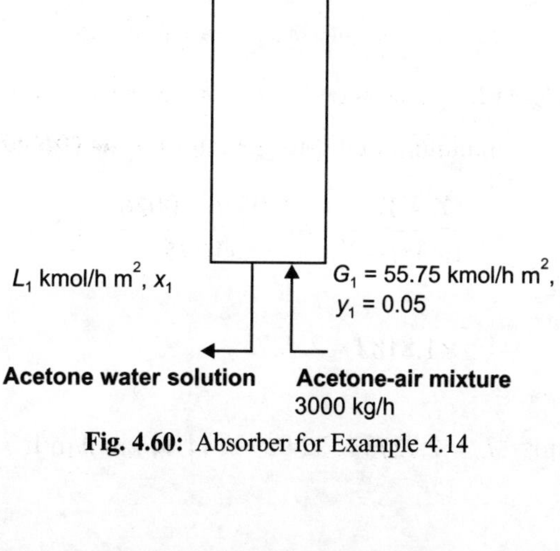

Fig. 4.60: Absorber for Example 4.14

The acetone-free air flow rate is

$$G_s = G_1(1-y_1) = 55.75(1-0.05) = 52.96 \text{ kmol/hm}^2$$

Since 95% of acetone in the entering gas is absorbed, the mole ratio of acetone in the exit gas is

$Y_2 = 0.05Y_1 = 0.05 \times 0.0526 = 0.0026$ and the mole fraction is $y_2 = \dfrac{Y_2}{1+Y_2} = \dfrac{0.0026}{1.0026} = 0.0026$

(a) Since the liquid is admitted at a rate 1.5 times the minimum, the minimum liquid rate permissible in the column is to be determined next. This requires plotting the equilibrium curve and the limiting operating line on mole ratio coordinates.

The equilibrium data is given in terms of mole fractions. These are converted into mole ratios by noting that

$$X = \frac{x}{1-x} \text{ and } Y = \frac{y}{1-y}$$

The following results are obtained.

X	0.0101	0.0204	0.0309	0.0417
Y	0.0204	0.0397	0.0584	0.0749

The equilibrium curve is drawn on mole ratios as shown in Fig. 4.61. The point A represents the concentrations at the top of the column. Coordinates of point A are, $X_2 = 0$, $Y_2 = 0.0026$. The limiting operating line AB is drawn tangential to the equilibrium curve to meet the horizontal line drawn at $Y_1 = 0.0526$ at point B. It is seen that the x-coordinate of point B is $X_1(\text{max}) = 0.0275$

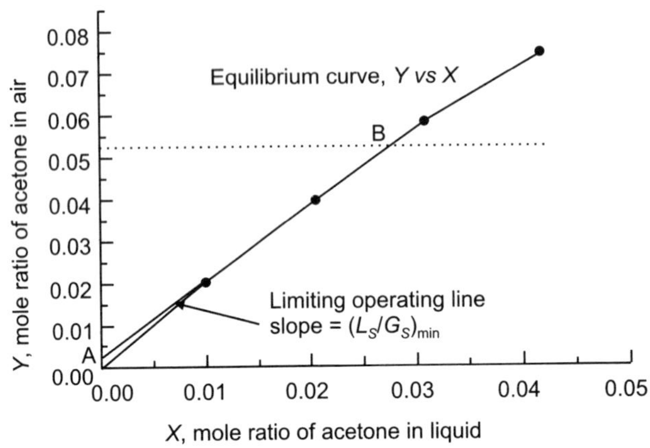

Fig. 4.61: Minimum solvent requirement in Example 4.14

The slope of line AB is the minimum liquid-to gas ratio for the column.

$$\left(\frac{L_s}{G_s}\right)_{\text{min}} = \frac{Y_1 - Y_2}{X_1(\text{max}) - X_2} = \frac{0.0526 - 0026}{0.0275} = 1.8182$$

$$\left(\frac{L_s}{G_s}\right)_{\text{actual}} = 1.5 \times 1.8182 = 2.7273$$

Since $G_s = 52.96 \text{ kmol/hm}^2$, $L_s = 2.7273 \times 52.96 = 144.44 \text{ kmol/m}^2\text{h}$.

Since water is pure, $L_s = L_2 = 144.44$ kmol/m² h. This is the rate of liquid admitted per unit area. The liquid is admitted at a total rate of

$$144.44 \times \frac{\pi}{4} \times 1.5^2 = 255.24 \text{ kmol/h} \text{ or } 255.24 \times 18 = 4594.4 \text{ kg/h} \qquad \textbf{(Answer)}$$

(b) The Operating line for the column is obtained by drawing a straight line from A with a slope 2.7273 to terminate at the horizontal drawn at Y_1. The x-coordinate of the upper terminus of this line is the exit concentration of the solution X_1. From the graph, $X_1 = 0.0183$.

(Alternatively, since the slope of the operating line is 2.7273, we have the following relationship for the slope:

$$\frac{Y_1 - Y_2}{X_1 - X_2} = \frac{0.0526 - 0.0026}{X_1 - 0} = 2.7273$$

so that $X_1 = 0.0183$ *mol acetone per mol water.*)

The mole fraction of acetone in the exit liquid $= x_1 = \frac{X_1}{1+X_1} = \frac{0.0183}{1.0183} = 0.0180$

The mole percent of acetone in the exit liquid $= \frac{X_1}{1+X_1} \times 100 = \frac{0.0183}{1.0183} \times 100 = 1.80\%$. **(Answer)**

(c) To find the number of transfer units by graphical area method, we have to plot the operating line on mole fraction coordinates. We know that the equation for the operating line on mole ratios is obtained by material balance as $G_s(Y - Y_2) = L_s(X - X_2)$ where X and Y are mole ratios of acetone in the liquid and the gas, respectively. The operating line is given as

$$Y = \frac{L_s}{G_s} X + \left(Y_2 - \frac{L_s}{G_s} X_2 \right) = 2.7273X + (0.0026 - 2.7273 \times 0) \text{ or}$$

$$Y = 0.0026 + 2.7273X$$

In terms of mole fractions the equation becomes

$$\frac{y}{1-y} = 0.0026 + 2.7273 \frac{x}{1-x} \qquad \text{(A)}$$

Since the gas undergoes a concentration change from 0.05 mole fraction to 0.0026, assume different values for y in the range 0.0026 to 0.05 and calculate x using Eq. (A). The following results are obtained:

x	0	0.0009	0.0027	0.0046	0.0065	0.0084	0.0103	0.0122	0.0141	0.0161	0.0180
y	0.0026	0.005	0.01	0.015	0.02	0.025	0.03	0.035	0.04	0.045	0.05

The operating line is plotted using the above values and the equilibrium curve is plotted with the data given as shown in Fig. 4.62.

The driving force $(y - y^*)$ is given by the vertical distance between the operating line and the equilibrium curve. For different y – values, (that is, for different points on the operating line), $(y - y^*)$ are determined from the graph.

The following table gives y^*, $y - y^*$, $1/(y - y^*)$ against y:

y	0.0026	0.01	0.02	0.03	0.04	0.05
y^*	0	0.0056	0.0129	0.0204	0.0273	0.0346
$y - y^*$	0.0026	0.0044	0.0071	0.0096	0.0127	0.0154
$1/(y - y^*)$	384.6	227.3	140.8	104.2	78.7	64.9

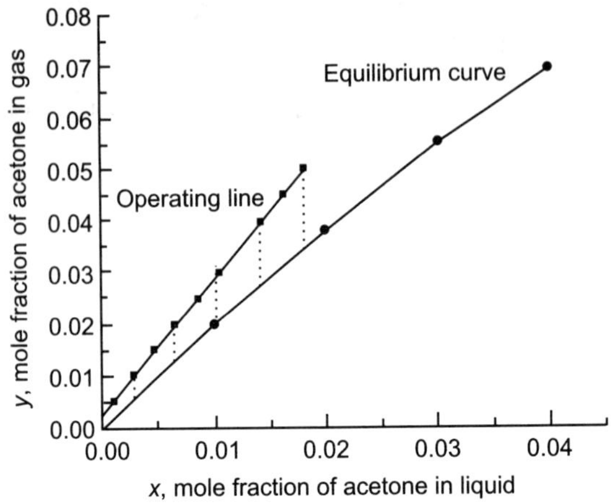

Fig. 4.62: Evaluation of driving force $(y - y^*)$ in the absorption column

A graph is plotted (Fig. 4.63) taking $\dfrac{1}{(y-y^*)}$ as ordinate against y as abscissa. The area under the curve between limits $y_2 = 0.0026$ and $y_1 = 0.05$ gives the integral $\int\limits_{0.0026}^{0.05} \dfrac{dy}{y-y^*}$. The area is found to be 6.96. The number of overall gas transfer units by Eq. (4.124) is

$$N_{tOG} = \int\limits_{0.0026}^{0.05} \frac{dy}{y-y^*} + \frac{1}{2}\ln\frac{1-y_2}{1-y_1}$$

$$N_{tOG} = 6.96 + \frac{1}{2}\ln\frac{1-0.0026}{1-0.05} = 6.98$$

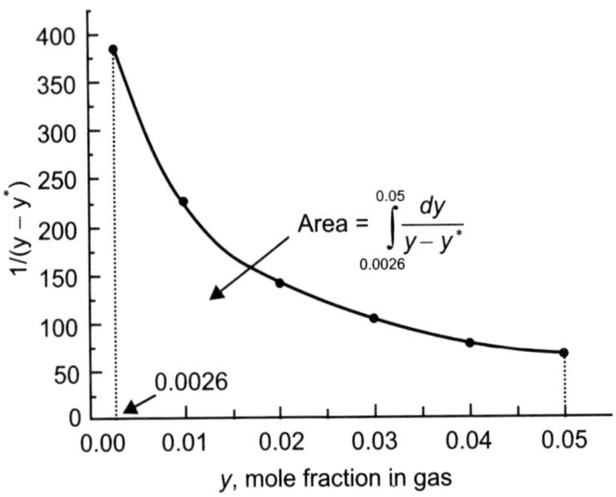

Fig. 4.63: Graphical integration for number of transfer units in Example 4.14

The height of overall gas transfer units is

$$H_{tOG} \approx \frac{G}{K_y a} = \frac{G}{K_G aP} \tag{B}$$

Here $$G = \frac{G_1 + G_2}{2}$$

with $G_1 = 55.75$ kmol/h m^2 and $G_2 = G_s(1 + X_2) = 52.96(1 + 0.0026) = 53.10$ kmol/h m^2

Therefore,

$$G = \frac{55.75 + 53.10}{2} = 54.43 \text{ kmol/h m}^2 \text{ and}$$

Using Eq. (B),

$$H_{tOG} = \frac{G}{K_G aP} = \frac{54.43}{1.1 \times 101.3} = 0.488 \text{ m}$$

Height of the column $Z = H_{tOG} \times N_{tOG} = 0.488 \times 6.98 = 3.41$ m **(Answer)**

Example 4.15: A countercurrent packed tower is used to remove benzene from an effluent gas containing 10% (mole) benzene by scrubbing with a mineral oil at 315 K and 101.3 kPa. The gas is admitted to the column at 1.5 m^3/h per square meter of tower cross section and the solvent enters at 2.5×10^{-2} kmol/m^2 h with a benzene content of 1% (mole). It is desired that the benzene content in the exit gas is to be reduced to 1 mole %. The height of the gas and liquid transfer units are respectively 0.6 m and 0.4 m and the equilibrium relationship for the system is $y = 0.263x$. Determine the height of the tower

(a) using graphical area method
(b) using log-mean driving force
(c) assuming dilute solutions and Henry's law
(d) by Baker's method

***Solution*:** The gas flow rate = 1.5 m^3/h per square meter of tower cross section at 315 K and 101.3 kPa.

$$G_1 = \frac{PV}{RT} = \frac{101.3 \times 1.5}{8.314 \times 315} = 0.058 \, \text{kmol/m}^2\text{h}$$

Since the gas contains 10% (mole) benzene $y_1 = 0.10$

Since the benzene content in the exit gas is 1 mole %, $y_2 = 0.01$

The solvent is admitted at a rate 2.5×10^{-2} kmol/m^2 h. $L_2 = 0.025$ kmol/m^{2}h.

Since the benzene content in the liquid entering is 1%, $x_2 = 0.01$

The given composition in mole fractions can be converted to mole ratios as follows.

$$Y_1 = \frac{y_1}{1 - y_1} = \frac{0.10}{0.90} = 0.1111$$

$$Y_2 = \frac{y_2}{1 - y_2} = \frac{0.01}{0.99} = 0.0101$$

$$X_2 = \frac{x_2}{1 - x_2} = \frac{0.01}{0.99} = 0.0101$$

The flow rate on benzene-free basis are

$$G_S = G_1(1 - y_1) = 0.058 \times 0.90 = 0.0522 \text{ kmol/m}^2\text{h}$$
$$L_S = L_2(1 - x_2) = 0.025 \times 0.99 = 0.02475 \, \text{kmol/m}^2\text{h}$$

A benzene balance for the column gives

$$G_S(Y_1 - Y_2) = L_S(X_1 - X_2)$$

$$0.0522(0.1111 - 0.0101) = 0.02475(X_1 - 0.0101)$$

Solving this equation gives

$$X_1 = 0.2231$$

Mole fraction of benzene in the exit liquid $= x_1 = \dfrac{X_1}{1+X_1} = \dfrac{0.2231}{1.2231} = 0.1824$

$$G_2 = G_S(1+Y_2) = 0.0522 \times 1.0101 = 0.0527 \text{ kmol/m}^2\text{h}$$

$$L_1 = L_s(1+X_1) = 0.02475 \times 1.2231 = 0.0303 \text{ kmol/m}^2\text{h}$$

The absorption tower is shown in Fig. 4.64.

The height of the packed tower is calculated using Eq. (4.119)

$$Z = H_{tOG} \times N_{tOG} \qquad \text{...(4.119)}$$

The height of the overall gas transfer unit is related to the heights of the gas and the liquid transfer units by Eq. (4.162)

$$H_{tOG} = H_{tG} + \frac{mG}{L} H_{tL}$$

The liquid flow L and gas flow G may be taken to be the average of the respective quantities at the ends of the column. The equilibrium distribution constant $m = 0.263$.

Thus $$L = \frac{L_1 + L_2}{2} = \frac{0.0303 + 0.025}{2} = 0.0277 \text{ kmol/m}^2\text{h and}$$

$$G = \frac{G_1 + G_2}{2} = \frac{0.0527 + 0.058}{2} = 0.0554 \text{ kmol/m}^2\text{h}$$

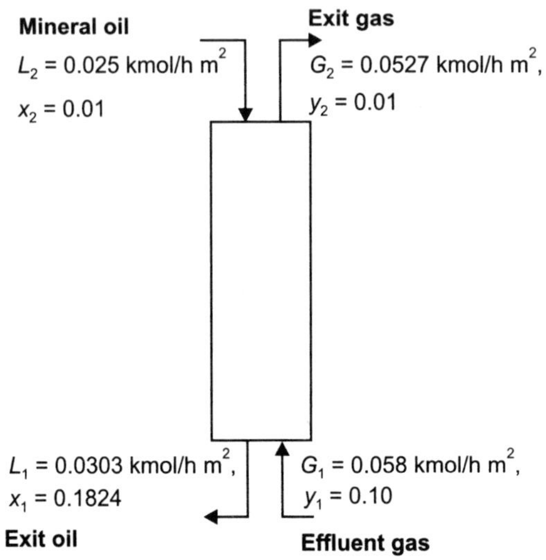

Fig. 4.64: The absorber in Example 4.15

$$H_{tOG} = H_{tG} + \frac{mG}{L} H_{tL} = 0.60 + \frac{0.263 \times 0.0554}{0.0277} \times 0.4 = 0.81 \text{ m}$$

The number of transfer unit N_{tOG} can be calculated using different methods.

(a) Graphical area method: N_{tOG} can be determined by Eq. (4.124)

$$N_{tOG} = \int_{y_2}^{y_1} \frac{dy}{y - y^*} + \frac{1}{2} \ln \frac{1 - y_2}{1 - y_1} \quad \ldots(4.124)$$

The first term in Eq. (4.124) is determined by graphical integration.

Equilibrium curve is plotted on mole fraction coordinates using the given relationship: $y^* = 0.263x$. Equilibrium curve is a straight line passing through the origin with slope 0.263.

To plot the operating line on mole fractions, the coordinates of different points on the line are to be estimated by material balance. We know that the equation for the operating line on mole ratios is obtained by material balance as

$G_S(Y - Y_2) = L_S(X - X_2)$ where X and Y are mole ratios of the diffusing component in the gas and the liquid, respectively. The operating line is given as

$$Y = \frac{L_S}{G_S} X + \left(Y_2 - \frac{L_S}{G_S} X_2 \right) = \frac{0.02475}{0.0522} X + \left(0.0101 - \frac{0.02475}{0.0522} \times 0.0101 \right)$$

$$Y = 0.4741X + 0.00531$$

Or the operating line written in terms of mole fraction is

$$\frac{y}{1 - y} = 0.4741 \frac{x}{1 - x} + 0.00531 \quad \text{(A)}$$

Take y values in the range 0.01 to 0.10 and calculate corresponding x using Eq. (A).

x	0.01	0.0309	0.0513	0.0712	0.0908	0.1099	01286	0.1469	0.1649	0.1824
y	0.01	0.02	0.03	0.04	0.05	0.06	0.07	0.08	0.09	0.10

The x and y values are the coordinates of different points on the operating line by joining which the operating line AB shown in Fig. 4.65 is obtained.

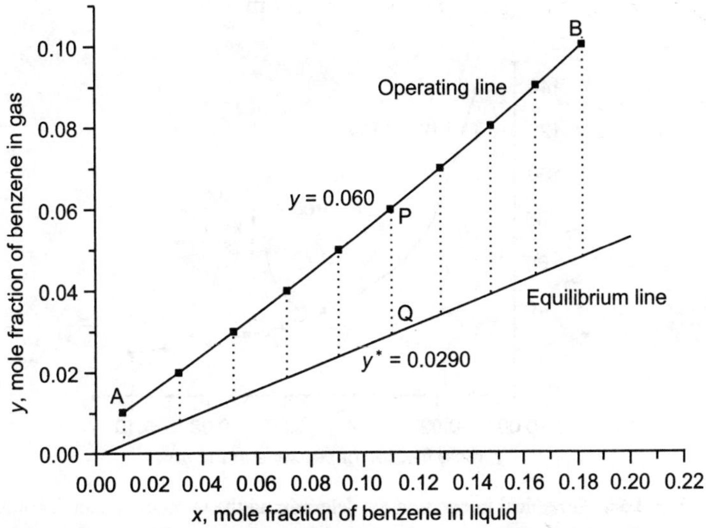

Fig. 4.65: Determination of driving force $y - y^*$ graphically for Example 4.15

The driving force $y - y^*$ at a given point in the column is given by the vertical distance between the operating line and the equilibrium curve. For example, consider a point such as point P on the operating line whose coordinates are (0.1099, 0.06). The composition of the gas at this point is $y = 0.06$ and the composition of the liquid phase in contact with the gas is $x = 0.1099$. The concentration of a gas phase which is in equilibrium with the liquid of composition x is given by y^* which is determined by moving vertically downward to point Q on the equilibrium line. $y^* = 0.0290$ from the Fig. 4.65. The driving force $y - y^* = 0.06 - 0.0290 = 0.031$.

The driving force $y - y^*$ are determined for various points on the operating line with various y – values beginning with y_2 at the top of the column to y_1 at the bottom. The results are given in the table below where y, y^*, $y - y^*$ and its reciprocal $(y - y^*)^{-1}$ are tabulated.

y	0.01	0.02	0.03	0.04	0.05	0.06	0.07	0.08	0.09	0.10
y^*	0.0027	0.0082	0.0134	0.0189	0.0238	0.0290	0.0339	0.0389	0.0434	0.0480
$y - y^*$	0.0073	0.0118	0.0166	0.0211	0.0262	0.0310	0.0361	0.0411	0.0466	0.0520
$1/(y - y^*)$	137.0	84.7	60.2	47.4	38.2	32.3	27.7	24.3	21.5	19.2

To calculate $\int_{0.01}^{0.10} \frac{dy}{y - y^*}$, a graph is plotted with $\frac{1}{y - y^*}$ as ordinate against y as abscissa. The curve is shown in the Fig. 4.66. The area under the curve between the limits $y = 0.01$ to $y = 0.10$ gives the integral $\int_{0.01}^{0.10} \frac{dy}{y - y^*}$. The area is found to be 4.14. Therefore, $\int_{0.01}^{0.10} \frac{dy}{y - y^*} = 4.14$

From Eq. (4.124), $N_{tOG} = \int_{y_2}^{y_1} \frac{dy}{y - y^*} + \frac{1}{2} \ln \frac{1 - y_2}{1 - y_1}$

Therefore, $N_{tOG} = 4.14 + \frac{1}{2} \ln \frac{1 - 0.01}{1 - 0.10} = 4.19$

The height of the packed section (Z) is the product of the height of transfer units (H_{tOG}) and the number of transfer units (N_{tOG}).

$$Z = H_{tOG} \times N_{tOG} = 0.81 \times 4.19 = 3.39 \text{ m}$$ **(Answer)**

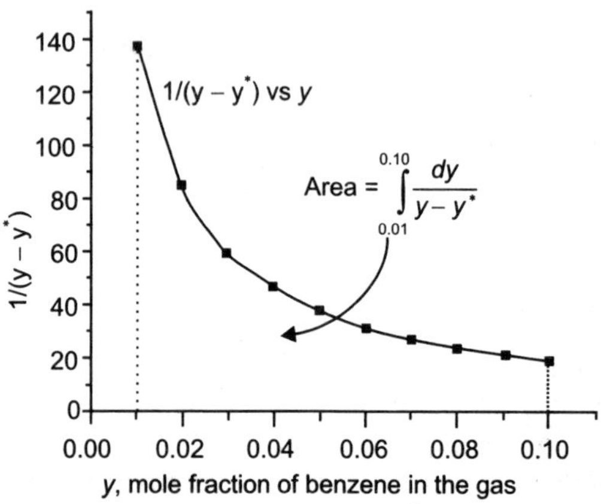

Fig. 4.66: Graphical integration for determining the number of transfer units

(b) The number of transfer units can be determined by a simplified method which makes use of the average driving force. The number of transfer units based on the gas phase is given by Eq. (4.137).

$$N_{tOG} = \frac{y_1 - y_2}{(y - y^*)_M}$$

where $(y - y^*)_M$ denotes the logarithmic mean of the concentration difference driving forces at the top of the column, $(y_2 - y_2^*)$ and at the bottom of the column, $(y_1 - y_1^*)$. Referring to the table where driving forces $y - y^*$ are tabulated, we see that $(y_1 = 0.10$ and $y_1^* = 0.0480)$ and $(y_2 = 0.01$ and $y_2^* = 0.0027)$. Therefore, the driving force at the top of the column $(y_2 - y_2^*) = 0.0073$ and the driving force at the bottom of the column $(y_1 - y_1^*) = 0.0520$. The log mean driving force

$$(y - y^*)_M = \frac{(y_1 - y_1^*) - (y_2 - y_2^*)}{\ln \frac{y_1 - y_1^*}{y_2 - y_2^*}} = \frac{0.0520 - 0.0073}{\ln \frac{0.0520}{0.0073}} = 0.0228$$

Using Eq. (4.137), $N_{tOG} = \dfrac{y_1 - y_2}{(y - y^*)_M} = \dfrac{0.10 - 0.01}{0.0228} = 3.95$

Height of the packed section is $Z = H_{tOG} \times N_{tOG} = 0.81 \times 3.95 = 3.2$ m **(Answer)**

(c) The number of transfer units assuming applicability of the Henry's law is given as

$$N_{tOG} = \frac{A}{A-1} \ln\left[\frac{y_1 - mx_2}{y_2 - mx_2}\left(1 - \frac{1}{A}\right) + \frac{1}{A}\right] \qquad \ldots(4.149)$$

The absorption factor A is the geometric mean of the absorption factors at the top and bottom of the column.

$$A = (A_1 A_2)^{0.5}$$

where $$A_1 \approx \frac{L_1}{mG_1} \quad \text{and} \quad A_2 \approx \frac{L_2}{mG_2}$$

Utilizing the values already available,

$$A_1 = \frac{L_1}{mG_1} = \frac{0.0303}{0.263 \times 0.058} = 1.9864$$

$$A_2 = \frac{L_2}{mG_2} = \frac{0.025}{0.263 \times 0.0527} = 1.8037$$

$$A = (A_1 A_2)^{0.5} = (1.9864 \times 1.8037)^{0.5} = 1.8928$$

Substituting in Eq. (4.149), we get

$$N_{tOG} = \frac{A}{A-1} \ln\left[\frac{y_1 - mx_2}{y_2 - mx_2}\left(1 - \frac{1}{A}\right) + \frac{1}{A}\right]$$

$$= \frac{1.8928}{1.8928 - 1} \ln\left[\frac{0.10 - 0.263(0.01)}{0.01 - 0.263(0.01)}\left(1 - \frac{1}{1.8928}\right) + \frac{1}{1.8928}\right] = 4.05$$

Height of the packed section is $Z = H_{tOG} \times N_{tOG} = 0.81 \times 4.05 = 3.28$ m **(Answer)**

(d) To find the number of transfer units by Baker's graphical method, the operating line (y vs x) and equilibrium curve (y^* vs x) are plotted using the coordinates already evaluated for the various points on the curves (Fig. 4.67). A construction line MN is drawn midway between the operating line and the equilibrium curve. The y', the ordinate of the point on the line MN is evaluated using y and y^* values for any given x as

$$y' = y^* + 0.5(y - y^*)$$

y' are tabulated below against x along with y and y^* values:

x	0.01	0.0309	0.0513	0.0712	0.0908	0.1099	01286	0.1469	0.1649	0.1824
y	0.01	0.02	0.03	0.04	0.05	0.06	0.07	0.08	0.09	0.10
y^*	0.0027	0.0082	0.0134	0.0189	0.0238	0.0290	0.0339	0.0389	0.0434	0.0480
y'	0.0064	0.0141	0.0217	0.0295	0.0369	0.0445	0.0520	0.0595	0.0667	0.0740

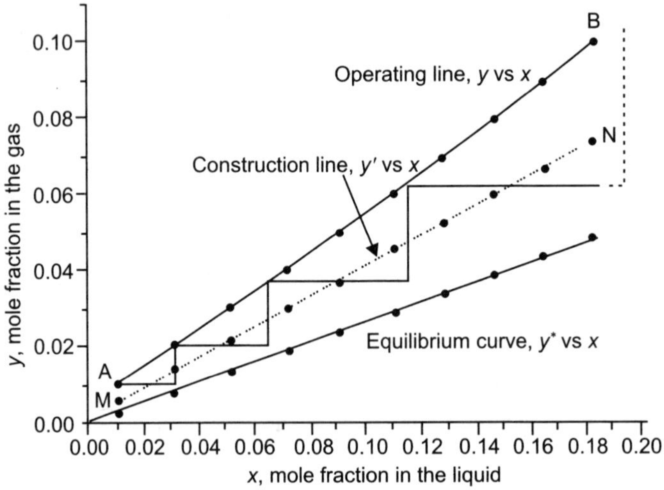

Fig. 4.67: Baker's method for Example 4.15

The curve y' versus x gives the curve MN in Fig. 4.67. Stepwise construction is carried out beginning with the lower end of the operating line A as discussed earlier. The integral number of triangles that can be constructed is 3 and the fractional number is evaluated to be 0.87. Therefore, the number of transfer units, $N_{tOG} = 3.87$. The height of packed section $Z = H_{tOG} \times N_{tOG} = 0.81 \times 3.87 = 3.13$ m **(Answer)**

The results of calculation of the height of the packed column by different methods are summarised in the following table:

Method	**Height of the column, *Z***
Graphical integration	3.39 m
Log-mean driving force	3.20 m
Henry's law analytical method	3.28 m
Baker's graphical construction	3.13 m

Example 4.16: It is required to produce a 10% (mole) ammonia in air mixture by stripping an aqueous 20 weight% ammonia solution with air at 101.3 kPa. The spent solution from the stripper should contain not more than 1 (weight)% NH_3. On the basis of 100 kmol/h of air entering the stripper, calculate the volume of packing required if $K_x a = 50$ kmol/m^{3}h mole fraction. The equilibrium data is given below:

$\bar{p}_{NH_3}$, kPa	1.6	2.4	4.3	6.7	9.3	22.1	30.3
g NH_3/100g water	2	3	5	7.5	10	20	25

***Solution*:** The stripper is represented in Fig. 4.68. Air is admitted at a rate 100 kmol/h. Let A be the area of cross section of the column. Then $G_1 = 100/A$ kmol/m^{2}h. Since air enters free of ammonia, $y_1 = Y_1 = 0$ and $G_S = G_1 = 100/A$. Since it is desired that air-ammonia mixture leaving the column contains 10% ammonia, $y_2 = 0.1$ and $Y_2 = y_2/(1-y_2) = 0.1111$. The exit gas rate, $G_2 = G_S(1+Y_2) = 1.1111 \times 100/A =$

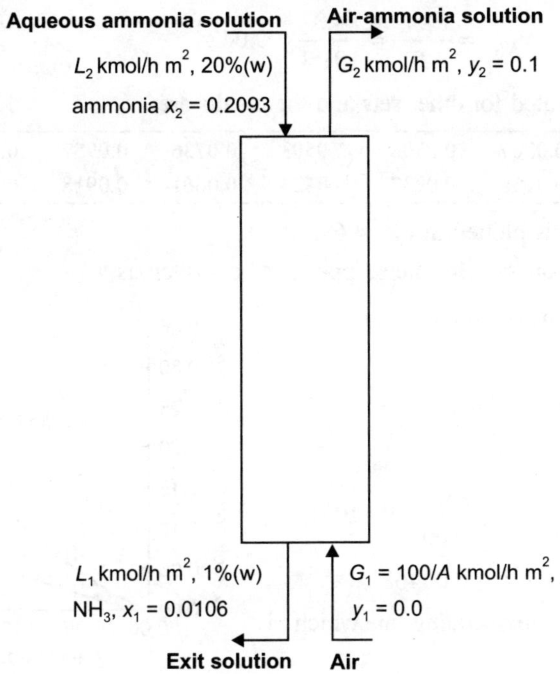

Fig. 4.68: Stripper in Example 4.16

111.11/A kmol/m^2 h. The solution admitted to the column contains 20% (weight) ammonia, the mole fraction x_2 is $x_2 = \dfrac{20/17}{20/17 + 80/18} = 0.2093$ and mole ratio $X_2 = 0.2093/(1 - 0.2093) = 0.2647$. Since the spent solution contain not more than 1% (weight) ammonia, $x_1 = \dfrac{1/17}{1/17 + 99/18} = 0.0106$ and $X_1 = 0.0106/(1 - 0.0106) = 0.0107$

The liquid flow rate can be estimated by material balance.

$$G_S(Y_2 - Y_1) = L_S(X_2 - X_1)$$

$$\frac{100}{A}(0.1111 - 0) = L_S(0.2647 - 0.0107)$$

On solving this equation, we get $L_S = 43.74/A$ kmol/m^{2}h.

$$L_2 = L_S(1 + X_2) = \frac{43.74}{A}(1.2647) = \frac{55.32}{A} \text{ kmol/m}^2\text{h}$$

$$L_1 = L_S(1 + X_1) = \frac{43.74}{A}(1.0107) = \frac{44.21}{A} \text{ kmol/m}^2\text{h}$$

The equilibrium data is expressed in terms of mole fractions. Consider for example, the data $\bar{p}_{NH_3} = 4.3$ kPa and liquid composition 5 g NH_3/100g water.

$$5 \text{ g NH}_3\text{/100g water} = \frac{5/17}{100/18} \frac{\text{mole ammonia}}{\text{mole water}} \equiv \frac{5/17}{5/17 + 100/18} = 0.0503 \frac{\text{mole ammonia}}{\text{mole solution}}$$

Or mole fraction in the liquid, $x_{NH_3} = 0.0503$

Mole fraction in the gas is $y_{NH_3} = \frac{\bar{p}_{NH_3}}{P} = \frac{4.3}{101.3} = 0.0424$

The calculation is repeated for other sets and the results are:

x	0.0	0.0207	0.0308	0.0503	0.0736	0.0957	0.1748	0.2093
y	0.0	0.0158	0.0237	0.0424	0.0661	0.0918	0.2182	0.2991

The equilibrium curve is plotted in Fig. 4.69.

The material balance equation for the stripper can be written as

$$G_S(Y - Y_1) = L_S(X - X_1)$$

which gives

$$Y = \frac{L_S}{G_S}X - \frac{L_S}{G_S}X_1 + Y_1$$

$$Y = \frac{43.74}{100}X - \frac{43.74}{100}(0.0107)$$

$$Y = 0.4374X - 0.00468$$

This is the equation for the operating line which in terms of mole fractions is

$$\frac{y}{1-y} = 0.4374\frac{x}{1-x} - 0.00468 \quad \text{(A)}$$

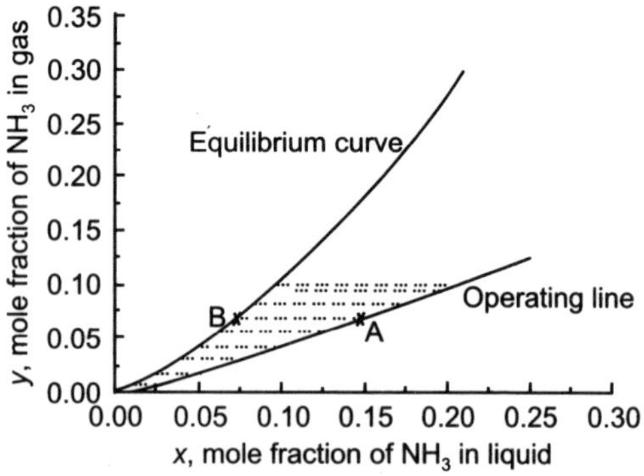

Fig. 4.69: Graphical determination of $x - x^*$ for the stripper in Example 4.16

To find points on the operating line, assume x in the range x_1 to x_2 and calculate y using Eq. (A).

x	0.0106	0.05	0.10	0.15	0.20	0.2093	0.25
y	0.0	0.0180	0.0421	0.0676	0.0948	0.10	0.1237

The operating line is plotted in Fig. 4.69.

The height of the packed section of the stripper may be estimated as $Z = H_{tOL} \times N_{tOL}$ where H_{tOL} and N_{tOL} are the height and number of overall liquid transfer units. The number of transfer units is given by the Eq. (4.154A) and the height of transfer unit by Eq. (4.152)

The height of transfer unit for stripper can be evaluated by

$$N_{tOL} = \int_{x_2}^{x_1} \frac{dx}{x - x^*} + \frac{1}{2}\ln\frac{1-x_1}{1-x_2} \quad \text{(B)}$$

The x-coordinate of a point on the operating line gives the concentration of the liquid x and y-coordinate gives the concentration of the gas phase y in contact with the liquid. For illustration, let us consider point A in Fig. 4.69. Its x-coordinate is 0.15. From this point move horizontally to point B on the equilibrium curve and the x-coordinate of B is x^*, the liquid-phase equilibrium equivalent of y. In this way, we can determine for various x, the corresponding x^*. The difference between x and x^* is the driving force based on the liquid side. The x and x^* so determined graphically and $x - x^*$, and $1/(x - x^*)$ are tabulated below.

x	0.0106	0.025	0.05	0.075	0.10	0.125	0.15	0.175	0.20	0.2093
x^*	0.0	0.0089	0.0245	0.0386	0.0503	0.0636	0.0750	0.0869	0.0962	0.0996
$x-x^*$	0.0106	0.0161	0.0255	0.0364	0.0497	0.0614	0.075	0.0881	0.1038	0.1097
$1/(x-x^*)$	94.3	62.1	39.2	27.5	20.1	16.3	13.3	11.4	9.6	9.1

A graph is plotted with $1/(x - x^*)$ on the y – axis against x on the abscissa. It is shown in Fig. 4.70.

The area between limits x_1 and x_2 is determined which gives $\int_{0.0106}^{0.2903} \frac{dx}{x - x^*}$.

From the graph $\int_{0.0106}^{0.2903} \frac{dx}{x - x^*} = 5.5$

The number of transfer units

$$N_{tOL} = \int_{x_1}^{x_2} \frac{dx}{x - x^*} + \frac{1}{2} \ln \frac{1 - x_1}{1 - x_2}$$

$$= 5.5 + \frac{1}{2} \ln \frac{1 - 0.0106}{1 - 0.2903} = 5.67$$

The height of overall liquid transfer units is

$$H_{tOL} = \frac{L}{K_x a (1 - x)_{*M}} \approx \frac{L}{K_x a}$$

Here L may be taken to be

$$\frac{L_1 + L_2}{2} = \frac{55.32 + 44.21}{2A} = \frac{49.77}{A} \text{ kmol/m}^2\text{h}$$

$$H_{tOL} = \frac{L}{K_x a} = \frac{49.77}{50A} = \frac{0.9953}{A} \text{ m}$$

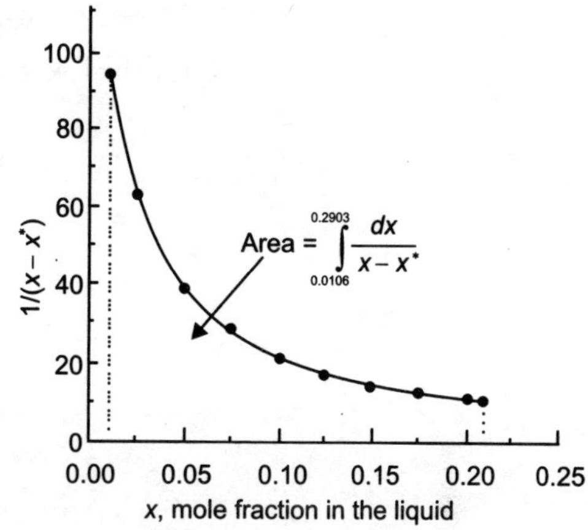

Fig. 4.70: Determination of number of transfer units in Example 4.16 by graphical integration

Since the height of the packed column is evaluated as

$$Z = H_{tOL} \times N_{tOL} = \frac{0.9953}{A} \times 5.67 = \frac{5.64}{A} \text{ m}$$

Since $Z = \frac{5.64}{A}$ m, the volume of packing $AZ = 5.64 \text{ m}^3$ **(Answer)**

Example 4.17: An air-ammonia mixture containing 20% (mole) NH_3 is scrubbed with water in a countercurrent packed tower to remove 95% ammonia in the entering gas. The gas enters the column at 1000 kg/h m^2 and the column is operated at 308 K and 101.3 kPa. Pure water is admitted at a rate 1.5 times the minimum rate. The equilibrium relation is $y = 0.75x$ where x and y are mole fractions. Using the same water rate as above and keeping all other conditions unchanged, determine the percent increase in the column height required to absorb 99% of the ammonia in the entering gas. Assume dilute solutions.

***Solution*:** The absorption column is illustrated in Fig. 4.71. Gas entering the column contains 20% ammonia. Therefore

$$y_1 = 0.20 \quad Y_1 = \frac{0.20}{1 - 0.20} = 0.25$$

Average molecular weight of the entering gas $= MW = 0.20 \times 17 + 0.80 \times 29 = 26.6$

The molar flow rate $G_1 = \frac{1000}{26.6} = 37.59 \text{ kmol/m}^2\text{h}$

$$G_s = G_1 (1 - 0.20) = 37.59 \times 0.80 = 30.08 \text{ kmol/m}^2\text{h}$$

Case(1) : Ammonia recovery is 95%

Since 95% of ammonia in the entering gas is absorbed, $Y_2 = 0.05 \times 0.25 = 0.0125$ and

$$y_2 = \frac{0.0125}{1+0.0125} = 0.0123$$

$$G_2 = G_s(1+Y_2) = 30.08 \times 1.0125 = 30.46 \text{ kmol/m}^2\text{h}$$

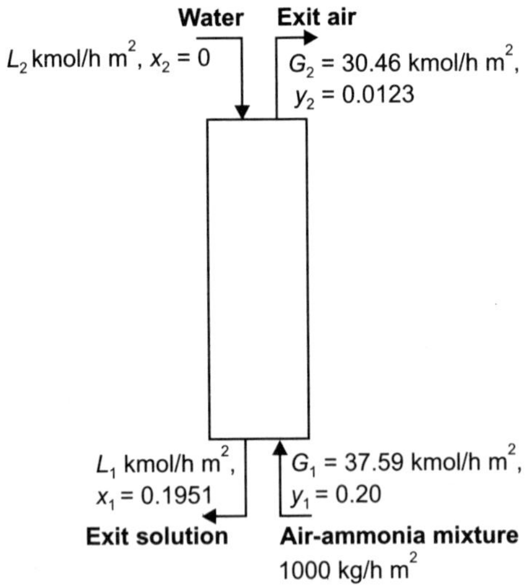

Fig. 4.71: Absorber for 95% ammonia recovery in Example 4.17

When the operating line is plotted on mole ratio coordinates, its slope is

$$\left(\frac{L_s}{G_s}\right)_{\min} = \frac{Y_1 - Y_2}{X_1(\max) - X_2} \tag{A}$$

Assuming that the equilibrium curve is a straight line, $X_1(\max)$ is the equilibrium value corresponding to the gas phase composition $Y_1 = 0.25$. Since the equilibrium relation is

$$y = 0.75x \text{ or } \frac{Y}{1+Y} = 0.75\frac{X}{1+X}, \; X_1 \text{ (max) is given by}$$

$$\frac{0.25}{1+0.25} = 0.75\frac{X_1(\max)}{1+X_1(\max)} \text{ or } X_1(\max) = 0.3636$$

(It can be seen that the result obtained by the graphical method is the same as this.)

Now using Eq. (A),

$$\left(\frac{L_s}{G_s}\right)_{\min} = \frac{Y_1 - Y_2}{X_1(\max) - X_2} = \frac{0.25 - 0.0125}{0.3636 - 0} = 0.6532$$

Since the liquid rate is 1.5 times the minimum,

$$\left(\frac{L_s}{G_s}\right) = \frac{Y_1 - Y_2}{X_1 - X_2} = \frac{0.25 - 0.0125}{X_1 - 0} = 1.5 \times 0.6532 = 0.9798$$

which gives $X_1 = 0.2424$ or the exit solution contains 0.2424 mole ammonia per mole solution.

$$x_1 = \frac{X_1}{1+X_1} = \frac{0.2424}{1.2424} = 0.1951$$

Since the equilibrium curve and the operating lines are approximately straight lines it is reasonable to calculate the number of transfer unit by the log-mean method.

$$N_{tOG} = \frac{y_1 - y_2}{(y - y^*)_M} \quad ...(4.137)$$

where $(y - y^*)_M$ denotes the logarithmic mean of the concentration difference driving forces at the top of the column, $(y_2 - y_2^*)$ and bottom of the column, $(y_1 - y_1^*)$.

Since the equilibrium relationship is $y = 0.75x$, $y_1^* = 0.75x_1 = 0.75 \times 0.1951 = 0.1463$ and $y_1 - y_1^* = 0.20 - 0.1463 = 0.0537$.

$$y_2^* = 0.75x_2 = 0.75 \times 0.0 = 0 \text{ and } y_2 - y_2^* = 0.0123 - 0 = 0.0123$$

$$(y - y^*)_M = \frac{(y_1 - y_1^*) - (y_2 - y_2^*)}{\ln \dfrac{y_1 - y_1^*}{y_2 - y_2^*}} = \frac{0.0537 - 0.0123}{\ln \dfrac{0.0537}{0.0123}} = 0.0281$$

Number of transfer units, $N_{tOG} = \dfrac{y_1 - y_2}{(y - y^*)_M} = \dfrac{0.20 - 0.0123}{0.0281} = 6.68$

Height of transfer unit, $H_{tOG} = \dfrac{G}{K_y a}$ where $G = \dfrac{G_1 + G_2}{2} = \dfrac{37.59 + 30.46}{2} = 34.03 \text{ kmol/m}^2\text{h}$.

$$H_{tOG} = \frac{G}{K_y a} = \frac{34.03}{K_y a} \text{ m}$$

The height of the column is $Z_1 = H_{tOG} \times N_{tOG} = \dfrac{34.03}{K_y a} \times 6.68 = \dfrac{227.32}{K_y a}$ m (C)

Case(2) : Ammonia recovery is 99%

Refer Fig. 4.72. Since 99% of ammonia in the entering gas is absorbed,

$$Y_2 = 0.01 \times 0.25 = 0.0025 \text{ and } y_2 = \frac{0.0025}{1 + 0.0025} = 0.0025$$

$$G_2 = G_s(1 + Y_2) = 30.08 \times 1.0025 = 30.16 \text{ kmol/m}^2\text{h}$$

Since the same quantity of water as in the previous case is used,

$$\left(\frac{L_s}{G_s}\right) = 0.9798 = \frac{Y_1 - Y_2}{X_1 - X_2} = \frac{0.25 - 0.0025}{X_1 - 0}$$

which gives the exit liquid concentration $X_1 = 0.2526$ mol ammonia per mol water. Therefore,

$$x_1 = 0.2526/1.2526 = 0.2017 \text{ and } y_1^* = 0.75x_1 = 0.75 \times 0.2017 = 0.1512$$

$$y_1 - y_1^* = 0.20 - 0.1512 = 0.0488.$$

$$y_2^* = 0.75x_2 = 0.75 \times 0.0 = 0 \text{ and } y_2 - y_2^* = 0.0025 - 0 = 0.0025$$

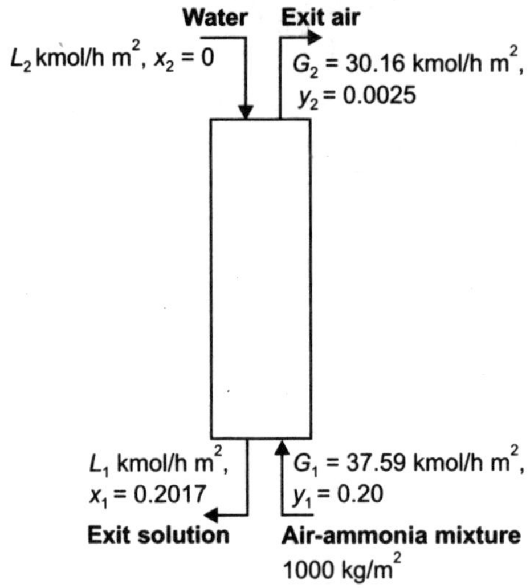

Fig. 4.72: Absorber for 99% ammonia recovery in Example 4.17

$$(y - y^*)_M = \frac{0.0488 - 0.0025}{\ln \dfrac{0.0488}{0.0025}} = 0.0156$$

Number of transfer units, $N_{tOG} = \dfrac{y_1 - y_2}{(y - y^*)_M} = \dfrac{0.20 - 0.0025}{0.0156} = 12.66$

Height of transfer unit, $H_{tOG} = \dfrac{G}{K_y a}$ where $G = \dfrac{G_1 + G_2}{2} = \dfrac{37.59 + 30.16}{2} = 33.88 \text{ kmol/m}^2\text{h}.$

$$H_{tOG} = \frac{G}{K_y a} = \frac{33.88}{K_y a} \text{ m}$$

The height of the column is $Z_2 = H_{tOG} \times N_{tOG} = \dfrac{33.88}{K_y a} \times 12.66 = \dfrac{428.92}{K_y a}$ m (D)

Dividing Eq. (D) by Eq. (C) and assuming that there is no substantial change in the overall mas transfer coefficient,

$$\frac{Z_2}{Z_1} = \frac{428.92}{227.32} = 1.887$$

To increase the percent recovery from 95% to 99%, the percent increase in the column height is

$$\frac{Z_2 - Z_1}{Z_1} \times 100 = \frac{1.887 - 1.0}{1.0} \times 100 = 88.7\%$$ **(Answer)**

Height of Packed Columns: General Method

Since the overall mass transfer coefficients depend upon the slope of the equilibrium curve (*see* Eq. (4.158)), the method based on the overall transfer unit is most suitable when the equilibrium curve is straight line in the concentration range of interest. The general method for the design of packed column should take into account the non linearity of the equilibrium relationship, the variation of gas and liquid flows through the column as well as the dependence of mass transfer coefficients on the gas and liquid flows and concentrations. In the following section we describe the general technique for determination of the height of the packed column for absorption.

Consider Eq. (4.113).

$$dz = \frac{G\,dy}{k_y a(1-y)(y-y_i)} \quad \ldots(4.113)$$

Multiply the numerator and denominator by $(1-y)_{iM}$, the logarithmic mean of the mole fraction of non diffusing components in the bulk and at the interface.

$$dz = \frac{G(1-y)_{iM}\,dy}{k_y a(1-y)_{iM}(1-y)(y-y_i)} \quad \ldots(4.176)$$

On integrating Eq. (4.176) between the top and bottom concentrations of the column, we get

$$\int_0^Z dz = \int_{y_2}^{y_1} \frac{G(1-y)_{iM}\,dy}{k_y a(1-y)_{iM}(1-y)(y-y_i)} \quad \ldots(4.177)$$

In Eq. (4.177), $k_y a(1-y)_{iM} = k'_y a$ where $k'_y a$ is the film coefficient of mass transfer. Replacing $k_y a$ which is a strong function of composition, by the film coefficient of mass transfer, Eq. (4.177) becomes

$$\int_0^Z dz = Z = \int_{y_2}^{y_1} \frac{G(1-y)_{iM}\,dy}{k'_y a(1-y)(y-y_i)} \quad \ldots(4.178)$$

The integral in Eq. (4.178) is to be determined graphically. Gas-phase concentration at the interface y_i is determined by the methods already explained in detail in Chapter 3. The operating line and the equilibrium curve are plotted on mole fraction coordinates. Let the ordinate of a point P on the operating line be y, the bulk concentration of the gas phase. For dilute solutions, the concentration y_i at the interface can be estimated graphically using the equation

$$\frac{y-y_i}{x-x_i} = -\frac{k_x a}{k_y a} \quad \ldots(4.179)$$

Draw the line from P with slope $-\frac{k_x a}{k_y a}$, which will intersect the equilibrium curve at Q whose coordinates are x_i and y_i. The mass transfer coefficients k_x and k_y are strong functions of concentration and therefore, use of Eq. (4.179) with constant values for k_x and k_y will lead to erroneous results when dealing with concentrated solutions. In terms of film coefficients, Eq. (4.179) can be written as:

$$\frac{y-y_i}{x-x_i} = -\frac{k_x a}{k_y a} = -\frac{k'_x a/(1-x)_{iM}}{k'_y a/(1-y)_{iM}} \quad \ldots(4.180)$$

Generally the empirical correlations on mass transfer coefficients provide the film coefficients directly. Though the film coefficients k'_x and k'_y do vary with liquid and gas flow rates, the use of average values for k'_x and k'_y will not introduce any significant error. Using Eq. (4.180), the interfacial compositions corresponding to the bulk compositions x and y can be estimated by trial and error. Assume interfacial compositions x_i and y_i and estimate $(1-x)_{iM}$ and $(1-y)_{iM}$ by the following equations:

$$(1-y)_{iM} = \frac{(1-y)-(1-y_i)}{\ln\dfrac{1-y}{1-y_i}} \quad \ldots(4.181)$$

$$(1-x)_{iM} = \frac{(1-x)-(1-x_i)}{\ln\dfrac{1-x}{1-x_i}} \qquad \text{...(4.182)}$$

Calculate slope of the driving force line $-\dfrac{k'_x a(1-y)_{iM}}{k'_y a(1-x)_{iM}}$ and draw a line with this slope from point P with coordinates (x, y) on the operating line as shown in Fig. 4.73. The intersection of this line with the equilibrium curve gives (x_i, y_i) which, if different from the assumed values, are used for finding the new estimate of $(1-x)_{iM}$ and $(1-y)_{iM}$ for the second trial. The calculations are repeated till a satisfactory estimate of interfacial concentrations is made. It is seen that generally not more than two trials are necessary for arriving at a satisfactory estimate of y_i.

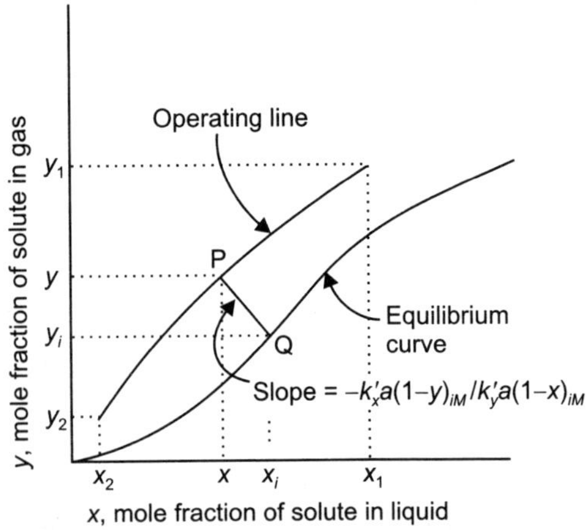

Fig. 4.73: Evaluation of the interfacial compositions

The interfacial composition y_i is substituted into Eq. (4.177) for determining the height of packed columns. A simplification is possible for Eq. (4.177), since the quantity $\dfrac{G}{k'_y a} = \dfrac{G}{k_y a(1-y)_{iM}}$ is found to remain approximately constant over the length of the tower. This quantity has the dimension of length and is taken outside the integral. It is designated as the height of a gas transfer unit represented by H_{tG}.

$$Z = \frac{G}{k_y a(1-y)_{iM}} \int_{y_2}^{y_1} \frac{(1-y)_{iM}\,dy}{(1-y)(y-y_i)}$$

$$= H_{tG} \int_{y_2}^{y_1} \frac{(1-y)_{iM}\,dy}{(1-y)(y-y_i)} \qquad \text{...(4.183)}$$

The integral $\displaystyle\int_{y_2}^{y_1} \frac{(1-y)_{iM}\,dy}{(1-y)(y-y_i)}$ is a dimensionless quantity called the number of gas transfer units and is denoted by N_{tG}. Thus

$$Z = H_{tG} N_{tG} \qquad \text{...(4.184)}$$

with

$$H_{tG} = \frac{G}{k_y a(1-y)_{iM}} = \frac{G}{k'_y a} \qquad \text{...(4.185)}$$

and

$$N_{tG} = \int_{y_2}^{y_1} \frac{(1-y)_{iM}\,dy}{(1-y)(y-y_i)} \qquad \text{...(4.186)}$$

Replacing the logarithmic mean $(1-y)_{iM}$ in Eq. (4.186) with arithmetic mean and proceeding as in the derivation of Eq. (4.124) we get

$$N_{tG}=\int_{y_2}^{y_1}\frac{[(1-y)+(1-y_i)]dy}{2(1-y)(y-y_i)}=\int_{y_2}^{y_1}\frac{dy}{y-y_i}+\frac{1}{2}\ln\frac{1-y_2}{1-y_1} \quad ...(4.187)$$

The first term in Eq. (4.187) is evaluated by graphical integration. Using the values of interfacial composition y_i estimated for given values of y, the driving force for mass transfer in the gas phase $y-y_i$ is calculated. This is repeated for different y values on the operating line and a plot of $1/(y-y_i)$ is drawn against y. The area under the curve between limits y_2 and y_1 gives the required integral as shown in Fig. 4.74.

In dealing with concentrated solutions, it will be necessary to account for the variation of the gas flow through the column and its effect on the mass transfer coefficients. In such cases, we have to use Eq. (4.178) for the determination of the height of the column.

$$Z=\int_{y_2}^{y_1}\frac{G(1-y)_{iM}\,dy}{k'_y a(1-y)(y-y_i)}$$

The determination of the height of the packed column using these general methods is illustrated in Examples 4.19 and 4.20.

[**Note:** To determine the interfacial compositions, we can use yet another technique. Equation (4.180) can be modified as

$$\frac{1-y_i}{1-y}=\left(\frac{1-x}{1-x_i}\right)^{k'_x a/k'_y a} \quad ...(4.188)$$

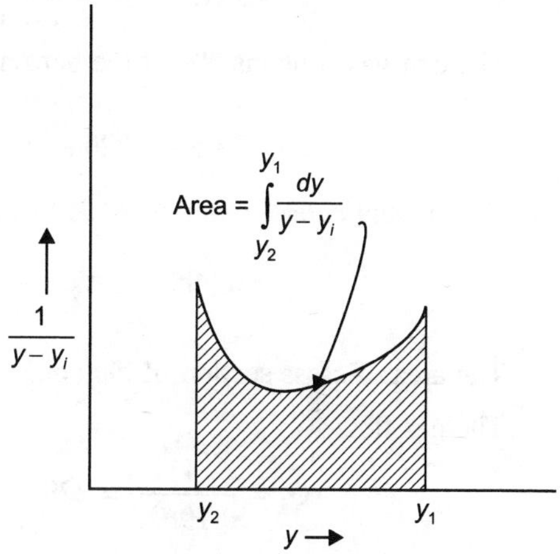

Fig. 4.74: Graphical integration for determining number of transfer units

For a point on the operating curve with coordinates (x, y) we can calculate the set of x_i and y_i values satisfying the above equation. Plot a curve through all these points which will intersect the equilibrium curve, the coordinates of the point of intersection being (x_i, y_i). Repeat the calculation for different points on the operating line.]

The height of the column can be determined using the liquid side coefficients in which case we use

$$Z=\int_{x_2}^{x_1}\frac{L(1-x)_{iM}\,dx}{k'_x a(1-x)(x_i-x)}$$

The height and number of transfer units based on the liquid side can be expressed as below:

$$Z=H_{tL}N_{tL} \quad ...(4.189)$$

with

$$H_{tL}=\frac{L}{k_x a(1-x)_{iM}}\approx\frac{L}{k'_x a} \quad ...(4.190)$$

and

$$N_{tL}=\int_{x_2}^{x_1}\frac{(1-x)_{iM}\,dx}{(1-x)(x_i-x)}=\int_{x_2}^{x_1}\frac{dx}{x_i-x}+\frac{1}{2}\ln\frac{1-x_2}{1-x_1} \quad ...(4.191)$$

Example 4.18: A 0.5 m diameter packed column is used for absorbing benzene from coal gas at 300K and 101.3 kPa. The gas containing 2 (mole) % benzene enters the column at a rate 45 kmol/h and it is scrubbed with absorbent oil to absorb 95% of the benzene. Oil containing 0.5 (mole) % benzene is used as the solvent at a rate 1.5 times the minimum. The equilibrium distribution of benzene between gas and liquid is given by $y = 0.125x$, where x and y are the mole fractions of benzene in the liquid and gas respectively. The film coefficients for mass transfer on the gas and liquid side are $k_y' a = 250$ kmol/m³h and $k_x' a = 50$ kmol/m³h. Calculate the packed height of the column

(a) assuming the gas flow is constant

(b) taking into account the variation of gas flow through the column.

Solution: The gas inlet concentration

$$y_1 = 0.02 \qquad Y_1 = \frac{0.02}{1-0.02} = 0.0204$$

The exit gas contains 5% of the benzene present in the entering gas

$$Y_2 = 0.05 \times 0.0204 = 0.00102 \qquad y_2 = \frac{0.00102}{1+0.00102} = 0.00102$$

The solvent enters with 0.5 (mole)% benzene.

$$x_2 = 0.005 \qquad X_2 = \frac{0.005}{1-0.005} = 0.00503$$

The area of cross section of the column is $A = \frac{\pi}{4}(0.50)^2 = 0.19635 \text{ m}^2$

The gas flow

$$G_1 = \frac{45}{0.19635} = 229.18 \text{ kmol/m}^2\text{h}$$

$$G_s = G_1(1-0.02) = 229.18 \times 0.98 = 224.60 \text{ kmol/m}^2\text{h}$$

$$G_2 = G_s(1+Y_2) = 224.60 \times 1.00102 = 224.83 \text{kmol/m}^2\text{h}$$

The concentrations and flow quantities are shown in the Fig. 4.75.

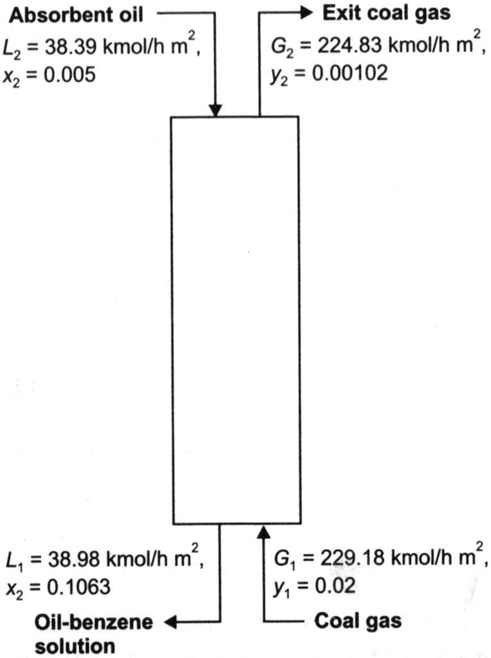

Fig. 4.75: Absorber for Example 4.18

The equilibrium relation is $y = 0.125x$, which is to be expressed in terms of the mole ratios, in order to evaluate minimum liquid-to-gas ratio.

$$\frac{Y}{1+Y} = 0.125\frac{X}{1+X} \tag{A}$$

By assuming different values of X calculate Y using Eq. (A):

X	0.0000	0.0250	0.0500	0.0750	0.1000	0.1250	0.1500	0.1750	0.2000	0.2250	0.2500
Y	0.0000	0.0031	0.0060	0.0088	0.0115	0.0141	0.0166	0.0190	0.0213	0.0235	0.0256

The equilibrium curve is drawn as shown in Fig. 4.76 and the limiting operating line AB is tangential to the equilibrium curve from point A whose coordinates are ($X_2 = 0.00503, Y_1 = 0.00102$). The coordinates of point B are noted [X_1 (max) = 0.1760, Y_2 = 0.0204]. The slope of the line AB is

$$\left(\frac{L_s}{G_s}\right)_{\min} = \frac{0.0204 - 0.00102}{0.1760 - 0.00503} = 0.1134$$

The liquid-to-gas ratio is 1.5 times the minimum.

$$\left(\frac{L_s}{G_s}\right) = 0.1134 \times 1.5 = 0.1701$$

The solute balance for the column is

$$(Y_1 - Y_2) = \left(\frac{L_s}{G_s}\right)(X_1 - X_2)$$

$$(0.0204 - 0.00102) = (0.1701)(X_1 - 0.00503)$$

On solving this equation, we get

$$X_1 = 0.1190 \quad x_1 = \frac{0.1190}{1.1190} = 0.1063$$

$$L_s = \left(\frac{L_s}{G_s}\right) \times G_s = 0.1701 \times 224.6 = 38.20 \text{ kmol/m}^2\text{h}$$

$$L_1 = L_s(1 + X_1) = 38.20 \times 1.0204 = 38.98 \text{ kmol/m}^2\text{h}$$

$$L_2 = L_s(1 + X_2) = 38.20 \times 1.00503 = 38.39 \text{ kmol/m}^2\text{h}$$

To calculate the number of transfer units graphically, the equilibrium curve and the operating line are to be drawn on mole fraction coordinates. Since the equilibrium relation is $y = 0.125x$, the equilibrium curve is a straight line with slope 0.125. Operating line on mole ratios is given by the equation

$$(Y - Y_2) = \left(\frac{L_s}{G_s}\right)(X - X_2) \text{ or}$$

$$(Y - 0.00102) = (0.1701)(X - 0.00503)$$

$$Y = 0.1701X + 0.00016 \tag{B}$$

This is converted into mole fractions

$$\frac{y}{1-y} = 0.1701\frac{x}{1-x} + 0.00016 \tag{C}$$

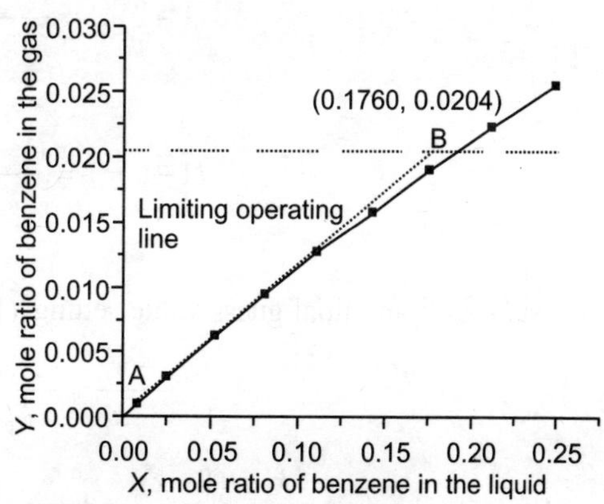

Fig. 4.76: Determination of minimum liquid-to-gas ratio for Example 4.18

Equation (C) is the equation for the operating line on mole fraction coordinates. Using Eq. (C), coordinates of various points on the operating line are evaluated as given in the table below:

x	0.005	0.0200	0.0400	0.0600	0.0800	0.1000	0.1063
y	0.00102	0.0036	0.0072	0.0109	0.0147	0.0187	0.02

The operating line (*AB*) is drawn through the above points as shown in Fig. 4.77. Having obtained the operating line and the equilibrium curve, the interfacial compositions can now be determined using Eq. (4.180)

$$\frac{y-y_i}{x-x_i}=-\frac{k_xa}{k_ya}=-\frac{k'_xa/(1-x)_{iM}}{k'_ya/(1-y)_{iM}} \quad \ldots(4.180)$$

Consider a point such as point *E* on the operating line whose coordinates are (x, y). The corresponding interfacial compositions are determined by drawing a straight line from point *E* with slope $-\frac{k'_x/(1-x)_{iM}}{k'_y/(1-y)_{iM}}$ to intersect the equilibrium curve at point *E*′. The coordinates of point *E*′ are (x_i, y_i). To evaluate $(1-x)_{iM}$ and $(1-y)_{iM}$ for using Eq. (4.180), we should know the interface compositions (x_i, y_i) because,

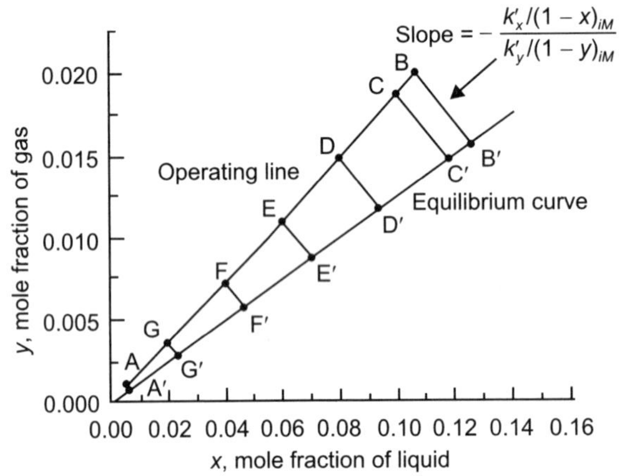

Fig. 4.77: Determination of interfacial composition, Example 4.18

$$(1-y)_{iM}=\frac{(1-y)-(1-y_i)}{\ln\frac{1-y}{1-y_i}} \quad \ldots(4.182)$$

$$(1-x)_{iM}=\frac{(1-x)-(1-x_i)}{\ln\frac{1-x}{1-x_i}} \quad \ldots(4.183)$$

We start with an initial guess value setting $(1-y)_{iM}=(1-y)$ and $(1-x)_{iM}=(1-x)$ so that

$$\frac{y-y_i}{x-x_i}=-\frac{k_xa}{k_ya}=-\frac{k'_xa/(1-x)}{k'_ya/(1-y)} \quad (A)$$

Evaluate slope as $-\frac{k'_xa/(1-x)}{k'_ya/(1-y)}$ and draw a straight line with this slope from point *E* to intersect the equilibrium curve at (x_i, y_i). Using these x_i and y_i values in Eqs. (4.182) and (4.183), we modify the slope as $-\dfrac{k'_x a/(1-x)_{iM}}{k'_y a/(1-y)_{iM}}$. If the slope now evaluated is not significantly different from the initial guess we can accept (x_i, y_i) as the interfacial compositions. Otherwise, draw the driving force line EE' with the new slope and obtain the new set of interface concentrations. The interfacial concentrations are to be determined for different points starting from point B (x_1, y_1). At point B, the bottom of the column, $x = x_1 = 0.1063$ and $y = y_1 = 0.02$. Let $(1-y)_{iM} = (1-y) = 0.98$ and $(1-x)_{iM} = (1-x) = 0.8937$

$$\text{Slope of the driving force line } BB' = -\frac{k'_x a/(1-x)_{iM}}{k'_y a/(1-y)_{iM}} = -\frac{k'_x a/(1-x)}{k'_y a/(1-y)} = -\frac{50}{250}\times\frac{0.98}{0.8937} = -0.2193$$

A straight line BB' is drawn with slope – 0.2193 which intersect the equilibrium curve at (0.1258, 0.01573). Assuming that $x_i = 0.1258$ and $y_i = 0.01573$, $1 - x_i = 0.8742$, $1 - y_i = 0.98427$ and we have

$$(1-y)_{iM} = \frac{(1-y)-(1-y_i)}{\ln\dfrac{1-y}{1-y_i}} = \frac{0.98-0.98427}{\ln\dfrac{0.98}{0.98427}} = 0.98214$$

$$(1-x)_{iM} = \frac{(1-x)-(1-x_i)}{\ln\dfrac{1-x}{1-x_i}} = \frac{0.8937-0.8742}{\ln\dfrac{0.8937}{0.8742}} = 0.8839$$

$$\text{The slope} = -\frac{k'_x/(1-x)_{iM}}{k'_y/(1-y)_{iM}} = -\frac{50}{250}\times\frac{0.98214}{0.8839} = -0.2222$$

Since the new slope is slightly different from the original guess, the driving force line is drawn with the new slope from point B to intersect at B' ($x_i = 0.1256$, $y_i = 0.0157$). Using these values for the interface concentrations, the slope of driving force line is again calculated. $1 - x_i = 0.8744$, $1 - y_i = 0.9843$.

$$(1-y)_{iM} = \frac{(1-y)-(1-y_i)}{\ln\dfrac{1-y}{1-y_i}} = \frac{0.98-0.9843}{\ln\dfrac{0.98}{0.9843}} = 0.98215$$

$$(1-x)_{iM} = \frac{(1-x)-(1-x_i)}{\ln\dfrac{1-x}{1-x_i}} = \frac{0.8937-0.8744}{\ln\dfrac{0.8937}{0.8744}} = 0.8840$$

$$\text{The slope} = -\frac{k'_x a/(1-x)_{iM}}{k'_y a/(1-y)_{iM}} = -\frac{50}{250}\times\frac{0.98215}{0.8840} = -0.2222$$

The slope is same as the one used in the preceding trial, and therefore, $x_i = 0.1256$, $y_i = 0.0157$ are accepted as the interfacial compositions corresponding to bulk compositions $x = 0.1063$, $y = 0.02$.

These calculations are carried out for different points such as, C, D, ….., A on the operating line; the first two rows in the following table give the y and y_i values so evaluated.

y	0.00102	0.0036	0.0072	0.0109	0.0147	0.0187	0.02
y_i	0.00078	0.00292	0.00583	0.00876	0.01172	0.01474	0.01570
$y-y_i$	0.00024	0.00068	0.00137	0.00214	0.00298	0.00396	0.00430
$(y-y_i)^{-1}$	4098.4	1466.3	728.3	467.7	335.8	252.7	232.8(a)

(a) The height of the column is determined using the equation

$$Z = H_{tG} N_{tG} \quad \ldots(4.184)$$

Assuming the mass transfer coefficient and gas flow rate remain constant, we can use the relatively simple method of calculation using

$$H_{tG} = \frac{G}{k'_y a} \quad \ldots(4.185)$$

and

$$N_{tG} = \int_{y_2}^{y_1} \frac{dy}{y - y_i} + \frac{1}{2} \ln \frac{1 - y_2}{1 - y_1} \quad \ldots(4.187)$$

To evaluate the integral in Eq. (4.187) graphically the $(y - y_i)^{-1}$ is plotted against y and the area is found between limits y_2 and y_1. The $(y - y_i)^{-1}$ values for different y values are also given in the above table.

A graph is drawn with $1/(y - y_i)$ against y as shown below (Fig. 4.78). The area is determined between limits $y_2 = 0.0102$ and $y_1 = 0.02$.

$$\text{Area} = \int_{0.00102}^{0.02} \frac{dy}{y - y_i} = 14.7$$

$$N_{tG} = \int_{y_2}^{y_1} \frac{dy}{y - y_i} + \frac{1}{2} \ln \frac{1 - y_2}{1 - y_1}$$

$$= 14.8 + \frac{1}{2} \ln \frac{1 - 0.00102}{1 - 0.02} = 14.71$$

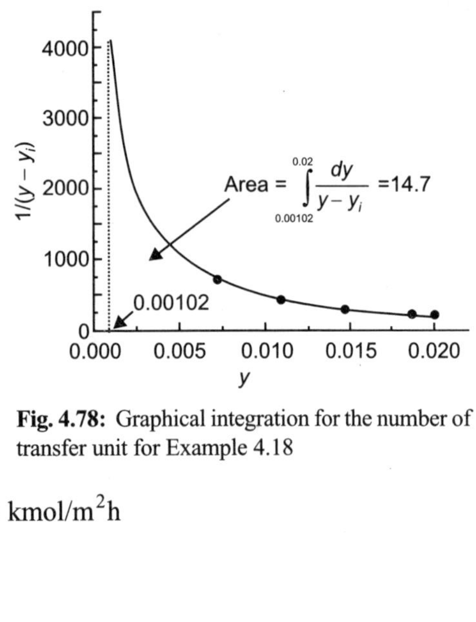

Fig. 4.78: Graphical integration for the number of transfer unit for Example 4.18

The height of transfer unit is now determined.

$$H_{tG} = \frac{G}{k'_y a}$$

Here

$$G = \frac{G_1 + G_2}{2} = \frac{229.18 + 224.83}{2} = 227.00 \text{ kmol/m}^2\text{h}$$

$$H_{tG} = \frac{G}{k'_y a} = \frac{227.00}{250} = 0.908 \text{ m}$$

$$Z = H_{tG} \times N_{tG} = 0.908 \times 14.71 = 13.36 \text{ m}$$ **(Answer)**

[**Note:** Since the graphical integration with $1/(y\text{-}y_i)$ versus y involves very large range for the ordinate, determination of area will be difficult. This can be overcome by plotting $y/(y–y_i)$ against log y and finding the area between limits log y_2 and log y_1. Thus we have the following values for $y/(y–y_i)$ against log y:

ln y	–6.8880	–5.6268	–4.9337	–4.5190	–4.2199	–3.9792	–3.9120
$y/(y–y_i)$	4.1803	5.2786	5.2440	5.0983	4.9363	4.7246	4.6554

$$N_{tG} = \int_{\ln y_2}^{\ln y_1} \frac{y\, d(\ln y)}{y - y_i} + \frac{1}{2} \ln \frac{1 - y_2}{1 - y_1} = 14.8 + \frac{1}{2} \ln \frac{1 - 0.00102}{1 - 0.02} = 14.71]$$

These are plotted as shown in Fig. 4.79.

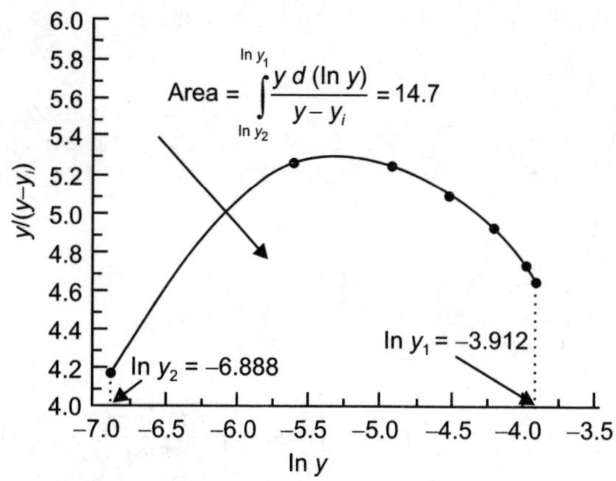

Fig. 4.79: Graphical integration for number of transfer units, Example 4.18

(b) We can take into account the variation in the gas flow along the height of the column and determine the height of the column by graphical integration directly by the general method which makes use of the Eq. (4.178).

$$Z = \int_{y_2}^{y_1} \frac{G(1-y)_{iM}\,dy}{k'_y a(1-y)(y-y_i)}$$

For different y values we have already determined the interface concentration y_i from which we can calculate $(1 - y_i)_{iM}$ as

$$(1-y)_{iM} = \frac{(1-y)-(1-y_i)}{\ln \dfrac{1-y}{1-y_i}}$$

For a given position in the column where the bulk gas concentration is y, the gas flow rate G is determined as

$$G = G_s(1+Y) = G_s\left(\frac{1}{1-y}\right)$$

in which the flow of non-diffusing gas G_s is constant and = 224.6 kmol/m^2 h. The mass transfer coefficient $k'_y a$ = 250 kmol/m^3 h (mole fraction). For any chosen y value, calculate

$$f(y) = \frac{G(1-y)_{iM}}{k'_y a(1-y)(y-y_i)}$$

by which the following results are obtained:

y	0.00102	0.0036	0.0072	0.0109	0.0147	0.01870	0.0200
y_i	0.00078	0.00292	0.00583	0.00876	0.01172	0.01474	0.01570
$1-y$	0.99898	0.9964	0.99280	0.98910	0.98530	0.9813	0.9800
$1-y_i$	0.99922	0.99708	0.99417	0.99124	0.98828	0.98526	0.98430
$y-y_i$	0.00024	0.00068	0.00137	0.00214	0.00298	0.00396	0.00430
$(1-y)_{iM}$	0.99910	0.99674	0.99349	0.99017	0.98679	0.98328	0.98215
G kmol/m^{2}h	224.830	225.410	226.23	227.080	227.950	228.88	229.180
$f(y)$	3686.18	1322.51	659.53	425.300	306.640	231.78	213.860

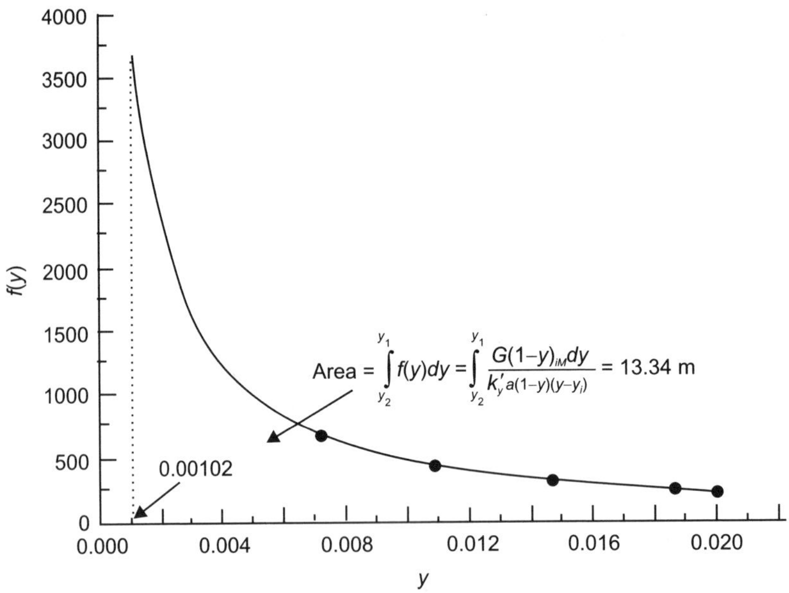

Fig. 4.80: Height of the column by general graphical method

To determine the integral in Eq. (4.178), a graph is plotted with $\dfrac{G(1-y)_{iM}}{k'_y a(1-y)(y-y_i)}$ as ordinate against y as abscissa. (Fig. 4.80). The area under the resulting curve between y_2 and y_1 is evaluated. The area represents the height of the column. From Fig. 4.80, the area is found to be 13.34 m. Therefore Z = 13.34 m. **(Answer)**

We can see that the height of the column obtained by taking into account the changes in gas flow is approximately the same as the height evaluated in part (a) assuming constant G. This is expected as solutions are dilute and the variation in the gas flow along the length of the absorber is negligible.

Example 4.19: Sulphur dioxide is absorbed from air by using pure water in a 350-mm diameter packed column operated at 101.3 kPa and 293 K. The gas enters the column at 2.5 kmol/h on SO_2 free basis and contains 20% SO_2 which is to be reduced to 1% by absorbing in pure water which enters the column at 150 kmol/h. The film coefficients for mass transfer are found to be

$$k'_y a = 0.0895(G')^{0.7}(L')^{0.25} \quad k'_x a = 0.6637(L')^{0.82}$$

where L' and G' are respectively, the liquid and gas flow, in kg/m²h and $k'_x a$ and $k'_y a$ are the liquid phase and gas phase volumetric mass transfer coefficients in kmol/m³ h (mole fraction). The equilibrium data for SO_2–water system is given below:

x	0	0.00006	0.00014	0.00028	0.00042	0.00056
$\bar{p}_{SO_2}$ kPa	0	0.0666	0.1599	0.4265	0.7731	1.1330
x	0.00084	0.00140	0.00197	0.00279	0.00420	0.00698
$\bar{p}_{SO_2}$ kPa	1.8794	3.4655	5.1983	7.8641	12.2626	21.4596

Determine the height of the packed column.

***Solution*:** The flow rates are divided by the cross sectional area of the column. The solute free gas flow per unit area of tower cross section is

$$G_s = \frac{2.5}{\frac{\pi}{4}(0.35)^2} = 25.98 \text{ kmol/m}^2\text{h}$$

The solute free liquid flow per unit area is

$$L_s = \frac{150}{\frac{\pi}{4}(0.35)^2} = 1559.07 \text{ kmol/m}^2\text{h}$$

The gas enters with 20% (mole) SO_2 and leaves with 1% SO_2.

$$y_1 = 0.20 \qquad Y_1 = \frac{0.20}{1-0.20} = 0.25$$

$$y_2 = 0.01 \qquad Y_2 = \frac{0.01}{1-0.01} = 0.0101$$

Since pure water is used for absorption,

$$x_2 = 0.0 \qquad X_2 = 0.0$$

The concentrations and flow rates are shown in Fig. 4.81. SO_2 balance over the column gives the exit concentration of the liquid:

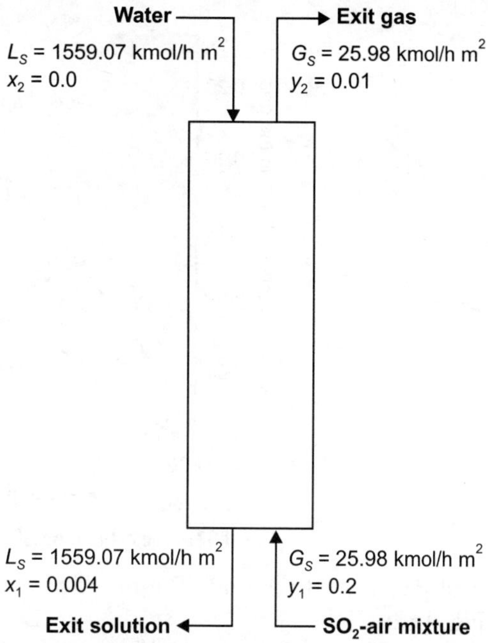

Fig. 4.81: Absorber in Example 4.19

$$G_s(Y_1 - Y_2) = L_s(X_1 - X_2)$$

$$25.98(0.25 - 0.0101) = 1559.07(X_1 - 0)$$

So that $X_1 = 0.004$ and $x_1 = \frac{0.004}{1+0.004} = 0.004$

Equation for the operating line in terms of mole fractions can be written as

$$G_s(Y - Y_2) = L_s(X - X_2)$$

Or
$$\frac{y}{1-y} = \frac{L_s}{G_s}\frac{x}{1-x} + \left(Y_2 - \frac{L_s}{G_s}X_2\right)$$

$$\frac{y}{1-y} = \frac{1559.07}{25.98}\frac{x}{1-x} + (0.0101 - 0)$$

$$\frac{y}{1-y} = 60.0104\frac{x}{1-x} + 0.0101 \qquad \text{(A)}$$

Equation (A) is used to evaluate the coordinates of say, four or five points on the operating line. This is done by assuming values for y in the range $y_2 = 0.01$ to $y_1 = 0.20$ and calculating the corresponding x. The results are indicated in the table below.

x	0	0.00071	0.00168	0.00276	0.004
y	0.01	0.05	0.1	0.15	0.2

A curve is drawn through the above points as shown by line AB in Fig. 4.82. The equilibrium curve also is plotted in the figure. To plot the equilibrium curve, the partial pressure of SO_2 in the equilibrium data is expressed as mole fraction of SO_2 using $y = \frac{\bar{p}}{P}$ where $\bar{p}$ is the partial pressure and P is the

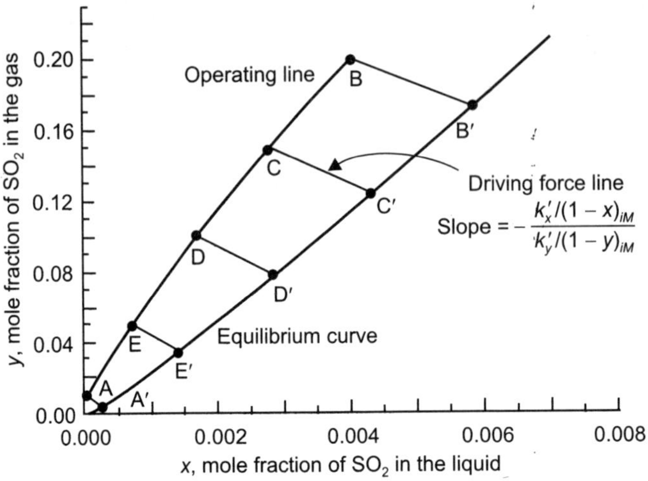

Fig. 4.82: The estimation of interfacial compositions for Example 4.19

total pressure (101.3 kPa). The modified equilibrium data is given below which is also plotted in Fig. 4.82:

x	0	0.00006	0.00014	0.00028	0.00042	0.00056
y	0	0.00066	0.00158	0.00421	0.00763	0.01118
x	0.00084	0.00140	0.00197	0.00279	0.00420	0.00698
y	0.01855	0.03421	0.05132	0.07763	0.12105	0.21184

Interfacial compositions for given bulk phase compositions can be determined graphically once the operating line and the equilibrium curve are plotted. Consider for example, the point C on the operating line. Its coordinates are $x = 0.00276$ and $y = 0.15$. These are the bulk compositions of the gas and liquid phases in contact at certain point in the column. The film coefficients for mass transfer $k'_x a$ and $k'_y a$ are to be calculated at this point using

$$k'_y a = 0.0895(G')^{0.7}(L')^{0.25} \quad \text{(B)}$$

$$k'_x a = 0.6637(L')^{0.82} \quad \text{(C)}$$

where L' and G' are respectively, the liquid or gas flow, in kg/m^2 h.

The mass flow rates are

$$L' = L_s \times (MW)_{\text{water}} + L_s X \times (MW)_{SO_2} = 1559.07\left(18 + 64.1\frac{x}{1-x}\right) \quad \text{(D)}$$

$$G' = G_s \times (MW)_{\text{Air}} + G_s Y \times (MW)_{SO_2} = 25.98\left(29 + 64.1\frac{y}{1-y}\right) \quad \text{(E)}$$

Substituting $x = 0.00276$ and $y = 0.15$ in Eqs. (D) and (E), we get $L' = 28339.85$ kg/m^{2}h and $G' = 1047.30$ kg/m^{2}h. Substitute L' and G' in Eqs. (B) and (C), we get

$$k'_x a = 0.6637(L')^{0.82} = 0.6637(28339.85)^{0.82} = 2971.24 \text{ kmol/m}^3\text{h (mole fraction)}$$

$$k'_y a = 0.0895(G')^{0.7}(L')^{0.25} = 0.0895(1047.30)^{0.7}(28339.85)^{0.25} = 151 \text{ kmol/m}^3\text{h (mole fraction)}$$

The driving force line is drawn from point C with slope $= -\dfrac{k'_x/(1-x)_{iM}}{k'_y/(1-y)_{iM}}$. Since the interface compositions are not known, we make an initial guess so that

$$\text{Slope} = -\frac{k'_x/(1-x)_{iM}}{k'_y/(1-y)_{iM}} \approx -\frac{k'_x/(1-x)}{k'_y/(1-y)} = -\frac{2971.24}{151.00} \times \frac{(1-0.15)}{(1-0.00276)} = -16.7718$$

A straight line is drawn with slope –16.7718 which is found to intersect the equilibrium curve at ($x_i = 0.0043$ $y_i = 0.1242$). With these values for x_i and y_i, the values of $(1-x)_{iM}$ and $(1-y)_{iM}$ are now found out.

$$(1-y)_{iM} = \frac{(1-y)-(1-y_i)}{\ln\dfrac{1-y}{1-y_i}} = \frac{0.85-0.8758}{\ln\dfrac{0.85}{0.8758}} = 0.8628$$

$$(1-x)_{iM} = \frac{(1-x)-(1-x_i)}{\ln\dfrac{1-x}{1-x_i}} = \frac{0.99724-0.9957}{\ln\dfrac{0.99724}{0.9957}} = 0.9965$$

$$\text{New slope} = -\frac{k'_x/(1-x)_{iM}}{k'_y/(1-y)_{iM}} = -\frac{2971.24}{151.00} \times \frac{0.8628}{0.9965} = -17.038$$

Since the modified slope is appreciably different from the previous value, the driving force line is redrawn with the new slope. It is found that the line CC' intersects the equilibrium curve at ($x_i = 0.00429$ $y_i = 0.12393$). It can be seen that there is no change in the slope when it is recalculated with the new interfacial compositions. Therefore the interfacial compositions are $x_i = 0.00429$ and $y_i = 0.12393$ corresponding to point C on the operating line with coordinates $x = 0.00276$ and $y = 0.15$.

The interfacial compositions are determined in a similar manner for other points A, B, D, and E. The results of these computations are summarised below:

x	y	G'	L'	k'_x	k'_y	x_i	y_i	Slope
0	0.01	770.24	28063.26	2947.441	121.4776	0.00028	0.00319	–24.1065
0.00071	0.05	841.07	28134.27	2953.555	129.2748	0.00139	0.03514	–21.8963
0.00168	0.1	938.46	28231.44	2961.917	139.6998	0.00281	0.07817	–19.354
0.00276	0.15	1047.30	28339.85	2971.24	150.9983	0.0043	0.12393	–17.0384
0.004	0.2	1169.75	28464.61	2981.962	163.3294	0.00582	0.17287	–14.9226

The height of the column is determined using Eq. (4.178) which takes into account the variations in mass transfer coefficient and gas flow with y.

$$Z = \int_{y_2}^{y_1} \frac{G(1-y)_{iM}\,dy}{k'_y a(1-y)(y-y_i)}$$

For a given position in the column where the bulk gas concentration is y, the gas flow rate G is determined as

$$G = G_s(1+Y) = G_s\left(\frac{1}{1-y}\right)$$

In which the flow of non-diffusing gas G_s is constant and = 25.98 kmol/m²h. The mass transfer coefficient $k'_y a = 0.0895(G')^{0.7}(L')^{0.25}$ are already evaluated for different values of y. Calculate $f(y)$ for given value of y where

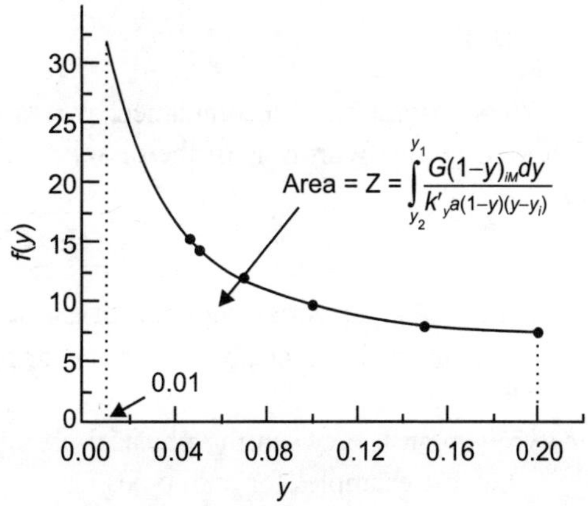

Fig. 4.83: Height of the column by the general graphical method (Example 4.19)

$$f(y) = \frac{G(1-y)_{iM}}{k'_y a(1-y)(y-y_i)}$$

The following results are obtained:

y	0.01	0.05	0.1	0.15	0.20
y_i	0.00319	0.03514	0.07817	0.12393	0.17287
$y-y_i$	0.00681	0.01486	0.02183	0.02607	0.02713
$(1-y)_{iM}$	0.99340	0.95741	0.91087	0.86297	0.81349
$k'_y a$	121.4776	129.2748	139.6998	150.9983	163.3294
G kmol/m^{2}h	26.2424	27.3474	28.8667	30.5647	32.475
$f(y)$	31.83	14.34	9.58	7.88	7.45

To determine the integral in Eq. (4.178), a graph is plotted with $f(y)$ as ordinate against y as abscissa (Fig. 4.83).

The area under the resulting curve between $y_2 = 0.01$ and $y_1 = 0.20$ is evaluated. The area represents the height of the column. From Fig. 4.83, the area is found to be 2.23 m. Therefore,

$$Z = 2.23 \text{ m}$$ **(Answer)**

4.8 ABSORPTION WITH CHEMICAL REACTION

Absorption accompanied by chemical reaction between the dissolved gas and the liquid solvent is attractive from the point of view of industrial applications because (i) the reaction results in the increase in capacity of the solvent for dissolving the solute, (ii) if the reaction at the gas-liquid interface is rapid, it may increase the mass transfer coefficient, (iii) reaction may increase the selectivity of the solvent to preferentially dissolve only certain components of the gas and (iv) the reaction may convert hazardous chemicals to safer compounds. The increased capacity of the solvent reduces the solvent circulation through the absorber. The reaction in the liquid reduces the equilibrium partial pressure of the solute over the solution thereby increasing the driving force for absorption. Absorption of an acid gas such as CO_2 in aqueous sodium hydroxide solution is accompanied by an irreversible neutralization reaction. Absorption of carbon dioxide and hydrogen sulphide gas in aqueous solutions of monoethanolamine or diethanolamine are accompanied by a reversible chemical reaction. If the reaction is irreversible, the number of transfer units can be determined by setting the equilibrium partial pressure equal to zero ($y^* = 0$), so that by Eq. (4.131), the number of transfer units is

$$N_{tOG} = \int_{y_2}^{y_1} \frac{dy}{y - y^*} = \int_{y_2}^{y_1} \frac{dy}{y} = \ln \frac{y_1}{y_2} \qquad \ldots(4.188)$$

When absorption is accompanied by a slow reaction, such as the absorption of chlorine in water followed by the hydrolysis of the dissolved chlorine, rate of absorption is reduced by the reaction.

REVIEW QUESTIONS

1. Give some industrial examples of gas absorption as a separation technique
2. Distinguish between absorption and stripping with reference to the mechanism of interphase mass transfer.
3. Distinguish between the physical absorption and absorption with chemical reaction with some suitable examples for both types.
4. Bring out the advantages and disadvantages of absorption involving chemical reaction over the physical absorption.

5. Absorption and distillation are two important gas-liquid contact operations. Bring out the essential difference between the two.
6. Describe the working of a tray tower with the help of a neat sketch.
7. Compare the different types of trays for stage-wise gas-liquid contact.
8. With reference to plate columns, explain the following: (a) Flooding (b) Entrainment (c) Priming (d) Weeping (e) Dumping (f) Coning.
9. "In tray towers, conditions leading to high tray efficiencies result in several operational difficulties." Elaborate.
10. Explain the influence of the liquid head and the gas velocity on tray efficiency.
11. What is the effect of liquid density, viscosity and gas solubility on the overall column efficiency?
12. Substantiate the statement that the Murphree plate efficiencies can even exceed 100%.
13. Explain why the Murphree point efficiency will always be less than the plate efficiency.
14. Show that when there is no horizontal liquid mixing over the plate, the Murphree plate efficiency (E_{ML}) and the point efficiency (E_{OL}) referred to the liquid can be related as $E_{ML} = 1 - e^{-N_{toL}}$ where N_{tOL} is the number of transfer units referred to the liquid.
15. Prove that the number of overall gas transfer unit in terms of mole ratios is given by

$$N_{tOG} = \int_{Y_2}^{Y_1} \frac{dY}{Y - Y^*} + \frac{1}{2}\ln\frac{1+Y_2}{1+Y_1}$$

16. Describe the relative advantages of structured packings over random packing.
17. Explain channelling in packed columns. How can it be prevented?
18. Distinguish between random packing and regular packing with suitable examples.
19. What are the most desirable properties of tower packings for gas absorption?
20. Draw neat sketches of different types of random packings in use for gas-liquid contact.
21. Describe the terms flooding and loading with reference to packed columns.
22. Compare the advantages and disadvantages of packed columns and tray columns.
23. Discuss the industrial application of sparged vessels (bubble columns).
24. What are the major limitations of bubble columns for gas absorption?
25. What are the major limitations of spray columns?
26. Explain why spray tower are preferred when absorption is gas-phase controlled.
27. Why should the absorption operations be carried out at low temperatures and high pressures?
28. Why absorption columns are generally provided with cooling coils?
29. What are the important characteristics of ideal solutions?
30. Mention few examples for ideal solutions.
31. When ideal solutions are formed, the solubilities of gases in liquids are independent of the solvent. Do you agree?
32. Why Henry's law constant increases with temperature?
33. How would you predict the solubility of a gas in a liquid if the Henrys law constant for the system is available?
34. Distinguish between Raoult's law and Henry's law.
35. What is the number of degrees of freedom when ammonia is absorbed from air-ammonia mixture in pure water?
36. What is the effect of temperature on solubility of gases in liquid?
37. Define absorption factor. What would be the value of absorption factor if the operating and equilibrium lines are straight parallel lines?

38. To determine minimum solvent for absorption graphically, it is convenient to use mole ratio coordinate system for plotting the equilibrium curve and the operating line. Why?
39. Actual quantity of solvent used in absorption columns should be above the minimum quantity. Why?
40. What is the physical significance of operating line and equilibrium curve in absorption?
41. Explain why the operating line is not a straight line when plotted on mole fraction coordinates.
42. It is the L/G ratio and not the absolute values of L and G that determines the height of an absorption column. Do you agree? Substantiate.
43. What is HETP? Why is HETP-based method abandoned as a design technique?
44. What are the assumptions used in deriving the Kremser equations for determining the number of theoretical plates?
45. Why is it not possible to achieve a good solute recovery in absorbers even by providing large number of plates, if the absorption factor is less than unity?
46. The mole fraction of solute **A** in the gas phase is reduced from 0.5 to 0.48 over a certain plate in an absorption column. The mole fraction of **A** in the liquid reaching a plate is 0.2 and that in the liquid leaving is 0.23. The equilibrium is represented by $y = 2x$. What is the plate efficiency?
47. Distinguish between the height equivalent of a theoretical plate and the height of a transfer unit.
48. How is the height of overall transfer unit related to the heights of individual gas and liquid phase transfer units?
49. Derive Eq. (4.188) which relates the interfacial compositions and individual mass transfer coefficients:

$$\frac{1-y_i}{1-y} = \left(\frac{1-x}{1-x_i}\right)^{k'_x a/k'_y a}$$

50. Absorption accompanied by a chemical reaction is sometimes advantageous for industrial applications. Explain.

MULTIPLE CHOICE QUESTIONS

4.1 Packed columns are
A. continuous contact equipment B. stage contact equipment
C. gas-solid contact equipment D. solid-liquid contact equipment

4.2 When do you prefer ring type distributors?
A. When the gas rate is low B. When the liquid rate is high
C. When the gas rate is low and the liquid rate is high
D. When the gas rate is high and the liquid rate is low

4.3 The uneven distribution of liquid over the tower cross section is known as
A. channelling B. flooding
C. loading D. wetting

4.4 Liquid redistributors are necessary to avoid
A. flooding B. channelling
C. weeping D. overheating

4.5 The tower packing should provide large free cross section for flow, so that
A. the bulk density is high B. the interfacial area is high
C. the pressure drop is high D. the frictional loss is reduced

4.6 A high interfacial area provided by packings leads to
A. lesser number of theoretical stages B. high Murphree plate efficiency
C. small packed height D. small number of gas transfer units

4.7 Inexpensive, relatively strong, poor wettability, unsuitable to handle certain organic solvents. These are the characteristics of

A. random packings B. ceramic packings
C. plastic packings D. structured packing

4.8 Which one of the following materials would you recommend as packing materials for use in a corrosive and elevated temperature condition?

A. Plastic packing B. Ceramic packing
C. Metal packing D. None of the above

4.9 All except one of the following characteristics refer to regular packing relative to random packing. Spot the odd one out.

A. Expensive B. Greater chances of channeling
C. Low pressure drop D. High flooding velocities

4.10 As the size of packing increases, the pressure drop through the column

A. increases B. decreases
C. unaffected D. cannot be generalized

4.11 The interfacial surface of packing

A. is more for random packing compared to regular packing
B. determines the height of packed section
C. determines the diameter of the column
D. determines the area of mass transfer and the pressure drop

4.12 For gas absorption, it is better to operate the packed column with a gas rate lying in the loading region, because

A. the pressure drop is less
B. the liquid holdup promotes interfacial contact
C. the liquid holdup is less
D. the liquid flow is obstructed

4.13 The loading and flooding point are affected by

A. gas and liquid rates B. gas rates only
C. liquid rates only D. the surface area of packing

4.14 At higher liquid rates, the flooding velocity is

A. dependent on the surface characteristics of packing
B. higher than that at low liquid rates
C. same as that at low liquid rates
D. lower than that at low liquid rates

4.15 Generally in absorption columns employing random packings, the packing sizes exceeding one-eighth of the tower diameter are not recommended. Why?

A. To avoid flooding B. To reduce cost of tower
C. To prevent channelling D. To ensure more free flow cross section

4.16 Why baffles are provided in mechanically agitated absorbers?

A. To increase interfacial area B. To ensure thorough mixing
C. To prevent vortex formation D. To produce smaller bubbles

4.17 Bubble columns and agitated vessels are generally equivalent to one theoretical stage. Why?

A. Because bubbles are small
B. Because the interfacial area is high
C. Because the height/diameter ratio is high
D. Because of liquid back-mixing

4.18 Spray towers can be used when absorption is gas phase controlled, because
- A. gas is well mixed and little mixing within liquid droplets
- B. liquid is well mixed and little mixing of the gas phase
- C. liquid drops provide large interfacial surface
- D. none of the above

4.19 Flooding in plate columns is the result of
- A. large gas pressure drop
- B. low gas rate
- C. low gas rate and high liquid rate
- D. entrainment of liquid

4.20 Which of the following occurs due to low liquid rate?
- A. Flooding
- B. Entrainment
- C. Coning
- D. Weeping

4.21 Liquid drains through perforations and do not flow through the downcomer. This condition is known as
- A. weeping
- B. dumping
- C. priming
- D. flooding

4.22 To prevent weeping in plate columns
- A. increase the gas rate
- B. decrease the gas rate
- C. increase the liquid rate
- D. increase turbulence

4.23 Which of the following is most suitable to achieve high tray efficiency?
- A. Low depth of liquid on the plate and high gas rate
- B. Low depth of liquid on the plate and low gas rate
- C. Deep liquid pool on the plate and high gas rate
- D. Deep liquid pool on the plate low gas rate

4.24 Which of the following trays are most suitable to handle satisfactorily a wide range of liquid and gas rates?
- A. Valve trays
- B. Bubble-cap trays
- C. Sieve trays
- D. Turbo-grid trays

4.25 The bubble-cap trays are not much in use now. Why?
- A. Because of high cost
- B. Because they cannot handle wide range of liquid flow rates
- C. Because they cannot handle wide range of gas flow rates
- D. Because of low efficiency

4.26 Which of the following is true?
- A. There is less chance of weeping in valve trays in comparison to sieve trays.
- B. Bubble cap plates have become more popular now because of their low cost.
- C. Sieve trays can operate over a wide range of flow rates than valve trays.
- D. Flooding condition does not exist in sieve trays

4.27 For gas-liquid systems, the deviation from equilibrium determines:
- A. The driving force for mass transfer
- B. The height of transfer unit
- C. Height equivalent of theoretical plate
- D. The liquid to gas ratio of the absorber

4.28 Which one of the following statements is wrong?
- A. Murphree point efficiency can never exceed 100%
- B. Murphree plate efficiency can never exceed 100%
- C. Overall plate efficiency can never exceed 100%
- D. There are situations when Murphree plate and point efficiencies will be equal.

4.29 Absorption equivalent to more than one or two theoretical plates is not achievable in
A. bubble columns
B. packed columns
C. plate columns
D. centrifugal absorber

4.30 When the headroom available for installing an absorption unit is small the following is preferred:
A. Bubble columns
B. Packed columns
C. Plate columns
D. Centrifugal absorber

4.31 Given the following statements, select the correct combination of true statements from the choices that follow this list.
P. Plate columns preferred when the operations involves liquids containing suspended solids.
Q. Packed towers are preferred if the liquids have a large foaming tendency
R. The pressure drop through packed towers is more than the pressure drop through plate columns designed for the same duty.
S. Packed columns are preferred when large temperature changes are involved in distillation operations
T. Packed towers are cheaper than plate towers if highly corrosive fluids must be handled.
A. T, S, P
B. P, Q, T
C. S, R, T
D. R, Q, S.

4.32 The solubility of a gas in a liquid generally
A. increases with temperature
B. decreases with temperature
C. decreases with pressure
D. not affected by temperature and pressure

4.33 Absorption operations are carried out at relatively low temperatures because
A. absorption of gas in liquid is exothermic
B. absorption of gas in liquid is endothermic
C. heating is costly and energy intensive
D. liquid available is at low temperatures

4.34 Partial pressure exerted by a sparingly soluble gas in liquid is .... that of a highly soluble gas
A. higher than
B. lower than
C. same as
D. higher or lower than

4.35 The solubility curve for highly soluble gas in liquid
A. lies closer to the liquid concentration axis
B. lies closer to the gas concentration axis
C. is concave downwards
D. is concave upwards

4.36 If an ideal solution is formed on dissolving a gas in the liquid, the solubility (expressed as mole fraction of solute) is
A. independent of the properties of the gas
B. independent of the properties of the solvent
C. independent of the temperature
D. independent of the temperature and properties of gas and liquid

4.37 For very dilute saturated solution of gas in a liquid, the solubility (expressed as mole fraction) is
A. independent of the temperature
B. independent of the pressure
C. inversely proportional to Henry's law constant
D. decreases with total pressure of the gas

4.38 Since absorption is exothermic process, Henry's law constant
A. increases with temperature
B. decreases with temperature
C. may increase or decrease with temperature
D. unaffected by change in pressure

4.39 A high value of Henry's law constant means
A. the gas solubility is high
B. the solubility is low
C. equilibrium partial pressure is low
D. none of the above

4.40 For high overall column efficiency
A. the liquid viscosity and the gas solubility are high
B. the liquid viscosity is low and the gas solubility is high.
C. the liquid viscosity and the gas solubility are low
D. the liquid viscosity is high and the gas solubility is low

4.41 The ratio of the number of theoretical plates to the number of actual plates is known as
A. overall stage efficiency
B. Murphree column efficiency
C. Murphree plate efficiency
D. Murphree point efficiency

4.42 The Murphree plate efficiency will exceed 100% if
A. the liquid travels in plug flow over the plate
B. the liquid over the plate is well mixed
C. the gas reaching the plate is of uniform concentration
D. the plate diameter is small

4.43 The Murphree tray efficiency and point efficiency are identical
A. for valve plates
B. if there is no liquid entrainment
C. if the liquid on the tray is well mixed
D. if the flooding and entrainment are eliminated

4.44 The absorption factor is defined as
A. L/mG
B. G/mL
C. mL/G
D. LG/m

4.45 The individual mass transfer coefficient (kmol/m^{2}s) for absorption of a solute from a gas mixture into a liquid solvent are $k_L = 4.5$ and $k_G = 1.5$. The slope of the equilibrium line is 3. Which of the following resistances is controlling?
A. liquid side
B. gas side
C. interfacial
D. both liquid and gas side

4.46 In the absorption of a solute gas from a mixture containing inerts in a solvent, it has been found that the overall gas transfer coefficient is nearly equal to the individual gas film transfer coefficient. It may therefore be concluded that:
A. the process is liquid film controlled
B. the gas is sparingly soluble in the solvent
C. transfer rate can be increased substantially by reducing the thickness of the liquid film
D. the transfer rate can be increased substantially by reducing the thickness of the gas film.

4.47 For absorbing a sparingly soluble gas in a liquid, mass transfer rate can be improved by
A. increasing the gas side mass transfer coefficient
B. increasing the liquid side mass transfer coefficient
C. decreasing the liquid side mass transfer coefficient
D. keeping the mass transfer coefficients constant

4.48 For stripping of a liquid in a countercurrent stripper, the operating line
A. lies above the equilibrium curve
B. lies below the equilibrium curve
C. can lie above or below the equilibrium curve
D. is always parallel to the equilibrium curve

4.49 Benzene in air-benzene mixture is to be reduced from 5.2 mol% in the feed to 0.5 mol% by contacting with wash oil in a countercurrent gas absorber. The inlet flow rate of air-benzene mixture is 10 mol/s while benzene-free wash oil comes in at 9.5 mol/s. If the equilibrium curve is given as $Y=X$ where X and Y are equilibrium mole ratio of benzene in oil ands air, respectively, the number of equilibrium stages required to achieve the above separation is

A. 12 B. 10
C. 8 D. 6

4.50 In a tray column separating a binary mixture with non-ideal stages, which one of the following statements is true?

A. Point efficiency can exceed 100%
B. Murphree efficiency cannot exceed 100%
C. Murphree efficiency can exceed 100%
D. Both Murphree and point efficiency can exceed 100%

4.51 A separation column for vapour-liquid contact processes 200 kmol/h of vapour. The flooding velocity is 3 m/s. If the column operates at 85% of flooding velocity and the downcomer area is 10% of the total cross sectional area, what is the diameter of the column? Average density of the vapour is 2 kg/m^3 and its molecular weight is 44.

A. 0.82 m B. 0.72 m
C. 0.78 m D. 1.0 m

4.52 Binary gas mixture of a solute and a carrier gas is treated in a countercurrent gas absorption column, containing ideal trays, using a solvent. Assume that the carrier gas is insoluble in the solvent and that the vapour pressure of the solvent is very low at the given conditions of the column. The gas and liquid are sufficiently dilute that L and G may be assumed to be constant throughout the column. The equilibrium relation is given by $y=mx$ where m is a positive constant. For any value of m, the change in liquid composition across a tray is independent of the tray location if

A. $L/G=1$ B. $mL/G=1$
C. $m/LG=1$ D. $L/mG=1$

4.53 A binary gas mixture of a solute and a carrier gas is treated in a countercurrent gas absorption column, containing ideal trays, using a solvent. Assume that the carrier gas is insoluble in the solvent and that the vapour pressure of the solvent is very low at the given conditions of the column. The gas and liquid are sufficiently dilute that L and G may be assumed to be constant throughout the column. The equilibrium relation is given by $y=mx$ where m is a positive constant. If $L/mG=1$, then the number of ideal stages in the column is given by

A. $N=\dfrac{x_0-x_N}{x_0-(y_{N+1}/m)}$ B. $N=\dfrac{x_0-(y_{N+1}/m)}{x_0-x_N}$
C. $N=\dfrac{x_N-x_0}{(y_{N+1}/m)-x_N}$ D. $N=\dfrac{(y_{N+1}/m)-x_N}{x_N-x_0}$

4.54 Acetone is to be removed from air in an isothermal dilute absorber using pure water as the solvent. The incoming air contains 5 mol% of acetone. The number of trays may be estimated by $N+2=6\log\dfrac{y_1}{y_2}$. For 98% recovery of acetone, the number of trays required is

A. 1 B. 8
C. 9 D. 10

4.55 When the slope of the gas-liquid solubility curve is very small (i.e. the curve is flat), the interphase mass transfer is

A. liquid-phase controlled B. gas-phase controlled
C. from gas to liquid D. from liquid to gas

4.56 For gas absorption, the height of transfer unit based on the gas phase is given by

A. $\frac{G}{K_y a}$ B. $\frac{K_y}{Ga}$

C. $\frac{Ga}{K_y}$ D. $\frac{L}{K_y G}$

4.57 The interfacial area per unit volume of dispersion, in a gas-liquid contactor, for fractional hold-up of gas = 0.1 and gas bubble diameter is 0.5 mm is given by (in m^2/m^3)

A. 500 B. 1200

C. 900 D. 800

4.58 H_2S is being absorbed in a gas absorption unit. The height of the transfer unit based on the overall mass transfer coefficient on the gas side is 0.4 m. The equilibrium data is given by $y = 1.5x$. The bulk concentration of H_2S has to be reduced from 0.05 to 0.001 mole fraction in the gas side. The height of the tower in meters corresponding to an operating line given by $y = 5x + 0.001$ is

A. 2.0 B. 1.56

C. 1.0 D. 0.56

4.59 If the operating line coincide with the equilibrium curve, which one of the following is not true for absorbers?

A. The solvent rate is minimum B. The number of plates is infinity

C. The L/G ratio is maximum D. The driving force becomes zero.

4.60 For absorbers with a given liquid to gas ratio,

A. the column diameter is fixed

B. the column height is fixed

C. the number of plates increases with liquid rate

D. the number of plates decreases with increase in liquid rate

4.61 In absorption, the ratio of the slope of the operating line to that of the equilibrium line is called

A. distribution coefficient B. Henry's law constant

C. absorption factor D. stripping factor

4.62 If the absorption factor A is less than 1

A. the gas leaving the column can never approach equilibrium with liquid entering

B. the gas leaving will be in equilibrium with liquid entering

C. the operating line and the equilibrium curve are straight parallel lines

D. Kremser equation is not applicable to the system

4.63 A large value of absorption factor results in

A. large number of plates B. large solvent throughput

C. low cost for solvent regeneration D. small diameter columns

4.64 HETP is defined as

A. height of a plate column achieving the concentration change equal to the driving force

B. height of a packed column achieving a concentration change equal to the driving force

C. height of a packed column achieving a concentration change as in theoretical plate

D. height equivalent of a transfer plate.

4.65 In one transfer unit, the gas undergoes a change in concentration

A. equal to the average driving force

B. equal to that in an ideal plate

C. that varies with the size and type packing material

D. which is estimated as $G/K_y a$

4.66 The height of a packed column (where N is the number of theoretical plates) is estimated by

A. $Z = HTU \times N$
B. $Z = HETP \times N$
C. $Z = HETP \times NTU$
D. None of the above

4.67 The height of a transfer unit is

A. the number of times the average driving force divides into the overall concentration change
B. a measure of the separation effectiveness of the packing material
C. increased if the rate of mass transfer increase
D. high if interfacial area of packing is large

4.68 NTU (number of transfer unit) = NTP (number of theoretical plates) if the absorption factor A is

A. zero
B. infinity
C. greater than unity
D. one

4.69 The equilibrium curve and the operating line are straight lines as shown below.

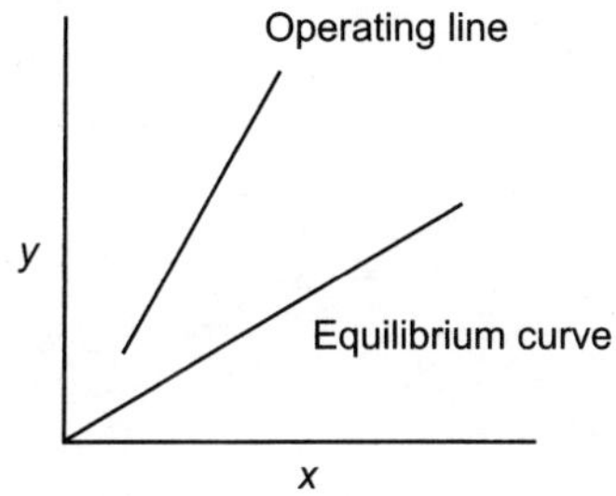

Which one of the following statements is true?

A. $N_{toG} = N$
B. $A < 1$
C. $N_{toG} > N$
D. $N > N_{toG}$

4.70 For gas-phase controlled systems,

A. $Z = H_{tG}$
B. $H_{tOG} = H_{tG}$
C. $H_{tOG} = H_{tOL}$
D. $Z = H_{tOG}$

4.71 For absorption of sparingly soluble gas in liquids,

A. height of the column is equal to the height of gas transfer unit
B. height of the overall gas transfer unit and overall liquid transfer unit are the same
C. height of the column is approximately the height of liquid transfer unit
D. height of overall liquid transfer unit is equal to the height of the liquid transfer unit.

4.72 What is turndown ratio in plate columns?

A. Ratio of area of perforations to the column cross section
B. Ratio of flooding velocity to operating velocity
C. Ratio of maximum to minimum gas rate
D. Ratio of liquid to gas flow in the column

4.73 The mole fraction of solute **A** in the gas phase is reduced from 0.5 to 0.48 on a certain plate in a column. The liquid reaching the plate has a mole fraction of **A** 0.2 and the Murphree plate efficiency referred to gas side is 50%. The equilibrium is represented by $y = 2x$. What is the concentration of **A** in the liquid leaving?

A. 66.67%
B. 40%
C. 50%
D. 23%

4.74 A gas mixture is in contact with a liquid. Component P in the gas mixture is highly soluble in the liquid. Possible concentration profile during absorption of P are shown in the figure. x = mole fraction of P in the bulk liquid, y = mole fraction of P in the bulk gas, x_i = mole fraction of P at

the interface in the liquid, y_i = mole fraction of P at the interface in gas. y^* = equilibrium gas phase mole fraction corresponding to x_i.
The correct profile is

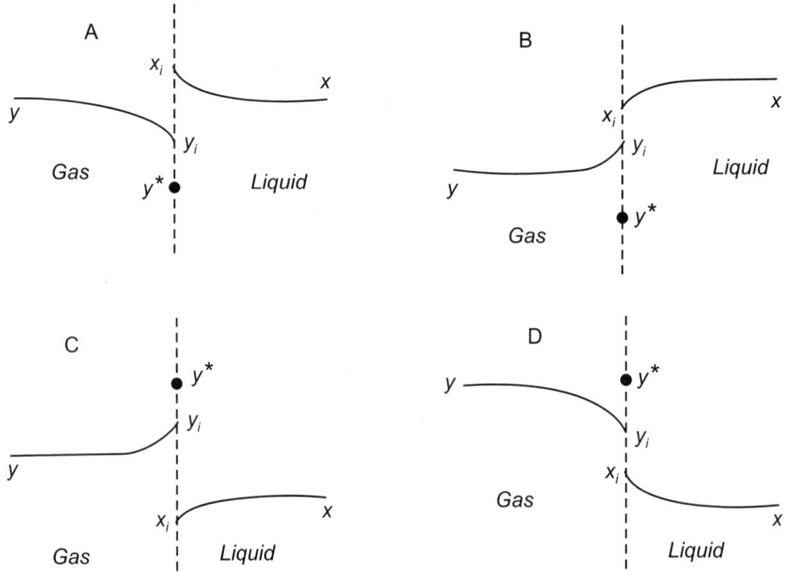

4.75 For which of the following combinations, does the absorption operation become gas-film controlled?

P. The solubility of the gas in the liquid is very high
Q. The solubility of the gas in the liquid is very low
R. The liquid-side mass transfer coefficient is much higher than the gas-side mass transfer coefficient
S. The liquid-side mass transfer coefficient is much lower than the gas-side mass transfer coefficient.

A. P & Q B. P & R
C. P & S D. Q & R

4.76 Water is used to absorb ammonia from a gas mixture, in a single stage contactor. The process is schematically represented in figure below:

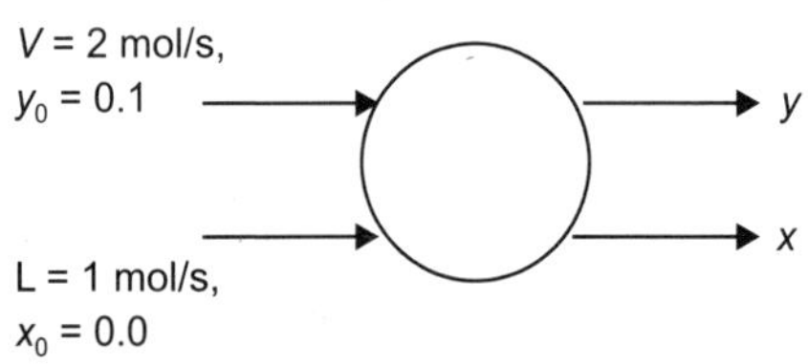

The molar gas and liquid flow rates, and the inlet mole fraction are given in the figure. Both the liquid and the gas phases are well mixed, and the equilibrium relation is given by $y^* = x$. If the stage is ideal, then the value of y is

A. 1/15 B. 1/10
C. 2/15 D. 1/6

4.77 Water is used to absorb ammonia from a gas mixture, in a single stage contactor. The process is schematically represented in figure below:

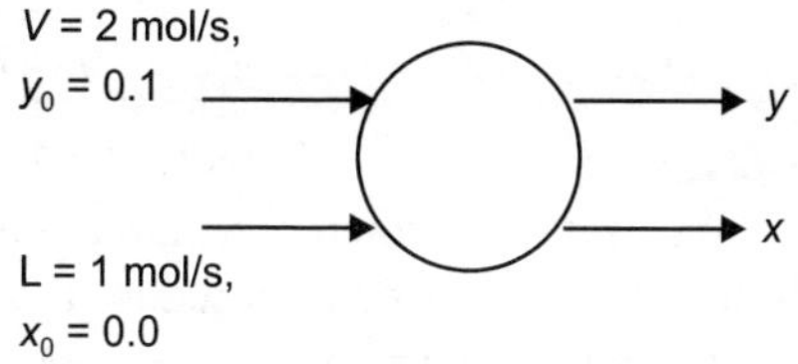

The molar gas and liquid flow rates, and the inlet mole fraction are given in the figure. Both the liquid and the gas phases are well mixed, and the equilibrium relation is given by $y^* = x$. If the stage efficiency is 50%, then the value of y is

A. 1/12 B. 1/6

C. 1/4 D 1/3

EXERCISES

Equilibrium

4.1 The Henry's law constant for oxygen in water at 298 K is 4.4×10^4 bar. Estimate the solubility of oxygen in water at 298 K for a partial pressure of oxygen of 0.25 bar and express it as g oxygen per 100 g water. (**Answer:** 1.01×10^{-3})

4.2 Calculate the concentration of nitrogen in water exposed to air at 298 K and 1 bar if Henry's law constant for nitrogen in water is 8.68×10^4 bar/mole fraction N_2 at this temperature. Assume that air is 79 percent nitrogen by volume. Express the result in moles of nitrogen per kg water

(**Answer:** 5.056×10^{-4} mol N_2/kg water)

4.3 Estimate the solubility of oxygen in water under normal atmospheric pressure and 298K when the solution is in equilibrium with air at 101.3 kPa. The Henry's law constant at 298 K is 4.437×10^6 kPa / mole fraction. Express the result in moles per cubic meter of water.

(**Answer:** 0.266 mol/m^3)

4.4 The Henry's law constant for CO_2 in water at 298 K is 1.66×10^3 bar. Determine the amount of carbon dioxide dissolved in one cubic meter of soda water if the carbonation of water is carried out at a CO_2 pressure of 2.5 bar. (**Answer:** 3.68 kg)

4.5 Calculate the composition of the liquid when a gas mixture containing 60% methane, 20% ethane, 8% propane, 6% *n*-Butane and 6% *n*-Pentane is in equilibrium with hydrocarbon oil at 200 kPa and 300K. Solubility of methane may be assumed negligible under these conditions. The vapour pressures at 300 K are: 43.6 bar for ethane, 10 bar for propane, 2.6 bar for *n*-Butane and 0.7 bar for *n*-Pentane.

(**Answer:** 75.78 mol% oil, 0.92% ethane, 1.6% propane, 4.6% butane, 17.1% pentane)

4.6 At 363 K, the vapour pressures of benzene and toluene are 136 kPa and 54 kPa, respectively. Determine the total pressures and weight composition of the vapour in equilibrium with a liquid mixture consisting of equal weights of the two components.

(**Answer:** 98.3 kPa, 71.5% benzene)

Material balance

4.7 Ammonia-air mixture containing 4.0% ammonia by volume is admitted to a countercurrent absorber at a rate of 50 m^3/h at 293 K and 101.3 kPa in order to reduce the ammonia content to

1.0%. If pure water at a rate 40 kg/h is used as the solvent, determine $(L_s/G_s)_{actual}$ and $(L_s/G_s)_{minimum}$. Equilibrium data for absorption of ammonia in water at 293 K are given below:

X, kmol NH_3/kmol water	0.01	0.02	0.03	0.04	0.05	0.06	0.07
Y, kmol NH_3/ kmol air	0.013	0.025	0.037	0.047	0.057	0.068	0.078

(**Answer:** 1.113 kmol/kmol, 0.91 kmol/kmol)

4.8 Ammonia-air mixture containing 4.0% ammonia by volume is admitted to a cocurrent absorber at a rate of 50 m^3/h at 293 K and 101.3 kPa in order to reduce the ammonia content to 1.0% using pure water as the solvent. (a) Determine the minimum (L_s/G_s) value (b) If the water used is 1.5 times the minimum, determine the actual water rate (c) determine the exit liquid composition for the conditions in part (b). The equilibrium data are

X, kmol NH_3/kmol water	0.01	0.02	0.03	0.04	0.05	0.06	0.07
Y, kmol NH_3/ kmol air	0.013	0.025	0.037	0.047	0.057	0.068	0.078

(**Note:** The minimum solvent requirement may be estimated as the flow rate corresponding to which the liquid leaves in equilibrium with the exit gas.)

(**Answer:** (a) 4.05 (b) 12.13 kmol/h (c) $X_2 = 0.\ 0052$)

4.9 Coal gas is scrubbed in a countercurrent absorber at a rate of 1000 m^3/h at 300 K and 101.3 kPa. The gas admitted to the tower contains 2% (volume) benzene, 95% of which is to be absorbed using wash oil (Molecular weight 260) containing 0.5% benzene. Solvent used is at a rate 1.5 times the minimum. The wash oil-benzene solution may be assumed to behave as ideal solution. The vapour pressure of benzene at 300 K is 13.8 kPa. Calculate (a) the oil circulation rate in kg/h and (b) the exit liquid composition expressed as mole% benzene.

(**Answer:** (a) 1944.3 kg/h of benzene-free wash oil, (b) 9.8%)

4.10 Wash oil is used for absorbing benzene from coal gas. The wash oil-benzene solution containing 10.0% benzene is to be regenerated in a stripper using superheated steam at 400 K and 101.3 kPa. The steam is supplied at a rate 1.5 times the minimum. The solution is admitted at a rate of 7.5 kmol/h. The regenerated solvent should contain not more than 0.5% benzene. The wash oil-benzene solution may be assumed to behave as ideal solution. The vapour pressure of benzene at 400 K is 350 kPa. Calculate (a) the rate of circulation of steam in kg/h and (b) the concentration of benzene in the exit gas stream expressed as mole%.

(**Answer:** (a) 42.77 kg/h (b) 23.15%)

Plate columns

4.11 Ammonia is to be removed from a 10% ammonia-air mixture by absorbing in water. The absorber is operated at 101.3 kPa and treats 300 kg/h of the gas entering the column and recover 95% of the ammonia. The water used is 20% more than the minimum required. Overall stage efficiency is 70%. Operation is assumed to take place isothermally. The equilibrium partial pressure for ammonia solution at the operating temperature is given below:

$\bar{p}_{NH_3}$, kPa	1.6	2.4	4.3	6.7	9.3
kg NH_3/100kg water	2	3	5	7.5	10

Determine (a) The minimum water rate, kg/h (b) The number of actual plates.

(**Answer:** (a) 161.75 kg/h (b) 9)

4.12 A counter current plate absorber is to be installed for scrubbing of air mixture containing 5% ammonia by volume. The scrubber is fed with water containing 0.002 mole NH_3 per mole of water. The scrubbing water flows at a rate of 1.0 mole water per mole air. It is necessary to absorb

85 percent of the ammonia present in the air. The equilibrium distribution constant is $K = 0.8 \dfrac{\text{mol NH}_3\text{/mol air}}{\text{mol NH}_3\text{/mol water}}$. Calculate (a) The concentration of ammonia in the outgoing liquid and (b) estimate the number of stages necessary for this operation.

(**Answer:** (a) 4.27 mole% NH_3 (b) 3.64)

4.13 A gas mixture containing 15% CO_2 and the balance inerts is admitted into an absorption column operated at 101.3 kPa and 298 K. Methanolamine solution containing 0.06 mol CO_2 per mol solution entering at 298 K is used as the solvent for absorption at a rate 1.2 times the minimum. The leaving gas contains 2% CO_2. The equilibrium partial pressure of carbon dioxide over aqueous solutions of methanolamine are:

$\bar{p}_{CO_2}$, kPa	0.746	1.706	3.865	7.464	20.66
x	0.058	0.060	0.062	0.064	0.068

Determine

(a) Minimum liquid to gas ratio

(b) The number of plates required.

(**Answer:** (a) 14.32 (b) 2.3)

4.14 Hydrogen sulphide is to be removed from a mixture of hydrogen sulphide and methane using monoethanolamine as solvent. The feed gas containing 20% H_2S is admitted to a plate absorber at a rate of 2000 m^3/h at 303 K and 101.3 kPa. The equilibrium relation $y = 1.1x$ is found to be applicable in the concentration range of interest. The solvent enters the column free of H_2S at a rate 1.2 times the minimum. If it is desired to remove 90% of the sulphide from the gas, determine:

(a) Minimum solvent flow rate, kmol/h

(b) The number of theoretical plates required by graphical method

(c) The number of theoretical plates using Kremser equation

(**Answer:** (a) 65.21 (b) 6.4 (c) 6.6)

4.15 Air containing 5% butane by volume is sent to a plate column for absorption of 95% of the butane. A heavy non volatile oil (Molecular weight 250 and specific gravity 0.9) is used as the solvent. The column is operated at 288 K and 101.3 kPa. The solvent is circulated at a rate 1.5 times the minimum. Raoult's law is found to be applicable for butane – oil solution. The vapour pressure of butane at 288 K is 194.5 kPa. If it is desired that 0.5 m^3/h of liquid butane is to be recovered by absorption, calculate:

(a) Solvent used in m^3/h (The density of liquid butane is 580 kg/m^3)

(b) The number of plates assuming a plate efficiency of 70%.

(**Answer:** (a) 77.11 (b) 8)

4.16 Acetone is absorbed from acetone – air mixture in a plate column using pure water admitted at a rate 30% in excess of the minimum. Acetone content in the entering gas is 20% and it is desired that 96% of it is to be absorbed. If the gas is admitted at a rate of 150 kmol/h and the equilibrium relationship is given by $Y = 1.5X$, where X and Y are mole ratios of acetone, determine:

(a) Water circulation rate, kg/h

(b) Number of theoretical plates

(**Answer:** (a) 4036 kg/h (b) 8.2)

4.17 An air-acetone mixture containing 12% acetone by volume is scrubbed with water in a valve tray column with an expected overall tray efficiency of 50% to recover 95% of the acetone in the feed stream. The column is to operate at 293 K and 101.3 kPa. The following equilibrium data may be used:

x	0.033	0.072	0.117	0.171
	4.0	8.4	11.4	13.7

Determine for 100 kmol/h of feed gas
(a) The minimum L_s/G_s
(b) The number of plates required for a solvent rate 1.25 times the minimum
(c) Concentration of acetone in the exit liquid

(**Answer:** (a) 1.06 (b) 18 (c) 8.86%)

4.18 It is desired to absorb acetone from a dilute mixture of acetone/air containing 1 mole% acetone by contacting it countercurrently with pure water in an absorber consisting of 2 theoretical stages. The total inlet gas flow rate is 30 kmol/h and that of water is 90 kmol/h. Under the operating conditions, the equilibrium relationship for acetone in gas-liquid is $y = 2x$. Estimate the mole fraction of acetone in the water stream leaving the absorber. (**Answer:** 0.0026)

4.19 The gas leaving the fermenter used for alcohol production consists of 98% CO_2 and 2% ethanol. 97% of the ethanol vapours are recovered by absorbing in pure water in a plate column operated at 303 K and 110 kPa. The gas is admitted to the column at a rate 180 kmol/h and water at a rate 1.3 times the minimum rate. The equilibrium can be represented by $y = 0.57x$. Determine (a) The liquid rate, kg/h (b) The number of theoretical plates required.

(**Answer:** (a) 2246.4 kg/h (b) 8.9)

4.20 A countercurrent multistage stripper is used to remove an impurity from a cream using pure steam. 100 kg/h of liquid cream containing 20 ppm (by weight) of impurity is fed to the stripper. It is desired to reduce the concentration in the cream to 1 ppm. Assume that the liquid cream does not evaporate and the steam does not condense. The equilibrium relation is $y = 10x$, where y and x are the ppm of impurity in steam and cream, respectively.
(a) Determine the minimum flow rate of steam required
(b) If the steam input to the stripper is 1.5 times the minimum rate, determine the required the number of theoretical stages.

(**Answer:** (a) 9.5 kg/h (b) 5.4)

4.21 The solvent oil used for absorbing benzene which leaves the absorber with 8 mol% benzene is to be steam stripped in a sieve-plate column at 101.3 kPa to recover 75% of the dissolved benzene. The steam leaving the stripper contains 3 mol% benzene. Equilibrium relationship is given by the Henry's law. The Henry's law constant is 50.65 kPa/mole fraction benzene. Determine:
(a) The number of theoretical plates
(b) Steam requirement for stripping 100 kmol oil
(c) The number of plates required if 85% recovery is desired

(**Answer:** (a) 2.8 (b) 194.4 (c) 3.8)

4.22 A mixture of paraffinic hydrocarbon (MW = 114) and a non-volatile organic compound (MW = 135) containing 10 weight% paraffin is subjected to steam stripping in a plate tower to reduce the paraffin concentration to 0.1%. Saturated steam at 373 K is used for stripping at a rate twice the minimum. How many theoretical stages are required? The vapour pressure of paraffin at 373 K is 53kPa. The tower is operated isothermally at 101.3 kPa. Assume that the system obeys Raoult's law. (**Answer:** 5.9)

4.23 An absorber is fed with a gas stream containing 0.05 mol benzene per mole of inert gas. The column is operated isothermally at 298 K. 90% of the benzene in the gas is absorbed in a hydrocarbon oil (Molecular weight 200) containing 0.01 mol benzene per mol oil. The liquid leaving the absorber which contains 0.16 mol benzene per mol oil is regenerated by stripping off the benzene using superheated steam at 101.3 kPa and 383 K in a plate column stripper. The liquid leaving the stripper will have a concentration of 0.01 mol benzene per mol oil. Equilibrium data at the respective operation temperatures are:

X, mol benzene/mol oil	0.04	0.08	0.12	0.16	0.20	0.24	0.28
Y, mol benzene per mol gas (298 K)	0.011	0.0215	0.032	0.042	0.0515	0.060	0.068
Y, mol benzene/mol steam (383 K)	0.10	0.21	0.33	0.47	0.62	0.795	1.05

Determine:

(a) The L_s/G_s ratio on the absorber

(b) The number of theoretical plates in the absorber

(c) Minimum steam requirement for stripping 100 kmol of oil.

(d) If the steam is used at a rate 20% in excess of the minimum, determine the number of theoretical stages in the stripper.

(**Answer:** (a) 0.3 (b) 10 (c) 29 kmol (d) 9.9)

4.24 An aqueous solution is stripped off ammonia at 303 K and 100 kPa in a plate column. At a particular plate in the column it is found that the air entering is essentially free of ammonia and the air-water ratio is 2 kmol/kmol. The solution entering the plate contains 0.1% ammonia. The murphree vapour efficiency is 80% and the Henry's law constant is 130 kPa/mole fraction at 303 K. Determine exit water composition. (**Answer:** 0.033% ammonia)

4.25 Stripping of ammonia is carried out at a pressure of 110 kPa. One cubic meter of water enters the system and the ratio of the molar flow rate of air and that of the water is 4. The inlet air and the inlet water have 0.1 and 1.0 mole percent ammonia respectively. The Murphree vapour plate efficiency for ammonia removal is 50% and the Henry's law constant for ammonia in water is 2.6075 kPa m^3/kmol. Determine the ammonia content in the (a) Exit water and (b) The exit air.

(**Answer:** (a) 0.33% (b) 0.27%)

Packed columns

4.26 3000 kg/m^2 h of ammonia-air mixture containing 5%(mole) ammonia is scrubbed with water to absorb 90% NH_3. Pure water at rate 2500 kg/m^{2}h is supplied at the top of the countercurrent absorber. The column is operated at 100 kPa. Equilibrium can be represented by $y = 0.95x$ and the overall mass transfer coefficient is $K_Ga = 0.65$ kmol/m^{3}h kPa. Find the height of packing required. (**Answer:** 6.75 m)

4.27 Ammonia-air mixture containing 6% ammonia by volume is admitted to a countercurrent packed tower in which 99% of ammonia gets absorbed in water. The absorber is operated isothermally at 293 K and 101.3 kPa. The gas flow rate is 4000 m^3/h at 293 K and 101.3 kPa and the liquid flow rate is twice the minimum. The velocity at flooding is 2 m/s and the operating velocity is 50% of the flooding velocity. The overall mass transfer coefficient K_ya is 150 kmol/m^{3}h (mole fraction). Equilibrium solubility of ammonia is

$\bar{p}_{NH_3}$, kPa	1.6	2.4	4.3	6.7
kg NH_3/100kg water	2	3	5	7.5

Determine the diameter and packed height of the column

(**Answer:** $D = 1.2$ m, $Z = 7.00$ m)

4.28 Acetone-air mixture containing 5% (mole) acetone is scrubbed with water in a countercurrent packed column of diameter 1.5 m in order to remove 95% of the acetone. The flow rates of entering gas and liquid are 3000 kg/h and 5000 kg/h respectively. The column is operated at 101.3 kPa pressure. The overall mass transfer coefficient K_Ga is 1.1kmol/m^{3}h kPa. Calculate the height of the absorber. Equilibrium data at the operating conditions are

x	0.01	0.02	0.03	0.04
y	0.02	0.0382	0.0552	0.0697

(**Answer:** 3.03 m)

4.29 A gas mixture is scrubbed in a countercurrent packed column of height 1.5 m to absorb solute **A** in a liquid solvent. It is observed that the recovery of **A** in the solvent is 85%. If it is desired to achieve 95% recovery of solute **A** using the same packing material, what should be the height of the bed? The equilibrium partial pressure of **A** over the solution is negligible. Assume that the solutions are dilute. (**Answer:** 2.37 m)

4.30 Ammonia is absorbed from an ammonia-air mixture containing 10% ammonia by scrubbing with pure water in a packed column. The tower is operated at 101.3 kPa and is to be designed to absorb 95% of ammonia in the gas. The gas is admitted at a rate of 4000 kg/h m^2 and the liquid at a rate 4500 kg/hm^2. The overall mass transfer coefficient is $K_G a = 3.5$ kmol/m^{3}h (kPa) and the equilibrium data at the operating temperature is:

mol NH_3/mol water	0.021	0.031	0.042	0.053	0.079	0.106	0.159	0.209
$\bar{p}_{NH_3}$, mmHg	12	18.2	24.9	31.7	50	69.6	114	166

Determine the height of the packed section by (a) Graphical method (b) Log-mean driving force.
(**Answer:** (a) 1.65 m (b) 1.66 m)

4.31 Ammonia-air mixture containing 5% ammonia by volume is admitted to a counter current absorber at a rate 2 kmol/h in order to reduce the ammonia content to 1.5% using pure water as the solvent admitted at a rate 40 kg/h. The column is 150-mm in diameter, is packed to a height 1.5 m and is operated at 293 K and 101.3 kPa. The equilibrium data are

X, kmol NH_3/kmol water	0.0164	0.0252	0.0349	0.0455	0.0722
Y, kmol NH_3/ kmol air	0.021	0.032	0.042	0.053	0.08

Assuming that the overall mass transfer coefficient is constant, determine $K_G\, a$.
(**Answer:** 1.87 kmol/m^{2}h kPa)

4.32 Ammonia content is reduced from an ammonia-air mixture from 10% (volume) to 0.5% (volume) using water as solvent in an experimental absorption unit operated at 101.3 kPa and 293 K. The experimental unit is made up of a 0.25 m diameter tower packed to a height of 2 m with 25-mm Raschig rings. The ammonia-free air flow rate is 100 m^3/h. Interfacial area of packing is 200m^2/m^3. The partial pressure of ammonia in equilibrium with solution at the bottom of the column is 1.2 kPa. Assuming that the equilibrium line and the operating line are straight lines, determine:

(a) The overall mass transfer coefficient K_G
(b) The number of overall gas transfer units
(c) The height of overall gas transfer unit

(**Answer:** (a) 7.25×10^{-3} kmol/m^2 h (kPa) (b) 3.28 (c) 0.61 m)

4.33 The effluent gas from a process industry is scrubbed with water in a 10-m high column to reduce the pollutant concentration from 2 to 0.5 percent. The gas flow rate is 50 kmol/m^2 h and water flow rate is 100 kmol/m^{2}h. Determine the concentration of pollutant in the exit gas stream if the same column is operated with double the liquid rate keeping the gas flow constant. Assume dilute solution and the Henry's law $y = 1.5x$ is applicable to represent the equilibrium. It is reported that the overall gas phase mass transfer coefficient varies with solute free liquid flow rate as $K_y a \propto L_S^{0.4}$. (**Answer:** 0.21%)

4.34 Acetone-air mixture containing 5% acetone by volume is admitted to a countercurrent packed tower in which 96% of acetone gets absorbed in water. The absorber is operated isothermally at 293 K and 101.3 kPa. The acetone-free air flow rate is 1500 m^3/h at standard temperature and pressure and the liquid flow rate is 3000 kg/h. Equilibrium solubility of acetone in water is given by $y = 1.70x$ where x and y are mole fractions of acetone. The tower packings provide an interfacial area of 200 m^2/m^3 and the overall mass transfer coefficient K_y is 0.4 kmol/m^2 h mole fraction. The flooding velocity is 1.6 m/s and the column is to be operated at a velocity is 25% lower than the flooding velocity. Assume dilute solutions. Determine:

(a) The diameter of the tower
(b) The height of packed section

(**Answer:** (a) 0.707 m (b) 15.2 m)

4.35 It is desired to absorb 99.5% of acetone from an acetone-air mixture containing 2.0 (volume)% acetone at 298 K and 100.0 kPa. The mass velocity of air leaving the absorber is 4000 kg/m^2 of tower cross section and water is used for absorption at a rate 6000 kg/h m^2. The absorption coefficient is related to the liquid rate as $K_G a = 0.083\sqrt{L}$ where K_G is kmol acetone/h m^3 (kPa) and L is kmol water/hm^2. Equilibrium relation is $\bar{p} = 180.0x$ where $\bar{p}$ is the partial pressure of acetone in kPa and x is mole fraction of acetone in the liquid. If the unit is to recover 300 kg acetone/h. Determine the height and diameter of the column.

(**Answer:** Z = 14.57 m, D = 1.53 m)

4.36 It is desired to absorb 95% acetone from an acetone air mixture containing 3 mol% acetone in a countercurrent packed tower. Water is used as the solvent at a rate of 7000 kg/hm^2. gas flow rate is 4000 kg/h m^2. The equilibrium solubility of acetone in water is given by $y = 2.53x$, where x and y are mole fractions in the liquid and gas, respectively. The heights of individual transfer units are 0.3 m and 0.5 m for the liquid and gas phase respectively. Calculate

(a) the height of the tower
(b) the minimum solvent rate, kg/h, that can be employed for achieving the desired separation

(**Answer:** (a) 7.00 m (b) 5728)

4.37 1000 kg/h of acetone-air mixture containing 3% acetone by volume is scrubbed with water in a countercurrent packed column to remove 95% of acetone. Pure water used is admitted at a rate 25% in excess of the minimum required. The equilibrium relation is $y = 2.53x$. The heights of individual transfer units are 0.3 m for liquid phase and 0.35 m for gas phase. Determine the height of the packed section of the column. (**Answer:** 5.23 m)

4.38 An inert gas contaminated with hydrogen sulphide is to be scrubbed with an aqueous solution of triethanolamine in a countercurrent packed column operated at 300 K and 101.3 kPa to reduce H_2S content from the initial concentration of 0.03 mole H_2S per mole of inert gas to one-hundredth of this value. Solvent enters the tower free of H_2S and leaves with a concentration of 0.013 mole H_2S per mole solvent. The flow of inert gas is 50.0 kmol/m^2 h. The equilibrium is described by the linear relation $Y = 2X$ where X and Y are mole ratios of H_2S. The overall mass transfer coefficient $K_y a$ = 125 kmol/m^3h (mole fraction). Determine (a) The rate at which solvent is introduced, kmol/m^2h and (b) the height of the absorber. (**Answer:** 114.23, 8.48 m).

4.39 Air containing 5 mole % methanol vapour is scrubbed with water in a countercurrent packed tower at 300 K and 100.0 kPa. The gas enters the column at a rate 2000 kg/m^2h and the water used is free of methanol and is admitted at a rate 500 kg/m^2 h. It is found that 95% recovery of methanol is achieved. The height of overall transfer unit based on the liquid side is 0.8 m. The equilibrium partial pressure in kPa of methanol over the aqueous solution is given by $\bar{p} = 28.4x$. Determine the height of the tower. (**Answer:** 2.8 m)

4.40 The gas leaving the fermenter used for production of ethyl alcohol consists of 98% CO_2 and 2% ethanol. 97% of the ethanol vapours are recovered by absorbing in pure water in a packed column operated at 303 K and 110 kPa. The gas is admitted to the column at a rate 200 kmol/h and water at a rate 1.5 times the minimum rate. The equilibrium can be represented by $y = 0.57x$. Determine (a) The water rate and (b) The packed height required if the height of the overall gas transfer unit is 0.5 m. (**Answer:** 2878 kg/h (b) 4.14 m)

4.41 A column, 1.5 m^2 in cross sectional area with a packed height of 4.0 m is used in an experimental study for absorbing SO_2 from air containing 2.0% (mole) SO_2. The mole fraction of SO_2 in the exit gas is 0.005. The gas flow rate is 6500 kg/h and the water flow rate is 2350 kg/min. Equilibrium is represented by $y = 40x$. Determine:
(a) The number of overall gas transfer units
(b) The height of overall gas transfer units
(c) The overall mass transfer coefficient $K_y a$ in kmol/m^{3}h (mole fraction)
(**Answer:** (a) 3.77 (b) 1.06 m (c) 135.33)

4.42 Sulphur dioxide is recovered from smelter gas by scrubbing with water in a packed column at 293 K and 100.0 kPa. The partial pressure of SO_2 in the gas entering and leaving the column are, respectively, 4 kPa and 1.2 kPa. The water enters the top of the column free of SO_2 at 1800 kmol/h and leaves the bottom with an SO_2 content of 0.0012 kmol SO_2 per kmol water. The tower cross sectional area is 1.5 m^2 and the overall absorption coefficient is $K_L a = 700$ kmol SO_2/m^{3}h (kmol SO_2/kmol water). Determine the height of the column. The equilibrium data for SO_2 and water at 293 K are:

kmol SO_2/kmol H_2O × 10^3	0.056	0.14	0.28	0.42	0.56	0.84	1.405	1.59
kmol SO_2/kmol inert gas × 10^3	0.7	1.6	4.3	7.9	11.6	19.4	36.3	42.0

(**Answer:** 4.35 m)

4.43 Ethylene oxide is recovered from its mixture with an inert gas by absorbing in water in a 1.5 m diameter packed column. The gas and liquid feed rates are, respectively, 1500 kmol/h and 2000 kmol/h. The individual mass transfer coefficients are estimated to be $k_y a = 3000$ kmol/m^{3}h (mole fraction) and $k_x a = 2500$ kmol/m^{3}h (mole fraction). Equilibrium is represented by $y = 0.86x$. Calculate:
(a) The overall volumetric mass transfer coefficient, $K_y a$
(b) The height of overall gas transfer unit
(**Answer:** (a) 1485.15 kmol/m^3 h mole fraction (b) 0.57 m)

4.44 The concentration of chlorine in the exit gas from a chlorinator is to be reduced from 10 mole% chlorine to 1 mole% by absorbing in water at 293 K and 101.3 kPa. The water rate employed is twice the minimum. The equilibrium data for chlorine in water at the operating temperature are (mole fraction chlorine)

x	0.0001	0.00015	0.0002	0.00025	0.0003	0.0004
y	0.006	0.012	0.024	0.04	0.06	0.102

Determine:
(a) The minimum water rate in kg/h
(b) The number of overall gas transfer unit for treating 100 kmol/h of exit gas from the chlorinator.
(**Answer:** (a) 4.14×10^5 kg/h (b) 2.73)

4.45 Acetone is to be recovered from an acetone-air mixture containing 5% acetone by volume at 290 K and 101.3 kPa. The gas mixture is admitted at a rate 200 m^3/h and the recovery is to be 98%. Water used is 25% in excess of the minimum. The maximum allowable gas superficial velocity is 300 m/h. It may be assumed that in the range of operation the equilibrium relation $Y = 1.75X$, where X and Y are moles of acetone per mole of acetone-free substance. The overall gas mass transfer coefficient is 50 kmol/m^3 h (mole fraction). Assume dilute solutions and determine:

(a) The minimum water rate, kg/h
(b) The actual water rate. kg/h
(c) The maximum concentration of acetone in the aqueous solution
(d) The number of theoretical stages
(e) The number of overall gas transfer units
(f) The height of packing

(**Answer:** (a) 246.5 (b) 308.1 (c) 2.92% (d) 11.6 (e) 12.75 (f) 3.14 m)

4.46 Equilibrium relationship for the system heptane-oil-air is given by $Y = 2X$ where Y and X are kg heptane/kg air and kg heptane/kg oil respectively. Oil containing 0.005 kg heptane/kg oil is being used as solvent for reducing the heptane content of air from 0.10 to 0.02 kg heptane/kg air in a countercurrent packed bed absorber. What column height is required to treat 1400 kg/hm^2 of empty tower cross section of pure air containing heptane if the overall gas mass transfer coefficient is 320 kg/h m^3 per unit gradient of Y. The oil rate employed is 5100 kg/hm^2. Solve analytically.

(**Answer:** 14.8 m)

4.47 It is required to produce a 10% (mole) ammonia in air mixture by stripping an aqueous 20 weight% ammonia solution with air at 101.3 kPa. The spent solution from the stripper should contain not more than 1 (weight)% NH_3. On the basis of 100 kmol/h of air entering the stripper, calculate the volume of packing required if $K_ya = 65$ kmol/m^{3}h mole fraction. The equilibrium data is given below:

$\bar{p}_{NH_3}$, kPa	1.6	2.4	4.3	6.7	9.3	22.1	30.3
g NH_3/100g water	2	3	5	7.5	10	20	25

(**Answer:** 4.5 m^3)

4.48 The CO_2 issuing out of a fermenter contains 0.01 mole fraction of ethanol which has to be reduced to 0.0001 mole fraction by scrubbing with water in a countercurrent packed tower. The gas flow rate is 200 kmol/h and may be assumed constant throughout the tower. The equilibrium mole fraction of ethanol in the gas phase is related to that in the liquid by $y = 1.07x$. Determine (a) the minimum liquid rate needed and (b) the number of overall gas transfer units needed at 1.5 times the minimum liquid rate. The entering liquid may be assumed to be free of ethanol.

(**Answer:** (a) 212 kmol/h (b)10.72)

4.49 Wash oil is used as the solvent for absorbing benzene from coal gas containing 4 volume% benzene. The benzene content is to be reduced to 0.05 volume% using an oil circulation at the rate 5 kg for each cubic meter of coal gas at STP. The oil enters the column free of benzene. The height of overall gas transfer unit is 1.2 m and the operating pressure is 101.3 kPa. The equilibrium data for the system are:

Weight % benzene in oil	0.05	0.1	0.50	1.00	2.00	3.00	3.32
Partial pressure of benzene in gas, kPa	0.013	0.033	0.20	0.53	1.33	3.33	4.05

Determine the height of the tower. (**Answer:** 10.36 m)

4.50 Water is used for absorbing a solute gas **A** from its mixture with air in a counter current packed tower operated at 298 K and 101.3 kPa. The gas fed to the column contains 5 mole% **A** and a 99% recovery is desired. The gas and liquid flow rates are 0.25 kmol/m^2 h and 1.2 kmol/m^2 h, respectively. The equilibrium relation is given by $y = 3.0x$. The individual liquid and gas mass transfer coefficients are 2.0 kmol/m^3 h mol fraction and 0.5 kmol/m^3 h. Assume that the absorber operates isothermally and the solutions are dilute. Determine:

(a) The number and height of overall gas transfer unit (b) The height of the column

(**Answer:** (a) 9.46, 0.85 m (b) 8.1 m)

4.51 A gas mixture consisting of 95% air and 5% benzene is scrubbed with an absorbing oil (MW = 230) to absorb 90 percent of entering benzene. The gas flow rate is 1000 m^3/h at 300 K and 101.3 kPa. The mass velocity of the gas is to be 4000 kg/m^2h. The oil enters the column at a rate 2000 kg/h containing 0.2% benzene by weight. The vapour pressure of benzene is 14 kPa. The height of overall gas transfer unit is 1.5 m. Assume that the Raoult's law is applicable. Determine the height and diameter of the packed column. (**Answer:** $D = 0.64$ m, $Z = 5.5$ m)

4.52 Air containing 2.0 (volume)% SO_2 is scrubbed with pure water in a packed column of 1.5 m^2 cross sectional area and 3.5 m height at a pressure of 101.3 kPa. Total gas flow rate is 6500 kg/h, the water flow rate is 2350 kg/min and the outlet gas concentration is $y = 0.005$. The equilibrium relation is $y = 40x$. Determine:

(a) Ratio of actual liquid to minimum liquid flow rates
(b) The number of gas transfer units and number of theoretical plates
(c) The height of overall gas transfer unit and HETP
(d) The overall mass transfer coefficient K_Ga

(**Answer:** (a) 1.2 (b) 3.76, 3.83 (c) 0.94, 0.91 (d) 1.52 kmol/m^3 h kPa

4.53 Ammonia-air mixture containing 40% (mole) ammonia is being absorbed in water in a countercurrent packed column at 101.3 kPa. The liquid to gas ratio employed is 5 kmol/kmol on ammonia-free basis. The recovery of ammonia is to be 95%.The equilibrium partial pressure for ammonia solution at the operating temperature is given below:

	1.6	2.4	4.3	6.7	9.3	22.1	30.3	40.5
kg NH_3/100 kg water	2	3	5	7.5	10	20	25	29.6

Calculate the number of overall gas transfer units graphically. (**Answer:** 3.34)

4.54 A solute **A** is absorbed in water using a packed column with the following terminal compositions (mole fraction **A**):

	Top	Bottom
x	0	0.08
y	0.001	0.009

If the equilibrium can be represented as $y = 0.06x$ and the height of gas transfer unit is 0.4 m and that of liquid transfer unit is 0.25 m, determine the height of packing. (**Answer:** 2 m)

4.55 0.47 m diameter packed column is used for absorbing benzene from coal gas at 300K and 101.3 kPa. The gas containing 2 (mole)% benzene enters the column at a rate 38.7 kmol/h and it is scrubbed with wash oil to absorb 95% of the benzene. Wash oil containing 0.5(mole)% benzene is used as the solvent at a rate 1.5 times the minimum. The equilibrium distribution of benzene between gas and liquid is given by $y = 0.125x$, where x and y are the mole fractions of benzene in the liquid and gas respectively. The film coefficients for mass transfer on the gas and liquid side are k'_y = 258.84 kmol/m^3h and k'_x = 49.57 kmol/m^3h. Calculate the packed height of the column

(a) assuming the gas flow is constant
(b) taking into account variation of gas flow through the column.

(**Answer:** (a) 12.83 m (b) 12.77 m)

4.56 A packed column absorber is used to reduce the ammonia content in the air-ammonia mixture from 4.0 (mole)% to 0.5% by absorbing in water under atmospheric conditions. The gas enters the column at 132 kmol/m^2 h and water at 155 kmol/m^2h. The film coefficients for mass transfer are $k'_y = 250$ kmol/m^3h and $k'_x = 571.8$ kmol/m^3h. Equilibrium data for ammonia-water system is given below:

x	0.0208	0.0258	0.0309	0.0405	0.0503
y	0.0158	0.0197	0.0239	0.0328	0.0416

Calculate the height of the column using individual gas-phase mass transfer coefficient and ignoring the variation of total gas flow through the column. (**Answer:** = 2.5 m)

4.57 A packed gas-liquid contactor employs certain Intallox saddles (38 mm). The gas flow rate is $G = 1.5$ kg/s, gas density $\rho_G = 1.5$ kg/m^3. The liquid flow rate $L = 30$ kg/s, and liquid density $\rho_L = 1000$ kg/m^3. The column is to be designed for a pressure drop of 5.6 kPa. Norton's correlation for this pressure drop is

$$K = \frac{13.1 G_w^2 F_p \left(\frac{\mu_L}{\rho_L}\right)^{0.1}}{\rho_G(\rho_L - \rho_G)}$$

where K = flow coefficient, μ_L is liquid viscosity, Ns/m^2, G_W is gas flow rate per unit cross sectional area, kg/m^2 s, F_p = packing factor = 170 m^{-1}. K is given by the equation:

$K = 0.62 - 0.5(m - 0.5)$ for $0.5 \le m \le 1.0$, where $m = \frac{L}{G}\sqrt{\frac{\rho_G}{\rho_L}}$.

Calculate the diameter of the contactor. (**Answer:** 1.3 m)

Chapter 5

Distillation

5.1 INTRODUCTION

Separation of liquid mixtures into component parts by distillation is one of the most challenging tasks in chemical processing industries. Distillation is an art which was practiced by the mankind from ancient times. Distillation for the production of alcohol for medicinal purposes and for making perfumes was known to the Chinese of the 30th century BC, the East Indians of the 25th century BC, the Egyptians of thc 20th century BC, the Greek of the 10th century BC, and the Romans of the 2nd century BC. Europeans came to know of the technique of distillation by the 6th century AD, after the invasion of Europe by the Arabs. Early distillation consisted of simple batch stills to produce ethanol. Later, kerosene was produced using the same method with crude oil heated in batch stills. Alchemists and monks progressively improved both the distillation technique and the equipment. The invention of the highly effective 'pure distillation' is credited to Arabic and Persian chemists in the Middle East during the 8th century. They practiced distillation processes for isolation and purification of chemical substances, isolation of natural esters for perfumes and for production of pure alcohol. Today distillation is by far one of the major operations in chemical and petroleum industries for separating liquid mixtures containing two or more constituents into component fractions of the desired purity. Separation of mixture of alcohol and water to pure components, separation of liquefied air into oxygen and nitrogen, distillation of crude oil into petrol, diesel, kerosene, lubricating oil, etc. are achieved by distillation.

In distillation, separation is achieved by applying heat. On boiling a liquid mixture, the components get differently distributed between the vapour and liquid phases. The vapour will be richer in more volatile components than the liquid with which the vapour is in equilibrium, as well as with the liquid from which the vapour is produced. Thus the original feed is separated into products of different compositions. It is this difference in compositions caused due to the difference in the boiling points or the volatilities of the constituents that forms the basis for the separation by distillation. If the vapour composition were the same as the composition of the liquid from which the vapours is produced, separation would not have been possible by distillation. The products of any desired purity can be produced by means of suitable distillation techniques or by repeated contacting of phases.

Distillation can be treated as a *direct separation method* as it employs no external agent other than heat to generate the second phase necessary for mass transfer. In other separation operations such as absorption, extraction, leaching, etc. foreign substances are introduced into the system to effect separation. We have seen in Chapter 4 that absorption makes use of a liquid solvent and stripping makes use of a gas as separating agents. Being a direct technique requiring no additional component to effect separation, the products of distillation require no further purification. Besides this, distillation offers the following advantages over other separation techniques (Smith R., 2005):

(i) Ability to separate mixtures with a wide range of throughputs.

(ii) Ability to separate mixtures with a wide range of concentrations; most of the alternative separation techniques can handle only relatively pure products.

(iii) Ability to produce high-purity products.

In absorption, as we have already seen, the mechanism of mass transfer is diffusion through stagnant gas, which means only certain components are diffusing while others are stagnant or non-diffusing. Consider distillation of a binary mixture comprising of the more volatile component (**A**) and less volatile component (**B**). Equimolar counter diffusion prevails under ideal circumstances, for each mole of **A** transferred from the liquid to the vapour phase, one mole of **B** gets transferred to the liquid from the vapour. Another important difference between distillation and absorption is that the latter is a low temperature separation operation whereas distillation occurs at the boiling point of the mixture.

Though evaporation also is carried out at the boiling temperature of the liquid, distillation is different from evaporation. Evaporation generally involves separation of a liquid solution consisting of a volatile solvent – usually water – and non volatile solutes. The vapour produced during evaporation is almost pure solvent, leaving behind the non-volatile solutes in the residue. In contrast, in distillation, all the constituent substances will be present, distributed differently, in the coexisting vapour and liquid phases. 'Distilled water' is produced from raw water by 'evaporation' and not by distillation, whereas absolute alcohol is produced from dilute alcohol solution by 'distillation'. Industries frequently combine evaporation with distillation to achieve the desired degree of separation economically. For example, aqueous glycerine can be concentrated to about 80% glycerine by evaporation, and purification beyond this concentration is achieved by distillation. Initially only water is removed as vapour from the dilute aqueous glycerine solution on boiling, whereas both glycerine and water appear in the vapour phase evolved from concentrated solutions.

Limitations

Distillation is a highly energy intensive operation. Since separation is achieved by applying heat, and not by an external agent, the cost of addition and removal of heat is one of the major operating costs in the plant and may even consume more than 50% of the plant's operating energy costs.

In processes utilizing additional substances as separating agents, there is always scope for improvement in the separation effectiveness by opting for a better agent. Thus we use hydrocarbon oils, and not water, for absorbing benzene vapours from the coke oven gases. However, no such choice is possible in distillation. We have to content with the concentration change brought about by the differential distribution of components between the vapour and the liquid phases generated by applying heat. In some cases, this difference in concentration between the equilibrium phases will be so small that separation by distillation becomes impractical. Distillation is not feasible or is uneconomical in the following cases:

(1) Separation of close-boiling liquids, e.g. separation of butadiene (BP 268.4 K) from butylenes (BP 267.2 K - 268.2 K).

(2) Separation of liquids of poor relative volatility, e.g. separation of acetic acid and water from dilute aqueous acetic acid.

(3) Separation of mixtures that form azeotropes.

(4) Separation where boiling points overlap, e.g. separation of aromatic and aliphatic hydrocarbons of the same boiling point range.

(5) Separation of mixtures of low molecular weights. Distillation columns have to be operated at excessively high pressures if vapour condensation is to be achieved by means of cooling water. Otherwise, a refrigerant will be needed as coolant in the overhead condenser. Both the above options are very expensive.

(6) Separation of heat sensitive materials. Such materials have to be distilled under vacuum; operation of columns under very high vacuum is not economical.

(7) Separation of mixtures containing components in very low concentration into pure products.

(8) Separation of mixtures containing condensable and non-condensable components.

The alternatives available for distillation such as solvent extraction, membrane separation and adsorption generally warrant higher investment cost. Therefore, distillation remains the ideal choice for separation in large-scale chemical and petrochemical plants. It is particularly well suited for high purity separations since any degree of separation can be achieved with fixed energy consumption by increasing the number of stages.

5.2 VAPOUR-LIQUID EQUILIBRIA

The difference in the compositions of the vapour and liquid phases, arising due to the differential distribution of the constituents of the mixture on boiling, is the basis of distillation operations. Knowledge of the relationship between the compositions of the phases in equilibrium and the dependence of the equilibrium compositions on temperature and pressure is very essential for better understanding of distillation operations. The design or simulation of a distillation column requires a quantitative estimate of the vapour-liquid equilibrium (VLE) properties of the mixture. Engineer should be familiar with the different equilibrium behaviours and their representation, the sources of equilibrium data, the method of predicting the VLE data in the absence of measured values and methods of extrapolating available data.

Consider a binary mixture consisting of components **A** and **B**. Let **A** be the more volatile component (MVC) in the mixture and **B** the less volatile component (LVC). It means that the vapour pressure exerted by component **A** in the pure state is greater than that exerted by pure **B**. The boiling point of pure **A** is lower than that of pure **B**. The composition of the liquid and vapour phases that exist in equilibrium are represented by the mole fractions of the more volatile component **A** in the phases. Thus x represents the mole fraction of **A** in the liquid and y represents the mole fraction of **A** in the vapour. Study of vapour-liquid equilibrium is intended to provide an insight into the interrelationship between these compositions and the operating conditions of temperature T and pressure P.

Gibb's Phase rule is used to determine the number of independent intensive variables among the group of T, P, x or y that are required to define the state of the system uniquely. It states that $F = C - \pi + 2$, where F is the number of degrees of freedom or the number of independent variables available to define the state uniquely, C is the number of components and π is the number of phases constituting the system. For the vapour-liquid equilibria involving a binary mixture, $C = 2$, $\pi = 2$ and therefore, $F = 2$. That is, two variables can be independently specified to describe the intensive state of the system uniquely. If the pressure is fixed, as is usually the case with distillation, only one variable, say, the liquid phase composition x, can be freely varied. The equilibrium temperature and the vapour phase compositions follow. At a given pressure and temperature, the compositions of the phases in equilibrium x and y are not independent. These observations can be represented graphically by means of the phase diagrams. The constant pressure vapour-liquid equilibria are usually represented by means of a $T - x - y$ diagram commonly known as the boiling point diagram or $x - y$ diagram commonly known as the equilibrium distribution diagram or simply, the equilibrium diagram.

The Boiling-Point Diagram

Figure 5.1 shows the boiling point diagram for the mixture consisting of components **A** and **B**. The upper curve is a plot of temperature versus vapour composition – *T* versus *y* – and is known as the dew-point curve. The lower curve is the plot of temperature versus liquid composition – *T* versus *x* – and is known as the bubble-point curve. Saturated liquid is represented by points lying on the bubble-point curve; saturated vapour by the points lying on the dew-point curve. Since the vapour is richer in the more volatile component, the dew-point curve lies above the bubble-point curve. The mixtures lying below the bubble-point curve are subcooled liquid and those above the dew-point curve are superheated vapour. The vapour and liquid phases coexisting in equilibrium are represented by points lying within the closed loop formed by the dew-point and bubble-point curves.

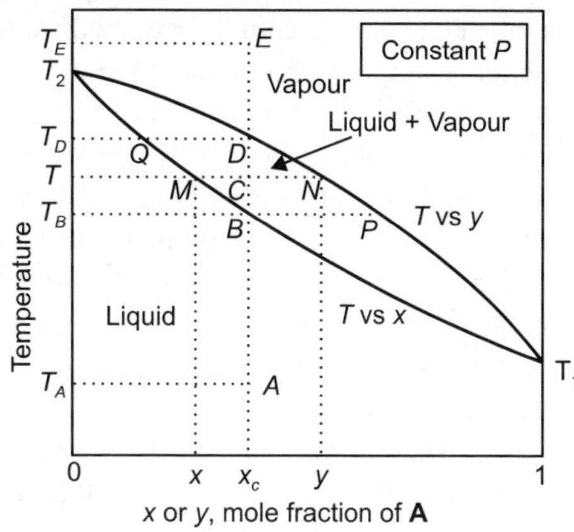

Fig. 5.1: Boiling-point diagram

To understand the significance of the boiling point diagrams in distillation studies, consider what happens when a liquid mixture represented by point *A* in the figure is heated at constant pressure in a closed container. The mixture temperature increases along the line *ABCDE*. The line is vertical because no material is removed from the system and the overall composition of the mixture is constant. Initially at point *A*, the mixture is a subcooled liquid. When its temperature reaches T_B, the mixture, becomes a saturated liquid and begins to boil. T_B is known as the bubble-point temperature of the mixture, as the *first bubble of vapour* appears at this temperature. The vapour so produced will be at the same temperature and pressure as the saturated liquid but will be rich in more volatile component; its condition is represented by point *P* on the upper curve shown in Fig. 5.1. *P* represents the saturated vapour in equilibrium with the liquid at the bubble point. If heating is continued, the temperature increases and the mixture enters the two phase region. The mixture can no longer exist as a single phase; it spontaneously separates into saturated vapour and liquid phases which are in equilibrium. For example, at temperature *T*, saturated liquid of composition *x* represented by point *M* and saturated vapour of composition *y* represented by point *N* coexist in equilibrium. The combined system of vapour and liquid is represented by point *C*. If heating is continued, more and more liquid gets vaporised and when the temperature reaches T_D, the last drop of the liquid is also vaporised and the mixture is entirely saturated vapour represented by point *D*. The temperature T_D at which the vaporisation is completed is called the dew point of the mixture. Further heating takes the mixture along the line *DE* into the superheated vapour region.

The changes occurring due to cooling of a superheated vapour can also be explained using the boiling point diagram. Suppose the original mixture were a superheated vapour represented by say, point *E* in Fig. 5.1. When it is cooled, the first drop of condensate is formed at the dew-point temperature T_D and the liquid so formed is represented by point *Q* on the *T*-*x* curve. The liquid at *Q* is richer in less volatile component **B** than the vapour at *D* with which it is in equilibrium. On further reduction of temperature, the mixture enters into the two-phase region as indicated by point *C* in the figure. At *C*, the mixture is made up of liquid and vapour phases which are in equilibrium with compositions *x* and *y*, respectively. Further cooling will bring down the temperature of the mixture to the bubble point T_B, at which the entire mixture is condensed to saturated liquid. Reducing the temperature further, takes the liquid to the subcooled state as at point *A*.

Pure liquids boil at the constant temperatures T_1 or T_2, the boiling points of pure **A** or pure **B**, as the case may be, till they are completely converted into vapour. In contrast, the boiling of a liquid mixture at a given pressure takes place over a range of temperatures starting with the bubble point and ending at the dew point. At $x = 1$, the liquid is pure **A** and when $x = 0$, the liquid is pure **B** as shown in the figure. At the bubble point, the entire mixture is in the liquid state and only infinitesimally small quantity of vapour exist, which is the richest vapour that can be produced under the stated conditions. At the dew point, the mixture is entirely saturated vapour and only an infinitesimally small quantity of liquid exists in equilibrium with the vapour, and this is the leanest liquid referred to more volatile component, that can be obtained from the given mixture.

For temperatures lying between the bubble point and the dew point, the relative amounts of the vapour and liquid that exist in equilibrium depends on how close the equilibrium temperature is to the dew point or to the bubble point. Consider the mixture at point C in the figure. Let C be the total moles of the mixture and x_c be the mole fraction of **A** in it. The mixture will separate into M moles of saturated liquid of composition x represented by point M in the figure and N moles of saturated vapour of composition y represented by point N in the figure. The line MN is one of the tie lines that connects the liquid and vapour phases in equilibrium. The tie line MN is divided by the mixture point C into two sections MC and NC and it can be shown that the ratio of MC to NC determines the ratio of the amounts of saturated vapour and saturated liquid coexisting in equilibrium. A total material balance and a component-**A** balance gives the following equations.

$$C = M + N \qquad \text{...(5.1)}$$

$$Cx_c = Mx + Ny \qquad \text{...(5.2)}$$

Eliminating C in Eq. (5.2) using Eq. (5.1), we get

$$(M + N)x_c = Mx + Ny \qquad \text{...(5.3)}$$

Equation (5.3) can be rearranged as

$$M(x_c - x) = N(y - x_c) \qquad \text{...(5.4)}$$

$$\frac{M}{N} = \frac{(y - x_c)}{(x_c - x)} \qquad \text{...(5.5)}$$

Referring to Fig. 5.1, the right-hand side of Eq. (5.5) is nothing but the ratio of the lengths of the line segments NC to MC. The left-hand side of Eq. (5.5) is the ratio of the quantity of the liquid to that of the vapour existing in equilibrium. Therefore,

$$\frac{M}{N} = \frac{\text{Moles of liquid}}{\text{Moles of vapour}} = \frac{\text{line } NC}{\text{line } MC} \qquad \text{...(5.6)}$$

Equation (5.6) is known as the inverse-lever-arm rule.

If the vaporisation is carried out in an open vessel and the vapour formed on heating the solution were removed from the system as soon as it is formed, the temperature of the residual liquid increases and the concentration of **A** in the liquid falls along the lower curve from point B to point Q. Similarly on condensation of a mixture, the composition of the residual vapour increases along the dew-point curve from point D to point P in the figure.

Equilibrium Diagram

The constant pressure vapour-liquid equilibrium can be represented as x–y diagram, also known as the equilibrium distribution diagram. In these plots, y, the mole fraction of more volatile component **A** in the vapour phase is the ordinate and x, the mole fraction of **A** in the liquid phase is the abscissa. A tie line such as the line MN in Fig. 5.1 provides one set of x and y values for locating the point such as P on the equilibrium curve in Fig. 5.2. Several such tie lines can be drawn on the boiling point diagram, each providing the necessary x and y values for plotting the equilibrium curve. In the figure, the dotted line is a 45° diagonal which acts as a reference line on which $y = x$. The closer the equilibrium curve to the diagonal, the smaller the difference between y and x and separation of the mixture by distillation will be difficult. From the point of view of easiness of separation by distillation, it is advantageous if the equilibrium curve is far removed from the diagonal. An equilibrium curve coinciding with the diagonal means that separation of the mixture by distillation is impossible.

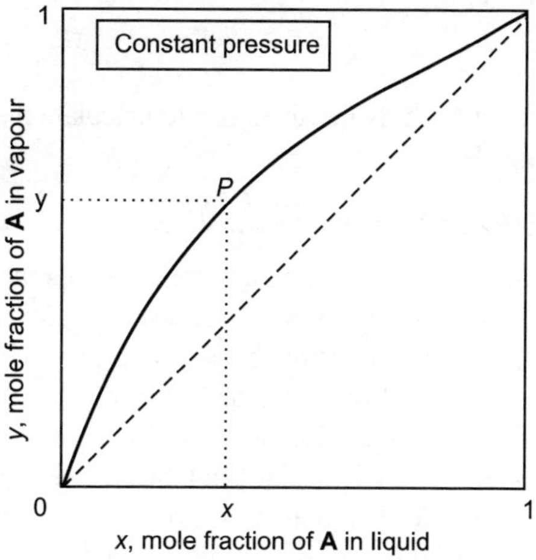

Fig. 5.2: The equilibrium diagram

Relative Volatility

Relative volatility provides a numerical measure of the separability of a mixture by distillation. It is defined as the ratio of volatility of more volatile component **A** to that of the less volatile component **B**. It is denoted by α_{AB} or simply α, and its numerical value is greater than 1.

$$\text{Relative volatility, } \alpha = \frac{\text{Volatility of } \mathbf{A}}{\text{Volatility of } \mathbf{B}} \qquad \text{...(5.7)}$$

Volatility of a constituent in a solution is measured as the ratio of its partial pressure in the vapour to its mole fraction in the liquid. For pure substances, the volatility is nothing but their vapour pressures. Thus,

$$\text{volatility of } \mathbf{A} = \frac{\bar{p}_A}{x}$$
$$\text{volatility of } \mathbf{B} = \frac{\bar{p}_B}{1-x} \qquad \text{...(5.8)}$$

Substituting Eq. (5.8) into Eq. (5.7), we get

$$\alpha = \frac{\bar{p}_A}{x} \bigg/ \frac{\bar{p}_B}{(1-x)} = \frac{\bar{p}_A}{\bar{p}_B} \bigg/ \frac{x}{(1-x)} \qquad \text{...(5.9)}$$

If the vapour phase is assumed to behave as an ideal gas, the ratio of the partial pressure is nothing but the ratio of the mole fractions in the vapour. Therefore,

$$\frac{\bar{p}_A}{\bar{p}_B} = \frac{y}{(1-y)} \qquad \text{...(5.10)}$$

Substituting Eq. (5.10) into Eq. (5.9), we get

$$\alpha = \frac{y}{(1-y)} \bigg/ \frac{x}{(1-x)} \qquad \text{...(5.11)}$$

Equation (5.11) means that the relative volatility is the ratio of the mole ratio of **A** to **B** in the vapour phase to the mole ratio of **A** to **B** in the liquid phase. The greater the value of α, the larger the distance between the equilibrium curve and the diagonal in Fig. 5.2, and the easier the separation of the components of the mixture by distillation. α varies with liquid composition x, and if $\alpha = 1$ at any x other than $x = 0$ or $x = 1$, separation by distillation is not possible.

If the relative volatility remains approximately constant over the entire concentration range, the $y - x$ equilibrium curve can be determined and plotted easily by utilizing the definition of relative volatility, Eq. (5.11).

$$\frac{y}{(1-y)} = \alpha \frac{x}{(1-x)} \qquad \text{...(5.12)}$$

Rearrangement of Eq. (5.12) leads to

$$y(1-x) = \alpha\, x(1-y) \qquad \text{...(5.13)}$$

$$y = \frac{\alpha x}{1 + (\alpha - 1)x} \qquad \ldots(5.14)$$

Equation (5.14) enables one to calculate the composition of the vapour y, given the composition of the liquid x.

Effect of Pressure on VLE

On the boiling point diagram, the temperatures corresponding to $x = 0$ and $x = 1$ are the boiling points of pure substances **B** and **A**, respectively. The boiling points of pure substances increase with pressure. This is true for the bubble and dew points of a mixture of given composition as well. Consequently, the boiling point diagrams at higher pressures will be above the low pressure boiling point diagrams as shown in Fig. 5.3a. Since the relative volatility decreases as pressure is increased, the closed loop formed by the dew point and the bubble-point curves become narrow at high pressures. P_3 is the *critical pressure* for component **A;** above this pressure, the looped curves are shorter. Figure 5.3b indicates the effect of pressure on the distribution diagram.

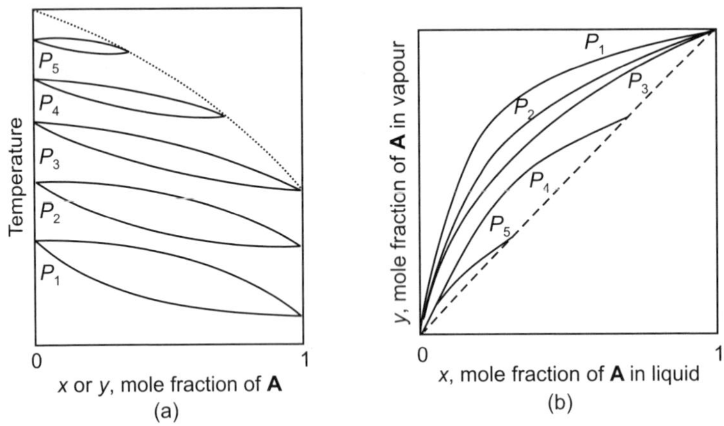

Fig. 5.3: Effect of pressure on (a) the boiling point diagram (b) the equilibrium diagram

Ideal Solutions

Normally the vapour pressure exerted by a constituent in solution will be different from the vapour pressure of the pure component. In solutions, the liquid surface which was occupied by molecules of a single component in the pure state is occupied by several components and consequently, the pressure exerted by individual components gets reduced. Also, the forces that the component molecules experience in solution are different from those in the pure state because of the difference in strengths of the unlike pair molecular interactions present in solutions and the like-pair interactions present in the pure state. However, if the components are chemically similar and the molecular interactions in the pure state and in solutions are of the same order of magnitude, the vapour pressure exerted will be a function only of the concentration in the liquid. Ideal solutions satisfy this requirement. If the liquid phase is an ideal solution and the vapour behaves as an ideal gas, the VLE data can be estimated easily without resorting to direct experimentation. Estimation of the vapour-liquid equilibrium data of ideal solutions requires only the vapour pressures of pure components. A solution conforming to the ideal behaviour has the following characteristics, all interrelated.

1. The components are chemically similar. This requirement implies that the average intermolecular forces in the pure state and in the solution are of approximately the same order of magnitude.

2. There is no volume change on mixing ($\Delta V = 0$) or the volume of the solution varies linearly with composition.
3. There is neither absorption nor evolution of heat on mixing the constituents that form an ideal solution ($\Delta H = 0$); that is, there is no temperature change on mixing.
4. The components in an ideal solution obey Raoult's law. *Raoult's law states that the partial pressure in the vapour in equilibrium with the liquid is directly proportional to the concentration in the liquid.* That is,

$$\bar{p}_i = x_i P_i^S \qquad \text{...(5.15)}$$

where $\bar{p}_i$ is the partial pressure of component i and x_i is its mole fraction in the liquid. P_i^S is the vapour pressure of pure i. Raoult's law requires that the total vapour pressure over an ideal solution is a linear function of its composition.

It should be understood that there exists no ideal solution in the strict sense of the term; but actual mixtures approach ideality as a limit. Ideality requires that the molecules of the constituents are similar in size, structure and chemical nature; only optical isomers of organic compounds meet these requirements. Thus a mixture of *ortho*-, *meta*-, and *para*-xylene conforms very closely to the ideal solution behaviour. Practically, adjacent or nearly adjacent members of the homologous series of organic compounds can be expected to form ideal solutions. Thus mixtures of benzene and toluene, *n*-octane and *n*-hexane, ethyl alcohol and propyl alcohol, acetone and acetonitrile, paraffin hydrocarbons in paraffin oils etc. can be treated as ideal solutions in engineering calculations.

Consider a binary mixture of components **A** and **B**. Assume that Raoult's law is applicable to the solution over the entire composition range. When the solution is in equilibrium with its vapour, the equilibrium partial pressure of the constituents can be expressed as

$$\bar{p}_A = x P_A^S \qquad \text{...(5.16)}$$

$$\bar{p}_B = (1 - x) P_B^S \qquad \text{...(5.17)}$$

where $\bar{p}_A$ and $\bar{p}_B$ are the partial pressures of components **A** and **B**, respectively. P_A^S and P_B^S are the vapour pressures of pure **A** and pure **B**, respectively. By Dalton's law, the total pressure of the vapour (P) is equal to sum of the partial pressures of the constituents. Therefore,

$$P = x P_A^S + (1 - x) P_B^S \qquad \text{...(5.18)}$$

which may be rearranged as

$$x = \frac{P - P_B^S}{P_A^S - P_B^S} \qquad \text{...(5.19)}$$

The mole fraction of a component in a gas phase, by Dalton's law, is the ratio of partial pressure to the total pressure.

$$y = \frac{\bar{p}_A}{P} \qquad \text{...(5.20)}$$

Substituting Eqs. (5.16) and (5.18) into Eq. (5.20), we get

$$y = \frac{x P_A^S}{P} = \frac{x P_A^S}{x P_A^S + (1 - x) P_B^S} \qquad \text{...(5.21)}$$

The boiling point diagram (T–x–y diagram) of ideal solutions can be constructed as explained below.

The total pressure (P) is fixed and is constant. Assume a temperature and get the vapour pressures of pure components at this temperature. Use Eq. (5.19) to calculate the liquid composition x. Use x in Eq. (5.21) and calculate y. These calculations are repeated for various values of equilibrium temperatures that lie between the boiling points of the pure components at the given pressure.

Alternatively, Eq. (5.18) can be used to calculate the equilibrium temperature for assumed values of x and use this x to calculate y using Eq. (5.21). In this method, determination of temperature using Eq. (5.18) involves a trial and error calculation.

P-x-y diagram

The equations presented above - Eqs. (5.16) through (5.21) - can be used to plot the P–x–y diagram of ideal solutions. P–x–y diagrams refer to constant temperatures just as T–x–y diagram refers to constant pressures. Equations (5.16) and (5.17) indicate that when the partial pressures over the solution at constant temperature are plotted against the mole fraction x, we would get two straight lines shown as dotted lines in Fig. 5.4.

We can rewrite Eq. (5.18) as

$$P = (P_A^s - P_B^s)x + P_B^s \qquad \ldots(5.22)$$

which indicates that the total pressure versus x will be a straight line as shown in the figure. Figure 5.4 also shows the total pressure versus y at constant temperature. The P–y curve lies below the P–x line as the vapour concentration y is greater than liquid concentration x.

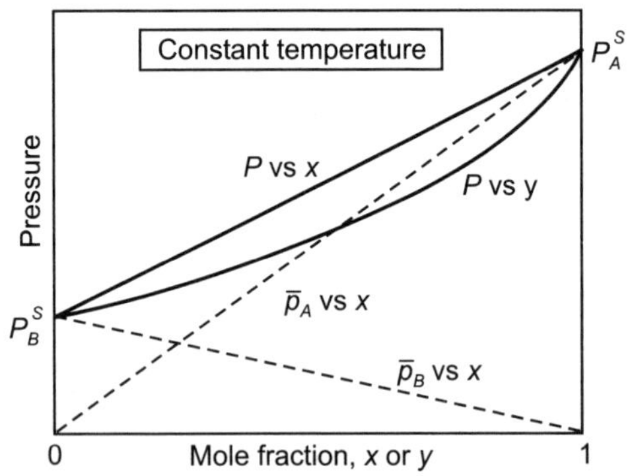

Fig. 5.4: The P–x–y diagram of ideal solutions

Example 5.1: Show that the relative volatility of ideal solution is equal to the ratio of the vapour pressures of the pure components.

***Solution*:** The relative volatility of component **A** with reference to component **B** is defined as the ratio of the volatility of **A** to the volatility of **B**. Volatility of a substance is the ratio of partial pressure to the mole fraction in the liquid. Thus by Eq. (5.9)

$$\alpha = \frac{\bar{p}_A}{x} \bigg/ \frac{\bar{p}_B}{(1-x)}$$

Since the components in ideal solutions obey Raoult's law,

$$\bar{p}_A = xP_A^S \qquad \ldots(5.16)$$

$$\bar{p}_B = (1-x)P_B^S \qquad \ldots(5.17)$$

Substituting Eq. (5.16) and (5.17) into Eq. (5.9), we get

$$\alpha = \frac{P_A^S}{P_B^S} \qquad \ldots(5.23)$$

The relative volatility of ideal solutions is calculated as the ratio of the vapour pressures of the pure components. Although the vapour pressures vary with temperature, it is reasonable to assume that for chemically similar components, this variation will be to the same extent for all components, so that the ratio of vapour pressures remains almost constant; and therefore, for ideal solutions α can be treated to

Table 5.1: Relative volatility of benzene-toluene solution estimated using Eq. (5.23)

T, K	353	358	363	368	373	378	383
P_A^S, kPa	100.5	117.0	135.5	156.2	179.3	204.9	233.3
P_B^S, kPa	38.6	45.8	54.0	63.3	73.8	85.7	99.1
α	2.60	2.56	2.51	2.47	2.43	2.39	2.35

be independent of temperature. For example, for benzene-toluene system, the relative volatility remains substantially constant over a wide range of temperature as shown in Table 5.1. The table lists the vapour pressures of benzene and toluene estimated using the Antoine equations and the relative volatility calculated using Eq. (5.23).

Example 5.2: At 365 K the vapour pressures of benzene (**A**) and toluene (**B**) are 143.5 kPa and 57.6 kPa, respectively. Determine the partial pressures and composition of the vapour in equilibrium with a liquid mixture consisting of equal weights of the two components.

***Solution*:** To calculate the molar composition of the liquid, let us consider 1 kg of the liquid as basis. It contains 0.5 kg benzene and 0.5 kg toluene. Or, 0.5/78 kmol benzene and 0.5/92 kmol toluene. Therefore, the mole fraction of benzene in the liquid is

$$x = \frac{0.5/78}{0.5/78 + 0.5/92} = 0.5412$$

Since the solutions obey Raoult's law [Eq. (5.16)], the partial pressures of benzene and toluene are evaluated as

$$\bar{p}_A = xP_A^S = 0.5412 \times 143.5 = 77.66 \text{ kPa}$$

$$\bar{p}_B = (1-x)P_B^S = 0.4588 \times 57.6 = 26.43 \text{ kPa}$$

The total pressure is the sum of partial pressures.

$$P = \bar{p}_A + \bar{p}_B = 77.66 + 26.43 = 104.09 \text{ kPa}$$

The mole fraction of toluene in the vapour is [Eq. (5.20)]:

$$y = \frac{\bar{p}_A}{P} = \frac{xP_A^S}{P} = \frac{77.66}{104.09} = 0.746$$

The mole fraction of toluene in the vapour is $1 - y = 1 - 0.746 = 0.254$

The vapour contains 74.6 mol% benzene and 25.4 mol% toluene at a total pressure of 104.09 kPa.

(Answer)

Example 5.3: Two liquids **A** and **B** form ideal solutions. The vapour pressures of pure **A** and **B** can be evaluated by the Antoine equations

$$\ln P_A^S = 14.5463 - \frac{2940.46}{T - 35.93}$$

$$\ln P_B^S = 14.2724 - \frac{2945.47}{T - 49.15}$$

where T is in K and P is in kPa.

(a) Construct the boiling point diagram and the equilibrium diagram for the system at a pressure of 101.3 kPa.

(b) Suggest the equilibrium relationship for the system using the average relative volatility.

***Solution*:** (a) The boiling point diagram is a plot of temperature on y-axis and x and y on x-axis. Assume temperatures between the boiling points of pure liquids and calculate the vapour pressure using Antoine equations. The boiling points of the pure liquids can be determined from Antoine equations by substituting for the vapour pressures, the total pressure 101.3 kPa.

$$\ln(101.3) = 14.5463 - \frac{2940.46}{T_A - 35.93}$$

$$\ln(101.3) = 14.2724 - \frac{2945.47}{T_B - 49.15}$$

where T_A and T_B are the boiling points of pure liquids **A** and **B**, respectively. Solving the above equations we get T_A = 332.1 K and T_B = 354.2 K. The vapour pressures are evaluated at different temperatures in the range 332.1 K to 354.2 K using Antoine equations.

Let T = 340 K. Then

$$\ln P_A^S = 14.5463 - \frac{2940.46}{340 - 35.93} \Rightarrow P_A^S = 131.1 \text{ kPa}$$

$$\ln P_B^S = 14.2724 - \frac{2945.47}{340 - 49.15} \Rightarrow P_B^S = 63.14 \text{ kPa}$$

The results are given below:

T, K	332.1	335	340	345	350	354.24
P_A^S, kPa	101.30	111.53	131.10	153.30	178.37	201.83
P_B^S, kPa	47.59	52.89	63.14	74.92	88.40	101.30

Composition of liquid x and vapour y under equilibrium can be calculated for each temperature using Eqs. (5.19) and (5.20). For example at T = 340 K,

$$x = \frac{P - P_B^S}{P_A^S - P_B^S} = \frac{101.3 - 63.14}{131.10 - 63.14} = 0.5615$$

$$y = \frac{xP_A^S}{P} = \frac{0.5615 \times 131.10}{101.3} = 0.7267$$

These calculations are repeated for other temperatures and the results are:

T, K	332.1	335	340	345	350	354.2
x	1.0	0.8256	0.5615	0.3365	0.1434	0.0
y	1.0	0.9089	0.7267	0.5093	0.2524	0.0

A graph is plotted with temperature as ordinate and x as abscissa which gives the bubble-point curve, and with temperature as ordinate against y as abscissa which gives the dew-point curve of the boiling point diagram. The boiling point curve is plotted in Fig. 5.5a and the equilibrium diagram in Fig. 5.5b.

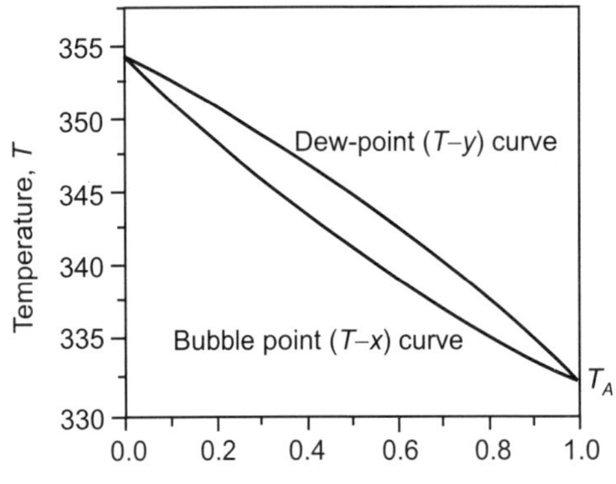

(a)

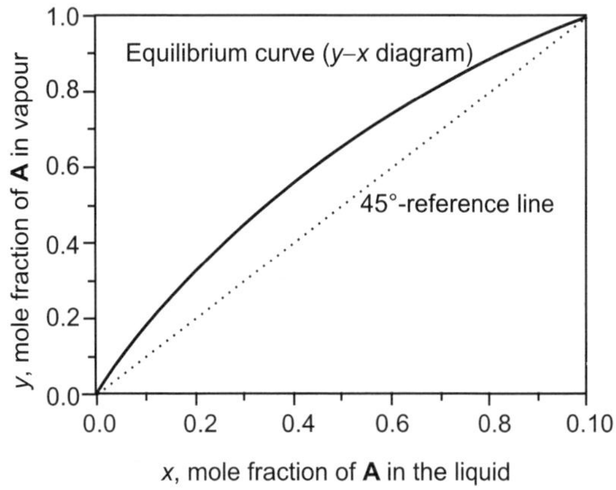

(b)

Fig. 5.5: (a) The boiling point diagram (b) Equilibrium diagram for Example 5.2

(b) For ideal solutions, the relative volatility may be taken to be the ratio of vapour pressures, (*see* Eq. (5.23)). Calculate relative volatility at different temperatures as $\alpha = \frac{P_A^S}{P_B^S}$. They are shown below in the table as last row.

T, K	332.1	335	340	345	350	354.24
P_A^S, kPa	101.30	111.53	131.10	153.30	178.37	201.83
P_B^S, kPa	47.59	52.89	63.14	74.92	88.40	101.30
α	2.129	2.109	2.076	2.046	2.018	1.992

The average of relative volatility values is found out to be $\alpha = 2.062$.

We have shown that in terms of α, the equilibrium relationship is

$$y = \frac{\alpha x}{1 + (\alpha - 1)x} \qquad \text{...(5.14)}$$

Substituting $\alpha = 2.062$ into the above, we have

$$y = \frac{2.062\,x}{1 + 1.062x} \qquad \textbf{(Answer)}$$

Equilibrium vapour composition y can be calculated using the above equation for assumed x-values:

x	1.0	0.8256	0.5615	0.3365	0.1434	0.0
y	1.0	0.9071	0.7253	0.5112	0.2566	0.0

Non-Ideal Solutions

Positive and negative deviations

Components in an ideal solution obey Raoult's law over the entire concentration range. Only very few systems approach ideal behaviour in the true sense of the term, although many can be treated as ideal solutions for engineering calculations. The non-ideal behaviour of liquid mixtures arises due to the dissimilarity among molecules. The dissimilarities may be due to the difference in the molecular structure or the molecular weights of the constituents. For such solutions, the plots of partial pressure against liquid composition will not be straight lines. The partial pressure curves will lie above or below the straight line that represents partial pressures in ideal solution. If the partial pressure is greater than that given by Raoult's law, the solution is said to exhibit positive deviation from ideality, and if it is less than the partial pressure given by Raoult's law, the solution is said to exhibit negative deviation from ideality. The total pressure versus liquid composition also will be above or below the Raoult's law line depending upon whether the solution exhibits positive or negative deviation from ideality. The general nature of the vapour pressure curves showing positive and negative deviation are shown in Fig. 5.6. These diagrams refer to constant temperature conditions. The uppermost curves give the total vapour pressure as function of liquid composition (P-x curve). The corresponding curves as a function of the vapour composition (P-y curve) lie below the P-x curves as the vapour is rich in the more volatile component.

Positive deviations generally result when the intermolecular forces between like molecules are stronger than those between unlike molecules. Heat is absorbed upon mixing two liquids which yield a solution exhibiting positive deviation from ideality. Examples are oxygen-nitrogen, ethanol-ethyl ether, water-ethanol, carbon disulphide-acetone, benzene-cyclohexane, etc. Negative deviation indicates stronger unlike-pair interaction than like-pair interaction. The formation of such solutions from the pure constituents is generally accompanied by an evolution of heat. Chloroform-ethyl ether, chloroform-benzene, hydrochloric acid-water, chloroform-acetone, phenol-cyclohexanol, etc. are examples of solutions exhibiting negative deviation.

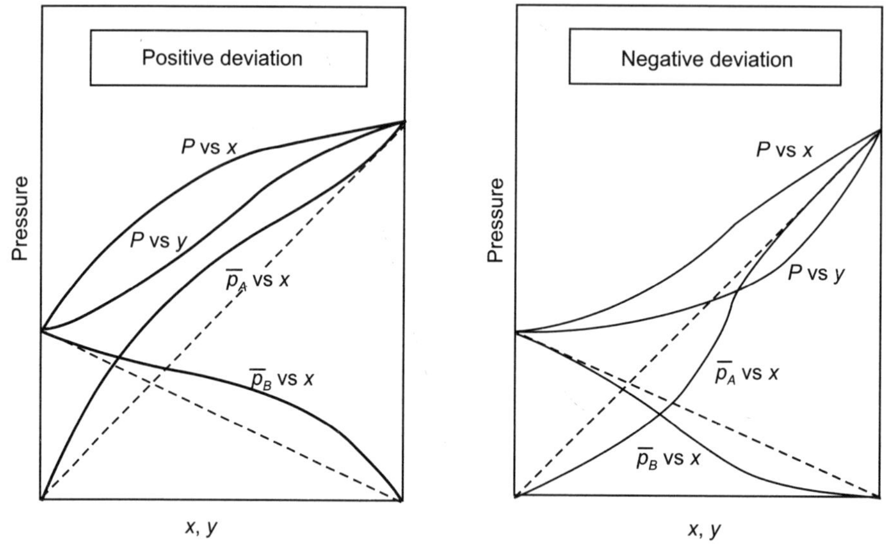

Fig. 5.6: Positive and negative deviation from ideal behaviour

Azeotropes

The ultimate in nonideal behaviour is exhibited by azeotropes. Azeotropes are constant boiling mixtures. The word azeotrope is derived from Greek words meaning 'boiling without changing.' When an azeotrope is boiled, the resulting vapour will have the same composition as the liquid from which it is produced. Whereas the equilibrium temperature of an ordinary solution varies from the bubble-point to the dew point, the boiling point of an azeotrope remains constant till the entire liquid is vaporised. The azeotropes are formed by solutions showing large positive or negative deviation from ideality. If the vapour pressures of the constituents of a solution are very close, then an appreciable positive deviation from ideality will lead to a maximum in the vapour pressure curve and negative deviations from ideality under the same conditions leads to a minimum in the vapour pressure curve. Even if the difference in the vapour pressures of the pure components is large, the chances for the occurrence of maximum or minimum in the vapour pressures should not be ruled out if the deviation from ideal behaviour is quite high. At the composition at which a maximum or a minimum occurs in the vapour pressure curve, a minimum or maximum, as the case may be, exists in the boiling point diagrams. The mixture is said to form an azeotrope at this composition under the given temperature and pressure. Azeotrope will distil without change in composition, because the vapour produced has the same composition as the liquid.

Minimum-boiling azeotropes. Solutions showing positive deviation from ideality in certain cases may lead to the formation of azeotropes of the minimum-boiling type. The *P-x-y*, *T-x-y* and *x-y* curves for the minimum-boiling azeotropes are shown in Fig. 5.7. In the boiling-point diagram, the liquid and vapour curves are tangents at point *M*, the point of azeotropism at this pressure. The temperature at *M* is the minimum temperature of boiling for the system. It is lower than the boiling points of either of the pure components. For all mixtures of composition less than *M*, the equilibrium vapour is richer in the more volatile component than the liquid. For all mixtures richer than *M*, the vapour is less rich than the liquid in terms of the more volatile component. A mixture of composition *M* boils producing a vapour of identical compositions and consequently, boiling occurs at constant temperature without change in composition. If solutions either at *P* or *Q* are boiled in an open vessel with continuous escape of vapours, the temperature and composition move along the lower curve away from *M* and towards the pure substances. Solutions like these cannot be distilled by usual distillation methods. Distillation of a solution which forms a minimum boiling azeotrope always gives a distillate product which is closer in composition to the azeotrope than the original feed. One of the most important azeotropes in this category is ethanol-water which forms azeotrope at 89.4 mol% ethanol at 351.4 K and 101.3 kPa. Other examples are benzene-ethanol (341.2 K, 55% benzene), carbon disulphide-acetone (312.5 K, 61% carbon disulphide), isopropyl ether-isopropyl alcohol (345.1 K, 39.3% alcohol) all at 101.3 kPa.

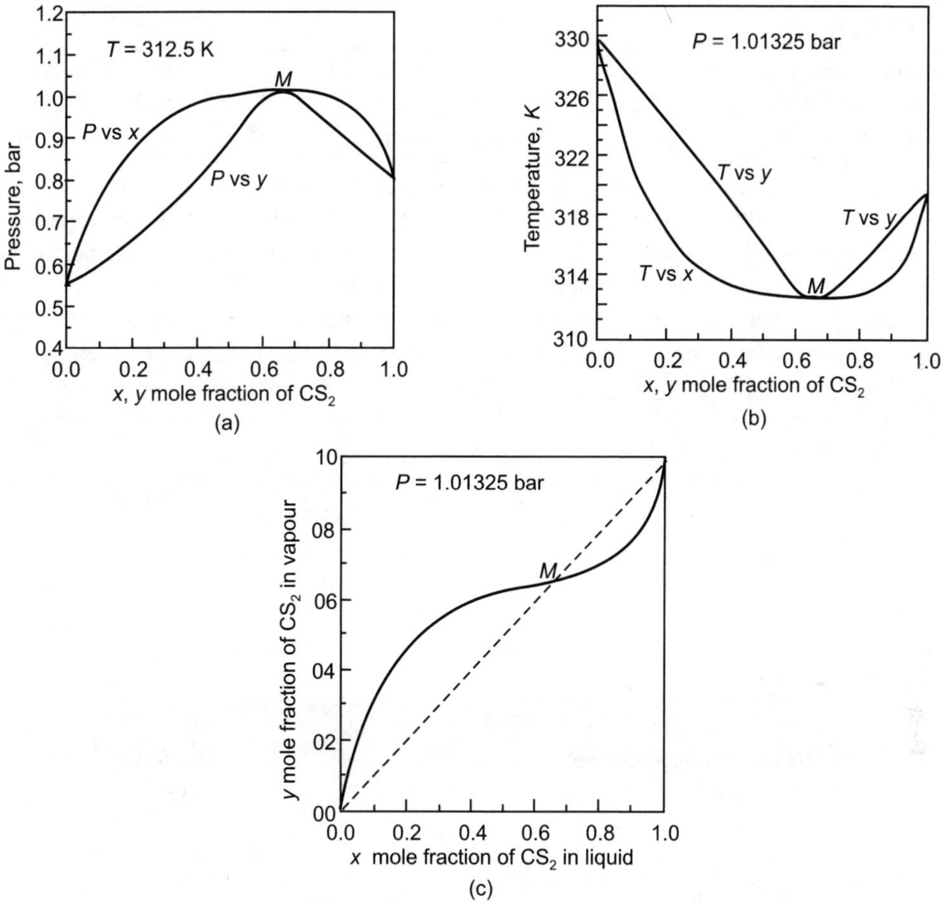

Fig. 5.7: Minimum boiling azeotrope (carbon disulphide – acetone system) (a) *P–x–y* diagram (b) *T–x–y* diagram (c) Equilibrium curve

Maximum-boiling azeotropes. When the total pressure of the system at equilibrium is less than the ideal value, the system is said to exhibit negative deviation from ideality. When the difference in vapour pressures of the pure components is not too large, and in addition, the negative deviations are large, the curve for total pressure against composition passes through a minimum. This condition results in a maximum in the boiling temperature and a condition of azeotropism as at point N in Fig. 5.8. The vapour is leaner in the more volatile component for liquids whose concentration is less than the azeotropic concentration. Solutions on either side of the azeotrope, if boiled in an open vessel with escape of vapours will ultimately leave a residual liquid of the azeotropic composition in the vessel. That is, when a solution that forms a maximum boiling azeotrope is distilled, the residue will be closer to the azeotrope than the original solution. Maximum-boiling azeotropes are less common than the minimum-boiling type. Hydrochloric acid-water system forms an azeotrope at 11.1% (mol) HCl at 383 K and 101.3 kPa. Other examples are chloroform-acetone (337.7 K and 65.5% chloroform), phenol-cyclohexanol (455.65 K, 90% phenol) all at 101.3 kPa.

Effect of pressure on azeotropes. The azeotropic composition shifts continuously with change in pressure or temperature. In some cases, changing the pressure may eliminate azeotropism from the system. Azeotropism disappears in ethanol-water system at pressures below 9.33 kPa. The table below illustrates the effect of pressure on this system. The last row gives the mole percent of alcohol in the azeotrope.

P, kPa	13.3	20.0	26.6	53.2	101.3	146.6	193.3
T, K	307.4	315.2	321	336	351.3	361.0	368.5
% (mol)	99.6	96.2	93.8	91.4	89.43	89.3	89.0

The effect of pressure on azeotropic composition is made use of in the separation of such mixtures by *pressure-swing distillation*. The pressure is adjusted to such a value that the azeotrope is closer to the 100% composition. For example by distilling ethanol-water solution at a pressure below 20 kPa, it is possible to obtain a distillate of composition as high as, say, 96%. If this product is subjected to further fractionation at atmospheric pressure it is possible to obtain alcohol of any desired composition, since the azeotropic composition of 89.43% at atmospheric pressure is already crossed.

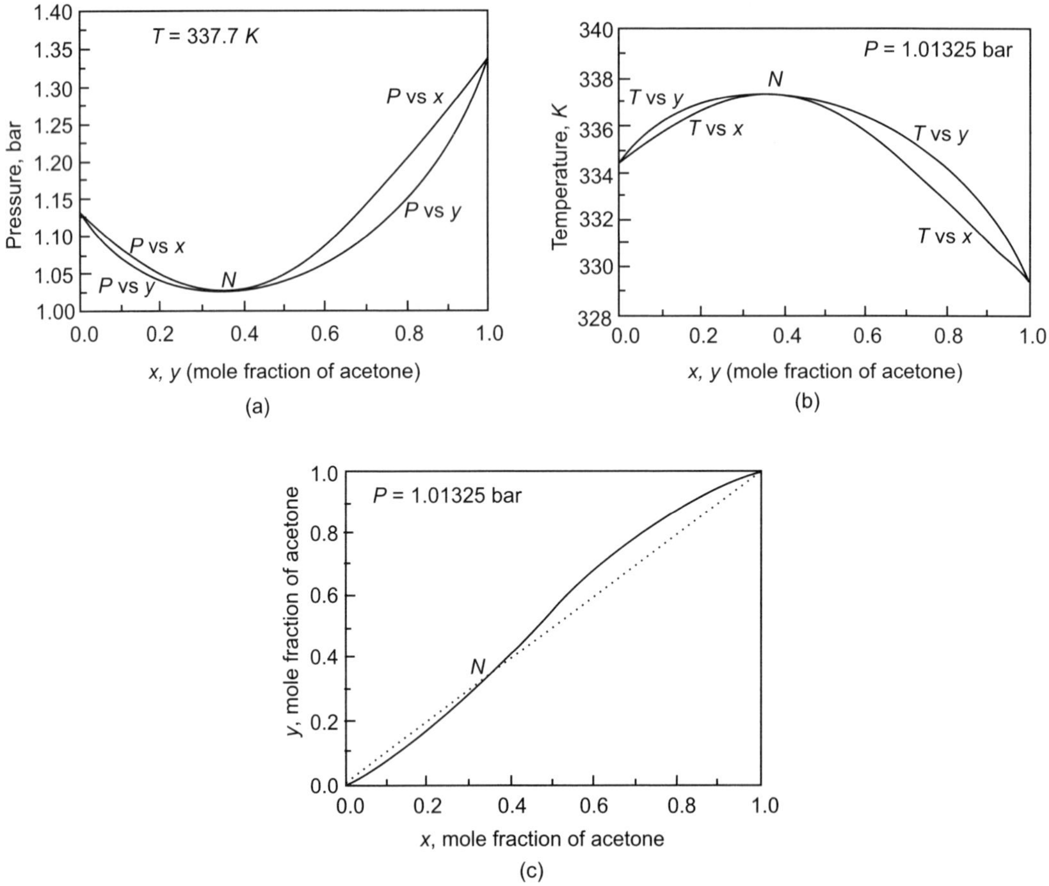

Fig. 5.8: Maximum-boiling azeotrope (acetone-chloroform system) (a) P–x–y diagram (b) T–x–y diagram (c) Equilibrium diagram

Partially miscible systems and heteroazeotropes

Figure 5.9 shows the temperature-composition diagram for a partially miscible system at constant total pressure. If the average composition of the mixture consisting of liquids **A** and **B** lies in the two-liquid region represented by the envelope *JLMK*, the mixture cannot exist as a single liquid, but separates into two saturated liquid phases one lying on the curve *JL* and the other lying on the curve *KM*. Thus if the liquids **A** and **B** are mixed in such proportion that the average mixture composition is represented by point *O*, it separates into two saturated liquids *N* and *P*. The saturated liquid *N* is rich in **B** and the saturated liquid *P* is rich in **A**. If the mixture lies in the liquid regions on the left of the solubility curve *JL* but below the curve *BL,* the solution is homogeneous liquid rich in **B**. Similarly, if it is on the right of the solubility curve *KM* but below *AM*, the solution is a homogeneous liquid rich in **A**. Referring to the Fig. 5.9a, if the mixture at *O* were heated, its temperature increases along the vertical line passing through *O*. When the temperature reaches T^*, vaporisation begins and the system now is made up of two saturated liquids represented by points *L* and *M* and a saturated vapour represented by point *C*. The temperature remains constant at T^* till one of the liquid phases disappears. The temperature T^*, at which the three phases coexist in equilibrium is called the three phase equilibrium temperature and according to phase rule, this temperature is fixed at a given total pressure. A binary system consisting of two liquid phases and a vapour phase is univariant according to the phase rule. By fixing the pressure, the system is completely defined. Points *L* and *M* represent the saturated liquid phases and point *C*, the vapour in equilibrium with these liquids. If the original liquid mixture had the average composition same as *C*, the mixture will boil without change of composition till the entire liquid is vaporised. A liquid mixture of composition *C* can be treated as a heteroazeotrope of components **A** and **B**.

At temperatures above T^*, one of the liquid phases disappears and the system can exist either as two phases (liquid and vapour) or as a single phase, depending on the overall composition. The liquid phase that disappears depends on whether the combined mixture composition is to the left or right of point *C*. For the mixture at *O*, the **B**-rich phase disappears. The system now consists of a saturated **A**-rich liquid represented by point *U* in the figure in equilibrium with saturated vapour represented by the point *V*. When the temperature is increased, the overall composition must still lie on the vertical line passing through *O*, but, the relative amounts and composition of the liquid and the vapour phases change. Above the temperature corresponding to point *X*, the mixture is entirely vapour.

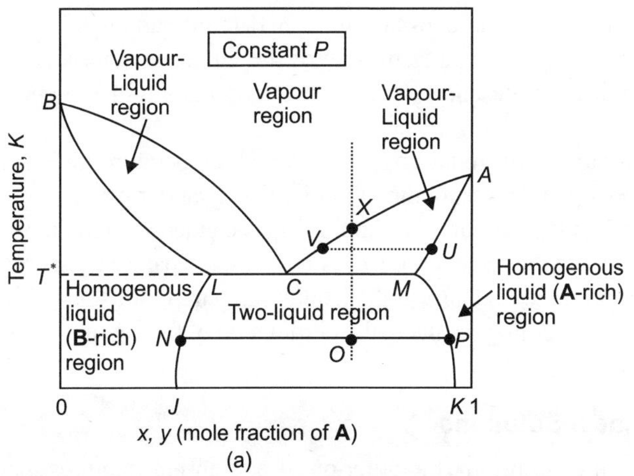

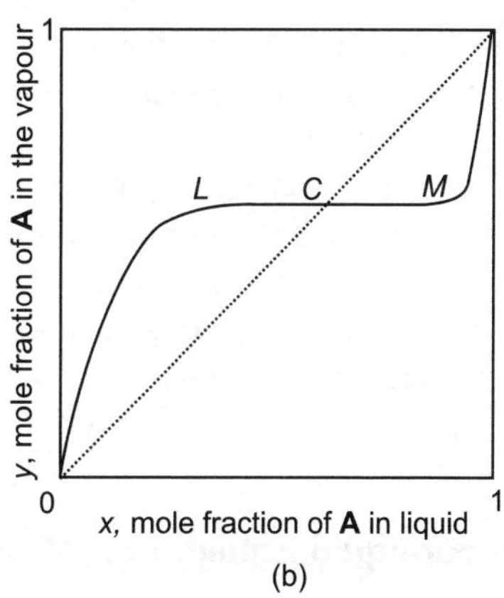

Fig. 5.9: (a) The *T-x-y* diagram of a partially miscible binary liquid mixture (b) The equilibrium diagram of a partially miscible binary liquid mixture

If the initial mixture were to the left of point C, the liquid remaining in equilibrium with the vapour would have been **B**-rich saturated liquid lying on the curve BL. The three-phase equilibrium temperature and hence the mutual solubility of **A** and **B** increase with pressure and at the *upper critical solution temperature* the two liquid phases become identical. For pressures above this critical condition, the three-phase equilibrium conditions do not exist.

Immiscible systems

The phase diagram for a completely immiscible system is shown in Fig. 5.10. This diagram is a special case of Fig. 5.9 if the **A**-rich and the **B**-rich phases are the pure liquids **A** and **B**, respectively. Consider a binary mixture of two immiscible liquids having an overall composition represented by point D. When this mixture is heated its temperature increases along the line DE. For all temperatures up to point E, each component exerts its full vapour pressure as its partial pressure in the vapour phase. When the temperature T^* corresponding to point E is reached, the sum of the vapour pressures becomes equal to the surrounding pressure and the liquid boils. Three phases, two pure liquids and one vapour phase coexist in equilibrium at this condition. T^* is therefore known as the *three phase temperature.*

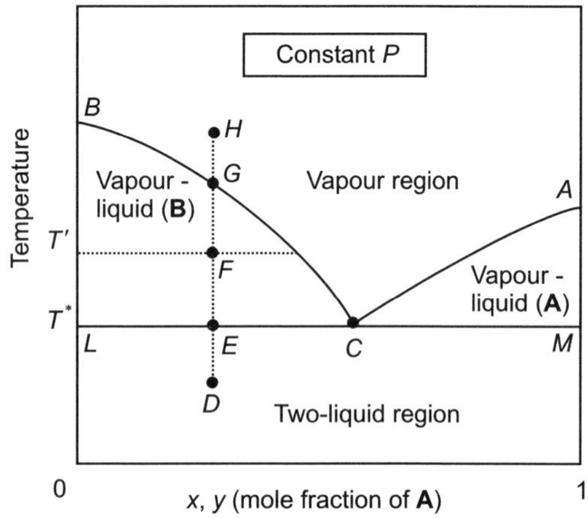

Fig. 5.10: Vapour-liquid equilibrium diagram of immiscible systems

The composition of the vapour in equilibrium with the pure liquids at the three-phase temperature is given by point C. Since there are three phases and two components present, according to the phase rule, the number of degrees of freedom is one as in the case with partially miscible systems. It means that when the pressure is fixed, the system is completely defined. The three-phase temperature and the equilibrium vapour composition get automatically fixed by specifying the pressure in such a way that the sum of the vapour pressures equals the surrounding pressure.

$$P = P_A^S + P_B^S \qquad \ldots(5.24)$$

On further addition of heat, the temperature remains constant at T^* and more vapour of the same composition as given by point C is formed. This continues till one of the liquids disappears and the system becomes a two-phase mixture, either vapour and liquid **A** or vapour and liquid **B**, depending upon the initial composition.

Now let us consider the cooling of a vapour of initial composition and temperature indicated by point H. When the temperature is lowered to that corresponding to point G the vapour pressure of pure **B** will be equal to the partial pressure of **B** in the vapour. Pure liquid **B** gets condensed and the vapour composition changes along the line GC. When the temperature T^* is reached, the partial pressure of **A** in the vapour will be equal to the vapour pressure of **A** and at this condition, pure liquids **A** and **B** and the vapour are present in equilibrium. Further cooling results in the elimination of the vapour phase and the system now consists of two immiscible liquids.

Vapour-liquid Equilibrium of Nonideal Solutions

When the vapour and liquid phases are in equilibrium, the criterion of equilibrium requires that the fugacities of the components in the liquid and vapour phases are the same for all components. For component i,

$$\bar{f}_i^L = \bar{f}_i^V \qquad \text{...(5.25)}$$

where $\bar{f}_i^L$ is the fugacity in the liquid phase and $\bar{f}_i^V$ is the fugacity in the vapour-phase. The fugacity in the vapour phase is the product of the vapour phase composition y_i, fugacity coefficient for the vapour, ϕ_i^V and the total pressure P.

$$\bar{f}_i^V = y_i \phi_i^V P \qquad \text{...(5.26)}$$

Fugacity in the liquid phase can be estimated as either the product of liquid phase composition x_i, fugacity coefficient in the liquid ϕ_i^L and the pressure P

$$\bar{f}_i^L = x_i \phi_i^L P \qquad \text{...(5.27)}$$

or as the product of the liquid phase composition, activity coefficient in the liquid γ_i, and the fugacity of the pure substance in the standard state f_i^0.

$$\bar{f}_i^L = x_i \gamma_i f_i^0 \qquad \text{...(5.28)}$$

Replacing the standard state fugacity by the saturation pressure of the pure component P_i^s, Eq. (5.28) becomes

$$\bar{f}_i^L = x_i \gamma_i P_i^s \qquad \text{...(5.29)}$$

Substituting in Eq. (5.25), the fugacity of vapour as defined by Eq. (5.26) and the fugacity of liquid as defined by Eq. (5.29), we get

$$y_i \phi_i^V P = x_i \gamma_i P_i^s \qquad \text{...(5.30)}$$

Instead of Eq. (5.29), if the fugacity of liquid as given by Eq. (5.27) is used in Eq. (5.25) along with Eq. (5.26) for fugacity of the vapour, we get

$$y_i \phi_i^V = x_i \phi_i^L \qquad \text{...(5.31)}$$

Equation (5.30) forms the basis for estimating the VLE of non ideal solutions using the *activity coefficient models* and Eq. (5.31) forms the basis of estimating VLE by the *equation of state approach*. These are discussed briefly in Appendix-A. In using Eq. (5.30), the liquid-phase activity coefficients can be estimated by any of the models described there. Estimation of the vapour-liquid equilibria using activity coefficients is useful when polar components are present in the system. For most substances at low pressures, ϕ_i^V can be assumed to be unity. If the pressure is very high, ϕ_i^V must be evaluated using an equation of state. The equations of state provide thermodynamic models for evaluating ϕ_i^V and ϕ_i^L in Eq. (5.31) from volumetric properties. The method has the advantage that the standard state fugacity f_i^0 need not be specified. Also, the continuity at the critical point is ensured, as the properties in the gas and liquid phases are evaluated using the same model. However, in practice, the equations of state approach is applied only to mixtures of nonpolar components. They are widely used to predict VLE of light hydrocarbon mixtures.

Vaporisation equilibrium constants: For muticomponent systems, it is convenient to express the phase equilibrium relations in terms of the vaporisation equilibrium constants or *K*-factors. It is defined as the ratio of the mole fraction in the vapour phase y to that in the liquid x or $K_i = (y_i / x_i)$.

Equation (5.30) can be rearranged as below:

$$K_i = \frac{y_i}{x_i} = \frac{\gamma_i P_i^s}{\phi_i^V P} \qquad \text{...(5.32)}$$

Similarly Eq. (5.31) gives

$$K_i = \frac{y_i}{x_i} = \frac{\phi_i^L}{\phi_i^V} \qquad \text{...(5.33)}$$

For components i and j in a multicomponent solution, the K-factors can be written as

$$K_i = \frac{y_i}{x_i} \qquad \text{...(5.34)}$$

$$K_j = \frac{y_j}{x_j} \qquad \text{...(5.35)}$$

We can define the relative volatility α_{ij} of component i with reference to component j by Eq. (5.11)

$$\alpha_{ij} = \frac{y_i/y_j}{x_i/x_j} \qquad \text{...(5.11)}$$

In terms of the K-factors, the above relation becomes

$$\alpha_{ij} = \frac{y_i/y_j}{x_i/x_j} = \frac{y_i/x_i}{y_j/x_j} = \frac{K_i}{K_j} \qquad \text{...(5.36)}$$

The relative volatility α for a binary solution can be treated as the ratio of vaporisation equilibrium constant of more volatile component to that of the less volatile component.

For ideal solutions, the activity coefficient γ_i is unity and therefore, Eq. (5.32) can be simplified. If the vapour phase is assumed to behave as an ideal gas, the vapour-phase fugacity coefficient ϕ_i^V, which is a measure of the departure of the gas from ideal gas behaviour, also equals unity and therefore, Eq. (5.32) becomes

$$K_i = \frac{y_i}{x_i} = \frac{P_i^s}{P} \qquad \text{...(5.37)}$$

which is nothing but Raoult's law written for the case where the vapour behaves as ideal gas.

$$y_i = \frac{P_i^s}{P} x_i \qquad \text{...(5.38)}$$

The same result was obtained earlier for ideal solutions. *See* Eq. (5.21). We see that for ideal solution, the vaporisation equilibrium constant is the ratio of vapour pressure to total pressure [Eq. (5.37)] and as such K-factors for ideal solutions depend only on the temperature and pressure and are readily correlated as a function of these two variables.

It is possible to express Eq. (5.32) in the following form:

$$K_i = \frac{P_i^s \phi_i^s \gamma_i}{P\phi_i^V} \exp\left(\frac{V_i^L (P - P_i^s)}{RT}\right) \qquad \text{...(5.39)}$$

where the exponent is known as *Poynting correction* which is approximately unity when pressure is low. The fugacity coefficient ϕ_i^s is the ratio of the fugacity of the component i under its saturation conditions to the saturation pressure of the substance and tends to unity if the vapour behaves as an ideal gas. The advantage of this equation is that it contains just the properties of the pure substances and therefore expresses K-values as functions of only the temperature and pressure, independent of the compositions of the liquid and vapour phases. Since ϕ_i^s and ϕ_i^V can be determined from equations of state or generalized correlations, it is possible to provide correlations for K-values of substances as functions of temperature and pressure. The DePriester nomographs provide such correlations for many hydrocarbons. Figure 5.11 gives the nomograph for light hydrocarbons in the high temperature range.

Bubble-point equilibria: The bubble-point temperature is the temperature at which the first bubble of vapour is produced from the liquid on heating at constant pressure. At the bubble-point, the liquid has the same composition as the original mixture. Therefore, in problems in which bubble-point temperature

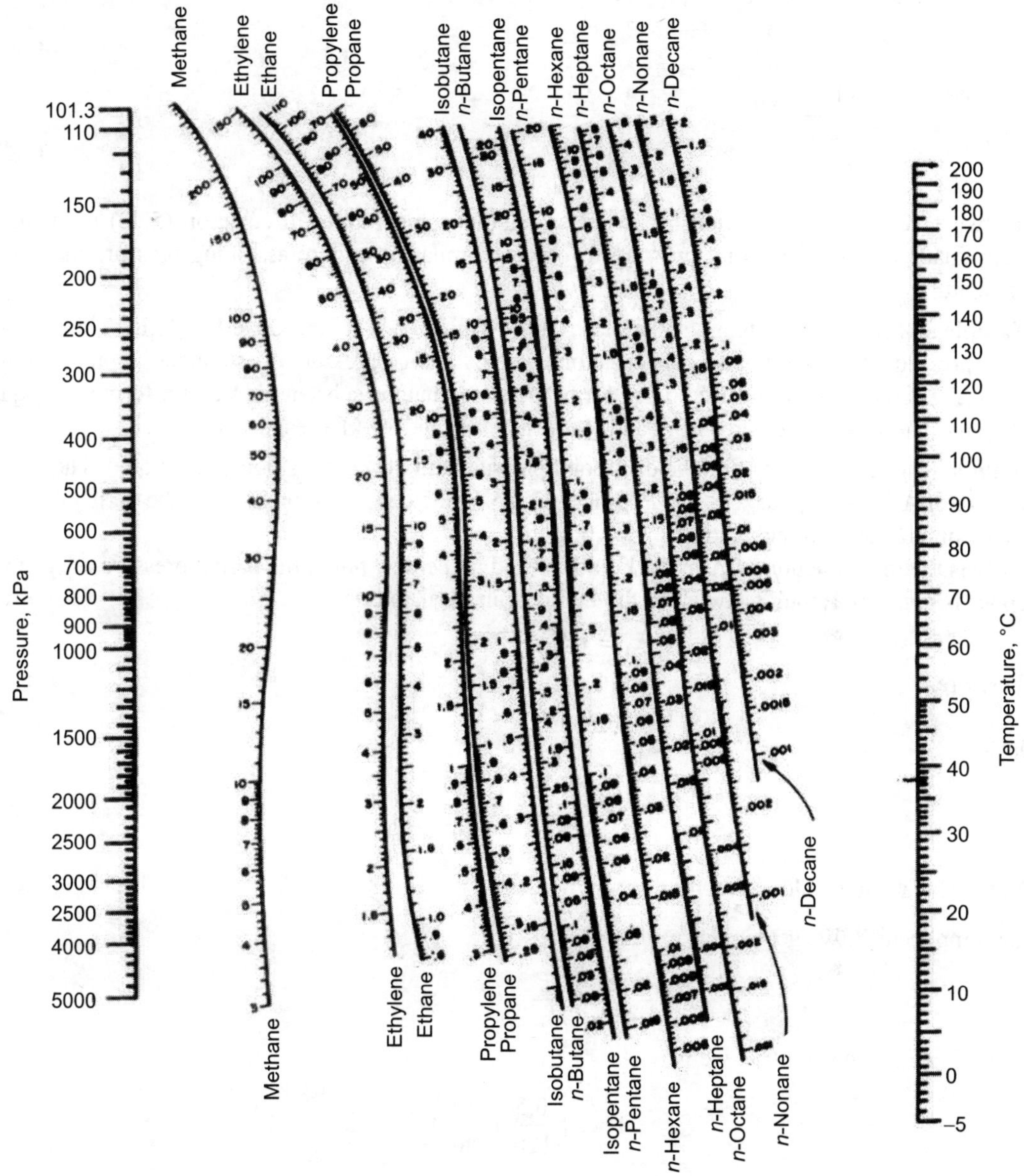

Fig. 5.11: Distribution coefficient K for light hydrocarbons

is to be determined, the x_i are known. Assume a temperature and get the K_i values at this temperature. Calculate y_i using $y_i = K_i x_i$. If the assumed temperature is the correct choice, then

$$\sum y_i = \sum K_i x_i = 1 \qquad \text{...(5.40)}$$

Otherwise, repeat the calculations with another temperature. To find the bubble-point pressure, a similar procedure as above is adopted, but by assuming various values of pressure until $\sum K_i x_i = 1$.

Dew-point equilibria: The dew-point temperature is the temperature at which the first drop of condensate is formed on cooling a vapour at constant pressure. The vapour in equilibrium with the liquid at the dew point has the same composition as the original mixture. In order to find the dew-point temperature, a temperature is assumed arbitrarily and K_i is determined. Then,

$$x_i = \frac{y_i}{K_i} \quad ...(5.41)$$

At the dew point,

$$\Sigma x_i = \Sigma \frac{y_i}{K_i} = 1 \quad ...(5.42)$$

Otherwise, repeat the calculation by assuming another temperature till equation (5.42) is satisfied. Determination of the dew-point pressure involves a similar procedure assuming pressure instead of temperature.

Example 5.4: A vapour containing 50 mol% **A** and 50 mol% **B** is at 300 K and 101.3 kPa. Tl is mixture is compressed isothermally until condensation occurs. At what pressure does condensation occur and what is the composition of the liquid that forms? Assume that the substances **A** and **B** form ideal liquid solutions. The vapour pressures of **A** and **B** are 120 kPa and 90 kPa respectively.

***Solution*:** The problem is to find the dew-point pressure and the composition of the liquid. The partial pressure of **A** in the vapour $= \overline{p}_A = yP$. Since $y = 0.5$, at the dew-point pressure P, the partial pressure of the component **A** in the vapour is $\overline{p}_A = 0.5P$.

If x is the mole fraction of **A** in the liquid formed at the dew point, the partial pressure of **A** in the vapour is given by Raoult's law since the liquid is an ideal solution.

$$\overline{p}_A = xP_A^S = 120x$$

Therefore,

$$120x = 0.5P$$

Or

$$x = \frac{0.5P}{120} \quad (A)$$

[**Note:** The above result is directly given by Eq. (5.41): $x_i = \frac{y_i}{K_i}$ with $K_i = \frac{P_A^S}{P}$ for ideal solutions.]

For component **B**, we get similar results:

$$1 - x = \frac{0.5P}{90} \quad (B)$$

Adding Eqs. (A) and (B), we get

$$1 = P\left(\frac{0.5}{120} + \frac{0.5}{90}\right)$$

solving which we get

$$P = 102.86 \text{ kPa.} \quad \textbf{(Answer)}$$

Substituting P =102.86 kPa in Eq. (A), we get $x = 0.4286$ **(Answer)**

Example 5.5: For the *n*-hexane – ethanol system the liquid phase activity coefficients are represented by Wilson equations (*See* Appendix-A):

$$\ln \gamma_A = -\ln (x_A + \Lambda_{AB}\, x_B) + x_B \left[\frac{\Lambda_{AB}}{x_A + \Lambda_{AB}\, x_B} - \frac{\Lambda_{BA}}{\Lambda_{BA}\, x_A + x_B}\right]$$

$$\ln \gamma_B = -\ln (x_B + \Lambda_{BA}\, x_A) - x_A \left[\frac{\Lambda_{AB}}{x_A + \Lambda_{AB}\, x_B} - \frac{\Lambda_{BA}}{\Lambda_{BA}\, x_A + x_B}\right]$$

The Wilson parameters Λ_{AB} and Λ_{BA} are found to be 0.281 and 0.041, respectively. Assuming the parameters to be temperature independent, calculate the composition of the vapour in equilibrium with a solution containing 30 mol% *n*-hexane and the equilibrium temperature. The Antoine constants are

n-Hexane	13.8216	2697.55	48.78
Ethyl alcohol	16.5092	3578.91	50.50

***Solution*:** For a solution containing 30 mol% *n*-hexane (A) and 70 mol% ethanol (B), $x_A = 0.30$, $x_B = 0.70$. The activity coefficients of this solution are evaluated using Wilson equations. The Wilson parameters are given. $\Lambda_{AB} = 0.281$ and $\Lambda_{BA} = 0.041$. γ_A is the activity coefficient of *n*-hexane and γ_B is the activity coefficient of ethanol.

$$\ln \gamma_A = -\ln(x_A + \Lambda_{AB}x_B) + x_B\left[\frac{\Lambda_{AB}}{x_A + \Lambda_{AB}x_B} - \frac{\Lambda_{BA}}{\Lambda_{BA}x_A + x_B}\right]$$

$$\ln \gamma_B = -\ln(x_B + \Lambda_{BA}x_A) - x_A\left[\frac{\Lambda_{AB}}{x_A + \Lambda_{AB}x_B} - \frac{\Lambda_{BA}}{\Lambda_{BA}x_A + x_B}\right]$$

Substituting the values in the Wilson equations, we get

$$\ln \gamma_A = -\ln(0.30 + 0.281 \times 0.70) + 0.70\left[\frac{0.281}{0.30 + 0.281 \times 0.70} - \frac{0.041}{0.041 \times 0.30 + 0.70}\right] = 1.05549$$

$$\ln \gamma_B = -\ln(0.70 + 0.041 \times 0.30) - 0.30\left[\frac{0.281}{0.30 + 0.281 \times 0.70} - \frac{0.041}{0.041 \times 0.30 + 0.70}\right] = 0.1868$$

$$\gamma_A = \exp(1.05549) = 2.8734$$

$$\gamma_B = \exp(0.1868) = 1.2054$$

The vapour-liquid equilibrium is expressed by Eq. (5.30).

$$y_i \phi_i^V P = x_i \gamma_i P_i^S \quad \text{...(5.30)}$$

Since the pressure is low, the vapour phase may be assumed to behave as ideal gas. Therefore, $\phi_i^V = 1$ and Eq. (5.30) reduces to $y_i P = x_i \gamma_i P_i^S$.

For hexane and ethanol, Eq. (5.30) gives

$$y_A P = x_A \gamma_A P_A^S \quad \text{(A)}$$

$$y_B P = x_B \gamma_B P_B^S \quad \text{(B)}$$

Here y_A and y_B are, respectively the mole fractions of hexane and ethanol in the vapour, x_A and x_B are those in the liquid, P_A^S and P_B^S are the vapour pressures of hexane and ethanol respectively. Adding Eq. (A) and Eq. (B), we get

$$(y_A + y_B)P = (x_A \gamma_A P_A^S + x_B \gamma_B P_B^S) \quad \text{(C)}$$

Since $y_A + y_B = 1$, Eq. (C) becomes

$$P = (x_A \gamma_A P_A^S + x_B \gamma_B P_B^S) \quad \text{(D)}$$

Substitute the known values in Eq. (D).

$$101.3 = (0.3 \times 2.8734 P_A^S + 0.70 \times 1.2054 P_B^S)$$

$$101.3 = (0.8620 P_A^S + 0.8438 P_B^S) \quad \text{(E)}$$

We have to get the temperature such that Eq. (A) is satisfied. This requires a trial-and-error calculation in which we assume arbitrarily a temperature, calculate P_A^S and P_B^S using the Antoine equations

$$\ln P_A^S = 13.8216 - \frac{2697.55}{T - 48.78}$$

$$\ln P_B^S = 16.5092 - \frac{3578.91}{T - 50.50}$$

and check whether Eq. (E) is satisfied. The results of the calculations for $T = 332$ K and $T = 333$ K are given below.

T	P_A^S	P_B^S	$(0.8620P_A^S + 0.8438P_B^S)$
332	73.4809	44.49994	100.8896
333	75.98506	46.54837	104.7766

We can see that the equilibrium temperature is between 332 K and 333 K.

At $T = 332.11$ K

$P_A^S = 73.7531$ kPa

$P_B^S = 44.7215$ kPa

$0.8620P_A^S + 0.8438P_B^S = 101.31$. The equilibrium temperature is 332.11 K.

From Eq. (A), $y_A = \dfrac{x_A\gamma_A P_A^S}{P} = \dfrac{0.30 \times 2.8734 \times 73.7531}{101.3} = 0.6276$

The mole fraction of n-hexane in the vapour = 0.6276

Thus $x = 0.30$, $y = 0.6276$, $T = 332.11$ K **(Answer)**

Example 5.6: At a pressure of 101.3 kPa, the acetone (1) - chloroform (2) system forms an azeotrope of composition 34.5 mol% acetone which boils at 337.7 K. The van Laar equations (*See* Appendix A) are given below which can be used to estimate the activity coefficients for the system:

$$\ln \gamma_1 = \frac{A_{12}x_2^2}{\left[(A_{12}/A_{21})x_1 + x_2\right]^2}$$

$$\ln \gamma_2 = \frac{A_{21}x_1^2}{\left[x_1 + (A_{21}/A_{12})x_2\right]^2}$$

Determine the constants in the van Laar equations. The vapour pressure of acetone at 337.7 K is 134.04 kPa and that of chloroform is 112.67 kPa.

***Solution*:** The vapour-liquid equilibrium is expressed by Eq. (5.30).

$$y_i\phi_i^V P = x_i\gamma_i P_i^S \quad \text{...(5.30)}$$

where γ_i is the activity coefficient of component i in the liquid. Since the pressure is low, the vapour phase may be assumed to behave as ideal gas. Therefore, $\phi_i^V = 1$ and Eq. (5.30) reduces to

$$y_i P = x_i\gamma_i P_i^S \quad \text{(A)}$$

The activity coefficient in Eq. (A) can be evaluated by the van Laar equations.

$$\ln \gamma_1 = \frac{A_{12}x_2^2}{\left[(A_{12}/A_{21})x_1 + x_2\right]^2}$$

$$\ln \gamma_2 = \frac{A_{21} x_1^2}{\left[x_1 + (A_{21}/A_{12})\, x_2\right]^2} \tag{B}$$

To use this equation for determining the activity coefficients, we should know the van Laar constants A_{12} and A_{21}. The constants can be evaluated if one set of γ_1 and γ_2 values are known. It is for finding these constants that we make use of the characteristics of the azeotropic state. For the azeotrope, the vapour and liquid have same composition or $x_i = y_i$. Therefore, Eq. (A) becomes $P = \gamma_i P_i^S$ at the azeotropic point. Therefore $\gamma_i = \dfrac{P}{P_i^S}$. Representing acetone as component 1 and chloroform as component 2, we have

$$\gamma_1 = \frac{P}{P_1^S} = \frac{101.3}{134.04} = 0.7557$$

$$\gamma_2 = \frac{P}{P_2^S} = \frac{101.3}{112.67} = 0.8991$$

with $x_1 = 0.345$ and $x_2 = 0.655$ at azeotropic point. Substituting these in van Laar equations we get

$$\ln (0.7557) = \frac{A_{12}\,(0.655)^2}{\left[(A_{12}/A_{21})\, 0.345 + 0.655\right]^2}$$

$$\ln (0.8991) = \frac{A_{21}\,(0.345)^2}{\left[\,0.345 + (A_{21}/A_{12})\, 0.655\right]^2}$$

These equations are solved simultaneously to get the van Laar constants:

$A_{12} = -0.8298$ and $A_{21} = -0.6060$ **(Answer)**

Example 5.7: A hydrocarbon mixture contains 20 mol% propane, 35 mol% *n*-butane, 30 mol% *n*-pentane and 15 mol% *n*-hexane at 2000 kPa. Calculate:

(a) The bubble-point temperature and composition of the liquid
(b) The dew-point temperature and the composition of the vapour

***Solution*:** (a) To find the bubble point of the mixture, a temperature is assumed, the K_i values at this temperature are obtained from standard nomographs. y_i is calculated using $y_i = K_i x_i$. If the assumed temperature is correct, then

$$\Sigma y_i = \Sigma K_i x_i = 1 \tag{5.40}$$

At $T = 388.8$ K, $\Sigma K_i x_i = 0.949$ and at $T = 394.3$ K, $\Sigma K_i x_i = 1.019$ as shown in the table below.

Component, i	*Mole fraction, x_i*	$T = 388.8$ K		$T = 394.3$ K		$T = 392.9$ K	
		K_i	$y_i = K_i x_i$	K_i	$y_i = K_i x_i$	K_i	$y_i = K_i x_i$
Propane	0.20	2.20	0.440	2.30	0.460	2.28	0.456
n-Butane	0.35	0.95	0.333	1.05	0.368	1.02	0.357
n-Pentane	0.30	0.46	0.138	0.50	0.150	0.49	0.147
n-Hexane	0.15	0.25	0.038	0.27	0.041	0.265	0.040
	$\Sigma y_i = \Sigma K_i x_i =$		0.949		1.019		1.000

It means that the bubble-point temperature lies between 388.8 K and 394.3 K. By interpolation, this temperature is found to be 392.9 K. The K– values at this temperature and $y_i = K_i x_i$ are shown in the table, in the last two columns. Since at 392.9 K, $\Sigma y_i = \Sigma K_i x_i = 1$, the bubble point is 392.9 K. The vapour-phase composition is given in the last column.

Vapour-phase in equilibrium with the given liquid contains 45.6% propane, 35.7% *n*-butane, 14.7% *n*-pentane and 4.0% *n*-hexane **(Answer)**

(b) To find the dew point, a temperature is assumed arbitrarily and K_i is determined. At the dew point,

$$\Sigma x_i = \Sigma \frac{y_i}{K_i} = 1.00 \qquad \ldots(5.42)$$

It is seen that

at temperature 422.1 K, $\Sigma x_i = \Sigma \frac{y_i}{K_i} = 1.031$

and at 427.6 K $\Sigma x_i = \Sigma \frac{y_i}{K_i} = 0.952$,

indicating that the dew point is in between these two temperatures. By interpolation it is found to be 424.3 K. The liquid compositions computed at this temperature add up to 1 as shown in the following table.

Component, i	*Mole fraction, y_i*	*T* = 422.1 K		*T* = 427.6 K		*T* = 424.3 K	
		K_i	$x_i = y_i/K_i$	K_i	$x_i = y_i/K_i$	K_i	$x_i = y_i/K_i$
Propane	0.20	3.00	0.067	3.20	0.063	3.08	0.065
n-Butane	0.35	1.48	0.236	1.60	0.219	1.53	0.229
n-Pentane	0.30	0.76	0.395	0.84	0.357	0.79	0.380
n-Hexane	0.15	0.45	0.333	0.48	0.313	0.46	0.326
	$\Sigma x_i = \Sigma \frac{y_i}{K_i} =$		1.031		0.952		1.000

The dew point is 424.3 K and the liquid in equilibrium at the dew point contains 6.5 mol% propane, 22.9 mol% *n*-butane, 38.0 mol% *n*-pentane and 32.6 mol% *n*-hexane. **(Answer)**

5.3 DISTILLATION METHODS: CLASSIFICATION

In distillation of binary solutions, the feed mixture is generally separated into at least two products: the first is a distillate product which is richer in more volatile component than the feed. The second product is called the bottoms or residue which is leaner in more volatile component than the original feed. The methods available for distillation can be classified into two broad categories. The first one is distillation without rectification and the second is fractionation. In the first method, the vapour produced by boiling the liquid is recovered as distillate product after condensation. No part of the condensate is returned back to the still for enriching the vapours. Flash vaporisation, differential distillation and simple steam distillation are examples for this type of distillation. In the second method, the vapours are condensed and a part of the condensate known as reflux is returned back to the still. The reflux flows in contact with the vapours, but in the opposite direction, thus enabling the vapours to get enriched in more volatile component due to mass transfer. This type of distillation is known as rectification, fractional distillation or simply fractionation.

5.4 FLASH VAPORISATION

When a saturated liquid at a relatively high pressure is subjected to a reduction in pressure, the liquid gets partially vaporised with components differently distributed between the vapour and liquid products. This is the principle behind flash vaporisation or flash distillation. Flash distillation is also called equilibrium distillation because, in an ideal flash distillation process, the vapour and liquid products are in equilibrium.

Flash distillation is essentially a single stage operation and can be carried out in batch or continuous fashion. It involves three steps. They are:

(1) Increasing the temperature and pressure of the feed to the desired level.

(2) Reducing the pressure across a valve.

(3) Separating the liquid and vapour products.

A typical continuous flash distillation unit is schematically represented in Fig. 5.12. The feed is heated under pressure by passing through a conventional tubular heat exchanger or through the tubes of a fuel fired furnace. The pressure is then reduced by passing across a valve or some other pressure reducing devices. The reduction in pressure causes the liquid to get partially vaporised. As the operation is adiabatic, the required heat of vaporisation is extracted from the sensible heat of the total material and as a result, the mixture gets cooled during flashing. The liquid and vapour formed are allowed to be in intimate contact for sufficiently long time so that equilibrium is established between the phases before they are separated in a liquid-vapour separator. The vapour product leaving the liquid-vapour separator may be taken to a condenser and the condensate collected as distillate. The liquid leaves at the bottom of the separator.

Unless the relative volatility is very large, only limited separation is achieved by flash vaporisation. Therefore, it is not as effective for the separation of binary mixtures as for the separation of multi-component solutions. Flash distillation is very common in oil refineries. Crude oil is first heated in pipe stills and the superheated liquid is then subjected to pressure reduction and flashed under equilibrium conditions. Liquid leaving absorption columns may be flashed to recover the dissolved solute; the liquid from a high pressure reactor may be flashed to vaporise a portion of the liquid. Even in cases where separation is achieved by means of other techniques, a 'pre-flash' is frequently employed in industries as an auxiliary operation to reduce the load on the separation units.

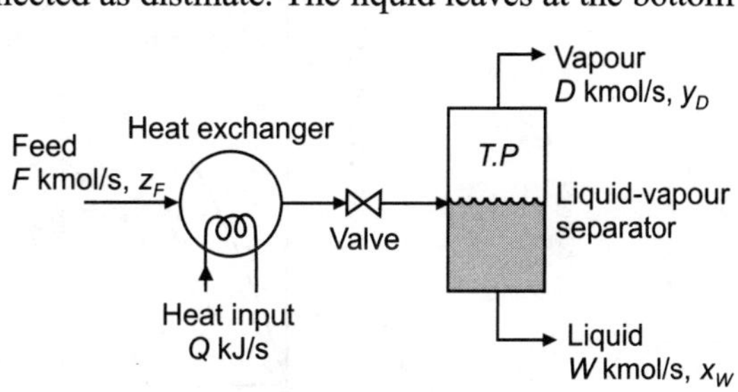

Fig. 5.12: Flash distillation

Consider a feed consisting of two components **A** and **B** subjected to flash distillation at a rate F kmol/s. z_F is the mole fraction of more volatile component (**A**) in the feed. A portion of the feed is vaporised and the vapour portion of concentration y_D mole fraction **A** is removed at the top at a rate D kmol/s. The liquid rate is W kmol/s and its composition is x_W. The temperature and pressure in the separator are T and P respectively, and under these conditions, x_W and y_D are the concentrations of the liquid and vapour phases in equilibrium.

The total material balance for the flash distillation unit can be written as

$$F = D + W \qquad \text{...(5.43)}$$

Component-**A** balance gives

$$Fz_F = Dy_D + Wx_W \qquad \text{...(5.44)}$$

Eliminating F from Equations (5.43) and (5.44), we get

$$(D + W)z_F = Dy_D + Wx_W \qquad \text{...(5.45)}$$

Equation (5.45) may be rearranged as

$$\frac{y_D - z_F}{x_W - z_F} = -\frac{W}{D} \qquad \text{...(5.46)}$$

Let f denote the fractional vaporisation, i.e. the fraction of the feed that has vaporised. Then $D = fF$ and $W = F - D = (1 - f)F$. Using these in Eq. (5.46), we get

$$\frac{y_D - z_F}{x_W - z_F} = -\frac{W}{D} = -\frac{(1-f)}{f} \qquad \text{...(5.47)}$$

Given the feed composition and the degree of vaporisation, Eq. (5.47) may be used to determine the composition of the vapour and liquid products, y_D and x_W. If the equilibrium relationship is known, then $y_D = f(x_W)$ and Eq. (5.47) can be solved analytically to determine the product compositions. Otherwise, the product compositions can be determined graphically.

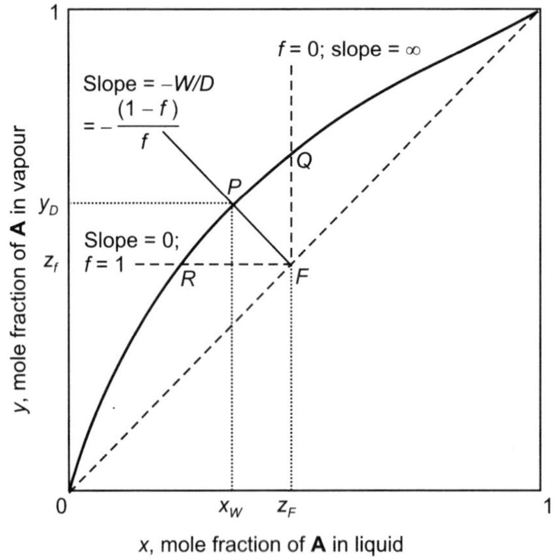

Fig. 5.13: Flash vaporisation represented on equilibrium diagram

Figure 5.13 shows the equilibrium curve plotted on *y*-*x* coordinates. The left-hand side of Eq. (5.47) is the slope of a straight line joining points (x_W, y_D) and (z_F, z_F). The point with coordinates (z_F, z_F) is marked *F* on the diagram and it lies on the 45° diagonal. Since x_W and y_D are concentrations of the liquid and vapour phases in equilibrium, the point *P*, having coordinates x_W and y_D should lie on the equilibrium diagram. Equation (5.47) means that a straight line drawn from point *F* with slope $-\frac{W}{D}$ or $-\frac{(1-f)}{f}$ will intersect the equilibrium curve at point *P*, the coordinates of which are x_W and y_D. The line *FP* is the *operating line* for flash vaporisation, which is a single stage operation. Thus, knowing the feed composition and fractional vaporisation, the compositions of the liquid and vapour products can be determined graphically. Alternatively, the amount of liquid to be vaporised from a given quantity of feed (or the extent of vaporisation) for producing a distillate of known composition y_D (or a liquid residue of known composition x_W) can be calculated from the slope of the operating line *PF*.

It is obvious from Eq. (5.47) that the compositions of the products depend on the extent of vaporisation, which in turn depends on the pressure and temperature of flashing. We have seen that at a given pressure, the two phases can exist in equilibrium at temperatures ranging from the bubble point to the dew point of the mixture. If flashing is done at the bubble point, the vapour produced, though infinitesimal in amount, is the richest in the more volatile component. Fractional vaporisation $f = 0$, slope of the operating line is infinity and the line is vertical as shown in Fig. 5.13. The intersection of this line with the equilibrium curve is *Q*, the ordinate of which gives the vapour composition under this condition and this is the richest vapour that can be produced under a given pressure. In a similar manner, at the dew point of the mixture, flashing results in infinitesimally small amounts of the liquid since $f = 1$. The slope of the operating line is zero and the line is horizontal as shown by *FR* in the figure. The leanest liquid is produced at the dew point as shown by the abscissa of point *R*. Depending upon the degree of vaporisation, the composition of the actual products will lie between the limits set by the vertical and horizontal operating lines.

The analysis presented here for flash vaporisation can be applied to partial condensation as well. Here the feed which is in the vapour state is cooled by passage through a heat exchanger whereby it is partially condensed and the vapour and liquid products which are in equilibrium, are separated.

Multicomponent solutions

The flash vaporisation is important in the separation of multicomponent solutions. The general flash vaporisation problem can be stated thus: Given a mixture of known overall composition z_i at temperature T and pressure P, what is the fraction that is vapour and what are the composition of the liquid and vapour phases in equilibrium?

The overall material balance for the system is

$$F = D + W \qquad \text{...(5.48)}$$

where F is the total number of moles of the initial mixture and D and W are the moles of vapour and liquid products in equilibrium. The relationship between the composition of the liquid and vapour phases in equilibrium is described by means of the vaporisation equilibrium constant K_i.

$$y_i = K_i x_i \qquad \text{...(5.49)}$$

The component-i balance for the system is

$$Fz_i = Dy_i + Wx_i \qquad \text{...(5.50)}$$

Using Eq. (5.49) x_i can be eliminated from Eq. (5.50) to get the following:

$$Fz_i = Dy_i + W\frac{y_i}{K_i} \qquad \text{...(5.51)}$$

$$y_i = \frac{F}{D}\left(\frac{z_i}{1 + W/(DK_i)}\right) \qquad \text{...(5.52)}$$

As $\quad \Sigma y_i = 1,$

$$\Sigma y_i = \Sigma \frac{F}{D}\left(\frac{z_i}{1 + W/(DK_i)}\right) = 1 \qquad \text{...(5.53)}$$

Or

$$\Sigma \frac{z_i}{1 + W/(DK_i)} = \frac{D}{F} \qquad \text{...(5.54)}$$

Equation (5.54) can be utilized in an iterative procedure to estimate T, P or the fraction of the initial mixture that is vaporised.

Alternatively, y_i can be eliminated from Eq. (5.50) which leads to

$$x_i = \frac{Fz_i}{K_i D + W} = \frac{F}{D}\left(\frac{z_i}{K_i + W/D}\right) \qquad \text{...(5.55)}$$

Since $\Sigma x_i = 1$, we have

$$\Sigma \frac{z_i}{K_i + W/D} = \frac{D}{F} \qquad \text{...(5.56)}$$

The above equation can also be used for the calculation of T, P or fractional vaporisation iteratively.

Example 5.8: A mixture containing 60 mol% **A** and the rest **B** is flash vaporised at 101.3 kPa until 40% of the original mixture is vaporised. If the relative volatility of the mixture is 2.5 and is expected to remain constant, determine the composition of the products.

***Solution*:** Equation (5.47) is the material balance equation for flash distillation:

$$\frac{y_D - z_F}{x_W - z_F} = -\frac{W}{D} = -\frac{(1-f)}{f} \qquad \ldots(5.47)$$

Here $z_F = 0.60, f = 0.40$ so that

$$\frac{y_D - 0.6}{x_W - 0.6} = -\frac{(1-0.4)}{0.4} = -\frac{3}{2} \qquad \text{(A)}$$

Since the products of flash distillation are in equilibrium, x_W and y_D are equilibrium compositions. The equilibrium relationship can be written in terms of relative volatility α, since α is given to be constant. By Eq. (5.14),

$$y = \frac{\alpha x}{1+(\alpha-1)x}$$

Substituting $\alpha = 2.5$ into the above we have

$$y = \frac{2.5\,x}{1+1.5x}$$

Since x_W and y_D are equilibrium compositions

$$y_D = \frac{2.5\,x_W}{1+1.5x_W} \qquad \text{(B)}$$

Substituting Eq. (B) into Eq. (A)

$$\frac{\dfrac{2.5\,x_W}{1+1.5x_W} - 0.6}{x_W - 0.6} = -\frac{3}{2} \qquad \text{(C)}$$

On solving Eq. (C) we get $x_W = 0.5155$.

Substituting $x_W = 0.5155$ into Eq. (B), $y_D = 0.7268$

Distillate contains 72.68 mol% **A** and the residue contains 51.55 mol% **A**. **(Answer)**

Example 5.9: A mixture consisting of 40 mol% benzene and the rest toluene is subjected to flash distillation at 101.3 kPa at a rate 100 kmol/h so that one-third of the feed mixture is vaporised. The equilibrium data for benzene-toluene system at 101.3 kPa:

T, K	383.8	379.6	375.5	371.8	368.4	365.7	362.7	360.2	357.8	355.6	354.4	353.3
x	0	0.1	0.2	0.3	0.4	0.5	0.6	0.7	0.8	0.9	0.95	1
y	0	0.205	0.369	0.502	0.618	0.706	0.789	0.849	0.907	0.955	0.978	1

Determine the following:

(a) Flow rates of distillate and residue.
(b) Composition of the products.
(c) Temperature of flash vaporisation.
(d) The fraction of the feed that is vaporised, if the residue composition is 25%.

***Solution*:** **Basis:** 100 kmol/h of feed. $F = 100$ kmol/h.

(a) When one-third of the feed mixture is vaporised, fractional vaporisation, $f = 1/3$

$$D = F \times 1/3 = 33.33 \text{ kmol/h} \qquad \textbf{(Answer)}$$

$$W = F - D = 100 - 33.33 = 66.67 \text{ kmol/h} \qquad \textbf{(Answer)}$$

(b) Feed consists of 40 mol% benzene and 60 mol% toluene. Therefore $z_F = 0.40$. We have Eq. (5.47) to describe the material balance for flash distillation.

$$\frac{y_D - z_F}{x_W - z_F} = -\frac{W}{D} = -\frac{(1-f)}{f} \qquad \ldots(5.47)$$

According to Eq. (5.47), a straight line with slope $-\frac{W}{D} = -\frac{(1-f)}{f}$ drawn from point (z_F, z_F) will intersect the equilibrium curve at (x_W, y_D) where x_W and y_D are, respectively, the compositions of the residue and distillate.

The equilibrium curve and the 45° diagonal are plotted and the point (z_F, z_F) is marked on the diagonal. The slope of the operating line is

$$-\frac{W}{D} = -\frac{(1-f)}{f} = -\frac{2/3}{1/3} = -2.0$$

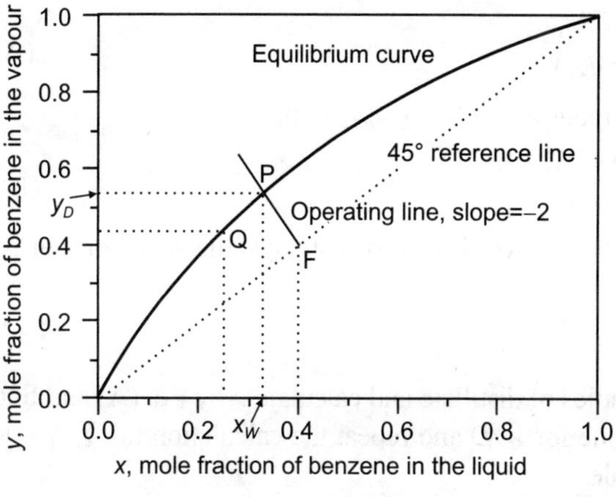

Fig. 5.14: Flash distillation

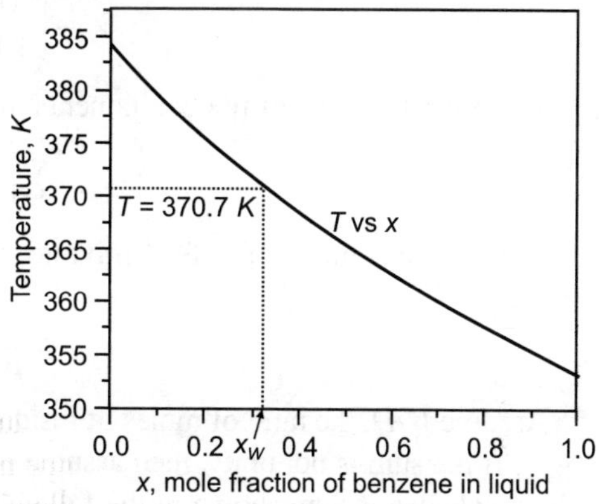

Fig. 5.15: Estimate of temperature of flash vaporisation in Example 5.9

Operating line with slope –2.0 is drawn from point F (0.4, 0.4) as shown in Fig. 5.14. The intersection of this line with the equilibrium curve is at P (0.330, 0.539).

Composition of the residue, $x_W = 0.330$ **(Answer)**

Composition of the distillate, $y_D = 0.539$ **(Answer)**

(c) Temperature of flash vaporisation is the temperature at which the liquid ($x = 0.330$) and the vapour ($y = 0.539$) exist in equilibrium. The temperature versus x curve is plotted as shown in Fig. 5.15. The temperature corresponding to $x = 0.33$ is determined. From the graph we see that the temperature of flash vaporisation is 370.7 K. **(Answer)**

(d) When the liquid residue composition is 25 mol% benzene, $x_W = 0.25$. Since the distillate is in equilibrium with liquid of composition 0.25, the coordinates of point Q lying on the equilibrium curve with $x = 0.25$ represent the product composition. y-coordinate of Q from the figure is $y_D = 0.431$. Slope of the new operating line FQ gives $-W/D$ from which the number of moles to be vaporised can be determined. Alternatively,

$$\frac{y_D - z_F}{x_W - z_F} = -\frac{W}{D} = -\frac{(1-f)}{f}$$

$$\frac{0.431 - 0.40}{0.25 - 0.40} = -0.2067 = -\frac{(1-f)}{f}$$

This is solved and the fractional vaporisation is determined as $f = 0.8287$ which means 82.87% of the feed is to be vaporised in order that the liquid residue composition is 25 mol% benzene.

(Answer)

Example 5.10: A mixture containing 40 mol% benzene, 35 mol% toluene and 25 mol% ethyl benzene is partially vaporised at 373 K and 90 kPa pressure. The vapour pressures at 373 K are: benzene = 179.06 kPa, toluene = 73.72 kPa, ethyl benzene = 34.02 kPa. Determine the fraction of the feed vaporised and composition of the vapour and liquid products.

***Solution*:** Equation (5.52) relates the composition of the vapour to the amounts of distillate and residue, the composition of the feed mixture and the *K*-factors.

$$y_i = \frac{F}{D}\left(\frac{z_i}{1 + W/(DK_i)}\right) \quad \text{...(5.52)}$$

Since $F = D + W, F/D = 1 + W/D$, Eq. (5.52) can be written as

$$y_i = \left(\frac{(1 + W/D)z_i}{1 + W/(DK_i)}\right) \quad \text{(A)}$$

The mole fractions of the components in the vapour add to 1, $\sum y_i = 1$, or

$$\sum y_i = \sum\left(\frac{(1 + W/D)z_i}{1 + W/(DK_i)}\right) = 1 \quad \text{(B)}$$

z_i is the composition of the feed mixture which are known. For ideal solution, the *K*-factor can be written as

$$K_i = \frac{P_i^s}{P} \quad \text{...(5.37)}$$

Assume *W*/*D*, the ratio of moles of residue to moles of distillate and calculate y_i by Eq. (A) and find $\sum y_i$. If the sum is not unity, then assume new value for *W*/*D* and repeat the calculation till $\sum y_i = 1$. These calculations are shown in the following table.

Component	z_i	P_i^S	$K_i = \frac{P_i^s}{P}$	$y_i = \left(\frac{(1+W/D)z_i}{1+W/(DK_i)}\right)$		
				W/*D* = 1.4	*W*/*D* = 1.3	*W*/*D* = 1.371
				y_i	y_i	y_i
Benzene	0.40	179.06	1.9896	0.5635	0.5564	0.5615
Toluene	0.35	73.72	0.8191	0.3101	0.3112	0.3104
Ethyl benzene	0.25	34.02	0.3780	0.1276	0.1295	0.1281
				1.0011	0.9971	1.0000

The ratio of the amount of residue to distillate *W*/*D* = 1.371

$$\frac{W}{D} + 1 = \frac{W + D}{D} = \frac{F}{D} = 1.371 + 1 = 2.371$$

$$\frac{D}{F} = \frac{1}{2.371} = 0.4218$$

Fraction of the feed vaporised = 0.4218. **(Answer)**

The vapour composition is (y_i) given in the last column of the above table. The liquid composition is calculated as $x_i = y_i/K_i$. The results are given below:

Component	$K_i = \dfrac{P_i^s}{P}$	y_i	$x_i = y_i/K_i$
Benzene	1.9896	0.5615	0.2822
Toluene	0.8191	0.3104	0.3790
Ethyl benzene	0.3780	0.1281	0.3388
		1.0000	1.0000

(Answer)

Example 5.11: A hydrocarbon mixture containing 20 mol% propane, 35 mol% *n*-butane, 30 mol% *n*-pentane and 15 mol% *n*-hexane is subjected to flash vaporisation at 2000 kPa. Calculate the temperature and the composition of the liquid and vapour products when 40 mol% of the initial mixture is vaporised.

***Solution*:** Equation (5.54) relates the fraction of the mixture of given composition that is in the vapour phase to the composition and the *K*-values.

$$\sum \frac{z_i}{1+W/(DK_i)} = \frac{D}{F} \quad \ldots(5.54)$$

Assume that $F = 100$ mol; since 40 mole percent of the mixture is vaporised, $D = 40$ mol and $W = 60$ mol. Equation (5.54) becomes

$$\sum \frac{z_i}{1+60/(40K_i)} = \frac{40}{100}$$

$$\sum \frac{z_i}{1+1.5/K_i} = 0.40 \quad \text{(A)}$$

A temperature is assumed and *K*-values determined at this temperature. The term on the left-side of Eq. (A) is evaluated for each component and their sum is determined. If the sum is 0.4, the assumed temperature can be taken as the equilibrium temperature. Otherwise, the calculations are repeated till the desired result is obtained.

At $T = 394.3$ K, $\dfrac{z_i}{1+1.5/K_i} = 0.363 < 0.4$; At $T = 422.1$ K, $\dfrac{z_i}{1+1.5/K_i} = 0.443 > 0.4$. The equilibrium temperature therefore lies between 394.3 K and 422.1 K.

At $T = 405.4$ K, it is found that $\dfrac{z_i}{1+1.5/K_i} = 0.40$; The equilibrium temperature is 405.4 K

(**Note:** A rough idea about the equilibrium temperature is derived from the knowledge of the bubble point and dew point of the mixture, which are already calculated in Example 5.7.)

Component, i	Mole fraction, z_i	$T = 394.3$ K		$T = 422.1$ K		$T = 405.4$ K	
		K_i	$\dfrac{z_i}{1+1.5/K_i}$	K_i	$\dfrac{z_i}{1+1.5/K_i}$	K_i	$\dfrac{z_i}{1+1.5/K_i}$
Propane	0.20	2.3	0.121	3	0.133	2.58	0.126
n-Butane	0.35	1.05	0.144	1.48	0.174	1.3	0.162
n-Pentane	0.30	0.5	0.075	0.76	0.101	0.6	0.086
n-Hexane	0.15	0.27	0.023	0.45	0.035	0.32	0.026
		$\sum \dfrac{z_i}{1+1.5/K_i} =$	0.363		0.443		0.400

Equation (5.52) can be used to calculate the vapour composition.

$$y_i = \frac{F}{D}\left(\frac{z_i}{1+W/(DK_i)}\right) \quad \text{...(5.52)}$$

$$y_i = \frac{100}{40}\left(\frac{z_i}{1+1.5/K_i}\right)$$

Multiplying the results in the last column by $F/D = 100/40 = 2.5$, we get the mole fraction of each component in the vapour.

The mole fraction in the liquid is $x_i = y_i/K_i$. Alternatively, it can be calculated from the component balance between the feed, vapour and the liquid.

$$Fz_i = Dy_i + Wx_i$$

$$x_i = \frac{Fz_i - Dy_i}{W} = \frac{100z_i - 40y_i}{60}$$

For example, the mole fraction of propane in the vapour is

$$y_i = 2.5\left(\frac{z_i}{1+1.5/K_i}\right) = 2.5 \times 0.126 = 0.315$$

Mole fraction of propane in the liquid is

$$x_i = \frac{100z_i - 40y_i}{60} = \frac{100 \times 0.20 - 40 \times 0.315}{60} = 0.123$$

The y_i and x_i are calculated for other components as well and the results are summarized below:

Component, i	Mole fraction, z_i	T = 405.4 K		$y_i = 2.5\left(\frac{z_i}{1+1.5/K_i}\right)$	$x_i = \frac{100z_i - 40y_i}{60}$
		K_i	$\frac{z_i}{1+1.5/K_i}$		
Propane	0.20	2.58	0.126	0.315	0.123
n-Butane	0.35	1.30	0.162	0.405	0.313
n-Pentane	0.30	0.60	0.086	0.215	0.357
n-Hexane	0.15	0.32	0.026	0.065	0..207
			0.400	1.000	1.000

The vapour product contains 31.5% propane, 40.5% *n*-butane, 21.5% *n*-pentane and 6.5% *n*-hexane. The liquid contains 12.3% propane, 31.3% *n*-butane, 35.7% *n*-pentane and 20.7% *n*-hexane. The equilibrium temperature = 405.4 K. **(Answer)**

5.5 SIMPLE DISTILLATION

Simple distillation is also known as differential distillation. Laboratory distillation carried out without reflux is an example of simple distillation. An important industrial application of differential distillation is the batch type of petroleum stills called topping stills. Simple distillation can be carried out either as batch operation or as continuous operation. A batch distillation unit is shown in Fig. 5.16. The mixture to be separated is charged into a still. The charge is heated by passing steam through the jackets or by other means. The vapours produced are withdrawn as rapidly as they form and are condensed by passing through a condenser. The condensate known as distillate is collected cumulatively in the receivers. The distillation is stopped when separation to the desired extent is achieved as determined by the average composition or the amount of the distillate collected. The liquid residue remaining in the still is richer in less volatile component and the distillate collected in the receivers is richer in the more volatile component.

Simple distillation is not very effective for sharp separation of the components. However, the method finds application when the components to be separated have widely different boiling points or when a sharp separation is not desired.

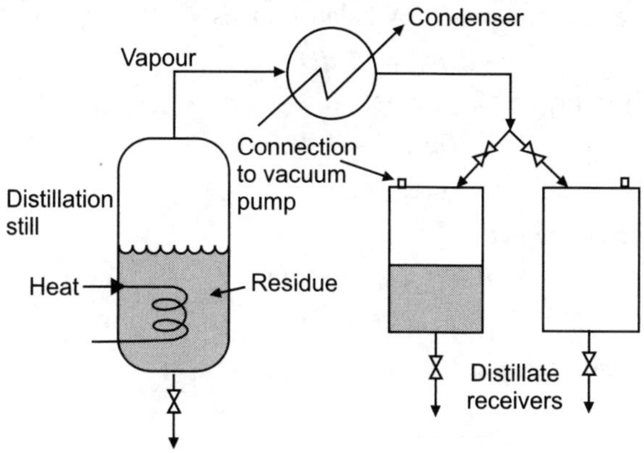

Fig. 5.16: Differential distillation

In simple distillation, the composition of the vapour is not constant; the concentration of the more volatile component in the vapours decreases continuously as distillation progresses. The vapours coming out first are the richest in the more volatile component than the vapours evolved subsequently. Since the vapours are removed and condensed as they are formed, and collected in the same receiver, the composition of the distillate progressively diminishes. Therefore, the composition of the distillate means the average of the compositions of the condensate collected over the given time interval. The liquid in the still gets leaner in more volatile component with progress of distillation; therefore the temperature of distillation is not constant but increases continuously. Instead of collecting the distillate in a single reservoir, it is possible to collect distillate batches or cuts of different composition in separate receivers as shown schematically in Fig. 5.16.

The essential difference between flash distillation and simple distillation is that in the former, the vapour and liquid products attain equilibrium compositions and the temperature remains constant, whereas in the latter, the vapour in equilibrium with the liquid residue in the still will have composition different from the average composition of the distillate. However, the vapours evolved in an infinitesimally short period during distillation will be in equilibrium with the liquid in the still and the process occurring in this short interval is equivalent to flash vaporisation. Therefore, simple distillation can be treated as a series of flashing operations, each producing infinitesimal quantities of vapours which are in equilibrium with the liquid. This model makes simple distillation amenable to analytical treatment as explained below.

Binary Distillation: The Rayleigh Equation

Let F be the moles of feed charged into the kettle and x_F be the mole fraction of more volatile component (**A**) in the feed. The mixture is heated and vapours produced are taken to the condenser as fast as they are formed and the condensate is collected as the distillate product. Finally there remains in the kettle, W moles of liquid with concentration of component **A** equal to x_W. At any given time, let the total moles of liquid in the still be n and x be the mole fraction of **A** in it. Let dD be the moles of vapour produced during a short time interval and let y^* be the concentration of **A** in the vapour. The basic assumption in deriving the Rayleigh equation is that the vapour produced at any instant is in equilibrium with the liquid in the still and therefore x and y^* are equilibrium compositions. Since the moles of distillate produced is equal to the moles of liquid vaporised, the total material balance can be written as

$$-dn = dD \qquad \text{...(5.57)}$$

The rate of disappearance of component-**A** from the liquid is equal to the rate at which **A** is added to the distillate. Therefore, the component-**A** balance gives

$$-d(nx) = y^* dD \qquad \text{...(5.58)}$$

Substituting Eq.(5.57) in Eq. (5.58), we get

$$-d(nx) = -y^* dn \qquad \text{...(5.59)}$$

Or

$$d(nx) = y^* dn \qquad \text{...(5.60)}$$

On expanding Eq. (5.60), we get

$$n\,dx + x\,dn = y^* dn \qquad \text{...(5.61)}$$

$$n\,dx = (y^* - x)dn \qquad \text{...(5.62)}$$

Equation (5.62) may be rearranged as

$$\frac{dn}{n} = \frac{dx}{y^* - x} \qquad \text{...(5.63)}$$

Equation (5.63) can be integrated knowing that initially $n = F$ and $x = x_F$ and finally $n = W$ and $x = x_W$.

$$\int_F^W \frac{dn}{n} = \int_{x_F}^{x_W} \frac{dx}{y^* - x} \qquad \text{...(5.64)}$$

That is,

$$\ln\frac{W}{F} = \int_{x_F}^{x_W} \frac{dx}{y^* - x} \qquad \text{...(5.65)}$$

Or

$$\ln\frac{F}{W} = \int_{x_W}^{x_F} \frac{dx}{y^* - x} \qquad \text{...(5.66)}$$

Equation (5.66) is known as the Rayleigh equation. The equation involves four variables F, W, x_F and x_W. Rayleigh equation can be used to determine any one of these variables knowing the other three. The integral in Eq. (5.66) can be evaluated graphically. Using the equilibrium data, the reciprocal of $(y^* - x)$ is plotted on the y-axis with x on the x-axis. The area under the resulting curve between limits x_W and x_F is the required integral (Fig. 5.17).

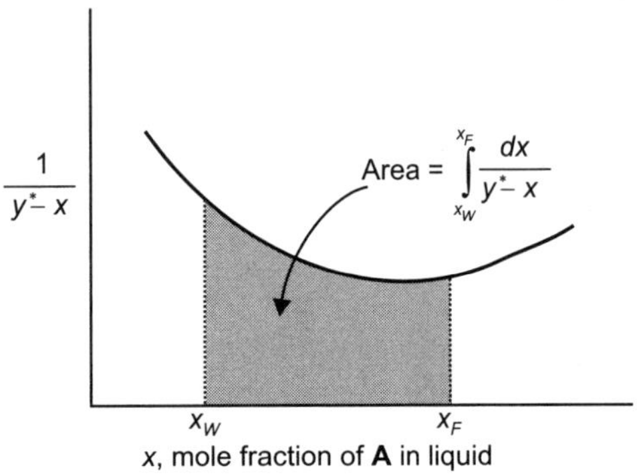

Fig. 5.17: Graphical integration for evaluating the integral in Rayleigh equation

The average distillate composition $y_{D,av}$ can be determined using the following material balance equations:

$$F = D + W \qquad \text{...(5.67)}$$

$$Fx_F = Dy_{D,av} + Wx_W \qquad \text{...(5.68)}$$

Special cases (a) *When the equilibrium curve is a straight line*: If the equilibrium relationship is linear over the concentration region of interest,

$$y^* = mx + c \qquad \text{...(5.69)}$$

integral in Eq. (5.66) can be determined analytically.

$$\ln\frac{F}{W}=\int_{x_W}^{x_F}\frac{dx}{y^*-x}=\int_{x_W}^{x_F}\frac{dx}{(m-1)x+c} \quad ...(5.70)$$

On integration and substituting the limits, we get

$$\ln\frac{F}{W}=\frac{1}{(m-1)}\ln\frac{(m-1)x_F+c}{(m-1)x_W+c} \quad ...(5.71)$$

Equation (5.71) can be rearranged as

$$\frac{F}{W}=\left[\frac{(m-1)x_F+c}{(m-1)x_W+c}\right]^{\frac{1}{m-1}}=\left[\frac{(mx_F+c)-x_F}{(mx_W+c)-x_W}\right]^{\frac{1}{m-1}} \quad ...(5.72)$$

Equation (5.72) may be equivalently written as

$$\frac{F}{W}=\left[\frac{y_F^*-x_F}{y_W^*-x_W}\right]^{\frac{1}{m-1}} \quad ...(5.73)$$

or as

$$\frac{y_F^*-x_F}{y_W^*-x_W}=\left(\frac{F}{W}\right)^{m-1} \quad ...(5.74)$$

y_F^* and y_W^* appearing in Eqs. (5.73) and (5.74) are concentration of fictitious vapour phases which are in equilibrium with the feed and residue, respectively.

(b) *When relative volatility is constant*: When an average relative volatility α is found to be valid for the system, then the equilibrium relationship according to Eq. (5.14) is

$$y^*=\frac{\alpha x}{1+(\alpha-1)x} \quad ...(5.14)$$

Substituting Eq. (5.14) in Eq. (5.66), we get

$$\ln\frac{F}{W}=\int_{x_W}^{x_F}\frac{dx}{\frac{\alpha x}{1+(\alpha-1)x}-x}=\int_{x_W}^{x_F}\frac{\left[(\alpha-1)x+1\right]dx}{(\alpha-1)x(1-x)} \quad ...(5.75)$$

Equation (5.75) may be further simplified as

$$\ln\frac{F}{W}=\frac{\alpha}{\alpha-1}\int_{x_W}^{x_F}\frac{dx}{1-x}+\frac{1}{\alpha-1}\int_{x_W}^{x_F}\frac{dx}{x} \quad ...(5.76)$$

$$\frac{F}{W}=\left(\frac{1-x_W}{1-x_F}\right)^{\frac{\alpha}{\alpha-1}}\left(\frac{x_F}{x_W}\right)^{\frac{1}{\alpha-1}} \quad ...(5.77)$$

$$\left[\frac{F(1-x_F)}{W(1-x_W)}\right]^{\alpha}=\frac{Fx_F}{Wx_W} \quad ...(5.78)$$

Let the moles of **A** in the feed and residue be represented by A_F and A_W, respectively, then

$$Fx_F=A_F \text{ and } Wx_W=A_W$$

Similarly, let B_F and B_W represent the moles of **B** in the feed and residue, respectively. Then,

$$F(1-x_F)=B_F \text{ and } W(1-x_W)=B_W$$

Equation (5.78) can now be written as

$$\left[\frac{B_F}{B_W}\right]^{\alpha}=\frac{A_F}{A_W} \qquad \text{...(5.79)}$$

Equations (5.78) and (5.79) are expressions for Rayleigh equations applicable for constant relative volatility systems.

For Rayleigh equation to be applicable, the vapour produced at all times should be in equilibrium with the liquid and the concentrations change continuously. These ideal requirements of a differential distillation can be attained in actual practice if the operation is carried out infinitely slowly, if the liquid entrainment is completely eliminated, and if there is no cooling and condensation of the vapours on their way to the condenser.

Differential Condensation

In differential condensation, a vapour feed is slowly condensed under equilibrium conditions and the condensate is withdrawn as rapidly as it is formed, leaving behind a residual distillate. Let F be the moles of vapour fed and y_F be its composition. After condensation, the product remaining uncondensed is D moles and y_D is its compositions. At any instant, let the vapour present is n moles and y be its composition. In a small time interval during condensation, if dn is the differential amount of liquid condensate formed which is also equal to the vapour disappeared due to condensation, we have similar to Eq. (5.60)

$$d(ny)=x^*\,dn \qquad \text{...(5.80)}$$

where x^* is the concentration of liquid formed which is assumed to be in equilibrium with vapour of composition y. On expanding Eq. (5.80), we get

$$n\,dy+y\,dn=x^*\,dn \qquad \text{...(5.81)}$$

$$-n\,dy=(y-x^*)dn \qquad \text{...(5.82)}$$

Equation (5.82) may be rearranged as

$$-\frac{dn}{n}=\frac{dy}{y-x^*} \qquad \text{...(5.83)}$$

Equation (5.83) can be integrated knowing that initially $n = F$ and $y = y_F$ and finally $n = D$ and $y = y_D$.

$$-\int_F^D \frac{dn}{n}=\int_{y_F}^{y_D}\frac{dy}{y-x^*} \qquad \text{...(5.84)}$$

That is,

$$\ln\frac{F}{D}=\int_{y_F}^{y_D}\frac{dy}{y-x^*} \qquad \text{...(5.85)}$$

Multicomponent Systems: Ideal Solutions

For ideal multicomponent solutions, Eq.(5.78) can be written for any two components. One of the components (say, component-k) is chosen as the reference substance and relative volatilities of the other n-1 constituents are expressed with reference to k. For each of the n-1 constituents, equations similar to Eq. (5.78) can be written. For example, for component-j, we have

$$\left[\frac{Fx_{k,F}}{Wx_{k,W}}\right]^{\alpha_{jk}} = \frac{Fx_{j,F}}{Wx_{j,W}} \quad \text{...(5.86)}$$

where $x_{k,F}$ and $x_{j,F}$ are mole fraction of components k and j in the feed and $x_{k,W}$ and $x_{j,W}$ are mole fraction of components k and j in the residue. α_{jk} is the relative volatility of j with reference to k. Since

$$\sum x_W = 1.0 \quad \text{...(5.87)}$$

equations (5.86) and (5.87) can be used to find the composition of the residual liquid solution.

Example 5.12: A liquid mixture containing 60 mol% acetone and 40 mol% water is differentially distilled at 101.3 kPa. The residue remaining in the still has a composition of 30.0 mol% acetone. The equilibrium data is given below:

x	0.10	0.15	0.20	0.30	0.40	0.50	0.60	0.70	0.80	0.90
y	0.755	0.798	0.815	0.830	0.839	0.849	0.859	0.874	0.898	0.935

Determine the amount of products obtained and the composition of the distillate.

***Solution*:** The Rayleigh equation [Eq. (5.66)] describes simple distillation:

$$\ln\frac{F}{W} = \int_{x_W}^{x_F} \frac{dx}{y^* - x} \quad \text{...(5.66)}$$

It relates the four variables in simple distillation: Feed quantity F and its concentration x_F and the residue quantity W and its concentration x_W.

Assume a basis of F = 100 kmol

Feed contains 60 mol% acetone; $x_F = 0.60$

The concentration of the residue is $x_W = 0.30$

The fourth variable, the quantity of residue is determined using the Rayleigh equation.

The integral $\int_{0.30}^{0.60} \frac{dx}{y^* - x}$ is found out by plotting $\frac{1}{y^* - x}$ on the y-axis against liquid composition x. y^* is the concentration of the vapour in equilibrium with the liquid of composition x.

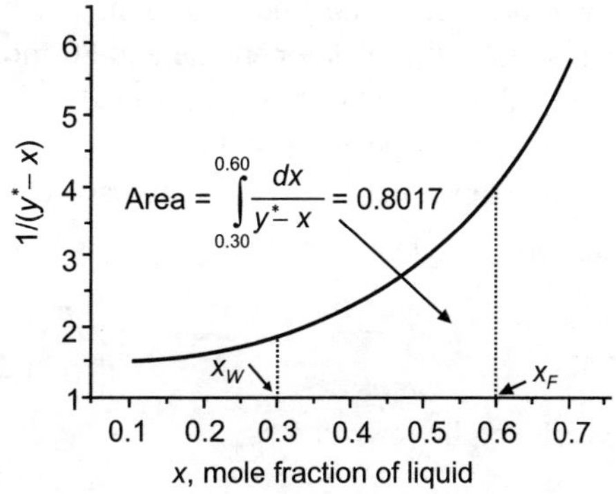

Fig. 5.18: Graphical integration for Example 5.12 on simple distillation

Using the equilibrium data $\frac{1}{y^* - x}$ values are calculated and the results are shown below:

x	0.1	0.15	0.2	0.3	0.4	0.5	0.6	0.7
y^*	0.755	0.798	0.815	0.83	0.839	0.849	0.859	0.874
$y^* - x$	0.655	0.648	0.615	0.53	0.439	0.349	0.259	0.174
$\frac{1}{y^* - x}$	1.5267	1.5432	1.6260	1.8868	2.2779	2.8653	3.8610	5.7471

The graph of $\frac{1}{y^* - x}$ versus x is shown in the Fig. 5.18. The area under the curve between the limits $x_W = 0.3$ and $x_F = 0.6$ is the required integral $\int_{0.30}^{0.60} \frac{dx}{y^* - x}$. The area is found out to be 0.8017.

$$\ln\frac{F}{W} = \int_{x_W}^{x_F} \frac{dx}{y^* - x} = 0.8017$$

Therefore, $\dfrac{F}{W} = \exp(0.8017) = 2.2293$

Amount of residue left in the still is $W = \dfrac{F}{2.2293} = \dfrac{100}{2.2293} = 44.857$ kmol

The amount of distillate is $D = F - W = 100 - 44.857 = 55.143$ kmol

Average composition of the distillate is obtained by material balance as explained earlier.

$$Fx_F = Dy_{D,av} + Wx_W \qquad \ldots(5.68)$$

Rearranging Eq. (5.68)

$$y_{D,av} = \frac{Fx_F - Wx_W}{D} \qquad \text{(A)}$$

Substituting the values in Eq. (A)

$$y_{D,av} = \frac{Fx_F - Wx_W}{D} = \frac{100 \times 0.60 - 44.857 \times 0.30}{55.143} = 0.844$$

Products: Distillate = 55.143 kmol, residue = 44.857 kmol, average distillate composition = 84.4 mol% acetone. **(Answer)**

Example 5.13: Ninety percent of the *n*-pentane is to be recovered by simple distillation from a mixture containing 20 mol% *n*-pentane and 80 mol% *n*-hexane. Determine

(a) the amount of hexane distilled
(b) the composition of the residue

Solve analytically using an average α value of 2.59.

***Solution*:** Rayleigh equation simplifies to the following form for systems with constant relative volatility.

$$\left[\frac{F(1-x_F)}{W(1-x_W)}\right]^{\alpha} = \frac{Fx_F}{Wx_W} \qquad \ldots(5.78)$$

which can be rewritten as:

$$\left[\frac{B_F}{B_W}\right]^{\alpha} = \frac{A_F}{A_W} \qquad \ldots(5.79)$$

Assume a basis of 100 kmol of feed containing 20 mol% *n*-pentane and 80 mol% *n*-hexane.

A_F = moles of **A** (*n*-pentane) in the feed = 20 kmol

B_F = moles of **B** (*n*-hexane) in the feed = 80 kmol

Since 90% of *n*-pentane in the feed is recovered in the distillate, only 10% remain with the residue.

A_W = moles of **A** (*n*-pentane) in the residue $= 0.1 \times 20 = 2$ kmol

Substituting in Eq. (5.79)

$$\left[\frac{80}{B_W}\right]^{2.59} = \frac{20}{2}$$

Solving this we get $B_W = 32.88$ kmol

$$W = A_W + B_W = 2 + 32.88 = 34.88 \text{ kmol and } D = F - W = 100 - 34.88 = 65.12 \text{ kmol}$$

(a) Amount of hexane **(B)** distilled = *D* – amount of pentane **(A)** in the distillate = 65.12 – 18 = 47.12 kmol **(Answer)**

(b) $x_W = \dfrac{\text{Amount of pentane in the residue}}{\text{Amount of the residue}} = \dfrac{A_W}{W} = \dfrac{2}{34.88} = 0.0573$

Amount of hexane distilled = 47.12 kmol

Mole fraction of pentane in the residue = 0.0573. **(Answer)**

Example 5.14: A mixture of *n*-heptane and *n*-octane containing 40 mol% *n*-heptane is subjected to simple distillation at 101.3 kPa. Equilibrium data at 101.3 kPa:

x	0	0.1571	0.312	0.487	0.655	1.0
y	0	0.279	0.492	0.674	0.810	1.0

(a) Determine the composition of the residue and distillate if half the feed is distilled off.

(b) Determine the composition of the residue and moles of the products if it is desired that the distillate product has an average composition of 57 mol% *n*-heptane.

***Solution*:** (a) Assume a basis of F = 100 kmol

Since 50% is distilled, the residue remaining is W = 100 – 50 = 50 kmol

Feed contains 40 mol% *n*-heptane; $x_F = 0.40$

The composition of the residue is determined using the Rayleigh equation.

$$\ln\frac{F}{W} = \int_{x_W}^{x_F} \frac{dx}{y^* - x}$$

$$\ln\frac{F}{W} = \ln\frac{100}{50} = 0.693$$

To find the integral $\int_{x_W}^{x_F} \frac{dx}{y^* - x}$, a graph is plotted with $\frac{1}{y^* - x}$ on the y-axis against liquid composition x. y^* is the concentration of the vapour in equilibrium with the liquid of composition x. Using the equilibrium data $\frac{1}{y^* - x}$ values are calculated and the results plotted as shown in Fig. 5.19.

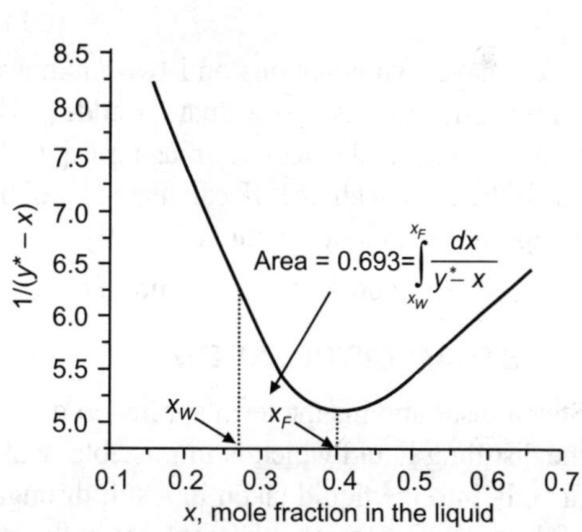

Fig. 5.19: Graphical integration for simple distillation example 5.14

x	0.1571	0.312	0.487	0.655
y^*	0.279	0.492	0.674	0.81
$y^* - x$	0.1219	0.18	0.187	0.155
$\frac{1}{y^* - x}$	8.2034	5.5556	5.3476	6.4516

The residue composition x_W is located on the graph in such a way that the area under the curve between the limits $x = x_W$ and $x = x_F = 0.4$ is equal to $\ln\frac{F}{W} = 0.693$. From the graph (Fig. 5.19) it is found that the residue composition is $x_W = 0.272$.

Average composition of the distillate is obtained by material balance as explained earlier.

$$F x_F = D y_{D,av} + W x_W \qquad \ldots(5.68)$$

Rearranging Eq. (5.68)

$$y_{D,av} = \frac{Fx_F - Wx_W}{D} \tag{A}$$

Substituting the values in Eq. (A)

$$y_{D,av} = \frac{Fx_F - Wx_W}{D} = \frac{100 \times 0.40 - 50 \times 0.272}{50} = 0.528$$

Composition of the residue = 27.2 mol% *n*-heptane and the distillate composition = 52.8 mol% *n*-heptane. **(Answer)**

(b) We have the Rayleigh equation which relates x_F, x_W, F and W.

$$\ln\frac{F}{W} = \int_{x_W}^{x_F} \frac{dx}{y^* - x} \qquad \ldots(5.66)$$

The material balance equation relates F, W, x_F, x_W and $y_{D,av}$.

$$Fx_F = (F - W)y_{D,av} + Wx_W \qquad \ldots(5.68)$$

Here, the average composition of the composited distillate $y_{D,av}$ is given to be 0.57.

$$F = 100 \text{ kmol}, x_F = 0.40$$

We have two equations and two unknown variables: W and x_W. Equations (5.66) and (5.68) are solved simultaneously to evaluate W and x_W. This requires a trial and error calculation. Assume arbitrarily a value for x_W and calculate W using Eq. (5.68) and also using Eq. (5.66) graphically. If the assumed one is the correct choice, W calculated by both equations should be the same. Otherwise the calculation is repeated with a new value of x_W. By a few such trials, it is found that

$x_W = 0.35$ and $W = 77.12$ kmol. $D = F - W = 100 - 77.12 = 22.88$ kmol **(Answer)**

5.6 STEAM DISTILLATION

Steam distillation is the term applied to distillation operations in which steam is admitted directly into the distilling liquid which is immiscible with water. In simple steam distillation, the steam is admitted directly into the liquid taken in a still through coils with numerous perforations, placed in the bottom of the still. If necessary, the contents in the still can be heated externally by passing steam through the jackets or by direct fire. The vapours evolved is taken through the overhead vapour line to a condenser, the condensate is collected in receivers and finally separated into pure liquids in gravity separators.

The major areas of applications of steam distillation are the following:

(a) Separation of high boiling materials from nonvolatile impurities. High boiling organic materials that decompose if boiled under atmospheric pressure, can be distilled at relatively lower temperature without the danger of decomposition, by steam distillation, e.g. separation of high-boiling heat sensitive essential oils from water soluble impurities.

(b) Removal of high boiling volatile impurities from still higher boiling substances, e.g. removal of free fatty acids which give undesirable odour and taste to edible vegetable oils.

(c) Distillation of material which cannot be distilled by direct heating because of the high boiling temperatures prevailing, even under vacuum. The poor heat transfer characteristics of the liquid will lead to local superheating on indirect heating.

Steam distillation is used to purify essential oils, perfumes, fine organic chemicals and such materials. It is also used to distil waxes, petroleum fractions, etc. at relatively low temperatures. Steam distillation can be combined with rectification in continuous fractionating columns as in the treatment of lubricating oils.

Steam distillation can be carried out in two different ways. In the first method, steam is allowed to condense in the still forming two immiscible liquid phases. In the second method, the formation of the second liquid phase by condensation of steam is prevented by indirect heating or by using highly superheated steam.

Consider the steam distillation in which steam condenses in the still forming two immiscible liquid phases. This will be the case if the charge in the still is not heated by indirect means and the steam supplied is not excessively superheated. Steam condenses when it is brought into contact with the feed; the latent heat liberated is used to raise the sensible heat of the feed and to provide the heat of vaporisation of the liquid distilled. In a mixture of two immiscible liquids, both components exert their full vapour pressures as if they are existing independently. The total pressure is the sum of partial pressures, which in this case is sum of their vapour pressures. The mixture boils when the sum of the vapour pressures equals the surrounding pressure P.

$$P = P_A^S + P_B^S$$

where P_A^S is the vapour pressure of pure **A** (water) and P_B^S is the vapour pressure of pure **B** (the material being distilled). It is obvious that the mixture will boil at a temperature lower than the boiling points of either pure **A** or pure **B**. Therefore, the high boiling organic material can be steam distilled under atmospheric pressure at a temperature well below 373 K without danger of decomposition, leaving behind non-volatile impurities in the still.

When distillation occurs, there are two liquid phases and one vapour phase in equilibrium and as described under the phase equilibrium behaviour of immiscible systems, the number of degrees of freedom is one. The system is univariant and the three-phase equilibrium temperature (or the distillation temperature) remains constant till the entire liquid is vaporised and removed. The composition of the vapour also remains constant which is given by the point C in Fig. 5.10. The distillation temperature can be determined by plotting $P_A^S + P_B^S$ as the ordinate against temperature as the abscissa. The temperature at which $P_A^S + P_B^S = P$ can be located in the graph which is the distillation temperature. Alternatively a Hausbrand type graph with P_B^S versus temperature and $P - P_A^S$ versus temperature can be plotted as shown in Fig. 5.20. The temperature corresponding to the point of intersection of the resulting curves is the distillation temperature. The figure refers to the steam distillation of chlorobenzene and it can be seen that for chlorobenzene, the distillation temperature at atmospheric pressure is 364 K.

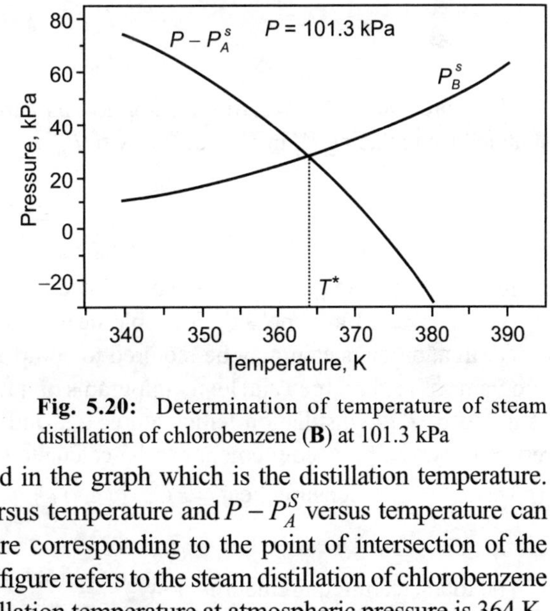

Fig. 5.20: Determination of temperature of steam distillation of chlorobenzene (**B**) at 101.3 kPa

Since the water and the liquid being distilled both exert their full vapour pressures as their partial pressures, the molar ratio of steam (**A**) to the liquid (**B**) in the vapour is given by

$$Y = \frac{n_A}{n_B} = \frac{P_A^S}{P_B^S} \qquad \text{...(5.88)}$$

where n_A and n_B are respectively, the number of moles of **A** and **B**, present in the vapour phase. The vapour phase composition is given by the mole fractions of water (y_A) and the distilling material (y_B) in the vapour.

$$y_A = \frac{n_A}{n_A + n_B} = \frac{P_A^S}{P_A^S + P_B^S} = \frac{P_A^S}{P} \qquad \text{...(5.89)}$$

$$y_B = (1 - y_A) = \frac{n_B}{n_A + n_B} = \frac{P_B^S}{P_A^S + P_B^S} = \frac{P_B^S}{P} \quad \ldots(5.90)$$

The rate of steam consumption is the kilograms of steam utilized for distillation of one kilogram of the liquid. In addition to providing the sensible heat of the feed and the latent heat of vaporisation of the liquid, steam also serves as the carrier vapour to carry the vapours of the liquid distilled to the condenser. The theoretical minimum requirement of steam as carrier vapour is given by

$$\left(\frac{w_A}{w_B}\right)_{\min} = \frac{P_A^S}{P_B^S} \times \frac{M_A}{M_B} \quad \ldots(5.91)$$

where w_A is the mass of **A** and w_B is the mass of **B** in the vapour and M_A and M_B are the molecular weights of water and the distilled material, respectively. In actual operation, the steam requirement will be higher than that given by Eq. (5.91). Due to imperfect mixing, steam may leave the still without itself getting saturated with the distillate vapours, and as a result, the partial pressure exerted by the material in the vapour will be less than the vapour pressure. This departure from equilibrium is measured by the term, the *vaporisation efficiency*. It is defined as the ratio of actual partial pressure ($\bar{P}_B$) to the vapour pressure of the material distilled and is denoted by η_v.

$$\eta_v = \frac{\bar{P}_B}{P_B^S} \quad \ldots(5.92)$$

Therefore, the actual steam consumption as carrier vapour per unit weight of liquid distilled is obtained by replacing P_B^S in Eq. (5.91) by $\eta_v P_B^S$.

$$\left(\frac{w_A}{w_B}\right)_{actual} = \frac{P_A^S}{\eta_v P_B^S} \times \frac{M_A}{M_B} \quad \ldots(5.93)$$

The total steam requirement can be evaluated by knowing the steam required as carrier vapour and the steam required to supply the sensible heat of the feed and the latent heat of vaporisation of the liquid. In addition, steam may be required to compensate for the heat loss from the unit. Let F be the kilograms of feed charged and let w_B kilograms of material **B** is vaporised. If T_F is the initial temperature of the feed, T the distillation temperature, C_{pF} and λ_F the specific heat and latent heat of the feed, respectively, the heat requirement can be calculated as:

$$\text{Sensible heat} = FC_{pF}(T - T_F) \quad \ldots(5.94)$$

$$\text{Latent heat} = w_B \lambda_F \quad \ldots(5.95)$$

The total steam requirement is now

$$W_A = \frac{1}{\lambda_s}\left[FC_{pF}(T - T_F) + w_B \lambda_F\right] + \frac{P_A^S}{\eta_v P_B^S} \times \frac{M_A}{M_B} \times w_B \quad \ldots(5.96)$$

where W_A is the total steam required in kilograms per w_B kilograms of material distilled, and λ_S is the latent heat of vaporisation of steam.

Since the distillation temperature under atmospheric pressure is less than 373 K, the boiling point of pure water, the vapour pressure of the high boiling liquid will be very small at this temperature; consequently, the steam consumption as carrier vapour alone will be very high according to Eq. (5.93). The major challenge in steam distillation is to reduce the cost of steam consumption. By carrying out steam distillation at a reduced pressure allowing formation of the two liquid phases in the distilling apparatus will not serve the purpose. The distillation temperature gets reduced by reducing the pressure;

however, at the new equilibrium temperature, both P_A^S and P_B^S are reduced and their ratio which determines the steam consumption may remain unchanged or the change may be insignificant. On the other hand, if the formation of two liquid layers is prevented by either indirect heating of the feed or by using highly superheated steam, the steam consumption can be considerably reduced.

Suppose that the liquid in the still is heated by indirect means to a temperature T' which is above the three-phase equilibrium temperature T^* shown in Fig. 5.10, and assume that the temperature is maintained constant at this value at pressure P. Steam will not condense; it only acts as the carrier vapour which mixes with and removes the vapours of the liquid. There are only one liquid phase (pure **B**) and one vapour phase (consisting of **A** and **B**) in equilibrium. According to the phase rule, the number of degrees of freedom is two and hence both the temperature and the pressure can be altered independently. That permits us to maintain the system at the lowest feasible pressure and heat it to the maximum allowable temperature level without the danger of thermal decomposition. Both these contribute to a reduction in the steam consumption. The pressure is determined by the cooling water temperature in the condenser. If the material being distilled is allowed to reach equilibrium with its vapour, the partial pressure of the distillate vapour is its vapour pressure at T'. Steam is supplied in such quantities that its partial pressure in the vapour is the difference between the existing total pressure P and the vapour pressure of the material, P_B^S. The minimum steam consumption is given by

$$\left(\frac{w_A}{w_B}\right)_{\min} = \frac{P - P_B^S}{P_B^S} \times \frac{M_A}{M_B} \qquad \ldots(5.97)$$

Allowing for the vaporisation efficiency, the actual steam consumption may be estimated as

$$\left(\frac{w_A}{w_B}\right)_{\text{actual}} = \frac{P - \eta_v P_B^S}{\eta_v P_B^S} \times \frac{M_A}{M_B} = \frac{M_A}{M_B}\left(\frac{P}{\eta_v P_B^S} - 1\right) \qquad \ldots(5.98)$$

We can see from Eq. (5.98) that both the pressure and the temperature influence the steam consumption. The amount of steam required per unit mass of distillate may be reduced by either raising the temperature of the liquid which increases P_B^S or by lowering the total pressure. The steam consumption vanishes as the total pressure approaches the vapour pressure of the liquid. When this happens, the process is equivalent to vacuum distillation. At a given pressure, the partial pressure of steam in the vapour may be varied at will by blowing more or less steam through the feed. As a result, the partial pressure of the liquid required for vaporisation can be changed as per our requirement and consequently, the vaporisation temperature also can be changed. The effect of changing the temperature and the pressure on the steam consumption is represented in Fig. 5.21 for the specific case of distillation of chlorobenzene assuming a vaporisation efficiency of 100%.

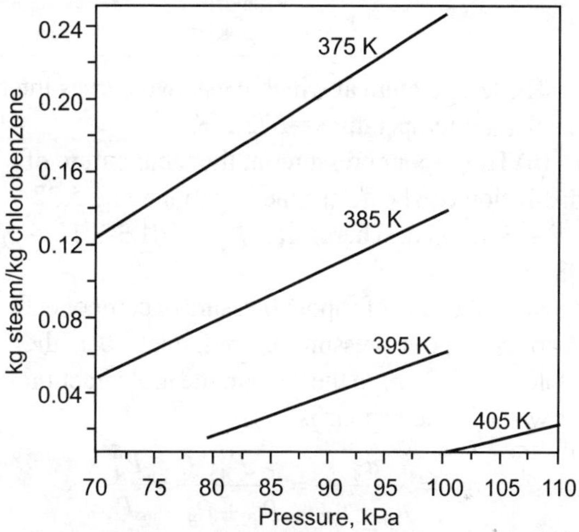

Fig. 5.21: Effect of changing temperature and pressure on steam consumption for the distillation of chlorobenzene

Any inert gas such as nitrogen, carbon dioxide, flue gas, etc., or condensable vapour which forms an immiscible liquid phase with the material to be vaporised, can be used in the place of steam and the operation will be governed by the same principles that we have discussed above. However, steam is preferred because: (i) it is readily available at low cost, (ii) it is immiscible with many high-boiling organic compounds, (iii) it can be removed easily from most systems by condensation and gravity separation and (iv) it can furnish the necessary heat of vaporisation under proper conditions.

Since distillation temperature can be brought down by reducing the pressure, vacuum distillation can be considered as an alternative to steam distillation. However, due to the difficulty in maintaining a very high vacuum and due to the unfavourable heat transfer characteristics of the liquid, steam distillation is preferred over vacuum distillation. For example, in the removal of free fatty acids from edible vegetable oils, even under a very high vacuum, vaporisation of fatty acids is not appreciable. Similarly, in the separation of aniline, the vacuum required is so high that it is practically very difficult to ensure an effective low temperature operation by vacuum distillation. Another advantage of steam distillation is that the injected steam keeps the charge thoroughly agitated and prevents local superheating arising from the poor heat transfer characteristics of the liquid.

Example 5.15: Ethyl aniline (MW 121.2) contaminated with small amounts of a nonvolatile impurity is purified by steam distillation at a total pressure of 101.3 kPa. Saturated steam at 101.3 kPa is passed through a charge consisting of 100 kg each of water and ethyl aniline taken in the still. Determine

(a) the distillation temperature
(b) the composition of the vapour
(c) the kilogram steam required for distilling one kilogram ethyl aniline

The vapour pressure of water and ethyl aniline are given below:

Temperature, K	353.8	369.2	372.3	386.4
P_A^S, kPa	48.5	87.7	98.3	163.3
P_B^S, kPa	1.33	2.67	3.04	5.33

***Solution*:** (a) P_A^S = vapour pressure of water; P_B^S = vapour pressure of ethyl aniline

To find the distillation temperature, P_B^S and $P - P_A^S$ are plotted against temperature as shown in Fig. 5.22.

Temperature, K	353.8	369.2	372.3	386.4
$101.3 - P^S_{water}$, kPa	52.8	13.6	3.0	-62.0
$P^S_{ethyl\,aniline}$, kPa	1.33	2.67	3.04	5.33

The temperature at which these two curves intersect is the distillation temperature. From the graph, distillation temperature = 372.2 K. **(Answer)**

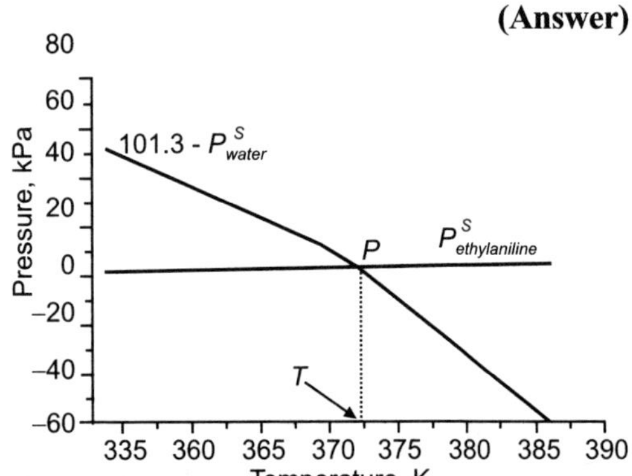

Fig. 5.22: Determination of temperature of steam distillation in Example 5.15

(b) The vapour pressures at the temperature of distillation can be determined from the Fig. 5.22. P_B^S = 3.12 kPa. Therefore, P_A^S = 101.3-3.12 = 98.18 kPa.

Since the ratio of vapour pressure of component **A** to the vapour pressure of component **B** is the mole ratio n_A to n_B in the vapour, the mole fraction of water in the vapour is

$$y_A = \frac{n_A}{n_A + n_B} = \frac{P_A^S}{P_A^S + P_B^S} = \frac{P_A^S}{P}$$

$$y_A = \frac{98.18}{101.3} = 0.9692$$

$$y_B = 1 - 0.9692 = 0.0308$$

The vapour consists of 96.92 mol% water and 3.08 mol% ethyl aniline. **(Answer)**

(c) Minimum steam consumption can be estimated as

$$\left(\frac{w_A}{w_B}\right) = \frac{P_A^S}{P_B^S} \times \frac{M_A}{M_B} \quad \text{...(5.91)}$$

$$\left(\frac{w_A}{w_B}\right) = \frac{98.18}{3.12} \times \frac{18}{121.2} = 4.67 \text{ kg steam per kg ethyl aniline distilled.}$$

Steam consumption = 4.67 kg steam per kg ethyl aniline distilled. **(Answer)**

Example 5.16: A batch of 1000 kg of turpentine (MW = 140) is to be purified by steam distillation using saturated steam at 101.3 kPa. The charge is initially at 298 K and the heat loss from the unit is negligible. The vaporisation efficiency is 0.8. Heat capacity of turpentine including the heat absorbed by the still is 2.09 kJ/kg K. Latent heat of vaporisation of turpentine = 298 kJ/kg. Latent heat of saturated steam at 101.3 kPa = 2256.9 kJ/kg. Vapour pressure data of turpentine **(B)** and water **(A)** are:

Temperature, K	323	333	343	353	363	373
P_A^S, kPa	12.4	19.9	31.1	47.3	70.1	101.3
P_B^S, kPa	2.5	3.5	5.0	7.5	11.8	17.5

Determine the distillation temperature, vapour composition and steam consumption if

(a) no additional heat is supplied

(b) the charge is heated to the temperature of the inlet steam by external source.

***Solution*:** (a) Since the vaporisation efficiency is 0.8, the temperature of distillation is the temperature at which

$$P = P_A^S + \eta_v P_B^S$$

$$101.3 = P_A^S + 0.8 P_B^S$$

P_A^S and $\eta_v P_B^S$ at each temperature are calculated and their sum $P_A^S + 0.8P_B^S$ is plotted against temperature as shown in Fig. 5.23.

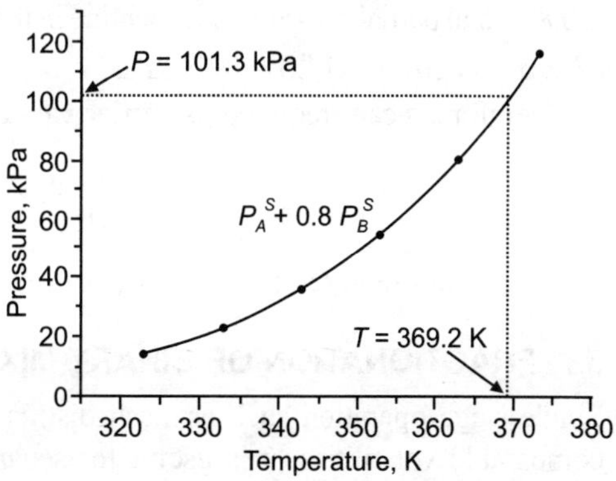

Fig. 5.23: Estimate of distillation temperature for Example 5.16

Temperature, K	323	333	343	353	363	373
P_{water}^S, kPa	12.4	19.9	31.1	47.3	70.1	101.3
$0.8P_{turpentine}^S$, kPa	2.0	2.8	4.0	6.0	9.44	14.0
$P_A^S + 0.8P_B^S$, kPa	14.4	22.7	35.1	53.3	79.54	115.3

The distillation temperature is the temperature at which $P_A^S + 0.8P_B^S = 101.3$ K. From the figure the distillation temperature is found to be 369.2 K. **(Answer)**

The vapour pressure P_A^S at 369.2 can be determined by plotting P_A^S against T or by interpolation P_A^S is found to be 89.1 kPa. Therefore, $0.8P_B^S = 101.3 - 89.1 = 12.2$ kPa.

The mole fraction of water vapour in the distillate is found out by Eq. (5.89) as

$$y_A = \frac{n_A}{n_A + n_B} = \frac{P_A^S}{P} = \frac{89.1}{101.3} = 0.8796$$

The vapour contains 87.96 mol% water vapour and the rest turpentine. **(Answer)**

Heat is required to raise the temperature of the feed to the distillation temperature and to provide the latent heat of vaporisation of turpentine.

Sensible heat required by the feed $= FC_{pF}(T - T_F) = 1000 \times 2.09 \times (369.2 - 298) = 1{,}48{,}808\,\text{kJ}$

Latent heat required by the feed $= F\lambda_F = 1000 \times 298 = 2{,}98{,}000\,\text{kJ}$

Total heat required = sensible heat + latent heat = 1,48,808 + 2,98,000 = 4,46,808 kJ

If λ_S is the latent heat of vaporisation of steam, the amount of steam required to supply the sensible heat and latent heat of the feed

$$= \frac{446808}{\lambda_S} = \frac{446808}{2256.9} = 197.97\text{ kg.}$$

In addition, heat is required to carry the vapours of the distillate. The steam required per kilogram of turpentine distilled is evaluated as $\left(\frac{w_A}{w_B}\right) = \frac{P_A^S}{\eta_v P_B^S} \times \frac{M_A}{M_B} = \frac{89.1}{12.2} \times \frac{18}{140} = 0.9390$ kg.

For 1000 kg of turpentine distilled, carrier steam required $= 0.939 \times 1000 = 939$ kg.

Total steam requirement = 197.97 + 939 = 1136.7 kg **(Answer)**

(b) When the heat requirement is met by indirect heating, steam is required only as carrier gas. The temperature of distillation is the temperature of steam supplied = 373 K.

At this temperature, the vapour pressure of turpentine = =17.5 kPa. Since the vaporisation efficiency is 0.8, actual partial pressure of turpentine in the vapour is $\eta_v P_B^S = 0.8 \times 17.5 = 14.0$ kPa. Partial pressure of steam is 101.3 – 14.0 = 87.3 kPa.

Therefore, steam required as carrier vapour per kg turpentine distilled is

$$\left(\frac{w_A}{w_B}\right)_{actual} = \frac{P - \eta_v P_B^S}{\eta_v P_B^S} \times \frac{M_A}{M_B} = \frac{87.3}{14.0} \times \frac{18}{140} = 0.8017$$

Total steam required for 1000 kg of turpentine $= 0.8017 \times 1000 = 801.7\,\text{kg}$. **(Answer)**

5.7 FRACTIONATION OF BINARY MIXTURES

Single stage operation such as flash distillation is not very effective for separating components of comparable volatilities; it is useful for separating components with widely different boiling points. Fractionation, also known as rectification, is a distillation technique by which substantially pure products can be separated even from mixtures with relative volatility very close to unity. In fractionation, the vapours evolved by boiling the liquid are condensed and part of the condensate called 'reflux' is brought into intimate countercurrent contact with the rising vapours. The operation may be carried out as a multistage contact operation in a plate tower or as a continuous contact operation in a packed column. The vapour gets enriched in more volatile component as it moves up in contact with the liquid reflux that is flowing down and consequently a distillate product is obtained at the top of the column which is very rich in more volatile component than the vapour obtained from the flash distillation. Similarly, the liquid stream flowing down the column is stripped off the more volatile component by contacting it with a vapour that is generated by the partial vaporisation of the liquid. The bottom product is therefore richer in less volatile component than the residual liquid resulting from a single stage flashing operation.

Principle of Fractionation

Consider a mixture of benzene and toluene containing 30 mol% benzene. Flash distillation of the mixture is represented on the boiling-point diagram in Fig. 5.24.

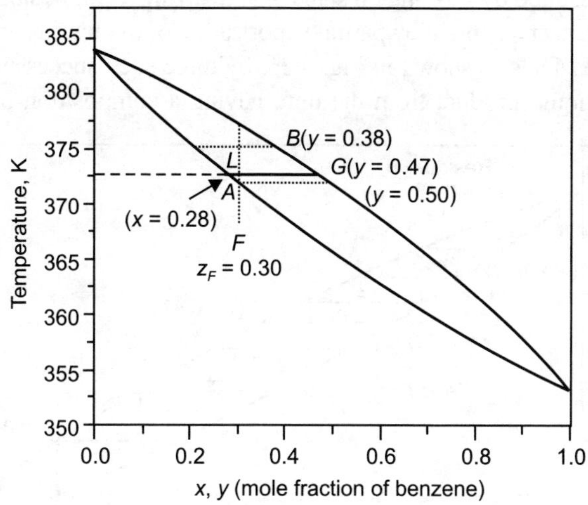

Fig. 5.24: Flash vaporisation on boiling point diagram

If the liquid is flashed at the bubble point of the feed (represented by point A in the figure), distillate will have a composition of 50 mol% benzene, which is the maximum concentration that the vapour can attain by flash vaporisation at 101.3 kPa. Though the richest vapour is produced at this condition, the quantity vaporised is infinitesimally small. The concentration of the distillate depends on the extent of vaporisation. If the extent of vaporisation is say, 50%, the concentration of the benzene in the distillate would be about 38 mol% as shown by point B in Fig. 5.24.

Let us now consider flash vaporisation of the mixture at 372.6 K so that about 10% of the feed is vaporised. Assume that the operation is carried out on a device designated as the 'stage' represented by m in Fig. 5.25. The distillate vapour has a composition of 47 mol% benzene and the residue 28 mol% benzene. The distillate and residue are represented by points G and L in Figs. 5.24 and 5.26. From 100 moles of a feed mixture containing 30 mol% benzene, we could produce about 10 moles of distillate containing 47% benzene by a single stage flashing operation.

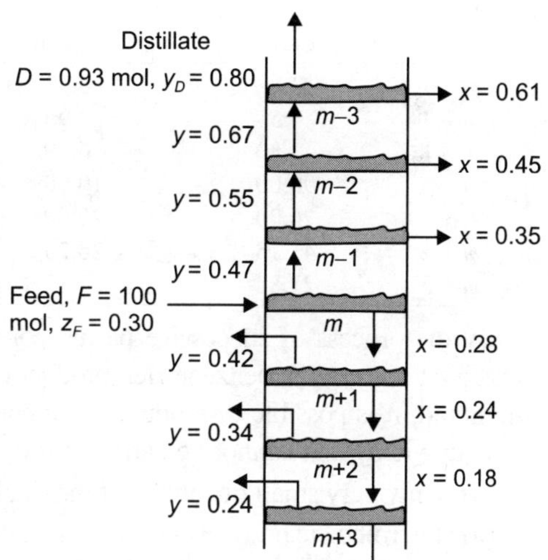

Fig. 5.25: Schematic representation of flash vaporisation and successive partial condensation of the resulting vapour and partial vaporisation of the resulting liquid

It is possible to obtain a vapour richer than G by partially condensing this vapour on stage $m - 1$. The figure shows the vapour resulting by the partial condensation on stage $m - 1$ as having a benzene content of 55% and its quantity will be about 6.5 moles. The concentrations of the vapour and liquid products depend on the degree of liquefaction of the vapour. The vapour leaving stage $m - 1$ is again partially condensed on stage $m - 2$ and then on stage $m - 3$ and so on. Partial condensation of the vapour on stage $m - 3$ produces about 0.93 moles of vapour having a composition of 80% benzene. So by successive condensation on a series of separate devices we get a distillate product which is richer than the product that would be obtained by a single-stage flashing. The residual liquid (L) that resulted by flashing on stage m and having a composition of about 28% benzene can be stripped off the benzene content by partial vaporisation of the liquids on stages $m + 1$, $m + 2$ and so on as shown in figure. Thus as shown in Fig. 5.26, by three such successive partial vaporisations it is possible to obtain a liquid product from the unit, having a composition of about 12% benzene in

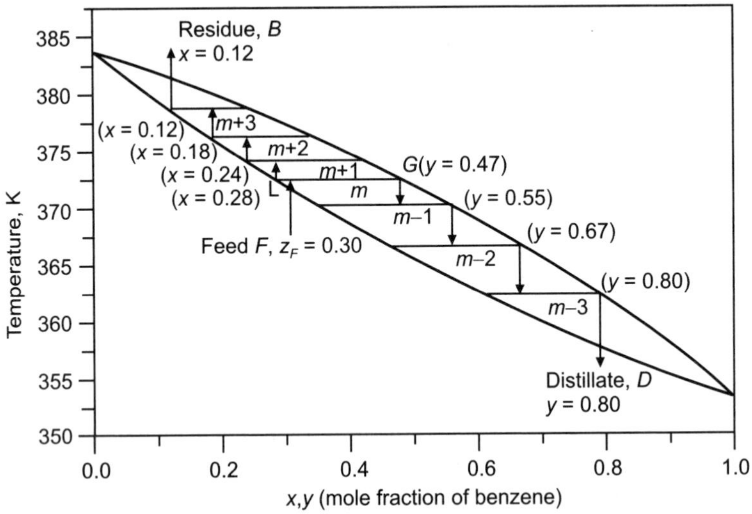

Fig. 5.26: Boiling point-diagram of benzene-toluene system showing the partial vaporisation of liquid and partial condensation of vapour

amounts approximately equal to 22 moles. The quantities, composition and temperature of operation of each stage in the chain are given in the following table.

Stage	Quantity, mol		Composition		Temperature
	Liquid	Vapour	x	y	K
$m - 3$	2.02	0.93	0.61	0.80	362.3
$m - 2$	3.55	2.95	0.45	0.67	367.0
$m - 1$	3.50	6.50	0.35	0.55	370.2
m	90.00	10.00	0.28	0.47	372.5
$m + 1$	70.00	20.00	0.24	0.42	374.1
$m + 2$	43.75	26.25	0.18	0.34	376.2
$m + 3$	21.88	21.88	0.12	0.24	378.7

By the process of successive partial condensation of the vapour generated in the first step, it is possible to produce a benzene-rich product of desired purity and by successive partial vaporisation of the liquid, it is possible to produce a toluene-rich product of desired purity. However, this process is quite unwieldy and cannot be carried out on a commercial scale because of the following reasons:

(i) Only very small quantities of the final products are obtained by this process

(ii) Intermediate products - both liquid and vapour - available at different stages are not subjected to enrichment and are not utilized.

(iii) Condensation of the vapour requires cooling by external means and vaporisation requires external heaters. This means that all the stages are to be maintained at their operating temperatures by external means.

A remarkable improvement can be achieved if the liquid stream resulting by the partial condensation of the vapours are returned back to the stage where the composition of the incoming stream is closer to the composition of the returned liquid. Assume that the process of successive partial vaporisation and condensation are carried out on a plate column and the feed mixture is admitted to the *m*th plate of the column. The scheme is represented on the boiling-point diagram in Fig. 5.27a and the process is represented schematically in Fig. 5.27b. The vapour leaving plate *m* is admitted to the plate above (*m*-1) where it is partially condensed. The condensate from plate *m* – 1 (having a benzene content of 35% in our example) is sent back to the plate *m* where the feed is admitted (with composition 30% benzene in the present example). In a similar manner the liquid leaving plate *m* – 2 ($x = 0.45$) is returned to plate *m* – 1 where it is brought into contact with vapour coming from plate *m* with almost similar composition ($y = 0.47$). The vapour reaching plate *m*-1 is at a higher temperature than the liquid reaching there. When these phases are brought into contact, the liquid gets partially vaporised and the vapour gets partially condensed. The heat required for vaporisation of the liquid is supplied by the heat liberated by condensation of the liquid. No external cooling or heating is required on any plate in the column. This process is repeated on other plates above the feed plate till a vapour of desired purity is produced. This vapour is condensed and a part of the condensate is withdrawn as distillate product and the other part returned back to the top plate as reflux. The enrichment of the vapour in more volatile component happens here because of transfer of the more volatile component (**A**) from the liquid to the vapour and less volatile component from the vapour to the liquid (**B**). The liquid phase required for the mass transfer is generated by the partial condensation of the vapour.

In a similar manner, the liquid flowing down the feed plate is stripped off the more volatile component by bringing the liquid in contact with a vapour phase. This is achieved by returning the vapour generated by partial vaporisation of the liquids on plate *m*+1 and *m*+2, etc. to the plate above them. The reboiler provided at the bottom of the column receives liquid from the bottommost plate, subjects it to partial vaporisation and permits the resulting liquid to be withdrawn as residue product and the vapour returned back to the column. The reboiler performs the same duty as a plate in the column and according to Fig. 5.27, reboiler acts as plate *m*+3. The liquid leaving is depleted off the more volatile component and is therefore richer in less volatile component. The scheme described here explains the principle of the process known as fractionation or rectification.

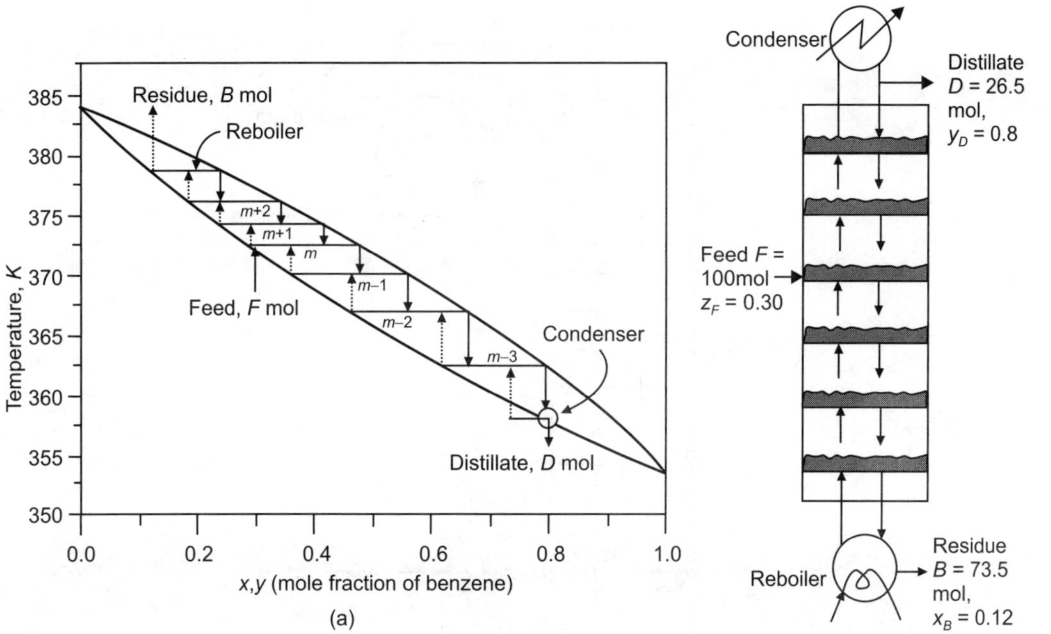

Fig. 5.27: (a) Principles of fractionation on boiling-point diagram (b) Schematic representation of fractionation

By this scheme, distillate product which is almost pure **A** and bottom product which is almost pure **B** can be produced in sufficiently large quantities from a given binary mixture of any composition. No devices are required to maintain the plates at the desired temperatures; in continuous operation, the plates will attain steady state temperature automatically and will remain at these temperatures, unless the operating conditions are altered. From 100 moles of the feed consisting of 30 mol% benzene and 70 mol% toluene, entering the column as saturated liquid and returning about 2.4 parts of liquid leaving the condenser back to the column for every part of distillate withdrawn as product, it is possible to obtain in the column equipped with just 6 ideal plates, a total condenser and a partial reboiler, 26.5 moles of a distillate containing 80 mol% benzene and 73.5 moles of a bottom product containing 88 mol% toluene.

Fractionation in Plate Columns

Fractional distillation can be carried out either as multi stage countercurrent operation in plate columns or as continuous contact operation in packed columns. The equipment for gas-liquid contact operations are discussed in Chapter 4 under Gas Absorption. Plates used for distillation may be the sieve plates, bubble-cap plates, valve plates or any of the proprietary type plates. The feed is admitted on the feed plate which is located generally in the middle of the column. The column is connected at the top to a condenser in which the overhead vapours are condensed. A portion of the condensate is recycled to the top of the column which is called the *reflux*. The ratio of the moles of liquid returned as reflux to the moles of distillate product withdrawn is designated as *reflux ratio*. It is the reflux that is responsible for the enrichment of the vapour rising from the feed plate and in its absence, the distillate composition will not be different from the composition of the vapour produced at the feed plate. The liquid leaving the bottommost plate is admitted to a reboiler. The heat supply to the reboiler is usually by steam. The vapour produced in the reboiler is returned to the column as *vapour boil up* and the liquid leaving the reboiler is the bottom product. The ratio of the moles of vapour sent back to the column to the moles of bottom product is designated as the *boil-up ratio*. The essential parts of a fractionating column are shown in Fig. 5.28.

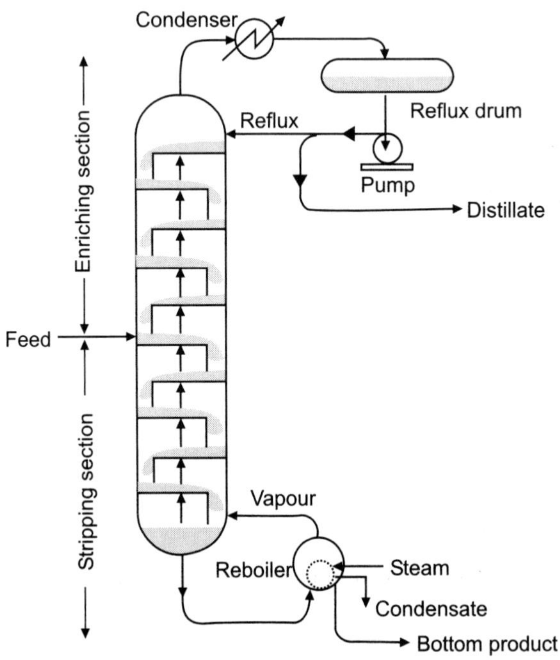

Fig. 5.28: Plate column for fractionation

Liquid and vapour phases are brought into contact on each plate. The vapour reaching a plate is the one originating from the plate below it and the liquid reaching a plate is the one originating from the plate above. Enrichment of the vapour occurs because it has a lower concentration of the more volatile component (**A**) than the vapour that would be in equilibrium with the liquid with which it is contacted. On each plate, the liquid gets depleted off or stripped off component **A** because of vaporisation and transfer of **A** to the vapour phase.

Assume all the plates in the column to be ideal plates and consider the *n*th plate in the column as shown in Fig. 5.29a. Vapour leaving plate $n+1$ reaches plate n with concentration y_{n+1} and liquid flowing down from plate $n-1$ reaches plate n with composition x_{n-1}. These liquid and vapour streams are represented by L_{n-1} and G_{n+1} on the boiling point diagram shown in Fig. 5.29b. The vapour is at its dew point and the liquid is at its bubble point. These streams are not in equilibrium; otherwise they should lie at the ends of one of the horizontal tie lines that can be drawn on the *T-x-y* diagram. When these streams at different temperatures are brought into contact, the saturated liquid gets partially vaporised and the saturated vapour gets partially condensed. The heat of vaporisation of the liquid is obtained from the heat liberated by the condensation of the vapour.

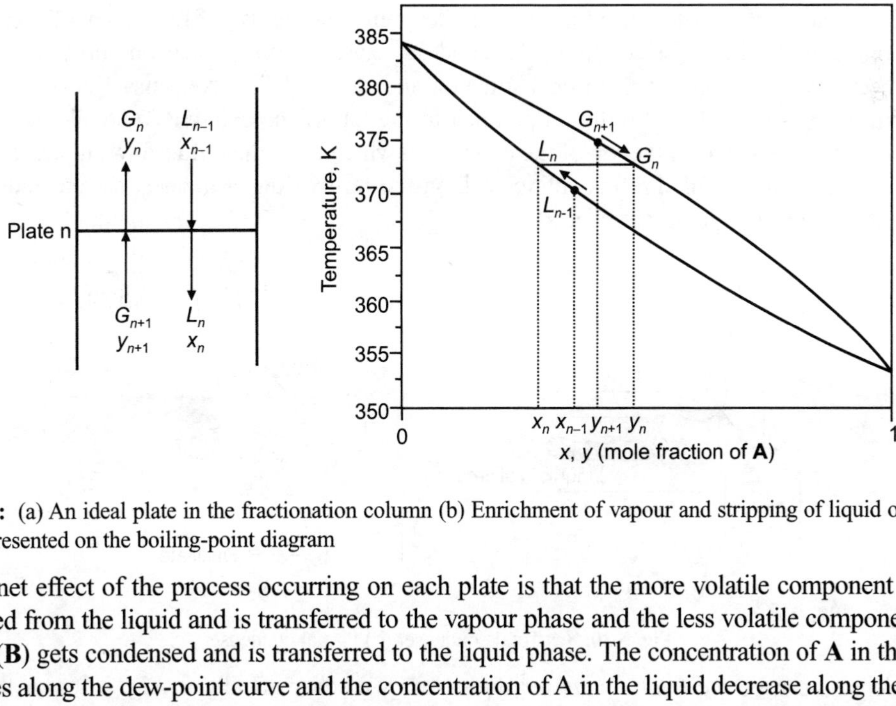

Fig. 5.29: (a) An ideal plate in the fractionation column (b) Enrichment of vapour and stripping of liquid on an ideal plate represented on the boiling-point diagram

The net effect of the process occurring on each plate is that the more volatile component (**A**) gets vaporised from the liquid and is transferred to the vapour phase and the less volatile component in the vapour (**B**) gets condensed and is transferred to the liquid phase. The concentration of **A** in the vapour increases along the dew-point curve and the concentration of A in the liquid decrease along the bubble-point curve till the concentrations reach the equilibrium values x_n and y_n shown by the points L_n and G_n. The vapour thus gets enriched in **A** and the liquid gets stripped off **A** on each plate in the column. If the molar latent heats of vaporisation of the components **A** and **B** are approximately equal, for every mole of **B** condensed and transferred from the gas to the liquid, one mole of **A** is vaporised and transferred from the liquid to the vapour phase. That is, under ideal conditions, equimolar counter diffusion occurs on all plates in the fractionating column. Since the concentration of more volatile component increases as we go up the column, the temperature on each plate decreases from bottom of the column to the top.

On the top plate, the liquid reflux is brought into contact with the vapour. Since the vapour reaching the top plate and the reflux are not in equilibrium, the vapour gets enriched in **A** by the mechanism described above. The reflux may sometimes get cooled below its bubble point; if it is so, on reaching the plate, it gets immediately heated to its bubble point by the vapour on the top plate. On all other plates the liquid and vapour are at their saturation temperatures. The liquid leaving the bottom plate is sent to a reboiler which produces the vapour that flow upward the column in such quantities that both the overhead product and the reflux required are obtained from it by condensation at the top. The steam is supplied to the reboiler for producing this vapour; the energy required in the reboiler represents the major cost of the process of separation by distillation.

The section of the column above the feed plate, including the condenser is designated as the *enriching section* as shown in Fig. 5.28. It is also called *absorption section or rectifying section.* The top product is enriched in more volatile component in this section. The portion of the column below the feed plate and including the feed plate is designated as the *stripping section* or the *exhausting section.* In this section, the liquid stream gets stripped off the more volatile component by mass transfer with the vapour flowing up.

Condenser

The vapour leaving the top plate is taken to a condenser which is usually a conventional tubular heat exchanger. Water is the coolant used in most applications; but the use of refrigerants for condensing highly volatile mixtures also is common. The condensers are generally arranged horizontally near the ground because of the convenience in construction and cleaning. The condensate is collected in the reflux drum from which liquid reflux is pumped to the top of the column. In smaller columns, the condenser can be placed near the top of the column, in which case, the reflux flows to the top plate by gravity. A special valve called reflux splitter will ensure reflux flow at the desired rate and no reflux accumulator and reflux pump are needed.

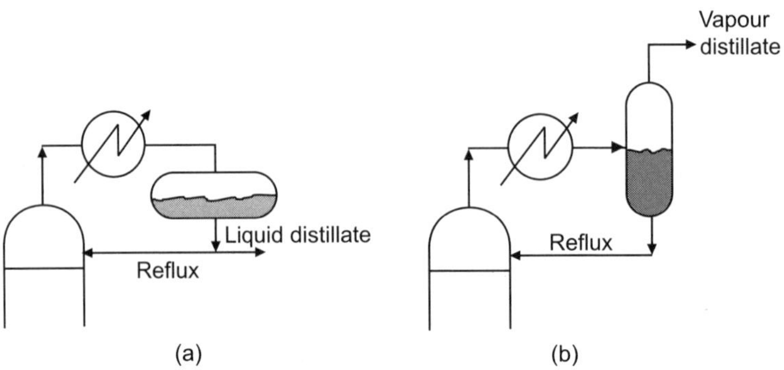

Fig. 5.30: (a) Total condenser (b) Partial condenser

Condensers can be the *total condenser* type as shown schematically in Fig. 5.30a or the *partial condenser* type as shown in Fig. 5.30b. In the total condenser, the entire vapour issuing from the top plate is condensed and the condensate is collected in the reflux accumulator. A portion of the condensate is pumped back to the column as reflux. The distillate product, the reflux and the vapour leaving the top plate, all have the same composition. If water is the coolant used, the pressure in the column should be such that a reasonably good temperature difference driving force between the vapour and the cooling water exists for rapid condensation of the vapour. The cost of fractionator increases with the operation pressure and if the condensation pressure exceeds about 2.52 MPa, as happens in the case of highly volatile distillates, refrigerants are often used as the condenser coolant.

In a partial condenser, only a fraction of the vapour is condensed and the residual vapours are collected as distillate. The distillate vapour can be assumed to be in equilibrium with the liquid reflux and in stage computations, the partial condenser can be treated as a theoretical plate. Partial condensers are used under the following situations: (i) When it is desired to withdraw the distillate product in the vapour state, complete condensation is not warranted. Only a portion of the vapour that is required to provide the liquid reflux is condensed and the uncondensed vapours form the distillate product. (ii) If complete condensation of the vapour necessitates an excessive column pressure, partial condenser can be used with lower operating pressure.

Reboilers

Reboilers attached to the bottom of the fractionation tower ensure vapour boilup through the column by receiving the liquid leaving the bottom plate and vaporising it either completely or partially. There are two broad types of reboilers. The *internal reboilers* are used in pilot plant work and small columns and the *external reboilers* in large-scale industrial applications. The internal reboilers can be of the jacketted type in which the bottom of the column is shaped as a jacketed kettle through which steam is circulated as the heating fluid. In the tubular type of internal reboilers, the heat exchanger tubes are built into the bottom of the column and steam or some heating fluid is passed through the tubes which are immersed in the pool of liquid at the bottom of the tower. Internal reboilers generally have small heat transfer surface and their vapour capacity is therefore limited. Also clean up of the tubes requires shutdown of the unit. In contrast, external reboilers shown in Fig. 5.31 can be arranged with spares which facilitate cleaning without disrupting the column operation.

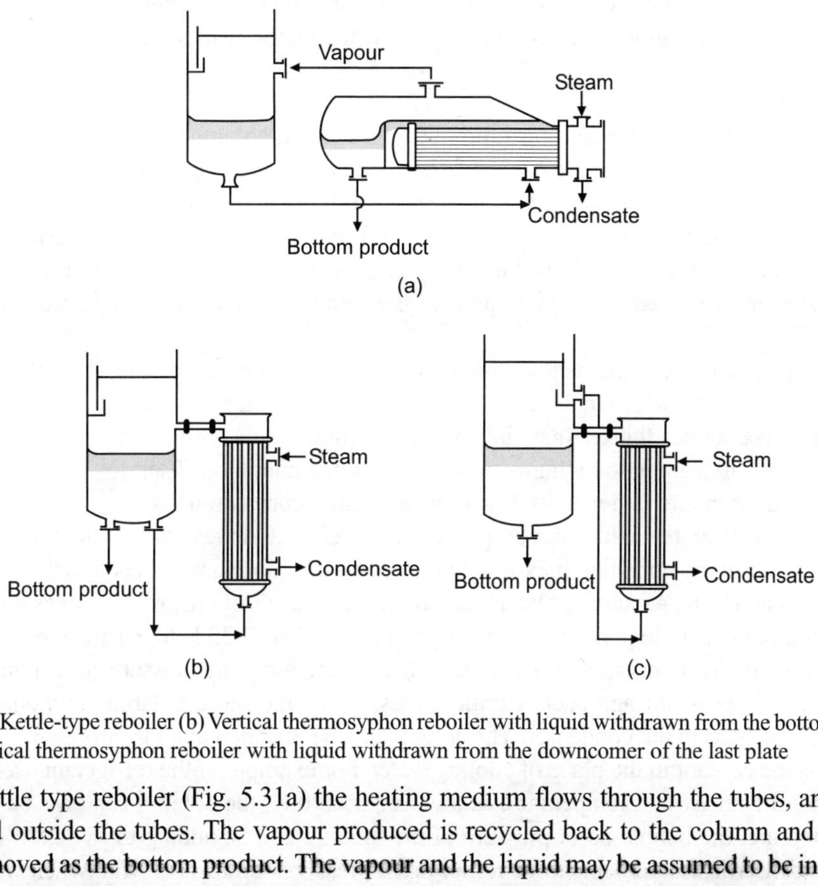

Fig. 5.31: (a) Kettle-type reboiler (b) Vertical thermosyphon reboiler with liquid withdrawn from the bottom sump of the tower (c) Vertical thermosyphon reboiler with liquid withdrawn from the downcomer of the last plate

In the kettle type reboiler (Fig. 5.31a) the heating medium flows through the tubes, and the liquid is circulated outside the tubes. The vapour produced is recycled back to the column and the residual liquid is removed as the bottom product. The vapour and the liquid may be assumed to be in equilibrium and therefore, the kettle-type reboiler acts as a theoretical plate. In the vertical thermosyphon reboiler

(Fig. 5.31b) the heating medium is outside the tubes and the liquid removed from the bottom sump is vaporised inside the tubes of the vertical heat exchanger. Total vaporisation is possible inside the tubes, but chances of fouling are more. Generally only partial vaporisation is carried out and therefore, a mixture of liquid and vapour leaves the heater tubes, which is then returned to the column. The bottom product is withdrawn from the sump at the bottom. Vapour enrichment occurs in the reboiler, but it is never equivalent to that achieved by a theoretical plate because of the mixing of the liquid leaving the heater tubes with the liquid leaving the last tray. The thermosyphon reboiler of the type shown in Fig. 5.31c is similar in construction, but the liquid for vaporisation is removed from the downcomer of the last plate whereas bottom product is withdrawn from the reservoir at the column bottom. Since the liquid leaving the heater is not contacted with the liquid from the last tray, this type of reboiler can be treated as equivalent to a theoretical plate.

Kettle-type reboilers are widely used in industries. Thermosyphon reboilers are advantageous when the bottom product contains heat sensitive materials, or if the bottom pressure is high or there is heavy fouling. Though inexpensive, they are not suitable for high viscosity liquids or vacuum operation. Steam, heat-transfer oils or other hot fluids can be used as heating medium. The choice of reboiler depends on the following:

(i) The properties such as viscosity, fouling tendencies, etc. of the liquid vaporised.
(ii) Heat sensitivity of the bottom product.
(iii) Operating pressure.
(iv) Temperature difference between the process fluid and the heating medium.
(v) The equipment layout with stress on the space available for head room.

Column Pressure

An important factor in the design and operation of fractionation columns is the choice of the column pressure. The operating pressure in the column is decided based on several considerations. The column is generally operated at a low pressure, because the separability is high if the pressure is low. At high pressures, the equilibrium curve is closer to the diagonal so that the vapour and liquid have very close values for the equilibrium compositions. Separation in this case is difficult; for given separation more number of stages are required than the column of comparable efficiency, but operated with a lower pressure.

Since the vapour volume to be handled increases with reduction in pressure, the column diameter correspondingly increases, which in turn, may result in high investment cost for fractionators. However, a high pressure necessitates thicker materials of construction for the column, more number of plates and a taller column, all leading to high investment cost for the operation. Therefore, the selection of an optimum operating pressure is generally based on economic considerations.

The type of condenser, the nature and properties of the reflux and the nature of the coolant used also influence the choice of the operating pressure. Suppose that the cooling water used as the coolant in the condenser leaves at 315 K. Assuming that a total condenser is used and the reflux leaves the condenser as saturated liquid, the bubble point of the reflux should be at least 325 K to ensure a reasonably good rate of condensation. The condenser pressure should be the bubble-point pressure corresponding to 325 K. However, if this pressure approaches critical pressure of the more volatile component, a lower pressure should be used in the condenser. The column can be operated at a lower pressure if we go for a refrigerant as the coolant in the place of cooling water. For example, if the refrigerant used is the one which can cool the reflux to a temperature as low as say, 270 K, the column can now be operated at a very low pressure, the bubble point pressure of the mixture corresponding to say, 280K.

A total condenser is recommended for reflux-drum pressures up to 1.48 MPa when water is used as the coolant. A partial condenser is appropriate from 1.48 MPa to 2.52 MPa. If pressure exceeds

2.52 MPa, a refrigerant may be used as condenser coolant. A condenser pressure drop from 0 to 14 kPa and a column pressure drop of 35 kPa (or to be more accurate, 0.7 kPa per tray) may be used for computing pressure at the bottom of the tower (Seader, 2006).

Certain mixtures which are thermally unstable, when subjected to distillation may undergo decomposition, polymerisation or other interactions when they attain certain critical temperatures. The design pressure is chosen such that the decomposition temperature is not reached in the column. Since the highest temperature is encountered in the bottom of the column, the thermal stability of the mixture at the reboiler temperature usually restrict the design pressure for the fractionating column.

Material and Energy Balances

A general procedure for design of fractionating columns involves calculation or specification of the following:

1. Distillation pressure
2. Number of theoretical plates required for a specified separation
3. Efficiency of the stages and the number of actual plates
4. Type and design of plates
5. Plate spacing
6. Feed-entry stage
7. Height of column
8. Column diameter
9. Type of condenser
10. Type of reboiler
11. The column internals

An important step in the analysis of the column performance is to study the material and energy quantities involved in the column. By conducting the material and energy balance, the flow rates and the composition of the streams and the heat requirement for fractionation can be evaluated. The cooling water circulation rate in the condenser, and the rate of supply of steam in the reboiler can also be determined by the material and energy balance.

The significance of the reflux ratio, the single most important quantity that influence the fractionator performance, can be explained with the help of the material and energy balance. The flow rate, composition and enthalpy of various streams in the distillation column are shown in Fig. 5.32. The flow rates of liquid streams in the enriching section are denoted by L and that in the stripping section are represented by $\overline{L}$, kmol/h. The flow rates of vapour streams in the enriching section are represented by G and those in the stripping section are represented by $\overline{G}$, kmol/h. The plates are numbered from the top to the bottom; the total number of plates in the column is N. The plate n stands for any plate in the enriching section and plate m stands for any plate in the stripping section. The suffix appearing with the symbols indicates the number of the plate from which the stream is originating. Thus L_n and $\overline{L}_m$ are the liquid flows in kmol/h leaving the plates n and m, respectively and G_{n+1} is the flow of vapour leaving plate $n + 1$ and $\overline{G}_{m+1}$ is the vapour flow leaving plate $m + 1$. Enthalpies of the liquid and vapour streams (in kJ/kmol) are represented by H_L and H_G, respectively. Feed is admitted at a rate F kmol/h, has a concentration z_F mole fraction **A** and enthalpy H_F kJ/kmol. The distillate is produced at a rate D kmol/h with concentration of **A**, x_D and enthalpy, H_D. Bottom product is withdrawn at a rate B kmol/h with concentration of **A**, x_B and enthalpy, H_B. Assume that a total condenser is used and the cooling water is supplied to remove the heat of condensation at rate Q_C kJ/h. Steam is used in the reboiler to supply heat at rate Q_B kJ/h.

Total material balance for the fractionating column gives

$$F = D + B \quad \text{...(5.99)}$$

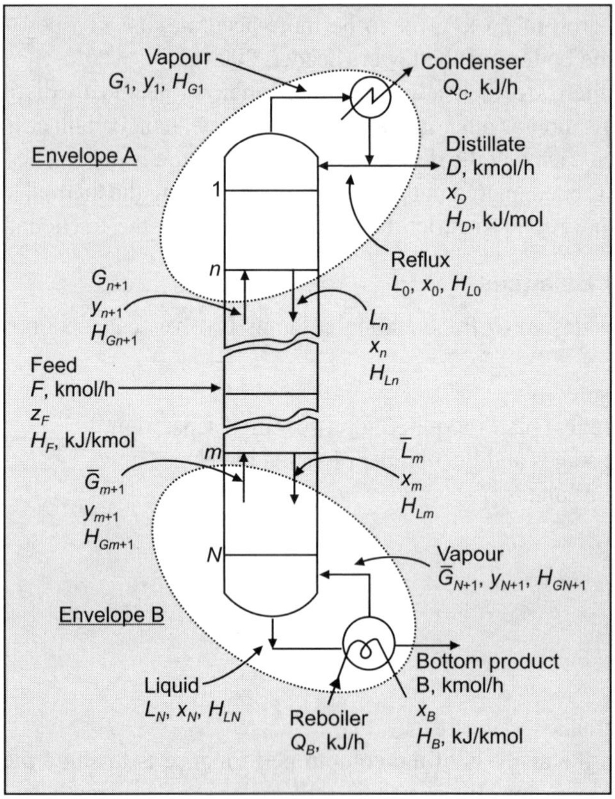

Fig. 5.32: Fractionation column showing the flow rate, composition and enthalpy of different streams

Component-**A** balance for the column is

$$Fz_F = Dx_D + Bx_B \quad \ldots(5.100)$$

Eliminating B from Eq. (5.100) using Eq. (5.99) and rearranging, we get

$$D = F\frac{z_F - x_B}{x_D - x_B} \quad \ldots(5.101)$$

Similarly, elimination of D from Eq. (5.100) gives the following equation.

$$B = F\frac{x_D - z_F}{x_D - x_B} \quad \ldots(5.102)$$

It can be shown that the net flow of the total material and the net flow of components remain constant in the enriching section and stripping section of the column regardless of the changes in the flow rates and compositions of the vapour and the liquid from plate to plate. A total material balance for the enriching section can be written by considering the section above plate n represented by the envelope **A** in Fig. 5.32.

$$G_{n+1} = D + L_n \quad \ldots(5.103)$$

Similarly a component-**A** balance over the envelope-A gives

$$G_{n+1}y_{n+1} = L_n x_n + Dx_D \quad \ldots(5.104)$$

The total material balance and component-**A** balance can be written for the condenser as follows.

$$G_1 = D + L_0 \quad \ldots(5.105)$$

$$G_1 y_1 = D x_D + L_0 x_0 \qquad \text{...(5.106)}$$

Equations (5.103) and (5.105) can be rearranged as:

$$G_{n+1} - L_n = D = G_1 - L_0 \qquad \text{...(5.107)}$$

In Eq. (5.107) $G_{n+1} - L_n$ is the net flow of material upward the column between plates n and $n + 1$ and $G_1 - L_0$ is the net flow of total materials above plate 1. Since, for a specified rate of distillate, D is constant, Eq. (5.107) means that the net flow of total materials is constant throughout the enriching section of the column. Similarly Eqs. (5.104) and (5.106) can be rearranged as

$$G_{n+1} y_{n+1} - L_n x_n = D x_D = G_1 y_1 - L_0 x_0 \qquad \text{...(5.108)}$$

This equation means that the net flow of component **A** upward the enriching section of the column is constant and is equal to $D x_D$ kmol/h.

Now consider envelope **B** in Fig. 5.24 which encompasses the stripping section in the column below plate m. Total material balance and the component **A** balance gives

$$\bar{L}_m - \bar{G}_{m+1} = \bar{L}_N - \bar{G}_{N+1} = B \qquad \text{...(5.109)}$$

$$\bar{L}_m x_m - \bar{G}_{m+1} y_{m+1} = \bar{L}_N x_N - \bar{G}_{N+1} y_{N+1} = B x_B \qquad \text{...(5.110)}$$

Equations (5.109) and (5.110) state that the net flow of total material and net flow of components downward the stripping section remain constant for a specified process.

An enthalpy balance can be written for the distillation column as a whole, taking into consideration that heat is brought to the column by the feed and by the steam supplied to the reboiler and the heat is carried away from the column by the distillate, the bottoms and by the coolant in the condenser. It is assumed that the heat loss from the column is negligible.

$$F H_F + Q_B = D H_D + B H_B + Q_C \qquad \text{...(5.113)}$$

Since the enthalpies H_F, H_D and H_B, which depend on the compositions and temperatures of the streams are fixed by the process, Eq. (5.113) means that only one of the remaining variables, either the condenser heat duty Q_C or the reboiler heat load Q_B can be varied as we wish, but not both. Fixing the rate of steam supply to the reboiler automatically fixes the rate of circulation of cooling water in the condenser, and *vice versa*.

Now consider a heat balance on the condenser alone.

$$G_1 H_{G1} = D H_D + L_0 H_{L0} + Q_C \qquad \text{...(5.114)}$$

Here G_1 is the kmol/h of vapour reaching the condenser and H_{G1} is its enthalpy. L_0 is the reflux flow rate in kmol/h and H_{L0} is its enthalpy. Since $G_1 = D + L_0$ [*See* Eq. (5.105)], we can write Eq. (5.114) as

$$Q_C = D(H_{G1} - H_D) + L_0 (H_{G1} - H_{L0}) \qquad \text{...(5.115)}$$

Assuming that a total condenser is used, the reflux and distillate have the same enthalpy or $H_D = H_{L0}$. Therefore Eq. (5.115) can be modified as

$$Q_C = D(H_{G1} - H_{L0}) + L_0 (H_{G1} - H_{L0}) = (D + L_0)(H_{G1} - H_{L0}) \qquad \text{...(5.116)}$$

If the reflux is not subcooled, the enthalpy difference $(H_{G1} - H_{L0})$ is the latent heat of vaporisation of the distillate, λ_0.

$$Q_C = (D + L_0)\lambda_0 \qquad \text{...(5.117)}$$

Since by definition, the reflux ratio (also known as external reflux ratio) R is $\frac{L_0}{D}$, Eq. (5.117) can be written in terms of the reflux ratio as

$$Q_C = D(R + 1)\lambda_0 \qquad \text{...(5.118)}$$

Equation (5.118) indicates that the reflux ratio is the single most important quantity that determines the performance of distillation columns. It is an operating variable that can be controlled at will by adjusting the split between the reflux and the overhead product or by changing the amount of vapour formed in the reboiler for a given rate of production of distillate. Once reflux ratio R is specified, the condenser duty is fixed. Since the condenser duty and the reboiler heat duty are not independent as is evident from Eq. (5.113), the reboiler heat duty also is fixed by specifying the reflux ratio.

5.8 McCABE-THIELE METHOD

Operating Lines for Distillation Columns

A component-**A** balance for the enriching section of the column gave

$$G_{n+1} y_{n+1} = D x_D + L_n x_n \quad \text{...(5.104)}$$

This can be modified as

$$y_{n+1} = \frac{L_n}{G_{n+1}} x_n + \frac{D x_D}{G_{n+1}} \quad \text{...(5.117)}$$

Similarly the component balance for the stripping section gives

$$\bar{L}_m x_m - \bar{G}_{m+1} y_{m+1} = B x_B \quad \text{...(5.110)}$$

which can be written as

$$y_{m+1} = \frac{\bar{L}_m}{\bar{G}_{m+1}} x_m - \frac{B x_B}{\bar{G}_{m+1}} \quad \text{...(5.118)}$$

Equation (5.117) is the material balance equation for the enriching section and Eq. (5.118) is that for the stripping section. On plotting these equations with mole fraction of **A** in the vapour (y) as the ordinate against mole fraction of **A** in the liquid (x) as the abscissa, we get the *operating lines* for the fractionator. Operating lines are graphical representations of material balance equations. The physical significance of the operating line is that the coordinates of any point on these lines represent the mole fractions of component **A** in the liquid and the vapour phases crossing each other between two adjacent plates. The slope of the enriching section operating line is $\frac{L_n}{G_{n+1}}$ and the slope of the stripping section operating line is $\frac{\bar{L}_m}{\bar{G}_{m+1}}$. Since the liquid flow and the vapour flow change with the column height, the operating lines need not be straight lines. The McCabe-Thiele graphical construction can be used to determine the number of theoretical plates if the equilibrium curve and the operating lines are drawn. The operating lines as given by Eqs. (5.117) and (5.118) are difficult to plot because the variation in the flow rates of the liquid and vapour as they move from plate to plate should be known. The graphical construction can be considerably simplified if the operating lines can be approximated as straight lines. The basic assumptions of the McCabe-Thiele method are intended to render the operating lines straight lines.

Constant Molar Overflow

McCabe-Thiele method assumes that the molar flow of liquid remains constant at each section of the column irrespective of the plate. Thus

$$L_n = L \quad \text{(for all } n \text{ in the rectifying section)}$$

$$\bar{L}_m = \bar{L} \quad \text{(for all } m \text{ in the stripping section)}$$

The above simplification is known as the assumption of *constant molar overflow*. Since $G_{n+1} = D + L_n$ [Eq. (5.103)], if the liquid flow is constant at L throughout the enriching section, the vapour flow also should be constant. This is true for the stripping section as well. Thus

$$G_n = G \quad \text{(for all } n \text{ in the rectifying section)}$$

$$\bar{G}_m = \bar{G} \quad \text{(for all } m \text{ in the stripping section)}$$

This is the condition of *constant molar vaporisation*. It means that the constant molar vaporisation follows from the assumption of constant molar overflow.

Under what conditions, the above assumptions represent the real picture in the column? Generally, the vapour and liquid flows in the column are influenced by the following factors:

1. The molar heat of vaporisation of the mixture.
2. The heat of mixing or solution.
3. Change in the sensible heat of the liquid and vapour due to temperature changes.
4. Heat losses from the column to the surroundings.

If the molar latent heats of vaporisation of the components **A** and **B** are almost equal, the latent heat of vaporisation of the mixture will be independent of composition. Then, if other heat effects are negligible, the heat required for vaporisation of one mole of the liquid will be the same as the heat released by the condensation of one mole of the vapour. As an example, let us consider the benzene-toluene system with pure component molar heat of vaporisation values 30800 kJ/kmol for benzene and 33362 kJ/kmol for toluene. It means that by utilizing the heat liberated by condensation of one mole of toluene, about 1.08 kmol of benzene can be vaporised. That is, for every mole of liquid vaporised on the plate, nearly one mole of vapour is condensed; the net effect is that there is no change in the molar flows of liquid or the vapour. The conditions of constant molar overflow or constant molar vaporisation are thus realized. Even if all the four heat effects listed above are present in appreciable quantities as to influence the flows in the column, constant molar overflow can still exist, if their net effect is zero.

The assumption of constant molar overflow should not be viewed as a serious limitation of McCabe-Thiele method, since in very few systems do the molar heats of vaporisation differ by more than 10 per cent. It is observed that substances with fairly close boiling points have almost identical molar latent heats. Solutions with a nearly constant relative volatility also exhibit nearly constant molar overflow (King, 1980).

Since the liquid and vapour flow rates remain constant in the rectifying section and the stripping section, the material balance equations for the enriching and stripping sections, Eq. (5.104) and Eq. (5.110), can be modified as

$$G y_{n+1} = L x_n + D x_D \qquad \text{...(5.119)}$$

$$\bar{G} y_{m+1} = \bar{L} x_m - B x_B \qquad \text{...(5.120)}$$

and the slope of the operating line $\frac{L_n}{G_{n+1}}$ in Eq. (5.117) can be replaced by $\frac{L}{G}$ and slope $\frac{\bar{L}_m}{\bar{G}_{m+1}}$ in Eq. (5.118) can be replaced by $\frac{\bar{L}}{\bar{G}}$ so that we have for the operating lines, the following equations.

Operating line for rectifying section:

$$y_{n+1} = \frac{L}{G} x_n + \frac{D x_D}{G} \qquad \text{...(5.121)}$$

Operating line for the stripping section:

$$y_{m+1} = \frac{\bar{L}}{\bar{G}} x_m - \frac{B x_B}{\bar{G}} \qquad \text{...(5.122)}$$

The operating lines represented by Eqs. (5.121) and (5.122) are straight lines and can be easily drawn if the coordinates of any two points on the lines or the coordinates of one point and the slope of the line, are known.

It should be noted that for many binary systems, the latent heat of vaporisation of components expressed on mass basis will be considerably different even for chemically similar components with very close boiling points. Even when the molar flow rates (kmol/h) remain constant, the flow rates evaluated on mass basis (kg/h) may vary throughout the column. The operating lines plotted on the mass basis will not be straight lines, and the McCabe-Thiele method is therefore not applicable for the estimation of the number of plates, on mass basis.

Rectifying Section

The operating line for the enriching section is given by Eq. (5.119)

$$y_{n+1} = \frac{L}{G} x_n + \frac{D x_D}{G} \quad \text{...(5.121)}$$

Equation (5.119) when plotted gives a straight line because the ratio $\frac{L}{G}$ which is the slope of the line is constant under constant molar overflow. Taking the material balance around the condenser, we see that

$$G = L + D \quad \text{...(5.123)}$$

Substituting Eq. (5.121) into Eq. (5.119), we get

$$y_{n+1} = \frac{L}{L+D} x_n + \frac{D x_D}{L+D} \quad \text{...(5.124)}$$

Dividing the numerator and the denominator of the terms on the right side of Eq. (5.122) by D and noting that the reflux ratio $R = \frac{L}{D}$, we can modify the equation for the enriching operating line as

$$y_{n+1} = \frac{L/D}{L/D+1} x_n + \frac{x_D}{L/D+1} \quad \text{...(5.125)}$$

$$y_{n+1} = \frac{R}{R+1} x_n + \frac{x_D}{R+1} \quad \text{...(5.126)}$$

The slope of the enriching operating line is $\frac{R}{R+1}$ and the line intersects the y-axis at $\frac{x_D}{R+1}$. Put $x_n = x_D$ in Eq. (5.124) and we see that $y_{n+1} = x_D$. The operating line passes through the point (x_D, x_D) which is a point on the diagonal of the equilibrium diagram. Knowing the distillate composition, the point (x_D, x_D) can be marked on the diagonal as represented by the point D in Fig. 5.33b. A straight line drawn through point D, making an intercept $\frac{x_D}{R+1}$ with y-axis is the enriching operating line. The liquid and vapour crossing each other anywhere in the enriching section have compositions such as x_n and y_{n+1} and these compositions can be represented by points lying on the operating line. Thus (x_0, y_1), (x_1, y_2), (x_2, y_3), etc., are points on the operating line for the enriching section.

Assume that a total condenser as shown in Fig. 5.33a is used. The vapour reaching the condenser, the reflux and the distillate have the same composition. Therefore, $y_1 = x_0 = x_D$. The liquid and the vapour above the top plate (plate 1) have compositions x_0 and y_1 and since these are equal to the distillate composition x_D, the point D represents conditions above plate 1. The liquid flowing down the plate 1 will be in equilibrium with the vapour leaving the plate. Intersection of the horizontal line drawn

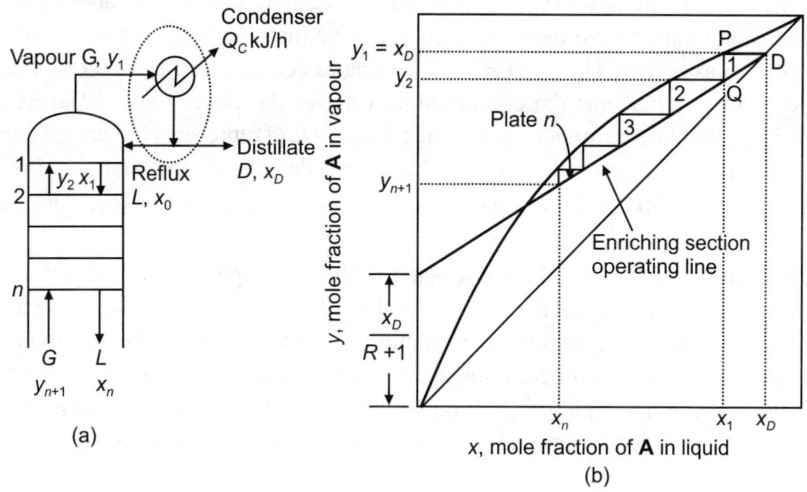

Fig. 5.33: (a) The enriching section with a total condenser (b) McCabe-Thiele construction for enriching section

from point D with the equilibrium curve will give the compositions of the vapour and the liquid leaving the plate 1. Thus point P has coordinates (x_1, y_1). Since x_1 and y_2 are the composition of the streams crossing each other between plates 1 and 2, a vertical line drawn at P intersect the operating line at Q (x_1, y_2). Thus the step (triangle) DPQ constructed in Fig. 5.33b represents plate 1 since by moving from point D to point Q, we have moved from the conditions above the plate 1 to conditions between plate 1 and plate 2. The next step that can be constructed between the operating line and the equilibrium curve represents plate 2 and so on. The construction is continued till we reach the feed plate, the location of which is determined by the method to be discussed in the later sections.

When a partial condenser is used, the liquid reflux does not have the same composition as the distillate (Fig. 5.34a). Partial condenser produces a vapour and liquid which may be assumed to be in equilibrium. The liquid of composition x_0 is returned to the column as reflux and the vapour of composition y_0 is withdrawn as the distillate. Since the distillate composition is specified by the process, $y_0 = x_D$.

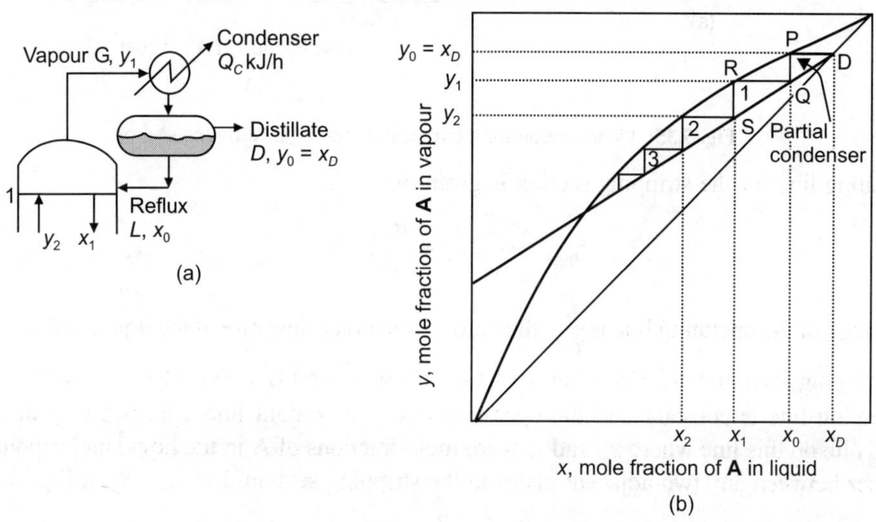

Fig. 5.34: (a) Partial condenser (b) Construction for theoretical plates for enriching section with a partial condenser

Here, the point D with coordinates (x_D, x_D) does not represent the conditions above plate 1. The point representing the conditions above the plate 1 should have the coordinates (x_0, y_1). Now consider the triangle DPQ drawn as before. The point P has coordinates (x_0, y_0) and the point Q has coordinates (x_0, y_1). Therefore point Q represents the conditions just above the plate 1 and as far as the column is concerned the operating line terminates at Q. The triangle DPQ represents the partial condenser which acts as a theoretical plate outside the column. Now consider the second triangle QRS shown in the figure. Point R has coordinates (x_1, y_1) and S has coordinates (x_1, y_2) and therefore QRS represents plate 1.

The number of triangles that can be constructed in the enriching section is equal to the number of theoretical plates in the enriching section when a total condenser is used, and the number of plates when a partial condenser is used is one less than the number of triangles that can be constructed. Though the partial condenser achieves enrichment equivalent to one theoretical plate outside the column, it is practically difficult to provide equilibrium condensation. Therefore, it is safer to ignore the enrichment brought out by the partial condenser and include an additional plate in the design of new columns.

Stripping Section

Plate m represents any general plate in the stripping section of the column as shown in Fig. 5.35a.

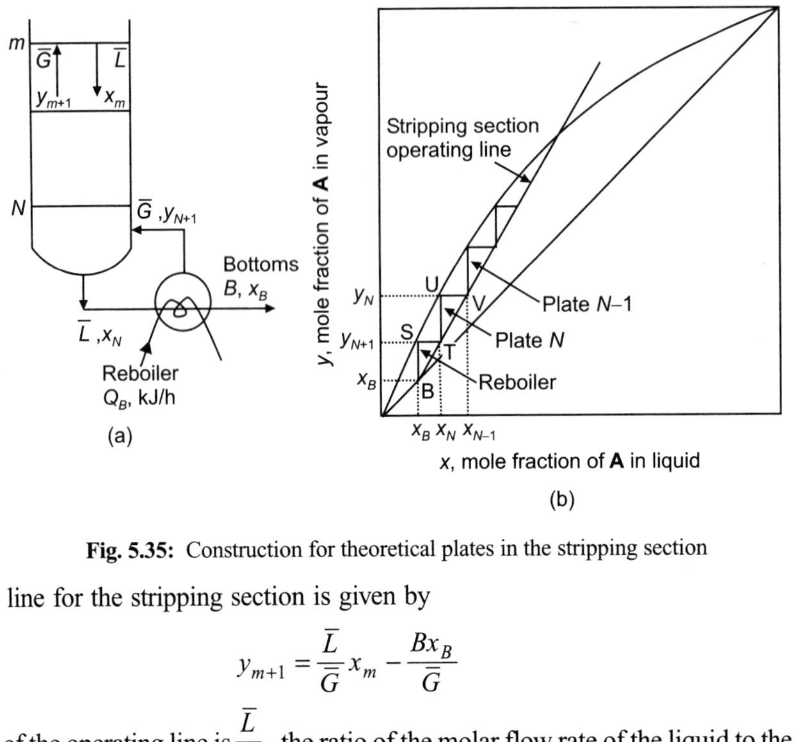

Fig. 5.35: Construction for theoretical plates in the stripping section

Operating line for the stripping section is given by

$$y_{m+1} = \frac{\bar{L}}{\bar{G}} x_m - \frac{B x_B}{\bar{G}} \quad \text{...(5.122)}$$

The slope of the operating line is $\frac{\bar{L}}{\bar{G}}$, the ratio of the molar flow rate of the liquid to the vapour flows in the stripping section. Under constant molar overflow, $\bar{L}$ and $\bar{G}$ are constant; therefore, the slope of the operating line is constant and the operating line is a straight line. The points with coordinates (x_m, y_{m+1}) lie on this line where x_m and y_{m+1} are mole fractions of **A** in the liquid and vapour that cross each other between any two adjacent plates in the stripping section. Let $x_m = x_B$ in Eq. (5.120). Then

$$y_{m+1} = \frac{\bar{L} - B}{\bar{G}} x_B = \frac{\bar{G}}{\bar{G}} x_B = x_B$$

It means that the stripping operating line pass through point (x_B, x_B) represented by point B on the diagonal of the equilibrium diagram. The point B can be located on the diagonal since the bottoms composition x_B is specified by the process. If the slope $\frac{\bar{L}}{\bar{G}}$ also is known, we can draw the operating line. This slope depends on the reflux ratio, quality as well as quantity of the feed entering the column, as will be shown later. In practice, the stripping operating line can be drawn easily by using the feed line as explained in the following sections.

In some cases, the vapour boilup ratio is specified instead of the reflux ratio. The vapour boilup, V, is defined as the moles of vapour returned to the column from the reboiler per mole of bottom product withdrawn. $V = \frac{\bar{G}}{B}$. Equation (5.120) can be written in terms of V. By material balance on the reboiler, $\bar{L} = \bar{G} + B$. Substitute this in Eq. (5.122) and divide the numerator and denominator of the terms on the right side by B.

$$y_{m+1} = \frac{\bar{G}+B}{\bar{G}}x_m - \frac{Bx_B}{\bar{G}} = \frac{\bar{G}/B+1}{\bar{G}/B}x_m - \frac{x_B}{\bar{G}/B}$$

$$y_{m+1} = \frac{V+1}{V}x_m - \frac{x_B}{V} \qquad \text{...(5.127)}$$

As the slope of the operating line for the stripping section is $\frac{V+1}{V}$ and as it passes through the point (x_B, x_B), the operating line can be drawn, if the vapour boilup ratio V is known.

Let the operating line be drawn as shown in Fig. 5.35b. Point B is the lower terminus of the line. The vertical line drawn at B intersects the equilibrium curve at S. For most of the common type of reboilers the vapour leaving the reboiler is in equilibrium with the bottom product of composition x_B, and hence the coordinates of S are (x_B, y_{N+1}). The intersection of the horizontal line drawn through S and the operating line is represented in the figure by point T and has coordinates (x_N, y_{N+1}). x_N is the composition of the liquid leaving and y_{N+1} is the composition of the vapour reaching the bottom plate. The triangle BST therefore represents the reboiler which acts as a theoretical plate outside the column. The triangle TUV shown in the diagram stands for plate N. The stepwise construction is continued till the feed plate is reached. The number of triangles so constructed less one will be equal to the number of plates in the stripping section.

Feed Quality and the Feed Line

Feed quality, *q* Vapour flow in the stripping section $\bar{G}$ and the liquid flow $\bar{L}$ will be different from the corresponding quantities G and L in the rectifying section because of the introduction of the feed on the feed plate. It is not the quantity of feed admitted alone that will determine the difference in the flows in the stripping section and rectifying section. The quantity as well as the quality of the feed influences the flow rates. By the feed quality we mean the physical state and thermal conditions at which the feed enters the column after it has been flashed adiabatically to the pressure at the feed plate. There are five possible states for the feed entering the column. It may enter the column as (1) a cold liquid, (2) as a saturated liquid, (3) as partially vaporised mixture, (4) as a saturated vapour or (5) as a superheated vapour. These feed conditions are illustrated in Fig. 5.36.

If the feed admitted is a saturated liquid (Fig. 5.36b), the entire feed join the liquid coming from the plate above the feed plate and flows down. The flow rate $\bar{L} = L + F$. The vapour flows in the column are unaffected by feed entry and $\bar{G} = G$. On the other hand if the feed is a saturated vapour (Fig. 5.36d), the entire feed joins the vapour reaching the feed plate from the plate below it and flows up as G in the enriching section. The liquid flow L and $\bar{L}$ remain the same. That is $G = \bar{G} + F$ and $\bar{L} = L$. If the feed is partially vaporised, it enters the column as a mixture of saturated liquid and saturated vapour (Fig. 5.36c). The liquid portion of the feed becomes part of $\bar{L}$ and vapour portion becomes part of G. Therefore, $\bar{L} > L$, $G > \bar{G}$. If the feed is a subcooled liquid, it has to be brought to the saturation temperature first. The sensible heat required for this is supplied by the condensation of a portion of the vapour reaching the feed plate, the quantity of which is dependent on the degree of subcooling of the liquid. The condensate so formed, the feed which is heated to its bubble point and the liquid flowing down the enriching section join together and flow down the stripping section as $\bar{L}$. Therefore $\bar{L} > L + F$ and because of the condensation of a portion of $\bar{G}$ on the feed plate, $G < \bar{G}$. If the feed is a superheated vapour (Fig. 5.36e), on reaching the fed plate it releases the sensible heat which is used up by a portion of the saturated liquid reaching there, for vaporisation. The feed which is cooled to saturated vapour state, the vapour coming from the stripping section and the vapour formed on the feed plate flow upward from the feed plate. The liquid flowing down the feed plate will be less by the amount vaporised on the plate using the sensible heat given out by the feed, which in turn is dependent on the degree of superheat of the feed. $G > \bar{G} + F$ and $\bar{L} < L$.

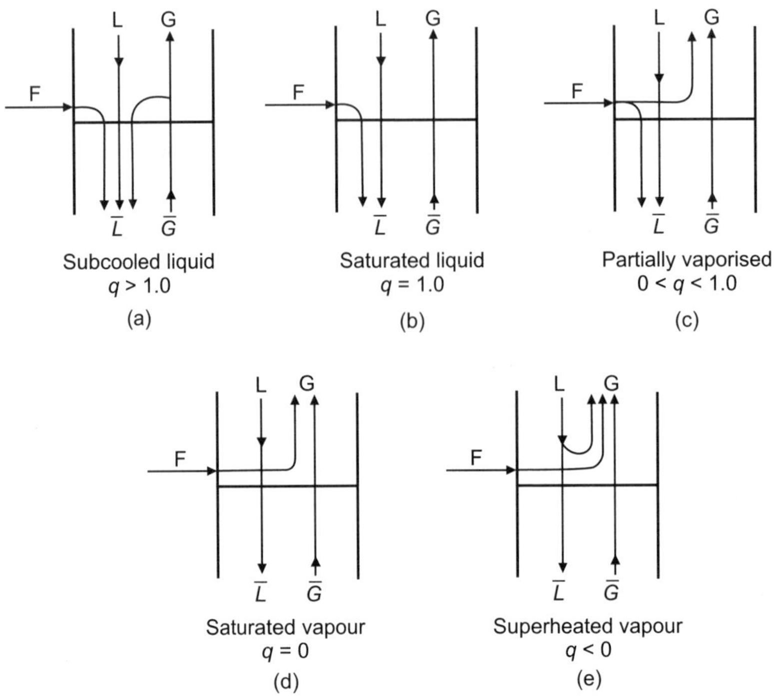

Fig. 5.36: Different thermal conditions of feed

How do we express all these feed conditions? The *feed quality factor* q takes care of all the five feed conditions. The simplest definition for q is that it is the fraction of saturated liquid present in the feed. Thus if feed is saturated liquid $q = 1.0$ and if it is saturated vapour $q = 0$. For a mixture of saturated liquid and vapour $0 \le q \le 1.0$, depending upon the extent of vaporisation of the feed. Can this definition of q provide q values for the other two feed conditions, the subcooled liquid and the superheated vapour?

The feed quality q can also be expressed as the increase in the liquid flow in the stripping section that results by the introduction of one mole of feed. That is,

$$q = \frac{\bar{L} - L}{F} \quad \text{...(5.128)}$$

For saturated liquid feed, $\bar{L} = L + F$ and therefore $q = 1.0$. For saturated vapour feed, $\bar{L} = L$ and therefore $q = 0$. If the feed is a mixture of liquid and vapour containing 40% liquid, $\bar{L} = L + 0.4F$ and therefore $q = 0.4$.

For subcooled liquid and superheated vapour, Eq. (5.128) can be combined with an energy balance on the feed plate to estimate q (Fig. 5.37). Assume that F kmol/h of feed with enthalpy H_F kJ/kmol is admitted to the feed plate. Saturated liquid L and saturated vapour $\bar{G}$ reach the plate and the saturated liquid $\bar{L}$ and saturated vapour G leave the plate. It is assumed that the temperature and concentration change on the feed plate is so small that the saturated liquid streams are assumed to have same enthalpy H_L and the saturated vapour, same enthalpy H_G.

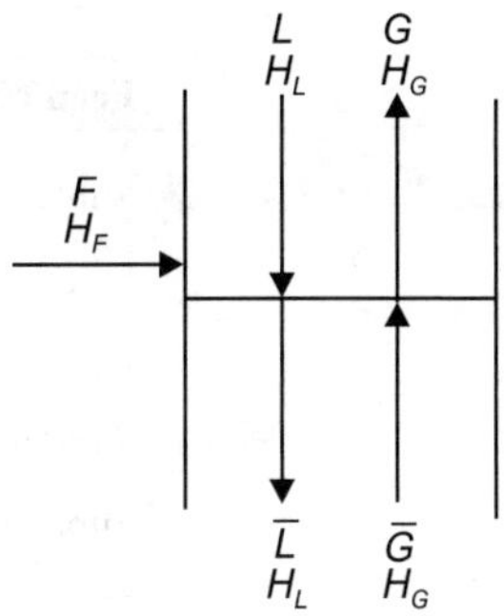

Fig. 5.37: Vapour and liquid flows at the feed plate

Total material balance on the feed plate is

$$F + \bar{G} + L = G + \bar{L} \qquad ...(5.129)$$

which can be rearranged as

$$\bar{G} - G = (\bar{L} - L) - F \qquad ...(5.130)$$

Enthalpy balance on the feed plate can be written as

$$FH_F + \bar{G}H_G + LH_L = GH_G + \bar{L}H_L \qquad ...(5.131)$$

Rearrange Eq. (5.131) and substitute Eq. (5.130) into it.

$$(\bar{L} - L)H_L = (\bar{G} - G)H_G + FH_F \qquad ...(5.132)$$

$$(\bar{L} - L)H_L = (\bar{L} - L)H_G + FH_F - FH_G \qquad ...(5.133)$$

$$(\bar{L} - L)(H_G - H_L) = F(H_G - H_F) \qquad ...(5.134)$$

The feed quality factor q is obtained from Eq. (5.128).

$$q = \frac{\bar{L} - L}{F} = \frac{H_G - H_F}{H_G - H_L} = \frac{H_G - H_F}{\lambda_v} \qquad ...(5.135)$$

where λ_v is the latent heat of vaporisation of the mixture.

$$q = \frac{\text{Heat required for converting one mole of the feed into a saturated vapour}}{\text{molar heat of vaporisation of the feed}} \qquad ...(5.136)$$

The feed quality q is the ratio of the heat required by the feed for converting it completely as saturated vapour to the latent heat of vaporisation of the mixture.

Assume that the feed is a subcooled liquid at temperature T_F and its heat capacity is C_{pL}. Let the bubble point of the feed at the operating pressure be T_b. The heat required for converting one mole of the feed into a saturated vapour = sensible heat required to raise the temperature of the feed to the bubble point + latent heat of vaporisation $= C_{pL}(T_b - T_F) + \lambda_v$. Therefore,

$$q = \frac{C_{pL}(T_b - T_F) + \lambda_v}{\lambda_v} = 1 + \frac{C_{pL}(T_b - T_F)}{\lambda_v} \qquad ...(5.137)$$

As is clear from Eq. (5.137), for subcooled liquid feed, $q > 1.0$.

If the feed is superheated to temperature T_F and if the dew point of the feed is T_d, the heat given out on cooling the vapour to its saturated state is $= C_{pV}(T_F - T_d)$ where C_{pV} is the heat capacity of the feed vapour. Therefore, $H_G - H_F$, in Eq. (5. 135) is $-C_{pV}(T_F - T_d)$ and

$$q = -\frac{C_{pV}(T_F - T_d)}{\lambda_v} \qquad ...(5.138)$$

Equation (5.138) indicates that for superheated vapour, $q < 0$.

The feed quality factors are summarised in Table 5.2

Table 5.2: Values of q for different feed conditions

Feed condition	Quality q	Remarks
Subcooled liquid	$q > 1.0$	$q = 1 + \frac{C_{pL}(T_b - T_F)}{\lambda_v}$
Saturated liquid	$q = 1.0$	
Partially vaporised	$0 \le q \le 1.0$	
Saturated vapour	$q = 0$	
Superheated vapour	$q < 0$	$q = -\frac{C_{pV}(T_F - T_d)}{\lambda_v}$

Feed line. Using the definition of q (Eq. 5.128), the liquid rate $\bar{L}$ can be written as

$$\bar{L} = L + qF \quad \text{...(5.139)}$$

Since the reflux ratio $R = L/D$, Eq. (5.139) becomes

$$\bar{L} = RD + qF \quad \text{...(5.140)}$$

By a material balance around the reboiler, the vapour rate $\bar{G}$ can be written as

$$\bar{G} = \bar{L} - B \quad \text{...(5.141)}$$

The overall material balance for the column gives $B = F - D$. Substitute this result and Eq. (5.140) into Eq. (5.141), we get

$$\bar{G} = (R+1)D + (q-1)F \quad \text{...(5.142)}$$

We have shown that [Eq. (5.120)] the slope of the stripping operating line is $\bar{L}/\bar{G}$. Using Eqs. (5.140) and (5.142)

$$\frac{\bar{L}}{\bar{G}} = \frac{RD + qF}{(R+1)D + (q-1)F} = \frac{R + q\dfrac{F}{D}}{(R+1) + (q-1)\dfrac{F}{D}} \quad \text{...(5.143)}$$

Slope of the enriching section operating line [by Eq. (5.126)] is

$$\frac{L}{G} = \frac{R}{(R+1)} \quad \text{...(5.144)}$$

The last two equation show that the slope of the operating line for the enriching section depends on the reflux ratio, whereas the slope of the stripping section operating line depends in addition, on the quality and quantity of the feed.

In the McCabe-Thiele construction, the point of intersection of the upper and lower operating lines (point P) is located first and the stripping section operating line is then drawn as a straight line passing through P and point B (x_B, x_B) on the diagonal. To find the intersection, let us consider the operating line equations for the enriching section and stripping section already developed. *See* equations (5.119) and (5.120).

$$Gy_{n+1} = Lx_n + Dx_D \quad \text{...(5.119)}$$

$$\bar{G}y_{m+1} = \bar{L}x_m - Bx_B \quad \text{...(5.120)}$$

Let (x', y') be the point of intersection of the operating lines. Then

$$Gy' = Lx' + Dx_D \quad \text{...(5.145)}$$

$$\bar{G}y' = \bar{L}x' - Bx_B \quad \text{...(5.146)}$$

Subtracting Eq. (5.145) from Eq. (5.146), we get

$$(\bar{G}-G)y' = (\bar{L}-L)x' - (Bx_B + Dx_D) \quad \text{...(5.147)}$$

Since $\bar{G}-G = (\bar{L}-L)-F$ (by Eq. (5.130)), and $(Bx_B + Dx_D) = Fz_F$ by the component-**A** balance on the fractionation column, Eq. (5.147) can be modified as

$$(\bar{L}-L-F)y' = (\bar{L}-L)x' - (Fz_F) \quad \text{...(5.148)}$$

Dividing by F on both sides of Eq. (5.148), we get

$$\left(\frac{\bar{L}-L}{F}-1\right)y' = \frac{\bar{L}-L}{F}x' - z_F \quad \text{...(5.149)}$$

Using Eq. (5.128) into Eq. (5.149), we get

$$(q-1)y' = qx' - z_F \quad \text{...(5.150)}$$

Or

$$y' = \frac{q}{q-1}x' - \frac{z_F}{q-1} \quad \text{...(5.151)}$$

Equation (5.151) is the equation of a straight line with slope $\frac{q}{q-1}$. The line is called the q line or the feed line. Put $x' = z_F$ in Eq. (5.151). We can see that $y' = z_F$. It means that the q-line passes through the point (z_F, z_F) on the diagonal of the equilibrium diagram. The significance of q-line is that it represents the locus of the point of intersection of the enriching operating line and the stripping operating line.

The slope of the q-line depends on the feed quality as shown in the Fig. 5.38. When the feed is saturated liquid, q = 1, the slope of q-line $q/(q-1)$ is infinity and the q-line is vertical. When the feed is saturated vapour, q = 0, the slope of q-line is zero and therefore q-line is a horizontal line as shown in the figure. When the feed is partially vaporised, slope of q-line is negative and it lies between the horizontal and vertical lines drawn through F (z_F, z_F). For example, if the feed contains 50% vapour and 50% liquid, q = ½. Slope of the q-line is – 1 and the line makes 45° with the horizontal at F.

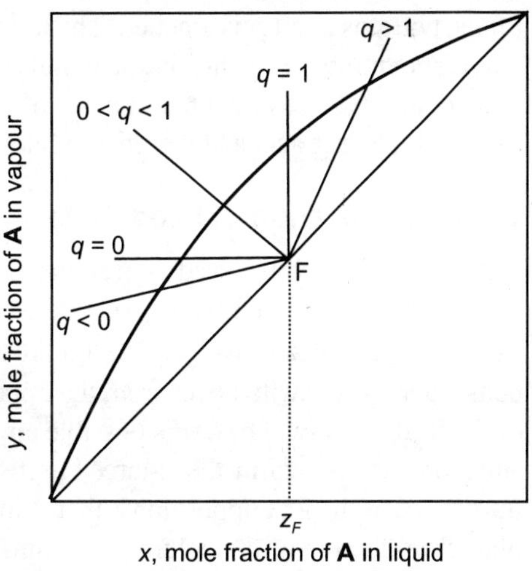

Fig. 5.38: Effect of feed quality on the feed line

Construction of Operating Lines

The fact that the lower operating line (the stripping operating line) passes through the intersection of the upper operating line (enriching operating line) and the q-line allows us to draw the lower operating line easily. The upper operating line is drawn from the knowledge of the reflux ratio and distillate composition. Distillate composition x_D fixes point D and the reflux ratio determines the intercept that the upper operating line makes with the y-axis. Draw the line passing through point D such that it makes an intercept $x_D/(R+1)$ with the y-axis. The q-line is drawn from point F (z_F, z_F) on the diagonal with a slope $q/(q-1)$. Its intersection with the upper operating line is denoted by P in Fig. 5.39.

The straight line passing through points P and B is the lower operating line. The construction of triangles which represents the theoretical plates in the column is initiated from point D. Stair-case construction is continued between the enriching operating line and the equilibrium curve till the point of intersection of the operating lines (point P) is reached. On crossing point P, further construction is

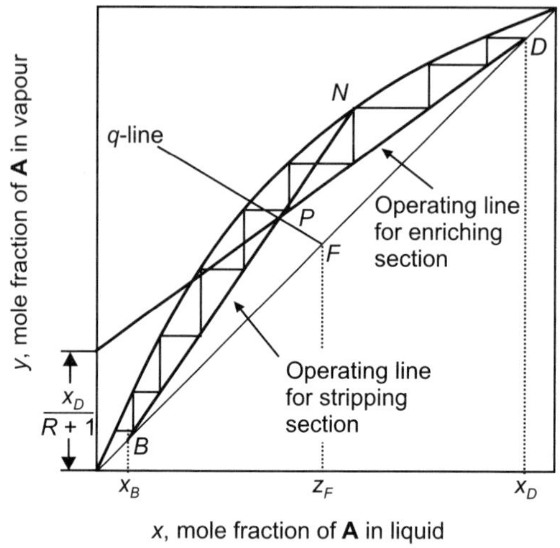

Fig. 5.39: Construction of operating line for the stripping section

carried out between the lower operating line and the equilibrium curve. Construction of steps is continued till the bottoms point B is reached. The triangle that straddles the operating line intersection P is the feed plate. The number of plates in the column is one less than the number of triangles if a total condenser is used, and two less than the number of triangles, if a partial condenser is used. The reboiler and the partial condenser acts as theoretical plates outside the column as described earlier.

Optimum Location of Feed Plate

The plate on which the feed is introduced is called the feed plate. The feed plate need not necessarily be the plate represented by the triangle which straddles the point of intersection of the operating lines. Introduction of feed changes the liquid to vapour flow ratio in the column, from L/G to $\overline{L}/\overline{G}$ and consequently we switch over from the upper operating line to the lower one at the point of introduction of the feed. This switch over is possible anywhere between point M and N according to Fig. 5.40. Three situations are shown in the figure. Figure 5.40a represents the early feed entry in which the feed is introduced at the first opportunity, that is immediately after passing point N. The feed plate is the fourth plate from the top of the column assuming that a total condenser is used. The switch over from upper operating line to the lower one is done at the fourth step and the triangle has its one corner on the upper operating line and the other corner on the lower operating line. We see that roughly 9.5 theoretical plates are required in the column in addition to a partial reboiler and total condenser. In Fig. 5.40b the feed introduction is delayed up to the seventh plate from the top. Upper operating line is used for stepping off stages up to the seventh stage and for further construction the lower operating line is used. We see that the number of theoretical plates in the column is 9. In contrast to both (a) and (b), in figure (c), the transition from the upper to the lower operating line is done immediately after P, the intersection of the operating lines, is crossed. We see that the number of theoretical plates required is 8. This particular choice for the feed plate location is the optimum feed plate location, as it always results in lesser number of theoretical stages in the column in comparison with delayed or early feed entry. In other words the greatest enrichment per plate is achieved if the feed plate is located at the optimum point. If the feed plate is located anywhere else, an unnecessarily large number of stages would be necessary. In the design of new columns, this practice is generally adhered to in locating the feed plate. However in the adaptation of an existing column, the provision of the feed entry nozzles in the column may restrict our choice of the feed plate. Since the number of theoretical plates in each section of the column is fixed in this case, the reflux ratio and the product compositions may be determined by trial and error to satisfy this condition.

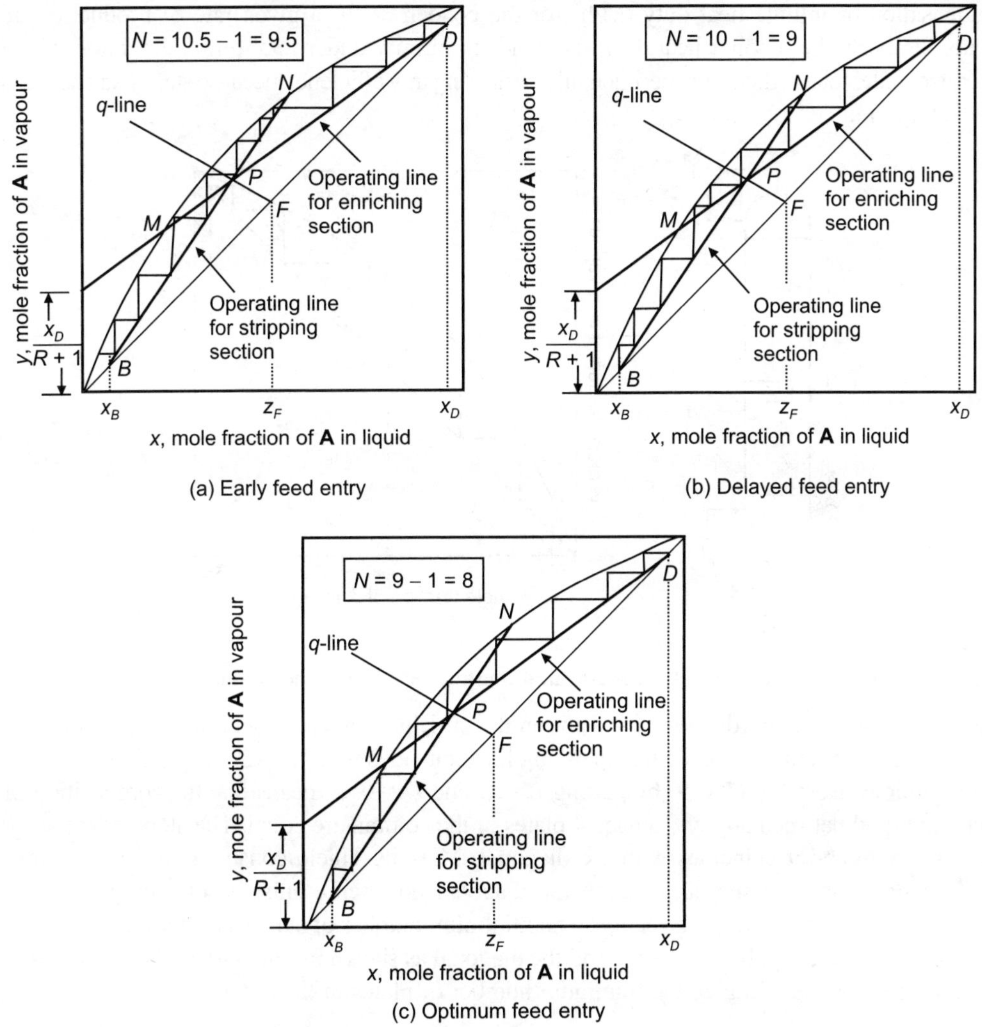

Fig. 5.40: Optimum location of the feed plate

Total Reflux and Minimum Number of Stages

The reflux ratio is a very important parameter in the design, operation and analysis of fractionating columns. The reflux ratio fixes the slope of the operating lines which, in turn determines the number of plates in the column. The slope of enriching operating line $L/G = R/(R + 1)$. Under normal operating conditions, $L < G$, the slope of the enriching line is less than unity. When the reflux ratio increases, the operating line slope increases and it moves toward the diagonal. In the limiting case when $R = \infty$, $L = G$, the slope of the operating line becomes equal to unity and the line coincides with the diagonal. This condition is known as the total reflux condition. For any specified separation, the lowest number of plates results, when the reflux ratio employed is the maximum. Total reflux condition is a limiting condition in the column operation. The entire liquid leaving the condenser is returned to the column as reflux and the entire liquid entering the reboiler is vaporised and returned to the column (Fig. 5.41a). There is neither continuous addition of the feed, nor withdrawal of the products. $F = D = B = 0$. Since under total reflux, $L/D = \infty$, the total reflux conditions may also be interpreted as requiring infinite rate of condensation or infinite heat duty (kJ/h) for the condenser for a given rate of production of the distillate. Similarly the reboiler heat duty also tends to infinity under total reflux condition. Columns are operated under total reflux for experimentally evaluating tray efficiency because steady state conditions can be achieved under total reflux.

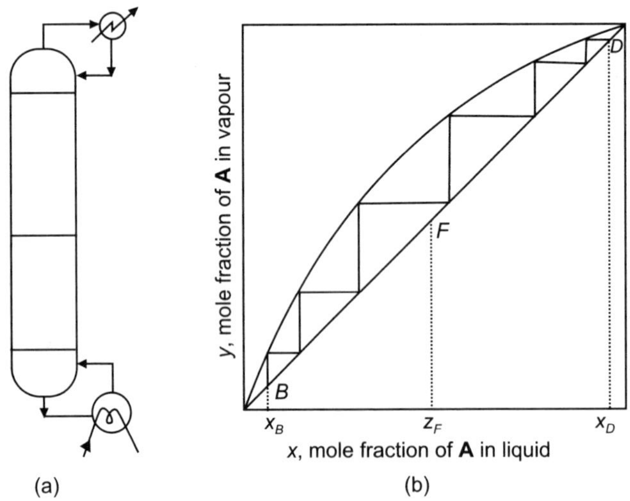

Fig. 5.41: (a) Total reflux condition (b) The graphical construction to determine the minimum number of stages

We have shown that coordinates of a point on the enriching line represent the compositions of the liquid and vapour x_n and y_{n+1}. Since the operating line coincides with the 45° diagonal, $y_{n+1} = x_n$. This condition follows from Eq. (5.119) by putting $L = G$ and $D = 0$. It means that the composition of the vapour and liquid between any two adjacent plates in the column are equal. The slope of the stripping line is also unity and it coincides with the diagonal. McCabe-Thiele construction for the minimum number of stages is very simple, because the diagonal now represents the operating lines for the enriching section and the stripping section. The minimum number of stages is determined by stepping off stages between the equilibrium curve and the diagonal as shown in Fig. 5.41b. The number of steps so constructed less one will give the minimum number of plates in the column.

Fenske Equation for Minimum Number of Stages

The minimum number of plates under total reflux conditions can be determined analytically from the terminal concentrations x_D and x_B using Fenske equation. This is applicable to systems for which the relative volatility remains constant and the equilibrium relationship can be expressed in terms of the relative volatility.

$$\frac{y}{1-y} = \frac{\alpha x}{1-x} \qquad \ldots(5.152)$$

For ideal solutions, the relative volatility is the ratio of the vapour pressures of the components which can be assumed to remain constant within a reasonable temperature range that is normally encountered in distillation columns. The small variation in α can be accommodated by taking the geometric mean of the α values at the top and bottom conditions of the column.

$$\alpha = (\alpha_1 \alpha_N)^{0.5} \qquad \text{...(5.153)}$$

where α_1 is the relative volatility of the overhead vapour and α_N, that of the liquid leaving the bottom plate. The vapour and liquid leaving plate $(n + 1)$, assuming it to be an ideal plate are in equilibrium and therefore by Eq. (5.152)

$$\frac{y_{n+1}}{1-y_{n+1}} = \frac{\alpha x_{n+1}}{1-x_{n+1}} \qquad \text{...(5.154)}$$

We have already shown that under total reflux, $y_{n+1} = x_n$. Therefore Eq. (5.154) can be modified as

$$\frac{x_n}{1-x_n} = \frac{\alpha x_{n+1}}{1-x_{n+1}} \qquad \text{...(5.155)}$$

Equation (5.155) is applicable to all theoretical plates including the reboiler. Put $n = 0$ in Eq. (5.155). Assuming a total condenser, $x_0 = x_D$ and we get

$$\frac{x_D}{1-x_D} = \frac{\alpha x_1}{1-x_1} \qquad \text{...(5.156)}$$

x_D is the mole fraction of **A** in the liquid reaching plate 1 and x_1 is the mole fraction of **A** in the liquid leaving plate 1. Put $n = 2$ in Eq. (5.155) and we get

$$\frac{x_1}{1-x_1} = \frac{\alpha x_2}{1-x_2} \qquad \text{...(5.157)}$$

Similar equations can be written for all the plates in the column. Let N_m be the number of plates in the column. For $n = N_m - 1$,

$$\frac{x_{N_m-1}}{1-x_{N_m-1}} = \frac{\alpha x_{N_m}}{1-x_{N_m}} \qquad \text{...(5.158)}$$

For the reboiler $(n = N_m)$ the liquid reaching it has composition x_{Nm} and the liquid leaving it has composition x_B,

$$\frac{x_{N_m}}{1-x_{N_m}} = \frac{\alpha x_B}{1-x_B} \qquad \text{...(5.159)}$$

Likewise we can write N_m+1 equations. Equating the product of the terms on the left side of all the above equations with the product of the terms on the right side we get

$$\frac{x_D}{1-x_D} = \frac{\alpha^{N_m+1} x_B}{1-x_B} \qquad \text{...(5.160)}$$

which may be rearranged into the following forms:

$$\alpha^{N_m+1} = \frac{x_D/1-x_D}{x_B/1-x_B} \qquad \text{...(5.161)}$$

$$N_m + 1 = \frac{\ln\left(\dfrac{x_D}{1-x_D} \Big/ \dfrac{x_B}{1-x_B}\right)}{\ln \alpha} \qquad \text{...(5.162)}$$

Equation (5.162) is known as the Fenske equation.

Minimum Reflux Ratio and Infinite Number of Stages

We have seen that the number of plates required for separating a feed of given composition and thermal state, into distillate and residue of specified compositions, is the minimum when the reflux ratio is maximum. If the reflux is reduced, the number of plates increases and a limiting condition is reached corresponding to which the reflux ratio is the minimum. An infinitely large number of plates is required for the separation under this condition. Operating reflux ratio in distillation columns should lie between the minimum reflux at which the number of plates is infinity and total reflux at which the number of plates is the minimum.

The operating lines coincide with the diagonal under total reflux conditions. Since no products are obtained under total reflux, the operating reflux ratio for a distillation column should be less than that in the total reflux condition. The line DP in Fig. 5.42a is a typical operating line for the enriching section and PB is the corresponding operating line for the stripping section for a reflux ratio which is less than the total reflux condition. Since the operating lines lie closer to the equilibrium curve, the triangles that can be constructed between the operating lines and the equilibrium curve and therefore, the number of plates required is considerably larger than the number required under total reflux. If the reflux ratio is reduced further, the operating lines move farther away from the diagonal towards the equilibrium curve and the number of plates is further increased. However, there is a limit to which reflux ratio can be reduced. Referring to Fig. 5.42a, we see that for a particular reflux ratio, the point of intersection of the q-line and the enriching operating line falls on the equilibrium curve as indicated by P'. The stripping section operating line also should pass through P'. We see that a *pinch or an invariant zone* occurs at the feed point and an infinite number of plates is required to achieve the desired separation. The reflux ratio corresponding to this limiting condition is the minimum reflux R_m. Any reflux ratio smaller than R_m can result in infinite number of plates, but the pinch point will be shifted away from the feed point. Thus the minimum reflux ratio can be treated as the largest reflux ratio that will lead to infinite number of plates in the column, for the separation of a feed of given composition and thermal quality, into products of specified composition. Employing a reflux ratio slightly above the minimum reflux will achieve the separation with a finite number of plates.

The minimum reflux ratio R_m can be determined by measuring the slope of the limiting operating line. Alternatively, it can be determined from the coordinates of the point of intersection P'. Let (x', y') be the coordinates of P'. The limiting operating line which determines the minimum reflux is DP' and its slope is $\frac{R_m}{R_m+1}$. Since DP' passes through (x', y') and (x_D, x_D), the slope can be expressed as

$$\frac{R_m}{R_m+1} = \frac{x_D - y'}{x_D - x'} \quad \ldots(5.163)$$

By rearranging Eq. (5.163), the minimum reflux ratio can be written as

$$R_m = \frac{x_D - y'}{y' - x'} \quad \ldots(5.164)$$

For highly non-ideal systems with an equilibrium curve of the type shown in Fig. 5.42b, the operating line drawn through the intersection of the q-line and the equilibrium curve will lie outside the curve and therefore cannot be treated as the limiting operating line. The operating line which determines the minimum reflux ratio in this case is tangential to the equilibrium curve and passes through the point P on the q-line as shown. The line joining P and B is the stripping operating line. The invariant zone or the pinch occurs somewhere above the feed point in the enriching section. R_m should be obtained from the slope of the line DP and cannot be calculated using Eq. (5.164).

Another special case in the evaluation of minimum reflux is for the non-ideal systems of the type shown in Fig. 5.42c. The equilibrium curve is concave upward in the lower concentration region. The limiting operating line is obtained by drawing the stripping section operating line from B tangential to the equilibrium curve to meet the q-line at P. The upper operating line is the line joining D to P. The slope of the line DP gives the $\frac{R_m}{R_m+1}$ from which minimum reflux ratio R_m can be evaluated. The invariant zone in this case is below the feed plate.

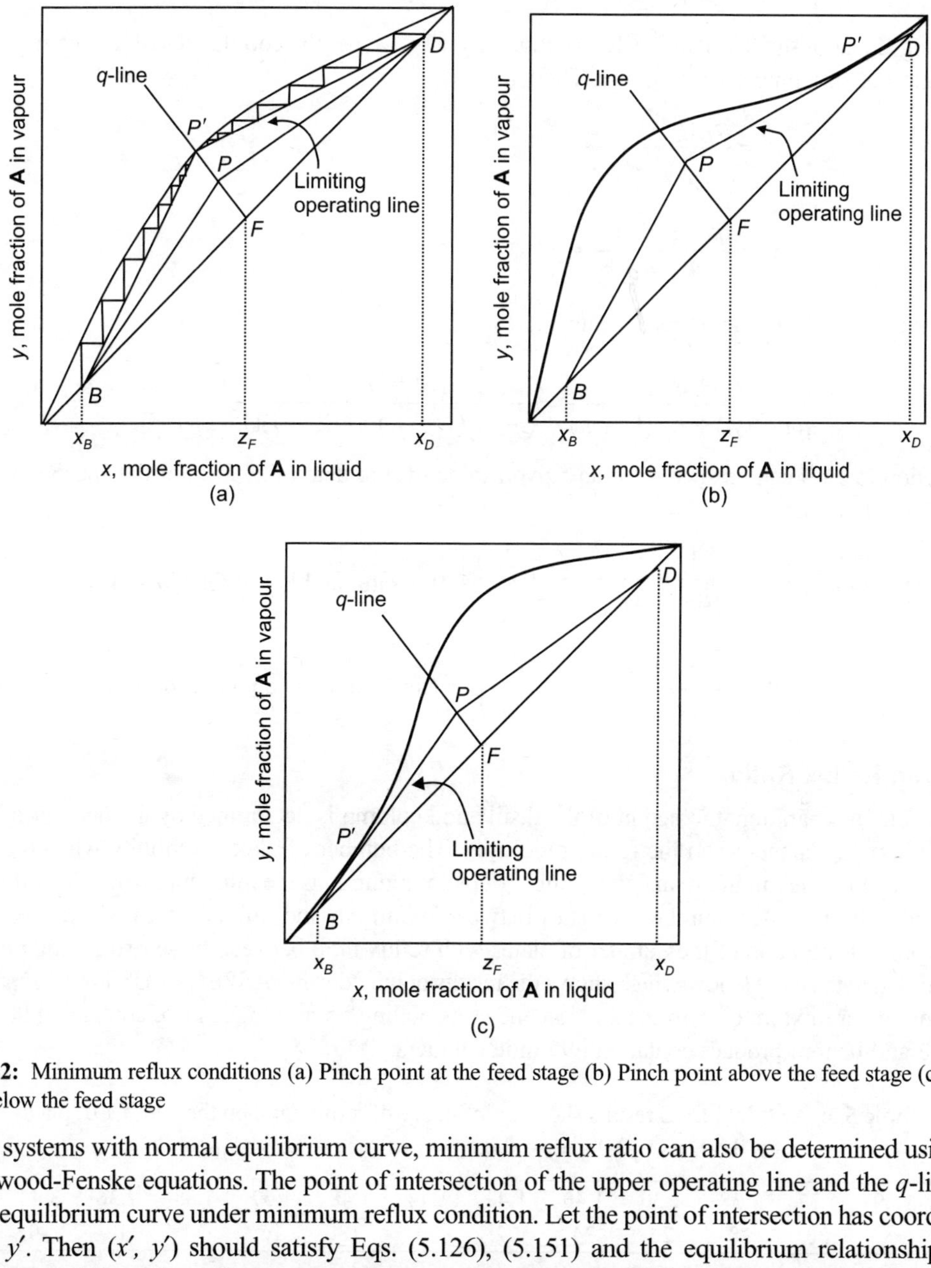

Fig. 5.42: Minimum reflux conditions (a) Pinch point at the feed stage (b) Pinch point above the feed stage (c) Pinch point below the feed stage

For systems with normal equilibrium curve, minimum reflux ratio can also be determined using the Underwood-Fenske equations. The point of intersection of the upper operating line and the q-line lies on the equilibrium curve under minimum reflux condition. Let the point of intersection has coordinates x' and y'. Then (x', y') should satisfy Eqs. (5.126), (5.151) and the equilibrium relationship. The equilibrium relationship can be expressed in terms of the relative volatility as given by Eq. (5.152). That is, we have

$$y' = \frac{R_m}{R_m+1}x' + \frac{x_D}{R_m+1} \qquad \text{...(5.126)}$$

$$y' = \frac{q}{q-1}x' - \frac{z_F}{q-1} \quad \ldots(5.151)$$

$$\frac{y'}{1-y'} = \frac{\alpha x'}{1-x'} \quad \ldots(5.152)$$

Solving Equations (5.126) and (5.151) simultaneously, we get the coordinates of the intersection of q–line and the enriching operating line. They are

$$x' = \frac{x_D(q-1) + z_F(R_m+1)}{R_m+q} \quad \ldots(5.165)$$

$$y' = \frac{qx_D + R_m z_F}{R_m+q} \quad \ldots(5.166)$$

Substitute Eq. (5.165) and (5.166) into Eq. (5.152).

$$\frac{qx_D + R_m z_F}{q(1-x_D) + R_m(1-z_F)} = \frac{\alpha\left[x_D(q-1) + z_F(R_m+1)\right]}{(q-1)(1-x_D) + (R_m+1)(1-z_F)} \quad \ldots(5.167)$$

Equation (5.167) can be solved for any given value of q to determine R_m. For the special case of $q = 0$ and $q = 1$, we get

$$R_m = \frac{1}{\alpha-1}\left[\frac{x_D}{z_F} - \frac{\alpha(1-x_D)}{1-z_F}\right] \text{ (for saturated liquid feed, } q=1) \quad \ldots(5.168)$$

$$R_m = \frac{1}{\alpha-1}\left(\frac{\alpha x_D}{z_F} - \frac{1-x_D}{1-z_F}\right) - 1 \text{ (for saturated vapour feed, } q=0) \quad \ldots(5.169)$$

Optimum Reflux Ratio

For a specified separation, the height of the distillation column is determined by the number of plates which in turn is dependent on the reflux ratio used. The number of plates is infinity when the reflux ratio employed is the minimum and the number of plates required is the minimum when the reflux ratio employed is infinity. Any number of plates between minimum and infinity will provide the desired separation. The variation of the number of plates with reflux ratio between these two limiting cases is as illustrated in the table below which shows the results obtained using ASPEN PLUS for the separation of an equimolar mixture of benzene and toluene at its boiling point into distillate containing 98 mol% benzene and bottom product containing 95 mol% toluene.

Table 5.3: ASPEN PLUS results showing influence of reflux ratio on the number of plates ($z_F = 0.50$, $x_D = 0.98$, $x_B = 0.05$, $q = 1.0$, $R_m = 1.29$, $N_{min} = 7.7$)

R	1.29	1.37	1.38	1.39	1.40	1.42	1.44	1.48	1.57	1.88	2.36	3.45	∞
N	∞	30	28	26	24	22	20	18	16	14	12	10	7.7

The results are also plotted in Fig. 5.43. The number of plates falls off rapidly with reflux ratio in the immediate vicinity of the minimum reflux. However, at high reflux ratio values, the reduction in the number of plates with reflux ratio is very small. The number of plates determines the height of the column. We see that there is considerable saving in the cost of equipment if the reflux ratio is increased from the minimum reflux ratio because of the reduction in the number of plates. However, at very high reflux ratio values the height is not influenced much and the saving on this account is insignificant.

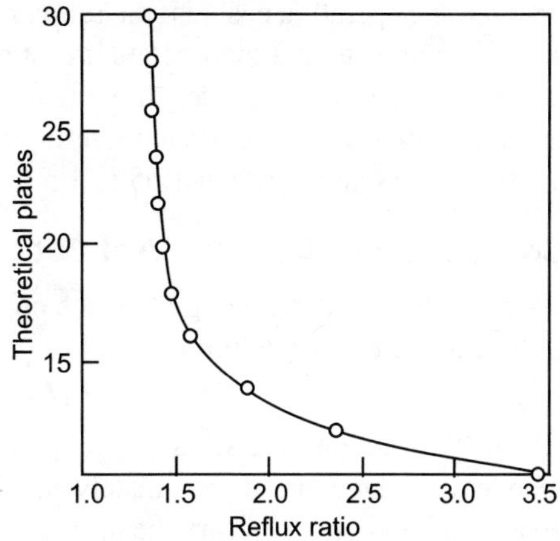

Fig. 5.43: Reflux ratio versus the number of theoretical plates

The fixed cost on the equipment varies with the height and diameter of the column. The diameter of the column is dependent on the vapour flow through the column, and the vapour flow varies with R. For example, for the feed entering at its boiling point, the vapour flow in the enriching section and stripping section are equal, $\bar{G} = G$. Since $G = L + D = (R + 1)D$, the vapour flow in the column and consequently, the diameter of the column increases steadily with reflux ratio. With a reflux ratio in the neighbourhood of the minimum reflux, the diameter of the column increases slowly, but the number of plates decreases rapidly, with the net result that initially the cost of the column decreases with the reflux ratio. At high reflux ratio values the column diameter increases steadily with increase of reflux ratio. However, the reduction in the number of plates due to the increase of reflux ratio in this region is very small. Here, the marginal saving in the cost of equipment due to the reduction in the number of plates is offset by the sharp increase in the diameter of the column with R. Thus the fixed cost on the equipment initially decreases and thereafter increases as reflux ratio is increased from the minimum reflux ratio as shown by curve (a) in Fig. 5.44.

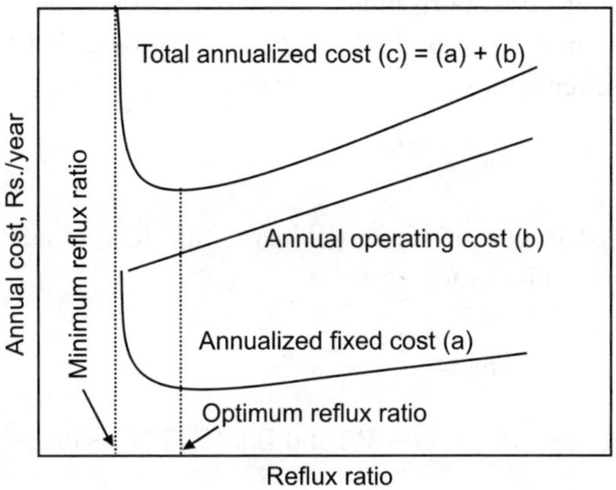

Fig. 5.44: Estimation of optimum reflux ratio

The cost of distillation operation is determined as sum of the capital cost of the equipment and the operating cost, the latter comprising mainly of the cost of steam supply in the reboiler, the cost of circulation of the cooling water and related pumping costs. Each one of these is influenced by reflux ratio. The cost of steam supplied to the reboiler is nearly eight times the cost of cooling water circulated in the condenser and therefore, the operating cost is dominated by steam consumption cost. The cooling water circulation rate depends on G, the amount of vapour condensed and G is directly proportional to $R + 1$ for given rate of production of the distillate, as shown above. The increase of liquid flow in the column due to increase in the reflux ratio also necessitates increased vaporisation in the reboiler. Thus the rate of steam supply also increases directly with reflux ratio. The operating cost of distillation columns is represented by curve (b) as shown in the figure which shows almost direct proportionality between the operating cost and the reflux ratio. The total cost of the distillation unit is the sum of fixed cost of equipment and the operating cost and is represented by the curve (c) in Fig. 5.44.

The total cost curve passes through a minimum as shown and the reflux ratio corresponding to the minimum total cost is the optimum reflux ratio for the column. In practice, the column is operated with a reflux ratio lying between 1.05 to 1.5 times the minimum reflux ratio.

Condenser and Reboiler Heat Duties

We have discussed the energy balance on the distillation column and have shown that in the absence of heat loss from the column,

$$FH_F + Q_B = DH_D + BH_B + Q_C \qquad \text{...(5.113)}$$

For the feed entering at its boiling point, $FH_F = DH_D + BH_B$ and $Q_B = Q_C$. The condenser and reboiler heat duties are approximately equal. In general, if total condenser is used, the amount of vapour condensed is G and therefore the heat removed at the condenser by the cooling medium is $G\lambda_D$ kJ/h, where λ_D is the molar latent heat of vaporisation of the distillate. If cooling water is circulated at a rate $\dot{m}_c$ kg/h, then

$$\dot{m}_c = \frac{G\lambda_D}{C_{pc}(T_o - T_i)} = \frac{(R+1)D\lambda_D}{C_{pc}(T_o - T_i)} \qquad \text{...(5.170)}$$

where C_{pc} is the heat capacity of cooling water (kJ/kg K) and $(T_o - T_i)$ is the rise in temperature of cooling water in the condenser.

Similarly, the rate of steam supply to the reboiler can be evaluated knowing the vapour boilup rate. Since the vapour produced at the reboiler is $\bar{G}$, the heat requirement for vaporisation is $\bar{G}\lambda_B$ kJ/h where λ_B is the molar latent heat of vaporisation at the reboiler conditions (kJ/kmol). This quantity of heat is liberated by the condensation of $\dot{m}_s$ kg/h of steam. Assuming that saturated steam is used, we can calculate the steam requirement as

$$\dot{m}_s = \frac{\bar{G}\lambda_B}{\lambda_s} \qquad \text{...(5.171)}$$

where λ_s is the heat of vaporisation of steam (kJ/kg). If the feed is admitted at the boiling point, $G = \bar{G} = (R+1)D$, Eq. (5.171) becomes

$$\dot{m}_s = \frac{(R+1)D\lambda_B}{\lambda_s} \qquad \text{...(5.172)}$$

In terms of the vapour boilup ratio V, $\bar{G} = VB$ and Eq. (5.171) becomes

$$\dot{m}_s = \frac{VB\lambda_B}{\lambda_s} \qquad \text{...(5.173)}$$

Summary of McCabe-Thiele Method

A typical problem for the determination of the number of theoretical plates may be stated thus: *A binary mixture of components* **A** *and* **B** *containing z_F mole fraction* **A** *admitted to a plate column with a thermal quality q is to be separated into a distillate containing x_D mole fraction* **A** *and a bottom product containing x_B mole fraction* **A**. *A reflux ratio R may be employed. A total/partial condenser may be used. Equilibrium data for the system may be taken from standard references. How many theoretical plates are required?*

The McCabe-Thiele graphical method for the determination of number of theoretical plates can be summarized as below:

1. Plot the equilibrium diagram and the 45° diagonal.
2. Mark the feed point $F\,(z_F, z_F)$, the distillate point $D\,(x_D, x_D)$ and the bottoms point $B\,(x_B, x_B)$ on the diagonal.
3. Draw the q-line with slope $\dfrac{q}{q-1}$ from the point F.
4. Draw the enriching operating line from point D in such a way that it makes an intercept $\dfrac{x_D}{R+1}$ with y-axis.
5. The intersection of the enriching operating line and the q-line is marked P and join P to B. The line PB is the stripping operating line.
6. Starting from D construct steps between the equilibrium curve and the enriching operating line till the operating line intersection P is crossed. After that continue construction of steps between the equilibrium curve and the stripping operating line till point B is reached.
7. Report the number of theoretical plates as:

N = The number of steps - 1 (if total condenser is used)

N = The number of steps - 2 (if partial condenser is used)

The step that straddles point P is the feed plate.

(**Note:** If the ratio R/R_m is given instead of R, the reflux ratio R is to be estimated after evaluating the minimum reflux R_m by the methods described earlier. If the feed is subcooled liquid or superheated vapour, the thermal quality q is to be calculated by the methods discussed earlier.)

Example 5.17: A plate column equipped with a total condenser and a kettle-type reboiler is used to separate 100 kmol/h of a benzene-toluene solution containing 50 mol% benzene into a distillate product containing 95 mol% benzene and bottom product containing 5 mol% benzene. The column is operated at 101.3 kPa. The feed is partially vaporised and is one-third vapour and two-thirds liquid. Determine the following:

(a) The minimum number of plates graphically and analytically using an average α of 2.39
(b) The minimum reflux ratio
(c) The number of plates if the operating reflux ratio is 30% in excess of the minimum.
(d) The feed plate location
(e) The number of plates required if the overall stage efficiency is 75%

The equilibrium data for benzene-toluene system at 101.3 kPa:

x	0	0.1	0.2	0.3	0.4	0.5	0.6	0.7	0.8	0.9	0.95	1
y	0	0.205	0.369	0.502	0.618	0.706	0.789	0.849	0.907	0.955	0.978	1

***Solution*:** Feed : 100 kmol/h of benzene(**A**)-toluene (**B**) mixture.

F = 100 kmol/h; z_F = mole fraction of benzene in the feed = 0.50, x_D = mole fraction of benzene in the distillate = 0.95, x_B = mole fraction of benzene in the bottoms = 0.05. Since the feed enters in partially vaporised condition containing two-thirds liquid, the feed quality q = 2/3.

(a) The equilibrium curve and the 45° reference line (the diagonal) are plotted as shown in Fig. 5.45. The minimum number of plates corresponds to the total reflux conditions. Under total reflux, $R = \infty$, the slope of the operating line is unity and the operating line coincides with the diagonal as shown in Fig. 5.45. In the figure, points B and D are points lying on the diagonal that can be located from the bottoms and distillate compositions. Point D has coordinates (x_D, x_D) or (0.95, 0.95) and point B has coordinates (x_B, x_B) or (0.05, 0.05). To find the number of plates, steps are constructed between the diagonal and the equilibrium diagram starting from point D and ending at point B. The number of steps = 7.2 from the figure. This includes the reboiler which acts as a theoretical plate. Therefore,

Minimum number of plates N_m = 7.2 – 1 = 6.2 **(Answer)**

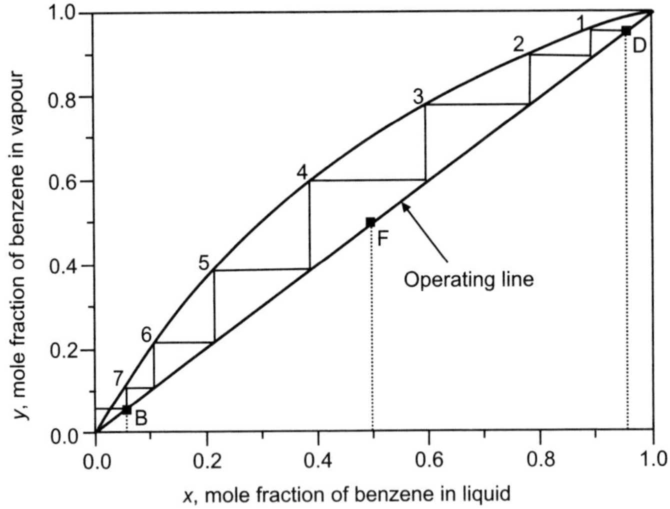

Fig. 5.45: McCabe-Thiele construction for Minimum number of plates, Example 5.17

Minimum number of Pates using Fenske equation

Assuming a constant α value of 2.39, Fenske equation can be used to find the minimum number of plates.

$$N_m + 1 = \frac{\ln\left(\dfrac{x_D}{1-x_D} \Big/ \dfrac{x_B}{1-x_B}\right)}{\ln \alpha}$$

Substitute $x_D = 0.95$, $x_B = 0.05$ and $\alpha = 2.39$ into the above equation:

$$N_m + 1 = \frac{\ln\left(\dfrac{0.95}{1-0.95} \Big/ \dfrac{0.05}{1-0.05}\right)}{\ln 2.39} = 6.8$$

$$N_m = 5.8 \qquad \textbf{(Answer)}$$

(Minimum number of plates = 5.8 by Fenske equation)

(b) The q-line is drawn at point F on the diagonal. F is marked knowing the feed composition z_F. The coordinates of F are $(z_F, z_F) = (0.5, 0.5)$. The slope of the q line is

$$\frac{q}{q-1} = \frac{2/3}{2/3-1} = -2$$

The q-line is drawn which intersects the equilibrium curve at Q as shown in Fig. 5.46. The coordinates of Q (x', y') are noted. ($x' = 0.4283$ and $y' = 0.6431$). The limiting operating line the slope of which is $\dfrac{R_m}{R_m+1}$ where R_m is the minimum reflux ratio passes through the point Q. The minimum reflux ratio can be calculated as

$$R_m = \frac{x_D - y'}{y' - x'} = \frac{0.95 - 0.6431}{0.6431 - 0.4283} = 1.4288$$

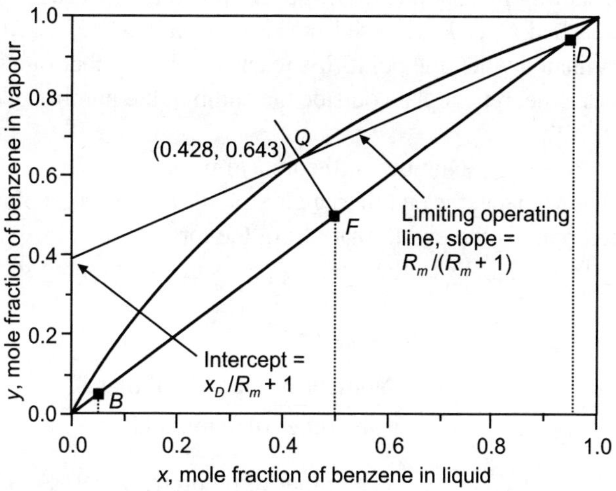

Fig. 5.46: Estimation of the minimum reflux ratio for Example 5.17

(R_m can be calculated also from the slope of the limiting operating line or from the intercept the limiting operating line makes with the y-axis)

The minimum reflux ratio $R_m = 1.4288$ **(Answer)**

(c) Operating reflux ratio is 30% in excess of R_m. $R = 1.3\ R_m = 1.3(1.4288) = 1.8574$.

To find the number of theoretical plates the equilibrium curve is plotted as shown in Fig. 5.47 along with the diagonal. Points F (0.5, 0.5), D (0.95, 0.95) and B (0.05, 0.05) are marked on the diagonal. The q-line is drawn at F with slope $q/(q - 1) = -2$. The enriching operating line intersects the y-axis at $\frac{x_D}{R+1} = \frac{0.95}{1.8574+1} = 0.3325$. The intersection is shown by point M in the figure. Line DM is the enriching operating line. This line intersects the q-line at point P. The stripping operating line passes through points P and B. The line BP is the stripping operating line.

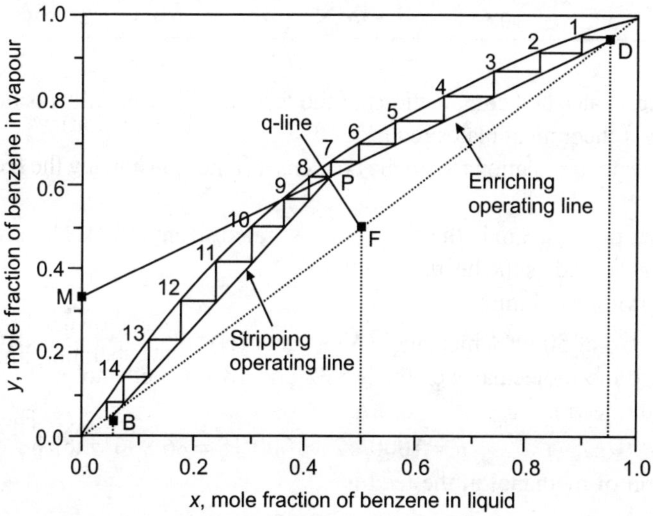

Fig. 5.47: Determination of number of theoretical plates, Example 5.17

The steps are drawn starting from point D. The upper operating line is used for construction of steps till the intersection P is reached. Once P is crossed, the lower operating line is used for further construction. The construction is terminated when the point B is reached. The number of steps is found to be 14.7. Since the reboiler acts as a theoretical plate outside the column, the number of theoretical plates in the column $N = 14.7 - 1 = 13.7$

Number of theoretical plates = 13.7. **(Answer)**

(d) From Fig. 5.47, it is seen that the eighth triangle straddles P, the point of intersection of operating lines. Therefore the feed plate is the eighth plate from the top. **(Answer)**

There are 7 plates in the enriching section and 6.7 plates in the stripping section which includes the feed plate. **(Answer)**

(e) The overall stage efficiency is defined as

$$E_O = \frac{\text{Number of theoretical plates}}{\text{Number of actual plates}}$$

$$\text{Therefore, the number of actual plates} = \frac{\text{Number of theoretical plates}}{E_O} = \frac{13.7}{0.75} = 18.3 \approx 19$$

Number of actual plates = 19. **(Answer)**

Example 5.18: 2000 kg/h of methanol-water solution containing 50 wt% methanol is to be fractionated in a plate column to recover 95% methanol in the feed as distillate of composition 96 mol%. The feed enters the column at 330 K. The column is provided with a total condenser and a kettle-type reboiler. A reflux ratio 1.5 times the minimum may be employed. The vapour velocity in the column is to be 0.6 m/s. The heat capacity of the feed solution is 3.852 kJ/kg K. The latent heat of vaporisation of methanol is 1046.7 kJ/kg and that of water is 2284 kJ/kg and may be assumed constant. Equilibrium data at 101.3 kPa (mole fraction methanol)

T, K	373.2	365.6	360.9	357.6	354.9	351.2	348.5
x	0.00	0.05	0.10	0.15	0.20	0.30	0.40
y	0.000	0.267	0.418	0.517	0.570	0.665	0.729
T, K	346.3	344.4	342.5	340.8	339.2	338.2	337.7
x	0.50	0.60	0.70	0.80	0.90	0.95	1.00
y	0.779	0.825	0.870	0.915	0.958	0.979	1.000

Determine:

(a) The molar flow rates and compositions of the feed, distillate and the residue
(b) The number of theoretical plates required.
(c) The cooling water circulation rate in the condenser if the water enters the condenser at 300 K and leaves at 325 K.
(d) The rate of supply of steam in the reboiler if saturated steam at 101.3 kPa (latent heat of steam = 2256.9 kJ/kg) is used as the heating medium
(e) The diameter of the column.

Solution: (a) 2000 kg/h of 50 wt% methanol (**A**) – water (**B**) solution contains 1000 kg methanol and 1000 kg water. Since the molecular weights are 32 and 18 for methanol and water respectively, the molar flow rate of the feed is

$$F = 1000/32 + 1000/18 = 86.806\,\text{kmol/h}$$

and the mole fraction of methanol in the feed is

$$z_F = \frac{1000/32}{1000/32 + 1000/18} = 0.36$$

The average molecular weight of the feed $= 0.36 \times 32 + 0.64 \times 18 = 23.04$

Since 95% of the methanol in the feed is recovered as distillate, distillate contains $0.95 \times \dfrac{1000}{32}$ = 29.688 kmol methanol. Since distillate composition is 96 mol% methanol, the amount of distillate is

$$D = \frac{29.688}{0.96} = 30.925\,\text{kmol/h}$$

The amount of bottoms is

$$B = F - D = 86.806 - 30.925 = 55.881\,\text{kmol/h}$$

By methanol balance,

$$Bx_B = Fz_F - Dx_D$$

$$x_B = \frac{Fz_F - Dx_D}{B} = \frac{86.806 \times 0.36 - 30.925 \times 0.96}{55.881} = 0.028$$

Thus

$$F = 86.806 \text{ kmol/h},\ D = 30.925 \text{ kmol/h and } B = 55.881 \text{ kmol/h}$$

$$z_F = 0.36,\ x_D = 0.96,\ x_B = 0.028$$ **(Answer)**

(b) Since the feed is a subcooled liquid, the quality of feed q is calculated using Eq. (5.137)

$q = 1 + \dfrac{C_{pL}(T_b - T_F)}{\lambda_v}$, where C_{pL} is the molar heat capacity of the feed, λ_v is the heat of vaporisation of the feed, and T_b and T_F are respectively, the bubble point temperature and the temperature of the feed. The feed enters the column at $T_F = 330$ K. The bubble point of the liquid with $x = 0.36$ is found to be 349.6 K from the $T - x - y$ data. That is T_b = 349.6 K. The molar heat capacity of the feed $C_{pL} = 3.852 \times 23.04 = 88.75$ kJ/kmol K. The molar latent heat of vaporisation of the feed solution is approximated as the mole fraction weighted average of the latent heats of the pure components.

$$\lambda_v = x \times 32 \times \lambda_A + (1 - x) \times 18 \times \lambda_B$$

$$\lambda_v = 0.36 \times 32 \times 1046.7 + (1 - 0.36) \times 18 \times 2284 = 38369.66 \text{ kJ/kmol.}$$

Substitute these in Eq. (5.137)

$$q = 1 + \frac{C_{pL}(T_b - T_F)}{\lambda_v} = 1 + \frac{88.75(349.6 - 330)}{38369.66} = 1.045$$

The equilibrium curve and the 45° reference line are plotted as shown in Fig. 5.48. Knowing z_F, x_D and x_B, the points F (0.36,0.36), D (0.96, 0.96) and B (0.028,0.028) are marked on the diagonal. The q-line is drawn at F with slope $\dfrac{q}{q-1} = \dfrac{1.045}{1.045 - 1} = 23.22$ to intersect the equilibrium curve at Q. The coordinates of Q (x', y') are read from the graph.

$$x' = 0.375,\ y' = 0.711$$

The minimum reflux ratio is now calculated.

$$R_m = \frac{x_D - y'}{y' - x'} = \frac{0.96 - 0.711}{0.711 - 0.375} = 0.7411$$

The operating reflux ratio $R = 1.5(R_m) = 1.5\ (0.7411) = 1.1116$.

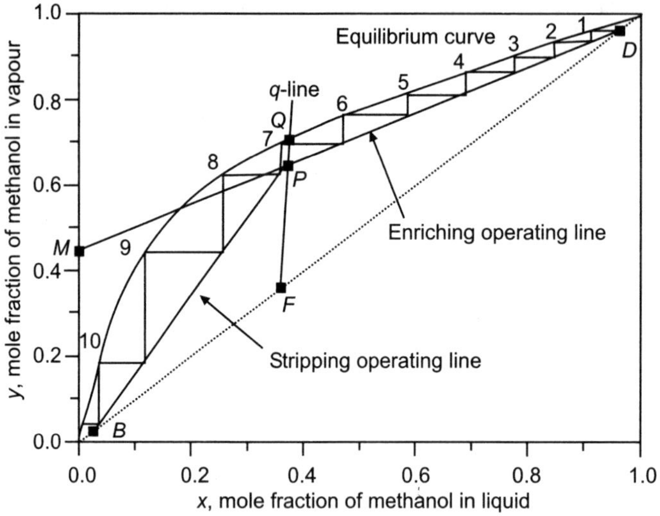

Fig. 5.48: McCabe-Thiele construction for Example 5.18

The enriching operating line meets the y-axis at $\dfrac{x_D}{R+1}=\dfrac{0.96}{1.1116+1}=0.4546$. Point M is marked at $y = 0.4546$ on the y-axis. Line DM is drawn, which is the enriching operating line. It intersects the q-line at P. Line PB is the stripping operating line. Stages are stepped off beginning from point D and terminating at B. The number of steps from the graph = 10.3.

The number of plates in the column = 10.3 – 1 = 9.3 **(Answer)**

(c) The cooling water is required to remove heat liberated on condensation of the vapour in the condenser. The heat liberated on condensation of the vapour $= Q_c = G\lambda_1$ where G is the amount of vapour reaching the condenser and λ_1 is the latent heat of vaporisation of the vapour leaving plate 1, which is same as λ_D, the latent heat of vaporisation of the distillate. The latent heat of vaporisation is estimated corresponding to the distillate composition, from the pure component latent heats. Since $x_D = 0.96$,

$$\lambda_D = x_D\lambda_A + (1-x_D)\lambda_B = 0.96\times(1046.7\times 32)+(1-0.96)\times(2284\times 18) = 33799.1 \text{ kJ/kmol}$$

By material balance around the condenser,

$$G = L + D = (R+1)D = (1.1116+1)30.925 = 65.301 \text{ kJ/kmol}$$

Let $\dot{m}_c$ be the cooling water circulation rate kg/h. Then by Eq. (5.170)

$$\dot{m}_c = \frac{G\lambda_D}{C_{pc}(T_o - T_i)} = \frac{65.301\times 33799.1}{4.1868(325-300)} = 21086.5 \text{ kg/h}$$

(It is assumed that heat capacity of cooling water = 4.1868 kJ/kg K)

Rate of circulation of cooling water = 21086.5 kg/h. **(Answer)**

(d) Rate of supply of steam to the reboiler is evaluated using Eq. (5.171)

$$\dot{m}_s = \frac{\bar{G}\lambda_B}{\lambda_s}$$

We have shown [Eq. (5.142)] by a total material balance on the feed plate that

$$\bar{G} = (R+1)D + (q-1)F = (1.1116+1)30.925 + (1.045-1)86.806 = 69.2075 \text{ kmol/h}$$

Let λ_W be the latent heat of vaporisation for the bottom product (Strictly speaking, it should be the latent heat of vaporisation of the liquid leaving the bottommost plate). Then

$\lambda_W = x_B\lambda_A + (1 - x_B)\lambda_B = 0.028(1046.7 \times 32) + 0.972(2284 \times 18) = 40898.71$ kJ/kmol.

The amount of steam required can be computed using Eq. (5.171).

$$\dot{m}_s = \frac{\bar{G}\lambda_W}{\lambda_s} = \frac{69.2075 \times 40898.71}{2256.9} = 1254.15\,\text{kg/h}.$$

Steam requirement is 1254.15 kg/h **(Answer)**

(e) The vapour flow rate at the top of the column, $G = 65.301$ kmol/h. The temperature of the vapour at the top of the column may be evaluated from the *T-x-y* data. It is found to be 339.2 K. The pressure is 101.3 kPa throughout. Volumetric flow rate corresponding to the conditions at the top of the column is evaluated as:

$$V_D = G \times 22.414 \times \frac{P_0}{P} \times \frac{T}{T_0} = 65.301 \times 22.414 \times \frac{339.2}{273.2} = 1817.25\,\text{m}^3/\text{h}$$

The vapour flow at the bottom $= \bar{G} = 69.2075$ kmol/h and the temperature of the vapour reaching the bottom plate from *T-x-y* diagram is 368.9 K. Volumetric flow rate for these conditions is

$$V_B = \bar{G} \times 22.414 \times \frac{P_0}{P} \times \frac{T}{T_0} = 69.2075 \times 22.414 \times \frac{368.9}{273.2} = 2094.60\,\text{m}^3/\text{h}$$

Since the volumetric flow of vapour at the bottom of the column (V_B) is higher than that at the top, the column diameter is based on the bottom condition. The vapour velocity is given to be 0.6 m/s.

V_B = Area of the tower $\times$ velocity $= \frac{\pi}{4}D^2(0.6 \times 3600) = 2094.60\ \text{m}^3/\text{h}$, where D is the diameter of the tower.

Therefore, the diameter $D = 1.11$ m. **(Answer)**

Example 5.19: A distillation column provided with the equivalent of 10 ideal plates, a total condenser and a reboiler is used to separate 100 kmol/h of an equimolar mixture of components **A** and **B**, the relative volatility of which is 2.5. The feed enters the column at its bubble point and it is desired that the distillate composition be 95 mol% **A**. The reflux ratio employed is 1.5 times the minimum. Determine (a) the composition of the residue and (b) kmol/h of distillate produced.

***Solution*:** The relative volatility is used to compute the equilibrium data.

$$y = \frac{\alpha x}{1 + (\alpha - 1)x} \qquad \ldots(5.14)$$

Substituting $\alpha = 2.50$, the equilibrium relationship becomes

$$y = \frac{2.5x}{1 + 1.5x} \qquad \text{(A)}$$

For different x-values, y is calculated from Eq. (A) and the results are:

x	0	0.1	0.2	0.3	0.4	0.5	0.6	0.7	0.8	0.9	1
y	0	0.217	0.385	0.517	0.625	0.714	0.789	0.854	0.909	0.957	1

The equilibrium curve and the reference line (diagonal) are drawn as shown in Fig. 5.49. The feed being at its bubble point, $q = 1$, and the q-line is vertical. The q-line intersects the equilibrium curve at Q whose coordinates are (x', y'). From the figure, ($x' = 0.5$, $y' = 0.7143$).

The limiting operating line, the slope of which determines the minimum reflux R_m passes through D and Q. R_m is evaluated by (Eq. 5.164) as

$$R_m = \frac{x_D - y'}{y' - x'} = \frac{0.95 - 0.7143}{0.7143 - 0.5} = 1.1$$

The operating reflux ratio $R = 1.5(R_m) = 1.65$

Intercept of the upper operating line with x-axis is $\frac{x_D}{R+1} = \frac{0.95}{2.65} = 0.3585$, and this determines the location of the point M in the figure. The upper operating line is DM which intersects the q-line at P.

The lower operating line can be drawn if the residue composition is known. Assign arbitrarily a value for x_B and mark B on the diagonal. Draw the lower operating line BP. Find the number of steps that can be constructed from point D to point B. If it exactly equals 11 (10 ideal plates in the column + reboiler), the assumed value of x_B is the right choice. Otherwise find the number of steps with a new value of x_B.

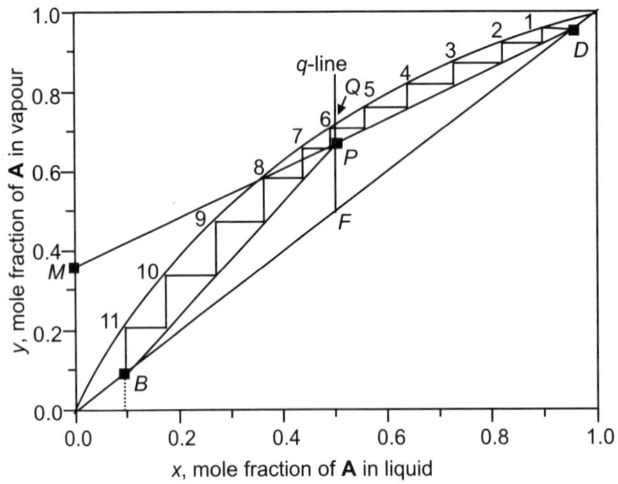

Fig. 5.49: McCabe-Thiele construction to find x_B when N is fixed

(a) The residue composition is found to be $x_B = 0.094$. **(Answer)**

Component-**A** balance for the column gives

$$D = \frac{F(z_F - x_B)}{(x_D - x_B)} = 100\frac{0.5 - 0.094}{0.95 - 0.094} = 47.43 \text{ kmol/h and } B = 100 - 47.43 = 52.57 \text{ kmol/h.}$$

(b) Amount of distillate product = 47.43 kmol/h. **(Answer)**

High-Purity Products

When nearly pure products are required, mole fraction $x_W \rightarrow 0$ and/or $x_D \rightarrow 1$; in the McCabe-Thiele method, a large number of stages are to be constructed and majority of them get crowded at the ends of the diagram. It becomes very difficult to determine the number of plates accurately. In such cases, the following alternatives are suggested.

(i) Plot the ends of the equilibrium curve separately on an expanded scale and carry out construction in the usual manner.

(ii) Plot the lower end of the equilibrium curve and the operating line on log-log graph sheet. The equilibrium curve will be a straight line of slope 45°, whereas the operating line is a curve on this graph. The latter can be plotted using the points calculated using the equation for the operating line for the stripping section. To plot the distillate end of the equilibrium diagram, the graph-sheet is turned upside down marking the coordinates as 1 – printed coordinates. The equilibrium curve is again a straight line and the operating line is to be plotted by calculating certain points on the curve using the operating line equation for the enriching section.

(iii) Use Kremser equation derived in Chapter 4 for the ends and the McCabe-Thiele method for the middle portion.

Incorporating Murphree Efficiency in McCabe-Thiele Construction

The number of plates determined by the McCabe-Thiele method is the number of theoretical plates because it is assumed that the liquid and the vapour leaving a plate are in equilibrium and the points representing their compositions lie on the equilibrium curve. However, it is very difficult to achieve equilibrium conditions in practice, because it requires infinite time of contact and perfect mixing between the liquid and the vapour. The plate efficiencies are defined to express the degree of departure of an actual plate from the ideal behaviour. For example, the Murphree vapour efficiency is the ratio of actual change in concentration of the vapour over a plate to the change in concentration of the vapour if it were in equilibrium with the liquid leaving the plate. Thus the ratio of the line lengths *PM*/*PR* in Fig. 5.50 is the Murphree vapour efficiency of the plate. If the efficiency of the plate is 70%, this ratio is 0.7. The point *M* which represents the actual effluent composition from the plate can be located at a height which is the efficiency times the vertical distance between the equilibrium curve and the operating line. The dashed line in the Fig. 5.50 is drawn through the points lying at a height 0.7 times the vertical distance, assuming that all the plates in the column have a vapour efficiency of 0.7. The number of actual plates in the column can be determined by constructing steps between the operating line and the *non-equilibrium curve* as shown in the figure. If the Murphree liquid efficiency is used, the non-equilibrium curve can be constructed in such a way that the horizontal distance between the equilibrium curve and the operating line is divided in proportion of the efficiency factor. A detailed discussion of the different type of efficiencies applicable to absorption columns were provided in Chapter 4, which are applicable to distillation columns as well.

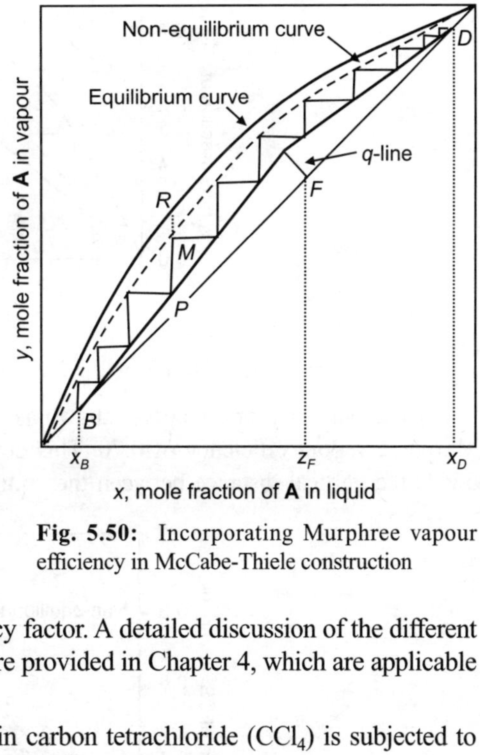

Fig. 5.50: Incorporating Murphree vapour efficiency in McCabe-Thiele construction

Example 5.20: A solution of carbon disulphide (CS_2) in carbon tetrachloride (CCl_4) is subjected to fractionation in a plate column. The feed contains 40 mol% CS_2 and it is 50% vaporised. It is desired to obtain a distillate containing 95 mol% CS_2 and bottoms product containing 5 mol% CS_2. The reflux ratio employed is 4. The Murphree vapour efficiency is 70%. Assume that the condenser used is total and the reboiler is kettle-type. The equilibrium data (mole fraction of CS_2) are

x	0.0	0.05	0.1	0.2	0.3	0.4	0.5	0.6	0.7	0.8	0.9	1.0
y	0.0	0.135	0.245	0.42	0.545	0.64	0.725	0.79	0.85	0.905	0.955	1.0

Using the McCabe-Thiele method, determine

(a) The number of theoretical plates

(b) The number of plates, taking the Murphree efficiency into account.

***Solution*:** $z_F = 0.40$; $x_D = 0.95$; $x_B = 0.05$

q = fraction of the feed that is vapour = 0.5; Slope of q-line $= \frac{q}{q-1} = \frac{0.5}{0.5-1} = -1.0$

$R = 4$; Intercept of the operating line with y-axis $= \frac{x_D}{R+1} = \frac{0.95}{4+1} = 0.19$

(a) Equilibrium curve is plotted in the Fig. 5.51. The q-line is plotted at F with slope = −1.0. Knowing x_D and the y-intercept, the enriching operating line (DP in the figure) and knowing x_B, the stripping operating line (PB in figure) are drawn. By the usual stair-case construction, the number of plates is determined to be 8.6 – 1 = 7.6. **(Answer)**

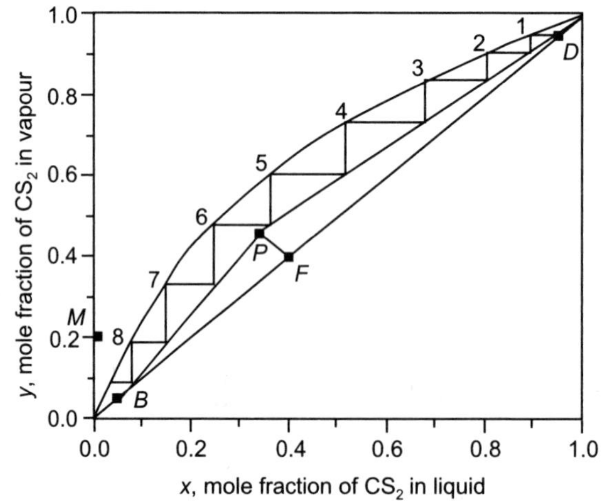

Fig. 5.51: Example 5.20: Determination of number of theoretical plates

(b) The non-equilibrium curve (shown as dotted line in Fig. 5.52) is drawn taking into account the Murphree vapour efficiency of 0.70. This curve is drawn such that it passes through the points that divide the vertical distance between the equilibrium curve and the operating line by 70%. Now the

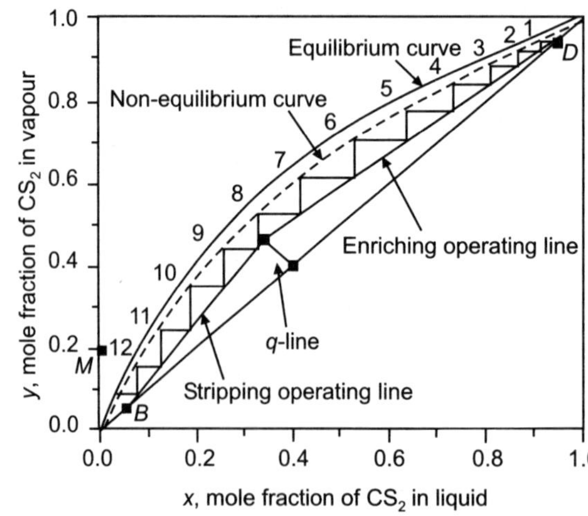

Fig. 5.52: Determination of number of plates in Example 5.20 taking the Murphree vapour efficiency into account

stair-case construction for plates is carried out as usual, but using non-equilibrium curve for the construction instead of the equilibrium curve. The number of plates is found to be

$$12.8 - 1 = 11.8 \quad \textbf{(Answer)}$$

Special Case 1: Cold Reflux

Normally the reflux enters the top plate as a saturated liquid. However, the reflux may enter the column as a subcooled liquid. When using a total condenser, this happens if (i) the bubble point of the reflux is significantly greater than the cooling water inlet temperature, (ii) if the condenser exit pressure is less than the pressure at the top tray, (iii) if there is significant heat loss from the reflux line *en route* to the tower. For a partial condenser, the subcooled reflux results due to the last two reasons. If the reflux is below its bubble point, the liquid flowing down from the top plate (L) will be greater than the reflux reaching the plate (L_0). Consider Fig. 5.53. The reflux on reaching the plate 1 gets heated to its bubble point. The sensible heat required for this purpose is taken from the latent heat of condensation of the vapour reaching plate 1. A portion of the saturated vapour reaching the plate condenses on the plate and adds to the reflux as it flows down. Thus L is greater than L_0 by the amount of vapour condensed which in turn depends on the degree of subcooling. Let ΔL be the increase in the liquid flow down the plate 1.

$$L = L_0 + \Delta L \qquad \text{...(5.174)}$$

Let the reflux enter at T_0 and let $T_{0,b}$ be its bubble point. The sensible heat required by the reflux to raise its temperature to the bubble point $= L_0 C_{pL}(T_{0,b} - T_0)$ where C_{pL} is the molar heat capacity of the reflux liquid (kJ/kmol K). Assume that the molar latent heat of condensation of the vapour at plate 1 to be λ_1 (kJ/kmol). Then

$$\Delta L \lambda_1 = L_0 C_{pL}(T_{0,b} - T_0) \qquad \text{...(5.175)}$$

Therefore

$$\Delta L = \frac{L_0 C_{pL}(T_{0,b} - T_0))}{\lambda_1} \qquad \text{...(5.176)}$$

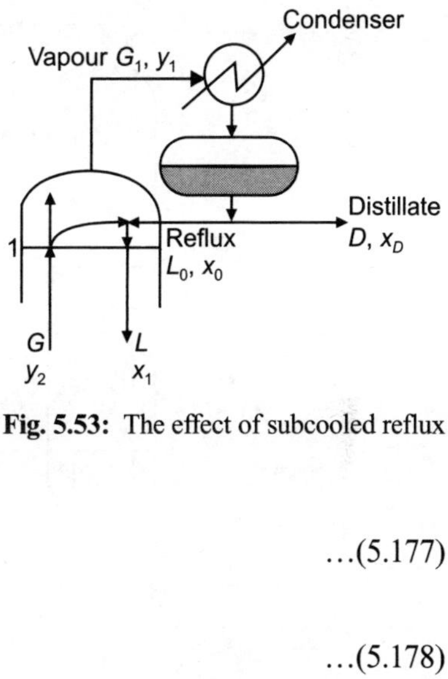

Fig. 5.53: The effect of subcooled reflux

Substituting Eq. (5.176) into Eq. (5.174), we get

$$L = L_0 + \frac{L_0 C_{pL}(T_{0,b} - T_0)}{\lambda_1} \qquad \text{...(5.177)}$$

$$L = L_0 \left[1 + \frac{C_{pL}(T_{0,b} - T_0)}{\lambda_1}\right] \qquad \text{...(5.178)}$$

Dividing throughout by D

$$\frac{L}{D} = \frac{L_0}{D}\left[1 + \frac{C_{pL}(T_{0,b} - T_0)}{\lambda_1}\right] \qquad \text{...(5.179)}$$

The ratio $\frac{L_0}{D}$ is the external reflux ratio or simply the reflux ratio (R) for the column whereas the ratio $\frac{L}{D}$ is called the internal reflux ratio and is denoted by R'. When the reflux enters at the bubble point, $T_{0,b} = T_0$ and $R = R'$. The internal reflux ratio R' is constant throughout the rest of the column and it should be used in the graphical construction of operating lines in place of R. The equation for the upper operating line (5.126) gets modified as

$$y_{n+1} = \frac{R'}{R'+1}x_n + \frac{x_D}{R'+1} \quad \text{...(5.180)}$$

Special Case 2: Open Steam

When an aqueous solution of a volatile substance is separated by distillation, the bottom product will be nearly pure water. Instead of sending the liquid leaving the bottom plate to the reboiler for producing vapours by partially vaporising the liquid, steam that is available in the plant can be used to serve the purpose of the vapour phase. The reboiler can be abandoned and steam can be directly sent to the column. The scheme is as shown in Fig. 5.54a.

Saturated or superheated steam is admitted to the column at a rate $\bar{G}$ kmol/h. Since no **A** is present in the steam supplied, $y_{N+1} = 0$. Component **A**-balance from the bottom of the column to the plate m in the stripping section gives

$$\bar{G}y_{m+1} = \bar{L}x_m - Bx_B \quad \text{...(5.181)}$$

$$y_{m+1} = \frac{\bar{L}}{\bar{G}}x_m - \frac{B}{\bar{G}}x_B \quad \text{...(5.182)}$$

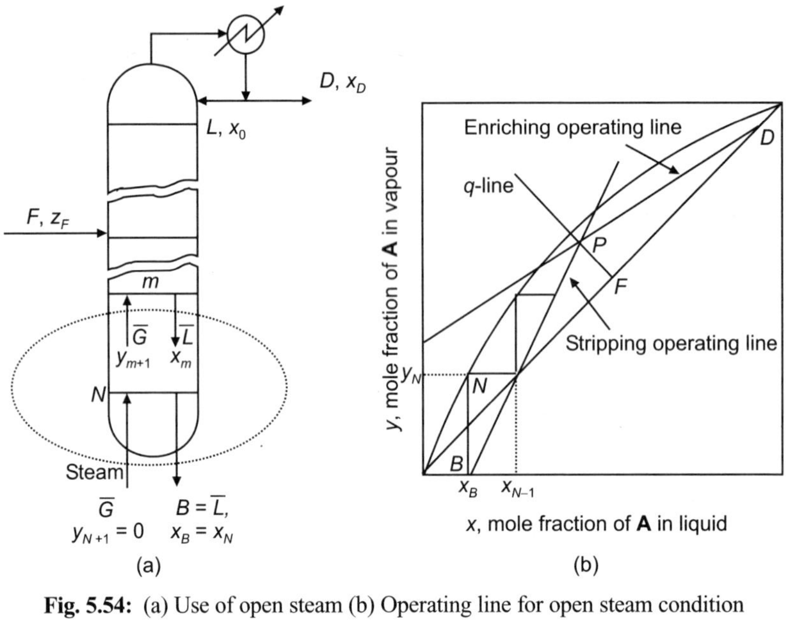

Fig. 5.54: (a) Use of open steam (b) Operating line for open steam condition

Equation (5.182) represents the operating line for stripping section of the column. Since $\bar{L} = B$, if we let $x_m = x_B$ in Eq. (5.182), we see that $y_{m+1} = 0$; or the operating line passes through the point $(x_B, 0)$ which lies on the x-axis.

The enriching operating line remains the same as that described by Eq. (5.126)

$$y_{n+1} = \frac{R}{R+1}x_n + \frac{x_D}{R+1} \quad \text{...(5.126)}$$

The stripping operating line can be drawn from point B $(x_B, 0)$; it passes through the intersection of the enriching and feed lines. Steps are constructed as before starting from point D. The construction is terminated when the point B on the x-axis is reached. The last triangle so constructed is the last plate N in the cascade.

The quantity and the composition of the residue can be computed by taking a material balance on the column.

$$F + \bar{G} = B + D \qquad ...(5.183)$$

$$Fz_F + \bar{G}y_{N+1} = Bx_B + Dx_D \qquad ...(5.184)$$

Combining Eqs. (5.183) and (5.184) and noting that $y_{N+1} = 0$, we see that

$$x_B = \frac{Fz_F - Dx_D}{B} = \frac{Fz_F - Dx_D}{F + \bar{G} - D} \qquad ...(5.185)$$

Saving in the cost of the reboiler and the cost of its maintenance make the open steam an attractive option. However, the bottoms get diluted and more number of plates would be needed in the stripping section than that required in a column with reboiler, for the specified rate of distillate. For illustration, let us assume that 100 kmol/h of an equimolar mixture of **A** and **B** (**B** in this case should be water) is subjected to fractionation to produce 50 kmol/h of a distillate containing 90% **A**. In the conventional column using the reboiler the bottoms will be 50 kmol/h and its concentration is $x_B = 0.1$. If open steam were used at a rate 100 kmol/h, other conditions remaining the same, the bottoms quantity increases to 150 kmol/h and its composition falls to 0.033. We see that the bottom composition falls and the operating line for the stripping section moves closer to the equilibrium curve. Consequently, more number of plates and hence a taller tower would be needed. Only an economic analysis can tell us whether using the open steam is attractive over the reboiler with its associated operating and maintenance costs.

Example 5.21: An aqueous solution of ethyl alcohol containing 12 mol% alcohol is sent to a distillation column operated at 101.3 kPa for producing a distillate containing 85 mol% alcohol. It is desired to have 90% recovery of alcohol as distillate. The feed enters the column at a rate 100 kmol/h, as saturated liquid. A total condenser is used and the reflux ratio employed is 3. No reboiler is used and direct steam at 101.3 kPa is admitted to the bottom plate. Determine:

(a) Steam consumption, kg/h.
(b) Composition of the bottoms.
(c) The number of theoretical plates.

Equilibrium data at 101.3 kPa (mole fraction of ethanol):

x	0.0190	0.0721	0.0966	0.1238	0.1661	0.2337	0.2608	0.3273
y	0.1700	0.3891	0.4375	0.4704	0.5089	0.5445	0.5580	0.5836
x	0.3965	0.5079	0.5198	0.5732	0.6763	0.7472	0.8943	
y	0.6122	0.6564	0.6599	0.6841	0.7385	0.7815	0.8943	

***Solution*:** Ethyl alcohol (**A**)-water (**B**)

$$F = 100 \text{ kmol/h}, \; z_F = 0.12, \; x_D = 0.85$$

Since 90% of the alcohol in the feed is recovered in the distillate,

$$Dx_D = 0.90Fz_F = 0.90 \times 100 \times 0.12 = 10.8 \text{ kmol/h}$$

$$\text{Therefore, } D = \frac{10.8}{x_D} = \frac{10.8}{0.85} = 12.71 \text{ kmol/h}$$

The reflux ratio $R = 3$;

$$\frac{L}{D} = 3; \text{ Therefore, } L = 3D = 38.12 \, \text{kmol/h}$$

Consider Fig. 5.55. Taking a total material balance around the condenser,

$$G = L + D = 38.12 + 12.71 = 50.83 \text{ kmol/h}$$

Since the feed is admitted as a saturated liquid the vapour flow in the stripping section and that in the enriching section are the same. Therefore, $\bar{G} = G = 50.83$ kmol/h.

By taking a material balance on the column (Fig. 5.55).

$$F + \bar{G} = B + D \qquad \text{...(5.183)}$$

$$B = F + \bar{G} - D = 100 + 50.83 - 12.71 = 138.12 \text{ kmol/h}$$

Since 10% of the alcohol in the feed should be present in the residue,

$$x_B = \frac{0.10 \times F z_F}{B} = \frac{1.2}{138.12} = 0.0087$$

(a) Steam consumption $= \bar{G} = 50.83$ kmol/h **(Answer)**

(b) The bottom product contains 0.87 mol% alcohol and the rest water **(Answer)**

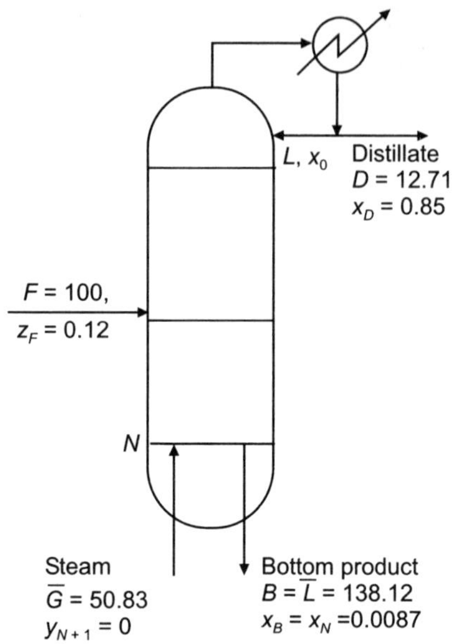

Fig. 5.55: Example 5.21 on open steam

(c) The equilibrium curve is plotted as shown in Fig. 5.56. The point F has coordinates (0.12, 0.12) and point D has coordinates (0.85, 0.85). The streams below the bottom plate have compositions x_B and y_{N+1}. Here, $x_B = 0.0087$ and $y_{N+1} = 0$. The point B which represents the lower terminus of the operating line lies on the x –axis as shown. The q-line is vertical as the feed is a saturated liquid. q-line is drawn at F. The enriching operating line meets the y-axis at M. The y-coordinate of M is

$$\frac{x_D}{R+1} = \frac{0.85}{4} = 0.2125$$

The enriching line passes through D and M. Its intersection with q-line is marked P. P is joined to B to get the operating line for stripping section.

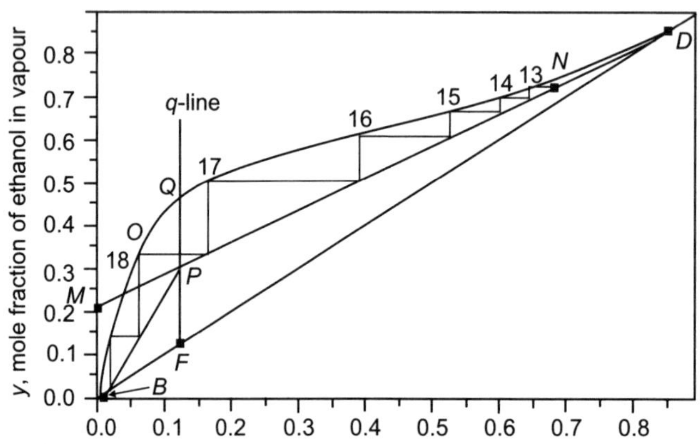

Fig. 5.56: The McCabe –Thiele construction for Example 5.21

The operating line for the enriching section being very close to the equilibrium curve near the distillate end of the diagram, stepping off of stages become very difficult in the ordinary plot covering the full concentration range. The higher concentration region is separately plotted on an expanded scale as shown in Fig. 5.57. Construction for stages is carried out in the usual manner. The construction below point N on the operating line is continued in the original diagram of Fig. 5.56.

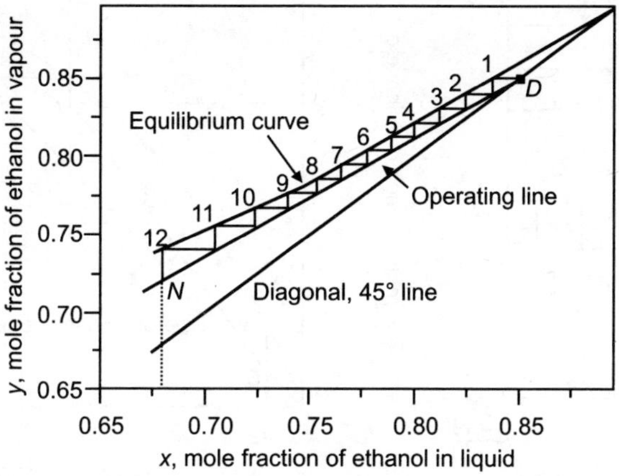

Fig. 5.57: McCabe-Thiele construction for the high-concentration region in Example 5.20

The stair-case construction of plates is terminated at B on the x-axis. The lower end of the diagram can also be drawn on a larger scale as shown in the Fig. 5.58.

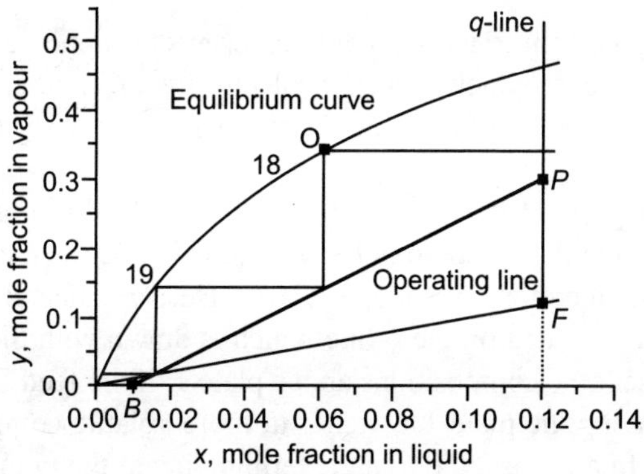

Fig. 5.58: McCabe-Thiele construction for lower end of the diagram

Total number of triangles drawn = Number of theoretical plates = 19.54 and the feed enters on the 18th plate from the top. **(Answer)**

Special Case 3: Stripping Column

In a stripping column (Fig. 5.59a), the feed enters as a saturated or a subcooled liquid at the top plate and the overhead vapour leaves the column without subjecting part of it to be returned as reflux after condensation. The tower is used to strip off the more volatile components from the liquid. The column is, in fact, similar to the stripping section of the usual fractionation tower. The operating line for the column is established by considering the material balance to the section of the column below plate m. We see that the operating line is described by Eq. (5.120) developed for the stripping section of the fractionation column.

$$\bar{G}y_{m+1} = \bar{L}x_m - Bx_B \qquad \ldots(5.120)$$

The operating line can be drawn with slope $\bar{L}/\bar{G}$, and the line passes through the point B (x_B, x_B) on the diagonal. We also know that the upper terminus of the operating line should be point (x_0, y_1).

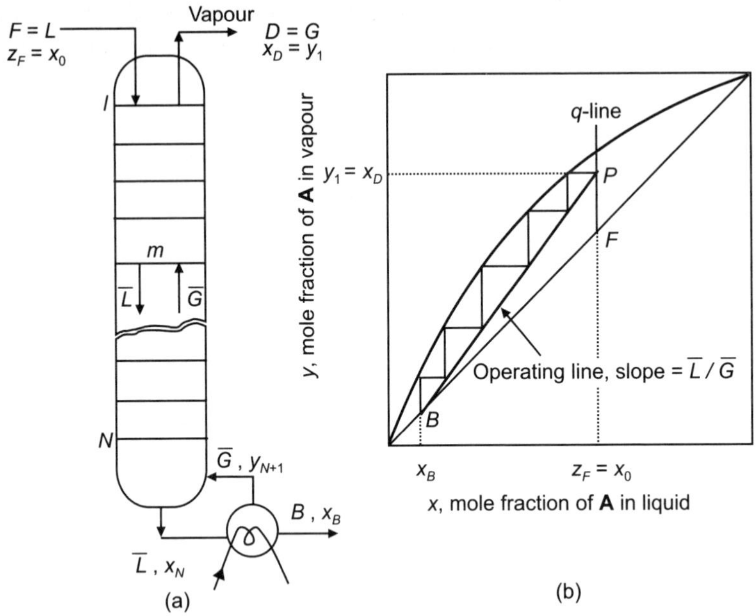

Fig. 5.59: (a) Stripping column (b) McCabe-Thiele construction for theoretical plates in the stripping column

Since $x_0 = z_F$ and $y_1 = x_D$, the point P in Fig. 5.59b represents the upper terminus of the operating line. It lies on the q-line for the feed, which is a vertical line for the saturated liquid feed as shown in Fig. 5.59b. We can draw the operating line using the slope $\bar{L}/\bar{G}$ knowing that it passes through points B and P.

The slope of the operating line $\bar{L}/\bar{G} = L/G = F/D$ when the feed is a saturated liquid. On the other hand, if the feed is a subcooled liquid, some of the vapour reaching the top plate condenses to heat the feed to its saturated state and therefore $\bar{L} > F$ and $\bar{G} > G$. Here $q > 1.0$ and the q-line for the feed is not vertical. The point P is to be located on the q-line which is drawn with the slope $q/(q-1)$ at F (z_F, z_F). The point P will not represent the compositions above plate 1, if the feed is subcooled liquid. It is also possible that the liquid leaving the plate 1 is richer in more volatile component than the feed if it is cooled to a greater extent. In any case, the triangle terminating at the point P represents the top plate.

Special Case 4: Enriching Column

In enriching columns the feed is admitted to the bottom of the column at plate N as a saturated or super heated vapour. The vapour rising up through the column is contacted with the liquid reflux as is done in the enriching section of the conventional fractionation column. A material balance from the top of the column up to plate n gives

$$Gy_{n+1} = Lx_n + Dx_D \qquad \ldots(5.119)$$

The equation can be rewritten as

$$y_{n+1} = \frac{R}{R+1}x_n + \frac{x_D}{R+1} \qquad \ldots(5.126)$$

This is the equation for the operating line of the enriching column shown in Fig. 5.60a. Its slope is $R/(R+1)$ and it passes through points (x_D, x_D) on the diagonal and (0, $x_D/(R+1)$) on the y-axis. The operating line intersects the q-line at point P. The coordinates of point P will represent the compositions of the liquid and vapour below plate N. If the feed is a saturated vapour, $F = \bar{G} = G$ and $B = \bar{L} = L$. However, if feed is a superheated vapour, some of the liquid reaching the plate N from the plate above it, is vaporised utilizing the sensible heat of the feed and therefore $G > \bar{G}$ and $L > \bar{L}$. Though the triangle terminating at P represents the last plate N in the column, the coordinates of point P are not the same as the composition of the liquid and vapour below plate N, because of the difference in the flows above and below the plate N. From Fig. 5.60b we can see that the vapour rising from plate N can be leaner in A than the feed, if the feed is superheated.

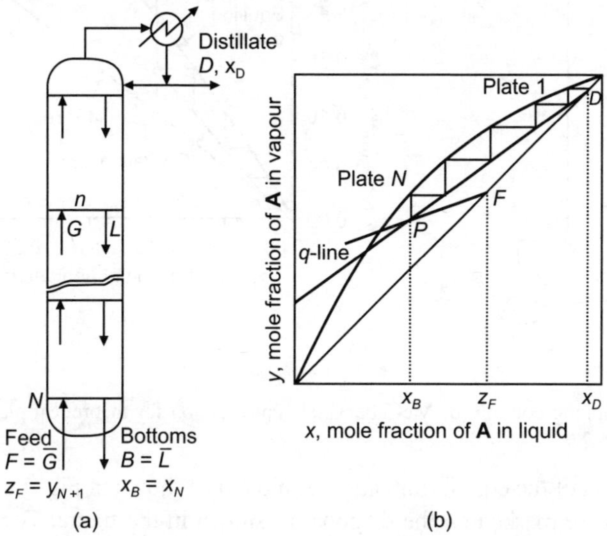

Fig. 5.60: (a) Enriching column (b) Construction for number of plates in enriching column with a superheated vapour feed

Example 5.22: A liquid mixture containing 10 mol% *n*-heptane and the rest *n*-octane is admitted to the top plate of a stripping tower. The liquid enters the column as a saturated liquid at 101.3 kPa. The bottoms product contains 98 mol% *n*-octane. For every 3 mol feed, 2 mol vapour is withdrawn as distillate product. Determine:

(a) The kmol/h of distillate and bottom product per 100 kmol/h of feed.
(b) The composition of the distillate.
(c) The number of theoretical plates.

The equilibrium data (mole fraction *n*-heptane) are:

x	0	0.012	0.039	0.067	0.097	0.284
y	0	0.025	0.078	0.131	0.184	0.459

***Solution*:** Basis: 100 kmol/h feed.

(a) $D = 100 \times \dfrac{2}{3} = 66.67$ kmol/h

Total material balance gives $B = F - D = 33.33$ kmol/h

Distillate = 66.67 kmol/h and bottoms = 33.33 kmol/h **(Answer)**

(b) Composition of the bottoms product: $x_B = 0.02$

A heptane balance gives

$$x_D = \frac{Fz_F - Bx_B}{D} = \frac{10 - 33.33 \times 0.02}{66.67} = 0.14$$

Mole fraction of heptane in the distillate = 0.14 **(Answer)**

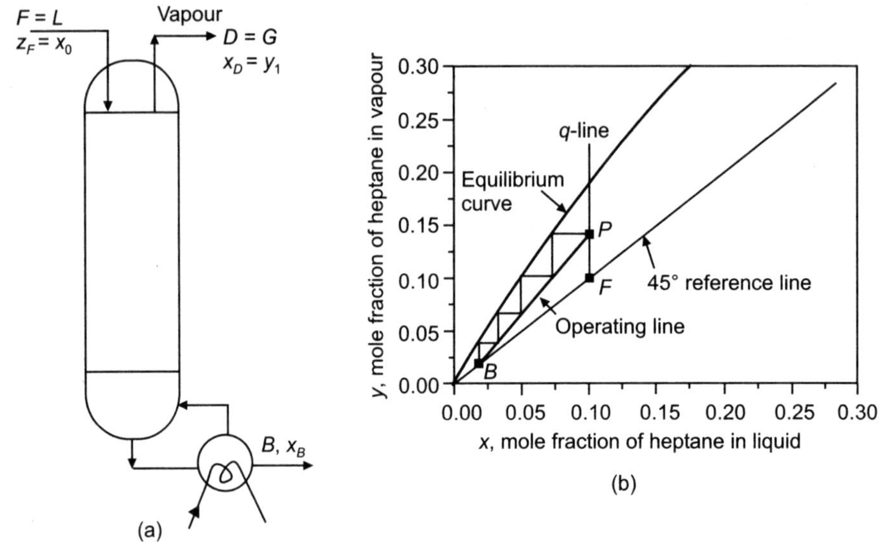

Fig. 5.61: (a) Stripping column (b) McCabe-Thiele construction for theoretical plates in the stripping column, Example 5.22

(c) The lower portion of the equilibrium curve is plotted as shown in Fig. 5.61b. The points B (0.02, 0.02) and F (0.1, 0.1) are marked on the diagonal as shown in the figure. The upper terminus of the operating line has coordinates (x_0, y_1) which is same as (z_F, x_D). Therefore point P (0.10, 0.14) can be marked on the diagram. Since the feed is a saturated liquid, the point P lies on the q-line which is a vertical line at point F (0.10, 0.10). The line PB is the operating line for the stripping column. Starting from P, steps are constructed as shown. The number of theoretical plates = 4 – 1 = 3. **(Answer)**

Example 5.23: An enriching column is fed with a feed consisting of 40 mol% benzene and 60 mol% toluene at 101.3 kPa. The feed enters at a rate 100 kmol/h as a saturated vapour and it is desired to produce a distillate containing 90 mol% benzene. Assume that a total condenser is used and the reflux ratio is 4.0 mol per mol of distillate. Determine:

(a) Molar rate of production of distillate and residue.
(b) Composition of the residue.
(c) Number of theoretical plates.

Equilibrium data (mole fraction of benzene) are:

x	0	0.1	0.2	0.3	0.4	0.5	0.6	0.7	0.8	0.9	0.95	1
y	0	0.205	0.369	0.502	0.618	0.706	0.789	0.849	0.907	0.955	0.978	1

***Solution*:** Consider the enriching column shown in Fig. 5.62a.

$$F = 100 \text{ kmol/h}; \; z_F = 0.40; \; x_D = 0.90; \; R = 4$$

(a) The material balance on the column is

$$F = D + B = D + L = D + RD = (R+1)D = 5D$$

$$D = \frac{F}{5} = 20 \text{ kmol/h}$$

$$B = F - D = 100 - 20 = 80 \text{ kmol/h}$$

Molar flow of distillate = 20 kmol/h and residue = 80 kmol/h. **(Answer)**

(b) $$x_B = \frac{Fz_F - Dx_D}{B} = \frac{100 \times 0.40 - 20 \times 0.90}{80} = 0.275$$ **(Answer)**

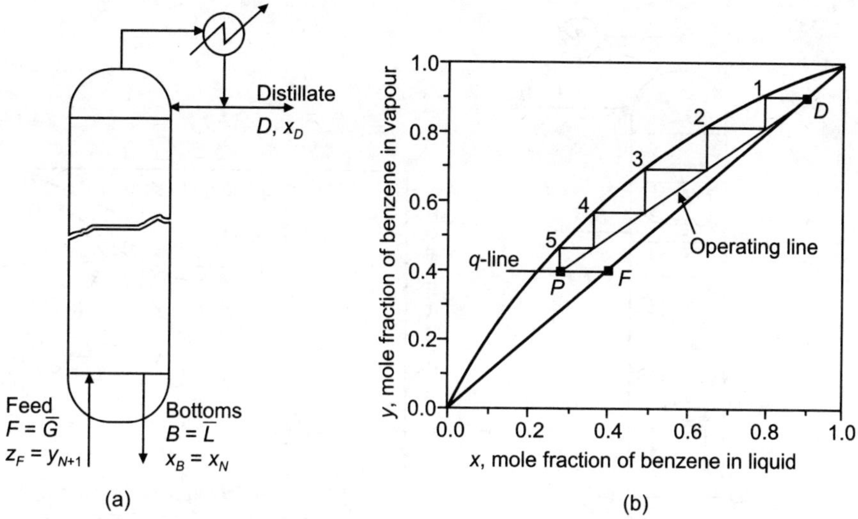

Fig. 5.62: (a) Enriching column (b) Construction for number of plates in enriching column, Example, 5.23

(c) Since the feed is a saturated vapour, $q = 0$ and q-line is a horizontal line as shown on the equilibrium diagram in Fig. 5.62b. The coordinates of the point P represent the composition of liquid and vapour below the bottom plate. Therefore P can be marked on the diagram with coordinates (x_B, z_F): (0.275, 0.4). Here, since the feed is a saturated vapour, the point P lies on the q-line and it is the lower terminus of the operating line. Number of plates is determined by the McCabe-Thiele staircase construction. We see that the number of plates = 5 from the figure. **(Answer)**

Special Case 5: Multiple Feeds and Side Stream

Side stream

When a side stream is withdrawn as shown in Fig. 5.63a, the internal flows of liquid and vapour below the plate from which it is withdrawn, will not be the same as those above the plate. Thus each side stream withdrawal leads to an additional operating line in the McCabe-Thiele construction. Let S kmol/h of liquid of composition x_S be removed as an intermediate product as shown in the figure. The liquid and vapour flows below the plate from which the side stream is removed are represented by L' and G' kmol/h respectively and those above this plate are L and G kmol/h.

Considering a plate i in the intermediate section, a component-**A** balance from top of the column to plate i gives

$$G'y_{i+1} = L'x_i + (Dx_D + Sx_S) \quad \text{...(5.186)}$$

Equation (5.186) is the equation for the operating line for the section of the column between the feed and the side stream. Equations (5.119) and (5.120) still represent the operating lines for the section above the side stream withdrawal and section below the feed plate, respectively.

$$Gy_{n+1} = Lx_n + Dx_D \quad \text{...(5.119)}$$

$$\bar{G}y_{m+1} = \bar{L}x_m - Bx_B \quad \text{...(5.120)}$$

The intersection of the operating line for the enriching section and the intermediate section is established considering Eq. (5.119) and (5.186). Let the point of intersection has coordinates (x', y'). Then

$$G'y' = L'x' + (Dx_D + Sx_S) \quad \text{...(5.187)}$$

$$Gy' = Lx' + Dx_D \quad \text{...(5.188)}$$

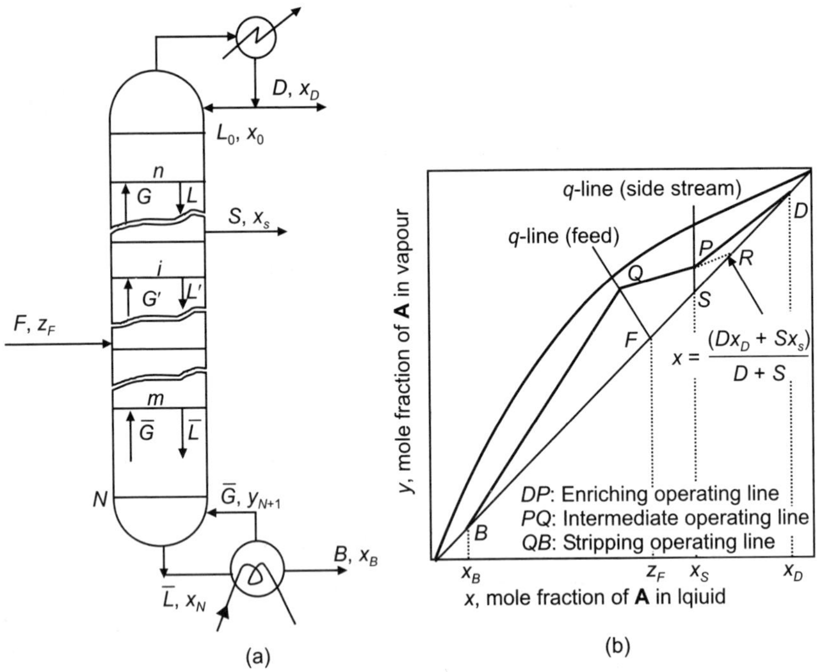

Fig. 5.63: (a) Column with one side stream (b) McCabe-Thiele construction for fractionation column in the presence of side stream

Subtracting Eq. (5.187) from Eq. (5.188), we get

$$(G - G')y' = (L - L')x' - Sx_S \quad \text{...(5.189)}$$

Since

$$(G - G') = (L - L') - S \quad \text{...(5.190)}$$

Eq. (5.190) may be rearranged as

$$\left[(L - L') - S\right]y' = (L - L')x' - Sx_S \quad \text{...(5.191)}$$

$$\left[\frac{L - L'}{S} - 1\right]y' = \frac{L - L'}{S}x' - x_S \quad \text{...(5.192)}$$

$$(q - 1)y' = qx' - x_S \quad \text{...(5.193)}$$

$$y' = \frac{q}{q-1}x - \frac{x_S}{q-1} \quad \text{...(5.194)}$$

where q' is the thermal quality of side stream. Thermal quality q' indicates by how much the liquid flow in the intermediate section is affected by the removal of one mole of side stream. The locus of intersection of the enriching operating line and the intermediate operating line is given by the q-line for the side stream.

Rearranging Eq. (5.186) we get

$$y_{i+1} = \frac{L'}{G'}x_i + \frac{(Dx_D + Sx_S)}{G'} \quad \text{...(5.195)}$$

Equation (5.195) is the equation for the operating line for the intermediate section. Its slope is $\frac{L'}{G'}$, the ratio of liquid to vapour flows in this section $L' = L - qS$ and $G = G' - (1 - q)S$, where q is the thermal quality of the side stream. Side streams can be withdrawn as saturated liquids or saturated vapour, but generally side streams are saturated liquids. For saturated liquid $q = 1.0$ and therefore $L' = L - S$ and $G' = G$. Therefore, L' and G' are known if L and G and thermal quality of the side stream are known. L and G are liquid and vapour flow in the enriching section and are related to the reflux ratio.

Since we know that the slope of the operating line for the intermediate section is $\frac{L'}{G'}$ and it passes through the intersection of the q-line and the enriching operating line P, the line can be drawn. Alternatively it is possible to find the intersection of the operating line with the diagonal. put $y_{i+1} = x_i$ in Eq. (5.195). We can see that the operating line intersect the diagonal at $x = \frac{(Dx_D + Sx_S)}{D+S}$. This point is marked R in Fig. 5.63b. The straight line passing through R and P is the operating line for the intermediate section. The intersection of the intermediate operating line with the q-line for the feed is marked Q in Fig. 5.63b. The line QB is the operating line for the stripping section as before. The number of plates required is determined by stepping off plates between the equilibrium curve and the operating lines. The step which represents the plate from which side stream is withdrawn should be at the intersection of the upper and the middle operating lines.

Multiple feed

When more than one feed stream of different compositions are admitted to the tower at different locations, three or more operating lines with different slopes exist. The Fig. 5.64a shows the column with two feeds: the upper one is a saturated liquid (feed 1) and the lower one a saturated vapour (feed 2). The McCabe-Thiele construction is shown in Fig. 5.64b. The operating line for the enriching section (the line DP) is drawn from the point D making an intercept $x_D/(R+1)$ with y-axis. Its intersection with the q-line for feed 1 is indicated by point P. The slope of the q-line for feed F_1 is $(q_1/q_1 - 1)$ and here

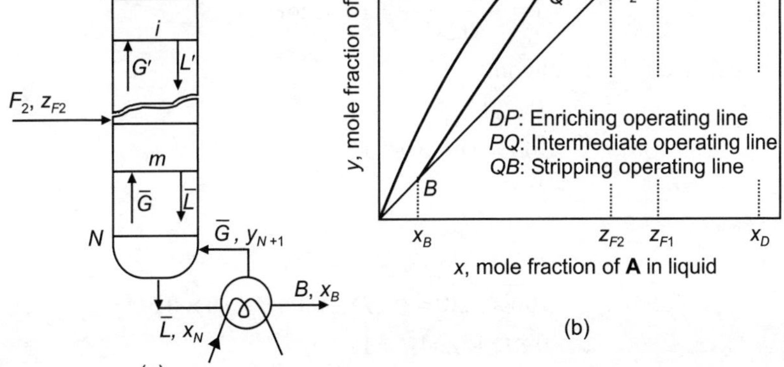

Fig. 5.64: (a) Fractionation column with multiple feed (b) McCabe-Thiele construction for distillation column employing multiple feeds

it is a vertical line drawn at z_{F1} on the diagonal, because feed 1 is assumed to be a saturated liquid. The slope of the intermediate operating line is L'/G', which can be evaluated knowing L, G and the thermal quality q_1 of the feed.

$$L' = L + q_1 F_1$$

$$G' = G - (1 - q_1)F_1$$

A line with slope L'/G' passing through P is the operating line for the middle section, the line PQ in the figure. The point Q is its intersection with the q-line for feed 2 drawn at point F_2 on the diagonal. The x-coordinate of F_2 is z_{F2} and slope of q-line is $q_2/(q_2 - 1)$. The line joining Q to B is the operating line for the stripping section. Its slope is $\bar{L}/\bar{G}$, the ratio of the liquid to the vapour flow in the stripping section. Triangles are constructed between the operating line and the equilibrium curve which gives the number of theoretical plates required in different sections of the column.

Example 5.24: One hundred kmol/h of an aqueous solution of ethyl alcohol containing 20 mol% alcohol is subjected to fractional distillation at 101.3 kPa to produce a distillate product containing 80 mol% alcohol and a residue containing 4 mol% alcohol. It is also desired that a side stream of composition 50 mol% alcohol is withdrawn so that 25% recovery of alcohol is achieved in the side stream. The feed enters as saturated liquid and a reflux ratio of 2 is used. Determine:

(a) The molar rate of products, kmol/h

(b) The number of theoretical plates required and the plate from which the side stream should be withdrawn.

Equilibrium data (mole fraction alcohol) are:

x	0.0190	0.0721	0.0966	0.1238	0.1661	0.2337	0.2608	0.3273
y	0.1700	0.3891	0.4375	0.4704	0.5089	0.5445	0.5580	0.5836
x	0.3965	0.5079	0.5198	0.5732	0.6763	0.7472	0.8943	
y	0.6122	0.6564	0.6599	0.6841	0.7385	0.7815	0.8943	

***Solution*:** Feed F = 100 kmol/h; $z_F = 0.20$

Distillate $x_D = 0.80$; Residue $x_B = 0.04$;

Side stream $x_S = 0.50$

Alcohol content in the feed = $Fz_F = 100 \times 0.20 = 20\ \text{kmol/h}$

Since 25% of alcohol in the feed is recovered in the side stream, $Sx_S = 0.25 \times 20$ where S is the side stream withdrawn (kmol/h).

$$S = \frac{0.25 \times 20}{0.50} = 10\ \text{kmol/h}$$

Total material balance for the column

$$F = D + B + S$$

$$100 = D + B + 10$$

Or

$$D + B = 90 \quad \text{(A)}$$

Alcohol balance

$$Fz_F = Dx_D + Bx_B + Sx_S$$

$$100 \times 0.20 = 0.8D + 0.04B + 10 \times 0.50$$

$$0.8D + 0.04B = 15 \quad \text{(B)}$$

Solving (A) and (B) we get

D = 15 kmol/h; B = 75 kmol/h

D = 15 kmol/h; B = 75 kmol/h; S = 10 kmol/h **(Answer)**

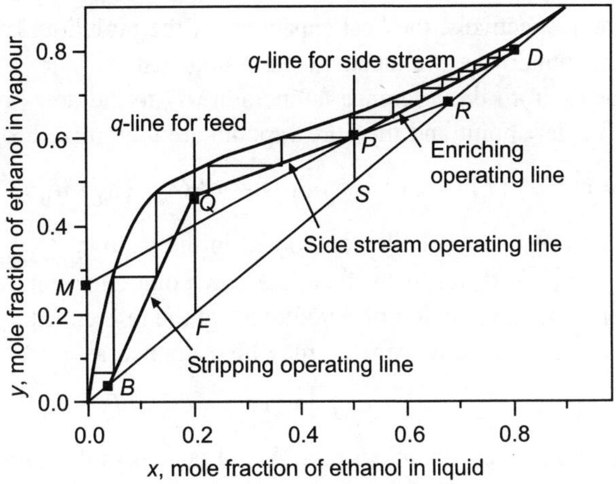

Fig. 5.65: Number of theoretical plates in the presence of side stream, Example 5.24

(b) The equilibrium curve is shown in Fig. 5.65. The points F (z_F, z_F), D (x_D, x_D), B (x_B, x_B) and S (x_S, x_S) are marked on the diagonal of the equilibrium curve knowing the concentrations z_F, x_D, x_B and x_S. Since the feed and the side streams are saturated liquids, their q-lines are vertical lines. The q-lines are erected at F and S as shown in the figure. The enriching operating line is drawn as usual knowing the y-intercept $\dfrac{x_D}{R+1} = \dfrac{0.80}{3} = 0.2667$. The enriching operating line DM intersects the q-line for the side stream at P. The intermediate operating line passes through point P and the point on the diagonal with coordinates $x = \dfrac{(Dx_D + Sx_S)}{D+S} = \dfrac{15 \times 0.8 + 10 \times 0.5}{15+10} = 0.68$. This latter point is shown by R in the figure. The line PR is joined and extended to meet the q-line of the feed at Q. The line PQ is the operating line for the intermediate section and BQ is the operating line for the stripping section. Now the stages are stepped off as shown. The number of steps = 12.25. Allowing for the partial reboiler, the number of plates in the column = 11.25. The side stream is to be withdrawn from the 8th plate from the top and the feed is to be introduced on the 11th plate. **(Answer)**

5.9 ENTHALPY-COMPOSITION DIAGRAMS

Enthalpy-composition diagrams give the relationship between enthalpy and composition of binary mixtures at constant pressure. Enthalpy in kJ/kmol (or kJ/kg) is taken as ordinate and mole fraction (or weight fraction) of more volatile components as abscissa. Of particular importance in distillation calculations are the enthalpies of saturated liquid and vapour streams which are shown as H_L versus x curve and H_G versus y curve in Fig. 5.66, where H_L is the enthalpy of saturated liquid and H_G is the enthalpy of saturated vapour. H_L is estimated as

$$H_L = C_L(T_b - T_0) + \Delta H_S \qquad \text{...(5.196)}$$

In Eq. (5.196), C_L is the heat capacity of the liquid mixture, T_b is the bubble point of the mixture, T_0 is the reference temperature at which the pure liquids are arbitrarily assigned zero enthalpy, and ΔH_S is the heat of mixing or heat of solution at the reference temperature T_0. ΔH_S is positive if heat is absorbed and negative if heat is evolved during the mixing. For ideal solutions, the heat capacity of the solution C_L may be evaluated as the composition weighted average of the heat capacities of the pure liquids.

$$C_L = xC_{A,L} + (1-x)C_{B,L} \quad \ldots(5.197)$$

where $C_{A,L}$ and $C_{B,L}$ are, respectively, the heat capacities of the pure liquids **A** and **B**.

The saturated vapour enthalpy is evaluated by assuming that the pure components **A** and **B** are heated separately as liquids from the reference temperature T_0 to the dew point temperature T_d, then allowed to vaporise at the dew point and the pure vapours are then mixed together. Thus

$$H_G = y\left[C_{A,L}\left(T_d - T_0\right) + \lambda_{A,d}\right] + (1-y)\left[C_{B,L}\left(T_d - T_0\right) + \lambda_{B,d}\right] \quad \ldots(5.198)$$

Here y is the mole fraction of more volatile component in the vapour, $\lambda_{A,d}$ and $\lambda_{B,d}$ are the heat of vaporisation of pure **A** and pure **B**, respectively, at the dew point temperature of the mixture.

Alternatively, if the heat of vaporisation of pure liquids at the reference temperature $\lambda_{A,0}$ and $\lambda_{B,0}$ are available, then the saturated vapour enthalpy may be estimated as

$$H_G = y\left[C_{A,G}(T_d - T_0) + \lambda_{A,0}\right] + (1-y)\left[C_{B,G}(T_d - T_0) + \lambda_{B,0}\right] \quad \ldots(5.199)$$

where $C_{A,G}$ and $C_{B,G}$ are the heat capacities of pure **A** and pure **B** in the vapour state.

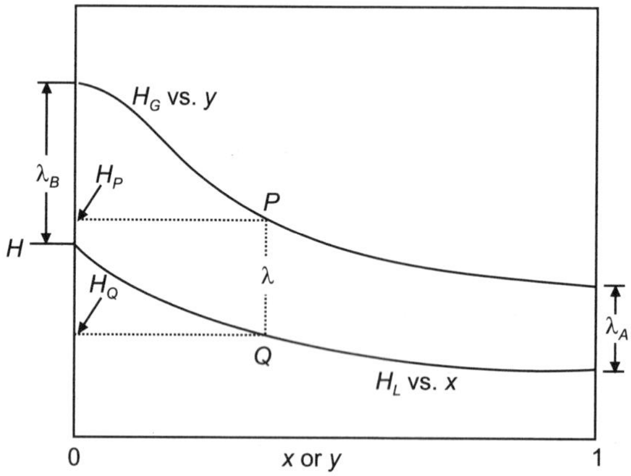

Fig. 5.66: Enthalpy-composition diagram

The upper curve in Fig. 5.66 is the enthalpy of saturated vapour versus composition (H_G versus y) and the lower curve is the enthalpy of saturated liquid versus composition (H_L versus x). The vertical distance between these two curves gives the heat of vaporisation of the mixture. For example, point P represents saturated vapour with enthalpy H_P and point Q represents saturated liquid of the same composition as the vapour P, with enthalpy H_Q; the distance PQ is H_P–H_Q, the heat of vaporisation λ of the mixture. At $x = 0$ and $x = 1$, these heat of vaporisation values represent the heats of vaporisation of pure liquids λ_B and λ_A, respectively. Below the H_L versus x curve, the solution is a subcooled liquid and above the H_G versus y curve, the mixture is superheated vapour. We can see that the saturation curves will be straight parallel lines if the latent heat of vaporisation of the components λ_B and λ_A are approximately equal, the heat of solution is negligible and the heat capacity of the solution varies linearly with composition.

Vapour-liquid Equilibrium on H-x-y Diagram

The equilibrium distribution curve is shown plotted below the enthalpy-composition diagram in Fig. 5.67. The liquid and vapour phases in equilibrium are represented by points such as M lying on the equilibrium curve. This point is equivalent to the line PQ on the enthalpy-composition diagram. P and Q represent the saturated liquid and vapour phases whose compositions are the same as the compositions of the phases represented by M on the equilibrium diagram. The line PQ which connects the two phases in equilibrium is known as a *tie line*. Several such tie lines can be drawn on the H-x-y diagram and each can be equivalently represented by a point on the x-y equilibrium distribution curve.

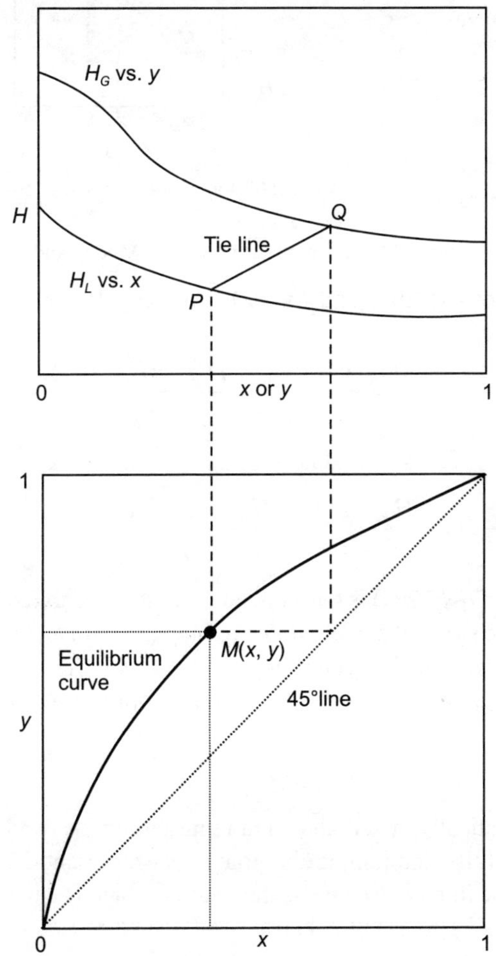

Fig. 5.67: Representation of equilibrium relationship on the H-x-y diagram

Adiabatic Mixing Process on H-x-y Diagram

Consider two streams, M kmol of stream M and N kmol of stream N mixed together adiabatically to yield P kmol of product P. The mixing process is shown in Fig. 5.68a.

Let the mole fraction of **A** in the stream M, N and P be x_M, x_N and x_P respectively, and let the enthalpies be H_M, H_N and H_P, respectively. Points M and N represent these streams on the enthalpy composition diagram as shown in Fig. 5.68c. The total material balance, component-**A** balance and enthalpy balance can be written as

$$M + N = P \quad \text{...(5.200)}$$

$$Mx_M + Nx_N = Px_P \quad \text{...(5.201)}$$

$$MH_M + NH_N = PH_P \quad \text{...(5.202)}$$

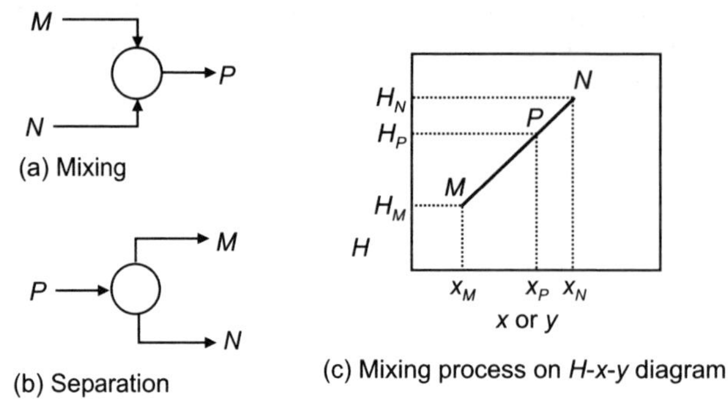

Fig. 5.68: (a) Mixing of two streams (b) Separation of streams (c) Mixing and separation on *H-x-y* diagram

Eliminate P from Equations (5.201) and (5.202) using Eq. (5.200) and rearrange the resulting equations.

$$\frac{M}{N} = \frac{x_N - x_P}{x_P - x_M} = \frac{H_N - H_P}{H_P - H_M} \qquad ...(5.203)$$

Equation (5.203) shows that

$$\frac{H_P - H_M}{x_P - x_M} = \frac{H_N - H_P}{x_N - x_P} \qquad ...(5.204)$$

The left hand side term of Eq. (5.204) is the slope of a straight line joining (x_P, H_P) and (x_M, H_M) and the right hand side term is the slope of a straight line joining (x_N, H_N) and (x_P, H_P). That is, Eq. (5.204) means that the slope of line *MP* and that of the line *PN* are equal. Since these lines have equal slopes and have one point in common, the points *P*, *M* and N are collinear. Also by Eq. (5.203)

$$\frac{M}{N} = \frac{\text{line NP}}{\text{line PM}} \qquad ...(5.205)$$

When two streams are adiabatically mixed, the point representing the resultant stream on the enthalpy composition diagram lies on the line joining the points that represent the original streams. Also the resultant point divides the line into two segments the ratio of which is equal to the ratio of the mass (or moles) of the original streams. This result is known as the *mixture rule* or the *lever-arm rule.*

The mixture rule can be extended to the subtraction or removal of one stream from another. Let stream M (x_M, H_M) be removed from stream P (x_P, H_P) under adiabatic conditions as shown in Fig. 5.68b. The resultant stream N should lie on the extension of the line joining M to P and N will be closer to the larger of the two streams.

$$P - M = N \qquad ...(5.206)$$

$$Px_P - Mx_M = Nx_N \qquad ...(5.207)$$

$$PH_P - MH_M = NH_N \qquad ...(5.208)$$

Eliminating N from the last two equations using Eq. (5.206), we get

$$\frac{M}{P} = \frac{x_N - x_P}{x_N - x_M} = \frac{H_N - H_P}{H_N - H_M} = \frac{\text{line PN}}{\text{line MN}} \qquad ...(5.209)$$

The resultant stream can be located on the extension of the line joining the initial two streams, if their amounts are known.

Flash Vaporisation on H-x-y Diagram

The flash vaporisation was discussed in detail under Section 5.4. An enthalpy balance for this operation can now be considered. Consider the process which is shown schematically in Fig. 5.69a.

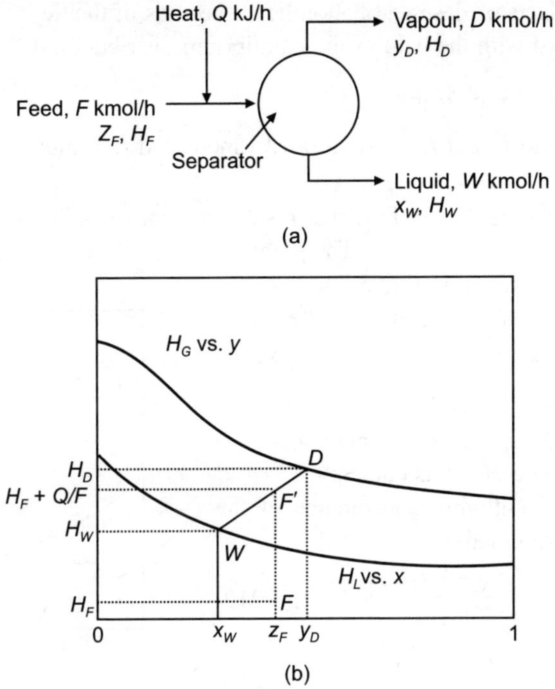

Fig. 5.69: (a) Flash distillation (b) Flash distillation on enthalpy-composition diagram

Feed mixture with enthalpy H_F and composition z_F is represented by point F in Fig. 5.69b. The mixture is heated by passing through a heat exchanger where it receives heat at a rate of Q kJ/h. Enthalpy increases by Q/F kJ/kmol and F' in the figure shows the condition of the feed leaving the heat exchanger. The composition of the feed is z_F and its enthalpy is now $H_F + Q/F$. The mixture being in the two-phase region, separates into a saturated liquid phase W (x_W, H_W) and a saturated vapour phase D (y_D, H_D) which are in equilibrium. Distillate product will lie on the H_G vs y-curve and the residue W lies on the H_L vs x-curve as shown.

The total material balance, the component-**A** balance and the enthalpy balance gives the following:

$$F = D + W \quad \text{...(5.210)}$$

$$Fz_F = Dy_D + Wx_W \quad \text{...(5.211)}$$

$$F(H_F + Q/F) = DH_D + WH_W \quad \text{...(5.212)}$$

Eliminating F from Eqs. (5.211) and (5.212) and rearranging we get

$$\frac{y_D - z_F}{x_w - z_F} = -\frac{W}{D} \quad \text{...(5.213)}$$

and

$$\frac{H_D - (H_F + Q/F)}{H_w - (H_F + Q/F)} = -\frac{W}{D} \quad \text{...(5.214)}$$

Equation (5.213) was developed earlier as the operating line on y-x coordinates for single-stage flash vaporisation. Combining equations (5.213) and (5.214), we see

$$\frac{y_D - z_F}{x_W - z_F} = \frac{H_D - (H_F + Q/F)}{H_w - (H_F + Q/F)} = -\frac{W}{D} \quad \ldots(5.215)$$

This is similar to Eq. (5.203) and by analogous reasoning, we can show that the points *D, F* and *W* are collinear and the products of flash distillation lie at the ends of the tie line that passes through F'. This tie line can be located with the help of the equilibrium distribution diagram.

Net Flow and the Difference Point

Consider two streams *K* and *L*. Let *K* be a saturated vapour and *L* a saturated liquid and let

$$K - L = \Delta \quad \ldots(5.216)$$

The stream Δ can be treated as the net flow that results on adiabatically removing saturated liquid *L* from a saturated vapour *K* on condensation. The points *K, L* and Δ are collinear and can be located on the extension of line *KL* as shown in Fig. 5.70.

The solute and enthalpy balances can be written as

$$Ky_K - Lx_L = \Delta x_\Delta \quad \ldots(5.217)$$

$$KH_K - LH_L = \Delta H_\Delta \quad \ldots(5.218)$$

where x_Δ is the mole fraction of component **A** in the fictitious net flow stream and H_Δ is its enthalpy.

The concentration and enthalpy coordinates of the difference point can be expressed as

$$x_\Delta = \frac{\text{Net moles of } K}{\text{Net total material}} = \frac{Kx_K - Lx_L}{K - L} \quad \ldots(5.219)$$

$$H_\Delta = \frac{\text{Net heat}}{\text{Net total material}} = \frac{KH_K - LH_L}{K - L} \quad \ldots(5.220)$$

The stream Δ can equally be treated as the one that results when the saturated liquid *Q* is adiabatically removed from the saturated vapour *P* or when the saturated liquid *N* is removed from the saturated vapour *M*, or by many such other combinations.

$$K - L = P - Q = M - N = \Delta \quad \ldots(5.221)$$

The net flow is constant and the straight line joining *P* and *Q* when extended will pass through the difference point Δ as also the straight line joining *M* and *N*. This is true for all pairs of streams the net flow of which is constant and equal to Δ. Thus any point such as Δ on the *H-x-y* diagram can be considered as the resultant of a process that adiabatically separates and removes a saturated liquid from a saturated vapour stream.

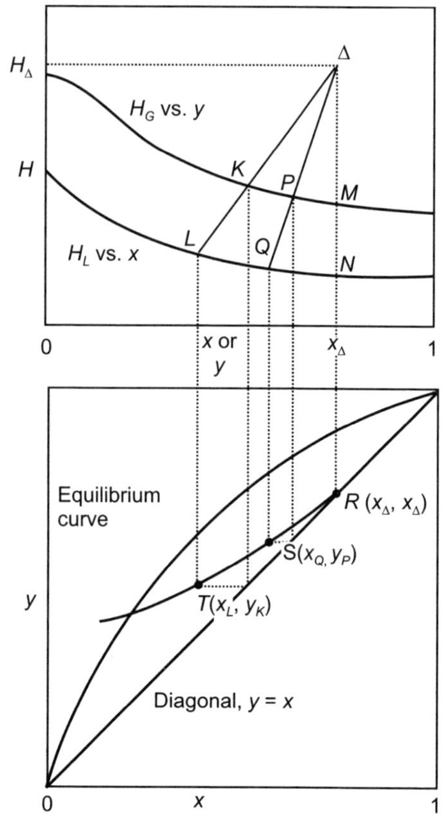

Fig. 5.70: Representation of net flow

The lines *KL, PQ* and *MN* are not tie lines since they don't connect phases in equilibrium. However, these lines can be projected on to the equilibrium diagram as shown in Fig. 5.70. Thus line *KL* on the *H-x-y* diagram is equivalently represented by point *T* on the *y-x* diagram with coordinates (x_L, y_K), the line *PQ* is represented by point *S* with coordinates (x_Q, y_P) and the vertical line *MN* is represented by point *R*. Point *R* lies on the 45° line with coordinates (x_Δ, x_Δ). Thus the net stream Δ on the *H-x-y* diagram is equivalent to the smooth curve *RST* drawn on the equilibrium coordinates. Any number of random lines can be drawn originating from Δ on the *H-x-y* diagram, intersecting the H_G-y and H_L-x saturation curves. These intersections when projected on to the *x-y* diagram, all lie on the curve *RST*.

If the difference point is located far removed from the saturation curves, the curve *RST* becomes steeper and ultimately coincides with the 45° diagonal when Δ is at infinity. Elimination of Δ from Eqs. (5.217) and (5.218) using Eq. (5.216) gives

$$\frac{L}{K} = \frac{x_\Delta - y_K}{x_\Delta - x_L} = \frac{\text{line } \Delta\text{K}}{\text{line } \Delta\text{L}} \quad \text{...(5.222)}$$

Similarly, it can be shown that

$$\frac{Q}{P} = \frac{\text{line } \Delta\text{P}}{\text{line } \Delta\text{Q}} \quad \text{...(5.223)}$$

$$\frac{N}{M} = \frac{\text{line } \Delta\text{M}}{\text{line } \Delta\text{N}} \quad \text{...(5.224)}$$

On the *y-x* coordinates, Eq. (5.222) is the slope of the straight line joining points *R* and *T* and the slope is equal to *L*/*K*, the ratio of the quantities of streams *L* and *K*. We can see from Eqs. (5.222), (5.223) and (5.224) that if the saturation curves H_G versus *y* and H_L versus *x* are straight parallel lines, then

$$\frac{L}{K} = \frac{Q}{P} = \frac{N}{M} \quad \text{...(5.225)}$$

This means that the curve *RST* on the *x-y* diagram will be a straight line if the enthalpy concentration curves are straight parallel lines. This condition is met when the molar latent heats of the components are equal, the heat of mixing is negligible and the heat capacities of the components are constant. These are the conditions which lead to the constant molar overflow and the constant molar vaporisation in the column. If these conditions are not applicable, the operating lines will be curved as shown by *RST* in Fig. 5.70.

5.10 PONCHON-SAVARIT METHOD

Material and Enthalpy Balances for the Fractionator

With reference to the Fig. 5.24, the total material balance, the component-**A** balance and the overall enthalpy balance for the fractionator can be written as

$$F = D + B \quad \text{...(5.226)}$$

$$Fz_F = Dx_D + Bx_B \quad \text{...(5.227)}$$

$$FH_F + Q_B = DH_D + BH_B + Q_C \quad \text{...(5.228)}$$

Here H_F, H_D, H_B are the enthalpies of the feed, the distillate and the residue, respectively in kJ/kmol and Q_C and Q_B are respectively, the condenser and reboiler heat loads in kJ/h. Equation (5.228) may be rearranged as

$$FH_F = D\left(H_D + \frac{Q_C}{D}\right) + B\left(H_B - \frac{Q_B}{B}\right) \quad \text{...(5.229)}$$

Let us designate $H_D + \dfrac{Q_C}{D}$ (the enthalpy of the distillate plus the condenser duty per kmol of the distillate) by H'_D, which then represents the net heat leaving the enriching section of the column per kmol of distillate product. Similarly, $H_B - \dfrac{Q_B}{B}$ (the enthalpy of the bottoms product less the reboiler heat duty per kmol of the bottoms product) is denoted by H'_B, which is the net heat leaving the stripping section of the column. Then Eq. (5.229) becomes

$$FH_F = DH'_D + BH'_B \qquad \text{...(5.230)}$$

The material balance and the enthalpy balance equations - Eqs. (5.226), (5.227) and (5.230) - can be combined together and expressed as

$$F = \Delta_D + \Delta_B \qquad \text{...(5.231)}$$

where Δ_D represents the quantity, composition and enthalpy of the net flow in the enriching section of the column and Δ_B represents the corresponding values of the net flow in the stripping section of the column. Thus quantity of Δ_D is D kmol/h, composition is x_D, and enthalpy is $H'_D = H_D + \frac{Q_C}{D}$ kJ/kmol. Δ_D is equivalent to D with respect to the amount and concentration, and not with respect to the enthalpy. The net flow in the stripping section Δ_B is B kmol/h in quantity, mole fraction x_B is its composition, and $H'_B = H_B - \frac{Q_B}{B}$ kJ/kmol is its enthalpy. Thus Δ_B is equivalent to B only with respect to the amount and concentration, and not with respect to the enthalpy.

Similar to the derivation of Eq. (5.203), we can show using Eqs. (5.226), (5.227) and (5.230) that

$$\frac{D}{B} = \frac{z_F - x_B}{x_D - z_F} = \frac{H_F - H'_B}{H'_D - H_F} \qquad \text{...(5.232)}$$

Equation (5.232) means that a straight line can be drawn connecting the lower difference point $\Delta_B(x_B, H'_B)$, the feed point $F(z_F, H_F)$ and the upper difference point $\Delta_D(x_D, H'_D)$. The line $\Delta_B F \Delta_D$ is designated as the *feed line*.

Knowing the composition and enthalpy of the feed, feed point F can be marked on the diagram as shown in Fig. 5.71. The point F will be on the lower curve (H-x curve) if the feed is a saturated liquid, on the upper curve (H-y curve) if the feed is saturated vapour, below the H-x curve if the feed is subcooled liquid and above the H-y curve if the feed is superheated vapour. If the feed is partially vaporised, F will be in the two-phase region bounded by the upper and lower saturation curves. Figure 5.71 shows that the feed as a saturated liquid. Similarly, the points representing the distillate (D) and bottoms product (B) are also shown in the figure as saturated liquids. We can locate Δ_D on the enthalpy-composition diagram knowing its concentration and enthalpy coordinates, that is, the distillate composition x_D and enthalpy $H_D + \frac{Q_C}{D}$. The latter is determined if the enthalpy of the distillate and the condenser heat duty Q_C are known. It will be shown in the following sections that Δ_D can be easily located on the H-x-y diagram knowing the reflux ratio R. After locating points F and Δ_D, draw line $\Delta_D F$ the extension of which intersects the vertical line drawn through B (x_B, H_B) at Δ_B.

Material and Enthalpy Balances for the Enriching Section

Consider plate n in the enriching section of the column (Fig. 5.72). The molar flow rate, mole fraction of **A** and enthalpy of the vapour reaching the plate n are, respectively G_{n+1}, y_{n+1} and $H_{G,n+1}$. A material and enthalpy balance around the enriching section give

$$G_{n+1} - L_n = D \qquad \text{...(5.233)}$$

$$G_{n+1}y_{n+1} - L_n x_n = Dx_D \qquad \text{...(5.234)}$$

$$G_{n+1}H_{G,n+1} - L_n H_{L,n} = D\left(H_D + \frac{Q_C}{D}\right) = DH'_D \qquad \text{...(5.235)}$$

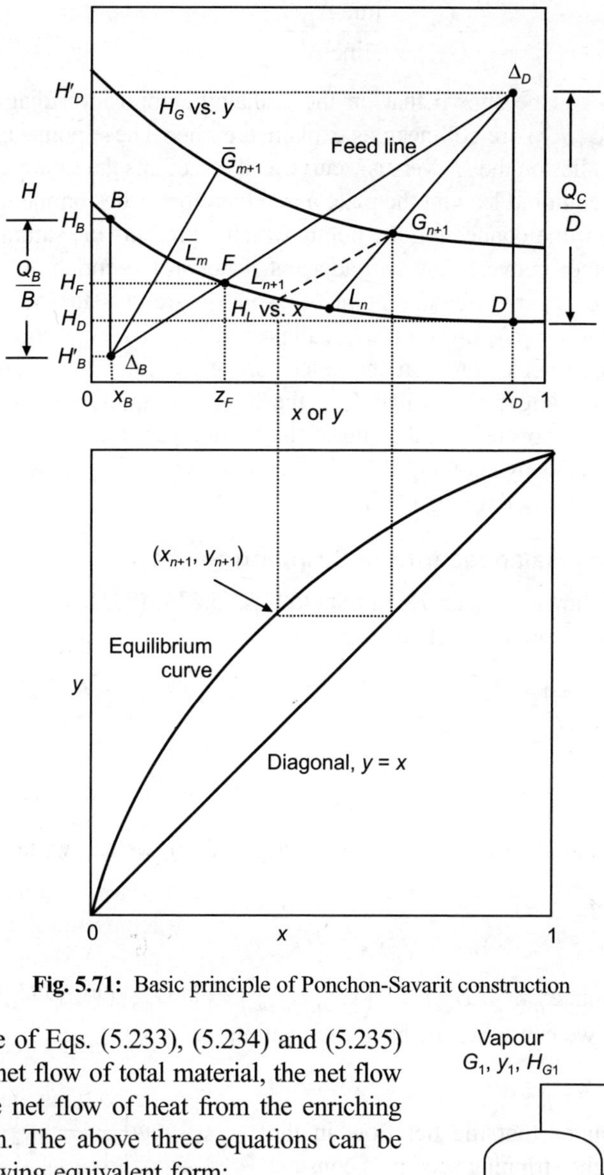

Fig. 5.71: Basic principle of Ponchon-Savarit construction

The right-hand side of Eqs. (5.233), (5.234) and (5.235) give respectively, the net flow of total material, the net flow of material **A** and the net flow of heat from the enriching section of the column. The above three equations can be combined in the following equivalent form:

$$G_{n+1} - L_n = \Delta_D \quad \text{...(5.236)}$$

Eliminating D from Eqs. (5.234) and (5.235) using Eq. (5.233), we get

$$\frac{L_n}{G_{n+1}} = \frac{x_D - y_{n+1}}{x_D - x_n} = \frac{H'_D - H_{G,n+1}}{H'_D - H_{L,n}} \quad \text{...(5.237)}$$

Using Eq. (5.237), the following relation can be written for determining geometrically the internal reflux ratio at any plate n. Referring to Fig. 5.71,

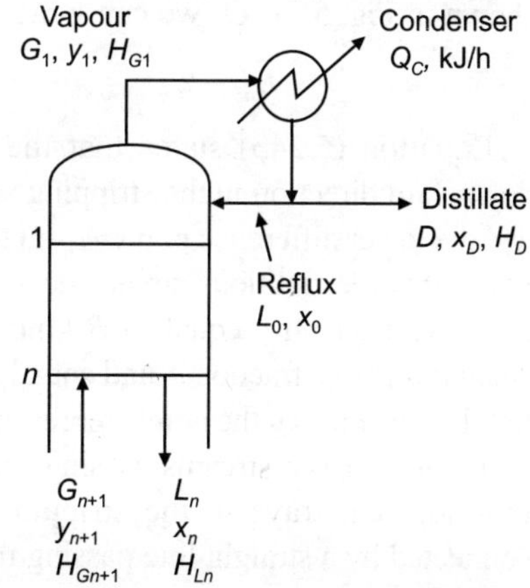

Fig. 5.72: Flow rates, concentration and enthalpies of streams in the enriching section

$$\frac{L_n}{G_{n+1}} = \frac{\text{line } \Delta_D G_{n+1}}{\text{line } \Delta_D L_n} \qquad \ldots(5.238)$$

Using Eq.(5.237), it can be shown that on the enthalpy-composition diagram, the points (y_{n+1}, $H_{G,n+1}$), (x_n, $H_{L,n}$) and (x_D, H'_D) are collinear, as explained earlier. These points are denoted as G_{n+1}, L_n and Δ_D in Fig. 5.71. G_{n+1} lies on the H_G versus y curve as it represents the saturated vapour reaching the plate n. L_n is the saturated liquid leaving the plate n and therefore it lies on the lower saturation curve, H_L versus x. The straight line connecting the points which represents the saturated liquid and vapour streams passing each other between any two adjacent plates in the enriching section of the column should pass through the upper difference point Δ_D which represents the net flow in the enriching section. Thus the lines L_0G_1, L_1G_2, L_2G_3, L_3G_4,.... all pass through Δ_D. As a corollary, if G_{n+1} is known, L_n can be located at the intersection that the extension of the line $\Delta_D G_{n+1}$ makes with the lower saturation curve. We have already seen that L_{n+1}, the liquid in equilibrium with the vapour G_{n+1} is determined on the H_L versus x curve at the end of the tie line passing through G_{n+1} as shown by the dotted line in the figure. Thus by knowing G_{n+1}, L_n can be determined by drawing the construction line and L_{n+1} can be determined by drawing the tie line.

Material and Enthalpy Balances for the Stripping Section

The stripping section is shown in Fig. 5.73. Similar to Eqs. (5.233)-(235), we can write the material and enthalpy balances for the stripping section as

$$\bar{L}_m - \bar{G}_{m+1} = B \qquad \ldots(5.239)$$

$$\bar{L}_m x_m - \bar{G}_{m+1} y_{m+1} = B x_B \qquad \ldots(5.240)$$

$$\bar{L}_m H_{L,m} - \bar{G}_{m+1} H_{G,m+1} = B\left(H_B - \frac{Q_B}{B}\right) = BH'_B \qquad \ldots(5.241)$$

By eliminating B from the last two equations using Eq. (5.239), we can write

$$\frac{\bar{L}_m}{\bar{G}_{m+1}} = \frac{y_{m+1} - x_B}{x_m - x_B} = \frac{H_{G,m+1} - H'_B}{H_{L,m} - H'_B} \qquad \ldots(5.242)$$

which shows that points $\Delta_B(x_B, H'_B)$, $\bar{L}_m(x_{L,m}, H_{L,m})$ and $\bar{G}_{m+1}(y_{m+1}, H_{G,m+1})$ are collinear as shown in Fig. 5.71. Or we can write the balance equations as

$$\bar{L}_m - \bar{G}_{m+1} = \Delta_B \qquad \ldots(5.243)$$

Equation (5.243) states that the net flow in the downward direction in the stripping section is constant and the lower difference point Δ_B on the H-x-y diagram represents the fictitious stream designated as the 'net flow' with quantity equal to B kmol/h, concentration equal to x_B mole fraction **A** and enthalpy equal to H'_B kJ/kmol. It means that the points representing the saturated liquid and vapour streams passing each other between two adjacent trays in the stripping section can be connected by a straight line passing through Δ_B. Thus the lines $\bar{G}_{N+1}\bar{L}_N$, $\bar{G}_N\bar{L}_{N-1}$, $\bar{G}_{N-1}\bar{L}_{N-2}$ ......, etc. when extended pass through the stripping section difference point Δ_B.

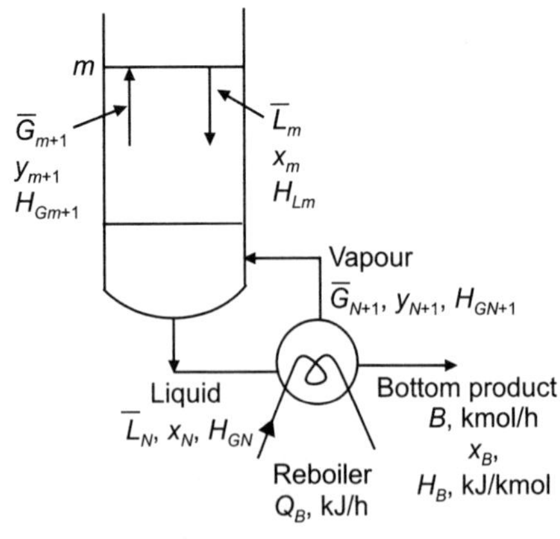

Fig. 5.73: Flow rates, concentration and enthalpies of streams in the stripping section

Graphical Construction

Assume that a total condenser is used. Since all the vapour leaving the top tray get condensed, the reflux L_0 and the distillate D have the same properties (i.e., same composition and enthalpy), they are both represented by a single point D (or L_0) as shown in Fig. 5.74a. Figure assumes that the reflux is at the bubble point, so that L_0 lies on the H_L versus x curve. The composition of the saturated vapour G_1 reaching the condenser is same as that of the reflux, i.e., $x_0 = x_D = y_1$. Therefore G_1 can be located on the H_G versus y curve and on the vertical line passing through D. We know that the enriching section difference point Δ_D should also fall on this vertical line because the concentration coordinate of Δ_D is x_D. The exact location of Δ_D can be fixed once its enthalpy coordinate H'_D is known, which in turn depends on the condenser heat duty.

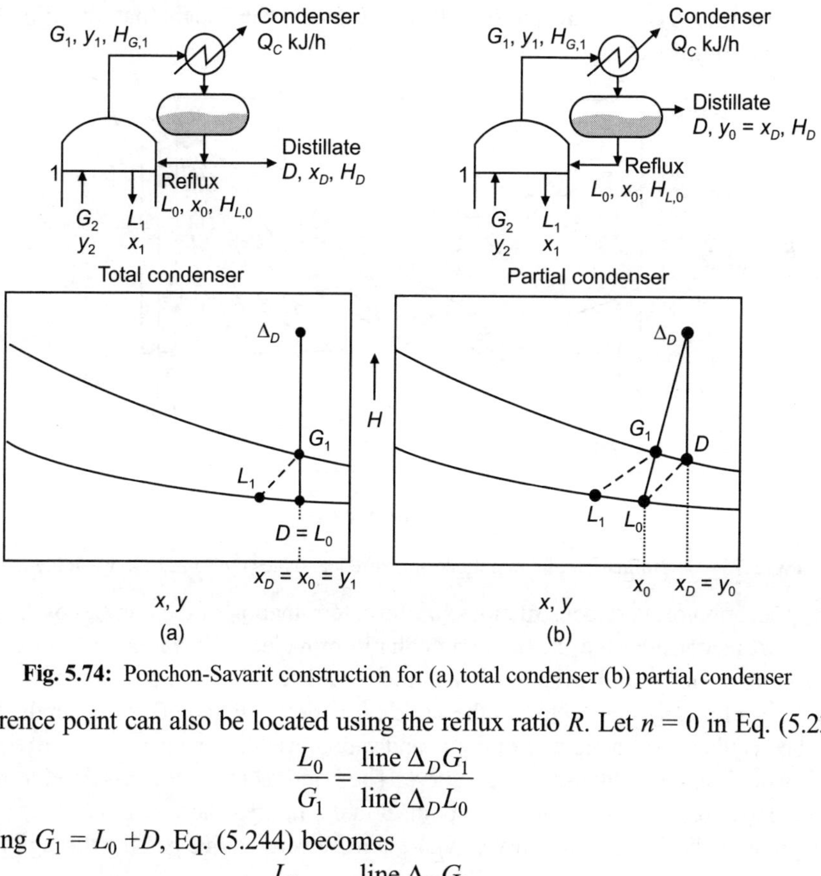

Fig. 5.74: Ponchon-Savarit construction for (a) total condenser (b) partial condenser

The difference point can also be located using the reflux ratio R. Let $n = 0$ in Eq. (5.238). We get

$$\frac{L_0}{G_1} = \frac{\text{line } \Delta_D G_1}{\text{line } \Delta_D L_0} \qquad \text{...(5.244)}$$

Substituting $G_1 = L_0 + D$, Eq. (5.244) becomes

$$\frac{L_0}{L_0 + D} = \frac{\text{line } \Delta_D G_1}{\text{line } \Delta_D L_0} \qquad \text{...(5.245)}$$

Dividing the terms in the numerator and denominator of the left hand side of Eq. (5.245) by D and noting that $R = L_0/D$, the above equation becomes

$$\frac{R}{R+1} = \frac{\text{line } \Delta_D G_1}{\text{line } \Delta_D L_0} \qquad \text{...(5.245 A)}$$

which can be written as

$$R = \frac{\text{line } \Delta_D G_1}{\text{line } \Delta_D L_0 - \text{line } \Delta_D G_1} = \frac{\text{line } \Delta_D G_1}{\text{line } L_0 G_1} \qquad \text{...(5.246)}$$

$$\text{line } \Delta_D G_1 = R(\text{line } L_0 G_1) \qquad \ldots(5.247)$$

Equation (5.247) means that Δ_D can be located above G_1 on the vertical line passing through D and G_1 at a height R times the distance L_0G_1. Condenser heat duty Q_C can then be computed from the enthalpy coordinate of Δ_D.

If a partial condenser as shown in Fig. 5.74b is used, the distillate is a saturated vapour and the reflux is the saturated liquid in equilibrium with the distillate. Let the distillate composition be denoted by y_0 $(= x_D)$. The point D representing the distillate lies on the H_G versus y curve as shown and the reflux L_0 should be on the H_L versus x curve at the end of the tie line passing through D. Since the points L_0, G_1 and Δ_D are collinear, intersection of the line joining L_0 and Δ_D intersects the upper saturation curve at G_1. If Δ_D is to be located using the reflux ratio, Eq. (5.247) can still be used but, this time the Δ_D has to be located on the vertical line passing through D by trial such that the line length $\Delta_D G_1$ is R times the line length L_0G_1.

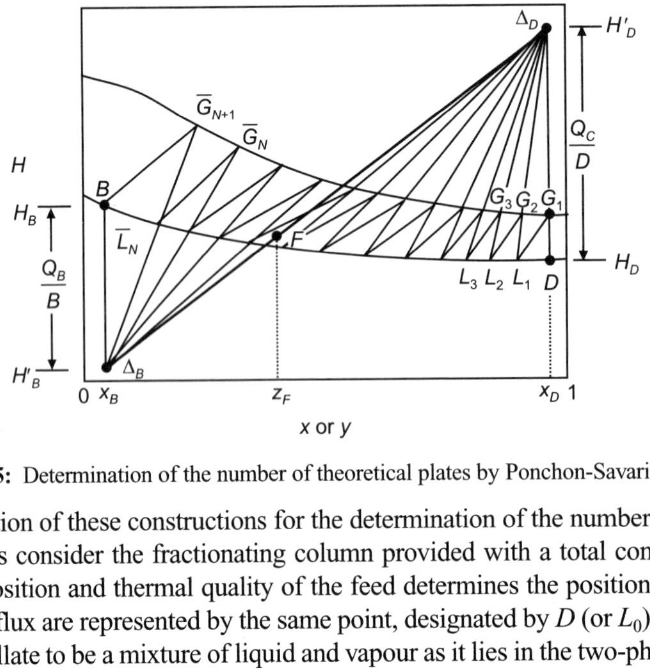

Fig. 5.75: Determination of the number of theoretical plates by Ponchon-Savarit method

Now the application of these constructions for the determination of the number of theoretical plates is quite clear. Let us consider the fractionating column provided with a total condenser and a partial reboiler. The composition and thermal quality of the feed determines the position of the feed point F. Distillate and the reflux are represented by the same point, designated by D (or L_0) in the figure. Figure 5.75 shows the distillate to be a mixture of liquid and vapour as it lies in the two-phase region. If it were subcooled, it would lie below the saturation curve. The bottoms product is assumed to be a saturated liquid with composition x_B and is denoted by point B in the figure. The vapour G_1 lies vertically above the distillate point on the H_G versus y curve. Δ_D lies on the vertical line passing through L_0 and G_1 and its exact location is determined either by knowing the condenser heat duty $\left(H'_D = H_D + \dfrac{Q_C}{D}\right)$ or by using the reflux ratio R [Eq. (5.247)]. To determine the stripping section difference point Δ_B, the line joining Δ_D and F is extended. It intersects the vertical line passing through B at Δ_B. The enthalpy coordinate of this point is used to evaluate the reboiler heat duty $\left(H'_B = H_B - \dfrac{Q_B}{B}\right)$.

The liquid leaves plate 1 is in equilibrium with G_1 and therefore a tie line drawn through G_1 will locate L_1 as shown. The tie lines can be easily drawn on the H-x-y diagram with the help of the distribution curve plotted just below the enthalpy-composition diagram (not shown in the figure). Since L_n, G_{n+1}, Δ_D are collinear, the line connecting L_1 and Δ_D will intersect the upper saturation curve at G_2.

Having obtained G_2, L_2 lies at the end of the tie line through G_2 and G_3 on the construction line connecting L_2 and Δ_D and so on. Here the tie lines G_1L_1, G_2L_2, G_3L_3, etc. represent theoretical plates 1, 2, 3, etc. in the column. Thus the number of tie lines drawn for the enriching section will be the number of theoretical plates in the column. However, when partial condenser is used and the distillate is a saturated vapour and the reflux is saturated liquid in equilibrium with the distillate, the first tie line drawn is DL_0 which stands for the partial condenser. The second tie line is L_1G_1 and it represents the plate 1. When a partial condenser is used the number of plates in the enriching section is the number of tie lines minus one.

The optimum feed plate location corresponds to the tie line which crosses the feed line $\Delta_B F \Delta_D$. In the Fig. 5.75, the 9th plate from the top of the column is the feed plate. When the tie line has crossed the feed line, construction lines are to be drawn through Δ_B instead of Δ_D to determine $\bar{G}$. Alternating tie line and construction line are drawn until we reach point B. The last tie line which passes through the point B connects the saturated liquid of composition x_B and the saturated vapour of composition y_{N+1}. These are the compositions of the effluent streams from the reboiler and therefore, the last tie line represents the reboiler and the last but one tie line is the bottommost plate N. Thus the number of plates in the stripping section is equal to the number of tie lines minus one.

If the feed is admitted at a plate earlier than the one at the optimum location, the transfer from the upper to the lower difference point for construction lines is made early, and if the feed entry is delayed beyond the optimum location, the transfer to the lower difference point also is delayed. Both these cases, will lead to a larger number of plates than that obtained when the feed entry is at the optimum location. The optimum location of the feed addition is that plate which corresponds to the tie line that crosses the feed line.

Aliter

We have shown that any point on the enthalpy-composition diagram can be considered as representing a net flow which results when a saturated liquid is adiabatically removed from a saturated vapour stream. The net flow can be equivalently represented by a curve such as the curve *RST* in Fig. 5.70. When Δ_D is considered as the net flow, the equivalent curve that is obtained on the *x-y* diagram is the operating line for the enriching section. This is because the line such as $G_{n+1}L_n$ that connects the phases that cross each other between two adjacent plates n and $n+1$, when extended pass through Δ_D on the *H-x-y* diagram and the projection of G_{n+1} and L_n to the equilibrium diagram has therefore the coordinates (x_n, y_{n+1}) which are the coordinates of points on the operating line on *x-y* coordinates. The operating line is drawn through the points with coordinates that represent the composition of the passing streams such as (x_n, y_{n+1}). To draw the operating curve, after fixing Δ_D, random lines are drawn from Δ_D to intersect the upper and lower saturation curves. The intersections of each such line gives rise to a point on the *x-y* coordinate diagram. A smooth curve is drawn through these points, which is the operating line for the enriching section. Similarly, the intersections of the random lines originating from Δ_B with the upper and lower saturation curves when projected on to the equilibrium curve give rise to the operating line for the stripping section of the column. After the operating lines are drawn, the McCabe-Thiele stair-case construction gives the number of theoretical plates. The construction is shown in the Fig. 5.76.

Summary of Ponchon-Savarit Method

The Ponchon-Savarit method for determining the number of theoretical plates in the fractionator for separating a feed mixture of given composition into the products of specified purity can be summarized as below:

1. Plot the enthalpy-composition diagram showing the saturated vapour and liquid enthalpy curves.
2. Mark points F (z_F, H_F), D (x_D, H_D) and B (x_B, H_B) on the diagram.

3. Point L_0 representing the reflux and D are identical if the condenser used is total. Mark point G_1 on the upper saturation curve vertically above D. If the condenser is partial, point D lies on the upper saturation curve. Mark point L_0 on the lower saturation curve at the end of the tie line passing through D.
4. Locate the difference point Δ_D on the vertical line passing through D using either the condenser heat duty or the reflux ratio R.
   (a) If condenser duty is known, Δ_D is located from its enthalpy coordinate which is
   $$H'_D = H_D + \frac{Q_C}{D}$$
   (b) If reflux ratio R is known, and the condenser is total, then Δ_D is located such that
   $$\Delta_D G_1 = R(G_1 L_0)$$
   (c) If R is known and the condenser is partial, a straight line is drawn from L_0 by trial which intersects the upper saturation curve at G_1 and meet the vertical line passing though D at Δ_D such that $\Delta_D G_1 = R(G_1 L_0)$
5. Connect Δ_D to F and extend it to meet the vertical line passing though B at Δ_B; the line $\Delta_D F \Delta_B$ is the feed line.
6. The tie line G_1L_1 is drawn and it represents the first plate in the column. The construction line $L_1 \Delta_D$ is drawn which intersects the upper saturation curve at G_2.

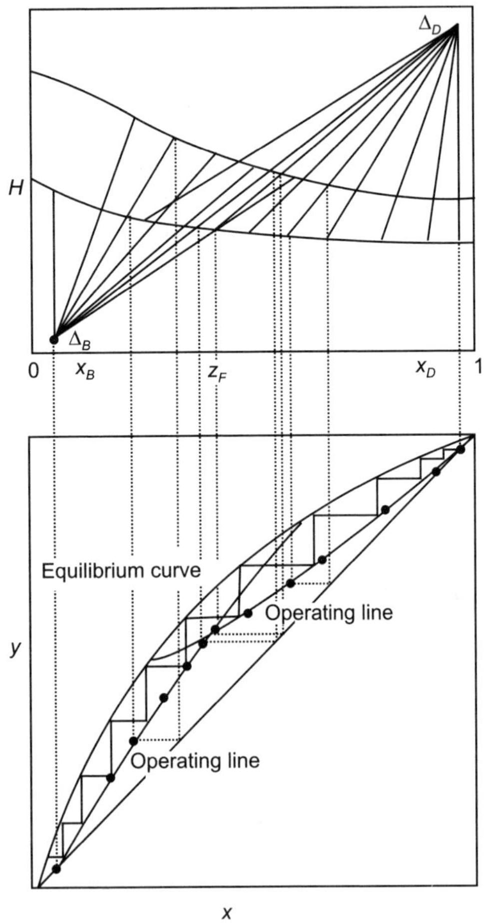

Fig. 5.76: Construction of operating line using *H-x-y* data

7. The construction of alternate tie line and the construction line is continued till the tie line crosses the feed line.
8. Continue the construction of alternate tie line and the construction line beyond the feed line, but the lower difference point is used for construction once the tie line crosses the feed line.
9. Stop the construction when the x-coordinate of the saturated liquid end of the tie line is less than or equal to x_B.
10. The number of theoretical plates in the column is the total number of tie lines constructed between point D to point B minus one if the condenser is total, minus two if the condenser is partial.

Total Reflux

As reflux ratio is increased, the condenser heat duty increases and the difference point Δ_D moves away from the saturation curves. Under total reflux condition, $R = \infty$, and the difference point Δ_D is at infinity. As the difference point for the stripping section Δ_B, F and Δ_D lie on the same line, under total reflux condition, Δ_B moves to $-\infty$. The construction lines G_1L_1, G_2L_2, G_3L_3, …etc become vertical lines. The number of theoretical plates is the minimum under total reflux conditions and this can be determined graphically as shown in the Fig. 5.77.

It may also be noted that since all the construction lines are vertical lines their projections on the equilibrium diagram fall on the diagonal. The operating lines for the total reflux condition, thus coincide with the diagonal of the equilibrium diagram.

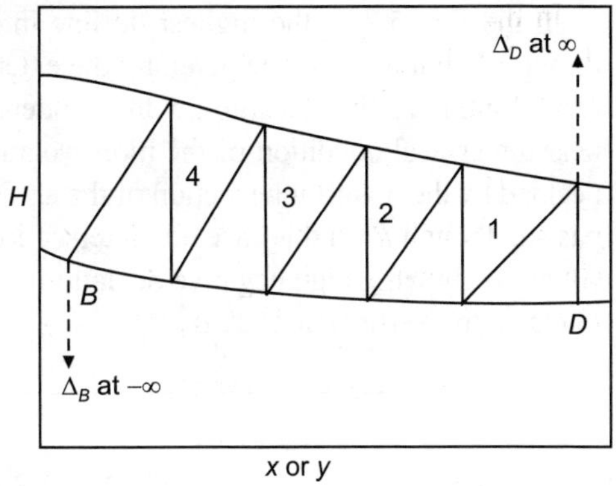

Fig. 5.77: Total reflux and minimum number of plates

Minimum Reflux Conditions

The reflux ratio determines the position of the difference points on the *H-x-y* diagram as given by Eq. (5.246)

$$R = \frac{\text{line } \Delta_D G_1}{\text{line } L_0 G_1} \qquad \ldots(5.246)$$

As the reflux ratio is reduced, the cost of circulation of the cooling water in the condenser and the cost of supply of the steam to the reboiler get reduced. It means that the lower the reflux ratio the smaller the cost of operation of the distillation column. That is, the operating cost is the lowest at the minimum reflux condition. However, at the minimum reflux, the number of plates required to accomplish the specified separation becomes infinitely large. The minimum reflux ratio may be treated as the maximum reflux ratio which will result in infinite number of plates for the specified separation because, for the reflux ratio values below the minimum, the separation requires infinite plates and the region of constant composition or the pinch point gets shifted from the feed plate, as explained elsewhere.

Geometrically, an infinite number of plates will result if the difference point Δ_D is located such that one of the tie lines for the system when extended passes through this point. The construction line and the tie line coincide and an infinite number of these lines will be required to cross this composition. To overcome this difficulty, the reflux ratio should be such that Δ_D lies above the highest intersection made by the extended tie lines with the vertical line passing through *D*. In general, for systems with normal equilibrium curves, that is, equilibrium curves which are concave downward throughout, the tie line passing through the feed point when extended intersects the vertical at *D* at the highest position. The tie line *PQ* in Fig. 5.78 provides the highest intersection at $\Delta_{D,m}$ the difference point corresponding to the minimum reflux. $\Delta_{D,m}$ is the lowest possible position for Δ_D as the construction line and the tie line will coincide at the feed point and a zone of constant composition is developed around the feed stage. $\Delta_{B,m}$ is determined on the extension of the line joining $\Delta_{D,m}$ and *F*.

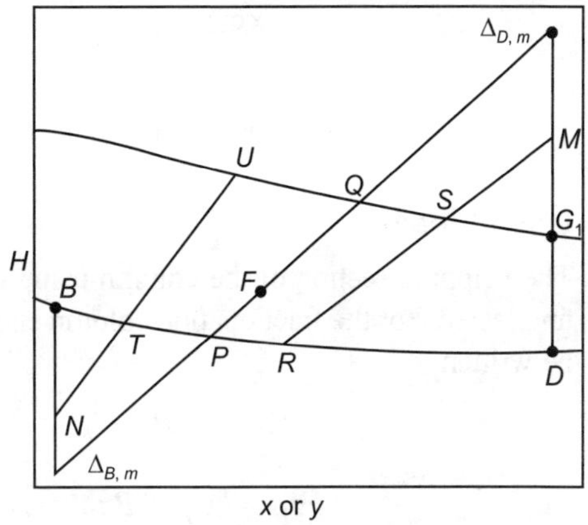

Fig. 5.78: Determination of the minimum reflux

$\Delta_{D,m}$ is located at the highest tie line intersection and the $\Delta_{B,m}$ at the lowest tie line intersection for the minimum reflux condition. Tie lines such as *RS* on the right side of *F* when extended intersects below $\Delta_{D,m}$ and all tie lines such as *TU* on the left side of *F* intersect above $\Delta_{B,m}$. If $\Delta_{D,m}$ were located at *M* in Fig. 5.78 or $\Delta_{B,m}$ located at *N* will also result in infinite plates, but will not correspond to the minimum reflux condition, because the minimum reflux is the maximum of all the reflux values giving infinite number of plates. The minimum condenser duty is determined from the *H*-coordinate of $\Delta_{D,m}$ and the minimum reboiler heat load from that of $\Delta_{B,m}$. The minimum reflux ratio can be computed using Eq. (5.246).

In the Fig. 5.79a, the highest tie line intersection is not the one provided by the tie line passing through F, but it is the one given by the extension of the tie line RS. This is true for the systems which exhibit high positive deviation with a tendency to form azeotrope. This will be the case for all systems near the critical condition of the more volatile component. In Fig. 5.79b, the minimum reflux ratio is decided by the lowest intersection of the extension of the stripping section tie lines with the vertical line passing though B; in this case, the intersection of the line TU. This will be the case for certain nonideal solutions showing high negative deviation. After locating $\Delta_{B,m}$ in this way, the extension of $\Delta_{B,m}F$ will intersect the vertical at D at $\Delta_{D,m}$.

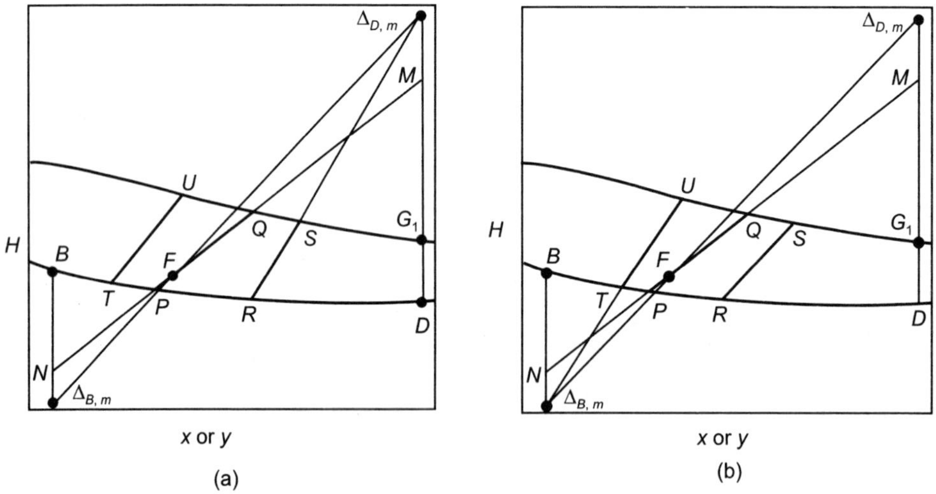

Fig. 5.79: Determination of minimum reflux ratio for highly nonideal systems

Open Steam

The stripping section of the column using open steam is shown in Fig. 5.80. The enriching section is unaffected. For the fractionation column employing open steam, the material and enthalpy balances can be written as

$$F = D + B - \bar{G}_{N+1} \quad \text{...(5.248)}$$

$$Fz_F = Dx_D + Bx_B - \bar{G}_{N+1}y_{N+1} \text{ with } y_{N+1} = 0 \quad \text{...(5.249)}$$

$$FH_F = D\left(H_D + \frac{Q_C}{D}\right) + (BH_B - \bar{G}_{N+1}H_{G,N+1}) \quad \text{...(5.250)}$$

$\bar{G}_{N+1}$ is the rate at which open steam is admitted and $H_{G,N+1}$ is the enthalpy of steam. y_{N+1} is the mole fraction of **A** in the steam which is zero. The point $\bar{G}_{N+1}$ therefore lies on the enthalpy axis of the H-x-y diagram as shown. It will be on the upper curve if saturated steam is used, and above the upper curve if superheated steam is used. The last equation can be modified as

$$FH_F = DH'_D + (B - \bar{G}_{N+1})H'_B \quad \text{...(5.251)}$$

where

$$H'_D = H_D + \frac{Q_C}{D} \quad \text{...(5.252)}$$

$$H'_B = \frac{BH_B - \bar{G}_{N+1}H_{G,N+1}}{B - \bar{G}_{N+1}} \qquad \ldots(5.253)$$

Equations (5.248), (5.249) and (5.251) can be jointly expressed as

$$F = \Delta_D + \Delta_B \qquad \ldots(5.254)$$

where Δ_D is the net flow for the enriching section and Δ_B is the net flow in the stripping section. The enriching section of the column is unaffected by using open steam and the difference point Δ_D has the usual meaning. The difference point for the stripping section which represents the net flow downward the stripping section of the column is designated by Δ_B with the following specifications:

Amount of net flow $= \Delta_B = B - \bar{G}_{N+1}$ ...(5.255)

Composition coordinate of $\Delta_B = x'_B = \dfrac{Bx_B}{B - \bar{G}_{N+1}}$ ...(5.256)

Enthalpy coordinate of net flow $\Delta_B = H'_B = \dfrac{BH_B - \bar{G}_{N+1}H_{G,N+1}}{B - \bar{G}_{N+1}}$ ...(5.257)

It can be easily proved that Δ_B, F and Δ_D are collinear.

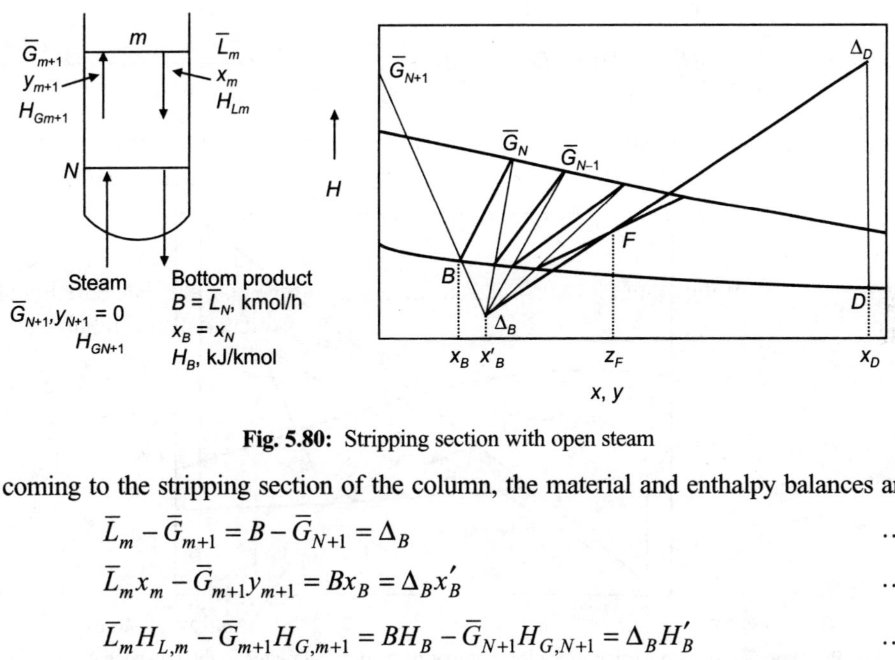

Fig. 5.80: Stripping section with open steam

Now coming to the stripping section of the column, the material and enthalpy balances are:

$$\bar{L}_m - \bar{G}_{m+1} = B - \bar{G}_{N+1} = \Delta_B \qquad \ldots(5.258)$$

$$\bar{L}_m x_m - \bar{G}_{m+1}y_{m+1} = Bx_B = \Delta_B x'_B \qquad \ldots(5.259)$$

$$\bar{L}_m H_{L,m} - \bar{G}_{m+1}H_{G,m+1} = BH_B - \bar{G}_{N+1}H_{G,N+1} = \Delta_B H'_B \qquad \ldots(5.260)$$

where Δ_B, x'_B and H'_B have meanings as indicated by Eqs. (5.255), (5.256) and (5.257). Equations (5.258)-(5.260) can be combined as

$$\bar{L}_m - \bar{G}_{m+1} = \Delta_B \qquad \ldots(5.261)$$

By eliminating Δ_B from Eqs. (5.259) and (5.260) using Eq. (5.258), we see that

$$\frac{\bar{L}_m}{\bar{G}_{m+1}} = \frac{y_{m+1} - x'_B}{x_m - x'_B} = \frac{H_{G,m+1} - H'_B}{H_{L,m} - H'_B} \qquad \ldots(5.262)$$

which implies that the points $\bar{L}_m$ $(x_{L,m}, H_{L,m})$, $\bar{G}_{m+1}(y_{L,m+1}, H_{G,m+1})$ and $\Delta_B(x'_B, H'_B)$ are collinear. This feature can be used to determine the number of plates.

Multiple Feeds and Side Streams

Introduction of additional feeds or withdrawal of side stream products will not introduce difficulties in the Ponchon-Savarit geometrical construction. Each addition of the feed or the withdrawal of the side stream will require an additional difference point in the graphical construction. For example, consider withdrawal of side stream as shown in Fig. 5.63. Taking material and enthalpy balance over the top section of the column which includes any plate i below the stage from which side stream is withdrawn, we get

$$G'_{i+1} - L'_i = D + S = \Delta_S \qquad \text{...(5.263)}$$

$$G'_{i+1}y_{i+1} - L'_i x_i = Dx_D + Sx_S = \Delta_S x'_S \qquad \text{...(5.264)}$$

$$G_{i+1}H_{G,i+1} - L_i H_{L,i} = D\left(H_D + \frac{Q_C}{D}\right) + SH_S = \Delta_S H'_s \qquad \text{...(5.265)}$$

where Δ_S is the additional difference point due to the side stream withdrawal. The concentration and enthalpy coordinates of Δ_S are given as

$$x'_S = \frac{Dx_D + Sx_S}{D+S} = \frac{Dx_D + Sx_S}{\Delta_S} \qquad \text{...(5.266)}$$

$$H'_S = \frac{DH'_D + SH_S}{D+S} = \frac{DH'_D + SH_S}{\Delta_S} \qquad \text{...(5.267)}$$

These difference points are shown in Fig. 5.81. A similar analysis can be carried out for multiple feeds also.

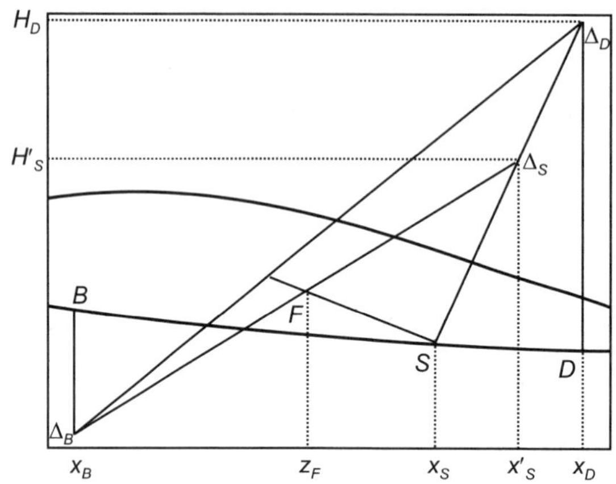

Fig. 5.81: Ponchon-Savartit construction for the number of theoretical plates in the presence of one side stream

Ponchon-Savarit Method versus McCabe-Thiele Method

The Ponchon-Savarit method is a theoretically sound rigorous method for the determination of the number of theoretical plates in a distillation column. It is free from simplifying assumptions. The McCabe-Thiele method assumes constant molar overflow which will be valid only when the column heat effects are negligible. The components should have almost equal molar heats of vaporisation; there should be no appreciable heats of mixing, no heat loss from the column to the surroundings, no sensible heat changes along the column height. For the separation of mixtures such as benzene-toluene, these assumptions will not introduce any significant error. However, for mixtures such as ammonia-water, there is appreciable heat of mixing and the Ponchon-Savarit method gives accurate results compared to the McCabe-Thiele method.

The exactness of the Ponchon-Savarit method is dependent on the accuracy of the enthalpy-composition data used for the computation. Measured data are not always available and they are usually computed using properties of the components. The accuracy of the Ponchon method will be lost if approximate data are used for enthalpy-composition calculations.

Ponchon-Savarit method is applicable for calculations either on mole or mass basis. McCabe-Thiele method is found to be generally applicable for mole basis calculations, because it is the molar latent heats and not latent heats expressed on mass basis that are found to be close to each other for chemically similar components. In short, the operating lines on mass basis will not be straight even for those systems for which straight lines are obtained on mole basis.

Ponchon-Savarit method gives the energy requirements in the column directly and presents a complete picture of what is happening in the column. Whereas this method gives the heat quantities such as condenser and reboiler duties directly, these quantities are hidden and need be computed separately in McCabe-Thiele method.

The Ponchon-Savarit method has now largely been superseded by rigorous computer simulation techniques applicable to binary as well as multicomponent distillation. Except for cases, where heat losses or heats of mixing are unusually large, the McCabe-Thiele method is found to be adequate for most applications and is widely used in practice.

Example 5.25: Explain how the constant molar overflow conditions follow from the Ponchon-Savarit construction.

***Solution*:** Assume that the enthalpy-composition diagram is as shown in Fig. 5.82. The saturation curves are assumed to be straight parallel lines. Two construction lines are shown in the enriching section. The first one connects the liquid and vapour phases passing each other below plate 1 and the second one connects the liquid and vapour phases below plate 2. These lines are shown as $\Delta_D G_2 L_1$ and $\Delta_D G_3 L_2$.

The internal reflux ratio was given by Eq. (5.238) as

$$\frac{L_n}{G_{n+1}} = \frac{\text{line } \Delta_D G_{n+1}}{\text{line } \Delta_D L_n} \quad \ldots(5.238)$$

For the passing streams below plate 1,

$$\frac{L_1}{G_2} = \frac{\text{line } \Delta_D G_2}{\text{line } \Delta_D L_1} \quad \ldots(5.268)$$

For the passing streams below plate 2,

$$\frac{L_2}{G_3} = \frac{\text{line } \Delta_D G_3}{\text{line } \Delta_D L_2} \quad \ldots(5.269)$$

Since the saturation curves are straight parallel lines, the triangles $\Delta_D G_3 G_2$ and $\Delta_D L_2 L_1$ are similar triangles and

$$\frac{\text{line } \Delta_D G_2}{\text{line } \Delta_D L_1} = \frac{\text{line } \Delta_D G_3}{\text{line } \Delta_D L_2} \quad \ldots(5.270)$$

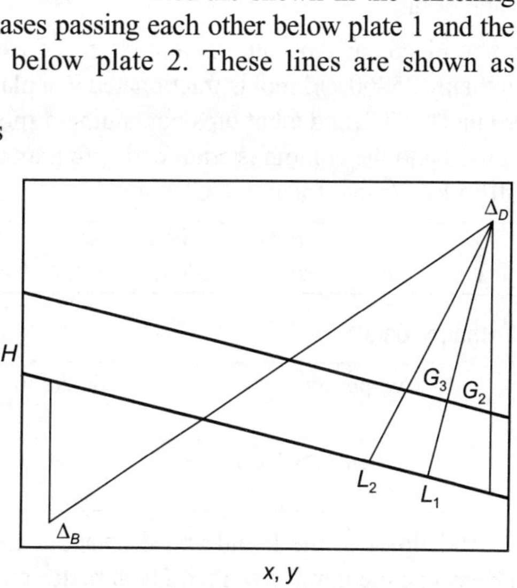

Fig. 5.82: Constant molar overflow condition on enthalpy-composition diagram

Combining Eqs. (5.268), (5.269) and (5.270),

$$\frac{L_1}{G_2} = \frac{L_2}{G_3} \quad \ldots(5.271)$$

Similarly it can be shown that

$$\frac{L_2}{G_3} = \frac{L_3}{G_4} \qquad \text{...(5.272)}$$

The above result can be extended to any plate in the enriching section so that we can conclude that the internal reflux ratio is constant and

$$\frac{L_1}{G_2} = \frac{L_2}{G_3} = \frac{L_3}{G_4} = \ldots \quad = \frac{L_n}{G_{n+1}} = \text{Constant} \qquad \text{...(5.273)}$$

For the enriching section of the column, we have the overall material balance which gives

$$G_{n+1} = L_n + D \qquad \text{...(5.274)}$$

Substituting Eq. (5.274) into Eq. (5.273), we get

$$\frac{L_n}{L_n + D} = \text{Constant} \qquad \text{...(5.275)}$$

For a given distillate rate D is constant so that Eq. (5.275) means

$$L_n = \text{Constant} \qquad \text{...(5.276)}$$

Equation (5.276) is the mathematical statement of the constant molar overflow condition

$$L_1 = L_2 = L_3 = \ldots.L_n = \text{Constant} \qquad \text{...(5.277)}$$

Substituting Eq. (5.274) in Eq. (5.277), it follows that

$$G_1 = G_2 = G_3 = \ldots.G_n = \text{Constant} \qquad \text{...(5.278)}$$

In a similar manner, we can show that the liquid and vapour flows in the stripping section of the column also remain constant. It is clear that the constant molar overflow and constant molar vaporisation conditions follow from the geometric condition which makes the saturation enthalpy curves straight parallel lines.

Example 5.26: An equimolar mixture of carbon tetrachloride (CCl_4) and toluene (C_7H_8) with an enthalpy 25460 kJ/kmol is fractionated in a plate column at 101.3 kPa to produce a distillate containing 90 mol% CCl_4 and a bottoms containing 5 mol% CCl_4. A reflux ratio 20% in excess of the minimum is used and the column is equipped with a total condenser and partial reboiler. The equilibrium data at 101.3 kPa (mole fraction CCl_4) are:

x	0	0.06	0.16	0.29	0.43	0.56	0.64	0.78	0.95	1.0
y	0	0.13	0.31	0.49	0.64	0.76	0.81	0.90	0.97	1.0

Enthalpy data:

Component	Saturation enthalpy, kJ/kg		Boiling point, K
	Liquid	Vapour	
Carbon tetrachloride	63.0	257.5	349.1
Toluene	130.0	460.0	383.6

Enthalpies of the liquid and the vapour may be assumed to be linear functions of compositions. Determine the number of theoretical plates required for the separation using Ponchon-Savarit method. Compare the results with that obtained by the McCabe-Thiele method assuming that the feed to be a mixture of equal amounts of the liquid and the vapour.

***Solution*:** Saturation enthalpies of pure components are as follows:

For carbon tetrachloride, $H_L = 63.0 \times 153.8 = 9689.4$ kJ/kmol, $H_G = 257.5 \times 153.8 = 39603.5$ kJ/kmol

For toluene, $H_L = 130 \times 92 = 11960$ kJ/kmol, $H_G = 460 \times 92 = 42320$ kJ/kmol

A graph is plotted with enthalpies H (kJ/kmol) on y-axis and composition in mole fraction on the x-axis. Since the enthalpies of saturated liquid and saturated vapour are assumed to vary linearly with composition, the saturation enthalpy curves are straight lines as shown in Fig. 5.83. H_L versus x is a straight line connecting the pure component saturated liquid enthalpies (11960 at $x = 0$ and 9689.4 at $x = 1.0$). Similarly H_G versus y curve is the straight line joining the pure component saturated vapour enthalpies (42320 at $y = 0$ and 39603.5 at $y = 1.0$). The equilibrium curve y versus x is also plotted in the same graph.

The composition and enthalpy of the feed are $z_F = 0.50$, $H_F = 25460$ kJ/kmol. The point F representing the feed is marked as shown in the Fig. 5.83a. The feed is partially vaporised and lies in the two-phase region. The distillate and bottoms are saturated liquids and lies on the H_L versus x curve. Point D with $x_D = 0.90$ and B with $x_B = 0.05$ are marked as shown in the figure. The difference point for the enriching section lies on the vertical line passing through D. The difference point $\Delta_{D,m}$ corresponds to the minimum reflux ratio. To find the difference point, the tie line passing through F is determined by trial. In Fig. 5.83a the tie line corresponding to point P on the equilibrium curve passes through the feed point. This tie line is extended to meet the vertical erected at D at $\Delta_{D,m}$. The coordinates of $\Delta_{D,m}$ are (0.90, 83193). The intersection of the vertical line at D with the H_G versus y curve gives G_1 whose coordinates are (0.90, 39918). The enthalpy of the distillate is found to be $H_D = 9804$ kJ/kmol from the graph.

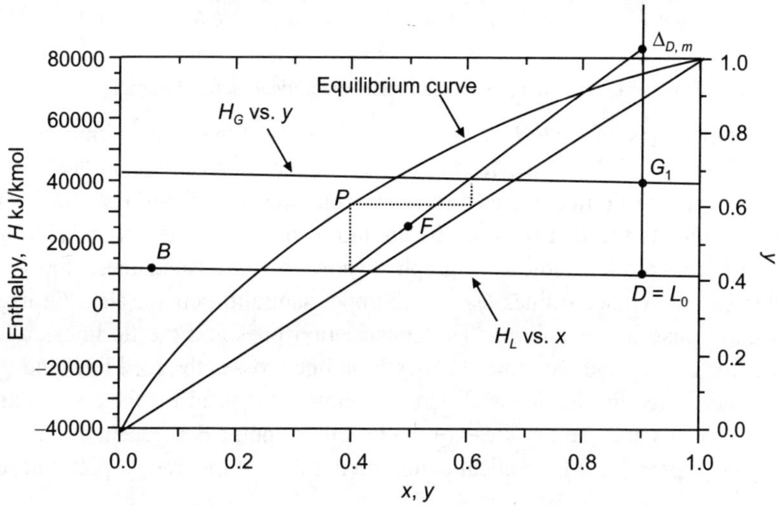

Fig. 5.83 (a) Determination of minimum reflux ratio for example 5.26

The minimum reflux ratio is computed as

$$R_m = \frac{\Delta_{D,m}G_1}{G_1D} = \frac{H'_{D,m} - H_{G,1}}{H_{G,1} - H_D}$$

$$= \frac{83193 - 39918}{39918 - 9804} = 1.437$$

Since the operating reflux ratio is 20% in excess of the minimum,

$$R = 1.2R_m = 1.2 \times 1.437 = 1.724$$

To locate the difference point Δ_D, we can use the equation $\Delta_D G_1 = R(G_1 D)$ or,

$$H'_D - H_{G,1} = R(H_{G,1} - H_D)$$

That is

$$H'_D = H_{G,1} + R(H_{G,1} - H_D) = 39918 + 1.724(39918 - 9804) = 91835 \text{ kJ/kmol}$$

where H'_D is the enthalpy coordinate of the difference point Δ_D. Now the Δ_D is located on the vertical line passing through the point D as shown in Fig. 5.83b

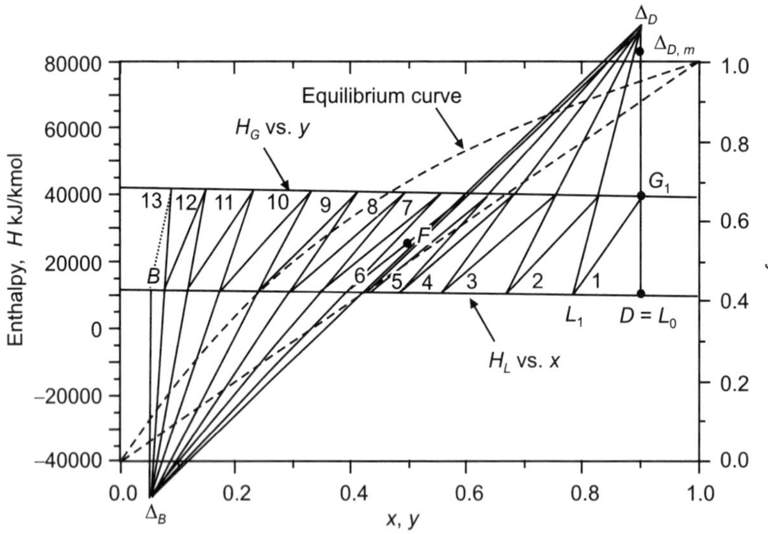

Fig. 5.83: (b) Determination of number of theoretical plates: Example 5.26

The difference point Δ_D is joined to F and extended to intersect the vertical line passing through the point B at Δ_B. The line $\Delta_D F \Delta_B$ is the feed line. The number of theoretical plates is determined as shown in the figure. The first tie line passes through G_1 and intersects the lower saturation curve at L_1. The tie line L_1G_1 represents the first plate in the column. This tie line is drawn with the help of the equilibrium curve which is plotted in the same graph as shown by the dashed lines. Now the construction line drawn joining L_1 and Δ_D determines G_2 on the upper saturation curve. Next tie line is L_2G_2 and represents the second plate in the column. The construction lines and the tie lines are drawn in this fashion till the tie line crosses the feed line. The sixth tie line crosses the feed line, and therefore sixth plate is the feed plate. Hereafter, the lower difference point Δ_B is used for the construction lines. The alternate construction lines and the tie lines are drawn till the point B is reached.

Since each tie line represents a theoretical plate, the number of theoretical plates according to Fig. 5.83b is 12.7.

(***Note*:** The fraction of stage may be computed as $\dfrac{x_{12} - x_B}{x_{12} - x_{13}}$).

Since the reboiler acts as a theoretical plate, the number of plates in the column is 11.7 **(Answer)**

McCabe-Thiele method also gives the same number of theoretical plates. This is to be expected because the saturation curves are straight lines and are almost parallel, the condition leading to constant molar overflow in the column.

Example 5.27: It is desired to fractionate an aqueous ammonia solution containing 30 mol% ammonia into products containing 99.5 mol% ammonia and bottoms containing 2 mol% ammonia in a plate column operated at 1250 kPa pressure. The solution enters the column as saturated liquid at the operating pressure. A total condenser is employed. Assume that the reflux is returned to the column at its bubble point. The following are the enthalpy-composition data and equilibrium data (mole fraction ammonia) at 1250 kPa:

Enthalpy-composition data:

x	0	.05	0.10	0.20	0.30	0.40	0.50	0.60
H_L, kJ/kmol	14555	12624	11027	7815	5234	2987	1390	424
H_G, kJ/kmol	49910	48943	48312	45415	43484	40904	38658	36410
x	0.70	0.80	0.90	0.95	1.0			
H_L, kJ/kmol	424	1056	1705	2022	2987			
H_G, kJ/kmol	34160	30950	28040	26100	21945			

Equilibrium data (mole fraction ammonia) are:

x	0	0.05	0.10	0.15	0.20	0.25	0.30	0.35	0.40
y	0	0.279	0.501	0.651	0.767	0.850	0.895	0.934	0.958
x	0.50	0.60	0.70	0.80	0.85	0.90	0.95	1.0	
y	0.982	0.994	0.996	0.998	0.9982	0.9996	0.9998	1.0	

Determine:

(a) The minimum number of plates under total reflux.
(b) The minimum reflux ratio.
(c) The number of theoretical plates when the reflux ratio used is 1 kmol/kmol product.
(d) The condenser duty and the reboiler heat load per 100 kmol/h of feed.

Solution: (a) The enthalpy-composition diagram is drawn as shown in Fig. 5.84. The feed is saturated liquid and $z_F = 0.30$; distillate is saturated liquid with $x_D = 0.995$. The bottoms composition is $x_B = 0.02$. Using these composition values points F, D and B are marked on the H_L versus x curve.

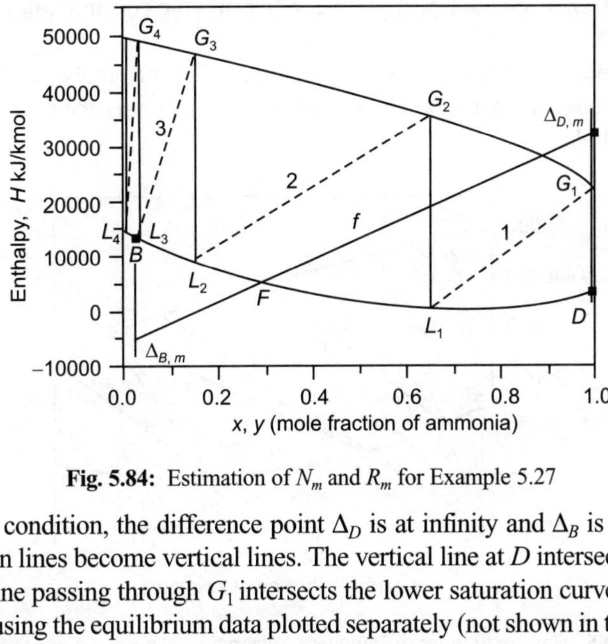

Fig. 5.84: Estimation of N_m and R_m for Example 5.27

Under total reflux condition, the difference point Δ_D is at infinity and Δ_B is at minus infinity. As a result, the construction lines become vertical lines. The vertical line at D intersects the upper saturation curve at G_1. The tie line passing through G_1 intersects the lower saturation curve at L_1. (*See* Fig. 5.84). The tie line is drawn using the equilibrium data plotted separately (not shown in the figure). The vertical erected at L_1 is the construction line which intersects the upper saturation curve at G_2. The tie line passing though the point G_2 gives L_2. In this way, the number of tie lines that can be drawn till the point B is reached is found out. It is found that the minimum number of plates = number of tie lines – 1 = 3.3 – 1 = 2.3 theoretical plates. (Here, the fractional stage can be estimated as the ratio of $x_3 - x_B$ to $x_3 - x_4$.)

Minimum number of plates = 2.3 **(Answer)**

(b) The tie line passing through the feed point F is determined using the equilibrium data. It is shown as f in the figure. This tie line is extended to intersect the vertical line passing through D at $\Delta_{D,m}$. Since $\Delta_{D,m}G_1 = R_m(G_1D)$,

$$R_m = \frac{\Delta_{D,m}G_1}{G_1D} = \frac{H'_D - H_{G,1}}{H_{G,1} - H_D} = \frac{32465 - 22329}{22329 - 2932} = 0.523$$

The minimum reflux ratio = 0.523 mol/mol product **(Answer)**

(c) The enthalpy composition diagram is drawn as shown in the Fig. 5.85 and the points F, D and B are marked on the lower saturation curve since the feed, distillate and the residue are saturated liquids. When $R = 1$, the location of the upper difference point Δ_D is determined as

$$\Delta_D G_1 = R(G_1D) = G_1D$$

$$H'_D - H_{G,1} = H_{G,1} - H_D$$

$H'_D - 22329 = 22329 - 2932$ or $H'_D = 41726$ kJ/kmol. Δ_D is located on the vertical line at D with the enthalpy coordinate of Δ_D is 41726 kJ/kmol. The line $\Delta_D F$ is extended to meet the vertical passing through B at Δ_B. The vapour G_1 reaching the condenser can be located vertically above D on the upper saturation curve, because the vapour and distillate have the same composition (total condenser). Plate 1 in the column is the tie line passing through G_1. L_1 is the saturated liquid leaving the plate 1 and it lies at the end of this tie line on the lower saturation curve. The construction line drawn from L_1 to Δ_D intersects the upper saturation curve at G_2. G_2L_2 is the second tie line and it represents the second plate from the top. Thus alternate construction line and the tie line are drawn using the upper difference point Δ_D till the feed line $\Delta_D F\ \Delta_B$ is crossed by a tie line. After that the construction is continued using the lower difference point Δ_B.

The number of tie lines = 6. Allowing for the partial reboiler, the number of theoretical plates is 5. The feed plate is represented in the diagram by the tie line which crosses the feed line. The second plate from the top is the feed plate. **(Answer)**

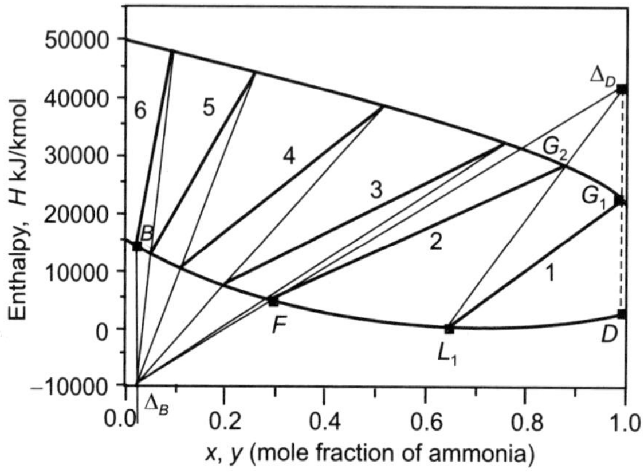

Fig. 5.85: Construction for the number of theoretical plates in Example 5.27

(d) The enthalpy coordinate of the upper difference point Δ_D is H'_D where,

$H'_D = H_D + \dfrac{Q_C}{D}$ where Q_C is the condenser heat duty.

From the figure, $H'_D = 32465$ kJ/kmol, $H_D = 2932$ kJ/kmol

$$D = F\frac{z_F - x_B}{x_D - x_B} = 100\frac{0.3 - 0.02}{0.995 - 0.02} = 28.72 \text{ kmol/h and } B = F - D = 100 - 28.72 = 71.28 \text{ kmol/h.}$$

$$32465 = 2932 + \frac{Q_C}{28.72}$$

$$Q_C = 8.48188 \times 10^5 \text{ kJ/h}$$ **(Answer)**

The enthalpy coordinate of the lower difference point Δ_B is H'_B where,

$H'_B = H_B - \frac{Q_B}{B}$ where Q_B is the reboiler heat load.

From the figure, $H'_B = -9182$ kJ/kmol and $H_B = 13771$ kJ/kmol

$$-9182 = 13771 - \frac{Q_B}{71.28}$$

$$Q_B = 16.3609 \times 10^5 \text{ kJ/h}$$ **(Answer)**

5.11 PACKED COLUMNS FOR DISTILLATION

Packed columns are generally limited to towers of small diameters. They are particularly useful for distillation of heat sensitive materials. Damage to such materials due to long exposure on the plates to high temperature can be minimized using packed columns, which provides smaller liquid hold up in comparison to tray towers. Also, when pressure drop through the tower is to be restricted as in the case

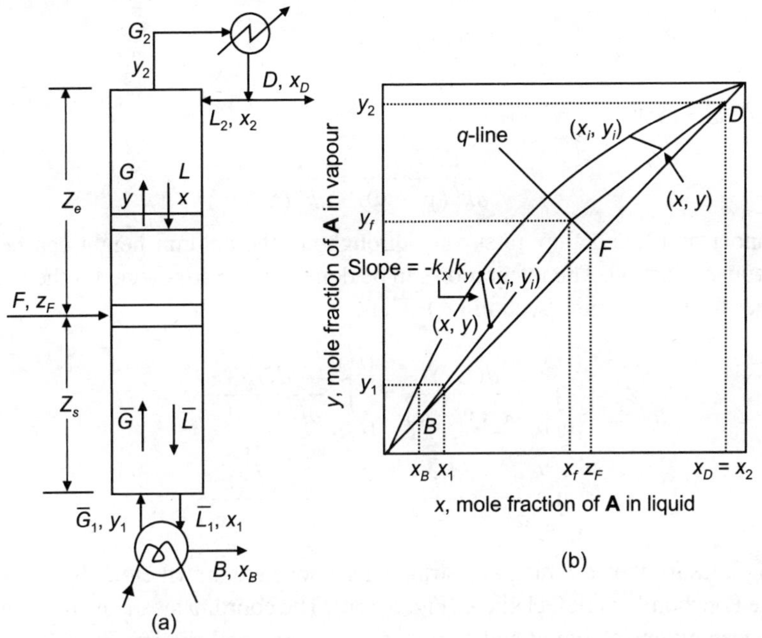

Fig. 5.86: (a) Packed column for distillation (b) Determination of the number of transfer units

of low pressure distillation, packed columns are preferred over tray towers. Packed columns are easy to construct, and can be made of relatively less expensive materials such as non-corrosive refractory, earthenware, glass and carbon as well as the usual metals used in plate column construction. Channelling is the major difficulty encountered in packed column operation and channelling leads to loss of efficiency of the gas-liquid contact. To minimize channelling, tall columns are often split into smaller sections of 3 to 4 m height, every section provided with redistributors.

In packed columns, the liquid flows over the surface of the packing material while the vapour flows up through the voids. Transfer occurs between the vapour and liquid by the mechanism approximating to equimolar counter diffusion. The mass flux at a height *z* units from the bottom of the tower (Fig. 5.86a) where the concentrations in the vapour and liquid are y and x respectively, can be written as

$$N_A = k_y'(y_i - y) = k_x'(x - x_i) \quad \text{...(5.301)}$$

where x_i and y_i are the concentrations at the interface and k_x' and k_y' are the liquid phase and gas phase mass transfer coefficients, characteristic of equimolar counter diffusion. Equation (5.301) can be rearranged as

$$\frac{y_i - y}{x_i - x} = -\frac{k_x'}{k_y'} \quad \text{...(5.302)}$$

Consider the rate of mass transfer at the section under consideration. Let dz be the thickness of the packed section and let a be the interfacial area per unit volume of packing. The mass flux at this section where the vapour flow rate is G kmol/h m^2 of tower cross section and the liquid flow is L kmol/h m^2 is

$$N_A = \frac{d(Gy)}{adz} = \frac{d(Lx)}{adz} \quad \text{...(5.303)}$$

Comparing Eq. (5.303) with Eq. (5.301) we get

$$k_y'(y_i - y) = \frac{d(Gy)}{adz} \quad \text{...(5.304)}$$

$$k_x'(x - x_i) = \frac{d(Lx)}{adz} \quad \text{...(5.305)}$$

Equations (5.304) and (5.305) can be rearranged as

$$dz = \frac{d(Gy)}{ak_y'(y_i - y)} = \frac{d(Lx)}{ak_x'(x - x_i)} \quad \text{...(5.306)}$$

Since the vapour and liquid flow rates vary throughout, the column height can be computed by integrating the above equation. The integration can be done in two parts—one for the enriching section (Z_e) and the other for the stripping section (Z_s). Thus,

$$Z_e = \int_{(Gy)_f}^{(Gy)_2} \frac{d(Gy)}{ak_y'(y_i - y)} = \int_{(Lx)_f}^{(Lx)_2} \frac{d(Lx)}{ak_x'(x - x_i)} \quad \text{...(5.307)}$$

$$Z_s = \int_{(Gy)_1}^{(Gy)_f} \frac{d(\bar{G}y)}{ak_y'(y_i - y)} = \int_{(Lx)_1}^{(Lx)_f} \frac{d(\bar{L}x)}{ak_x'(x - x_i)} \quad \text{...(5.308)}$$

The operating lines for the enriching and stripping sections are drawn either by the McCabe-Thiele method or by the Ponchon-Savarit technique (Fig. 5.86b). The coordinates of any point on the operating line give the concentrations of liquid and vapour at any horizontal section in the column. From any point on the operating line, a straight line with slope $-\frac{k_x'}{k_y'}$ is drawn which intersects the equilibrium curve at (x_i, y_i). The height z_e can be determined by plotting Gy against $\frac{1}{ak_y'(y_i - y)}$ or by plotting Lx

against $\frac{1}{ak'_x(x - x_i)}$ and by determining the area under the curve between the appropriate limits. The choice of x_f and y_f as the coordinates of the intersection of the operating lines results in shorter columns in comparison with the results obtained by other possible choices for feed entries.

The method described takes into account the variations in G, L, x, y, k'_x, k'_y and the interfacial area a along the length of the column. On the other hand, if the simplifying assumptions are valid, G and L remain constant in each section and if the mass transfer coefficients also are constant, Eq. (5.307) becomes

$$z_e = \frac{G}{k'_y a} \int_{y_f}^{y_2} \frac{dy}{(y_i - y)} = H_{tG} \times N_{tG} \quad ...(5.309)$$

or

$$z_e = \frac{L}{k'_x a} \int_{x_f}^{x_2} \frac{dx}{(x - x_i)} = H_{tL} \times N_{tL} \quad ...(5.310)$$

Similar equation can be written for z_s. In the above equations, the height of the gas transfer units $H_{tG} = \frac{G}{k'_y a}$ and the height of the liquid transfer units $H_{tL} = \frac{L}{k'_x a}$ are assumed constant. It should be noted that even if the molar flow rates remain constant, the mass flow rates may vary because of the change in the molecular weights with concentration. The assumption of constant height of transfer unit becomes invalid because the interfacial area and mass transfer coefficients vary with mass flow rates of the phases.

The use of the overall coefficients and the overall height of transfer units is not recommended for computing the height of the packed distillation columns because, the equilibrium curve is generally curved for all sections of the column. If the equilibrium curve is a straight line, the following equations may be used for estimating the heights.

$$z_e = \frac{G}{K'_y a} \int_{y_f}^{y_2} \frac{dy}{(y^* - y)} = H_{toG} \times N_{toG} \quad ...(5.311)$$

or

$$z_e = \frac{L}{K'_x a} \int_{x_f}^{x_2} \frac{dx}{(x - x^*)} = H_{toL} \times N_{toL} \quad ...(5.312)$$

In equations (5.311) and (5.312), $(y^* - y)$ and $(x - x^*)$ are the overall driving forces for mass transfer in terms of the vapour compositions and the liquid compositions, respectively.

Example 5.28: One hundred kmol/h of a binary mixture consisting of 40 mol% **A** and 60 mol% **B** is distilled in a packed column of diameter 1.5 m to produce a distillate containing 75 mol% **A** and a residue containing 5 mol% **A**. The column is equipped with a total condenser and a partial reboiler. The feed enters as saturated liquid and the reflux ratio is 50% in excess of the minimum. Determine the height of the packed column.

The gas side and liquid side mass transfer coefficients $\left(\frac{\text{kmol}}{\text{m}^3\text{h (mole fraction)}}\right)$ are estimated to be $k'_x a = 700$ and $k'_y a = 200$ for the enriching section and $k'_x a = 1500$ and $k'_y a = 300$ for the stripping section. The equilibrium data (mole fraction **A**) are:

x	0	0.1	0.2	0.3	0.4	0.5	0.6	0.7	0.8	0.9	1.0
y	0	0.289	0.450	0.549	0.616	0.664	0.704	0.742	0.788	0.859	1.0

***Solution*:** Feed rate $F = 100$ kmol/h; $z_F = 0.40$; $x_D = 0.75$; $x_B = 0.05$

$$D = F\frac{z_F - x_B}{x_D - x_B} = 100\frac{0.40 - 0.05}{0.75 - 0.05} = 50 \text{ kmol/h}, \ B = F - D = 50 \text{ kmol/h}$$

Since the feed enters as saturated liquid, $q = 1$ and the q-line is vertical as shown in Fig. 5.87. The limiting operating line passes through the intersection of q-line and the equilibrium curve, point P, in Fig. 5.87. It intersects the y-axis at $x_D/(R_m+1)$ where R_m is the minimum reflux ratio. The intercept is found to be 0.463 so that $0.463 = \dfrac{0.75}{R_m + 1}$. Solving this, we get $R_m = 0.62$. Reflux ratio $R = 1.5\,R_m$ or $R = 0.93$. (Alternatively, $R_m = \dfrac{x_D - y'}{y' - x'}$ where x' and y' are the coordinates of the point of intersection of the q-line and the equilibrium curve.)

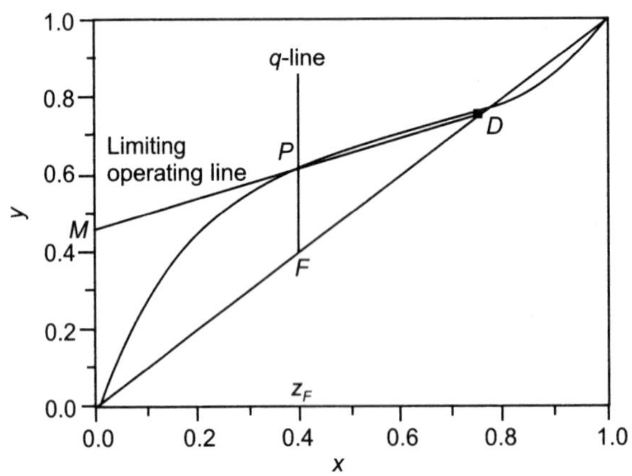

Fig. 5.87: Minimum reflux ratio for Example 5.28

For the enriching section, assuming constant molar overflow, the vapour and the liquid flow are

$$L = RD = 0.93 \times 50 = 46.5 \text{ kmol/h}$$

$$G = L + D = 96.5 \text{ kmol/h}$$

For the stripping section, the corresponding values are:

$$\bar{L} = L + qF = 46.5 + 100 = 146.5 \text{ kmol/h}$$

$$\bar{G} = \bar{L} - B = 146.5 - 50 = 96.5 \text{ kmol/h}$$

The flow rates are to be expressed per unit area of tower cross section for estimating the height of transfer units. Area of cross section of the tower $= \dfrac{\pi}{4}(1.5)^2 = 1.767 \text{ m}^2$. Therefore,

$$L = 46.5/1.767 = 26.316 \text{ kmol/m}^2\text{h and } G = 96.5/1.767 = 54.612 \text{ kmol/m}^2\text{h}$$

$$\bar{L} = 146.5 / 1.767 = 82.909 \text{ kmol/m}^2\text{h and } \bar{G} = 96.5/1.767 = 54.612 \text{ kmol/m}^2\text{h}$$

Assuming constant molar overflow, the height of the enriching section is

$$z_e = \frac{G}{k'_y a}\int_{y_f}^{y_2} \frac{dy}{(y_i - y)} = H_{tG} \times N_{tG} \qquad \text{...(5.309)}$$

where $$H_{tG} = \frac{G}{k'_y a} \text{ and } N_{tG} = \int_{y_f}^{y_2} \frac{dy}{(y_i - y)}$$

The height of the gas transfer units is

$$H_{tG} = \frac{G}{k'_y a} = \frac{54.612}{200} = 0.273 \text{ m}$$

To find the number of gas transfer units, interfacial concentration y_i is to be obtained for a given bulk composition y. The operating line for the enriching section is drawn from point D such that it makes an intercept with y-axis equal to $\frac{x_D}{R+1} = \frac{0.75}{0.93+1} = 0.389$. It intersects the q-line at point P whose coordinates are found to be (0.4, 0.579). The stripping operating line is the line joining P to B. Since the reboiler brings about a concentration change equal to that of an ideal plate, a theoretical stage is stepped off from point B as shown in Fig. 5.88. The point C (0.108, 0.144) is the lower terminus of the stripping operating line as far as the packed section of the column is concerned.

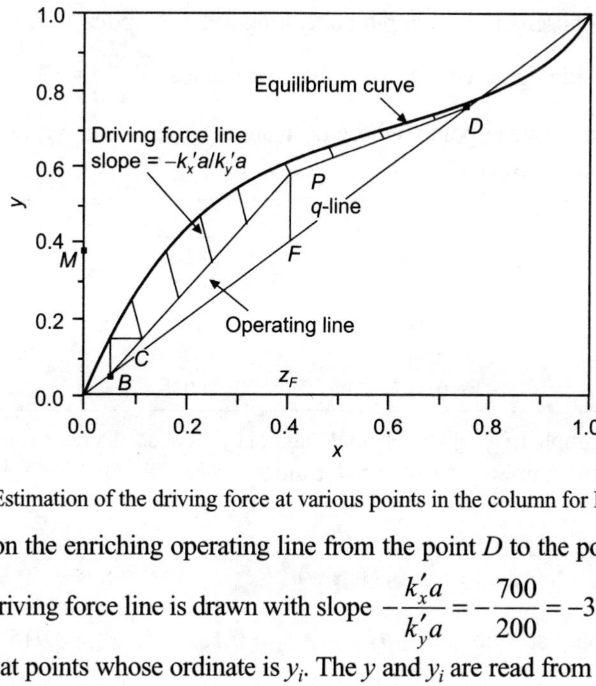

Fig. 5.88: Estimation of the driving force at various points in the column for Example 5.28

Points are chosen on the enriching operating line from the point D to the point P. From each point (with ordinate y) the driving force line is drawn with slope $-\frac{k'_x a}{k'_y a} = -\frac{700}{200} = -3.5$. These lines intersect the equilibrium curve at points whose ordinate is y_i. The y and y_i are read from the figure and $1/(y_i - y)$ computed as given below:

y	y_i	$y_i - y$	$1/(y_i - y)$
0.579	0.609	0.03	33.33
0.620	0.655	0.035	28.57
0.670	0.697	0.027	37.04
0.720	0.734	0.014	71.43
0.750	0.764	0.014	71.43

A graph is drawn with $1/(y_i - y)$ as ordinate against y as abscissa. The area is found out from the graph, which is the number of gas transfer unit.

$$N_{tG} = \int_{y_f}^{y_2} \frac{dy}{(y_i - y)} = \int_{0.579}^{0.75} \frac{dy}{(y_i - y)} = 7.764$$

The height of enriching section is, $z_e = H_{tG} \times N_{tG} = 0.273 \times 7.764 = 2.12$ m **(Answer)**

In a similar manner, for stripping section,

$$z_e = \frac{\bar{G}}{k'_y a} \int_{y_1}^{y_f} \frac{dy}{(y_i - y)} = H_{tG} \times N_{tG}$$

where $$H_{tG} = \frac{\bar{G}}{k'_y a} \text{ and } N_{tG} = \int_{y_1}^{y_f} \frac{dy}{(y_i - y)}$$

The height of gas transfer unit is

$$H_{tG} = \frac{\bar{G}}{k'_y a} = \frac{54.612}{300} = 0.182 \text{ m}$$

Different points are chosen on the stripping operating line from point C to point P. From each point (with ordinate y), the driving force line is drawn with slope $-\frac{k'_x a}{k'_y a} = -\frac{1500}{300} = -5.0$. These lines intersect the equilibrium curve at points whose ordinate is y_i. The y and y_i are read from the figure and $1/(y_i - y)$ computed as given below:

y	y_i	$y_i - y$	$1/(y_i - y)$
0.144	0.254	0.11	9.09
0.250	0.380	0.13	7.69
0.350	0.476	0.126	7.94
0.450	0.548	0.098	10.20
0.579	0.615	0.036	27.78

The area under the graph $1/(y_i - y)$ as ordinate against y as abscissa between limits 0.144 and 0.579 is found out which is the number of gas transfer units.

$$N_{tG} = \int_{y1}^{y_f} \frac{dy}{(y_i - y)} = \int_{0.144}^{0.579} \frac{dy}{(y_i - y)} = 5.028$$

The height of stripping section is, $z_s = H_{tG} \times N_{tG} = 0.182 \times 5.028 = 0.915$ m **(Answer)**

Total height of the packed column = $z_e + z_s$ = 2.12 + 0.915 = 3.035 m **(Answer)**

5.12 FRACTIONATION OF PARTIALLY MISCIBLE MIXTURES

The vapour-liquid equilibrium diagram of two partially miscible liquids **A** and **B** are shown in the Fig. 5.89. The solubility of **A** in **B** is $x_{0,1}$ and that of **B** in **A** is $x_{0,2}$. The vapour phase with composition lying between y' and y'' in the figure, on condensation separates into two liquid phases one rich in **B** with mole fraction of **A** equal to $x_{0,1}$ and the other rich in **A** with mole fraction of **A** of $x_{0,2}$.

The fractionation of partially miscible system requires two columns arranged as shown in Fig. 5.90. The feed is admitted to the column 1. The vapour leaving the top of the column 1 has composition $y_{D,1}$ and the bottoms product is almost pure **B**. The liquid obtained on condensation of the overhead vapours from the column 1 has an average composition lying in the two-phase region represented by point P in the figure. When this liquid is allowed to separate, two saturated liquid phases with composition $x_{0,1}$ and $x_{0,2}$ are obtained. The first liquid having composition $x_{0,1}$ is returned to the column 1 as reflux and the second having composition $x_{0,2}$ is subjected to stripping in column 2. It is possible to obtain almost pure **A** as the bottoms product from the column 2. The vapour emerging from the top of column 2 has composition $y_{D,2}$ which on condensation and separation gives two immiscible liquid phases, with composition $x_{0,1}$ and $x_{0,2}$, same as those obtained by condensation of vapours from column 1. Thus, it is possible to condense the vapours from both columns in a single condenser as shown in Fig. 5.90. The upper terminus of the operating line for column 1 is ($x_{0,1}$, $y_{D,1}$) and the lower terminus of the operating line for column 2 is ($x_{0,2}$, $y_{D,2}$). The number of theoretical plates can be determined by the usual McCabe-Thiele graphical construction or by the Ponchon-Savarit method.

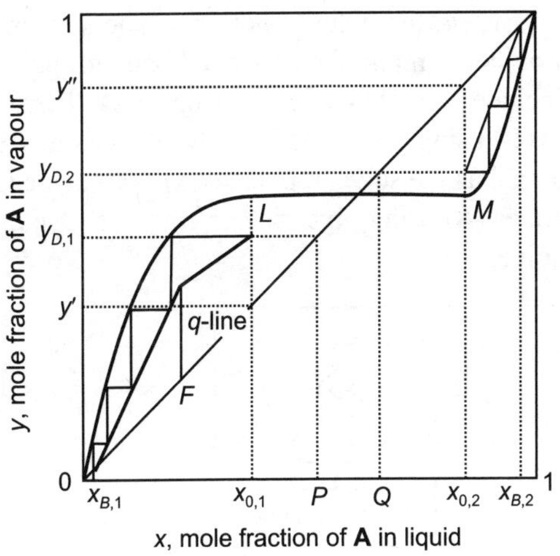

Fig. 5.89: McCabe-Thiele construction for separation of partially miscible systems

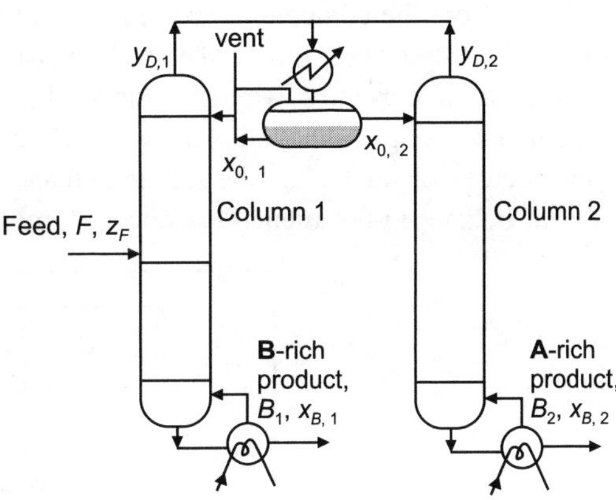

Fig. 5.90: Fractionation of partially miscible system

5.13 AZEOTROPIC AND EXTRACTIVE DISTILLATION

Not all liquid mixtures can be separated by ordinary fractional distillation. When the components of the system have low relative volatilities ($1.00 < \alpha < 1.05$), separation becomes difficult and expensive as a large number of trays are required for obtaining high purity products. The column would have to be operated with a high reflux ratio which leads to high heat requirements and towers of large cross sectional area. Both equipment and utilities costs thus increase markedly and the operation becomes uneconomical. If the system forms an azeotrope, a different problem arises – the azeotrope composition limits the separation, and for a better separation this azeotrope must be bypassed in some way. Azeotropic and extractive distillations are the two techniques employed for the fractionation of binary mixtures with poor relative volatility or which form azeotropes, by adding a third component to facilitate separation.

Extractive Distillation

Extractive distillation is defined as the distillation in the presence of a miscible nonvolatile component, the *solvent*, added to a binary mixture which is difficult or impossible to separate by ordinary means. The solvent alters the relative volatility of the original components, but forms no azeotrope with them. Extractive distillation is widely used in chemical and petrochemical industries for separating azeotropic, close-boiling, or low relative volatility mixtures.

Extractive distillation is effective because the solvent is specially chosen to interact differently with the components of the original mixture, thereby altering their relative volatilities. The extent of departure from the ideality between the solvent and the components will be different for the two original constituents of the mixture. This difference is responsible for the separation by extractive distillation. On adding a solvent to the binary mixture, that component which exhibits larger extent of nonideality with the solvent than the other component can be separated easily as the overhead vapour. The solvent is continuously added at the top of the column to ensure that an appreciable amount is present in the liquid phase on all the trays below. However, solvent cannot be admitted on the top plate because a few stages must be provided above the solvent entry point to reduce the solvent concentration in the overhead vapours. The mixture to be separated is added through the second feed point further down the column (Fig. 5.91). In extractive distillation column, the component having the greater volatility (**A**), not necessarily the component having the lowest boiling point, is taken overhead as a relatively pure distillate. The other component (**B**) leaves with the solvent via the column bottoms. The solvent is separated from the component (**B**) in a second distillation column and then recycled back to the first column. Industrial applications of extractive distillation include separation of cyclohexane (**A**) – benzene (**B**) using aniline as the solvent, isooctane (**A**) – toluene (**B**) using phenol as the solvent, acetone (**A**)-methanol(**B**) using aniline, ethylene glycol, water or butanol as the solvent, butene (**A**) – *n*-butane (**B**) using acetone or furfural as the solvent, ethanol (**A**) –water (**B**) using glycerine or ethylene glycol as solvent, butane(**A**)-butadiene(**B**) using furfural as the solvent, etc.

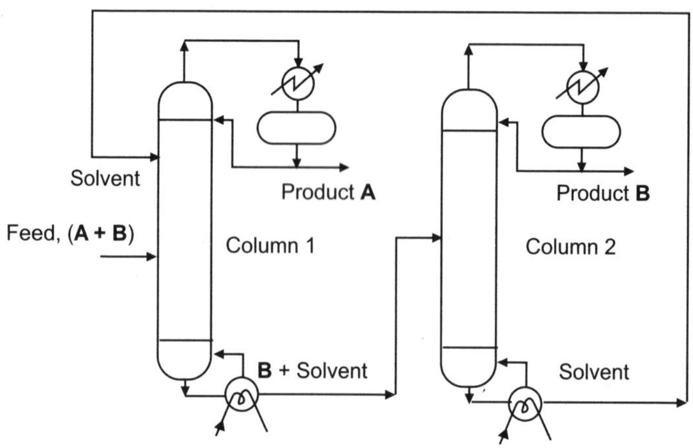

Fig. 5.91: Extractive distillation

The solvent used for extractive distillation has the following characteristics:

1. High selectivity to enhance the relative volatility of the system.
2. High capacity so that the solvent requirement per unit weight of the feed is low.
3. Low volatility to minimize the vaporisation and the loss of the solvent with the distillate.
4. Solubility should be such that it remain soluble in the feed components and should not lead to the formation of two phases.
5. Should not form azeotrope with the original constituents.
6. Should be easily separable from the bottom product.
7. Inexpensive and readily available.
8. Stable at the temperature of the distillation and solvent separation.
9. Non reactive with the components in the feed mixture.

10. Low latent heat of vaporisation to reduce the cost of heat.
11. Low freezing point to facilitate storage and outdoor handling.
12. Low viscosity to provide high tray efficiencies.
13. Nontoxic and noncorrosive towards common materials of construction.

Generally the solvent chosen will be chemically similar to one of the components in the original mixture so that the other component can easily be separated in the extractive distillation column. When both the components are chemically similar, a solvent of an entirely different chemical nature will be necessary.

Azeotropic Distillation

Separation by the azeotropic distillation is accomplished by the addition of a third component, called the *entrainer*, which forms a minimum boiling azeotrope with at least one of the components of the feed solution.

Consider for example, the case where a binary mixture of **A** and **B** with poor relative volatility is distilled in the presence of an entrainer **E** which forms a low-boiling heteroazeotrope with component **A**. The entrainer is added in sufficiently large quantities to form the azeotrope with the entire **A** in the mixture. The azeotrope is then vaporised and removed as the overhead product while the bottom product is almost pure **B**. The distillate vapours on condensation separates into two immiscible liquid layers, one rich in **A** but saturated with **E**, and the other rich in **E** but saturated with **A**. The **E**-rich layer is recycled back to the column as reflux and the **A**-rich layer is subjected to further purification in a smaller column as shown in Fig. 5.92.

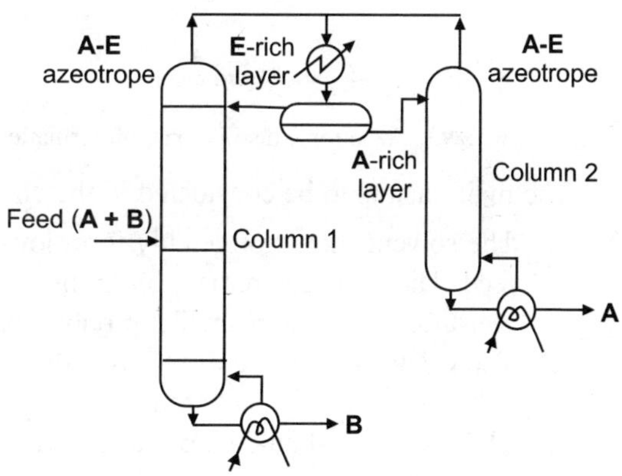

Fig. 5.92: Azeotropic distillation with formation of heteroazeotrope

An example of this type of azeotropic distillation is the separation of water (**A**) - acetic acid (**B**) solution using butyl acetate (**E**) as entrainer. Butyl acetate forms a heteroazeotrope with water, which boils at 363.4 K at 101.3 kPa. The normal boiling points of acetic acid and water are, respectively, 391.3 K and 373 K. Separation is achieved in columns with relatively few number of trays, but since the entrainer also is to be vaporised, the heat load for the operation is generally very high.

The addition of the entrainer may lead to the formation of a ternary azeotrope which constitutes the overhead product from the azeotropic distillation column. For example, the dehydration of ethanol-water system is achieved in this fashion using benzene as the entrainer (Fig. 5.93).

Benzene forms a minimum boiling azeotrope with ethanol and water (53.9 mol% benzene, 23.3 mol% water and 22.8 mol% ethanol), which boils at a lower temperature (338.1 K) than the ethanol-water binary azeotrope (BP = 351.4K) and also contains a higher ratio of water to ethanol. The ternary azeotrope is removed from the primary column as the distillate and pure alcohol as the bottoms product. The distillate vapour on condensation separates into two layers. The upper one, rich in benzene, is returned as reflux. The lower layer, rich in water, is subjected to distillation in a secondary column. The ternary azeotrope leaves this column as overhead vapours which is combined with the vapours leaving the primary column and sent to the condenser and separator. The aqueous alcohol solution leaving the secondary column as the bottoms product is further separated into alcohol-water binary azeotrope as the overhead product and pure water as the bottoms in a third column as shown in the figure. The binary azeotrope is recycled and joins the feed to the primary column.

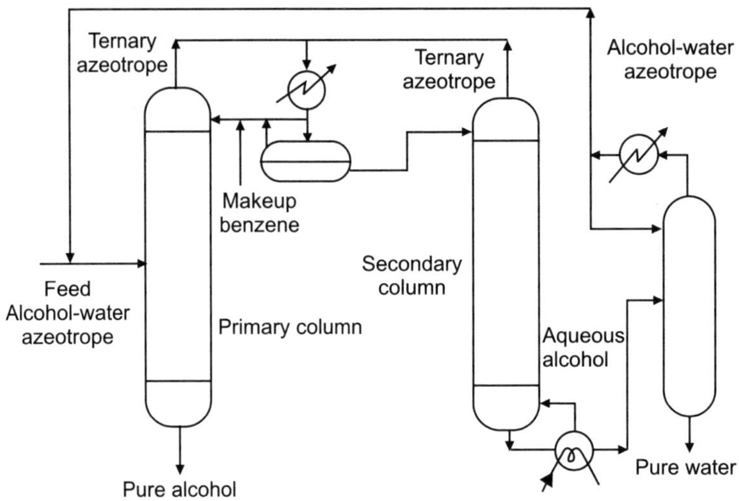

Fig. 5.93: Azeotropic distillation with formation of ternary azeotrope (Ethanol-water-benzene system)

The main factors to be considered in the choice of entrainers are:

1. The solvent should preferably form low-boiling azeotrope with one of the constituents of the feed. The heat requirement of the process can be reduced if the azeotrope is formed with the constituent present in smaller proportions.
2. The volatility of the new azeotrope should be high so that it is easily separated from the primary column as the overhead product.
3. The azeotrope should be of heterogeneous-liquid type so that easy recovery of the entrainer is possible.

In addition, the properties 7 to 13 listed in the previous sections on the choice of solvent for extractive distillation, are applicable for the choice of entrainer for the azeotropic distillation as well.

Extractive distillation offers a greater flexibility in the choice for the added solvent where as the entrainer in the azeotropic distillation has to form an azeotrope with the constituents and therefore, the choice is quite restricted. The extractive and azeotropic distillation employ a foreign substance to effect separation. Since the substance added introduces undesired impurity in the products, they have to be removed. But it is impossible to achieve 100 percent removal of the added solvent or the entrainer. The cost of recovery of the solvent sometimes can be very large. Even though the loss of the solvent is only about 0.1 percent of the solvent circulation rate, the loss is appreciable since the solvent feed rate is usually in the range 3 to 4 times the feed rate.

5.14 PRESSURE-SWING DISTILLATION

Pressure-swing distillation depends on the effect of the pressure on the azeotropic composition. Azeotropic composition can be altered by changing the pressure of distillation. For example, ethanol-water forms an azeotrope of composition 89.43 mol% ethanol at 101.3 kPa. When the pressure is reduced to say, 20 kPa, the azeotropic composition shifts to 96.2 mol% ethanol. When the azeotropic composition shifts by more than 5 mol% or more over moderate changes in pressure, pure components can be separated from the mixtures by operating two distillation columns at different pressures. This process is known as the pressure-swing distillation.

Figure 5.94 shows the *T-x-y* diagram of the **A-B** binary at two pressures P_1 and P_2. The azeotrope shifts towards pure **A** as the pressure is reduced. Suppose that a feed solution F of composition z_F is to

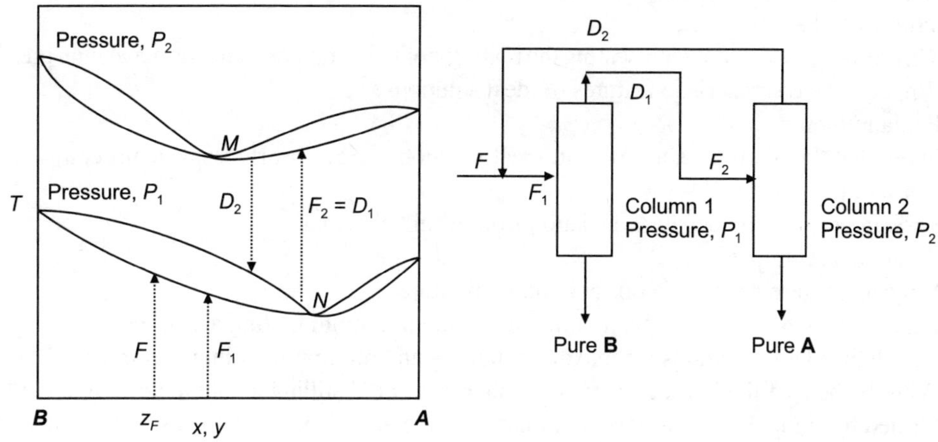

Fig. 5.94: Pressure swing distillation

be separated by distillation at pressure P_2. It is not possible to obtain pure **A** at this pressure, because the distillate composition can only approach the azeotropic composition at pressure P_2. This composition can be "jumped" if the feed is admitted first to a column which is operated at pressure P_1 which is lower than P_2. Pure **B** is withdrawn as bottom product from this column and the distillate D_1 with composition closer to point N, the azeotrope corresponding to pressure P_1, is sent to the second column which is operated at pressure P_2. Since the composition of D_1 is on the right side of the azeotrope M at pressure P_2, it is possible to obtain pure **A** as the bottom product in column 2. The overhead vapour D_2 which has composition closer to the azeotrope at P_2 is admitted to the first column along with fresh feed. Thus we see that pure **A** is obtained from column 2 and pure **B** from column 1.

REVIEW QUESTIONS

1. Why is distillation treated as a direct separation method?
2. Distinguish between distillation and evaporation
3. In what ways absorption and distillation are different?
4. What are the limitations of distillation as a separation technique?
5. What do you mean by the number of degrees of freedom? How many degrees of freedom are there in the distillation of a binary mixture?
6. What are boiling point diagrams?
7. What do you mean by equilibrium distribution curve?
8. At a given pressure, a liquid mixture boils at a constant temperature known as the boiling point of the mixture, till the entire liquid gets vaporised. Do you agree?
9. What are the bubble point and the dew point of a binary liquid?
10. With reference to the boiling point diagram, what is a tie line?
11. Under vapour-liquid equilibrium, the vapour is at its dew point and the liquid is at its bubble point. Do you agree?
12. Describe the inverse lever-arm-rule.
13. If the equilibrium curve lies very close to the diagonal, what does that imply?
14. Define relative volatility.
15. Show that relative volatility of ideal solution is the ratio of the vapour pressures of pure components.

16. Vapour-liquid-equilibrium of a constant relative volatility system can be easily evaluated in terms of α. Explain
17. Why is the looped curve representing the boiling-point diagram become shorter at high pressures?
18. What are the characteristic features of ideal solutions?
19. Explain Raoult's law.
20. Show that the total pressure over an ideal solution varies linearly with liquid composition at constant temperature.
21. Explain positive and negative deviation from ideal behaviour.
22. What are azeotropes?
23. Why are azeotropes called constant boiling mixtures?
24. Explain with examples the maximum boiling and minimum boiling azeotropes?
25. Sketch the phase diagrams for maximum boiling and minimum boiling azeotropes.
26. With the help of the boiling diagram, explain why on distilling a solution of any composition formed by components that form minimum-boiling azeotrope we get a distillate product that is closer to the azeotrope.
27. With the help of the boiling-point diagram explain why on distilling a solution of any composition formed by components that form maximum-boiling azeotrope, the residue is closer to the azeotrope than the original feed.
28. What is pressure swing distillation?
29. Describe the principle behind the activity coefficient model approach and the equations of state approach for evaluating the vapour-liquid equilibria.
30. How is fractionation different from other distillation techniques?
31. Why is the flash distillation known as equilibrium distillation?
32. Explain the relationship between the compositions of the products of flash distillation and the extent of vaporisation.
33. What are the major industrial applications of flash distillation?
34. For the flash distillation at a given pressure, temperature is automatically fixed and cannot be changed. Do you agree?
35. How is the vaporisation equilibrium constant useful in the analysis of the flash vaporisation of multicomponent mixtures?
36. In the simple distillation, the liquid residue in the still is not in equilibrium with the distillate product. Do you agree? Substantiate your answer.
37. Simple distillation is also known as the differential distillation. Explain.
38. Differentiate between the simple distillation and the flash distillation.
39. Simple distillation can be considered as the net result of a series of flash vaporisations. Do you agree?
40. State Rayleigh equation for simple distillation.
41. Does the distilling temperature increase or decrease with progress in the differential distillation?
42. For simple distillation, how would you estimate the distillate composition given the quantity and composition of the feed and quantity of the distillate?
43. An equimolar mixture with relative volatility 2 is subjected to differential distillation so that one-half of the feed is vaporised. What is the composition of the residue?
44. What are the major areas of applications of steam distillation?
45. High boiling heat sensitive materials can be distilled without danger of decomposition by steam distillation. Why?
46. Steam distillation can be carried out in two different ways. Explain.
47. What are the advantages of steam distillation over vacuum distillation?

48. What is the advantage of steam distillation carried out with indirect heating over the other type in which heating is done by condensing the steam?
49. Why is the temperature of steam distillation lower than the boiling point of water?
50. How is distillation temperature evaluated in type 1 steam distillation?
51. What is Hausbrand chart?
52. In steam distillation how is the steam consumption related to the vapour pressures of the components?
53. What is vaporisation efficiency in steam distillation?
54. How do you evaluate the total steam consumption in type 1 steam distillation?
55. To reduce steam consumption, type 2 distillation with indirect heating is preferred over type 1. Explain.
56. Why is steam distillation preferred to distill liquids of poor heat transfer characteristics?
57. When steam condenses in the distillation still, the pressure and temperature of distillation cannot be independently varied. Why?
58. Do the amounts of liquid and water present in the still influence the composition of the vapour produced in steam distillation?
59. In the absence of two immiscible liquid phases in the still, the temperature of steam distillation can be varied by passing more or less steam through the charge. Do you agree? Why?
60. What is reflux? Define reflux ratio. What is the role of the reflux in fractionation?
61. What is boilup ratio? Why is boilup necessary in fractionation columns?
62. Equimolar counter diffusion exists in distillation of binary mixtures. Do you agree?
63. What are the enriching section and the stripping section in fractionation columns?
64. What are the factors that govern the choice of distillation column pressure?
65. When do you go for a partial condenser in distillation columns?
66. What are the major types of reboilers used in distillation?
67. What are internal reboilers? Why are they not preferred for industrial applications?
68. Why thermosyphon reboilers which receive the liquid for vaporisation from the bottom sump of distillation columns cannot be considered as equivalent to one theoretical plate?
69. What are the factors that determine the choice of reboilers?
70. Show that the net flows in distillation columns are constant.
71. Can you fix the condenser duty and the reboiler heat load in distillation columns independently?
72. Show that by fixing the reflux ratio in the distillation columns the condenser duty and reboiler heat load get automatically fixed.
73. What are the McCabe-Thiele assumptions?
74. What are the factors that are responsible for change in the flow rates of vapour and liquid streams in columns?
75. Justify that the constant molar overflow in the fractionation columns also lead to constant molar vaporisation.
76. What is the slope of the enriching operating line in the McCabe-Thiele method?
77. Do you agree that the slope of the stripping operating line is fixed by specifying the reflux ratio?
78. Can we use McCabe-Thiele method on mass basis for determining the number of plates?
79. Justify that the partial condenser acts as a theoretical plate outside the column.
80. What is the criterion of locating feed plate in McCabe-Thiele method?
81. What do you mean by the feed quality factor q?
82. Explain the different feed conditions and their respective q-values.
83. In what way the q-value is helpful in the graphical method for determining the number of plates?

84. Under total reflux condition what is the slope of the operating line? What is the value of R under total reflux?
85. What is Fenske equation? When is it applicable?
86. Explain how the minimum reflux ratio is determined for different type of VLE systems.
87. What is pinch point? How is it eliminated?
88. Show that for the high-purity products, the minimum reflux ratio R_m can be approximated as $R_m = \frac{1}{(\alpha - 1)z_F}$.
89. Explain how the optimum reflux ratio for the column is estimated.
90. With increase in the reflux ratio, the cost of the column first decreases and then increases. Justify.
91. What is the effect of cold reflux on the design and operation of the distillation column?
92. When is open steam used in distillation and what are its advantages and disadvantages?
93. Describe how the saturation curves H_L vs x and H_G vs. y are established on enthalpy-composition diagrams.
94. Under what conditions, the saturation enthalpy curves are straight parallel lines?
95. Explain how the tie lines are drawn on enthalpy-composition diagram.
96. What is the lever-arm rule for adiabatic mixing of two streams?
97. What is the significance of difference points on the *H-x-y* diagram?
98. The difference points on the *H-x-y* diagram can be equivalently represented by a line on the *x-y* diagram. Justify. When will this line be a straight line?
99. How do you locate the upper difference point Δ_D on the enthalpy-composition diagram?
100. Given the reflux ratio how do you compute the condenser and reboiler heat duties by the Ponchon-Savarit method?
101. What is the geometric interpretation of the internal reflux ratio in the Ponchon-Savarit construction?
102. How do you determine the optimum location of the feed plate by the Ponchon-Savarit method?
103. How would you obtain the operating lines using the Ponchon-Savarit construction?
104. How is the minimum reflux ratio predicted using the Ponchon-Savarit technique?
105. What features of the *H-x-y* diagram make the constant molar overflow conditions valid?
106. When is packed columns preferred over plate columns for distillation and why?
107. What is the major operational difficulty in using the packed columns for distillation?
108. For estimating the height of the packed columns used for distillation, the method based on the overall transfer units is not preferred. Why?

MULTIPLE CHOICE QUESTIONS

5.1 Distillation is a direct separation technique because
- A. no foreign substance is used as separating agent
- B. separation makes use of differential distribution of components between phases
- C. there is no interphase mass transfer
- D. it is an energy intensive operation

5.2 In distillation operation we make use of the following as separating agent:
- A. Solvent
- B. Condenser
- C. Heat
- D. Trays

5.3 Under ideal situation, distillation involves
- A. diffusion through stagnant gas
- B. equimolar counter diffusion
- C. diffusion in liquid phase only
- D. none of the above

5.4 The major disadvantage of distillation as a separation technique is that
A. it does not make use of external agents
B. it cannot be carried out batch-wise
C. high purity separation is not possible
D. it is energy intensive

5.5 Distillation makes use of which one of the following to effect separation?
A. Difference in volatility of the components
B. Difference in boiling points of the components
C. Difference in compositions of the vapour and liquid phases produced on boiling
D. All the above

5.6 On boiling a liquid mixture, the vapour produced is rich in
A. high-boiling components
B. less volatile components
C. low boiling components
D. component with relative volatility unity

5.7 When vapour and liquid coexist in equilibrium, the dew point of the vapour is
A. equal to the bubble-point of the liquid
B. above the bubble point of the liquid
C. below the bubble point of the liquid
D. not related to the bubble point

5.8 Which one of the following is called boiling point diagram?
A. *P-x-y* diagram
B. *x-y* diagram
C. *T-x-y* diagram
D. *P-T* diagram

5.9 At a given pressure, the number of degrees of freedom for the vapour and liquid phases of a binary mixture in equilibrium is
A. two
B. three
C. one
D. zero

5.10 For ideal liquid phase and ideal vapour phase, the relative volatility α_{AB} of component **A** (vapour pressure = P_A^S) relative to component **B** (P_B^S) is equal to
A. $\dfrac{P_A^S}{P}$
B. $\dfrac{P_B^S}{P}$
C. $\dfrac{P_A^S}{P_B^S}$
D. $\dfrac{P_B^S}{P_A^S}$

5.11 The relative volatility $\alpha = \dfrac{y/1-y}{x/1-x}$ is applicable to
A. all solutions whether ideal or non ideal
B. only ideal solutions
C. solutions exhibiting positive deviation
D. minimum boiling azeotropes

5.12 Which one of the following cannot be treated as the relative volatility of a component in a binary mixture?
A. The ratio of vapour pressure of pure **A** to vapour pressure of pure **B**.
B. The ratio of moles of **A** in the vapour to that in the liquid in equilibrium with the vapour
C. The ratio of mole ratio of **A** to **B** in the vapour to mole ratio of **A** to **B** in the liquid
D. The ratio of vaporisation equilibrium constants K_A/K_B

5.13 Relative volatility for a binary system
A. decrease with increase in pressure
B. increase with increase in pressure

C. increase with increase in temperature at constant pressure
D. has no significance in distillation process

5.14 In binary distillation, easier separation of components is achieved when
A. $\alpha >> 1$ B. $\alpha << 0$
C. $\alpha = 1$ D. None of these

5.15 Relative volatility α for a binary system is a measure of separability by distillation. If $\alpha = 1$ separation by distillation is
A. maximum possible B. least possible
C. impossible D. optimum possible

5.16 Boiling point of a binary liquid solution
A. lies between the boiling points of pure components
B. lies between the bubble point and dew point of the mixture
C. lies above the boiling points of pure components
D. is above the dew point but below the bubble point of the mixture.

5.17 In the boiling point diagram, it is seen that the point which represents the composition of the combined mixture of liquid and vapour phases in equilibrium divides the tie line in the ratio 3:1 and the point lies closer to the dew point curve, the mixture contains
A. 33.33% vapour, 66.67% liquid B. 33.33% liquid, 66.67% vapour
C. 25% liquid, 75% vapour D. 25% vapour, 75% liquid

5.18 A system with large negative deviation from ideal behaviour is
A. Acetaldehyde-carbon disulphide B. Ethyl alcohol-chloroform
C. Acetone- chloroform D. None of the above

5.19 In a binary mixture, the negative deviation occurs when
A. unlike molecule interaction is stronger
B. enthalpy of mixing is negative
C. volume of mixing is negative
D. all of the above

5.20 A mixture of two liquids has boiling points lower than either of them when it
A. forms ideal solution B. shows positive deviation
C. shows negative deviation D. is saturated

5.21 The ratio of moles of **A** in the vapour to that in the liquid in equilibrium with the vapour is called
A. relative volatility B. vaporisation equilibrium constant
C. separation factor D. volatility of component **A**

5.22 Which one of the following statements is not true with regard to azeotropes?
A. They are formed from solutions exhibiting positive deviation from ideal behaviour
B. They are constant boiling mixtures.
C. The vapour and liquid resulting on boiling an azeotrope have the same compositions
D. Components exhibiting deviation from ideal behaviour do not form azeotropes

5.23 On distillation, if the composition of the distillate product is closer to the azeotrope, the solution
A. Forms a minimum boiling azeotrope
B. Forms a maximum boiling azeotrope
C. Shows negative deviation from ideal behaviour
D. None of the above

5.24 Components **A** and **B** are known to form azeotrope. If a mixture of **A** and **B** of any composition on distillation gives a residue product closer in composition to the azeotrope, then the azeotrope is
A. not constant boiling B. maximum boiling
C. minimum boiling D. heteroazeotrope

5.25 The system which does not form azeotrope at 1 bar is
A. ethanol-water B. benzene-toluene
C. HCl-water D. acetone-chloroform

5.26 An example of minimum boiling azeotrope at 1 bar is
A. ethanol-water B. benzene-toluene
C. HCl-water D. acetone-chloroform

5.27 An example of maximum boiling azeotrope at 1 atm is
A. ethanol-water B. benzene-ethanol
C. carbon disulphide-acetone D. acetone-chloroform

5.28 At the azeotropic composition of a binary mixture, relative volatility is
A. 0 B. ∞
C. 1 D. <1

5.29 An ideal solution containing 40% **A** and 60% **B** is in equilibrium with its vapour. The vapour pressures of the pure liquids at the equilibrium temperature are 80 kPa (**A**) and 40 kPa (**B**). The vapour composition is
A. 80% **A** B. 67%**A**
C. 57% **A** D. 40% **A**

5.30 If $y^* = x$, the relative volatility for a binary system is always
A. 0 B. ∞
C. 1 D. 1, except at $x = 0$ or 1

5.31 Azeotropes
A. are mixtures with constant boiling point at the specified pressure
B. obeys Raoult's law at all temperatures
C. are mixtures of partially miscible liquids
D. none of the above

5.32 Which one of the following statements is not true with reference to minimum boiling azeotropes?
A. There is a maximum on the vapour pressure curve
B. The solution exhibits positive deviation from ideal behaviour
C. With regard to forces of molecular interaction, the like-pair interactions are stronger than unlike pair interactions.
D. The boiling point of the azeotrope lies between the bubble point and the dew point

5.33 In a binary liquid solution of components **A** and **B**, if **A** exhibits positive deviation from Raoult's law, then component **B**
A. exhibits positive deviation Raoult's law
B. exhibits negative deviation from Raoult's law
C. obeys Raoult's law
D. may exhibit either positive or negative deviation from Raoult's law

5.34 In flash vaporisation of a binary mixture at a given pressure,
A. the temperature is constant and cannot be varied
B. the temperature can be varied over a range
C. the temperature is above the dew point of the mixture
D. the temperature should be above the boiling points of pure components

5.35 If the distillate and residue have equilibrium composition, the separation achieved is by
A. flash distillation B. rectification
C. differential distillation D. multistage operation

5.36 In flash distillation of a binary mixture, the richest vapour is obtained
A. at the dew point of the mixture
B. at the bubble point of the mixture
C. at the boiling point of the mixture
D. at the three-phase equilibrium temperature

5.37 During flashing operation, the temperature falls because
A. the operation is adiabatic
B. of Joule-Thomson expansion
C. the mixture is passed through a condenser
D. the heat is removed by cooling coils

5.38 Flash distillation operation is suitable for separating components which
A. boils at very close temperatures
B. boils at widely different temperature
C. form minimum boiling azeotrope
D. form maximum boiling azeotrope

5.39 The flash vaporisation process is carried out by
A. reducing the pressure on the feed stream
B. cooling the feed at constant pressure
C. increasing the pressure on the feed stream
D. none of the above

5.40 The composition of the products in flash distillation depend upon
A. the fraction of the feed vaporised.
B. the temperature of distillation
C. the separator pressure
D. all the above

5.41 The vapour pressures of benzene and toluene are 3 and 4/3 atmospheres respectively. The liquid product obtained on flash distillation contains 0.4 moles of benzene and 0.6 moles of toluene. Assuming that the products are in equilibrium, the vapour phase mole fraction of benzene is
A. 0.4
B. 0.6
C. 0.8
D. 0.2

5.42 One mole feed of a binary mixture of a given composition is flash vaporised at a fixed P and T. If Raoult's law is obeyed, then changing the feed composition would affect
A. the product composition but not the fraction vaporised.
B. the product composition as well as the fraction vaporised.
C. fraction vaporised but not the product composition.
D. neither the product composition nor the fraction vaporised.

5.43 If the extent of flash vaporisation is 50%, the slope of the operating line is
A. 1/2
B. –1/2
C. 1
D. –1

5.44 A flash distillation drum (Fig. (i)) is used to separate methanol-water mixture. The mole fraction of methanol in the feed is 0.5 and the feed flow rate is 1000 kmol/h. The feed is preheated in a

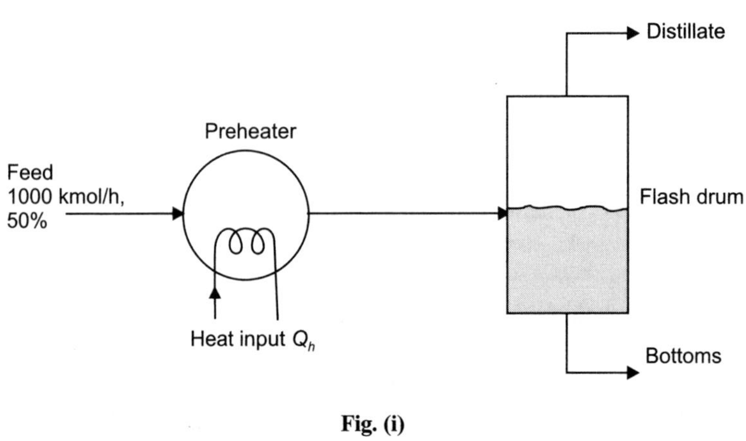

Fig. (i)

heater with heat duty Q_h and is subsequently flashed in the drum. The flash drum can be assumed to be an equilibrium stage operating adiabatically. The equilibrium relationship between the mole fractions of methanol in the vapour and the liquid phases is $y^* = 4x$. The ratio of distillate to feed flow rate is 0.5

The mole fraction of methanol in the distillate is

A. 0.2 B. 0.7
C. 0.8 D. 0.9

5.45 A flash distillation drum (Fig. (i)) is used to separate methanol-water mixture. The mole fraction of methanol in the feed is 0.5 and the feed flow rate is 1000 kmol/h. The feed is preheated in a heater with heat duty Q_h kJ/h and is subsequently flashed in the drum. The flash drum can be assumed to be an equilibrium stage operating adiabatically. The equilibrium relationship between the mole fractions of methanol in the vapour and the liquid phases is $y^* = 4x$. The ratio of distillate to feed flow rate is 0.5. If the enthalpy of the distillate with reference to the feed is 3000 kJ/kmol, and the enthalpy of the bottoms with reference to the feed is –1000 kJ/kmol, the heat duty of the preheater in kJ/h is

A. -2×10^6 B. -1×10^6
C. 1×10^6 D. 2×10^6

5.46 A binary hydrocarbon liquid mixture of **A** and **B** ($K_A = 1.5$) containing 60 mole% **A** is flash vapourised. If 40% of the feed is vaporised, the mole fraction of **A** in the bottom is

A. 0.25 B. 0.5
C. 0.75 D. 0.85

5.47 An equimolar mixture of **A** and **B** is flash distilled continuously at a feed rate of 100 kmol/h, such that the liquid product contains 40 mol% **A**. If the relative volatility is 6, then the vapour product, in kmol/h, is

A. 10 B. 20
C. 25 D. 45

5.48 Spot the statement that is not true with respect to simple distillation.

A. The distillate and residue composition change with time.
B. The distillate and residue have equilibrium compositions.
C. The distillate vapours produced at any instant is in equilibrium with the liquid.
D. The temperature of distillation increases continuously.

5.49 A binary liquid mixture containing 50% mole **A** is differentially distilled so that 70% of the feed is vaporised. If the equilibrium is represented by $y^* = 2x$ in the region of interest, what would be the molar concentration of **A** in the residue?

A. 10.3% B. 12.2%
C. 15.0% D. 5.3%

5.50 100 moles of a binary mixture F containing 60 mole% **A** and 40 mole% **B** is treated in a batch distillation still. After 1 hour, 70 moles of distillate D is collected leaving behind the residue W. Relative volatility α is 2. The governing equation is

$$\log \frac{F x_F}{W x_w} = \alpha \log \frac{F(1-x_F)}{W(1-x_W)}$$

The average mole fraction of **A** in the distillate is

A. 0.43 B. 0.61
C. 0.69 D. 0.73

5.51 An equimolar mixture of **A** and **B** with constant relative volatility is subjected to simple distillation so that 50% of the feed is distilled off. If the residue has a composition of 30% **A**, then the relative volatility of the system is

A. 2.4 B. 2.0
C. 3.4 D. 4.3

5.52 An equimolar mixture of **A** and **B** with relative volatility 2.5 is subjected to simple distillation so that 50% of the feed is distilled off. The composition of the residue is nearly

A. 30% **A** B. 35% **A**
C. 40% **A** D. 45% **A**

5.53 Rayleigh equation is used for estimating product composition in

A. Flash distillation B. Simple distillation
C. Steam distillation D. Fractionation

5.54 Rayleigh equation relates

A. F, W, x_F, x_W B. x_F, x_W, y_D, F
C. F, D, y_D, x_F D. x, y^*, x_W, x_F

5.55 The number of degrees of freedom in steam distillation allowing condensation of steam in the still is

A. one B. two
C. three D. zero

5.56 Steam distillation reduces the boiling point of the material because

A. of the reduction in the total pressure
B. the liquid boils when sum of the vapour pressures of water and the material is equal to the total pressure
C. the partial pressure of the material depend upon the composition of the immiscible liquid mixture
D. steam agitates the liquid and brings down the boiling point.

5.57 With reference to steam distillation, the ratio of actual partial pressure of the distillate vapour to the vapour pressure at the distillation temperature is called

A. relative saturation B. thermal efficiency
C. vaporisation efficiency D. vaporisation equilibrium constant K

5.58 If there is condensation of carrier steam in the steam distillation of high boiling organic materials

A. both temperature and pressure can be independently specified
B. either the temperature or pressure must be specified
C. the temperature is always more than 373K at 1atm pressure
D. the temperature is always equal to 373K at 1atm pressure

5.59 A mixture of two immiscible liquids containing 25 mol water (**A**) and 75 mol liquid **B** is in equilibrium with its vapour. The vapour pressures of the components are 90 kPa for water and 10 kPa for the liquid. The mole percent of water in the vapour is

A. 7.5 B. 22.5
C. 25 D. 90

5.60 In steam distillation with indirect heating, which of the following is recommended to reduce steam consumption?

A. Maintain low temperature and high pressure
B. Maintain high temperature and low pressure
C. Allow formation of two immiscible phases
D. None of the above

5.61 For steam distillation, though several gases can be used in its place, steam is preferred because of all but one of the following reasons. Spot the incorrect statement.
A. Water is available at low cost.
B. Water is generally immiscible with high-boiling organic materials.
C. Water vapour can be easily condensed and removed from the distillate.
D. Latent heat of vaporisation of water is low.

5.62 In steam distillation, the
A. temperature is 373 K.
B. temperature is more than 373 K.
C. product must be immiscible with water.
D. the temperature is higher than the boiling point of either component.

5.63 Components having widely different boiling points in a binary mixture can be separated using
A. extractive distillation
B. azeotropic distillation
C. steam distillation
D. simple distillation

5.64 Benzene and water may be considered immiscible. A mixture of benzene (20 kg) and water (80 kg) is taken in a vessel and boiled. It boils at 342 K and 101.3 kPa pressure. At this temperature, vapour pressure of benzene is 71.18 kPa and that of water is 30.12 kPa. What is the concentration of benzene in the vapour in mass percent.
A. 70
B. 91
C. 20
D. 80

5.65 Steam distillation is preferred for separating
A. low boiling mixtures
B. high boiling mixtures
C. high volatile liquids
D. miscible liquids

5.66 In steam distillation employing indirect external heating (type II distillation), we can reduce steam consumption by
A. reducing the temperature of distillation.
B. increasing the pressure.
C. increasing the temperature of distillation.
D. increasing the temperature and pressure.

5.67 The temperature at which a mixture of water and toluene with normal boiling points 373 K and 383.6 K respectively, will exert a pressure of one standard atmosphere is
A. less than 373 K
B. 373 K
C. between 373 K and 383.6 K
D. 383.6 K

5.68 At a total pressure of one standard atmosphere exerted by the vapours of water and toluene, the mole fraction of water x_w in the vapour phase satisfies
A. $0 < x_w < 0.5$
B. $x_w = 0.5$
C. $0.5 < x_w < 1.0$
D. $x_w = 1.0$

5.69 Fractionation is different from flash distillation and simple distillation because, in fractionation
A. the distillate and residue are not in equilibrium.
B. the distillate vapours are not condensed.
C. a part of the condensed vapours are recycled.
D. the distillate and residue are not subjected to rectification.

5.70 An ideal plate is defined as one where
A. the vapour and liquid leaving streams are in equilibrium.
B. the vapour and liquid entering streams are in equilibrium.
C. the vapour leaving stream is in equilibrium with the liquid entering stream.
D. the vapour entering stream is in equilibrium with the liquid leaving stream.

5.71 Reflux ratio in distillation column is defined as the ratio of

A. feed to distillate.
B. reflux to distillate.
C. reflux to feed.
D. reflux to bottom product.

5.72 Which of the following is most important in the separation of a liquid mixture by distillation?

A. Selectivity of the solvent
B. Relative volatility of the system
C. Mutual solubility of the components
D. Diffusion of **A** through stagnant **B**

5.73 Reflux ratio (external reflux ratio) is defined as the ratio of

A. moles of vapour recycled to moles of bottoms
B. moles of liquid recycled to moles of distillate
C. moles of feed introduced to moles of distillate
D. molar flow of liquid to molar flow of vapour

5.74 Increasing the reflux ratio

A. decreases the reboiler load
B. decreases the vapour boilup
C. increases the number of plates
D. increases the vapour boilup

5.75 In distilling heat sensitive liquids, the operating pressure is decided by

A. cooling water inlet temperature
B. temperature of the feed
C. temperature at the reboiler
D. bubble point of the distillate

5.76 Which one of the following reboilers cannot be treated as equivalent to a theoretical plate in the McCabe-Thiele construction?

A. Kettle-type reboiler
B. Thermosyphon reboilers with liquid withdrawn from the bottom sump of the column
C. Thermosyphon reboilers with liquid withdrawn from the downcomer of the bottom plate
D. Tubular heat exchanger type internal reboilers

5.77 Regarding partial condensers, one of the following statements is not correct.

A. Used when vapour distillate is desired
B. Used when total condensation requires very high pressure
C. Used when total condensation requires operation under vacuum
D. Acts as a theoretical plate

5.78 When the liquid shows heavy fouling tendency, the reboiler preferred is

A. Kettle-type reboiler
B. Thermosyphon vertical reboiler
C. Internal tubular heat exchanger type
D. Jacketted type reboiler

5.79 Which one of the following statements is incorrect with reference to reflux ratio R?

A. By increasing R, the liquid to vapour flow ratio (L/G) is unaffected
B. By increasing R, the cooling water circulation is increased
C. By increasing R, the steam supply to the reboiler is increased
D. By increasing R, the column diameter increases

5.80 The slope of the rectifying operating line is

A. $\frac{x_D}{R+1}$
B. $\frac{R}{R+1}$
C. $\frac{q}{q-1}$
D. $\frac{x_D - y'}{y' - x'}$

5.81 In the McCabe Thiele diagram for binary distillation, vertical feed line represents .... feed and horizontal feed line represent .... feed.

A. saturated vapour, saturated liquid B. saturated liquid, saturated vapour
C. saturated feed, saturated vapour D. none of the above

5.82 In the McCabe-Thiele method for fractionating columns, if R = reflux ratio, q = feed quality, F = feed rate, D = distillate rate the quantity, $\dfrac{R+qF/D}{(R+1)+(q-1)F/D}$ represents

A. condenser heat duty B. reboiler heat load
C. slope of the rectifying operating line D. slope of the stripping operating line

5.83 In the McCabe-Thiele construction for theoretical plates for a column using a kettle type reboiler and a total condenser, the number of steps constructed between the operating lines and the equilibrium curve is found to be N. Then the number of theoretical plates in the column is

A. $N-2$ B. $N-1$
C. N D. $N+1$

5.84 Three feed lines are plotted in Fig. (ii). They refer to

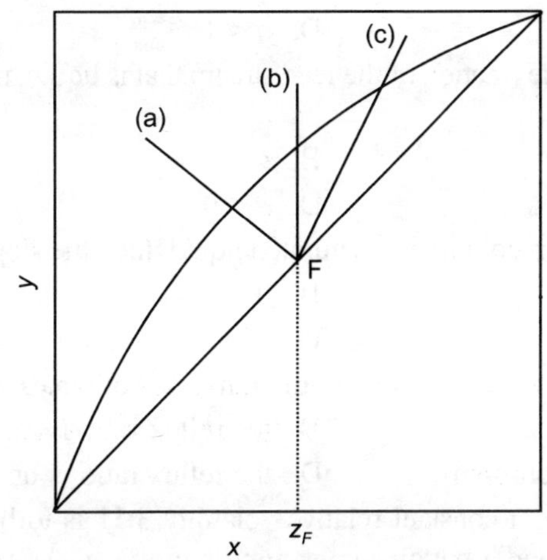

Fig. (ii)

A. (a) saturated liquid (b) saturated vapour (c) subcooled liquid
B. (a) partially vaporised (b) saturated liquid (c) superheated vapour
C. (a) partially vaporised (b) saturated liquid (c) subcooled liquid
D. (a) partially vaporised (b) saturated vapour (c) superheated vapour

5.85 The q-factor of feed can be evaluated as (λ_v = latent heat of vaporisation of the feed, H_G = enthalpy of saturated vapour feed, H_L = enthalpy of saturated liquid feed, H_F = enthalpy of feed)

A. $q=\dfrac{H_G-H_L}{\lambda_v}$ B. $q=\dfrac{\lambda_v}{H_G-H_F}$

C. $q=\dfrac{H_L-H_F}{\lambda_v}$ D. $q=\dfrac{H_G-H_F}{\lambda_v}$

5.86 The slope of the operating line for the stripping section of the distillation column is

A. 0 B. ∞
C. <1 D. >1

5.87 The slope of the operating line for the rectifying section of the distillation column is

A. ∞ B. 0

C. >1 D. <1

5.88 If q is the fraction of the feed that is liquid, then the slope of the feed line in distillation operation is

A. $-\dfrac{(1-q)}{q}$ B. $-\dfrac{q}{1-q}$

C. $-q$ D. $-\dfrac{q}{q-1}$

5.89 For a two-phase feed, where 80% of the feed is vaporised under column conditions, the feed line slope in the McCabe-Thiele method for distillation column design is

A. –1/4 B. +1/4

C. 4 D. –4

5.90 If q is interpreted to be the fraction of the feed stream that is liquid, then for saturated vapour feed to a distillation column

A. $q = 1$ B. $q = 0$

C. $q > 1$ D. $q < 1$

5.91 If q is interpreted to be the fraction of the feed stream that is liquid, then for saturated liquid feed to a distillation column

A. $q = 1$ B. $q = 0$

C. $q > 1$ D. $q < 0$

5.92 If the feed to a distillation column is a saturated liquid, then the slope of the feed line is

A. ∞ B. 0

C. >1 D. <1

5.93 The fractionating column requires the minimum number of plates when

A. the reflux ratio is maximum B. the reflux ratio is minimum

C. the reflux ratio is optimum D. the reflux ratio is unity

5.94 A binary mixture having a constant relative volatility 3.51 is to be separated into a distillate containing 99 mol% **A** and a bottoms containing 5 mol% **A**. Assuming a total condenser, the minimum number of plates required in the column is

A. 10 B. 6

C. 5 D. 4

5.95 An equimolar mixture having a constant relative volatility 2.5 is to be separated into a distillate containing 98 mol% **A**. If the feed is saturated liquid, the minimum reflux ratio for the column is

A. 2.44 B. 1.24

C. 1.45 D. 3.21

5.96 Which of the following methods assume constant molal vaporisation and overflow?

A. McCabe-Thiele method B. Ponchon-Savarit method

C. Enthalpy concentration method D. Plate absorption method

5.97 McCabe-Thiele method

A. uses molal units for material and energy balance

B. uses weight fractions to express liquid and vapour compositions

C. can use any type of units

D. is more accurate than Ponchon–Savarit method

5.98 Fenske equation is used to calculate the minimum number of theoretical stages required to obtain a given separation by distillation. This equation is valid for

A. total reflux
B. constant molal overflow
C. constant relative volatility
D. all of the above

5.99 In distillation operation, the heat removed in the condenser

A. remains unaffected with change in reflux ratio
B. increases with increase in reflux ratio
C. decrease with increase in reflux ratio
D. and the heat required in reboiler decreases with increase in reflux ratio

5.100 In a distillation operation, the reflux ratio may vary between

A. 0 and 1
B. 0 and ∞
C. –1 and 1
D. minimum reflux ratio and infinity

5.101 In a distillation, operation total reflux requires

A. minimum number of plates
B. infinite number of plates
C. minimum reboiler and condenser load
D. continuous flow of fresh feed

5.102 Most distillation columns are designed for reflux ratio between

A. 3 to 5 times R_m
B. 1.2 to 1.7 times R_m
C. 2 to 10 times R_m
D. 0.2 to 0.7 times R_m

where R_{min} is the minimum reflux ratio

5.103 What will be the minimum number of plates required in a distillation column to separate an equimolar mixture of **A** and **B** into a distillate containing 99% **A** and a bottom product containing 98% **B**? Assume a constant relative volatility of 2 and apply Fenske equation.

A. 5
B. 12
C. 9
D. 28

5.104 For obtaining a given separation in distillation column, the minimum number of theoretical stages is obtained with

A. minimum reflux ratio
B. optimum reflux ratio
C. total reflux
D. no reflux

5.105 At very high reflux ratios, increasing the reflux ratio results in

A. large increase in column diameter and small decrease in column height
B. large increase in column diameter and column height
C. large reduction in column diameter and small decrease in column height
D. decrease of column diameter and column height

5.106 Compared to saturated reflux, subcooled reflux leads to

A. increase in the number of plates in the column
B. increase of column pressure
C. lesser number of plates in the column
D. no change in the height and diameter of the column.

5.107 Which of the following is correct regarding the slopes of the enriching operating line (EOL) and stripping operating line (SOL)

A. EOL >1; SOL >1
B. EOL < 1; SOL >1
C. EOL >1; SOL < 1
D. No such generalization is possible

5.108 Minimum number of ideal stages are required in fractionating column when the reflux ratio is equal to

A. minimum reflux ratio
B. optimum reflux ratio
C. zero
D. infinity

5.109 In distillation under minimum reflux conditions, the number of theoretical stages would be

A. one
B. minimum
C. infinite
D. none of these

5.110 In a distillation column where q is defined as the moles of liquid flow in the stripping section per mole of feed introduced, for saturated liquid feed

A. $q > 1$
B. $q < 1$
C. $q = 1$
D. $q = 0$

5.111 In the McCabe-Thiele diagram, if the x-coordinate of the point of intersection of the q-line and the equilibrium curve is greater than the x-coordinate of the feed point, then the feed is

A. superheated vapour
B. liquid below the bubble point
C. saturated vapour
D. saturated liquid.

5.112 If open steam is used instead of a reboiler supplying vapour to the distillation column, for a given rate of production of the distillate.

A. more number of plates required in the stripping and enriching sections
B. more number of plates required in the stripping section
C. lesser number of plates required in the stripping section
D. lesser number of plates required in the stripping and enriching sections

5.113 An infinite number of plates are required

A. under total reflux
B. under minimum reflux
C. if feed is at its boiling point
D. if the feed is superheated vapour

5.114 If the slope of the q-line is finite and greater than 1, the feed is

A. superheated vapour
B. saturated vapour
C. saturated liquid
D. subcooled liquid

5.115 If q is interpreted to be the fraction of the feed stream that is liquid, then for superheated vapour feed to a distillation column

A. $q > 1$
B. $0 < q < 1$
C. $q = 0$
D. $q < 0$

5.116 In a distillation column, the minimum reflux ratio

A. corresponds to the maximum ratio which will require an infinite number of trays for the desired separation
B. corresponds to the maximum reboiler heat load for the separation desired
C. corresponds to the maximum cooling load for the separation desired
D. is always zero

5.117 A saturated vapour is fed to a distillation column at 180 kmol/h. Both the rectifying and stripping sections of the column operate at 60% of their respective flooding velocities. The flooding velocity of the rectifying section is twice that of the stripping section. The assumptions of constant molar overflow and constant molar vapour density throughout the column are valid. If the boil-up rate is 60 kmol/h, then the relationship between the diameters of the rectifying section (d_r) and stripping section (d_s) is

A. $d_r = \sqrt{2}d_s$
B. $d_r = \sqrt{3}d_s$
C. $d_r = 2d_s$
D. $d_r = 3d_s$

5.118 A binary distillation column separates 100 mol/h of a feed mixture into distillate D and residue W. The McCabe-Thiele diagram for this process is given below [Fig. (iii)]. The relative volatility for the binary system is constant at 2.4.

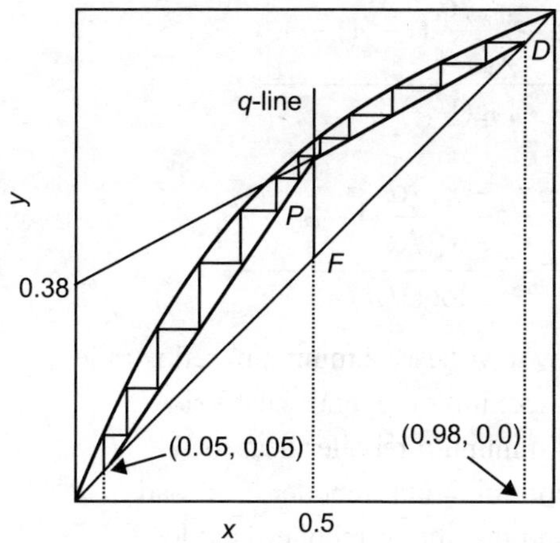

Fig. (iii)

The distillate and residue flow rates (in mol/h) are

A. $D = 48.4$, $W = 51.6$ B. $D = 51.6$, $W = 48.4$

C. $D = 54.7$, $W = 45.3$ D. $D = 45.3$, $W = 54.7$

5.119 For the above case, the ratio of liquid to vapour molar flow rates in the rectifying section is

A. 0.61 B. 1.00

C. 1.55 D. 1.80

5.120 The minimum number of theoretical stages (inclusive of reboiler) for this process is

A. 5.2 B. 6.1

C. 7.8 D. infinite

5.121 In a binary distillation column, if the feed contains 40 mole% vapour, the q-line will have a slope of

A. 1.5 B. –0.6

C. –1.5 D. 0.6

5.122 In a distillation operation, it is desired to have a very high purity bottom product. Initially a kettle type reboiler is used at the bottom of the column and the following analytical equation is used to obtain the equilibrium trays in the exhausting section of the column

$$N - m + 1 = \frac{\log\left[\dfrac{x_m - x_w/\alpha}{x_w - x_w/\alpha}(1-\bar{A}) + \bar{A}\right]}{\log(1/\bar{A})}$$

where x_m is the composition of the liquid leaving tray m. Tray m is the last equilibrium tray obtained by a McCabe-Thiele graph of the exhausting section. If the kettle type reboiler is replaced by a thermo-syphon reboiler, the analytical equation, for the exhausting section will be

A. $$N - m + 1 = \frac{\log\left[\dfrac{x_m - x_w/\alpha}{x_w - x_w/\alpha}(1-\bar{A}) + \bar{A}\right]}{\log(1/\bar{A})}$$

B. $$N + 1 = \frac{\log\left[\dfrac{x_m - x_w/\alpha}{x_w - x_w/\alpha}(1-\bar{A}) + \bar{A}\right]}{\log(1/\bar{A})}$$

C. $N - m = \dfrac{\log\left[\dfrac{x_m - x_w/\alpha}{x_w - x_w/\alpha}(1-\bar{A}) + \bar{A}\right]}{\log(1/\bar{A})}$

D. $N - m + 2 = \dfrac{\log\left[\dfrac{x_m - x_w/\alpha}{x_w - x_w/\alpha}(1-\bar{A}) + \bar{A}\right]}{\log(1/\bar{A})}$

5.123 When a multistage tray tower uses a minimum reflux ratio, it implies
A. infinite trays and maximum reboiler heat load
B. infinite trays and minimum reboiler heat load
C. minimum trays and minimum reboiler heat load
D. minimum trays and maximum reboiler heat load

5.124 In binary distillation, the separation of the components is easier if the relative volatility is
A. $\alpha >> 1$ B. independent of composition
C. constant D. none of these

5.125 The feed to a binary distillation column has 40 mol% vapour and 60 mol% liquid. Then the slope of the q-line in the McCabe-Thiele plot is
A. –1.5 B. –0.6
C. 0.6 D. 1.5

5.126 An equimolar mixture of four hydrocarbons (1, 2, 3, 4) is to be separated into high purity individual components using a sequence of simple distillation columns (one overhead and one bottom stream). Four possible schemes are shown below [*See* Fig. (iv)].

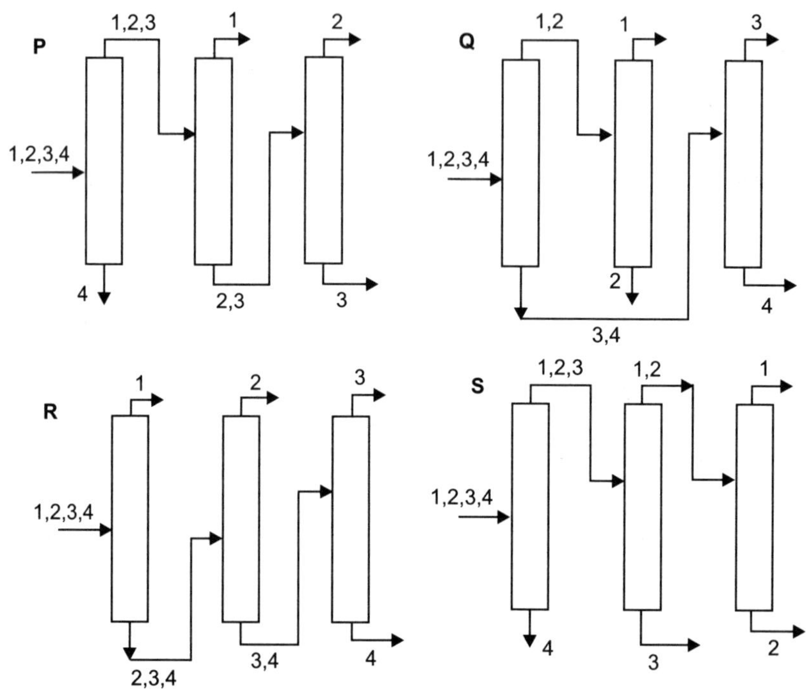

Fig. (iv)

Component	K_i
1.	6
2.	3
3.	2.5
4.	1.1

Using the K_i values given above, the optimal Scheme is

A. P B. Q
C. R D. S

5.127 In a distillation operation, what is the effect of temperature of the reflux stream (given below) on the condenser and reboiler loads?

Reflux conditions:

(i) Reflux stream is completely liquid and is at its bubble point.
(ii) Reflux stream is below its bubble point.

A. Condenser and reboiler loads are the same in both the cases.
B. Reboiler load is the same in both the cases but condenser load is higher in case (ii).
C. Condenser load is the same in both the cases but reboiler load is higher in case (ii).
D. Both condenser and reboiler loads are higher in case (ii) as compared to case (i).

5.128 A distillation column with *N* plates is being operated under normal conditions. At some point in time the operation is shifted to total reflux conditions (i.e., no product and residue are being withdrawn and feed to the column is stopped). At the new steady state

A. composition of vapours and that of liquid do not vary throughout the column
B. reboiler load and condenser load are minimum.
C. the top and bottom compositions are unchanged with and without total reflux
D. the top and bottom compositions correspond to the maximum enrichment achievable

5.129 An aqueous solution of methanol is to be distilled in a tray column. High pressure steam is available as a source of heat. For a given reflux ratio and overhead composition, two options are being explored; (i) a reboiler is used and (ii) no reboiler is used but steam is fed directly to the bottom of the column. As compared to option (i), in option (ii)

A. less number of trays are required.
B. composition of the residue remains unchanged.
C. more number of trays are required but the residue composition remains unchanged.
D. more number of trays required and the residue composition is more dilute in methanol.

5.130 A distillation column at a pilot plant is scaled up by 3 times for industrial use at steady state. After scale up

A. the number of theoretical trays increases by 3 times.
B. the minimum reflux ratio is increased by 3 times.
C. the feed flow rate and product flow rates are increased by 3 times.
D. the feed composition and product composition are increased by 3 times.

5.131 A separation column for vapour liquid contact processes 200 kmol/h of vapour. The flooding velocity is 3 m/s. If the column operates at 85% of the flooding velocity and the downcomer area is 10% of the total cross sectional area, what is the diameter of the column? Average density of vapour is 2 kg/m^3 and its molecular weight is 44.

A. 0.82 m B. 0.72 m
C. 0.78 m D. 1 m

5.132 For a fixed number of ideal stages in a distillation column, as the reflux ratio is increased, the difference in composition between the top and bottom product streams

A. increases B. decreases
C. remains unaffected D. passes through a maximum

5.133 In a distillation operation, what is the effect of the temperature of the reflux stream (given below) on the condenser and reboiler loads?

Reflux conditions

1. Reflux stream is completely liquid and is at its bubble point
2. Reflux stream is below its bubble point

A. Condenser and reboiler loads are the same in both cases
B. Reboiler load is the same in both cases but condenser load is higher in case (2)
C. Condenser load is the same in both cases but the reboiler load is lower in case (2).
D. The internal reflux ratio is higher in case (2) as compared to case (1).

5.134 In distillation column design, the McCabe-Thiele procedure is inadequate and Ponchon-Savarit procedure is needed when

A. saturated feed is not used
B. an azeotrope is formed
C. the difference in the latent heats of vaporisation of the components is very large
D. a total condenser is used

5.135 Minimum reflux ratio in a distillation column results in

A. Optimum number of trays B. Minimum reboiler size
C. Maximum condenser size D. Minimum number of trays

5.136 In distillation column sizing calculations by short cut methods, match the following

P Underwood's Equation	1. Number of real trays
Q Fenske's equation	2. Column diameter
R Gilliland's equation	3. Minimum number of ideal trays
S Vapour velocity at flooding	4. Actual number of ideal trays
	5. Minimum reflux ratio
	6. Tray efficiency

A. P-1, Q-3, R-4, S-6 B. P-2, Q-5, R-1, S-3
C. P-5, Q-3, R-6, S-2 D. P-5, Q-3, R-4, S-2

5.137 A distillation column separates 10000 kg/h of a benzene toluene mixture as shown in Fig. (v). In the figure z_F, x_D, x_B represent the weight fraction of benzene in the feed, distillate, and residue, respectively.

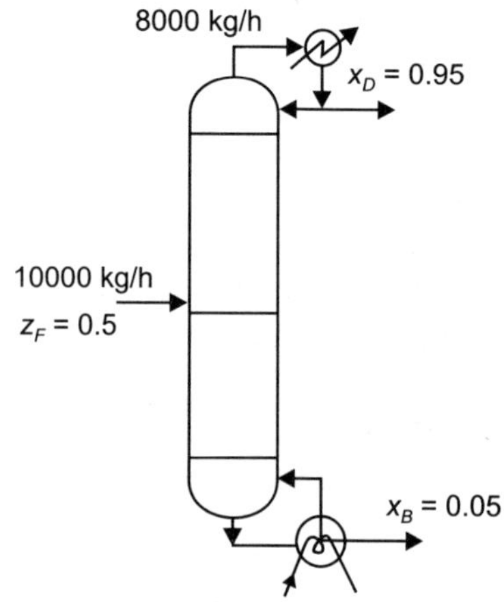

Fig. (v)

The reflux ratio is

A. 0.5
B. 0.6
C. 1.0
D. 2.0

5.138 The feed is admitted to a distillation column at a rate of 200 kmol/h and is 60% vaporised. The distillate is withdrawn at 120 kmol/h with 0.95 mole fraction of the more volatile component. Steam consumption (latent heat = 2200 kJ/kg) in the reboiler is 1500 kg/h. Latent heat of column liquid may be assumed to be 33000 kJ/kmol. What is the operating reflux ratio for the column?

A. 1.667
B. 0. 600
C. 0.833
D. 1.20

5.139 The feed is admitted to a distillation column at a rate of 200 kmol/h and is 60% vaporised. The distillate is withdrawn at 120 kmol/h with 0.95 mole fraction of the more volatile component. Steam consumption (latent heat = 2200 kJ/kg) in the reboiler is 1500 kg/h. Latent heat of column liquid may be assumed to be 33000 kJ/kmol. The mole fraction of component A in the liquid leaving a plate in the enriching section is found to be 0.7. What is the mole fraction of A in the vapour coming to that plate?

A. 0.93
B. 0.84
C. 0.70
D. 0.60

5.140 A binary feed mixture containing equimolar quantities of components *S* and *T* is to be distilled in a fractionating tower at atmospheric pressure. The distillate contains 96 mol% S. The *q*-line intersects the equilibrium curve at $x' = 0.46$ and $y' = 0.66$, where x' and y' are mole fractions. Assume that the McCabe-Thiele method is applicable and the relative volatility is constant. The minimum reflux ratio is

A. 1.6
B. 1.5
C. 0.66
D. 0.6

5.141 A binary feed mixture containing equimolar quantities of components S and T is to be distilled in a fractionating tower at atmospheric pressure. The distillate contains 96 mol% S. The *q*-line intersects the equilibrium curve at $x' = 0.46$ and $y' = 0.66$, where x' and y' are mole fractions. Assume that the McCabe-Thiele method is applicable and the relative volatility is constant. The feed is

A. at dew point
B. at bubble point
C. superheated vapour
D. partially vaporised.

5.142 On the enthalpy-composition diagram, the saturated vapour and liquid enthalpy curves are straight parallel lines, if

A. the latent heats of components are equal
B. the heat of solution is negligible
C. the heat capacities are constant
D. all the above are true

5.143 A tie line on the *H-x-y* diagram connects

A. the liquid and vapour phases passing each other between two adjacent plates
B. the liquid and vapour phases leaving a plate
C. the feed point and the difference points
D. the feed point *F*, distillate point *D* and the bottoms point *B*.

5.144 The saturated liquid and vapour streams below plate *n* are L_n and G_{n+1}, respectively. The difference point for the enriching section is Δ_D [Fig. (vi)].

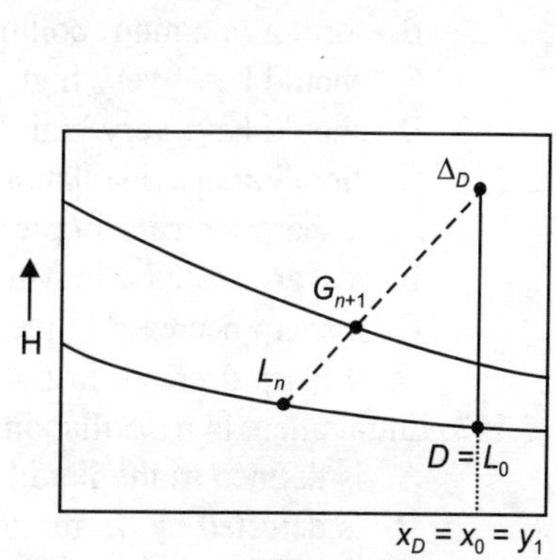

Fig. (vi)

The internal reflux ratio $\frac{L_n}{G_{n+1}}$ is

A. $\frac{\Delta_D G_{n+1}}{\Delta_D L_n}$ B. $\frac{\Delta_D L_n}{\Delta_D G_{n+1}}$

C. $\frac{L_n G_{n+1}}{\Delta_D L_n}$ D. $\frac{L_n G_{n+1}}{\Delta_D G_{n+1}}$

5.145 Under total reflux conditions, one of the following statements is wrong.
A. The condenser heat duty is infinity.
B. Condenser and reboiler heat duties are infinity.
C. Number of plates is the maximum.
D. There is no continuous addition of feed or withdrawal of products.

5.146 In most cases, the minimum reflux ratio in Ponchon-Savarit method is determined by
A. positioning the difference points at infinity
B. the intersection of the extension of the tie line passing through the feed with $x = x_D$
C. the lowest tie line intersection with $x = x_D$
D. the highest tie line intersection with $x = x_B$

5.147 Solvent used in extractive distillation
A. alters the relative volatility of the original substance
B. is of low volatility
C. must form no azeotrope with the original substance
D. all of the above

5.148 Alcohol is dehydrated using
A. extractive distillation B. azeotropic distillation
C. steam distillation D. molecular distillation

5.149 In extractive distillation, the solvent is
A. added to alter the relative volatility of the mixture
B. of high volatility
C. is present in overhead stream
D. is of high viscosity to provide high tray efficiency

5.150 Entrainer used in azeotropic distillation of binary mixture
A. form no azeotrope with the original substance
B. form a minimum boiling azeotrope with one or both of the feed components
C. should be of very high viscosity to provide high tray efficiency
D. should have very high latent heat of vaporisation

5.151 The flooding in a distillation column is detected by
A. a sharp increase in pressure drop
B. a sharp increase in Murphree plate efficiency
C. a sharp decrease in pressure drop
D. a sharp decrease in liquid hold up in the column

5.152 Entrainment in a distillation column
A. is defined as the liquid carried with the vapour from one tray to the tray above
B. is detected by decrease in Murphree plate efficiency
C. may be detrimental since it can carry non-volatile constituents upward
D. all of the above

5.153 When gas velocity is low, the liquid drains through the perforations on the sieve tray. This condition is known as

A. loading B. flooding
C. weeping D. channelling

5.154 Inside the distillation column

A. the pressure decreases gradually from bottom to the top of the column
B. the higher temperature is at the top of the column
C. the pressure increases gradually from bottom to the top of the column
D. the pressure remains constant

5.155 The minimum residence time for the liquid in the downspout of a distillation column is

A. 8 minutes B. 1 minute
C. 8 seconds D. 80 seconds

5.156 For a given type of tray at flooding the superficial velocity of gas v_F is related to fluid densities by

A. $u_F = C_F\left(\dfrac{\rho_L - \rho_G}{\rho_G}\right)^{3/2}$ B. $u_F = C_F\left(\dfrac{\rho_L - \rho_G}{\rho_G}\right)^{1/2}$

C. $u_F = C_F\left(\dfrac{\rho_L - \rho_G}{\rho_L}\right)^{1/2}$ D. $u_F = C_F\left(\dfrac{\rho_G}{\rho_L}\right)^{1/2}$

5.157 The recommended tray spacing for distillation tower of diameters up to 4 ft is

A. 36 inches B. 30 inches
C. 15 inches D. 18 to 20 inches

5.158 In atmospheric distillation tower, the pressure drop in a tray is about

A. 0.07–0.12 psi B. 0.15–0.20 psi
C. 1–2 mm Hg D. none of the above

5.159 To ensure reasonably uniform distribution of liquid flow on single pass trays in the distillation column, a weir length used is about … ……percent of the tower diameter

A. 25 to 30 B. 60 to 75
C. 15 to 20 D. 5 to 15

5.160 In distillation column, the number of bubble caps per tray primarily depends on the

A. allowable liquid velocity B. allowable gas velocity
C. allowable gas and liquid velocities D. feed composition

5.161 Channelling is most severe in towers which contain

A. random packing with crushed solids B. stacked packing
C. dumped packing of regular units D. all of the above

5.162 The diameter of the distillation column is decided by

A. the depth of the downcomer B. number of sieve perforations
C. allowable vapour velocity D. number of theoretical stages

5.163 Inside a distillation column

A. the driving force for the vapour flow is the pressure drop
B. the driving force for the liquid flow is its weight
C. the liquid and its vapour are always at their bubble point and dew point respectively
D. all of the above

5.164 The flooding velocity in a plate column, operating at 1 atm pressure is 3 m/s. If the column is operated at 2 atm pressure, under otherwise identical conditions, the flooding velocity will be

A. $3/\sqrt{2}$ B. $3/2$
C. 1 D. $3/4$

5.165 Murphree efficiency is concerned with the performance of separation

A. at a specific location on a plate B. in the entire column
C. on a single plate D. none of the above

5.166 Murphree plate efficiency (E_{MV}) is defined as

A. $E_{MV} = \dfrac{y_n - y_{n+1}}{y_n^* - y_{n+1}}$ B. $E_{MV} = \dfrac{y_n - y_{n+1}}{y_n - y_{n+1}^*}$

C. $E_{MV} = \dfrac{x_n - x_{n+1}}{x_n^* - x_{n+1}}$ D. $E_{MV} = \dfrac{y_n - y_{n+1}}{y_n^* - y_{n-1}}$

5.167 For the nth tray (counted from the bottom of a distillation column), the Murphree tray efficiency is given by

A. $\dfrac{y_{n+1} - y_n}{y_n^* - y_{n-1}}$ B. $\dfrac{y_n - y_{n-1}}{y_n^* - y_{n-1}}$

C. $\dfrac{y_{n-1} - y_n}{y_{n+1} - y_n}$ D. $\dfrac{y_n^* - y_{n-1}}{y_n^* - y_{n+1}}$

5.168 Benzene-toluene mixture ($\alpha = 2.5$) is being fractionated in a plate column. Under total reflux conditions, the liquid from two adjacent plates analyzed 0.4 and 0.5 mole fraction benzene. Determine the Murphree vapour efficiency of the lower plate.

A. 70% B. 53.2%
C. 44.4% D. 92.0%

5.169 In small columns, local efficiency is ....to Murphree efficiency; in large column local efficiency is ....Murphree efficiency

A. equal to, less than B. equal to, more than
C. less than, equal to D. less than, more than

5.170 In batch distillation with constant reflux, overhead product composition

A. increase with time B. decreases with time
C. does not vary with time D. none of the above

5.171 The reason for preferring packed towers over plate towers in distillation practice is that packed tower operation gives

A. low pressure drop and high hold up
B. high pressure drop and low hold up
C. low pressure drop and low hold up
D. high pressure drop and high hold up

EXERCISES

VLE of ideal solutions

5.1 At 303 K the vapour pressures of benzene (**A**) and toluene (**B**) are 15.75 kPa and 4.89 kPa respectively. Determine the partial pressures and weight composition of the vapour in equilibrium with a liquid mixture consisting of equal weights of the two components.

(**Answer:** $\bar{p}_{benzene} = 8.52$ kPa, $\bar{p}_{toluene} = 2.24$ kPa, 76.3 (wt)% benzene)

5.2 *n*-Heptane and toluene form ideal solution. At 373 K their vapour pressures are 106 and 74 kPa respectively. Determine the composition of the liquid and vapour in equilibrium at 373 K and 101.3 kPa. (**Answer:** $x = 0.853, y = 0.893$)

5.3 An equimolar mixture of benzene and toluene is contained in a piston cylinder arrangement at a temperature T. What is the pressure below which this mixture will exist as a vapour phase alone? At the given T, the vapour pressures of benzene and toluene are 203.9 and 85.31 kPa respectively. Assume that Raoult's law is valid. **(Answer:** 120.29 kPa)

5.4 An equimolar solution of benzene and toluene is totally evaporated at a constant temperature of 363 K. At this temperature, the vapour pressures of benzene and toluene are 135.4 and 54 kPa respectively. What are the pressures at the beginning and at the end of the vaporisation process? **(Answer:** 94.7 kPa, 77.21 kPa)

5.5 Two substances **A** and **B** are known to form ideal liquid solutions. A vapour mixture containing 50 mol% **A** and 50 mol% **B** is at 311 K and 101.3 kPa. This mixture is compressed isothermally until condensation occurs. At what pressure does condensation occur and what is the composition of the liquid that forms? The vapour pressures of **A** and **B** are 142 kPa and 122 kPa respectively. **(Answer:** 131.24 kPa, $x = 0.462$)

5.6 The vapour pressures of components **A** and **B** can be evaluated by the Antoine equations

$$\ln P_A^S = 14.5463 - \frac{2940.46}{T - 35.93}$$

$$\ln P_B^S = 14.2724 - \frac{2945.47}{T - 49.15}$$

where T is in K and P is in kPa. Assuming that the solutions formed by these are ideal, calculate the bubble point and dew point temperature of a mixture containing 40 mol% **A** at 65 kPa. **(Answer:** 330.4 K, 334 K)

5.7 The vapour pressures of components **A** and **B** which form ideal solutions can be evaluated by the Antoine equations

$$\ln P_A^S = 14.5463 - \frac{2940.46}{T - 35.93}$$

$$\ln P_B^S = 14.2724 - \frac{2945.47}{T - 49.15}$$

where T is in K and P is in kPa. Calculate the bubble point and dew point pressures of a mixture containing 40 mol% **A** at 327 K. **(Answer:** 57.63 kPa, 50.09 kPa)

5.8 The vapour pressures of components **A** and **B** can be evaluated by the Antoine equations

$$\ln P_A^S = 14.5463 - \frac{2940.46}{T - 35.93}$$

$$\ln P_B^S = 14.2724 - \frac{2945.47}{T - 49.15}$$

where T is in K and P is in kPa. Assuming that the solutions formed by these are ideal, calculate the fraction of the system that is liquid and the composition of the liquid and vapour in equilibrium at 327 K and 65 kPa when the overall composition of the system is 70 mole percent A.

(Answer: 20% of the feed is liquid, $x = 0.5608$, $y = 0.7344$)

5.9 The binary system of components **A** and **B** conforms closely to Raoult's law. Using the vapour pressure data given below prepare the following

(a) Plot the *P-x-y* diagram at 323 K

(b) Plot the boiling point diagram at 53.32 kPa

T, K	311.45	315	319	323	327	331	335.33
P_A^S, kPa	53.32	61.09	70.91	81.97	94.36	108.2	124.95
P_B^S, kPa	21.25	24.61	28.90	33.79	39.35	45.62	53.32

5.10 Mixtures of *n*-heptane (**A**) and *n*-octane (**B**) are expected to behave ideally. The total pressure over the system is 101.3 kPa. Using the vapour pressure data given below

(a) Construct the boiling point diagram and

(b) Construct the Equilibrium diagram

(c) Deduce an equation for the equilibrium diagram using an arithmetic average a value.

T, K	371.4	378	383	388	393	398.6
P_A^S, kPa	101.3	125.3	140.0	160.0	179.9	205.3
P_B^S, kPa	44.4	55.6	64.5	74.8	86.6	101.3

5.11 Assuming the Raoult's law to be valid for the system benzene (**A**)-ethyl benzene (**B**) and the vapour pressures are given by the Antoine equations

$$\ln P_A^S = 13.8858 - \frac{2788.51}{T - 52.36}$$

$$\ln P_B^S = 14.0045 - \frac{3279.47}{T - 60.00}$$

where P is in kPa and T is in K. Construct the *T-x-y* diagram at 101.3 kPa.

5.12 Plot the boiling point diagram and equilibrium diagram for benzene-toluene system at 101.3 kPa pressure assuming that the relative volatility is constant and using the relative volatility evaluated at the average of the boiling points of the pure liquids. The Antoine constants are:

	A	*B*	*C*
Benzene	13.8858	2788.51	52.36
Toluene	13.9987	3096.52	53.67

VLE of nonideal solutions

5.13 Under atmospheric pressure the **A** and **B** forms an azeotrope which boils at 340 K and contains 34.5 mol% **A**. The vapour pressure of acetone at 340 K is 131.5 kPa and that of **B** is 112.7 kPa. Assume that the van Laar equations are applicable.

(a) Calculate the composition of the vapour in equilibrium with a liquid analysing 10.0 mol% A at a temperature 340 K.

(b) What is the total pressure at this condition?

(**Answer:** (a) 6.7 mol% (b) 107.5 kPa)

5.14 Ethyl alcohol and hexane form an azeotrope at 33.2 mol% ethanol. It boils at 331.9 K at 101.3 kPa. At 331.9 K the vapour pressures are 44.25 kPa for ethanol and 72.24 kPa for hexane. Determine

(a) The van Laar constants.

(b) The vapour composition for a solution containing 50 mol% hexane boiling at 331.9 K

(c) The total pressure for the conditions in part (b)

(**Answer:** (a) 1.6625, 2.7475 (b) 67.81 mol% (c) 101.56 kPa)

5.15 From vapour-liquid equilibrium measurements for ethanol-benzene system at 318 K and 40.25 kPa it is found that the vapour in equilibrium with a liquid containing 38.4 mol% benzene contained 56.6 mol% benzene. At 318 K, the vapour pressures of ethanol and benzene are 22.9 and 29.6 kPa respectively. Determine the composition of the vapour and total pressure in equilibrium with a liquid containing 50 mol% benzene. Assume that van Laar equation is applicable for the system. (**Answer:** $y = 0.5884$, 40.78 kPa)

5.16 Using the van Laar method, calculate the vapour-liquid equilibrium compositions for acetone (**A**)-chloroform (**B**) system at a pressure of 101.3 kPa. At this pressure the system forms an azeotrope of composition 65.5 mol% chloroform which boils at 337.7 K. The vapour pressures of the pure components are given below.

T, K	318	323	333	343	353
P_A^S, kPa	66.4	79.67	112.6	155.6	210.5
P_B^S, kPa	57.4	68.8	96.7	133	179.1

The normal boiling points of acetone and chloroform are respectively 329.8 K and 334.4 K.

5.17 Using the van Laar method, calculate the vapour-liquid equilibrium compositions for acetone (**A**)-chloroform (**B**) system at a pressure of 101.3 kPa. At this pressure the system forms an azeotrope of composition 65.5 mol% chloroform which boils at 337.7 K. The vapour pressures of the pure components are given by the Antoine equations:

$$\ln P_A^S = 14.39155 - \frac{2795.817}{T - 43.798}$$

$$\ln P_B^S = 13.9985 - \frac{2696.2663}{T - 46.968}$$

5.18 For the binary system n-pentanol (**A**)-n-hexane (**B**), determine the activity coefficients at 313 K in an equimolal mixture given the Wilson parameters as follows:

$$a_{AB} = 7194.18 \text{ J/mol}, \; a_{BA} = 697.52 \text{ J/mol}$$
$$v_A = 109.2 \times 10^{-6} \text{ m}^3\text{/mol}, \; v_B = 132.5 \times 10^{-6} \text{m}^3\text{/mol}.$$

(**Answer:** $\gamma_A = 1.355, \gamma_B = 1.682$)

5.19 For the n-hexane (**A**) – ethanol (**B**) system the liquid phase activity coefficients are represented by the Wilson equations:

$$\ln \gamma_A = -\ln(x_A + \Lambda_{AB}\, x_B) + x_B\left[\frac{\Lambda_{AB}}{x_A + \Lambda_{AB}\, x_B} - \frac{\Lambda_{BA}}{\Lambda_{BA}\, x_A + x_B}\right]$$

$$\ln \gamma_B = -\ln(x_B + \Lambda_{BA}\, x_A) - x_A\left[\frac{\Lambda_{AB}}{x_A + \Lambda_{AB}\, x_B} - \frac{\Lambda_{BA}}{\Lambda_{BA}\, x_A + x_B}\right]$$

The Wilson parameters Λ_{AB} and Λ_{BA} are found to be 0.281 and 0.041, respectively. Assuming the parameters to be temperature independent, calculate the composition of the vapour in equilibrium with a solution containing 70 mol% n-hexane and the equilibrium temperature. The Antoine constants are

n-Hexane (**A**)	13.8216	2697.55	48.78
Ethyl alcohol (**B**)	16.5092	3578.91	50.50

(**Answer:** $x = 0.70$, $y = 0.6734$, $T = 331.09$ K)

5.20 The Wilson parameters for acetone (**A**)-methanol (**B**) are obtained from the following values:

$$a_{AB} = -712.51 \text{ J/mol}, \; a_{BA} = 2487.71 \text{ J/mol}$$
$$v_A = 74.05 \times 10^{-6} \text{ m}^3\text{/mol}, \; v_B = 40.73 \times 10^{-6} \text{ m}^3\text{/mol}.$$

The vapour pressures are given by the Antoine equations:

$$\ln P_A^S = 14.39155 - \frac{2795.817}{T - 43.798}$$

$$\ln P_B^S = 16.5725 - \frac{3626.55}{T - 34.29}$$

Calculate the total pressure and vapour composition in equilibrium with a liquid containing 40 mol% acetone at 333 K.

(**Answer:** 111.5 kPa, $y = 0.5$)

5.21 For the 2-propanol (**A**)-water (**B**) system, the following Wilson parameters are reported.

$$a_{AB} = 1833.74 \text{ J/mol}, a_{BA} = 5183.26 \text{ J/mol},$$
$$v_A = 76.92 \times 10^{-6} \text{ m}^3\text{/mol}, v_B = 18.07 \times 10^{-6} \text{m}^3\text{/mol}.$$

The vapour pressures can be calculated by the Antoine equations, which are given below:

$$\ln P_A^S = 16.678 - \frac{3640.20}{T - 53.54}$$

$$\ln P_B^S = 16.26205 - \frac{3799.887}{T - 46.854}$$

where T is in K and the vapour pressures are in kPa. Calculate for an equimolar solution at 350 K

(a) The bubble point pressure and vapour composition

(b) The dew point pressure and liquid composition

(**Answer:** (a) 86.13 kPa, $y = 0.6182$ (b) 77.28 kPa, $x = 0.2322$)

5.22 Wilson equation may be used to predict the activity coefficients of methanol (**A**)-benzene (**B**) system with following parameters.

$$\Lambda_{AB} = 2.1952 \exp\left[-\frac{862.2043}{T}\right]$$

$$\Lambda_{BA} = 0.4555 \exp\left[-\frac{94.177}{T}\right]$$

The Antoine Constants are

	A	B	C
Methanol	16.5725	3626.55	34.29
Benzene	13.8858	2788.51	52.36

Determine (a) The bubble-point temperature and equilibrium vapour composition of a solution containing 20% (mole) methanol at 101.3 kPa and (b) The dew point temperature and equilibrium liquid composition of a vapour containing 40% (mole) methanol at 101.3 kPa.

(**Answer:** (a) 332.87 K, $y = 0.54$ (b) 338.5, $x = 0.0582$)

5.23 Calculate and plot the boiling point diagram and equilibrium diagram for the mixture of benzene (**A**) and 4-methyl-2-pentanol (**B**) at 101.3 kPa. Treat the system as consisting of non-ideal liquid and ideal vapour and neglect Poynting correction. Antoine constants are

	A	B	C
Benzene	13.8858	2788.51	52.36
4-methyl-2-pentanol	17.4813	5007.85	15.38

where the Antoine equation is $\ln P^s = A - \dfrac{B}{T - C}$ (with P in kPa and T in K). The Wilson parameters are

$$\Lambda_{AB} = 1.4241 \exp\left[-\frac{107.918}{T}\right]$$

$$\Lambda_{BA} = 0.7022 \exp\left[-\frac{183.26}{T}\right]$$

5.24 Calculate and plot the boiling point diagram and equilibrium diagram for the acetone-water system at 101.3 kPa. Treat the system as consisting of non-ideal liquid and ideal vapour and neglect Poynting correction. Antoine constants are

	A	B	C
Water	16.26205	3799.887	46.854
Acetone	14.39155	2795.817	43.198

where the Antoine equation is $\ln P^s = A - \frac{B}{T-C}$ (with P in kPa and T in K). The Wilson parameters are

$$\Lambda_{AB} = 0.2440 \exp\left[-\frac{246.4360}{T}\right]$$

$$\Lambda_{BA} = 4.0980 \exp\left[-\frac{716.5117}{T}\right]$$

Bubble-point and dew-point calculation of multicomponent systems

5.25 A mixture contains 45 mol% methanol (**A**), 30 mol% ethanol (**B**) and the rest n-propanol (**C**). Liquid solution may be assumed to be ideal and perfect gas law is valid for the vapour phase. Calculate at a total pressure of 101.3 kPa

(a) The bubble-point temperature and the vapour composition

(b) The dew-point temperature and the liquid composition.

The vapour pressures of the pure liquids are given below:

Temperature, K	333	343	353	363
P_A^S, kPa	81.97	133.29	186.61	266.58
P_B^S, kPa	49.32	73.31	106.63	166.61
P_C^S, kPa	39.32	62.65	93.30	133.29

(**Answer:** (a) $T = 344.6$ K, $y_A = 0.610$, $y_B = 0.226$, $y_C = 0.162$
(b) $T = 347.5$ K, $x_A = 0.297$, $x_B = 0.357$, $x_C = 0.345$)

5.26 A solution of hydrocarbons contains n-propane 5.0%, n-butane 30.0%, n-pentane 40.0% and n-hexane 25.0%. Compute the bubble-point and the dew-point at 350 kPa. Take K-values from the DePriester nomographs. (**Answer:** 329.7 K, 356.9 K)

5.27 The bottom product from a distillation column has the following composition:
i-butane : 3.19%, n-Butane : 79.93%, i-Pentane : 10.40% and n-Pentane : 6.48%.
Estimate the temperature of the mixture if the bottom pressure is 700 kPa. (**Answer:** 341 K)

5.28 Assuming ideal liquid solutions determine the bubble point temperatures and equilibrium vapour composition of mixture containing 35% (mole) n-hexane(**A**), 40% (mole) n-heptane(**B**), and 25% (mole) n-Octane(**C**) at 101.3 kPa.

Temperature, K	290	300	310	320	330	340	350	360	370
P_A^S, kPa	14.26	22.86	34.38	51.63	72.45	96.78	136.8	177.25	222.6
P_B^S, kPa	4.53	7.18	11.56	19.13	25.08	37.73	51.63	73.8	99.65
P_C^S, kPa	1.17	2.09	3.82	6.47	9.61	14.26	21.93	30.87	45.18

Repeat the calculations using K-factors taken from the DePriester nomographs.

(**Answer:** 360.9K, $y = 0.6274, 0.3007, 0.0720$; 362.1K, $y = 0.616, 0.298, 0.092$)

5.29 Calculate the bubble point temperature and dew point temperature of a mixture containing 10% ethane, 20% propane, 55% n-butane and 15% *n*-pentane at 1450 kPa. Use *K*-values taken from nomographs. **(Answer:**334 K, 370.7 K)

5.30 A distillation column is charged with a feed containing 6% (mole) propane, 65% (mole) isobutene, 26% (mole) *n*-butane and 3% (mole) *n*-pentane. The overhead product should contain not more than 5% of *n*-butane in the feed and the bottom product should contain not more than 5% of the isobutene in the feed. Assume complete separation of propane and pentane. Assuming bubble point condensate, an accumulator temperature of 311 K, and a feed at its bubble point, calculate
(a) Accumulator pressure
(b) Top Vapour temperature
(c) Reboiler temperature

(Answer: (a) 608 kPa (b) 313.8 kPa (c) 332.6 K)

Flash distillation

5.31 A liquid mixture containing 65 mol% benzene and 35 mol% toluene is subjected to flash vaporisation at 363 K and 101.3 kPa. The vapour pressure of benzene at this temperature is 136.09 kPa and the vapour pressure of toluene is 54.21 kPa. Flash vaporisation is essentially an equilibrium stage operation. Calculate
(a) The exit liquid composition (b) The exit vapour composition
(c) The mole percent of the feed that is vaporised.

(Answer: (a) 0.575 (b) 0.7726 (c) 37.9%)

5.32 A vapour containing 30 mol% benzene and the rest toluene is partially condensed at 101.3 kPa and 373.2 K. Determine the fraction of the vapour liquefied and composition of the liquid and vapour produced. The vapour pressure data as follows:

T, K	361	363	367	371	373	377
P^S (Benzene), kPa	128.4	136	152.6	170.5	180	200.4
P^S (Toluene), kPa	50.8	54.2	61.6	69.8	74.2	83.6

(Answer: $f = 0.25$, $x_W = 0.2507$, $y_D = 0.4467$)

5.33 An equimolar mixture of *n*-heptane (**A**) and *n*-octane (**B**) is continuously flash vaporised at atmospheric pressure to vaporise 60% of the feed. Determine:
(a) Composition of the products
(b) Temperature in the separator
(c) The percent of the feed to be vaporised to produce a distillate containing 65 mol% heptane.
The mole fraction of heptane in the liquid and vapour in equilibrium at 101.3 kPa is as follows:

T, K	398.8	393	388	383	378	371.6
x	0	0.1571	0.312	0.487	0.655	1.0
y	0	0.279	0.492	0.674	0.810	1.0

(Answer: (a) $x_W = 0.388$, $y_D = 0.572$ (b) 385.7 K (c) 18.92%)

5.34 An equimolar mixture of benzene and toluene is subjected to flash distillation at 101.3 kPa. Determine the composition of the products and temperature in the separator for a fractional vaporisation of 40%.

T, K	383.8	379.6	375.5	371.8	368.4	365.7	362.7	360.2	357.8	355.6	354.4	353.3
x	0	0.1	0.2	0.3	0.4	0.5	0.6	0.7	0.8	0.9	0.95	1
y	0	0.205	0.369	0.502	0.618	0.706	0.789	0.849	0.907	0.955	0.978	1

(Answer: $x_W = 0.41$, $y_D = 0.63$ (b) 368.2 K)

5.35 One hundred kmol of a mixture of *n*-pentane and *n*-heptane containing 60 mol% pentane is partially vaporised at 101.3 kPa in order to produce 45 kmol distillate. The mole fraction of *n*-pentane in the liquid and vapour in equilibrium at 101.3 kPa are given below:

x	0	0.059	0.145	0.254	0.398	0.594	0.867	1.00
y	0	0.271	0.521	0.701	0.836	0.925	0.984	1.00

Calculate the composition of the products. (**Answer:** $x_W = 0.40$, $y_D = 0.84$)

5.36 A feed consisting of 50 mol% **A** and 50 mol% **B** is flash vaporised at 13.3 kPa. If the residue contained 60 mol% **B** and the feed rate is 20 kmol/h, determine

(a) The composition of the distillate

(b) Flow rates of distillate and residue

Equilibrium data at 13.3 kPa is as follows:

x	0	0.1	0.2	0.3	0.5	0.6	0.8	0.9	1.0
y	0	0.2	0.35	0.47	0.70	0.78	0.91	0.96	1.0

(**Answer:** (a) $y_D = 0.584$ (b) $D = 10.87$ kmol/h $W = 9.13$ kmol/h)

5.37 A solution has the following composition in mol%: ethane 0.25%, propane 25.00%, isobutane 18.5%, *n*-butane 56.0%, isopentane 0.25%. For a pressure of 10 bars, calculate

(a) The bubble-point temperature

(b) The dew-point temperature

(c) The composition of the liquid and vapour when 40% of the mixture is vaporised.

(**Answer:** (a) 330 K (b) 340.6 K (c) $x_i = 0.0009, 0.1745, 0.1874, 0.6335, 0.0038$; $y_i = 0.0048, 0.3630, 0.1818, 0.4498, 0.0013$)

5.38 A mixture containing 45 mol% acetone, 35 mol% acetonitrile and 20 mol% nitromethane is partially vaporised at 353.2 K and 110 kPa pressure. Determine the fraction of the feed vaporised and composition of the vapour and liquid products. The vapour pressures are given by the Antoine equations $\ln P^s = A - \dfrac{B}{T - C}$ where pressure is in kPa and temperature in K, with following constants:

Material	A	B	C
Acetone	14.39155	2795.817	43.198
Acetonitrile	14.2724	2945.47	49.2
Nitromethane	14.2043	2972.64	64.2

(**Answer:** 0.8291, $x_i = 0.2508, 0.3854, 0.3638$ $y_i = 0.4911, 0.3427, 0.1662$)

5.39 One hundred kmol/h of a liquid mixture containing 60 mol% benzene, 25 mol% toluene and 15 mol% *o*-xylene is subjected to flash vaporisation at 101.3 kPa and 373 K. Determine the amount of distillate and residue products and their compositions. Assume that the solution obeys the Raoult's law. The vapour pressures of benzene, toluene and xylene are, respectively, 183 kPa, 73 kPa and 27 kPa.

(**Answer:** $D = 69.25$ kmol/h, $y_i = 0.697, 0.223, 0.080$; $W = 30.75$ kmol/h, $x_i = 0.386, 0.310, 0.304$)

5.40 One hundred kmol/h of a hydrocarbon mixture containing 20 mol% *n*-butane, 50 mol% *n*-pentane, and 30 mol% *n*-hexane is subjected to flash vaporisation at 395 K and a pressure of 1013 kPa. Find the composition of the products and their flow rates. The vaporisation equilibrium constant values are 2.13, 1.10 and 0.59 for butane, pentane and hexane, respectively.

(**Answer:** $D = 70.4$ kmol/h, $y_i = 0.2372, 0.5139, 0.2489$; $W = 29.6$ kmol/h, $x_i = 0.1114, 0.4672, 0.4219$)

5.41 A liquid mixture consisting of 20% **A**, 50% **B**, 30% **C** is to be flashed at 500 kPa so that 50% of the feed is vaporised.

The vapour pressures are given by the Antoine equations $\ln P^s = A - \dfrac{B}{T - C}$ where pressure is in kPa and temperature in K, with following constants:

Material	A	B	C
A	16.5092	3578.91	50.50
B	15.81647	3310.416	74.737
C	16.6780	3640.2	53.54

Assuming ideal solution behavior, determine:

(a) The temperature of the mixture

(b) The bubble point and dew point of the mixture

(**Answer:** (a) 409.8 K (b) 408.4 K, 411K)

5.42 A vapour mixture containing 15% ethane, 20% propane, 60% isobutane and the rest n-butane is subjected to partial condensation so that 75% of the vapour is condensed. If the condenser temperature is 300 K determine the pressure. (**Answer:** 758 kPa)

5.43 A 100 kmol/h feed consisting of 10 mol% propane, 20 mol% n-butane, 30 mol% n-pentane and 40 mol% n-hexane is at 690 kPa and 365 K. What fraction of the mixture is liquid? If the liquid and vapour phases are allowed to separate what are their compositions? Take K-values from the DePriester nomographs.

(**Answer:** 0.884 fraction, $y_i = 0.307, 0.318, 0.229, 0.147$,
$x_i = 0.073, 0.185, 0.309, 0.433$)

5.44 One hundred kmol of a feed composed of 25 mol% n-butane, 40 mol% n-pentane, 35 mol% n-hexane are flashed at 390 K, so that 80% of the hexane is to be recovered in the residue, calculate (a) The pressure of operation (b) Compositions of the liquid and vapour.

(**Answer:** (a) 793 kPa (b) $y_i = 0.39, 0.40, 0.21$, $x_i = 0.18, 0.40, 0.42$)

5.45 One hundred kmol/h of an equimolar mixture of ethane, propane, n-butane, and n-pentane is subjected to a flash vaporisation at 340 K and 1410 kPa. Determine the quantities and compositions of liquid and vapour produced. Take K values as 4.0, 1.61, 0.60 and 0.24 for ethane, propane, butane and pentane respectively.

(**Answer:** $D = 48$ kmol/h, $y_i = 0.409, 0.311, 0.186, 0.094$;
$W = 52$ kmol/h, $x_i = 0.102, 0.193, 0.310, 0.395$)

5.46 A hydrocarbon mixture contains 25 mol% propane, 40 mol% n-butane and 35 mol% n-pentane at 1450 kPa. Assume ideal solution behaviour and calculate

(a) The bubble-point temperature and composition of the vapour

(b) The dew-point temperature and the composition of the liquid

(c) The temperature and the composition of the liquid and vapour in equilibrium when 45 mol% of the initial mixture is vaporised. Take K_i from nomographs.

(**Answer:** (a) 361 K, 0.53, 0.34, 0.13 (b) 387 K, 0.088, 0.320, 0.593
(c) 374.6 K, $y_i = 0.3697, 0.4130, 0.2173$; $x_i = 0.1521, 0.3894, 0.4586$)

Simple/Differential distillation

5.47 A vapour consists of components **A** and **B** which forms ideal solutions in the liquid state. If the vapour is differentially condensed, show that

$$\ln\frac{B_F}{B_D} = \alpha\ln\frac{A_F}{A_D}$$

where A_F and A_D are moles of **A** in the feed and distillate, respectively, and B_F and B_D are moles of **B** in the feed and distillate, respectively.

5.48 Ninety percent of the *n*-butane is to be recovered by simple distillation from a mixture containing 20 mol% *n*-butane and 80 mol% *n*-pentane. Determine for 100 kmol of feed
(a) the amount of pentane distilled
(b) the composition of the residue
Solve analytically using an average α value of 3.5

(**Answer:** (a) 38.56 kmol (b) 4.6% butane)

5.49 A mixture of *n*-heptane and *n*-octane containing 40 mol% *n*-heptane is differentially distilled so that a residue containing 30 mol% heptane remains in the still. If an average α value of 2.16 can be used to represent the equilibrium, determine the fraction of the feed vaporised and the composition of the distillate analytically. (**Answer:** 0.4144, $y_D = 0.5413$)

5.50 A mixture of 50 mol% **A** and 50 mol% **B** is subjected to differential distillation till one half of the liquid is vaporised. If the average volatility of the system is 2.5, calculate the composition of the liquid in the still and the overhead product. (**Answer:** $x_W = 0.346$, $y_D = 0.654$)

5.51 A liquid mixture consisting of acetonitrile-nitromethane is differentially distilled in a batch still at 70 kPa. The feed is 65 mol% acetonitrile. Distillation is continued till the liquid residue contains 35 mol% acetonitrile. The equilibrium is represented in terms of the mole fractions of acetonitrile by:

$$y = 1.05x + 0.13 \quad \text{for} \quad 0.3 \le x \le 0.52$$

$$y = 0.77x + 0.2756 \quad \text{for} \quad 0.52 \le x \le 0.8$$

Determine the average composition of the distillate.

(**Answer:** $y_D = 0.6946$)

5.52 An equimolar mixture of *n*-heptane and *n*-octane is subjected to simple distillation so that the liquid residue has a composition 30 mol% heptane. Determine
(a) the fraction of feed vaporised
(b) the composition of the distillate
Equilibrium data for *n*-heptane and *n*-octane at 101.3 kPa:

x	0	0.1571	0.312	0.487	0.655	1.0
y	0	0.279	0.492	0.674	0.810	1.0

(**Answer:** (a) 0.677 (b) 0.595)

5.53 A mixture of 40 mol% isopropanol in water is to be distilled at 101.3 kPa by simple batch distillation until 70 mol% of the charge has been distilled. What will be the composition of the liquid remaining in the still and of the distillate? The equilibrium data are:

x	0	0.0118	0.0841	0.1978	0.3496	0.4525	0.6794	0.7693	0.9442	1.00
y	0	0.2195	0.4620	0.5242	0.5686	0.5926	0.6821	0.7421	0.9160	1.00

(**Answer:** $x_W = 0.06$, $y_D = 0.5457$)

5.54 Five kmol of a mixture containing 30 mol% naphthalene and 70 mol% dipropylene glycol is differentially distilled at 13.3 kPa. Equilibrium data at 13.3 kPa in terms of mole fractions of naphthalene are given below:

x	0.084	0.116	0.28	0.506	0.687	0.806	0.88
y	0.223	0.411	0.629	0.748	0.802	0.844	0.88

(a) Determine the composition of the liquid residue if one-third of the feed is distilled off.
(b) Determine the composition of the residue and the amount of distillate produced if it is desired to produce a distillate of average composition 55 mol% naphthalene.

(**Answer:** (a) 0.16 (b) $x_W = 0.12$, $D = 2.116$ kmol)

5.55 One hundred kilograms of an aqueous solution containing 20 mol% phenol is subjected to differential distillation at 35 kPa so that the residue has composition 80 mol% phenol. Determine
(a) The amount of feed vaporised in kg
(b) The composition of the distillate in mol%

Equilibrium data at 35 kPa in terms of the weight fractions of water is

x'	0.015	0.050	0.069	0.077	0.196	0.284	0.397	0.830	0.900	0.934	0.957
y'	0.410	0.797	0.828	0.845	0.899	0.911	0.912	0.919	0.928	0.942	0.956

(**Answer:** (a) 45.94 kg (b) 97.69% water)

5.56 A liquid mixture containing 70 mol% acetone and 30 mol% acetic acid is subjected to differential distillation at 101.3 kPa. The equilibrium data in terms of the mole fractions of acetone at 101.3 kPa:

x	0.05	0.10	0.20	0.30	0.40	0.50	0.60	0.70	0.80	0.90
y	0.162	0.306	0.557	0.725	0.840	0.912	0.947	0.969	0.984	0.993

(a) What fraction of the charge must be vaporised to leave a liquid containing 30 mol% acetone?
(b) What will be the average composition of the distillate?

(**Answer:** (a) 0.65 (b) $y_D = 0.915$)

5.57 A liquid mixture containing 60 mol% acetone and 40 mol% water is differentially distilled at 101.3 kPa to vaporise 40% of the feed. The equilibrium data is given below:

x	0.10	0.15	0.20	0.30	0.40	0.50	0.60	0.70	0.80	0.90
y	0.755	0.798	0.815	0.830	0.839	0.849	0.859	0.874	0.898	0.935

Determine the composition of the products. (**Answer:** $x_W = 0.43$ and $y_D = 0.855$)

5.58 One hundred moles of a solution containing 40 mol% benzene and 60 mol% toluene is differentially distilled at atmospheric pressure to give a distillate product having 55 mol% benzene. Calculate the composition and amount of residual liquid. Equilibrium data at 101.3 kPa in terms of mole fractions of benzene:

x	0	0.1	0.2	0.3	0.4	0.5	0.6	0.7	0.8	0.9	1
y	0	0.205	0.369	0.502	0.618	0.706	0.789	0.849	0.907	0.955	1

(**Answer:** $x_W = 0.27$, $W = 53.46$ kmol)

5.59 One thousand kilograms of ethyl alcohol-water solution containing 50 weight% alcohol is subjected to simple distillation. If the residue contained 5 weight% alcohol, Determine:
(a) The composition of the distillate
(b) Quantities of the distillate and residue

The equilibrium data in terms of the weight fractions of alcohol at atmospheric pressure is given below:

x'	0.025	0.05	0.10	0.20	0.30	0.40	0.50	0.55	0.60
y'	0.225	0.36	0.516	0.655	0.71	0.74	0.767	0.778	0.789

(**Answer:** (a) $y_D = 0.7018$ (b) $W = 309.6$ kg, $D = 690.4$ kg)

5.60 A feed consisting of 50 mol% **A** and the rest **B** is differentially distilled. If it is desired that the distillate vapour produced at any instant should contain not less than 75 mol% **A**, determine

(a) The fraction of the feed vaporised
(b) The composition of the composited distillate
The equilibrium data is given by $y = 1.2x + 0.3$ in the range $0.3 \le x \le 0.8$

(**Answer:** (a) 0.2758 (b) 82.82% A)

5.61 A still is charged with 100 kmol of an equimolar mixture of benzene and toluene. The mixture is subjected to differential distillation such that vaporisation occurs at a constant rate of 10 kmol/h. Determine the following at the end of 5 h after commencement of distillation
(a) the distillation temperature
(b) the residue composition
(c) the composition of the vapour evolved
(d) the average composition of the distillate collected
The equilibrium data in terms of mole fraction of benzene are:

T, K	383.8	379.6	375.5	371.8	368.4	365.7	362.7	360.2	357.8	355.6	354.4	353.3
x	0	0.1	0.2	0.3	0.4	0.5	0.6	0.7	0.8	0.9	0.95	1
y	0	0.205	0.369	0.502	0.618	0.706	0.789	0.849	0.907	0.955	0.978	1

(**Answer:** (a) 370 K (b) $x_W = 0.35$ (c) $y = 0.58$ (d) $y_D = 0.65$)

5.62 A steam jacketed kettle is used for differential distillation of a liquid mixture consisting of 40 mol% **A** and 60 mol% **B**. Vapours are withdrawn and condensed as fast as they are formed. The kettle is initially charged with 30 moles of the feed liquid. Feed mixture at a rate 15 kmol/h is continuously admitted to the kettle and the vapour is produced and withdrawn at a rate 15 kmol/h so that the total molar content in the kettle remains constant.
(a) How much time will it take to produce a vapour of composition 50 mol% **A**?
(b) What will be the average composition of the distillate collected during the above time interval?
(c) If the vapour is produced at a rate 18 kmol/h instead of the desired 15 kmol/h, how long will it take to produce vapour containing 50% **A**?
Assume that the system has a constant relative volatility of 2.5

(**Answer:** (a) 1.461 h (b) $y_D = 0.556$ (c) 1.07 h)

5.63 A liquid hydrocarbon mixture containing 25 mol% *n*-butane, 30 mol% *n*-pentane and 45 mol% *n*-hexane is subjected to a simple batch distillation at a pressure of 1000 kPa. Determine the composition of the residue when half of the butane has been removed. Assume that the relative volatilities of butane and pentane with respect to hexane are constant at 3.5 and 1.85, respectively.

(**Answer:** butane = 0.178, pentane = 0.2962, hexane = 0.5258)

5.64 A liquid containing 45 mol% benzene, 30 mol% toluene and 25 mol% o-xylene is differentially distilled at 101.3 kPa. Determine the composition of the distillate and the residue when 30% of the charge has been distilled. Take the vapour pressures of benzene, toluene and xylene to be 183 kPa, 73 kPa and 27 kPa, respectively at the average temperature. Assume ideal solution behaviour.

(**Answer:** $x_i = 0.1004, 0.3405, 0.5592$; $y_i = 0.60, 0.2826, 0.1174$)

Steam distillation

5.65 A liquid **A** of molecular weight 150 is insoluble in water and is to be purified by steam distillation at a total pressure of 101.3 kPa. The vapour pressure of **A** at the temperature of distillation is 3.5 kPa. How many kg of steam is required theoretically to distil one kg of **A**? (**Answer:** 3.35 kg)

5.66 Dimethylaniline is distilled with steam at 90 kPa to free it from the non-volatile impurities. Assuming it to be completely immiscible with water, determine
(a) The distillation temperature

(b) The composition of the vapour produced.
The vapour pressure data follows:

T, K	343.2	374.8
P^S (aniline), kPa	1.333	5.332
P^S (water), kPa	31.19	106.91

Plot of $\ln P^S$ versus $1/T$ may be assumed linear.

(**Answer:** (a) 368.9 K, (b) $y_A = 0.9537$)

5.67 A mixture containing chlorobenzene and water is being distilled. The liquids may be assumed immiscible. The distillation is carried out under such condition that the two liquid layers are present in the still. The vapour pressure data follow:

Clorobenzene: $$\ln P^s = 13.9926 - \frac{3295.12}{T - 55.60}$$

Water: $$\ln P^s = 16.26205 - \frac{3799.887}{T - 46.854}$$

Determine the temperature of distillation and the composition of the vapour if the pressure is 40 kPa.

(**Answer:** 341 K, $y_A = 0.71$)

5.68 A high boiling organic liquid is purified from non-volatile impurities by allowing it to mix with steam directly at a total pressure of 95 kPa. The vapour pressure data are as follows:

Temperature, K	353	373
Vapour Pressure of water, kPa	47.98	101.3
Vapour pressure of liquid, kPa	2.67	5.33

Assume that water and the organic liquid are immiscible and the impurities do not affect the vaporisation characteristics. The vapour pressures vary linearly with temperature. Calculate under three-phase equilibrium

(a) the equilibrium temperature and
(b) the composition of the resulting vapour.

(**Answer:** (a) 368.85 K (b) $y_A = 0.9497$)

5.69 A mixture of two high-boiling organic acids is to be separated from nonvolatile impurities by steam distillation at 373 K under a total pressure of 20 kPa. The organic acid mixture contains 30 mol% low boiling acid and 70 mol% high boiling acid and at 373 K the vapour pressures of the two acids are 2.7 kPa and 1.1 kPa, respectively. The acid mixture obeys Raoult's law and are immiscible with water. Determine the kilograms of steam required per kmol of acid recovered.

(**Answer:** 209.5 kg)

5.70 A 1000-kg batch of nitrobenzene containing nonvolatile impurities is charged into a steam jacketed kettle for purification by steam distillation at a pressure of 101.3 kPa. The charge is initially at 300K. A liquid water level is maintained in the kettle by introducing water at 300 K continuously. The contents in the kettle are heated by passing saturated steam at 134 kPa through the jacket. Determine:

(a) Distillation temperature
(b) Quantity of water vaporised
(c) Quantity of steam condensed in the kettle

The heat capacity of nitrobenzene is 1.382 kJ kg K and the latent heat is 331kJ/kg. The heat capacity of water is 4.1868 kJ/kg K and latent heat of vaporisation of water is 2257 kJ/kg. Latent heat of vaporisation steam at 134 kPa is 2235 kJ/kg. Neglect the heat required to bring the still to the distilling temperature. The vapour pressures are given by the Antoine equations $\ln P^s = A - \dfrac{B}{T-C}$ where pressure is in kPa and temperature in K, with following constants:

Material	A	B	C
Water	16.26205	3799.887	46.854
Nitrobenzene	14.3692	4021.689	71.417

(**Answer:** (a) 372.42 K (b) 5244 kg (c) 6176 kg)

5.71 A batch of 2500 kg of impure turpentine (MW=140) is being subjected to steam distillation at 101.3 kPa. The feed containing 85 (weight)% turpentine and 15% water is fed to the still at 300K. Saturated steam is supplied to the still at 125 kPa. The vaporisation efficiency is 70%. Heat loss from the still to the surroundings is estimated to be one-tenth of the usefully consumed heat. Determine:

(a) The composition of the vapour
(b) The steam consumption.

The vapour pressure data:

Temperature, K	323	333	343	353	363	373
P^S_{water}, kPa	12.4	19.9	31.1	47.3	70.1	101.3
$P^S_{\text{turpentine}}$, kPa	2.5	3.5	5.0	7.5	11.8	17.5

Latent heat of vaporisation of turpentine = 298 kJ/kg, heat capacity of turpentine = 2 kJ/kg K. Temperature of saturation of steam at 125 kPa = 379.15 K. Enthalpy of saturated steam at 125 kPa = 2685.2 kJ/kg, enthalpy of saturated steam at 101.3 kPa = 2676.0 kJ/kg,

(**Answer:** (a) $y_A = 0.8983$ (b) 2904.24 kg)

5.72 Glycerine is purified by distillation with superheated steam at 503 K and under a vacuum of 75 kPa in a still with provision for external heating. The vaporisation efficiency 0.75. The crude glycerine is fed in at the distillation temperature. The vapour pressure (kPa) of glycerine is

$$\ln P^s = 12.18032 - \frac{2385.6225}{T - 245.103}$$

(a) Determine the amount of steam withdrawn with 1000 kg of glycerin in the distillate.
(b) How will the steam consumption and vapour composition change if the vacuum is changed to 85 kPa?

(**Answer:** (a) 170.7 kg (b) 31.395 kg, water content decreases from 46.59% to 13.83%)

5.73 Stearic acid (MW = 284) is being purified by steam distillation at a pressure of 10 kPa. Superheated steam is admitted to the vessel and leaves at 500 K, 70% saturated with the acid. Temperature of the feed is 295 K. There is no heat loss to the surroundings. The stearic acid has the following properties:

Melting point: 337 K; heat capacity (kJ/kg K): 1.68 (for solid), 2.31 (for liquid); latent heat (kJ/kg): 200 (fusion), 277 (vaporisation); vapour pressure at 500 K is 2.5 kPa.
Enthalpy of steam at 510 kPa and 500 K is 3489 kJ/kg.
Determine the heat requirement for the process per kilogram of acid distilled.

(**Answer:** 1939.16 kJ/kg)

Fractionation: Material and energy balances, reflux ratio, plate efficiency

5.74 A continuous rectification column is to separate a binary mixture of **A** and **B**. The rate of distillate production is 100 kmol/h and contains 98 mol% **A**. The mole fraction of **A** in the liquid and in the vapour, x and y respectively leaving two adjacent plates in the enriching section are

x	0.65	0.58
y	0.82	0.78

Assuming that the latent heat of vaporisation are the same for all mixtures, and the feed is saturated liquid, calculate (a) the reflux ratio (b) the vapour rate in the stripping section in kmol/h.

(**Answer:** (a) 1.5385 (b) 253.85 kmol/h)

5.75 A distillation column is fed with a binary mixture at 350 kmol/h. It is desired to produce a distillate containing 97 mol% **A** at a rate 150 kmol/h and residue containing 2 mol% **A**. The reflux ratio employed is 3.5. A vapour sample leaving a plate in the stripping section analyzed 0.33 mole fraction **A** while the mole fraction of **A** in the liquid coming to that plate is 0.25. Assuming constant molar overflow, determine (a) the feed quality (b) the vapour boilup ratio.

(**Answer:** (a) 0.7143 (b) 2.875)

5.76 A mixture consisting of 40 mol% **A** and 60 mol% **B** is distilled in a plate column equipped with a total condenser at the top. A test run gave the following compositions (mole fraction of **A**) for liquid and vapour streams taken from three adjacent plates in the enriching section of the column.

Plate	x	y
$n-1$	0.68	0.795
n	0.60	0.740
$n+1$	0.51	0.679

Determine:

(a) The reflux ratio

(b) The distillate composition

(**Answer:** (a)3.2163 (b) 0.9331)

5.77 The feed is admitted to a distillation column at a rate of 200 kmol/h and is 60% vaporised. The distillate is withdrawn at 120 kmol/h with 0.95 mole fraction of the more volatile component. Steam consumption (latent heat = 2200 kJ/kg) in the reboiler is 1500 kg/h. Latent heat of the column liquid may be assumed to be 33000 kJ/kmol. Calculate

(a) the operating reflux ratio and

(b) the composition of vapour entering that plate from which liquid overflow contains 0.7 mole fraction of the more volatile component (**Answer:** (a) 0.833 (b) 0.8364)

5.78 A laboratory distillation column consists of 5 plates, a total condenser a perfect reboiler and is operated at atmospheric pressure at total reflux on a benzene toluene mixture. The relative volatility of benzene to toluene for this mixture is 2.5. Samples of the liquid from the condenser and reboiler analyse 90 mol% benzene and 10 mol% benzene, respectively. Calculate the average plate efficiency. (**Answer:** 76.0%)

5.79 The following figures represent the equilibrium conditions in mole fraction of benzene toluene mixture at their boiling points

Liquids(x):	0.51	0.38	0.26	0.15
Vapour (y):	0.72	0.60	0.45	0.30

If the liquid composition on four adjacent plates are 0.18, 0.28, 0.41, 0.5 under conditions of total reflux, determine the Murphree vapour efficiencies of the plates.

(**Answer:** 60.6%, 66.7%, 40.9%)

Fractionation: McCabe-Thiele method

Number of plates not specified

5.80 A mixture containing 40 mole% **A** and 60 mole% **B** is to be separated in a distillation tower. The concentration of **A** in the distillate is 95% of **A** and 96% is to be recovered as distillate product.

The feed is half vapour and the reflux ratio used is 4. The relative volatility of **A** to **B** is 2. Total condenser is employed.

(a) How many theoretical plates are required in each section?

(b) Determine the minimum number of stages graphically and analytically.

(Answer: (a) 8, 7 (b) 8.8 (graphical), 8.4 (analytical)

5.81 A solution containing 40 mol% *n*-heptane and the rest methyl cyclohexane is to be fractionated to produce a distillate containing 95 mol% heptane. The feed enters at a rate of 100 kmol/h at its boiling point and it is desired that the recovery of heptane in the distillate be 94% of that in the feed. The column is provided with a total condenser and a kettle-type reboiler. Assume relative volatility to be constant at 2.0.

Evaluate:

(a) The composition of the residue.

(b) The molar flow rates of the products.

(c) The minimum number of stages (i)using Fenske equation (ii) graphically.

(d) The minimum reflux ratio (i) graphically (ii) using Underwood-Fenske equation.

(e) The number of theoretical plates when a reflux ratio 50% in excess of the minimum is used.

(Answer: (a) 0.0397 (b) $D = 39.58$ kmol/h, $B = 60.42$ kmol/h (c) (i) and (ii) 7.8 (d) (i) 2.2089 (ii) 2.2083 (e) 16.2)

5.82 A saturated feed containing 40 mol% chloroform and 60 mol% benzene is to be separated continuously. The distillate should contain 95 mol% chloroform and residue not more than 5 mol% chloroform. The column is provided with a total condenser and partial reboiler. The vapour-liquid equilibrium data for the system benzene-chloroform is represented by the following equation

$$y = 1.29x + 0.25x^2 - 0.54x^3$$

where x and y are the mole fractions of chloroform in liquid and vapour phases respectively.

Calculate:

(a) The minimum number of trays for this separation.

(b) The actual number of trays if the reflux ratio is twice the minimum

(Answer: (a) 11(b) 18.7)

5.83 It is desired to fractionate a given mixture of a binary volatile liquid with an initial composition 30 mol% **A** into an overhead product containing 97 mol% **A**. Loss of **A** in the bottoms should not be more than 5% of that entering the system. Take the relative volatility of the system to be 4.8. If the feed is cold liquid requiring 0.1 kmol additional condensation per mole of feed at the feed plate, and if a total condenser is employed determine:

(a) The number of theoretical plates at twice the minimum reflux.

(b) The number of plates under total reflux (i) analytically (ii) graphically.

(Answer: (a) 7 (b) (i) 4 (ii) 3.7)

5.84 A solution containing 30 mol% benzene, 70 mol% toluene is subjected to continuous fractionation to produce a distillate containing 90 mol% benzene and bottoms product containing 95 mol% toluene. The equilibrium data (mole fraction benzene) are:

T, K	383.8	379.6	375.5	371.8	368.4	365.7	362.7	360.2	357.8	355.6	354.4	353.3
x	0	0.1	0.2	0.3	0.4	0.5	0.6	0.7	0.8	0.9	0.95	1
y	0	0.205	0.369	0.502	0.618	0.706	0.789	0.849	0.907	0.955	0.978	1

The latent heat of vaporisation of the mixture may be assumed constant at 3×10^4 kJ/kmol. Assume that the reflux ratio used is 5 and a total condenser is employed. Determine the no of theoretical plates and the feed plate location if

(a) the feed is liquid at 283 K and specific heat of the feed solution is 1.85 kJ/kg K.
(b) the feed is saturated vapour

(**Answer:** (a) 7, 4th from top (b) 9.8, 7th from top)

5.85 A feed consisting of 40 mol% heptane and 60 mol% ethyl benzene is distilled to produce a distillate containing 95 mol% heptane and a bottoms product containing 96 mol% ethyl benzene. Reflux ratio employed is 2.8. Determine the number of theoretical plates if

(a) the feed is saturated liquid
(b) the feed is 40% vapour and 60% liquid
(c) the feed is saturated vapour

Equilibrium data (mole fraction of heptane) are:

x	0	0.08	0.185	0.251	0.335	0.487	0.651	0.788	0.914	1.00
y	0	0.233	0.428	0.514	0.608	0.729	0.834	0.904	0.963	1.00

(**Answer:** (a) 7 (b) 8 (c) 10)

5.86 A distillation column is fed with a mixture containing 35 mole% benzene and 65 mole% toluene as saturated liquid at 101.3kPa. The distillate is to contain 95 mole% benzene and the bottom products from the column contain 5 mole% benzene. The reflux ratio used is 3.0 mol per mol of product. Assuming that the plates used has an overall efficiency of 60% find the number of plates required and the position of the feed plate.

Equilibrium data (mole fraction benzene) are:

x	0	0.1	0.2	0.3	0.4	0.5	0.6	0.7	0.8	0.9	0.95	1
y	0	0.205	0.369	0.502	0.618	0.706	0.789	0.849	0.907	0.955	0.978	1

(**Answer:** 19, 11th plate from the top)

5.87 A solution containing methanol and 1,4-dioxane (MW = 88) is separated by distillation in a column operated at 101.3 kPa. The feed contains 3000 kg/h methanol and 7000 kg/h dioxane as a saturated liquid mixture. The equilibrium data (mole fraction of methanol) are:

T, K	374.3	368.1	362.8	355.4	350.1	346.3	343.9
x	0	0.050	0.100	0.200	0.300	0.400	0.500
y	0	0.212	0.365	0.542	0.657	0.735	0.779
T, K	342	340.7	339.4	338.3	337.9	337.7	
x	0.600	0.700	0.800	0.900	0.950	1.00	
y	0.818	0.850	0.892	0.947	0.974	1.00	

A reflux ratio of 2.5 may be employed. The distillate is to recover 98% of the methanol and should have a purity of 99 mol%. The average heat of vaporisation of the mixture is 35000 kJ/kmol.

Determine:

(a) The flow rates of the product streams, kmol/h
(b) Composition of the residue and fractional recovery of dioxane in the residue
(c) The heating and cooling requirements, kJ/h
(d) The minimum reflux ratio and minimum boilup ratio
(e) The actual boil-up ratio
(f) The minimum number of plates
(g) The internal flow rates: $L, \bar{L}, G, \bar{G}$ (kmol/h)

(h) The number of theoretical plates and the location of feed plate

(**Answer:** (a) $D = 92.8$, 80.5 (b) 0.0233, 0.988 (c) 1.1368×10^7 (d) 0.7677, 2.04 (e) 4.0348 (f) 7 (g) $L = 232$, $\bar{L} = 405.3$, $G = \bar{G} = 324.8$ (h) 11, 9th from top)

5.88 A mixture of benzene and toluene containing 40% (mol) benzene is to be separated to give a product of 90 mol% benzene at the top and a bottom product containing not more than 10 mol% benzene. The feed enters the column at its bubble point, the vapour is totally condensed, and a reflux ratio of 3 moles per mole of product is employed. Determine:

(a) The number of theoretical plates needed
(b) The feed plate location
(c) The minimum number of plates required using
  (i) graphical construction
  (ii) Fenske equation assuming average α value of 2.4
(d) The minimum reflux ratio using
  (i) graphical method
  (ii) Underwood – Fenske equation.

The following equilibrium data (mole fraction benzene) may be used.

x	0	0.1	0.2	0.3	0.4	0.5	0.6	0.7	0.8	0.9	0.95	1
y	0	0.205	0.369	0.502	0.618	0.706	0.789	0.849	0.907	0.955	0.978	1

(**Answer:** (a) 6.7 (b) 4th from top (c) (i) 4 (ii) 4.02 (d) (i) 1.2934 (ii) 1.3214)

5.89 One hundred kmol/h of a feed mixture consisting of 60 mol% benzene and the rest toluene is admitted to a fractionating column to produce a distillate containing 95 mol% benzene and a bottoms product containing 4 mol% benzene. The feed is partially vaporised and enters the column with a vapour fraction equal to the distillate to feed ratio in the column. The column is provided with a partial condenser and kettle-type reboiler. The operating reflux ratio is 1.3 times the minimum. The pressure is 101.3 kPa and the equilibrium data (mole fraction benzene) are:

x	0	0.1	0.2	0.3	0.4	0.5	0.6	0.7	0.8	0.9	0.95	1
y	0	0.205	0.369	0.502	0.618	0.706	0.789	0.849	0.907	0.955	0.978	1

Determine:

(a) Minimum number of stages using graphical and analytical method ($\alpha = 2.4$)
(b) Minimum reflux ratio
(c) The number of theoretical plates required in the column

(**Answer:** (a) 4.99, 4.99 (b) 1.3024 (c) 11.4)

5.90 A continuous fractionating column is to be designed to separate 30000 kg/h of a mixture 40% (weight) benzene and 60% (weight) toluene into an overhead product containing 97% (weight) benzene and bottom product containing 98% (weight) toluene. A reflux ratio 2.5 times the minimum may be used. The feed enters at 293K while its bubble point at 101.3 kPa is 367K. The latent heats of vaporisation of benzene and toluene are respectively, 30.912 kJ/mol and 33.452 kJ/mol. The specific heat of the feed is 1.85 kJ/kg K. Steam supplied to the reboiler is saturated at 150 kPa. A total condenser is used in which the cooling water enters at 300K and leaves at 320K. Overall stage efficiency of the column is 80%. Equilibrium data in mole fraction of benzene.

T, K	383.8	379.6	375.5	371.8	368.4	365.7	362.7	360.2	357.8	355.6	354.4	353.3
x	0	0.1	0.2	0.3	0.4	0.5	0.6	0.7	0.8	0.9	0.95	1
y	0	0.205	0.369	0.502	0.618	0.706	0.789	0.849	0.907	0.955	0.978	1

Latent heat of vaporisation of steam at 150 kPa is 2267 kJ/kg

Determine the following:

(a) The minimum number of plates
(b) The minimum reflux ratio
(c) The number of ideal plates and the feed plate location
(d) Number of actual plates
(e) Steam requirement, kg/h
(f) Rate of cooling water circulation.

(**Answer:** (a) 7.8 (b) 1.2632 (c) 12, 6 (d) 15 (e) 11250.3 (f) 2.3557×10^5 kg/h)

5.91 A continuous fractionating column separates 5000 kg/h of a liquid mixture of 25 weight% CS_2 and the rest chloroform into a distillate containing 90 weight% CS_2 and a residue of 97 weight% chloroform. Overall stage efficiency is 70% and the reflux ratio employed is 3. The feed enters at 290 K and the specific heat of the feed solution is 1.7 kJ/kg K. The column is provided with a total condenser and partial reboiler. Assume a constant latent heat of vaporisation of 25900 kJ/kmol for CS_2 and chloroform.

Equilibrium data:

T, K	349.9	348.1	346.3	343.5	341.8	337	332.5	328.5	325.5	323.6	321.7	319.5
x	0	.0296	.0615	.1106	.1435	.2585	.3908	.5318	.6630	.7574	.8604	1.0
y	0	.0823	.1555	.266	.3325	.4950	.6340	.7470	.8290	.8780	.9320	1.0

Determine:

(a) The product rates, kmol/h
(b) The number of theoretical plates
(c) The number of actual plates

(**Answer:** (a) $D = 16.03$, $B = 31.80$ kmol/h (b) 7.6 (c) 11)

5.92 An equimolar solution of chloroform ($CHCl_3$) and carbon disulphide (CS_2) is to be continuously fractionated at 101.3 kPa into a distillate containing 95 mol% CS_2 and a residue of 3 mol% CS_2. The feed is 30 mol% vapour before entering the tower. A total condenser is used.

Determine:

(a) Minimum reflux ratio
(b) Minimum number of theoretical plates
(c) The Number of theoretical plates at a reflux ratio twice the minimum

Equilibrium data (mole fraction CS_2) are:

x	0	.0296	.0615	.1106	.1435	.2585	.3908	.5318	.6630	.7574	.8604	1.0
y	0	.0823	.1555	.266	.3325	.4950	.6340	.7470	.8290	.8780	.9320	1.0

(**Answer:** (a) 1.2386 (b) 5.8 (c) 9)

5.93 A continuous fractionating column is used to separate a mixture containing 70 mol% n-heptane and the rest n-octane into products of 97 mol% purity. The feed is supplied to the column as saturated liquid at a rate of 5000 kg/h. The column is to operate at a pressure of 101.3 kPa with a vapour velocity of 0.6 m/s. The reflux ratio used is twice the minimum. The bubble point at the top of the column may be taken as 372 K.

Equilibrium data (mole fraction n-heptane) are:

x	1.0	0.92	0.82	0.69	0.57	0.46	0.32	0.22	0.13	0
y	1.0	0.96	0.91	0.83	0.74	0.65	0.50	0.37	0.24	0

Determine:

(a) Minimum reflux ratio

(b) Number of theoretical plates

(c) Diameter of the column

(**Answer:** (a) 1.139 (b) 12 (c) 1.42 m)

5.94 A solution containing 40 weight% methanol and 60 weight% water at 300K is admitted to a distillation column at a rate 5000 kg/h after being preheated by heat exchange with the residue leaving the tower. The overhead product should contain 95 weight% methanol and the bottom product contains 3 weight% methanol. The distillate is to be totally condensed to a liquid and the reflux returned at the bubble point. The reflux ratio used is 1.5 times the minimum. The residue leaves the system at 310K. Equilibrium data:

Equilibrium data (mole fraction methanol) are:

T, K	373.2	360.9	354.9	351.2	348.5	346.3	344.4	342.5	340.8	339.2	337.7
x	0	0.1	0.2	0.3	0.4	0.5	0.6	0.7	0.8	0.9	1.0
y	0	0.417	0.579	0.669	0.729	0.78	0.825	0.871	0.915	0.959	1.0

Heat capacity of feed: 3.852 kJ/kg K, heat capacity of residue: 4.179 kJ/kg K, latent heat of vaporisation : 1047 kJ/kg (methanol), latent heat of vaporisation: 2284 kJ/kg (water).

Determine:

(a) The minimum reflux ratio

(b) The minimum number of theoretical trays

(c) The number of theoretical trays

(d) The liquid and vapour quantities inside the tower

(e) The condenser and reboiler heat loads

(**Answer:** (a) 0.7264 (b) 3.75 (c) 8 (d) $L = 71.16, \bar{L} = 306.4, G = 136.44, \bar{G} = 142.5$
(e) $Q_C = 4.66 \times 10^6$ kJ/h, $Q_B = 5.84 \times 10^6$ kJ/h)

5.95 An ethanol-water solution containing 40 weight% ethanol is to be separated in a continuous fractionating column operating at atmospheric pressure. The feed is at 310 K and is admitted at rate 20000 kg/h. The column will be equipped with a reboiler and a partial condenser. The top product is to contain 90 weight% ethanol and the bottom product is to contain 5 weight% ethanol. The reflux ratio employed is 50% in excess of the minimum. Determine:

(a) The minimum reflux ratio

(b) The minimum condenser duty

(c) The condenser duty

(d) The reboiler duty

(e) The number of ideal plates

Equilibrium data (mole fraction of ethanol) are:

T, K	373.2	368.7	359.9	357.3	355.5	353.9	352.9	351.94	351.35
x	0	.019	.0966	.1661	.2608	.3965	.5198	.6763	.8943
y	0	.17	.4375	.5089	.558	.6122	.6599	.7385	.8943

Heat capacity of ethanol: 3.465 kJ/kg K, Heat capacity of water: 4.187 kJ/kg K, Latent heat of vaporisation : 860 kJ/kg (ethanol), Latent heat of vaporisation : 2284 kJ/kg (water).

(**Answer:** (a) 0.8577 (b) 1.5335×10^7 kJ/h (c) 1.8875×10^7 kJ/h
(d) 2.3083×10^7 kJ/h (e) 9.8)

Number of plates specified

5.96 One thousand kmol/h of a solution containing 30 mol% *n*-hexane and the rest *n*-octane is subjected to fractionation in a column consisting of one theoretical plate, a kettle type reboiler and a partial condenser. The feed at its bubble point is admitted to and the bottoms product is withdrawn from the reboiler. The distillate contains 80 mol% hexane. The reflux ratio employed is 2 mol per mol distillate. Determine:

(a) Composition of the bottoms
(b) Quantity of distillate and bottoms, kmol/h
(c) The composition of the bottoms, if the feed were introduced at the plate instead of the reboiler

The equilibrium data (mole fraction of hexane) are:

T	399.1	388.5	379.7	372.4	366.1	360.7	356.0	351.9	348.2	345.0	342.0
x	0	0.1	0.2	0.3	0.4	0.5	0.6	0.7	0.8	0.9	1.0
y	0	0.324	0.535	0.676	0.773	0.842	0.892	0.930	0.959	0.982	1.0

(**Answer:** (a) 0.142 (b) $D = 240.12$, $B = 759.88$ (c) 0.1)

5.97 A fractionating column consisting of two theoretical plates, a total condenser and a partial reboiler is fed with a solution containing 55 mol% benzene and the rest chlorobenzene. The feed is admitted to the bottom plate as saturated liquid. Reflux ratio is provided to maintain the L/G ratio of 0.5 above the feed plate. Reboiler is supplied with steam to produce vapour at the rate of 0.9 mol per mol feed. Equilibrium data (mole fraction benzene) are:

x	0	0.1	0.2	0.3	0.4	0.5	0.6	0.7	0.8	1.0
y	0	0.314	0.508	0.640	0.734	0.806	0.862	0.905	0.943	1.0

Determine the compositions of the distillate and bottoms product

(**Answer:** $x_D = 0.9$, $x_B = 0.27$)

5.98 An equimolar mixture of **A** and **B** is to be separated into an overhead product containing 96 mol% **A** using a fractionating column consisting of 10 theoretical plates and equipped wit a partial reboiler and total condenser. The feed enters the column at its bubble point and the relative volatility of **A** to **B** is 2. The reflux ratio employed is 50% in excess of the minimum. Determine:

(a) The minimum reflux ratio
(b) The composition of the residue
(c) The yield of distillate per mole of feed.

(**Answer:** (a) 1.7594 (b) 0.293 (c) 0.31)

5.99 A feed consisting of 40 mol% methanol and 60 mol% water is admitted at its bubble point into a fractionation column consisting of seven equilibrium stages, a reboiler and total condenser. The feed enters the column as saturated liquid. 95% of the methanol is to be recovered in a liquid distillate containing 98 mol% methanol. What will be the necessary rate of vapour production in the reboiler in kmol/h, if the feed is admitted at rate 100 kmol/h? Equilibrium data (mole fraction methanol) are:

x	0	0.1	0.2	0.3	0.4	0.5	0.6	0.7	0.8	0.9	1.0
y	0	0.417	0.579	0.669	0.729	0.78	0.825	0.871	0.915	0.959	1.0

(**Answer:** 122.66)

Number of plates incorporating Murphree efficiency

5.100 A feed containing 60 kmol/s methanol and 60 kmol/s water at its saturation temperature is to be distilled at 101.3 kPa pressure to produce a distillate containing 98 mol% methanol. 95% of the methanol is to be recovered in the distillate. A reflux ratio of twice the minimum reflux rate must be employed. Assume an average Murphree vapour efficiency E_{MG} of 0.75. Find the number of actual plates given the following equilibrium data (mole fraction of methanol):

x	0	0.1	0.2	0.3	0.4	0.5	0.6	0.7	0.8	0.9	1.0
y	0	0.417	0.579	0.669	0.729	0.78	0.825	0.871	0.915	0.959	1.0

(**Answer:** 13 plates)

5.101 An equimolar solution of acetone and isopropanol is subjected to fractional distillation in a plate column to produce a distillate containing 80 mol% acetone and bottoms containing 25 mol% acetone. The Murphree vapour efficiency is found to be 50%. The reflux ratio used is 0.5. The feed is admitted at its bubble point. Column is provided with a total condenser and partial reboiler. Determine the number of plates.

The equilibrium data (mole fraction of acetone) are:

x	0	0.026	0.054	0.117	0.207	0.297	0.341	0.440	0.520
y	0	0.089	0.174	0.315	0.456	0.557	0.601	0.687	0.743
x	0.639	0.746	0.803	0.865	0.902	0.925	0.957	1.0	
y	0.815	0.870	0.894	0.923	0.942	0.955	0.974	1.0	

(**Answer:** 10 plates)

5.102 A plant must distil a mixture containing 50 mol% methanol and 50 mol% water. The overhead product must contain 97 mol% methanol and a bottom product containing 3 mol% methanol. The feed is admitted as saturated liquid and the reflux ratio used is twice the minimum. Calculate the number of plates using a total condenser and a still assuming an average Murphree plate efficiency of 70%.

Equilibrium data:

x	0	0.1	0.2	0.3	0.4	0.5	0.6	0.7	0.8	0.9	1.0
y	0	0.417	0.579	0.669	0.729	0.78	0.825	0.871	0.915	0.959	1.0

(**Answer:** 12 plates)

Stripping and enriching columns, open steam

5.103 It is desired to strip off the more volatile component **A** using open steam using a multistage countercurrent column. The feed composition is 20 mol% **A** and it enters the column with a quality of 0.8. If the feed rate is 100 kmol/h and the reject composition is not to exceed 0.5 mole% **A**, determine:

(a) The minimum steam requirement

(b) The number of ideal stages if 1.5 times the minimum steam rate is used

Equilibrium data (mole fraction **A**):

x	0.05	0.075	0.1	0.15	0.20
y	0.08	0.145	0.375	0.55	0.62

(**Answer:** (a) 46.38 kmol/h (b) 7)

5.104 A mixture containing 2 mol% ethanol and the rest water is to be stripped off in a plate column to a bottom product containing not more than 0.01 mol% ethanol. Steam admitted through the open coil in the liquid on the bottom plate is to be used as the source of vapour. The feed is at its boiling point. The steam flow is to be 0.2 mole per mole of feed. For dilute ethanol water system, the equilibrium line is straight and is given by $y = 9.0x$. Determine the number of plates graphically and using the Kremser equation. (**Answer:** 7.5, 7.64)

5.105 An aqueous solution of a valuable solvent containing 15 mol% solvent is to be steam stripped to recover 95% of all solvent entering the system. If for every mol of feed admitted there is an additional condensation of 0.2 mol and the Murphree stage efficiency based on vapour compositions is 70% determine the number of actual stages if the steam rate used is 1.3 times the minimum. Equilibrium data for solvent:

x	0.025	0.05	0.075	0.10	0.125	0.15	0.175	0.20
y	0.035	0.065	0.132	0.23	0.285	0.32	0.35	0.38

(**Answer:** 8.6)

5.106 A liquid mixture at its boiling point and containing 70 mol% benzene and 30 mol% toluene is fed at a rate of 400 kmol/h to a stripping tower at 101.3 kPa. The bottom product is to be 60 kmol/h containing only 10 mol% benzene. Calculate the kmol/h overhead vapour, its composition and the number of theoretical trays required. Equilibrium data:

x	0	0.1	0.2	0.3	0.4	0.5	0.6	0.7	0.8	0.9	0.95	1
y	0	0.205	0.369	0.502	0.618	0.706	0.789	0.849	0.907	0.955	0.978	1

(**Answer:** 340, 0.8059, 4.5)

5.107 An enriching tower is fed 100 kmol/h of a saturated vapour feed containing 40 mole% benzene and 60 mole% toluene at 101.3kPa. The distillate is to contain 90 mole% benzene. The reflux ratio is 4. Calculate:

(a) The rate of production of distillate and residue, kmol/h and their compositions
(b) The number of plates

Equilibrium data (mole fraction benzene) are:

x	0	0.1	0.2	0.3	0.4	0.5	0.6	0.7	0.8	0.9	0.95	1
y	0	0.205	0.369	0.502	0.618	0.706	0.789	0.849	0.907	0.955	0.978	1

(**Answer:** (a) $D = 20$, $B = 80$, $x_D = 0.9$, $x_B = 0.275$, (b) 5)

5.108 An aqueous solution of ethanol at 300K containing 40weight% ethanol is to be stripped off in a plate column to a product containing 4 weight% ethanol. The distillate is the condensed overhead vapours. Calculate:

(a) Minimum heat required per kg bottom product for the required operation
(b) Minimum ratio of vapour leaving the still to the bottom product
(c) Composition of the distillate product for conditions in part (b)
(d) Composition of distillate and number of actual plates if the vapour boil-up ratio is 1.5 times the minimum and overall stage efficiency is 0.6
(e) Heat load in the reboiler.

Heat capacity of ethanol: 3.465 kJ/kg K, latent heat of vaporisation of ethanol: 860 kJ/kg, latent heat of vaporisation of water: 2284 kJ/kg.
Equilibrium data (mole fraction ethanol) are:

T, K	373.2	368.7	359.9	357.3	355.5	353.9	352.9	351.94	351.35
x	0	0.019	0.0966	0.1661	0.2608	0.3965	0.5198	0.6763	0.8943
y	0	0.17	0.4375	0.5089	0.558	0.6122	0.6599	0.7385	0.8943

(**Answer:** (a) 2.69×10^3 kJ/h (b)1.2127 (c) 0.5495 (d) xD=0.333.2, 2 (e) 4.036×10^3 kJ/h)

Side streams and multiple feeds

5.109 A mixture of water and ethyl alcohol containing 15 mol% ethanol is continuously distilled in a plate column to give a product containing 80 mol% ethanol and a residue containing 2 mol%

ethanol. It is proposed to withdraw 25% of alcohol in the entering stream as a side stream containing 50 mol% alcohol. Determine the number of theoretical plates required and the plate from which the side stream should be withdrawn if the feed is liquid at the bubble point and a reflux ratio of 2 is used.

Equilibrium data (mole fraction ethanol) are:

x	0	0.019	0.0966	0.1661	0.2608	0.3965	0.5198	0.6763	0.8943
y	0	0.17	0.4375	0.5089	0.558	0.6122	0.6599	0.7385	0.8943

(**Answer:** 11, 7th plate)

5.110 A solvent **A** is to be recovered by distillation from its water solution. It is necessary to produce an overhead product containing 95 mol% **A** and to recover 95% of the **A** in the feed. The feed is available at the plant site in two streams one containing 40 mol% **A** and the other containing 60 mol% **A**. Each stream will provide 50 kmol/h of component **A**, and each will be fed into the column as saturated liquids. The vapour boilup is provided through open steam. The reflux ratio used is 30% in excess of the minimum. Total condenser is used. The overall plate efficiency is 70%. The relative volatility may be assumed to be constant at 3.0. Determine:

(a) The composition of the bottoms product
(b) The number of plates required.

(**Answer:** (a) 0.0174 (b) 22)

5.111 A fractionation column is fed with a 100 kmol/h of a solution containing 75 mol% methanol and 25 mol% ethanol. The feed is saturated liquid at the column operating pressure of 101.3 kPa. The relative volatility of the system may be assumed to be constant at 1.56. The distillate contains 96 mol% methanol and bottoms product is 95 mol% ethanol. It is desired to withdraw a side stream containing 80 mol% ethanol at a rate of 15 kmol/h from the stripping section. If the reflux ratio employed is 20% in excess of the minimum and the condenser used is total, determine the number of theoretical plates required. (**Answer:** 25)

5.112 One hundred kmol of an equimolar mixture of **A** and **B** with a relative volatility of 2.3 is to be separated into a distillate product with 98.5 mol% **A**, a bottoms product with 2 mol% **A**, and an intermediate liquid product containing 80 mol% **A** with 40% of **A** in the feed recovered. The feed enters the column at its bubble point. Determine:

(a) Amounts of each product
(b) Minimum reflux ratio
(c) How much greater is the minimum reflux rate because of the withdrawal of side stream?
(d) If the reflux ratio employed is 3, find the number of theoretical plates.

(**Answer:** (a) $D = 29.53$, $B = 45.47$, $S = 25.00$ (b) 2.7511 (c) 15.47 kmol (d) 21)

5.113 Two liquid streams made up of benzene and toluene, one containing 40 mol% benzene and the other containing 70 mol% benzene are available in equal amounts at their bubble points as feed streams to a distillation column. It is desired to produce a distillate containing 98 mol% benzene and residue containing 97 mol% toluene. Reflux ratio employed is 2.5 mol/mol distillate. Determine:

(a) The number of theoretical plates if the streams are combined and admitted to the column as saturated liquid at a single point.
(b) The number of plates if the feeds enter as saturated liquids at two different plates.

Equilibrium data (mole fraction benzene) are:

x	0	0.1	0.2	0.3	0.4	0.5	0.6	0.7	0.8	0.9	0.95	1
y	0	0.205	0.369	0.502	0.618	0.706	0.789	0.849	0.907	0.955	0.978	1

(**Answer:** (a) 12 (b) 11.5)

Enthalpy-composition diagram

5.114 A continuous stream of a binary liquid mixture, containing z_F mole fraction of **A** represented by point Q on the enthalpy concentration diagram is fed to a flash distillation unit. The products out of the unit are a vapour stream (y mole fraction of **A**) and a liquid stream (x mole fraction of **A**) which lie at the two ends of the tie line passing through point P which represents the enthalpy and composition of the combined mixture of vapour and liquid.

(a) Obtain an expression for the ratio of vapour flow rate to feed flow rate in terms of compositions.

(b) Show that the heat added per mole of feed is given by the distance PQ.

5.115 A liquid mixture containing 50 mol% each of benzene and toluene at 313 K is to be continuously flash vaporised to vaporise 60% of the feed. The residual liquid product contains 35 mol% benzene. The enthalpies of the feed, liquid product and vapour product are respectively 5, 2 and 30 kJ/mol. Represent the process on H-x-y diagram and calculate the heat added in kJ per kmol of vapour product. **(Answer:** 23000 kJ)

5.116 Construct the enthalpy-composition diagram for the system benzene-toluene at 101.3 kPa. The T-x-y data for the system (mole fraction of benzene) are:

T, K	383.8	379.6	375.5	371.8	368.4	365.7	362.7	360.2	357.8	355.6	354.4	353.3
x	0	0.1	0.2	0.3	0.4	0.5	0.6	0.7	0.8	0.9	0.95	1
y	0	0.205	0.369	0.502	0.618	0.706	0.789	0.849	0.907	0.955	0.978	1

Enthalpy data:

Component	Enthalpy of vaporisation, kJ/kmol	Specific heat at constant pressure, kJ/kmol K		Boiling point, K
		Liquid	Vapour	
Benzene	30,815	138.2	96.3	353.3
Toluene	33,327	167.5	138.2	383.8

(**Answer:**

x	0	0.1	0.2	0.3	0.4	0.5	0.6	0.7	0.8	0.9	1.0
$H_L \times 10^{-3}$ kJ/kmol	18.53	17.51	16.54	15.65	14.83	14.14	13.42	12.79	12.19	11.63	11.07
$H_G \times 10^{-3}$ kJ/kmol	51.85	50.88	49.93	48.90	47.94	46.94	45.95	44.98	43.98	42.95	41.88

Fractionation: Ponchon-Savarit method

5.117 Explain the method of estimating the number of theoretical plates in the presence of multiple feeds by Ponchon-Savarit method.

5.118 One thousand kg/h of an alcohol-water solution containing 50 weight% alcohol is separated into a distillate containing 85 weight% alcohol and a bottoms containing 3 weight% alcohol in a plate column at 101.3 kPa. A total condenser is employed and the reflux ratio used is 1.5. The feed enters the column as saturated liquid. The equilibrium data (weight fraction ethanol) are:

x'	0	0.02	0.05	0.10	0.20	0.30	0.40
y'	0	0.192	0.377	0.527	0.656	0.713	0.746
x'	0.50	0.60	0.70	0.80	0.90	0.94	0.96
y'	0.771	0.794	0.822	0.858	0.912	0.942	0.959

The enthalpy data (weight fraction ethanol)are:

x'	0	0.10	0.30	0.50	0.70	0.90	1.0
H_L, kJ/kg	419	372	314	286	258	225	207
H_G, kJ/kg	2675	2517	2193	1870	1544	1223	1064

Determine:

(a) The amounts of distillate and residue, kg/h
(b) The number of theoretical plates
(c) Condenser and reboiler loads, kJ/h

(**Answer:** (a) $D = 573.17$ kg/h, $B = 426.83$ kg/h
(b) 3.8, (c) 1.5289×10^6 kJ/h, 1.5440×10^6 kJ/h)

5.119 A methanol-water solution containing 30 mol% methanol enters a distillation column as saturated liquid. It is continuously distilled to produce a distillate containing 95 mol% methanol and a residue containing 4% methanol. Distillate is totally condensed and the reflux is at its bubble point. The equilibrium data (mole fraction methanol) are:

x	0.00	0.05	0.10	0.15	0.20	0.30	0.40
y	0.000	0.267	0.418	0.517	0.570	0.665	0.729
x	0.50	0.60	0.70	0.80	0.90	0.95	1.00
y	0.779	0.825	0.870	0.915	0.958	0.979	1.000

Enthalpy data:

Component	Enthalpy, kJ/kmol	
	Liquid	Vapour
Methanol	5000	40000
Water	7500	48000

Determine:

(a) The minimum reflux ratio
(b) The number of plates required for a reflux ratio 2

(**Answer:** (a) 0.86 (b) 5.7)

5.120 One thousand kg/h of a solution containing 40 mol% *n*-heptane and the rest ethyl benzene is fractionated at 101.3 kPa in a plate column equipped with a partial condenser and kettle-type reboiler. It is desired that the distillate should recover 97% of heptane and 4% of ethyl benzene in the feed. The feed enters the column at its bubble point. Determine:

(a) Product rates, kg/h and their composition
(b) Minimum reflux
(c) Minimum number of plates at total reflux
(d) Number of plates at a reflux ratio of 2 mol/mol distillate.
(e) Condenser and reboiler duties

Equilibrium data (mole fraction *n*-heptane) are:

x	0	0.08	0.18	0.25	0.49	0.65	0.79	0.91	1.0
y	0	0.28	0.43	0.51	0.73	0.83	0.90	0.96	1.0

The enthalpy (mole fraction heptane) data:

x	0	0.08	0.18	0.25	0.49	0.65	0.79	0.91	1.0
$H_L \times 10^{-3}$ kJ/kmol	24.3	24.1	23.2	22.8	22.05	21.75	21.7	21.6	21.4
$H_G \times 10^{-3}$ kJ/kmol	61.2	59.6	58.5	58.1	56.5	55.2	54.4	53.8	53.3

(**Answer:** (a) $D = 399.1$ kg/h, $B = 600.9$ kg/h $x_D = 0.9418$, $x_B = 0.0204$
(b) 1.278 (c) 4 (d) 8.5 (e) 2.568×10^5 kJ/h, 3.91×10^5 kJ/h)

5.121 It is desired to fractionate an aqueous ammonia solution containing 30 mol% ammonia into products containing 99.0 mol% ammonia and bottoms containing 2 mol% ammonia in a plate column operated at 1250 kPa pressure. The solution enters the column as saturated liquid at the operating pressure. A total condenser is employed. Open steam saturated at 1250 kPa is admitted below the bottom plate. Assume that the reflux is returned to the column at its bubble point. A reflux ratio of 1 kmol/kmol product is used. The following are the enthalpy-composition data and equilibrium data (mole fraction ammonia) at 1250 kPa:

Enthalpy-composition data:

x	0	.05	0.10	0.20	0.30	0.40	0.50	0.60
H_L, kJ/kmol	14555	12624	11027	7815	5234	2987	1390	424
H_G, kJ/kmol	49910	48943	48312	45415	43484	40904	38658	36410
x	0.70	0.80	0.90	0.95	1.0			
H_L, kJ/kmol	424	1056	1705	2022	2987			
H_G, kJ/kmol	34160	30950	28040	26100	21945			

Equilibrium data:

x	0	0.05	0.10	0.15	0.20	0.25	0.30	0.35	0.40
y	0	0.279	0.501	0.651	0.767	0.850	0.895	0.934	0.958
x	0.50	0.60	0.70	0.80	0.85	0.90	0.95	1.0	
y	0.982	0.994	0.996	0.998	0.9982	0.9996	0.9998	1.0	

Determine:

(a) The minimum reflux ratio

(b) The number of theoretical plates when the reflux ratio used is 1 kmol/kmol product.

(c) The condenser duty per 100 kmol/h of feed.

(**Answer:** (a) 0.459 (b) 5.6 (c) 1.1483×10^6 kJ/h)

5.122 An equimolar mixture of carbon tetrachloride (CCl_4) and toluene (C_7H_8) is fractionated in a plate column to produce a distillate containing 95 mol% CCl_4 and a bottoms containing 5 mol% CCl_4. The feed is a saturated liquid and the reflux ratio is 1.6 mol per mol of product. The equilibrium data at 101.3 kPa (mole fraction CCl_4) are:

x	0	0.06	0.16	0.29	0.43	0.56	0.64	0.78	0.95	1.0
y	0	0.13	0.31	0.49	0.64	0.76	0.81	0.90	0.97	1.0

Enthalpy data:

Component	Saturation enthalpy, kJ/kg		Boiling point, K
	Liquid	Vapour	
Carbon tetrachloride	63.0	257.5	349.1
Toluene	130.0	460	383.6

Enthalpies of liquid and vapour may be assumed to be linear functions of compositions. Determine the number of theoretical plates required for the separation.

(**Answer:** 13)

5.123 A binary mixture of **A** and **B** containing 35 mol% **A** entering the column as saturated liquid is separated in to a distillate containing 95 mol% **A**. It is specified that not more than 5% of **A** in the entering feed be lost in the residue. Assume that a total condenser is used. Feed is admitted at a rate 100 kmol/h. The enthalpy composition data are:

x	0	0.1	0.2	0.3	0.4	0.5	0.6	0.7	0.8	0.9	1.0
$H_L \times 10^{-3}$ kJ/kmol	29.4	16.8	12.6	13.7	14.7	15.0	16.8	18.9	20.0	20.2	20.6
$H_G \times 10^{-3}$ kJ/kmol	244	248	252	254	250	242	229	210	193	168	141

The equilibrium data (mole fraction of A) are:

x	0	0.1	0.2	0.3	0.4	0.5	0.6	0.7	0.8	0.9	1.0
y	0	0.275	0.45	0.56	0.685	0.775	0.85	0.90	0.945	0.98	1.0

Determine:

(a) The minimum reflux ratio

(b) The number of ideal plates at twice the minimum reflux

(c) The condenser and reboiler heat duties

(**Answer:** (a) 2.4664 (b) 7 (c) 2.7169×10^7 kJ/h, 2.8556×10^7 kJ/h)

5.124 Construct the enthalpy-composition diagram and determine the number of theoretical plates for separating a mixture of 35 mol% benzene and the rest toluene entering a plate column as saturated liquid at 101.3 kPa. The distillate contains 95 mol% benzene and the residue 10 mol% benzene. The reflux ratio used is 1.5 times the minimum and a total condenser is used for condensing the vapours leaving the top plate. The *T-x-y* data for the system (mole fraction of benzene) and the enthalpy data are given below.

T, K	383.8	379.6	375.5	371.8	368.4	365.7	362.7	360.2	357.8	355.6	354.4	353.3
x	0	0.1	0.2	0.3	0.4	0.5	0.6	0.7	0.8	0.9	0.95	1
y	0	0.205	0.369	0.502	0.618	0.706	0.789	0.849	0.907	0.955	0.978	1

Enthalpy data:

Component	Enthalpy of vaporisation, kJ/kmol	Specific heat at constant pressure, kJ/kmol K		Boiling point, K
		Liquid	Vapour	
Benzene	30,815	138.2	96.3	353.3
Toluene	33,327	167.5	138.2	383.8

(**Answer:** 9.4)

5.125 A methanol-water solution containing 30 mol% methanol enters a distillation column as saturated liquid. It is continuously distilled at a rate 100 kmol/h to produce a distillate containing 97.5 mol% methanol and a residue containing 3% methanol. Distillate is totally condensed and the reflux is at its bubble point. Determine:

(a) The minimum reflux ratio

(b) The number of plates if the reflux ratio is 1.5 times the minimum.

(c) The condenser duty and reboiler heat load, kJ/h

x	0.10	0.20	0.30	0.40	0.50	0.60	0.70	0.80	0.90	0.95	1.00
y	0.418	0.570	0.665	0.729	0.779	0.825	0.870	0.915	0.958	0.979	1.000

The equilibrium data (mole fraction methanol) are:

Enthalpy data:

x	H_G, kJ/kmol	H_L, kJ/kmol	x	H_G, kJ/kmol	H_L, kJ/kmol
0	48195	7536	0.50	44124	5745
0.05	47730	7141	0.60	43380	5606
0.10	47311	6862	0.70	42589	5513
0.15	46892	6629	0.80	41821	5420
0.20	46520	6420	0.90	41124	5327
0.30	45683	6094	1.00	40449	5234
0.40	44915	5908			

Answer: (a) 0.96 (b) 9.5 (c) 2.4696×10^6 kJ/h, 2.5290×10^6 kJ/h)

5.126 An equimolar mixture of carbon tetrachloride (CCl_4) and toluene (C_7H_8) is fractionated in a plate column to produce a distillate containing 95 mol% CCl_4 and a bottoms containing 5 mol% CCl_4. A side steam containing 80 mol% CCl_4 at its boiling point is with drawn at a rate 10% of feed rate. The feed is a saturated liquid and reflux ratio of 1.6 mol per mol product. The equilibrium data at 101.3 kPa (mole fraction CCl_4) are:

x	0	0.06	0.16	0.29	0.43	0.56	0.64	0.78	0.95	1.0
y	0	0.13	0.31	0.49	0.64	0.76	0.81	0.90	0.97	1.0

Enthalpy data:

Component	Saturation enthalpy, kJ/kmol		Boiling point, K
	Liquid	Vapour	
Caron tetrachloride	63.0	257.5	349.1
Toluene	130.0	460	383.6

Enthalpies of liquid and vapour may be assumed to be linear functions of compositions. Determine:

(a) Quantities of distillate and residue as percentages of the feed rate.

(b) Number of theoretical plates.

(**Answer:** (a) $D = 51.67$, $B = 38.33$, $S = 10$ (b) 17)

Packed columns for distillation

5.127 A fractionation column packed to a height of 3 m. The charge consisting of components **A** and **B** is admitted to the reboiler. The samples of distillate and residue collected when the column is operated under total reflux condition analyzed and gave $x_D = 0.70$ and $x_B = 0.3$. Calculate HETP if the equilibrium data (mole fraction **A**) are:

x	0	0.1	0.2	0.3	0.4	0.5	0.6	0.7	0.8	0.9	1.0
y	0	0.11	0.22	0.325	0.426	0.526	0.625	0.720	0.815	0.91	1.0

(**Answer:** 206.9 mm)

5.128 An aqueous solution of ethanol containing 10 mol% ethanol is admitted to a packed distillation column at its bubble point. The distillate is to contain 80 mol% ethanol and the bottoms contain not more than 1 mol% ethanol. The reflux ratio is 50% in excess of the minimum. Assuming that the height of overall gas transfer unit is 0.4 m determine the height of packed column given the following equilibrium data (mole fraction of ethanol):

x	0.00	0.05	0.10	0.15	0.20	0.30	0.40
y	0.000	0.267	0.418	0.517	0.570	0.665	0.729
x	0.50	0.60	0.70	0.80	0.90	0.95	1.00
y	0.779	0.825	0.870	0.915	0.958	0.979	1.000

(**Answer:** 3.28 m)

5.129 An equimolar solution consisting of components **A** and **B** is admitted to a packed column for fractionation at 101.3 kPa. The feed is liquid at its bubble point and it is admitted at a rate 60 kmol/m^2 h. The distillate and bottoms product composition are 75 mol% **A** and 5 mol% **A**, respectively. A total condenser is used and the reflux ratio used is 1 kmol/kmol distillate. The individual gas side and liquid side mass transfer coefficients for the enriching section are 200 and 600 kmol/m^3 h, respectively and those for the stripping section are 300 and 1200 kmol/m^3 h respectively. Assume that the variation of gas and liquid flow rates along the length of the column is negligible. Determine the height of the packed section of the column.
Equilibrium data (mole fraction **A**) are:

x	0.00	0.05	0.10	0.15	0.20	0.30	0.40
y	0.000	0.267	0.418	0.517	0.570	0.665	0.729
x	0.50	0.60	0.70	0.80	0.90	0.95	1.00
y	0.779	0.825	0.870	0.915	0.958	0.979	1.000

(**Answer:** 0.84 m)

Chapter 6

Humidification Operations

Humidification and dehumidification are operations in which a pure liquid is in contact with an inert gas phase which is insoluble in the liquid. In humidification, the liquid vaporises and the vapours are transferred to the gas phase, whereas in dehumidification, the vapours are condensed and transferred from the gas phase to the liquid phase. In majority of these operations, the liquid phase is pure water and the gas is air. Since the liquid phase is pure, there are no concentration gradients and no resistance to mass transfer exists in the liquid phase; the mass transfer in these operations is confined to the gas phase. In this respect, the process is simpler than the gas absorption and the distillation, which are operations in which mass transfer occurs in both liquid and gas phases. However, the analysis of the process is complex because it is not the mass transfer alone that controls the rate; it is the simultaneous transfer of mass and heat that governs the rate of humidification and dehumidification operations.

It is true that in almost all separation operations involving mass transfer, the transfer of mass is accompanied by the transfer of heat as well. Heat transfer is of minor significance as a rate limiting mechanism in operations such as distillation, gas absorption involving dilute solutions, extraction, leaching, etc. In contrast, humidification and dehumidification are operations in which heat transfer significantly influences the rate of mass transfer.

Applications

1. **Liquid cooling:** Liquid can be cooled by bringing the hot liquid into direct contact with the gas phase. Cooling occurs by the transfer of sensible heat and by evaporation. Cooling towers are used for cooling the water recycled from various processing units and power generating stations in the process plant. Cooling towers operate with the atmospheric air as the coolant for process water and the operation is a typical humidification operation.
2. **Gas cooling:** Gas can be cooled by direct contact with cold liquid. If the presence of the vapours in the gas is not objectionable, the gas cooling by humidification offers substantial saving in the cost of equipment and maintenance. The gas cooling by humidification is free from the fouling problems associated with the gas cooling using heat exchangers.
3. **Adiabatic humidification:** There are certain processes that require gas of controlled humidity and temperature. For example, in drying of certain solids, humidified air will be necessary to control the rate of drying and to avoid case hardening. The gas humidity can be increased by adiabatic humidification.

4. **Dehumidification:** Adiabatic dehumidification of the gas is achieved by the direct contact with a cold liquid. Dehumidification is also achieved under non-adiabatic condition by passing the gas-vapour mixture in contact with refrigerated pipes. The application of dehumidification include:
   (a) *Solvent recovery*: The recovery of the solvent vapours from the gas is achieved by dehumidification, e.g. recovery of carbon tetrachloride, trichloroethylene, benzene, methanol, etc. are achieved by dehumidification.
   (b) *Gas drying*: Moisture present in the gases can be removed by dehumidification. Presence of moisture in the chlorine gas leads to severe corrosion of the steel equipment handling the wet gas. The gas is dried by removing the chlorine by dehumidification. Moisture present in the gas admitted to the sulphuric acid converters can be removed by a dehumidification operation.
   (c) *Air-conditioning*: Apart from houses and buildings, the air in certain plant areas in process industries have to be maintained at the desired temperature and humidity. Air conditioning is the process of maintaining the water content in the air at some desired level. Dehumidification is an important operation in the air conditioning systems.
5. **Evaporative cooling:** This is essentially a non-adiabatic operation in which the gas or the liquid to be cooled is passed through the heat exchanger tubes and water is allowed to flow as a thin film around the tubes. The film of water is cooled by vaporisation and transfer of sensible heat to the air passed in contact with the liquid.
6. **Water desalination:** The humidification-dehumidification processes offer a technique for desalination of water. The process claims several advantages such as the flexibility in capacity, moderate costs, etc. over the conventional methods of desalination such as evaporation, distillation and reverse osmosis.

6.1 VAPOUR-LIQUID EQUILIBRIUM FOR PURE SUBSTANCES

Humidification operations involve gas-liquid interactions, more frequently, air-water interactions. Since a pure liquid is in contact with a gas phase and vaporisation or condensation occurs till equilibrium is established between the gas and the liquid, it is necessary to know the vapour-liquid equilibrium characteristics of pure substances. Consider a liquid **A** (say, water) in contact with a non-condensable gas **B** (say, air) in a closed container at a constant temperature and pressure. Initially the air is dry so that the partial pressure of water vapour in the air is zero. When the dry air comes in contact with water, vaporisation occurs and the air picks up moisture from the liquid. As a result, the partial pressure of water in the gas increases. This continues till equilibrium is established between the liquid and the gas. When equilibrium between the liquid and the gas is finally established, the partial pressure of water vapour in the gas stream attains a value that remains constant unless the equilibrium is disturbed. Even if the liquid and gas are allowed to be in contact for an infinitely long period, the equilibrium partial pressure exerted by the vapour in the gas will not change as long as the temperature is not altered. At this state, the gas is said to be *saturated* with the vapours of the liquid. The partial pressure exerted by the vapour under saturation is equal to the *vapour pressure* of the liquid at the prevailing temperature. If the gas is *unsaturated*, the partial pressure of the vapour will be less than the vapour pressure of the liquid at that temperature.

At moderate pressures, the vapour pressure of a substance is independent of pressure. The equilibrium partial pressure of substance **A** in the gas phase is, therefore, a unique function of temperature and is identical to the vapour pressure curve of substance **A** as shown in Fig. 6.1. Figure shows the vapour pressure of water as a function of temperature. At a given temperature, water exerts a definite vapour pressure; the vapour pressure increases with increase in temperature. If the water at this temperature is in equilibrium with an insoluble gas such as air, the partial pressure exerted by the water vapour in the gas phase will be equal to the vapour pressure. For example, at 340 K, the vapour pressure of water is

27.34 kPa. If the air is in contact with liquid water at 340 K, the partial pressure exerted by the water in the air under equilibrium will be 27.34 kPa. When moist air with a partial pressure of water less than 27.34 kPa is contacted with water at 340 K, water vaporises and the air gets humidified.

The vapour pressure versus temperature plot serves as the phase equilibrium curve for the humidification and the dehumidification operations. The equilibrium can also be represented in the conventional way by plotting the concentrations expressed as mole fractions. The mole fraction in the gas phase y is calculated using Dalton's law:

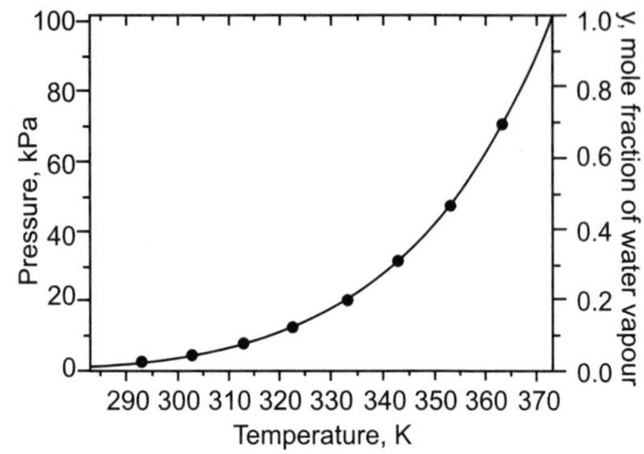

Fig. 6.1: Vapour pressure of pure water versus temperature and equilibria for the system air-water at 101.3 kPa

$$y = \frac{\bar{p}_A}{P} = \frac{P_A^S}{P} \quad \text{...(6.1)}$$

The liquid phase being pure **A**, mole fraction of the liquid cannot be used to represent the equilibrium; temperature of the liquid is plotted as the x-coordinate of the equilibrium curve in the place of mole fraction x. The y versus T curve for the air-water vapour system is also shown in Fig. 6.1.

Example 6.1: The vapour pressure of acetone at 300 K is 33.26 kPa. For a mixture of nitrogen gas saturated with the vapours of acetone at 300 K and 101.3 kPa, calculate:

(a) The mole percent of acetone in the mixture

(b) The percent composition by weight

(c) The kg of vapour per m^3 of the mixture

***Solution*:** Since the nitrogen gas is saturated with acetone vapours, the partial pressure of acetone in the gas at 300 K, $\bar{p}_A = P_A^S$ = the vapour pressure of acetone at 300 K = 33.26 kPa.

(a) The mole fraction of acetone $= \dfrac{\text{Partial pressure}}{\text{Total pressure}} = \dfrac{\bar{p}_A}{P} = \dfrac{P_A^S}{P} = \dfrac{33.26}{101.3} = 0.3283$

Mole percent = mole fraction × 100 = 32.83% **(Answer)**

(b) Molecular weight of acetone = 58, molecular weight of nitrogen = 28.0

One kmol of the gas mixture contains 0.3283 kmol acetone, 0.6717 kmol of nitrogen. That is, one kmol of the gas mixture contains 0.3283 × 58 = 19.04 kg acetone and 0.6717 × 28 = 18.81 kg nitrogen. Therefore, the composition in weight percent is:

$$\text{Acetone} = \frac{19.04}{(19.04 + 18.81)} \times 100 = 50.30\%$$

$$\text{Nitrogen} = \frac{18.81}{(19.04 + 18.81)} \times 100 = 49.70\%$$ **(Answer)**

(c) Volume of one kmol gas at 300 K and 101.3 kPa =

$$V_0 \times \frac{P_0}{P} \times \frac{T}{T_0} = 22.4 \times \frac{101.3}{101.3} \times \frac{300}{273.15} = 24.60 \text{ m}^3$$

Kilograms of vapour present in one kmol of gas = 19.04 kg

$$\text{Concentration of the vapour} = \frac{19.04}{24.60} = 0.7740 \text{ kg/m}^3$$ **(Answer)**

6.2 DEFINITIONS

Humidification involves contact between a pure liquid (**A**) and a gas phase consisting of an insoluble gas (**B**) and the vapours of the liquid (**A**). Certain special terms and definitions are used in the study of humidification operations, particularly those involving water (**A**)-air (**B**) interactions.

Humidity

The term *humidity* is used to indicate the concentration of the vapour in a gas-vapour mixture and is defined as the amount of vapour present per unit quantity of vapour-free gas. For air-water vapour system, humidity refers to the amount of water vapour present per unit quantity of dry air. Humidity can be expressed on mole basis or on mass basis. The *molal humidity* is defined as the moles of vapour present per mole of vapour-free gas. Let there be n_A moles of vapour and n_B moles of dry gas in a given sample of the gas-vapour system. Then,

$$\text{Molal humidity, } Y = \frac{\text{moles of vapour}}{\text{moles of dry gas}} = \frac{n_A}{n_B} \quad \ldots(6.2)$$

Assuming that the gas phase behaves as an ideal gas, the following equations can be used to determine the number of moles of the constituents.

$$n_A = \frac{\bar{p}_A V}{RT}$$

$$n_B = \frac{\bar{p}_B V}{RT} \quad \ldots(6.3)$$

where V is the volume of the mixture and T is the temperature. Combining Eqs. (6.2) and (6.3), we get

$$\text{Molal humidity, } Y = \frac{n_A}{n_B} = \frac{\bar{p}_A}{\bar{p}_B} = \frac{\bar{p}_A}{P - \bar{p}_A} \quad \ldots(6.4)$$

Equation (6.4) means that the molal humidity is equal to the ratio of the partial pressure of the vapour to the partial pressure of vapour-free gas.

If the gas were saturated with the vapours, the molal humidity is the *molal humidity at saturation* denoted as Y_S. It is equal to the moles of vapour per mole of dry gas under saturation. Since, at saturation, the partial pressure of the vapour is equal to the vapour pressure, the molal saturation humidity can be written as

$$Y_S = \frac{P_A^S}{P - P_A^S} \quad \ldots(6.5)$$

where P_A^S is the vapour pressure of the liquid **A**.

The *absolute humidity* is defined as the mass of vapour present per unit mass of vapour-free gas. For the air-water vapour system it is the kilograms of water present per kg of dry air. If m_A and m_B are the weights of the vapour and vapour-free gas, then

$$\text{Absolute humidity, } Y' = \frac{m_A}{m_B} \quad \ldots(6.6)$$

The absolute humidity is related to the molal humidity as:

$$Y' = \frac{m_A}{m_B} = \frac{n_A}{n_B} \times \frac{M_A}{M_B} = Y \times \frac{M_A}{M_B} \quad \ldots(6.7)$$

where M_A and M_B are the molecular weights of **A** and **B**, respectively. Combining with Eq. (6.4), the absolute humidity can be expressed in terms of the partial pressures of **A** and **B** in the gas phase.

$$Y' = \frac{\bar{p}_A}{P - \bar{p}_A} \times \frac{M_A}{M_B} \quad \ldots(6.8)$$

Noting that for air, the average molecular weight is approximately 29 and the molecular weight of water is 18, the absolute humidity of the air-water vapour system can be related to the partial pressure as

$$Y' = \frac{\bar{p}_A}{P - \bar{p}_A} \times \frac{18}{29} \qquad \ldots(6.9)$$

The *absolute saturation humidity* (Y_S') denotes the kilograms of water vapour that one kg dry air can hold at saturated conditions. Analogous to Eq. (6.5), we can write

$$Y_S' = \frac{P_A^S}{P - P_A^S} \times \frac{M_A}{M_B} \qquad \ldots(6.10)$$

At the boiling point of the liquid, the vapour pressure becomes equal to the total pressure so that the denominator in Eq. (6.10) reduces to zero. It means that as the boiling point of the liquid is approached, the saturation humidity tends to infinity. We can see from Eqs. (6.5) and (6.10) that at a given temperature, the molal humidity at saturation is independent of the nature of the gas and depends only upon the pressure, whereas the absolute saturation humidity depends upon the characteristics of the gas as well.

Example 6.2: A mixture of acetone vapour and nitrogen gas at 101.3 kPa and 300 K contains acetone vapour to the extent that it exerts a partial pressure of 20 kPa. The vapour pressure of acetone is given by the Antoine equation

$$\ln P^s = 14.39155 - \frac{2795.817}{T - 43.198}$$

where the pressure is in kPa and temperature is in K. Determine the following:

(a) The molal humidity
(b) The absolute humidity
(c) The molal saturation humidity
(d) Absolute saturation humidity

***Solution*:** (a) Molal humidity is the ratio of moles of acetone to moles of dry nitrogen. It is given by Eq. (6.4).

$$Y = \frac{\bar{p}_A}{P - \bar{p}_A} = \frac{20}{101.3 - 20} = 0.2460 \ \frac{\text{moles acetone}}{\text{mole nitrogen}} \qquad \textbf{(Answer)}$$

(b) Absolute humidity is the ratio of kilograms of acetone to kg of nitrogen. It is obtained by multiplying the molal humidity by the ratio of the molecular weights of acetone and nitrogen.

$$Y' = 0.2460 \times \frac{58}{28} = 0.5096 \ \frac{\text{kg acetone}}{\text{kg nitrogen}} \qquad \textbf{(Answer)}$$

(c) Vapour pressure of acetone at 300 K, from Antoine equation

$$\ln P^s = 14.39155 - \frac{2795.817}{300 - 43.198}$$

$$P^s = 33.265 \text{ kPa}$$

$$\text{Saturation humidity,} \quad Y_S = \frac{P^S}{P - P^S} = \frac{33.265}{101.3 - 33.265} = 0.4889 \ \frac{\text{moles of acetone}}{\text{mole of nitrogen}} \qquad \textbf{(Answer)}$$

(d) Absolute saturation humidity $Y_S' = 0.4889 \times \frac{58}{28} = 1.0128 \ \frac{\text{kg acetone}}{\text{kg nitrogen}}$ **(Answer)**

Relative Humidity and Percent Humidity

The degree of unsaturation of a gas-vapour mixture can be expressed either as the percent relative saturation or as the percent saturation. The *percent relative saturation* or *the percent relative humidity* is the ratio of the actual partial pressure exerted by the vapour to its vapour pressure at the same temperature, expressed as percentage. Thus

$$\text{Relative saturation (relative humidity)} = \frac{\bar{p}_A}{P_A^S} \times 100\% \qquad \text{...(6.11)}$$

The *percent saturation* or the *percent humidity* is the ratio of the moles of vapour present per mole of dry gas in the existing conditions to the moles of vapour per mole of dry gas if the gas were saturated, expressed as percentage. It is the same as the ratio of the kg of vapour present per kg of the dry gas at the given temperature to the kg of the vapour associated with one kg dry gas under saturated conditions at the same temperature.

$$\text{Percent saturation (percent humidity)} = \frac{(n_A/n_B)}{(n_A/n_B)_{\text{saturation}}} \times 100\% \qquad \text{...(6.12)}$$

The numerator in Eq. (6.12) is the molal humidity and the denominator is the molal humidity at saturation. Therefore, the percent humidity can be defined as the ratio of the humidity of the mixture to the humidity in the saturated state.

$$\text{Percent saturation} = \frac{Y}{Y_S} \times 100\% = \frac{Y'}{Y_S'} \times 100\% \qquad \text{...(6.13)}$$

Suppose that the percent saturation is 60%. It means that the vapour content of the mixture per one kg of vapour-free gas is only 60% of the vapour that will be present with one kg dry gas if the gas were saturated with the vapours at the same temperature. In contrast, the relative saturation of 60 percent means that the partial pressure of the vapour in the gas phase is sixty percent of the vapour pressure. The relationship between percent saturation and percent relative saturation can be derived by combining Eqs. (6.4), (6.5) and (6.13).

$$\text{P.S} = \frac{\bar{p}_A}{P_A^S} \times \frac{P - P_A^S}{P - \bar{p}_A} \times 100\% \qquad \text{...(6.14)}$$

Equation (6.14) can be rewritten using Eq. (6.11) as

$$\text{P.S} = \text{R.S} \times \frac{P - P_A^S}{P - \bar{p}_A} \qquad \text{...(6.15)}$$

where P.S and R.S are the abbreviations for percent saturation and relative saturation, respectively. Since the ratio $(P - P_A^S)/(P - \bar{p}_A)$ in Eq. (6.15) is always less than 1, percent saturation is less than relative saturation. That is, P.S < R.S

Example 6.3: A mixture of acetone vapour and nitrogen gas at 101.3 kPa and 300 K contains acetone vapour to the extent that it exerts a partial pressure of 20 kPa. The vapour pressure of acetone at 300 K is 33.26 kPa. Determine the relative saturation and the percent saturation of the mixture.

***Solution*:** The relative saturation is the ratio of the partial pressure of acetone to the vapour pressure expressed as percentage, i.e.,

$$\text{R.S.} = \frac{\bar{p}_A}{P_A^S} \times 100 = \frac{20}{33.26} \times 100 = 60.13\% \qquad \textbf{(Answer)}$$

The percent saturation is given by Eq. (6.13).

Here, the molal humidity, $$Y = \frac{\bar{p}_A}{P - \bar{p}_A} = \frac{20}{101.3 - 20} = 0.2460 \frac{\text{kmol acetone}}{\text{kmol nitrogen}}$$

The molal humidity at saturation, $$Y_S = \frac{p_A^S}{P - p_A^S} = \frac{33.26}{101.3 - 33.26} = 0.4888 \frac{\text{kmol acetone}}{\text{kmol nitrogen}}$$

The percent saturation P.S. $= \dfrac{Y}{Y_s} \times 100 = \dfrac{0.2460}{0.4888} \times 100 = 50.33\%$ **(Answer)**

Example 6.4: Air contains 0.015 kg water vapour per cubic metre of the mixture at 303 K and 101.3 kPa. Determine:

(a) The partial pressure of water vapour
(b) The relative saturation
(c) The absolute humidity of the air
(d) The percent saturation
(e) The temperature to which the mixture be heated so that its percent saturation becomes 10%

The vapour pressure of water (in kPa) is approximated by the Antoine equation as

$$\ln P^s = 16.26205 - \frac{3799.887}{T - 46.854}$$

***Solution*:** The total number of moles that one cubic metre of air at 303 K and 101.3 kPa contains can be calculated using the ideal gas equation:

$$N = \frac{V}{22.414} \times \frac{P}{P_0} \times \frac{T_0}{T} = \frac{1}{22.414} \times \frac{101.3}{101.3} \times \frac{273.15}{303} = 0.0402 \text{ kmol}$$

0.015 kg water = 0.015/18 = 8.33×10^{-4} kmol

(a) Mole fraction of water vapour $= \dfrac{n_W}{N} = \dfrac{8.33 \times 10^{-4}}{0.0402} = 0.0207$

Partial pressure of water vapour = mole fraction × total pressure

$= 0.0207 \times 101.3 = 2.0969$ kPa **(Answer)**

(b) The vapour pressure at 303 K is calculated using Antoine equation.

$$\ln P^S = 16.26205 - \frac{3799.887}{303 - 46.854}$$

Therefore, $P^S = 4.167$ kPa

Relative saturation is the ratio of partial pressure to vapour pressure.

$$\text{R.S.} = \frac{2.0969}{4.167} \times 100 = 50.32\%$$ **(Answer)**

(c) Absolute humidity can be evaluated by Eq. (6.9)

$$Y' = \frac{\bar{p}_A}{P - \bar{p}_A} \times \frac{18}{29} = \frac{2.0969}{101.3 - 2.0969} \times \frac{18}{29}$$

$= 0.01312$ kg water/kg dry air **(Answer)**

(d) Saturation absolute humidity is

$$Y'_S = \frac{p_A^S}{P - p_A^S} \times \frac{18}{29} = \frac{4.167}{101.3 - 4.167} \times \frac{18}{29} = 0.0266 \text{ kg water/kg dry air}$$

Percent saturation $= \text{P.S.} = \dfrac{Y'}{Y'_S} \times 100 = \dfrac{0.01312}{0.0266} \times 100 = 49.32\%$ **(Answer)**

(e) When the mixture is heated at constant pressure, its humidity (Y') will not change. However, saturation humidity at the new temperature will be different from the earlier value. Since the percent saturation at the new temperature is given to be 10%, we have

$$\text{P.S.} = \frac{Y'}{Y'_S} \times 100 = 10.0\% \qquad Y'_S = \frac{Y'}{0.1} = \frac{0.01312}{0.1} = 0.1312 \text{ kg water/kg dry air.}$$

But, by Eq. (6.10), $Y'_S = \dfrac{P_A^S}{101.3 - P_A^S} \times \dfrac{18}{29} = 0.1312$ kg water/kg dry air.

Therefore, $P_A^S = 17.68$ kPa. The temperature corresponding to this vapour pressure is calculated using the Antoine equation. This temperature is 330.6 K. **(Answer)**

Dew Point

Dew point is the lowest temperature to which a vapour-gas mixture can be cooled at constant pressure without condensation, out of contact with the liquid. On cooling a gas to its dew point, the gas gets saturated with the vapours, without any additional vapour input. Stated in another way, the dew point is that temperature at which the vapour pressure of the liquid and the partial pressure of the vapour in the vapour-gas mixture are the same.

Consider an air-water vapour system at 101.3 kPa and 313.15 K with a partial pressure of water, say, 1.7 kPa. At this condition, the humidity of the air can be estimated to be $Y' = \dfrac{\bar{p}_A}{P - \bar{p}_A} \times \dfrac{18}{29}$ $= \dfrac{1.7}{101.3 - 1.7} \times \dfrac{18}{29} =$ 0.0106 kg water per kg dry air. The vapour pressures of water at different temperatures are listed in Table 6.1.

Table 6.1: Vapour pressures of water at different temperatures

T, K	283.15	288.15	293.15	298.15	303.15	308.15	313.15
P_A^S, kPa	1.23	1.70	2.34	3.17	4.24	5.62	7.36

Since the vapour pressure of water at 313.15 K is 7.36 kPa, the relative humidity is $\dfrac{1.7}{7.36} \times 100 =$ 23.1%. If this mixture is cooled to say, 303.15 K, its humidity remains at 0.0106 kg water per kg dry air, since the partial pressure of water is still 1.7 kPa. However, since the vapour pressure at the new temperature is 4.24 kPa, the relative saturation becomes $\dfrac{1.7}{4.24} \times 100 = 40.1\%$. On cooling further to say, 293.15 K, the relative humidity increases to 72.6%. If the cooling is continued, eventually the air becomes saturated (R.S. = 100%) at a temperature of 288.15 K, which is the dew point of the given air-water vapour mixture. The vapour pressure of water at this temperature equals the partial pressure of the water in the original air. If cooling is continued to temperature below the dew-point, the water vapour condenses and the humidity decreases, air remaining saturated at all temperatures below the dew-point. Suppose that the air is cooled to, say, 283.15 K, the partial pressure reduces to 1.23 kPa, the vapour pressure at 283.15 K. This corresponds to a humidity of 0.0076 kg water vapour per kg dry air. Since the initial humidity was 0.0106 kg water per kg dry air, cooling the air by 5 K below the dew point, results in condensation of 0.003 kg water vapour per every kg of dry air.

Since the dew point is that temperature at which the saturation humidity will be equal to the humidity of the given air-water vapour system, knowing the dew point and the vapour pressure of water as a function of the temperature, the humidity of the air is readily evaluated. However, to establish the state of the given mixture uniquely one more specification such as the temperature, the relative saturation or the percent saturation also is necessary.

Dew point of air can be determined experimentally by cooling a well polished metal surface in the presence of air and measuring the temperature at which, the water vapour begins to condense and a fog appears on the metallic surface. As the dew point is a measure of the humidity, this is one of the methods for measuring atmospheric humidity.

Example 6.5: Air-water vapour mixture at 325 K and 120 kPa has a dew point of 293 K. The vapour pressure of water is given by the Antoine equation (P in kPa, T in K)

$$\ln P^s = 16.26205 - \frac{3799.887}{T - 46.854}$$

Determine:

(a) The absolute humidity of air
(b) The saturation humidity
(c) The relative saturation and percent saturation
(d) The dew point if the total pressure is reduced to 100 kPa.

***Solution*:** The partial pressure of water vapour is equal to the vapour pressure at the dew point of 293 K. The vapour pressure at 293 K is calculated from Antoine equation:

$$P^S = \exp\left(16.26205 - \frac{3799.887}{293 - 46.854}\right) = 2.28 \text{ kPa.}$$

(a) Partial pressure of water $\bar{p}_A$ = vapour pressure at the dew point = 2.28 kPa
Absolute humidity:

$$Y' = \frac{\bar{p}_A}{P - \bar{p}_A} \times \frac{M_A}{M_B} = \frac{2.28}{120 - 2.28} \times \frac{18}{29} = 0.012 \text{ kg water/kg dry air.}$$ **(Answer)**

(b) The vapour pressure of water at 325 K

$$= P_A^S = \exp\left(16.26205 - \frac{3799.887}{325 - 46.854}\right) = 13.47 \text{ kPa.}$$

Saturation humidity:

$$Y_S' = \frac{P_A^S}{P - P_A^S} \times \frac{M_A}{M_B} = \frac{13.47}{120 - 13.47} \times \frac{18}{29} = 0.0785 \text{ kg water/kg dry air.}$$ **(Answer)**

$$\text{(c) Relative humidity} = \frac{\bar{p}_A}{P_A^S} \times 100 = \frac{2.28}{13.47} \times 100 = 16.93\%$$

$$\text{Percent humidity} = \frac{Y'}{Y_S'} \times 100 = \frac{0.012}{0.0785} \times 100 = 15.29\%$$ **(Answer)**

(d) The mole fraction of water vapour in the air is equal to the ratio of partial pressure to total pressure.

$$y = \frac{\bar{p}_A}{P} = \frac{2.28}{120} = 0.019$$

Since the humidity remains the same, the partial pressure of water at 100 kPa = mole fraction × total pressure = 0.019×100 = 1.9 kPa. Using the Antoine equation, the temperature at which the vapour pressure of water is 1.9 kPa is found out, which is the dew point at 100 kPa.

Dew point = 290.12 K. **(Answer)**

Dry-bulb Temperature

In psychrometry, the temperature of the gas vapour-mixture that is measured by an ordinary thermometer or other temperature measuring devices placed in the gas is known as the dry-bulb temperature.

Humid Volume

The volume of one kg dry gas and the accompanying vapour is known as the humid volume of the vapour-gas mixture. Assuming the ideal gas law to be applicable,

$$V_H = V_0 \times \frac{P_0}{P} \times \frac{T}{T_0} \qquad ...(6.16)$$

where V_H is the humid volume and V_0 is the volume occupied by one kg of the dry gas and the accompanying vapour at the standard temperature and pressure.

$$V_0 = \left(\frac{1}{M_B} + \frac{Y'}{M_A}\right) \times 22.414 \qquad ...(6.17)$$

where V_0 is in m^3. For the gas at the dry-bulb temperature T, pressure P (kPa) and having humidity Y', the humid volume is calculated using Eqs. (6.16) and (6.17) as

$$V_H = \left(\frac{1}{M_B} + \frac{Y'}{M_A}\right) \times 22.414 \times \frac{101.325}{P} \times \frac{T}{273.15} \qquad ...(6.18)$$

V_H is the humid volume in m^3/kg dry air. M_A and M_B are the molecular weights of the vapour and the dry gas, respectively. For the air-water vapour system, Eq. (6.18) gives

$$V_H = 8.314\left(\frac{1}{29} + \frac{Y'}{18}\right) \times \frac{T}{P} \qquad ...(6.19)$$

where P is in kPa and V_H is in m^3/kg dry air. Given the humidity of the air, the humid volume at any temperature can be calculated using this equation.

Humid Heat

Humid heat (C_S) is the amount of heat required to raise the temperature of one kg of dry gas and the accompanying vapour by one degree. Thus it represents the heat capacity of the gas-vapour system at constant pressure. If the specific heat of the vapour is C_A (kJ/kg K) and that of the dry gas is C_B (kJ/kg K), then the humid heat is

$$C_S = C_B + Y'C_A \qquad ...(6.20)$$

For the air-water vapour system, C_A = 1.884 kJ/kg K and C_B = 1.005 kJ/kg K.

Knowing the humid heat of the air-water system, the heat requirement in the operations such as drying and humidification can be easily calculated as

$$Q = G_s C_s \Delta T \qquad ...(6.21)$$

where Q is the quantity of sensible heat required for heating without phase change G_S kg of dry air and the accompanying vapour by (ΔT) K.

Example 6.6: Air at 101.3 kPa and 328 K has a volume of 100 m^3 and is 10% saturated with water vapour. The vapour pressure of water at 328 K is 15.7 kPa. Determine:

(a) The humid volume of the sample
(b) The humid heat
(c) The amount of heat to be supplied to heat the given quantity of air to 370 K.

***Solution*:** Since the vapour pressure of water at 328 K is 15.7 kPa, the saturation humidity of the air is

$$Y'_S = \frac{p_A^S}{P - p_A^S} \times \frac{18}{29} = \frac{15.7}{101.3 - 15.7} \times \frac{18}{29} = 0.1138 \text{ kg water/kg dry air}$$

Since the air is 10% saturated, the humidity

$$Y' = 0.1 \times Y'_S = 0.10 \times 0.1138 = 0.0114 \text{ kg water/kg dry air}$$

(a) Eq. (6.19) is used to calculate the humid volume.

$$V_H = 8.314\left(\frac{1}{29}+\frac{Y'}{18}\right)\times\frac{T}{P}$$

$$V_H = 8.314\left(\frac{1}{29}+\frac{0.0114}{18}\right)\times\frac{328}{101.3} = 0.9453 \text{ m}^3\text{/kg dry air} \qquad \textbf{(Answer)}$$

(b) The humid heat $C_S = C_B + Y'C_A$

$$C_S = 1.005 + 0.0114(1.884) = 1.0265 \text{ kJ/(kg dry air)K.} \qquad \textbf{(Answer)}$$

(c) Since the humid volume is 0.9453 m³/kg dry air, 100 m³ of the given air contains 100/0.9453 = 105.79 kg dry air with its accompanying vapour. The heat required for heating 105.79 kg dry air and accompanying vapour from 328 to 370 K is given by Eq. (6.21).

$$Q = G_S C_S \Delta T = 105.79 \times 1.0265 \times (370 - 328) = 4560.92 \text{ kJ.} \qquad \textbf{(Answer)}$$

Specific Enthalpy

Enthalpy of the gas-vapour system is calculated relative to an arbitrarily chosen reference state. For the air-water vapour system, the reference temperature chosen for enthalpy calculation is generally, 273.15 K at which water exists as saturated liquid. If the humidity is Y' and the dry-bulb temperature is T, the specific enthalpy of the humid air is the heat required for bringing one kg of dry air and Y' kg of the liquid water from the initial temperature T_0 to the final state at T at constant pressure. Thus the enthalpy H' of the air is

$$H' = C_B(T - T_0) + Y'\left[\lambda_0 + C_A(T - T_0)\right] \qquad \text{...(6.22)}$$

where λ_0 is the latent heat of vaporisation of the water at the reference temperature T_0. Latent heat of vaporisation of the water at 273.2 K (λ_0) can be taken to be 2502.3 kJ/kg. Using Eq. (6.20) for the humid heat, Eq. (6.22) can be modified as

$$H' = C_S(T - T_0) + Y'\lambda_0 \qquad \text{...(6.23)}$$

Equation (6.23) reveals that the enthalpy can be kept constant in a process in which the humidity is increased and the temperature is reduced, as happens in the adiabatic humidification operations to be discussed later.

Example 6.7: Determine the specific enthalpy of air at 101.3 kPa and 328 K, 10% saturated with water vapour. The vapour pressure of water at 328 K is 15.7 kPa.

***Solution*:** The humidity of air at 328 K with percent saturation of 10% was found to be 0.0114 kg water vapour per kg dry air in Example 6.6.

$$Y' = 0.0114 \text{ kg water/kg dry air}$$

Enthalpy can be calculated using Eq. (6.23)

$$H' = C_S(T_G - T_0) + Y'\lambda_0$$

where $C_S = C_B + Y'C_A = 1.005 + 0.0114(1.884) = 1.0265$ kJ/kg dry air K.

$T = 328$ K, $T_0 = 273.15$ K, $\lambda_0 = 2502.3$ kJ/kg.

Substituting in Eq. (6.23) we get

$$H' = 1.0265(328 - 273.15) + 0.0114(2502.3) = 84.83 \text{ kJ/kg dry air} \qquad \textbf{(Answer)}$$

6.3 HUMIDITY CHART

The humidity charts usually refer to the air-water vapour systems, although charts for any vapour-gas combination can be constructed. Humidity chart for the air-water system at atmospheric pressure is shown in Fig. 6.2. The ordinate represents the absolute humidity Y' (kg water vapour per kg dry air)

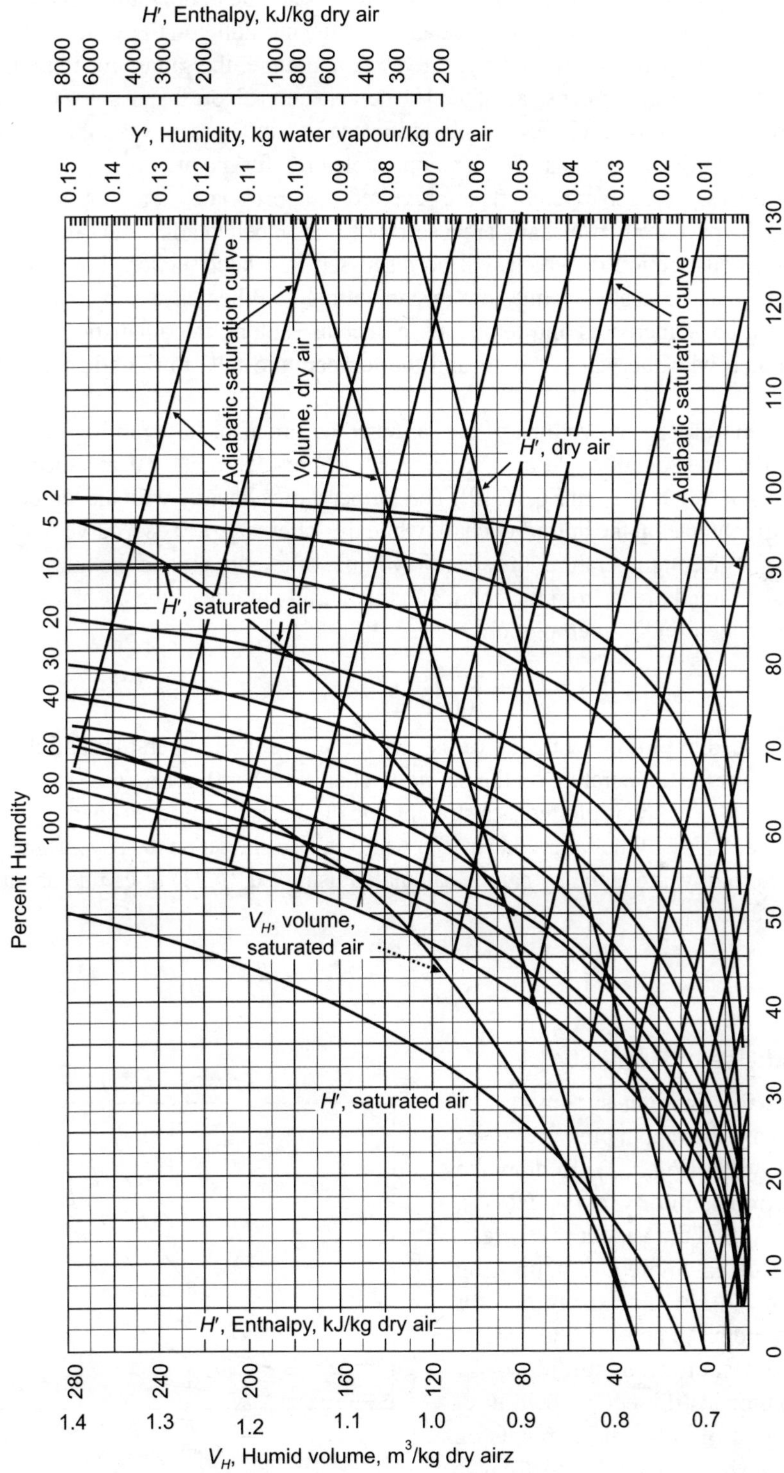

Fig. 6.2: Psychrometric chart for air-water vapour system at 1 std atmosphere

and the abscissa represents the dry-bulb temperature. The absolute saturation humidity at a given temperature can be read from the 100 percent saturation curve. Take the temperature on the horizontal axis and move vertically up to the 100 percent curve and read the humidity on the vertical axis. To find the humidity of air of say, 50 percent saturation at a given temperature, the same procedure is adopted but using the 50 percent saturation curve instead of the 100 percent curve. Note that in some psychrometric charts, the percent relative saturation curves are plotted instead of the percent saturation curves.

One limitation of humidity charts is that they are applicable only for the pressure for which they are constructed. For other pressures, separate charts are required or appropriate correction terms are to be used. Also plotted on the psychrometric charts are a series of lines labelled adiabatic saturation curves, the properties and the construction of which will be discussed in a later section. For most practical calculations, these lines can be treated as lines of constant enthalpy. When air is subjected to adiabatic humidification, the humidity increases and temperature decreases along the adiabatic cooling curves till the air gets saturated with the water vapour and the temperature falls to the adiabatic saturation temperature.

The volume of the dry air and the volume of the saturated air are also shown in Fig. 6.2. These are plotted using Eq. (6.19). The specific volume of the dry air (V_D) results for $Y' = 0$ and specific volume of saturated air (V_s) results for $Y' = Y_s'$ in Eq. (6.19). The former varies linearly with the temperature and is represented by a straight line on the humidity chart, while the latter is shown as a curve that is slightly concave upwards. For a given temperature, the specific volume of the dry air and the humid volume of the saturated air can be directly read from the chart, and the humid volume of the air of known percent humidity calculated using a linear interpolation.

$$V_H = V_D + \frac{\text{P.S}}{100}(V_S - V_D) \qquad \text{...(6.24)}$$

Equation (6.23) is used to plot the enthalpy of the dry air versus the temperature and the enthalpy of the saturated air versus the temperature on the humidity chart. Put $Y' = 0$ in Eq. (6.23) and we get the enthalpy of the dry air (H_D') which is plotted as a straight line on the psychrometric chart. For $Y' = Y_s'$ in Eq. (6.23), we get the enthalpy of the saturated air (H_S') which is also shown on the psychrometric chart. For air of the given humidity, the enthalpy can be calculated using Eq. (6.23) or can be obtained from the chart by interpolation:

$$H' = H_D' + \frac{\text{P.S}}{100}(H_S' - H_D') \qquad \text{...(6.25)}$$

Construction and Use of Humidity Chart

The humidity charts for air-water system can be constructed knowing just the vapour pressure versus temperature data for water (Fig. 6.3). To draw the 100 percent saturation humidity curve (*BCD*), assume a temperature, find the vapour pressure at that temperature and calculate the saturation humidity using Eq. (6.10). For example, for temperature corresponding to point *A*, the vertical height *AC* gives the saturation humidity. By repeating the calculation at different tempera-tures below the boiling point of water, other points on the saturation curve can be obtained. A smooth curve drawn through these points establishes the 100 percent

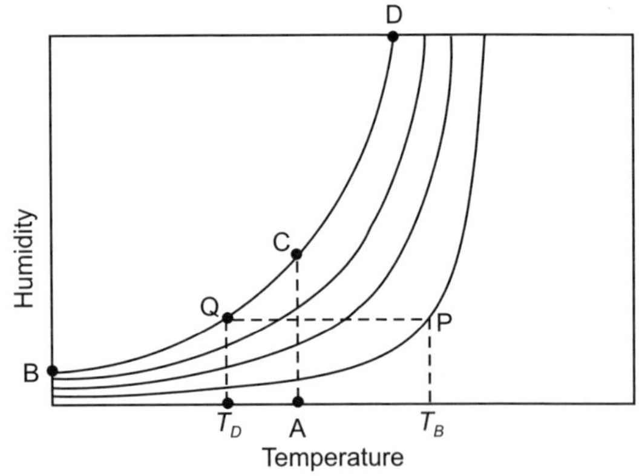

Fig. 6.3: Construction of psychrometric chart

saturation curve. Saturation humidity at the boiling point is infinity and therefore the curve rises asymptotically to infinity near the boiling point. Other percent saturation curves are drawn by dividing the vertical distance between the 100% saturation curve and the temperature axis equally into as many divisions as required.

Specifying the dry-bulb temperature and the percent humidity determines the state of the air-water vapour system uniquely. For example, when the dry-bulb temperature is T_B and the percent saturation is 25%, the state of the air-water vapour system is represented by the point P in Fig. 6.3. The dew point of the air at point P is obtained by moving horizontally to point Q on the 100 percent curve and by reading the temperature T_D on the horizontal axis. If the dew point of the air, say T_D is known, the humidity will be the same as that at Q. Then the state of the system can be anywhere on the horizontal line passing through Q. Though by specifying the dew point, the humidity is known, to specify the state completely, one more property should be specified in addition to the dew point.

Example 6.8: An air-water vapour sample at 101.3 kPa has a dry-bulb temperature of 330.7 K and is 20% saturated with water vapour. Using the psychrometric chart, determine the following:

(a) The absolute humidity, kg water vapour per kg dry air.
(b) The partial pressure of water vapour
(c) The absolute saturation humidity at 330.7 K
(d) The vapour pressure of water at 330.7 K
(e) The percent relative saturation
(f) The dew point of the system

***Solution*:** (a) Point P in Fig. 6.4 corresponds to 330.7 K and 20% saturation. The y-coordinate of this point is read from the chart. This is the humidity at this condition. This is equal to 0.0265 kg water vapour/kg dry air.

$$Y' = 0.0265 \text{ kg water per kg dry air.}$$ **(Answer)**

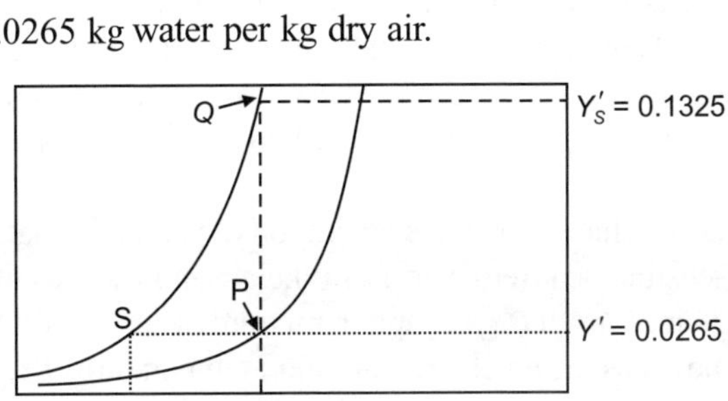

Fig. 6.4: Solution of Example 6.8

(b) Partial pressure is calculated using Eq. (6.8)

$$Y' = \frac{\bar{p}_A}{P - \bar{p}_A} \times \frac{18}{29} = 0.0265$$

$0.04269\left(P - \bar{p}_A\right) = \bar{p}_A$ where P is 101.3 kPa. Solving this, we get $\bar{p}_A = 4.15$ kPa **(Answer)**

(c) The saturation humidity at 330.7 K is obtained directly from the psychrometric chart by reading the y-coordinate of point Q which is the point on the 100% saturation curve corresponding to temperature 330.7 K. This is found to be 0.1325 kg water per kg dry air.

$$Y'_s = 0.1325 \text{ kg water per kg dry air.}$$ **(Answer)**

(d) Equation (6.10) can be used to calculate vapour pressure,

$$Y'_s = \frac{P_A^S}{P - P_A^S} \times \frac{M_A}{M_B} = \frac{P_A^S}{P - P_A^S} \times \frac{18}{29}$$

$$0.1325 = \frac{P_A^S}{101.3 - P_A^S} \times 0.6207$$

Solving this we get

$$P_A^S = 17.8 \text{ kPa}$$ **(Answer)**

(e) The percent relative saturation is obtained using Eq. (6.11)

$$\text{R.S.} = \frac{\bar{p}_A}{P_A^S} \times 100\ \% = \frac{4.15}{17.8} \times 100\% = 23.3\%$$ **(Answer)**

(f) The dew point is obtained by moving from the point P horizontally to point S on the saturation curve and noting its temperature coordinate.

The dew point is found to be 302.7 K. **(Answer)**

(The vapour pressure at the dew point should be equal to the partial pressure of water in the air which is obtained as 4.15 in part (b). Using steam tables, the vapour pressure at 302.7 K is found to be 4.13 kPa)

6.4 ADIABATIC SATURATION TEMPERATURE

Adiabatic saturation temperature is a thermodynamic property of the moist air and is defined as the temperature that the air stream would attain if it were allowed to become saturated adiabatically. For adiabatic humidification, an unsaturated air-water vapour mixture is passed through a thermally insulated and sufficiently large chamber. Figure 6.5 shows an adiabatic humidification chamber. Water is sprayed into the unsaturated air-water vapour mixture. The operation is adiabatic; that is, there is no addition to or removal of heat from the chamber. The quantity of unsaturated gas mixture is small in comparison with the liquid with which it is in contact so that the temperature as well as the humidity of the air changes during the process. Water temperature falls due to evaporative cooling and in continuous operation, water attains the equilibrium temperature which is maintained constant by recirculation. The loss of water by evaporation is compensated by the addition of make up water which is also at the equilibrium temperature.

Humidification of the gas is achieved by the evaporation of water. The latent heat of vaporisation of water is to be met by the sensible heat removed from the air and as a result, the air gets cooled. The equilibrium temperature attained by the gas-vapour mixture when it is adiabatically humidified in contact with the liquid is known as the adiabatic saturation temperature, T_{as}, and the humidity attained by the gas under equilibrium is the saturation humidity at adiabatic saturation temperature, Y'_{as}.

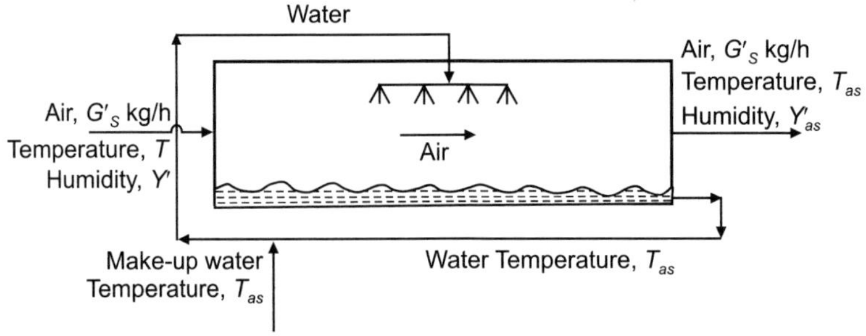

Fig. 6.5: Adiabatic saturation chamber

Let G'_S be the quantity of the dry air circulated, kg/h. The air gets cooled from its initial temperature T to the adiabatic saturation temperature T_{as} and its humidity increases from the initial value Y' to the final value Y'_{as} .The sensible heat removed from the gas stream is $G'_S C_S (T - T_{as})$ where C_S is the humid heat of the air. The quantity of water vaporised is $G'_S (Y'_{as} - Y')$ and the heat required for vaporisation is $G_S'(Y_{as}' - Y')\lambda_{as}$ where λ_{as} is the latent heat of vaporisation of water (kJ/kg) at the adiabatic saturation temperature. The heat balance at the equilibrium conditions requires that, the sensible heat lost by the gas should be equal to the heat of vaporisation of water, so that

$$G_S'C_S(T - T_{as}) = G_S'(Y_{as}' - Y')\lambda_{as} \quad \text{...(6.26)}$$

Equation (6.26) can be rearranged as

$$T_{as} = T - \frac{\lambda_{as}}{C_S}(Y_{as}' - Y') \quad \text{...(6.27)}$$

Eq. (6.27) gives the adiabatic saturation temperature of a gas-vapour mixture at the dry-bulb temperature T and humidity Y'. Equation (6.27) can be rearranged as

$$(T - T_{as}) = \frac{\lambda_{as}}{C_S}(Y_{as}' - Y') \quad \text{...(6.28)}$$

$$(Y_{as}' - Y') = -\frac{C_S}{\lambda_{as}}(T_{as} - T) \quad \text{...(6.29)}$$

On the humidity chart, Eq. (6.29) can be plotted as a curve passing through the point (T, Y') which represents the initial state of the given air (point P in Fig. 6.6) and (T_{as}, Y'_{as}) which lies on the saturation curve (point S). Point S represents the state of the saturated air subjected to adiabatic humidification. The temperature-humidity history of the gas undergoing adiabatic saturation is represented by a point lying on the curve PS, the slope of which is $-\frac{C_S}{\lambda_{as}}$. Since, by definition, the humid heat C_S is equal to $C_B + Y'C_A$, it varies with the humidity. Therefore, the slope is not a constant and the curve PS is not generally a straight line. The curve PS in Fig. 6.6 is known as the *adiabatic saturation curve or* the *adiabatic cooling line*. To make the interpolation easier, generally the coordinates of the humidity charts are manipulated and the adiabatic cooling lines are made straight and parallel. The air-water vapour mixture at any given state has an adiabatic saturation temperature, which the air attains under equilibrium, when it is subjected to adiabatic humidification. Due to the finite size of the saturation chamber, it may not be practicable to attain equilibrium and generally the air leaving the chamber will be unsaturated as represented by a point such as the point R on the cooling line below the 100 percent saturation curve in Fig. 6.6.

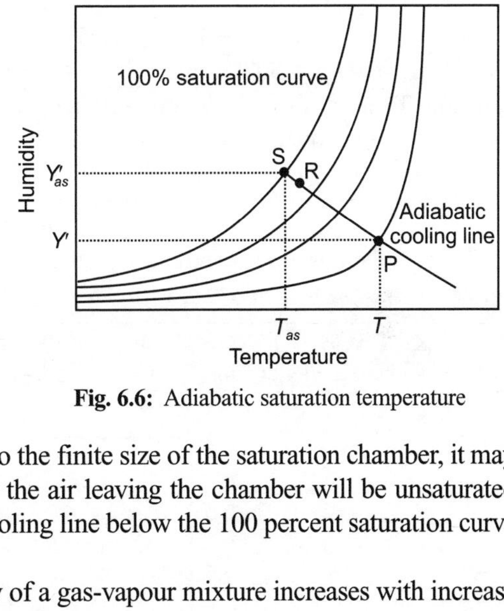

Fig. 6.6: Adiabatic saturation temperature

Refering to Eq. (6.23), we can see that the enthalpy of a gas-vapour mixture increases with increase in the humidity and decreases with decrease in the temperature. It is therefore quite natural to expect that in adiabatic saturation where the humidity of the gas increases and temperature decreases, the enthalpy would remain constant; and it is found to be true and the adiabatic saturation process is essentially a constant enthalpy process.

Example 6.9: The vapour pressure of benzene varies with temperature as

$$\ln P^S = 13.8858 - \frac{2788.51}{T - 52.36}$$

Specific heat of the benzene vapour is 1.2 kJ/kg K and that of the dry air is 1.005 kJ/kg K. The latent heat of vaporisation of benzene at 300 K is 433 kJ/kg. Calculate and plot:

(a) The saturation curve for the benzene-air system at 101.3 kPa for the temperature range 280 K to 310 K with temperature increments of 10 K.

(b) The adiabatic cooling line for the adiabatic saturation temperature of 300 K.

Solution: (a) The vapour pressures at different temperatures are calculated using the Antoine equation. The saturation humidity values are determined at each temperature using Eq. (6.10).

$$Y'_S = \frac{P_A^S}{P - P_A^S} \times \frac{M_A}{M_B} = \frac{P_A^S}{P - P_A^S} \times \frac{78}{29}$$

The results are tabulated below.

T, K	280	290	300	310
P_A^S (kPa)	5.14	8.60	13.81	21.38
Y'_S, kg benzene/kg dry air	0.144	0.250	0.425	0.720

A graph is plotted with Y'_S on the y-axis and temperature on the x-axis as shown in Fig. 6.7.

(Answer)

(b) For a given point on the saturation curve, an adiabatic cooling line passing through it can be drawn using Eq. (6.27) which can be written as

$$T = T_{as} + \frac{\lambda_{as}}{C_S}(Y'_{as} - Y') \qquad \text{...(A)}$$

Here $C_S = C_B + Y' C_A = 1.005 + 1.2Y'$ and $\lambda_{as} = 433$ kJ/kg K. To draw the adiabatic cooling line, for 300 K, put $T_{as} = 300$ K and $Y'_{as} = 0.425$ in Eq.(A) and find the sets of T and Y' that satisfy the equation. To make the calculations simpler, assume Y' (Y' should be lower than Y'_{as}) and calculate T that satisfies the equation. For example, let us assume $Y' = 0.30$ kg benzene/kg dry air. Then

$$T = T_{as} + \frac{\lambda_{as}(Y'_{as} - Y')}{C_s} = 300 + \frac{433(0.425 - Y')}{1.005 + 1.2Y'} = 300 + \frac{433(0.425 - 0.30)}{1.005 + 1.2 \times 0.30} = 339.65 \text{ K}$$

The calculations are carried out for other humidity values and the results are as shown in the table below:

T, K	300	307.29	314.88	322.79	331.04	339.65
Y' kg benzene/kg dry air	0.425	0.40	0.375	0.35	0.325	0.30

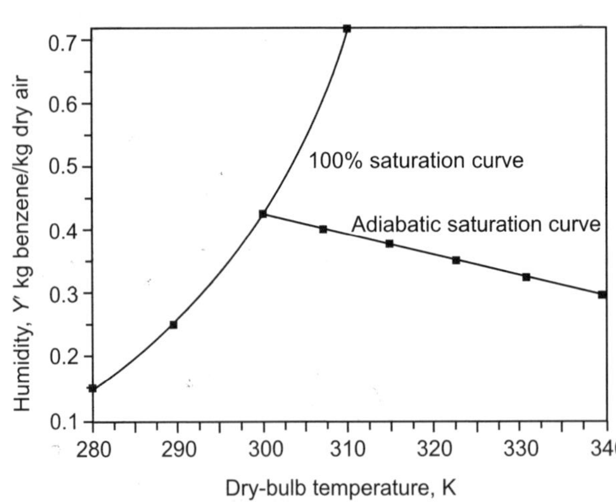

Fig. 6.7: Construction of 100% saturation curve and adiabatic cooling line for Example 6.9

The above points are marked and a smooth curve is drawn joining these points as shown in Fig. 6.7, which gives the adiabatic cooling line for the adiabatic saturation temperature of 300 K. The curve passes through (T_{as}, Y'_{as}) on the 100% saturation curve. In a similar way, the curves for other adiabatic saturation temperatures also can be plotted. **(Answer)**

Example 6.10: Air at 355.7 K has a dew point temperature of 313.2 K. Using psychrometric chart, determine:

(a) The humidity
(b) The percent saturation
(c) Adiabatic saturation temperature
(d) The air is subjected to adiabatic humidification. If the air leaves the humidifier 80% saturated with water vapour what are the humidity and temperature of the exit air?

***Solution*:** (a) Point *P* in Fig. 6.8 represents the air with dew point 313.2 K and dry-bulb temperature 355.7 K. Corresponding to this point on the humidity chart, the humidity can be read on the vertical axis as $Y' = 0.0485$ kg water per kg dry air **(Answer)**

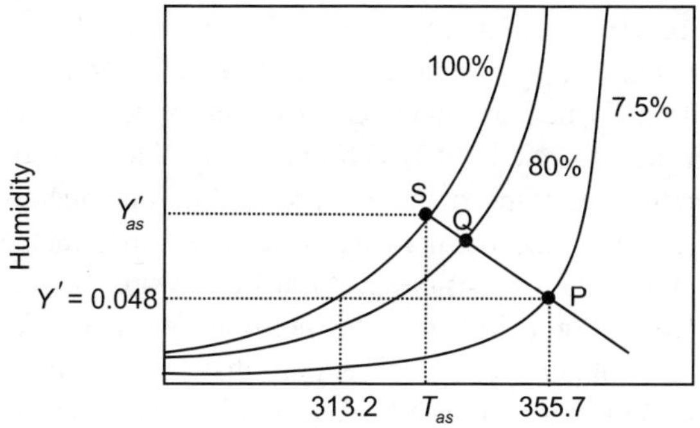

Fig. 6.8: Application of humidity chart for Example 6.10

(b) The percent saturation can be read from the psychrometric chart to be 7.5% **(Answer)**

(c) The adiabatic saturation temperature can be determined by moving along the adiabatic cooling line passing through point *P*. The temperature coordinate of the point *S* on the 100% saturation curve gives the adiabatic saturation temperature. From the humidity chart, $T_{as} = 318.2$ K. **(Answer)**

(d) Since the air leaves the humidifier 80% saturated with water vapour, intersection of the adiabatic cooling line with the 80% saturation curve gives the condition of the exit air from the humidifier. Point *Q* in Fig. 6.8 gives the state of the exit air. The humidity and temperature of the exit air are obtained from the psychrometric chart. They are $T = 321.7$ K; $Y' = 0.0635$ kg water per kg dry air. **(Answer)**

6.5 WET-BULB TEMPERATURE

The wet-bulb temperature is the steady-state temperature attained by a small quantity of the liquid evaporating into a large quantity of unsaturated gas-vapour stream and is a measure of the humidity of the gas-vapour system. The wet-bulb temperature of a gas stream is measured by a wet-bulb thermometer, as the one shown in Fig. 6.9. The bulb of the thermometer used to measure the humidity of the air-water vapour system is usually covered by a wet wick the end of which is immersed in water. The air is passed in such large quantities that its temperature and humidity are not changed by the evaporation of water.

The film of air which is in contact with the bulb of the thermometer will be saturated with the water vapour which exerts a partial pressure equal to the vapour pressure of water. The partial pressure of water vapour in the air passed across the thermometer is less than the vapour pressure unless the air is saturated. This results in the vaporisation of water and the diffusion of the vapour into the air stream.

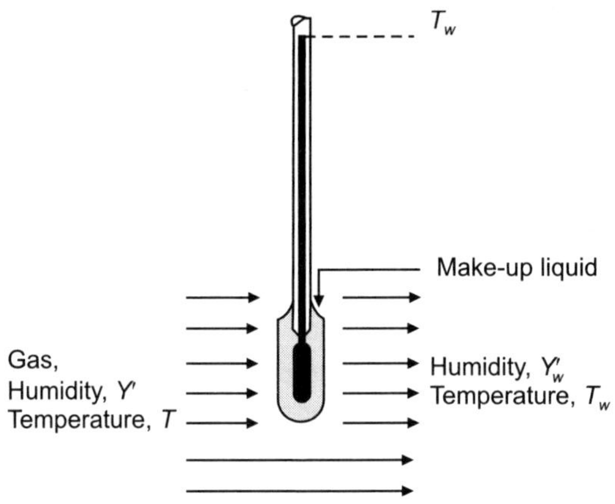

Fig. 6.9: Wet-bulb thermometer

The heat of vaporisation of water is initially supplied by the water film surrounding the bulb and consequently the water cools down, lowering the temperature indicated by the thermometer. When the temperature of the water film falls, heat transfer occurs from the bulk of the air to the thermometer bulb. Thus two processes occur, the directions of which are opposite to each other. First, the transfer of sensible heat from the air to the thermometer and the second, the simultaneous transfer of heat and mass in the opposite direction, that is, from the thermometer to the air stream. Eventually a dynamic equilibrium will be established at which the rate of sensible heat transfer towards the thermometer and the rate at which the latent heat of vaporisation is lost from the thermometer are equal. Under this condition, the thermometer indicates a constant and steady temperature that remains unaltered with further passage of air. The temperature indicated under this steady state condition is known as the wet-bulb temperature of the air.

Let the air passed across the thermometer bulb be at temperature T, humidity Y' and let the wet-bulb temperature of the air be T_W. At steady state, the rate of transfer of sensible heat from the bulk air to the film covering the thermometer bulb (Q_S) is,

$$Q_s = h_G A(T - T_W) \qquad \text{...(6.30)}$$

Here h_G is the film coefficient for heat transfer between the air and the water and A is the area of heat transfer.

The rate at which the heat is transferred from the liquid surrounding the thermometer bulb to the bulk air due to vaporisation (Q_V) is,

$$Q_v = N_A M_A A \lambda_W \qquad \text{...(6.31)}$$

where N_A is the molar flux of vaporisation of liquid (kmol/m^2 h), A is the area of mass transfer (m^2), M_A is the molecular weight of liquid and λ_W is the latent heat of vaporisation of liquid at the wet-bulb temperature, kJ/kg. The molar flux of vaporisation N_A can be written as the product of mass transfer coefficient k_y and the difference between the mole fractions of water vapour in the bulk air and the air in contact with the liquid. Since the concentration of the water vapour in the gas is very small, the bulk flow effect is negligible and the rate of vaporisation can be written as

$$N_A = k_y(y_W - y) \qquad \text{...(6.32)}$$

For small values of mole fraction y, we can write $y = \dfrac{Y}{1+Y} \approx Y$ where Y is the molal humidity. The mole fraction in Eq. (6.32) can be replaced by absolute humidity as

$$y \approx Y = Y' \times \frac{M_B}{M_A} \qquad \text{...(6.33)}$$

so that Eq. (6.32) becomes

$$N_A = k_y \frac{M_B}{M_A}(Y'_W - Y') \qquad \text{...(6.34)}$$

Here M_B is the molecular weight of the dry gas. Substituting Eq. (6.34) into Eq. (6.31), we get

$$Q_v = k_y M_B A \lambda_W (Y'_W - Y') \qquad \text{...(6.35)}$$

Under steady state, the rate of sensible heat transfer from the gas to the liquid surrounding the thermometer bulb is equal to the rate of removal of the heat from the bulb due to vaporisation. Using Eqs. (6.30) and (6.35) we get

$$h_G A(T - T_W) = k_y M_B A \lambda_W (Y'_W - Y') \qquad \text{...(6.36)}$$

It is assumed that the wick covering the thermometer bulb is completely wet and no dry spots are present, so that the area for heat transfer and mass transfer are the same. Equation (6.36) can be written as

$$T_W = T - \frac{k_y M_B}{h_G} \lambda_W (Y'_W - Y') \qquad \text{...(6.37)}$$

The wet-bulb temperature can be calculated using Eq. (6.37). Equation (6.37) can now be put in the following form.

$$(Y' - Y'_W) = -\frac{h_G}{k_y M_B \lambda_W}(T - T_W) \qquad \text{...(6.38)}$$

Given the wet-bulb temperature and the dry-bulb temperature, Eq. (6.38) can be used to evaluate the humidity of the air. The difference between the dry-bulb and wet-bulb temperature $(T - T_W)$ is known as the *wet-bulb depression* and is a measure of the humidity of the air stream. If the air passed across the wet-bulb thermometer is saturated, there will not be any vaporisation and consequent lowering of temperature; the dry-bulb and wet-bulb temperature will be the same and the wet-bulb depression will be zero. The higher the wet-bulb depression, the larger the degree of unsaturation of the air.

The ratio of heat transfer coefficient to mass transfer coefficient $\frac{h_G}{k_y M_B}$, is known as the *psychrometric ratio*. The values of h_G and k_y can be estimated independently for the particular shape of the wetted surface using standard correlations developed from the analogy between heat and mass transfer. Psychrometric ratio can be related to the fluid properties expressed by the Lewis number (*Le*) for the fluid.

$$Le = \frac{\alpha}{D_{AB}} = \frac{Sc}{Pr}$$

where α and D_{AB} are the thermal diffusivity and mass diffusivity of the fluid respectively, and *Sc* and *Pr* are the Schmidt number and Prandtl number, respectively. Consider the Chilton-Colburn analogy [refer (Eq. 3.112)].

$$St_H Pr^{2/3} = St_D Sc^{2/3}$$

where St_H, St_D are the Stanton number for heat and mass transfer, respectively, and *Pr* and *Sc* are the Prandtl number and Schmidt number, respectively.

$$\frac{h_G}{C_S G} Pr^{2/3} = \frac{k_y M}{G} Sc^{2/3}$$

Assuming that $M \approx M_B$, the psychrometric ratio becomes

$$\frac{h_G}{k_y M_B} = C_S \left(\frac{Sc}{Pr}\right)^{2/3}$$

Or
$$\frac{h_G}{k_y M_B} = C_S (Le)^{2/3} \quad \text{...(6.39)}$$

Depending upon the geometry of the wetted surface, the values of the exponent of *Le* in Eq. (6.39) differ. A modification suggested based on the experimental evidence on vapour-gas systems is

$$\frac{h_G}{k_y M_B} = C_S (Le)^{0.567}$$

For the air-vapour systems, assuming $C_S = C_B$, and with $Pr = 0.707$ for air, the above relation gives

$$\frac{h_G}{k_y M_B} = 1.223\, Sc^{0.567} \quad \text{...(6.40)}$$

where the psychrometric ratio is in kJ/kg K. For the air-water vapour systems, this further simplifies to $\frac{h_G}{k_y M_B} = 0.950$ kJ/kg K, but the experimental value is found to be slightly higher and equal to 1.09 kJ/kg K.

Lewis Relation

Consider Eq. (6.38).

$$(Y' - Y'_W) = -\frac{h_G}{k_y M_B \lambda_W}(T - T_W)$$

On the psychrometric chart, the point (T_w, Y'_W) lies on the 100% saturation curve. The curve with slope $-\frac{h_G}{k_y M_B \lambda_W}$ passing through this point is the *wet-bulb temperature line* or the *psychrometric line* for the given wet-bulb temperature, T_W. The air-water system with the dry-bulb temperature (T) and the humidity (Y') that satisfy Eq. (6.38) for the specified wet-bulb temperature can be represented by a point lying on the psychrometric line. Several such lines can be drawn, each one corresponds to a given wet-bulb temperature.

The slope of the psychrometric line is $-\frac{h_G}{k_y M_B \lambda_W}$. We have seen that the slope of the adiabatic cooling line is $-\frac{C_S}{\lambda_{as}}$. In general, the psychrometric line and the adiabatic cooling line passing through a given point on the 100% saturation curve are distinct lines and the relationship between them depends on the relative magnitudes of the psychrometric ratio $\frac{h_G}{k_y M_B}$ and the humid heat C_S. For mixtures of air and organic vapours, the psychrometric lines are steeper than the adiabatic cooling lines and therefore, the wet-bulb temperature is higher than the adiabatic saturation temperature.

The adiabatic saturation temperature and the wet-bulb temperature will be identical if the psychrometric ratio and the humid heat are equal. For the air-water system this is found to be approximately true. According to the Lewis relation, the psychrometric ratio is approximately equal to the humid heat (C_s) for the air-water vapour system.

$$\frac{h_G}{k_y M_B} = C_S \quad \text{...(6.41)}$$

It can be seen that the Lewis relation follows from Eq. (6.39) for the case when the Lewis number for the system is unity. Lewis number is found to be approximately unity for the air-water vapour systems. Consequently, the psychrometric lines and the adiabatic cooling lines for air-water vapour system are identical.

The humidity and other properties of air can be determined using humidity charts if the wet-bulb temperature and the dry-bulb temperature are known. Take the wet-bulb temperature on the temperature axis (Fig. 6.10) and move vertically to the point N on the 100 percent saturation curve. Move along the psychrometric line passing through N downward to its intersection with the vertical line (point M) at the given dry-bulb temperature. This point (M) determines the state of the air-water system and the humidity and other properties corresponding to this point can be read from the chart directly.

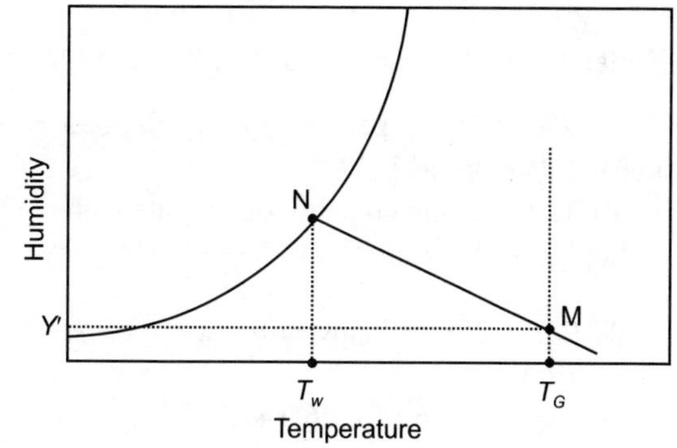

Fig. 6.10: Determination of humidity using wet-bulb temperature

Example 6.11: The wet-bulb and dry-bulb temperature of air are 313 K and 333 K, respectively. Determine using the psychrometric chart:

(a) Absolute humidity
(b) Percent humidity
(c) Humid volume
(d) Enthalpy of wet air.

***Solution*:**

(a) Obtain point A on the humidity chart such that the psychrometric line at the WBT 313 K intersects the vertical at the DBT 333 K at point A (Fig. 6.11). The humidity at point A is read from the chart and is equal to 0.04 kg water per kg dry air. **(Answer)**

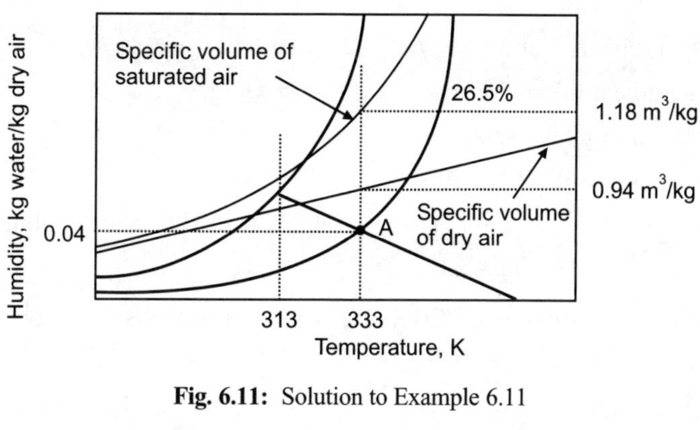

Fig. 6.11: Solution to Example 6.11

(b) The percent humidity at point A is 26.5% (from the chart) **(Answer)**

(c) At 333 K the volume of saturated air = 1.18 m^3/kg dry air. (V_S)

Volume of dry air at 333 K = 0.94 m^3/kg dry air. (V_D)

By Eq. (6.24),

$$\text{humid volume} = V_H = V_D + \frac{\text{P.S}}{100}(V_S - V_D)$$

$$= 0.94 + 0.265(1.18 - 0.94) = 1.0036 \text{ m}^3/\text{kg dry air}$$ **(Answer)**

[**Note:** Eq. (6.19) can be used to calculate the humid volume. $V_H = 8.314\left(\frac{1}{29}+\frac{Y'}{18}\right)\times\frac{T}{P}$. Here, $Y' = 0.04$ kg water per kg dry air, $T = 333$ K and $P = 101.3$ kPa. Substituting the values, $V_H = 1.0032$ m^3/kg dry air]

(d) At 333 K the enthalpy of saturated air = 470 kJ/kg dry air (H'_S). **(Answer)**

Enthalpy of dry air at 333 K = 60 kJ/kg dry air (H'_D).

By Eq. (6.25), enthalpy of the wet air $= H' = H'_D + \frac{\text{P.S}}{100}(H'_s - H'_D)$

$$60 + 0.265(470 - 60) = 168.65 \text{ kJ/kg dry air} \quad \textbf{(Answer)}$$

[**Note:** Enthalpy of humid air can be calculated by Eq. (6.23).

$$H' = C_s(T - T_0) + Y'\lambda_0$$

$C_s = 1.005 + 0.04 \times 1.884 = 1.08036$ kJ/kg K; $T_0 = 273.15$ K; $\lambda_0 = 2502.3$ kJ/kg; $T = 333$ K; $Y' = 0.04$ kg water per kg dry air. Substituting the values in Eq. (6.23), we get $H' = 164.75$ kJ/kg dry air]

Example 6.12: Dry air at 101.3 kPa is blown across the bulb of a mercury thermometer. The bulb is covered with a wick. The wick is immersed in an organic liquid (MW = 58). The reading of the thermometer is 290 K. At this temperature, the vapour pressure of the liquid is 10.0 kPa. Find the air temperature given that the psychrometric ratio is 2 kJ/kg K and the latent heat of vaporisation of the liquid is 425 kJ/kg.

***Solution*:** Eq. (6.37) is used to evaluate the temperature.

$$T_W = T - \frac{\lambda_W (Y'_W - Y')}{(h_G/k_y M_B)} \quad \ldots(6.37)$$

Wet-bulb temperature $T_W = 290$ K;

$$P_W^S = 10 \text{ kPa. } Y'_W = \frac{P_W^S}{P - P_W^S}\times\frac{M_A}{M_B} = \frac{10}{101.3 - 10}\times\frac{58}{29} = 0.2191 \text{ kg vapour/kg dry air;}$$

Psychrometric ratio $= h_G/k_Y = 2$ kJ/kg K;

Latent heat of vaporisation $= \lambda_W = 425$ kJ/kg.

Since the air is free of organic vapour, $Y' = 0$.

Substituting these values, Eq. (6.37) gives

$$T = T_W + \frac{\lambda_W (Y'_W - Y')}{(h_G/k_y M_B)} = 290 + \frac{425 \times 0.2191}{2} = 336.5 \text{ K} \quad \textbf{(Answer)}$$

Example 6.13: A benzene-air mixture at 320 K and 101.3 kPa contains 0.05 kg benzene vapour per kg dry air. Antoine equation for vapour pressure of benzene is

$$\ln P^S = 13.8858 - \frac{2788.51}{T - 52.36}$$

The specific heat of dry air is 1.005 kJ/kg K and the specific heat of benzene vapour is 1.2 kJ/kg K. The psychrometric ratio for air-benzene system may be evaluated by the empirical equation

$$\frac{h_G}{k_y M_B} = 1.223\,\text{Sc}^{0.567}$$

Estimate:

(a) The adiabatic saturation temperature
(b) The wet-bulb temperature.

The latent heat of vaporisation of benzene in the range of interest may be taken to be constant at 430 kJ/kg. The diffusivity of benzene-air pair is 1.007×10^{-5} m^2/s at 320 K. Viscosity and density of air at 320 K are 1.07×10^{-5} kg/m s and 1.104 kg/m^3, respectively.

Solution: Dry-bulb temperature $T = 320$ K. Humidity $Y' = 0.05$ kg benzene/kg dry air

$$C_S = C_B + Y'C_A = 1.005 + 1.2 \times 0.05 = 1.065 \text{ kJ/kg K.}$$

(a) The adiabatic saturation temperature is estimated using Eq. (6.27)

$$T_{as} = T - \frac{\lambda_{as}}{C_S}(Y'_{as} - Y')$$

Neglecting the change in the latent heat of vaporisation of benzene with temperature,

$$T_{as} = 320 - \frac{430}{1.065}(Y'_{as} - 0.05) \tag{A}$$

Since Y'_{as} is the saturation humidity at temperature T_{as}, the adiabatic saturation temperature is determined by solving Eq. (A) by trial. Assume a temperature, calculate the vapour pressure of benzene at this temperature using the Antoine equation and use the vapour pressure to calculate the saturation humidity by

$$Y'_{as} = \frac{P_A^S}{P - P_A^S} \times \frac{M_A}{M_B} = \frac{P_A^S}{P - P_A^S} \times \frac{78}{29} \tag{B}$$

Substituting Y'_{as} in Eq. (A), calculate T_{as} and check whether the assumed and calculated temperatures are equal. It is found that at $T_{as} = 280.507$ K, Eq. (A) is satisfied. Therefore, the adiabatic saturation temperature = 280.507 K. **(Answer)**

(b) The wet-bulb temperature is calculated using Eq. (6.37)

$$T_W = T - \frac{k_y M_B}{h_G}\lambda_W (Y'_W - Y')$$

The psychrometric ratio $\dfrac{h_G}{k_y M_B}$ is evaluated using

$$\frac{h_G}{k_y M_B} = 1.223\, Sc^{0.567}$$

where Schmidt number $Sc = \dfrac{\mu}{\rho D_{AB}} = \dfrac{1.07 \times 10^{-5}}{1.104 \times 1.007 \times 10^{-5}} = 0.9625.$

$$\frac{h_G}{k_y M_B} = 1.223(0.9625)^{0.567} = 1.197 \text{kJ/kg K}$$

Substituting in Eq. (6.37), we get

$$T_W = 320 - \frac{430}{1.197}(Y'_W - 0.05) \tag{C}$$

Equation (C) is solved for T_W by trial. Assume a temperature, calculate the vapour pressure at this temperature, and using the vapour pressure, calculate the saturation humidity. Substitute the value in Eq. (C) and calculate T_W. Check the calculated temperature with the assumed one. The calculations are repeated till both are equal. It is found that $T_W = 281.574$ K. **(Answer)**

6.6 COOLING TOWERS

Cooling of the process streams and the condensation of the vapours are important operations in the process industries; water is the commonly used coolant in majority of such operations. Huge quantities of water are to be circulated through the plant cooling systems. For example, a typical large refinery processing 40,000 metric tonnes of crude oil per day requires cooling water circulation at a rate of about 80,000 cubic metre per hour. Using such large quantities of water as once-through coolant is uneconomical. Also, discarding such large amounts of warm water leaving the process units to the ocean, lake or river may create thermal pollution and is harmful to fish and other aquatic organisms. Therefore, the recovery and reuse of water after cooling are also important activities in large process industries, power plants and air conditioning units. Cooling of water on such a large scale is accomplished in cooling towers. The cooling towers are large diameter columns with special types of packing designed to give good gas-liquid contact. The towers vary in size from small units to very large hyperboloid structures about 180 m tall and 100 m in diameter.

Water reaches the cooling towers from different units in the plant, absorbing heat while performing the cooling duty. The warm water is taken to the top of the tower, distributed over the packing by spray nozzles or a grid of notched troughs or pipes. The water trickles down over the packings and the air is made to flow up or across the tower in direct contact with water. Since warm water is in contact with unsaturated air, part of the liquid evaporates and the liquid temperature drops. Thus, water gets cooled and the air gets humidified and heated. The cooled water is collected in a cold water basin below the packed section from which it is pumped back to the process for performing cooling duties. Cooling towers offer the following advantages over ordinary tubular heat exchangers for water cooling.

1. Heat exchangers are made of costly materials of construction; the cooling towers are built of inexpensive materials such as wood, brick, concrete, asbestos-board, etc.
2. Use of inexpensive coolant such as atmospheric air necessitates very large heat transfer area and consequently very expensive heat exchangers, because of the poor heat transfer coefficient of air. In cooling towers, the air is brought into direct contact with the water, which results in the evaporative cooling of the water.
3. Water can be cooled to temperatures as low as the wet-bulb temperature of the air admitted to the cooling tower. Even when the air is warmer than the water, water can be cooled by evaporation, provided the wet-bulb temperature is below the temperature of water. In contrast, the temperature of the water leaving a heat exchanger can never be brought down below the dry-bulb temperature of the incoming air.

Based on the method of generating and maintaining the air flow through the column, cooling towers are classified into two broad types: *natural draft* and *mechanical draft* cooling towers. Mechanical draft cooling towers rely on the power-driven fans to draw or force the air through the tower. In natural draft cooling towers the required air flow is created by the buoyancy effect resulting from the difference in the densities between the hot and humid air inside the tower and the cold denser ambient air outside. The mechanical draft or the natural draft cooling towers can be either of the *cross flow type* or the *counterflow type* depending upon the direction of flow and the nature of contact between the air and the water. In both types, the warm water is sprayed at the top of the column and flows downwards. In the counterflow cooling towers the air flow and the water flow are in the opposite directions; air travels upward through the packing also called *fill*, opposite to the downward motion of the water. In the cross flow type, the air flow is in the direction perpendicular to the flow of water; the air moves horizontally through the fill as the water moves downward. The counterflow tower occupies less floor space than a cross flow tower, but is taller for a given capacity. The principal advantage of the cross flow tower are the low pressure drop in relations to its capacity and the lower fan power requirement leading to lower energy costs.

Natural Draft Cooling Towers

For large applications as in power plants, the natural draft cooling towers are often used. Natural draft is provided by the displacement of the warm air inside the tower by the cooler outside air. The warm moist air is less dense than the relatively dry and cold outside air at the same pressure. This moist air buoyancy produces a current of air through the tower. Because the pressure difference created by the density effect is approximately proportional to the height of the tower, tall chimneys are needed to facilitate the flow of air.

Natural draft cooling towers consist of an empty shell constructed of ferro-concrete. The lower portion amounting to approximately 10 to 12% of the total height is occupied by the packings. Commonly, wood or plastic grid packings are employed, on to which the water is sprayed by means of distributor sprays. The empty upper portion serves merely to increase the draft. The atmospheric conditions, the temperature and the quantity of the water will influence the quantity, the velocity and the flow pattern of the air through the tower and hence affect the transfer coefficients between the gas and the liquid. One of the prime considerations in the design, therefore, is to construct a tower such that the resistance to the air flow is low. Hence the packings and distributors are arranged in open formation.

In the counterflow type, air enters the tower through open rings around the bottom, flows up through the packing materials over which the water is distributed, and finally goes up through the tower chimney. The cross-flow type natural draft towers have a fill configuration in which, the air flows perpendicular to the water falling downward. Figure 6.12a shows schematically the crossflow natural draft cooling tower and Fig. 6.12b the counterflow natural draft cooling tower.

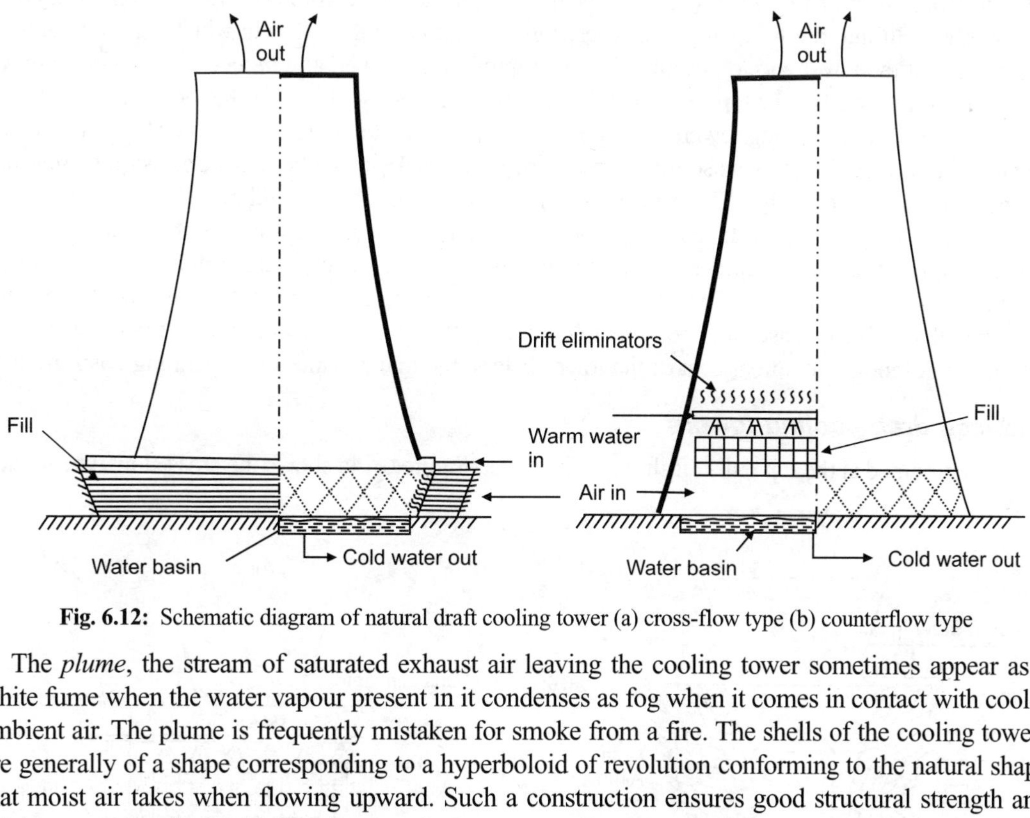

Fig. 6.12: Schematic diagram of natural draft cooling tower (a) cross-flow type (b) counterflow type

The *plume*, the stream of saturated exhaust air leaving the cooling tower sometimes appear as a white fume when the water vapour present in it condenses as fog when it comes in contact with cooler ambient air. The plume is frequently mistaken for smoke from a fire. The shells of the cooling towers are generally of a shape corresponding to a hyperboloid of revolution conforming to the natural shape that moist air takes when flowing upward. Such a construction ensures good structural strength and minimum usage of material. The increase in the cross section towards the top causes a reduction in the outlet velocity with a consequent recovery of kinetic energy into pressure energy. The hyperboloid shape also aids in accelerating the upward convective air flow, resulting in improved cooling efficiency. The natural draft towers are generally operated at a pressure difference of 50 N/m^2, and a mean air velocity typically in the range 1-2 m/s above the packing. The flow of air through a natural draft cooling tower is augmented by means of fan in the *hybrid type* or *fan assisted natural draft towers*.

Atmospheric cooling towers

In the atmospheric towers (Fig. 6.13) the air circulation is by the natural wind movement through the tower.

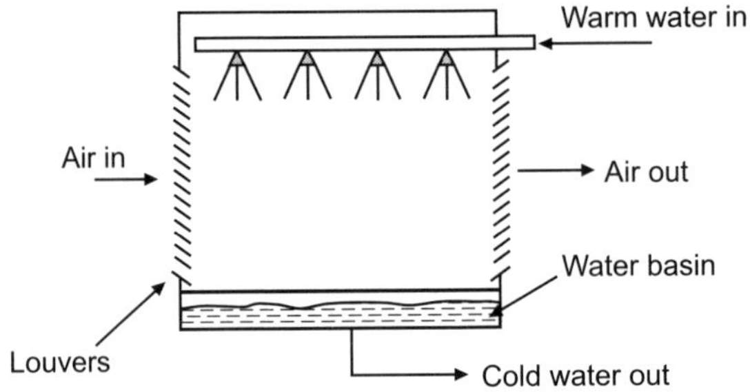

Fig. 6.13: Schematic diagram of atmospheric cooling tower

Mechanical Draft Cooling Towers

The *mechanical draft towers*, use the power driven fan motors to force or draw the air through the tower. They fall into the following two types: *forced draft cooling towers* in which fan is mounted at the bottom of the tower and forces air movement through the tower, and *induced draft cooling towers* with the fan mounted at the top of the tower that drives the moist air out at the top. The advantage of the mechanical draft cooling towers is that the air velocity can be increased appreciably above that in the natural draft cooling towers so that a greater depth of packing can be used. Corrosion is prevented by the construction entirely of inert material such as redwood, plastics and porcelain. The tower will extend only up to the top of the packing unless atmospheric conditions are such that a chimney must be provided in order to prevent recirculation of the moist air. The mechanical draft towers are generally confined to small installations. They are preferred when the water must be cooled to as low a temperature as is possible. In some cases it is possible to cool the water to within 1 K of the wet-bulb temperature of the air. Although the initial cost of the tower is less, the maintenance and operating costs are high.

Induced draft cooling towers

The fan mounted at the top pulls air through the tower. Figure 6.14a shows the cross-flow type induced

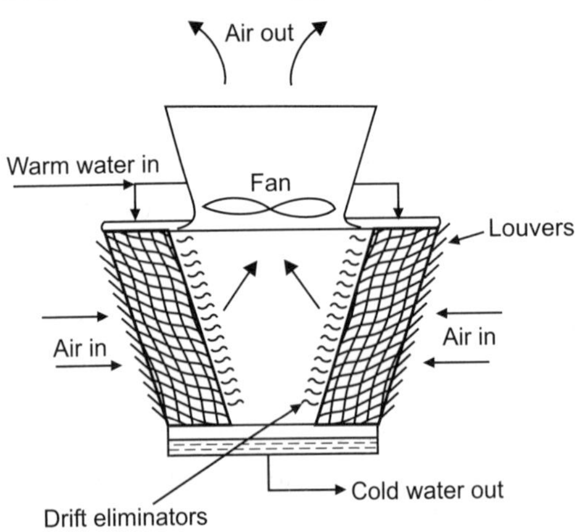

Fig. 6.14: (a) Induced draft cooling towers cross-flow type

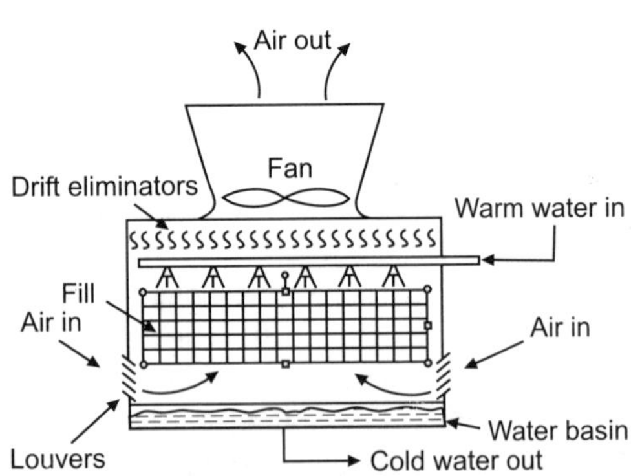

Fig. 6.14: (b) Induced draft cooling towers counterflow type

draft tower and Fig. 6.14b the counterflow type induced draft tower. In the former, the air passes horizontally through the inclined beds of packings or fill, while the water flows downward. Slanted louvers prevent the escape of water drops to the outside. Angled baffles called the *drift eliminators* capture most of the droplets that are entrained in the exit air. In the counterflow arrangement, water trickles down through a layer of fill and the air enters below the fill and flow up in the direction opposite to that of water. The drift eliminators are provided above the packings to capture the physical entrainment of water droplets with air. The counterflow arrangement is found to be more efficient than the cross flow type and provides a closer temperature approach.

Since the fan is at the air exit, the discharge velocity of the air is high and therefore the danger of the recirculation of the hot humid air back into the column through the air intake is eliminated. Also a uniform distribution of the air over the tower cross section is achieved. The fan is often located at the throat of a venturi-shaped cylinder that promotes smooth flow of the air to the fan and gives some pressure recovery in the expanding section.

Forced draft cooling towers

A forced draft tower is schematically represented in Fig. 6.15. The fan mounted at the base of the tower forces air through the tower with high entry and low discharge air velocities. The forced draft towers are smaller in size and can be installed in confined spaces and even indoor in some situations.

Since the fan and drive are outside the tower, the inspection, maintenance and repair are convenient. Being away from the hot and humid exit point, the fan is not exposed to severe conditions of corrosions. However, due to the low discharge velocity of the air the forced draft towers are subjected to severe recirculation of the hot humid air to the air intake which will result in poor performance of the tower. Also being on the air intake, the fan is more susceptible to complications such as icing in freezing weather conditions. Another disadvantage is that a forced draft design typically requires more motor horsepower than the induced draft design of equivalent capacity.

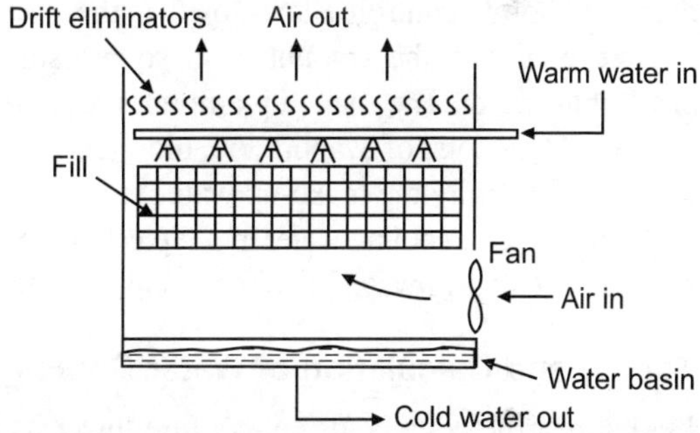

Fig. 6.15: Schematic of a forced draft cooling tower

Packings

Cooling tower packings called *fill* provide large interfacial area between the air and the water phases for heating of the air and evaporation of the water. There are two types of packing generally employed: the *splash fill* and the *film fill*. The *splash fill* are materials placed to interrupt the water flow and cause splashing. In older towers, several levels of redwood or cypress slats were used as the horizontal splash elements. Water falling on the slats would splash and the droplets would be intercepted by the next layer of the packing. Thus a cascade of many small droplets provides a large combined area. In modern towers using splash-type fill, V-shaped PVC bars are used as splash elements, instead of wooden slats. The pressure drop is low because the air flow is parallel to the slats or bars and the open structure makes the inspection and cleaning relatively easy. Splash fill is not recommended for counter flow towers.

The *film fill* or *cellular fill* is the most common type of packings in the new installations. These are composed of thin corrugated sheets of plastics upon which water spreads as a thin film. Generally film fill provides more transfer area per unit volume than splash type packing. The plastic sheets are spaced typically about 18-25 mm to permit high flow rates of the air and the water with only a moderate pressure drop. Since redistribution of water cannot occur naturally as in splash fill, it is essential that in towers packed with film fill, a good distribution of water is maintained at the top.

Make-up Water Requirement

Make-up water is generally required to compensate for the loss of water in the tower. Evaporation, blow-down, blow-out and drift are the common causes for loss of water in the cooling tower systems. The loss of water by evaporation during the cooling is small. For cooling 100 kg water by say, 12 K, the heat to be removed is approximately 5000 kJ. Utilizing this heat, about 2 kg of water can be evaporated because the latent heat of vaporisation of water is approximately 2500 kJ/kg. Thus, a range of 12 K in the tower means an evaporation loss of only 2% of the water circulated. *Drift* also causes water loss in the tower. Drift is the loss of water as droplets that are carried out of the cooling tower with the exhaust air. The drift rate is reduced by employing baffle-like devices, called drift eliminators, through which the air must travel after leaving the fill and spray zones of the tower. The drift is generally only 0.2% of the recirculation rate for induced draft towers. *Blow-out* is the loss of water as droplets blown out of the cooling tower by wind, generally at the air inlet openings. Water may also be lost, in the absence of wind, through splashing or misting. Devices such as wind screens, louvers, splash deflectors and water diverters are used to limit these losses.

Since it is the pure water that is lost by evaporation, the concentration of the dissolved minerals and other solids in circulating water tend to increase, unless some means such as *blow-down* is provided for dissolved solids control. Blow-down is the portion of the circulating water flow that is removed, in order to maintain the amount of dissolved solids and other impurities at an acceptable level in the circulating water. The make-up amount required in the cooling tower to maintain a steady water level must equal the total of evaporation, blow-down, drift, and other water losses such as wind blow out and leakage. Generally the make-up water requirement will be about 5% of the water circulation rate. The circulation rate of cooling water in a typical 700 MW coal-fired power plant amounts to about 72,000 m^3/h and the circulating water requires a supply water make-up rate of about 3,600 m^3/h.

Theory and Calculation of Water Cooling Towers

Let us treat the countercurrent cooling tower as a packed column in which water is admitted at the top at a rate L' kg/h m^2 of tower cross section and flow downwards and the air is admitted at the bottom and flow upward at a rate G'_S kg dry air/h m^2 of tower cross section as shown in Fig. 6.16. The temperatures of the water and the air are represented by T_L and T_G, respectively. The enthalpy of the air is represented by H' kJ/kg dry air. The conditions at the bottom of the column are represented by the suffix 1 and those at the top by the suffix 2. The humidity of the air entering the column is Y'_1 kg water per kg dry air and that of air leaving the tower is Y'_2 kg water per kg dry air.

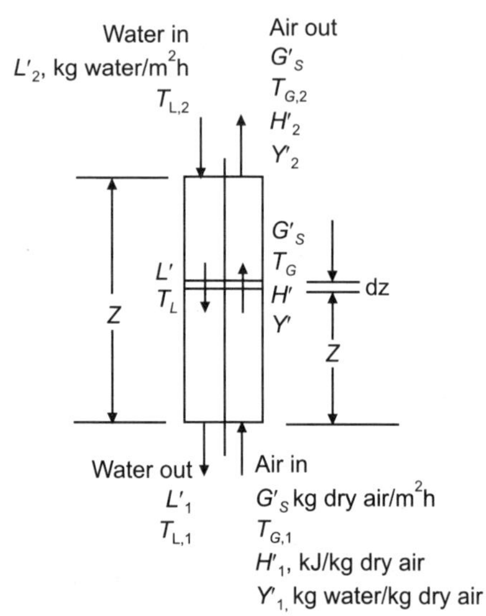

Fig. 6.16: Air and water flows in the counter-flow cooling tower

Temperature and concentration profile

Consider the conditions at the top of the cooling tower where the warm water is in contact with the relatively cold air. The temperature of the water (T_L) is greater than the temperature of the air (T_G) and there is transfer of sensible heat from the liquid side to the interface and from the interface to the bulk air. The temperature gradient is set up as shown in Fig. 6.17a. The air-water interface is at temperature T_i and the humidity of air at the interface Y_i' is the saturation humidity at T_i. As long as the interface temperature is above the dew point of the air, Y_i' is greater than the humidity of the air, Y', and the humidity gradient in the gas is from the interface to the bulk air, just as the temperature gradient. Evaporation takes place followed by the transfer of the water vapour from the interface to the bulk air, the vapour carrying with it the latent heat of vaporisation.

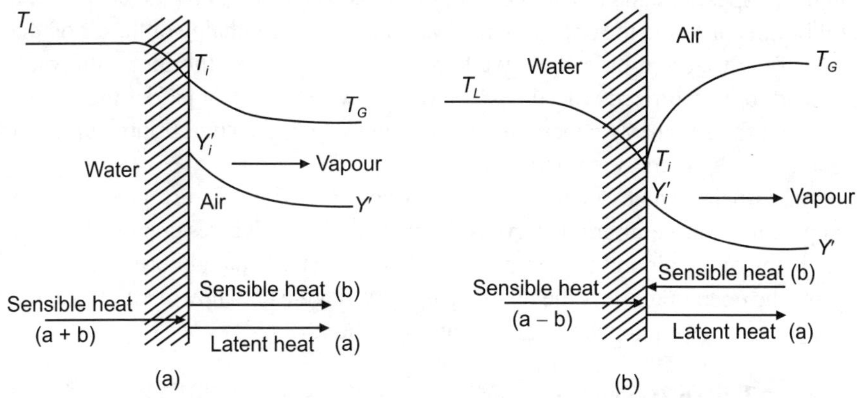

Fig. 6.17: Temperature and humidity profile at (a) the top of cooling tower (b) bottom of the cooling tower

The transfer of both the sensible heat and the latent heat take place in the direction from the interface to the bulk gas. The sensible heat removed from the water is equal to the sum of the heat transferred from the interface to the air as latent heat and sensible heat. Due to this combined effect of mass transfer and heat transfer, cooling of water takes place at a faster rate than that would result in the absence of evaporation and mass transfer. We can see that the mass transfer is confined to the gas side alone as no concentration gradient exists in the liquid phase which is pure water. However, heat transfer occurs on both the gas and liquid side.

The temperature profile at the bottom of the countercurrent cooling tower will be different from that at the top depending upon whether the water temperature is above or below the dry-bulb temperature of the air. Figure 6.17b shows the case when the water temperature is below the dry-bulb temperature of the air, but above its wet-bulb temperature. The interface being at a temperature above that of the water, the temperature profile assumes a V-shape as shown in the figure. The temperature gradient on the liquid side is from the bulk liquid to the interface and cooling of water occurs by transfer of heat to the interface. Since the dry-bulb temperature of air is above the interface temperature, a temperature gradient exists from the bulk air to the interface, and sensible heat transfer occurs from the air to the interface. Since the air at the interface is saturated with the water vapour, the humidity at the interface is still greater than the air humidity and therefore vaporisation occurs at the air-water interface followed by mass transfer. The transfer of latent heat by the vapours is from the interface to the air and it is in the direction opposite to that of the transfer of sensible heat in the air. The sensible heat transferred from the water now equal the latent heat removed by the vapours less the sensible heat transferred from the air to the interface.

Equilibrium curve and the operating lines

In the analysis of absorption and distillation operations, the equilibrium curve and the operating lines were plotted with concentration of the gas and the liquid phases as the coordinates. In the cooling towers, the liquid phase is pure water and no concentration gradient exists there and therefore no mass transfer occurs in the liquid. However, water gets cooled because of the combined effect of the heat and the mass transfer occurring on the gas side. Therefore, it is the liquid temperature and not the concentration on the liquid side that can be used to represent the changes occurring in the water. Since the humidity and the temperature of the air change, the property that can be used to represent the effect of the simultaneous transfer of heat and mass on the gas side is its enthalpy which by definition is,

$$H' = C_S(T_G - T_0) + Y'\lambda_0 \quad \text{...(6.42)}$$

and

$$dH' = C_S dT_G + \lambda_0 dY' \quad \text{...(6.43)}$$

The change in the temperature and the humidity of the air is taken care of by the change in its enthalpy. Thus, the equilibrium curve and the operating lines are plotted with enthalpy of the air on the *y*-axis and the temperature of the water on the *x*-axis. We know that the air in equilibrium with water at a given temperature is said to be saturated with the water vapour which exerts a partial pressure equal to the vapour pressure. On the enthalpy-temperature coordinate system, the equilibrium curve is therefore, a plot of saturation enthalpy versus temperature.

The saturation enthalpy values of the air-water system at 101.3 kPa are given in Table 6.2 as a function of temperature. The equilibrium curve is shown in Fig. 6.18. It gives graphically, the relationship between the enthalpy of the air and the temperature of the water when equilibrium is attained between them. In contrast, the operating line is the relationship between the enthalpy of air (H') and temperature of the water (T_L) which are in contact at any location in the tower.

Table 6.2: Saturation enthalpy of air-water system at 101.3 kPa

T, K	283.15	288.15	293.15	298.15	303.15	308.15	313.15	318.15	323.15
H', kJ/kg	29.3	42.0	57.4	76.2	99.6	128.9	165.8	213.0	273.8

Consider a differential section in the tower of thickness dz as shown in Fig. 6.16. The decrease in the sensible heat of water leads to the changes in the temperature and humidity of air, and consequently the enthalpy of the air changes, as explained earlier. The enthalpy balance for the differential section relates the changes in the temperature of the water and the enthalpy of the air as

$$G_S' \, dH' = L'C_L \, dT_L \quad \text{...(6.44)}$$

dH' is the rise in the enthalpy of the gas, and dT_L is the fall in the temperature of water. C_L is the heat capacity of the water which is assumed to be constant. Equation (6.44) can be rearranged as

$$\frac{dH'}{dT_L} = \frac{L'C_L}{G_S'} \quad \text{...(6.45)}$$

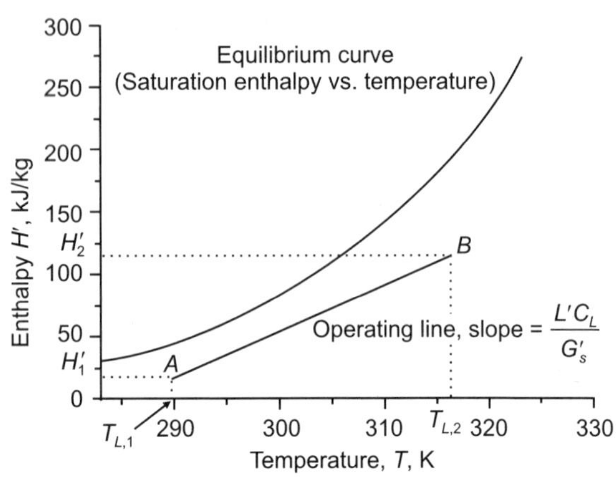

Fig. 6.18: The equilibrium curve and the operating lines of counterflow cooling towers

Left-hand side term of Eq. (6.45) is the slope of the H' versus T_L curve, where H' is the enthalpy of the air and T_L is the temperature of the water which is in contact with the air. This curve is the operating line for the cooling tower. Here G_S' is the flow rate of the dry air through the tower and is constant. Since the quantity of the water vaporised is negligibly small in comparison with the quantity circulated, L' can also be treated as a constant. That means the slope of H' versus T_L is constant or the operating line is a straight line. Integration of Eq. (6.45) gives

$$\frac{H_2' - H_1'}{T_{L,2} - T_{L,1}} = \frac{L'C_L}{G_S'} \quad \text{...(6.46)}$$

The operating line for the cooling tower is a straight line joining $(T_{L,1}, H'_1)$ and $(T_{L,2}, H'_2)$ where $T_{L,1}$ is the temperature of the cooling water and H'_1 is the enthalpy of the air at the bottom of the cooling tower and $T_{L,2}$ is the temperature of the water and H'_2 is the enthalpy of the air at the top of the tower. The line *AB* in Fig. 6.18 is the operating line.

In cooling tower design, generally the inlet and exit temperature of the cooling water ($T_{L,1}$ and $T_{L,2}$) and the inlet air enthalpy (H'_1) are known initially. Therefore, to draw the operating line, the lower terminus of the line $A(T_{L,1}, H'_1)$ is first located and a straight line of slope $\dfrac{L'C_L}{G'_S}$ is drawn to terminate at point *B* lying on the vertical line passing though $T_{L,2}$, the water inlet temperature.

Minimum air flow

For the design of cooling towers, the gas flow rate is to be set by the designer. The lower the gas rate, the smaller the cost of its circulation and the operating cost of the tower. However, there is a minimum allowable rate for the gas flow which is to be obtained from the slope of the limiting operating line.

Consider Fig. 6.19. The lower terminus *A* of the operating line is fixed because its coordinates are the exit water temperature and the inlet air enthalpy ($T_{L,1}$ and H'_1). The upper terminus *B* will be on the vertical line at $T = T_{L,2}$. Since the liquid circulation rate is specified, the slope of the operating line depends on the gas flow rate employed. The limiting operating line is the one which anywhere touches the equilibrium curve; the driving force reduces to zero where the operating line touches the equilibrium curve and the tower becomes infinitely tall to accomplish a given temperature change in the liquid. The slope of the limiting operating line is the maximum and this corresponds to the minimum gas flow rate. The upper terminus of the limiting operating line either coincides with the equilibrium curve as shown by point *B′* in the Fig. 6.19 or it falls on the tangent drawn to the equilibrium curve from the point *A*. *AB′* is the limiting operating line with slope $\dfrac{L'C_L}{G'_{S,\min}}$. From the slope of the limiting operating line, the minimum permissible air circulation rate is determined. The actual supply should be greater than the minimum; often a value lying between 1.3 to 1.5 times the minimum is employed as the actual rate.

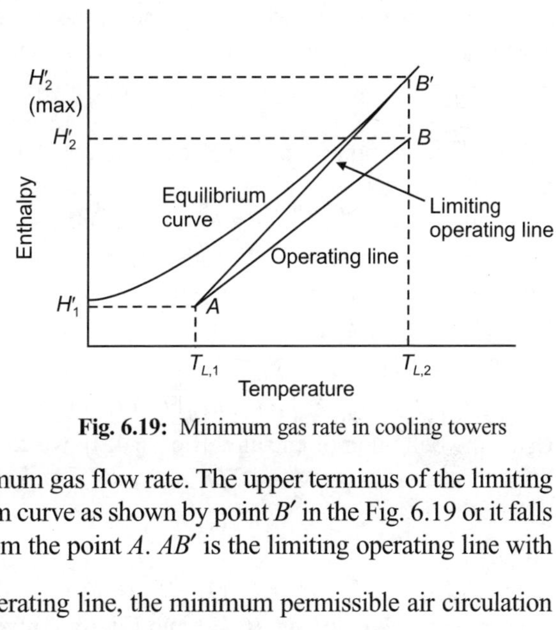

Fig. 6.19: Minimum gas rate in cooling towers

Height of the tower

Consider the differential section of thickness *dz* at a height *z* units from the bottom of the column as shown in Fig. 6.16. Let the area of cross section of the tower be unity and let *a* be the interfacial area per unit volume of packing, m²/m³ for both heat transfer and mass transfer. The analysis presented here refers to the top of the column, but is equally valid for the bottom conditions as well.

First let us consider the heat and mass transfer occurring from the air-water interface to the bulk air. The sensible heat transfer from the interface to the bulk air results in the increase in temperature of air by dT_G. Thus

$$G'_S C_S dT_G = h_G a(T_i - T_G)dz \qquad \text{...(6.47)}$$

where h_G is the film heat transfer coefficient for the gas side, T_i is the interface temperature and adz is the interfacial area available for heat transfer at the differential section of the packed tower. Since the Lewis relation is applicable for the air water system,

$$\frac{h_G}{k_y M_B} = C_S \qquad \text{...(6.41)}$$

where k_y is the mass transfer coefficient with mole fraction difference as the driving force, M_B is the molecular weight of the dry air and C_S is the humid heat of the air-water vapour system, Eq. (6.47) becomes

$$G_S' C_S dT_G = k_y a M_B C_S (T_i - T_G) dz \qquad \text{...(6.48)}$$

Now consider the mass transfer from the interface to the air. The interface is saturated with the water vapour and the humidity at the interface Y_i' is greater than the humidity of the bulk air Y'. The vaporisation of the water and the mass transfer from the interface is governed by the following equation.

$$G_S' dY' = k_y a (y_i - y) M_A dz \qquad \text{...(6.49)}$$

The right-side of Eq. (6.49) includes M_A, the molecular weight of the water, in order to express the rate as mass of water transferred instead of moles. Since the concentration of water vapour in the air is very small, the following simplification is possible

$$y = \frac{Y}{1+Y} \approx Y = Y' \times \frac{M_B}{M_A} \qquad \text{...(6.50)}$$

and the mole fraction y can be replaced in terms of the absolute humidity Y'. Substituting Eq. (6.50) into Eq. (6.49), we get

$$G_S' dY' = k_y a M_B (Y_i' - Y') dz \qquad \text{...(6.51)}$$

Multiply both sides by λ_0, the latent heat of vaporisation of water at T_0, where T_0 is the reference temperature chosen for estimating the enthalpy.

$$G_S' \lambda_0 dY' = k_y a M_B \lambda_0 (Y_i' - Y') dz \qquad \text{...(6.52)}$$

Add Eq. (6.48) to Eq. (6.52). The result is

$$G_S' \left(C_S dT_G + \lambda_0 dY'\right) = k_y a M_B \left[C_S (T_i - T_G) + \lambda_0 (Y_i' - Y')\right] dz \qquad \text{...(6.53)}$$

Using the definition of enthalpy [Eq. (6.42)], we can write

$$H_i' = C_S (T_i - T_0) + Y_i \lambda_0 \qquad \text{...(6.54)}$$

$$H' = C_S (T_G - T_0) + Y' \lambda_0 \qquad \text{...(6.55)}$$

So that

$$H_i' - H' = C_S (T_i - T_G) + \lambda_0 \left(Y_i' - Y'\right) \qquad \text{...(6.56)}$$

Substituting Eq. (6.56) and (6.43) into Eq. (6.53) we get

$$G_S' dH' = k_y a M_B \left(H_i' - H'\right) dz \qquad \text{...(6.57)}$$

Equation (6.57) states that the rate of change of the gas enthalpy at a given point in the column is the product of the enthalpy difference driving force from the interface to the bulk air, the gas film mass transfer coefficient and the interfacial area of mass transfer. Equation (6.57) forms the basis for the rigorous method for the design of cooling tower based on the film coefficient of mass transfer. Before we discuss this method in detail, let us consider a simplified method based on the overall coefficient.

Height of the Tower Using Overall Mass-transfer Coefficient

Writing the change in the gas enthalpy as the product of the overall gas-phase mass transfer coefficient, the overall enthalpy difference driving force and the area of mass transfer, Eq. (6.57) takes the following form:

$$G_S' dH' = K_Y a \left(H'^* - H'\right) dZ \qquad \text{...(6.58)}$$

Here H'^* is the enthalpy of the air in equilibrium with the water (that is, the enthalpy of the saturated air at temperature T_L) and K_Y is an overall gas phase mass transfer coefficient. The overall mass transfer coefficient may be related to the individual coefficients as

$$\frac{1}{K_Y a} = \frac{1}{k_y a M_B} + \frac{m}{h_G a} \quad \text{...(6.59)}$$

where m is the slope of the equilibrium curve. Equation (6.58) may be rearranged as

$$dZ = \frac{G'_S}{K_Y a} \frac{dH'}{H'^* - H'} \quad \text{...(6.60)}$$

On integrating Eq. (6.60), we get

$$Z = \frac{G'_S}{K_Y a} \int_{H'_1}^{H'_2} \frac{dH'}{H'^* - H'} \quad \text{...(6.61)}$$

Designating $\frac{G'_S}{K_Y a}$ as the height of overall gas-enthalpy transfer units (H_{tOG}) and $\int_{H'_1}^{H'_2} \frac{dH'}{H'^* - H'}$ as the number of the overall gas-enthalpy transfer units (N_{tOG}), we can write Eq. (6.61) as

$$Z = H_{tOG} \times N_{tOG} \quad \text{...(6.62)}$$

The number of transfer units (N_{tOG}) can be determined by graphical integration. The equilibrium curve and the operating lines are first plotted on H' versus T coordinates. For a given point on the operating line such as point C in Fig. 6.20, the ordinate is the enthalpy of the air and the abscissa is the temperature of the water in contact with the air at some location in the tower. The y-coordinate of the point D located on the equilibrium curve vertically above the point C is the enthalpy of the air saturated at the temperature of the water with which the air is contacted, or the ordinate of the point D is the enthalpy of the air in equilibrium with the water which is denoted by H'^*. Therefore, the vertical distance CD is the overall enthalpy difference driving force $H'^* - H'$.

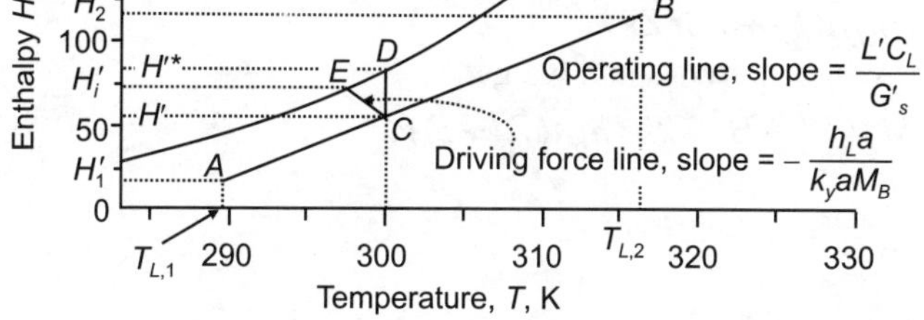

Fig. 6.20: Determination of overall enthalpy difference driving force

Choose different points on the operating line and for each such points the ordinate gives H' and the vertical distance between the equilibrium curve and the operating line gives $H'^* - H'$. Plot $\frac{1}{H'^* - H'}$ as the ordinate against H' as the abscissa. The area under the curve between the limits H'_2 and H'_1

gives $\int_{H_1'}^{H_2'} \frac{dH'}{H'^* - H'}$, the number of transfer units N_{tOG}. The height of the overall gas-enthalpy transfer units (H_{tOG}) is readily calculated from $H_{tOG} = \frac{G_S'}{K_Y a}$, and the product of H_{tOG} and N_{tOG} gives the height of the packed section of the tower. It is observed that the height of overall gas-enthalpy transfer unit under normal air flow rates lies in the range 0.6 – 1 m for cellular fill and in the range 3 – 6 m for wood slats.

The method of calculating the tower height using the overall coefficient can be expected to be reliable only if the saturation enthalpy versus temperature curve is straight line in the region of interest. The use of the overall driving force is justified if the liquid side film coefficient of heat transfer is very high (i.e. infinite) so that there is no temperature gradient in the liquid side and the air-water interface temperature is identical to the liquid temperature.

Height of the Tower Using Gas-film Mass-transfer Coefficient

Equation (6.57) can be put into the following form:

$$dZ = \frac{G_S'}{k_y a M_B} \frac{dH'}{(H_i' - H')} \quad \text{...(6.63)}$$

Integration of Eq. (6.63) gives

$$Z = \frac{G_S'}{k_y a M_B} \int_{H_2'}^{H_1'} \frac{dH'}{(H_i' - H')} = H_{tG} \times N_{tG} \quad \text{...(6.64)}$$

where

$$H_{tG} = \text{Height of gas-enthalpy transfer unit} = \frac{G_S'}{k_y a M_B} \quad \text{...(6.65)}$$

$$N_{tG} = \text{Number of gas-enthalpy transfer unit} = \int_{H_2'}^{H_1'} \frac{dH'}{(H_i' - H')} \quad \text{...(6.66)}$$

In order to determine the number of transfer units by Eq. (6.66), we should first find the enthalpy at the air-water interface (H_i'). Consider the rate of heat transfer from the bulk liquid to the interface:

$$L'C_L dT_L = h_L a(T_L - T_i)dZ \quad \text{...(6.67)}$$

where h_L is the liquid film heat transfer coefficient. Substituting Eq. (6.44) into Eq. (6.67), we get

$$G_S' dH' = h_L a(T_L - T_i)dZ \quad \text{...(6.68)}$$

Compare Eq. (6.57) with Eq. (6.68) and we get

$$k_y a M_B (H_i' - H')dZ = h_L a(T_L - T_i)dZ \quad \text{...(6.69)}$$

which means

$$\frac{H_i' - H'}{T_i - T_L} = -\frac{h_L a}{k_y a M_B} \quad \text{...(6.70)}$$

According to Eq. (6.70), the slope of the straight line joining points (T_L, H') and (T_i, H_i') is $-\frac{h_L a}{k_y a M_B}$. The point (T_L, H') lies on the operating line. Since the air at the interface is assumed to be saturated with the water vapour, H_i' is the saturation enthalpy at the interface temperature T_i. The point (T_i, H_i') therefore, lies on the equilibrium curve. To find the temperature and enthalpy at the interface corresponding to the bulk liquid temperature T_L and the air humidity H', locate the point with coordinates (T_L, H') on the operating line and draw a straight line (*the driving force line*) from this point with a slope $-\dfrac{h_L a}{k_y a M_B}$. The line will intersect the equilibrium curve at (T_i, H_i').

To evaluate the number of gas-enthalpy transfer units, for different H' values on the operating line ranging from H_2' to H_1', the corresponding H_i' values are evaluated by drawing the driving force line with slope $-\dfrac{h_L a}{k_y a M_B}$ and finding the enthalpy coordinate of their intersection with the equilibrium curve. For illustration, the driving force line drawn from the point C in Fig. 6.20 intersects the equilibrium curve at E and the y-coordinate of the point E is H_i'. For each H' value, calculate the reciprocal of the driving force $\dfrac{1}{(H_i' - H')}$. Plot a graph with the reciprocal of the driving force on the y-axis and H' on the x-axis. The area under the graph between the limits H_2' and H_1' is the number of transfer units N_{tG}.

It is clear that the driving force line will be vertical if the resistance to heat transfer on the liquid side is negligible or the liquid side heat transfer coefficient tends to infinity. The use of the overall driving force and the overall coefficient for determining the height of the cooling tower is justified under such situations.

Change in the Temperature and Humidity of Air

The methods discussed above do not yield any information on the variation of the temperature and humidity of the air flowing through the column. For predicting the variation of the condition of air through the column, divide Eq. (6.57) by Eq. (6.47) and rearrange to give

$$\frac{dH'}{dT_G} = \frac{k_y a M_B C_S}{h_G a} \frac{H_i' - H'}{T_i - T_G} \qquad \ldots(6.71)$$

Since by the Lewis relation, $\dfrac{k_y a M_B C_S}{h_G a} = 1$, Eq. (6.71) simplifies as

$$\frac{dH'}{dT_G} = \frac{H_i' - H'}{T_i - T_G} \qquad \ldots(6.72)$$

Equation (6.72) means that the rate of change of the enthalpy of the air with the air dry-bulb temperature is equal to the slope of the straight line joining (T_i, H_i') and (T_G, H'). Referring to Fig. 6.21, the coordinates of the point M are the enthalpy and temperature of air entering the cooling tower. This point lies on the horizontal line passing though A, the bottom terminus of the operating line whose coordinates are the temperature of water leaving the column and the enthalpy of the air entering the column. The interfacial conditions at the bottom of the column corresponding to A are given by the point N which is located by drawing the driving force line AN as explained earlier. By Eq. (6.72), the slope of the line MN represents the slope of the H' versus T_G curve. The enthalpy and temperature of air varies along this curve. The slope is assumed to remain constant over small increments, say, up to point P in Fig. 6.21. A line parallel to the

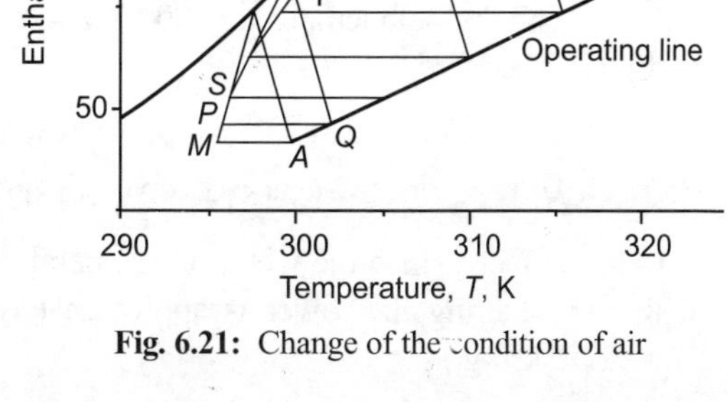

Fig. 6.21: Change of the condition of air

temperature axis and passing through P intersects the operating line at Q. The enthalpy coordinate of Q is the same as that of P and the temperature coordinate of Q is the temperature of water in contact with the air. The interfacial conditions corresponding to the bulk conditions represented by the point Q are obtained by drawing the line QR parallel to the line AN. The new direction along which the enthalpy and temperature of air in the tower change, is given by the line PR. The slope is assumed not to change for a small length, say, till the point S is reached. At S, the new slope of the H' versus T_G curve is determined by drawing the horizontal line at S to intersect the operating line and repeating the constructions explained above. A smooth curve drawn through M, P, S, etc. will represent the H' versus T_G curve. The final point on this curve will have as its coordinates, the temperature and enthalpy of the air leaving the column.

Design Considerations

Referring to Fig. 6.21, we can say for sure that so long as the point A representing the lower terminus of the operating line lies below the equilibrium curve, there is a positive enthalpy difference driving force for transfer of enthalpy from the water to the air and there is consequent cooling of water. The coordinates of point A are the water exit temperature ($T_{L,1}$) and the air inlet enthalpy (H_1'). Enthalpy of the air entering the tower is approximately equal to the saturation enthalpy at the wet-bulb temperature of the air. As long as the wet-bulb temperature of the air admitted to the column is below the exit temperature of water, the operating line will lie below the equilibrium curve and cooling of water is possible. Therefore, water can be cooled even if the dry-bulb temperature of the air with which it is contacted is above the temperature of water, provided its wet-bulb temperature is lower than the water exit temperature. It is also possible to cool water using saturated air if its temperature is below that of water. The difference between the exit water temperature and the entering-air wet-bulb temperature, $T_{L,1}$-$T_{W,1}$, is designated as the *wet-bulb temperature approach*. In the design of cooling towers, a wet-bulb temperature approach lying between 2.5 K and 8 K is commonly used. To account for the variation of the wet-bulb temperature with climatic conditions, the maximum wet-bulb temperature at the location which is never exceeded more than 5% of the time on the average, is used in the design.

Example 6.14: Water enters a forced draft counterflow cooling tower at a rate 1000 kg/m^{2}h at 318 K and leaves at 298 K. The air enters at 300 K with a wet-bulb temperature of 293 K. The heat transfer resistance in the liquid phase is negligible. The overall coefficient $K_Y a$ is 1200 kg/m^{3}h. Determine:

(a) The minimum air circulation rate, kg/m^{2}h

(b) The tower height, if the air rate used is 1.5 times the minimum.

The saturation enthalpy of air versus temperature data are as follows:

Temperature, K	283	288	293	298	303	308	313	318	323
Enthalpy, kJ/kg	29.1	41.8	57.2	76.0	99.4	128.7	165.7	212.8	273.6

***Solution*:** The cooling tower is schematically represented in Fig. 6.22.

The enthalpy of the air entering the tower (H_1') is determined using the psychrometric chart. The humidity of the air with dry-bulb temperature 300 K and wet-bulb temperature 293 K is 0.0118 kg water vapour per kg dry air. Therefore,

$$C_S = 1.005 + 1.884Y' = 1.005 + 1.884(0.0118) = 1.0272 \text{ kJ/kg K}$$

$$H_1' = C_S(T_G - 273.15) + Y'\lambda_0 = 1.0272(300 - 273.15) + 0.0118 \times 2502.3 = 57.1 \text{kJ/kg}$$

(Alternatively, since the wet-bulb temperature of the air entering the column is 293 K, the enthalpy of the air entering the tower is approximately equal to the saturation enthalpy at 293 K. That is, $H_1' = 57.2$ kJ/kg.)

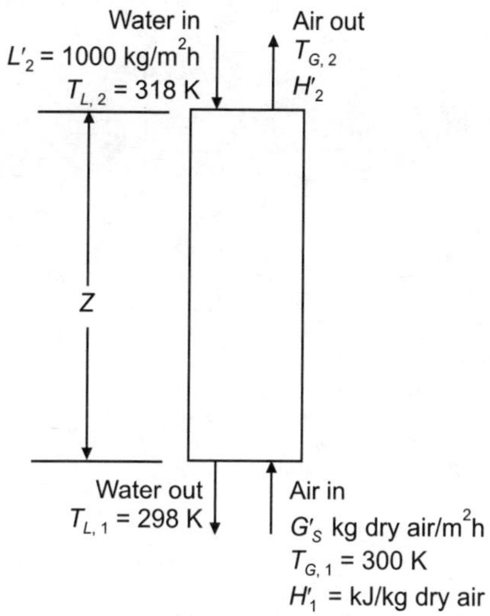

Fig. 6.22: Cooling tower in Example 6.14

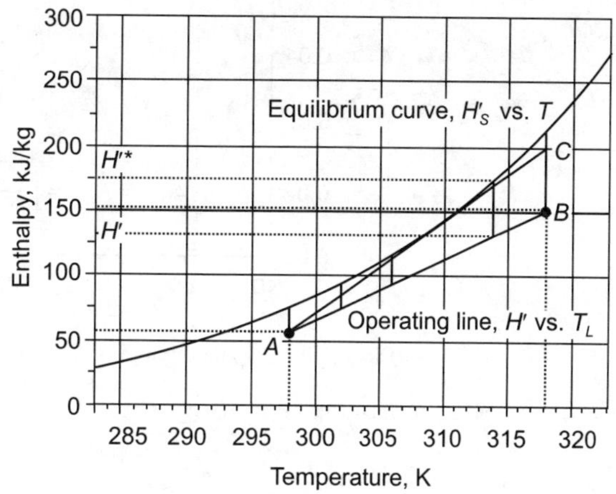

Fig. 6.23: Equilibrium curve and operating line for Example 6.14

The equilibrium curve is plotted with the enthalpy of the saturated air on the *y*-axis and the temperature on the *x*-axis as shown in Fig. 6.23. The lower terminus of the operating line is shown by point *A* whose coordinates are $T_{L,1}$ and H'_1. Here, the exit temperature of the cooling water, $T_{L,1} = 298$ K, and $H'_1 =$ 57.1. The upper terminus of the operating line *B* has coordinates H'_2, the enthalpy of the air leaving the tower and $T_{L,2}$, the temperature of the water entering the tower. Here $T_{L,2} = 318$ K. Point *B* should lie on the vertical line erected at $T = 318$ K. The exact location of *B* depends on the slope of the operating line which depends on the gas rate employed, and is equal to $\dfrac{L'C_L}{G'_S}$.

(a) For the minimum gas rate, the operating line just touches the equilibrium curve as shown by the limiting operating line, *AC*. The slope of the limiting operating line is $\dfrac{L'C_L}{G'_{S,\min}}$ where

$$\frac{L'C_L}{G'_{S,\,\min}} = \frac{H'_2(\max) - H'_1}{T_{L,2} - T_{L,1}}$$

$H'_2(\max)$ is the enthalpy coordinate of point *C* in Fig. 6.23. $H'_2(\max) = 200$ kJ/kg.

$$\frac{L'C_L}{G'_{S,\min}} = \frac{200 - 57.1}{318 - 298} = 7.145 \frac{\text{kJ}}{\text{kgK}} \qquad \text{(A)}$$

L' is the cooling water circulation rate. $L' = 1000$ kg/m^{2}h.

C_L is the heat capacity of cooling water, $C_L = 4.1868$ kJ/kg K.

Substituting these values in (A), we get,

$$\frac{L'C_L}{G'_{S,\min}} = \frac{1000 \times 4.1868}{G'_{S,\min}} = 7.145 \frac{\text{kJ}}{\text{kgK}}, \text{ or } G'_{S,\min} = 585.98 \text{ kg/m}^2\text{h}.$$

The minimum air circulation rate = 585.98 kg dry air /m^{2}h **(Answer)**

(b) Since the air rate used is 1.5 times the minimum, the actual gas rate is

$$G'_S = 1.5 \times 585.98 = 878.96 \text{ kg/m}^2\text{h}.$$

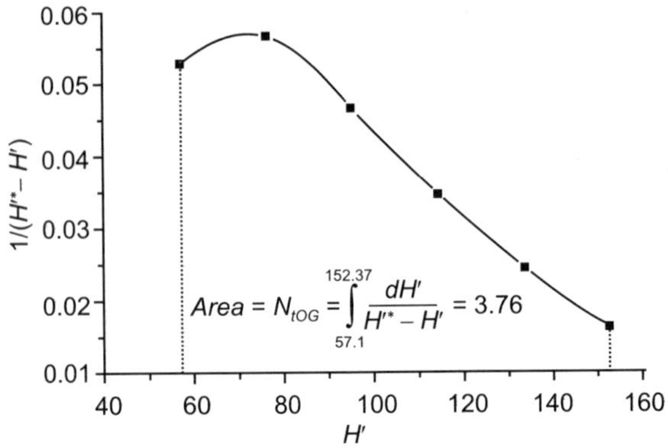

Fig. 6.24: Graphical integration to find the number of transfer units

Slope of the actual operating line AB is

$$\frac{L'C_L}{G'_S} = \frac{7.145}{1.5} = 4.763 \frac{\text{kJ}}{\text{kgK}}$$

The line AB is drawn from point A with slope 4.763 to terminate at point B on the vertical line at $T = 318$ K.

Since the liquid side offers no resistance to heat transfer, the height of the packed section can be estimated using Eq. (6.62)

$$Z = H_{tOG} \times N_{tOG}$$

where H_{tOG} is the height of the overall enthalpy transfer unit and N_{tOG} is the number of overall transfer units.

$$H_{tOG} = \frac{G'_S}{K_Y a} \text{ and } N_{tOG} = \int_{H'_1}^{H'_2} \frac{dH'}{H'^* - H'}$$

$$H_{tOG} = \frac{G'_S}{K_Y a} = \frac{878.96}{1200} = 0.7325 \text{ m}$$

To find the number of transfer units by graphical integration, $\frac{1}{H'^* - H'}$ is plotted on the y-axis and H' on the x-axis. For any given temperature, H' is the enthalpy coordinate of the point on the operating line and H'^* is the enthalpy coordinate of the point on the equilibrium curve. Thus the vertical distance between the operating line and the equilibrium curve gives $H'^* - H'$, the overall driving force for enthalpy transfer. The driving force is obtained from Fig. 6.23 by measuring the vertical distance between the equilibrium curve and the operating line at various points on the operating line. The reciprocal of the driving force is equal to $\frac{1}{H'^* - H'}$. The following table gives the result of such calculation.

T	H'	H'^*	$H'^* - H'$	$\frac{1}{H'^* - H'}$
298	57.10	76.00	18.90	0.0529
302	76.15	93.80	17.65	0.0567
306	95.21	116.60	21.39	0.0468
310	114.26	143.24	28.98	0.0345
314	133.31	174.04	40.73	0.0246
318	152.37	212.80	60.43	0.0165

A graph is drawn with H' on the x-axis and $\frac{1}{H'^{*}-H'}$ on the y-axis as shown in Fig. 6.24. The area under the curve between limits $H_1' = 57.1$ and $H_2' = 152.37$ gives the number of transfer units.

$$N_{tOG} = \int_{57.1}^{152.37} \frac{dH'}{H'^{*}-H'} = 3.76$$

The height of the packed section of the column is

$$Z = H_{tOG} \times N_{tOG} = 0.7325 \times 3.76 = 2.75 \text{ m}$$ **(Answer)**

Example 6.15: Water is to be cooled from 318 K to 303 K in an induced draft counterflow cooling tower at a rate 6000 kg/m² h. The air is admitted at a rate 5000 kg dry air /m²h. The 5% wet-bulb temperature for the location is 298 K. The estimated mass transfer coefficient $k_y a$ is 150 kmol/m³ h. Determine the height of the packed section given that the ratio $\frac{h_L a}{k_y a M_B} = 20 \text{ kJ/kg K}$.

The saturation enthalpy of air versus temperature data are as follows:

Temperature, K	283	288	293	298	303	308	313	318	323
Enthalpy, kJ/kg	29.1	41.8	57.2	76.0	99.4	128.7	165.7	212.8	273.6

***Solution*:** The liquid and air flow rates are $L' = 6000$ kg/m² h and $G_S' = 5000$ kg/m² h.

The water inlet temperature, $T_{L,2} = 318$ K.

Water exit temperature, $T_{L,1} = 303$ K.

Enthalpy of inlet air, H_1' = Enthalpy of saturated air at the wet-bulb temperature, 298 K = 76.0 kJ/kg.

Enthalpy of air leaving the tower (H_2') is obtained by the enthalpy balance over the tower.

$$G_S'(H_2' - H_1') = L' C_L (T_{L,2} - T_{L,1})$$

$$5000(H_2' - 76.0) = 6000 \times 4.1868(318 - 303)$$

On solving this equation we get

$$H_2' = 151.36 \text{ kJ/kg}$$

The saturation enthalpy versus temperature curve is drawn which is the equilibrium curve for the cooling tower as shown in Fig. 6.25. Operating line is the straight line joining point A (303,76) and point B (318,151.36) on this diagram.

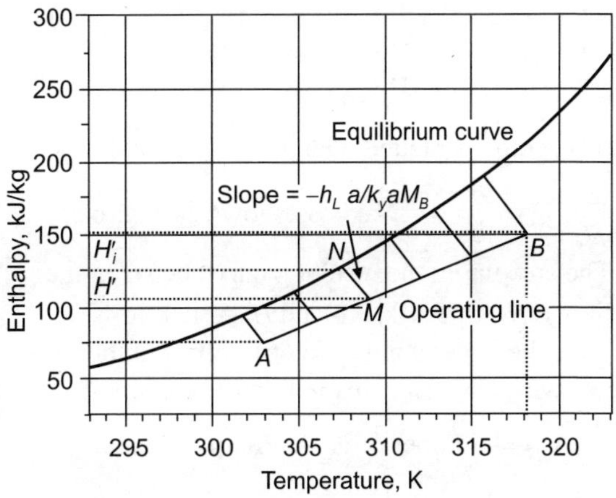

Fig. 6.25: Equilibrium curve and operating line for Example 6.15

Height of the tower is determined using Eq. (6.64)

$$Z = H_{tG} \times N_{tG}$$

where H_{tG} = Height of gas-enthalpy transfer unit $= \frac{G_S'}{k_y a M_B}$ ...(6.65)

N_{tG} = Number of gas-enthalpy transfer unit $= \int_{H_2'}^{H_1'} \frac{dH'}{(H_i' - H')}$ ...(6.66)

Since $k_y a = 150$ kmol/m³h, the height of transfer unit,

$$H_{tG} = \frac{G'_S}{k_y a M_B} = \frac{5000}{150 \times 29} = 1.149 \text{ m.}$$

To find the number of transfer units, N_{tG} by graphical integration, $\frac{1}{(H'_i - H')}$ is to be plotted against H', where H' is the enthalpy of the air in contact with water at a given section in the column and H'_i is the enthalpy of the air at the interface at that section. The enthalpy at the interface is related to bulk enthalpy as

$$\frac{H'_i - H'}{T_i - T_L} = -\frac{h_L a}{k_y a M_B} \qquad \ldots(6.70)$$

Choose a point such as the point M on the operating line, the ordinate of which is H' and draw the driving force line with slope $-\frac{h_L a}{k_y a M_B}$ through it which will intersect the equilibrium curve at point N the ordinate of which is H'_i. The reciprocal of the difference $H'_i - H'$ is calculated. This is repeated for different points on the operating line. The results are summarized in the following table.

T	H'	H'_i	$H'_i - H'$	$\frac{1}{H'_i - H'}$
303	76.00	93.80	17.80	0.0529
306	91.07	109.10	18.03	0.0567
309	106.10	125.50	19.40	0.0468
312	121.20	145.30	24.10	0.0345
315	136.30	165.50	29.20	0.0246
318	151.36	189.13	37.77	0.0165

A graph is plotted with H' on the x-axis and $\frac{1}{H'_i - H'}$ on the y-axis as shown in Fig. 6.26. The area under the resulting curve between the limits $H'_1 = 76.0$ kJ/kg and $H'_2 = 151.36$ kJ/kg gives the integral in Eq. (6.66) or the number of transfer units. The area is found to be 2.973.

N_{tG} = Number of gas-enthalpy transfer unit

$$= \int_{H'_2}^{H'_1} \frac{dH'}{(H'_i - H')} = 2.973$$

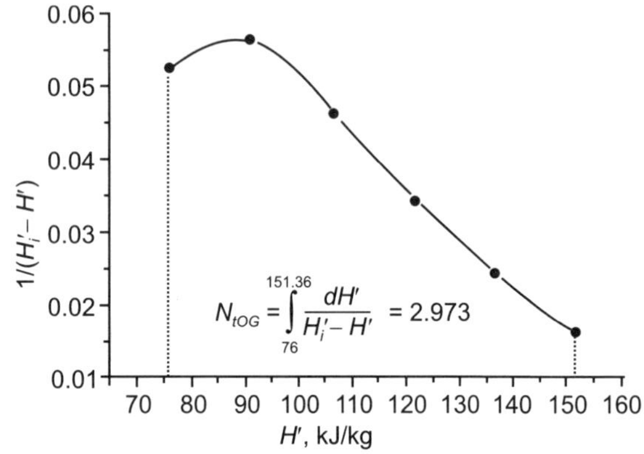

Fig. 6.26: Graphical integration to find the number of transfer units in Example 6.15

The height of the packed section is

$$Z = H_{tG} \times N_{tG} = 1.149 \times 2.973 = 3.42 \text{ m.}$$

(Answer)

Example 6.16: The cooling tower in Example 6.15 is supplied with water at a total circulation rate of 6×10^5 kg/h. Air is admitted at a rate of 5×10^5 kg dry air/h with a humidity of 0.01 kg/kg dry air. Assume that the air leaving the tower is saturated with the water vapour. It is desired that the circulating water should contain hardness not more than 2000 ppm dissolved solids. The make-up water contains 500 ppm hardness. The windage losses in the tower is estimated to be 0.2% of the circulation rate. A small amount of water is deliberately discarded (blow down) to keep the hardness at the toleration limit. Determine the make-up water requirement.

Solution: The cooling tower in Example 6.16 can be represented as in Fig. 6.27.

The exit air enthalpy is $H_2' = 151.36$ kJ/kg (Example 6.15). Assuming that the air-leaving the tower is saturated with the water vapour, H_2' is the saturation enthalpy at the temperature of the air leaving. Using the humidity charts, the temperature at which the saturation enthalpy is 151.36 kJ/kg is found to be 311 K. Therefore, the exit temperature of air = 311 K. Saturation humidity at 311 K = 0.0434 kg water/kg dry air from the psychrometric charts. Therefore, $Y_2' = 0.0434$ kg/kg dry air. The humidity of the air entering the column is $Y_1' =$ 0.01 kg/kg dry air.

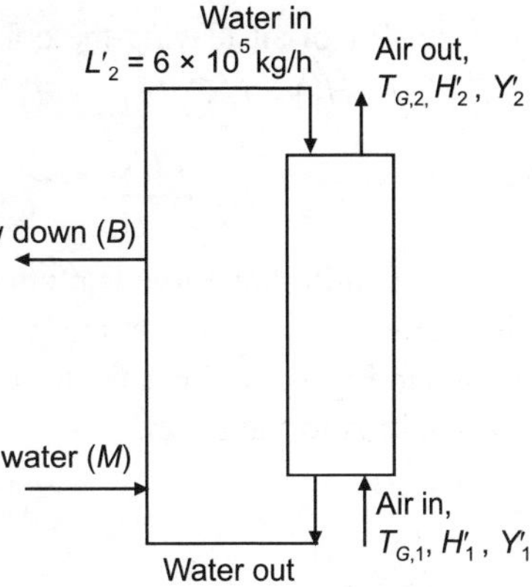

Fig. 6.27: Cooling tower in example 6.16

Let E be the evaporation loss in kg/h. Then

$$E = G_S(Y_2' - Y_1') = 5 \times 10^5 (0.0434 - 0.01)$$
$$= 1.67 \times 10^4 \text{ kg/h}$$

Let W be the windage loss and B be the blow down in kg/h. Since W is 0.2% of the circulation rate,

$$W = 0.002 \times 6 \times 10^5 = 1200 \text{ kg/h}.$$

Then a total material balance gives

$$M = E + B + W \tag{A}$$

Since water lost by evaporation is free of hardness, a hardness balance gives

$$Mx_M = (B + W)x_C \tag{B}$$

where x_M and x_C are the weight fraction of hardness in the make up water and the circulating water, respectively. Eliminate M in Eq. (B) using (A).

$$(E + B + W)x_M = (B + W)x_C$$

$$B = E\frac{x_M}{x_C - x_M} - W \tag{C}$$

$$B = 1.67 \times 10^4 \frac{500}{2000 - 500} - 1200 = 4366.67 \text{ kg/h}$$

Substituting in (A), we get

$$M = 1.67 \times 10^4 + 4366.67 + 1200 = 22{,}266.67 \text{ kg/h}$$

The make-up water required is 22,266.67 kg/h. This is $\frac{22266.67}{600000} \times 100 = 3.7\%$ of the circulating water. **(Answer)**

Example 6.17: Water is to be cooled in a small packed column from 323 to 295 K by means of air flowing countercurrently. The rate of flow of liquid is 1.5 kg/ m² s and the flow rate of the air is 5.4 kg/ m² s. The air enters at a temperature of 295 K with humidity 60%. The resistance to transfer is confined to the gas side. Determine the condition of the air leaving at the top of the column.

The saturation enthalpy of air versus temperature data are as follows:

Temperature, K	283	288	293	298	303	308	313	318	323
Enthalpy, kJ/kg	29.1	41.8	57.2	76.0	99.4	128.7	165.7	212.8	273.6

Solution: The water circulation rate $L' = 1.5$ kg/m² s and the air circulation rate $G_S' = 5.4$ kg/m² s. The inlet water temperature is $T_{L,2} = 323$ K and the exit water temperature is $T_{L,1} = 295$ K. The air enters at 295 K with 60% humidity. Therefore humidity of entering air $Y_1' = 0.60\, Y_S'$ where the saturation humidity Y_S' is read from humidity chart.

$$Y_1' = 0.6Y_S' = 0.6 \times 0.0166 = 0.01 \text{ kg water/kg dry air}.$$

The humid heat of the air entering the tower, $C_S = 1.005 + 0.01 \times 1.884 = 1.0238$ kJ/kg

The enthalpy of the entering air $H_1' = 1.0238(295 - 273.15) + 2502.3 \times 0.01 = 47.4$ kJ/kg

Enthalpy of air leaving the column is obtained from an enthalpy balance for the tower.

$$G_S'(H_2' - H_1') = L'C_L(T_{L,2} - T_{L,1})$$

$$H_2' = H_1' + \frac{L'C_L}{G_S'}(T_{L,2} - T_{L,1}) = 47.4 + \frac{1.5 \times 4.1868}{5.4}(323 - 295) = 80 \text{ kJ/kg} \quad \textbf{(Answer)}$$

The equilibrium curve is plotted on the enthalpy versus temperature coordinates and the operating line is plotted as a straight line connecting point A (295 K, 47.4 kJ/kg) and point B (323 K, 80 kJ/kg) as shown in Fig. 6.28. Since the air is admitted at 295 K, the point A also represents the initial state of the air admitted to the tower.

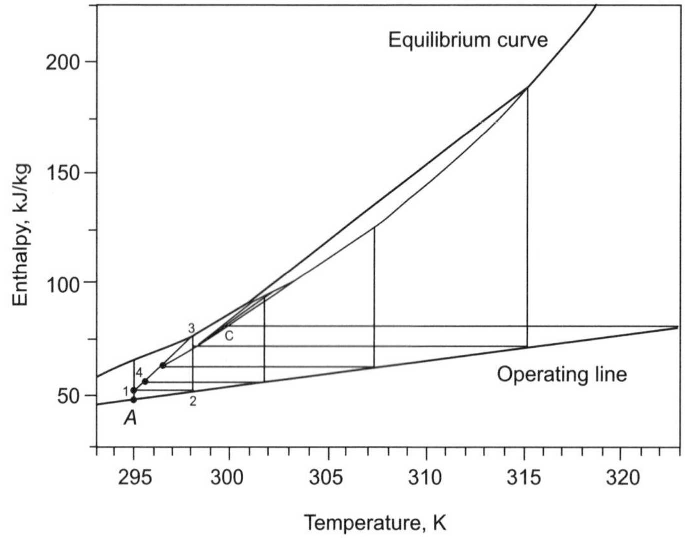

Fig. 6.28: Construction to trace the change in the condition of air in cooling towers

The rate of change of enthalpy of the air with its temperature is given by Eq. (6.72). From the point A, a vertical line is drawn to meet the equilibrium curve. For a small distance, say, up to point 1, the enthalpy and temperature change along this line. A horizontal line drawn at point 1 meets the operating line at point 2. A vertical line is erected at 2 which intersects the equilibrium curve at 3. The new direction in which the enthalpy and temperature change is given by the line 1-3. At point 4 on this line, the new direction for the change is determined by drawing a horizontal line passing through point 4 and by erecting a vertical at its intersection with the equilibrium curve. The construction is continued till we reach point C lying on the horizontal line passing thorough B. At point C, the enthalpy is the enthalpy of exit air H_2'

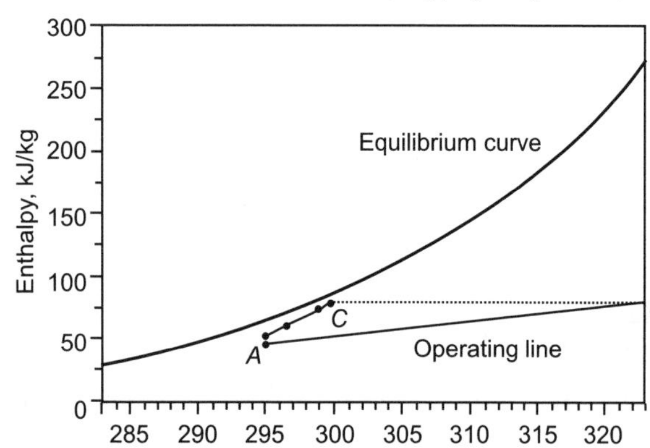

Fig. 6.29: Enthalpy – temperature history of the air in the cooling tower of example, 6.17

and the temperature is $T_{G,2}$. Figure 6.29 shows the result of the graphical construction in which the curve *AC* represents the enthalpy-temperature history of the air in the cooling tower.

Temperature of air leaving: 300 K **(Answer)**

6.7 ADIABATIC HUMIDIFICATION

Air humidification is used for preparing the air of desired humidity and temperature by bringing unsaturated air in direct contact with water. Assume that the air is initially at a dry-bulb temperature $T_{G,1}$ and humidity Y_1' and it is desired to produce the air having the humidity Y_2' at temperature $T_{G,2}$. The process can be represented on the psychrometric chart as shown in Fig. 6.30. Point *A* represents the initial state of the air and point *B* the final state. In the adiabatic humidification operation using recirculation of water, the temperature of water is brought to and maintained constant at the adiabatic saturation temperature of the air. The small quantity of the water lost by the evaporation is compensated by adding make-up water also at the adiabatic saturation temperature. The air is first preheated at constant humidity Y_1' to a temperature T_G such that the adiabatic saturation temperature of the air at condition *C* and the dew point of the air at the final state *B* are the same. The air at *C* is subjected to adiabatic humidification in an apparatus such as the spray chamber which are horizontal spray towers as shown in Fig. 6.31.

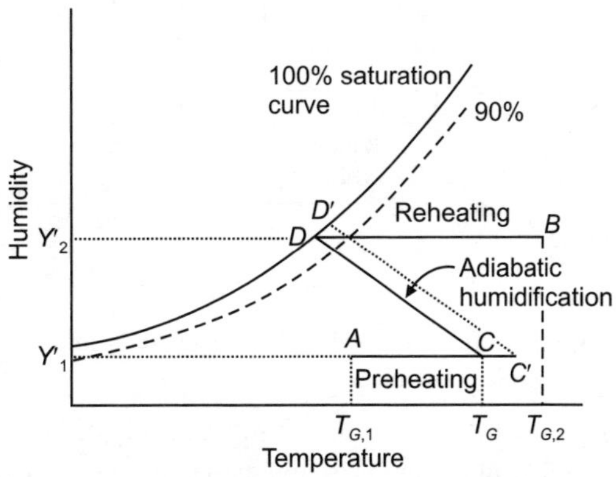

Fig. 6.30: Adiabatic humidification on psychrometric chart

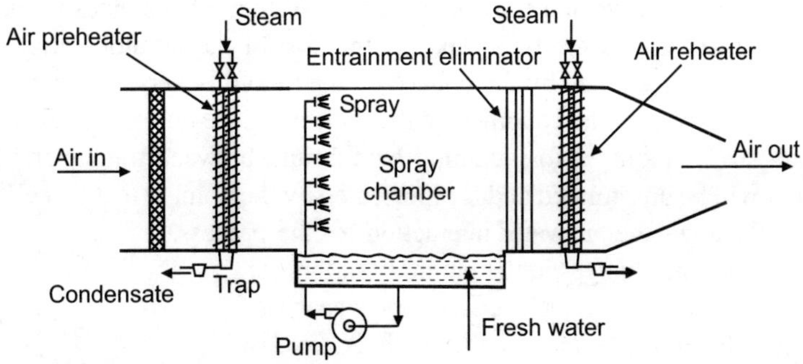

Fig. 6.31: Schematic sketch of spray chamber

Water is sprayed onto the air, and the air gets humidified and cooled. Heat transfer surfaces provided ahead of the chamber act as the air preheaters and those that follow the chamber act as air reheaters. If the chamber is large and there is sufficient contact between the air and water to attain equilibrium, the air leaving the humidification chamber will be saturated as represented by the point *D* in Fig. 6.30. The air leaving the humidification chamber is reheated at constant humidity Y_2' to the desired final temperature $T_{G,2}$. Frequently it is necessary to give allowance for incomplete humidification in the chamber. For example, if only 90% saturation is possible in the adiabatic humidification step, it is necessary to preheat the air to a temperature slightly above the temperature corresponding to point *C* so that air leaving the spray chamber at 90% saturated state still has the desired final humidity Y_2'. The line *C′D′* will represent the adiabatic humidification step in this case.

The temperature and the humidity of the streams in the adiabatic humidification chamber are shown in Fig. 6.32. The air-flow rate is G_S' kg dry air per hour. The air enters the chamber with humidity Y_1' kg

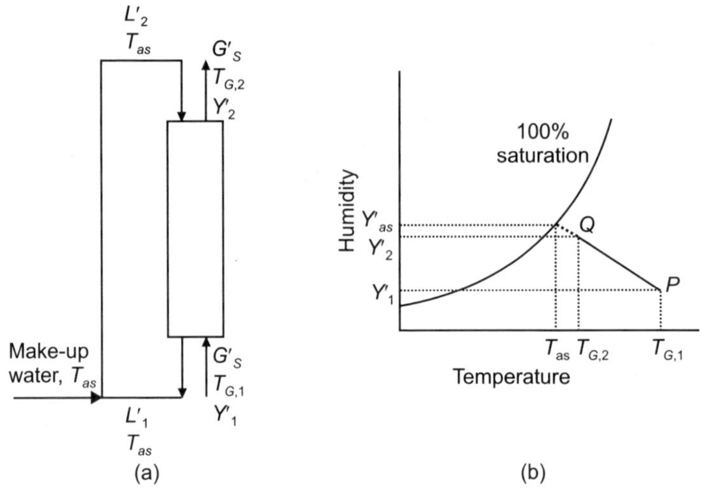

Fig. 6.32: (a) The gas and the liquid streams in adiabatic humidification chamber (b) Temperature –humidity relationship in adiabatic humidification

water per kg dry air and temperature $T_{G,1}$. The air leaves with a humidity Y_2' and temperature $T_{G,2}$. The water temperature is maintained constant at the adiabatic saturation temperature of the air by continuous recirculation and by the addition of make-up water, which also is at the adiabatic saturation temperature T_{as}.

The adiabatic humidification being essentially a constant enthalpy process, the air enthalpy remains constant during the process. The liquid temperature is also constant in the process using recirculation of the water. Therefore, the plot of the enthalpy versus temperature of water cannot be used for representing the equilibrium curve and the operating line. Rather, the changes in the temperature and the humidity of the gas can be used for estimating the size of the humidification chamber. The air entering the humidifier is represented by the point P on the humidity versus temperature diagram (Fig. 6.32b) and the point Q represents the condition of the air leaving the humidifier. Since a humidification chamber of infinite size is required for attaining equilibrium between the air and the water, the air leaving the chamber will be unsaturated and is represented by the point Q that lie below the equilibrium curve. Figure 6.33 illustrates the air-water interaction for the process.

First, we will consider the changes in the humidity of the gas stream and use the mass transfer from the interface for the analysis. The rate of mass transfer from the interface to the bulk air is

$$G_S'\,dY' = k_y a M_B (Y_i' - Y')dZ \qquad \ldots(6.73)$$

The humidity at the interface is the saturation humidity at the adiabatic saturation temperature, T_{as}. This is represented by Y_{as}' so that Eq. (6.73) becomes

$$G_S'\,dY' = k_y a M_B (Y_{as}' - Y')dZ \qquad \ldots(6.74)$$

Equation (6.74) can be written as

$$dZ = \frac{G_S'}{k_y a M_B}\frac{dY'}{(Y_{as}' - Y')} \qquad \ldots(6.75)$$

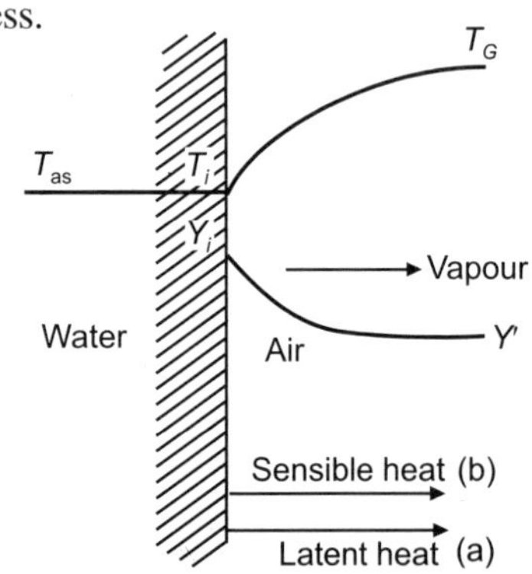

Fig. 6.33: Air-water interaction in adiabatic humidification

Integration of Eq. (6.75) gives

$$Z = \frac{G_S'}{k_y a M_B}\int_{Y_1'}^{Y_2'}\frac{dY'}{(Y_{as}' - Y')} \qquad \ldots(6.76)$$

Since Y_{as}' is constant, Eq. (6.76) is amenable to analytical integration.

$$Z = \frac{G'_S}{k_y a M_B} \ln \frac{Y'_{as} - Y'_1}{Y'_{as} - Y'_2} = H_{tG} \times N_{tG} \qquad \ldots(6.77)$$

where $H_{tG} = \dfrac{G'_S}{k_y a M_B}$ is the height of the gas transfer unit and

$$N_{tG} = \ln \frac{Y'_{as} - Y'_1}{Y'_{as} - Y'_2} = \frac{Y'_2 - Y'_1}{(\Delta Y')_m}$$ is the number of the gas transfer unit.

Here $(\Delta Y')_m$ is the logarithmic mean of the driving forces at the two ends of the chamber $Y'_{as} - Y'_1$ and $Y'_{as} - Y'_2$.

$$(\Delta Y')_m = \frac{(Y'_{as} - Y'_1) - (Y'_{as} - Y'_2)}{\ln \dfrac{Y'_{as} - Y'_1}{Y'_{as} - Y'_2}} \qquad \ldots(6.78)$$

The heat transfer rate equations for the gas phase can also be used for the design of the adiabatic humidification chamber in a similar way. Consider Eq. (6.47).

$$G'_S C_S dT_G = h_G a (T_i - T_G) dZ \qquad \ldots(6.47)$$

Since no temperature gradient exists in the liquid, the interface is at the liquid temperature which is equal to the adiabatic saturation temperature of the air. Replacing the interface temperature T_i in Eq. (6.47) with T_{as}, and rearranging, we get

$$G'_S C_S dT_G = h_G a (T_{as} - T_G) dZ \qquad \ldots(6.79)$$

$$dZ = \frac{G'_S C_S}{h_G a} \frac{dT_G}{T_{as} - T_G} \qquad \ldots(6.80)$$

Integrate Eq. (6.80) between limits $T_{G,1}$ and $T_{G,2}$. The result is

$$Z = \frac{G'_S C_S}{h_G a} \int_{T_{G1}}^{T_{G2}} \frac{dT_G}{T_{as} - T_G} = \frac{G'_S C_S}{h_G a} \ln \frac{T_{as} - T_{G,1}}{T_{as} - T_{G,2}} \qquad \ldots(6.81)$$

Writing Z as the product of the height and the number of transfer units, we have

$$Z = H_{tG} \times N_{tG} \qquad \ldots(6.82)$$

where $H_{tG} = \dfrac{G'_S C_S}{h_G a}$ is the height of the transfer unit and $N_{tG} = \displaystyle\int_{T_{G1}}^{T_{G2}} \frac{dT_G}{T_{as} - T_G} = \ln \frac{T_{as} - T_{G1}}{T_{as} - T_{G2}}$ is the number of transfer units. The number of transfer units can be expressed in terms of the logarithmic mean temperature difference driving force $(\Delta T)_{lm}$

$$(\Delta T)_{lm} = \frac{(T_{as} - T_{G,1}) - (T_{as} - T_{G,2})}{\ln \dfrac{T_{as} - T_{G,1}}{T_{as} - T_{G,2}}} \qquad \ldots(6.83)$$

The number of transfer units now becomes

$$N_{tG} = \ln \frac{T_{as} - T_{G1}}{T_{as} - T_{G2}} = \frac{T_{G,2} - T_{G,1}}{(\Delta T)_{lm}} \qquad \ldots(6.84)$$

and by substituting Eq. (6.84) into Eq. (6.81), we get the length of the chamber as

$$Z = \frac{G'_S C_S}{h_G a} \frac{(T_{G2} - T_{G1})}{(\Delta T)_{lm}} \qquad \ldots(6.85)$$

We can see that Eq. (6.85) is analogous to

$$Q = h_G A(\Delta T)_{lm} \qquad \ldots(6.86)$$

where $Q = G_S' C_S (T_{G2} - T_{G1})$ is the total heat transferred to the gas stream in unit time, $A = aZ$ is the area of heat transfer and $(\Delta T)_{lm}$ is the logarithmic mean temperature difference (*LMTD*) between the inlet and exit to the humidification chamber.

The approximate size of the humidification chamber when humidification is carried out adiabatically with circulating water, can be determined using either the integrated form of the mass transfer rate equation [Eq. (6.77)] or its heat transfer counterpart [Eq. (6.81)].

The design of humidifier in which the temperature of the water changes along the length of the exchanger is quite complex and we have to use the gas enthalpy versus liquid temperature coordinates for representing the changes occurring in the tower as we do in the analysis of the cooling towers. But here, since the exit water temperature and the exit air enthalpy are not known, it is not possible to plot the operating line, though we know its slope $\frac{L'C_L}{G_S'}$. The problem is solved by trial by assuming arbitrarily a water exit temperature, which will fix the upper terminus of the operating line. Now the operating line is drawn and the enthalpy and temperature of the exit air are determined using the method discussed under *cooling towers*; using the temperature and enthalpy of the air, its humidity is determined. The calculation has to be repeated by assuming a new exit water temperature and calculating the humidity of the exit air again, if the humidity calculated and that specified for the exit air are not the same. Using the exit water temperature calculated by trial, the size of the vessel is determined as explained earlier under the section on the cooling towers, with obvious modifications.

Example 6.18: Five thousand kg dry air/m^2 h at 338 K having a humidity of 0.012 kg water vapour/kg dry air is subjected to adiabatic humidification in a horizontal spray chamber. The water temperature is maintained constant by continuous recirculation. The air leaving the chamber is 90% saturated. The heat transfer coefficient $h_G a$ may be taken to be 1500 W/m^3 K. Determine the length of the spray chamber.

***Solution*:** The adiabatic humidification process can be represented on the humidity chart as shown in Fig. 6.34.

The air flow rate is $G_S' = 5000$ kg dry air/m^2 h. The humidity and temperature of the air admitted to the chamber are $Y_1' = 0.012$ kg water/kg dry air and $T_{G,1} = 338$ K. The state of the air entering the chamber is represented by point A. The adiabatic cooling line passing through A intersects the 90% saturation curve at point B, the coordinates of which are the humidity and temperature of the air leaving the chamber. Humidity $Y_2' = 0.0265$ kg/kg dry air and temperature $T_{G,2} = 304.8$ K. The water temperature is maintained constant at the adiabatic saturation temperature, which from the psychrometric chart is the temperature coordinate of point C. Adiabatic saturation temperature $T_{as} = 303.2$ K

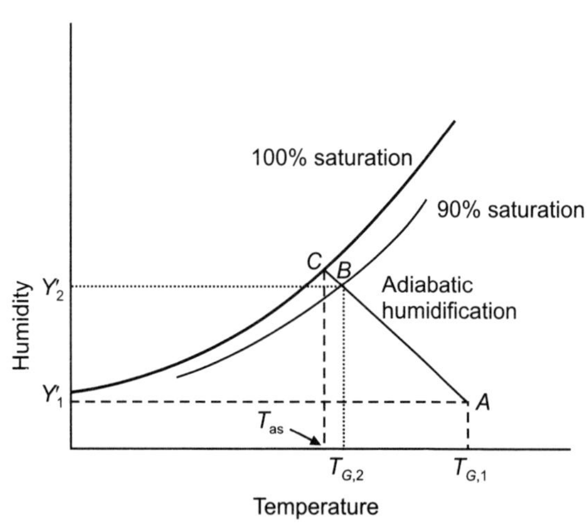

Fig. 6.34: Adiabatic humidification process in Example 6.18

The length of the spray chamber is given by Eq. (6.81).

$Z = \dfrac{G'_S C_S}{h_G a} \ln \dfrac{T_{as} - T_{G,1}}{T_{as} - T_{G,2}}$ where C_S is the humid heat of the air, kJ/kg K and the heat transfer coefficient $h_G a$ = 1500 W/m³ K.

Humid heat of the inlet air C_{S1} = 1.005 + 0.012(1.884) = 1.0276 kJ/kg K

Humid heat of the exit air C_{S2} = 1.005 + 0.0265(1.884) = 1.0549 kJ/kg K

Average humid heat, C_S = 0.5 (1.0276+1.0549) = 1.0413 kJ/kg K

Substituting the values in Eq. (6.81)

$$Z = \frac{5000 \times 1.0413}{1500 \times 10^{-3} \times 3600} \ln \frac{303.2 - 338}{303.2 - 304.8} = 2.97 \text{ m}$$

Length of the chamber = 2.97 m **(Answer)**

6.8 DEHUMIDIFICATION

Air conditioning requirements of plant areas and buildings have made dehumidification operations very important industrially. Removal of the moisture from the air for reuse as, for example, the removal of water vapour from hot air leaving a drier, is another important application of dehumidification. Dehumidification is done by bringing the air in contact with water, the temperature of which is below the dew point of the air. Dehumidification can be accomplished in a spray chamber as the one described for adiabatic humidification, but in this case the temperature of water should be low, well below the dew point of the air. It can also be done by passing the air outside a bank of finned tubes through the inside of which cold water is circulated. Water vapour condenses on the cold outer surface of the tube walls forming a film of condensate with which the air is in direct contact. In both these cases, the air-water interaction can be represented as in Fig. 6.35.

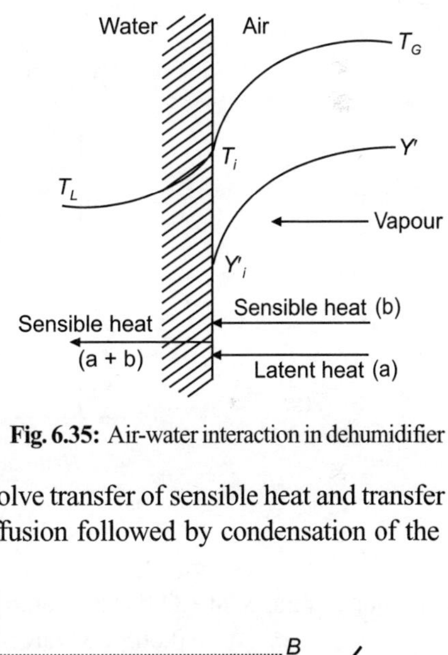

Fig. 6.35: Air-water interaction in dehumidifier

The simultaneous cooling and dehumidification of a gas involve transfer of sensible heat and transfer of latent heat to the interface, the latter resulting from the diffusion followed by condensation of the water vapour. Here Y' is greater than Y'_i and therefore, water vapour diffuses to the interface. Moisture can be removed from unsaturated air without first bringing the bulk gas to saturation. The temperature and humidity gradients are in the opposite direction to those in the upper portion of the cooling towers, but in other respects the process and the method of calculations are identical to those of cooling towers.

The operating line on the gas-enthalpy versus liquid temperature graph will be above the equilibrium curve as shown in Fig. 6.36 and the driving force is $H' - H'^*$. The height of the packed section is given by

$$Z = \frac{G'_S}{k_y a M_B} \int_{H'_2}^{H'_1} \frac{dH'}{H' - H'^*} \qquad \ldots(6.87)$$

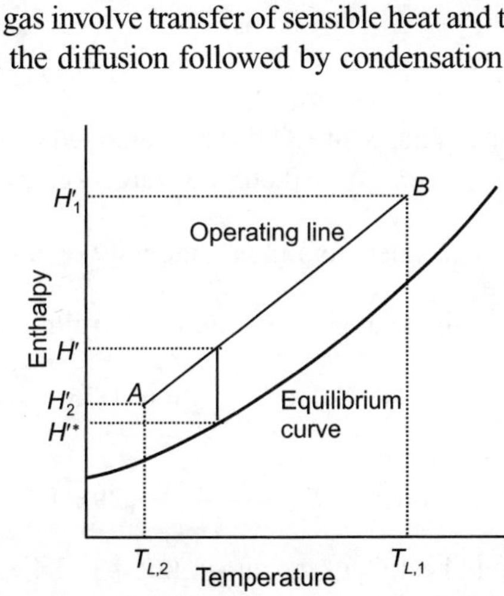

Fig. 6.36: Equilibrium curve and the operating line for dehumidification column

The integral in Eq. (6.87) can be determined by graphical integration by plotting $\frac{1}{H' - H'^*}$ against H', as described in the design of cooling towers.

Example 6.19: Air, 80% saturated with water vapour enters a dehumidifier at 322 K and leaves at a wet-bulb temperature of 300 K. Water enters the column at 294 K. The air flow rate is 720 m^3/m^2 h measured at standard conditions. The water rate is 1.5 times the minimum rate. The ratio of heat transfer coefficient to mass transfer coefficient h_L/k_yM_B has been found to be 50 kJ/kg K. The mass transfer coefficient k_yaM_B = 1260 kg/m^3 h. Determine:

(a) Water flow rate
(b) Exit temperature of water
(c) Height of the tower

The saturation enthalpy of air versus temperature data are as follows:

Temperature, K	283	288	293	298	303	308	313	318	323
Enthalpy, kJ/kg	29.1	41.8	57.2	76.0	99.4	128.7	165.7	212.8	273.6

***Solution*:** Figure 6.37 shows the flow rates, temperature and enthalpy of air and water streams in the dehumidifier.

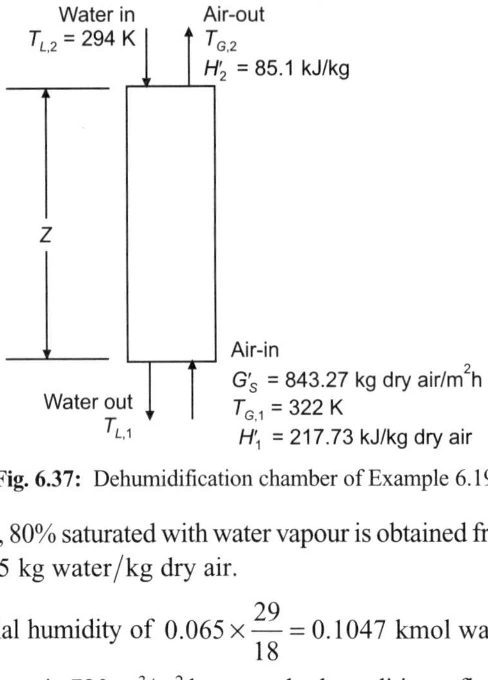

Fig. 6.37: Dehumidification chamber of Example 6.19

Humidity of the air at 322 K, 80% saturated with water vapour is obtained from the psychrometric chart.

$$Y'_1 = 0.065 \text{ kg water/kg dry air.}$$

(This is equivalent to a molal humidity of $0.065 \times \frac{29}{18} = 0.1047$ kmol water/kmol dry air.)

Since the total air flow rate is 720 m^3/m^2 h at standard conditions, flow rate of dry air is

$$720 \times \frac{1}{1.1047} = 651.76 \text{ m}^3 \text{ dry air/m}^2 \text{ h}$$

This corresponds to $G'_S = \frac{651.76}{22.414} \times 29 = 843.27$ kg dry air/m^2 h.

Humid heat C_S of the air $= 1.005 + 1.884 \times 0.065 = 1.1275$ kJ/kg K.

Enthalpy of the air entering the tower,

$$H'_1 = 1.1275 \times (322 - 273.15) + 0.065 \times 2502.3 = 17.73 \text{ kJ/kg.}$$

Since the air leaves at a wet-bulb temperature of 300 K, its enthalpy is approximately equal to the enthalpy of saturated air at 300 K. $H_2' = 85.1\,\text{kJ/kg}$ (from the humidity chart).

The enthalpy of the saturated air is plotted against the temperature as shown in the Fig. 6.38 which gives the equilibrium curve for the system. The operating line is a straight line with one terminus at A $(T_{L,2}, H_2')$ and the other at B $(T_{L,1}, H_1')$. The slope of the operating line is $\frac{L'C_L}{G_S'}$ and it depends on the quantity of water circulated. Therefore, the upper terminus B of the operating line can lie on the horizontal line drawn at $H_1' = 217.73$ kJ/kg, the exact location of which depends on the temperature of the water leaving the dehumidifier $(T_{L,1})$ which in turn, depends on the amount of water circulated. For minimum rate of liquid circulation, the operating line meets the equilibrium curve at point C and AC is the limiting operating line with minimum slope $\left(\frac{L'C_L}{G_S'}\right)_{\min}$ and the exit temperature of water is $T_{L,1}$ (max).

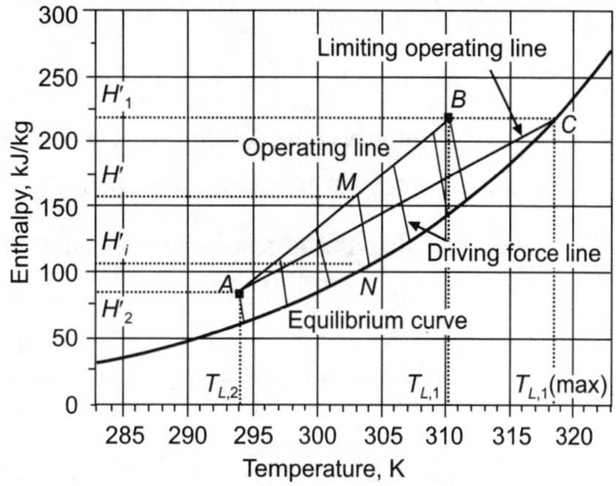

Fig. 6.38: Determination of number of transfer units of dehumidifier in Example 6.19

From the Fig. 6.38, $T_{L,1}\,(\max) = 318.4$ K.

(a) The slope of the line

$$AC = \left(\frac{L'C_L}{G_S'}\right)_{\min} = \frac{H_1' - H_2'}{T_{L,1} - T_{L,2}\ (\max)} = \frac{217.73 - 85.1}{318.4 - 294} = 5.4357\,\text{kJ/kg K}.$$

$$\text{The minimum water circulation rate} = L'_{\min} = 5.4357 \times \frac{G_S'}{C_L}$$

$$= 5.4357 \times \frac{843.27}{4.1868} = 1094.81\,\text{kg/m}^2\text{h}.$$

Since the actual rate is 1.5 times the minimum, the actual rate of water circulation

$$= 1.5 \times 1094.81 = 1642.22\,\text{kg/m}^2\text{h}.$$ **(Answer)**

(b) Slope of the actual operating line $= \frac{L'C_L}{G_S'} = 1.5\left(\frac{L'C_L}{G_S'}\right)_{\min} = 8.1535\,\text{kJ/kg K}.$

The actual operating line is drawn from the point A with slope = 8.1535 kJ/kg K, which will intersect the horizontal line drawn at $H_1' = 217.73$ kJ/kg at point B. The temperature coordinate of point B is the actual water exit temperature, $T_{L,1}$.

Exit temperature of water, $T_{L,1} = 310.27$ K **(Answer)**

[Alternatively, since $\frac{L'C_L}{G_S'} = \frac{H_1' - H_2'}{T_{L,1} - T_{L,2}}, 8.1535 = \frac{217.73 - 85.1}{T_{L,1} - 294}$ so that $T_{L,1} = 310.27$ K]

(c) The height of the dehumidification column is given by Eq. (6.87) modified to account for the resistance offered by the liquid phase as

$$Z = \frac{G_S'}{k_y a M_B} \int_{H_2'}^{H_1'} \frac{dH'}{H' - H_i'}$$

To find the integral $\int_{H_2'}^{H_1'} \frac{dH'}{H' - H_i'}$, points such as point M in Fig. 6.38 is chosen on the operating line. Its enthalpy coordinate is H'. The straight line drawn from the point M with slope $= -\frac{h_L a}{k_y a M_B}$ $= -50\,\text{kJ/kg K}$ intersects the equilibrium curve at point N, the enthalpy coordinate of which is H_i'. This is repeated for different points on the operating line from point A to point B. For each such point, $\frac{1}{H' - H_i'}$ is calculated. The results are tabulated in the table given below:

T	H'	H_i'	$H' - H_i'$	$\frac{1}{H' - H_i'}$
204	85.10	62.32	22.78	0.0439
297	109.56	74.34	35.22	0.0284
300	134.02	88.91	45.11	0.0222
303	158.48	105.50	52.98	0.0189
306	182.93	123.32	59.61	0.0168
309	207.39	144.91	62.48	0.0160
310.3	217.70	154.92	62.78	0.0159

A graph is plotted with $\frac{1}{H' - H_i'}$ on the y-axis and H' on the x-axis. The resulting graph is shown in Fig. 6.39.

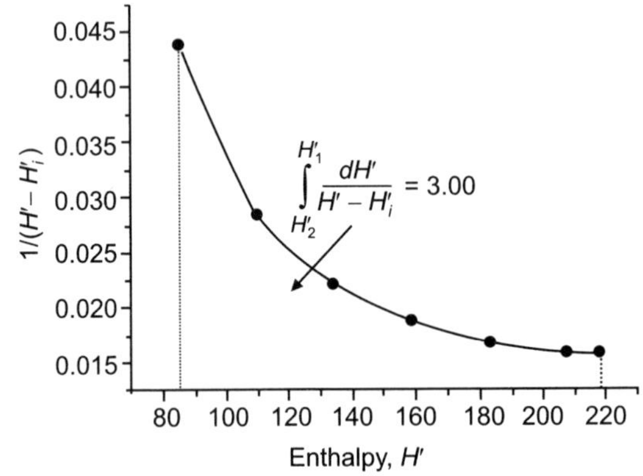

Fig. 6.39: Graphical integration to find the number of transfer units in Example 6.19

The area under the curve between the limits $H_2' = 85.10$ kJ/kg and $H_1' = 217.7$ kJ/kg. Therefore,

$$\int_{H_2'}^{H_1'} \frac{dH'}{H' - H_i'} = 3.0.$$

$$\text{Now } Z = \frac{G_S'}{k_y a M_B} \int_{H_2'}^{H_1'} \frac{dH'}{H' - H_i'} = \frac{843.27}{1260} \times 3.0 = 2.00 \text{ m}$$

Height of the dehumidification chamber = 2.00 m. **(Answer)**

REVIEW QUESTIONS

1. What are the major applications of humidification operations?
2. What do you mean by a saturated gas-vapour system?
3. Define the terms absolute humidity and molal absolute humidity. How are they related to partial pressures?
4. Explain why the molal humidity at saturation is the same for all gases at a given temperature, whereas the absolute humidity at saturation is different.
5. Distinguish between relative humidity and percent humidity. Show that the percent humidity can never exceed relative saturation.
6. Define dew point in at least three ways.
7. Describe a simple method for measuring the dew point of atmospheric air.
8. Explain how the knowledge of the dew point of a gas-vapour system helps us to evaluate its humidity.
9. What is humid volume? How can it be calculated for air-water system knowing the humidity and the dry-bulb temperature? How is humid volume determined using psychrometric charts?
10. Define humid heat and specific enthalpy. How are they related?
11. What is the temperature that you read on the psychrometric chart when you move horizontally from a given point on it to the 100% saturation curve?
12. What is the temperature that you read on the psychrometric chart when you move from a given point along the sloping line passing through it to the 100% saturation curve?
13. Explain how the enthalpy of air-water system at a given temperature is estimated using the psychrometric chart if the humidity is known.
14. Explain adiabatic saturation temperature.
15. Explain the significance of adiabatic cooling lines on the psychrometric charts.
16. Describe wet-bulb temperature.
17. Distinguish between the wet-bulb and the adiabatic saturation temperature.
18. What are psychrometric lines? Distinguish between the adiabatic cooling lines and the psychrometric lines.
19. What is Lewis relation?
20. Define wet-bulb depression? How does it indicate the degree of unsaturation of the gas-vapour mixture?
21. What is psychrometric ratio? How is this related to properties of the gas-vapour system?
22. Explain why the psychrometric lines and adiabatic cooling lines are not shown separately on the humidity charts of air-water vapour mixtures.
23. Why cooling towers are required in process industry?
24. What are the advantages of cooling towers over heat exchangers?
25. Describe how cooling towers are classified.
26. Describe the merits and demerits of counterflow and crossflow type cooling towers
27. Why are tall chimneys needed in natural draft cooling towers?
28. Compare the advantages and disadvantages of the natural draft and the mechanical draft cooling towers.
29. What are the roles of drift eliminators and louveres in cooling towers?
30. Compare the merits and demerits of forced draft and induced draft cooling towers.
31. Distinguish between splash fill and film fill used in cooling towers.
32. Explain blow-down and blow out with reference to cooling towers.
33. Explain why the temperature profile at the bottom of the counterflow cooling tower different from that at the top.

34. Explain the equilibrium curve and the operating line for cooling tower analysis
35. Under what circumstances the use of overall driving force and overall coefficient for estimating the height of cooling tower justified?
36. How would you estimate the variation of the air temperature along the length of the cooling tower?
37. In cooling towers water can be cooled to a temperature lower than the dry-bulb temperature of the entering air. Do you agree? Substantiate your answer.
38. What is wet-bulb temperature approach? Discuss its significance in the design of cooling towers.
39. Explain why the air enthalpy versus water temperature cannot be used for representing the operating line and equilibrium curve for adiabatic humidification with re-circulating water.
40. Explain the significance of air preheating and reheating steps in adiabatic humidification process using spray chambers.
41. How do you determine the temperature to which air should be preheated before it is subjected to adiabatic humidification?
42. Explain why the operating line for the dehumidification columns lies above the equilibrium curve.

MULTIPLE CHOICE QUESTIONS

6.1 At a specified temperature, the molal humidity at saturation
- A. depends on pressure but is independent of the properties of the gas
- B. independent of pressure and properties of the gas
- C. depends on the properties of the gas but is independent of the pressure
- D. depends on properties of the gas and the total pressure.

6.2 At what temperature, the saturation humidity value is infinity?
- A. At the triple point temperature of the liquid
- B. At the critical temperature of the liquid
- C. At the boiling point of the liquid
- D. At the dew point of the system.

6.3 Which one of the following statements is true with regards to the percent saturation (P.S) and relative saturation (R.S)?
- A. P.S = R.S
- B. P.S > R.S
- C. P.S < R.S
- D. P.S and R.S are not interrelated.

6.4 Which of the following statements is true with respect to the dew point of a vapour-gas mixture?
- A. The dew point is independent of the humidity of the mixture
- B. The vapour pressure at the dew point is equal to the partial pressure of the vapour
- C. Dew point is the temperature attained when the mixture is humidified to saturation
- D. Saturation humidity of the mixture is equal to the saturation humidity at the dew point.

6.5 What will happen if an air-water vapour sample is cooled below the dew-point?
- A. Its humidity will increase
- B. Its percent saturation will increase
- C. Its percent saturation will decrease
- D. Its humidity will decrease.

6.6 The wet-bulb temperature of an unsaturated gas-vapour system
- A. is the same as its dew point
- B. is always greater than the dew point
- C. is always less than the dew point
- D. none of the above

6.7 The dew point of air at constant temperature and pressure .... with increase in humidity
- A. does not change
- B. increases
- C. decreases
- D. may increase or decrease

6.8 The relative humidity of air can decrease even if the humidity is increased if the
A. temperature rises
B. pressure rises
C. temperature decreases
D. air is adiabatically humidified

6.9 Choose the right statement.
A. The vapour pressure at the wet-bulb temperature is equal to the existing partial pressure of the vapour in the vapour-gas mixture
B. The existing partial pressure of the vapour is less than the vapour pressure at the wet-bulb temperature.
C. Wet-bulb and dry-bulb temperatures can never be equal
D. The lower the wet-bulb depression, the lower the percent saturation of the vapour-gas mixture.

6.10 The enthalpy of a gas-vapour mixture can be increased by
A. increasing the humidity and decreasing the temperature
B. increasing the humidity and temperature
C. decreasing the humidity and increasing the temperature
D. decreasing both temperature and humidity.

6.11 Psychrometric ratio is defined as

A. $\dfrac{L'C_L}{G'_S}$ B. $\dfrac{h_L a}{k_y a M_B}$

C. $\dfrac{G'_S}{k_y a M_B}$ D. $\dfrac{h_G}{k_y M_B}$

6.12 Adiabatic saturation temperature of a vapour-gas mixture is the
A. steady-state temperature attained by the liquid evaporating into the gas
B. temperature to which the gas can be cooled at constant pressure without condensation
C. temperature attained by the gas when it is humidified to saturation in contact with the liquid at a constant temperature
D. temperature at which the existing vapour content is sufficient to saturate the gas.

6.13 An air-water vapour mixture has a dry-bulb temperature of 303 K and wet-bulb temperature of 290 K. The mixture is heated at constant pressure to 330 K. The wet-bulb temperature will be
A. constant at 290 K B. less than 290 K
C. greater than 290 K D. equal to 303 K.

6.14 The Lewis number is equal to
A. *Pr/Sc* B. *Sc/Pr*
C. *Pr Sc* D. $(Pr\ Sc)^{0.8}$

6.15 The Lewis number of a gas-vapour mixture is unity when mass diffusivity is equal to
A. kinematic viscosity B. thermal diffusivity
C. thermal conductivity D. absolute viscosity

6.16 An air-water vapour mixture has a dry-bulb temperature of 333 K and a dew-point temperature of 313 K. The total pressure is 101.3 kPa and the vapour pressure of water at 313 K is 7.3 kPa and that at 333 K is 19.91 kPa. The humidity of air expressed as kg water vapour per kg dry air is
A. 0.048 B. 0.079
C. 0.122 D. 0.152

6.17 Lewis relation states that

A. $\dfrac{h_L a}{k_y a M_B} = 1$ B. $\dfrac{h_G}{k_y M_B} = 1$

C. $\dfrac{G'_S}{k_y a M_B} = 1$ D. $\dfrac{h_G}{k_y M_B C_S} = 1$

6.18 An air-water vapour mixture has a dry-bulb temperature of 333 K and a dew-point temperature of 313 K. The total pressure is 101.3 kPa and the vapour pressure of water at 313 K is 7.3 kPa and that at 333 K is 19.91 kPa. The wet-bulb temperature T_w for the above mixture would be

A. less than 313 K B. 313 K

C. $313 < T_w < 333$ D. 333 K

6.19 At certain temperature, the saturation humidity of air-water vapour system is found to be 0.070 kg water vapour/kg dry air at 101.3 kPa. What is the vapour pressure of water at this temperature?

A. 3.9 kPa B. 10.3 kPa

C. 6.2 kPa D. 13.1 kPa

6.20 For the air-water system under ambient conditions, the adiabatic saturation temperature and wet-bulb temperature are nearly equal, because

A. water has a high latent heat of vaporisation

B. Lewis number of the system is close to unity

C. Schmidt number is equal to Lewis number

D. solubility of air in water is negligible.

6.21 Air at 318 K and 101.3 kPa has a dew point of 295.K. What is the partial pressure of water vapour in the air if the vapour pressure of water in kPa is given by the Antoine equation:

$$\ln P^S = 16.26205 - \frac{3799.887}{T - 46.854}$$

A. 5.28 kPa B. 9.72 kPa

C. 2.58 kPa D. 4.52 kPa

6.22 The enthalpy of dry air and saturated air are respectively, 60 kJ/kg dry air and 470 kJ/kg dry air at 333 K. What is the enthalpy of air if the dry-bulb temperature is 333 K and dew point is 310 K. The vapour pressure of water is given by

$$\ln P^S = 16.26205 - \frac{3799.887}{T - 46.854}$$

A. 350 kJ/kg B. 170 kJ/kg

C. 215 kJ/kg D. 440 kJ/kg

6.23 The following plot gives the saturation humidity (Y) versus temperature (T). Line joining (T_1, Y_1) and (T_2, Y_2) is a constant enthalpy line.

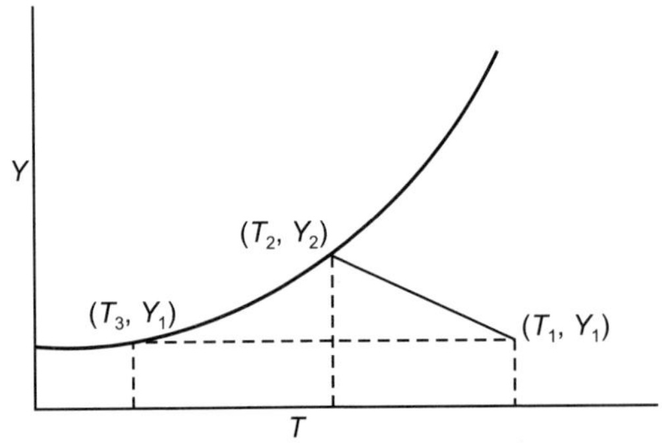

Choose the correct one from the following alternatives:
A. T_1 = Dew-point temperature, T_2 = Dry-bulb temperature, T_3 = Wet-bulb temperature
B. T_1 = Dew-point temperature T_2 = Wet-bulb temperature, T_3 = Dry-bulb temperature
C. T_1 = Wet-bulb temperature, T_2 = Dry-bulb temperature, T_3 = Dew-point temperature
D. T_1 = Dry-bulb temperature, T_2 = Wet-bulb temperature, T_3 = Dew-point temperature

6.24 The adiabatic saturation temperature of air-benzene system at 310 K is found to be 285 K at 101.3 kPa. The vapour pressure of benzene at 285 K is 6.7 kPa. The humid heat of the system may be taken to be constant at 1.155 kJ/kg K and the latent heat of vaporisation of benzene at 285 K is 435.4 kJ/kg. The benzene content in the air at 310 K is ...... kg benzene vapour per kg dry air.
A. 0.05 B. 0.20
C. 0.12 D. 0.25

6.25 It is desired to produce air at 308 K and 30% saturated with water vapour from an air supply of 30% saturated air at 298 K. Pressure is constant at 101.3 kPa. Suggest a suitable method for achieving this.
A. Heating at constant pressure from 298 K to 308 K.
B. Heating at constant pressure from 298 K to 308 K followed by adiabatic humidification
C. Adiabatic humidification followed by heating at constant pressure.
D. Adiabatic dehumidification

6.26 It is required that the air leaving an adiabatic spray chamber contains 0.02 kg water/kg dry air and is 30% saturated with water vapour. If the ambient air is preheated to 342 K before sending into the spray chamber, what is the humidity of the ambient air (kg/kg)?
A. 0.01 B. 0.02
C. 0.015 D. 0.005

6.27 If the percent humidity of air (303 K, 100 kPa) is 24% and the saturation pressure of water vapour at that temperature is 4 kPa, the percent relative humidity and the absolute humidity of air are
A. 25.2, 0.0062 B. 25, 0.0035
C. 20.7, 0.0055 D. 18.2, 0.0035

6.28 The absolute humidity of air at 101.3 kPa is measured to be 0.02 kg water per kg dry air. Determine the partial pressure of water vapour in the air?
A. 1.99 kPa B. 2.55 kPa
C. 3.16 kPa D. 3.87 kPa

6.29 If the specific heats of a gas and vapour are 0.2 kJ/kg K and 1.8 kJ/kg K respectively, and the humidity is 0.02, the humid heat in kJ/kg K is
A. 1.84 B. 0.236
C. 0.017 D. 2.0

6.30 It is required that the air leaving an adiabatic spray chamber contains 0.02 kg water/kg dry air and is 30% saturated with water vapour. If the ambient air at 308 K with 30% saturation is preheated before sending into the spray chamber, to what temperature this air need be preheated?
A. 319 K B. 342 K
C. 368 K D. No preheating is required.

6.31 At 298 K and 90% relative humidity, water evaporates from the surface of the lake at a rate of 1.0 kg/m^2 h. The relative humidity that will lead to an evaporation rate of 3.0 kg/m^2 h, with other conditions remaining the same, is
A. 30% B. 50%
C. 60% D. 70%

6.32 In cooling towers water can be cooled using ambient air which is warmer than water if
A. the air is saturated with water vapour.
B. if the dew point of air is below the water temperature
C. if the wet-bulb temperature is below the temperature of water
D. if the wet-bulb temperature is above the temperature of water.

6.33 Which one of the following is true for cooling towers?
A. The exit water temperature should be above the dry-bulb temperature of inlet air
B. The exit water temperature should be above the wet-bulb temperature of inlet air
C. The exit water temperature is always below the wet-bulb temperature of inlet air
D. The exit water temperature is not dependent on dry-bulb or wet-bulb temperature of inlet air as evaporative cooling occurs.

6.34 If air circulation in the cooling tower is by means of a fan provided above the packing, the tower belongs to the
A. induced-draft type B. forced draft type
C. natural draft type D. cocurrent flow type

6.35 The change in water temperature from inlet to exit of the cooling tower is known as
A. wet-bulb depression B. wet-bulb temperature approach
C. range D. LMTD

6.36 By reducing the wet-bulb temperature approach, the height of cooling tower
A. increases B. decreases
C. remains unaffected D. cannot be predicted

6.37 For very large cooling duties, the cooling tower preferred is
A. induced draft tower B. natural draft tower
C. forced draft tower D. atmospheric towers

6.38 Which one of the following types of cooling towers is more susceptible to air recirculation?
A. Induced draft tower B. Natural draft tower
C. Forced draft tower D. Atmospheric towers

6.39 Louvers are used in cooling towers to
A. prevent drift B. prevent blow-down
C. prevent blow-out D. prevent deterioration of packing

6.40 In cooling towers, blow-down is done
A. to prevent windage losses
B. to prevent damage of packing materials
C. to prevent algae growth on the packing surface
D. to prevent accumulation of dissolved solids

6.41 The difference between the exit water temperature in the cooling tower and the wet-bulb temperature of inlet air is called
A. wet-bulb depression B. wet-bulb temperature approach
C. Lewis number D. psychrometric ratio

6.42 With reference to cooling towers, the wet-bulb temperature approach is defined as the difference between the
A. dry-bulb and wet-bulb temperature of the incoming air
B. temperatures of inlet water and inlet air
C. wet-bulb temperature of inlet air and the exit water temperature.
D. wet-bulb temperature of exit air and inlet water temperature

6.43 With respect to the air-water interaction in cooling towers, which one of the following statements is not correct?

A. Sensible heat transfer in the liquid is always from the bulk liquid to the interface
B. Sensible heat transfer in the gas is always from the interface to the bulk air
C. Latent heat transfer is from the interface to the bulk air.
D. There is no concentration gradient in the liquid

6.44 The slope of the operating line for counterflow cooling tower is

A. $\dfrac{L'C_L}{G'_S}$ B. $\dfrac{G'_S C_S}{L'}$

C. $\dfrac{G'_S}{L'C_L}$ D. $-\dfrac{h_L a}{k_y a M_B}$

6.45 In dehumidification operation, the temperature of water brought in contact with the air should be

A. Below the dew-point of air
B. Below the wet-bulb temperature of air
C. Below the dry-bulb temperature of air
D. Between wet-bulb temperature and dew-point temperature of air.

6.46 Windage loss in induced-draft cooling towers is roughly _______ of the recirculation rate.

A. 3.0–5.0% B. 0.1–0.3%
C. 1.0–3.0% D. above 5.0%

6.47 The common types of packing in modern counterflow cooling towers are

A. Wooden slats B. Corrugated sheets of plastics
C. Ceramic packings D. Coal or charcoal

EXERCISES

Humidity and saturation

6.1 Benzene vapours exert a partial pressure of 15 kPa in a mixture of benzene and nitrogen gas at 105 kPa and 330 K. The vapour pressure of benzene is given by the Antoine equation

$$\ln P^s = 13.8858 - \frac{2788.51}{T - 52.36}$$

Determine the following

(a) The mole fraction of benzene in the mixture
(b) The weight fraction of benzene in the mixture
(c) The molal humidity
(d) The absolute humidity
(e) The molal saturation humidity
(f) Absolute saturation humidity
(g) The mass of benzene in 100 m^3 of the mixture
(h) Percent humidity
(i) Percent relative humidity

(**Answer:** (a) 0.1429 (b) 0.3172 (c) 0.1667 mol/mol (d) 0.4643 kg/kg (e) 0.7988 mol/mol (f) 2.2252 kg/kg (g) 42.64 kg (h) 20.87% (i) 32.17%)

6.2 Moist air contains 0.025 kg water vapour per cubic meter of the mixture at 313 K and 101.3 kPa. Calculate

(a) The partial pressure of water vapour
(b) The relative saturation
(c) The absolute humidity of the air
(d) The percent saturation
(e) The temperature to which the mixture be heated so that its percent saturation becomes 10%

The vapour pressure of water (in kPa) is given by the Antoine equation as

$$\ln P^S = 16.26205 - \frac{3799.887}{T - 46.854}$$

(**Answer:** (a) 3.615 kPa (b) 49.7% (c) 0.023 kg/kg (d) 47.82% (e) 340.22K)

6.3 Determine the dew point of a mixture of acetone vapour and air containing 8.3% acetone by volume if the temperature and pressure are 300 K and 100 kPa. The vapour pressure of acetone (kPa) is given by the Antoine equation with constants, $A = 14.39155$, $B = 2795.817$ and $C = 43.198$.

(**Answer:** 271 K)

6.4 The dew point of a mixture of benzene vapour and dry air at 303 K and 100 kPa is found to be 288 K. The vapour pressure of benzene at 288 K is 7.8 kPa and at 303 K is 15.8 kPa. Determine

(a) The absolute humidity and molal absolute humidity of the mixture
(b) The composition of the mixture in volume percent and weight percent
(c) The relative saturation and the percent saturation

(**Answer:** (a) 0.2275 kg/kg 0.0846 mol/mol (b) Benzene : 7.8% (v), 18.53%(w) (c) RS = 49.37% PS = 45.08%)

6.5 One cubic metre of a gas at STP is to be dried from a dew-point of 300 K to a dew point of 285 K. The vapour pressure of water is 1.4 kPa at 285 K and 3.56 kPa at 300 K.

(a) How much water must be removed?
(b) What will be the volume of the gas at STP after drying?

(**Answer:** (a) 0.0174 kg (b) 0.9773 m^3)

6.6 Air at 298 K, 55% saturated with water vapour which is initially at 100 kPa is compressed to 1000 kPa and cooled to a temperature so that 90% of the water vapour is condensed. Calculate:

(a) The final temperature of the air
(b) The volume of final air per cubic metre of original air

The vapour pressure of water (kPa) is given by the Antoine equation as

$$\ln P^S = 16.26205 - \frac{3799.887}{T - 46.854}$$

(**Answer:** (a) 288.9 K (b) 0.0955 m^3)

6.7 Hydrogen saturated with the vapours of water at 101.3 kPa and 305 K is passed through a cooler in which the temperature of the gas is reduced to 280 K. A part of the vapour condenses as a result of this operation. The gas leaving the cooler is heated to 293 K. Calculate:

(a) The weight of water condensed per kg of moisture-free hydrogen
(b) The percent humidity of the gas in the final state.

The vapour pressure of water (in kPa) may be estimated using the equation

$$\ln P^S = 16.26205 - \frac{3799.887}{T - 46.854}$$

(**Answer:** (a) 0.3467 kg (b) 41.73%)

6.8 A mixture of nitrogen and methanol has a dew-point of 313 K. The temperature and pressure of the mixture are 353 K and 200 kPa, respectively. The vapour pressure of methanol (kPa) is

$$\ln P^s = 16.5725 - \frac{3626.55}{T - 34.29}$$

Calculate

(a) Percent saturation of the mixture
(b) kg alcohol per kg nitrogen
(c) kg alcohol per m^3 of mixture
(d) Dew point of the mixture if the total pressure were 100 kPa
(e) Pressure required to condense 60 % of the alcohol from the mixture at 300 K.

(**Answer:** (a) 2.34% (b) 0.244 kg (c) 0.4657 kg (d) 298.9 K (e) 236.6 kPa)

6.9 A mixture of acetone vapour and nitrogen gas at 300 K has a dew-point of 283 K. It enters an evaporator where it picks up acetone vapour. The mixture leaving the evaporator is at 298 K and has a dew point of 293 K. The pressure is kept constant at 100 kPa. Determine

(a) The absolute humidity of the mixture entering and leaving the evaporator
(b) The kg of acetone evaporated per kg of nitrogen entering the evaporator
(c) The kg acetone evaporated per 100 m^3 of gas entering the evaporator
(d) The volume of gas leaving the evaporator per 100 m^3 of gas entering the unit

The vapour pressure of acetone at 283 K = 15.37 kPa, at 293 K = 24.52 kPa.

(**Answer:** (a) 0.3762 kg/kg, 0.6729 kg/kg (b) 0.2967 kg (c) 28.2 kg (d) 111.4 m^3)

6.10 Air at a rate of 25 m^3 per minute is supplied to a drier in which it removes moisture at a rate of 0.45 kg per minute. The air enters the drier at 333 K and 101.3 kPa, 10% saturated with water vapour and leaves at 310 K and 100 kPa. The vapour pressure of water at 310 K = 6.3 kPa and at 333 K = 19.9 kPa.

(a) What is the percent humidity of air leaving the drier?
(b) At what rate air leaves the drier in m^3 per minute?

(**Answer:** (a) 78.05% (b) 24.22 m^3/min)

6.11 A room of volume 100 m^3 is to be maintained at 300 K and 60 % relative humidity. For this fresh air from the outside saturated at 260 K is to be admitted to the room at such a rate that the air is completely renewed each hour. The pressure is 100 kPa. Vapour pressure of water = 3.56 kPa at 300 K and 0.199 kPa at 260 K. Calculate the weight of water evaporated in one hour.

(**Answer:** 1.65 kg/h)

6.12 A hydrocarbon fuel is burned with dry air. Determine the dew point of the flue gases at 533 K and 100 kPa if the dry flue gases analysed 13.0 % CO_2, 0.05 %CO, 2.5% O_2 and 84.45% N_2. The vapour pressure of water is given by the Antoine equation with $A = 16.26205$, $B = 3799.887$ and $C = 46.854$ when pressure is in kPa and temperature in K. (**Answer:** 322.9 K)

Humidity chart

6.13 The vapour pressure of *n*-hexane in kPa is given by

$$\ln P^s = 13.8216 - \frac{2697.55}{T - 48.78}$$

Plot the 100%, 75%, 50% and 25% saturation lines for hexane in nitrogen at 100 kPa.

6.14 An air-water vapour sample at 101.3 kPa has a dry-bulb temperature of 333 K and an absolute humidity of 0.01 kg water vapour per kg dry air. Using the psychrometric chart determine the following:

(a) The percent saturation

(b) The partial pressure of water vapour
(c) The absolute saturation humidity at 333 K
(d) The vapour pressure of water at 333 K
(e) The percent relative saturation
(f) The dew point of the system
(g) The humid heat
(h) The wet-bulb temperature

(**Answer:** (a) 6.6% (b)1.6050 kPa (c) 0.152 kg/kg (d) 19.93 kPa (e) 8.05% (f) 289.2 K (g) 1.0238 kJ/kg K (h) 301K)

6.15 The wet-bulb and dry-bulb temperature of air are 313 K and 333 K, respectively. Determine using psychrometric chart:
(a) Absolute humidity
(b) Molal humidity
(c) The percent humidity
(d) The humid volume
(e) The dew point
(f) The enthalpy of wet air.

(**Answer:** (a) 0.04 kg/kg (b) 0.064 mol/mol (c) 26.5% (d) 1.0032 m^3/kg (e) 310.2 K (f) 160 kJ/kg)

6.16 The dry-bulb temperature and the dew point of air are 343 K and 308 K, respectively. Estimate using psychrometric chart:
(a) Absolute humidity
(b) Molal humidity
(c) The percent humidity
(d) The humid volume
(e) The wet-bulb temperature
(f) The enthalpy of wet air

(**Answer:** (a) 0.035 kg/kg (b) 0.0563 mol/mol (c) 12.5% (d) 1.0255 m^3/kg (e) 313 K (f) 162.39 kJ/kg)

6.17 Moist air at 313 K and 50 % saturation is heated to 362 K at constant pressure of 101.3 kPa. Using the psychrometric chart, determine the heat added per cubic meter of initial air.

(**Answer:** 56.6 kJ)

6.18 An air stream at 101.3 kPa and 333 K is 25 % saturated with water vapour. Determine the humid volume and enthalpy of the air. If the air is adiabatically humidified to 70 % saturation, what will be its enthalpy in the final state? (**Answer:** 0.9986 m^3/kg, 157 kJ/kg, 156.9 kJ/kg)

6.19 Construct a psychrometric chart for acetone-air system in the temperature range 260 K to 320 K which includes the following details.
(a) 25%, 50%, 75% and 100% saturation curves
(b) Specific volume of dry air and humid volume of saturated gas
(c) Enthalpy of dry and saturated gas in kJ/kg dry air referred to liquid acetone and dry air at 260 K.
(d) Adiabatic cooling line for an adiabatic saturation temperature of 300 K
(e) Psychrometric line for a wet-bulb temperature of 300 K.

The vapour pressure (kPa) of acetone is given by the Antoine equation with constants, $A = 14.39155$, $B = 2795.817$ and $C = 43.198$. The specific heat of acetone vapour and air are 1.47 kJ/kg K and 1.005 kJ/kg K, respectively. Latent heat of vaporisation of acetone at 300 K = 544 kJ/kg and the psychrometric ratio for air-acetone system = 1.7 kJ/kg K.

6.20 Air with a dry-bulb temperature of 363 K and wet-bulb temperature of 308 K is passed over a wet material in a drier. The rate of evaporation is found to be 0.015 kg water per kg dry air passed through the drier. Assuming the process to be adiabatic, calculate using psychrometric chart

(a) The absolute humidity, molal humidity and percent saturation of air entering the drier.

(b) The dry-bulb temperature, the wet-bulb temperature, the absolute humidity and the percent saturation of air leaving the drier.

(**Answer:** (a) 0.0125 kg/kg, 0.0201 mol/mol, 0.89% (b) 329 K, 308 K, 0.0275 kg/kg, 22.5%)

6.21 Air at 365 K, 101.3 kPa and a dew-point of 283 K is supplied to a drier which operates adiabatically. Determine using psychrometric charts

(a) The minimum temperature to which the air gets cooled in the drier

(b) On the basis of 10 m^3 of air entering the drier, the maximum quantity of water evaporated in the drier

(c) The volume of air leaving the drier for the conditions in part (b)

(**Answer:** (a) 306 K (b) 0.2395 kg (c) 9.586 m^3)

6.22 Air is supplied to a drier at a dry-bulb temperature of 305.5 K and a wet-bulb temperature of 298 K. It is heated to 363 K before admitting to the drier. Drier is operated adiabatically. Air leaves the drier fully saturated. Using psychrometric chart, determine

(a) The humidity, percent humidity and the dew point of the initial air

(b) The heat to be supplied in the heater per 100 m^3 of initial air

(c) Water evaporated per 100 m^3 of air entering the drier

(d) The temperature of air leaving the drier.

(**Answer:** (a) 0.0175 kg/kg, 55%, 295 K (b) 6713.77 kJ (c) 2.53 kg (d) 310.7 K)

6.23 Wet material is spread on two trays stacked one above the other. Air containing 0.005 kg water/kg dry air is heated to 323 K and passed over the lower tray. It leaves the tray at 60% humidity and is then reheated to 323 K and passed over the upper tray after which the air leaves the drier, 60% saturated. On the assumption that material on each tray has reached the wet-bulb temperature and that heat losses from the drier can be neglected, determine:

(a) The temperature of the material on the trays

(b) The amount of water removed in kg/h if 1000 m^3/h of moist air leaves the drier

(c) The temperature to which the air is to be preheated, if the final condition of air is to be achieved in the first tray itself.

(**Answer:** (a) 295.7 K, 301 K (b) 18.87 kg (c) 343 K)

6.24 It is required to evaporate 100 kg/h of water from a wet solid. The drying is carried out in an adiabatic drier. Air is available at 101.3 kPa and 295 K at 70% humidity. Its temperature is raised by passing through a preheater to such a value that the air leaving the drier is at 311.7 K and 90% saturated. Calculate:

(a) The adiabatic saturation temperature of air leaving the drier.

(b) The temperature of air leaving the preheater

(c) The rate of circulation of moist air, kg/h

(d) The volume of moist air before preheating.

(**Answer:** (a) 310.7 K (b) 375.7 K (c) 3614 kg/h (d) 3039 m^3)

Wet-bulb temperature and adiabatic saturation temperature

6.25 For an air-water vapour mixture dry-bulb temperature 335 K and wet-bulb temperature of 308 K determine the humidity of air at 101.3 kPa. Latent heat of vaporisation of water at 308 K is 2419 kJ/kg and psychrometric ratio is 0.950 kJ/kg K. Vapour pressure of water at 308 K = 5.62 kPa.

(**Answer:** 0.0259 kg/kg)

6.26 An adiabatic drier is used to dry a wet material. The drying air enters at 380.7 K and 101.3 kPa with a dew-point of 298 K. Measurements show that 2.25 kg water is evaporated per 100 m^3 of wet inlet air. Calculate using psychrometric chart

(a) Humidity of air entering the drier
(b) The exit air humidity and percent humidity
(c) The exit air wet-bulb and dry-bulb temperatures
(d) The volume of exit air per 100 m^3 of inlet wet air.

(**Answer:** (a) 0.02 kg/kg (b) 0.045 kg/kg, 55% (c) 313.2 K, 322 K (d) 87.88 m^3)

6.27 A mixture of air and benzene vapour containing 0.7 kg benzene per kg dry air is to be cooled from 343 K to 288 K in a tubular cooler condenser. The specific heat of dry air is 1.005 kJ/kg K and the specific heat of benzene vapour is 1.1 kJ/kg K. The vapour pressure of benzene in kPa is given as

$$\ln P^S = 13.8858 - \frac{2788.51}{T - 52.36}$$

The psychrometric ratio and the latent heat of vaporisation of benzene in the range of interest may be taken to be constant at 1.7 kJ/kg K and 250 kJ/kg, respectively. The pressure is 101.3 kPa. Determine

(a) The wet-bulb temperature of the inlet stream
(b) The humidity of the exit stream
(c) The total amount of heat to be transferred per kg dry air

(**Answer:** (a) 314.09 K (b) 0.224 kg/kg (c) 216.65 kJ)

6.28 The vapour pressure of benzene varies with temperature as

$$\ln P^S = 13.8858 - \frac{2788.51}{T - 52.36}$$

The latent heat of vaporisation of benzene at 293 K is 435.4 kJ/kg and the psychrometric ratio is 1.717 kJ/kg K. Calculate and plot:

(a) The saturation curve for the benzene-air system at 101.3 kPa for the temperature range 283 K to 313 K with temperature increments of 10 K.
(b) The psychrometric line for the wet-bulb temperature of 293 K.

6.29 A toluene-air mixture at 330 K and 101.3 kPa contains 0.05 kg toluene vapour per kg dry air. Antoine equation for vapour pressure of toluene is

$$\ln P^S(\text{kPa}) = 13.9987 - \frac{3096.52}{T - 53.67}$$

The specific heats of toluene vapour and air are, respectively, 1.256 kJ/kg K and 1.005 kJ/kg K. Assume a constant value of 404 kJ/kg for latent heat of vaporisation of toluene.

(a) Estimate the adiabatic saturation temperature
(b) If the psychrometric ratio for toluene-air system is 1.842 kJ/kg K, estimate the wet-bulb temperature.

(**Answer:** (a) 299.7 K, 304 K)

6.30 The vapour pressure and latent heat of vaporisation of carbon tetrachloride as function of temperature is given below:

T, K	278	300	322	333	344	350
P^S, kPa	5.6	16.4	39.3	57.3	84.7	101.3
λ, kJ/kg	215.7	206.3	197.5	193.1	188.6	186.4

The specific heat of dry air and that of CCl_4 vapour may be assumed constant. They are respectively, 1.005 kJ/kg K and 0.942 kJ/kg K. For a pressure of 101.3 kPa

(a) Calculate the adiabatic saturation temperature for dry air at 350 K
(b) Calculate the adiabatic saturation temperature for air at 350 K and having a humidity 1.5 kg vapour per kg dry air. (**Answer:** (a) 280.6 K (b) 311 K)

Cooling towers

Note: In all the exercise problems that follow, the saturation enthalpy of air versus temperature data are as follow, unless otherwise specified:

Temperature, K	283	288	293	298	303	308	313	318	323	328
Enthalpy, kJ/kg	29.1	41.8	57.2	76.0	99.4	128.7	165.7	212.8	273.6	354.2

6.31 In a certain cooling tower operation, air enters at 303.2 K with a humidity 0.01 kg water/kg dry air and leaves at 313.2 K with a partial pressure of water 6.5 kPa. A 5 K wet-bulb temperature approach is to be provided. The column is working with equal rates (kg/h) for cooling water and dry air circulation. The total pressure is 100 kPa. Calculate:
(a) Relative humidity of air-water vapour at the inlet and exit of the tower
(b) Humidity of the air at the exit of the tower
(c) Specific enthalpy of air at the inlet and exit of the tower.
(d) The temperature range for water
(e) The inlet and exit temperature of the water
(f) The amount of water lost by evaporation per kg of dry air circulated.
Vapour pressure (in kPa) of water is given by

$$\ln P^s = 16.26205 - \frac{3799.887}{T - 46.854}$$

(**Answer:** (a) 38%, 88% (b) 0.0431 kg/kg (c) 55.79 kJ/kg, 151.35 kJ/kg (d) 22.82 K (e) 321 K, 298.2 K (f) 0.0331 kg)

6.32 A forced draft counter flow cooling tower cools water from 318 K to 298 K using air at a dry-bulb temperature of 297 K and a wet-bulb temperature of 293 K. The height of overall gas enthalpy transfer unit is 0.5 m. If the air supplied is 30% in excess of the minimum find the height of the tower. Neglect the heat transfer resistance of the liquid phase. (**Answer:** 2.62 m)

6.33 Calculate the height of a tower for cooling water from 328 K to 293 K by means of air flowing countercurrently. The liquid flows at rate of 1000 kg/m^{2}h and the air rate is 1000 m^3/m^{2}h. The air enters at a temperature of 293 K with a percent humidity of 20%. Assume that the whole of resistance to heat and mass transfer lie within the gas phase and the mass transfer coefficient K_ya can be taken as 30 kmol/m^{3}h. (**Answer:**3.2 m)

6.34 Water is to be cooled from 318 K to 303 K in a packed counterflow tower at a rate 1.5 kg/m^2 s. Air enters the tower at rate 1.5 kg dry air/m^2 s. The air dry-bulb temperature is 303 K and wet-bulb temperature is 298 K. The mass transfer coefficient k_ya is estimated to be 1.2×10^{-2} kmol/m^{3}s. Assume that heat transfer resistance in the liquid is negligible. Calculate the height of the packed tower. (**Answer:** 7.2 m)

6.35 A counterflow cooling tower is to be designed to cool 10000 kg/m^{2}h of water from 317K to 300 K using air entering at 300 K with a wet-bulb temperature of 295 K. The overall mass transfer coefficient K_ya is 6000 kg/m^{3}h. Determine:
(a) The minimum air rate that can be used
(b) Calculate the height of the tower if the air flow rate is 7000 kg/m^2 h.
(**Answer:** (a) 5409 kg/m^{2}h (b) 6.2 m)

6.36 During the test of a cooling tower, water was cooled from 305 K to 293 K. Water flow rate was 3.8 kg/m^2 s and the air rate was 3.2 kg/m^2 s. Air was available at dry-bulb temperature of 291 K and wet-bulb temperature of 286 K. If the height of packed section is 2 m and specific surface area of packing is 7.0 m^2/m^3, determine:

(a) The number of transfer units

(b) The overall mass transfer coefficient **(Answer:** (a) 3.984 (b) 0.91 kg/m^2 s)

6.37 During the test of a cooling tower, water was cooled from 306 K to 296 K. Water flow rate was 3 kg/m^2 s. Air was available at dry-bulb temperature of 292 K and wet-bulb temperature of 288 K. The air leaving the tower was saturated with water vapour and its temperature was 301 K. If the height of packed section is 3.0 m and specific surface area of packing is 7.0 m^2/m^3, determine:

(a) The air mass velocity, kg/m^2s

(b) The overall mass transfer coefficient, kg/m^2s

(Answer: (a) 2.62 kg/m^2 s (b) 0.2418 check)

6.38 Water is to be cooled from 318 K to 303 K in an induced draft counterflow cooling tower at a rate 5000 kg/h. The air is admitted at a rate 5000 kg dry air /m^2 h. The 5% wet-bulb temperature for the location is 298 K. The estimated mass transfer coefficient $k_y a$ is 30 kmol/m^3 h. Determine the height of the packed section if

(a) the ratio $\dfrac{h_L a}{k_y a M_B} = \infty$ (b) the ratio $\dfrac{h_L a}{k_y a M_B} = 20\,\text{kJ/kg K}$

(Answer: (a) 9.6 m (b) 12.79 m)

6.39 Water is to be cooled from 318 K to 303 K in a packed counterflow tower at a rate 1.5 kg/m^2 s. Air enters the tower at rate 1.5 kg dry air/m^2 s. The air dry-bulb temperature is 303 K and wet-bulb temperature is 298 K. The mass transfer coefficient $k_y a$ is estimated to be 1.2 × 10^{-2} kmol/m^3s and $\dfrac{h_L a}{k_y a M_B} = 40\,\text{kJ/kg K}$. Calculate the height of the packed tower.

(Answer: 8.8 m)

6.40 Water is cooled from 318 k to 295 K in a forced draft, counterflow cooling tower. The ambient air is having a dry-bulb temperature of 305 K and wet-bulb temperature is 288 K. The ratio $\dfrac{h_L a}{k_y a} = 20\,\text{kJ/kg K}$ and height of transfer unit is 5 m. A liquid rate of 7500 kg/h m^2 of tower cross section will be used. Determine:

(a) The minimum gas rate

(b) The height of the tower if the gas rate is twice the minimum.

(Answer: (a) 4787 kg/m^2 h (b) 12.91 m)

6.41 Water is to be cooled in a small packed column from 323 to 295 K by means of air flowing countercurrently. The rate of flow of liquid is 1.5 kg/ m^2 s and the flow rate of dry air is 3.0 m^3/m^2 s. Air enters at a temperature of 295 K and humidity 60%. The resistance to transfer is confined to the gas side and the mass transfer coefficient is 2.3kg/m^2 s. Determine:

(a) The height of the column

(b) The condition of air leaving at the top of the column.

(Answer: (a) 1.59 m (b) saturated at 302.4 K)

6.42 Water is to be cooled in a small packed column from 323 to 295 K by means of air flowing countercurrently. The rate of flow of liquid is 1.5 kg/ m^2 s and the flow rate of dry air is 5.4 kg/m^2 s. Air enters at a temperature of 295 K and humidity 60%. The resistance to transfer is confined to the gas side. Determine the condition of the air leaving at the top of the column.

(Answer: 299.5 K)

6.43 A cooling tower cools water from 328 K to 293 K by means of air flowing countercurrently. The liquid flows at rate of 1000 kg/m^2h and the dry air rate is 2500 m^3/m^2h. The dry air enters at a temperature of 293 K with a percent humidity of 20%. Assuming that the whole of resistance to heat and mass transfer lie within the gas phase calculate the condition of the air leaving the tower. (**Answer:** $T = 298.9$ K, $Y' = 0.0197$ kg /kg, 93% saturated)

6.44 Water is cooled at a rate of 60 kg/s from 315 to 300 K in a forced draft cooling tower. The evaporation loss in the tower is expected to be 1% for every 5 K drop in the temperature of water. Assuming the windage loss to be 0.1% of the circulation, estimate the blow down and make up water requirement if the maximum allowable concentration of the dissolved solids in water is three times the concentration in the makeup? (**Answer:** 0.84 kg/s, 2.7 kg/s)

6.45 Water is cooled from 318 K to 298 K in an induced draft cooling tower at a rate 50,000 kg/h. Air is available at 297 K dry-bulb temperature and 293 K wet-bulb temperature at a rate 50% in excess of the minimum. The air circulation rate is 7000 kg dry air/m^2 h. For the packing to be used the mass transfer coefficient $k_y a M_B$ is expected to be 3000 kg/m^3 h. Calculate:

(a) The height and diameter of the tower

(b) The make-up water required, kg/s, assuming the air leaving is saturated and the windage and other losses are negligible.

(**Answer:** (a) 8.67 m, 2.84 m (b) 1459.06 kg/h check)

6.46 It is desired that the circulating water in Exercise 6.45 should contain hardness not more than 2000 ppm dissolved solids. The make-up water contains 500 ppm hardness. The windage losses in the tower is 0.2 % of the circulation rate. Estimate the make-up water requirement if a small amount of water is deliberately discarded to keep the hardness at the toleration limit.

(**Answer:** 1945.41 kg/h)

6.47 An induced draft cooling tower is designed to have a capacity to cool 125kg/s of cooling water from 316 K to 303 K when the available air has a wet-bulb temperature of 297 K. A test on the tower when operated at full fan capacity gave the following results:

Water rate = 125kg/s, inlet temperature of water = 316 K. Outlet water temperature = 299 K, Inlet air dry-bulb temperature is 297 K and wet-bulb temperature = 288 K. Outlet air, saturated at temperature = 311 K.

Determine the fan capacity, m^3/s

Can the tower be expected to meet the design specifications?

(**Answer:** 68.57 m^3/s , can meet the specification)

6.48 A counterflow induced-draft cooling tower operates with inlet and exit water temperature of 314 K and 303 K when the air has dry-bulb and wet-bulb temperatures, respectively, of 305 K and 298 K. The tower has 1.22 m of stacked plastic fill, and the flow rates are $G_S' = 2.7$ kg dry air/m^2 s and liquid rate is 3 kg/m^2 s.

(a) Determine the number of transfer units, the height of a transfer unit based on the overall gas-phase driving force, and the temperature approach (b) If the cooling load remains the same but the air temperature drops to 294 K with a wet-bulb temperature of 289 K, predict the water temperature and the temperature approach

(**Answer:** (a) 1.6457, 0.7413 m, 5 K (b) 308.5 K – 297.5 K, 19.5 K approach)

6.49 A counter flow cooling tower with 1m of cellular fill was designed for an approach of 5 K and a cooling range of 10 K when the we-bulb temperature is 298 K. The tower cross section is 100 m^2 and the air and water flow rates are 250 m^3/s (at 305 K) and 400 kg/s. (a) What are the inlet air enthalpy and the changes in enthalpy across the tower? (b) Calculate the number of transfer units and height of transfer units for these conditions. (c) How much more fill would be needed if the wet-bulb temperature was 301K and the desired approach was 3 K?

(**Answer:** (a) inlet $H' = 76$ kJ/kg, change in $H' = 57.82$ kJ/kg
(b) 2.2723, 0.44 m (c) additional 0.31 m)

Humidification

6.50 Air at 313 K dry-bulb temperature and 295 K wet-bulb temperature is admitted to a spray chamber where water is sprayed into it adiabatically. If the exit air is at 300 K, determine using psychrometric chart the quantity of water added per kg dry air. (**Answer:** 0.0055 kg)

6.51 An air stream at 101.3 kPa and 333 K is 26.5% saturated with water vapour. It is sent to a spray chamber where it is adiabatically humidified. The temperature of the air leaving the chamber is 5 K above the adiabatic saturation temperature. Determine
(a) The absolute humidity of the initial air
(b) The wet-bulb temperature of the air
(c) The dew-point of the initial air
(d) The percent saturation of the air leaving the spray chamber
(e) The humidity and temperature of the air leaving the chamber
(f) The quantity of water picked up by 100 m^3 of wet air admitted to the chamber.
(**Answer:** (a) 0.04 kg/kg (b) 313.2 K (c) 310.2 K (d) 70% (e) 0.046 kg/kg, 318.2 K (f) 0.5981 kg)

6.52 It is desired to maintain the temperature of air entering a building at 308 K and humidity at 60 %. This is achieved by humidifying air available at 305 K with 65% percent humidity to saturation adiabatically in a spray chamber and by heating the gas leaving the chamber to the desired final condition. Assume the pressure to be atmospheric. Using psychrometric chart, determine
(a) The temperature of air leaving the spray chamber
(b) The amount of water picked up by one kg dry air passing through the chamber.
(**Answer:** (a) 299K (b) 0.004 kg)

6.53 It is desired that the air entering a building be at 308 K and 30% saturated with water vapour. This is achieved by passing air at 298 K and 30% saturated, through a series of water sprays where it is cooled and saturated. The air emerging from the spray chamber is then heated to 308 K. The water and air leave the chamber at the same temperature. Using psychrometric chart, calculate:
(a) The temperature of water
(b) Humidity of air entering the building
(c) The amount of water evaporated per 1000 m^3 of air entering the chamber.
(**Answer:** (a) 288.5 K (b) 0.01 kg/kg (c) 4.697 kg)

6.54 Air at 101.3 kPa has a dry-bulb temperature of 303 K and a wet-bulb temperature of 288 K. It is desired to heat the air in a preheater and then spray water into it adiabatically so that final air will be saturated at 303 K. Using psychrometric chart, determine
(a) The humidity of the initial air
(b) The percent humidity of the initial air
(c) The temperature to which the air is to be heated in the preheater
(d) The heat to be supplied per kg of dry air.
(**Answer:** (a) 0.004 kg/kg (b) 17% (c) 357 K (d) 54.68 kJ)

6.55 Seven thousand kg of dry air per hour at 338 K having a humidity of 0.012 kg water vapour per kg dry air is subjected to adiabatic humidification in a spray chamber. The water temperature is maintained constant by continuous recirculation. The air leaving the chamber is 90% saturated. The heat transfer coefficient $h_G a$ may be taken to be 1500 W/m^3 K. Determine:
(a) The temperature of the air leaving the humidifier.
(b) The adiabatic saturation temperature of air subjected to humidification
(c) The volume of spray chamber.
(**Answer:** (a) 304.7 K (b) 303.2 K (c) 4.24 m^3)

6.56 It is desired to produce air at 20% humidity and 325.7 K for certain process at a rate 5000 kg dry air per hour from the air available at the plant at 20% humidity and 293 K. This is achieved in three steps. Air is first heated then subjected to adiabatic humidification in a spray chamber followed by reheating to 325.7 K. Assuming that the air leaving the spray chamber is 90% saturated with water vapour, determine:

(a) The temperature of the air leaving the preheater
(b) The temperature of air leaving the spray chamber
(c) The adiabatic saturation temperature of the air in the humidifier
(d) The heat required for preheating and reheating

(**Answer:** (a) 340.5 K (b) 299.8 K (c) 298.0 K (d) 2.398×10^5 kJ, 1.348×10^5 kJ)

6.57 Air at a dry-bulb temperature of 338 K and with humidity 0.0115 kg water vapour per kg dry air is cooled and humidified at a rate 4 kg/s in a spray chamber to a dry-bulb temperature of 313 K. The spray chamber is 2 m long and 2 m^2 in cross section. If another chamber of the same dimensions is added in series to the existing chamber, what will be the conditions of the air leaving the second chamber? (**Answer:** $T = 305.9$ K, $Y_2 = 0.0256$ kg/kg) and check

Dehumidification

6.58 Air with a dew-point of 294 K and a percent humidity of 40 % enters a dehumidifier and leaves at 287 K and with a dew-point of 285 K. How much water is removed per kg dry air? How much heat is removed per kg dry air? Solve using psychrometric chart.

(**Answer:** 0.0069 kg, 41.25 kJ)

6.59 Air is to be cooled and dehumidified by countercurrent contact with water in a packed column. The air enters the column at a rate of 1000 kg dry air per hour with a dry-bulb temperature of 305 K and wet-bulb temperature of 303 K. Water is admitted at rate 725 kg/h with inlet temperature 291 K. Water exit temperature is 297 K. Assuming that the liquid phase resistance to mass transfer is negligible, determine:

(a) The maximum water rate that can be used in the column
(b) The number of transfer unit.

(**Answer:** (a) 1906.7 kg/h (b) 0.627)

6.60 Air 80% saturated with water vapour enters a 2 m diameter dehumidifier at 322 K and leaves at a wet-bulb temperature of 300 K. Water enters the column at 294 K. The air flow rate is 1700 m^3/h measured at standard conditions. The water rate is 1.5 times the minimum rate. The ratio of heat transfer coefficient to mass transfer coefficient $h_L/k_y M_B$ has been found to be 50 kJ/kg K. Assuming the height of transfer unit to be 0.5 m, determine:

(a) Exit temperature of water
(b) Height of the tower
(c) Water flow rate

(**Answer:** (a) 310.27 K, (b) 1.5 m (c) 1234.13 kg/m^2 h)

6.61 Air at 328 K, 70% saturated with water vapour is to be dehumidified at a rate of 3400 m^3/h to a moisture content of 0.046 kg per kg dry air. Cooling water enters the column at 296 K and leaves at 305 K at a rate 6800 kg/h. The gas mass velocity is 6000 kg dry air/ m^2 h. The heat transfer coefficients at these conditions were reported to be $h_G a = 6$ kW/m^3 K and $h_L a = 30$ kW/m^3 K. Determine the height and diameter of the tower. (**Answer:** $Z = 0.385$ m, $D = 0.83$ m)

Chapter 7

Liquid–liquid Extraction

Liquid–liquid extraction, also known as solvent extraction, is a technique for the separation of the components of a liquid mixture by contacting with an immiscible or partially miscible liquid solvent. Separation is due to the differential distribution of components between the two insoluble phases. This uneven distribution is the result of the difference in the solubility of the components, which in turn, is associated with the differences in their chemical structure. Acetone can be extracted from a solution containing acetone and water using trichloroethane (TCE) as the solvent. An aqueous solution of acetic acid can be separated by extracting with ethyl acetate as the solvent. Processes using two solvents for extraction, known as the double solvent extraction or fractional extraction also, are not uncommon. For example, a mixture of acetone and acetic acid can be separated by extraction with an insoluble mixture of chloroform and water, whereupon the acid preferentially dissolves in the water and acetone preferentially dissolves in the chloroform.

The large-scale industrial application of extraction began in the early 1930s with the recovery of sulphur and aromatic compounds from kerosene by extraction using liquid sulphur dioxide at 260 to 267 K. Liquid extraction is now utilized by industries for a wide variety of applications. These include extraction of lubricating oils, decaffeination of coffee, recovery of homogeneous catalysts, manufacture of penicillin, protein recovery from natural substances, recovery of metals such as copper from ammoniacal leach liquors, recovery of uranium and plutonium from spent fuel elements, separation of fatty acids from vegetable oils, phenol removal from aqueous waste water and extraction of high-boiling constituents such as phosphoric acid, boric acid, and sodium hydroxide from aqueous streams.

In its simplest form the liquid–liquid extraction can be represented as shown schematically in Fig. 7.1. Here, a binary mixture consisting of solute (**C**) and carrier liquid (**A**) is extracted using a solvent (**B**). The products that are separated are the solvent-rich phase called the *extract* and the carrier-rich phase called the *raffinate*. If the carrier liquid **A** and the solvent **B** are partially miscible, both the extract and raffinate phases will be ternary mixtures consisting of components **A, B** and **C**. The extract will be rich in **B**, the solvent used and the raffinate will be rich in **A**, the carrier liquid.

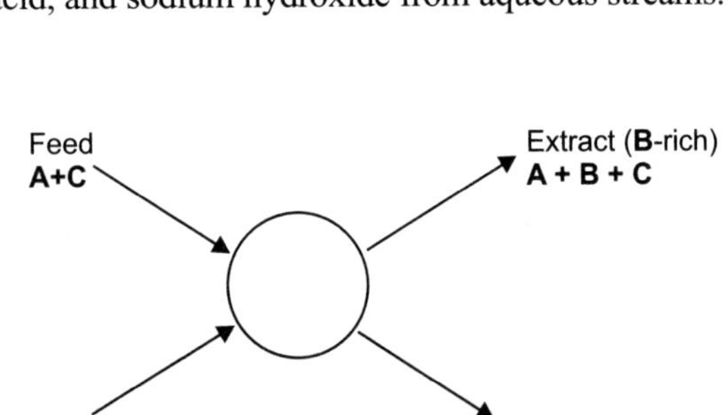

Fig. 7.1: Extraction operation

Consider an operation in which 100 kg of the feed solution containing 25% by weight acetone in water being treated with 50 kg of pure trichloroethane at 298 K. Acetone shows a preference to dissolve in the TCE than in the water and as a result, some of the acetone in the feed, but relatively little water gets transferred to the solvent. After attaining the equilibrium, the aqueous phase and the solvent phase are separated from each other. The mutual solubility of the water and the TCE being very small, quantity of water present in the solvent phase and quantity of TCE present in the aqueous phase will be very small. The solvent present in both the phases can be removed by evaporation or distillation, whereupon about 10.2 kg solvent-free extract product containing 95% acetone and 89.8 kg raffinate product containing 17% acetone are produced. The process is represented as shown in Fig. 7.2.

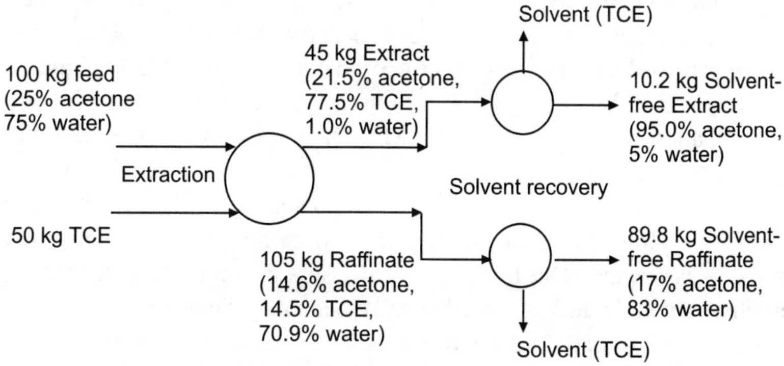

Fig. 7.2: Extraction of acetone from aqueous solution using trichloroethane as solvent

The acetone to water ratio in the products are widely different and also different from that in the original feed; we see that extraction with a solvent leads to separation of the components. In this example, about 40% of the acetone in the feed is recovered as a 95% solution from the original feed containing 25% acetone. The recovery can be substantially improved by repeated contacting in a multistage operation.

Since the separation by extraction is achieved by creating two insoluble phases on introducing the solvent, the products of extraction are inevitably contaminated by a 'foreign' substance (the solvent) and require in most cases an additional treatment for its removal. The raffinate and extract products should be subjected to post extraction treatments such as evaporation or distillation for removal of the solvent, not only for their purification but also for the recovery and reuse of costly solvent.

7.1 IMPORTANCE OF EXTRACTION

Extraction offers several advantages over the direct separation methods such as distillation, and evaporation. In distillation, separation of components of a liquid mixture is made possible by application of heat whereas the separation in extraction is achieved by the addition of a solvent. In extraction, separation is achieved according to the chemical type of the constituents whereas in distillation separation is achieved utilizing the difference in the volatilities. Since the second phase necessary for mass transfer is generated by the partial vaporisation of the mixture, frequently the composition of the phases will not be much different and therefore, the separability will be poor in distillation. Extraction offers greater flexibility in the choice of the operating conditions; by the proper choice of the type and amounts of solvents the effectiveness of separation can be improved significantly.

Extraction can be considered as a less expensive alternative means of separation when the distillation is either uneconomical or impossible as in the following areas:

1. *Close boiling liquids*: Distillation of mixtures of close boiling liquids is quite expensive, if not impossible. Extraction, which makes use of the difference in the solubility of the components

rather than their volatilities, can be used to separate such mixtures. Eg., separation of butadiene (bp: 268.45 K) from butylenes (bp: 267.2 K - 268.2 K) by extraction with cuprammonium acetate solution.

2. *Liquids of poor relative volatility*: Distillation of mixtures of certain liquids, in spite of the wide difference in their boiling points, is quite expensive either because of the poor relative volatility or because of the high heat of vaporisation. For example, for the separation of dilute aqueous solution of acetic acid by distillation, large quantities of water have to be vaporised and removed as distillate. The heat of vaporisation of the water being very high, the heat requirement for the process will be quite large. Separation of acetic acid by treating the aqueous solution with a suitable solvent such as ethyl acetate provides a less expensive alternative means.
3. *High boiling mixtures*: To avoid the danger of decomposition at high temperatures, distillation of mixtures of high boiling liquids has to be carried out under very high vacuum; this is generally a very costly operation. Liquid–liquid extraction can be used as a substitute for the highly expensive vacuum distillation, e.g. separation of long-chain fatty acids and vitamins from naturally occurring oils by extraction using liquid propane.
4. *Components present in traces*: Removal of components present in small concentrations, such as the colouring materials present in the tallow or hormones present in the animal oil.
5. *Heat sensitive substances*: Due to the chemical destruction of the substances at high temperatures, heat sensitive materials such as penicillin and other antibiotics cannot be recovered by distillation of fermentation mixtures. Extraction, which is often carried out at or near the ambient temperatures, is a useful technique for the separation of thermally unstable natural mixtures or pharmaceutical substances. For example, penicillin is recovered from the fermentation broth by extracting with suitable organic solvents, such as butyl acetate.
6. *Azeotropes*: Azeotropes are constant boiling mixtures and cannot be separated by ordinary distillation. However, they may be separated by extraction. For example, a mixture of methylethyl ketone and water which forms an azeotrope at 89.0% by weight ketone (bp: 346.65 K at 101.3 kPa) can be separated by extracting the ketone into a solvent such as trichloroethane. Alternatively, the extraction of water into aqueous calcium chloride brine can also be used for this separation.
7. *Liquids in the same boiling range*: Petroleum products with different chemical structures, but with overlapping boiling points cannot be separated by distillation because, for separation, distillation makes use of the difference in the volatilities of the components. Such mixtures can be separated by extraction, which makes use of the difference in the solubility caused by the distinct chemical characteristics of the components. Thus aromatic hydrocarbons are separated from paraffin hydrocarbons in the same boiling range by extraction. Solvents such as liquid sulphur dioxide, sulpholane, furfural or diethylene glycol are used for this purpose. Lube oil fractions can be freed off aromatic components, which if present in the oil lead to poor viscosity-temperature characteristics, by extraction with solvents such as phenol, furfural, or methyl pyrrolidone. Distillation cannot be used for this separation as the boiling points of the aromatics and the desired components of the oil overlap. Aromatic components such as the benzene, toluene and xylenes are recovered from the catalytic reformate by the extraction using high-boiling point polar solvents like sulpholane ($C_4H_8SO_2$) followed by distillation.

In addition to the above, extraction finds applications also as substitute for expensive evaporation, fractional crystallisation and separation by certain chemical methods. Separation of metals like tantalum-columbium, copper-iron, uranium-vanadium, etc. can be achieved by liquid–liquid extraction. Extraction is successfully employed for the purification of uranium fuel and for the recovery of spent fuel elements in the nuclear power industry. Super critical extraction is a recently developed method which uses super critical or near super critical fluids as solvents.

Unlike distillation, extraction is not a direct separation method and has therefore certain limitations. Contamination of the products of extraction with the added solvent which is definitely a "foreign" substance, is inevitable, and as a result, the products require additional treatment for the recovery or removal of the solvent. The recovery in many cases is to be achieved by distillation and the cost of recovery of solvent may constitute a substantial portion of the cost of separation by extraction and the combined operation may even be more expensive than the separation by ordinary distillation. Makeup solvent is required to compensate for the loss of solvent. Relatively large amounts of capital will be tied up if it is required to maintain in the plant, large inventories of the solvent. Since the solvent should be immiscible with the feed, the feed and the solvent necessarily be chemically dissimilar materials and extraction vessels have to frequently handle liquids of widely different corrosion properties. This restricts the choice of materials of construction that resist corrosion or may lead to costly materials of construction.

7.2 LIQUID EQUILIBRIA

Liquid–liquid extraction involves distribution of the components between two immiscible or partially miscible liquids. In its simplest application, the feed solution which is a mixture of solute **C** and a carrier liquid **A** is extracted with a solvent **B**. The extract will be rich in solvent **B** and the raffinate will be rich in the carrier **A**. The component **C** gets distributed between these two phases. If **A** and **B** are completely insoluble, the extract and raffinate phases will be binary mixtures, the former consisting of only **B** and **C**, and the latter of only **A** and **C**. The equilibrium distribution of **C** between the raffinate and extract phases can be very well represented on the rectangular coordinates, with concentration of solute in the extract on the *y*-axis and the concentration of solute in the raffinate on the *x*-axis as shown in Fig. 7.3.

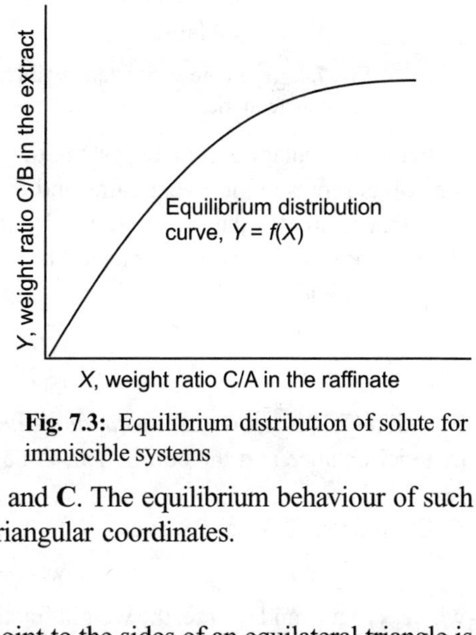

Fig. 7.3: Equilibrium distribution of solute for immiscible systems

On the other hand, if the carrier **A** and the solvent **B** are partially miscible, the extract and raffinate phases are ternary liquid mixtures consisting of components **A**, **B** and **C**. The equilibrium behaviour of such mixtures are conveniently represented on the equilateral triangular coordinates.

Equilateral Triangular Coordinates

The sum of the perpendicular distances from any interior point to the sides of an equilateral triangle is equal to the altitude of the triangle. If the altitude of the triangle is set to represent the 100% composition (weight fraction unity) of a ternary mixture, the perpendicular distances can be made to represent the percentages (or the weight fractions) of the three constituents of the mixture. The vertices of the triangle represent the pure components. Thus in Fig. 7.4a, the altitude *CD* which represents a weight fraction of unity is divided into 10 equal parts by horizontal lines that intersect the altitude, each horizontal line representing the increment of 0.1 in the weight fraction of **C** as we move from the base *AB* of the triangle to the vertex *C*. Weight fraction of component **C** is zero on the side *AB* of the triangle and is unity at the vertex *C*. A binary mixture consisting of **A** and **B** will be represented by a point on the side *AB*. The vertex *B* represents the pure component **B** and the vertex *A*, the pure component **A**. Consider a point such as *P* within the triangle. The line length *PS*, the perpendicular distance to the base *AB*, is the weight fraction of **C** in the mixture, the line length *PQ*, the weight fraction of **A** in the mixture, and the distance *PR*, the weight fraction of **B** in the mixture. Denoting the weight fractions of components

A, **B** and **C** by x_1, x_2 and x_3, respectively, the composition of the ternary mixture represented by the point P in Fig. 7.4a is $x_1 = 0.20$, $x_2 = 0.30$, and $x_3 = 0.50$ or the mixture at P consists of 20 (weight)% **A**, 30 (weight)% **B** and 50 (weight)% **C**.

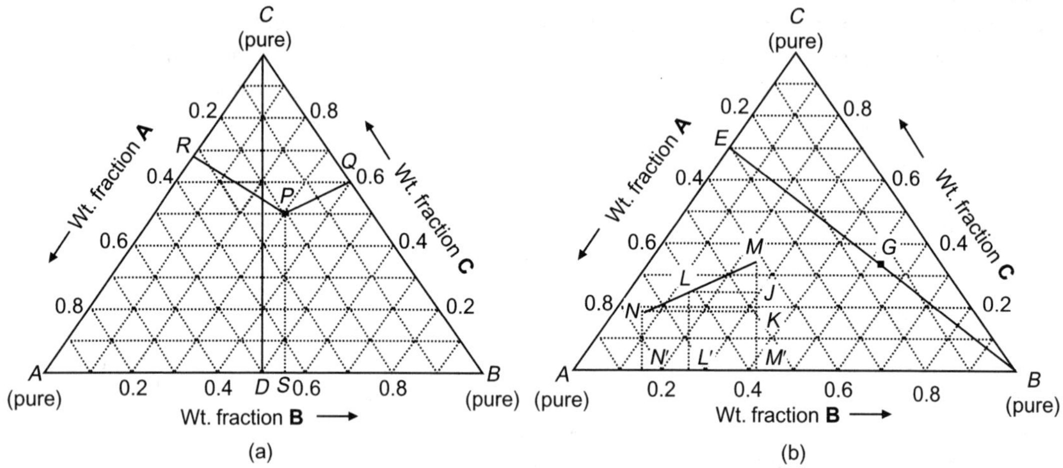

Fig. 7.4: (a) Representing ternary mixture on equilateral triangle

Fig. 7.4: (b) Mixture rule on triangular coordinates

In the triangular coordinate system, the mixing of two streams can be represented by the line joining the points representing the streams and the resultant stream can be represented by a point on the line such that it divides the line in the ratio of the quantities of the original streams. Consider Fig. 7.4b. When M kg of a stream represented by M and N kg of stream at N are mixed together the resultant mixture L divides the line NM in such a way that

$$\frac{\overline{NL}}{\overline{LM}} = \frac{M}{N} \qquad \text{...(7.1)}$$

Equation (7.1) is known as the *mixture rule* in extraction. This can be proved by considering the overall material balance and the component-**C** balance for the mixing process. The total material balance is

$$N + M = L \qquad \text{...(7.2)}$$

Component-**C** balance gives

$$Nx_N + Mx_M = Lx_L \qquad \text{...(7.3)}$$

where x_L, x_M and x_N are, the weight fractions of component **C** in the mixtures *L, M* and *N*, respectively. Eliminating L from Eq. (7.3) using Eq. (7.2), we get

$$Nx_N + Mx_M = (N + M)x_L \qquad \text{...(7.4)}$$

which may be rearranged as

$$\frac{M}{N} = \frac{x_L - x_N}{x_M - x_L} \qquad \text{...(7.5)}$$

From Fig. 7.4b, it is clear that $x_L - x_N$ is the length of the line *JK* and $x_M - x_L$ is the length *MJ*.

Using the properties of similar triangles, we have $\frac{\overline{JK}}{\overline{MJ}} = \frac{\overline{LN}}{\overline{ML}}$. Substituting these results in Eq. (7.5), we get

$$\frac{M}{N} = \frac{\overline{JK}}{\overline{MJ}} = \frac{\overline{NL}}{\overline{LM}} \qquad \text{...(7.6)}$$

The mixture rule also known as the *lever-arm rule*, can be used to determine geometrically, the location of the resultant of mixing two or more streams on the triangular coordinate system.

The mixture rule can be extended to determine the location of the product stream when a mixture is removed from another. When a certain quantity of a ternary mixture is subtracted from another, the resultant mixture will be located closer to the largest of the original streams on the extension of the straight line joining them. For example when M kg of a ternary mixture represented by the point M is removed from L kg of mixture at the point L, the resultant mixture will be located on the extension of the line ML and will be closer to the mixture at L.

Now consider a binary mixture such as the one represented by E in Fig. 7.4b. Point E represents a mixture consisting of 70 weight% C and 30 weight% **A**. When pure solvent **B** is added to this, the resultant product G will be a ternary mixture, which by the mixture rule should lie on the line joining E and B, the exact location being dependent on the quantities of the mixture E and the solvent **B** mixed together. For all mixture lying on the line EB, the ratio of **A** to **C** is constant at 3:7, but the amount of **B** varies. Nearer the product G is to the vertex B, the richer is the mixture in the solvent **B**. If all the solvent is removed from the mixture at G, the resultant product will be the binary mixture E which lies on the extension of the line GB. Similarly, for mixtures lying on a straight line passing through C, the weight ratio of **A** to **B** will be constant, and for the line passing through A, the weight ratio of **B** to **C** will be constant.

One Liquid Pair Partially Miscible

For the ternary systems consisting of components **A, B** and **C**, three binary pairs are involved. The most common type of the liquid–liquid system in extraction is the Type I system in which two such pairs are completely miscible and one pair is partially miscible. A few examples of such systems are listed in the table below:

A	B	C
Water	Ethyl acetate	Acetic acid
Water	Trichloroethane	Acetone
Water	Chloroform	Acetone
Water	Benzene	Ethanol
Water	Trichloroethane	Methyl ethyl ketone
Water	Furfural	Ethylene glycol
Water	Chlorobenzne	Pyridine

Liquid **C** dissolves completely in **A** and **B**, while **A** and **B** are only partially miscible with each other.

At a given temperature, the miscibility limits of liquid **B** in liquid **A** is represented by the point K and that of liquid **A** in liquid **B** is represented by the point L. Thus when two pure liquids **A** and **B** are mixed together and allowed to settle, the mixture will separate into a **A**-rich phase, represented by K (which is about 92.3% **A** and 7.7% **B** in Fig. 7.5) and a **B**-rich phase represented by L (which is about 95% **B** and 5%**A**). Assume that **A** and **B** are mixed together in such proportion that the composition of the resultant mixture lies in-between K and L, say, for example at J. This mixture cannot exist as a homogeneous liquid, and will separate spontaneously into mixtures K and L, the relative quantities of which are dependent, according to the mixture rule [Eq. (7.1)], on the amounts of **A** and **B** mixed.

Now consider a ternary system obtained by mixing the three liquids in such proportions that the combined mixture is represented by M in Fig. 7.5. The mixture cannot exist as a homogeneous phase. It separates into the **B**-rich phase called the extract denoted by E in the figure and a raffinate phase, rich in the carrier liquid **A**, denoted by R in the figure. The line joining R and E is called a tie line since it joins the two phases which are in equilibrium. The mutual solubility of **A** and **B** is altered by adding the component **C** to the binary mixture of **A** and **B**. The ratio of **A** to **B** in the raffinate phase R is not equal to that in the binary K. Similarly the ratio of **A** to **B** in the extract E and the **B**-rich solution L are different indicating the influence of **C** on the solubility of **A** and **B**.

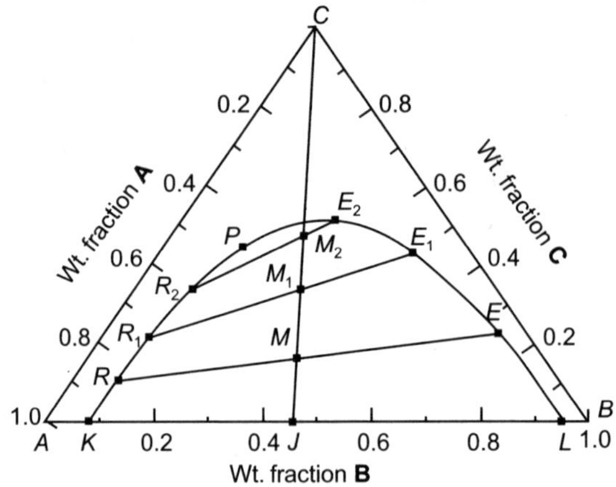

Fig. 7.5: Liquid–liquid equilibrium of Type I system

The relative quantities of the raffinate and extract phases are determined by the ratio in which the mixture point M divides the tie line. For another ternary mixture denoted by M_1, the equilibrium raffinate and extract phases lie at the ends of the tie line R_1E_1 passing through M_1. Similarly R_2E_2 and several other tie lines can be drawn. The last of the tie lines is the shortest and is represented by the point P which is called the *plait point*. The **A**-rich and **B**-rich solubility curves merge, the raffinate and extract phases have identical compositions and the phases become indistinguishable at the plait point. The mutual solubility of **A** and **B** in the extract and raffinate phases become identical at the plait point. Now a smooth curve can be drawn through the points K, R, R_1, R_2, …., P, …. , E_2, E_1, E, L which is called the binodal solubility curve for the ternary system.

The solubility curve separates the area within the triangle into two regions: the heterogeneous region and the homogeneous region. The dome-shaped region bounded by the solubility curve is the heterogeneous area within which the mixture cannot exist as a homogeneous liquid. The mixture in this region separates into raffinate and extract equilibrium phases that are represented by points lying on the solubility curve. Outside the dome, the mixture is a homogeneous liquid; no mixture in this region can separate into raffinate and extract phases. It is the heterogeneous region that is important from the extraction view point. By the phase rule the number of degrees of freedom in the homogeneous region is $F = C - \pi + 2 = 3 - 1 + 2 = 4$, which means that at definite temperature and pressure, there are two degrees of freedom, or concentrations of two components must be specified to define the state of the system. In the heterogeneous region $F = C - \pi + 2 = 3 - 2 + 2 = 3$, which means that at given temperature and pressure, the system is univariant and is uniquely determined by specifying concentration of only one component. For example, by specifying the weight fraction of **C** in the raffinate phase, the phase R is determined on the solubility curve and the extract E in equilibrium with it is determined at the end of the tie line passing through R that lies on the extract side of the solubility curve.

Distribution Curve

The liquid–liquid equilibrium can be represented by means of distribution curves as well. The distribution curve shows graphically the relationship between the weight fraction of solute in the extract phase (y) taken on the y-axis and the weight fraction of the solute in the raffinate phase (x) taken on the x-axis. Each tie line on the triangular coordinate system gives rise to a point on the distribution curve as shown in Fig. 7.6. If the tie line slopes upwards as shown in this figure, y will be greater than x and the distribution curve will lie above the diagonal. The ratio of the weight fraction of the solute in the extract to that in the raffinate is called the *distribution coefficient, m*.

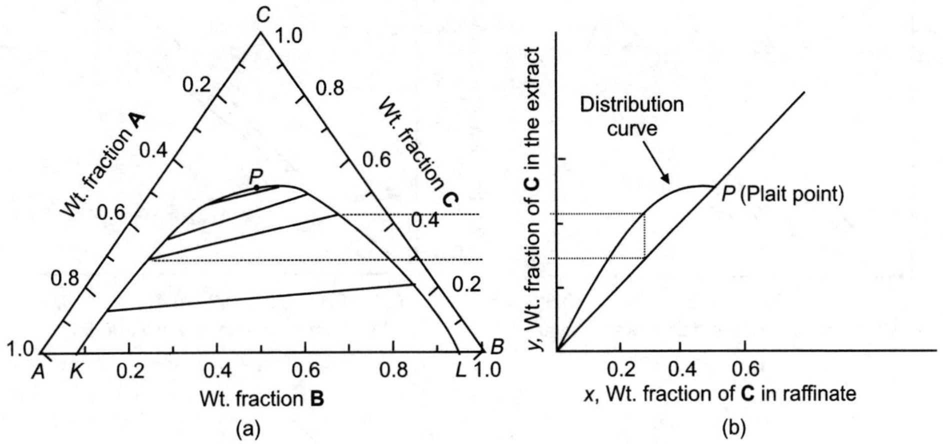

Fig. 7.6: (a) Equilateral triangular diagram (b) Distribution curve

Thus,
$$m = \frac{y}{x} \quad \text{...(7.7)}$$

Figure 7.7b shows the distribution curve of acetone (**C**) – water (**A**)- monochlorobenzene (**B**) system in which the distribution of acetone at equilibrium favours the solvent, rather than water. The distribution coefficient is greater than unity and the distribution curve lies above the 45° reference line.

The tie line slope need not always be positive. For example, for the system acetic acid (**C**) – water (**A**) – isopropyl ether (**B**) system, the tie line slopes downward and the distribution coefficient is less than one. It means that at equilibrium the distribution of acetic acid favours the aqueous phase rather than the solvent phase. The distribution curve lies below the diagonal in such cases as shown in Fig. 7.7a. It should be noted that this is not a limitation for extraction and acetic acid can still be separated by extraction from its solution with water using the isopropyl ether solvent as will be explained later.

For certain systems, there is a reversal of slope for the tie lines after certain concentrations. Pyridine (**C**) – water (**A**) – chlorobenzene (**B**), pyridine (**C**) – water (**A**) – benzene (**B**), isopropanol (**C**) – water (**A**) – benzene (**B**) are examples of such systems. These are known as "solutropes". The distribution curve lies above the diagonal up to certain concentration and below the diagonal after that. For solutropes there is one tie line that is horizontal on the triangular coordinate system and this corresponds to the point of intersection of the distribution curve and the diagonal. Figure 7.7c shows the distribution curve of pyridine – water – chlorobenzene system.

Selectivity

Selectivity or separation factor (β) is analogous to the relative volatility in distillation. Selectivity of a solvent **B** for **C** reflects the ability of the solvent **B** to dissolve **C** in preference to **A** when a mixture of **A** and **C** is treated with **B**. It is defined as the ratio of **C** to **A** in the extract phase to the ratio of **C** to **A** in the raffinate phase at equilibrium. For separation by extraction to be possible, its value should be different from unity. Thus,

$$\beta = \frac{\text{weight fraction of C in extract/weight fraction of A in extract}}{\text{weight fraction of C in raffinate/weight fraction of A in raffinate}} \quad \text{...(7.8)}$$

$$\beta = \frac{\text{weight fraction of C in extract}}{\text{weight fraction of C in raffinate}} \times \frac{\text{weight fraction of A in raffinate}}{\text{weight fraction of A in extract}} \quad \text{...(7.9)}$$

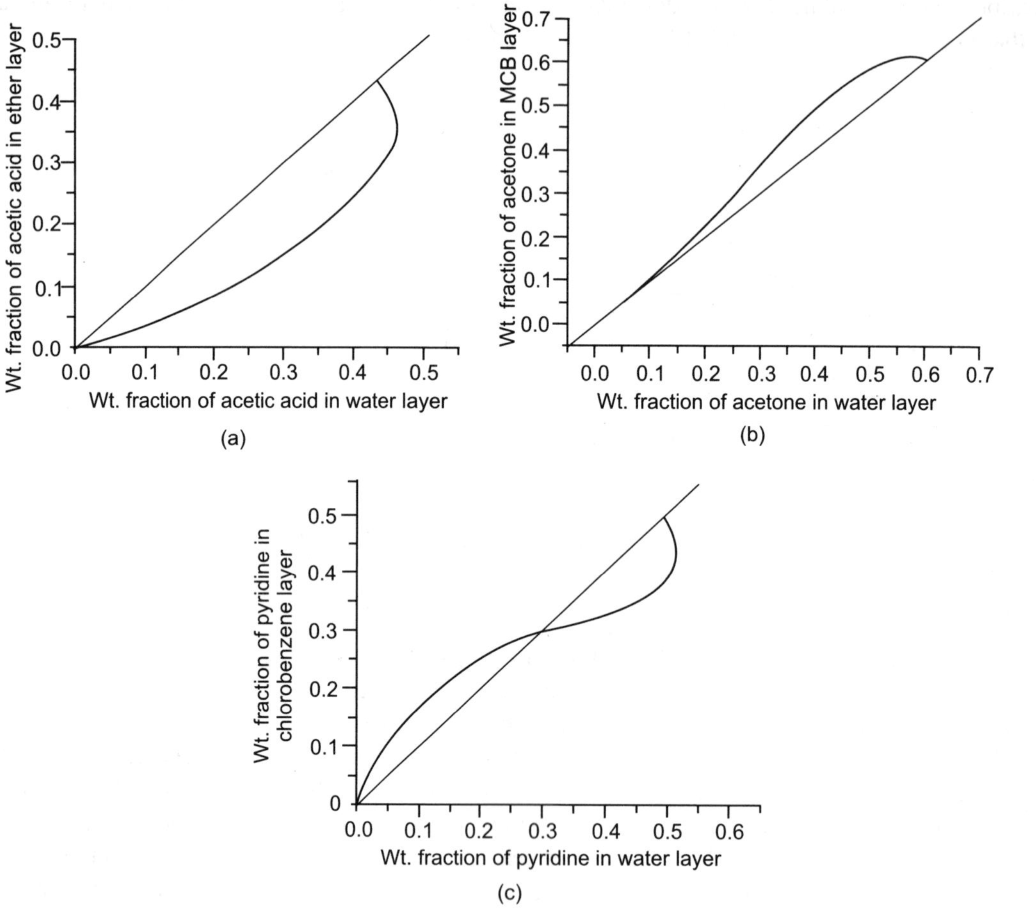

Fig. 7.7: Distribution curves for (a) Acetic acid-water-Isopropyl ether system (b) Acetone-water-monochlorobenzene system (c) Pyridine-water-chlorobenzene system

The first ratio on the left-hand-side of Eq. (7.9) is the distribution coefficient, m, by definition.

$$\beta = m \times \frac{\text{weight fraction of A in raffinate}}{\text{weight fraction of A in extract}} \qquad \ldots(7.10)$$

As the second ratio on the right side of Eq. (7.10) will always be greater than one, even if the distribution coefficient is less than one, the selectivity need not be. Unless the mutual solubility of **A** and **B** is very high, systems with unfavourable distribution coefficient ($m < 1$) and solutropes also will give selectivity greater than unity because of the high value of the second ratio in Eq. (7.10).

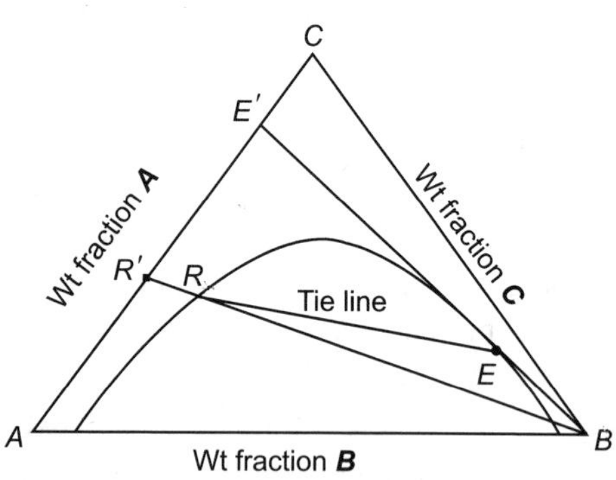

Fig. 7.8: Geometrical interpretation of selectivity

Figure 7.8 shows a system with distribution coefficient less than unity. RE is one of the tie lines for the system. When the solvent is completely removed from the raffinate, we get the solvent-free raffinate R' lying on the side AC of the triangle. By the mixture rule, for subtraction of one stream from another, R' should lie on the extension of line BR. The ratio of **C** to **A** in the mixture R' is equal to the ratio of the length of the lines $R'A$ to $R'C$. Similarly E' is the solvent-free extract resulting when all the solvent present in it is removed. The ratio of **C** to **A** in the extract is the ratio of the line lengths $E'A$ to $E'C$. Thus geometrically, the selectivity is

$$\beta = \frac{\overline{E'A}/\overline{E'C}}{R'A/R'C} \qquad ...(7.11)$$

which we can see is greater than one, by inspection of Fig. 7.8. That means, in spite of the distribution coefficient being unfavourable, the selectivity of solvent **B** for extraction of solute **C** is quite good. The farther the points R' and E' are from each other, as will be the case when tie lines slope upwards, the greater the selectivity of the solvent towards the solute.

Selectivity usually varies with the solute concentration and it will be unity at the plait point. In some systems, the selectivity changes from values greater than unity to fractional values, with selectivity equal to unity at certain solute concentrations. Such systems are analogous to azeotropes in distillation. For completely immiscible systems, the selectivity is infinity.

Effect of Temperature and pressure

The mutual solubility of **A** and **B** increases with temperature. The points K and L in Fig. 7.9 move closer to each other as the temperature is increased and the area of heterogeneity shrinks at elevated temperatures. At the *critical solution temperature* of the binary pair AB, **A** and **B** are completely soluble. Above this temperature, extraction cannot be carried out. The effect of temperature on the ternary systems of the Type I is depicted in Fig. 7.9. When the critical solution temperature T_4 is reached, the **A**-rich and **B**-rich phases become identical and the heterogeneous area disappears.

Effect of pressure on the liquid–liquid equilibrium under normal operating pressures is so small that it is usually ignored.

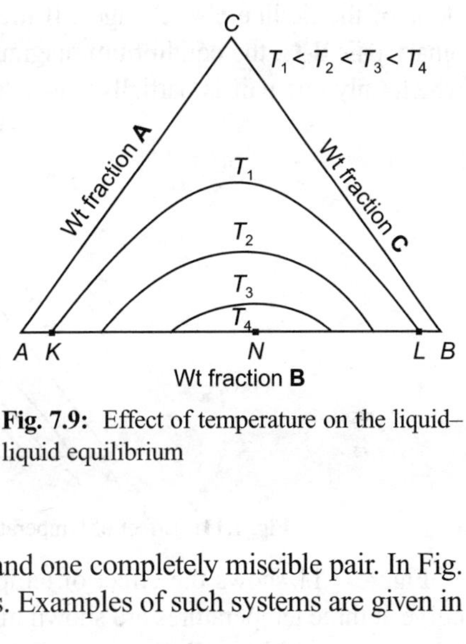

Fig. 7.9: Effect of temperature on the liquid–liquid equilibrium

Two Liquid Pairs Partially Soluble

The Type II systems form two partially miscible binary pairs and one completely miscible pair. In Fig. 7.10, the binary pairs **A-B** and **B-C** are partially miscible pairs. Examples of such systems are given in the table below:

A	B	C
Water	Ether	Ethyl alcohol
Isooctane	Furfural	Hexane
Water	Chlorobenznene	Methyl ethylketone
Hexane	Aniline	Methyl cyclopentane
Methanol	Isooctane	Nitrobenzene

A typical equilibrium diagram of the Type II system is shown in Fig. 7.10a and the corresponding distribution curve in Fig. 7.10b. The curve KRT is the raffinate side of the binodal curve and UEL is the extract side of the binodal curve. The region bounded by these two curves and the sides AB and BC of the triangle forms the region of heterogeneity and any mixture such as M within this region splits into the raffinate R lying on the curve KRT and an extract E lying on the curve UEL. RE is one of the tie lines for the system which when projected to the y versus x coordinate system on the right, we get the point P on the distribution curve.

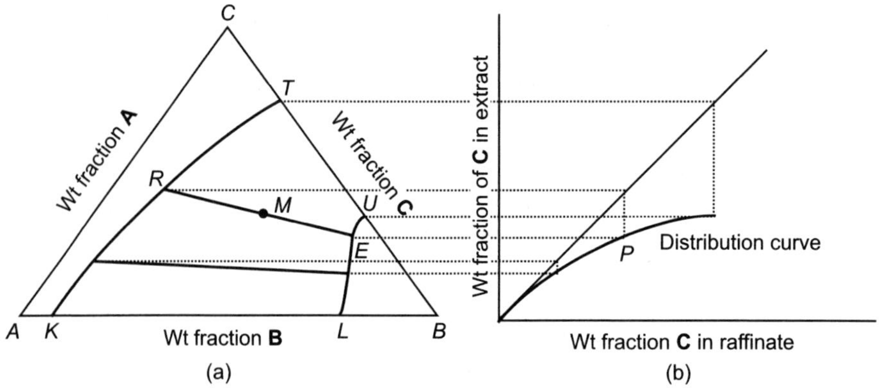

Fig. 7.10: (a) Liquid equilibrium diagram of Type II system (b) The Distribution curve for Type II system

At higher temperatures, the mutual solubility of **A** and **B** as well as that of **B** and **C** increase and the slope of the tie line also changes. If the temperature is above the critical solution temperature of the binary pair **B-C**, the equilibrium diagram of type II system will resemble that of the Type I system in which only one pair is partially miscible.

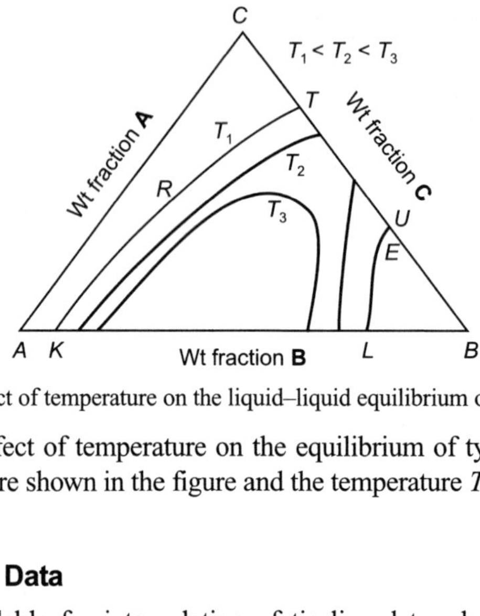

Fig. 7.11: Effect of temperature on the liquid–liquid equilibrium of Type II system

Figure 7.11 shows the effect of temperature on the equilibrium of type II systems. The solubility curve at three temperatures are shown in the figure and the temperature T_3 is above the critical solution temperature of binary **B-C**.

Interpolation of Tie Line Data

Several techniques are available for interpolation of tie line data when only a limited number of experimental results are available. The distribution curve can be used for interpolation as explained earlier. A geometrical method for interpolation is illustrated in Fig. 7.12. Consider the tie line RE. From R a line parallel to the side BC and from E a line parallel to the side AC are drawn to intersect at point M. Similarly the tie line R_1E_1 is used to get the point M_1 and tie line R_2E_2 gives the point M_2. A smooth curve is drawn connecting the points M_1, M, M_2, etc. which gives the tie-line correlation curve or conjugation curve, M_1MM_2P as shown in Fig. 7.12. The curve passes through the plait point P. This curve can be used to interpolate tie line data. From any point on this curve draw a line parallel to the side BC of the triangle which intersects the solubility curve at the raffinate end of the tie line and the line drawn parallel to the side AC intersects the solubility curve at the extract end of the tie line.

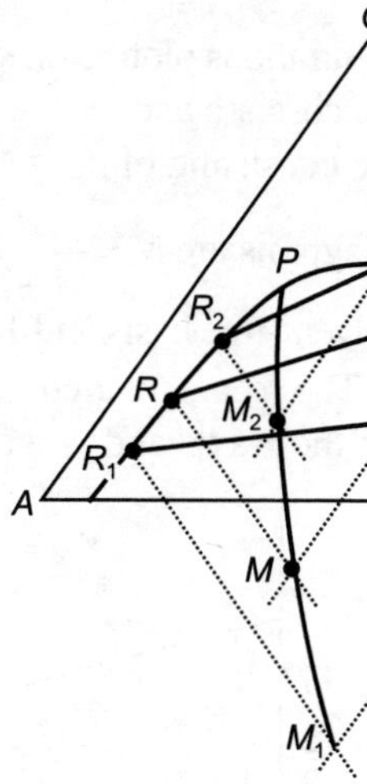

Fig. 7.12: Graphical construction for interpolation of tie line

Rectangular Coordinates

In equilateral triangle, the coordinate scales should be equal and therefore, it is difficult to expand one axis in relation to the other. Graphical constructions for the determination of number of stages usually get crowded and interpolation becomes difficult. This problem is overcome using the rectangular coordinate system in which it is possible to expand one concentration scale relative to another.

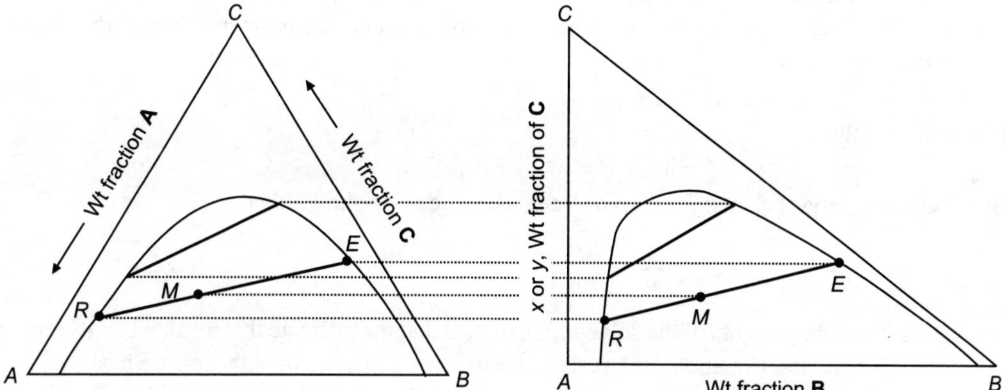

Fig. 7.13: Relation of equilateral triangular coordinates and the rectangular coordinates

The ternary equilibrium diagram can be represented on rectangular coordinate system as shown in Fig. 7.13. The weight fraction of solvent **B** is plotted as abscissa against the weight fraction of solute **C** (x and y). The rectangular coordinate system is, in fact a right-triangle system in which the weight fraction of the third component is obtained by subtracting the sum of the weight fractions of **B** and **C** from one. The mixture rule (Eq. 7.6) is applicable on right triangular system as well.

$$\frac{R}{E} = \frac{\overline{EM}}{\overline{MR}} \quad \ldots(7.6)$$

The Janecke Diagrams

These are rectangular coordinate systems with concentrations plotted on solvent-free basis. The weight fraction of solute on solvent free basis is taken as the abscissa and the weight fraction of the solvent on solvent-free basis as the ordinate. Thus, in a mixture consisting of A kg component **A**, B kg solvent **B** and C kg solute **C**, the coordinate of the Janecke diagrams are $N = \dfrac{B}{A+C}$ and X or $Y = \dfrac{C}{A+C}$, X is the weight fraction of solute **C** in the raffinate on solvent-free basis and Y is that in the extract, N is the weight fraction of the solvent on solvent-free basis. The rectangular coordinate system on solvent-free basis is also amenable for using different scales for the x-axis and y- axis.

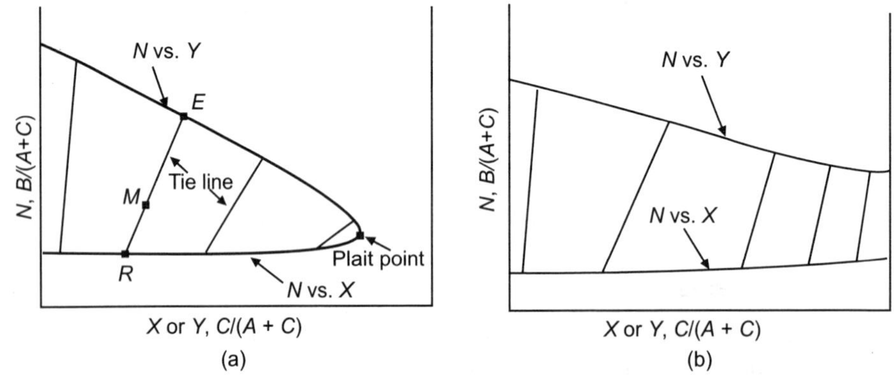

Fig. 7.14: Rectangular coordinates, solvent- free basis (Janecke type) for (a) one pair partially miscible system (b) Two pairs partially miscible system

Let M' be the weight of $A+C$ in the mixture of total weight M represented by point M in Fig. 7.14a. Let N_M and X_M be the weight fraction of solvent and solute on **B**-free basis in the mixture. Then the weight of **B** in the mixture is

$$B = M'N_M \quad \text{...(7.12)}$$

and the total weight is

$$M = M' + B = M'(1 + N_M) \quad \text{...(7.13)}$$

By taking a material balance on solvent-free basis, it can be shown that

$$\frac{E'}{R'} = \frac{\overline{RM}}{\overline{ME}} \quad \text{...(7.14)}$$

where E' and R' are the weights of the solvent free extract and the raffinate that result when the mixture M is allowed to separate. Equation 7.14 is the *inverse-lever-arm rule* on solvent-free basis.

Example 7.1: A ternary liquid mixture consisting of 22 kg pyridine, 33 kg water and 45 kg chlorobenzene is allowed to separate into the equilibrium raffinate and extract phases. The raffinate phase weighed 40.13 kg and analysed 18.9 weight% pyridine, 0.4 weight% chlorobenzene and 80.7 weight% water.

(a) Determine the quantity and composition of the extract phase by material balance.
(b) Estimate the selectivity of chlorobenzene for extraction of pyridine from water
(c) Verify the applicability of the mixture rule (inverse-lever-arm rule) for this separation.

***Solution*:** *Basis*: 100 kg original mixture. Here, **A** = water, **B** = chlorobenzene, **C** = pyridine.

$$M = 100 \text{ kg; Raffinate } R = 40.13 \text{ kg.}$$

(a) Consider the total material balance for the separation.

$$E = M - R = 100 - 40.13 = 59.87 \text{ kg}$$

The original mixture M contains 22 kg **C**. By a pyridine balance, the amount of pyridine present in the extract = pyridine in the original mixture – pyridine in the raffinate

$$= 22 - 40.13 \times 0.189 = 14.42 \text{ kg.}$$

Similarly, the chlorobenzene present in the extract = $45 - 40.13 \times 0.004 = 44.84$ kg

Amount of water present in the extract is obtained by subtracting the sum of the weights of pyridine and chlorobenzene from the total weight of the extract.

Thus, water present in the extract $= 59.87 - (14.42 + 44.84) = 0.61$ kg

The composition of the extract phase is computed as given below: **(Answer)**

Component	Weight, kg	Weight fraction
Pyridine	14.42	$\frac{14.42}{59.87} = 0.24$
Chlorobenzene	44.84	$\frac{44.84}{59.87} = 0.75$
Water	0.61	$\frac{0.61}{59.87} = 0.01$
Total	59.87	1.00

(b) The composition of the raffinate and extract phases that separate from the given mixture are given below:

Component	Raffinate phase, weight%	Extract phase, weight%
Pyridine (**C**)	18.9	24.0
Chlorobenzene (**B**)	0.4	75.0
Water (**A**)	80.7	1.0

Selectivity of the solvent **B** for solute **C** is defined as

$$\beta = \frac{\text{weight fraction of } \mathbf{C} \text{ in extract/weight fraction of } \mathbf{A} \text{ in extract}}{\text{weight fraction of } \mathbf{C} \text{ in raffinate/weight fraction of } \mathbf{A} \text{ in raffinate}}$$

$$\beta = \frac{\text{(weight of pyridine/weight of water) in the extract}}{\text{(weight of pyridine/weight of water) in the raffinate}}$$

$$\beta = \frac{24.0/1.0}{18.9/80.7} = 102.5$$ **(Answer)**

Since the C/A ratio in the extract and raffinate phases vary with composition, the selectivity will also vary with composition.

(c) The raffinate and extract phases are marked on the equilateral triangle as shown in Fig. 7.15 by points R and E, respectively. Using the original composition of the mixture, the point M is also marked. It can be seen that points R, M and E are collinear. The ratio of the line lengths is found to be

$$\frac{\overline{RM}}{\overline{EM}} = 1.492$$

The ratio of weights of extract and raffinate is

$$\frac{E}{R} = \frac{59.87}{40.13} = 1.492$$

Thus we see that

$$\frac{\overline{RM}}{\overline{EM}} = \frac{E}{R}$$

or the mixture rule is satisfied. **(Answer)**

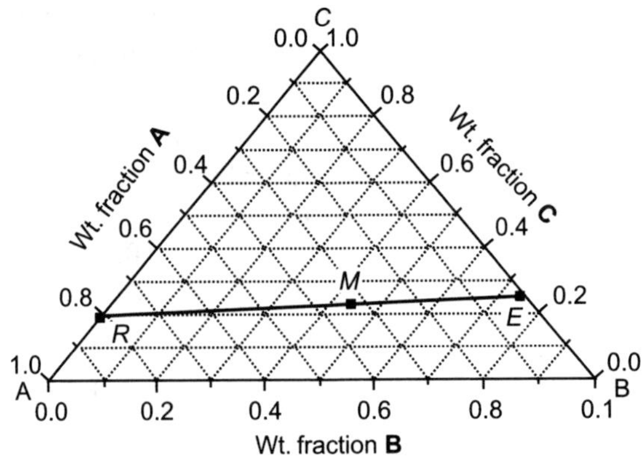

Fig. 7.15: Illustration of the mixture rule for Example 7.1

Example 7.2: Construct a triangular phase diagram for a ternary system made up of liquids **A**, **B** and **C** from the experimental data given in the table below. Also plot the distribution diagram for the system. Draw the tie line for the raffinate of composition 25 weight% **C**. What is the composition of the **B**-rich phase in equilibrium with this raffinate?

The ternary equilibrium data in weight% are:

Raffinate phase, weight%			Extract phase, weight%		
A	**B**	**C**	**A**	**B**	**C**
97	3	0	4	96	0
86	4	10	5	90	5
75	5	20	6	80	14
62	8	30	8	70	22
48	12	40	12	56	32
40	15	45	14	48	38
30	20	50	21	32	47
22	30	48	22	30	48

***Solution*:** The equilibrium data are converted to weight fractions and the equilibrium curve for the system is plotted on the equilateral triangular coordinate system as shown in Fig. 7.16a. The vertices *A*, *B* and *C* of the triangle represent the pure components **A**, **B** and **C** respectively. Raffinate side of the

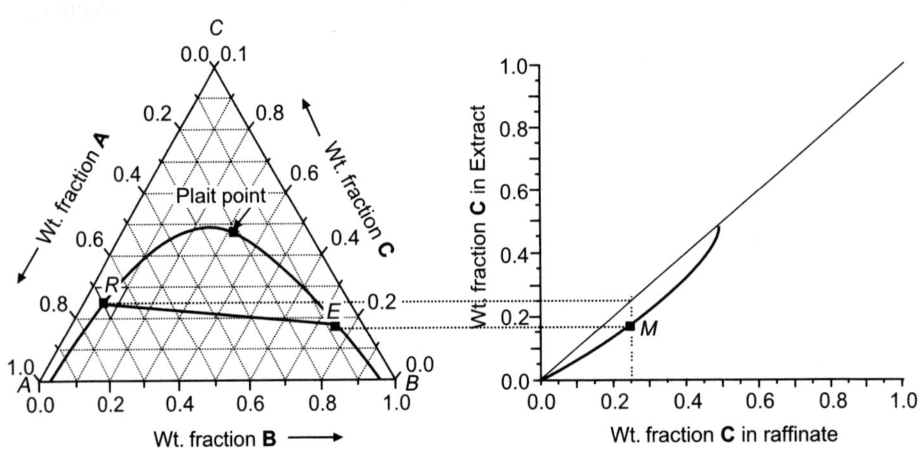

Fig. 7.16: (a) Equilibrium curve for Example 7.2 (b) Distribution curve for Example 7.2

binodal curve is obtained when the raffinate phase concentrations are plotted on the triangle and the extract side is obtained when the corresponding equilibrium extract compositions are plotted. The raffinate side and the extract side of the binodal curve meet at the plait point where the phases are identical.

The distribution curve shown in Fig. 7.16b is obtained when the weight fraction of the solute **C** in the extract is taken on the *y*-axis and the weight fraction of the solute **C** in the raffinate is taken on the *x*-axis. The distribution curve meets the 45° reference line at the plait point concentration.

To obtain the tie line passing through the raffinate phase having a concentration of 25 weight% **C**, the point *R* is located on the raffinate side of the binodal curve at weight fraction of **C** = 0.25. The figure shows the distribution curve drawn by the side of the triangular phase diagram choosing identical scales for its ordinate and the altitude of the triangle. The point *M* on the distribution diagram of Fig. 7.16b corresponds to the desired tie line in Fig. 7.16a.The *x*- and *y*- coordinates of *M* give the weight fractions of **C** in the raffinate and extract phases in equilibrium, x_R and y_E, respectively. Then by moving from *M* horizontally to the extract side of the binodal curve, we can locate *E*, the extract phase in equilibrium with the raffinate *R*. The line joining *R* and *E* is the required tie line. Alternatively, a conjugation curve can be constructed to facilitate interpolation of tie lines, as described earlier. From the raffinate end of a tie line, a line parallel to the side *BC* is drawn and from the extract end of the tie line, a line parallel to the side *AC* is drawn. The locus of the point of intersection of these lines provides the conjugation curve as shown in Fig. 7.17. To obtain the tie line passing through point *R* on the binodal curve, a line is drawn parallel to the side *BC* which intersects the conjugation curve at *N* as shown in the figure. Now the line drawn parallel to the side *AC* of the triangle will intersect the extract side of the binodal curve at *E*, which is the extract end of the desired tie line.

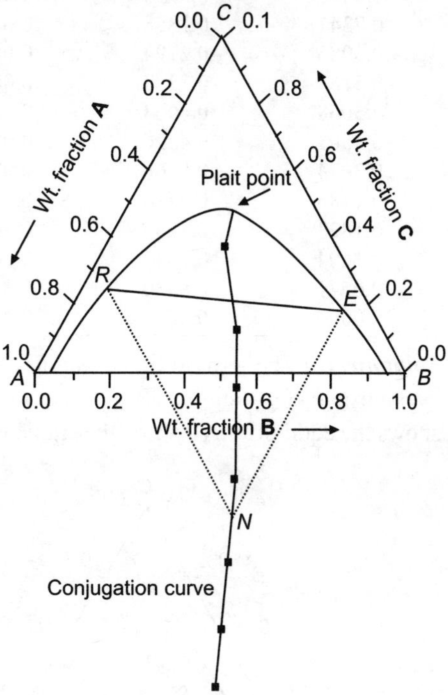

Fig. 7.17: Conjugate line for tie line interpolation for Example 7.2

The weight fractions of the components in the the **B**-rich phase in equilibrium with the raffinate *R* is obtained from the coordinates of pւint *E*. They are:

A = 0.072, **B** = 0.751 and **C** = 0.177

Therefore, the extract contains 17.7% solute (**C**), 75.1% solvent (**B**) and 7.2% carrier (**A**).

(Answer)

Example 7.3: The ternary equilibrium data of the type-II system aniline (**B**) – methylcyclohexane (**C**) – *n*-heptane (**A**) are given below. Use the data to construct

(a) The equilibrium diagram on triangular coordinates

(b) The Janecke diagram.

(c) On the Janecke diagram draw the tie line passing through the raffinate of composition 0.50 kg methylcyclohexane per kg of (methylcyclohexane + *n*-heptane) and estimate the composition in weight percent of the extract phase in equilibrium with this raffinate.

Raffinate phase, weight fraction			Extract phase, weight fraction		
n-Heptane	MCH	Aniline	*n*-Heptane	MCH	Aniline
0.9260	0.0000	0.0740	0.0620	0.0000	0.9380
0.8315	0.0914	0.0772	0.0599	0.0080	0.9321
0.7341	0.1858	0.0800	0.0530	0.0270	0.9200
0.6985	0.2194	0.0821	0.0511	0.0300	0.9190
0.5768	0.3373	0.0859	0.0450	0.0461	0.9089
0.5068	0.4063	0.0869	0.0400	0.0600	0.9000
0.4505	0.4596	0.0900	0.0360	0.0740	0.8900
0.3074	0.5966	0.0960	0.0298	0.0980	0.8723
0.2286	0.6714	0.1000	0.0211	0.1132	0.8658
0.1823	0.7158	0.1019	0.0160	0.1269	0.8571
0.1604	0.7357	0.1039	0.0139	0.1310	0.8551
0.0541	0.8330	0.1128	0.0062	0.1559	0.8379
0.0000	0.8811	0.1189	0.0000	0.1689	0.8311

Solution: (a) The ternary liquid–liquid equilibrium diagram for the system *n*-heptane (**A**) – methylcyclohexane (**C**) – aniline (**B**) – system is plotted on triangular coordinates in Fig. 7.18. Figure 7.18a shows the equilibrium curve on the equilateral triangle and Fig. 7.18b on the right-triangular coordinates.

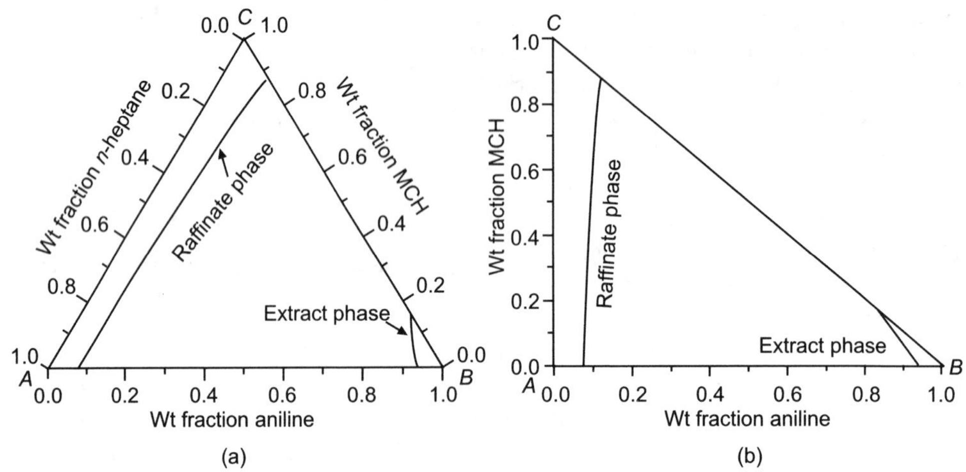

Fig. 7.18: Ternary equilibrium diagram for the system methylcyclohexane (MCH) – aniline – *n*-heptane (a) on equilateral triangle (b) on right triangle. (Example 7.3)

(b) The *x*-coordinate of the Janecke diagrams is the weight fraction of the solute on a solvent-free basis (*X* or *Y*) and the *y*-coordinate is the weight fraction of the solvent also on a solvent-free basis (*N*). These coordinates are calculated from the given equilibrium data. Calculation is illustrated below using the third set of data.

Raffinate: **C** = 0.1858, **A** = 0.7341, **B** = 0.08

Extract: **C** = 0.027, **A** = 0.053, **B** = 0.92

Raffinate:

$$\text{Weight fraction of solute, } X = \frac{C}{A+C} = \frac{0.1858}{0.7341+0.1858} = 0.202$$

$$\text{Weight fraction of solvent, } N = \frac{B}{A+C} = \frac{0.08}{0.7341+0.1858} = 0.087$$

Extract:

$$\text{Weight fraction of solute, } Y = \frac{C}{A+C} = \frac{0.027}{0.027+0.053} = 0.3375$$

$$\text{Weight fraction of solvent, } N = \frac{B}{A+C} = \frac{0.92}{0.027+0.053} = 11.5$$

The calculations are repeated for other sets and the results are tabulated below:

X	N	Y	N
0.0000	0.0799	0.0000	15.1290
0.0990	0.0836	0.1178	13.7275
0.2020	0.0870	0.3375	11.5000
0.2390	0.0894	0.3699	11.3317
0.3690	0.0940	0.5060	9.9769
0.4450	0.0952	0.6000	9.0000
0.5050	0.0989	0.6727	8.0909
0.6600	0.1062	0.7668	6.8255
0.7460	0.1111	0.8429	6.4468
0.7970	0.1135	0.8880	5.9979
0.8210	0.1159	0.9041	5.9013
0.9390	0.1272	0.9618	5.1690
1.0000	0.1349	1.0000	4.9207

The Janecke diagram is plotted with N on the y-axis and X (or Y) on the x-axis as shown in Fig. 7.19a. The tie lines are shown as dotted lines in the figure.

(c) The distribution curve is plotted by taking the weight fraction of the solute in the extract on solvent-free basis (Y) as the ordinate against the weight fraction of the solute on a solvent-free basis in the raffinate (X) as the abscissa. The curve is plotted in Fig. 7.19b just below the N versus X (or Y) curve. To draw the tie line passing thorough the raffinate phase with coordinate $X = 0.50$, mark point M on the N versus X curve at $X = 0.5$ a shown in Fig. 7.19a. The point P on the distribution curve vertically below M will have its x-coordinate $X = 0.5$. The point L on the Janecke diagram obtained as shown in Fig. 7.19a will represent the extract in equilibrium with M. ML is the desired tie line. The coordinates of L are found out from the figure. They are

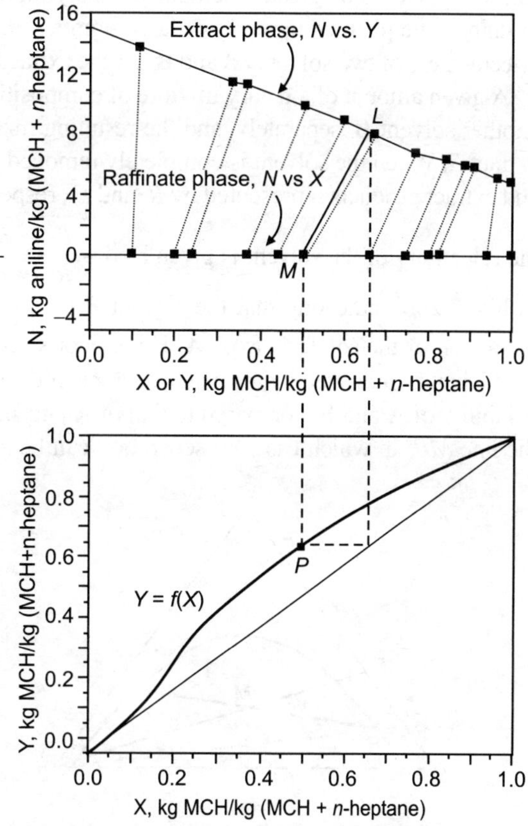

Fig. 7.19: (a) Janecke diagram for the system n-heptane (**A**) – aniline (**B**) – methylcyclohexane (**C**) (b) The distribution curve on solvent-free coordinates, for Example 7.3

$$Y = 0.657 \text{ and } N = 8.25$$

The extract in equilibrium with the raffinate of composition $X = 0.5$ kg MCH/kg (n-heptane + MCH) contains $Y = 0.657$ kg MCH/kg (n-heptane + MCH) and $N = 8.25$ kg aniline/kg (n-heptane + MCH).

That is, for every 1 kg (n-heptane + MCH) in the extract there are 0.657 kg MCH and 8.25 kg aniline. The weight percent of the constituents of the extract can be calculated as given below:

Component	Weight	Weight%
Methylcyclohexane	0.657	(0.657/9.25)×100 = 7.10
n-Heptane	1.0 – 0.657 = 0.343 kg	(0.343/9.25)×100 = 3.71
Aniline	8.25 kg	(100 – 7.10 – 3.71) = 89.19
Total	9.25 kg	100

Extract phase contains 7.1% methylcyclohexane, 3.71% n-heptane and 89.19% aniline. **(Answer)**

7.3 CHOICE OF SOLVENT

The desirable characteristics of a good extraction solvent are listed below.

1. *Selectivity*: The solvent should have good selectivity for the solute to be extracted; that is, it should dissolve the maximum amount of the solute (**C**) and the minimum amount of the carrier (**A**). The selectivity should be greater than unity. No separation is possible if the selectivity is unity. If it is close to unity large plant equipments or large number of extraction stages will be required. In Fig. 7.20 the effectiveness of two solvents **B** and **B′** for the extraction of solute **C** from its solution with **A** is explored.

A given amount of a binary mixture of composition represented by F is treated with a solvent **B** and another solvent **B′** separately, and the resulting mixture M is allowed to separate into raffinate R and extract E. When the solvent is completely removed from the products, we get the solvent free raffinate and extract products represented by R' and E', respectively. It was shown (Eq. 7.11) that geometrically, the selectivity of the solvent is given by $\beta = \dfrac{\overline{E'A} / \overline{E'C}}{\overline{R'A} / \overline{R'C}}$. Clearly this ratio is greater in Fig. 7.20a than in Fig. 7.20b, indicating that the solvent **B** is more selective than solvent **B′** and is preferred for the extraction of the solute **C** from **A**. Here, poor selectivity of **B′** is due to the unfavourable distribution coefficient as indicated by the negative slope of the tie line RE in Fig. 7.20b, and due to the high mutual solubility of **A** and **B′** compared to that of **A** and **B**. It may also happen that the tie line coincides with the line MB' in which case, no separation can be achieved by the extraction.

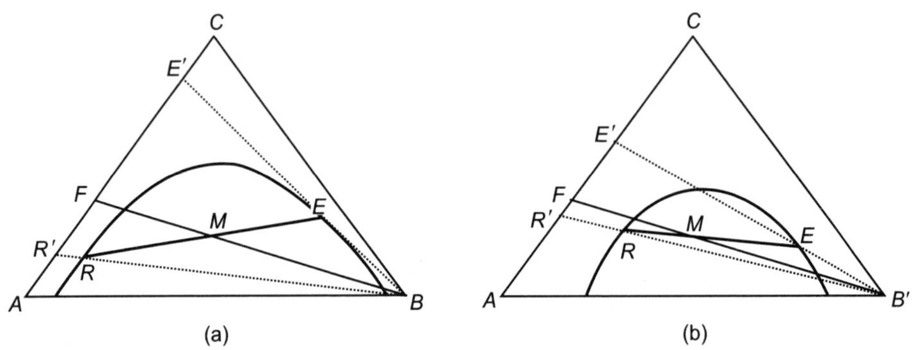

Fig. 7.20: (a) Solvent with high selectivity (b) Solvent with poor selectivity

2. *Distribution coefficient*: It is already shown that a a favourable distribution coefficient ($m > 1$) will lead to high selectivity values. However, it is not essential that the distribution coefficient values should be greater than one for extraction to be feasible. The larger the distribution coefficient values, the smaller the quantity of solvent required for extraction. The value of the distribution coefficient can be altered by changing the pH of the solution when the extracted solute is ionisable, or by maintaining the pH at the desired values by adding a buffer when the solute extracted is an acidic or basic substance.

3. *Capacity*: To reduce the solvent requirement for extraction, it is essential that the solvent should be capable of dissolving the solute in large quantities. High selectivity indicates that the ratio of **C** to **A** in the extract is high compared to that in the raffinate, but the high capacity indicates that the quantity of the solute that unit weight of the solvent can dissolve is high. A good solvent should therefore give not only large value of m, but also large possible concentration y of the solute in the extract.

4. *Solvent insolubility*: The solvent should be insoluble in the carrier **A**. The lower the mutual solubility, the larger will be the area of heterogeneity. In Fig. 7.21, the solvent **B′** is more soluble in **A** than **B**. Since the mixtures falling in the area outside the binodal solubility curve cannot be separated by extraction, only feed mixtures whose compositions fall in the range *A* to *F* can be treated. Thus the range of concentrations that can be handled is larger with solvent **B,** which compared to **B′**, is relatively more insoluble in **A**.

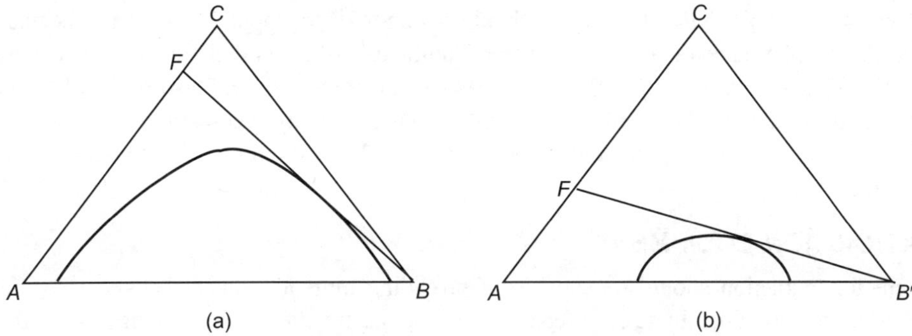

Fig. 7.21: Effect of mutual solubility of **A** and **B** on extraction

The mutual solubility of **A** and **B** also influence the selectivity. Referring to Eq. (7.10), we can see that the selectivity is high if the ratio $\dfrac{\text{weight fraction of A in raffinate}}{\text{weight fraction of A in extract}}$ is high. This ratio will be high if the solvent is less soluble in **A**.

In addition, the solvent recovery from a highly insoluble system is quite simple and for completely immiscible systems, the raffinate product is free of the solvent and requires no further treatment for solvent recovery.

5. *Solvent recoverability*: Recovery of solvent from the extraction products is important to permit reuse of the solvent and to prevent the contamination of the products. If distillation is the method employed for recovery, the properties such as volatility, heat of vaporisation, thermal stability, the chance of formation of azeotrope with the solute, etc. should be considered when a solvent for extraction is selected.

6. *Density*: The density difference between the solvent-rich and the carrier-rich phases should be as large as possible to permit rapid coalescence and disengagement of phases, and to achieve a high capacity in the extractor.

7. *Interfacial Tension*: High interfacial tension between the immiscible phases helps rapid coalescence and easy disengagement of phases, but for dispersion of one phase into the other, the energy required will be large. With low interfacial tension between the phases, dispersion is easier but disengagement of phases will be difficult. Generally, solvent giving high interfacial tension is preferred since the dispersion can be achieved by mechanical means as well.

8. *Chemical stability and reactivity*: The solvent should not undergo any chemical reaction with the components in the feed and materials of construction of the equipment. It should be chemically stable and should be unaffected by the oxidizing agents, should not undergo polymerization and air oxidation, should not be affected by the repeated heating in the solvent recovery operations.

9. *Viscosity*: Use of low viscosity solvent leads to low power requirements for pumping and agitation, rapid settling of dispersions and high solute mass transfer rates.

10. *Corrosiveness*: The solvent should not be corrosive to common materials of construction.

11. *Vapour pressure*: The vapour pressure should be low so that storage and extraction are possible at atmospheric conditions and loss of the solvent by vaporisation can be minimized. However, the vapour pressure should not be very low which may make the recovery by distillation difficult. Taking advantage of the high selectivity and easy recovery, liquid propane and liquid sulphur dioxide are used as solvents in petroleum refining, in spite of their high volatility.

12. *Wettabilty*: The solvent should have desirable wetting properties with the extractor surfaces.

13. *Freezing point*: To facilitate easy handling and storage in cold weather, the freezing point of the solvent should be low.

14. *Toxicity and flammability*: The solvent should be non-toxic for applications such as the manufacture of foodstuffs and pharmaceuticals. Even for the manufacture of general chemicals, the solvent should be non-toxic for safety reasons. For the same reasons, the solvent should also be non-flammable.

15. *Cost*: To maintain the large solvent inventory that is frequently necessary in extraction operations and to provide the makeup requirement at low cost, the solvent should be inexpensive and should be readily available.

7.4 EXTRACTION EQUIPMENT

Equipment for extraction should be capable of providing intimate contact between the immiscible liquid phases and should facilitate easy separation of the phases after mass exchange. Since the mixing of the two insoluble phases and subsequent separation after mass transfer are usually more difficult compared to those in the gas-liquid contact operations, the equipment used for absorption, distillation, etc. are ineffective as extractors unless the density difference between the phases is very high and the interfacial tension is very small. Specialized equipment have been designed and a wide variety of them are in use depending upon the properties of the phases being contacted. Extraction equipment can be classified into the following groups:

1. Mixer-settler assemblies
2. Static columns
   - (i) Spray columns
   - (ii) Packed columns
   - (iii) Sieve-plate columns
3. Mechanically agitated columns
   - (i) Rotary-impeller columns
   - (ii) Pulse columns
   - (iii) Reciprocating plate columns
   - (iv) Rotating-disk contactors
4. Centrifugal contactors

Mixer-settler contactors and sieve-plate columns are stage-wise contact equipment in which the liquids are brought into contact, and then separated into raffinate and extract phases in a series of stages. Majority of the extraction equipment are differential extractors in which the concentration change in the liquids is brought about by bringing them into contact continuously with separation of the phases allowed only at the exits.

Mixer-Settlers

The simplest type of extraction equipment is the one in which the liquids are mixed thoroughly in one vessel (the mixer) and the phases are allowed to settle and separate in another vessel (the settler).

Mixers are either agitated vessels or line mixers. The transfer of the solute from one phase into the other takes place in the mixer. Settlers are gravity settlers or centrifuges. The operations may be carried out in batch or continuous fashion.

Agitated vessels are usually vertical cylindrical mixing tanks provided with baffles. The mixing and dispersion is achieved either mechanically by means of the rotating impellers immersed in the liquid, or by gas agitation. Propellers, turbine or paddle type impellers provide mechanical agitation. Either the vapour generated by boiling the liquids being contacted, or the permanent gases such as air admitted through spargers at the bottom of the vessel may be used for gas agitation. One notable advantage of the agitated vessels for extraction is that they can handle liquids of any viscosity. Figure 7.22a shows an agitated vessel-gravity settler combination for extraction.

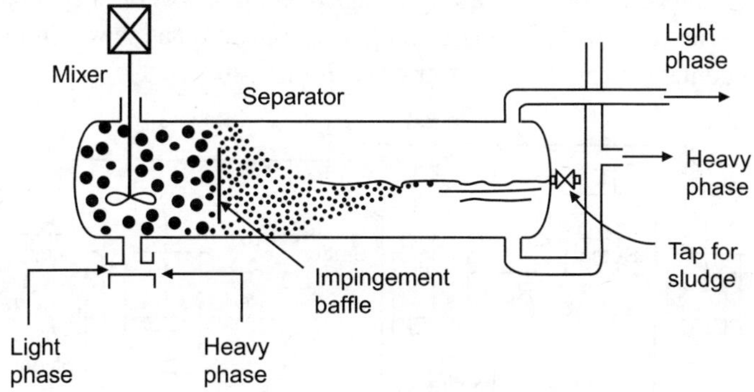

Fig. 7.22: (a) Single-stage mixer settler

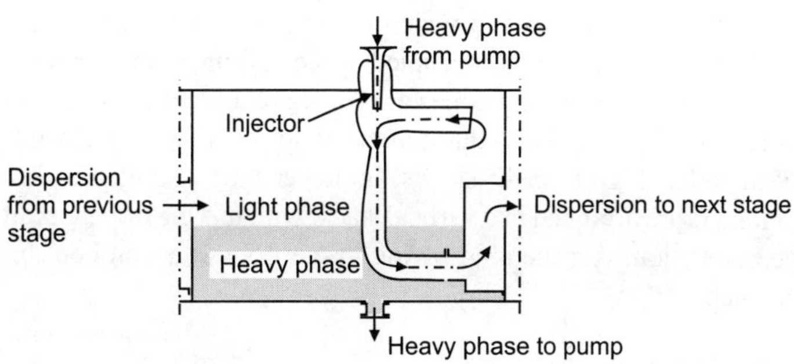

Fig. 7.22: (b) Injection mixer and settler

Jet mixers, injectors, orifices, mixing nozzles, etc. are line mixers or flow mixers through which the liquids to be contacted are passed. The energy required for the mixing and dispersing is derived from the pressure drop resulting from the flow. Line mixers are used only for continuous or semibatch operation. Since they have small volumes, the line mixers provide a very small holding time for mass transfer. Their application is limited to cases where dispersion is not difficult and the equilibrium is rapidly attained, as in the case of liquids of low interfacial tension and viscosity. Figure 7.22b shows an injector mixer where the mixer is incorporated into the settling chamber.

For settling, no separate vessels are needed for batch operation, as the mixture can be settled in the mixing vessel itself. For continuous operation, however, separate gravity settlers or centrifuges are used for settling. Centrifuges are machines that are rotated rapidly and that permit fast and continuous sedimentation of dispersion even when the density difference between the liquids is low. Being costly units compared to gravity settlers, centrifuges are used only where low holding time is essential or where phase separation is difficult. The simplest gravity settler is usually a horizontal vessel as the one shown in Fig. 7.22a, in which the dispersion issuing from the agitated mixing vessel is passed to the settler. The impingement baffle shown is either a slotted or perforated plate which protects the vessel contents from the disturbance by the jet of the entering dispersions. The tap at the downstream end of the settler facilitates periodical removal of the scum or sludge of the dust particles and other solids that collect at the interface.

The small drops of fine dispersion may be made to coalesce by passing through a coalescer. The coalescers are usually mats, beds, or layers of porous or fibrous solids, by passing through which, the drops get enlarged in size and settle fast.

Mixer-Settler Cascades

A mixer and settler combination constitute a single stage operation. Several such mixer settler assemblies can be connected in series to form a continuous multistage operation as shown in Fig. 7.23. These units can be operated in countercurrent, cocurrent or cross-flow fashion.

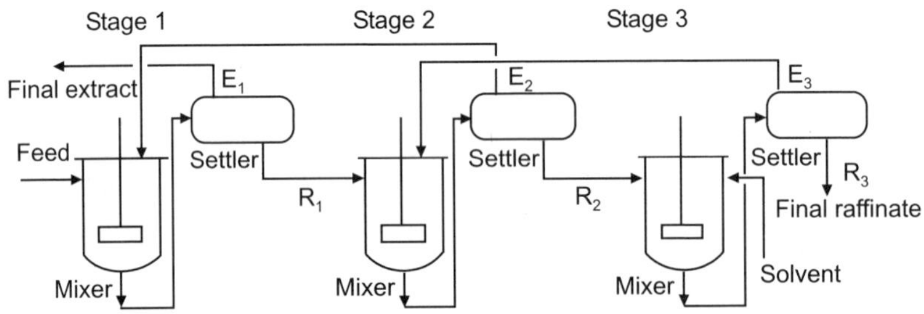

Fig. 7.23: Three-stage countercurrent mixer-settler cascade

Figure 7.24 shows two different ways in which the mixer settler assemblies can be arranged for multistage operation in order to reduce the occupied space and inter-stage piping. Figure 7.24a is the box type in which the mixing and settling zones are separated by plates. The impellers pump as well as mix by drawing the heavy liquid upward through the hollow shaft and discharging it at a higher level through the hollow impeller. Figure 7.24b shows the tower type arrangement in which a battery of settlers and external mixers are arranged in a vertical stack, with mixing impellers attached to a common stack. For multistage operation, considerable saving in the floor area can be achieved by the vertical column-type arrangement.

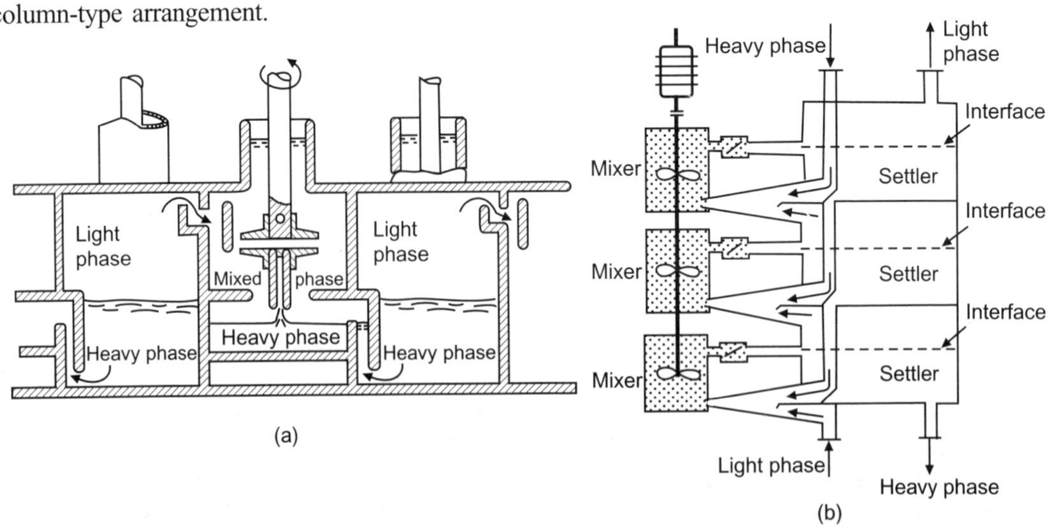

Fig. 7.24: (a) Pump-mix mixer settler (b) Vertical arrangement of a column of settlers and external mixers

Since the height of the mixer-settler units is small, the headroom requirement is low. Their performance is not affected by the presence of suspended solids. The stages are independent and can be added to or removed from the cascade as required, and high efficiency of extraction can be attained per stage. The large floor area required and the high holdup of valuable or hazardous solvent because of the high volumes of the equipment, formation of stable emulsions by severe mixing and the need for pumping one or both phases between the tanks are the disadvantages of mixer-settler units. The requirement of power for the agitation equipment at every stage is costly in terms of investment, operation and maintenance.

Static Extraction Columns

These are extraction columns with no moving parts to facilitate flow and mixing of the phases. The liquids flow countercurrently through the column due to the difference in their densities. Spray columns, packed columns and tray columns belong to this class. These columns are available in large diameters and can handle high production rates. The operation is simple as there are no moving parts and associated seals. The equivalent of a large number of theoretical stages can be achieved in columns which require less floor space compared to the mixer-settler units of the same capacity. However, the mass transfer efficiency of static columns is very low compared to those of mechanically agitated extractors. In general, the applications of the static columns are limited to such operations which require not more than three to five equilibrium stages and to liquids with low viscosities (less than 5 cP) and low to moderate interfacial tensions (0.003 to 0.02 N/m). Static columns find wide application as extractors in petroleum industries.

Spray columns

The simplest type of static column is the spray column. These are empty vessels with provision for introducing the liquids either as the dispersed or as the continuous phase. Because of the large free volumes available for flow, they can be built to handle large volumes of the liquids and are not influenced by the presence of suspended solids. Spray columns are best suited for handling liquids that usually foul or clog other equipment. Figure 7.25a shows a spray column with light liquid dispersed. The heavy liquid is introduced at the top without producing high local velocities. The bottom end is

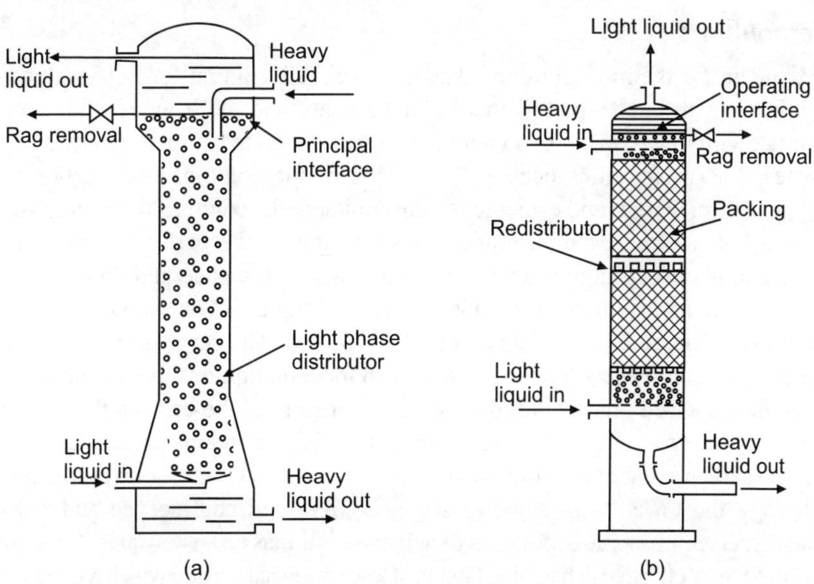

Fig. 7.25: Static column extractors (a) Spray column (b) Packed column

flared gradually to reduce the velocity of the continuous phase so that the formation of the bubbles at the distributor is not affected. This arrangement ensures high holdup for the dispersed phase, for the flow regime at flooding.

A serious limitation of spray columns is the problem of back mixing and therefore, the mass transfer performance of even columns with height exceeding 10 m, is not better than that of a single equilibrium stage. Axial dispersion is so serious in columns with high diameter to length ratio that the continuous phase gets completely mixed. Because of the low efficiency and therefore large size, this type of column is used only for special purposes such as operations in which rapid irreversible chemical reactions occur; e.g., the neutralization of waste acids.

Packed columns

Packed column consists of a cylindrical tower filled with packing similar to those employed for gas-liquid contact operations such as absorption or distillation. The role of the packing is primarily to reduce axial mixing. By breaking up large drops, packings help to increase the interfacial area and by distorting the droplet shape they promote mixing in the drops, both contributing to improved rates of mass transfer. A material that is preferentially wet by the continuous phase is considered to be ideal for the packing, because it prevents the coalescence of dispersed liquid within the packed volume.

Either *random* or *structured packings* can be employed in the extraction columns. The random packings include ceramic packings such as Raschig rings, Berl saddles, Intalox saddles, Lessing rings and metallic and plastic packings such as Pall rings, Hy-Pak and IMTP packings. Though many types of structured packings are in use, modular assemblies in the form of woven wire mesh or crimped sheets arranged in layers at specific angles are extensively used in the extraction columns. Distribution of the dispersed phase at the inlet and the wettablity of the fillings influence the efficiency of the column. It is customary to arrange redistributers for the liquids at intervals of about 3 to 5 m in tall towers as shown in Fig. 7.25b. This is done to prevent excessive back mixing and channelling of the liquids. It may be necessary to provide tall unpacked section at the dispersed-liquid-exit end for coalescing and maintaining a liquid–liquid interface. Packed towers may be employed when extraction equivalent to around 5-10 equilibrium stages suffice. They are not satisfactory at interfacial tensions above 0.01 N/m.

Sieve-plate columns

A schematic diagram for the most common type of sieve plate column for liquid–liquid extraction is shown in Fig. 7.25c with light liquid as the dispersed phase and the heavy liquid as the continuous phase. Sieve plate columns consist of cylindrical towers separated into different compartments by perforated plates. The perforations occupy 15 to 25% of the column cross section with diameter ranging from 3 to 8 mm. Design and construction are similar to the columns used for gas-liquid contact; but, without overflow weirs. The light liquid flows up through the perforations on the plate as the dispersed phase and the heavy liquid flows horizontally across the plate and down through the liquid downcomer as the continuous phase. The bubbles rising through the continuous phase coalesce into a layer and accumulates beneath each plate and get redispersed. The same design, but with Fig. 7.25c turned upside down will represent the tower used when the light liquid is the continuous phase and the heavy liquid is the dispersed phase, with the light liquid rising up the column through the upcomers.

Plate towers are free from the vertical recirculation of the continuous phase that is unavoidable in spray towers. The axial mixing of the continuous phase is confined to the region between trays and does not spread through the tower from stage to stage. By repeated coalescence and redispersion into droplets of the dispersed phase, it is possible to achieve high mass transfer rates in the plate columns. They are found to be most effective in handling liquids of low interfacial tension which requires no mechanical agitation for good dispersion. In terms of the liquid handling capacity and extraction efficiency, the

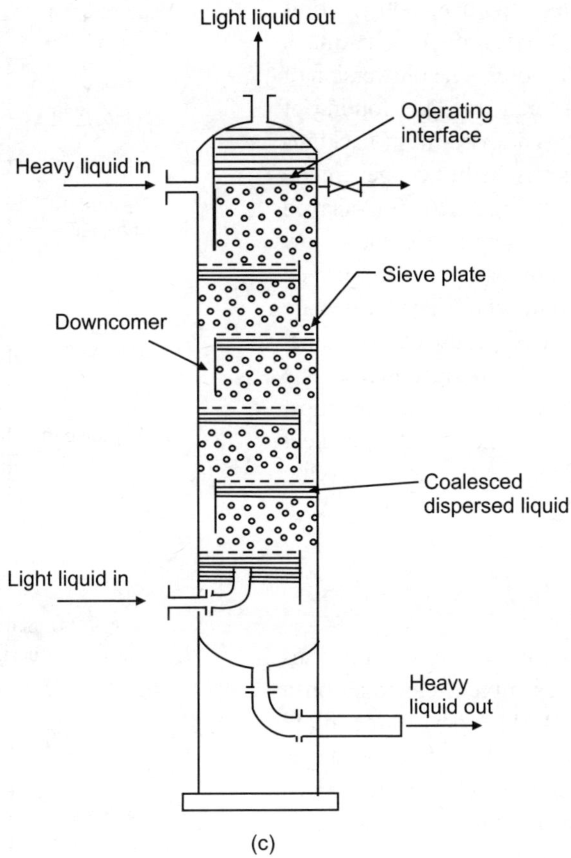

Fig. 7.25: (c) Sieve-plate column

plate columns are found to be superior to other unagitated columns. The overall efficiencies of sieve trays lie typically between 10 and 30 percent. Sieve plate columns work well when the density difference between the phases is appreciable; else it will be difficult to overcome the pressure drop at the plates.

Agitated Extraction Columns

The mass transfer efficiency of a static extraction column is quite low especially for systems with moderate to high interfacial tensions. Efficiency can be improved by agitating the dispersion within the column, mechanically or hydraulically. Agitation helps to control the drop size and the dispersed-phase hold-up. Agitated extraction columns can be further classified as rotary-impeller columns, the pulse columns, the reciprocating plate columns and the rotating-disc contactor (RDC).

Rotary-impeller columns

The mass transfer efficiency can be improved independent of the properties of the liquids by agitating the vessel contents mechanically by means of rotating impellers. Important extractors in this category are Scheibel extractor and Kunhi extractor.

Scheibel extractors

These are ideally suited for liquids of high interfacial tension and viscosity and for applications that require large number of extraction stages. The older designs consisted of alternate compartments agitated with centrally located impellers, with the other compartments packed with open woven wire mesh providing about 97 to 98% voids. The old version of the Scheibel columns suffered from the fouling of these wire-mesh packing. The wire-mesh packings are not used in the new designs. In the new-generation Scheibel column shown in Fig. 7.26, ring-shaped baffles are arranged immediately above and below the impeller, and shallow compartments are formed by stator rings attached to the shell. The baffles are held in position by the tie rods. Because of the internal baffling which controls the mixing patterns, the Scheibel columns have proved to be one of the most efficient extractors in terms of the height equivalent of a theoretical stage. Columns of widely varying sizes with diameters ranging from 75 mm to well above 2 m are in use.

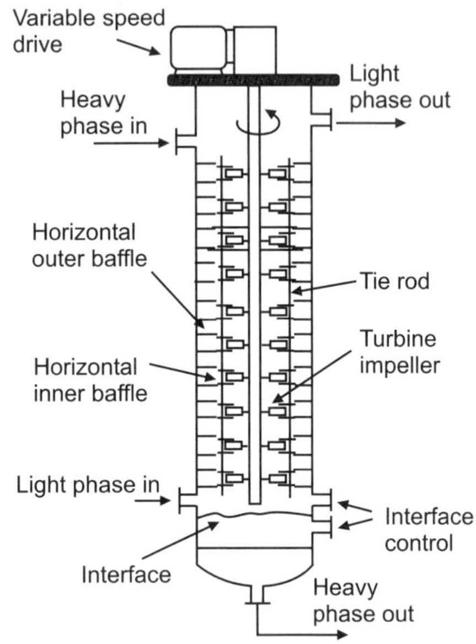

Fig. 7.26: Scheibel extractor (Third generation design)

Kuhni column

Like the Scheibel column, the Kuhni column uses closed turbine impellers as mixing elements on a central shaft (Fig. 7.27). The perforated plate partitions are used to separate the extraction stages into different compartments which help to reduce the back mixing between stages. The impellers promote in each compartment, the circulation action shown in Fig. 7.27b.

The Oldshue-Rushton Column (not shown) with flat-blade-disc turbine impellers to disperse and mix the liquids and horizontal plates fixed to the walls, is another popular rotary agitated column in use. Another novel device is the Graesser Raining-Bucket contactor which is a horizontal rotary agitated vessel used primarily for extraction processes involving liquids of small density difference, low interfacial tension and liquids with a tendency to form emulsions.

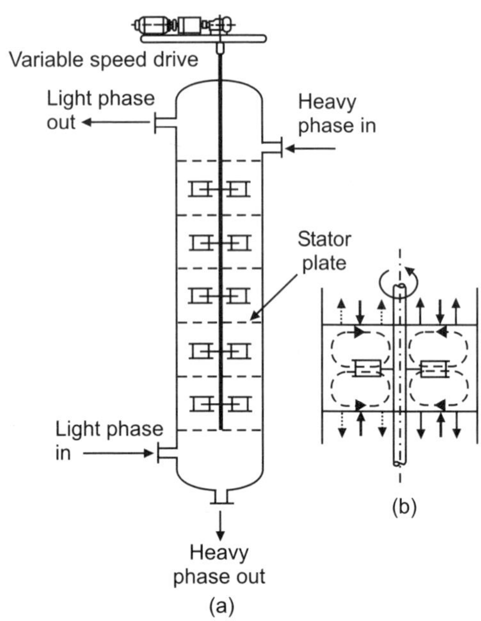

Fig. 7.27: (a) Kuhni column extractor (b) Liquid-circulation pattern in the shell compartment

Pulsed-liquid columns

The pulsed extraction columns are plate or packed columns in which a rapid reciprocating motion of relatively short amplitude is applied to the liquid contents to give improved rates of extraction. Several methods are used for providing the pulsing action. In Fig. 7.28, the pulsing action is achieved using plunger pump from which the valves have been removed, connected to the plate column extractor as shown.

The plates occupy the whole cross section of the column as shown in the figure and hence both liquids have to move through the perforations on the plate. Externally pulsing the liquid to impart mechanical agitation allows for a sealed agitated extraction column with no moving parts and bearings in contact with the liquid. This feature permits the use of these columns for special applications involving highly corrosive or dangerously radioactive liquids as in nuclear energy applications. For example, the extraction and separation of metals from the solutions in atomic energy operations can be done by pulsed columns. However, the energy required is generally more for moving the entire column of liquid than the energy required for moving the plates of a reciprocating column, discussed below.

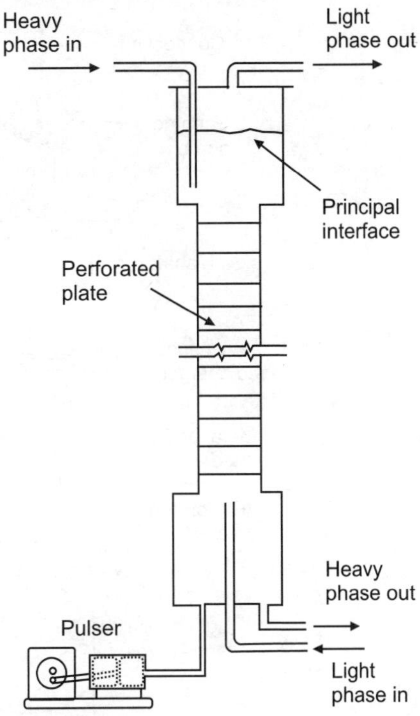

Fig. 7.28: The liquid-pulsed column

Reciprocating plate columns

In these types of extractors, agitation of the dispersion is achieved by means of reciprocating plates. Compared to pulsed columns, this mode consumes much less power and provides equally good extraction efficiency. The holes of the reciprocating plates are larger than those of pulsed columns. This generally results in a more uniform drop size distribution, because the shear forces get more evenly distributed over the entire cross section of the column. Reciprocating plate columns have a wide turndown range and are well suited to systems with moderate interfacial tension. Because of their high open-area design, they can even handle slurries of solids.

Karr reciprocating plate column shown in Fig. 7.29 is a popular example for reciprocating plate columns. It uses dual flow plates with 50 to 60% open area and has no downcomer. The high open area of Karr column permits operation with relatively high throughput. The plates are mounted on a central shaft that moves up and down approximately two times per second through a stroke length of typically up to 50 mm. In large diameter columns, the mass transfer efficiency decreases due to the axial mixing and to reduce the effect, doughnut shaped baffle plates may be added at intervals within the plate stack. Karr columns can handle corrosive liquids since the plates can be fabricated from non metallic materials.

Rotating-disc contactor

Rotating-disc contactors (RDC) are, perhaps, the world's most widely used liquid–liquid extraction device thanks to its geometrical simplicity and effectiveness. The column is divided into several compartments by means of stator rings fixed to the walls. Mechanical agitation is provided by means of a series of flat, circular disks, arranged on a central shaft which is rotated by means of variable-speed motor. The diameter of the disks is less than the diameter of the stator-rings openings as shown in Fig. 7.30. Tower of sizes ranging from 150 mm diameter laboratory scale units to over 3 m diameter columns have been constructed and are in use. RDC is suitable for systems with moderate viscosities (up to 100 cP) and for liquids that tend to foul easily. Originally used in furfural extraction of petroleum lubricating oils, RDC now finds extensive applications in almost all petroleum extraction operations including propane deasphalting and naphtha sweetening, by-product coke, food, metal-separation, and organic chemical industries.

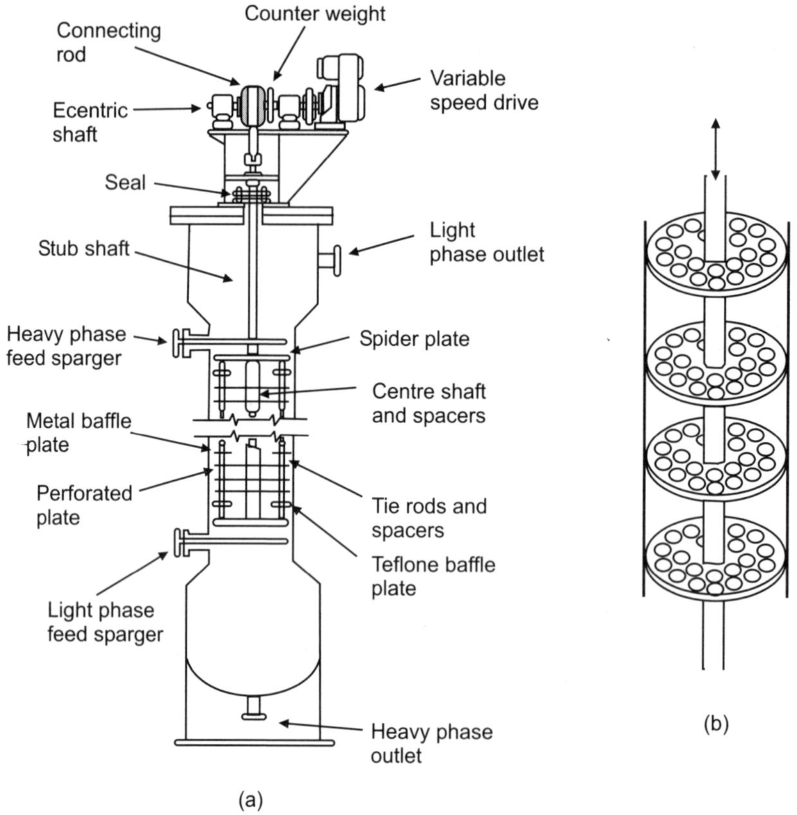

Fig. 7.29: (a) Karr reciprocating plate column (b) Perforated plate assembly of Karr column

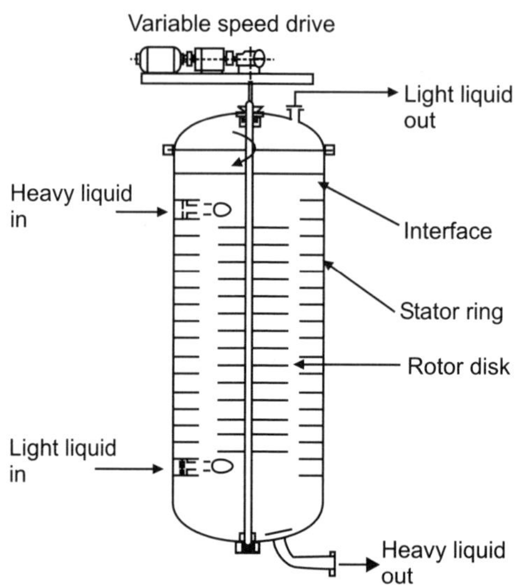

Fig. 7.30: Rotating-disc contactor

RDCs are not recommended for systems with low interfacial tension, low density differences and for feeds that exhibit tendencies to form stable emulsions. The large hold-up and long residence times in the rotating disk contactors may be harmful to certain unstable substances. For applications that require the equivalent of not more than a few theoretical stages, the conventional mixer-settler units will be cheaper than the RDCs.

Asymmetrical-Rotating-Disc-Contactor (ARDC) is a further development of the rotating-disk contactor. In these columns, the shaft with the rotating discs is asymmetric to the axis of the column.

Centrifugal Extractors

It is not possible to attain large liquid velocities in ordinary extractors since counterflow is maintained by the force of gravity which is determined by the density difference between liquids. In centrifugal extractors, the counterflow of phases is achieved by means of centrifugal forces generated by the rapidly rotating drum by which the heavier phase is transported to the periphery and the lighter phase to the centre. Large centrifugal force results in high fluid velocities that lead to reduced extractor volumes.

Centrifugal extractors can handle systems that emulsify easily, that have small density differences between the phases, or large interfacial tensions. Extraction reaches steady state quickly resulting in short residence times in the extractor. This feature makes them ideal for handling materials that undergo product degradation on long exposure in the equipment. Large capacities and small hold-ups are the other major attractions of centrifugal extractors. These features lead to small extractor volumes for high flow capacities and hence reduced inventory of expensive solvent. The disadvantages of the centrifugal extractors are the relatively high costs for the equipment, operation and maintenance. The number of theoretical stages that can be achieved per machine is also limited.

A Podbielniak extractor shown in Fig. 7.31, is a classic example of a centrifugal extractor designed for multistage performance. The body of the extractor is a horizontal cylindrical drum containing concentric perforated cylinders. The liquids are introduced through the horizontal rotating shaft with

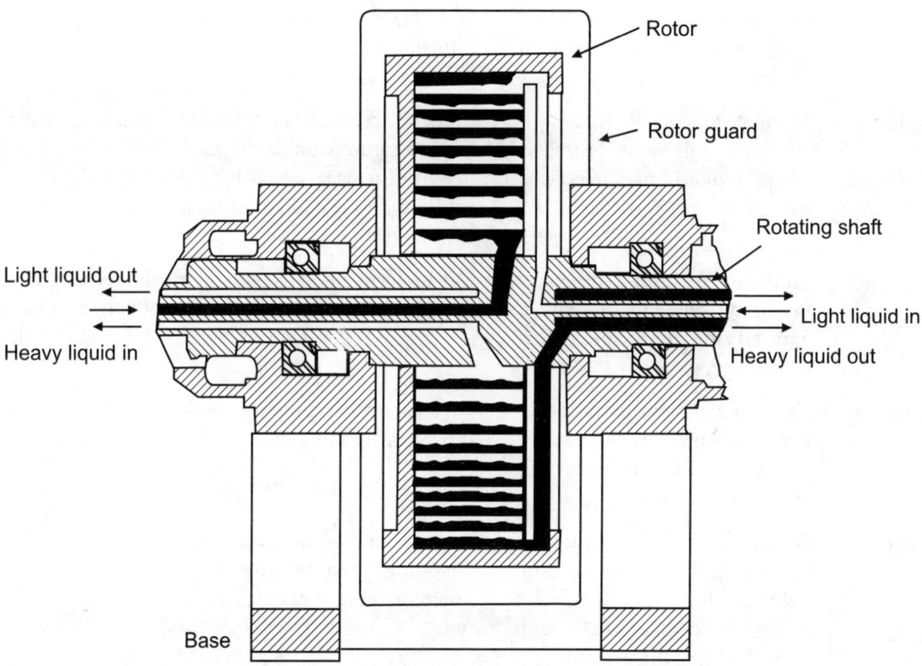

Fig. 7.31 Podbielniak extractor

the help of special metallic seals. The light liquid is fed internally to the drum periphery and the heavy liquid to the axis of the drum. Rotation with speed ranging up to several thousand revolutions per minute causes radial counterflow of the liquids which then flow out through the shaft. Materials of construction include steel, stainless steel, Hastelloy, and other corrosion resistant alloys. Podbielniak extractors are used in all phases of the pharmaceutical manufacture, petroleum processing, extraction of uranium from ore leach liquors, and for clarification and phase separation work. Since they provide very low hold up of liquid per stage, they are particularly useful for extraction of antibiotics, such as penicillin and the like, for which multistage extraction and phase separation must be done rapidly to avoid chemical destruction of the product under the conditions of the extraction.

Comparison and Selection of Extractors

Table 7.1 gives a comparison of the advantages and disadvantages of different extractors and Fig. 7.32 is a rough guide for selection of extractors.

Table 7.1: Comparison of extractors

Types of extractor	Advantages	Disadvantages	Industrial application
Mixer-settlers	Good contacting- handle wide flow ratios- low head-room- high efficiency- many stages available- reliable scale up- long residence time- handle high viscosity liquids and suspended solids	Large hold-up- high power cost- high investment- large floor area- inter-stage pumping	Petrochemical, Nuclear, Fertilizer, Metallurgical
Static extraction columns	Low initial cost and operating cost- simple construction- handle high production rate- suited best to systems with low interfacial tensions	Limited throughput for small density difference- cannot handle high flow ratio- high headroom, difficult scale up	Petrochemical, Chemical, Food
Rotary agitated column	Moderate to high efficiency- low operating cost- good flexibility- handle liquids of high interfacial tension	Moderate to high capital cost- limited throughput with small density difference- cannot handle emulsifying systems and high flow ratios	Petrochemical, Chemical, Pharmaceutical, Metallurgical, Fertilizer, Food
Reciprocating plate column	Moderate to high efficiency- flexibility- suited to systems of emulsifying tendencies- low operating cost	Cannot handle liquids with high interfacial tension	Petrochemical, Chemical, Pharmaceutical, Metallurgical, Food
Pulsed columns	No internal moving parts- moderate to high efficiency- useful for highly corrosive and toxic feeds- moderate production rate	High energy consumption- back mixing is present	Nuclear, Petrochemical, Metallurgical
Centrifugal extractors	Allow short contact time for unstable solutes- low hold-up volumes- less floor space and headroom - handle systems with low density difference and tendency to emulsify- low inventory of solvent	High initial cost- high operating cost- high maintenance cost- limited number of stages in single unit	Petrochemical, Chemical, Pharmaceutical, Nuclear

Selection of Extractors

The following rough guide may be used for selection of equipment for extraction.

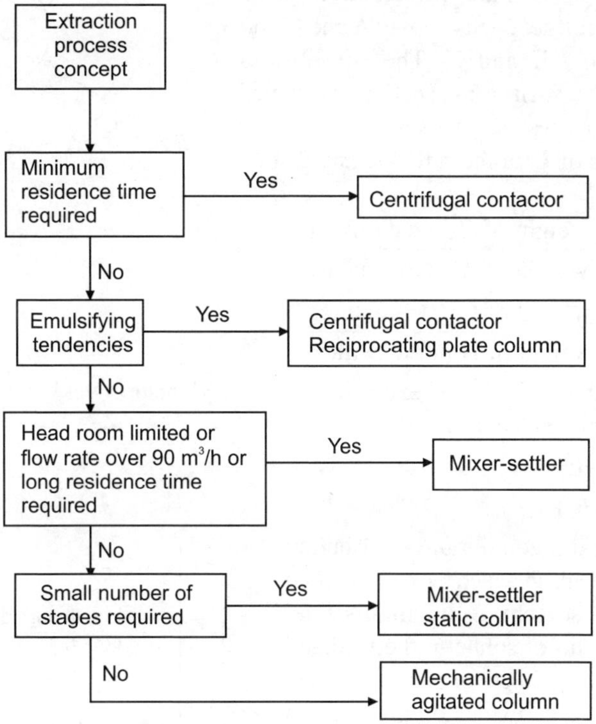

Fig. 7.32: Decision guide for selection of extractors

7.5 IMMISCIBLE SYSTEMS: STAGE CONTACT

Extraction processes may be either stage contact or continuous contact operations. Stage contact operations are carried out in equipment such as mixer-settlers in which the immiscible phases are brought into intimate contact, the phases are allowed to approach equilibrium by transfer of mass between them, and finally the phases are allowed to separate. The changes in the concentration of the phases occur only intermittently, whereas in continuous contact equipment, the phases undergo concentration change over the entire effective length of the unit, continuously. Stage contact operations can be carried out as single stage, multistage cross-current, multistage countercurrent operations or countercurrent operation with reflux. We will consider first the extraction processes involving immiscible systems, the treatment of which is relatively simple compared to the partially miscible systems.

Single-Stage Extraction

Extraction involves a minimum of three components, **A**, **B** and **C**. The solute **C** is extracted from the feed consisting of components **A** and **C** using solvent **B**. Figure 7.33 represents schematically a single-stage extraction process, operated batch wise. F kg of a feed containing x_F weight fraction solute is treated with S kg of solvent. For continuous operation, F and S are the feed and solvent rates, respectively, kg/h.

Now consider a ternary mixture consisting of components **A**, **B** and **C** where the carrier liquid **A** and the solvent **B** are completely insoluble, for all concentrations of **C** in the mixture. Let the weight of carrier **A** present in the feed be A kg and the weight of component **B** present in the solvent be B kg. The entire **A** present in the feed should leave with the raffinate and the entire **B** present in the solvent should leave with the extract. Thus when the mixture separates as raffinate and extract phases, the former contains only **A** and **C** and the latter contains only **B** and **C**. The equilibrium distribution of solute **C** in such cases is better represented on rectangular coordinate system X versus Y where X and Y are the weight ratios of **C** in the raffinate and extract phases, respectively.

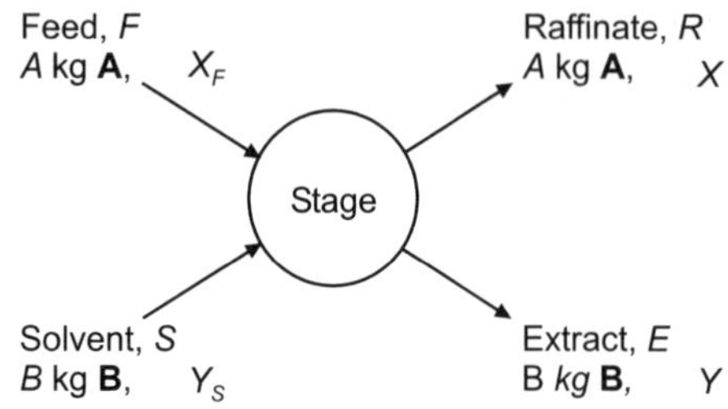

Fig. 7.33: Single-stage extraction

$$X = \frac{\text{weight of } \mathbf{C} \text{ in the raffinate}}{\text{weight of } \mathbf{A} \text{ in the raffinate}} \quad \text{...(7.15)}$$

$$Y = \frac{\text{weight of } \mathbf{C} \text{ in the extract}}{\text{weight of } \mathbf{B} \text{ in the extract}} \quad \text{...(7.16)}$$

The solute content in the raffinate and extract are AX and BY respectively and the weights of raffinate and extract streams can be calculated as

$$R = A(1+X) \quad \text{...(7.17)}$$

$$E = B(1+Y) \quad \text{...(7.18)}$$

Figure 7.34 shows the equilibrium distribution diagram for ternary systems in which the carrier liquid **A** and the extraction solvent **B** are immiscible. Denoting the weight ratio of solute in the feed and solvent by X_F and Y_S, respectively, the solute-**C** balance for the single stage extraction process can be written as

$$AX_F + BY_S = AX + BY \quad \text{...(7.19)}$$

Equation (7.19) can be rearranged as

$$\frac{Y - Y_S}{X - X_F} = -\frac{A}{B} \quad \text{...(7.20)}$$

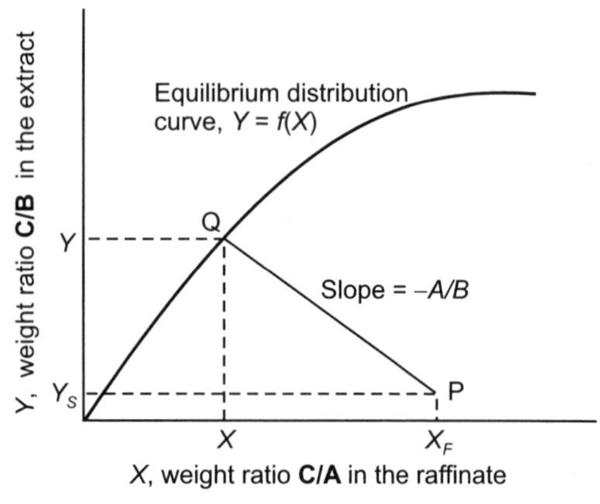

Fig. 7.34: Single-stage extraction represented on equilibrium distribution coordinates for immiscible systems

Equation (7.20) means that the slope of the straight line joining the points representing the input streams and the effluent streams from a single stage extraction process is equal to $-A/B$. Point P in Fig. 7.34 has coordinates (X_F, Y_S). Since the effluent streams from an ideal stage are in equilibrium, the point Q with coordinates representing the composition of the product streams (X, Y) lies on the distribution curve as shown in Fig. 7.34. Since the slope of the line PQ is $-A/B$ according to Eq. (7.20), knowing the quantities and compositions of the feed and solvent treated, the product compositions and their quantities can be estimated.

Multistage Crosscurrent Extraction

The multistage crosscurrent extraction is used to achieve a recovery of solute above that obtainable in a single stage extraction using the same quantity of solvent. The feed is contacted with the solvent in the first stage and the raffinate leaving the stage is again contacted with new batches of the solvent in subsequent stages. Figure 7.35 shows a three-stage crosscurrent unit; the final raffinate product is withdrawn from the third stage. The extract from each stage can be withdrawn separately or can be combined to form a composite extract product.

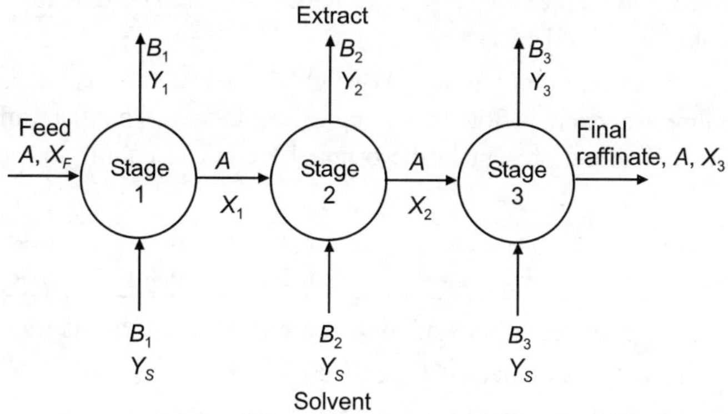

Fig. 7.35: Crosscurrent extraction, immiscible systems

The feed is a binary mixture consisting of A kg of carrier liquid **A** and the concentration of solute **C** in the feed is X_F kg **C** per kg **A**. The feed is admitted to the first stage where it is mixed with the solvent containing B_1 kg **B**. Let the weight ratio of **C** in the solvent be Y_S kg **C** per kg **B**. Since **A** and **B** are immiscible, the raffinate leaving the stage contains the entire **A** present in the feed. Let the weight ratio of **C** in the raffinate be X_1. Similarly, the entire **B** present in the solvent added will leave with the extract. Let the weight ratio of **C** in the extract from stage 1 be Y_1. The solute balance for stage 1 is

$$A(X_F - X_1) = B_1(Y_1 - Y_S) \quad \text{...(7.21)}$$

$$\frac{(Y_S - Y_1)}{(X_F - X_1)} = -\frac{A}{B_1} \quad \text{...(7.22)}$$

Figure 7.36 shows the distribution curve for the system on Y versus X coordinates. Point M_1 represents the concentration of the feed and solvent to the first stage, (X_F, Y_S). Assuming the stages to be ideal, the effluent streams from stage 1 will be in equilibrium. Therefore point N_1 representing the compositions of the effluent streams (X_1, Y_1) should lie on the distribution curve as shown. Equation (7.22) states that the slope of the straight line joining points M_1 (X_F, Y_S) and N_1 (X_1, Y_1) is $-A/B_1$. Thus knowing the feed composition (X_F) and the solvent purity (Y_S) and the quantities of the feed and the solvent, the operating line M_1N_1 can be drawn which intersects the distribution curve at (X_1, Y_1).

For any stage n in the cascade, similar analysis gives

$$\frac{(Y_S - Y_n)}{(X_{n-1} - X_n)} = -\frac{A}{B_n} \quad \text{...(7.23)}$$

For second stage (n = 2), the feed is the raffinate leaving stage 1 with composition (X_1) and solvent contains B_2 kg **B** with concentration of solute Y_S. The effluent streams have compositions X_2 and Y_2, so that Eq. (7.23) gives

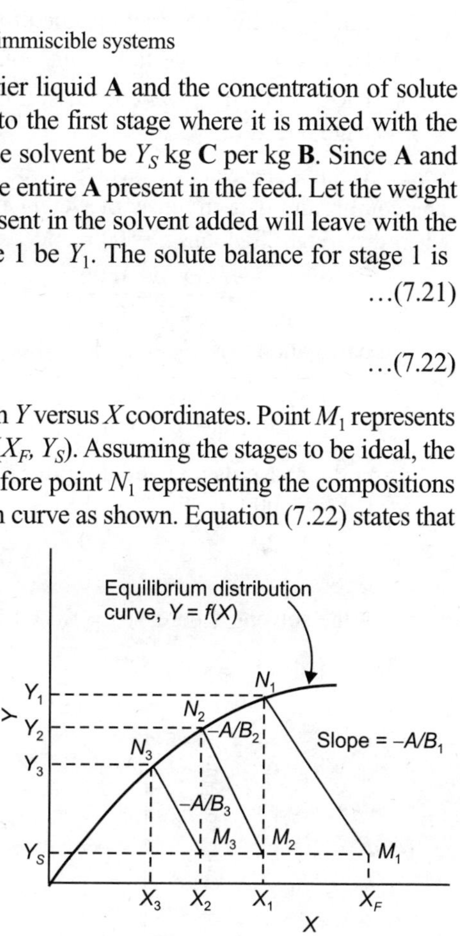

Fig. 7.36: Graphical construction for crosscurrent extraction

$$\frac{(Y_S - Y_2)}{(X_1 - X_2)} = -\frac{A}{B_2} \quad \text{...(7.24)}$$

Point M_2 has coordinates (X_1, Y_S) and N_2 has coordinates (X_2, Y_2) and the operating line has slope $-A/B_2$ as shown in Fig. 7.36. It can be seen that the operating lines M_1N_1 and M_2N_2 will be parallel if equal quantities of solvent are used in each stage (i.e., $B_1 = B_2$).

The construction is repeated for the third stage. For a three-stage unit, coordinates of the point N_3 gives the composition of the final raffinate and extract.

Example 7.4: An aqueous solution containing 20 (weight)% of a valuable solute is subjected to liquid–liquid extraction in a three-stage cross flow unit using a pure solvent. The final raffinate should contain not more than 10% of the solute present in the original feed. The equilibrium data under operating conditions are

Weight fraction of solute in aqueous layer	0.06	0.12	0.18	0.24
Weight fraction of solute in extract layer	0.24	0.40	0.56	0.73

Selectivity is infinity. If equal quantities of solvent are used in all the stages, determine:

(a) Total quantity of solvent used per 1000 kg feed

(b) Total quantity of extract obtained for conditions in part (a)

(c) The average composition of the combined extract

(d) The quantity of solvent required if the same recovery of the solute is to be obtained by a single-stage operation.

***Solution*:** Since the selectivity is given to be infinity, the solvent **B** and the carrier liquid **A** are immiscible. The equilibrium data in terms of weight fractions are converted into weight ratios. For the first set, when the weight fractions $x = 0.06$ and $y = 0.24$, the weight ratios are

$$X = \frac{x}{1-x} = \frac{0.06}{0.94} = 0.0638 \text{ and } Y = \frac{y}{1-y} = \frac{0.24}{0.76} = 0.3158$$

The calculations are repeated for other sets and the results are:

X, kg solute/kg carrier	0.0638	0.1364	0.2195	0.3158
Y, kg solute/kg solvent	0.3158	0.6667	1.2727	2.7037

Using the above data the equilibrium distribution curve is plotted as shown in Fig. 7.37a. Feed, $F = 1000$ kg. Since the feed is 20% solute,

$$X_F = \frac{x_F}{1-x_F} = \frac{0.20}{0.80} = 0.25$$

Since the solvent is pure, $y_S = Y_S = 0$.

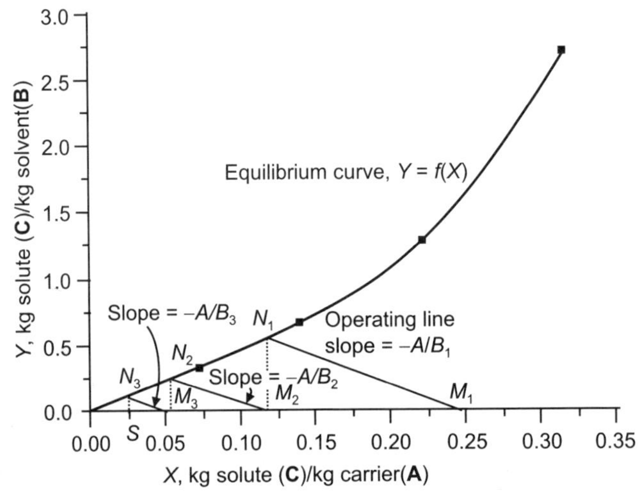

Fig. 7.37: (a) Equilibrium distribution curve and the crosscurrent operating lines for Example 7.4

Since the raffinate leaving the third stage contains only 10% of the solute in the feed,

$$X_3 = 0.10 \times 0.25 = 0.025.$$

The quantity of carrier **A** present in the feed $= A = F(1 - x_F) = 1000 \times (1 - 0.20) = 800$ kg.

The operating line for the first stage is M_1N_1. The coordinates of M_1 are $(X_F, Y_S) = (0.25, 0)$. The slope of the line is $-A/B_1$. Since the quantity of solvent used is not known, the operating line is drawn with an arbitrarily chosen slope. The line intersects the equilibrium curve at N_1, whose coordinates are the effluent compositions in stage 1, (X_1, Y_1). The line M_2N_2 is the operating line for stage 2. The point M_2 has coordinates (X_1, Ys) and M_2 is located on the x-axis as shown. Since the solvent used in all stages are equal, $B_1 = B_2 = B_3$ and therefore, the operating line M_2N_2 has the same slope as M_1N_1. A line is drawn parallel to M_1N_1 through M_2 which intersects the equilibrium curve at N_2. The coordinates of N_2 are (X_2, Y_2). Now M_3 is located at (X_2, Y_S) and the operating line for the third stage M_3N_3 is drawn as explained for second stage. It intersects the equilibrium curve at (X_3, Y_3). The solvent quantity used and hence the slope of the operating line are the desired quantities, if the final raffinate composition X_3 is equal to 0.025. Otherwise chose a different value for the solvent quantity and repeat the procedure, till the final effluent composition $X_3 = 0.025$. By a few such trials it can be seen that

$$-\frac{A}{B_1} = -\frac{A}{B_2} = -\frac{A}{B_3} = -4.24$$

$$B_1 = B_2 = B_3 = \frac{A}{4.24} = \frac{800}{4.24} = 188.68 \text{ kg}$$

(a) Therefore total quantity of the solvent required, $S = B = 188.68 \times 3 = 566.04$ kg **(Answer)**

(b) Total quantity of extract $= E = E_1 + E_2 + E_3 = B_1(1 + Y_1) + B_2(1 + Y_2) + B_3(1 + Y_3)$ where E_1, E_2, E_3 are the extracts leaving stages 1, 2 and 3 respectively and Y_1, Y_2, Y_3 are the respective solute weight ratios. Since $B_1 = B_2 = B_3$, the above equation simplifies as

$$E = B(3 + Y_1 + Y_2 + Y_3)$$

From Fig. 7.37a, $Y_1 = 0.568$, $Y_2 = 0.263$, $Y_3 = 0.123$

Therefore, total extract obtained, $E = 188.68(3 + 0.568 + 0.263 + 0.123) = 746.04$ kg **(Answer)**

[*Alternatively, the total extract can be computed by subtracting the weight of raffinate from the sum of the weights of feed and solvent supplied. That is,*

$$E = F + S - R = F + B - A(1 + X_3) = 1000 + 566.04 - 800(1 + 0.025) = 746.04 \text{ kg.}]$$

(c) The average composition of the combined extract can be calculated as

$$y = \frac{\text{weight of solute in the extract}}{\text{total weight of extract}} = \frac{B(Y_1 + Y_2 + Y_3)}{E}$$

$$= \frac{188.68(0.568 + 0.263 + 0.123)}{746.04} = 0.2413$$

The combined extract contains 24.13% solute **(Answer)**

(d) If the same recovery is achieved by a single-stage process, the operating line is drawn from M_1 to terminate at N_3 in Fig. 7.37a. In Fig. 7.37b the line MN represents the operating line of the single stage process. The input streams have compositions $X_F = 0.25$, $Y_S = 0$ and the exit streams have compositions $X = 0.025$, $Y = 0.123$. Thus the slope of the operating line is

$$-\frac{A}{B} = \frac{0.123 - 0}{0.025 - 0.25} = -0.5467$$

Since $A = 800$ kg,

$$B = \frac{800}{0.5467} = 1463.3 \text{ kg}$$ **(Answer)**

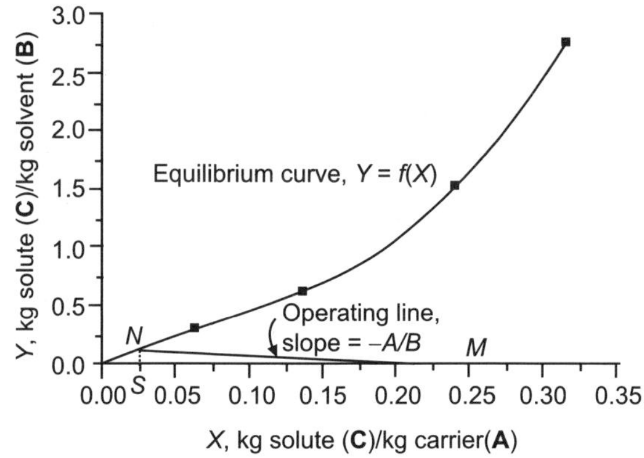

Fig. 7.37: (b) The operating line for the single stage operation in Example 7.4

Thus for the same percent recovery of solute, the solvent required in the single stage operation is more than two-and-half times that required for a three stage crosscurrent unit.

Multistage Countercurrent Extraction

A countercurrent multiple stage cascade consists of a number of extraction stages connected in series as shown in Fig. 7.38. The feed is admitted at one end of the cascade and the solvent at the other end and they flow in opposite directions as shown. The raffinate, leanest in the solute content, is in contact with the fresh solvent in the last stage and the fresh feed containing mostly easily separable solute, is in the first stage in contact with the extract containing substantial concentration of the solute. For the same extent of extraction as in a crosscurrent unit, a continuous countercurrent unit using the same quantity of solvent requires less number of stages; with the same number of stages as in the crosscurrent unit, the solvent required will be less in the countercurrent unit.

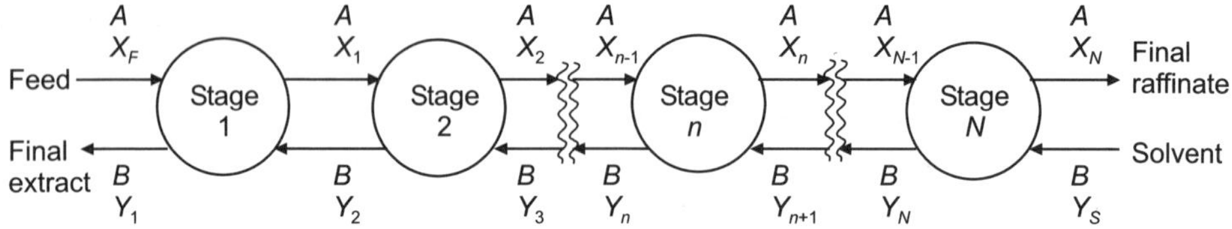

Fig. 7.38: Countercurrent extraction using immiscible solvents

An N-stage countercurrent cascade using immiscible solvents is shown in Fig. 7.38. The feed admitted to the first stage contains A kg component-**A** and the concentration of **C** in the feed is X_F kg **C** per kg **A**. All the raffinate streams contain the same quantity of **A** as in the feed; the solute content in the raffinate changes from X_F in the feed to X_N in the final raffinate. Amounts of **B** in the extract streams are constant and equal to B kg and the solute content in the solvent admitted to the last stage is Y_S kg **C** per kg **B**. The **C** concentration of solute in the extracts progressively increases and finally the extract product leaves stage 1 with concentration of solute Y_1.

A solute balance for the entire cascade gives

$$A(X_F - X_N) = B(Y_1 - Y_S) \qquad \text{...(7.25)}$$

which can be rearranged as

$$\frac{(Y_1 - Y_S)}{(X_F - X_N)} = \frac{A}{B} \qquad \text{...(7.26)}$$

Figure 7.39 shows the distribution curve plotted on Y versus X coordinates. Point N in the figure has coordinates (X_F, Y_1) and point M has coordinates (X_N, Y_S). According to Eq. (7.26), the slope of the straight line joining M and N is equal to the ratio of raffinate to extract flow rates on solute free basis. The line MN is the *operating line* for the countercurrent extraction process.

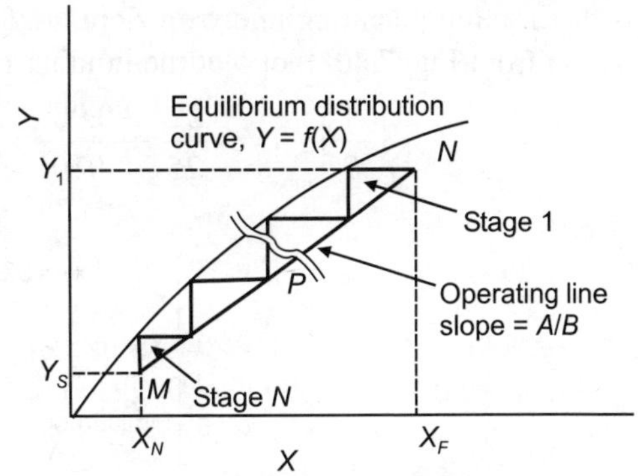

Fig. 7.39: Graphical construction for theoretical stages in countercurrent extraction cascade

Now considering a solute balance from stage 1 through n, we get

$$A(X_F - X_n) = B(Y_1 - Y_{n+1}) \quad \ldots(7.27)$$

Or
$$\frac{(Y_1 - Y_{n+1})}{(X_F - X_n)} = \frac{A}{B} \quad \ldots(7.28)$$

This is the equation for a straight line passing through $N(X_F, Y_1)$ and $P(X_n, Y_{n+1})$ with slope A/B. This line has the same slope as the operating line MN and has a point in common. Therefore, P should lie on the operating line MN. The point P represents the compositions of the passing streams between any two adjacent stages; hence all the passing streams between two adjacent stages can be represented by the points lying on the operating line MN.

Having drawn the equilibrium distribution curve and the operating line, the number of stages can be determined by stepping off stages starting with point N as shown in Fig. 7.39.

Example 7.5: A solution containing 20 weight% acetone in water is to be extracted at a rate 1000 kg/h with monochlorobenzne (MCB) containing 0.5% acetone by countercurrent extraction process. Monochlorobenzene and water are assumed immiscible in the operating conditions. The equilibrium data for the system is given below:

kg acetone/kg water	0.0258	0.0739	0.1605	0.2670
kg acetone/kg MCB	0.0288	0.0704	0.1560	0.2370

If it is desired that the final raffinate contains 1% acetone, calculate:

(a) Minimum solvent rate

(b) Number of stages required if the solvent is used at a rate 1.25 times the minimum.

***Solution*:** The equilibrium curve is plotted using the given data directly.

X	0.0258	0.0739	0.1605	0.2670
Y	0.0288	0.0704	0.1560	0.2370

The feed rate is 1000 kg/h and its concentration is 20 weight% acetone.

$F = 1000$, $x_F = 0.20$

Therefore, $A = F(1 - x_F) = 1000(1 - 0.20) = 800\,\text{kg/h}$ and $X_F = \dfrac{x_F}{1 - x_F} = \dfrac{0.20}{0.80} = 0.25$

Since the solvent contains 0.5% acetone, $y_S = 0.005$, $Y_S = \dfrac{y_S}{1 - y_S} = \dfrac{0.005}{0.995} = 0.0050$

The final raffinate contains 1% acetone. Therefore, $x_N = 0.01$, $X_N = \dfrac{0.01}{0.99} = 0.0101$

(a) The operating line is drawn from point M whose coordinates are $(X_N, Y_S) = (0.0101, 0.005)$. The operating line terminates on the vertical line at $X_F = 0.25$. For minimum solvent rate, the operating line has the maximum slope. The dashed line drawn from M to intersect the equilibrium curve at point P

is the limiting operating line with slope $(A/B)_{max}$. The coordinates of point P are found to be (0.25, 0.224) from Fig. 7.40. Slope of the limiting operating line

$$\left(\frac{A}{B}\right)_{max} = \frac{0.224 - 0.005}{0.25 - 0.0101} = 0.9129$$

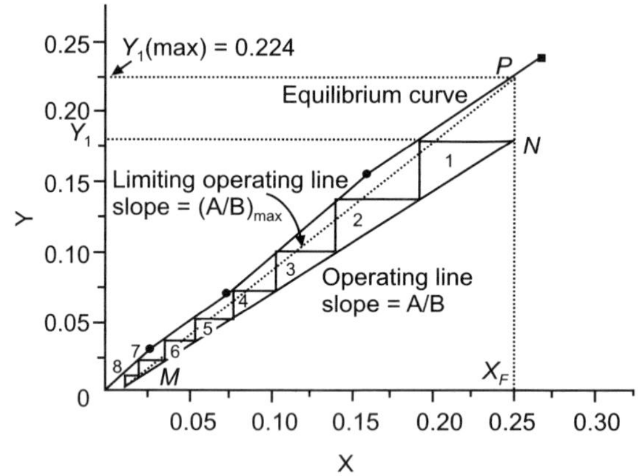

Fig. 7.40: Determination of number of stages for Example 7.5

The maximum slope corresponds to the minimum solvent rate. Therefore, the minimum solvent rate

$$B_{min} = \frac{A}{(A/B)_{max}} = \frac{800}{0.9129} = 876.3\,\text{kg}$$

Minimum solvent rate $S_{min} = B_{min}(1 + Y_S) = 876.3(1.005) = 880.7$ kg **(Answer)**

(b) For a solvent rate 1.25 times the minimum, slope of the operating line is

$$\frac{A}{B} = \frac{(A/B)_{max}}{1.25} = \frac{0.9129}{1.25} = 0.7303$$

The operating line *MN* is drawn with this slope as shown in the figure. *N* should lie on the vertical line passing through $X = X_F$.

Stages are stepped off between the operating line and the equilibrium curve as shown, starting with the upper terminus *N* of the operating line. The number of stages = 8 from Fig. 7.40. **(Answer)**

Analytical Method for Multistage Countercurrent Extraction

We have seen that the operating line for countercurrent operations plotted on weight ratio coordinates is a straight line with slope *A/B*. If the distribution coefficient *m* defined as

$$m = \frac{Y}{X} \quad \text{...(7.29)}$$

is constant for the system, the equilibrium curve also will be a straight line as shown in Fig. 7.41. For such cases, Kremser equations similar to Eqs. (4.81)-(4.83) derived for gas absorption can be developed for countercurrent extraction as well. The number of theoretical stages is given by

$$N = \frac{\log\left[\left(\frac{X_F - Y_S/m}{X_N - Y_S/m}\right)\left(1 - \frac{1}{\varepsilon}\right) + \frac{1}{\varepsilon}\right]}{\log(\varepsilon)} \quad \text{...(7.30)}$$

where $\varepsilon = \frac{mB}{A}$ is the extraction factor.

If ε = 1,

$$N = \frac{X_F - X_N}{X_N - Y_S/m} \quad \ldots(7.31)$$

Equation (7.30) can be written in the following analogous form which can be used for finding the composition of the unknown streams.

$$\frac{(X_F - X_N)}{(X_F - Y_S/m)} = \frac{\varepsilon^{N+1} - \varepsilon}{\varepsilon^{N+1} - 1} \quad \ldots(7.32)$$

The distribution coefficient is found to be constant for very dilute systems. For such systems, Kremser equations can very well be written in terms of the weight fractions and total weights replacing the quantities expressed on solute-free basis in Eqs. (7.30).

$$N = \frac{\log\left[\left(\frac{x_F - y_S/m}{x_N - y_S/m}\right)\left(1 - \frac{1}{\varepsilon}\right) + \frac{1}{\varepsilon}\right]}{\log(\varepsilon)} \quad \ldots(7.33)$$

with $\varepsilon = \dfrac{mE}{R}$.

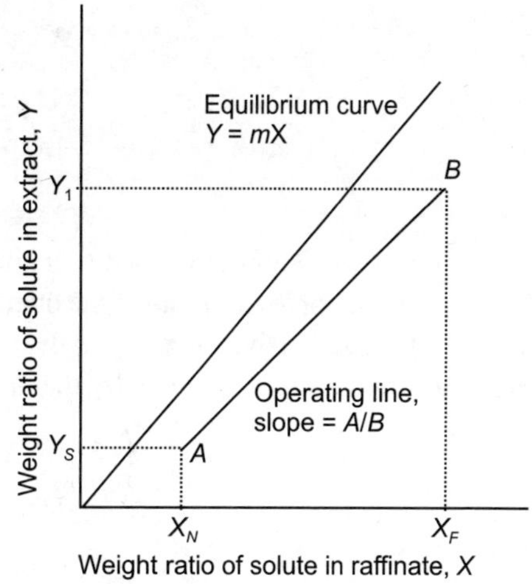

Fig. 7.41: Condition for using Kremser equation for estimating the number of stages in countercurrent extraction

Example 7.6: An aqueous solution containing 1.0 weight% nicotine in water is extracted using pure kerosene stream in a countercurrent multistage unit. The water and kerosene are essentially immiscible in each other. It is desired to reduce the nicotine content in the exit water to 0.10 weight%. The equilibrium data are as follows.

Weight fraction nicotine in water	0.001010	0.00246	0.005	0.00746	0.00988	0.0202
Weight fraction, nicotine in kerosene	0.000806	0.001959	0.00454	0.00682	0.00904	0.0185

Determine the number of stages required (a) graphically and (b) by Kremser equation, if a solvent rate 1.25 times the minimum is used.

***Solution*:** The equilibrium data are converted into weight ratios:

X, kg nicotine/kg water	0.001011	0.002466	0.005025	0.007516	0.009979	0.020616
Y, kg nicotine/kg kerosene	0.000807	0.001963	0.004561	0.006867	0.009122	0.018849

(a) The equilibrium curve is plotted as shown in Fig. 7.42

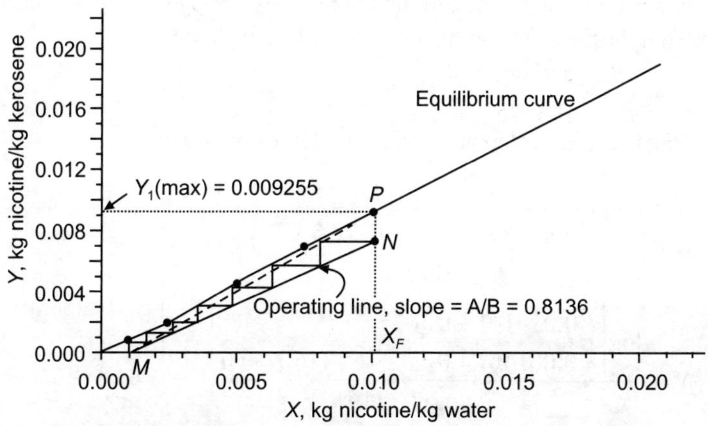

Fig. 7.42: Determination of number of stages for Example 7.6 graphically

$$X_F = \frac{0.01}{1-0.01} = 0.0101$$

$$X_N = \frac{0.001}{1-0.001} = 0.001$$

$$Y_S = 0$$

The lower terminus of the operating line has coordinates (X_N, Y_S). Thus M is marked with coordinates (0.001, 0). The upper terminus of the operating line lies on the vertical line at $X = X_F = 0.0101$, the exact location depends on the amount of solvent used. For the minimum solvent, the operating line intersects the equilibrium curve as shown by the dashed line MP. The slope of the limiting operating line MP is

$$\left(\frac{A}{B}\right)_{\max} = \frac{0.00926-0}{0.0101-0.001} = 1.017$$

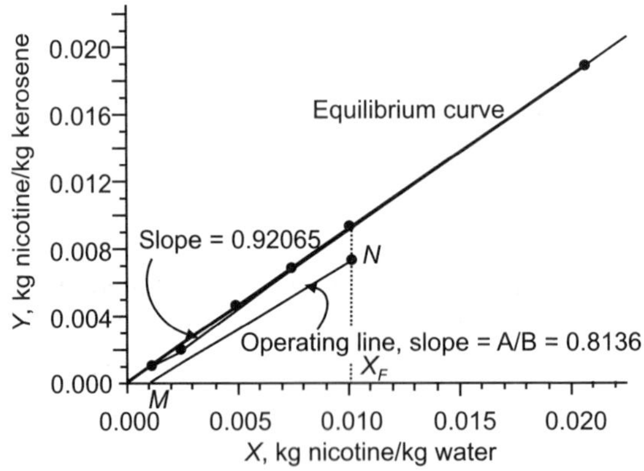

Fig. 7.43: Equilibrium curve and the operating line represented as straight lines for example 7.6

Since the solvent used is 1.25 times the minimum,

$$\frac{A}{B} = \frac{1.017}{1.25} = 0.8136$$

The operating line MN is drawn with this slope and the stages stepped off as described earlier. The number of stages is found to be 6.7. **(Answer)**

(b) The equilibrium curve can be approximated as a straight line with slope $m = 0.92065$ (Fig. 7.43). The Kremser equation can be used to determine the number of stages if the equilibrium curve and the operating lines are straight lines. The extraction factor is calculated as

$$\varepsilon = \frac{mB}{A} = \frac{0.92065}{0.8136} = 1.1316$$

The number of stages can be calculated using the Kremser equation:

$$N = \frac{\log\left[\left(\frac{X_F - Y_S/m}{X_N - Y_S/m}\right)\left(1-\frac{1}{\varepsilon}\right)+\frac{1}{\varepsilon}\right]}{\log(\varepsilon)}$$

$$N = \frac{\log\left[\left(\frac{0.010101-0}{0.001001-0}\right)\left(1-\frac{1}{1.1316}\right)+\frac{1}{1.1316}\right]}{\log(1.1316)} = 5.8$$

The number of stages by Kremser equation = 5.8. **(Answer)**

The difference between the number of stages calculated by the Kremser equation and the graphical method can be attributed to the appreciable deviation of the actual equilibrium from the linear behaviour.

7.6 PARTIALLY MISCIBLE SYSTEMS: STAGE CONTACT

Single-Stage Extraction

Figure 7.44 represents schematically a continuous single-stage extraction process. F kg/h of a feed containing x_F weight fraction solute is treated with S kg/h of solvent. For batch operation, F and S are kg of the feed and solvent being mixed. If the solvent used is the one recovered from a previous extraction operation, it may not be pure **B**, but may contain small amounts of **A** and **C**. Let y_S be the weight fraction of **C** in the solvent. The solute **C** gets distributed between the partially miscible liquids **A** and **B** giving an **A**-rich phase called the raffinate and a **B**-rich phase called the extract. Let R be the kg/h of raffinate and E be the kg/h of extract that separate as the extraction products. Assuming the stage to be an equilibrium stage or a theoretical stage, the raffinate and extract products leaving the unit are in equilibrium and x, the weight fraction of **C** in the raffinate and y, the weight fraction of **C** in the extract are equilibrium compositions.

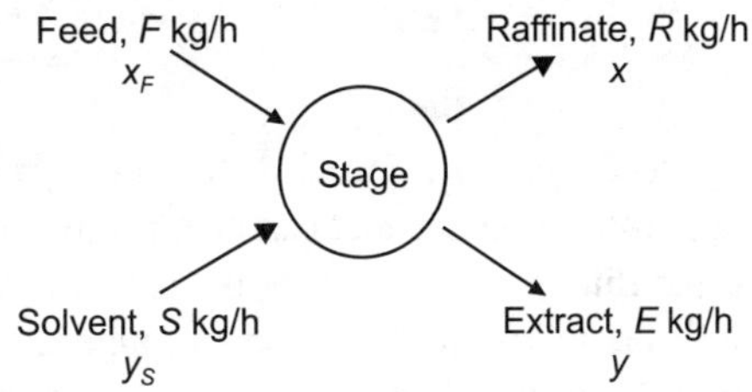

Fig. 7.44: Single-stage extraction

The operation can be represented on the triangular diagram as shown in Fig. 7.45a. Feed is a binary mixture of **A** and **C** and is represented by point F on the side AC of the triangle. Solvent is represented by point S. The point S will coincide with the vertex B of the triangle, if the solvent used were free of solute.

On mixing F kg feed with S kg solvent, M kg of mixture will result, represented by the point M in the figure. If M falls in the two-phase region, the mixture cannot exist as a homogeneous phase; when allowed to settle it will separate into the extract and the raffinate phases, the amounts and compositions of which can be readily evaluated graphically. The analysis presented here is applicable to the phase diagram plotted on equilateral triangle as well as on the right triangle.

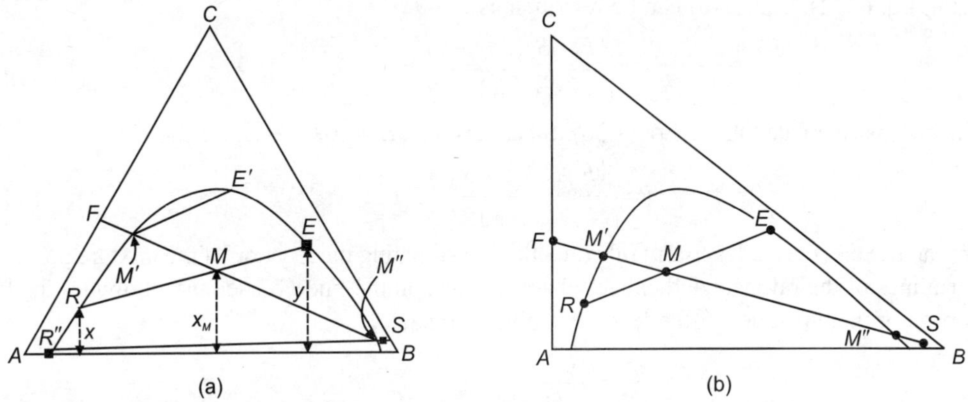

Fig. 7.45: (a) Single-stage extraction process on equilateral triangular coordinate system. (b) Single stage extraction process on right triangular coordinate system

The overall material balance gives

$$F + S = M = R + E \qquad \ldots(7.34)$$

Component-**C** balance gives

$$Fx_F + Sy_S = Mx_M \qquad \ldots(7.35)$$

where x_M is the weight fraction of solute in the mixture M. Combining with Eq. (7.34), Eq. (7.35) can be rearranged as

$$x_M = \frac{Fx_F + Sy_S}{F + S} \quad \text{...(7.36)}$$

By mixture rule, M should lie on the line joining F and S. Using the x_M value calculated using Eq. (7.36), the point representing the combined mixture M can be located on the line FS as shown in Fig. 7.45. Alternatively, if the location of point M is fixed by specifying the concentration of the raffinate or extract product, the amount of solvent required to treat a given quantity of feed can be determined using Eq. (7.36), which can be rearranged as

$$S = \frac{F(x_F - x_M)}{(x_M - y_S)} \quad \text{...(7.37)}$$

According to Eq. (7.34), the point M should lie on the line joining R and E as well. R and E are equilibrium phases that can be represented by the points lying on the binodal solubility curve; R lies on the raffinate side and E on the extract side of the solubility curve as shown in Fig. 7.45. Thus the line RE is a tie line which passes through the point representing the combined mixture M. Identify the tie line RE by trial, with the help of a distribution diagram or a conjugation curve for tie line correlation described elsewhere. The vertical distance from R and E to the base AB of the triangle are the weight fraction of solute in the raffinate (x) and extract (y), respectively. After reading x and y from the graph, the solute balance can be used to estimate the amounts of R and E produced.

$$Mx_M = Rx + Ey \quad \text{...(7.38)}$$

Noting that $R = M - E$, Eq. (7.38) becomes

$$M(x_M - x) = E(y - x) \quad \text{...(7.39)}$$

which may be rearranged as

$$E = \frac{M(x_M - x)}{(y - x)} \quad \text{...(7.40)}$$

Using Eq. (7.34), Eq. (7.40) can be rewritten as

$$E = \frac{(F + S)(x_M - x)}{(y - x)} \quad \text{...(7.41)}$$

in which x_M is calculated using Eq. (7.36). Since $R = M - E = F + S - E$, we can also write

$$R = \frac{(F + S)(y - x_M)}{(y - x)} \quad \text{...(7.42)}$$

We can use the inverse-lever-arm rule directly for estimating the location of the mixture point M and the quantities of the raffinate and extract phases. M lies on the line FS such that it divides the line in proportion of the quantities of the feed and solvent contacted.

$$\frac{F}{S} = \frac{\overline{SM}}{\overline{FM}} \quad \text{...(7.43)}$$

Draw the tie line RE passing through the point M by trial as described above. The ratio of the amount of the raffinate to the amount of the extract is given as

$$\frac{R}{E} = \frac{\overline{EM}}{\overline{RM}} \quad \text{...(7.44)}$$

Equation (7.44) in combination with the material balance Eq. (7.34) can be used to determine the amounts of the raffinate and the extract products R and E.

Limitations on the amounts of solvent

Separation by extraction is not possible if the mixture point M lies outside the dome shaped area in the figure that represents the two-phase region. The addition of solvent S to the feed F can be followed on the line joining F and S. The minimum quantity of the solvent required for single-stage extraction is the quantity when the mixture point falls at the intersection of the line FS with the raffinate side of the solubility curve as shown by M' in Fig. 7.45. The minimum solvent quantity S_{min}, is given by Eq. (7.37) as

$$S_{\min} = \frac{F(x_F - x_{M'})}{(x_{M'} - y_S)} \qquad \text{...(7.45)}$$

or by Eq. (7.43) as

$$\frac{F}{S_{\min}} = \frac{\overline{SM'}}{\overline{FM'}} \qquad \text{...(7.46)}$$

The richest extract E' but infinitesimal in amount is obtained corresponding to the minimum solvent. If the solvent quantity is less than that given by the above equations, M will lie in the homogeneous area and separation by extraction is not possible. In a similar manner, there is a maximum for the solvent quantity that can effect separation in a single stage operation and it corresponds to point M'' lying on the extract side of the solubility curve. Solvent in excess of that corresponding to M'' will make the mixture homogeneous and no extraction will be possible. An infinitesimal amount of raffinate R'' would result by this process and that is the raffinate with the lowest concentration possible in a single stage operation using the solvent of the given purity.

Example 7.7: Five hundred kg of a solution containing 35 weight% acetic acid in water is subjected to single-stage extraction with an equal weight of pure isopropyl ether.

Equilibrium data are given in the table below:

Water rich phase, weight%			Ether rich phase, weight%		
Acetic acid	Water	Isopropyl ether	Acetic acid	Water	Isopropyl ether
1.41	97.1	1.49	0.37	0.73	98.9
2.89	95.5	1.61	0.79	0.81	98.4
6.42	91.7	1.88	1.93	0.97	97.1
13.3	84.4	2.3	4.82	1.88	93.3
25.5	71.1	3.4	11.4	3.9	84.7
36.7	58.9	4.4	21.6	6.9	71.5
45.3	45.1	9.6	31.1	10.8	58.1
46.4	37.1	16.5	36.2	15.1	48.7

Determine:

(a) The composition of the raffinate and extract products

(b) The amounts of raffinate and extract products

(c) The percent recovery of acetic acid

(d) The minimum quantity of solvent required for extraction.

***Solution*:** For graphical solution, the equilibrium curve can be plotted either on the equilateral triangular coordinates or on the right triangular coordinates. In Fig. 7.46, the equilibrium data are plotted on right triangular coordinates, as these are more convenient for graphical construction of stages.

Feed, $F = 500$ kg, $x_F = 0.35$

Solvent, $S = 500$ kg, $y_S = 0$

The point F represents the feed and the point S represents the solvent. Since the solvent used is pure, S coincides with the vertex B of the triangle. The mixture M that results when 500 kg feed is mixed with 500 kg solvent lies on the line FS. The solute concentration in this mixture is evaluated by Eq. (7.36)

$$x_M = \frac{Fx_F + Sy_S}{F+S} = \frac{500\times 0.35 + 500\times 0}{(500+500)} = 0.175$$

Using this coordinate, M is located on the line FS as shown in Fig. 7.46. (Alternatively, M can be located knowing that $\frac{\overline{FM}}{\overline{MS}} = \frac{S}{F} = \frac{500}{500} = 1$ or $\overline{FM} = \overline{MS}$)

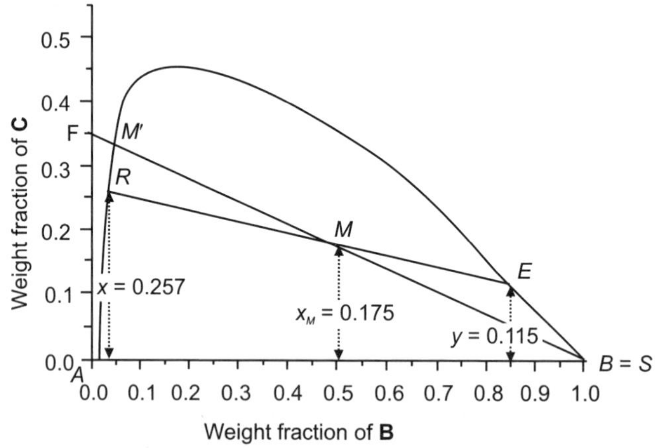

Fig. 7.46: Single stage extraction involving acetic acid-water-isopropyl ether system (Example 7.7)

The mixture M separates into raffinate (R) and extract (E) phases, which can be connected by a tie line passing through M. This tie line is obtained by trial with the help of the distribution diagram or conjugation curve for tie line interpolation. (Not shown in the figure). The tie line intersects the raffinate side of the solubility curve at R and the extract side at E.

(a) The coordinates of R and E are read from the figure. The x-coordinate is the weight fraction of solvent (isopropyl ether) and y-coordinate is the weight fraction of solute (acetic acid). The concentrations are:

Raffinate product:

Acetic acid = 0.257, isopropyl ether = 0.034 and water = (1– 0.257 – 0.034) = 0.709

Extract product:

Acetic acid = 0.115, isopropyl ether = 0.843 and water = (1– 0.115 – 0.843) = 0.042

Constituent	Raffinate, weight%	Extract, weight%
Acetic acid	25.7	11.5
Isopropylether	3.4	84.3
Water	70.9	4.2

(Answer)

(b) The quantities of the raffinate phase are calculated using Eq. (7.42).

$$R = \frac{(F+S)(y-x_M)}{(y-x)}$$

We have $x = 0.257$, $y = 0.115$, $x_M = 0.175$. Substituting the values, we get

$$R = \frac{(500+500)(0.115-0.175)}{(0.115-0.257)} = 422.5 \text{ kg}$$ **(Answer)**

The amount of extract can be calculated by the total material balance:

$$E = F + S - R = 1000 - 422.5 = 577.5 \text{ kg}$$ **(Answer)**

(c) The percent recovery of acetic acid is calculated as:

$$\text{percent recovery} = \frac{\text{solute extracted}}{\text{solute in the feed}} \times 100 = \frac{Ey}{Fx_F} \times 100$$

$$= \frac{577.5 \times 0.115}{500 \times 0.35} \times 100 = 37.95\% \qquad \textbf{(Answer)}$$

(d) The minimum quantity of solvent corresponds to the mixture point lying at M', the intersection of the line FS with the raffinate side of the solubility curve. The weight fraction of solute in this mixture is read from the graph to be $x_{M'} = 0.335$. The minimum quantity of solvent is calculated using Eq. (7.45)

$$S_{\min} = \frac{F(x_F - x_{M'})}{(x_{M'} - y_S)} = \frac{500(0.35 - 0.335)}{(0.335 - 0)} = 22.39 \text{ kg} \qquad \textbf{(Answer)}$$

Crosscurrent Multistage Operation

In the multistage crosscurrent extraction, the feed is contacted with the fresh solvent in the first stage and in all other stages, the raffinate from the immediately preceding stage is contacted with new batches of fresh solvent as shown in Fig. 7.47a. The solute recovery in a multistage crosscurrent unit is substantially greater than that achieved with the same total quantity of solvent in a single stage operation. Fig. 7.47a shows a three stage crosscurrent extraction unit and Fig. 7.47b shows the n th stage of a multistage crosscurrent unit. As is clear from the figure, the final raffinate product is obtained from the last stage and the extract leaving each stage can be combined to give an extract product of average composition y which can be estimated as

$$y = \frac{E_1 y_1 + E_2 y_2 + E_3 y_3 + \ldots.}{E_1 + E_2 + E_3 + \ldots\ldots} \qquad \ldots(7.47)$$

where E_i is the extract leaving the ith stage and y_i is the weight fraction of solute in it.

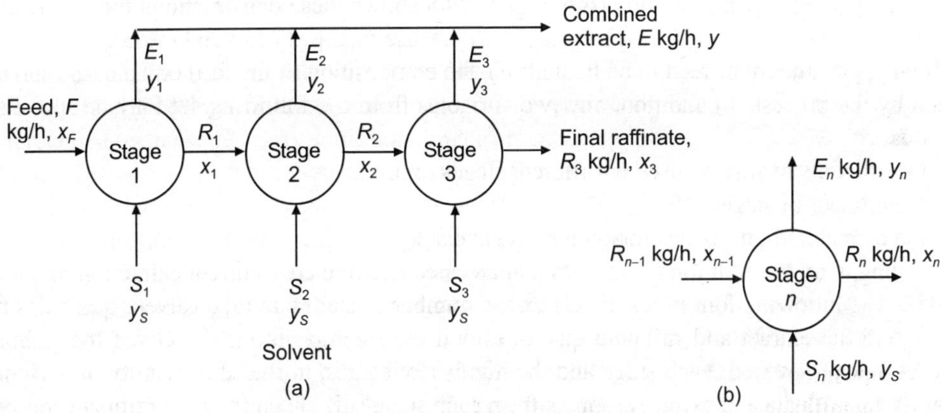

Fig. 7.47: (a) Three stage crosscurrent extraction (b) n th stage in a multistage crosscurrent unit

The quantity and composition of the extract and raffinate phases leaving each stage can be estimated graphically by the method described for single stage extraction. Thus, for the nth stage, we can write the total and solute balances as follows:

$$R_{n-1} + S_n = M_n = R_n + E_n \qquad \ldots(7.48)$$

$$R_{n-1}x_{n-1} + S_n y_S = M_n x_{M,n} = R_n x_n + E_n y_n \qquad \ldots(7.49)$$

Equations (7.48) and (7.49) are analogous to Eq. (7.34) and (7.35) written for the single stage operation. On rearranging these equations we get

$$x_{M,n} = \frac{R_{n-1}x_{n-1} + S_n y_S}{R_{n-1} + S_n} \qquad \text{...(7.50)}$$

where $x_{M,n}$ is the weight fraction of solute in the mixture M_n that results when R_{n-1} kg feed (raffinate from n-1st stage) and S_n kg solvent are mixed together in the nth stage. Knowing the compositions and quantities of the input streams, the composition of the combined mixture in the nth stage can be computed using Eq. (7.50) and the point M_n can be located on the line $R_{n-1}S$. Alternatively, M_n can be located graphically by the inverse-lever-arm principle:

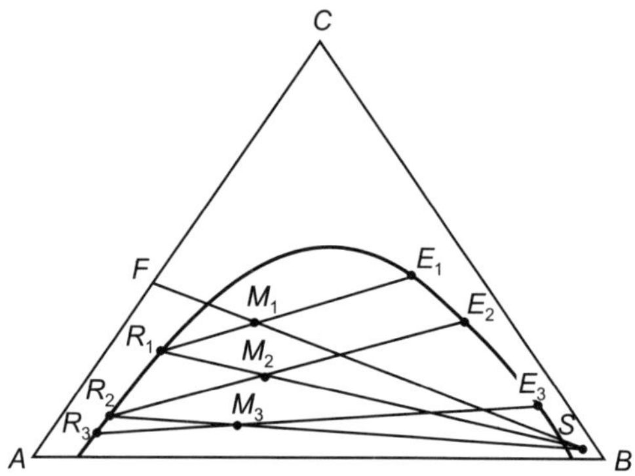

Fig. 7.48: Construction of stages for three-stage crosscurrent extraction operation

$$\frac{R_{n-1}}{S_n} = \frac{\overline{S_n M_n}}{\overline{R_{n-1} M_n}} \qquad \text{...(7.51)}$$

Corresponding to Eqs. (7.41) and (7.42), we can derive from Eqs.(7.48) and (7.49), the following:

$$E_n = \frac{(R_{n-1} + S_n)(x_{M,n} - x_n)}{(y_n - x_n)} \qquad \text{...(7.52)}$$

$$R_n = \frac{(R_{n-1} + S_n)(y_n - x_{M,n})}{(y_n - x_n)} \qquad \text{...(7.53)}$$

Construct the tie line passing through M_n and obtain E_n and R_n on the binodal curve at the opposite ends of the tie line. Read the solute concentration in the raffinate (x_n) and the extract (y_n) phases from the diagram. These values are substituted in Eqs. (7.52) and (7.53) to estimate the quantities of extract and raffinate phases leaving the nth stage. Figure 7.48 shows these constructions for n = 1, 2 and 3. Remember that when n = 1, $R_{n-1} = F$ and $x_{n-1} = x_F$ for use in Eqs. (7.52) and (7.53).

Normally, the amount of feed to be treated (F), the composition of the feed (x_F) and the solvent (y_S) are fixed by the process. In addition, any two variables from the following set may be specified in a given situation:

- the amounts of solvent used in different stages (S_n)
- the number of stages (N)
- the extract and raffinate compositions (y_n and x_n).

Depending upon the additional variables that are specified, the crosscurrent extraction problems fall into one of the following four types: (i) Given the number of stages and the solvent quantities in each stage, to find the extract and raffinate quantities and their compositions (ii) Given the raffinate (or extract) compositions from each stage and the number of stages, to find the quantity of solvent used, amount of the raffinate and extract products from each stage (iii) Given the final raffinate composition and the solvent used in each stage, to find the number of stages and the quantity of the raffinate and quantity and composition of the extract product. (iv) Given the number of stages, the solvent quantity in each stage, and the final raffinate composition, to find the quantity of the raffinate and quantity and composition of the extract product. The last two types of problems require trial-and-error solutions.

Example 7.8: One thousand kg of mixture containing 40 weight% acetone in water is extracted using methyl isobutyl ketone (MIK) as solvent. (a) What fraction of the acetone can be extracted in a single-stage process using 1000 kg pure MIK? (b) What fraction of the acetone could be extracted in a two-stage crosscurrent unit using 500 kg solvent in each stage? The miscibility data and the tie line data for the acetone-water-MIK system are given below:

Miscibility data, weight fraction				Tie line data, weight fraction of acetone	
MIK	Acetone	Water		Raffinate	Extract
0.98	0	0.02		0	0
0.932	0.046	0.022		0.025	0.045
0.773	0.1895	0.0375		0.055	0.1
0.71	0.244	0.046		0.075	0.135
0.655	0.289	0.056		0.1	0.175
0.547	0.376	0.077		0.125	0.213
0.462	0.432	0.106		0.15	0.255
0.124	0.427	0.449		0.175	0.282
0.0501	0.309	0.6409		0.2	0.312
0.0323	0.209	0.7587		0.225	0.34
0.0212	0.0373	0.9415		0.25	0.365
0.022	0	0.978		0.26	0.375

***Solution*:** The miscibility data is plotted on the right-triangular coordinates to get the binodal solubility curve, shown in Fig. 7.49. The tie line data can be plotted separately to get the distribution curve (not shown in the figure).

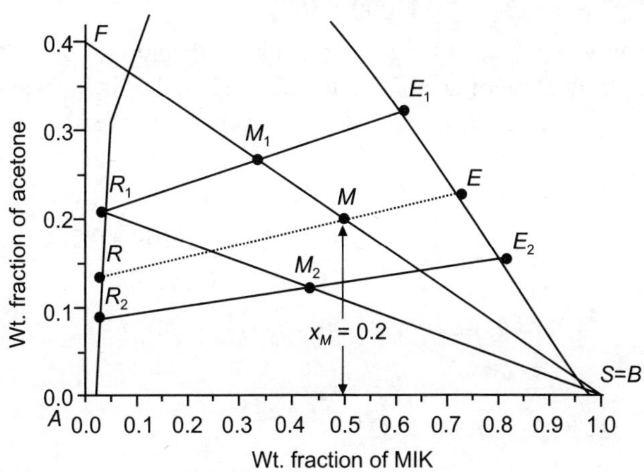

Fig. 7.49: Crosscurrent extraction involving acetone-water-MIK system (Example 7.8)

(a) Single stage extraction:

$F = 1000$ kg, $x_F = 0.40$

Solvent $= 1000$ kg, $y_S = 0$

The feed is represented by F and it lies on the vertical axis as shown. The solvent S being pure, lies at the vertex B of the triangle. When 1000 kg feed containing 40% acetone, is mixed with 1000 kg pure solvent, the mixture lies on the line FS at a height x_M, where

$$x_M = \frac{Fx_F + Sy_S}{F+S} = \frac{1000 \times 0.40 + 1000 \times 0}{(1000+1000)} = 0.20$$

Using the distribution curve, the tie line RE passing through M is determined and the coordinates of R and E are read from the figure. The weight fraction of acetone in the raffinate is found to be $x = 0.1357$ and the weight fraction of acetone in the extract is found to be $y = 0.2311$. The quantities of the raffinate and extract phases are calculated using Eq. (7.42).

$$R = \frac{(F+S)(y-x_M)}{(y-x)}$$

$$R = \frac{(1000+1000)(0.2311-0.20)}{(0.2311-0.1357)} = 652\,\text{kg}$$

The amount of extract can be calculated by the total material balance:

$$E = F + S - R = 2000 - 652 = 1348 \text{ kg}$$

$$\text{The fraction of acetone recovered} = \frac{\text{acetone in the extract}}{\text{acetone in the feed}} = \frac{Ey}{Fx_F} = \frac{1348 \times 0.2311}{1000 \times 0.40} = 0.779$$

(Answer)

(b) Crosscurrent extraction:

Stage 1

Feed, $F = 1000$ kg, $x_F = 0.40$

Solvent, $S = 500$ kg, $y_S = 0$

The mixture point M_1 is located at x_{M1} on the line FS where

$$x_{M_1} = \frac{Fx_F + Sy_S}{F+S} = \frac{1000 \times 0.40 + 500 \times 0}{(1000+500)} = 0.2667$$

The tie line passing through M_1 determines the quantity and composition of the raffinate R_1 and the extract E_1. x_1 is the weight fraction of acetone in the raffinate and y_1 is the weight fraction of acetone in the extract. The y-coordinates of R_1 and E_1 are read from the figure and they are $x_1 = 0.2082$ and $y_1 = 0.3209$

The raffinate and extract leaving the first stage are

$$R_1 = \frac{(F+S_1)(y_1 - x_{M,1})}{(y_1 - x_1)} = \frac{(1000+500)(0.3209-0.2667)}{(0.3209-0.2082)} = 721.83 \text{ kg}$$

$$E_1 = (F + S_1 - R_1) = 1000 + 500 - 721.83 = 778.17 \text{ kg}$$

Stage 2

Feed = Raffinate $R_1 = 721.83$ kg, $x_1 = 0.2082$

Solvent, $S_2 = 500$ kg, $y_S = 0$

The mixture point M_2 is located at $x_{M,2}$ on the line R_1S where

$$x_{M_2} = \frac{R_1x_1 + S_1y_S}{R_1 + S_1} = \frac{721.83 \times 0.2082 + 500 \times 0}{(721.83+500)} = 0.123$$

The tie line passing through M_2 determines the quantity and composition of the raffinate R_2 and extract E_2. x_2 is the weight fraction of acetone in the raffinate and y_2 is the weight fraction of acetone in the extract. The y-coordinates of R_2 and E_2 are read from the figure and they are $x_2 = 0.0872$ and $y_2 = 0.1558$

The raffinate and extract leaving the second stage are

$$R_2 = \frac{(R_1+S_2)(y_2 - x_{M,2})}{(y_2 - x_2)} = \frac{(721.83+500)(0.1558-0.123)}{(0.1558-0.0872)} = 584.2 \text{ kg}$$

$$E_2 = (R_1 + S_2 - R_2) = 721.83 + 500 - 584.2 = 637.63 \text{ kg}$$

$$\text{Fraction of acetone extracted} = \frac{Fx_F - R_2x_2}{Fx_F} = \frac{400 - 584.2 \times 0.0872}{400} = 0.873$$

(Answer)

Multistage Countercurrent Extraction

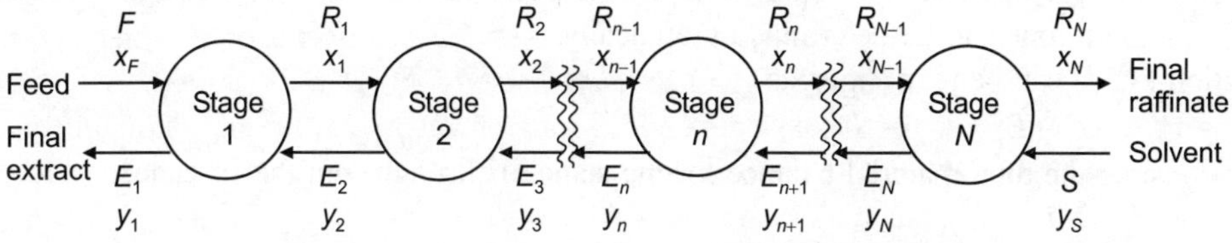

Fig. 7.50: Multistage countercurrent extraction process

F kg/h of a feed solution containing x_F weight fraction solute (component-**C**) is admitted to the first stage and S kg/h of the solvent containing y_S weight fraction solute is admitted to the Nth stage where N is the number of stages in the cascade. The final extract product leaves the first stage at a rate E_1 kg/h containing y_1 weight fraction solute. The final raffinate leaves the Nth stage with concentration of solute x_N. Any general stage in the cascade can be represented by stage n in which the streams entering are $R_{n\text{-}1}$ with concentration of solute $x_{n\text{-}1}$ and extract E_{n+1} with concentration y_{n+1} and the streams leaving are R_n and E_n with concentration x_n and y_n respectively as shown in Fig. 7.50.

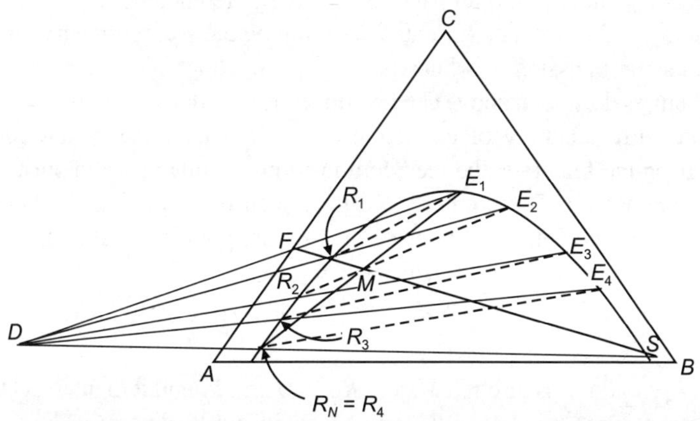

Fig. 7.51: Graphical construction for the number of stages in countercurrent extraction

Assuming the stages to be ideal, E_n and R_n are in equilibrium, which can be represented by points lying on the solubility curve connected by means of tie line.

Graphical method for the design of countercurrent multistage process can be followed on equilateral triangular phase diagram or the right triangular phase diagram (Fig. 7.51). The point F and S representing the feed and solvent can be located on the diagram knowing their compositions x_F and y_S. Considering the entire cascade, the overall material balance is

$$F + S = M = R_N + E_1 \qquad \text{...(7.54)}$$

Here M is the combined mixture that results when F kg feed and S kg solvent are mixed together. M is a fictitious stream because F and S are not brought into contact directly. The component-**C** balance for the cascade as a whole is

$$Fx_F + Sy_S = Mx_M = R_N x_N + E_1 y_1 \qquad \text{...(7.55)}$$

By mixture rule, the point M should lie on the line joining F and S such that

$$x_M = \frac{Fx_F + Sy_S}{F + S} \qquad \text{...(7.56)}$$

Since R_N and E_1 are effluents from equilibrium stages, they should lie on the solubility curve. Therefore, if the final raffinate composition x_N is known, R_N can be located on the solubility curve and by the mixture rule, E_1 lies on the extension of the line R_NM. Instead of x_N, if y_1 is specified, R_N can be located on the extension of the line E_1M. Thus knowing any three of the four terminal streams R_N, E_1, F and S, the fourth one can be determined graphically.

Equation (7.54) can be rearranged as

$$F - E_1 = R_N - S \qquad \text{...(7.57)}$$

Now consider the total material balance for the stages 1 through n in the cascade.

$$F + E_{n+1} = R_n + E_1 \qquad \text{...(7.58)}$$

which on rearrangement gives

$$F - E_1 = R_n - E_{n+1} \qquad \text{...(7.59)}$$

$R_n - E_{n+1}$ represents the net flow or the difference between the flow rates of the phases passing each other between any two adjacent stages n and n+1. Let us represent the net flow by D.

Combining Eqs. (7.57) and (7.59), we get

$$F - E_1 = R_N - S = R_n - E_{n+1} = D \qquad \text{...(7.60)}$$

Using the mixture-rule, the point D which represents the net flow can be located on the diagram at the intersection of the extension of the lines E_1F and SR_N as shown. Equation (7.60) also means that the line joining the points R_n and E_{n+1} should pass through the difference point D. Alternatively, by fixing the point D, we can locate the raffinate (or extract) stream on the solubility curve if the location of the extract (or raffinate) stream passing it is known.

The geometric features discussed above can be utilized for the determination of the number of stages required in a countercurrent battery of extraction stages. Let us consider the determination of the number of stages, given the feed rate, the feed composition, the quantity and purity of the solvent and the final raffinate composition. The points F, S and R_N can be located since their compositions are known. Using Eq. (7.56), calculate x_M and locate the mixture point M on the line FS. Alternatively, M can be located using the inverse lever-arm-rule which gives

$$\frac{F}{S} = \frac{\overline{SM}}{\overline{FM}} \qquad \text{...(7.61)}$$

Since the line R_NE_1 should pass through M, join R_N to M and extend it to intersect the solubility curve at E_1. If instead of x_N, y_1 is specified, then E_1 can be located and R_N determined by joining E_1 to M and extending it to intersect the solubility curve at R_N.

Having located the four terminal streams F, S, E_1 and R_N as described above, determine the difference point D at the intersection of the extension of the line E_1F and SR_N as shown in Fig. 7.51. Point D may lie on the solvent side or the feed side of the diagram depending on the solvent-to-feed ratio and the slope of the tie lines. Now, E_1 *is* the extract leaving the first stage which will be in equilibrium with the raffinate R_1 leaving the stage. R_1 can be determined at the end of the tie line passing through E_1. The tie line R_1E_1 is shown in the figure as a dashed line and it represents the first stage of the countercurrent unit. To draw the tie lines, either the conjugate curve or the distribution curve may be used as explained earlier. Since R_1 and E_2 are the two passing streams between stages 1 and 2, they should lie on the straight line passing through the difference point D which can be treated as equivalent to the operating line. Draw a straight line connecting D to R_1 and extend it to the extract side of the solubility curve to locate E_2. The dashed line E_2R_2 in the figure is the tie line connecting the phases leaving the second stage. The construction is continued by alternating between the tie line and the operating line till the final raffinate point R_N is reached or it is crossed by the tie line. The last tie line passing though R_N is the final stage in the countercurrent unit. Figure 7.51 shows four ideal stages. The method described here is known as the Hunter-Nash graphical method (Hunter & Nash, 1934).

Generally, the feed rate (F), feed composition (x_F) and the solvent purity (y_S) are fixed by the process and the remaining variables – the number of stages (N), the solvent rate (S), the compositions of the final raffinate and extract (x_N and y_1) – are to be specified by the designer. If any two of these variables are specified in addition to those already fixed by the process, the remaining variables are not independent; these can be estimated by the method described above. Depending upon which of these variables are specified, four types of problems are encountered in countercurrent extraction. They are: (i) Given the solvent rate (S) and the product composition (x_N or y_1) determine N and the remaining variables (ii) Given the product compositions (x_N and y_1), find N and S (iii) Given the solvent circulation rate (S) and the number of stages (N), find the product compositions (iv) Given the number of stages (N) and the composition of the final raffinate or extract (x_N or y_1), find the solvent circulation rate (S). The method has already been discussed for determining the number of stages for the problems of the first type. This can be extended to the second type as well with obvious modifications. The last two require trial-and-error solutions assuming arbitrary values for the product compositions or the amounts of solvent and determining the number of stages. If the number of stages so determined matches with the specified value, the assumed compositions are acceptable; otherwise change the compositions till the given number of stages and the number of tie lines that can be constructed are exactly equal.

To find the ratio of the raffinate to extract flow rates at any point in the cascade, consider the total material balance and solute balance for the stages 1 through n in the cascade.

$$R_n - E_{n+1} = D \qquad \ldots(7.62)$$

$$R_n x_n - E_{n+1} y_{n+1} = D x_D \qquad \ldots(7.63)$$

Here D is a fictitious stream representing the net flow between two adjacent stages and x_D is the weight fraction of the solute in the net flow. Eliminating D from Eq. (7.63) using Eq. (7.62), we get

$$\frac{R_n}{E_{n+1}} = \frac{y_{n+1} - x_D}{x_n - x_D} = \frac{\overline{DE_{n+1}}}{\overline{DR_n}} \qquad \ldots(7.64)$$

The ratio of the raffinate flow to the extract flow between two adjacent stages n and n+1 can be computed from the ratio of the line lengths as indicated by Eq. (7.64).

Minimum and maximum solvent-to-feed ratio

In order that the extraction be possible with a finite number of stages, the solvent used for extraction should be above the minimum permissible amount. If the operating line drawn from the difference point coincides with any of the tie lines, an infinite number of stages will be required to effect separation. The difference point, therefore, should be located in such a way that none of the tie lines coincide with the operating lines. Consider Fig. 7.52 which shows five tie lines labelled as a, b, c, d and e. Each one of these when extended meets the extension of the line SR_N at points marked 1,2,3,4 and 5 respectively. If the difference point D were located at any of these points, the operating lines will coincide with the tie lines and will result in infinite number of stages for carrying out the separation. The minimum solvent-to-feed ratio corresponds to the highest amount of solvent that gives rise to infinite number of stages. According to Fig. 7.52, the minimum solvent corresponds to the location D_{min} for the difference point. To avoid the pinch point and infinite number of stages, the actual operating difference point should be located away from D_{min} towards the left. To locate D_{min}, there is no shortcut other than finding the intersection with SR_N of the extension of all the tie lines below the one passing through the feed point and determining the one farthest from S. But in general, the tie line which when extended passes through the feed point F will determine the location of D_{min} as shown in the figure.

Corresponding to the minimum amount of solvent S_{min}, the final extract will be the richest in the solute and can be determined by extending $D_{min}F$ to intersect the extract side of the solubility curve at $E_{1,m}$. Knowing the four terminal compositions, the mixture point M corresponding to S_{min} is determined at the intersection of the lines FS and $R_NE_{1,m}$. Let the weight fraction of the solute in the mixture be x_M. A total material balance and a solute balance for the whole cascade lead to

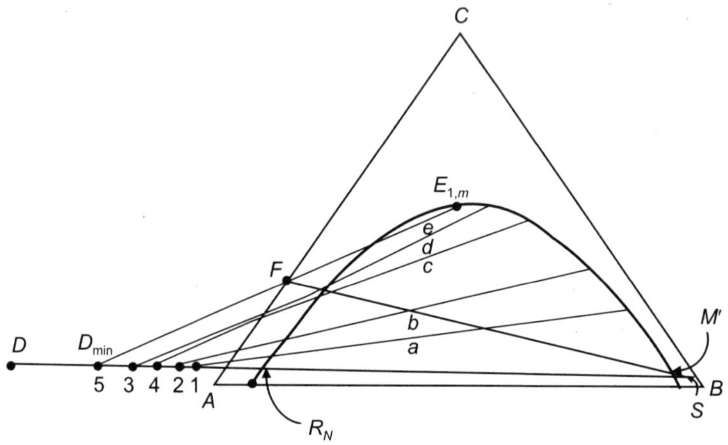

Fig. 7.52: Minimum solvent-to-feed ratio for countercurrent multistage extraction

$$F + S_{\min} = M \qquad ...(7.65)$$

$$Fx_F + S_{\min} y_S = Mx_M \qquad ...(7.66)$$

Eliminating M from Eq. (7.66) using Eq. (7.65) and rearranging the resulting equation, we get

$$S_{\min} = \frac{F(x_F - x_M)}{(x_M - y_S)} \qquad ...(7.67)$$

When the tie line slopes downward, the difference point will be located on the solvent side, in which case, the minimum difference point is given by the tie line intersection that lies closest to S as shown in Fig. 7.53. The tie lines a, b and c intersect the line R_NS at points 1, 2 and 3. The difference point D_{min} corresponding to the minimum amount of the solvent has to be located at point 3. Actual operating difference point D lies toward the left of D_{min}.

The addition of the solvent to the feed should form mixture M that falls within the heterogeneous area. If the amount of solvent added can completely dissolve the feed, the mixture point M will fall outside the heterogeneous area and extraction will not be possible. Thus, there is a maximum value for the solvent-to-feed ratio that can be used for extraction which corresponds to the mixture point coinciding with the intersection of the line FS with the extract side of the binodal curve. This point is represented by point M' in Fig. 7.52.

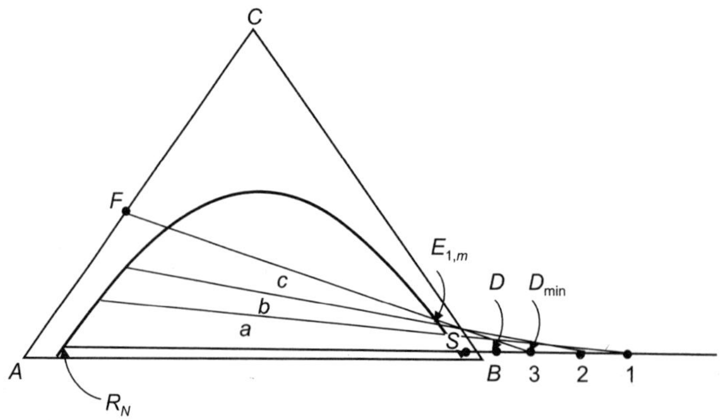

Fig. 7.53: Determination of the minimum solvent-to-feed ratio when the tie line slope is negative

Number of stages using the distribution diagram

A McCabe-Thiele type construction can also be used for finding the number of stages in countercurrent extraction after establishing the terminal concentrations and the difference point D on the triangular phase diagram, as described earlier. This method will be convenient when a large number of stages have to be accommodated in the triangular diagram. Using the tie line data, the distribution curve is plotted as shown in Fig. 7.54b with weight fraction of solute in the extract (y) on the y-axis and the weight fraction of solute in the raffinate (x) on the x-axis.

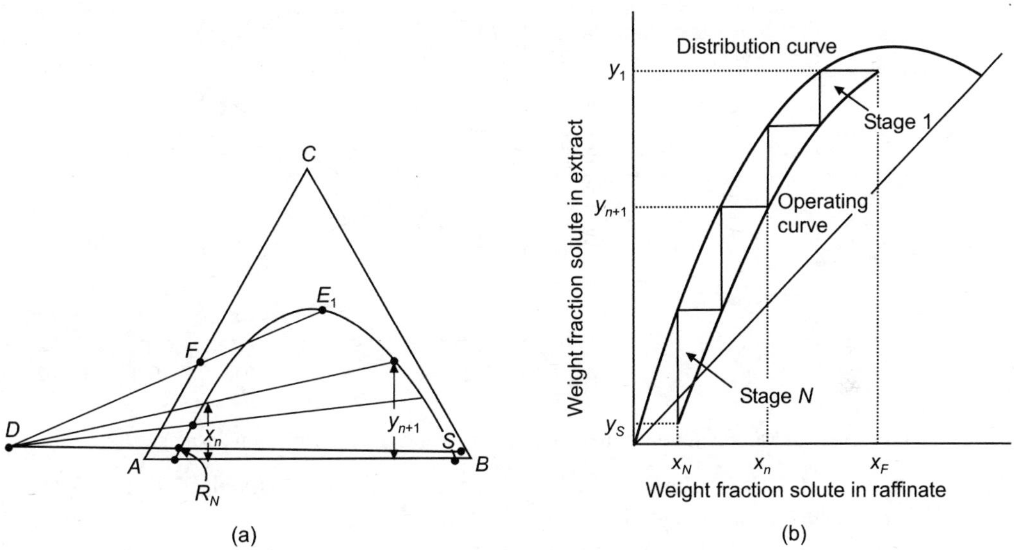

Fig. 7.54: Countercurrent extraction stages on the distribution diagram

Random lines are drawn originating from the difference point which will intersect the raffinate side of the solubility curve at x_n and the extract side at y_{n+1}. The point with coordinates (x_n, y_{n+1}) should lie on the operating line in Fig. 7.54b. Thus the operating line is obtained by drawing few such random lines from the difference point and finding the coordinates of their intersections (x_n, y_{n+1}). The upper terminus of the operating curve is (x_F, y_1) and the lower terminus is (x_N, y_S). Starting from the upper terminus, stages are stepped off as shown in the Fig. 7.54b.

Example 7.9: One thousand kilograms per hour of an aqueous solution containing 20.0 weight% acetone is being extracted in a countercurrent multistage unit using pure methyl isobutyl ketone solvent so that the final raffinate contains 5.0 weight% acetone. Use the Equilibrium data given in Example 7.8. Calculate:

(a) The minimum solvent that can be used.

(b) The number of theoretical stages if a solvent rate 25% in excess of the minimum is used.

***Solution*:** (a) The miscibility data is plotted on right triangular coordinates with weight fraction of MIK on the x-axis and weight fraction of acetone on the y-axis to get the binodal curve, as shown in Fig. 7.55a. The distribution curve may also be plotted using the tie line data to assist in the interpolation of tie lines. (Distribution curve is not shown in the figure).

Feed = 1000 kg, $x_F = 0.20$. Point F in the figure represents the feed. Since the final raffinate is to contain 5% acetone, $x_N = 0.05$ and the point R_N is marked on the raffinate side of the binodal curve with acetone content of 0.05. Since the solvent is pure, the point S coincides with the vertex B of the triangle.

The minimum solvent corresponds to the tie line lying below the feed which when extended intersects the line SR_N farthest from S. It is found that the tie line passing through the feed point is the one which decides the minimum solvent. With the help of the distribution curve or the conjugation curve, the tie line which when extended pass through the feed F, is found out. This tie line intersects the line SR_N at D_{min}, the difference point that corresponds to the minimum solvent. The tie line intersects the extract side of the binodal curve at $E_{1,m}$, which represents the extract phase that results if the minimum quantity of solvent is used for extraction.

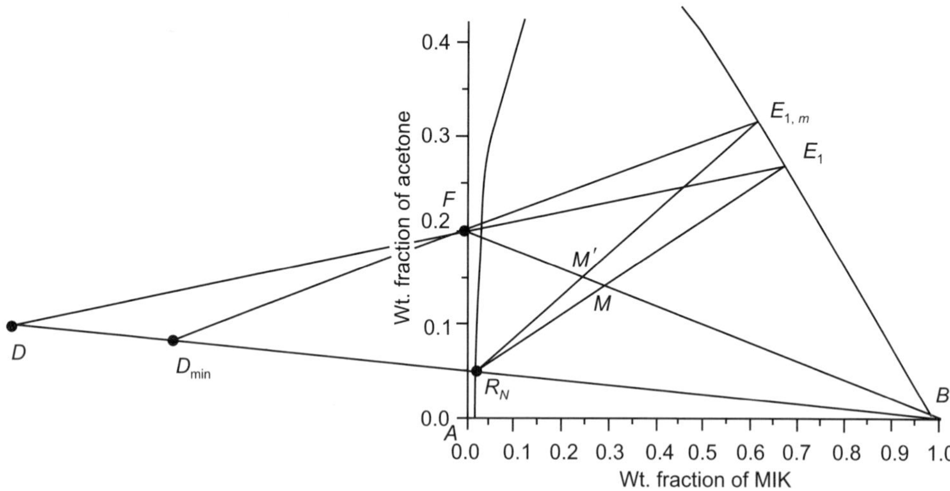

Fig. 7.55: (a) Minimum solvent requirement for countercurrent extraction (Example 7.9)

Join $E_{1,m}$ to R_N and S to F. The intersection of these two lines is marked M' in Fig. 7.55a. The y-coordinate of M' is the solute content in the mixture formed, if the minimum quantity of solvent were used for extraction. Thus $x'_M = 0.1508$ from the figure. Equation (7.67) is used to calculate the minimum solvent requirement.

$$S_{min} = \frac{F(x_F - x_{M'})}{(x_{M'} - y_S)} = \frac{1000(0.20 - 0.1508)}{(0.1508 - 0)} = 326.3 \text{ kg} \qquad \textbf{(Answer)}$$

(b) When the solvent used is 25% in excess of the minimum, $S = 1.25 \times 326.3 = 407.88$ kg

The mixture point M lies on the line FS such that

$$x_M = \frac{Fx_F + Sy_S}{F + S} = \frac{1000 \times 0.20 - 407.88 \times 0}{1000 + 407.88} = 0.1421$$

The line R_NM is extended to meet the extract side of the solubility curve at E_1, the point representing the final extract.

The construction for estimating the number of stages is shown in Fig. 7.55b. The intersection of the extensions of the lines E_1F and SR_N is marked D, the difference point. The tie line passing through E_1 is determined using the tie line data and R_1 is located at the raffinate end of the tie line. Tie line R_1E_1 represents the first stage in the countercurrent cascade. The difference point D is joined to R_1 and extended to meet the extract side at E_2. The tie line E_2R_2 is drawn. The construction of alternate operating line and tie line is repeated till the final raffinate point is reached.

The figure shows that more than three stages are required for the desired separation. The fraction of the stage can be evaluated as

$$\frac{x_3 - x_N}{x_3 - x_4} = \frac{0.0824 - 0.05}{0.0824 - 0.0380} = 0.7$$

where x_3, x_4, x_N are the weight fractions of solute in R_3, R_4 and R_N, respectively. The number of stages required = 3.7 **(Answer)**

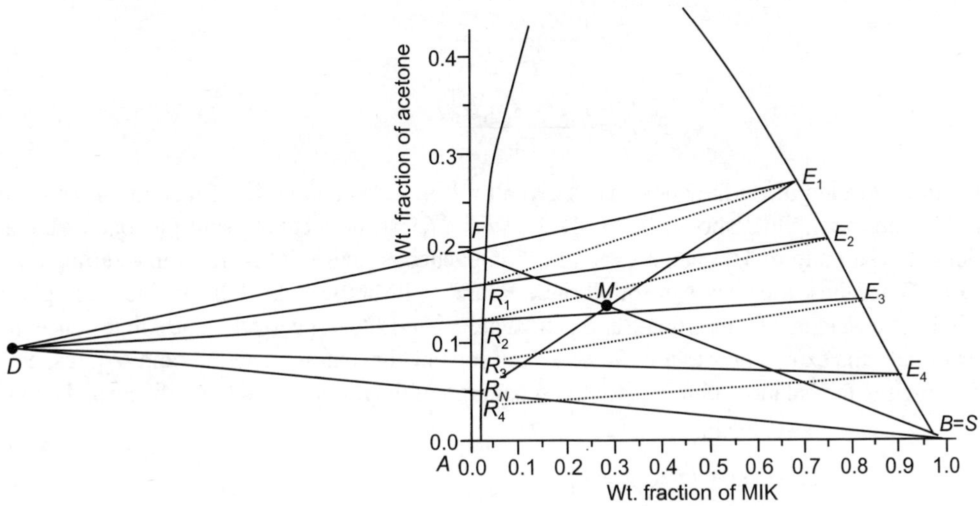

Fig. 7.55: (b) Determination of the number of stages of countercurrent extraction (Example 7.9)

Janecke Diagrams for Computation of Theoretical Stages

The extraction processes can be analyzed using the Janecke diagrams also. The multistage crosscurrent extraction and countercurrent extraction processes are discussed below using the Janecke diagrams.

Crosscurrent multistage contact

Consider the multistage crosscurrent unit shown in Fig. 7.56. F kg/h feed of composition x_F weight fraction C is extracted with S_1 kg solvent of composition y_S in the first stage. On solvent free basis the quantities treated are F' kg/h of feed and S_1' kg/h of solvent. The concentrations of **B** and **C** in the feed on solvent-free basis are N_F and X_F, respectively and concentrations in the solvent are N_S and Y_S, respectively. F and S can be represented in the N versus X coordinate system as shown in Fig. 7.56.

If the solvent used is pure, $N_S = \infty$ and the solvent point S will be located at infinity, which means the lines originating from S and intersecting the solubility curve will be vertical lines. Since the feed entering stage 1 is free of solvent **B**, $N_F = 0$ and F will be on the x-axis as shown. On solvent free basis, the following balance equations can be written:

A+C balance:

$$F' + S_1' = M_1' \quad \ldots(7.68)$$

C balance: $$F'X_F + S_1'Y_S = M_1'X_{M1} \quad \ldots(7.69)$$

B balance: $$F'N_F + S_1'N_S = M_1'N_{M1} \quad \ldots(7.70)$$

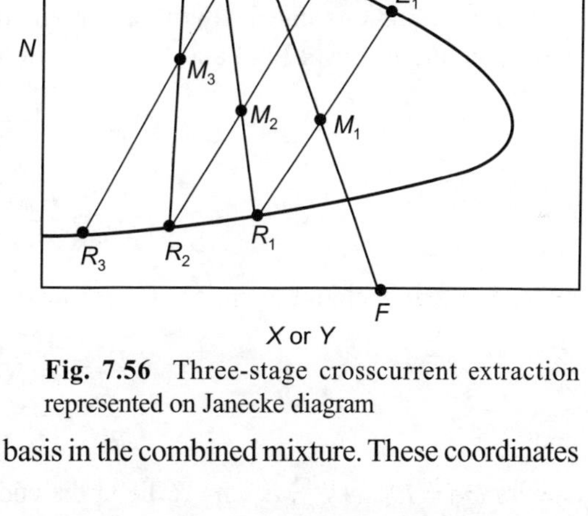

Fig. 7.56 Three-stage crosscurrent extraction represented on Janecke diagram

Here M_1' is the weight of **A+C** in the combined mixture of total weight M_1 and N_{M1} and X_{M1} are the weight fractions of the solvent and the solute on **B**-free basis in the combined mixture. These coordinates of M_1 are estimated using Eqs. (7.68) to (7.70) as

$$X_{M,1} = \frac{F'X_F + S_1'Y_S}{F' + S_1'} \qquad ...(7.71)$$

$$N_{M,1} = \frac{F'N_F + S_1'N_S}{F' + S_1'} \qquad ...(7.72)$$

M_1 can be located using the coordinates given by Eq. (7.71) and (7.72). Since the mixture rule is applicable and the mixture should lie on the line FS, M_1 can be located using just one of the above coordinates. The mixture M_1 cannot exist as a homogeneous phase, it separates into raffinate R_1 and extract E_1. Since these are phases in equilibrium, R_1 and E_1 lie at the ends of the tie line passing through M_1. The **B**-free weights of the raffinate and extract are R_1' and E_1', respectively. The solute concentration in the raffinate and extract phases are X_1 and Y_1 and solvent concentration are N_{R1} and N_{E1}, respectively.

We can write the balance equations for the separation of the mixture into the raffinate and extract as:

A+C balance: $M_1' = R_1' + E_1'$ ...(7.73)

C balance: $M_1'X_{M1} = R_1'X_1 + E_1'Y_1$ ...(7.74)

B balance: $M_1'N_{M1} = R_1'N_{R1} + E_1'N_{E1}$ ...(7.75)

Eliminating R_1' from Eq. (7.74) using Eq. (7.73), we get

$$E_1' = \frac{M_1'(X_{M1} - X_1)}{(Y_1 - X_1)} \qquad ...(7.76)$$

where $M_1' = F' + S_1'$ by Eq. (7.68). R_1' can be determined by substituting E_1' in Eq. (7.73). The total weights of the raffinate and extract products can be calculated from their **B**-free weights as

$$R = R_1'(1 + N_{R1}) \qquad ...(7.77)$$

$$E = E_1'(1 + N_{E1}) \qquad ...(7.78)$$

The inverse-lever-arm rule is applicable on solvent free basis and the following relations also may be used for estimating amounts of different streams.

$$\frac{F'}{S_1'} = \frac{\overline{SM_1}}{\overline{FM_1}} \qquad ...(7.79)$$

$$\frac{R_1'}{E_1'} = \frac{\overline{E_1M_1}}{\overline{R_1M_1}} \qquad ...(7.80)$$

Similar equations can be written for all the crosscurrent stages. For the nth stage, we have, corresponding to Eqs. (7.71) and (7.72), the following:

$$X_{Mn} = \frac{R_{n-1}'X_{Rn-1} + S_n'Y_S}{R_{n-1}' + S_n'} \qquad ...(7.81)$$

$$N_{Mn} = \frac{R_{n-1}'N_{Rn-1} + S_n'N_S}{R_{n-1}' + S_n'} \qquad ...(7.82)$$

which locate the mixture point M_n. Corresponding to Eq. (7.76), the extract leaving stage n is given by

$$E_n' = \frac{M_n'(X_{Mn} - X_n)}{(Y_n - X_x)} \qquad ...(7.83)$$

and $R_n' = M_n' - E_n'$ ...(7.84)

where $M_n' = R_{n-1}' + S_n'$. E_n and R_n lie at the ends of the tie line passing through M_n.

Countercurrent extraction

Consider the multistage countercurrent extraction unit shown in Fig. 7.50. On solvent-free basis the material balance can be written as

$$F' + S' = M' = R'_N + E'_1 \quad \text{...(7.85)}$$

The component-**C** balance is

$$F'X_F + S'Y_S = M'X_M = R'_N X_N + E'_1 Y_1 \quad \text{...(7.86)}$$

Here M is the combined mixture that results when F kg feed and S kg solvent are mixed together. F', S', E'_1 and R'_N are the B-free rates of feed, solvent, final extract and final raffinate, respectively. X_F, Y_S, Y_1 and X_N are the weight fractions of solute **C** on solvent-free basis. The point M representing the mixture M should lie on the line joining F and S such that

$$X_M = \frac{F'X_F + S'Y_S}{F' + S'} \quad \text{...(7.87)}$$

$$N_M = \frac{F'N_F + S'N_S}{F' + S'} \quad \text{...(7.88)}$$

Of the four terminal streams, F, S, R_N and E_1, if the specifications permit locating any three on the Janecke diagram, the fourth one can be located using the result that the line FS and R_NE_1 should pass through M.

Having located F, S, R_N and E_1, the difference point D can be determined. Considering the **A**+**C** balance for the first n stages in the cascade, we get

$$F' + E'_{n+1} = R'_n + E'_1 \quad \text{...(7.89)}$$

Rearranging Eq. (7.85), and (7.89) we get

$$F' - E'_1 = R'_N - S' = R'_n - E'_{n+1} = D' \quad \text{...(7.90)}$$

Equation (7.90) means that the difference between the rates of the raffinate and extract flows between any two adjacent stages is constant and can be represented by D' which represents the net flow in the cascade towards the right. It can be located on the N-X-Y diagram at the intersection of the extension of lines E_1F and SR_N.

To determine the number of stages graphically, proceed as follows. Draw the tie line passing through E_1 and determine R_1 on the raffinate side of the solubility curve. The tie line R_1E_1 is shown in Fig. 7.57 as a dashed line. Now $R'_1 - E'_2 = D'$ and therefore points D, R_1 and E_2 are collinear. E_2 is thus located by joining the difference point D to R_1 and extending to E_2 which lies on the extract side of the solubility curve as shown. Having obtained E_2, R_2 is determined at the end of the tie line passing through E_2. Draw the operating line and continue the construction by alternating between the tie line and the operating line till the raffinate end of a tie line coincides with or crosses

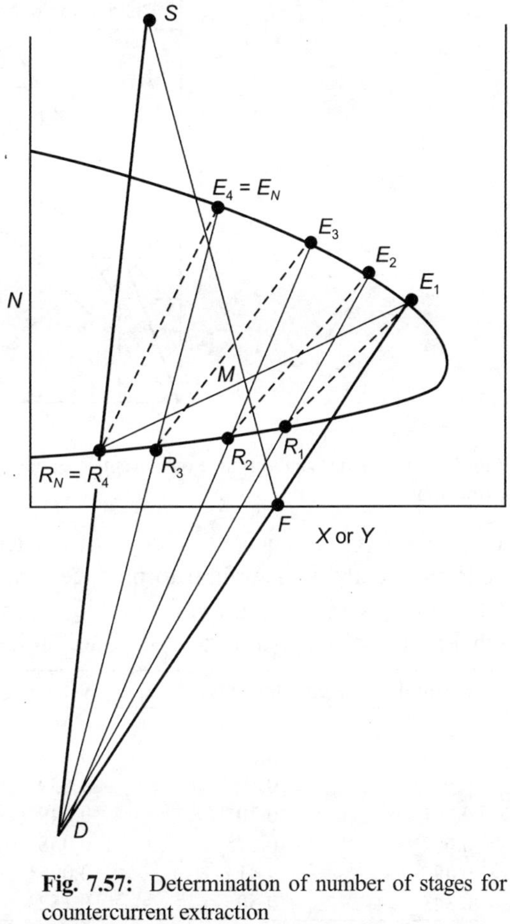

Fig. 7.57: Determination of number of stages for countercurrent extraction

the final raffinate point R_N. The number of tie lines drawn between E_1 and R_N is the number of equilibrium stages required for extraction.

Now consider the total material balance and solute balance for the stages 1 through n in the cascade.

$$R'_n - E'_{n+1} = D' \quad \ldots(7.91)$$

$$R'_n X_n - E'_{n+1} Y_{n+1} = D' X_D \quad \ldots(7.92)$$

Here X_D is the weight fraction of solute in the net flow on **B**-free basis. Eliminating D' from Eq. (7.92) using Eq. (7.91), we get

$$\frac{R'_n}{E'_{n+1}} = \frac{Y_{n+1} - X_D}{X_n - X_D} = \frac{\overline{DE}_{n+1}}{\overline{DR}_n} \quad \ldots(7.93)$$

The ratio of the raffinate flow to the extract flow between two adjacent stages n and n+1 can be computed from the ratio of line lengths as indicated by Eq. (7.93).

The number of stages in countercurrent extraction can also be determined graphically using the distribution diagram plotted on solvent free basis. This is obtained by projecting the tie lines in Fig. 7.57 to Y versus X coordinate system as shown in Fig. 7.58. The operating curve is drawn by drawing random lines from point D and noting the coordinates of their intersection with the solubility curve as discussed earlier for triangular coordinate system.

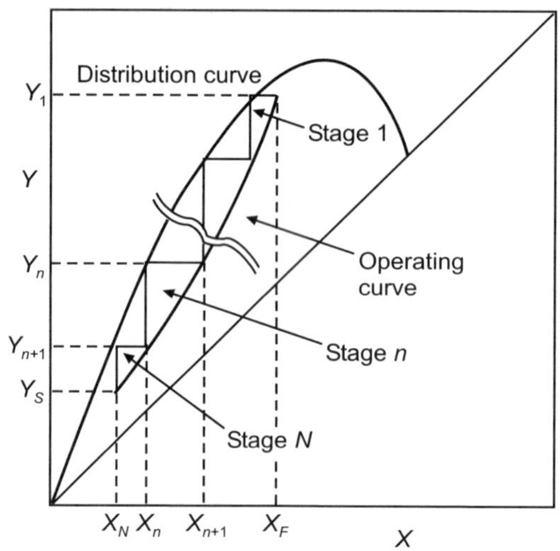

Fig. 7.58: Use of distribution curve on solvent-free coordinates for determination of the number of stages in countercurrent extraction

Example 7.10: An aqueous solution of ethylene glycol containing 30 weight% glycol is extracted countercurrently with furfural to produce a raffinate product containing 5% glycol on solvent-free basis. Feed is admitted at a rate 1000 kg/h. Equilibrium solubility data (in weight fractions) for the ethylene glycol-water-furfural system at 298 K are:

Furfural	Ethylene glycol	Water	Furfural	Ethylene glycol	Water
0.95	0	0.05	0.275	0.525	0.20
0.903	0.052	0.045	0.139	0.475	0.386
0.861	0.10	0.039	0.11	0.40	0.49
0.751	0.20	0.049	0.097	0.30	0.603
0.667	0.275	0.058	0.084	0.15	0.766
0.49	0.415	0.095	0.077	0	0.923
0.343	0.505	0.152			

The tie line data for the system are:

Raffinate phase			Extract phase		
Furfural	Glycol	water	Furfural	Glycol	Water
0.08	0.073	0.847	0.926	0.025	0.049
0.083	0.15	0.767	0.904	0.052	0.044
0.088	0.2	0.712	0.894	0.062	0.044
0.097	0.3	0.603	0.882	0.075	0.043
0.11	0.4	0.49	0.86	0.1	0.04
0.139	0.475	0.386	0.803	0.15	0.047
0.251	0.515	0.234	0.751	0.2	0.049
0.277	0.525	0.198	0.665	0.275	0.06
0.414	0.415	0.171	0.414	0.415	0.171

Determine:

(a) The minimum amount of solvent required

(b) The number of theoretical stages when pure solvent is used at a rate 4000 kg/h

***Solution*:** The equilibrium solubility data is converted into solvent free coordinates, N and X. For illustration, consider the third set of data:

Furfural (**B**) = 0.861, Ethylene glycol (**C**) = 0.10, Water (**A**) = 0.039

$$X = \frac{\text{weight of } \mathbf{C}}{\text{weight of } (\mathbf{A}+\mathbf{C})} = \frac{0.10}{(0.10+0.039)} = 0.7194$$

$$N = \frac{\text{weight of } \mathbf{B}}{\text{weight of } (\mathbf{A}+\mathbf{C})} = \frac{0.861}{(0.10+0.039)} = 6.1942$$

These calculations are repeated for other sets of data and the results are:

X	0.0	0.536	0.719	0.803	0.826	0.814	0.769	0.724	0.552	0.449	0.332	0.164	0.000
N	19.00	9.309	6.194	3.016	2.003	0.961	0.522	0.379	0.161	0.124	0.107	0.092	0.083

The weight fraction of solvent (N) is taken on the y-axis and the weight fraction of the solute (X or Y) is taken on the x-axis to plot the Janecke diagram shown in Fig. 7.59.

The tie line data are used to find the solute weight fractions on solvent – free basis in the equilibrium raffinate and extract phases. Consider the fourth set of tie line data.

Raffinate phase: Furfural (**B**) = 0.088, Ethylene glycol (**C**) = 0.20, Water (**A**) = 0.712

Extract phase: Furfural (**B**) = 0.894, Ethylene glycol (**C**) = 0.062, Water (**A**) = 0.044

$$X = \frac{C}{(A+C)} \text{ in the raffinate} = \frac{0.20}{(0.20+0.712)} = 0.219$$

$$Y = \frac{C}{(A+C)} \text{ in the extract} = \frac{0.062}{(0.062+0.044)} = 0.585$$

Results for other sets give

X	0.0	0.079	0.164	0.219	0.332	0.449	0.552	0.688	0.726	0.708
Y	0.0	0.338	0.542	0.585	0.636	0.714	0.761	0.803	0.821	0.708

The tie line data are plotted to give the distribution curve that is used for interpolation of the tie lines. This plot is not shown in the figure.

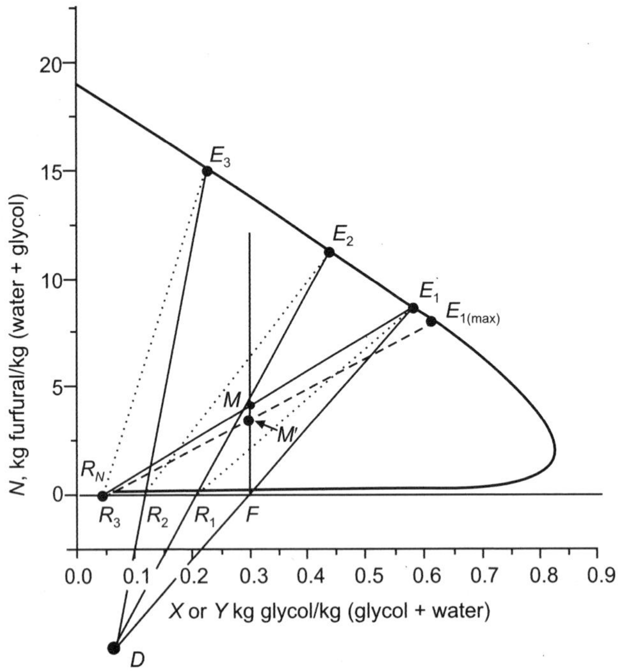

Fig. 7.59: The Janecke diagram for determining the number of stages in Example 7.10

(a) The feed is 1000 kg. $F = 1000$ kg $= F'$ (because, the feed is free of solvent)

$$X_F = \frac{C}{(A+C)} \text{ in the feed} = \frac{300}{1000} = 0.3$$

$$N_F = \frac{B}{(A+C)} \text{ in the feed} = \frac{0.0}{1.0} = 0.0$$

The feed point F is marked at (0.3, 0) as shown. The solvent being pure, its N-coordinate is ∞ and its x-coordinate is zero. The lines passing through the solvent point on this diagram will be vertical lines.

The final raffinate contains 5% glycol on solvent-free basis. Therefore R_N can be located on the raffinate side of the solubility curve at $x = x_N =$ 0.05. The tie line passing through the feed point is extended to meet the vertical line through R_N (the line SR_N should be vertical as S lies at infinity). The meeting point is $D'_{\min}$ which is the difference point corresponding to the minimum solvent. E_1(max) is the extract phase that will result if the quantity of solvent used is the minimum. E_1(max) is joined to R_N (shown as dashed line) and the coordinates of the intersection of R_NE_1(max) with the vertical at F (which is in fact, the line FS) are read from the figure. Thus N-coordinate of $M' = 3.507$ from the figure. Knowing that the solvent used is pure,

$$3.507 = \frac{B}{A+C} \text{ in the mixture } M' = \frac{B \text{ in } F+S}{(A+C) \text{ in } F+S} = \frac{S_{\min}}{1000}$$

Therefore, $S_{\min} = 3507$ kg **(Answer)**

(b) When the solvent used is $S = S' = 4000$ kg, the mixture point M can be located on the vertical line through F at height N_M, where

$$N_M = \frac{B}{A+C} \text{ in the mixture } M = \frac{B \text{ in } (F+S)}{(A+C) \text{ in } (F+S)} = \frac{4000}{1000} = 4$$

The line R_NM is extended to meet the solubility curve at E_1 as shown. The difference point D is located at the intersection of E_1F and the vertical passing through R_N as shown in Fig. 7.59.

To find the number of stages, the tie line E_1R_1 is drawn first, which gives the first stage and the operating line DR_1 is extended to meet the solubility curve at E_2. The tie line E_2R_2 is constructed which gives the second stage. The construction of the operating line and the tie line is repeated till the final raffinate R_N is reached. We can see that the number of tie lines that can be drawn = 3 and therefore, the number of stages = 3. **(Answer)**

7.7 EXTRACTION WITH REFLUX

In the countercurrent extraction process described above, the maximum solute concentration achievable for the extract product is equal to the concentration of the liquid in equilibrium with the feed entering. By providing a reflux, it is possible to enrich the product beyond this limiting concentration and thereby a sharper separation can be achieved. Figure 7.60 shows a countercurrent multistage extraction unit with extract reflux. Feed is admitted at an intermediate stage *f* in the cascade. All stages towards the left of *f* constitute the extract enriching section in which a general stage is represented by the stage *m*. The stages towards the right of *f* including the last stage *N* constitute the raffinate stripping section. Stage *n* represents a general stage in the stripping section. Extract E_1 leaving stage 1 is subjected to solvent recovery and a portion of the solvent-free extract R_0 is recycled as the reflux to stage 1 and the other part E is withdrawn as the final extract product. Pure component-**B** is recovered as the solvent from extract E_1.

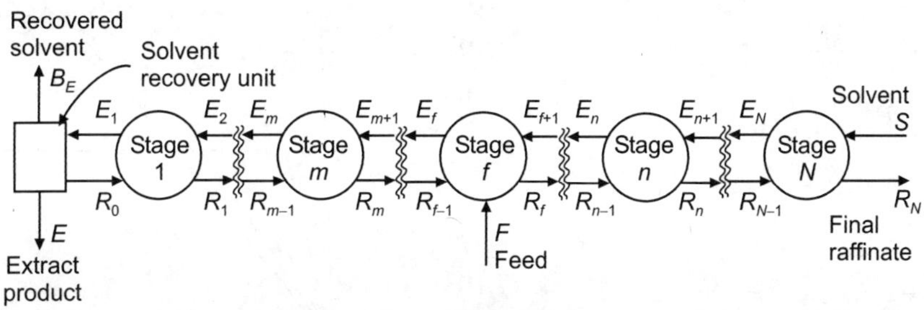

Fig. 7.60: Multistage countercurrent extraction with reflux

The Janecke diagram is found to be more convenient than the triangular diagram for graphical construction, as the latter frequently gets crowded and hence interpolation becomes difficult if the number of stages is large. The construction on the Janecke diagram is exactly similar to those on the enthalpy-composition diagram used in the Ponchon-Savarit method for distillation. The flow rates on solvent-free basis are represented by a prime on the symbol. The net flow outward in the enriching section is represented on the diagram by a fictitious stream D', which is the difference point or the operating point for the extract enriching section. The net flow outward the raffinate stripping section is represented by the stream D'', which is the difference point or operating point for the stripping section. The difference points can be located as described below.

The extract E_1 being the effluent from an ideal stage, it can be located on the upper solubility curve as shown in Fig. 7.61. The extract product E and the reflux R_0 are parts of the same stream and can be represented by a single point as shown. The point will lie on the horizontal axis if the stream is free of the solvent. Since the solute content of the reflux, the extract product and the extract leaving the stage 1 are the same, the points E_1 and R_0 lie on a vertical line as shown.

Now consider the enriching section and take a material balance around the first m stages.

A+C balance: $$E'_{m+1} - R'_m = E' \qquad \ldots(7.94)$$

It is assumed that the solvent recovered is B_E and no **A** or **C** is present in this stream. If all the solvent in E_1 were recovered, then $E = E'$

C balance: $$E'_{m+1}Y_{m+1} - R'_m X_m = E'Y_E \quad \text{...(7.95)}$$

B balance: $$E'_{m+1}N_{E,m+1} - R'_m N_{R,m} = E'N_E + B_E \quad \text{...(7.96)}$$

If the entire solvent in the extract E_1 were recovered as B_E, then $N_E = 0$, and the first term on the right vanishes. The above equations give the net flow of material, net flow of solute **C** and solvent **B**, respectively and all these are constant irrespective of the stage m. In short, we can write

$$E'_{m+1} - R'_m = D' \quad \text{...(7.97)}$$

D' is the net flow which is represented with coordinates ($X_{D'}$, $N_{D'}$) in Fig. 7.61 where

$$X_{D'} = Y_E \quad \text{...(7.98)}$$

$$N_{D'} = N_E + \frac{B_E}{E'} \quad \text{...(7.99)}$$

Knowing the coordinates as given by Eqs. (7.98) and (7.99) we can locate D' on the vertical passing through E_1. It can also be located if the external reflux ratio R_0/E is known, as described below.

When $m = 0$, Eqs. (7.94), (7.95) and (7.96) become

$$E'_1 - R'_0 = E' \quad \text{...(7.100)}$$

$$E'_1 Y_1 - R'_0 X_0 = E'Y_E \quad \text{...(7.101)}$$

$$E'_1 N_{E,1} - R'_0 N_{R,0} = E'N_E + B_E \quad \text{...(7.102)}$$

Substituting Eq. (7.100) into Eq. (7.102) and simplifying using Eq. (7.99), we get

$$\frac{R'_0}{E'} = \frac{(E'N_E + B_E) - N_{E,1}}{N_{E,1} - N_{R,0}} = \frac{N_{D'} - N_{E,1}}{N_{E,1} - N_{R,0}} \quad \text{...(7.103)}$$

The external reflux ratio defined as the ratio of the amount of extract recycled to the amount withdrawn as product, is

$$r = \frac{R_0}{E} = \frac{R'_0}{E'} = \frac{\overline{D'E_1}}{\overline{E_1R_0}} \quad \text{...(7.104)}$$

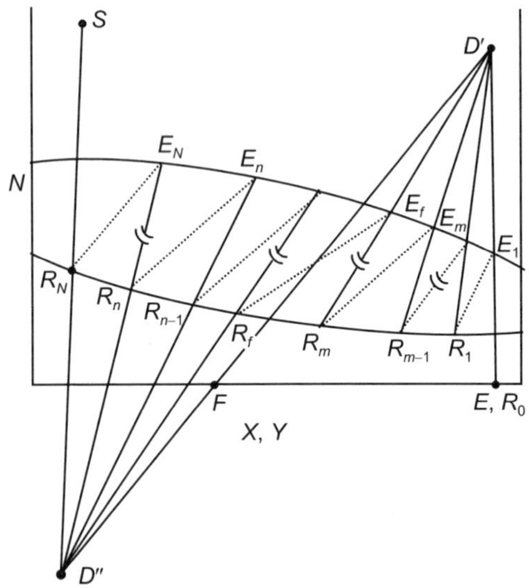

Fig. 7.61: Graphical construction to determine the number of stages for countercurrent extraction with reflux

Knowing the location of E_1 and R_0 on the Janecke diagram, the difference point for the enriching section can be located on the vertical line passing through E_1 at a height $r\overline{E_1R_0}$ from E_1. On the Janecke diagram, the points S and R_N represent the solvent and the final raffinate stream. Since the final raffinate leaves the last stage N in equilibrium with the extract from that stage, R_N should lie on the raffinate side of the solubility curve as shown in Fig. 7.61.

The stage n represents a general stage in the raffinate stripping section, a solvent-free material balance from stage n through stage N, the last stage in the cascade, we get

$$R'_{n-1} - E'_n = D'' = R'_N - S' \quad \text{...(7.105)}$$

Here D'' is the net flow outward the stripping section, which, by Eq. (7.105) is constant. On the Janecke diagram, the point D'' represents the difference point or the operating point for the stripping section.

Consider the solvent-free material balance for the countercurrent cascade as a whole.

$$F' + S' = R'_N + E' \quad \text{...(7.106)}$$

Rearranging Eq.(7.106), and utilizing the results of Eq. (7.105 and (7.97), we get

$$F' = E' + R'_N - S' = D' + D'' \qquad \ldots(7.107)$$

According to the mixture rule, the line joining the difference points D' and D'' should pass through F.

Now the construction for estimating the number of stages is straight forward. From E_1 draw the tie line which intersects the solubility curve at R_1. The points R_1, E_2 and D' are collinear according to Eq. (7.95). Therefore, the intersection of the line joining R_1 to D' with the extract side of the solubility curve gives E_2. Continue the construction alternating between tie line and the operating line till the tie line crosses the feed line $D'FD''$. The tie line that crosses the feed line represents the stage to which the feed is admitted. The construction of the operating lines and the tie lines is continued beyond the feed stage, but once the feed line is crossed, the operating line should be drawn from the difference point D'' instead of from D'. The number of tie lines drawn up to the point R_N gives the number of stages.

Limiting Reflux Ratios

The number of stages required for providing a given extent of extraction is finite and it increases with decrease in the reflux ratio, provided the reflux ratio employed is above the minimum permissible value. For minimum reflux ratio, the number of stages required for the specified separation will be infinity. In fact, the minimum reflux ratio is the highest reflux ratio that necessitates infinite number of stages for separation. In the graphical method described, we can see that an infinite number of stages will result if the tie line and the operating line coincide, which will be the case if the difference points lie at the intersection of the tie lines. The difference point that leads to infinite number of stages and lies farthest from $N = 0$ axis will determine the minimum reflux ratio for the counter-current cascade. In the Fig. 7.62 the dashed line p is a tie line which when extended passes through the feed point F. To determine the minimum reflux, extend all tie lines that lie towards the right of p to intersect the vertical passing through E_1 and if any tie line intersection is above that given by p, that intersection will determine the location of D'_{min}. Also extend the tie lies that lie towards the left of p to intersect the line SR_N. If any such tie line intersection lies below that given by the line p, the one farthest from the horizontal axis will determine D''_{min} from which the location of D'_{min} is to be determined. In Fig. 7.62 is shown D''_{min} located in this way, and D'_{min} established by extending the line $D''_{min}\,F$ to intersect the vertical line passing through E_1, because we know that the points D', D'' and F are collinear. After locating the difference point D'_{min}, the minimum reflux ratio r_m can be computed using Eq. (7.104). For most systems, the tie line passing through the feed point F will determine the location of the minimum difference point, in which case D'_{min} and D''_{min} will be at the ends of the line p in the figure.

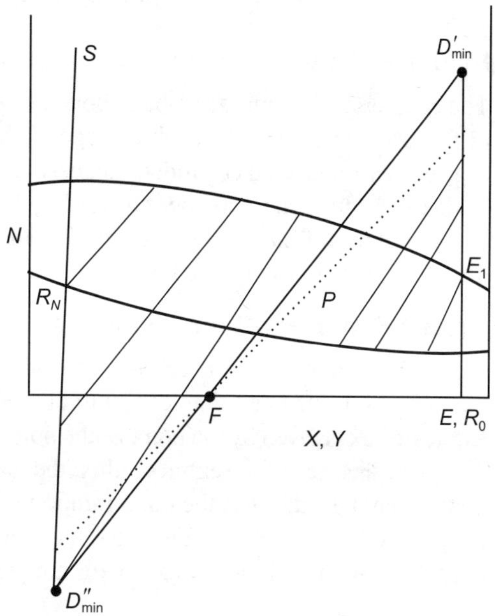

Fig. 7.62: Determination of minimum reflux ratio for countercurrent extraction

The larger the reflux ratio employed, the farther the location of the difference point D' from the horizontal axis. Consequently the number of stages required for a given separation decreases. At infinite reflux ratio, the difference point will be at infinity or $N_{D'} = \infty$; the operating lines radiating from D' will all be vertical lines and the number of stages required will be the minimum. Infinite reflux ratio means the total reflux condition in which there is no continuous addition of feed; there is no product withdrawal and all the solvent recovered is fed to the last stage as solvent. The determination of the minimum number of stages under total reflux conditions is as illustrated in Fig. 7.63.

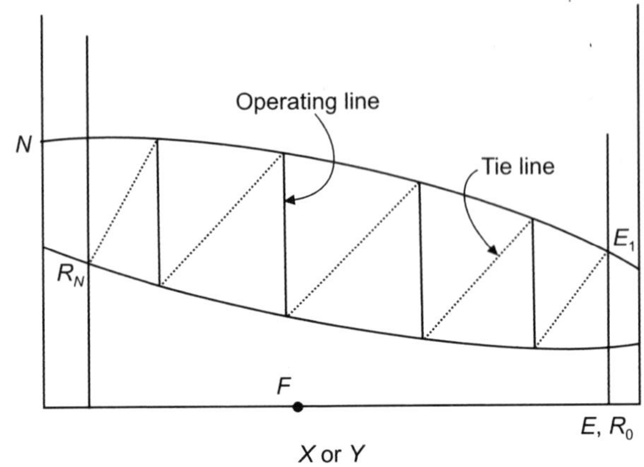

Fig. 7.63: Determination of minimum number of stages at total reflux

Raffinate Reflux

The countercurrent unit described above utilizes only the extract reflux. It is possible to return a part of the raffinate product R_N back to stage N, thus providing a raffinate reflux. It can be shown that for a given feed and extract product, since material balance fixes the composition and amount of the raffinate product, the net flow outward and hence the difference point D'' is not influenced by the quantity of the raffinate reflux. The result is that the number of stages in the cascade is the same whether a raffinate reflux is employed or not.

Extract Reflux and Type-I Systems

Figure 7.64 shows the Janecke diagram of systems with one pair of liquids partially miscible. The maximum solvent-free solute concentration of the final extract leaving the countercurrent cascade with extract reflux is given by the point E, the point farthest towards the right on the binodal curve. We know that in the absence of the extract reflux, the maximum solute concentration that can be achieved for the extract leaving stage 1 is the concentration of the liquid which is in equilibrium with the feed. If this happens to be close to E in Fig. 7.64, then the use of the extract reflux offers no significant advantage over the system with no reflux; this is frequently the case with type I systems. However, for type II

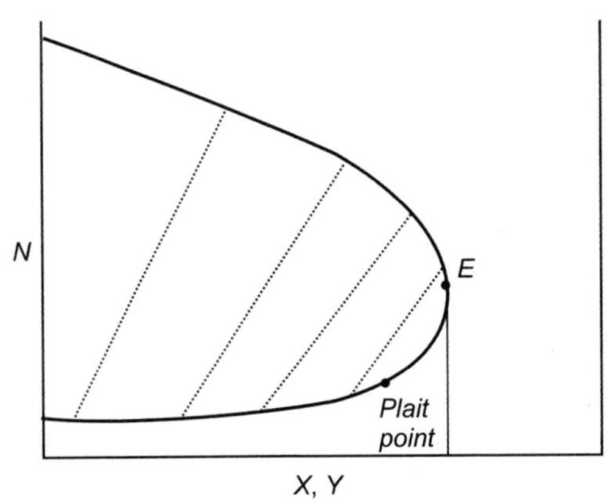

Fig. 7.64: Maximum solute concentration achievable for type I systems in countercurrent extraction

systems no such limit exists to the concentration of solute in the final extract product and therefore, the recovery can be as high as desired.

Example 7.11: A solution containing 50 weight% *n*-heptane and 50 weight% methylcyclohexane (MCH) is to be separated by extraction into products containing 90.0% MCH and 10.0% MCH, both on solvent-free basis. Aniline will be used as the solvent. Assume that the extract product and the reflux are free of solvent. Equilibrium data in weight fractions are:

Raffinate phase, weight fraction			Extract phase, weight fraction		
n-Heptane	MCH	Aniline	*n*-Heptane	MCH	Aniline
0.9260	0.0000	0.0740	0.0620	0.0000	0.9380
0.8315	0.0914	0.0772	0.0599	0.0080	0.9321
0.7341	0.1858	0.0800	0.0530	0.0270	0.9200
0.6985	0.2194	0.0821	0.0511	0.0300	0.9190
0.5768	0.3373	0.0859	0.0450	0.0461	0.9089
0.5068	0.4063	0.0869	0.0400	0.0600	0.9000
0.4505	0.4596	0.0900	0.0360	0.0740	0.8900
0.3074	0.5966	0.0960	0.0298	0.0980	0.8723
0.2286	0.6714	0.1000	0.0211	0.1132	0.8658
0.1823	0.7158	0.1019	0.0160	0.1269	0.8571
0.1604	0.7357	0.1039	0.0139	0.1310	0.8551
0.0541	0.8330	0.1128	0.0062	0.1559	0.8379
0.0000	0.8811	0.1189	0.0000	0.1689	0.8311

Determine:

(a) The minimum number of theoretical stages
(b) Minimum extract reflux ratio
(c) The number of theoretical stages for a reflux ratio of 5.0

***Solution*:** **A** = *n*-Heptane, **B** = Aniline, **C** = Methylcyclohexane

The equilibrium data are converted into the corresponding *N* versus *X* (or *Y*) data:

$$N = \frac{B}{A+C},\ X = \frac{A}{A+C} \text{ in the raffinate, } Y = \frac{A}{A+C} \text{ in the extract}$$

For the third set of equilibrium data, the raffinate phase contains

A = 0.7341 kg, **C** = 0.1858 kg and **B** = 0. 0.08 kg. Therefore,

$$X = \frac{C}{A+C} = \frac{0.1858}{0.7341+0.1858} = 0.202 \text{ kg MCH/kg (Heptane + MCH)}$$

and
$$N = \frac{B}{A+C} = \frac{0.08}{0.7341+0.1858} = 0.087 \text{ kg aniline/kg (MCH + Heptane)}$$

For the corresponding extract phase,

A = 00530, **C** = 0.0270, **B** = 0.92. Therefore,

$$Y = \frac{C}{A+C} = \frac{0.0270}{0.0530+0.0270} = 0.3375 \text{ kg MCH/kg (Heptane + MCH)}$$

and
$$N = \frac{B}{A+C} = \frac{0.92}{0.0530+0.0270} = 11.5 \text{ kg aniline/kg (MCH+Heptane)}$$

Similar calculations for all other data sets give the following results:

Raffinate phase		Extract phase	
X	N	Y	N
0	0.0799	0	15.12
0.099	0.0836	0.118	13.72
0.202	0.087	0.338	11.5
0.239	0.0894	0.370	11.34
0.369	0.094	0.506	9.98
0.445	0.0952	0.600	9.00
0.505	0.0989	0.673	8.09
0.660	0.1062	0.767	6.83
0.746	0.1111	0.843	6.45
0.797	0.1135	0.888	6.00
0.821	0.116	0.904	5.90
0.939	0.1272	0.962	5.17
1.00	0.135	1.00	4.92

The Janecke diagram is plotted with N on the y-axis and X or Y on the x-axis as shown in Fig. 7.65a. Point F represents the feed at $X = X_F = 0.50$ and $N_F = 0$. The final raffinate is at R_N on the solubility curve at $X = X_N = 0.10$. The extract product E and the raffinate R_0 lie on the x-axis at $X = 0.90$ as they are free of solvent. The extract E_1 can be located at the intersection of the vertical line passing through R_0 with the extract side of the solubility curve.

(a) The minimum number of equilibrium stages corresponds to the total reflux condition at which the difference point which represents the net flow is at infinity and the operating lines are vertical. The graphical construction for finding the number of stages is as follows: Draw the tie line E_1R_1 as explained earlier. The vertical line passing through R_1 intersects the upper solubility curve at E_2. E_2R_2 is the second tie line and R_2E_3 is the vertical line between the lower and the upper solubility curves. The construction is continued till the final raffinate R_N is reached. The number of tie lines gives the number of ideal stages. From the figure, the minimum number of stages = 7.7. **(Answer)**

(b) The tie line which when extended passes through the feed point F intersects the vertical line at E_1 at $D'_{\min}$ which is the difference point that determines the minimum reflux r_m. The minimum reflux can be calculated using

$$r_m = \frac{\overline{D'_{\min}E_1}}{\overline{E_1R_0}} = \frac{N_{D'_{\min}} - N_{E,1}}{N_{E,1} - N_{R,0}}$$

The N-coordinates of $D'_{\min}$, E_1 and R_0 are read from the figure. $N_{D'_{\min}} = 19.982$, $N_{E,1} = 5.888$, $N_{R,0} = 0$. Substituting the values, the minimum reflux ratio is

$$r_m = \frac{N_{D'_{\min}} - N_{E,1}}{N_{E,1} - N_{R,0}} = \frac{19.982 - 5.888}{5.888 - 0} = 2.394$$ **(Answer)**

(c) When the reflux ratio $r = 5$, we have $r = \dfrac{\overline{D'E_1}}{\overline{E_1R_0}} = 5$. Since the line length $\overline{E_1R_0}$ is available from the graph, D' can be located on the vertical line through E_1 at a height 5 times E_1R_0 above E_1. (Alternatively, the N-coordinate $N_{D'}$ of the difference point can be evaluated using the relation

$$N_{D''} = N_{E,1} + 5(N_{E,1} - N_{R,0}) = 5.888 + 5(5.888) = 35.33$$

The feed line $D'FD''$ is drawn by extending the line $D'F$ to meet the vertical line passing through RN at D'', the difference point for the raffinate stripping section of the cascade. The number of stages is determined by alternating the construction of the tie lines and operating lines as explained earlier,

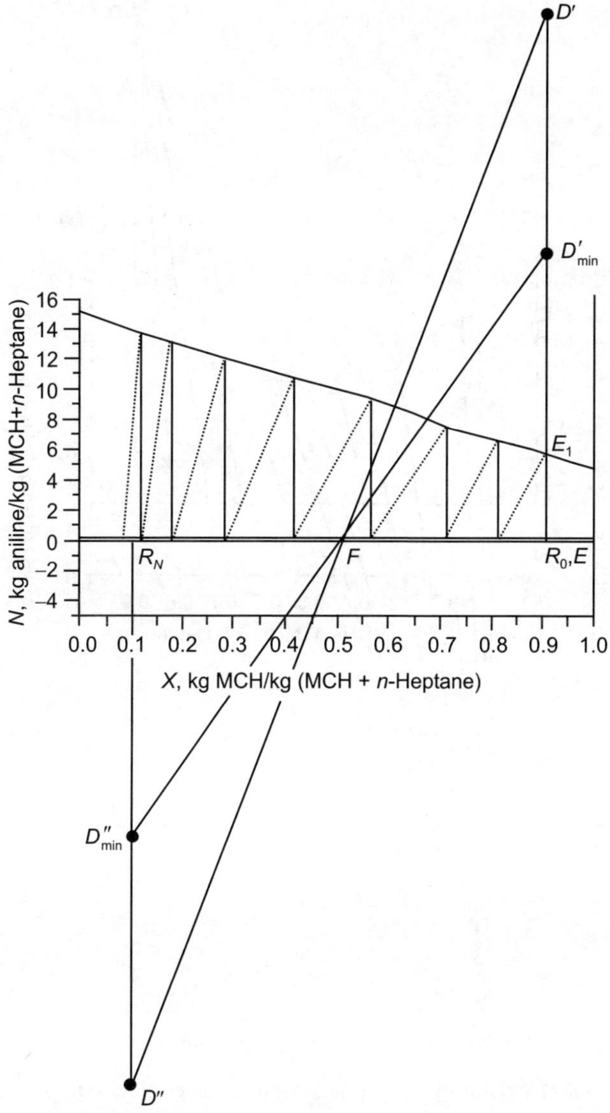

Fig. 7.65: (a) Graphical construction for minimum number of stages and minimum reflux ratio for example 7.11

starting from the point E_1. The operating lines are drawn using the upper difference point D' till the tie line crosses the feed line, and by using the lower difference point D'' after crossing the feed line. The construction is terminated when the final raffinate R_N is reached.

The number of stages $N = 13$. **(Answer)**

7.8 CONTINUOUS CONTACT EXTRACTORS

In stage contact equipment, the feed solution and the solvent are brought into contact in separate stages with intermediate settling and the phases are not in continuous contact. Instead, if the heavy liquid is made to flow down the equipment such as a hollow vertical tower by gravity and the light liquid is admitted at the bottom as a fine dispersion and flow in the direction opposite to that of the heavy phase, transfer of the solute and hence the concentration change of the phases occur continuously along the

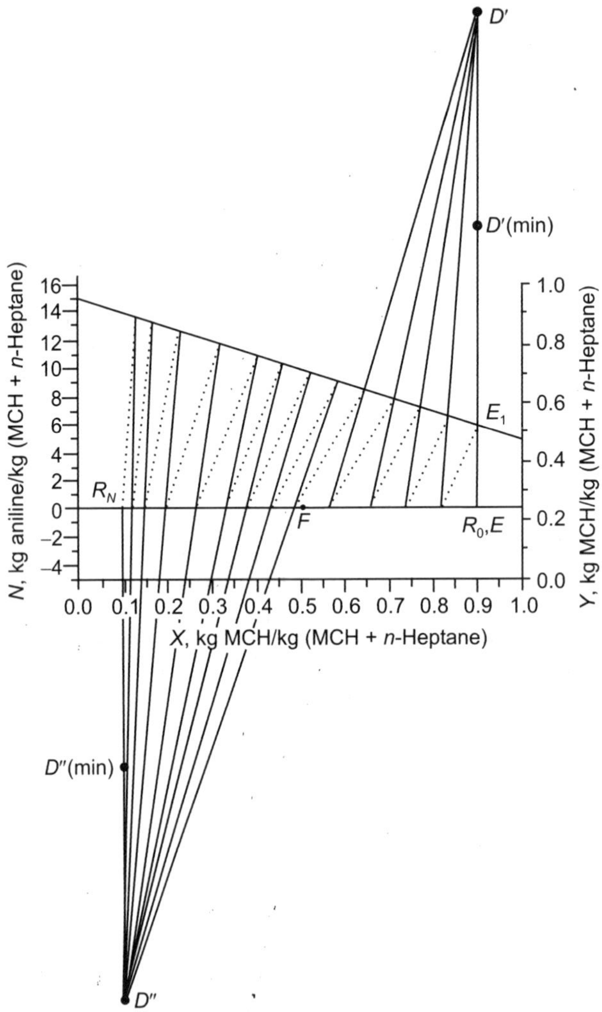

Fig. 7.65: (b) Graphical construction for finding the number of stages for Example 7.11

length of the device. The spray column is a typical continuous contact equipment. The packed column extractors provided with packing materials which increase the interfacial area of mass transfer, columns provided with baffles, perforated trays, etc. to promote turbulence and influence the flow characteristics to improve the mass transfer rate, are all continuous contact extractors.

The earlier method of designing continuous contact equipment was based on the concept of height equivalent to a theoretical stage (HETS). It is the height of a continuous contact equipment that gives the same extent of extraction as an ideal stage. HETS is an experimentally determined quantity. The height of the tower is the product of HETS and the number of theoretical stages required for the given separation. The method is not theoretically sound and is practically inconvenient because HETS is influenced by a variety of factors that include type of system, rates of flow, concentration of the contacting phases, type of tower, type and size of packing, etc. The HETS method is now abandoned in favour of the method based on the transfer units.

Consider unit area of cross section of a continuous contact equipment of height Z as shown in Fig. 7.66a. The raffinate and extract flow rates are R kmol/m^2 h and E kmol/m^2 h respectively and the mole fraction of solute in the raffinate and extract phases are represented by x and y, respectively. The conditions at the end where the raffinate phase is admitted are represented by suffix 1 and those at the end where the extract is admitted by suffix 2. It is assumed that the carrier liquid **A** and the solvent **B** are insoluble and only the solute **C** is transferred between the phases. The change in the mutual solubility of **A** and **B** due to transfer of **C** is ignored in the following analysis.

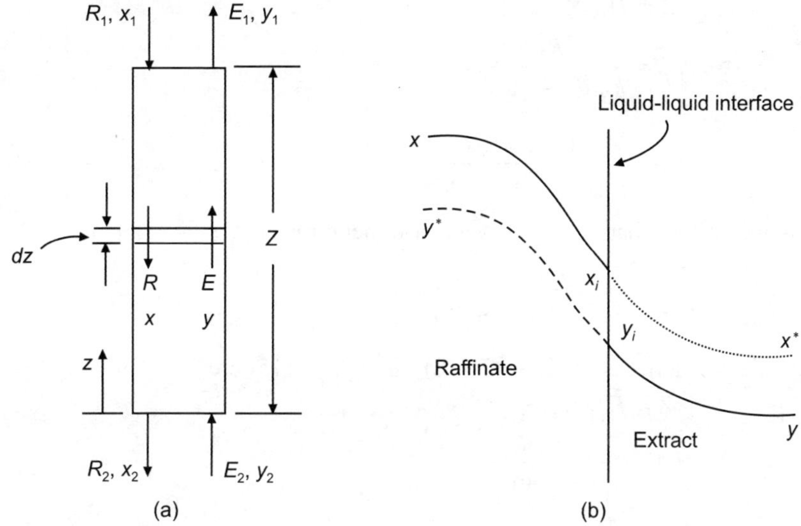

Fig. 7.66: (a) Schematic representation of a continuous contact unit for extraction (b) Concentration change across liquid–liquid interface

Consider a differential section in the equipment of thickness dz at a distance z from the bottom, where the raffinate and extract flows are R and E and the mole fraction of the solute in these streams are x and y, respectively. Figure 7.66b represents schematically the concentration profile at the liquid–liquid interface. Analogous to Eq. (4.112), we can show that the rate of mass transfer at this section is

$$k_R(x - x_i) = \frac{R\,dx}{(1-x)\,a\,dz} \quad \ldots(7.108)$$

where k_R is the raffinate phase mass transfer coefficient and a is the interfacial area for mass transfer per unit volume of packing. The concentrations at the raffinate and extract sides of the interface are represented by x_i and y_i, respectively. Introducing the term $(1-x)_{iM}$ into Eq. (7.108) and rearranging, we get

$$dz = \frac{R}{k_R a(1-x)_{iM}} \times \frac{(1-x)_{iM}\,dx}{(1-x)(x-x_i)} \quad \ldots(7.109)$$

where $(1-x)_{iM}$ is the logarithmic mean of $(1-x)$ and $(1-x_i)$.

$$(1-x)_{iM} = \frac{(1-x)-(1-x_i)}{\ln\dfrac{(1-x)}{(1-x_i)}} \quad \ldots(7.110)$$

Since the quantity $k_R(1-x)_{iM}$ is expected to remain constant along the length of the equipment, Eq. (7.109) can be integrated as

$$Z = \frac{R}{k_R a(1-x)_{iM}} \int_{x_2}^{x_1} \frac{(1-x)_{iM}\,dx}{(1-x)(x-x_i)} \quad \ldots(7.111)$$

Assuming $\frac{(1-x)_{iM}}{(1-x)} \approx 1$, the integral in Eq. (7.111) is approximately the number of times the average driving force for mass transfer divides into the concentration difference $(x_1 - x_2)$, or it represents the number of raffinate transfer units, designated as N_{tR}. The quantity outside the integral has the dimensions of length and is designated as the height of the raffinate transfer units, H_{tR}.

$$Z = H_{tR} \times N_{tR} \qquad \text{...(7.112)}$$

$$H_{tR} = \frac{R}{k_R a(1-x)_{iM}} \qquad \text{...(7.113)}$$

$$N_{tR} = \int_{x_2}^{x_1} \frac{(1-x)_{iM}\,dx}{(1-x)(x-x_i)} \qquad \text{...(7.114)}$$

Using the simplification that $(1-x)_{iM}$ is the arithmetic mean of $(1-x)$ and $(1-x_i)$, we get

$$N_{tR} = \int_{x_2}^{x_1} \frac{dx}{(x-x_i)} + \frac{1}{2}\ln\frac{1-x_2}{1-x_1} \qquad \text{...(7.115)}$$

The extract-side counterparts for Eqs. (7.112) through (7.115) are

$$Z = H_{tE} \times N_{tE} \qquad \text{...(7.116)}$$

$$H_{tE} = \frac{E}{k_E a(1-y)_{iM}} \qquad \text{...(7.117)}$$

$$N_{tR} = \int_{y_2}^{y_1} \frac{(1-y)_{iM}\,dy}{(1-y)(y_i-y)} \approx \int_{y_2}^{y_1} \frac{dy}{(y_i-y)} + \frac{1}{2}\ln\frac{1-y_2}{1-y_1} \qquad \text{...(7.118)}$$

For using the above expressions to find the size of the equipment, we should know the variation of the equilibrium interfacial concentrations along the length of the equipment. The interfacial concentrations are related to the raffinate and extract side mass transfer coefficients. If the coefficients are known as a function of composition, the interfacial concentrations can be estimated. But, only very rarely we get sufficient information on these coefficients for estimating the interfacial compositions accurately. For practical reasons, therefore, it is necessary to deal with the overall transfer units employing the overall mass transfer coefficients, K_R or K_E and the overall driving force for mass transfer $(x - x^*)$ or $(y^* - y)$.

$$Z = H_{toR} \times N_{tOR} \qquad \text{...(7.119)}$$

where N_{tOR} is the number of overall transfer unit and H_{tOR} is the height of the overall transfer unit based on the raffinate side.

$$N_{tOR} = \int_{x_2}^{x_1} \frac{(1-x)_{*M}\,dx}{(1-x)(x-x^*)} \qquad \text{...(7.120)}$$

$$H_{tOR} = \frac{R}{K_R a(1-x)_{*M}} \qquad \text{...(7.121)}$$

By replacing the logarithmic mean concentration difference $(1-x)_{*M}$ with the corresponding arithmetic mean, Eq. (7.120) can be simplified as

$$N_{tOR} = \int_{x_2}^{x_1} \frac{dx}{(x-x^*)} + \frac{1}{2}\ln\frac{1-x_2}{1-x_1} \qquad \text{...(7.122)}$$

For dilute solutions a further simplification is possible since $(1-x)_{*M} \approx 1$ in Eq. (7.121) and the second term in Eq. (7.122) vanishes. Then,

$$H_{tOR} = \frac{R}{K_R a} \quad \text{...(7.123)}$$

$$N_{tOR} = \frac{x_1 - x_2}{(x - x^*)_{LM}} \quad \text{...(7.124)}$$

where $(x - x^*)_{LM}$ is the logarithmic mean of the concentration difference $(x_1 - x_1^*)$ at the top and $(x_2 - x_2^*)$ at the bottom. In deriving Eq. (7.124) it is further assumed that the equilibrium curve and the operating lines are straight lines in the concentration range of interest.

Referred to the extract side, we can develop the following equations for the height and number of the overall transfer units.

$$Z = H_{toE} \times N_{tOE} \quad \text{...(7.125)}$$

$$N_{tOE} = \int_{y_2}^{y_1} \frac{(1-y)_{*M}\, dy}{(1-y)(y^* - y)} \approx \int_{y_2}^{y_1} \frac{dy}{(y^* - y)} + \frac{1}{2}\ln\frac{1-y_2}{1-y_1} \quad \text{...(7.126)}$$

$$H_{tOE} = \frac{E}{K_E a(1-y)_{*M}} \quad \text{...(7.127)}$$

For dilute solutions and straight equilibrium curve and operating lines,

$$H_{tOE} = \frac{E}{K_E a} \quad \text{...(7.128)}$$

$$N_{tOE} = \frac{y_1 - y_2}{(y^* - y)_{LM}} \quad \text{...(7.129)}$$

where $(y^* - y)_{LM}$ is the logarithmic mean of $(y_1^* - y_1)$ and $(y_1^* - y_1)$

Example 7.12: A valuable solute is being extracted in a packed column using a solvent immiscible with the carrier liquid of the feed. The equilibrium relationship for the distribution of solute between the immiscible phases is $y = 0.06x$ where x and y are the mole fractions of solute in the raffinate and extract phases, respectively. The terminal conditions are:

x	0.001	0.08
y	0.0	0.003

Determine:

(a) The number of transfer units based on the extract phase

(b) The height of packing, if the height of the transfer unit based on the extract side is 0.4 m.

***Solution*:** (a) The number of transfer units can be determined using the expression

$$N_{tOE} = \frac{y_1 - y_2}{(y^* - y)_{LM}}$$

where $(y^* - y)_{LM}$ is the logarithmic mean of $(y_1^* - y_1)$ and $(y_2^* - y_2)$. The terminal concentrations, are shown in Fig. 7.67.

$y_1^* = mx_1 = 0.06 \times 0.08 = 0.0048$.

Therefore, $y_1^* - y_1 = 0.0048 - 0.003 = 0.0018$

$y_2^* = mx_2 = 0.06 \times 0.001 = 0.00006$.

Therefore, $y_2^* - y_2 = 0.00006 - 0.0 = 0.00006$

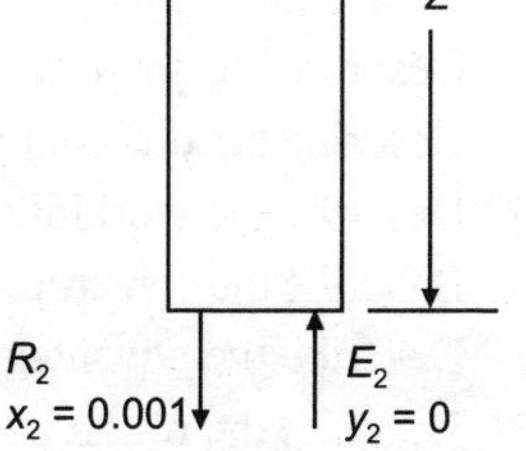

Fig. 7.67: Packed column in Example 7.12

$$(y^*-y)_{LM} = \frac{(y_1^*-y_1)-(y_2^*-y_2)}{\ln\dfrac{y_1^*-y_1}{y_2^*-y_2}} = \frac{0.0018-0.00006}{\ln\dfrac{0.0018}{0.00006}} = 0.00051$$

Therefore, the number of transfer units based on extract side

$$N_{tOE} = \frac{y_1-y_2}{(y^*-y)_{LM}} = \frac{0.003}{0.00051} = 5.88$$ **(Answer)**

(b) The height of the tower $Z = H_{tOE}N_{tOE}$ where H_{tOE} is the height of the extract side transfer unit. Then $Z = N_{tOE}H_{tOE} = 5.88 \times 0.4 = 2.35$ m.

Height of the tower = 2.35 m **(Answer)**

REVIEW QUESTIONS

1. Define the liquid–liquid extraction process.
2. List the major industrial applications of extraction.
3. What is fractional extraction?
4. Explain the terms raffinate and extract with a suitable example.
5. What are the advantages of extraction over distillation for the separation of liquid mixtures?
6. What are the major areas of applications of extraction as a separation technique?
7. Distillation is a direct technique for separation, whereas extraction is not. Why?
8. Why triangular diagrams are necessary for representing the liquid–liquid equilibrium?
9. Describe the mixture rule on triangular coordinate system.
10. Describe the type I and type II systems in extraction with examples.
11. What are solutropic systems?
12. Define the selectivity and the distribution coefficient.
13. Explain how selectivity is related to distribution coefficient.
14. What do you understand by the plait point?
15. Describe the phase rule as applied to liquid–liquid extraction.
16. What is the effect of temperature and pressure on the liquid–liquid equilibria?
17. Explain the methods for interpolating tie line data.
18. For graphical computation of the number of stages, the right triangular coordinates are preferred over equilateral triangle. Explain.
19. Explain the Janecke diagrams.
20. Describe the desirable properties of an extraction solvent.
21. Describe the classification of extraction equipment.
22. Describe the advantages and disadvantages of mixer-settlers for extraction.
23. Describe the different types of static extraction columns and list their relative merits and demerits.
24. Describe the working of Scheibel extractors and Kunhi extractors.
25. Describe the construction, working and advantages of rotating-disc contactors for liquid extraction.
26. Describe the advantages of centrifugal extractors.
27. Describe the rough guideline for selection of extractors for specific applications.
28. Explain how the inverse-lever-arm rule can be used for the single-stage extraction calculations.
29. How are the minimum and maximum solvent quantities for single stage extraction estimated?

30. Explain the advantages of multistage cross-current extraction over single-stage operations.
31. What are the advantages of multistage countercurrent extraction over cross-current extraction?
32. Describe the geometrical construction for finding the number of stages for specified recovery of solute by countercurrent extraction, from a feed solution of known rate and composition.
33. How is the minimum solvent requirement for multistage countercurrent extraction estimated?
34. Is there a maximum for the solvent quantity that can be used in countercurrent extraction?
35. How is the minimum solvent requirement and number of stages for countercurrent extraction determined graphically when the solvent **B** is immiscible with carrier liquid **A**?
36. What is extraction factor? How is it useful in estimating the number of stages in extraction operations?
37. Discuss the importance of extract reflux in countercurrent extraction?
38. Explain how the graphical construction on the Janecke diagram for countercurrent extraction with reflux, is similar to the Ponchon-Savarit method for distillation.
39. Explain how the minimum reflux ratio for countercurrent extraction is estimated graphically.
40. It is not advantageous to provide raffinate reflux for countercurrent extraction. Why?
41. Providing the reflux usually give no significant enrichment in extraction of the type-I systems, in comparison to the type-II systems. Why?
42. Define HETS. What are its limitations in the design of continuous contact extractors?

MULTIPLE CHOICE QUESTIONS

7.1 The most important characteristic of the liquids which determines the success of the liquid–liquid extraction is

A. the difference in the density
B. the difference in the volatility
C. the difference in the solubility
D. the difference in the viscosity

7.2 Extraction is preferred method of separation for recovering penicillin from fermentation broth because

A. extraction is a low temperature operation
B. penicillin is an antibiotic and hence should be separated by extraction
C. penicillin is non volatile
D. extraction is less expensive compared to distillation

7.3 Distillation is a direct technique for separation, whereas extraction is not. Why?

A. Because the products of extraction operation require further purification
B. Because separation by extraction is achieved at relatively low temperatures
C. Because extraction requires make up solvent
D. Because solvent used for extraction is partially miscible with the feed

7.4 Referred to the equilateral triangular coordinates used for representing liquid equilibria, which of the following is not true?

A. The vertices represent pure components.
B. The binary mixture can be represented by a point lying on the side of the triangle.
C. The ternary mixture always lie in the dome shaped region formed by the binodal curve.
D. The perpendicular distances from a point inside the triangle to the sides give the concentration of the components in the ternary mixture

7.5 What is the composition of the ternary mixture represented by point *P* in the following figure?

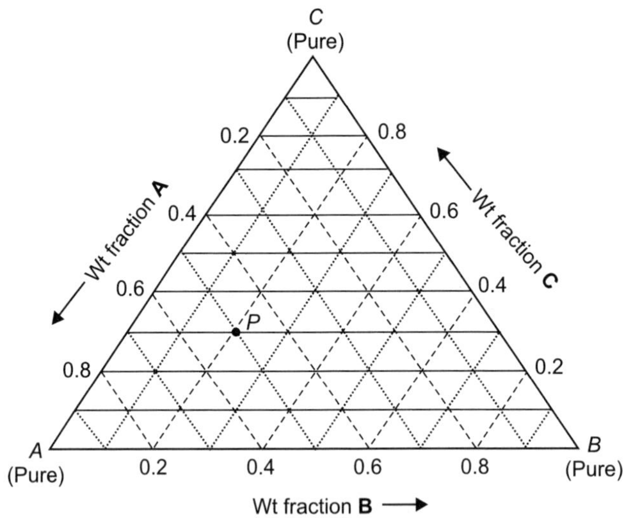

A. 20%A, 30%B, 50%C B. 30%A, 50%B, 20%C

C. 50%A, 20%B, 30%C D. 50%A, 30%B, 20%C

7.6 Points *P*, *Q* and *R* represent ternary mixtures on the equilateral triangle in such a way that the line *PQ* is divided by *R* in the ratio *PR*:*RQ* = 1:2; If *Q* = 50 kg, what is the weight of the mixture R?

A. 50 kg B. 100 kg

C. 150 kg D. 200 kg

7.7 In the triangular diagram represented below for batch separation process, a stream *F* is mixed with a solvent **B** to produce products *R* and *E*. Substance **A** is the carrier liquid and **C** is the solute being extracted. The amounts of **B** and *E* are 1 kg and 1.2 kg, respectively. The length *FM* is 3.1 and the length *FB* is 8.5 units on the figure. The ratio *R*/*E* is estimated to be:

A. 1.285 B. 2

C. 0.751 D. 2.5

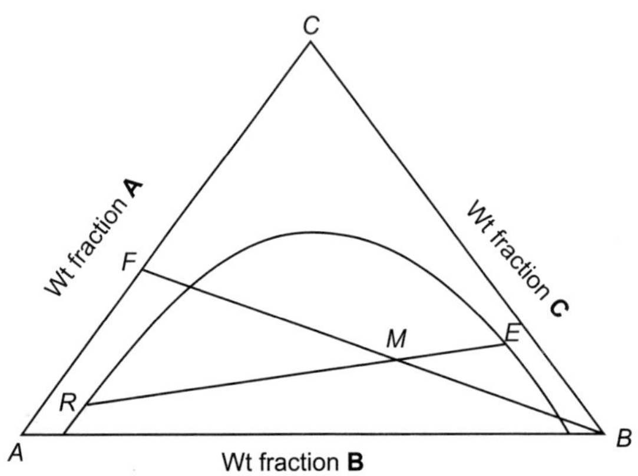

7.8 The solutropic systems are those in which

A. the direction of tie line slope remain unchanged

B. the direction of tie line slope changes with one tie line being horizontal

C. all tie lines are horizontal

D. none of the above.

7.9 In extraction, which of the following can be treated as equivalent to heat in distillation?

A. Raffinate B. Extract

C. Solvent D. Selectivity

7.10 If the distribution coefficient is less than unity,

A. separation by extraction is not possible

B. weight fraction of solute in the raffinate is greater than that in extract.

C. selectivity will be less than unity

D. slope of the distribution curve will be negative

7.11 Which of the following is not true with regard to the plait point in liquid equilibria?

A. The raffinate and extract phases are indistinguishable at the plait point

B. The plait point lies at the peak of the binodal solubility curve

C. The tie line merges to a single point at the plait point

D. The distribution coefficient is unity at the plait point

7.12 For a ternary system, at the plait point

A. The density difference between phases vanishes

B. The distribution coefficient is unity

C. The selectivity is unity

D. All of the above

7.13 In the liquid extraction involving ternary systems, the number of degrees of freedom at constant temperature and pressure is

A. four B. three

C. two D. one

7.14 In the extraction of solute **C** from a mixture of **A** and **C** using solvent **B**, if **A** and **B** are completely immiscible, the selectivity of solvent **B** for solute **C** is

A. infinity B. zero

C. unity D. negative

7.15 The selectivity at the plait point is

A. infinity B. zero

C. unity D. negative

7.16 One hundred kg of a solution containing 50 weight% **A** and 50 weight% **C** is mixed with a 75 kg solvent S containing 2 weight% **C** and the rest **B**. 90 kg raffinate of composition 54% **A** and 6% **B** is separated. Determine the selectivity of solvent **B** for **C**.

A. 10 B. 12

C. 8 D. 15

7.17 In the extraction of solute **C** from a mixture of **A** and **C**, pure solvent **B** which is partially miscible with **A** is added to the feed and the immiscible phases are separated. The raffinate phase analysed 80% **A**, 12% **B** and 8% **C** on weight basis and the extract phase contains 78% **B**. If the distribution coefficient on weight basis is 1.5, what is the selectivity of solvent **B** to **C**?

A. 12 B. 8

C. 1 D. None of the above.

7.18 The selectivity in extraction

A. is unaffected by change in temperature

B. increases with increase in temperature

C. decreases with increase in temperature

D. may increase or decrease with temperature.

7.19 For extraction involving type-I systems, the temperature should be

A. above the critical temperatures
B. above the critical solution temperature
C. below the critical solution temperature
D. between critical temperature and critical solution temperature

7.20 A mixture consists of 30 kg **A**, 50 kg **B** and 20 kg **C**. (**A** = carrier solvent in the feed, **B** = extraction solvent, **C** = solute extracted). On the Janecke diagrams the coordinates of the mixture point are:

A. $x = 0.4$, $N = 1$ B. $x = 0.4$, $N = 0.60$
C. $x = 0.25$, $N = 0.625$ D. $x = 0.20$, $N = 0.50$

7.21 At high .... the type-II system may exhibit type-I behaviour.

A. temperature B. pressure
C. solute concentration D. none of the above

7.22 A solvent with low interfacial tension leads to

A. easy coalescence and rapid disengagement of phases
B. formation of stable emulsions difficult to settle
C. easy separation by gravity
D. high rate of mass transfer

7.23 Which one of the following is not desirable for an extraction solvent ?

A. High selectivity for the solute B. High solubility with carrier liquid
C. Low viscosity D. High capacity for dissolving the solute.

7.24 500 kg feed containing 40% **C** is extracted with 400 kg pure solvent **B**. 600 kg extract is formed analyzing 25% **C**. The distribution coefficient corresponding to this concentration is

A. 1.67 B. 1.25
C. 0.25 D. 1.5

7.25 It is desired to reduce the concentration of pyridine in 500 kg of aqueous solution from 20 weight% to 5 weight% in a single batch extraction using chlorobenzene as solvent. Equilibrium compositions in terms of weight percent of pyridine-water-chlorobenzene are (5, 95, 0) and (11, 0, 89). The amount of pure solvent required in kg for the operation is

A. 607 B. 639
C. 931 D. 1501

7.26 An ideal single-stage extraction process is used to treat 100 mol/s of an organic feed solution. The solute concentration in the solution is to be reduced from 0.5 mol% to 0.1 mol%. A pure solvent B is used. To reduce the solvent requirement by half for the same operation

A. add one more ideal cocurrent stage
B. use another pure solvent whose partition coefficient is twice that of B
C. use solvent B containing 0.02 mole fraction of solute.
D. double the residence time of the solvent B in the contactor.

7.27 In liquid–liquid extraction 10 kg of a solution containing 2 kg of solute **C** and 8 kg of solvent A is brought into contact with 10 kg of solvent **B**. Solvents **A** and **B** are completely immiscible. The equilibrium relation is $Y = 0.9X$ where Y is kg **C**/kg **B** and X is kg **C** by kg **A**. Choose the correct answer.

A. The entire amount of **C** is transferred to solvent **B**
B. Less than 2 kg but more than 1 kg of **C** is transferred to **B**
C. Less than 1 kg of **C** is transferred to **B**
D. No amount of C is transferred to B

7.28 In countercurrent extraction, the maximum concentration of solute in the final extract is
A. that of liquid in equilibrium with the feed
B. less than that of liquid in equilibrium with the feed
C. Can be increased by using raffinate reflux
D. Cannot be increased by using extract reflux

7.29 For a given degree of separation using a given quantity of solvent, the number of stages required in continuous countercurrent extraction is
A. less than the number of stages required in crosscurrent extraction
B. more than the number of stages in crosscurrent extraction
C. the same as that required for crosscurrent extraction
D. equal to or greater than that in crosscurrent extraction

7.30 The number of stages required for a continuous countercurrent unit for a given extent of extraction
A. remains constant with change in solvent rate
B. remains constant with change in reflux ratio
C. will increase by increasing the solvent rate and reflux ratio
D. decreases with increase in solvent rate and reflux ratio

7.31 A 5-stage countercurrent mixer-settler cascade achieves 90% recovery of solute in certain extraction operation. If the same recovery is to be achieved even on removing one of the stages from the cascade, which one of the following is to be accepted?
A. Reduce the solvent rate
B. Increase the solvent rate
C. Use the same solvent rate
D. Not possible to achieve the recovery

7.32 A countercurrent extraction column is designed to remove 99% of solute **C** from a solution of solvent **A** and solute **C** using pure solvent **B**. The initial concentration of solute in the solution of **A** + **C** is 20 wt%, and the total flow of solution is 1000 kg/h. The equilibrium relationship is $Y = 0.50\,X$, where X is mass of **C**/mass of **A** and Y = mass of **C**/mass of **B**. The minimum flow rate of solvent **B** required (in kg/h) is
A. 1454 B. 1584
C. 1676 D. 1874

7.33 A countercurrent extraction column is designed to remove 99% of solute **C** from a solution of solvent **A** and solute **C** using pure solvent **B**. The initial concentration of solute in the solution of **A** + **C** is 20 wt%, and the total flow of solution is 1000 kg/h. The equilibrium relationship is $Y = 0.50\,X$, where X is mass of C/mass of A and Y = mass of **C**/mass of **B**. If the flow rate of **B** is 2400 kg/h, then the theoretical number of stages in the column, using Kremser equation (adjusted to the next integer) is
A. 5 B. 9
C. 11 D. 13

7.34 In a single stage extraction process, 10 kg of pure solvent S (containing no solute **C**) is mixed with 30 kg of feed F containing **C** at a mass fraction $x_F = 0.2$. The mixture splits into an extract phase E and a raffinate phase R, containing **C** at $y_E = 0.5$ and $x_R = 0.05$ respectively. The total mass of the extract phase is (in kg)
A 6.89 B. 8.89
C. 10 D. 8.25

7.35 Compound **C** is extracted from an equimolar solution of **A** + **C** into a pure solvent **B**. A cocurrent unit is used for the liquid–liquid extraction. The inlet rate of the solution containing **C** is 200 moles of **A**/h m² and the solvent flow rate is 400 moles of **B** /h m². The equilibrium data is represented by $X = 3\ Y^2$, where Y is in moles of **C**/moles of **B** and X is in moles of **C**/moles of **A**. The maximum percentage extraction achieved in the unit is

A. 25% B. 50%
C. 67% D. 90%

7.36 For handling liquids of any viscosity, the extractors preferred are:

A. Mixer-settlers B. Packed columns
C. Spray columns D. Sieve-plate columns

7.37 Extractors most suitable for handling the liquids that show a tendency to foul or clog are:

A. Mixer-settlers B. Packed columns
C. Spray columns D. Sieve-plate columns

7.38 Axial dispersion is a serious limitation of

A. mixer-settlers B. packed columns
C. spray columns D. sieve-plate columns

7.39 For highly corrosive or radioactive liquids, the extractors preferred are:

A. Pulsed columns B. Rotary impeller columns
C. Tray columns D. Mixer-settlers

7.40 Which of the following extractors can handle slurries of solids?

A. Mixer-settlers B. Plate towers
C. Packed columns D. Karr extractor

7.41 For systems that emulsify easily or have small density differences or large interfacial tensions, the most suitable extractors are:

A. Mixer-settlers B. Centrifugal extractors
C. Static columns D. Pulse columns

7.42 Which of the following is preferred for antibiotics and other pharmaceuticals because of their low holdup?

A. Scheibel extractor B. Karr extractor
C. Podbielniak extractor D. Rotating disc contactor

EXERCISES

Liquid equilibrium

7.1 One hundred kg of a mixture consisting of water, acetic acid and isopropyl ether containing 10% acetic acid by weight separates into a raffinate of composition $x_C = 0.133$, $x_A = 0.844$, $x_B = 0.023$ and an extract of composition $y_C = 0.048$, $y_A = 0.019$, $y_B = 0.933$.Determine the amounts of extract and raffinate.

(**Answer:** Raffinate: 61.18 kg, Extract = 38.82 kg)

7.2 One hundred kg of a mixture consisting of 30.0 weight% **A**, 30.0 weight% **B** and 40 weight% **C** is allowed to separate into an **A**-rich phase and a **B**-rich phase. The **A**-rich phase weighed 30 kg and analysed 80.0 weight% **A**, 10.0 weight% **B** and 10.0 weight% **C**. Using the mixture rule determine graphically the composition of the **B**-rich phase and compare with the result obtained by solute balance.

(**Answer:** A = 8.57%, B = 38.57%, C = 52.86%)

7.3 Construct a triangular phase diagram for a ternary system made up of liquids **A**, **B** and **C** from the experimental data given in the table below. Also plot the distribution diagram for the system. What is the composition of the **B**-rich phase in equilibrium with a raffinate solution containing 25% by weight **C**?

Raffinate phase, weight%			Extract phase, weight%		
A	**B**	**C**	**A**	**B**	**C**
97	3	0	4	96	0
86	4	10	5	90	5
75	5	20	6	80	14
62	8	30	8	70	22
48	12	40	12	56	32
40	15	45	14	48	38
30	20	50	21	32	47
22	30	48	22	30	48

(**Answer:** 18%**C**, 75%**B** and 7%**A**)

7.4 A mixture consisting of 10 kg acetic acid (**C**), 50 kg water (**A**) and 40 kg isopropyl ether (**B**) is allowed to separate into raffinate and extract phases which are in equilibrium. The equilibrium data for the system Acetic acid – water – Isopropyl ether at 298 K, 101.3 kPa.

Water-rich phase, weight%			Ether-rich phase, weight%		
Acetic acid	Water	Isopropyl ether	acetic acid	Water	Isopropyl ether
1.41	97.1	1.49	0.37	0.73	98.9
2.89	95.5	1.61	0.79	0.81	98.4
6.42	91.7	1.88	1.93	0.97	97.1
13.3	84.4	2.3	4.82	1.88	93.3
25.5	71.1	3.4	11.4	3.9	84.7
36.7	58.9	4.4	21.6	6.9	71.5
45.3	45.1	9.6	31.1	10.8	58.1
46.4	37.1	16.5	36.2	15.1	48.7

Determine the compositions of the equilibrium phases.

(**Answer:** Raffinate: 84.4% water, 13.3% acetic acid and 2.3% isopropyl ether
Extract: 1.88% water, 4.82% acetic acid, 93.3% isopropyl ether)

7.5 Construct a triangular phase diagram for the system water-isopropyl ether- acetic acid at 298 K from the experimental data given in Exercise 7.4. Use the diagram to determine: (a) the concentration of water and ether in an aqueous layer containing 20% by mass acetic acid. (b) The composition of ether layer in equilibrium with it (c) The minimum amount of acetic acid required to make the partially miscible system consisting of 4 kg ether and 6 kg water, homogeneous.

(**Answer:** (a) water = 77.1%, Ether = 2.9% (b) Water = 2.9%,
ether = 88.8%, acetic acid = 8.3% (c) 8.1 kg)

7.6 An aqueous solution of pyridine is treated with chlorobenzene for extraction of pyridine. If the extract and raffinate phases have the following compositions at equilibrium, calculate the selectivity of the solvent for pyridine.

Raffinate, weight%		Extract weight%	
Pyridine	Chlorobenzene	Pyridine	Chlorobenzene
11.1	0.24	19.0	79.9
45.0	4.2	35.1	61.0

(**Answer:** 136.7, 10.2)

7.7 Ternary equilibrium data for the system pyridine-water-chlorobenzene are given below in weight percent:

Extract phase			Raffinate phase		
Water	Chlorobenzene	Pyridine	Water	Chlorobenzene	Pyridine
0.05	99.95	0	99.92	0.08	0
0.67	88.28	11.05	94.82	0.16	5.02
1.15	79.9	18.95	88.71	0.24	11.05
1.62	74.28	24.1	80.72	0.38	18.9
2.25	69.15	28.6	73.92	0.58	25.5
2.87	65.58	31.55	62.05	1.85	36.1
3.95	61	35.05	50.87	4.18	44.95
6.4	53	40.6	37.9	8.9	53.2
13.2	37.8	49	13.2	37.8	49

Compute the selectivity of chlorobenzene for pyridine for each of the tie line data and plot the selectivity as a function of the solute concentration in the raffinate phase.

(Answer:

x	0	0.0502	0.1105	0.189	0.255	0.361	0.4495	0.532	0.49
β	∞	311.52	132.29	63.54	36.85	18.90	10.04	4.52	1

Immiscible systems: Crosscurrent extraction

7.8 Show that in a crosscurrent extraction unit where the liquids **A** and **B** are immiscible, the number of cross-current stages is given by

$$N = \frac{\log(X_N/X_F)}{\log(A/(A+Bm))}$$

where X and Y are the mass ratio of solute present in the raffinate and extract and m is the distribution coefficient given by $Y = mX$.

7.9 Nicotine in water solution containing 1% nicotine is to be extracted with kerosene at 293 K. Water and kerosene are essentially insoluble. Equilibrium data expressed as kg nicotine/kg liquid are as follows:

Weight fraction nicotine in water	0.001010	0.00246	0.005	0.00746	0.00988	0.0202
Weight fraction, nicotine in kerosene	0.000806	0.001959	0.00454	0.00682	0.00904	0.0185

(a) Determine the percentage extraction of nicotine if 100 kg of feed solution is extracted in a single-stage operation using 150 kg solvent. (b) What would be the percentage extraction if three theoretical stages are used with 50 kg solvent in each stage?

(Answer: (a) 57.23% (b) 67.43%)

7.10 An aqueous solution of a valuable solute with an initial concentration of 15 weight% of solute is subjected to liquid–liquid extraction in a three-stage cross flow unit using a pure solvent. The final raffinate should contain not more than 10% of the solute present in the original feed. The equilibrium data under operating conditions are

Weight fraction of solute in aqueous layer	0.06	0.12	0.18	0.24
Weight fraction of solute in extract layer	0.24	0.40	0.56	0.73

Selectivity is infinity. If equal quantities of solvent are used in all the stages, determine:

(a) Total quantity of solvent used per 1000 kg feed.

(b) Total quantity of extract obtained in part (a).

(c) The average composition of the combined extract.

(Answer: (a) 616 kg (b) 751 kg (c) 18% solute)

7.11 Benzene is used as solvent to recover the dioxane from its aqueous solution by liquid extraction. The equilibrium distribution of dioxane between water and benzene at 298 K is given below:

Dioxane in water layer, weight%	5.1	18.9	25.2
Dioxane in benzene layer, weight%	5.2	22.5	32.0

1000 kg of 25 weight% by weight aqueous solution of dioxane is to be treated with benzene to recover 90% of the dioxane. Calculate the quantity of solvent to be used in two stage crosscurrent cascade system if dioxane-free solvent is used in equal quantities in both stages.

(**Answer:** 2949.1 kg)

7.12 A 10 weight% solution of acetic acid in water is to be extracted with isopropyl ether which is nearly immiscible with water. How many crosscurrent stages will be required to reduce the concentration of acid to 2% if the ratio of the solution to solvent in each stage is 0.5. The following equilibrium data may be used.

kg acid/kg water	0.0197	0.04	0.0827	0.1290	0.1790
kg acid/kg ether	0.00493	0.0103	0.0227	0.0379	0.0558

(**Answer:** 4 stages)

7.13 A solution of 5% acetaldehyde in toluene is to be extracted with water in a 4-stage crosscurrent unit using 25 kg water in each stage for 100 kg feed. The equilibrium relation is $Y = 2.2X$ where Y is kg acetaldehyde/kg water and X is kg acetaldehyde per kg toluene. Determine:

(a) The percent recovery of acetaldehyde.

(b) The final raffinate concentration.

(**Answer:** (a) 83.85% (b) 0.84% acetaldehyde, 99.16% toluene)

7.14 A binary solution containing 10% by weight solute **C** dissolved in carrier **A** is fed to a three-stage countercurrent extraction unit at a rate of 1000 kg/h. Pure solvent **B** is used for extraction at rate 800 kg/h. Determine:

(a) The composition of the final raffinate.

(b) The composition of the final raffinate, if the same total quantity of solvent is used in equal amounts in a three stage crosscurrent unit. The equilibrium data are:

kg **C**/kg **A**	0.05	0.10	0.15
kg **C**/kg **B**	0.069	0.159	0.258

(**Answer:** (a) 1.85% C (b) 3.7% C)

Immiscible systems: Countercurrent extraction

7.15 One thousand kg/h of nicotine-water solution containing 1.5 weight% nicotine is extracted with 2000 kg/h kerosene solution containing 0.05 weight% nicotine in a countercurrent multistage unit to reduce the nicotine content to 10% of the original value. Calculate the number of stages required both graphically and using the Kremser equations.

Weight fraction, nicotine in water	0.001010	0.00246	0.005	0.00746	0.00988	0.0202
Weight fraction, nicotine in kerosene	0.000806	0.001959	0.00454	0.00682	0.00904	0.0185

(**Answer:** Graphical: 3.6, Kremser: 3.2)

7.16 One thousand kg/h of an aqueous solution of dioxane containing 25 weight% dioxane is subjected to continuous countercurrent extraction with benzene to remove 95% of the dioxane. Equilibrium distribution of dioxane between benzene and kerosene are as given below:

Weight% of dioxane in water	5.1	18.9	25.2
Weight% dioxane in benznene	5.2	22.5	32.0

Determine:

(a) Minimum solvent requirement in kg/h.

(b) The number of stages required if 900 kg/h of solvent is used for extraction.

(**Answer:** (a) 562 kg (b) 5.75)

7.17 An aqueous solution having a solute concentration of 0.1 kg/kg water is treated at a rate of 990 kg/h with 600 kg/h of pure immiscible organic solvent in a 2-stage mixer-settler system. If the organic and aqueous streams leaving each stage are in equilibrium such that $Y = 2X$, calculate the kg solute extracted per kg solvent if countercurrent operation is employed. Solve graphically and by using Kremser equation. (X = kg solute in aqueous phase/kg water, Y = kg solute in organic phase/kg solvent). (**Answer:** 83.16% (graphical), 73.2% (Kremser))

7.18 To recover an antibiotic from its solution with amyl acetate, pure water is used at a rate of 0.45 kg per kg of acetate in a countercurrent unit. If the distribution coefficient is 6.67 for this separation, how many ideal stages would be needed for 98% recovery of the antibiotic? Use the Kremser equation. (**Answer:** 3.2 stages)

7.19 An aqueous solution containing 1.0 weight% nicotine in water is extracted at a rate of 100 kg/h using a kerosene stream containing 0.05 weight% nicotine in countercurrent multistage unit. The water and kerosene are essentially immiscible in each other. It is desired to reduce the nicotine content in the exit water to 0.10 weight%. The equilibrium data are as follows:.

Weight fraction nicotine in water	0.001010	0.00246	0.005	0.00746	0.00988	0.0202
Weight fraction, nicotine in kerosene	0.000806	0.001959	0.00454	0.00682	0.00904	0.0185

Determine:

(a) The minimum solvent required.

(b) The number of stages required graphically and by Kremser equation if a solvent rate 1.25 times the minimum rate is used.

(**Answer:** (a) 103.6 kg (b) 8 (graphical), 7.95 (Kremser))

7.20 A solution containing 18 weight% acetone in water is to be extracted at a rate 1000 kg/h with monochlorobenzne containing 0.5% acetone by countercurrent extraction process. Monochlorobenzene and water are immiscible in the operating condition. The equilibrium data for the system is given below:

Kg acetone/kg water	0.0258	0.0739	0.1605	0.2670
Kg acetone/kg MCB	0.0288	0.0704	0.1560	0.2370

In order to obtain raffinate with 1% acetone content, calculate (a) Minimum solvent rate (b) Number of stages required with solvent rate 1.2 times the minimum.

(**Answer:** (a) 885.1 kg (b) 9)

7.21 Determine the number of theoretical stages required to extract nicotine from 100 kg/h of nicotine-kerosene solution containing 1.5 weight% nicotine using pure water at rate of 1.5 times the minimum in a countercurrent multistage unit, if it is desired to remove 90% of the nicotine. The equilibrium distribution of nicotine in water and kerosene are as follows:

Weight fraction nicotine in water	0.001010	0.00246	0.005	0.00746	0.00988	0.0202
Weight fraction, nicotine in kerosene	0.000806	0.001959	0.00454	0.00682	0.00904	0.0185

(**Answer:** 4 stages)

7.22 One thousand kg/h of an organic product containing 5.0 weight% of a water-soluble impurity is extracted with water to remove the impurity. Assume that water and the organic product are immiscible. It is found that on extraction with an equal amount of water in a single stage process, 90% of the impurity is extracted. Determine:

(a) The amount of water needed for a 90% removal of impurity in a two-stage countercurrent extraction unit. (b) The fraction of impurity removed, if a water-to-feed ratio of unity on mass basis is employed in a two-stage countercurrent unit.

(**Answer:** (a) 282.65 kg (b) 98.9%)

Partially miscible systems: Single-stage extraction

7.23 Five hundred kg of a solution containing 35 weight% acetic acid in water is subjected to single-stage extraction with an equal weight of pure isopropyl ether. Determine:

(a) The composition of the raffinate and extract products.
(b) The amounts of raffinate and extract products.
(c) The percent recovery of acetic acid.

Equilibrium data are given below:

Water-rich phase, weight%			Ether-rich phase, weight%		
Acetic acid	Water	Isopropyl ether	Acetic acid	Water	Isopropyl ether
1.41	97.1	1.49	0.37	0.73	98.9
2.89	95.5	1.61	0.79	0.81	98.4
6.42	91.7	1.88	1.93	0.97	97.1
13.3	84.4	2.3	4.82	1.88	93.3
25.5	71.1	3.4	11.4	3.9	84.7
36.7	58.9	4.4	21.6	6.9	71.5
45.3	45.1	9.6	31.1	10.8	58.1
46.4	37.1	16.5	36.2	15.1	48.7

(**Answer:** (a) $x = 25.1\%$ acetic acid, $y = 11.6\%$ acetic acid
(b) $R = 437$ kg, $E = 563$ kg (c) 37.3%)

7.24 A mixture consisting of, 40 kg acetic acid (**C**), 110 kg water (**A**) and 50 kg isopropyl ether (**B**) is allowed to separate into raffinate and extract phases which are in equilibrium. Determine the amounts and compositions of the raffinate and extract layers. Use the equilibrium data given in Exercise 7.23.

(**Answer:** Raffinate: 149.6 kg, $x_C = 0.232$, $x_B = 0.032$, $x_A = 0.736$
Extract: 50.4 kg, $y_C = 0.105$, $y_B = 0.860$, $y_A = 0.035$)

7.25 Three hundred kg of a mixture M of unknown composition containing water, acetic acid and isopropyl ether is mixed with 1000 kg of another mixture containing 20% acetic acid, 15% water, and 65% isopropyl ether. The resulting raffinate layer weighs 510 kg and contains 25% acetic acid, 71.6% water, and 3.4% isopropyl ether. All percentages are weight%. Use the equilibrium data given in Exercise 7.23. Determine:

(a) The original composition of the mixture M.
(b) The composition of the extract.

(**Answer:** (a) Acetic acid = 5.1%, water = 81.2%, Isopropyl ether = 13.7%
(b) Acetic acid = 11.1%, water = 3.6%, Isopropyl ether = 85.3%)

7.26 One hundred kg of a solution consisting of 50 weight% trimethylamine (TMA) and 50 weight% benzene is extracted with water in a single stage operation to produce a solvent-free extract containing 80% TMA. Determine:

(a) The composition of saturated extract and raffinate products.
(b) The quantities of saturated raffinate and extract products.
(c) The quantity of solvent free extract product.
(d) The quantity of solvent required for extraction.

The equilibrium miscibility data for the system are

Trimethylamine, weight%	Extract, weight%		Raffinate, weight%	
	Water	Benzene	Water	Benzene
5	94.6	0.4	0	95
10	89.4	0.6	0	90
15	84	1	1	84
20	78	2	2	78
25	72	3	4	71
30	66.4	3.6	7	63
35	58	7	15	50
40	47	13	34	26

The tie line data are

Weight% TMA in raffinate	31	14.5	9	6.8	3.5
Weight% TMA in extract	39.5	21.5	13	8.3	4

(**Answer:** (a) Raffinate: 28.2% TMA, 5.9% water; Extract: 36.5% TMA, 54.6% water, (b) R = 63.83 kg, E = 87.67 kg, (c) E′ = 39.8 kg, (d) S = 51.5 kg)

7.27 One thousand kg of a mixture of liquids **A** and **C** containing 30% **C** is extracted with a solvent **B** to recover component **C**. The equilibrium data at the temperature of extraction are:

Raffinate phase			Extract phase		
A	**B**	**C**	**A**	**B**	**C**
97	3	0	4	96	0
86	4	10	5	90	5
75	5	20	6	80	14
62	8	30	8	70	22
48	12	40	12	56	32
40	15	45	14	48	38
30	20	50	21	32	47
22	30	48	22	30	48

Calculate (a) the minimum amount of solvent (b) the maximum amount of solvent (c) the weight of solvent-free extract and raffinate and the percentage of **C** extracted in a single stage extraction with 750 kg solvent (d) If it is desired to produce a solvent free extract product containing 60.0% **C** determine the quantity of solvent to be used and the composition of the raffinate product on solvent-free basis.

(**Answer:** (a) 83 kg (b) 15300 kg (c) $R' = 820$ kg, $E' = 180$ kg, 41.7% (d) $S = 2871$ kg, 12.4% **C**, 87.6% **A**)

7.28 Determine the amounts and compositions of the raffinate and extract products that result from a single-stage extraction operation when a feed mixture weighing 1000 kg and containing 30 weight% acetone and 70 weight% water is treated with 500 kg methylisobutyl ketone (MIK). Solubility data are:

MIK	Acetone	Water	MIK	Acetone	Water
98	0	2	46.2	43.2	10.6
93.2	4.6	2.2	12.4	42.7	44.9
77.3	18.95	3.75	5.01	30.9	64.09
71	24.4	4.6	3.23	20.9	75.87
65.5	28.9	5.6	2.12	3.73	94.15
54.7	37.6	7.7	2.2	0	97.8

Tie line data:

Weight% acetone, water phase	2.5	5.5	7.5	10	12.5	15	17.5	20	22.5	25	26
Weight% of acetone MIK phase	4.5	10	13.5	17.5	21.3	25.5	28.2	31.2	34	36.5	37.5

(**Answer:** Raffinate: 814.1 kg; Acetone: 15.13%, MIK: 3.03%, water: 81.84%; Extract: 685.92 kg; Acetone: 25.78%, MIK: 69.24%, water: 4.98%)

7.29 One hundred kg of a solution of acetic acid in water containing 30% acid is to be extracted with isopropyl ether at 293 K. Use the equilibrium data given in Exercise 7.23. Calculate (a) the minimum quantity of solvent (b) The composition and quantities of the extract and raffinate products when the solution is treated with 100 kg solvent (c)The amounts of solvent-free extract and raffinate products for part (b).

(**Answer:** (a) 4.3 kg (b) Raffinate: 88.8 kg, 21.9% acetic acid, 3.1% ether; Extract: 111.2 kg, 9.49% acetic acid, 87.2% ether (c) R′ = 86.0 kg E′ = 14.0 kg)

Partially miscible systems: Crosscurrent extraction

7.30 A mixture containing 40 weight% acetone in water is extracted with an equal amount of MIK containing 1% acetone. (a) What fraction of the acetone can be extracted in a single-stage process? (b) What fraction of acetone could be extracted in a two-stage crosscurrent unit using equal quantities in both stages? Use the equilibrium data given in Exercise 7.28.

(**Answer:** (a) 79.1% (b) 88.5%)

7.31 It is desired to reduce the pyridine concentration of 2000 kg of an aqueous solution from 50 to 10% in a single batch extraction with chlorobenzene. The equilibrium data in weight percent are given below:

Extract phase			Raffinate phase		
Water	Chlorobenzene	Pyridine	Water	Chlorobenzene	Pyridine
0.05	99.95	0	99.92	0.08	0
0.67	88.28	11.05	94.82	0.16	5.02
1.15	79.9	18.95	88.71	0.24	11.05
1.62	74.28	24.1	80.72	0.38	18.9
2.25	69.15	28.6	73.92	0.58	25.5
2.87	65.58	31.55	62.05	1.85	36.1
3.95	61	35.05	50.87	4.18	44.95
6.4	53	40.6	37.9	8.9	53.2
13.2	37.8	49	13.2	37.8	49

Determine:

(a) The quantity of solvent required.
(b) The amount of raffinate and extract products.
(c) The composition of raffinate and extract products.

(**Answer:** (a) 4200 kg (b) $R = 1063.9$ kg, $E = 5135.7$ kg (c) Raffinate: 10% pyridine, 20% chlorobenzene; Extract: 17.4% pyridine, 81.7% chlorobenzene)

7.32 One thousand kg of a mixture of liquids **A** and **C** containing 30 weight% **C** is extracted twice with a total quantity of 750 kg **B** in two equal amounts. Use the equilibrium data given in Exercise 7.27. Determine the composition of final raffinate and percent of **C** extracted.

(**Answer:** Raffinate 860 kg, 19.26% **C**, 5.02% **B**; Percent extraction: 43.9%)

7.33 Calculate the amount of raffinate product, acetic acid concentration in it and the percent recovery of acetic acid if 1000 kg of a feed solution containing 25 weight% acid in isopropyl ether is extracted using pure water in

(a) a single stage process using 1000 kg of water.
(b) a four-stage crosscurrent unit using 250 kg fresh water in each stage.
Also calculate weight% of acetic acid in the combined extract in part (b).
Use the equilibrium data given in Exercise 7.23.

(**Answer:** (a) 759 kg, 16.13% acetic acid, recovery = 51%
(b) 690 kg, 2.3% acetic acid, recovery = 93.7%)

7.34 A pyridine-water solution containing 50% pyridine is to be extracted in a multistage crosscurrent unit with chlorobenzene as solvent to reduce the pyridine content to less than 10% in the final raffinate. Feed is admitted at a rate 2000 kg/h to the first stage and the quantity of solvent admitted to each stage equals the amount of feed to that stage. Use the equilibrium data given in Exercise 7.31. Determine:
(a) The number of theoretical stages required
(b) The concentration of the final raffinate
(c) The quantity of solvent, kg/h

(**Answer:** (a) 2 stages, (b) 7.12% pyridine, (c) 3199 kg)

Partially miscible systems: Countercurrent extraction

7.35 The equilibrium solubility data of acetone-water-1,1,2-dichloroethane system are given below. An aqueous solution containing 40 weight% acetone is treated with an equal weight of pure trichloroethane to produce a final raffinate containing 10% acetone on solvent-free basis in a countercurrent unit. The equilibrium data in weight% are given below:

Water layer, weight%			Trichloroethane layer, weight%		
Trichloreethane	Water	Acetone	Trichloreethane	Water	Acetone
0.52	93.52	5.96	90.93	0.32	8.75
0.73	82.23	17.04	73.76	1.1	25.14
1.02	72.06	26.92	59.21	2.27	38.52
1.17	67.95	30.88	53.92	3.11	42.97
1.6	62.67	35.73	47.53	4.26	48.21
2.1	57	40.9	40	6.05	53.95
3.75	50.2	46.05	33.7	8.9	57.4
6.52	41.7	51.78	26.26	13.4	60.34

Determine (a) The final extract composition (b) The location of the difference point analytically. (c) Derive an equation for the operating line passing through the difference point and the saturated raffinate containing 25% acetone on the right triangular coordinate system.

(**Answer:** (a) 24.65% acetone, 74.4% TCE, (b) The coordinates of the difference point are: ($x_C = 2.8222$, $x_B = -0.1822$), (c) $y = -0.1567x + 0.2602$)

7.36 Aqueous solution containing 30 weight% acetic acid is treated with an equal weight of pure solvent isopropyl ether in a countercurrent multistage extraction system. The feed is admitted at a rate 400 kg/h. If it is desired that the final raffinate should contain not more than 5 weight% acetic acid determine the composition and amounts of the extract and raffinate products. Use the equilibrium data given in Exercise 7.23.

(**Answer:** $E_1 = 533.3$ kg, 20% acetic acid, 6.4% water, 73.6% isopropyl ether;
$R_N = 266.7$ kg, 5% acetic acid, 93.2% water, 1.8% isopropyl ether)

7.37 An aqueous solution containing 30 weight% acetic acid is extracted at a rate of 200 kg/h using 400 kg/h of pure isopropyl ether in countercurrent multistage extraction unit. Calculate the number of stages required if the acetic acid concentration in the final raffinate is to be 7.5 weight%. Use the equilibrium data given in Exercise 7.23. (**Answer:** 4 stages)

7.38 One thousand kg/h of an aqueous solution containing 30.0 weight% acetic acid is being extracted in a countercurrent multistage unit using pure isopropyl ether solvent so that the final raffinate contains 2.0 weight% acid. Use the equilibrium data given in Exercise 7.23. Calculate:

(a) The minimum solvent that can be used.

(b) The number of theoretical stages if the solvent is used at rate of 2500 kg/h.

(**Answer:** (a) 1586 kg, (b) 7 stages)

7.39 Acetic acid is extracted from an aqueous feed containing 25 weight% acetic acid using pure isopropyl ether in a multistage countercurrent unit so that the acid concentration in the exit aqueous phase is reduced to 6.0 weight%. Use the equilibrium data given in Exercise 7.23. If 1000 kg/h of the feed is treated with 3000 kg/h of the solvent, calculate:

(a) The composition of the final raffinate and extract products

(b) The flow rates of final raffinate and extract products

(c) The number of theoretical stages

(**Answer:** (a) Raffinate: acetic acid = 6%, Isopropyl ether = 1.97, water = 92.03%,
Extract: acetic acid = 6.33%, Isopropyl ether = 91.50%, water = 2.17%,
(b) Raffinate = 970 kg, Extract = 3030 kg, (c) 3.0)

7.40 One thousand kg/h of a mixture containing 30% **C** in **A** is extracted countercurrently with 1500 kg/h of solvent **B** so as to reduce the concentration of **C** in the solvent-free final raffinate product to 10%. Determine the number of stages. Use the equilibrium data given in Exercise 7.27.

(**Answer:** 2.1)

7.41 Alcohol is extracted from an aqueous solution using ether in a multistage countercurrent unit. The feed contains 40% alcohol by weight and the solvent-free raffinate should contain not more than 5% alcohol. The feed is admitted at a rate 500 kg/h and the ether at a rate 600 kg/h. Equilibrium data for the system water-alcohol-ether are:

Water phase, weight fraction			Ether phase, weight fraction		
Alcohol	Ether	Water	Alcohol	Ether	Water
0	0.075	0.925	0	0.775	0.225
0.1	0.077	0.823	0.09	0.74	0.17
0.2	0.09	0.71	0.175	0.705	0.12
0.31	0.095	0.595	0.25	0.67	0.08
0.44	0.118	0.442	0.29	0.66	0.05
0.53	0.15	0.32	0.31	0.655	0.035
0.645	0.195	0.16	0.33	0.651	0.019
0.75	0.25	0	0.35	0.65	0

Calculate:

(a) The composition and flow rate of final raffinate.

(b) Composition and flow rate of final extract.

(c) The number of stages required.

(**Answer:** (a) 260 kg, 6.7% alcohol, (b) 840 kg, 21.7% alcohol, (c) 3.4)

7.42 One thousand kg/h of an aqueous solution containing 40% acetone is to be extracted in a countercurrent multistage unit with 1,1,2 trichloroethane at 298K to reduce the acetone content to 10%. The equilibrium data for the system trichloroethane-water-acetone are given in Exercise 7.35. Determine: (a) The minimum solvent rate (b) The number of stages required at a solvent rate 30% in excess of the minimum. (**Answer:** (a) 251.6 kg, (b) 5 stages)

7.43 A pyridine-water solution containing 50% pyridine is to be extracted at the rate of 5000 kg/h in a multistage countercurrent unit with chlorobenzene as solvent to reduce the pyridine content to

2% in the final raffinate. Use the equilibrium data given in Exercise 7.31. Determine: (a) The minimum solvent rate required (b) The number of stages required if solvent is used at rate of 5100 kg/h. (c) The quantities of the saturated extract and raffinate phases.

(**Answer:** (a) 3741 kg/h (b) 3 stages (c) $R = 2336.5$ kg/h, $E = 7763.5$ kg/h)

7.44 It is desired to reduce the acetic acid content in the aqueous solution from 30 weight% to less than 12.0 weight% using isopropyl ether as solvent in a 5-stage countercurrent unit. Determine the solvent-to-feed ratio and the composition of the extract. Use the equilibrium data given in Exercise 7.23. (**Answer:** $S/F = 1.37$, Extract: 13.5% acetic acid, 82.1% ether)

7.45 It is desired to reduce the acetic acid content in the aqueous solution from 30 weight% to 6.5 weight% using isopropyl ether as solvent in a 3-stage countercurrent unit. Determine the solvent-to-feed ratio and the composition of the extract. Use the equilibrium data given in Exercise 7.23.

(**Answer:** $S/F = 2.89$, Extract: 8% acetic acid, 89.2% ether)

Countercurrent extraction with reflux

7.46 Diethylene glycol (**B**) is used as solvent for separating by extraction a solution containing 50% ethyl benzene (**A**) and 50% styrene (**C**). The feed is admitted at a rate of 1000 kg/h and the concentration of styrene in the products are 10% and 90%. Equilibrium data are:

Raffinate phase, weight fraction			Extract phase, weight fraction		
Ethyl benzene	Glycol	Styrene	Ethyl benzene	Glycol	Styrene
99.33	0.67	0.00	10.40	89.60	0.00
90.56	0.81	8.63	9.84	88.52	1.64
80.42	0.93	18.66	9.31	87.20	3.50
70.49	1.00	28.51	8.72	85.80	5.48
60.93	1.09	37.98	8.07	84.47	7.45
53.55	1.20	45.25	7.36	83.39	9.25
52.96	1.20	45.84	7.31	83.19	9.50
43.29	1.39	55.32	6.32	81.68	12.00
42.11	1.39	56.51	6.07	81.38	12.55
21.51	1.80	76.69	3.74	77.63	18.64
0.00	2.50	97.50	0.00	72.90	27.10

Determine:

(a) The minimum number of theoretical stages.

(b) The minimum extract reflux ratio.

(c) The number of theoretical stages at a reflux ratio of 10 kg reflux per kg of extract product.

(**Answer:** (a) 10.5 stages, (b) 5.0, (c) 16 stages)

7.47 A countercurrent extraction unit employing extract reflux is used to extract a feed mixture consisting of 50 weight% methylcyclopentane (MCP, **C**) and 50 weight% *n*-hexane (**A**) admitted to the cascade at rate 1000 kg/h using pure aniline (**B**) as the solvent. The final extract is to contain 95% **C** on a solvent-free basis and the final raffinate is to contain 5% **C** on solvent-free basis and a the reflux ratio used is 8 kg reflux per kg product. Determine:

(a) The number of stages.

(b) The minimum reflux ratio.

Raffinate phase			Extract phase		
n-Hexane	Aniline	MCP	n-Hexane	Aniline	MCP
0.9279	0.0721	0.0000	0.0744	0.9256	0.0000
0.8261	0.0765	0.0973	0.0708	0.9165	0.0127

(Contd.)

Raffinate phase			Extract phase		
n-Hexane	Aniline	MCP	n-Hexane	Aniline	MCP
0.7239	0.0810	0.1951	0.0643	0.9074	0.0282
0.6203	0.0866	0.2932	0.0588	0.8974	0.0438
0.5152	0.0932	0.3916	0.0524	0.8856	0.0621
0.4076	0.1021	0.4903	0.0449	0.8709	0.0842
0.2976	0.1131	0.5893	0.0357	0.8534	0.1109
0.1853	0.1263	0.6884	0.0245	0.8313	0.1443
0.0685	0.1439	0.7876	0.0104	0.7996	0.1900
0.0000	0.1559	0.8441	0.0000	0.7762	0.2238

(**Answer:** (a) 28 stages, (b) 6.4)

7.48 A countercurrent extraction unit employing extract reflux is used to extract a feed mixture consisting of 50 weight% methylcyclohexane (MCH, **C**) and 50 weight% *n*-heptane (**A**) admitted to the cascade at rate 1000 kg/h using pure aniline (**B**) as the solvent admitted at a rate 15000 kg/h. The final extract is to contain 90% **C** on a solvent-free basis and the final raffinate is to contain 10% **C** on solvent-free basis. Determine:

(a) The reflux ratio.
(b) The number of stages and the location of the feed stage.
(c) The minimum number of stages.
(d) The minimum reflux ratio.

Raffinate phase, weight fraction			Extract phase, weight fraction		
n-Heptane	MCH	Aniline	n-Heptane	MCH	Aniline
0.9260	0.0000	0.0740	0.0620	0.0000	0.9380
0.8315	0.0914	0.0772	0.0599	0.0080	0.9321
0.7341	0.1858	0.0800	0.0530	0.0270	0.9200
0.6985	0.2194	0.0821	0.0511	0.0300	0.9190
0.5768	0.3373	0.0859	0.0450	0.0461	0.9089
0.5068	0.4063	0.0869	0.0400	0.0600	0.9000
0.4505	0.4596	0.0900	0.0360	0.0740	0.8900
0.3074	0.5966	0.0960	0.0298	0.0980	0.8723
0.2286	0.6714	0.1000	0.0211	0.1132	0.8658
0.1823	0.7158	0.1019	0.0160	0.1269	0.8571
0.1604	0.7357	0.1039	0.0139	0.1310	0.8551
0.0541	0.8330	0.1128	0.0062	0.1559	0.8379
0.0000	0.8811	0.1189	0.0000	0.1689	0.8311

(**Answer:** (a) 4.09 (b) 15, 6th stage (c) 7.7 (d) 2.4)

7.49 A solution containing 50 weight% n-heptane and 50 weight% methylcyclohexane is to be separated by extraction into products containing 92.5% MCH and 7.5% MCH, both on solvent-free basis. Aniline will be used as the solvent. Equilibrium data are given in Exercise 7.48. Determine:

(a) The minimum number of theoretical stages.
(b) Minimum extract reflux ratio.
(c) The number of theoretical stages for a reflux ratio of 7.0.

(**Answer:** (a) 10.3 (b) 2.72 (c) 15 stages)

7.50 Two liquids **C** and **A** are to be separated by extracting with solvent **B**. The feed contains 35% **C** and 65% **A** on a solvent-free basis and is a saturated raffinate. The following data represent the equilibrium between the two liquid phases at 368 K.

Extract layer			Raffinate layer		
C,%	**A,%**	**B,%**	**C,%**	**A,%**	**B,%**
0	7.0	93.0	0	92.0	8.0
1.0	6.1	92.9	9.0	81.7	9.3
1.8	5.5	92.7	14.9	75.0	10.1
3.7	4.4	91.9	25.3	63.0	11.7
6.2	3.3	90.5	35.0	51.5	13.5
9.2	2.4	88.4	42.0	41.0	17.0
13.0	1.8	85.2	48.1	29.3	22.6
18.3	1.8	79.9	52.0	20.0	28.0
24.5	3.0	72.5	47.1	12.9	40.0
31.2	5.6	63.2	31.2	5.6	63.2

The extract product contains 83% **C** and 17% **A** and a raffinate containing 10% **C** and 90% **A**, both on a solvent-free basis. Determine:

(a) The minimum reflux ratio.

(b) The number of ideal stages at a reflux ratio twice the minimum.

(**Answer:** (a) 3.3 (b) 10 stages)

7.51 An aqueous solution of ethylene glycol containing 30 weight% glycol is extracted with furfural to produce an extract product containing 80% glycol and a raffinate product containing 5% glycol both on solvent-free basis. Feed is admitted at a rate 1000 kg/h. A perfect solvent separator is used so that extract product and the reflux are free of solvent. Equilibrium solubility data for the ethylene glycol-water-furfural system at 298 K in weight fractions are:

Fufural	Ethylene glycol	Water	Fufural	Ethylene glycol	Water
0.95	0	0.05	0.275	0.525	0.20
0.903	0.052	0.045	0.139	0.475	0.386
0.861	0.10	0.039	0.11	0.40	0.49
0.751	0.20	0.049	0.097	0.30	0.603
0.667	0.275	0.058	0.084	0.15	0.766
0.49	0.415	0.095	0.077	0	0.923
0.343	0.505	0.152			

The tie line data for the system are:

Raffinate phase , weight fraction			Extract phase, weight fraction		
Furfural	Glycol	water	Furfural	Glycol	Water
0.08	0.073	0.847	0.926	0.025	0.049
0.083	0.15	0.767	0.904	0.052	0.044
0.088	0.2	0.712	0.894	0.062	0.044
0.097	0.3	0.603	0.882	0.075	0.043
0.11	0.4	0.49	0.86	0.1	0.04
0.139	0.475	0.386	0.803	0.15	0.047
0.251	0.515	0.234	0.751	0.2	0.049
0.277	0.525	0.198	0.665	0.275	0.06
0.414	0.415	0.171	0.414	0.415	0.171

Determine:

(a) The minimum number of theoretical stages

(b) Minimum extract reflux ratio

(c) The number of theoretical stages for a reflux ratio of 5.0

(**Answer:** (a) 3.75 (b) 4.12 (c) 6.8)

Continuous contact extraction

7.52 A lab-size countercurrent packed column of height 1.5 m is used to absorb acetic acid from an aqueous solution using benzene as solvent. Assume that benzene and water are immiscible. The acid concentration in the aqueous phase is to be reduced from 0.70 kmol/m^3 to 0.69 kmol/m^3. The acid concentration in the benzene phase are measured to be 0.004 kmol/m^3 at inlet and 0.015 kmol/m^3 at outlet. The equilibrium relationship is $C_E = 0.025C_R$. If the solvent is admitted at a rate is 1.5 × 10^{-3} m^3/m^2 s, determine:

(a) The overall transfer coefficient

(b) The height of the transfer unit (**Answer:** (a) 1.706 × 10^{-3} s^{-1} (b) 0.879 m)

7.53 Acetic acid is extracted counter currently in a packed column of height 3.0 m from an aqueous solution containing 4.0 weight% acid using isopropyl ether which is assumed to be immiscible with water. The feed is admitted at a rate 0.5 kg/m^{2}s and the solvent at a rate 1.0 kg/m^2 s. The concentration of acid in the ether phase leaving the column is 1.0 weight%. The equilibrium relationship is given by $Y = 0.3\,X$ where Y is kg acid/kg ether and X is kg acid per kg water. Calculate:

(a) The number of overall transfer units based on the raffinate phase, and

(b) The overall extraction coefficient based on the raffinate phase, kg/m^3 s.

(**Answer:** (a) 1.46 (b) 0.243 kg/m^{3}s)

7.54 A laboratory spray column of height 1.2 m is used to extract acetic acid from an aqueous solution using MIBK, which is assumed to be completely immiscible with water. The feed containing 0.70 kmol acid/m^3 enters the top of the column at a rate of 1.5×10^{-3} m^3/s m^2 of tower cross section and the solvent at the bottom at a rate of 2.5 × 10^{-3} m^3/s m^2 of tower cross section free of acetic acid and leaves with a concentration of 0.2 kmol/m^3. The equilibrium relationship under these conditions may be taken as $C_E = 0.55C_R$ where the concentrations are in kmol/m^3.Calculate:

(a) The overall extraction coefficient based on the driving force in the ketone phase

(b) The overall HTU based on the ketone phase. (**Answer:** (a) 2.16 × 10^{-3}s^{-1} (b) 1.16 m)

7.55 A valuable solute is being extracted in a packed column using a solvent immiscible with the carrier liquid in the feed. The equilibrium relationship for the distribution of solute between the immiscible phases is $y = 0.06x$ where x and y are the mole fractions of solute in the raffinate and extract phases, respectively. The terminal conditions are:

x	0.001	0.08
y	0.0	0.003

Determine:

(c) The number of transfer units based on the raffinate phase.

(d) The height of packing, if the height of the transfer unit based on the raffinate side is 0.25 m.

(**Answer:** (a) 9.27 (b) 2.32 m)

Chapter 8

Leaching

Leaching is the process by which the soluble constituents of a solid feed are separated by dissolving in liquid solvents. The process is also known as solid–liquid extraction. The soluble constituents exist within the insoluble solid material in the solid or liquid state. They exist in chemical combination with the insoluble matrix, physically adsorbed or mechanically bound on the pore structure. On leaching, the solutes get dissolved in the solvent and the insoluble solids are separated from the resulting solution. When the slurry consisting of the insoluble solids suspended in the solution is settled and separated, a thickened sludge called the 'underflow' in which all the insolubles are present and a clear 'overflow' solution in which the solutes are present, are obtained. The desired product of a leaching operation can be either the solute recovered in the overflow or the insoluble solids present in the underflow. Separation of oil from oilseeds and leaching of gold from gold ores are examples in which the objective is the recovery of the soluble components from the inert insoluble materials. However, in some cases, the purification of the insoluble solid matrix by removing the soluble constituents by contacting with solvents becomes the objective of leaching. An example is the removal of the soluble contaminants from insoluble pigments.

Earlier, the term leaching was used to describe only the processes involving the percolation of liquid through a fixed bed of solids. But the principles involved in the solid–liquid operations conventionally referred to by different names such as lixiviation, extraction, decoction, elutriation, and washing are now treated and analysed as leaching operations. Lixiviation originally referred to leaching of alkali from wood ash, decoction to leaching using solvents at their boiling points and elutriation to washing of soluble materials from the surface of insoluble solids. The term extraction was originally used for the extraction of oil from seeds and cakes by mass transfer and/or by mechanical means. Washing operations such as the washing of precipitates to free off the adhering mother liquor can also be carried out and analysed as leaching processes.

8.1 INDUSTRIAL APPLICATIONS

Metallurgical industries: Useful minerals occur in nature in combination with large proportions of undesired constituents; in the majority of cases, leaching is the process employed for the recovery of the minerals. Leaching is employed in metallurgical industries on a large scale for the recovery of metals such as copper, gold, aluminium, cobalt, manganese, nickel, zinc, etc. Copper is preferentially dissolved from its ore by leaching with sulphuric acid and gold by leaching the ore with sodium cyanide solution.

Natural organic products: Since many naturally occurring products can be easily separated from seeds, roots, fruits, leaves, stem and tree barks by leaching with suitable solvents, leaching operations find wide application in the food, pharmaceutical and biological processing industries. *Edible vegetable oils* are extracted from the seeds by leaching with organic solvents such as hexane, acetone and ether. Thus soyabeans, coconuts, cottonseeds, palm, sunflower seeds, rice bran etc. can be treated with solvent to recover the oils. When the oil content of the seeds is relatively high, it can be recovered to some extent by compressing the solid by a process known as *expression*. The solvent-extracted oilseed meal generally contains high quality residual protein suitable for further processing into food grade powders or food protein extenders. *Extraction of sugar* from sugar beets is achieved by means of hot water. Water-leaching process has now largely replaced the mechanical expression of sugarcane juice using multiple roller mills which is a highly energy and capital intensive operation. *Tannin* is extracted from tree barks using water. *Tea and coffee* are prepared domestically and industrially by leaching operation. For the production of soluble 'instant' coffee, ground roasted coffee is leached with fresh water; water-leaching of the tea leaves gives the soluble tea. The *removal of caffeine* from green coffee beans is achieved by means of water extraction. *Supercritical extraction* using CO_2 is now widely used in the food industries for a wide range of applications, as in the decaffeining of coffee and tea and in the preparation of the unique flavour extracts from the fruits and leaves of plants.

Washing operations: Primary leaching operations, filtration and precipitation always produce solids, the liquid adhering to which must be washed either to recover the solution if the soluble constituent is the desired material or to avoid contamination of the solid if it is the desired product. Washing of caustic soda liquor from the precipitated calcium carbonate following the reaction between soda ash and lime is an example of washing operation in which chemical precipitates are treated to remove the adhering mother liquor. These washing operations are also treated as leaching as mentioned earlier.

8.2 RATE OF LEACHING

In leaching, a solid feed is brought into contact with a liquid solvent. The following general processes occur when the feed solid and the leaching solvent are contacted:

1. The solvent is transferred from the bulk liquid to the solid surface.
2. The solvent diffuses into the solid particles.
3. The solute dissolves in the solvent.
4. The solute diffuses through the solid-solvent mixture to the surface of the solid particle.
5. The solute is transferred from the surface to the bulk solution.

One or more of these steps can limit the rate of leaching. In general, the first two are not the rate controlling steps as they are known to be occurring quite rapidly. By a proper choice of the leaching solvent, the dissolution of the solute in the solvent may also take place rapidly, and it may also be not the rate controlling step in leaching. The rate of diffusion of the solute through the solution to the surface of the particle is a complex function of several factors such as the particle size,. shape and the internal structure of the particle. If the solute is present in the solution contained within the pores of the inert solid, the diffusion through the porous solid can be described by means of an *effective diffusivity*. The use of a constant effective diffusivity is impractical for describing the diffusion process in the case of leaching of substances of natural or biological origin, in which the part of the solute transfer occurs by diffusion through the cell walls.

When the material is being dissolved from the surfaces of the solids as in the washing processes, the rate of leaching can be best described as the product of the mass transfer coefficient, interfacial area of mass transfer, and the concentration difference.

$$N' = k_L A(C_S - C_b) \qquad \ldots(8.1)$$

where N' is the rate of mass transfer, kg/s, k_L is the mass transfer coefficient m/s, A is the interfacial area provided by the particle surface m^2, C_S the saturation solubility of the solute in the liquid (kg/m^3) and C_b is the solubility of the solute in the bulk liquid at any time t s. The equation is also applicable to the general leaching situation if the diffusion within the solid is very rapid compared to the diffusion from the surface.

Let V m^3 be the volume of the solution which is constant for a batch process. The rate of leaching can be equated to the rate of accumulation of the solute in the solution so that Eq. (8.1) becomes

$$V\frac{dC_b}{dt} = k_L A(C_S - C_b) \quad \text{...(8.2)}$$

Integrating between limits $t = 0$ and $C_b = C_0$ to $t = t$ and $C_b = C_b$,

$$\int_{C_0}^{C_b} \frac{dC_b}{(C_S - C_b)} = \frac{Ak_L}{V}\int_0^t dt \quad \text{...(8.3)}$$

$$\ln\frac{C_S - C_0}{C_S - C_b} = \frac{Ak_L}{V}t \quad \text{...(8.4)}$$

$$\frac{C_S - C_b}{C_S - C_0} = e^{-\left(\frac{Ak_L}{V}\right)t} \quad \text{...(8.5)}$$

If pure solvent is used initially, $C_0 = 0$ and

$$1 - \frac{C_b}{C_S} = e^{-\left(\frac{Ak_L}{V}\right)t} \quad \text{...(8.6)}$$

Or

$$C_b = C_S\left(1 - e^{-\left(\frac{Ak_L}{V}\right)t}\right) \quad \text{...(8.7)}$$

Equation (8.7) shows that the solution approaches the saturation condition exponentially. The interfacial area will not remain constant during leaching. It tends to increase during leaching and when the solute constitute the major portion of the solid, the complete disintegration of the solid may occur on extraction, leading to a considerable increase in the area of mass transfer. However, the advantage of the increased area on the rate of leaching is lost by the adverse effect of the fine particles on the free flow of the solvent.

Factors Affecting Rate of Leaching

The selection of equipment for extraction process is influenced largely by the factors that limit the extraction rate. The four important factors that determine the rate of leaching are: the particle size, the solvent, the temperature of leaching and the agitation.

Particle size and preparation of feed

The particle size is important in deciding the rate of leaching, particularly when the rate is controlled by the diffusion of the solute through the porous structure of the solid. The material should be small in size so that the diffusion path is small. The particle size also determines the interfacial area of mass transfer. For smaller particles, the interfacial area between the solid and the liquid is greater and therefore the rate of mass transfer is high. The presence of very fine powder in the solid phase should be avoided, as very fine particles will wedge in the interstices of the solid and block the free passage of the liquid and hinder leaching. Also, if the particle size is very small, separation of the particles from the liquid and drainage of the solid residue are difficult. As far as possible, uniform particle size is preferred for the feed solid so that all the particles take approximately the same time for extraction.

In the preparation of the feed for leaching, it is important to consider the nature of the material being leached, whether the solid is, say, of inorganic origin or of vegetable origin, the proportion of the soluble constituent in the solid, its distribution in the solid, and the original particle size.

Inorganic and organic materials: If the solute being extracted is surrounded by a matrix of inert insoluble material, leaching involves the diffusion of the solvent into the solid, dissolution of the solute and finally diffusion of the solution back to the bulk liquid. In such cases, crushing and grinding of the solid, as practised in many hydrometallurgical operations, greatly improves the rate of leaching. However, it is necessary to limit the extent of crushing as it is an expensive operation. In certain situations, for diffusion to occur, it is essential that the material be crushed to very fine powders. This is the case with extraction of gold which is present in the ore scattered as fine particles throughout a matrix of impervious quartzite. On the other hand, if the soluble fraction is present as a solid solution or is uniformly distributed throughout the whole solid, grinding may not be necessary at all. On dissolution of the solute in the solvent, small channels are formed which go deeper into the solid with passage of additional solvent. If the soluble material is adsorbed on the surface or present in the solution adhering to the solid, grinding is not required; mere washing will serve the purpose as in the washing of chemical precipitates to separate the adhering mother liquor.

Materials of vegetable and animal origin: In materials of biological origin, for recovery of the solutes that are present within the cells, the cell walls offer additional resistance to mass transfer unless they are ruptured. But it is impractical and in some cases, undesirable, to grind the material to sizes small enough to rupture the cell walls. In the leaching of the sugar from the sugar beets, for example, the rupture of the cell walls permits easy passage into the solvent of not only sugar, but the undesired constituents such as colloidal and albuminous material. In such cases, instead of grinding to reduce the particle size and break the cell walls, the materials are subjected to one or more of the different processes such as slicing, rolling, flaking, drying, etc. Thus by slicing the sugar beets into thin wedge-shaped slices called cossettes, the rate of leaching is improved as the distance the solvent has to diffuse to reach the cell walls is reduced. In the recovery of pharmaceuticals from plant materials such as roots, leaves and stems, feed preparation includes drying which helps in the rupture of the cell walls, thereby releasing the desired solutes to the direct action of the solvent. Seeds and beans like soyabean are subjected to rolling or flaking to produce particles in the size range 0.1 mm to 0.5 mm, after tempering to a certain moisture content to prevent disintegration. Thin flakes provide very small diffusion path for the oil, which leads to a high rate of leaching.

Solvent

The solvent chosen should be selective and should offer high solubility towards the components to be leached, to minimize the solvent needed. Viscosity should be low in order that the solvent be circulated freely through the leaching equipment. The increase of viscosity of the solution with increase in the concentration of the solute may lead to fall in the leaching rate with progress of leaching. The solvent selected should be capable of producing an extract unimpaired by the presence of the solvent and should be chemically stable under the process conditions. Low vapour pressure, ease and economy of the recovery from the extracted solution, low density, low surface tension and low toxicity and flammability are other desirable characteristics of good leaching solvent.

Temperature

The higher the temperature, the greater will be the rate of leaching because of two reasons. First, the solubility of the solute is high at higher temperature so that the driving force for mass transfer is high. Second, the diffusivity of the solute is increased and viscosity of the solvent is lowered with rise in temperature. These factors lead to high rates of leaching at high temperatures. The upper limit for temperature, however, is determined by external factors such as the deterioration of the material that may set in at high temperature. For example, in leaching of sugar from sugar beets at high temperatures, the enzyme action may come into play with consequent chemical deterioration of the product. In some cases, temperature sensitivity of the materials of construction to corrosion or erosion attack may also be significant in choosing the operating temperature.

Agitation

If the diffusion of the solute from the surface of the particle to the bulk of the solution is the rate-controlling factor in leaching, the agitation of the fluid greatly increases the rate of leaching. Agitation also helps in preventing sedimentation of the solid and thereby more effective use of the interfacial surface can be achieved.

Example 8.1: 500 kg of a solid containing 30% solute and the rest insolubles is dissolved in 100 m^3 of water. The solubility of solute in water is 3.0 kg/m^3. How long it will take the solute to completely dissolve in water? In a pilot-plant study with a sample of the same solids, it took 30 s to attain 80% saturation when the solute was dissolved from the inert solid in 1 m^3 of water.

***Solution*:** Equation (8.7) gives

$$\frac{C_b}{C_S} = 1 - e^{-\left(\frac{Ak_L}{V}\right)t}$$

It took 30 s for attaining 80% saturation when the sample was dissolved in 1 m^3 of water. At 80% saturation, $C_b = 0.8\ C_S$ and therefore,

$$1 - e^{-(Ak_L/V)t} = 0.8$$

Or

$$e^{-\left(\frac{Ak_L}{V}\right)t} = 0.2$$

Therefore,

$$-\left(\frac{Ak_L}{V}\right)t = \ln(0.2) = -1.60943$$

Substituting $V = 1\ m^3$ and $t = 30$ s in the above,

$$Ak_L = 0.05365\ m^3/s$$

When 500 kg solid containing 30% solute is dissolved in 100 m^3 of water, the concentration C_b is

$$C_b = \frac{500 \times 0.3}{100} = 1.5\ kg/m^3$$

Substituting $C_b = 1.5$ kg/m^3, $C_S = 3.0$ kg/m^3, $Ak_L = 0.05365\ m^3/s$, $V = 100\ m^3$, in Eq. (8.7)

$1.5 = 3.0\left(1 - e^{-(0.05365/100)\,t}\right)$ which when solved gives the time for complete dissolution as

$$t = 1292\ s = 21.5\ min$$ **(Answer)**

8.3 CLASSIFICATION OF LEACHING METHODS

Leaching operations generally fall into two broad classes depending upon the method of contacting of the liquid and solid phases. They are *percolation* and *immersion*. In percolation, the liquid is allowed to flow through an open permeable bed of solids. The bed of solid may be either stationary or moving. The second method employs immersion of the solid in the liquid with agitation of the mixture and is generally used for handling impermeable solids and solids that disintegrate on leaching. Generally the percolation is used for leaching coarse particles and immersion for fine particles. Though very rapid and complete leaching is achieved by the immersion method, grinding of the large lumps of solids to fine particles and agitation and settling are costly. Also, the settled solids retain considerable amounts of solution and large quantities of the solvent are required for washing the solids to free the solute. Consequently, the immersion method produces dilute extract. On the other hand, the percolation methods in which relatively coarser particles are subjected to leaching, the operation is slow and incomplete. However, the mass of solution retained by the solids and hence the quantity of solvent required for the washing are less and therefore, the final extract produced is relatively concentrated.

Figure 8.1 shows a classification of leaching equipment on the basis of the method of contacting of the liquid and solid phases. Leaching operations, whether be percolation or immersion, can be carried out either batch-wise or continuously. Though both the stage contact and the continuous contact equipment are common for leaching, generally the countercurrent multistage operation is common with immersion type leaching operations.

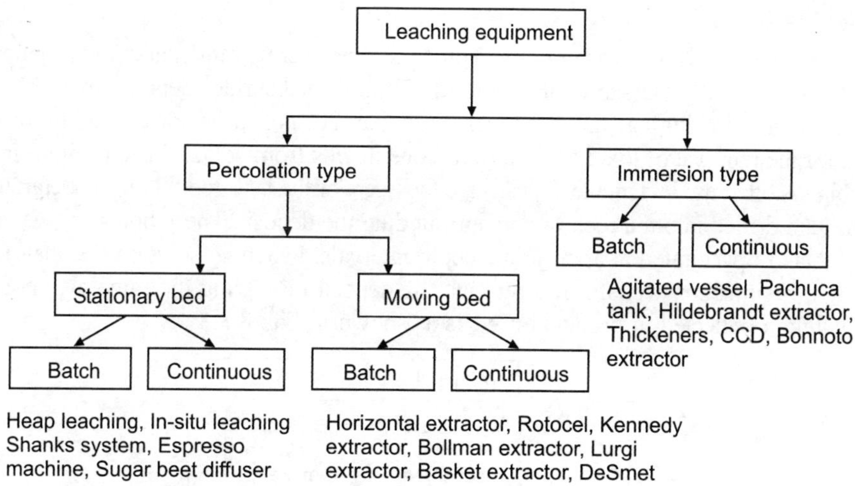

Fig. 8.1: Classification of leaching techniques and equipment

8.4 FIXED BED PERCOLATION EQUIPMENT

Heap Leaching

Heap leaching is a process where the solid is in a stationary heap and the solvent percolates through the solid. It is widely applied for the recovery of copper and precious metals such as gold from their ores. Heap leaching is the least expensive method of leaching. The heap is constructed on impervious polymeric membrane as shown in Fig. 8.2. In leaching of gold from low-grade oxidised gold ores, the

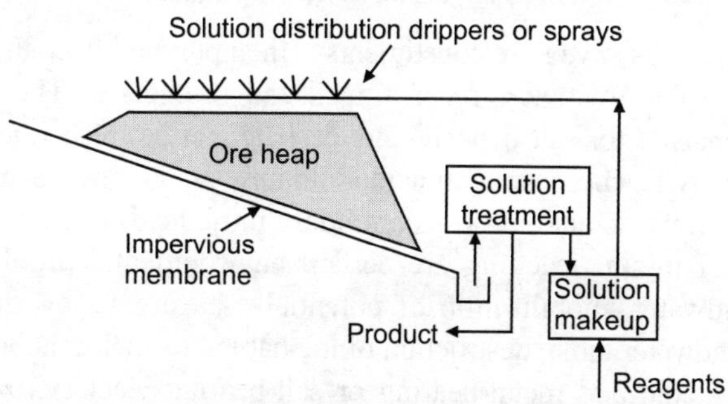

Fig. 8.2: Heap leaching for copper or precious metals

heap is made of ore crushed into small chunks. The heap can also be made of agglomerates formed from the fine powder produced by crushing of the ore. The solvent for leaching is distributed over the heap and it percolates through the material. Dilute alkaline solution of sodium cyanide is the solvent used for gold extraction, which is sprayed over the heap through solution drippers or sprays as shown in the figure. The solution containing gold-cyanide complex is collected in 'pregnant pond'. The solution is recycled back to the top of the heap after the gold is recovered from it.

Heap leaching of very low-grade gold ores and many oxide copper ores for which mineral value do not warrant expensive crushing or grinding, are conducted on run-of-mine material. In heap leaching of uranium, the ore is built into piles on a pattern of perforated drain pipes arranged on an impervious clay base. Heaps, trapezoidal in cross section, typically about 120 m wide at the base, 6 to 8 m tall and about 800 m long are common for leaching of uranium.

In-Situ Leaching

In-situ leaching also called *solution mining*, is a process of recovering minerals such as copper, uranium, etc. and water-soluble salts such as sodium chloride, from the subsurface deposits, without resorting to conventional mining. In-situ leaching produces more than 20% of the world's uranium production. It is also used for the removal of toxic or hazardous constituents from contaminated soil or groundwater. In-situ leaching is often used when the deposits are too deep, or the beds too thin for underground mining.

Boreholes are drilled into the rock or soil surrounding the deposit. These holes or "wells" are lined with tubing that are perforated at appropriate depth intervals. Leaching solution (leachate) is pumped down through the injection wells, flows through the deposit dissolving the minerals or salts and the 'pregnant' solution collected in production wells as shown in Fig. 8.3.

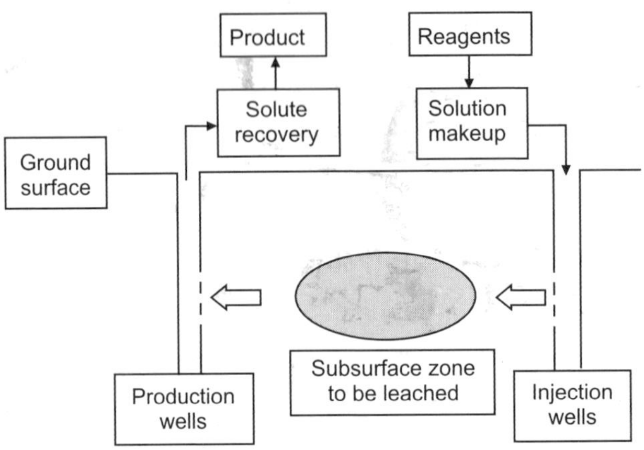

Fig. 8.3: In-situ leaching operation

The solution bearing the dissolved ore constituents is then pumped from the production wells to the surface and treated for solute recovery, reconstituted and re-injected. The leachate solution varies according to the ore deposit—for salt deposits the leachate can be fresh water in which salts readily dissolve. Copper is usually leached using an acid (sulphuric acid or hydrochloric acid). For uranium ores, an acid leachate (sulphuric acid, or less commonly nitric acid) or sodium bicarbonate are used.

The disadvantages of in-situ leaching are certain environmental problems associated with it. Acidification of groundwater, mobilisation of potentially hazardous or radioactive heavy metals, disturbance of the groundwater table, destruction of the habitat of rock-inhabiting organisms, bacteria, etc, potential spills of acidic and metal-bearing or salt-bearing leachates upon the surface are the environmental issues caused by the in-situ leaching processes.

Extraction Battery

Extraction batteries operating under the Shanks system are formed by interconnecting the fixed bed extractors as shown in Fig. 8.4. They are used for a variety of leaching applications such as the extraction of tannin from wood bark, sugar from sugar beets, water soluble substances from coffee, tea and spices and sodium nitrate from ore, etc. The individual extractors are either completely enclosed columns that can be operated at elevated pressures and temperatures or open vessels that are operated at atmospheric pressures. Although not shown in the figure, each extractor is equipped with provisions for solids loading–unloading and solvent feed and extract discharge. Heat exchangers are used to adjust the temperature of the solvent and pumps are used to transport the liquid from column to column. The system using closed tanks operated at high pressures is known as the *diffusion battery*.

In Fig. 8.4a, a series of four tanks containing the solids to be leached are in operation for quite some time. At the given instant, let one tank (tank 4), which contains solids which has undergone almost complete leaching, is isolated and emptied for refilling with fresh solids. Of the remaining three tanks, tank 3 is the one that is most recently charged with the solids and therefore contains the freshest solid. Tank 1 contains solids that have been subjected to leaching for more number of times. Fresh solvent is admitted to tank 1 which percolates through the bed. The extract solution leaving this tank is pumped to tank 2 and then to tank 3. The final extract is withdrawn from tank 3 which contains the fresh solid. At the end of this cycle, the extraction of solids in tank 1 is completed, and it is taken off-line for emptying and refilling of solids as shown in Fig. 8.4b. Fresh solvent is now admitted to tank 2 containing the most spent-up solid, the extract drawn off from tank 2 is percolated through tank 3 and finally through tank 4 which now contains the most recently charged solid. The final extract is drawn off from tank 4.

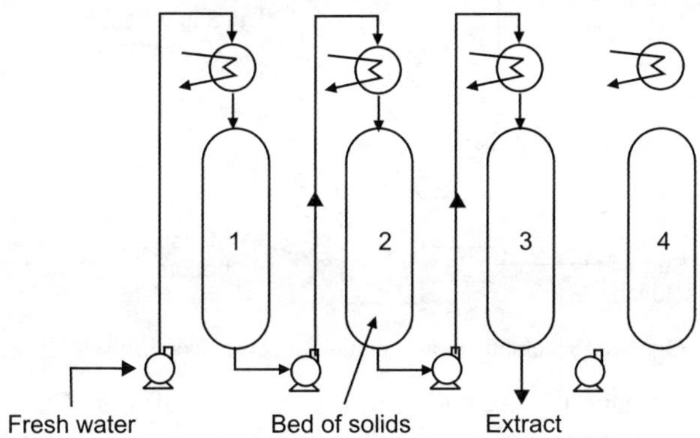

(a)

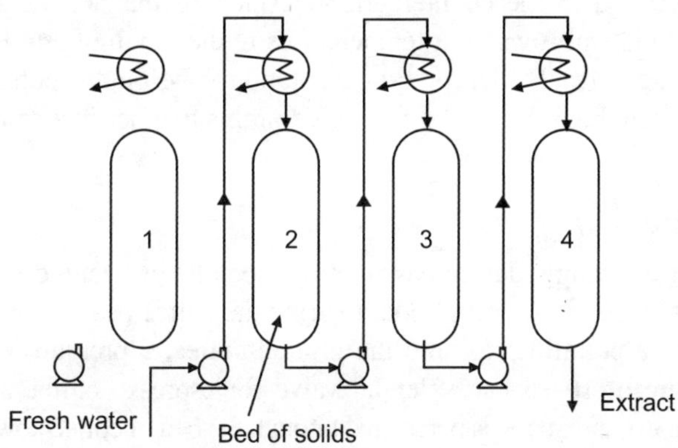

(b)

Fig. 8.4: Extraction battery

The operation is continued with most spent solids discharged periodically and fresh solids loaded into the tank vacated by the spent solid, and admitting the fresh solvent to the tank containing solids that have undergone the highest number of leaching contacts. Thus the extraction battery employing the Shanks system achieves a high concentration of solute in the solvent as in a true countercurrent multistage extraction, without physically moving the solids from tank to tank. The extraction battery can be operated with any number of tanks, and stages from 6 to 16 are the most common.

Diffusers

Diffusers are closed percolation vessels for leaching at high pressures and temperatures. They are used in the sugar industry, for extraction of tannin from wood barks, pharmaceuticals from barks and seeds, etc. The liquid is made to percolate through the bed of solids by gravity. When the pressure drop for flow of liquid is too high for gravity flow, pumping of the liquid through the bed of solid will be necessary. Closed vessels prevent evaporation losses when volatile solvents are used for leaching and also permit leaching operation above the boiling temperature of the solvent.

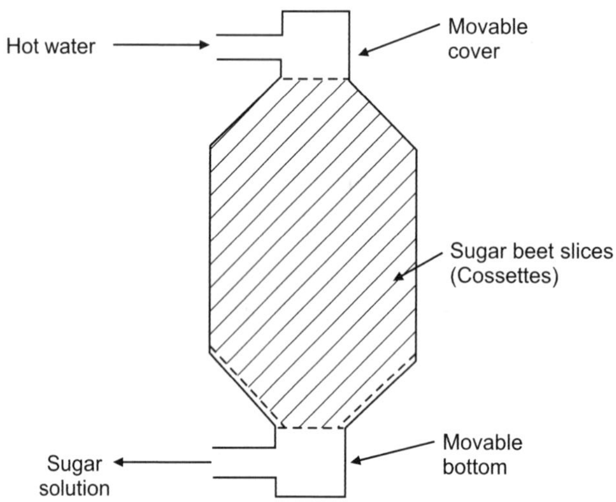

Fig. 8.5: Schematic representation of sugar-beet diffuser

Figure 8.5 shows the schematic representation of a sugar-beet diffuser. The sugar-beet slices (the cossettes) can be dumped into the diffuser by removing the top cover. Several diffusers can be arranged in a battery and operated in the countercurrent fashion of the Shanks system. Heaters are placed between the vessels and warm water at temperatures in the range of 344 K to 350 K is admitted into the bed to leach out the sugar. The leached sugar solution flows out the bottom into the next tank. About 95% to 98% of sugar in the beets can be recovered in this manner in a solution containing about 12% sugar.

Espresso Machine

Espresso machine is a leaching device based on the percolation method which is used to leach coffee from the ground coffee beans. A burr grinder to grind the coffee beans to powder of size typically from 250 to 750 microns, a portafilter to hold the ground coffee, a pressure vessel which receives water pumped at a pressure of 10-15 bar, solenoid valve for espresso coffee are its main parts. Water at temperature between 361 and 365 K is percolated through the bed of coffee powder for about 20 to 30 s. The high pressure facilitates pressure infusion of water into the particles that leads to rapid extraction of coffee. Espresso machine has the advantages of providing (1) short time interval between grinding and leaching (2) leaching operation with a short residence time (3) water at controlled temperature and pressure (4) very fine powder for leaching. All these factors contribute to extraction of favourable flavour-chemicals keeping the dissolution of unfavourable chemicals that give bitterness to the coffee to the minimum.

8.5 MOVING BED PERCOLATION EQUIPMENT

Bollmann Extractor

Bollmann extractor shown in Fig. 8.6 is a vertical basket type extractor used for leaching oil from seeds. Extractors typically of height 14 m and which can handle about 2000 to 20000 kg/h of flaky solids are used for this purpose. Buckets with perforated bottoms are attached to a chain conveyor which moves in the vertical direction. The baskets receive fresh solids feed from the top and begins their descending motion when a partially enriched solvent known as *half-miscella* is introduced into the baskets. The half-miscella is pumped from the bottom of the column of ascending buckets shown on the left side of the figure. The solution percolates through the bed of solids and drains to the basket below, achieving parallel flow of the contacting phases during the downward movement of the baskets. The strong solution of oil in the solvent known as *full miscella* is collected at the bottom of the extractor and is removed. During the upward movement of the baskets the solids come in contact with fresh solvent admitted at the top. Thus a countercurrent contact between the solid and the solvent is achieved during the ascending motion of the solids. The weak oil-solvent solution collected at the left-hand sump is recycled as half-miscella as explained earlier. A short drainage time is provided before the baskets are emptied and the solids are dumped at the top and the spent solids are carried away by means of paddle conveyors. Because the solids are un-agitated and the final miscella moves cocurrently, the Bollmann extractor permits the use of thin flakes while producing extracts of good clarity. However, it

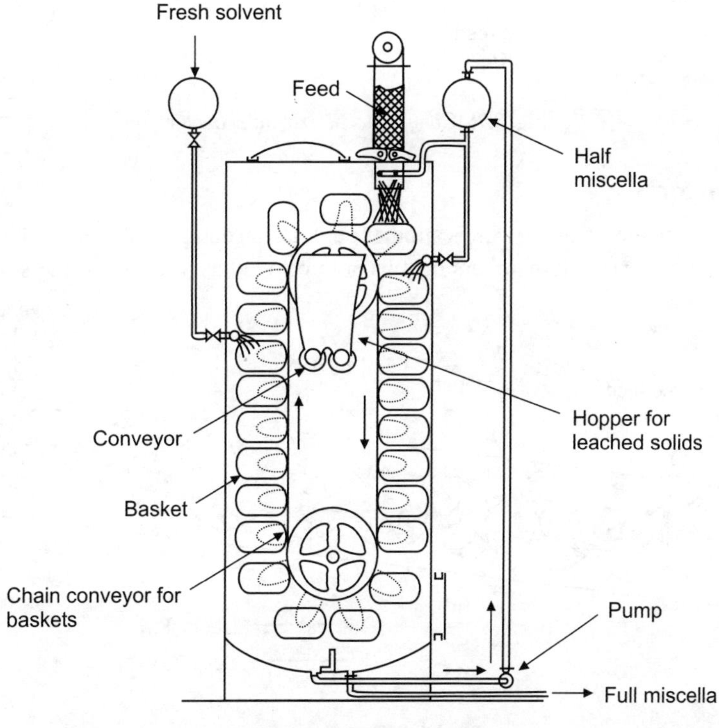

Fig. 8.6: Bollmann extractor

permits counterflow only partially and suffers from the problem of channelling. The Bollmann extractor has low stage efficiency and is being replaced in the oil industry by more efficient extractors such as the horizontal basket type.

Continuous Horizontal Extractor

The horizontal extractor shown schematically in Fig. 8.7 is a variant of Bollman extractor. The solids are charged into perforated baskets attached to a horizontal chain conveyor. The solvent is percolated through the bed of solids.

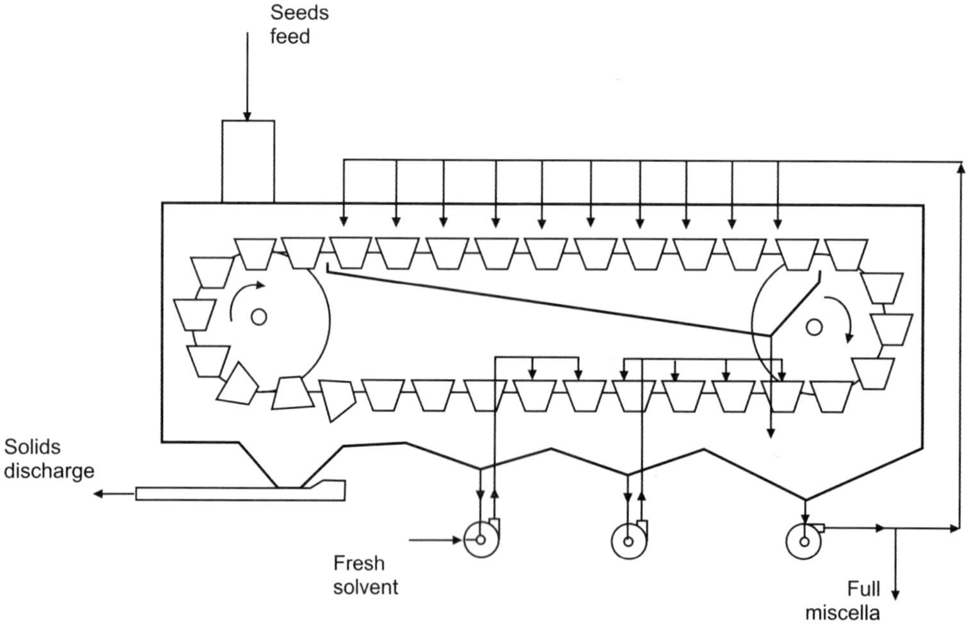

Fig. 8.7: Continuous horizontal extractor

Kennedy Extractor

The Kennedy extractor shown in Fig. 8.8 is another percolation equipment with the bed of solid moving through the solvent. A series of chambers or tubs are arranged as shown and the solids to be leached

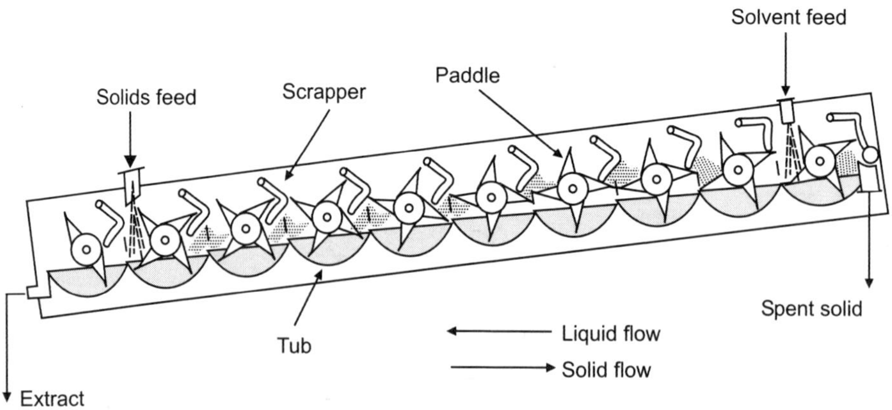

Fig. 8.8: Kennedy extractor

are transported in succession from chamber to chamber by slow impellers. The perforations provided in the impellers facilitates drainage of the solution from the bed of solids emerging from the solution when the impeller lifts the solids above the liquid level. Each impeller is scraped to disengage the solids which are then dumped into the adjacent chamber on the right as shown in the figure. Since the solvent is introduced at the upper end of the extractor, a perfect countercurrent contact is achieved. Since the solids are subjected to mechanical action somewhat more severe than other type of percolators, the Kennedy extractor is not preferred for leaching fragile materials such as flaked oil seeds.

The Kennedy extractor requires relatively less headroom and was originally used for leaching tannin from wood bark. It is now employed for leaching oil from seeds as well as for chemical leaching applications.

Rotocel

Rotocel extractor also called Carousal extractor is a moving bed percolation type continuous multistage equipment closely simulating countercurrent operation under the Shanks system. The Rotocel consists of a circular rotor slowly revolving above a stationary compartmented tank. The rotor is divided into a number of walled annular compartments called cells, the unit depicted schematically in Fig. 8.9 contains 18 such cells.

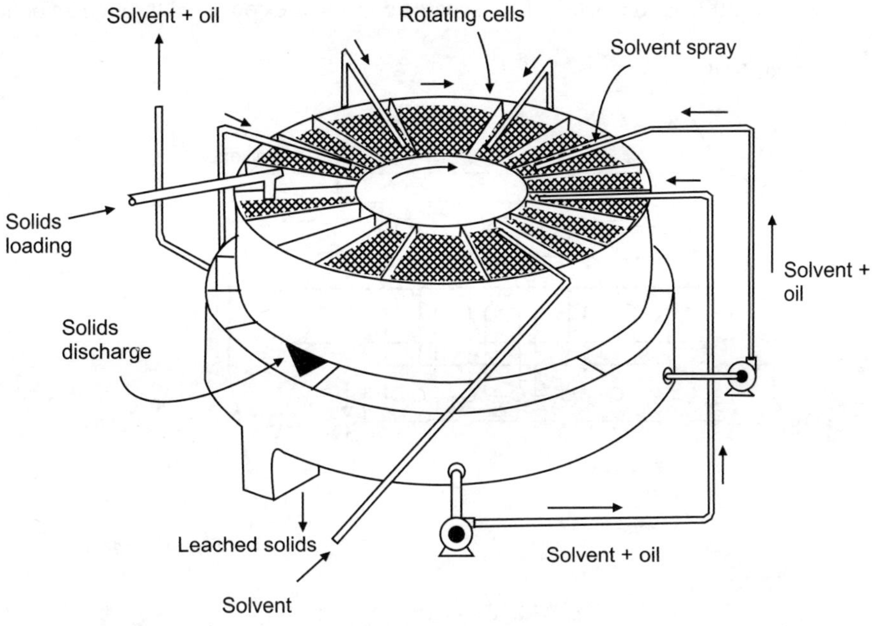

Fig. 8.9: Rotocel

The cells hold the solids and allow the drainage of the liquid through the perforated bottoms. As the rotor revolves, each cell passes successively through the solid feed area, a series of solvent sprays, a final spray and drainage area, and a solids discharge area. After completing nearly one revolution, the contents of each cell are drained and automatically dumped into one of the compartments of the static tank, from where they are continuously conveyed away. The empty cells move to the feed location to receive the next load of solids. Fresh solvent is supplied just ahead of the drainage and the discharge area. The solvent from each spray percolates through the bed and are drained to specific compartments in the bottom tank through the perforations on the supporting screen. The solvent-oil solution from these compartments is pumped to the cell preceding the cell from which they are withdrawn. A true countercurrent operation is realized in this manner and the strong solution is drawn off from the compartment under the freshest solids and is taken for treatment for solvent recovery.

The Rotocel can handle flaked seeds and beans. Typical units have diameters from 3.4-11.3 m, height 6.4-7.3 m, with bed depths of about 1.8-3.0 m. A widely used variant of the Rotocel is the French Stationary Basket-Extractor in which the cells are kept stationary and the solids-feed and solids-discharge zone rotate with periodic switching of the solvent feed and discharge connections.

Continuous Perforated-Belt Extractor

Continuous perforated-belt extractors are widely used for processing sugar cane, sugar beets, oil seeds and apples. The Lurgi extractor and deSmet extractor belong to this class and the latter is shown in Fig. 8.10. The feed solids are admitted through a hopper on to a slow moving perforated belt, driven by motorised sprockets at either end. The units may be as long as 7 to 40 m and with belt width ranging from 0.5 to 10 m. The depth of solids on the bed range from 0.8 to 2.5 m and it is adjusted by the damper at the outlet of the feed hopper.

Fresh solvent is admitted at the discharge point of the solids. The liquid percolates through the bed and are collected in the compartments below the perforated belt. The solution is drawn off from these compartments and sprayed over the bed of solids, thus ensuring true countercurrent multistage operation. The solids are scraped off from the belt at the discharge end and are conveyed through an outlet hopper.

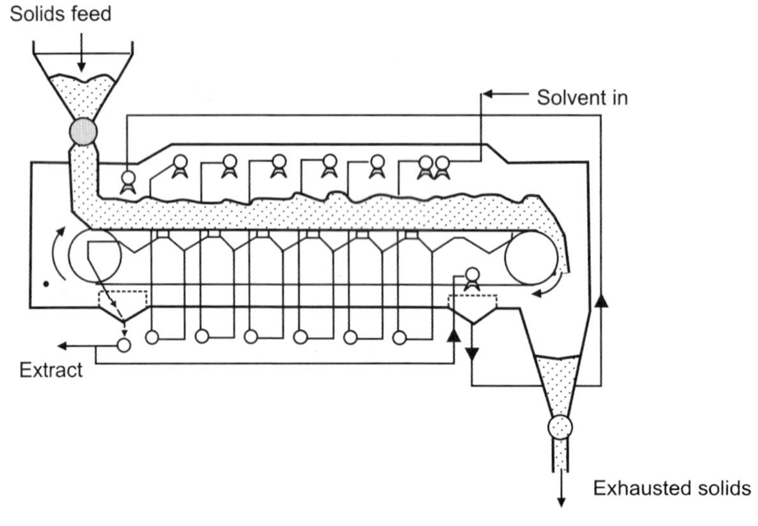

Fig. 8.10: Perforated-belt extractor

8.6 IMMERSION TYPE BATCH EXTRACTORS

Leaching is slow and incomplete in percolation-type leaching equipment, because of the channelling that usually occur in these equipment. Channelling can be eliminated in immersion type or solids-dispersion-type equipment by stirring. Equipment for batch leaching of fine solids in liquid suspension is now confined to batch tanks agitated by coaxial impellers (turbines, paddles, or propellers). The leached solids must be separated from the extract by external filters, centrifuges or thickeners. The difficulty of the physical separation of the solids from the leach liquor is the main limitation of batch stirred tanks. Added to this is the problem that the batch vessels generally face: the separation achieved is never more than that obtained with a single equilibrium stage.

Pachuca Tank

Pachuca tanks (Fig. 8.11) are air-agitated vessels widely used in metallurgical industry until the 1960s for leaching finely divided materials smaller than 0.1 mm in diameter. These are built of wood, metal or concrete and may be lined with inert non-corrosive metals such as lead. The solvent and solids are placed in the tank and the agitation is accomplished by the air lift. Air is admitted through a circular tube placed concentric with the tank. The solid liquid suspension is kept in circulation by the air admitted continuously. When leaching is completed, the air flow is cut off and the solids are allowed to settle. The clear solution is withdrawn by siphoning from the top of the tank. The settled solids are drawn off as a sludge from the bottom of the tank. The discharge of solids is facilitated by loosening the sludge by passing air. Since the agitation by mechanical means is more economical than agitation by the air-lift, impeller agitated tanks are now preferred over Pachuca tanks.

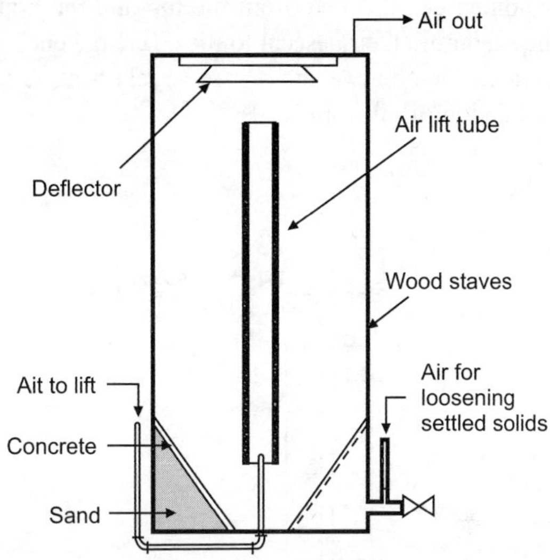

Fig. 8.11: Pachuca tank

Mechanically Agitated Vessels

Figure 8.12 shows mechanically agitated tanks of special construction which are used for leach-ing coarse solids. These tanks are extensively used for leaching of vegetable oil from the seeds. They can be used singly or can be arranged in series for the operation as a countercurrent battery. Figure 8.12a shows a closed cylindrical vessel arranged vertically and fitted with power-driven paddles or stirrers on a vertical shaft. They are also provided with false bottom for drainage of the solution when leaching is completed. Figure 8.12b shows a horizontal tank with stirrers arranged on a horizontal shaft.

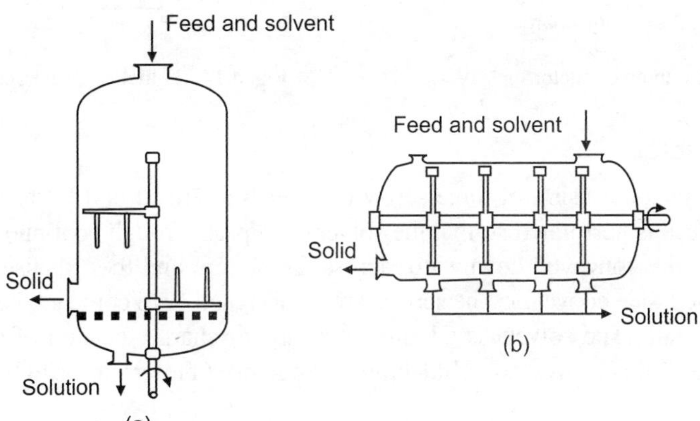

Fig. 8.12: Stirred vessels for batch leaching

8.7 IMMERSION-TYPE CONTINUOUS EXTRACTORS

Vertical Plate Extractor

The Bonnotto extractor shown in Fig. 8.13 is an example of a vertical plate extractor used for extraction of oil from the seeds. The vertical column is divided into a number of cylindrical compartments by means of horizontal circular plates. Each plate is provided with a radial opening to allow the passage of solids to the plate below. The openings on adjacent plates are staggered at 180°.

Radial blades are attached to a vertical shaft and each plate is wiped off by the blade. The solids are admitted continuously from the top and the liquid from the bottom. The solids get dispersed in the liquid during their descent to the discharge end. At the discharge end screw conveyors carry the solids away. Bonnotto extractors have largely been displaced by horizontal extractors or tray percolators for extraction of oil from seeds.

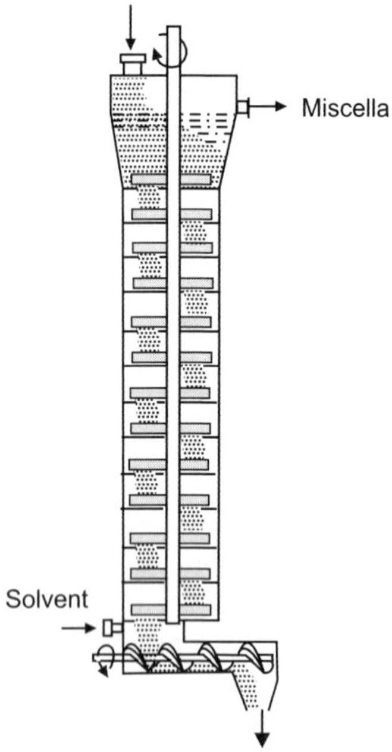

Fig. 8.13: Bonnoto extractor

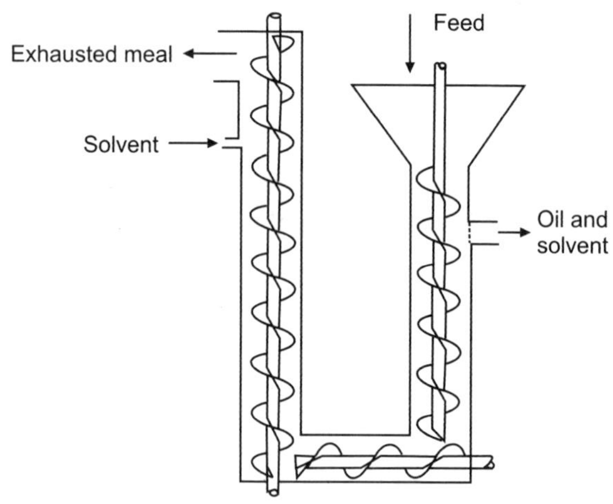

Fig. 8.14: Hildebrandt extractor

Hildebrandt Extractor

The Hildebrandt extractor consists of three screw conveyors arranged in a U-shape as shown in Fig. 8.14. The helix surface is perforated so that the solvent can pass through continuously. The solids are charged at the top right, conveyed downward, across the bottom and the exhausted meal discharged from the top of the left-side conveyor. The screws are so designed as to compact the solids during their passage through the unit. The solvent is admitted near the discharge location of the solid and flows countercurrent to the solids movement. Hildebrandt extractor is preferred for light permeable solids.

Continuous Agitated Vessels

Impeller agitated stirred tanks operating either singly or in series can be used for leaching of fine solids, which can be readily suspended in liquids. The tanks should be designed taking care to see that no solids accumulate in the tank. Stirred tanks are cylindrical vessels equipped with vertical baffles to prevent swirling and to make effective utilization of the energy supplied to the agitator. Mechanical agitation of the solid-liquid suspension is generally achieved by means of turbine-type agitators.

The Dorr agitator shown in Fig. 8.15 widely used in the metallurgical as well as chemical leaching applications, utilizes both the air-lift principle and the mechanical raking of solids. It consists of a

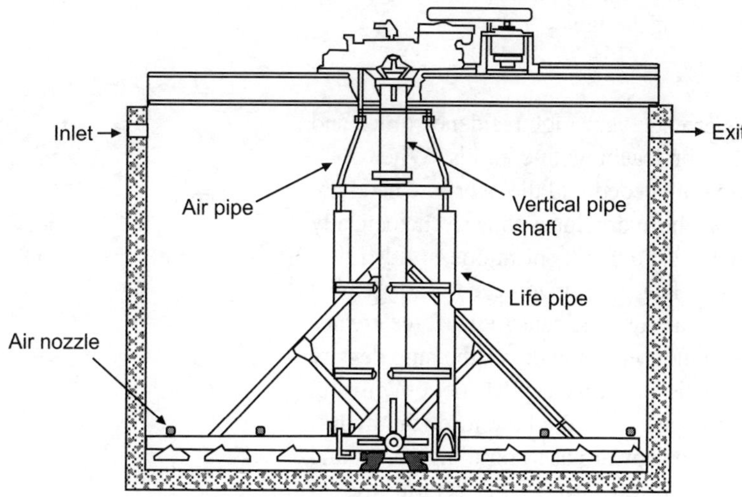

Fig. 8.15: Dorr agitator

cylindrical tank fitted with a central air-lift inside a hollow shaft which slowly rotates. Rakes fitted to the bottom of the shaft drag the solid material to the centre as it settles which is then picked up by the air lift. At the upper end of the shaft the airlift discharges into a perforated launder which distributes the suspension evenly over the surface of the liquid in the vessel.

Continuous Gravity Thickeners

The continuous gravity thickener (Fig. 8.16) consists of a large-diameter shallow tank with a flat or slightly conical bottom. Thickeners as large as 100 m in diameter and 3.5 m high have been constructed and are in use. The tank is equipped with a slowly revolving rake to move the thickened sludge to the central discharge. In large thickeners the rakes revolve at about 2 rpm. The rake arms can be raised automatically to lift the rakes off the bottom in the case of overload due to the interruption in the sludge discharge, and automatically lowered to the bottom when the overload is removed. The feed enters near the centre by means of a feed launder that discharges into a feed well at the top. Settling and sedimentation of the solid particles occur by gravity due to the difference in the densities of the solid and the liquid.

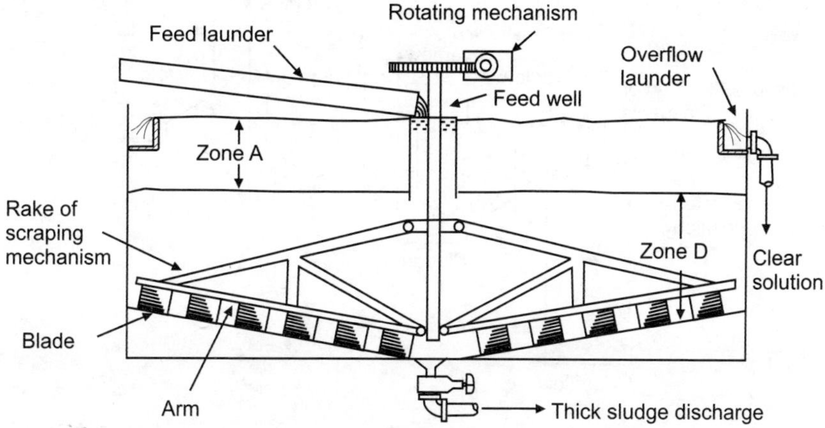

Fig. 8.16: Continuous gravity thickener

The clear supernatant solution is withdrawn through the overflow launder or weir situated around the upper, inner periphery of the tank.

Hydroclones

Gravity thickeners provide very high residence times and therefore, are ideal for leaching fine solids. When long residence times are not needed and the overflow need not be very clear solution, the hydroclones shown schematically in Fig. 8.17 can be used for continuous solid-liquid separation operation. The pressurized feed slurry enters tangentially to the separator. The centrifugal force creates downward spiralling motion which drives the high density solids to the wall, which is conical in shape, downward. The solids are discharged from the bottom of the hydroclone. The liquid moves upward as a spiralling vortex and are discharged from the top through the central vortex-finder pipe.

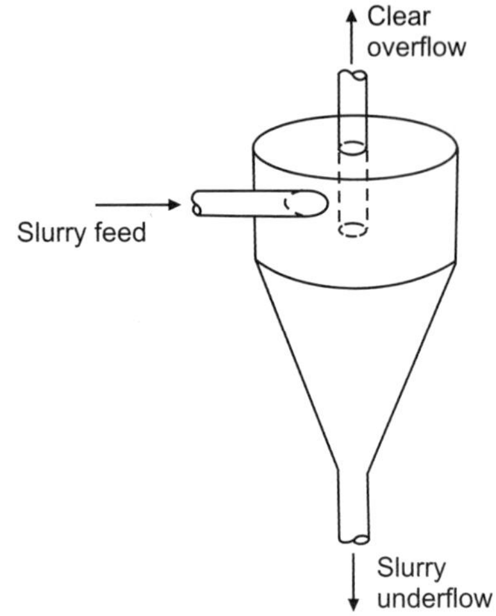

Fig. 8.17: Hydroclone

Continuous Countercurrent Decantation System (CCD)

When an appreciable amount of solution is still associated with the residual solids after separation, it is generally desirable to pass it through a battery of washers, arranged to give countercurrent flow of solids and the solvent. Classifiers are used when relatively coarse solids are to be washed and thickeners are used when fine solids are to be treated. The gravity thickeners or hydroclones can be interconnected to form a continuous countercurrent decanter cascade (CCD). Countercurrent washing of the solids to reduce the solute content in the liquid adhering to the solids, is a common practice in industries. Since the leaching of small particles containing very soluble solutes is completed very rapidly, frequently it will be necessary to subject the underflow sludge leaving such processes to countercurrent washing. This is true for solids formed by a chemical reaction in a solution.

A CCD system made up of gravity thickeners is shown in Fig. 8.18. The feed solids are admitted to the first thickener along with clear supernatant solution drawn off from tank 2. Thorough mixing of the

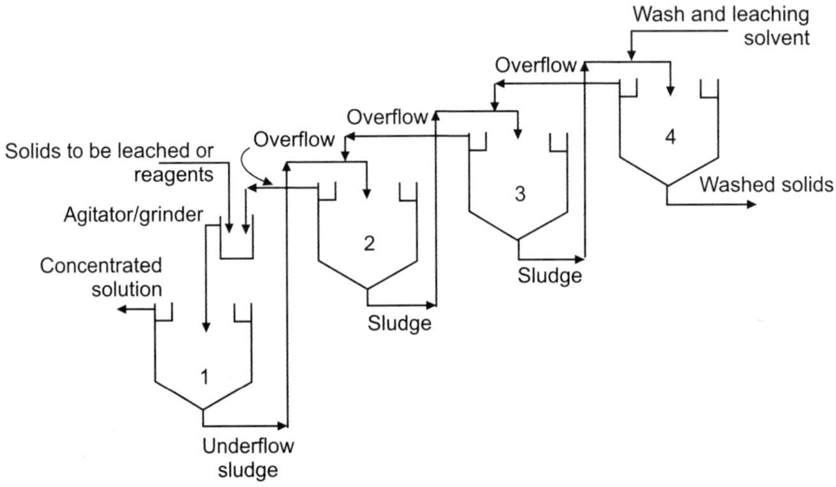

Fig. 8.18: Continuous countercurrent decanter

solids and the solution, followed by separation of the thickened sludge from the bottom of the tank and clear supernatant liquid from the top of the tank, occur in the thickener. In continuous countercurrent washing, each thickener is fed with thickened sludge leaving the preceding vessel and clear solution leaving the succeeding vessel. Thus thickener 2 receives solid sludge from tank 1 and clear solution from tank 3, and so on. Fresh solvent is admitted to the last stage from which the final underflow sludge is withdrawn. The final underflow sludge can be subjected to filtration, if the valuable solutes adhering to the solids are to be recovered or if the solids are to be freed of undesired solute materials.

8.8 LEACHING EQUILIBRIUM

Leaching involves bringing the solid feed and the liquid solvent into contact so that the soluble constituents dissolve in the liquid and separating the solution so formed from the insoluble solid residue. Since it is practically impossible to make a sharp separation of the solid from the suspension of solid and liquid, the settled solids inevitably carry some solution adhering with them. The separated clear solution is called the 'overflow' and the settled solid sludge and the accompanying solution is called the 'underflow'.

The leaching stage is a mechanical device in which the solid feed and the liquid solvent are brought into contact, and after dissolution of the solute, the underflow and the overflow are separated. An ideal leaching stage is the one which satisfies the following requirements:

1. The solvent is used in such large quantities that the solute is completely dissolved in the liquid.
2. No solute is preferentially adsorbed by the inert solid.
3. The composition of the solution, including that of any liquid retained within the pores of the inert solid is uniform throughout.
4. Concentration of the solute in the clear solution (overflow) and the solution associated with the settled solid (underflow) are the same.
5. No inert solid is present in the overflow in the dissolved or in the suspended form.

Since the solute in the feed gets completely dissolved in the liquid, 100% recovery of the solute would have been attained, had the separation of the resultant solution and the inert solid been perfect. The stage would have been an ideal equilibrium stage with an efficiency of 100% and a single equilibrium stage would have achieved complete separation of the solute. The mechanical separation of the solids from the solution by settling, filtration or drainage, can never be completed to 100% perfection and the solution retained by the solids carry some solute. The practical equilibrium curves take into account our inability to effect perfect physical separation of solids from the solution. These are established by studying the settling characteristics of the slurry consisting of the inert solids suspended in a solution of the soluble constituents in the leaching solvent under the conditions of temperature, pressure and time of settling identical with those in the actual equipment.

The leaching process involves three constituents: Solute (**C**), insolubles (**B**) and solvent (**A**). Soluble constituent (**C**) present in the feed in association with the insoluble material (**B**) is dissolved in liquid solvent (**A**). Thus, in the extraction of oil from seeds using hexane, the constituents are **A** = hexane, **B** = insoluble cake, and **C** = oil. The graphical method of solving leaching problem makes use of the right-triangular-coordinate system and/or rectangular coordinate system for representing the leaching equilibrium. We will be using the rectangular coordinate system because in the triangular diagrams, the graphical constructions to determine the number of stages frequently fall into one corner of the triangle.

The equilibrium curve is plotted on $N-x-y$ coordinates. N represents the concentration of the inert solid in the ternary mixture defined as

$$N = \frac{\text{mass of } \mathbf{B}}{\text{mass of } (\mathbf{A}+\mathbf{C})} \qquad \ldots(8.8)$$

The weight fraction of solute in the underflow on **B**-free basis is denoted by x and that in the overflow by y.

$$x \text{ or } y = \frac{\text{mass of } \mathbf{C}}{\text{mass of } (\mathbf{A}+\mathbf{C})} \quad \ldots(8.9)$$

If the feed stream is made up of components **B** and **C** and no **A** is present as is generally the case with fresh feed, it can be characterised by the coordinates $x = 1.0$ and N is equal to the weight ratio of inert to solute (weight of **B**/weight of **C**). If the solvent used is pure, no **B** and **C** is present and therefore, $y = 0$, $N = 0$. Consider, for example, 100 kg of a seed containing 25 weight % oil being leached with 225 kg of n-hexane. The unsettled slurry M consists of 75 kg inert insoluble cake (**B**) suspended in 250 kg oil-hexane solution (**A**+**C**) containing 25 kg oil (**C**) dissolved in it. For this suspension, N and x are given as

$$N = \frac{\text{mass of } \mathbf{B}}{\text{mass of } (\mathbf{A}+\mathbf{C})} = \frac{75}{250} = 0.30$$

$$x = \frac{\text{mass of } \mathbf{C}}{\text{mass of } (\mathbf{A}+\mathbf{C})} = \frac{25}{250} = 0.10$$

On the rectangular coordinate system, the weight fraction of inerts (N) is taken on the y-axis and the weight fraction of solute (x or y) on the x-axis. Thus the mixture M can be represented with coordinates ($x = 0.10$, $N = 0.3$) as shown in Fig. 8.19.

When this mixture is allowed to settle and the insoluble phases are separated, the clear solution called the overflow and represented by V and the settled sludge called the underflow, represented by L will result. It is impossible to separate from the suspension, the insoluble solids alone by settling, filtration or by drainage. Some solution with the same concentration as the concentration of the overflow is inevitably carried away by the underflow solids. The amount of solution associated with unit weight of the inert in the underflow depends upon the length of time allowed for settling. Let us assume that for the time allowed in our experiment for settling, for every 3 kg of solids, 2 kg solution is removed in the underflow. Then the underflow will consist of 75 kg inert solids **B** and 50 kg solution (**A**+**C**) and the overflow will be 200 kg solution (**A**+**C**). Since the same solution obtained by dissolving 25 kg oil in 225 kg solvent hexane is present in the overflow and the underflow streams, both these streams are represented on the vertical line at $x = y = 0.10$. Since the overflow contains no solid, the overflow V lies on the x-axis as shown in Fig. 8.19. The N-coordinate for the underflow is calculated as

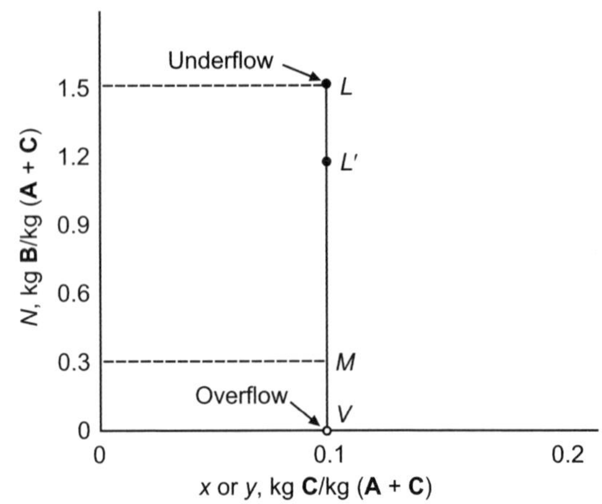

Fig. 8.19: Streams and concentrations in leaching

$$N = \frac{\text{mass of } \mathbf{B}}{\text{mass of } (\mathbf{A}+\mathbf{C})} = \frac{75}{50} = \frac{3}{2} = 1.5$$

Thus point L with coordinate ($x = 0.1$, $N = 1.5$) represents the underflow in Fig. 8.19. For ideal leaching stage, the underflow and the overflow should lie on the vertical tie line as the solution present in both have the same concentration of solute. However, the exact location of L on this vertical line depends upon the amount of solution adhering with unit weight of solid, which in turn, depends on the length of time allowed for settling. If the length of time allowed for settling of the original solution were less, the underflow will contain more solution associated with the solid and N would be less. The underflow in this case will be represented by a point such as L' in Fig. 8.19.

Constant and Variable Underflow

In majority of the leaching operations, particularly in washing operations, uniformity of solution concentration is attained rapidly on contacting the solvent stream with the feed. The lowered stage efficiency due to the incomplete recovery of solute in such cases is attributed to the incomplete drainage and settling of the slurry. The concentration of insoluble solid in the underflow sludge, settled for a fixed time depends on the viscosity and the relative density of the liquid in which the solid is suspended, which in turn, depends on the concentration of solute in the slurry. Experimental data on the settling characteristics plotted showing the variation of the composition of the thickened sludge with composition of the solution or the plot of N versus x, provide the practical equilibrium curve for analysing leaching operations. The data should be obtained under conditions of time, temperature and concentrations same as those prevailing in the proposed equipment or process for leaching.

If the density and viscosity of the solution do not change appreciably with concentration of the solution, the mass of the solution retained by the settled solids is constant, independent of the concentration of solution. N is constant or the locus of underflow (N versus x curve) is a straight horizontal line as shown in Fig. 8.20a. This is the *constant underflow condition* in leaching. Since the overflow carries no solid **B** in dissolved or in suspended form, the $N = 0$ axis (the x-axis) represents the locus of overflow (N versus y). The tie line connecting the underflow (L) and overflow (V) from each stage is vertical as the solute concentration in the overflow and underflow liquid are uniform.

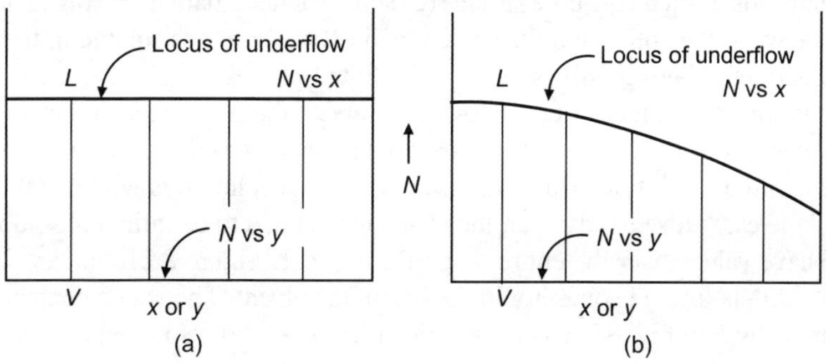

Fig. 8.20: (a) Constant underflow (b) Variable underflow

If the amount of solution retained by unit mass of solid in the underflow varies with the concentration of the solution, the *variable underflow condition* as shown in the Fig. 8.20b results. The viscosity and density of the solution increase with solute concentration; consequently, the rate of settling decreases with increase in solute content of the solution. More solution will be drawn off with the settled solids and N decreases with increase in x. The locus of overflow is represented by the x-axis since the clear solution is free of inert solids. The tie lines are vertical as the concentration of solution is uniform throughout.

Different Types of Equilibrium Curves

Equilibrium characteristics of leaching process may vary from the two ideal cases discussed above. Some typical non ideal cases are shown in Fig. 8.21. Figure 8.21a shows a variable underflow condition with nonuniform concentration of solute in the overflow solution and the solution drawn off by the underflow solids leading to nonvertical tie lines. The ratio of the solute to the solvent in the underflow is more than that of the overflow. This situation may arise when (1) the insoluble solids retain some

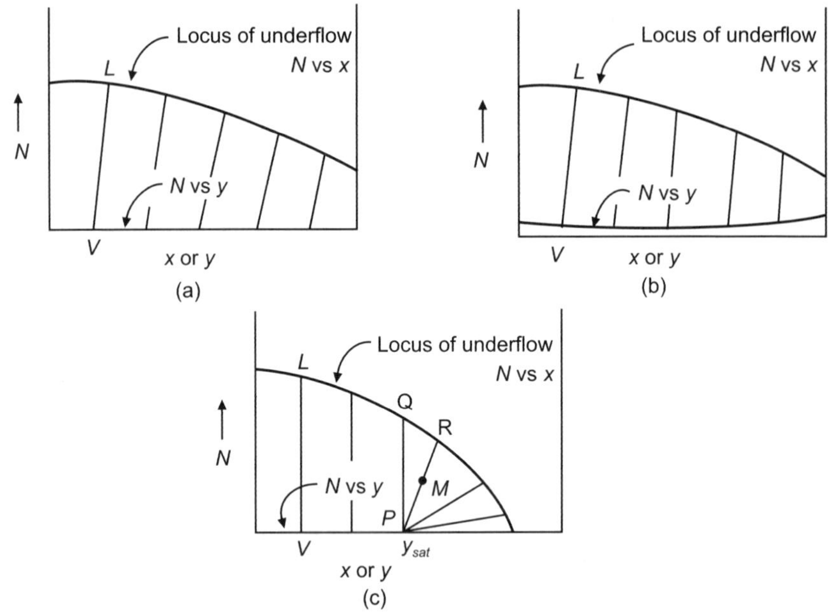

Fig. 8.21: Typical nonideal leaching equilibrium curves

solute preferentially adsorbed on the pore surface (2) some solute remain undissolved due to insufficient time of contact between the solids and the solvent and (3) the solute is soluble in the inert solids and gets unequally distributed between the solid and the liquid phases.

In Fig. 8.21b is shown, the locus of overflow lying above the $N = 0$ axis, indicating that inert solid is either partly soluble in solvent **A**, or does not settle to give an absolutely clear solution and hence is partly withdrawn with the solution. This will also occur if the solution is withdrawn before settling is completed. In all the cases discussed so far, the solute is assumed to be infinitely soluble in the solvent and the x and y have values over the entire range from 0 to 1. Figure 8.21c however, shows the case when the solute exhibits limited solubility in the leaching solvent. The solute content of the overflow solution represented by P in Fig. 8.21c is y_{sat} and this is the maximum solution concentration attainable by dissolving the solute **C** in the solvent **A** and the leaching process will not produce a clear solution stronger than y_{sat}. This saturation limit is not reached if the mixture M lies towards the left of line PQ in the figure. The phases separated lie at the ends of the vertical tie line such as VL, which means the solute concentration in the overflow and the underflow are the same. If the mixture lies towards the right of PQ as shown by point M in the figure, the clear solution separated is the saturated solution of **C** in **A** represented by the point P and the underflow represented by the point R will contain some undissolved solute also. The slope of the tie line and the location of the underflow stream depends upon the composition of the original mixture M subjected to settling.

Example 8.2: The settling characteristics of a slurry consisting of $CaCO_3$ suspended in a solution of NaOH and water is given below which indicate that the amount of the solution entrained with the solids vary with solution concentration:

x, weight fraction NaOH in the solution	0	0.02	0.04	0.06	0.08	0.10
Entrainment, kg solution /kg $CaCO_3$	1.496	1.553	1.640	1.760	1.911	2.083

Assume that there is no preferential adsorption of NaOH on the solid. A slurry consisting of 10% $CaCO_3$, 5% NaOH and 85% water is allowed to settle. (a) Plot the practical equilibrium curve on

N-x-y coordinates. (b) Locate the original mixture (*M*), underflow (*L*) and overflow (*V*) streams and determine the concentration of these streams.

***Solution*:** (a) The components present are:

$$\mathbf{A} = \text{water}, \mathbf{B} = CaCO_3, \mathbf{C} = NaOH.$$

The given data on the settling characteristics of the slurry consisting of $CaCO_3$, NaOH and water give the locus of underflow N vs. x. Here x is the weight fraction of NaOH on $CaCO_3$-free basis in the underflow and these are given in the first row of the table of settling data. Since N is the ratio of mass of inerts to mass of solution (ratio of mass of **B** to mass of **C**+**A**) in the underflow, the reciprocal of the values in the second row of the table of data gives N. Thus we get:

x, kg NaOH/kg solution	0	0.020	0.040	0.060	0.080	0.10
N, kg $CaCO_3$/kg solution	0.6684	0.6439	0.6098	0.5682	0.5233	0.4801

N versus x is plotted to get the locus of underflow as shown in Fig. 8.22. Assuming that the overflow is free of insolubles, the locus of overflow (N vs. y) lies on the x-axis. Thus Fig. 8.22 represents the equilibrium curve for the system.

(b) 100 kg of the original mixture (*M*) contains water (**A**) = 85 kg, $CaCO_3$ (**B**) = 10 kg and NaOH (**C**) = 5 kg. The mixture can be represented by the point M in Fig. 8.22 with following values for its coordinates N_M and x_M.

$$N_M = \frac{\text{mass of }\mathbf{B}}{\text{mass of }(\mathbf{A}+\mathbf{C})} = \frac{10}{90} = 0.1111$$

$$x_M = \frac{\text{mass of }\mathbf{C}}{\text{mass of }(\mathbf{A}+\mathbf{C})} = \frac{5}{90} = 0.0556$$

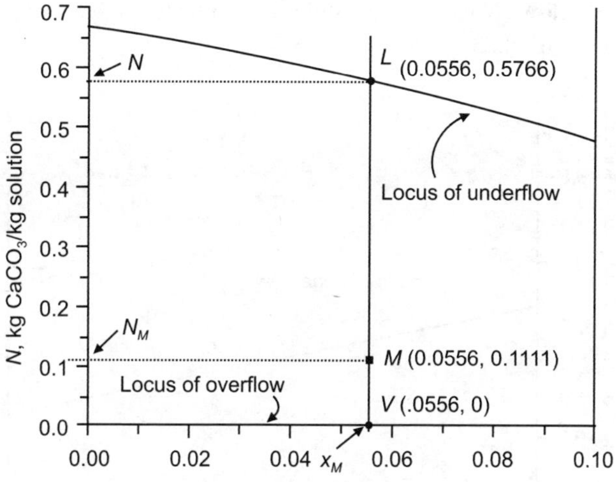

Fig. 8.22: Equilibrium curve for Example 8.2

The mixture when allowed to settle, separates into two streams: the clear solution or the overflow represented by point V and the settled solids or the underflow represented by point L. Since the entire solute (NaOH) is dissolved in water, and there is no preferential adsorption of solute on the solid, the concentration of NaOH in the solution associated with the underflow and the overflow are equal and they are also equal to the concentration of NaOH in the original mixture M. Thus a vertical line drawn through M is the tie line connecting the underflow and overflow streams. The tie line intersects the locus of underflow at L and the locus of overflow (the $N = 0$ axis) at V.

The coordinates of the underflow stream are read from the graph. The underflow (*L)* has the following coordinates: $N = 0.5766$ and $x = 0.0556$. It means that the underflow stream contains 0.5766 kg **B** and 0.0556 kg **C** per 1 kg (**A**+**C**). Or, the percent composition of the underflow can be computed as:

Component	Weight, kg	Weight %
NaOH	0.0556	(0.0556/1.5766) × 100 = 3.53%
$CaCO_3$	0.5766	(0.5766/1.5766) × 100 = 36.57%
Water	1 – 0.0556 = 0.9444	(0.9444/1.5766) × 100 = 59.90%
Total	1.5766	100%

The overflow (V) has the following coordinates: $N = 0, y = 0.0556$. Therefore the percent composition of the solution is 5.56% NaOH and 100 – 5.56 = 94.44% water. **(Answer)**

Example 8.3: In leaching of copper sulphate from roasted copper ore using pure water as the solvent, it is found that the solids in the underflow sludge retain solution such that the weight ratio of inert solids to water retained is 1:2 for all solution concentration. Plot the equilibrium curve on *N-x-y* coordinates.

***Solution*:** One kilogram inert solids (**B**) in the underflow retains 2 kg water (**A**) and the solute dissolved in it. If x is the concentration of solute in the solution, for 2 kg **A** in the underflow, there are $\frac{2}{1-x}$ kg solution. Thus the underflow contains 1 kg inert solids (**B**) which retains $\frac{2}{1-x}$ kg solution. Since N is defined as the ratio of mass of **B** to mass of solution retained,

$$N = \frac{B}{A+C} = \frac{1}{2/(1-x)} = 0.5(1-x)$$

Thus the locus of underflow (N vs x) is a straight line with slope –0.5 that makes an intercept with N-axis of 0.5 as shown in Fig. 8.23

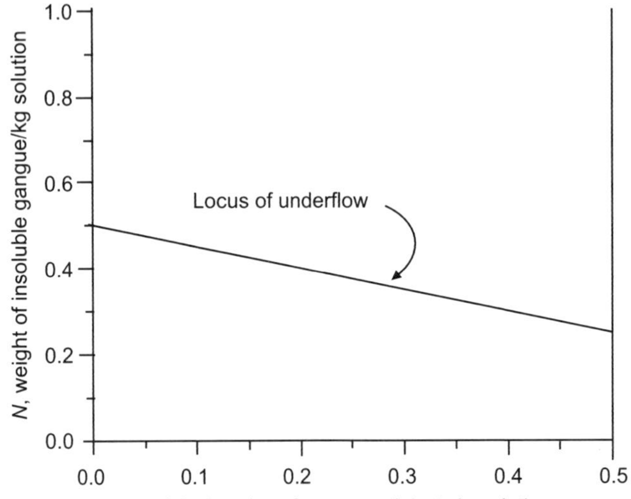

Fig. 8.23: Locus of underflow when solvent retained is constant. (Example 8.3)

8.9 CALCULATIONS FOR LEACHING STAGES

Leaching can be carried out as single stage or multistage cocurrent or countercurrent operation. Though the dissolution of the solute in the solvent is completed in a single stage, the underflow sludge is subjected to additional washing in multistage operation to reduce the solute content further. The extent of leaching, i.e. the amount of the soluble substance leached is to be estimated knowing the initial solute content of the solid, the amount of the solvent and the concentration of the solute in it, the number of stages and the method, whether batch or continuous countercurrent. The number of stages required to reduce the solute content of the solid to some specified value is estimated knowing the amount and solute concentration of the leaching solvent.

Single-Stage Leaching

Consider that the leaching is carried out in devices such as mixer-thickeners which represent one leaching stage as shown by the circle in Fig. 8.24. Feed solids are brought into contact with the liquid solvent and after dissolution of the solute in the liquid, the immiscible phases – the underflow and the overflow – are separated. Assuming that the inert solids are insoluble in the liquid, the underflow contains the entire insoluble matter present in the feed, and the solution retained by them. The clear solution drawn off as the overflow is free of inert material. The settling characteristics plotted in Fig. 8.25 are established for the system keeping the conditions of temperature, concentration and time of settling, same as those in the equipment.

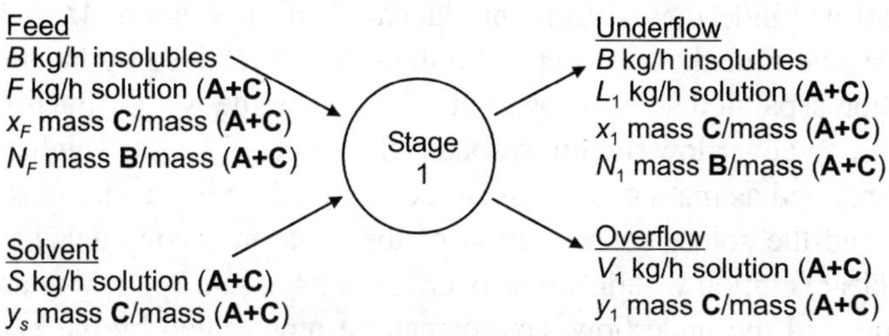

Fig. 8.24: Single stage leaching

Let the feed contain B kg/h of insoluble solids and F kg/h of solute (**C**) and solvent (**A**). N_F is the weight fraction of insolubles and x_F the weight fraction of solute in the feed on **B**-free basis. Solvent is admitted at a rate S kg/h containing y_S weight fraction solute. Knowing the coordinates N_F and x_F the feed is located at point F on the leaching $N-x-y$ diagram of Fig. 8.25. The solvent point S lies on the x-axis, as $N = 0$ for the solvent. The addition of the solvent to the feed produces a mixture M_1 which can be located in the figure, by calculating its coordinates x_{M1} and N_{M1} by material balance or by using the mixture rule.

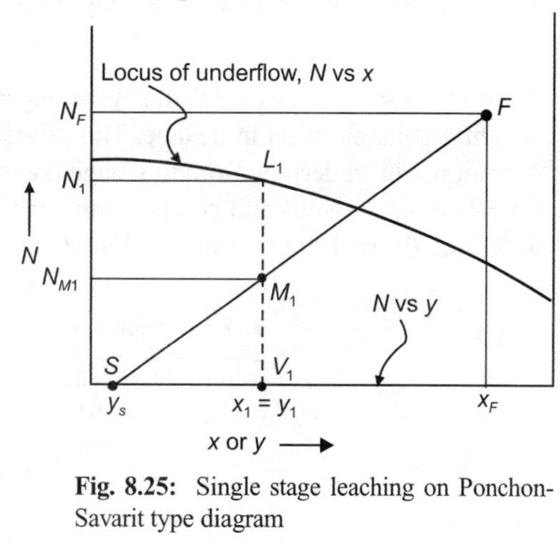

Fig. 8.25: Single stage leaching on Ponchon-Savarit type diagram

(**A**+**C**) - balance

$$F + S = M_1 \qquad ...(8.10)$$

C - balance

$$Fx_F + Sy_S = M_1 x_{M1} \qquad ...(8.11)$$

B – balance

$$FN_F = B = M_1 N_{M1} \qquad ...(8.12)$$

The coordinates x_{M1} and N_{M1} are calculated combining Eq. (8.10) with Eqs. (8.11) and (8.12).

$$x_{M1} = \frac{Fx_F + Sy_S}{M_1} = \frac{Fx_F + Sy_S}{F + S} \quad \text{...(8.13)}$$

$$N_{M1} = \frac{B}{M_1} = \frac{B}{F + S} \quad \text{...(8.14)}$$

The coordinates calculated using Eqs. (8.13) and (8.14) can be used to locate M_1 on the diagram. M_1 can be located by the mixture rule as well, which was shown to be applicable for similar cases in fractionation and liquid-liquid extraction. By the mixture rule, the point M_1 lies on the straight line joining F and S in such a way that

$$\frac{F}{S} = \frac{\overline{M_1S}}{\overline{M_1F}} \quad \text{...(8.15)}$$

The overflow and the underflow streams produced when the mixture M_1 is allowed to settle and separate can now be determined easily. Let L_1 be the solution retained by the solids in the underflow and V_1 be the solution separated as overflow. Let x_1 and y_1 be the weight fraction of solute on **B**-free basis in the underflow and overflow streams respectively and let N_1 be the weight fraction of inert solids in the underflow expressed as mass of **B**/mass of (**A**+**C**). Under ideal conditions, the entire solute has gone into solution and the solute concentration of the solution carried away with the solids in the underflow and the clear solution withdrawn as overflow are the same or $x_1 = y_1$ and they are also equal to x_{M1}. The overflow and the underflow streams can be represented by the vertical tie line passing through M_1. The overflow V_1 lies on the $N = 0$ axis and the underflow L_1 lies on the locus of the underflow as shown in Fig. 8.25. The N-coordinate of L_1 is N_1 which can be read from the graph. Now, since the entire insoluble solid in the feed will leave with the underflow, we have

$$B = FN_F = L_1N_1 \quad \text{...(8.16)}$$

Equation (8.16) can be used to calculate L_1, the solution retained by the solids in the settled sludge. The overflow V_1 is obtained by writing the solution (**A**+**C**) balance for the complete process.

$$F + S = M_1 = V_1 + L_1 \quad \text{...(8.17)}$$

Example 8.4: A slurry of flaked soyabeans consists of 75 kg inert solids suspended in 25 kg of a 10 weight% solution of oil in hexane. This slurry is contacted with 100 kg pure hexane in a single stage operation. The underflow from this stage contains 2 kg solution for every 3 kg insoluble solids present. Calculate the amounts and compositions of the overflow and the underflow leaving the stage.

Solution: Basis: 100 kg of the feed treated. The leaching operation is represented in Fig. 8.26

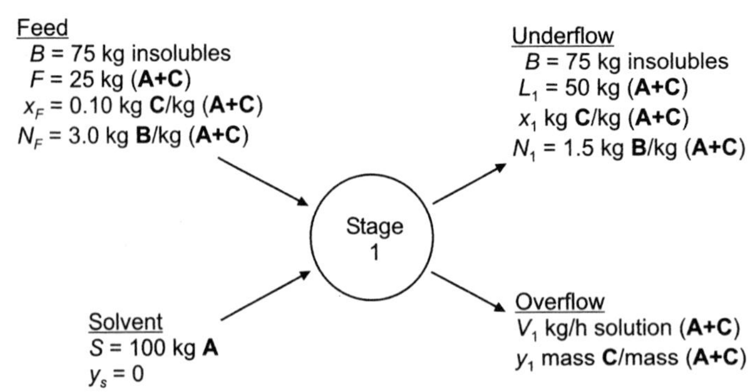

Fig. 8.26: Single-stage leaching (Example 8.4)

The feed consists of 75 kg inert solids (**B**) suspended in 25 kg of solution containing 10 % by weight oil (**C**) dissolved in hexane (**A**). Thus F = 25 kg (**A**+**C**) and **B** = 75 kg. The composition of the feed on **B**-free basis:

$$x_F = \frac{\text{kg oil in the feed}}{\text{kg (oil + hexane)}} = 0.10$$

$$N_F = \frac{\text{kg inert solids in the feed}}{\text{kg (oil + hexane)}} = \frac{75}{25} = 3.0$$

When 100 kg solvent hexane is added to the feed, the resulting mixture (M_1) has the following composition:

Oil $= Fx_F + Sy_S = 25 \times 0.1 + 0 = 2.5$ kg

Hexane $= F(1 - x_F) + S(1 - y_S) = 25 \times 0.9 + 100 = 122.5$ kg

Inert solids $= B = FN_F = 25 \times 3 = 75.0$ kg

$$x_{M1} = \frac{\text{kg oil in the mixture}}{\text{kg (oil + hexane)}} = \frac{2.5}{125} = 0.02$$

$$N_{M1} = \frac{\text{kg inert solids in the mixture}}{\text{kg (oil + hexane)}} = \frac{75}{125} = 0.60$$

The mixture (M_1) when allowed to settle, the underflow and overflow are separated.

Underflow: The underflow carries all the insoluble solids and the solution retained by the solids. Since the solution retained by the solids is 2 kg for every 3 kg insoluble solids, the underflow consists of:

Insolubles = 75 kg

Solution retained, $L_1 = 50$ kg

Since the solution retained has the same composition as the solution in the mixture M_1, the fraction of oil in the underflow solution, $x_1 = x_{M1} = 0.02$. The fraction of insolubles = 75/50 = 3/2.

The composition of the underflow sludge can also be expressed as mass percentages of the constituents.

Constituents	Weight, kg	Weight %
Oil	50 × 0.02 =1.0 kg	(1.0/125.0) × 100 = 0.8%
Hexane	50 – 1.0 = 49.0 kg	(49.0/125.0) × 100 = 39.2%
Inert solids	75.0 kg	(75.0/125.0) × 100 = 60.0%
Total	125.0 kg	100.0%

Overflow: The amount of solution withdrawn as clear solution overflow (V_1) is the difference between the amount of solution present in the mixture M_1 and the solution retained by the underflow solids (L_1). Or

$$V_1 = M_1 - L_1 = (F + S) - L_1 = 125 - 50 = 75 \text{ kg}$$

The concentration of oil in the overflow is the same as that of the original mixture (M_1) and the underflow solution (L_1). Thus

$$y_1 = x_1 = x_{M1} = 0.02$$

Thus overflow consist of 2.0% by weight of oil and 98.0% by weight of hexane.

Graphical representation: On the Ponchon-Savarit type diagram, the process can be represented as shown in Fig. 8.27. Point F represents the feed and is marked on the diagram with coordinates ($x_F = 0.10$, $N_F = 3.0$). Solvent S is represented by the origin of the coordinate system since the solvent is pure. When 25 kg of feed (F) on insolubles-free basis is mixed with 100 kg solvent S, the resultant mixture M_1 is located on the line joining F and S in such a way that

$$\frac{\overline{SM_1}}{\overline{M_1F}} = \frac{F}{S} = \frac{25}{100}$$

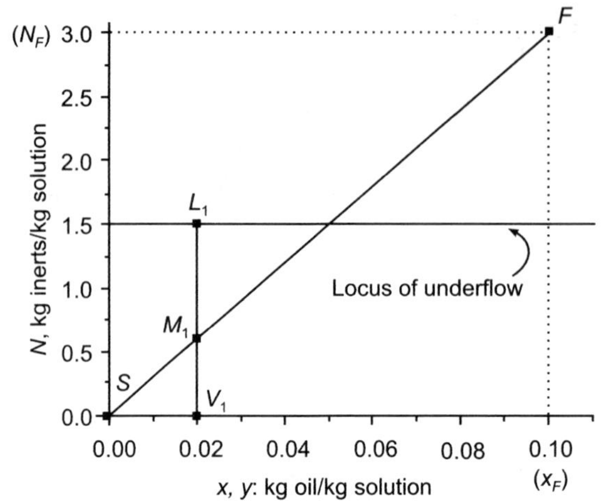

Fig. 8.27: Graphical representation of single-stage leaching of Example 8.4

M_1 can also be located knowing its coordinates (x_{M1}= 0.02, N_{M1} = 0.60), which were calculated earlier. The vertical tie line is drawn passing through M_1, which intersects the locus of underflow at L_1 and the $N = 0$ axis at V_1. The amounts of V_1 and L_1 are related as

$$\frac{\overline{V_1M_1}}{\overline{M_1L_1}} = \frac{L_1}{V_1}$$

and
$$V_1 + L_1 = M_1 = F + S$$

Alternatively, the coordinates of L_1 (x = 0.02, N = 1.5) and V_1 (y = 0.02, N = 0) can be used to calculate V_1 and L_1 through material balance.

(**Note:** The present problem can be solved both analytically and graphically. However, for variable underflow conditions, a graphical method is inevitable for solving leaching problems.)

Multistage Crosscurrent Leaching

Figure 8.28 shows a three stage crosscurrent leaching unit. Multistage crosscurrent leaching can be treated as an extension of the single-stage leaching. The analysis provided in the previous section for single stage leaching can be used to obtain the underflow (L_1) and overflow (V_1) leaving the first stage of the crosscurrent unit. The underflow from the stage 1 serves as the feed to the second stage, and so on. In the n th stage, L_{n-1} kg underflow solution leaving the preceding stage is mixed with S_n kg of solvent, the mixture M_n results, which when allowed to separate, yields the products L_n and V_n, the underflow and overflow from stage n. L_n and V_n lie at the ends of the vertical tie line passing through

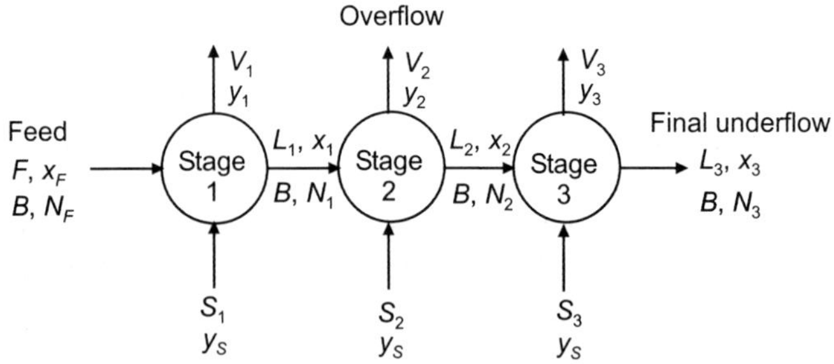

Fig. 8.28: Three-stage crosscurrent leaching unit

M_n. The N-coordinate of L_n gives N_n, which can be used to compute the quantities of the effluent streams from stage n as

$$L_n = \frac{B}{N_n} \qquad \ldots(8.18)$$

$$V_n = M_n - L_n = S_n + L_{n-1} - L_n \qquad \ldots(8.19)$$

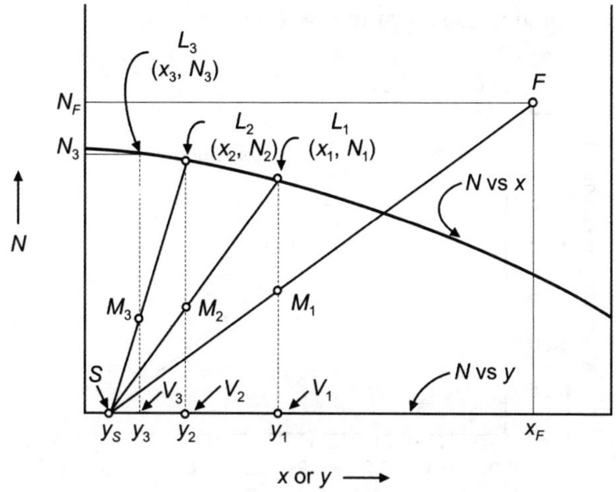

Fig. 8.29: Determination of the quantities and compositions in cross-current leaching

The final leached solids is removed from the last stage N (stage 3 in Fig. 8.29) with L_N (= L_3) kg solution retained by B kg insoluble solids.

Example 8.5: One-thousand kg of roasted copper ore containing 20 % $CuSO_4$, 70% gangue, and 10% water is to be extracted by washing it three times with 1500 kg batches of fresh water. The settling characteristics of the slurry have been found to be as follows:

x, wt fraction $CuSO_4$ in solution	0	0.1	0.2	0.3	0.4	0.5	0.6
kg solution retained/kg gangue	0.8	0.833	0.864	0.905	0.960	1.017	1.076

Determine (a) The percent composition by weight of the underflow sludge (b) The percent recovery of copper sulphate.

***Solution*:** The three stage crosscurrent unit is shown in Fig. 8.30.

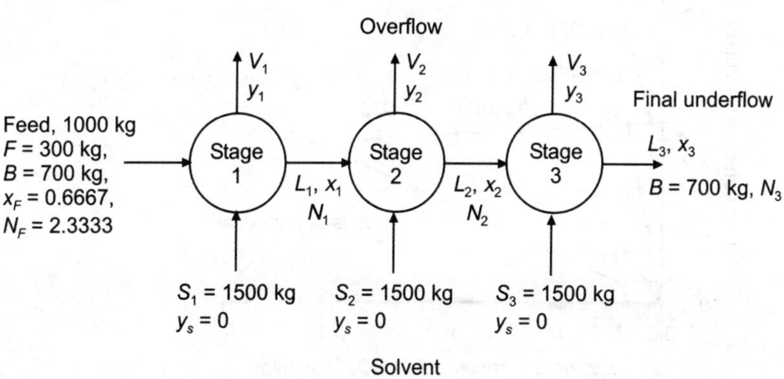

Fig. 8.30: Crosscurrent leaching (Example 8.5)

The settling data gives the locus of underflow. The reciprocal of the values in the second row of the table of data gives N. Thus, when the entrainment is 0.8 kg solution retained per kg inert, $N = \dfrac{1}{0.8} = 1.25$

The x versus N data are tabulated below:

x, wt fraction $CuSO_4$ in solution	0	0.1	0.2	0.3	0.4	0.5	0.6
N, kg gangue/kg solution	1.25	1.20	1.157	1.105	1.042	0.983	0.929

The equilibrium curve is plotted as shown in Fig. 8.31.

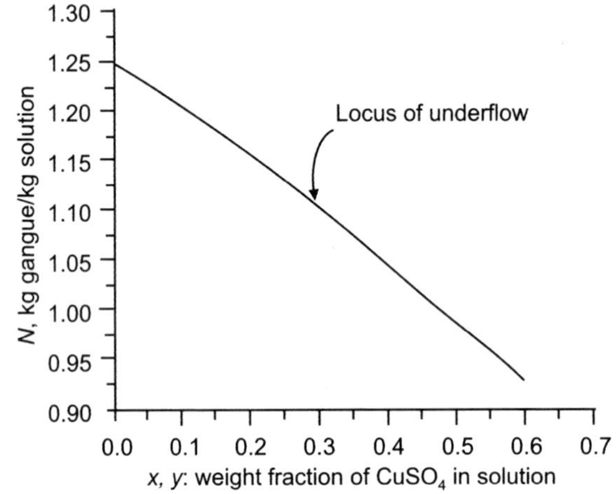

Fig. 8.31: Representation of the leaching equilibrium for Example 8.5

Basis: 1000 kg feed treated.

Stage-1: The feed consists of copper sulphate (**C**) = 200 kg; water (**A**) = 100 kg; gangue (**B**) = 700 kg

The feed is represented by point F with coordinates:

$$x_F = \frac{200}{300} = 0.6667 \quad N_F = \frac{700}{300} = 2.3333$$

In stage-1, 1500 kg solvent (S) is mixed with 1000 kg feed. Figure 8.32 gives the equilibrium curve on a slightly expanded scale. The resultant mixture can be represented by the point M_1 that lies on the

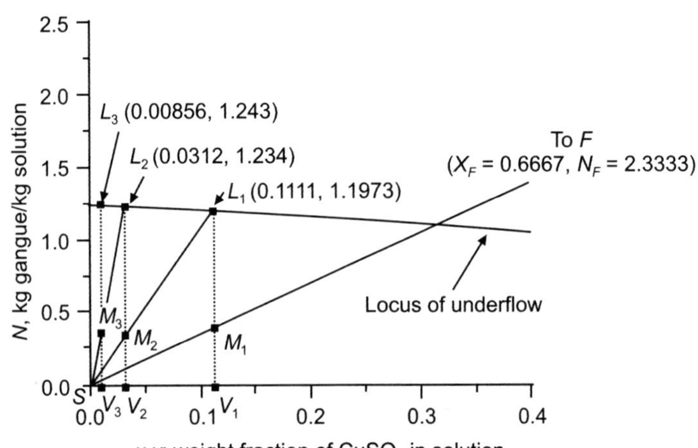

Fig. 8.32: Crosscurrent leaching: Example 8.5

line SF. (F is not shown in Fig. 8.32). The weight fraction of $CuSO_4$ on inert-free basis is calculated by Eq. (8.13)

$$x_{M1} = \frac{Fx_F + Sy_S}{F+S} = \frac{200}{300+1500} = 0.1111$$

Using this coordinate, M_1 can be located on the line FS as shown. A vertical tie line is drawn through M_1 which intersects the locus of underflow at L_1. The N-coordinate of the underflow is noted from the graph. $N_1 = 1.1973$. It means 1 kg solution is carried away with 1.1973 kg insoluble gangue. Therefore, solution leaving with the underflow sludge from stage 1 is

$$L_1 = \frac{700}{1.1973} = 584.65 \text{ kg}.$$

Stage-2: Underflow from stage-1 (L_1) is the feed for the second stage, which is mixed with 1500 kg solvent. The resultant mixture is represented by M_2 on the line joining L_1 and S at x_{M2}, where

$$x_{M2} = \frac{L_1 x_1 + Sy_S}{L_1 + S} = \frac{584.65 \times 0.1111 + 0}{584.65 + 1500} = 0.0312$$

The vertical tie line passing through M_2 is drawn which intersects the locus of underflow at L_2. Its coordinates are $x_2 = x_{M2}$ and $N_2 = 1.234$ read from the graph. Now

$$L_2 = \frac{700}{1.234} = 567.26 \text{ kg}$$

Stage-3: For stage-3, L_2 kg feed and 1500 kg solvent are mixed together and the mixture M_3 is located with x-coordinate x_{M3} on the line joining L_2 and S. x_{M3} is calculated as

$$x_{M3} = \frac{L_2 x_2 + Sy_S}{L_2 + S} = \frac{567.26 \times 0.0312 + 0}{567.26 + 1500} = 0.00856$$

The vertical tie line drawn through M_3 intersects the locus of underflow at L_3. The solute concentration of underflow is $x_3 = x_{M3} = 0.00856$. Its N-coordinate is 1.244 from graph. The quantity of underflow leaving stage-3,

$$L_3 = \frac{700}{1.244} = 562.7 \text{ kg}.$$

(a) The final underflow sludge contains 700 kg inert solids, 562.7 kg solution retained by the solids whose composition is 0.856% $CuSO_4$. The percent composition of the underflow sludge is

Constituents	Weight, kg	Weight %
$CuSO_4$	562.7 × 0.008567= 4.8 kg	(4.8/1262.7) × 100 = 0.38%
Water	562.7 × (1 – 0.008567) = 557.9 kg	(557.9/1262.7) × 100 = 44.18%
Gangue	700 kg	(700/1262.7) × 100 = 55.44%
Total	1262.7 kg	100.0%

(Answer)

(b) Since the underflow sludge contains 4.8 kg, $CuSO_4$ recovered is (200 – 4.8) = 195.2 kg.

$$\text{Percent recovery} = \frac{195.2}{200} \times 100 = 97.6\%.$$

(Answer)

Example 8.6: Black ash containing 45 weight% Na_2CO_3 and the rest water-insolubles is extracted with water such that 96% of soda ash in it is recovered. The underflow carries with it 15 kg water with every 100 kg insolubles. Determine the total quantity of water used and the concentration of the combined extract if the leaching has been carried out for a feed rate of 1000 kg/h of black ash in a two-stage cross-flow cascade system where equal amounts of water are added in both the stages.

***Solution*:** 1000 kg/h black ash as feed.

The feed admitted to stage 1 is $F = 450$ kg, $B = 550$ kg.

$$x_F = \frac{\text{kg } \mathbf{C}}{\text{kg } (\mathbf{A}+\mathbf{C})} = \frac{450}{450} = 1$$

$$N_F = \frac{\text{kg } \mathbf{B}}{\text{kg } (\mathbf{A}+\mathbf{C})} = \frac{550}{450} = 1.2222$$

Since there is 96% of solute recovery, the final underflow contains 4% of the soda ash present in the feed. Therefore, soda ash in the final underflow is $450 \times 0.04 = 18$ kg. The insoluble solids retain 15 kg water for every 100 kg solids. Since 550 kg insoluble solids are present in the underflow, water present in the final underflow $= \frac{15}{100} \times 550 = 82.5$ kg.

The underflow solution leaving stage-2 is $L_2 = 82.5 + 18 = 100.5$ kg.

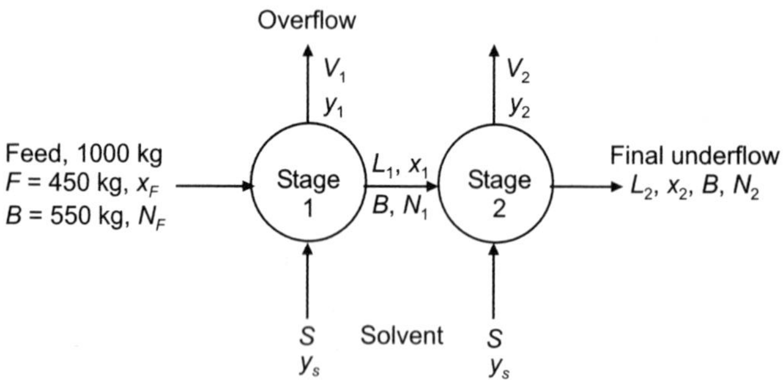

Fig. 8.33: Two-stage crosscurrent leaching of Example 8.6

Solute concentration in the final underflow is

$$x_2 = \frac{\text{kg } \mathbf{C} \text{ in the underflow}}{\text{kg } (\mathbf{C}+\mathbf{A}) \text{ in the underflow}} = \frac{18}{18+82.5} = 0.1791 \tag{A}$$

Let S kg solvent be added in both stages. Consider stage 1. Let the concentration of the solution in the first stage be x_{M1}. Then by Eq. (8.13)

$$x_{M1} = \frac{Fx_F + Sy_S}{M_1} = \frac{Fx_F + Sy_S}{F+S} = \frac{450}{450+S}$$

The concentration of solution leaving with the underflow and overflow are the same as x_{M1}. Therefore,

$$x_1 = y_1 = x_{M1} = \frac{450}{450+S} \tag{B}$$

The underflow carries 15 kg water with every 100 kg insolubles. The amount of water present in the underflow solution is, therefore, $\frac{15}{100} \times 550$ kg. This should be equal to $L_1(1 - x_1)$, where L_1 is the total weight of solution retained by the solids in the underflow and $(1 - x_1)$ is the weight fraction of water in the underflow. Therefore,

$$L_1(1-x_1) = L_1 \frac{S}{450+S} = \frac{15}{100} \times 550 = 82.5$$

Or $$L_1 = 82.5\frac{450+S}{S}\text{ kg} \tag{C}$$

Now consider the second stage. L_1 kg underflow is mixed with S kg solvent. The solute concentration of the resultant mixture M_2 is

$$x_{M2} = \frac{L_1x_1 + Sy_S}{M_2} = \frac{L_1x_1 + Sy_S}{L_1 + S} \tag{D}$$

Substituting (B) and (C) in Eq. (D), we get

$$x_{M2} = \frac{82.5\times 450/S}{\frac{82.5\times(450+S)}{S}+S} = \frac{37125}{37125+82.5S+S^2} \tag{E}$$

Since $x_{M2} = x_2$, Eq. (E) can be combined with Eq. (A) as:

$$\frac{37125}{37125+82.5S+S^2} = 0.1791 \tag{F}$$

Equation (F) is solved for S.

$$S = 373.3\text{ kg}$$

Since both stages use the same quantity of solvent, total water used is = $2S$ = 746.6 kg. **(Answer)**

Substituting S = 373.3 kg in Eq. (C)

$$L_1 = 82.5\frac{450+373.3}{373.3} = 181.95\text{ kg}$$

Now the overflow from stage-1, V_1 is obtained by material balance on stage 1.

$$V_1 = F + S - L_1 = 450 + 373.3 - 181.95 = 641.35\text{ kg}$$

The overflow from stage-2 V_2 is obtained by material balance on stage 2.

$$V_2 = L_1 + S - L_2 = 181.95 + 373.3 - 100.5 = 454.75\text{ kg}$$

Since 96% of the soda ash is present in the combined overflow, the soda ash concentration in the combined extract $= \frac{0.96\times 450}{V_1+V_2} = \frac{432}{(641.35+454.75)}\times 100 = 39.41\%$ **(Answer)**

Multistage Countercurrent Leaching

Ponchon-Savarit graphical method

A countercurrent multistage leaching cascade is shown in Fig. 8.34. The stages are numbered starting with the stage-1 in which fresh feed is admitted. F kg/h is the rate of addition of the feed on **B**-free basis and B kg/h is the insoluble solids in the feed. Solvent at a rate S kg/h containing y_S weight fraction solute is admitted in the last stage N and flows countercurrently to the flow of the solids.

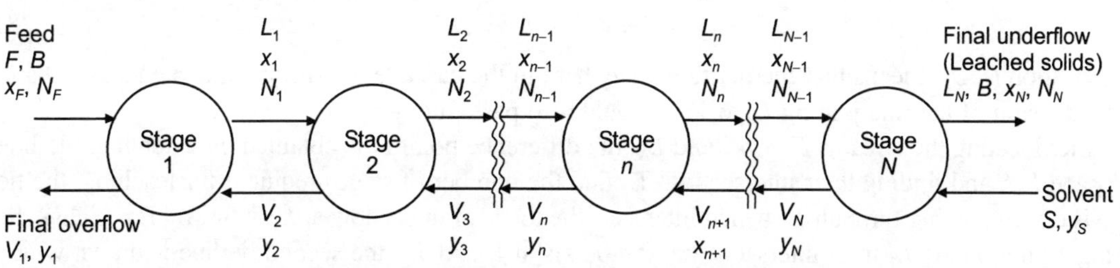

Fig. 8.34: Countercurrent leaching operation

Concentrated leach liquor (final overflow) is withdrawn from stage 1 at a rate V_1 kg/h and contains y_1 weight fraction solute **C** and the leached solids (final underflow) is removed from stage N. The final underflow contains B kg/h insolubles which carry L_N kg solution of concentration x_N weight fraction solute. The ratio of insolubles to solution in the underflow is N_N.

Knowing any three of the four terminal streams, F, S, L_N, and V_1, the fourth can be located on the $N-x-y$ coordinate system graphically. Consider the following overall material balances for the entire cascade.

(**A**+**C**) balance:

$$F + S = M = V_1 + L_N \qquad ...(8.20)$$

Here M is a fictitious stream that would result if the feed and solvent are brought into direct contact.

Solute – **C** balance:

$$Fx_F + Sy_S = Mx_M = V_1 y_1 + L_N x_N \qquad ...(8.21)$$

Insoluble solid **B** – balance:

$$FN_F = MN_M = L_N N_N = B \qquad ...(8.22)$$

The coordinates x_M and N_M are the solute and inert concentrations in the fictitious mixture. These are calculated by rearranging Eqs. (8.21) and (8.22)

$$x_M = \frac{Fx_F + Sy_S}{F + S} \qquad ...(8.23)$$

$$N_M = \frac{FN_F}{F + S} = \frac{B}{F + S} \qquad ...(8.24)$$

Knowing the quantities and compositions of the feed and the solvent, the mixture M can be located on the N–x–y coordinates system either by using Eqs. (8.23) and (8.24) or by mixture rule which can be written as

$$\frac{F}{S} = \frac{\overline{MS}}{\overline{MF}} \qquad ...(8.25)$$

By Eq. (8.20) the line joining V_1 and L_N should pass through M. Thus knowing V_1, L_N can be located on the extension of the line V_1M or knowing L_N, V_1 is also located in a similar manner. It should be remembered that V_1 should lie on the locus of overflow (the $N = 0$ axis) and L_N lies on the locus of underflow.

Equation (8.20) can be rearranged as

$$V_1 - F = S - L_N = D \qquad ...(8.26)$$

where D is the net flow towards the left. By using the lever-arm-rule the point D can be located at the intersection of the extensions of the line joining F to V_1 and L_N to S. Now consider the solution balance from stage 1 through stage n.

$$F + V_{n+1} = V_1 + L_n \qquad ...(8.27)$$

$$V_1 - F = V_{n+1} - L_n = D \qquad ...(8.28)$$

Equation (8.28) means that the net flow anywhere in the cascade is constant and is equal to D and the extension of the line joining L_n to V_{n+1} should also pass through D.

After locating the streams F, S, V_1 and L_N, the difference point D is obtained by extending the line FV_1 and L_NS and finding their intersection. To find the number of stages required for leaching, the tie line is drawn passing through V_1 which intersects the locus of underflow at L_1. Now, by Eq. (8.28), the straight line joining L_1 to D intersects the $N = 0$ axis at V_2. At V_2, the second tie line is drawn which locates L_2 on the locus of underflow as shown in Fig. 8.35. Now V_3 is determined by drawing the operating line through L_2D and the construction continued till the Nth tie line passes through the final underflow L_N in the Fig. 8.35. In the figure, the fourth tie line passes through L_N, and therefore the number of stages = 4.

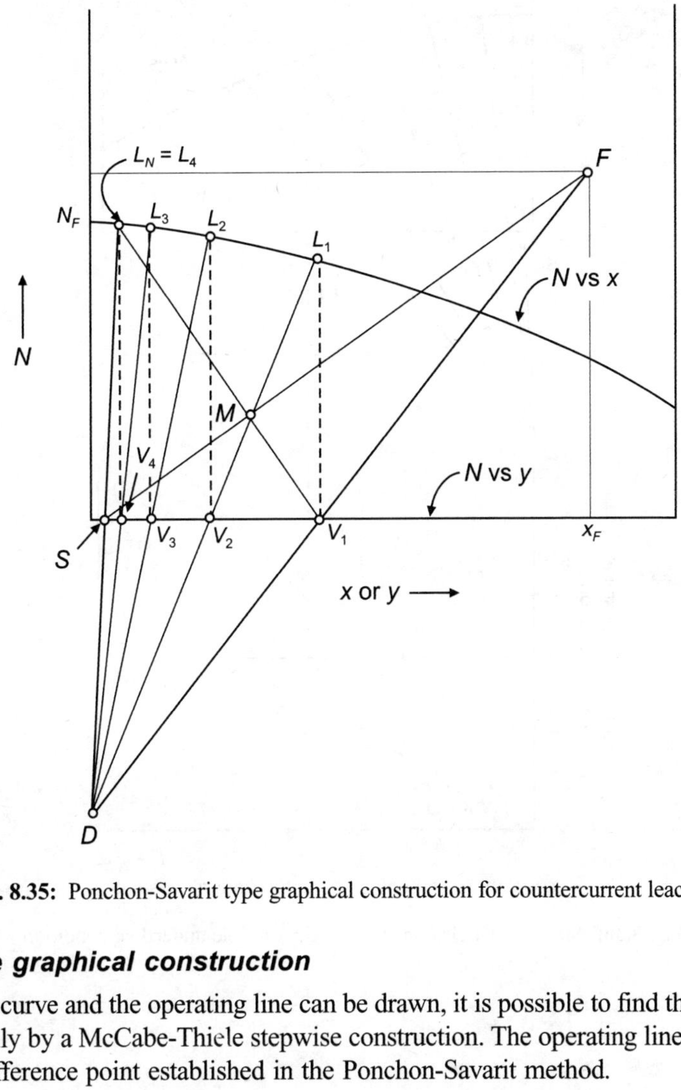

Fig. 8.35: Ponchon-Savarit type graphical construction for countercurrent leaching

McCabe-Thiele graphical construction

If the equilibrium curve and the operating line can be drawn, it is possible to find the number of stages required graphically by a McCabe-Thiele stepwise construction. The operating line can be drawn with the help of the difference point established in the Ponchon-Savarit method.

The intersections of the random lines drawn originating from the difference point D with the locus of underflow and overflow are projected to the y-x graph in the lower portion of Fig. 8.36. For example, the intersections of the line e with the locus of overflow and underflow in the upper diagram give the point e on the lower diagram. Similarly points a, b, c, d, f, etc. are also established and a smooth curve is drawn as shown from a to f which gives the operating curve. The number of stages is determined by the usual stepwise construction.

For constant underflow condition, the operating line is a straight line with slope L/V as will be shown in the next section, where L is the amount of solution retained by the solids in the underflow and V is the amount of overflow solution. The operating line can then be drawn (Fig. 8.37) knowing the concentrations at the two ends of the cascade or concentration at one end and the L/V ratio.

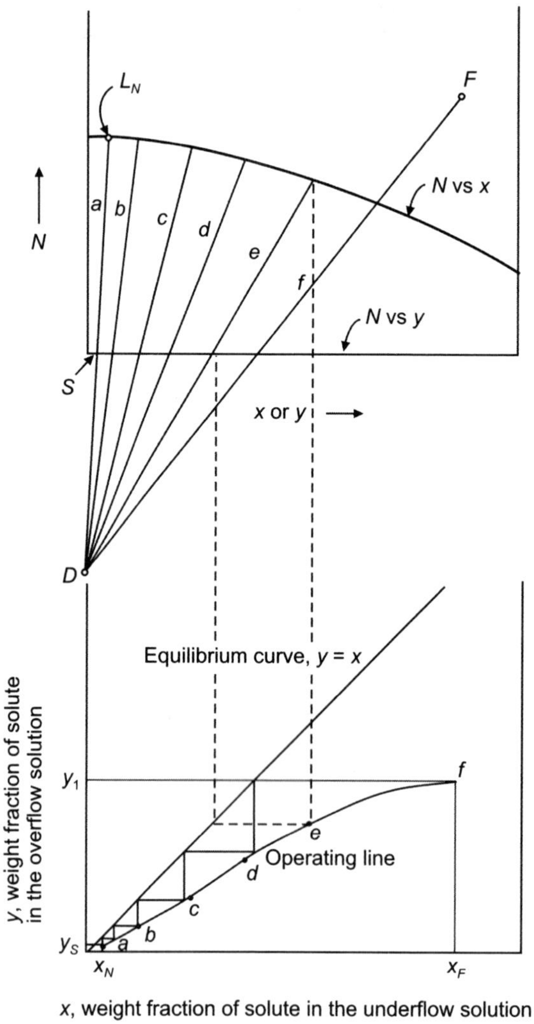

Fig. 8.36: McCabe-Thiele construction for variable underflow condition

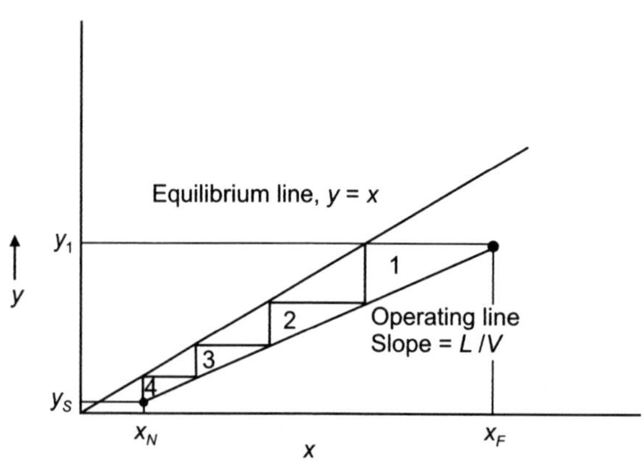

Fig. 8.37: McCabe-Thiele graphical construction for leaching stages for constant underflow conditions

Analytical method for constant underflow

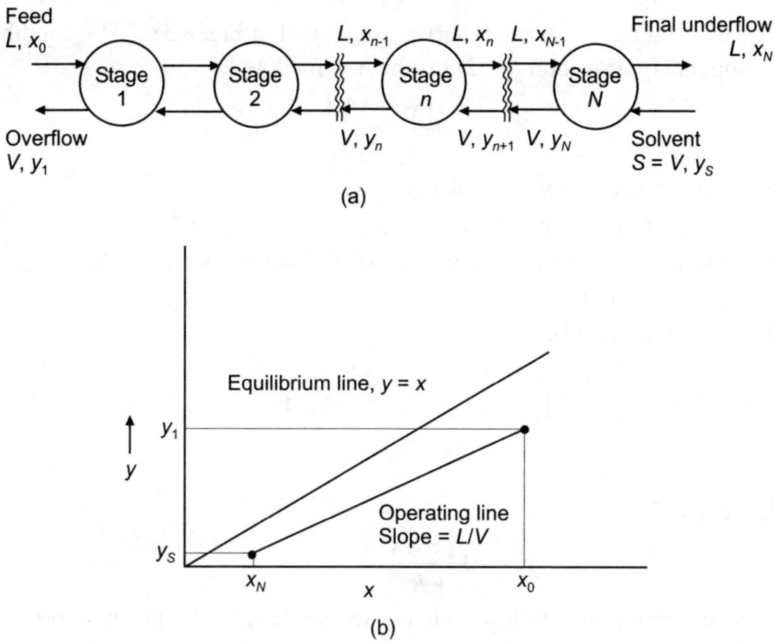

Fig. 8.38: (a) Continuous countercurrent system with N washing stages (b) Equilibrium curve and operating lines for constant underflow condition

Figure 8.38a shows a countercurrent cascade with N stages. It is assumed that the solution retained in the underflow remains constant at L kg/h throughout the cascade. If underflow is constant, so is the overflow. The overflow leaving each washing stage is denoted by V kg/h.

Considering a solute balance from stage 1 through n, we get

$$Lx_0 + Vy_{n+1} = Lx_n + Vy_1 \qquad \ldots(8.29)$$

which can be rearranged as

$$y_{n+1} = \frac{L}{V}x_n + \frac{Vy_1 - Lx_0}{V} \qquad \ldots(8.30)$$

This is the equation for the operating line for the washing cascade, when plotted on the y-x coordinates.

Let us denote the ratio of the weight of the solution in the overflow to the weight of the solution in the underflow by R. That is, $R = \frac{V}{L}$. For ideal leaching, it is assumed that the solute gets completely dissolved in the liquid and there is no preferential adsorption of the solute. The equilibrium relationship is $y = x$, because the concentration of overflow solution and the solution removed with the underflow streams are the same. Since the concentration of solute in the overflow (y) and that in the underflow (x) are equal under ideal conditions, R is also the ratio of the weights of solute in the overflow to that in the underflow.

$$R = \frac{\text{Weight of solution in the overflow}}{\text{Weight of solution in the underflow}} = \frac{\text{Weight of solute in the overflow}}{\text{Weight of solute in the underflow}} = \frac{V}{L} \qquad \ldots(8.31)$$

The slope of the operating line which passes through the terminal points (x_0, y_1) and (x_N, y_S) is $\frac{L}{V}$ which is given by

$$\frac{L}{V} = \frac{(y_1 - y_S)}{(x_0 - x_N)} = \frac{1}{R} \qquad \ldots(8.32)$$

The operating line and the equilibrium curves are plotted in Fig. 8.38b. The equilibrium curve is a straight line of slope unity. Equation (8.32) can be modified as

$$R = \frac{V}{L} = \frac{(x_0 - x_N)}{(y_1 - y_S)} \qquad \ldots(8.33)$$

For stage n in the cascade, the solute balance gives

$$L(x_{n-1} - x_n) = V(y_n - y_{n+1}) \qquad \ldots(8.34)$$

Since the concentration of liquid in the underflow and the overflow are the same,

$$y_n = x_n \quad \text{and} \quad y_{n-1} = x_{n-1} \qquad \ldots(8.35)$$

Substituting these in Eq. (8.34), we get

$$L(y_{n-1} - y_n) = V(y_n - y_{n+1}) \qquad \ldots(8.36)$$

Substituting $R = V/L$, and rearranging Eq. (8.36), we get

$$y_n = \frac{Ry_{n+1} + y_{n-1}}{1+R} \qquad \ldots(8.37)$$

Put $n = 1$ in Eq. (8.37)

$$y_1 = \frac{Ry_2 + y_0}{1+R} \qquad \ldots(8.38)$$

where y_0 is the concentration of overflow which is in equilibrium with the underflow of concentration x_0 entering stage-1.

For $n = 2$, Eq. (8.37) and Eq. (8.38) yield,

$$y_2 = \frac{Ry_3 + y_1}{1+R} = \frac{Ry_3}{1+R} + \frac{Ry_2 + y_0}{(1+R)^2} = \frac{R(1+R)y_3 + y_0}{1+R+R^2} \qquad \ldots(8.39)$$

Similarly for $n = 3$,

$$y_3 = \frac{R(1+R+R^2)y_4 + y_0}{1+R+R^2+R^3} \qquad \ldots(8.40)$$

And for $n = n$,

$$y_n = \frac{R(1+R+R^2+\ldots+R^{n-1})y_4 + y_0}{1+R+R^2+R^3+\ldots+R^n} \qquad \ldots(8.41)$$

Noting that

$$1+R+R^2+R^3+\ldots+R^n = \frac{1-R^{n+1}}{1-R} \qquad \ldots(8.42)$$

Eq. (8.41) can be written as

$$y_n = \frac{R(1-R^n)y_{n+1} + (1-R)y_0}{1-R^{n+1}} \qquad \ldots(8.43)$$

Replacing n with N, the total number of washing stages, and noting that $y_{N+1} = y_S$, Eq. (8.43) becomes

$$y_N = \frac{R(1-R^N)y_S + (1-R)y_0}{1-R^{N+1}} \qquad \ldots(8.44)$$

Now consider the solute balance from stage 1 through N.

$$L(x_0 - x_N) = V(y_1 - y_S) \qquad \ldots(8.45)$$

Substituting $R = V/L$ we get

$$R(y_1 - y_S) = (x_0 - x_N) \qquad \ldots(8.46)$$

At equilibrium, the underflow and overflow solution leaving a stage have the same composition, $x_N = y_N$ and $x_0 = y_0$. Using these, rearrange Eq. (8.46):

$$R(y_1 - y_S) = (y_0 - y_N) \quad \text{...(8.47)}$$

Or

$$y_N = y_0 - R(y_1 - y_S) \quad \text{...(8.48)}$$

Comparing Eqs. (8.44) and (8.48), we can show that

$$R^{N+1}(y_1 - y_S) = R^N(y_0 - y_S) + (y_1 - y_0) \quad \text{...(8.49)}$$

Substituting Eq. (8.46) into Eq. (8.49), we get

$$R^N(y_S - y_N) = (y_1 - y_0) \quad \text{...(8.50)}$$

which may be rearranged as

$$R^N = \frac{(y_1 - y_0)}{(y_S - x_N)} \quad \text{...(8.51)}$$

Or

$$N = \frac{\log\left(\frac{(y_1 - y_0)}{(y_S - x_N)}\right)}{\log R} \quad \text{...(8.52)}$$

Substituting for R using Eq. (8.33), Eq. (8.52) becomes

$$N = \frac{\log\left(\frac{(y_0 - y_1)}{(x_N - y_S)}\right)}{\log\left(\frac{(y_0 - x_N)}{(y_1 - y_S)}\right)} \quad \text{...(8.53)}$$

The above equation for the number of stages can be written in terms of R and f, where f is the fraction of solute entering stage-1 that remain unextracted. For the special case that the solvent used is pure, put $y_S = 0$ and $y_0 = x_0$ in Eq. (8.44):

$$y_N = \frac{(1 - R)x_0}{1 - R^{N+1}} \quad \text{...(8.54)}$$

Since $y_N = x_N$, Eq. (8.54) can be rewritten as

$$x_N = \frac{(1 - R)x_0}{1 - R^{N+1}} \quad \text{...(8.55)}$$

Let f denotes the fraction of solute in the feed that remains unextracted and hence leaves with the final underflow sludge, then

$$f = \frac{x_N}{x_0} \quad \text{...(8.56)}$$

and Eq. (8.55) can be modified as

$$R^{N+1} = 1 + (R - 1)\frac{1}{f} \quad \text{...(8.57)}$$

$$N = \frac{\log\left(1 + (R - 1)\frac{1}{f}\right)}{\log R} - 1 \quad \text{...(8.58)}$$

Equation (8.53) is used to determine the number of stages required when the amount of solution retained by the solids in the underflow is constant. The same equation can be used to determine the number of stages when the amount of solvent retained by the solids in the underflow is constant, with the weight fractions x_N, y_0, y_1 and y_S replaced by X_N, Y_0, Y_1 and Y_S, the respective weight ratios C/A (ratio of the weight of solute to weight of solvent) in the liquid.

$$N = \frac{\log\left(\dfrac{(Y_0 - Y_1)}{(X_N - Y_S)}\right)}{\log\left(\dfrac{(Y_0 - X_N)}{(Y_1 - Y_S)}\right)} \quad \text{...(8.59)}$$

The above equations are applicable to cases when the ratio of the weight of solution in the overflow (V) to that in the underflow remains constant throughout the cascade. This is true for many washing systems where the underflow leaving the primary leaching unit becomes the feed to stage-1 of the washing system. The primary leaching stage (feed stage) cannot be included in the above analysis as the ratio of the overflow V_0 to the underflow F in this stage is different from R which is the ratio of the overflow to the underflow in the rest of the cascade. The total number of stages required will be the sum of the stages required in the washing system plus the number of stages in the primary leaching unit.

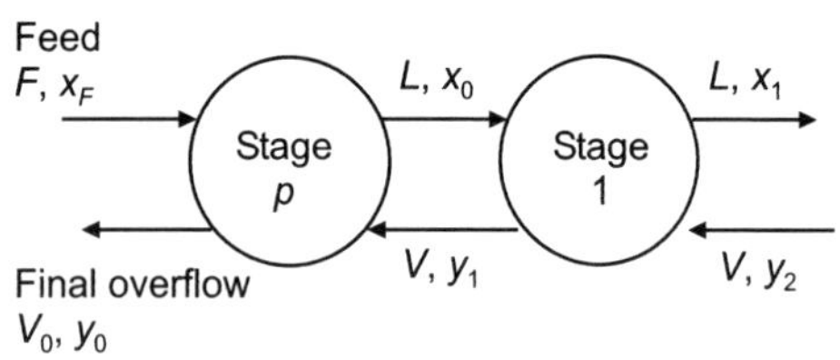

Fig. 8.39: The primary leaching unit preceding the washing stages

When the washing system uses the underflow leaving the primary leaching stage as the feed, in determining the number of stages using Eq. (8.53), y_0 in the equation is the concentration of the final overflow leaving the primary leaching unit and y_1 is the concentration of the solute in the overflow leaving stage-1 which is to be determined by a solute balance over stage p:

$$y_1 = \frac{(V_0 + L)y_0 - Fx_F}{V} \quad \text{...(8.60)}$$

Example 8.7: Sliced sugar beets having a composition of 25 weight% water, 35 weight% sugar, and 40 weight% inert material have to be extracted at a rate of 10,000 kg/h in a continuous countercurrent extraction plant using hot water as the solvent. The concentrated extract must contain 30 weight% sugar and the recovery of sugar is to be 90%. The settling characteristics is such that 1 kg inert material retains 3 kg solution at all solution concentrations. Determine:

(a) The quantity of solvent supplied.

(b) The number of stages required by (i) Ponchon-Savarit method, (ii) McCabe-Thiele graphical method and (iii) Analytical method.

***Solution*:** Basis: 10,000 kg/h sugar beets treated (Fig. 8.40).

Feed = 10,000 kg/h

Sugar (**C**) = 3500 kg; water (**A**) = 2500 kg; inert solids (**B**) = 4000 kg.

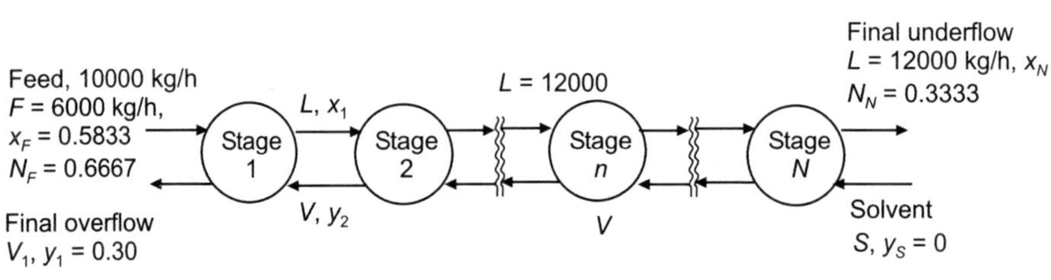

Fig. 8.40: Countercurrent leaching (Example 8.7)

$$x_F = \frac{C}{A+C} = \frac{3500}{3500+2500} = 0.5833$$

$$N_F = \frac{B}{A+C} = \frac{4000}{3500+2500} = 0.6667$$

Solvent, S kg/h; $y_s = 0$

Final Underflow:

The underflow contains all the inert solids in the feed. B = 4000 kg. Each kg inert solids retain 3 kg solution. Therefore, solution in the underflow, $L_N = 3 \times 4000 = 12000$ kg. Since 90% of the sugar present in the feed is leached, remaining 10% will be present in the underflow. Therefore, sugar present in the underflow = $0.10 \times 3500 = 350\,\text{kg}$.

$$\text{Solute concentration in the underflow} = x_N = \frac{350}{12000} = 0.0292$$

Final Overflow:

The overflow contains 90% of the sugar in the feed. That is, sugar present in the overflow = 3500 × 0.9 = 3150 kg. Since, $V_1 \times y_1 = 3150$ and the overflow concentration y_1 is given to be 0.3, the overflow rate is

$$V_1 = \frac{3150}{0.3} = 10500 \text{ kg/h}$$

(a) Now an overall material balance from stages 1 through N gives

$$S = V_1 + L_N - F = 10500 + 12000 - 6000 = 16500 \text{ kg/h}$$ **(Answer)**

(b)

(i) *Ponchon-Savarit method*

The equilibrium diagram is plotted in Fig. 8.41 with constant underflow locus at $N = 1/3$. The terminal streams F (0.5833, 0.6667), S (0, 0), L_N (0.0292, 0.3333) and V_1 (0.3, 0) are marked as shown.

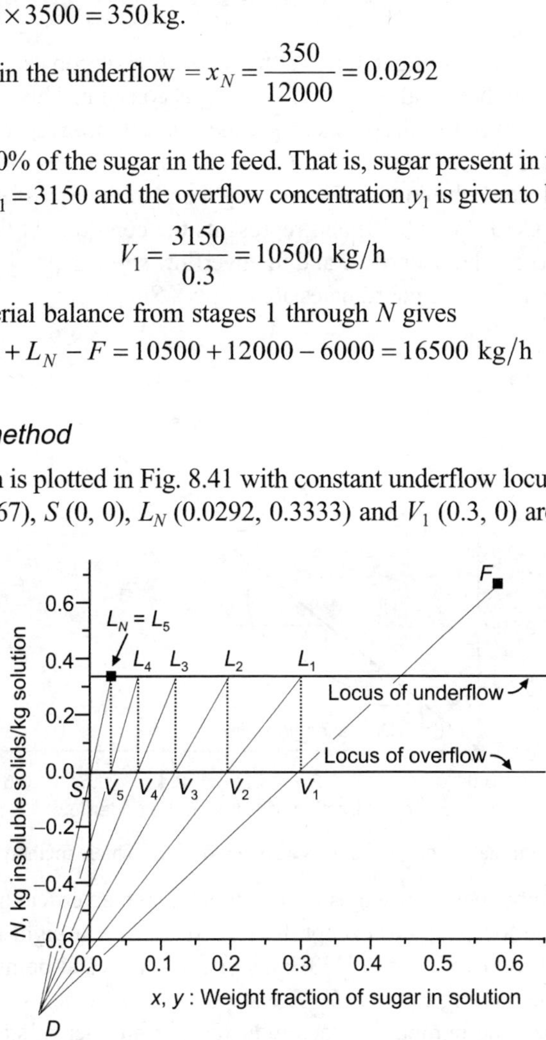

Fig. 8.41: Determination of the number of stages by Ponchon-Savarit method (Example 8.7)

The difference point D is located at the intersection of the extensions of lines FV_1 and L_NS. Vertical tie line V_1L_1 is drawn through V_1. L_1 lies at the intersection of the tie line with the locus of underflow. The line L_1D is drawn which intersects the $N = 0$ axis at V_2. The vertical tie line drawn at V_2 meets the locus of underflow at L_2. The construction is repeated till the underflow end of the tie line coincides with L_N or crosses it. The number of tie lines drawn equals the number of ideal stages, N. Here, $N = 5$.

The number of stages required = 5. **(Answer)**

(ii) *The McCabe-Thiele method*

In the McCabe-Thiele method of graphical construction of number of stages, the equilibrium curve and operating line are plotted on y versus x coordinates, where y is the concentration of solute in the overflow solution and x is the concentration of the solute in the underflow solution. Since the equilibrium relation is $y = x$, the equilibrium curve is a straight line of slope unity on this coordinates. Equation (8.30) is the equation of the operating line:

$$y_{n+1} = \frac{L}{V}x_n + \frac{Vy_1 - Lx_0}{V} \quad \text{...(8.30)}$$

The operating line will be a straight line, if the slope L/V is constant. For constant underflow condition L is constant throughout and so is V so that L/V is constant. This is true for all stages except the stage where the feed is introduced. At this stage, the L and V are respectively, F and V_1.

The equilibrium curve is plotted as shown in Fig. 8.42. The point A is marked with coordinates ($x_N = 0.0292$, $y_S = 0$) and represents the conditions at the right end of the cascade shown in Fig. 8.40. Point B has coordinates ($x_F = 0.5833$, $y_1 = 0.30$) and represents the condition at the left end of the cascade where the feed is introduced. The underflow and the overflow streams passing between stages 1 and 2 can be represented by a point whose coordinates are (x_1 , y_2). Since $x_1 = y_1 = 0.3$, the point representing

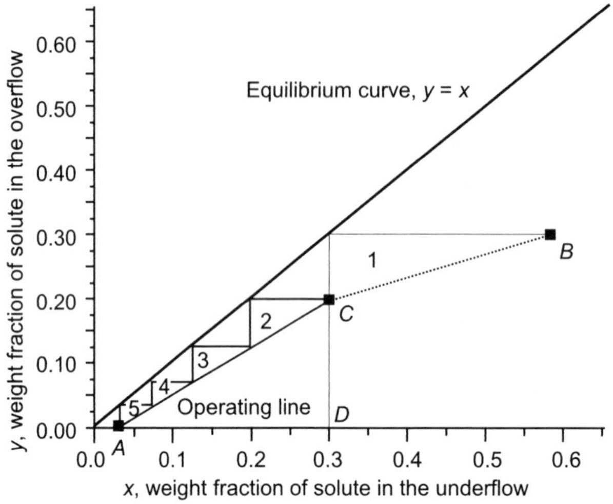

Fig. 8.42: Determination of number of stages by McCabe-Thiele method (Example, 8.7)

the streams passing each other between stages 1 and 2 will lie on the vertical line erected at point D with $x = 0.3$. The operating line for all stages except the feed stage, is a straight line originating at point A and has a slope = L/V = 12000/16500 = 0.7273. This line terminates at point C whose coordinates are (x_1, y_2) on the vertical line passing through D. The steps are constructed starting from B by the usual McCabe-Thiele technique. The number of steps indicates the number of stages.

Here number of stages = 5. **(Answer)**

(iii) *Analytical method*

Analytical method for the determination of number of stages is applicable if the ratio L/V is constant. The total number of stages for which this ratio is constant can be computed using Eq. (8.53). The feed stage is not included in the number of stages so determined. The countercurrent leaching system can be represented as shown in Fig. 8.43:

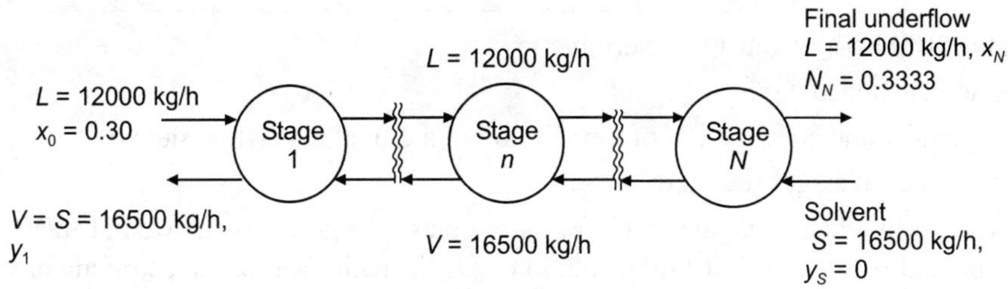

Fig. 8.43: Analytical solution of Example 8.7

The number of stages is given by Eq. (8.53):

$$N = \frac{\log\left(\dfrac{(y_0 - y_1)}{(x_N - y_S)}\right)}{\log\left(\dfrac{(y_0 - x_N)}{(y_1 - y_S)}\right)}$$

where N is the number of stages excluding the stage in which the feed is admitted. Here, $x_N = 0.0292$; $y_S = 0$. Since the weight fraction of the solute in the overflow and the underflow solution leaving the feed stage are equal, $y_0 = x_0 = 0.30$. $L = 12000$ kg/h, $V = S = 16500$ kg/h, $V_0 = 10500$ kg/h.

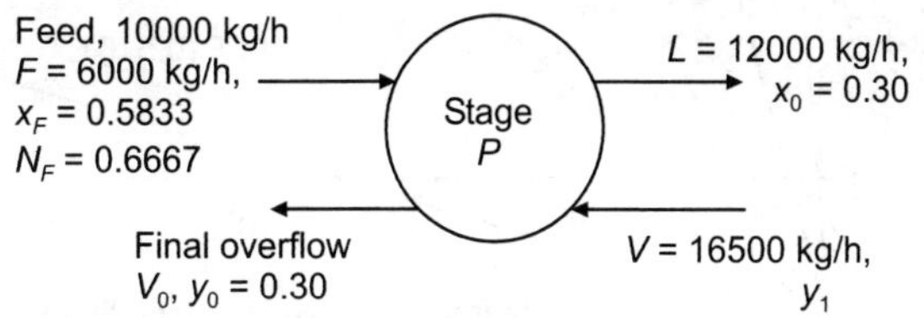

Fig. 8.44: Feed stage (primary leaching stage) of Example 8.7

A solute balance on the feed stage (Fig. 8.44) gives

$$Fx_F + Vy_1 = Lx_0 + V_0 y_0$$

$$y_1 = \frac{Lx_0 + V_0 y_0 - Fx_F}{V} = \frac{12000 \times 0.3 + 10500 \times 0.3 - 6000 \times 0.5833}{16500} = 0.1970$$

Substituting the values in Eq. (8.53), we get

$$N = \frac{\log\left(\dfrac{(0.30 - 0.1970)}{(0.0292 - 0)}\right)}{\log\left(\dfrac{(0.30 - 0.0292)}{(0.1970 - 0)}\right)} = 3.96 \approx 4$$

Thus the number of stages required including the feed stage = 4 + 1 = 5 **(Answer)**

Example 8.8: A countercurrent extraction battery is used to extract caustic soda adhering to the $CaCO_3$ sludge from the reaction

$$Na_2CO_3 + CaO + H_2O \rightarrow CaCO_3 + 2NaOH$$

The products leaving the reaction vessel contains 3 kg of water with every 4 kg of $CaCO_3$, and no excess reactants are present. It is desired to recover 98 percent of the NaOH. The final overflow contains 12% by weight NaOH. The underflow from each stage carries solution, the weight of which varies with the concentration as

wt % NaOH in solution	0.0	5.0	10.0	15.0	20.0
Entrainment, kg solution/ kg $CaCO_3$	1.50	1.75	2.20	2.70	3.60

For treating 100 kg/h of $CaCO_3$, determine:

(a) The solvent added, kg/h

(b) The quantity and composition of the solution in the final underflow sludge

(c) The number of stages required.

Solution: The settling data is tabulated below as x versus N, where x is the weight fraction of NaOH in the solution and N is the kg $CaCO_3$/kg solution. N is the reciprocal of the entrainment values given.

x, weight fraction NaOH	0.00	0.05	0.10	0.15	0.20
N, kg $CaCO_3$/kg solution	0.6667	0.5714	0.4545	0.3704	0.2778

The equilibrium curve is plotted as shown in Fig. 8.45

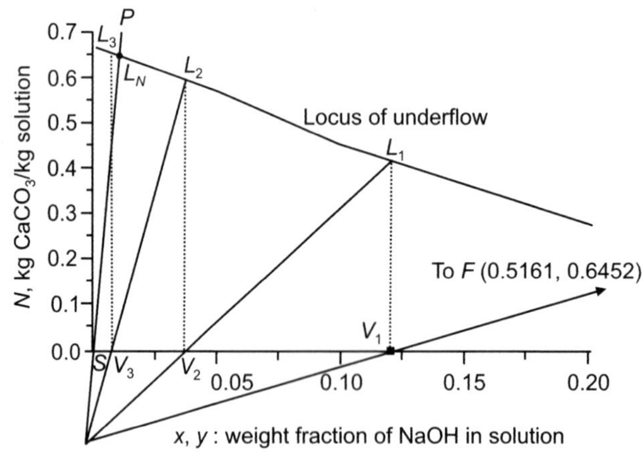

Fig. 8.45: Determination of the number of stages for Example 8.8

Since $CaCO_3$ and NaOH are the products of the following reaction

$$Na_2CO_3 + CaO + H_2O \rightarrow CaCO_3 + 2NaOH$$

with every kmol (100 kg) of $CaCO_3$ in the reaction product, there will be 2 kmol (80 kg) NaOH.

Therefore, the reaction product contains 100 kg $CaCO_3$, 80 kg NaOH and 3/4 × 100 kg water.

Thus the feed to the leaching unit consists of water (**A**) = 75 kg, $CaCO_3$ (**B**) = 100 kg, NaOH = 80 kg.

$$x_F = \frac{80}{80+75} = 0.5161$$

$$N_F = \frac{100}{80+75} = 0.6452$$

Feed: The feed point F is marked with the coordinates (x_F, N_F). (Not shown in Fig. 8.45.)

Solvent: Origin of the N versus x coordinate system represents the solvent S.

Overflow: Since the overflow contains 12% weight NaOH and is free of $CaCO_3$, the point V_1 representing the overflow is marked on the $N = 0$ axis at $y_1 = 0.12$.

Since 98% of the NaOH present in the feed is recovered by leaching, the overflow contains $0.98 \times 80 = 78.4$ kg. Since this constitutes 12% of the overflow,

$$V_1 = \frac{78.4}{0.12} = 653.33 \text{ kg}$$

Underflow: Underflow retains (80 – 78.4) = 1.6 kg unextracted NaOH and 100 kg $CaCO_3$. That is the inert/solute (**B/C**) ratio on weight basis in the underflow is 100/1.6 = 62.5. A straight line *SP* is drawn through the origin of the *N* versus *x* coordinates with slope 62.5. For all mixtures on this line the weight ratio of **B** to **C** is constant and is equal to 62.5. Since the **B/C** ratio in the underflow L_N is 62.5 and the underflow should lie on the locus of underflow, the intersection of the line *SP* and the locus of underflow gives the location for L_N. Its coordinates are noted from the figure: $x_N = 0.01035$, $N_N = 0.6470$. The solution retained by the solids in the underflow is

$$L_N = \frac{B}{N_N} = \frac{100}{0.6470} = 154.6 \text{ kg}$$

(a) To find the quantity of solvent, take the total material balance over the entire cascade.

$$S = V_1 + L_N - F = 653.33 + 154.6 - 155 = 652.9 \text{ kg.} \qquad \textbf{(Answer)}$$

(b) The final underflow sludge contains:

$CaCO_3$ (**B**) = 100 kg

Water (**A**) = $L_N(1 - x_N) = 154.6 - 1.6 = 153$ kg

NaOH (**C**) = 1.6 kg

The quantity of solution in the underflow sludge = L_N = 154.6 kg **(Answer)**

Weight fraction of NaOH in the underflow solution = x_N = 0.01035 or 1.035% **(Answer)**

(c) The points *F*, V_1, L_N and *S* are marked as shown in Fig. 8.45 and the difference point *D* is determined at the intersection of the extensions of the lines FV_1 and L_NS. The vertical tie line drawn at V_1 intersects the locus of underflow at L_1. The line L_1D intersects the $N = 0$ axis at V_2. The tie line at V_2 intersects the locus of underflow at L_2 ($x_2 = 0.0373$). L_2 is joined to *D* which intersects the *x*-axis at V_3 and the tie line erected at V_3 determines L_3 ($x_3 = 0.0067$). Since x_3 is less than $x_N = 0.01035$, it is clear that more than 2 but less than 3 stages are required.

$$\text{The number of stages} = 2 + \frac{0.0373 - 0.01035}{0.0373 - 0.0067} = 2.9 \qquad \textbf{(Answer)}$$

8.10 SUPER-CRITICAL FLUID EXTRACTION

Food processing uses leaching and liquid-liquid extraction for the recovery or removal of desired components from plant or animal materials. Supercritical fluid extraction (SFE) technology is an alternative to these conventional mode of separations using organic solvents. The technology uses more environment-friendly solvents such as supercritical carbondioxide, ethanol, water, etc. SFE is used in a variety of products including natural foods and flavours, pharmaceuticals, polymers, etc. Specific applications include decaffeination of coffee and tea, extraction of hop components for beer flavouring and extraction of essential oils, oleoresins, flavours and aromas from natural sources, removal of residual organic solvents or impurities like pesticides from final active compounds or food ingredients.

Principle of *SFE*

When a gas or liquid is heated past its critical point (Fig. 8.46), it enters a phase called 'supercritical phase' and the substance is referred to as a 'supercritical fluid'. At the supercritical state of a fluid, its properties such as the density, viscosity, heat capacity, diffusivity, thermal conductivity are affected in such a way that the ability of the substance to penetrate and extract the solute from the source materials is enhanced. The relative density of a SCF is in the range 0.1 to 1 compared to the density of 1 for liquids and 0.001 for gases. The relative viscosity is 0.1 to 1 compared with 1 for liquids and 0.01 for gases. The relative diffusivity is 10 to 100 compared with 1 for liquids and 10^4 for gases. The liquid-like density of a supercritical fluid solvent provides its high solvent power and the gas-like viscosity and diffusivity with zero surface tension enhance its transport properties. The SCF can penetrate into porous solid materials more effectively than liquid solvents resulting in faster mass transfer and extractions.

The most desirable solvent for SCF extraction of natural products for foods and medicines is CO_2. It is inert, relatively non-toxic, inexpensive, easily available, odourless, tasteless, environment friendly with convenient critical parameters (T_C = 304.2 K. P_C = 7.38 MPa). Many other solvents have been commercially applied for the processing of heavy hydrocarbons from petroleum fractions, recovery and purification of lube oils, coal liquefaction, polymer processing, critical point drying and cleaning of microelectronics materials. Carbon dioxide is not a very good solvent for high molecular weight and polar compounds. To increase the solubility of such compounds in supercritical CO_2, small amounts of polar or nonpolar co-solvents called modifiers may be added. For example, addition of small amounts (3.5 mol%) of methanol to CO_2 increases the solubility of cholesterol. Ethanol is the modifier of choice in the food industry, because it is non-toxic and is safe.

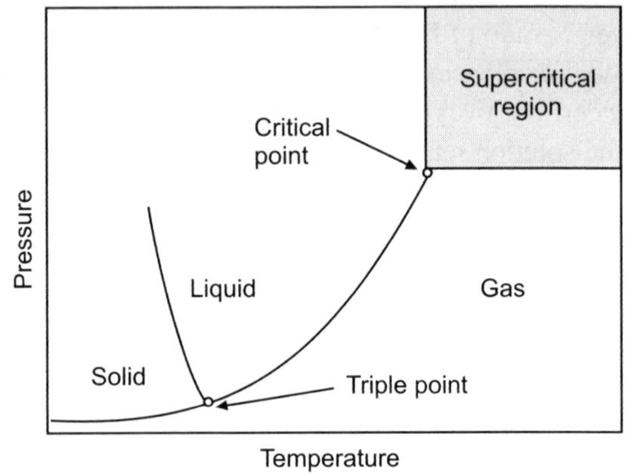

Fig. 8.46: *P-T* behaviour of pure fluid

The advantages of using supercritical fluid extraction are the high purity extracts, absence of residual solvents, single step processing, reduced operating costs, selective fractionation, faster separation, environmental acceptability and physiological compatibility. The oxygen-free operating system prevents oxidation, the low temperatures minimise thermal degradation of sensitive materials and the extract product is microbially sterile, as micro-organisms and their spores are not soluble in the SCF-CO_2.

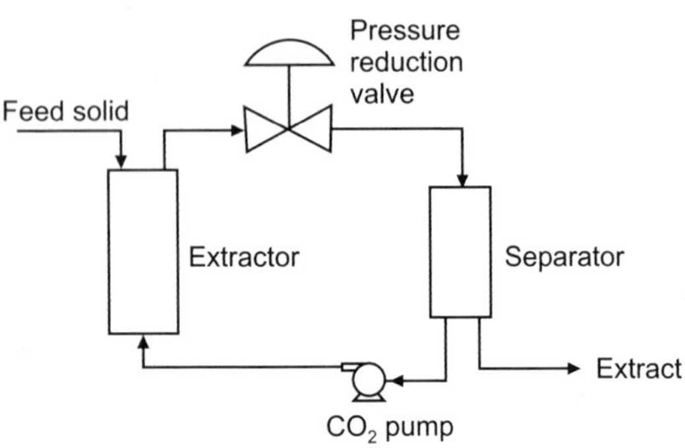

Fig. 8.47: Schematic diagram of supercritical fluid extraction process

Figure 8.47 shows schematically the supercritical fluid extraction process. The extractor consists of a high pressure vessel of volume ranging from typically, 0.01 – 1.0 m^3 into which the feed solid is admitted. The CO_2 is fed into the extractor through a high pressure pump. The extract-laden CO_2 is sent through a pressure reduction valve into a separator. The pressure is reduced by throttling through a needle valve or orifice. Heat exchangers may be required to maintain the temperature and prevent the excessive cooling at the throttling valve. At the reduced temperature and pressure in the separator, the fluid becomes a gas, the solute precipitates and is separated. The CO_2 is recycled to the extractor through the high pressure pump.

In the decaffeination of coffee, the beans are first soaked in water and charged into the extractor maintained at high pressure. The supercritical fluid CO_2 is admitted to the extractor and caffein in the beans are transferred to the solvent. The caffeine – CO_2 solution is then sent to a separate scrubbing vessel, also maintained at a high pressure where the caffeine is transferred from the CO_2 to the water. At the end of the batch cycle, caffeine is separated from the water by depressurizing and distilling. The caffeine content in the beans can be reduced to 0.02 percent from the initial value lying between 0.7 and 3.0 percent.

Most commercial SCF extraction processes are operated batch-wise, though semicontinuous operations are also practiced. The recent trend in SCF extraction process is to operate the extractor at very high pressure and temperature conditions to extract almost all the desired species and to subject the resulting extract to pass through a series of extractors each operated at different pressure-temperature conditions. In this way, specific species are allowed to precipitate individually in different separators, thus achieving a sort of fractional separation.

REVIEW QUESTIONS

1. Define leaching.
2. Cite examples where the objective of leaching is (a) the recovery of soluble constituent (b) the purification of the inert insoluble solid by removing the soluble impurities.
3. Discuss the major industrial applications of leaching.
4. What are the rate controlling processes in leaching operations?
5. List the four factors that determine the rate of leaching.
6. Why is high rate of leaching obtained if the particle sizes are small?
7. It is sometimes better to avoid the presence of very fine powder in the feed for leaching. Why?
8. Describe the significance of crushing and grinding in the preparation of feed for leaching.
9. Why crushing and grinding are not recommended in preparing materials of vegetable origin for leaching?
10. What are the factors to be considered in the choice of solvent for leaching?
11. In what way the temperature of operation influence the rate of leaching?
12. Distinguish between the percolation and the immersion methods of leaching. List the advantages and disadvantages of both.
13. Explain heap leaching with a schematic diagram.
14. What are the major applications of in-situ leaching? What environmental problems are associated with in-situ leaching?
15. Describe the Shanks system of continuous countercurrent extraction using the extraction battery consisting of fixed bed extractors.
16. Describe the working of the Espresso machine for making coffee.
17. With the help of a schematic diagram describe the working of a Bollmann extractor.
18. Explain how a true countercurrent operation is realized in Rotocel.
19. What are the advantages of perforated-belt extractors over the Bollmann extractors and their horizontal variants?
20. Explain the construction, working and applications of Pachuca tank. Why are the impeller agitated tanks preferred over Pachuca tanks?
21. Describe the advantages and drawbacks of Bonnotto extractors and Hildebrandt extractors.
22. Explain with the help of neat schematic sketch the working of Continuous Countercurrent Decantation system (CCD).
23. What are the conditions to be satisfied for a leaching stage to be ideal?
24. In leaching, even if a stage is ideal, the stage efficiency will be less than 100%. Why?
25. Determine the N and x coordinates' of the mixture that results when the seeds containing 25% oil and the rest insolubles are mixed with an equal weight of hexane.
26. Determine the percent composition of the mixture whose x and N coordinates are respectively, 0.25 and 0.667.
27. Under what conditions the constant underflow and variable underflow conditions occur in leaching?

28. Under what circumstances the locus of overflow lie above the x-axis?
29. In representing the leaching equilibrium, is it essential that the tie lines are always vertical? Substantiate your answer.
30. Derive an equation for the locus of underflow if it is known that the solids retain 0.8 kg solvent per kg inert solids.
31. Describe the Ponchon-Savarit and McCabe-Thiele graphical methods for determining the number of stages for countercurrent leaching.
32. Derive the Kremser type equation for determining number of stages for constant underflow conditions.
33. In a countercurrent washing system the weight ratio of overflow to underflow solution remains constant at 2. If it is desired that the soluble constituent of the washed solids leaving the system contains only 1/30th of the amount present in the feed, determine analytically the number of stages required.
34. What are the advantages of supercritical fluid extraction over the conventional methods of leaching? Give some examples for application of SCFE on a commercial scale.
35. What is the principle of SCFE? Why is CO_2 a preferred solvent for SCFE? What is the role of modifiers in SCFE?
36. With a schematic diagram, explain the supercritical fluid extraction process.

MULTIPLE CHOICE QUESTIONS

8.1 The process of leaching alkali from wood ash is also known as

A. decoction B. elution
C. lixiviation D. extraction

8.2 Decoction is a

A. leaching operation B. liquid-liquid extraction
C. drying operation D. adsorption operation

8.3 A suitable solvent for leaching copper from roasted copper ore is

A. cyanide solution B. sulphuric acid
C. hot water D. dilute ethanol

8.4 Which of the following is not a leaching operation?

A. Recovery of gold from ores using cyanide solution
B. Washing of caustic soda from precipitated calcium carbonate
C. Tannin extraction from tea barks
D. Expression of sugar cane juice from sugarcane

8.5 In leaching, if the particle size is small, all but one of the following will be true. Spot the odd one out.

A. Length of diffusion path is small.
B. Interfacial area of mass transfer is small
C. Separation of solid from the liquid and the drainage of the solid residue are difficult.
D. Relatively high rate of mass transfer

8.6 For preparing feed for leaching, crushing and grinding of solid are essential if

A. the soluble constituents are surrounded by inert insoluble material
B. the soluble constituents are contained within the cell walls of the plant structure
C. the soluble constituents are present as a solid solution
D. the soluble materials are adsorbed on the surface of the solid

8.7 For leaching of materials of vegetable origin, which of the following is not recommended for feed preparation?

A. Drying
B. Grinding
C. Flaking
D. Slicing

8.8 From the point of view of solvent viscosity and solute diffusivity it is preferable to carry out leaching

A. at low temperature
B. at high temperature
C. with agitation of the contents of the leaching equipment
D. with highly selective solvent

8.9 The temperature of leaching should be as high as possible because of

A. the high solubility of the solute in the solvent
B. the high diffusivity of the solute in the solution
C. low viscosity of the liquid
D. all the above reasons

8.10 Agitation of the contents of the leaching equipment will be desirable if

A. the rate controlling step is the diffusion within the solid
B. low viscosity of solvent and high solute diffusivity are necessary for high rate of leaching
C. leaching is to be conducted at high temperatures
D. the rate is controlled by the diffusion of the solute from the surface of the particle to the bulk.

8.11 The rate of leaching is increased at increased temperature due to

A. increased liquid viscosity and decreased diffusivity
B. increased liquid viscosity and increased diffusivity
C. decreased liquid viscosity and decreased diffusivity
D. decreased liquid viscosity and increased diffusivity

8.12 Which one of the following is an immersion type leaching vessel?

A. Bollmann extractor
B. Bonnoto extractor
C. Rotocel
D. Kennedy extractor

8.13 Which one of the following is a percolation type leaching equipment?

A. Bonnoto extractor
B. Pachuca tank
C. Kennedy extractor
D. Hildebrandt extractor

8.14 Gold is recovered from low-grade gold ores of very low mineral value by

A. in-situ leaching
B. heap leaching
C. moving bed percolation
D. continuous countercurrent decantation

8.15 Without subjecting to conventional mining, copper, uranium, salts etc can be recovered from subsurface deposits by

A. heap leaching
B. in situ leaching
C. extraction battery
D. continuous countercurrent decantation.

8.16 Which one of the following statements regarding Shank's system for leaching is incorrect?

A. Multistage countercurrent system with solid and liquid moving in opposite directions
B. Multistage countercurrent operation without physically moving the solids
C. Consists of number of tanks acting as fixed bed percolators with liquid flowing from tank to tank.
D. Diffusion cells and extraction battery operate under this scheme

8.17 An example of a percolation equipment with the bed of solids moving through the solvent which is flowing in the opposite direction is

A. Bollmann extractor B. Rotocel

C. Kennedy extractor D. Lurgi extractor

8.18 Rotocel is

A. fixed bed percolation type continuous equipment

B. moving bed percolation type batch extractor

C. continuous countercurrent immersion type multistage extractor

D. moving bed percolation type continuous multistage equipment

8.19 The main disadvantage of percolation type leaching equipment is

A. channelling and incomplete leaching

B. the inability to operate as continuous unit

C. the difficulty of separation of solid and liquid after extraction

D. the inability to handle coarse solids

8.20 Bollmann extractor is

A. continuous hybrid flow multistage sprayed percolator.

B. continuous countercurrent multistage percolator

C. continuous countercurrent horizontal multistage extractor

D. continuous hybrid flow solids dispersed multistage extractor

8.21 The .... consists of three screw conveyors arranged in a U shape.

A. Bollmann extractor B. Hildebrandt extractor

C. Kennedy extractor D. Rotocel

8.22 When the solid can be ground to very fine powders, the leaching is better carried out using

A. fixed bed percolators B. moving bed percolators

C. agitated vessels D. diffusers

8.23 One of the following extractors simulates closely the Shank's system of continuous extraction. Identify.

A. Bollmann extractor B. Dorr agitator

C. Hildebrandt extractor D. Rotocel

8.24 For an ideal leaching stage the tie lines are not vertical if

A. the solute is completely dissolved in the solvent.

B. there is preferential adsorption on the inert solids

C. no insolubles are carried away by the overflow solution

D. concentration of solute in the solution is uniform everywhere.

8.25 If the viscosity and density of the solution change significantly with change in its composition, the leaching equilibrium is characterized by

A. the tie lines which slope towards the right on N-x-y plot.

B. the tie lines which are vertical

C. a variable locus of underflow

D. a constant underflow condition

8.26 The amount of solution carried away by the inert solids in the underflow sludge does not depend on

A. the solute concentration in the liquid

B. the density of the solution

C. viscosity of the solution

D. diffusivity of the components

8.27 A constant underflow condition in leaching results if

A. the concentration of solute in the underflow and overflow are constant

B. the settling and separation of underflow solids and overflow solution are carried out for a fixed duration.

C. the rate of settling of the slurry is dependent on the viscosity and density of solution

D. the viscosity and density are independent of the composition of the solution

8.28 The seeds containing 25% oil and the rest insolubles are mixed with hexane. The composition of the resulting mixture if the seeds and solvent are mixed in equal amounts by weight is:

A. $x = 0.2$; $N = 0.6$ B. $x = 0.25$; $N = 0.75$

C. $x = 0.2$; $N = 3.0$ D. $x = 0.25$; $N = 0.6$

8.29 A mixture consisting of insoluble solids **B** suspended in a solution of **A** and **C** has the composition $x = 0.25$ and $N = 2/3$. The weight percent of **C** in the mixture is

A. 25.0 B. 67.67

C. 15.0 D. 40.0

8.30 If the solids retain 0.8 kg solvent per kg inert solids, the locus of underflow can be represented by

A. $N = 0.8(1 - x)$ B. $N = 1.25$

C. $N = 1.25x$ D. $N = 1.25 - 1.25x$

8.31 The settling characteristics of a slurry consisting of solids **B** suspended in a solution of **C** in **A** is given by the relation $N = 2(1 - x)$ where N is the weight ratio of inerts to solution retained by the settled solids and x is the solute concentration in the solution retained by the solids. A feed consisting 25% **C** and 75% **B** is subjected to leaching with solvent **A** in a countercurrent unit so that 90% of the solute is recovered in the final overflow. What is the weight percent of **C** in the solution retained by the solids?

A. 2.5% B. 22.5%

C. 6.25% D. 3.23%

8.32 One hundred kg each of a solid feed containing 25% solute (**C**) and 75% insolubles (**B**) is subjected to single-stage leaching using one hundred kg of a pure solvent (**A**). After settling 75 kg of clear solution is withdrawn as overflow. What is the inert to solution weight ratio in the underflow?

A. 1.0 B. 1.5

C. 2.0 D. 0.67

8.33 In a countercurrent washing system the weight ratio of overflow to underflow solution remains constant at 2. If it is desired that the soluble constituent of the washed solids leaving the system contains only 1/30th of that present in the feed, how many stages are required? (Do not use graph papers)

A. 2 B. 3

C. 4 D. 5

8.34 60 tons/day of oil sand (25 weight% oil and 75 weight% sand) is to be extracted with 40 tons/day of naphtha in a continuous countercurrent extraction battery. The final extract from the battery is to contain 40 weight% oil and 60 weight% naphtha and the underflow from each unit is expected to consist of 35 weight% solution and 65 weight% sand if the overall efficiency of the battery is 50%. How many stages will be required?

A. 3 B. 4

C. 5 D. 6

EXERCISES

Rate of settling

8.1 Five hundred kg of a solid containing 28% solute and the rest insolubles is dissolved in 100 m^3 of water. The solubility of solute in water is 2.5 kg/m^3. How long it will take the solute to completely dissolve in water? In a pilot plant study with a sample of same solids, it took 100s to attain 75% saturation when the solute was dissolved from the inert solid in 1 m^3 of water.

(**Answer:** 1.0 h 39 min)

Single-stage leaching

8.2 One hundred kg flaked soyabeans containing 25% oil by weight is leached with 100 kg fresh hexane solvent. If the underflow solids retain 2 kg solution with every 3 kg of insoluble solids, calculate the amounts and composition of overflow and underflow from the single stage operation.

[**Answer:** Underflow = 125 kg ($x = 0.2$, $N = 1.5$) Overflow = 75 kg solution ($y = 0.2$, $N = 0$)]

8.3 Water is to be used in a single equilibrium stage to dissolve 1000 kg/h of soda ash from 3750 kg/h of a solid, where the balance is an insoluble oxide. If 3000 kg/h of pure water is used, and the underflow from the stage retains 1 kg solution with every 2 kg insolubles, calculate the amounts and compositions of the overflow and the underflow leaving the stage.

[(**Answer:** Underflow = 4125 kg ($x = 0.25$, $N = 2.0$)
Overflow = 2625 kg solution ($y = 0.25$, $N = 0$)]

8.4 Fat is extracted from beef containing 60% water, 22% fat and 18% insoluble solids using isopropyl alcohol in a single stage leaching unit. Alcohol is supplied at a rate 5 times the rate of supply of feed on mass basis. The solution of isopropyl alcohol in the water present in the feed acts as the leaching solvent in which all fat is soluble. If 10% by weight of the total solution is lost with the solids leaving the stage, determine the concentration of fat in the overflow and the fat content in the solids on solvent free basis.

(**Answer:** Fat concentration in the overflow is 3.78%,
fat content in the underflow = 10.9% on solvent-free basis)

8.5 Feed solids at a rate 7500 kg/h, containing 2000 kg/h soda ash and the rest insoluble oxides is leached with 6000 kg/h water in a single stage operation. The underflow solids retain solution such that the ratio of solvent to inert solids in the underflow is 2:3. Compute the flow rates and compositions of the overflow and underflow.

(**Answer:** Underflow = 10389 kg ($x = 0.25$, $N = 1.125$)
Overflow = 2333 kg solution ($y = 0.25$, $N = 0$))

8.6 A slurry of flaked soyabeans consists of 75 kg inert solids suspended in 25 kg of a 10 weight% solution of oil in hexane. This slurry is contacted with 100 kg pure hexane in a single stage. The underflow from this stage contains 2 kg solution for every 3 kg insoluble solids present. Calculate the amounts and compositions of the overflow and the underflow leaving the stage.

(**Answer:** Underflow = 125 kg ($x = 0.02$, $N = 1.5$) Overflow = 75 kg ($y = 0.02$, $N = 0$)

8.7 10,000 kg/h of soyabean flakes containing 20% oil is extracted with 10,000 kg/h of hexane in a single stage leaching unit. It is found that it is impossible to reduce the oil content of the solid below 0.5 weight% as some oil remain with the solids without dissolving in the solvent. The settling characteristics of the solids are given as:

N, mass of insolubles /mass of solution	2.13	2.03	1.86	1.63	1.38	1.13
x, mass of oil/ mass of solution	0.0	0.2	0.4	0.6	0.8	1.0

Determine the composition and flow rates of the underflow and overflow, and the percent recovery of oil.

(**Answer:** Underflow 11971.84 kg ($x = 0.1724$, $N = 2.0142$)
Overflow = 8028.16 kg ($y = 0.1638$, $N = 0$), Recovery = 65.77%)

Crosscurrent leaching

8.8 One thousand kg of roasted copper ore containing 10 % $CuSO_4$, 85% gangue, and 5% water is to be extracted by washing it three times with 1500 kg batches of fresh water. It has been found that the solids retain 0.8 kg solution per kg gangue. Determine the percent composition by weight of the underflow sludge. (**Answer:** 0.26% $CuSO_4$, 44.18% water, 55.56% gangue)

8.9 A settling tank contains a sediment and 7000 kg of a solution in which 2000 kg of NaOH are dissolved. After settling, 6000 kg of the clear solution is withdrawn and replaced by 6000 kg of pure water and the suspension agitated. After settling, 6000 kg of clear solution is withdrawn, fresh water added and the process repeated once more. The clear solutions withdrawn are collected together which form the combined extract. Determine:

(a) The amount of NaOH remaining in the sediment
(b) The percent extraction of NaOH
(c) The concentration of the NaOH in the combined extract.

(**Answer:** (a) 5.8 kg, (b) 99.71% (c) 11% NaOH)

8.10 A mineral containing 20% elemental sulphur is to be leached with hot gas oil, in which the sulphur is soluble to the extent of 10% by weight. The solvent will be repeatedly pumped over the batch of ground mineral, using 1.5 kg fresh solvent per kg mineral. After no further solution of sulphur is obtained, the liquid will be drained and replaced with a fresh batch of 1.5 kg oil per kg original mineral, and the operation is repeated. On drainage the solid material retain the solution to the extent of one-tenth the weight of undissolved solid (sulphur and gangue). No preferential adsorption takes place. Determine the amount of sulphur unextracted and the sulphur concentration of the composited leach liquor.

(**Answer:** 1.03% of the original sulphur, 6.3% sulphur)

8.11 A slurry consisting of 10% $CaCO_3$, 7.2% NaOH and 82.8% water is obtained is settled and the clear solution of NaOH in water is withdrawn in the first stage of the three-stage crosscurrent leaching system. Fresh water is used for washing the underflow solids in the remaining stages, water being supplied in quantities equal to the mass of overflow removed in the preceding stage. What percentage of NaOH is lost unrecoverd in the final sludge? The settling characteristics of the slurry is given below which indicates preferential adsorption of NaOH on the solid:

x	0.0917	0.0762	0.0608	0.0452	0.0295	0.0204	0.01435	0.01015
y	0.0900	0.0700	0.0473	0.0330	0.0208	0.01187	0.00710	0.00450
N	0.495	0.525	0.568	0.600	0.620	0.650	0.659	0.666

(x = weight fraction NaOH in the settled sludge, y = weight fraction of NaOH in the overflow solution, N = kg $CaCO_3$/ kg solution). (**Answer:** 1.8%)

Countercurrent leaching (Constant underflow)

8.12 It is desired that 98% of soda ash is recovered from 1000 kg/h of a feed containing 20 weight% Na_2CO_3 and the rest insoluble solids in a countercurrent multistage unit consisting of one ideal leaching stage followed by a multistage washing system. The final extract should contain 20 weight% Na_2CO_3. The underflow from each stage contains 0.5 kg solution per kg insoluble solids. Determine:

(a) The quantity of solvent required
(b) The number of ideal stages required.

(**Answer:** (a) 1180 kg (b) 3.41 stages, both analytical and graphical)

8.13 Determine the number of ideal stages required if oil seeds containing 25% oil is extracted with hexane in a countercurrent plant to recover 95% of the oil in the seeds. The concentrated extract contains 50 weight% oil and the underflow from each stage contains 1 kg solution with every 2 kg insoluble matter. (**Answer:** 5 stages)

8.14 Oil is extracted from meal using benzene as solvent in a countercurrent extraction battery. The hourly feed of oil-meal to the unit contains 800 kg oil, 2000 kg insoluble meal and 50 kg benzene. The solvent is admitted at a rate 1500 kg/h and contains 20 kg oil. The solids leaving the last stage as final underflow contain 100 kg/h unextracted oil. The mass of solution retained by the solids is constant at 0.55 kg solution/kg solids. Determine: (a) The concentration of final overflow (b) Concentration of solution retained by the exhausted meal (c) The mass of solution in the final underflow solids (d) The mass of overflow (e) The number of stages required.

(**Answer:** (a) 57.6% oil (b) 9.1% oil (c) 1100 kg (d) 1250 kg (e) 4.15)

8.15 Two thousand kg/h of meal containing 55 weight% oil is to be extracted in a countercurrent multiple contact system by 2500 kg/h of *n*-hexane containing 1% oil. After extraction the meal contains 0.8 kg solution per kg oil-free meal. The overflow contains 0.005 kg fine meal particles per kg solution. How many ideal stages are required to recover 98% of the oil?

(**Answer:** 2.4 stages)

8.16 Sliced sugar beets containing 45% water, 15% sugar and 40% inert pulp are treated in a countercurrent multiple contact leaching system at a rate 10,000 kg/h, with fresh water as the solvent to recover 97% of the sugar in the feed. The strong solution leaving the system is to contain 10% weight sugar. Determine the number of stages and quantity of solvent required if each kg of dry pulp contains 3 kg of solution. (**Answer:** 5.7 stages, 20550 kg)

8.17 A five-stage countercurrent extraction battery is used to extract caustic soda adhering to the $CaCO_3$ sludge from the reaction

$$Na_2CO_3 + CaO + H_2O \rightarrow CaCO_3 + 2NaOH$$

The products leaving the reaction vessel contains 0.6 kg of water with every kg of $CaCO_3$ and no excess reactants are present in the products. It is desired to recover 97 percent of the NaOH by leaching with pure water. The underflow from each stage carries for each kg of $CaCO_3$, 1.5 kg solution when flowing to the next stage. Determine: (a) Kilograms of water required per kg of $CaCO_3$ treated (b) The concentration of solution leaving each stage.

(**Answer:** (a) 2.53 kg, (b) 32%, 17.5%, 9.2%, 4.4% and 1.6% NaOH)

8.18 A solid feed, 150 kg/h consisting of 30% Na_2CO_3 and the rest insoluble ash is to be leached and washed in a two-stage countercurrent system with 360 kg/h of water. Experiment show that the underflow sludge from each stage contains 2 kg solution per kg insoluble ash. Determine:

(a) The percent recovery of carbonate in the final extract
(b) The amount of additional carbonate that will be recovered, if a third stage is added in series.

(**Answer:** (a) 71.59%, (b) 5.58 kg)

8.19 Fifty thousand kg/h of flaked soyabean containing 20% weight oil is extracted in a countercurrent system consisting of a primary leaching stage and three washing stages. All stages are assumed to be ideal. Pure hexane is the solvent used and it is admitted at the same rate as the feed solid. Experiments have shown that the oil-free meal leaving with the underflow from each stage carries 0.8 kg solution.

(a) Determine the percent recovery of oil in the extract.

(b) If the leaching occurs in the first three stages with one-third of solute leached in each stage and the last stage alone acts as the washing stage, determine the percent recovery of oil in the final extract. (**Answer:** (a) 88.5% (b) 79.1%)

8.20 In the process for the manufacture of sugar from sugarcane, the cane which contains 21% fibre, 15% sugar, and 64% water is first subjected to milling which removes some juice and produces the fibrous residue called bagasse containing 30% fibre. The bagasse after shredding is subjected to multistage countercurrent leaching with water to extract 99% of the remaining sugar. The bagasse carries a constant amount of 1.25 kg solution per kilogram fibre. If it is desired that the concentration of sugar in the final extract must be 10%, calculate:

(a) The water required per 1000 kg/h of original cane.
(b) The number of ideal extraction stages
(c) The final sugar content if the extract is mixed with the juice first pressed out of the cane.

(**Answer:** (a) 693.2 kg/h (b) 3.6 (c) 12.2%)

8.21 Aluminium sulphate (MW = 342.12) is to be produced by the action of sulphuric acid on bauxite

$$Al_2O_3 + 3H_2SO_4 \rightarrow Al_2(SO_4)_3 + 3H_2O$$

A sample of bauxite containing 25% Al_2O_3 and the rest insolubles is leached with stoichiometric amount of aqueous acid containing 50% H_2SO_4. Assume the reaction is complete. The slurry obtained is to be washed in multistage countercurrent battery in order to recover 95% $Al_2(SO_4)_3$, formed as the concentrated leached solution of 20 % concentration using pure water. Each kg inert solid carry with it 2.5 kg of solution in the underflow. For a basis of 100 kg of aluminium sulphate produced by the reaction. determine the rate of wash water required and the number of ideal stages required for this operation. (**Answer:** 497 kg, 3.2)

8.22 A countercurrent washing system is used to remove oil associated with steel wool using chloroform. 1000 kg/day of steel wool containing 15% weight oil is being treated so that the finished product should not contain more than 0.2% oil by weight on solvent-free basis. The weight of solution retained by a kg of steel wool is 1.25 kg. The overflow solution which contains 3.5% oil is subjected to rectification to recover chloroform. The solvent recovered from the overflow and underflow is recycled to the washing system. The loss of chloroform is compensated by adding fresh make up solvent. The fresh solvent constitutes 10% by weight of the recycled solvent and the total solvent stream contains 0.06% oil. Determine:

(a) The quantity of fresh chloroform required per day
(b) The number of washing stages required.

(**Answer:** (a) 515 kg (b) 3)

8.23 The sludge consisting of 5% $CaCO_3$, 0.1% NaOH and the balance water is fed at a rate of 40000 kg/h to two thickeners in series in which it is washed countercurrently with 8000 kg/h of water. The thickened pulp leaving each thickener contains 20% solids and that from the last thickener is filtered and dried. Determine the mass percent of NaOH in the final dried $CaCO_3$ product. (**Answer:** 0.19%)

8.24 Aluminium sulphate (MW = 342.12) is to be produced by the action of sulphuric acid (MW = 98.08) on bauxite in a series of agitators, with a cascade of continuous thickeners to wash the insoluble mud free of aluminium sulphate.

$$Al_2O_3 + 3H_2SO_4 \rightarrow Al_2(SO_4)_3 + 3H_2O$$

The reaction agitators are fed with 25000 kg bauxite per day containing 50% Al_2O_3 (MW = 101.94) and the rest insolubles, the theoretical quantity of aqueous acid containing 60% H_2SO_4, and the overflow from the second thickener. Assume the reaction is complete. The strong product solution is to contain 25% $Al_2(SO_4)_3$, and no more than 2% of the $Al_2(SO_4)_3$ produced is to be lost in the washed mud. The last thickener is to be fed with pure wash water. The underflow from each thickener will contain 4 kg liquid per kg insoluble solid. There is no preferential adsorption of solute. Calculate the number of thickeners required and the amount of wash water required per day. (**Answer:** 3.3 stages, 141,822 kg.)

Countercurrent leaching (Constant underflow solvent)

8.25 Copper sulphate is to be leached from roasted copper ore using pure water as solvent in a multistage countercurrent extractor. In one hour, a feed consisting of 10000 kg of inert solids, 1200 kg of copper sulphate and 500 kg of water is admitted. It is desired that the extract composition should be 90% water and 10% $CuSO_4$ by weight. Inert solids retain solution such that the weight ratio of inert solids to water retained in each stage is 1:2. Determine the number of stages if 90% of $CuSO_4$ in the feed is to be recovered by (a) Ponchon-Savarit method (b) McCabe-Thiele method. (**Answer:** 6 stages)

8.26 Copper sulphate is to be leached from roasted copper ore using pure water as solvent in a multistage countercurrent extractor. In one hour, a feed consisting of 10000 kg of inert solids, 1200 kg of copper sulphate and 500 kg of water is admitted. It is desired that the extract composition should be 90% water and 10% $CuSO_4$ by weight. Inert solids retain solution such that the weight ratio of inert solids to water retained in each stage is 1:2. Determine the number of stages if 98% of $CuSO_4$ in the feed is to be recovered. (**Answer:** 9 stages)

8.27 Soluble materials from ground roasted coffee is extracted with water in a countercurrent unit. The feed to the leaching system contains 7% soluble solids, 3% water and 90% insoluble solids. The final extract should contain 0.1 kg solubles per kg water and spent coffee grounds should contain not more than 0.005 kg solubles per kg dry inert solids. The underflow sludge from all stages carry 1 kg water per kg inert insoluble solids. Determine:

(a) Kilograms of water circulated per 100 kg roasted coffee

(b) The number of stages required.

(**Answer:** (a) 152.5 kg (b) 5.1)

8.28 Black ash containing 45 weight% Na_2CO_3 and the rest water-insolubles is extracted with water such that 96% of soda ash in it is recovered. The underflow carries with it 15 kg water with every 100 kg insolubles. Determine the total quantity of water used and the concentration of the extract if the leaching has been carried out in a two-stage countercurrent leaching unit, with a feed rate of 1000 kg/h black ash. (**Answer:** 412 kg, 56.73% Na_2CO_3)

8.29 Waxed paper containing 20 weight% soluble wax and 80 weight% insoluble pulp is admitted at a rate of 5000 kg/day to a primary leaching unit where the wax gets completely dissolved in the solvent kerosene. The wax content in the solution adhering with the insoluble pulp is further reduced by washing the sludge with kerosene in a continuous countercurrent washing system. The final underflow sludge is to contain 0.2 kg wax per 100 kg pulp. The kerosene admitted to the last stage contains 0.05 kg wax per 100 kg kerosene. The kerosene flows countercurrent to the solids and leaves the primary leaching unit as the final extract with a concentration 5 kg wax per 100 kg kerosene. The underflow from each washing stage contains 2 kg kerosene per kg pulp. Determine the number of stages graphically and analytically. (**Answer:** 4)

8.30 Roasted zinc ore is treated with sulphuric acid to produce a slurry of water-soluble zinc sulphate and an insoluble residue called gangue. The slurry on decantation gave 20000 kg/h sludge containing 5 wt% water, 10 wt% zinc sulphate and the balance as gangue. This sludge is to be washed with water in a continuous countercurrent washing system to produce an extract containing 10 wt% zinc sulphate in water, with a 95% recovery of zinc sulphate. Assume that the underflow from each washing stage contains two parts by weight of water per every part of gangue. Determine the number of stages required.

(**Answer:** 7.2 stages (Analytical, McCabe), 7.8 (Ponchon-Savarit))

8.31 A countercurrent extraction battery is used to extract caustic soda adhering to the $CaCO_3$ sludge from the reaction

$$Na_2CO_3 + CaO + H_2O \rightarrow CaCO_3 + 2NaOH$$

A feed consisting of 2000 kg/h of Na_2CO_3 in water is admitted to the reaction vessel. The products leaving the reaction vessel contains 10 kg of water with every kg of $CaCO_3$ is fed through a series of three continuous countercurrent thickeners. It is desired that the final product contains on drying only 1 percent NaOH. The underflow from each stage carries 3 parts by mass of water with every part of $CaCO_3$. Determine the rate at which water is admitted for leaching.

(**Answer:** 18304 kg/h)

8.32 One thousand and five hundred kg/h of dry sea-shore sand containing 1 percent by mass of salt is to be washed with 1500 kg/h of fresh water running countercurrently to the sand through two classifiers in series. The sand discharged from each classifier contains one part by mass of water for every two of sand. Determine:

(a) The percentage of salt in the final dried sand
(b) Water rate required in a single classifier in order to wash the sand to the same extent as in (a).

(**Answer:** (a) 0.247% (b) 3030.3 kg/h)

8.33 One thousand kg/h of barium sulphide (MW = 169.4) dissolved in 2500 kg/h water together with stoichiometric quantities of sodium carbonate (MW=106) are fed to a reaction vessel where the following reaction occurs:

$$BaS + Na_2CO_3 \rightarrow BaCO_3 + Na_2S$$

The products of the reaction are washed with water in a three-stage countercurrent washing system to free the barium carbonate (MW =197.4) of the adhering sodium sulphide (MW = 78). The concentrated overflow solution is to contain 10% sodium sulphide. The underflow sludge from all thickeners contains two parts water with every part of barium carbonate by mass. Determine:

(a) The quantity and composition of the dried barium carbonate precipitate
(b) The rate at which water is to be supplied to the third thickener
(c) The percent recovery of sodium sulphide in the final extract

(**Answer:** (a) 1221.4 kg, 4.6% Na_2S (b) 3482.1 kg (c) 88.12%)

Countercurrent leaching (Variable underflow)

8.34 Seeds containing 20 % by weight oil are extracted in a countercurrent plant and 95% of the oil is to be recovered in a solution containing 50% of oil. It has been found that the amount of solution removed in the underflow in association with every kg of insoluble matter is given by

$$\frac{1}{N} = 0.7 + 0.5y + 3y^2$$

where y is mass fraction of solute in the overflow solution. If the seeds are extracted with fresh solvent, how many ideal stages are required?

(**Answer:** 10.8 stages)

8.35 Oil is extracted from meal using benzene as solvent in a countercurrent extraction battery. The hourly feed of oil-meal to the unit contains 500 kg oil, 1000 kg insoluble meal and 50 kg benzene. The solvent is admitted at a rate 800 kg/h and contains 12 kg oil. The solids leaving the last stage as final underflow contain 75 kg/h unextracted oil. The mass of solution retained by the solids from a stage varies with the concentration of the solution as:

x, kg oil/kg solution	0.0	0.1	0.2	0.3	0.4	0.5	0.6	0.7
N^{-1}, kg solution/ kg solid	0.500	0.505	0.515	0.530	0.550	0.571	0.595	0.620

Determine: (a) The concentration of final overflow (b) Concentration of solution retained by the exhausted meal (c) The mass of solution in the final underflow solids (d) The mass of overflow (e) The number of stages required.

(**Answer:** (a) 0.521 (b) 0.148 (c) 511.3 kg (d) 838.7 kg (e) 2.7)

8.36 Oil is extracted from soyabean flakes containing 20% oil using hexane as solvent which is introduced countercurrently at a rate of 0.50 kg per 1 kg flakes. The solid residue contains 0.01 kg oil per kg oil-free meal. The solvent introduced is free of oil. The settling characteristics of the solids are given as:

x, kg oil/kg solution	0	0.1	0.2	0.3	0.4	0.5	0.6	0.7
N, Kg solids/ kg solution	2.0	1.980	1.942	1.887	1.818	1.751	1.680	1.613

Determine: (a) The concentration of final overflow (b) Concentration of solution retained by the exhausted meal (c) The mass of solution in the final underflow solids (d) The mass of overflow (e) The number of stages required.

(**Answer:** (a) $y_1 = 0.64$ (b) $x_N = 0.019$ (c) 40 kg (d) 30 kg (e) 10]

8.37 A countercurrent extraction battery is used to extract caustic soda adhering to the $CaCO_3$ sludge from the reaction

$$Na_2CO_3 + CaO + H_2O \rightarrow CaCO_3 + 2NaOH$$

The products leaving the reaction vessel contains 0.6 kg of water with every kg of $CaCO_3$, but no excess reactants are present. It is desired to recover 97 percent of the NaOH. The final overflow contains 12% by weight NaOH. The underflow from each stage carries solution the weight of which varies with the concentration as

Wt % NaOH in solution	0.0	5.0	10.0	15.0	20.0
N^{-1}, kg solution/ kg $CaCO_3$	1.50	1.75	2.20	2.70	3.60

Determine for treating 100 kg/h of $CaCO_3$:

(a) The solvent added, kg/h

(b) The quantity and composition of the solution in the final underflow sludge

(c) The number of stages required.

(**Answer:** (a) 664 kg (b) 157 kg, 1.53% NaOH (c) 2.65)

8.38 Oil is extracted from halibut livers using pure ether as solvent in a countercurrent extraction battery. The charge consisting of 100 kg oil-free liver and 0.0352 m^3 oil is admitted to recover 95% of the oil in an extract containing 0.65 m^3 oil per m^3 of extract. The underflow concentration varies with solution concentration as given in the following table:

m^3 solution retained/kg exhausted livers $\times 10^4$	2.92	3.51	4.17	4.84	5.67	6.76	8.26	10.01
Solution concentration, m^3 oil/ m^3 solution	0	0.1	0.2	0.3	0.4	0.5	0.6	0.68

Determine:

(a) The solvent requirement in m^3

(b) The number of stages.

(**Answer:** (a) 0.0485 m^3 (b) 6.8)

8.39 Oil is extracted from halibut livers using pure ether as solvent in a countercurrent extraction battery. The charge containing 0.40 kg oil/kg exhausted livers is admitted to recover 95% of the oil. The underflow concentration varies with solution concentration as given in the following table:

Concentration of solution, kg oil/kg solution	0	0.1	0.2	0.3	0.4	0.5	0.6	0.67
Entrainment, kg solution/kg oil-free liver	0.28	0.34	0.40	0.47	0.55	0.66	0.80	0.96

Determine the number of stages required if 50 kg ether is used per 100 kg untreated solids.

(Answer 3.7)

8.40 Oil is extracted from halibut livers using pure ether as solvent in a countercurrent extraction battery. The charge containing 25.0 weight% oil is to be extracted with pure ethyl ether to recover 98% of the oil. The feed is admitted at a rate 1000 kg fresh livers per hour. The final overflow solution is to contain 70 weight% oil. The underflow concentration varies with solution concentration as given in the following table:

x, kg oil/kg solution	0	0.2	0.4	0.6	0.81
N, kg inert solid/kg solution	4.88	3.50	2.47	1.67	1.39

Determine:

(a) The amount of solvent required
(b) The amount of overflow solution
(c) The quantity and composition of underflow sludge
(d) The number of stages

(**Answer:** (a) 260.7 kg (b) 350 kg (c) 160.7 solution and 750 kg exhausted meal: 0.5% oil, 17.1% ether, 82.4% exhausted liver) (d) 8)

8.41 Titanium dioxide (TiO_2) pigment is freed of soluble impurities by washing with pure water in a countercurrent multistage system consisting of several thickeners operated in series. The mass rate of water admitted is the same as the impure pigment. The feed contains 50 weight% TiO_2, 20 weight% soluble salts and 30 weight% water and it is desired that the final product leaving the unit when dried yields 2×10^5 kg/h of 99.9% by weight of TiO_2 pigment. Determine the number of stages, the quantity and composition of overflow if

(a) the underflow from each stage contains 0.4 kg solution/kg TiO_2.
(b) the solution retained by the solids in the underflow varies with concentration as

x, kg solute/kg solution	0.0	0.1	0.2	0.3	0.4	0.5	0.6
N^{-1}, kg solution/ kg TiO_2	0.30	0.32	0.34	0.36	0.38	0.40	0.42

(**Answer:** (a) 3.5, 5.195×10^5 kg/h, 15.35% solubles (b) 3, 5.395×10^5 kg/h, 14.78% solubles)

8.42 Oil is extracted from soyabean flakes containing 25% oil using pure hexane as solvent which is introduced countercurrently at the same mass flow rate as that of the flakes. The solid residue contains 0.5% oil on solvent free basis. The overflow from all stages except the first is free of solids. The solution leaving the first stage contains 10% of the insoluble solids in the feed. The settling characteristics of the solids are given as:

Wt% oil in solution	0	20	30
kg solution retained per kg insoluble solid	0.58	0.66	0.70

Determine:

(a) Concentration of oil in the clear overflow solution
(b) The number of stages required.

(**Answer:** (a) 28.76% (b) 4.6)

8.43 Oil is extracted from soyabean flakes containing 25% oil using hexane as solvent. Both the feed and pure oil-free solvent are introduced countercurrently at a rate of 1000 kg/h. The solid residue contains 0.8 kg oil per 100 kg inert solids. The settling characteristics of the solids are given as:

Weight% oil in solution	0	20	30
kg inert solids/kg solution retained	1.73	1.52	1.43

Calculate:

(a) The flow rates of the effluent streams
(b) The percent compositions of the effluent streams and
(c) The number of stages.

(**Answer:** (a) Underflow: 750 kg/h inerts + 437.3 kg/h solution, Overflow: 812.7 kg/h solution (b) Underflow: inerts = 63.17%, hexane: 36.33%, oil = 0.5%; Overflow: 30.1% oil, 69.9% hexane (c) 4.3)

8.44 A multistage countercurrent system is used to leach protein from seaweed using hot water. The seaweed slurry consisting of 50% solids, 3% proteins, and 47% water is treated at a rate of 100 kg/h and hot water at a rate of 125 kg/h. It is desired that the maximum concentration of protein in the spent seaweed to be 0.2% on solids-free basis. Assume that the entire protein goes into solution with water. The settling characteristics of the sludge is given by the experimental data in the table below:

Weight fraction in underflow sludge		
Water	Solution protein	Solids
0.542	0.026	0.432
0.564	0.019	0.417
0.586	0.013	0.401
0.5954	0.0066	0.398
0.5994	0.0036	0.397
0.6028	0.0012	0.396

Determine:

(a) The percentage of protein in the strong leach liquor
(b) The number of stages required.

(**Answer:** (a) 2.9% (b) 4.7)

8.45 Sodium hydroxide is produced at a rat of 400 kg/h by the reaction of stoichiometric amounts of slaked lime $Ca(OH)_2$ and soda ash Na_2CO_3. The precipitated $CaCO_3$ is washed with water in a countercurrent multistage washing system consisting of a series of thickeners. Pure water is admitted to the last thickener and the final leach liquor containing 10% NaOH is removed from the reaction agitator. The settling characteristics of the slurry are given below which indicates preferential adsorption of NaOH on the solid:

x	0.0917	0.0762	0.0608	0.0452	0.0295	0.0204	0.01435	0.01015	0
y	0.0900	0.0700	0.0473	0.0330	0.0208	0.01187	0.00710	0.00450	0
N	0.495	0.525	0.568	0.600	0.620	0.650	0.659	0.666	0.682

(x = weight fraction NaOH in the settled sludge, y = weight fraction of NaOH in the overflow solution N = kg $CaCO_3$ /kg solution)

Determine:

(a) The number of thickeners required if the loss of NaOH in the underflow sludge is to be only 1% of that in the feed.
(b) The rate at which water is to be supplied. (**Answer:** (a) 3.8 stages (b) 4303 kg)

Chapter 9

Drying

Drying is the operation by which moisture is removed in relatively small quantities from solids or solid-like materials by evaporation. Most solid materials require drying at some stage during their manufacture. In process industries, drying generally follows operations such as evaporation, filtration or crystallisation. It is one of the last steps and is carried out just ahead of packaging and despatch of the product. Drying is different from the removal of water from wet solids by mechanical means such as squeezing, pressing, centrifugation, decantation, filtration, etc. Though dehumidification, adsorption and evaporation processes can be employed for the removal of moisture, our definition of 'drying' exclude them from treating as 'drying operations'. In dehumidification, water vapour present in a gas stream is removed by condensation; adsorption employs solid adsorbent such as silica gel or activated alumina to remove water vapour from gas streams. Drying differs from evaporation operation in which water is removed by vaporisation from solutions, in the following respects:

1. In evaporation, water is vaporised in large quantities from a solution whereas in drying, water is removed from solids in relatively smaller quantities.
2. Evaporation occurs at the boiling point of the solution whereas drying can be carried out at relatively low temperatures and the moisture is removed by vaporisation at temperatures well below the boiling point of water.
3. The vapour leaving an evaporator is generally pure water vapour unless it is contaminated by unavoidable leakages in the unit. On the other hand, drying operations frequently employ carrier gas such as air or flue gas for removing water vapour.

The liquid content of the dried product depends on the type of the materials. Dried table salt contains 0.5% water, coal about 4% and many food products about 5%. For products such as paper, timber, pharmaceuticals, solid catalysts, etc. drying is an integral part of the process, whereas for products such as soap, detergents, dyestuffs, fertilizers, coal, salt, sugar, milk, etc. removal of moisture is important for various reasons. Reducing the moisture content reduces the cost of transport of the material. The product must be made suitable for subsequent processing or sale. Material should have a certain moisture content for processing, pelletizing or moulding. It is highly desirable that the materials have easy-to-handle properties like the 'free flowing' nature of salt. Most pharmaceuticals and fine chemicals require drying to a suitable moisture content to facilitate satisfactory packaging. Removal of moisture from the process gas streams such as coal gas, gaseous fuels and benzene is necessary as moisture present in the gas streams may lead to corrosion of process equipment.

Micro-organisms responsible for food spoilage cannot survive, grow and multiply in the absence of water and drying is therefore used as a preservation technique for food and biological materials. Enzymatic action that cause certain chemical changes in food and biological materials can be prevented by reducing the moisture content by drying. It is also necessary to remove moisture from food stuffs for preserving their flavour and nutritional values.

Removal of liquid contaminants such as benzene and organic solvents from solids can also be treated as 'drying' and the theory and equipment discussed here for the removal of water from the solids are applicable to such operations as well, with necessary modifications.

9.1 DRYING EQUILIBRIUM

Since the quality of the dried product largely depends on the operating conditions prevailing in the dryer, the choice of the drying conditions is very important in the design and operation of dryers. Contamination of the pharmaceutical products, damage to the crystalline solids, checking or warping of the wood, shrinkage as in the drying of paper, loss of flavour in the drying of fruits, case hardening as in the case of clay and soap, etc. are some of the difficulties that occur during drying and these should be avoided by proper choice of the conditions of drying.

Considering the behaviour on drying, the solids may be classified into two broad categories:

Category 1: *Granular or crystalline solids*. Inorganic solids generally belong to this class. Moisture is contained within the interstices of particles or in shallow open surface pores. The solid is relatively unaffected by the presence of the liquid and by the drying process. These materials can be dried rapidly to very low moisture contents. The drying conditions can be chosen based on our convenience and economic advantage, as the final properties and appearance of the dried product are not affected by the choice. Examples of this class of materials are crushed rock, titanium dioxide, sand, catalysts, zinc sulphate monohydrate and sodium phosphates.

Category 2: *Amorphous, fibrous or gel like materials*. Organic solids of plant and animal origin which include wood, leather, soap, cotton, starch, wool, etc. belong to this class. Moisture is either dissolved by such materials or it exists as trapped in the fibres or very fine pores of the solid. Since moisture is present as an integral part of the solid structure, the moisture removal affects the structure. The surface layers dry more rapidly than the interior of the solid. If the rate of moisture removal is very high, the difference in moisture content through the solid, lead to warping or cracking. *Case hardening*, that is, the formation of an impervious shell of partly dry solid which prevents moisture removal from the interior, is another problem associated with the drying of certain amorphous materials. The drying conditions should be chosen with care as these determine the properties of the dried product.

Before we consider the drying behaviour and the drying equilibrium, we should be familiar with the method of expressing the moisture concentration in the solid.

Moisture Content

Generally the moisture content of a solid is expressed as weight percent on a *wet basis*, which is the kg of water contained in 100 kg of the wet solid. Thus

$$\text{Weight percent (wet basis)} = \frac{\text{weight of water in the solid}}{\text{weight of wet solid}} \times 100$$

The weight fraction denoted by w is the amount of water present in unit weight of the wet solid. Therefore, the weight percent on wet basis = $w \times 100$.

In the design and analysis of drying operations, it is more convenient to express the moisture concentration on *dry basis* because, only the moisture is exchanged between the solid and the gas phases, and the mass of dry solid and the mass of dry gas undergo no change. The weight percent of moisture on dry basis is the kilograms of moisture associated with 100 kg dry solid. Moisture content on dry basis is denoted by X, which is defined as the weight of water in kg associated with one kg dry solid. X is in fact the weight ratio of water in the wet solid.

$$X = \frac{\text{weight of water in the solid}}{\text{weight of dry solid}}$$

Weight percent of water on dry basis is then $X \times 100$.

If X is the moisture content on dry basis, 1 kg dry solid is associated with X kg water or $1 + X$ kg wet solid contains X kg moisture.

$$w = \frac{\text{weight of water in the solid}}{\text{weight of wet solid}} = \frac{X}{1+X}$$

Example 9.1: A wet solid weighs 50 kg and contains 10 kg moisture. Express the moisture concentration on wet and dry basis.

***Solution*:** 10 kg moisture is present in 50 kg wet solid. The moisture content on the wet basis, is defined as the kg moisture present per one kg of wet solid.

$$\text{Moisture content on wet basis, } w = \frac{\text{weight of water in the solid}}{\text{weight of wet solid}} = \frac{10}{50} = 0.2\frac{\text{kg water}}{\text{kg wet solid}} \quad \textbf{(Answer)}$$

Expressed as percent, the moisture content on the wet basis is 20% water.

Moisture content on dry basis is denoted by X and is defined as kg moisture present per kg dry solid. Since 10 kg water is associated with 40 kg dry solids,

$$X = \frac{\text{weight of water in the solid}}{\text{weight of dry solid}} = \frac{10}{40} = 0.25\frac{\text{kg moisture}}{\text{kg dry solid}} \quad \textbf{(Answer)}$$

Or, the solids contain 25% moisture on dry basis.

The weight percent on the wet basis is calculated as

$$\text{Weight percent (wet basis)} = \frac{\text{weight of water in the solid}}{\text{weight of wet solid}} \times 100 = \frac{X}{1+X} \times 100 = \frac{0.25}{1.25} \times 100 = 20\%$$

Equilibrium Moisture

Consider a material being dried in contact with air passed across its surface. Assume that the air is passed in sufficiently large quantities that the humidity and temperature of the air are not altered due to the heat and mass exchange with the wet solid. The moisture in the solid evaporates and diffuses to the air stream and the solid gets dried. This will continue till the moisture in the solid comes in equilibrium with the water vapour present in the air. The drying will come to a stop though the material still contains significant quantities of moisture. The moisture in the solid at this condition is the *equilibrium moisture content* and is denoted by X^*. The partial pressure exerted by the moisture in the solid under this condition equals the partial pressure of water vapour in the air. The solid material cannot be dried to a moisture content lower than X^* by further exposure to the same gas for an indefinitely long period.

The equilibrium moisture content for a given material is a function of the temperature, pressure and the relative humidity of the air in contact with the solid. If the wet material were contacted with air of lower relative humidity at the same temperature as above, the equilibrium moisture content X^* of the solid will take a lower value. Figure 9.1 shows the equilibrium moisture content of wood at 298 K plotted as a function of the percent relative humidity (RH) of air. Percent relative humidity indicates the concentration of moisture in the air and is defined as the ratio of the partial pressure of water vapour to the vapour pressure at the same temperature expressed as percentage. From Fig. 9.1, it can be seen that the equilibrium moisture content of wood in contact with air of 60% RH at 298 K is 0.113 kg water/kg dry solid and that with air of 40% RH is 0.075 kg water/kg dry solid. Since at equilibrium, the moisture in the solid exerts a partial pressure which is equal to the partial pressure of water vapour in the air, Fig. 9.1 may be treated as the plot of the moisture content of the solid versus the ratio of partial pressure exerted by the moisture in the solid at equilibrium to the vapour pressure of pure water at the given temperature.

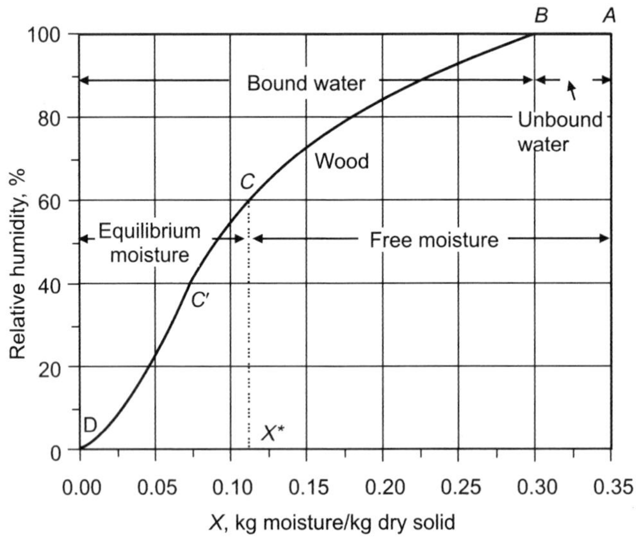

Fig. 9.1: Equilibrium moisture content of wood at 298 K

Suppose that the wood initially contains 35% moisture on dry basis and the initial condition is represented by point *A* in the figure. The moisture present in the solid at this condition exerts a partial pressure which is equal to the vapour pressure of water at the given temperature. It is brought in contact with the air of 60% relative humidity. The partial pressure of water in the air is only 60% of the vapour pressure of the water at 298 K. Since the partial pressure of water in the air is less than that exerted by water in the wood, moisture is evaporated and transferred to the gas stream. The moisture content of the solid decreases along the curve *ABCD*. However, when the moisture content falls to that corresponding to point *C*, the vapour pressure exerted by the moisture in the solid equals the partial pressure of water vapour in the air. The moisture content of the solid at this condition, 0.113 kg moisture/kg dry wood, is X^*, the equilibrium moisture content of wood when contacted with air of 60% RH. It is impossible to dry wood to a lower moisture content by contacting with air at 298 K and 60% RH. To do so, air of a lower relative saturation should be brought in contact with the solid. For example, by using 40% RH air, we can dry the wood to $X = 0.075$ kg moisture/kg dry solid which is the equilibrium moisture content X^* corresponding to the air of relative humidity 40%, represented by point C' in the figure.

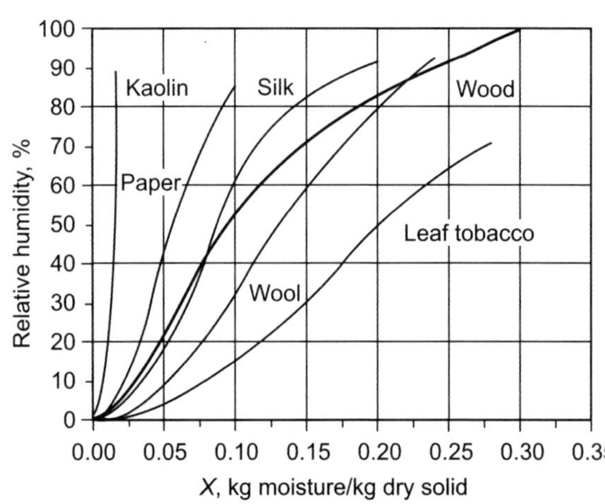

Fig. 9.2: Equilibrium moisture content of solids at 298 K

Different materials exhibit different equilibrium behaviour as shown in Fig. 9.2. For any given percent relative saturation, the equilibrium moisture content varies greatly with the type of materials. Amorphous and fibrous materials of biological origin usually show large equilibrium moisture content (e.g. wool, leather and wood), whereas granular or crystalline solids and inorganic materials show relatively low equilibrium moisture content (e.g. sand, china clay, glass wool, kaoline). To dry a material to the bone dry condition, perfectly dry air (RH = 0) has to be brought in contact with the wet material.

Equilibrium moisture content depends on the temperature and it decreases somewhat with increase in temperature. For example, for raw cotton in contact with air of RH = 50%, X^* falls from 7.3 kg water/kg dry solid at 311 K to 5.3 kg water/kg dry solid at 366.5 K. For a given species of solid, the equilibrium moisture may depend upon the particle size or specific surface, if moisture is physically adsorbed.

Free Moisture

Equilibrium moisture content sets the limit to which a material can be dried under specified set of conditions. *Free moisture* is the moisture in excess of the equilibrium moisture content X^*, that can be removed in any drying operation, as shown in Fig. 9.1. Exposing the material which is in equilibrium with the gas of a given humidity and temperature for even an indefinitely long period will not cause an additional loss of moisture from the solid. For example, in drying of the wood containing initially 35 weight% moisture on dry basis ($X = 0.35$) using air of 60% RH at 298 K,

$$\text{the free moisture} = X - X^* = 0.35 - 0.113 = 0.237 \text{ kg/kg dry solid}$$

That is, 23.7 kg water is removed from 100 kg of dry solids, which is nearly 68% of the water originally present in the wet solid.

Bound and Unbound Water

Moisture in the solids may be the *bound moisture*, the *unbound moisture* or both. Referring to the drying equilibrium curve for wood shown in Fig. 9.1, we see that the moisture in the solid exerts the full vapour pressure of water for all concentration greater than that indicated by point B in the figure. Such moisture, called the unbound moisture, behaves as free water, uninfluenced by the solid. It is the unbound moisture that is removed first in any drying operation when wet materials containing sufficiently large amounts of water are subjected to drying. The removal of unbound water from a wet solid occurs by the same mechanism as the evaporation of water from a free water surface. It is evident from Fig. 9.1 that only unbound moisture can be removed when the solid is in contact with saturated air.

Bound moisture exerts a partial pressure less than the vapour pressure of pure water. For moisture concentrations less than that shown by point B in Fig. 9.1, the partial pressure exerted by the moisture in the solid is less than the vapour pressure of pure water. The bound water can remain in equilibrium with the unsaturated air. It can exist in the solid under several conditions. The water held by fine capillaries in the solid or contained within the cell walls of biological materials, water in physical and chemical combination with the natural organic materials, water present as a liquid solution of soluble constituents of the solid, etc. constitute bound moisture. Lowering of the vapour pressure may be due to the concave curvature of the surface when water is held by fine capillaries or due to dissolution of soluble constituents when water is contained within the cell or fibre walls. Solids which contain bound water are called *hygroscopic substances*. The moisture content corresponding to point B in Fig. 9.1 is the highest moisture content of the solid that can exist in equilibrium with moist unsaturated air, and for cellular materials such as wood, it is often referred to as the *fibre saturation point*.

Hysteresis

The solid may attain the equilibrium moisture content either by sorption (condensation) or by desorption (evaporation) of water. If the original moisture content of the solid which was exposed to an air stream is less than the equilibrium moisture content X^*, the material will gain moisture from the air till the moisture concentration in the solid reaches the equilibrium value. In this case, the material gains water by sorption. In drying of a wet material on the other hand, the material loses moisture by desorption. Generally, the equilibrium moisture content attained by a solid is independent of the direction of approach to the equilibrium state. However, the equilibrium moisture content of certain materials depends upon the direction in which the equilibrium is attained. If the material exhibits different equilibrium moisture contents for sorption and desorption, the material is said to exhibit hysteresis.

A possible explanation for hysteresis can be provided by assuming that the solid consists of necked capillaries such as the ones shown in Fig. 9.3. The material attains equilibrium by desorption in Fig. 9.3a and by adsorption in Fig. 9.3b. The similarity in the curvature of the liquid meniscus leads to identical vapour pressure of the liquid even when the amount of moisture contained in the pores are different. Figure 9.4 shows the equilibrium behaviour of a solid exhibiting hysteresis. Sulphite pulp is an example of a material exhibiting hysteresis. For a given temperature and relative humidity of air, the equilibrium moisture content reached by desorption (X_d^*) is generally greater than that reached by adsorption (X_a^*). The dry material gains moisture along the adsorption curve and wet material looses moisture along the desorption curve.

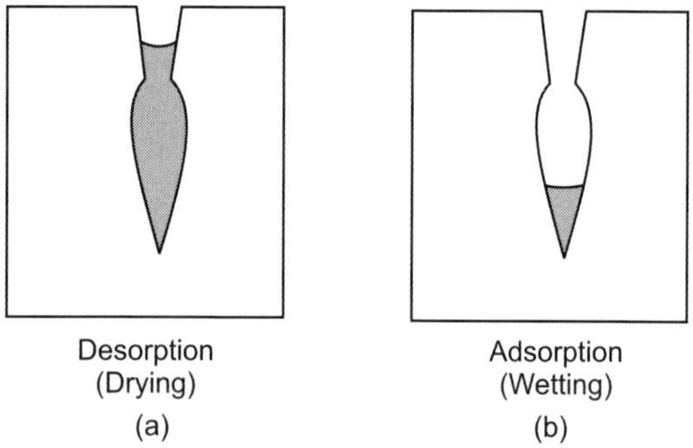

Fig. 9.3: Relationship between hysteresis and pore structure

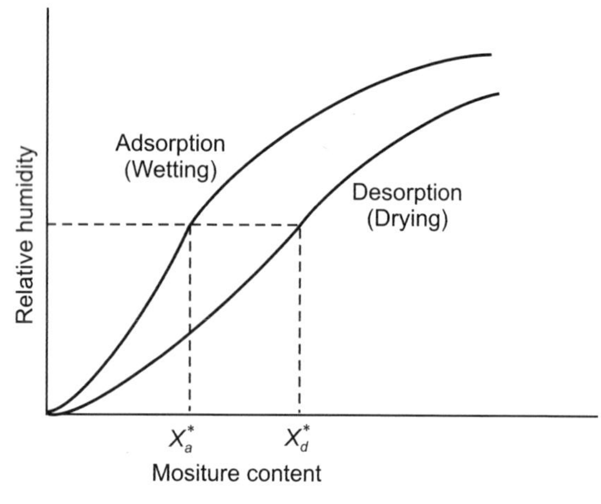

Fig. 9.4: Hysteresis in drying

To prevent the moisture regain of the dried product when exposed to moist air, it would be necessary to store the material in air-tight containers, immediately after drying. For material showing hysteresis behaviour, if the dried product is likely to be exposed to air of certain relative humidity, drying the material to a moisture content below the adsorption equilibrium value, is of no use.

Soluble Solids

Water of hydration of inorganic crystals is an example of the bound moisture held chemically by the solids. The hydrated form obtained on drying depends on the humidity as well as the temperature of the drying gas. Consider for example, a solution of $CuSO_4$ exposed to the air at 298 K. For partial pressure of water vapour in the air exceeding 3.08 kPa, the salt remains completely dissolved in the water (Fig. 9.5). If the partial pressure of water vapour in the air is between 1.04 kPa and 3.08 kPa, $CuSO_4.5H_2O$ is produced. At 298 K, the stable hydrate is the pentahydrate of $CuSO_4$. If the partial pressure is 3.08 kPa, a mixture of pentahydrate and saturated solution can coexist and at a partial pressure of 1.04 kPa, a mixture of penta and trihydrates of any proportion can coexist. For a partial pressure of water vapour in the air between 110 Pa and 750 Pa, it is the monohydrate ($CuSO_4.H_2O$) that is formed and for partial pressures below 110 Pa, dry anhydrous $CuSO_4$ crystals are formed on exposure to air at 298 K. The equilibrium moisture content falls to zero if bone-dry air is used.

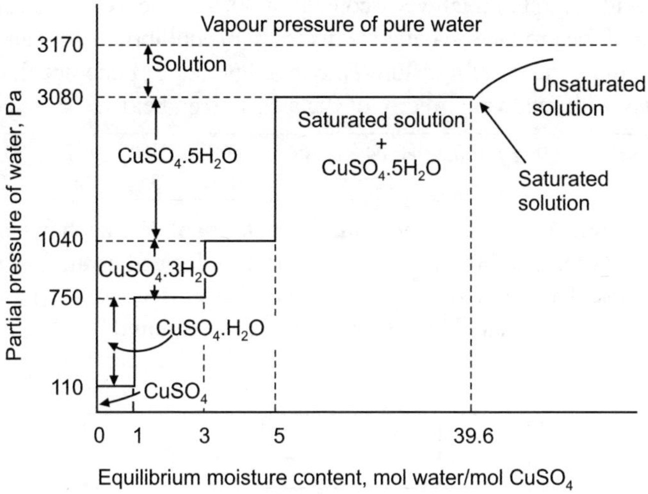

Fig. 9.5: Equilibrium moisture content of $CuSO_4$ at 298 K

Figure 9.6 shows the equilibrium moisture content of sodium nitrate at 298 K. Consider an unsaturated solution of $NaNO_3$ containing, say, 2 kg water per kg $NaNO_3$ represented by point *A* in the figure. Water in the solution exerts a partial pressure, which is less than the vapour pressure of pure water at 298 K. Suppose this solution is exposed to air containing water vapour with a partial pressure of say, 2 kPa. Moisture evaporates and the concentration of water in the solution falls along the curve *AB*. At point *B*, the solution becomes saturated. The partial pressure exerted by the water in the solution is about 2.35 kPa. On further exposure of this solution with relatively dry air, evaporation of water

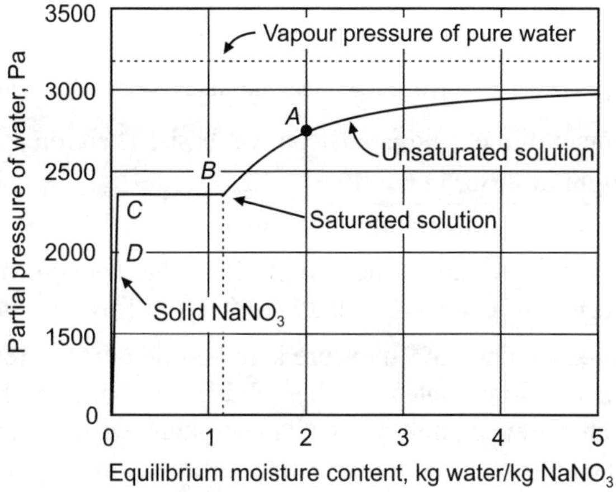

Fig. 9.6: Equilibrium moisture content of $NaNO_3$ at 298 K (not to scale)

continues along the line *BC* with crystallisation of anhydrous $NaNO_3$. The partial pressure remains constant on this line where the saturated solution and solid crystals coexist. At point *C* crystallisation is completed and the solid is almost free of water, but for the small traces adsorbed on the surface. On further contact with air with partial pressure of water 2 kPa, the moisture content falls to the equilibrium value corresponding to point *D*. We see that the solids which are soluble in water show insignificant equilibrium moisture content when contacted with air in which the partial pressure of moisture is less than the vapour pressure exerted by a saturated solution of the solid in water.

Example 9.2: Fibres with an initial total water content of 20% on wet basis are dried in a tunnel dryer operating at 101.3 kPa. If the products leaving the dryer are in equilibrium with air of relative humidity of 50% at 298 K, determine the kg of moisture evaporated per kg of bone-dry fibres. The equilibrium moisture content varies with relative humidity of the air as given below:

X, kg moisture/kg dry solid	0	0.02	0.04	0.06	0.08	0.10	0.12	0.14	0.16	0.18
Relative humidity,%	0	3	12	24	38	55	69	80	90	97

***Solution*:** The moisture content versus relative humidity data are plotted as shown in Fig. 9.7. The point A on the curve represents the equilibrium condition and it is located at the horizontal line drawn at relative humidity of 50%. The abscissa of point A is read from the graph which is the equilibrium moisture content of the solid corresponding to air of 50% relative humidity. $X^* = 0.094$ kg water/kg dry solid.

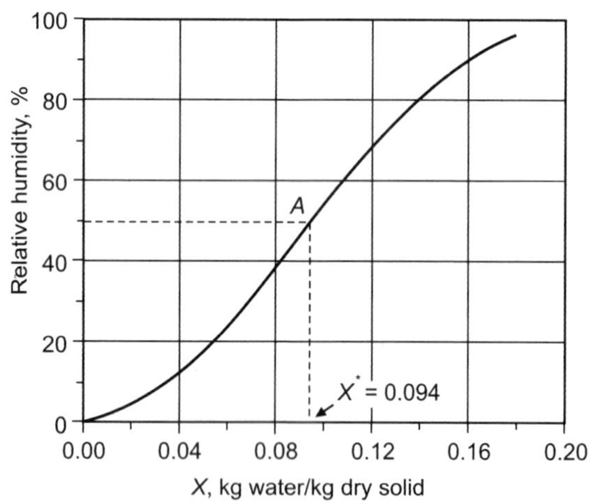

Fig. 9.7: Determination of equilibrium moisture content (Example 9.2)

The initial moisture content of the solid is 20% on wet basis. Therefore,

$$X = \frac{\text{weight of water in the fibre}}{\text{weight of dry fibre}} = \frac{20}{80} = 0.25$$

When the solid is dried from $X = 0.25$ to the equilibrium moisture content $X^* = 0.094$, the water removed per kilogram of dry fibre is $X - X^* = 0.25 - 0.094 = 0.156$ kg water. **(Answer)**

Example 9.3: A wet solid containing 33% moisture is to be dried in a current of air of 50% relative humidity. The equilibrium moisture content of the solid is 10% and the highest concentration of moisture in the solid that can exist in equilibrium with the unsaturated air is 28%. All moisture contents are on dry basis. Determine the free moisture, bound moisture and unbound moisture in the solid.

***Solution*:** The drying equilibrium curve is shown in Fig. 9.8. The initial moisture content in the solid is $X_1 = 0.33$ kg moisture/kg dry solid, and is represented by point A. The equilibrium moisture content $X^* = 0.10$ kg moisture/kg dry solid.

The free moisture content is the moisture present in the solid in excess of the equilibrium moisture content. Free moisture $= X_1 - X^* = 0.33 - 0.10 = 0.23$ kg moisture/kg dry solid. **(Answer)**

The unbound moisture in the solid exerts the full vapour pressure of water, whereas the partial pressure exerted by bound moisture in the solid is less than the vapour pressure. Therefore, the highest concentration of moisture in the solid that can exist in equilibrium with the unsaturated air represents the concentration of the bound moisture in the solid. Therefore, out of the 33 kg moisture present per 100 kg dry solid, 28 kg moisture is bound moisture and the rest, i.e. 33 – 28 = 5 kg is unbound moisture. **(Answer)**

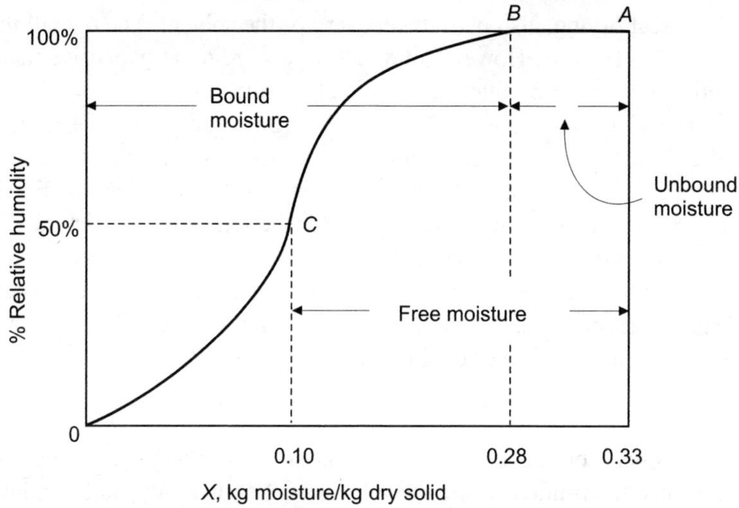

Fig. 9.8: Equilibrium curve for Example 9.3

9.2 CLASSIFICATION OF DRYERS

Dryers may be classified in several ways. Classification of dryers based on the conditions of the solids such as static, moving, fluidized or dispersed is important. Dryers can be distinguished as cocurrent, countercurrent and cross flow dryers based on the gas flow pattern in relation to the movement of the solids. When there is significant gas flow, dryers employ different methods of gas–solid contacting such as cross circulation, through circulation, impingement, etc. The classification based on the method of transfer of the heat required for evaporating the water from the solid, the method of operation whether batch or continuous, and the type and the nature of the feed and product materials are discussed below.

Mode of Heat Transfer

Dryers can be classified as direct, indirect, radiative and dielectric, based on the mode of transfer of heat to the material. In direct dryers, also known as convective or adiabatic dryers, heat transfer is achieved by the convection from hot gases passed in direct contact with the solids. The gas, besides supplying the necessary heat, acts as the carrier that sweeps away the moisture evaporated. Drying temperatures as high as 1000 K can be employed in direct dryers. It is easy to control the temperature of drying at desired levels and prevent heating beyond specified limits, a feature very important in drying temperature-sensitive materials. Direct dryers are less costly compared to indirect dryers. The main drawback of direct dryers is the dusting problem associated with the drying of very small particles. Since the entire heat requirement for drying is met by the gas, large quantities are to be circulated, necessitating very large and expensive dust recovery equipment. The overall thermal efficiency is low due to the loss of heat with the exhaust gas. Further, complications may arise if the solid reacts chemically with the oxygen in the drying gas.

In indirect dryers, the heating medium is contained in tubes and jackets and heat is transferred to the wet material through the walls by conduction or by radiation from a hot surface. A sweep gas, if necessary, may be provided to help remove the evaporated moisture. Indirect dryers can be operated under reduced pressure to bring down the temperature at which moisture is evaporated. Surface temperatures may range from temperature below the freezing point of moisture in the case of freeze dryers to above 800 K as in the case of dryers using combustion products as the heating medium. Since drying can be done in an inert atmosphere, oxidation of materials and explosion hazards are eliminated in indirect dryers. Indirect drying, also permits recovery of the solvent. The overall thermal efficiency is higher than that of direct dryers. However, indirect dryers are more expensive than the direct type. The danger of overheating of the products, also cannot be ruled out.

Infrared dryers depend on the transfer of radiant energy to evaporate the moisture. The radiant energy is supplied electrically by infrared lamps, by electric resistance elements, or by incandescent refractories heated by gas. The last method offers the additional benefit of convection heating. Infrared dryers are not widely used in the chemical industries for the removal of moisture.

Dielectric dryers, also known as the radio frequency dryers or microwave dryers generate heat within the solids and hence offer better scope for drying of massive geometric objects such as wood, sponge-rubber shapes, and ceramics. Power costs involved in the dielectric dryers are generally much higher than the fuel costs in the conventional methods.

Mode of Operation

Based on the mode of operation, the dryers may be classified as *batch dryers* and *continuous dryers*. Batch dryers are operated intermittently or cyclically and under unsteady-state conditions. The dryer is charged with the material, which remains in the dryer until dry, whereupon the dryer is emptied and recharged with fresh batch. Continuous dryers, in contrast, operate under steady state with the material moving continuously through the dryer.

Batch dryers are employed profitably by industries in which the overall process is predominantly batch. It is advantageous to go for batch drying when valuable materials, the total cost of which is little influenced by the cost of drying, are to be dried. An advantage of the batch dryer is that a single dryer usually can handle different materials. In batch drying, humidity can be altered during drying cycle. However, batch drying is a relatively expensive operation and is limited to low through-puts, usually 250 kg/h or less. These are preferred for operations that require very long drying times and are generally used for pilot plants and process development work.

Continuous dryers are useful for high evaporation rates and for high throughputs, generally 1000 kg/h or higher. The size of dryer is small relative to the quantity of product handled. The advantages of continuous dryers include the ease of their integration into the continuous chemical manufacture without intermediate storage, lower unit cost of drying and uniformity of the moisture content of the product.

Continuous adiabatic dryers may employ countercurrent, cocurrent or cross-flow of the gas and solid streams. In *countercurrent dryers*, the hot gas is admitted at the end where the dry solids are discharged. Therefore, relatively dry solids containing the last traces of bound moisture that are difficult to remove, are in contact with the gas at the highest temperature. This facilitates rapid removal of such moisture. At the same time, the possibility of damage of the dried product exists as it may get heated to temperatures as high as the inlet temperature of the gas. Thus, heat sensitive material cannot be dried in countercurrent dryers. Since the discharged products carry away a significant fraction of the heat energy contained in the inlet gas, the thermal efficiency of countercurrent dryer is low.

Parallel current adiabatic dryers operate with gas and solids streams flowing in the same direction. When the unbound moisture is removed, the solid is in contact with the gas at the highest temperature. Still, the temperature of the solid will not exceed the wet-bulb temperature of the gas, as will be shown in a later section. The wet-bulb temperature will be well below the boiling point of water, even if the dry-bulb temperature of the gas is very high. By the time the unbound moisture is completely removed, the gas temperature also may have fallen to a lower value that can cause no damage to the solid. Even heat sensitive materials can be dried with exposure to gas at high temperature, without the danger of damage to the materials. And also, by proper control of the gas flow rate, the moisture content of the solids leaving the parallel current dryer can be kept under control, if the product must not be completely dried.

Form of Feed and Product

Commercial dryers handle solid feed that comes in variety of forms. The feed may be granular solids, filter cakes, pastes, slabs, films, sheets, slurries, and solutions. No single design can handle such wide variety of materials. The dryers are specifically designed for particular feed and product forms. The physical form of the substance and the diverse methods of handling them have perhaps the greatest influence on the type of dryer used.

Solids Handling in Dryers

The different flow pattern of the solids and the type of contact between solid and gas stream within the dryer are important in understanding the principles involved in adiabatic drying of solids.

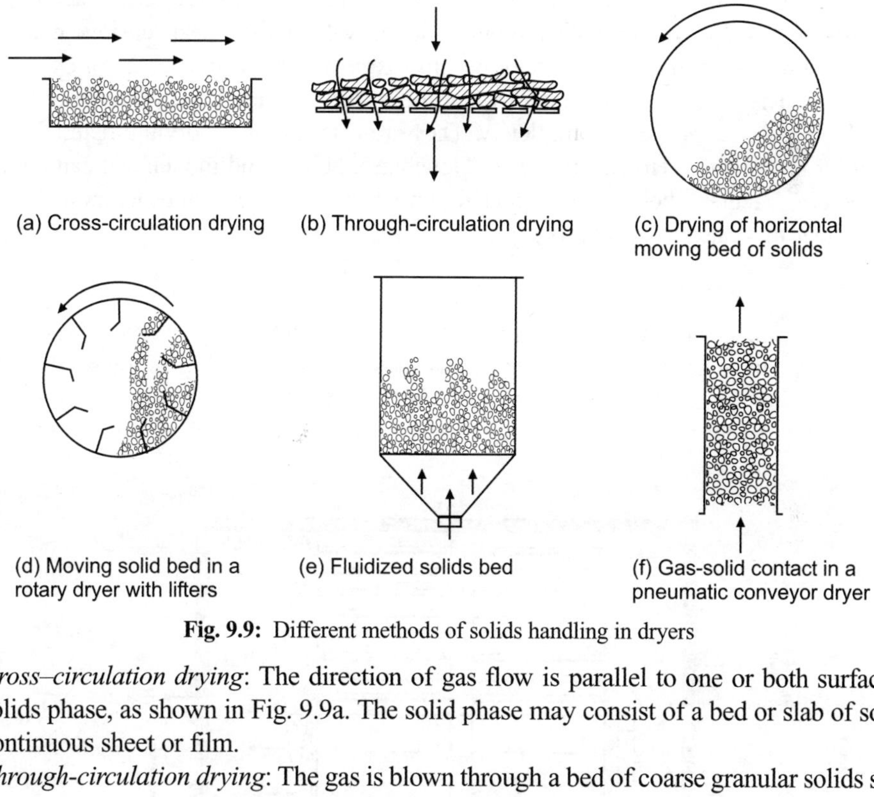

(a) Cross-circulation drying (b) Through-circulation drying (c) Drying of horizontal moving bed of solids

(d) Moving solid bed in a rotary dryer with lifters (e) Fluidized solids bed (f) Gas-solid contact in a pneumatic conveyor dryer

Fig. 9.9: Different methods of solids handling in dryers

1. *Cross–circulation drying*: The direction of gas flow is parallel to one or both surfaces of the solids phase, as shown in Fig. 9.9a. The solid phase may consist of a bed or slab of solids, or a continuous sheet or film.
2. *Through-circulation drying*: The gas is blown through a bed of coarse granular solids supported on a screen. The gas penetrates and flows through the interstices among the solids, circulating more or less freely around the individual particles, as shown in Fig. 9.9b.
3. *Moving bed of solids*: The solid particles are separated from one another and flow as a slightly expanded bed of solids by either mechanical agitation or gravity force. Usually the flow is downward under the force of gravity (Fig. 9.9c). Upward motion by mechanical lifting as in rotary dryer is also possible where the solids are showered downward through a slowly moving gas stream. (Fig. 9.9d).
4. *Fluidized bed*: The gas flows through the interstices among the particles and the resulting drag forces keep the bed of solids in an expanded state as in Fig. 9.9e. Gas forms the continuous phase and there is entrainment of fine particles in the gas.

5. *Pneumatic conveyor drying*: The gas phase and solid phase both flow in the same direction. The solids are completely entrained in the gas stream and are pneumatically conveyed to a mechanical separator.

9.3 BATCH DRYERS

Batch dryers are preferred when the overall process is batch and the throughput is low. They are characterised by long drying times ranging from 6 to 40 h, and they operate under unsteady-state conditions with the air temperature, humidity, material temperature and the moisture content changing continually with time at a given location in the dryer. Batch dryers may be classified into direct heat and indirect heat batch dryers.

In *direct heat batch dryers*, the dryer is charged with the wet material and the drying gas is passed in direct contact with the solids. One of the major difficulties of direct heat batch dryers is the nonuniformity of moisture content of the dried product resulting due to the nonuniform gas movement inside the dryer. Tray dryers and shelf dryers are typical examples of direct heat batch dryers.

The *indirect batch dryers* find application in drying of valuable and heat sensitive materials at low temperatures, or materials that decompose or become explosive in an atmosphere of oxygen prevalent in direct drying. They are especially useful when the material is dried of expensive or toxic organic vapours, which must be recovered completely. They are also useful for drying materials that tend to agglomerate if not agitated. Hot water, steam, Dowtherm, hot oil and molten salt can be used as the heating medium. Vacuum shelf dryers, agitated pan dryers, rotary vacuum dryers and batch freeze dryers fall under this category of dryers.

Tray Dryers

Tray dryers, also called cabinet, compartment or shelf dryers are the simplest and oldest of the drying equipment. They can be used to dry pasty materials such as wet filter cake, lumpy solids and all forms of particulate solids. Drying of lumber, sheet materials, painted metal objects and ceramics are also possible with tray dryers. The direct-heat tray dryers can be of two types: the cross-flow type and the through-circulation type.

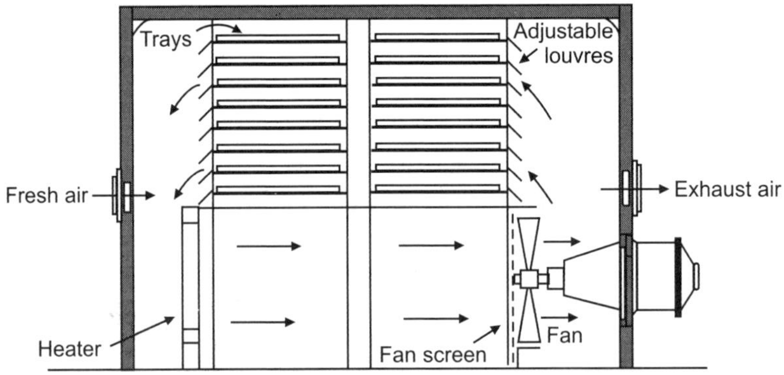

Fig. 9.10: Tray dryer (cross-circulation)

In cross-circulation type tray dryer, the material to be dried is spread on trays with solid bottom. The trays are stacked on shelves enclosed in a cabinet. Heated air is blown across and between the trays as shown in Fig. 9.10. The air is heated by passing through annular finned tube heat exchanger with steam condensing inside the tubes, but, gas or electric heating also can be used. The trays typically of area from 0.5 to 1 m^2 with a solids depth ranging from 10 to 150 mm are common, with ten, twenty or more trays arranged in the cabinet. To ensure high heat transfer coefficients and to eliminate air pockets, the air is blown across the surface of the solids at a high rate usually ranging from 1 to 10 m/s. Air is circulated by propeller or centrifugal fans. Adjustable louvres are used to ensure uniform distribution of air through all trays.

In through-circulation type tray dryer, the wet material is placed on trays with screen bottom. Heated air is directed to flow through the solids as shown in Fig. 9.11. These types of dryers are used for drying loosely packed granular or crystalline solids with appreciable voids, or solids with discrete shapes such as briquettes, noodles, pellets, etc. Pasty materials such as those resulting from precipitation can be transformed to these shapes by subjecting to preforming treatments such as extrusion, granulation and briquetting. Though there is considerable saving in the drying time compared to cross-circulation drying, quantities that can be dried per batch is smaller.

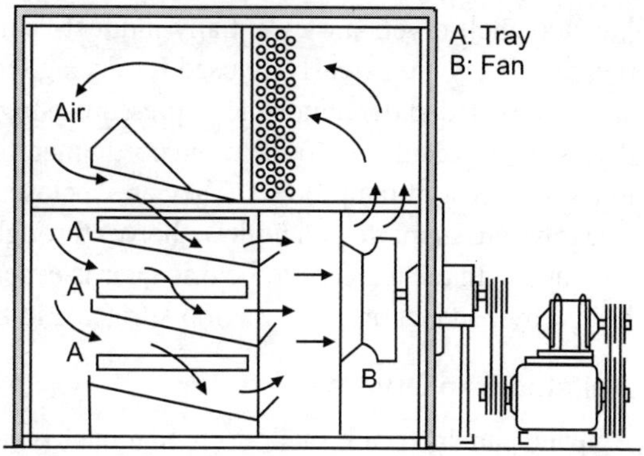

Fig. 9.11: Through-circulation tray dryer

The truck-tray dryers are modification of the stationary tray type explained earlier, with a view to reduce the time between the drying cycles for unloading and loading of solids on the trays. The racks are mounted on truck wheels so that at the beginning of the drying cycle the trucks are pushed in to the drying chamber with fresh batch of solids taken on the trays and at the end of the drying cycle, the trucks can be pulled out for unloading the dried product and recharging with the fresh feed. A double-truck tray dryer is illustrated in Fig. 9.12.

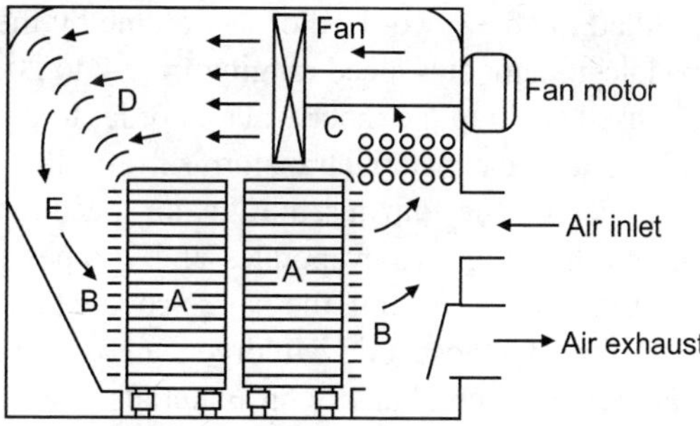

A: Trays, B: Adjustable air-blast nozzles, C: Fin heater, D: Turning vanes, E: Plenum chamber

Fig. 9.12: Double-truck tray dryer

Tray dryers are cheap to construct and easy to maintain and the drying conditions can be altered and controlled at will with the progress of drying. These features make the tray dryers most attractive for laboratory and pilot-plant scale operations. Tray dryers are ideal if the process warrants long drying cycle extending to several hours and even to days and for multi-product plants such as pharmaceuticals and speciality chemicals. For the latter, strict batch identity and thorough cleaning between batches are essential.

Non-uniform air movement is one of the most serious problems in the operation of tray dryers. This results in non uniformity in the moisture content of the finished product. By maintaining the circulation of hot air at a very high rate, this problem can be overcome to a great extent. To conserve heat, air recirculation is generally employed. Generally, 80 to 95% of the circulating air is recycled, the proportion of which is controlled by providing dampers within the dryer. Recirculation results in increase of the moisture content of the air in the dryer considerably above that of the outside fresh air. To achieve rapid drying, the temperature of the drying air should be maintained high. This necessitates thorough insulation of the drying chamber to prevent loss of heat and to avoid condensation of moisture on the walls of the chamber. High operating cost, low heat economy, high labour cost, etc. are the other limitations of the tray dryers. The energy cost is very high as the steam consumption is generally above 2.5 kg, and in some cases as high as 10 kg per kg of water evaporated. Thus tray dryers are most suited for drying valuable products like pharmaceuticals and dyes where the cost of drying contributes only a minor fraction of the total cost of the product.

Vacuum Shelf Dryers

Vacuum shelf dryers are essentially indirect-heat tray dryers that are operated under sub-atmospheric pressure. They are extensively used for drying temperature sensitive or easily oxidisable materials that cannot be dried under atmospheric pressure using direct heat in the presence of air. These are useful for drying pharmaceutical products and costly materials for which of cost of drying is negligible compared to the unit cost of the product. They are enclosed in vacuum tight cabinets made of cast iron or steel. The trays are supported on hollow shelves through which steam or hot water is circulated. After loading the trays with the material to be dried, air is evacuated from the cabinet. The materials on the trays are heated by conduction from the top surface and by radiation from the bottom surface of the shelves.

Agitated Pan Dryers

Agitated pan dryers are shallow circular pans typically 1 to 3 m in diameter and 0.6 m to 0.9 m in height with flat bottom and vertical sides. They are operated as either atmospheric dryers or vacuum dryers. The bottom and walls of the pans are jacketed, through which hot water or steam is circulated. The pan is filled to about two-third of its volume by the wet material. The contents of the pan are agitated by paddle agitator at a speed ranging from 2 to 20 rpm. The agitator carries spring loaded scrapers which are used to scrap the inner walls in order to prevent the cake build up on the walls. With the removal of moisture, the material transforms from a thin liquid into a sticky paste and finally to granular product form. To cope up with this, the agitator speed can be adjusted with the progress of drying. In atmospheric dryers, the moisture evaporates and escapes to the atmosphere through the opening at the top. If operated under vacuum, the pan is covered and the moisture is removed by vacuum pump.

Agitated atmospheric pan dryers are useful for drying pastes or slurries in small batches. They are particularly useful for drying materials that tend to agglomerate on drying. They provide the ideal choice if the process demands evaporation, crystallisation and drying to be accomplished in a single unit. However, these are not recommended for materials that transform into balls, suffer case-hardening or undergo particle size degradation on drying.

Rotating Batch Vacuum Dryers

Slurries or pastes can be dried under vacuum by means of a batch rotary vacuum dryer shown in Fig. 9.13. It consists of a stationary cylindrical shell, mounted horizontally with hot water, steam or Dowtherm

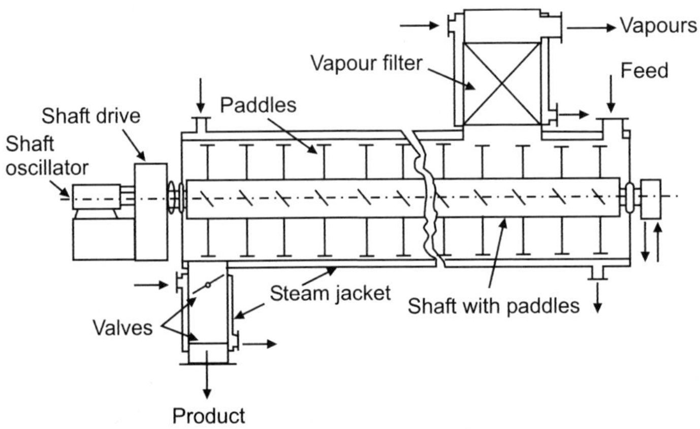

Fig. 9.13: Paddle–agitated cylinder dryer

circulated through the jacket surrounding the shell. The solids being dried is agitated by means of a paddle agitator as shown in Fig. 9.13 or by a double continuous spiral agitator mounted on a revolving central shaft that passes through the ends of the cylindrical shell. The outer blades of the agitator are very close to the walls leaving a clearance of only 3 to 6 mm, which prevent cake build up on the inner walls, by scraping. The feed is admitted through a port at the top and the discharge of the dried product is made through one or more discharge nozzles at the bottom of the cylinder. Vaporised moisture passes through the opening at the top to a condenser and noncondensable gases removed by a vacuum pump. The dryer is very suitable for materials which are heat sensitive, as low temperature can be maintained by high vacuum. Another advantage is the low energy consumption compared to other batch dryers due to high differential temperatures.

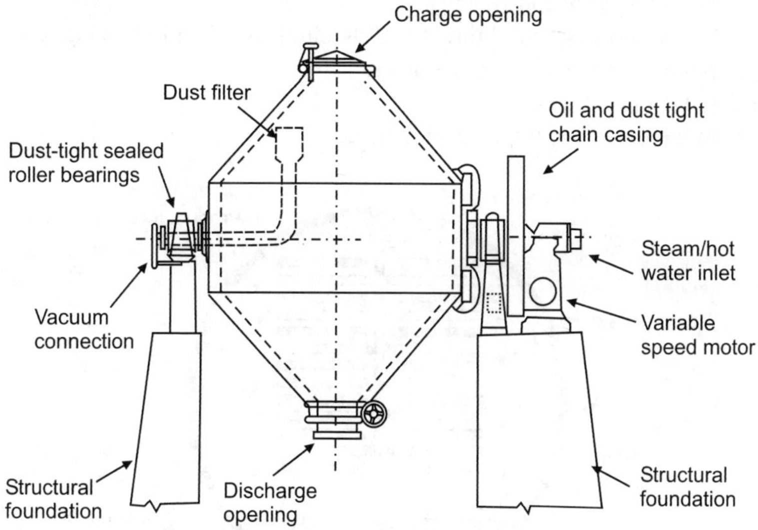

Fig. 9.14: Double-cone vacuum dryer

A rotating double-cone vacuum dryer shown in Fig. 9.14 is another type of a rotating batch dryer commonly used for drying applications related to pharmaceuticals and fine chemicals industries. The dryer consists of a jacketed rotating double cone and is characterised by mild agitation of the wet material by the gentle rotation of the dryer. The double cone dryer gives less attrition and therefore these are recommended for solids that undergo particle size degradation when the mechanically agitated dryers are used. The material is charged through openings at the top. The dryer volume generally ranges from 0.13 to 16 m^3 and the solids occupy up to 70% of the volume. The sloping walls of the dryer facilitate rapid emptying of the dryer even when it is in the stationary position. The major limitation of this type of dryer is the absence of provisions to prevent cake build up and the formation of lumps and balls.

Freeze Dryers

Freeze drying is used for removing moisture from materials that cannot be heated even to moderate temperatures, and therefore cannot be dried by conventional methods. Freeze drying is generally applied for drying pharmaceuticals, vegetables, premium coffee, fruits, meat, fish and beverages. In freeze drying, the moisture present in the solid is first frozen by cooling and then sublimed by controlled heating to prevent the moisture temperature rising to the melting point. Freeze drying differs from the conventional drying in that when the ice is sublimed only water vapour is removed from the interior of the solids causing no displacement to the dissolved substances such as sugar, salt and the acids. In conventional drying on the other hand, when the solution is transported from the interior to the surface of the solid by the capillary action and subsequent vaporisation at the surface, the dissolved solids gets displaced and deposited at the surface.

The advantages of freeze drying over the usual methods of drying are:

1. Freeze drying helps to preserve the original flavour, aroma, colour, texture and shape of the solids, all very essential in the drying of foods and beverages. For example, it is important that the instant coffee retains the original aroma and fruits like strawberries retain their colour, vegetables retain their texture and shape.
2. Results in little shrinkage of the solid which leads to instant and excellent rehydration, important in the drying of food stuffs.
3. Negligible loss of volatile constituents.
4. Low temperature and absence of direct heating eliminate chemical changes and oxidation.
5. No displacement of the soluble components.
6. Case hardening is prevented.
7. Cross contamination of the products eliminated.

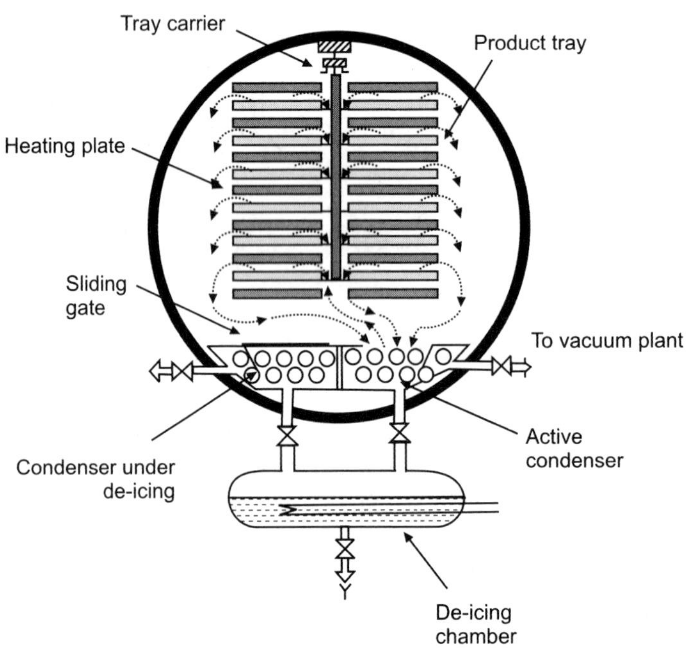

Fig. 9.15: Cross-section of a batch freeze dryer

For small quantities of biological and pharmaceutical materials, freeze drying is generally conducted batch-wise on trays. Cross-section view of a batch freeze dryer is shown in Fig. 9.15. The material to be dried is first frozen by exposure to very cold air and the frozen product is taken on trays enclosed in a vacuum chamber. For easy transport and rapid loading and unloading, the trays are carried on tray-trollies suspended from an overhead rail system as shown. The trays are placed between the heating plates through which the heating medium is circulated and the solids are subjected to controlled heating. Water or thermal oil is used as the heating medium. The solids are heated by radiation from the plates. When loading is completed, the freeze dryer is closed and vacuum applied.

Freeze drying of foods take place at a vacuum of 40 to 130 Pa corresponding to the sublimation temperature from 243 to 256 K. The operating vacuum should be reached quickly to avoid the risk of product melting. For the same reason, the heating plate is cooled to 298 K. When the desired vacuum is attained, the temperature of the heating plate is raised so that the maximum drying temperature is reached quickly. Approximately 75 to 80% of the free water gets sublimed. By the end of the sublimation drying, the product surface temperature must have risen to the maximum allowable drying temperature. Then the temperature of the heating plates is lowered gradually. The subsequent removal of the moisture is due to desorption drying which is continued till the residual moisture in the solid falls to 2 to 3 percent of the original. The sublimed ice is de-sublimed on condenser tubes through which a refrigerant is circulated. The continuous de-icing system is used to reduce the ice layer on the condenser to a maximum of 6 to 8 mm.

9.4 CONTINUOUS DRYERS

Continuous dryers are characterised by the continuous movement of the solids through the dryer. In direct heat continuous dryers the solids move through the dryer in direct contact with the moving gas stream. In adiabatic dryers, there is neither supply of heat within the dryer nor loss of heat to the surroundings. The moisture in the solids absorbs heat for vaporisation and consequently the gas lose sensible heat and cools down. In contrast, if heat is supplied within the dryer the gas temperature can be maintained constant. Direct heat dryers can be operated as countercurrent dryers or parallel current dryers. Tunnel dryers, band dryers, turbo dryers, direct heat rotary dryers, fluidised bed dryers, spouted bed dryers, spray dryers and flash dryers are examples of direct heat continuous dryers. Indirect dryers are preferred when the solids should be dried at high temperature but out of contact with hot flue gas. Indirect heat rotary dryer, thin film dryer, drum dryer, etc. are examples of indirect heat continuous dryers.

Tunnel Dryers

The simplest and perhaps the oldest equipment for continuous operation is the tunnel dryer, which is an adaptation of the batch truck dryer for continuous operation. Tunnel dryers are direct dryers consisting of a long drying chamber through which wheeled trucks loaded with solids to be dried move progressively from one end to the other. Trucks move on tracks or monorails and usually conveyed mechanically using chain drives connecting to the bottom of each truck. Hot air flows either countercurrent (Fig. 9.16a) or cocurrent (Fig. 9.16b) to the wet solids. For low temperature operation, air is heated by the steam condensing inside the finned tube heat exchanger. For high temperature drying and where purity and cleanliness of the product is not a serious concern, heating of the material can be done by flue gases

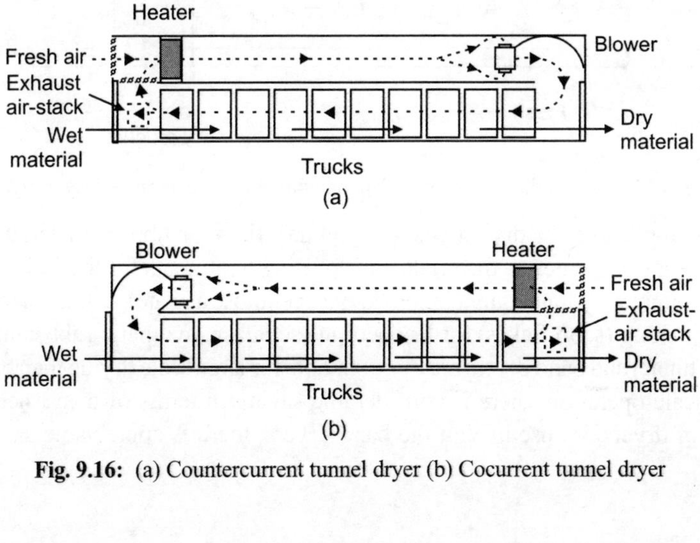

Fig. 9.16: (a) Countercurrent tunnel dryer (b) Cocurrent tunnel dryer

produced by the combustion of fuels. When a truck containing the dried product is removed at the discharge end of the tunnel, one filled with wet solid is introduced at the feed end. The loading and unloading of the trucks being carried out manually outside the tunnel, tunnel dryers are in fact, semi-continuous considering the overall operation. A typical tunnel dryer is over 30-m long and can accommodate around 15 trucks.

Tunnel dryers are used for drying any type of particulate materials, crystals, filter cake, pastes, and the like. They are used for drying large solid objects like refractory shapes and for drying and baking of enamelled pieces. Continuous sheets of wet cloth, yarn and hides may be dried in modified versions of the tunnel dryer. Skeins of wet yarn may be suspended from poles which move through the dryer, hides may be dried by hanging from the frames attached to a conveyor chain moving through the tunnel.

Band or Conveyor Dryers

Band dryers are also known as conveyor dryers, screen dryers or belt dryers. These are continuous direct-heat through-circulation dryers, in which the hot gas is passed through a permeable bed of wet solids. The solid is spread upon a moving endless conveyor which passes through the long drying chamber. The conveyors are, typically 2 m wide and 4 to 50 m long and moves with a very low velocity, typically, around 18 m/h. The conveyor is made up of perforated steel plates or woven wire screens. The solid forms a permeable bed on the conveyor to a depth of typically, from 25 to 150 mm. Air, heated by means of heating coils is directed to flow through the bed by means of fans. For moderate drying temperatures, the gas is heated by condensing steam in finned heating coils. In any case, temperature above 600 K is not advised as it affects adversely, the lubrication of the conveyor, chain and the roller drives. In large dryers, the drying chambers are made into separate sections, each provided with its own heater and fan. The sectional view and longitudinal view of a multi-section band dryer are shown in Figs. 9.17 and 9.18. The air flow is directed upward through the bed at the inlet sections and downward through the bed at the exit, thereby providing more uniform moisture concentration in the solid. Gas rate used is, typically between 0.5 to 1.5 m/s.

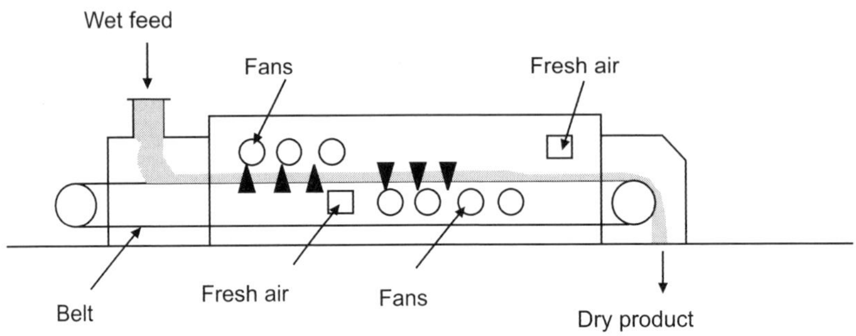

Fig. 9.17: Longitudinal view of continuous through-circulation conveyor dryer

Band dryers are applicable for drying coarse, granular, flaky or fibrous materials directly without any pre-treatment. For other types of materials, pre-forming treatments for the feed such as scoring on filter (for starch, aluminium hydrate), granulation (kaolin, cellulose acetate), extrusion (calcium carbonate, titanium dioxide), flaking (soap flakes), or briquetting (soda ash, synthetic rubber) may be necessary.

The continuous tunnel and band dryers are more suitable for large scale drying than batch compartment dryers. For large scale operation, there is considerable saving in terms of investment and installation when these types of dryers are used. With the band dryers, there is appreciable saving in the cost of labour as well.

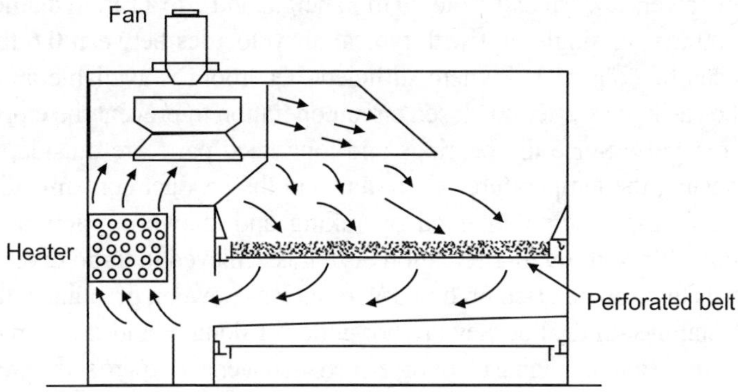

Fig. 9.18: Sectional view of perforated belt or band conveyor dryer

Turbo Dryer (Tower Dryer)

Turbo-type tower dryers are direct-heat continuous dryers which can handle from thick slurries to fine powders. The dryer consists of annular shelves mounted one above the other on a central shaft rotating at a speed of around 1 rpm. The wet material is fed through the roof, and falls on the top shelf as it rotates. The material is spread on the trays by means of stationary levellers. On completion of one revolution of the tray, the solid is scrapped off through radial slots on the trays by means of suitably placed arms or wipers, on to the tray below. The dried material is removed from the last shelf at the bottom. Hot air or gas can be admitted either countercurrent or cocurrent to the flow of solids. Figure 9.19 shows a turbo dryer in which the air flows upwards countercurrent to the flow of solids and by means of the turbine type fans attached to the central shaft, the air is allowed to flow across the material on the shelf. The gas is heated by a row of steam pipes. One or two trays at the bottom are used to cool

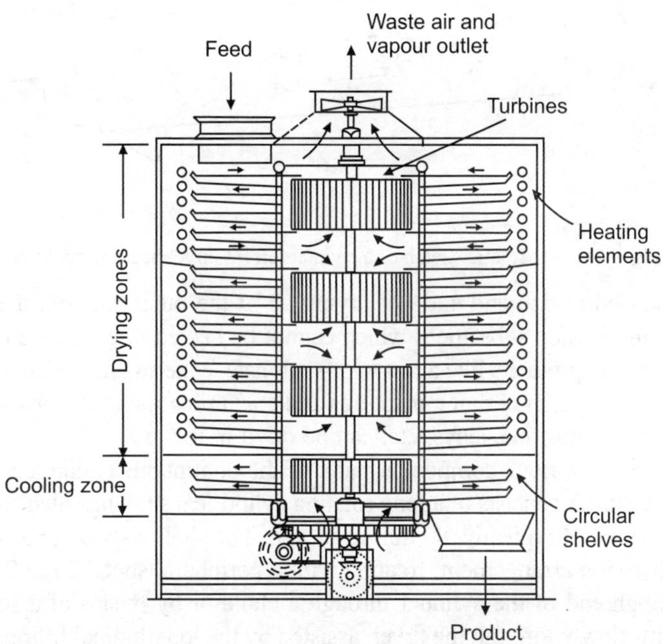

Fig. 9.19: Turbo-tray dryer

the dried product. The towers are typically 2 to 20 m in height and 2 to 11 m in diameter. The shelf area ranges from 0.7 to 200 m^2 per single unit with typical air velocities between 0.6 to 2.5 m/s.

The turbo dryers can be constructed where sufficient headroom is available and require less floor area. The dryer can be easily converted to closed circuit operation to prevent the emission of fumes and to facilitate solvent recovery. Since the bearings and lubricated parts are outside, and the drives are located under the housing, the temperature constraints and the product contamination are eliminated. The cross-circulation drying is supplemented by raking and showering action, and therefore, an evaporation rate comparable to through-circulation drying is achieved in turbo dryers. The turbo dryers can handle all materials that can be dried on tray dryers such as powder, granular solids, heavy sludges, pastes, beads and crystalline solids, but they are not good for doughy and tachy materials and fibrous materials which tend to mat. Before sending to the dryer, certain materials require some preliminary treatments such as granulation (for filter-press cake), scoring (of cakes leaving the rotary filter fed with thixotropic materials), extrusion (of pastes), etc. Calcium hypochlorite, urea, calcium chloride, sodium chloride, antibiotics, antioxidants and water-soluble polymers are some materials that are dried in turbo dryers.

Direct Heat Rotary Dryers

Rotary dryers are ideal for large scale continuous drying of materials of relatively small particle sizes if breakage of particles is not a concern. They handle granular, flaked or crystalline free flowing materials. Rotary dryers fall into three broad classes based on the mode of transfer of heat: Direct heat, indirect heat and direct-indirect heat rotary dryers. A direct-heat countercurrent hot air rotary dryer is schematically represented as shown in Fig. 9.20.

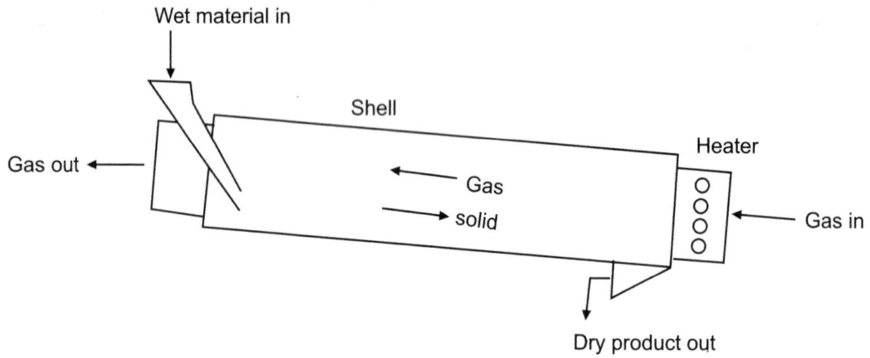

Fig. 9.20: Schematic sketch of a countercurrent direct heat rotary dryer

Wet solid is introduced at one end and hot drying air at the other end and they flow in opposite directions. Hot air is used when substances which cannot be heated to excessive temperature such as certain crystalline chemical products like ammonium sulphate, cane sugar, etc. are dried. Instead of air, flue gas can be used for drying, if high temperatures and contamination of the product are allowed. For example, minerals, sand, limestone, clays, etc. can be dried in this way.

The arrangement of the various component parts of the continuous countercurrent rotary dryer is shown in Fig. 9.21. The dryer consists of a long rotating cylindrical shell installed slightly inclined with the horizontal with a slope of typically, 80 mm/m or less. The shell rests on two sets of rollers and is driven by a gear and pinion arrangement, rotating with a peripheral speed 20 to 25 m/min. Wet solid is introduced at the high end of the cylinder through a chute or by means of a screw conveyor. The solids move by gravity slowly through the dryer, assisted by the longitudinal lifting flights mounted on the walls of the shell that extend over the full length of the dryer. The flights carry the solids and shower it down in a moving curtain through the dryer, thus providing better exposure of the solids to the gas

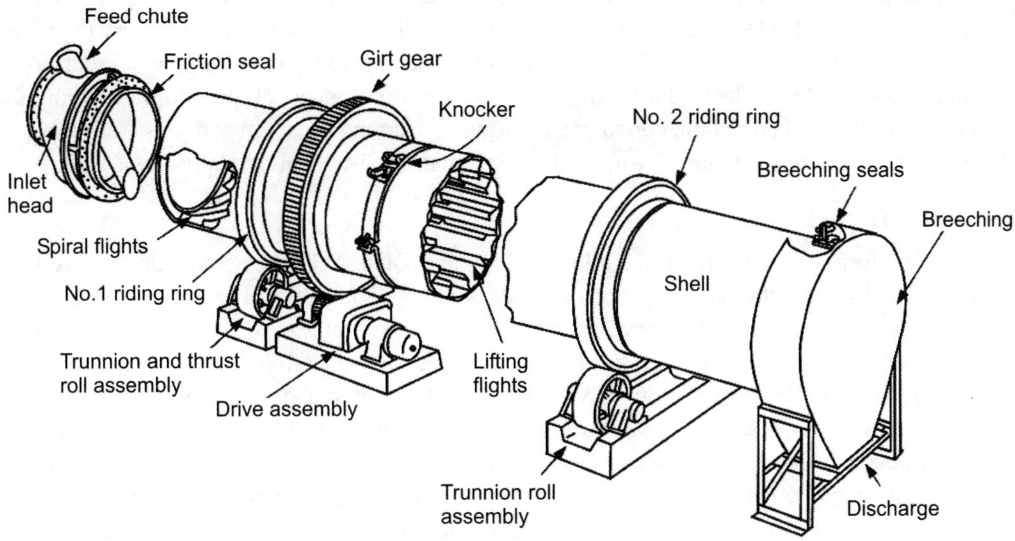

Fig. 9.21: Schematic sketch of component arrangement of a countercurrent direct heat rotary dryer

flowing countercurrently. For free flowing materials, radial flights with a 90° or 45° lip as the one shown in Fig. 9.22 is employed. For sticky materials, flat radial flights can be employed.

90° lip flights   Flat radial flights   45° lip flights

Fig. 9.22: Direct-heat rotary dryer flight arrangements

At the feed end of the dryer, short spiral flights are provided over a small length ahead of the principal longitudinal flights, to facilitate the initial forward motion. Knockers may be provided on the outside of the shell to prevent the solids sticking on to the inner walls of the shell. The dried product is discharged from the lower end through a screw conveyor. Rotary dryers of diameters typically from 0.5 to 6 m and length from 1.5 m to 50 m are common. About 10 to 20% of the shell volume is occupied by the solids. Residence time in the dryer varies from 300 s to 2 h.

Preheating of the drying air is achieved by passing air over extended surface pipes that is heated by the steam condensing on the inside. The fan at the stack helps to draw the air through the dryer and keeps the system under slight vacuum which prevents the escape of dust-laden gases to the atmosphere. Use of fan for gas circulation permits better control of the gas flow than that is possible with natural draft. The gas velocity is decided by the dusting characteristics of the feed and lie in the range 0.15 to 3 m/s. A dust collector of the cyclone, filter or washing type may be provided between the fan and the gas exit to recover the fine particles from the discharge gas. By using a blower for gas circulation, it is possible to maintain the pressure in the dryer near atmospheric, thus preventing the entry of cool air into the dryer from the outside. If flue gas is used as the drying medium in the place of air, a furnace burning coal, gas or oil will replace the air heater. In this case, if the gas exit temperature is high, it can be discharged through a stack and the gas circulation through the dryer can be achieved by the natural draft. The inlet gas temperatures usually range from 390 to 450 K for steam-heated air and 820 to 1075 K for flue gas from a furnace.

Materials that can be dried using flue gas without the fear of contamination, but are heat sensitive and therefore cannot be heated to high temperatures can be dried using parallel flow rotary dryers. For solids such as gypsum, iron pyrites, and organic materials such as peat and alfalfa, cocurrent flow of the gas and solids can be employed for drying. Since the gas flow can assist forward movement of the solids, the cocurrent dryer can be built horizontal. Rapid surface drying in cocurrent dryers is advantageous for materials that stick to the walls and for materials containing water of crystallisation.

Roto-Louvre Dryers

This is a special form of rotary dryer that allows through circulation drying without any showering of the solid, thus eliminating the dusting problems associated with conventional rotary dryers. The Roto-Louvre dryer is a through-circulation dryer suitable for coarse granular solids which do not offer high resistance to gas flow and solids that do not require intimate gas contacting for drying. It consists of a rotating double walled cylinder fitted with louvres which hold the solids and permit the passage of the gas.

As shown in Fig. 9.23, hot air is blown through the louvres up through the bed of solids. The solids bed move continuously through the cylinder. Constant turn over of the bed ensures uniform gas contacting for heat and mass transfer. The annular gas passage behind louvres is partitioned to permit the contacting air to enter only to those louvres which are underneath the bed of solids. Since there is significant gas pressure drop, to maintain an internal pressure close to the atmospheric, gas blowers are often provided both at the inlet and at the outlet. As the rates of heat and mass transfer are high, the dryer is smaller in size than the direct dryers used for similar drying duties. The diameter of the dryer lies in the range 0.8 to 3.6 m and length between 2.5 and 11 m. Hot gas temperature between 400 and 865 K may be used.

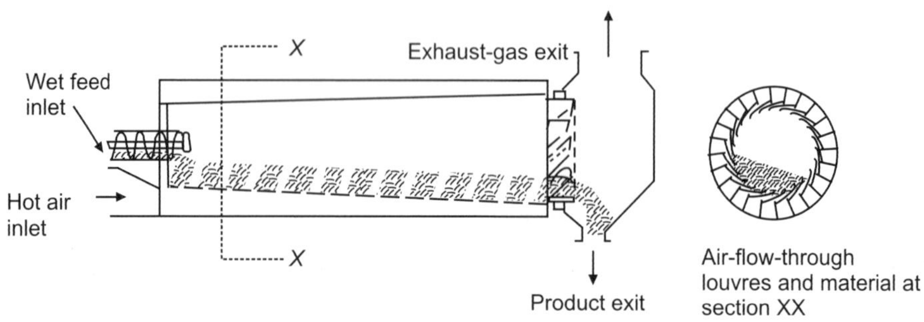

Fig. 9.23: Roto-louvre through circulation dryer

Indirect Heat Rotary Dryers

In indirect heat rotary dryers the drying air or gas is not in direct contact with the solids. Rather, heat is supplied by conduction through the walls of the tubes, shell or jacket through which the heating agent such as condensing steam is passed. Indirect-heat steam tube rotary dryers are applicable to dry high-moisture organic compounds and materials that undergo undesirable breakage, dust formation, contamination or product damage on drying in contact with the air or flue gas. For example, white pigments are to be dried at high temperature, out of contact with the flue gas. Inorganic crystals, silica, mica, flotation concentrates, pigments, filter cakes, precipitated calcium carbonate, citrus pulp, cellulose acetate and starch are dried in indirect rotary dryers. Also used for drying materials that should not be heated to excessively high temperature, such as cattle feed, feathers, brewer's grains, etc.

Indirect steam-tube rotary dryer consists of a rotating cylindrical shell that houses one or two rows of longitudinal concentric tubes that run along the full length of the shell as shown in Fig. 9.24. The tubes may be finned or unfinned and they rotate with the cylinder. Wet solid is admitted at one end through a chute or solids conveyor. The heating tubes give a gentle solids-lifting action. Lifting flights

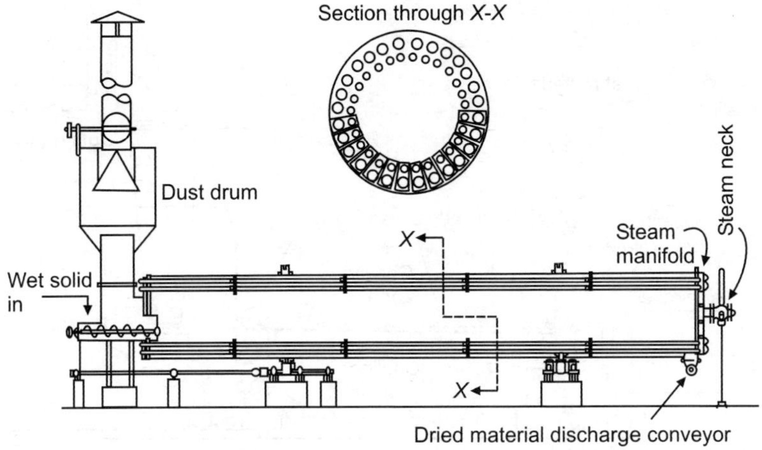

Fig. 9.24: Indirect-heat steam-tube rotary dryer

may or may not be provided. Steam is admitted at one end of the dryer through the central revolving steam manifold. The condensate is removed continuously by gravity through the steam joint, to a condensate receiver by means of the lifters in the condensate side of the manifold. The evaporation of the moisture occurs at its boiling point and the evaporated moisture is swept by a gentle current of air. The moisture-laden gas is removed at the feed end, usually to the atmosphere, through a natural draft stack and settling chamber or wet scrubber. The dried products leave through suitable ports at the end of the dryer. Steam tube rotary dryer size ranges typically, from 1 m to 2.5 m in diameter and 5 m to 25 m in length. The tubes of diameter from 60 mm to 115 mm are used for carrying condensing steam.

Direct-Indirect Heat Rotary Dryers

In direct-indirect rotary dryers hot flue gas enters the inner cylinder at 925 to 1250 K and transfers heat through the walls to the solids admitted in the annular space. In the central tube, the gas flow is cocurrent to the solids moving in the annular space. The gas cools to around 475 to 750 K by the time it leaves the central tube and enters the annular space. In the annular space, the gas flows countercurrently and in direct contact with the solids. Lifting flights are provided on the outer walls of the inner pipe and inner walls of the outer pipe to assist the solids movement. The gas is finally discharged at 330 to 340 K. These dryers are economical especially when the fuel costs are high. In this type of dryers, a large percentage of the moisture is removed at high temperatures. The heat recovery in direct-indirect heat rotary dryers is better than that in single flow direct dryers.

Fluidized Bed Dryers

Fluidized bed dryers (Fig. 9.25) are employed by a variety of industries which include heavy mining, fine chemicals and food and pharmaceuticals, for drying the free flowing moist materials with a narrow particle size distribution. The materials such as limestone, coal, sand, granular fertilizers, salt and polyvinyl chloride are examples of the products that are dried in fluidized bed dryers. The dryer consists of a cylindrical or rectangular fluidizing chamber. Hot gas is admitted through a perforated distributor plate at the bottom of the chamber and wet solid is admitted at the top from a bin or by a screw conveyor. Chamber with large cross sectional area and small bed height is preferred to ensure the required residence time for the solids in the bed. This will help in keeping to the minimum, the cost on the blower used for providing the gas pressure that is required for maintaining the pressure drops across the distributor plate and the bed. The gas velocity which is about twice the minimum fluidization

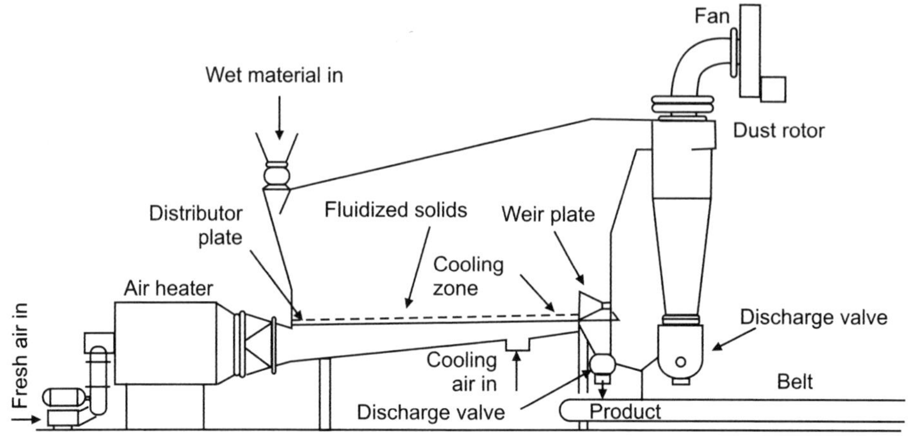

Fig. 9.25: Fluidized-bed dryer

velocity may be used to keep the bed in stable bubbling fluidization condition. Hot gases passed through the bed transfer to the solid, the heat for evaporation of the moisture and also keep the solid in circulation, thereby ensuring intense contact between the gas and the solid. The dust-laden gases that leave at the top of the chamber pass to a cyclone for dust removal. Sufficient height is provided for the chamber above the bed to allow the larger dust particles to settle back to the bed. The fluidized bed heights are typically in the range 0.15 m to 0.6 m, whereas the height of the chamber above the bed may be around 1.8 m. The solids are removed from the chamber through the overflow duct near the bottom and carried away by means of a conveyor.

Fluidized bed dryers are suitable for any gas temperature and can be used for batch as well as continuous operation. Mixing and heat transfer occurs very rapidly. Solid residence times in the dryer are controllable from seconds to hours. Typical residence time for the solid in the fluidized bed dryers ranges from 30 to 120 s when only the surface moisture is removed and from 0.25 to 0.5 h when the drying involves internal diffusion of the moisture. Short-circuiting of the solids from the feed distributor to the discharge duct may lead to undesired residence time distribution and consequent non-uniformity in the moisture contents of the dried product. Multiple fluidized beds is recommended if uniformity in moisture content is essential.

Spouted-Bed Dryers

Coarse solids that are too large for drying in fluidized-bed dryers, but are of uniform size and low density, are dried in spouted bed dryers. They find particular use in drying grains when the required drying time is more than just a few minutes. The solids are introduced at the bottom of the drying chamber from where they are carried by the drying gas up through the central draft tube as shown in Fig. 9.26.

The gas loses its velocity when it exits from the draft tube into an expanding conical section. The solids fall down through the annular downcomer where they come in contact with a countercurrent stream of hot gas. A portion of the solids is

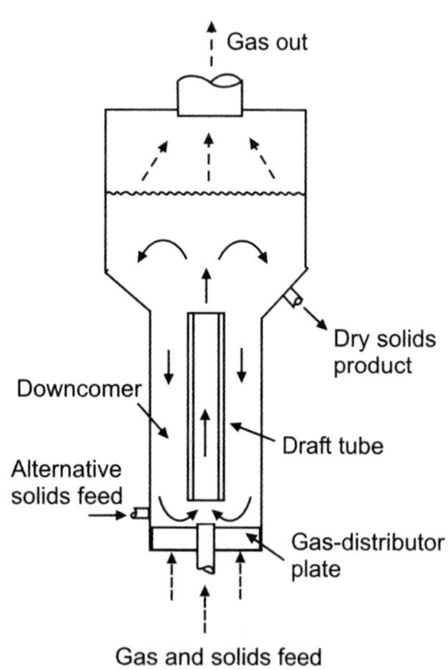

Fig. 9.26: Spouted-bed dryer

discharged from the conical section at the top. The gas leaves at the top after passing through a calming section that allows heavy particles to settle back.

Pneumatic-Conveyor Dryers

Drying is to be carried out with a very short time of contact between the solid and the gas when heat-sensitive, flammable or explosive materials are dried. Pneumatic conveyor dryers, also known as the flash dryer are suitable for such applications. These dryers are ideally suited when a paste or wet cake is to be partially dried with a very short contact time. For the pneumatic conveyor dryer to be most effective, the materials should be granular and free flowing when dispersed in the gas stream.

Figure 9.27 shows a pneumatic conveyor dryer which combines drying and pulverization in a single unit. Such a system produces a fine dry product in one step, eliminating the additional handling required when drying and grinding are accomplished separately. The wet solid feed is introduced into a paddle-mixer cum conveyor where it is mixed with a recycled fraction of the dried product for conditioning the feed. The de-agglomeration or disintegration of the wet solids is done in the hammer

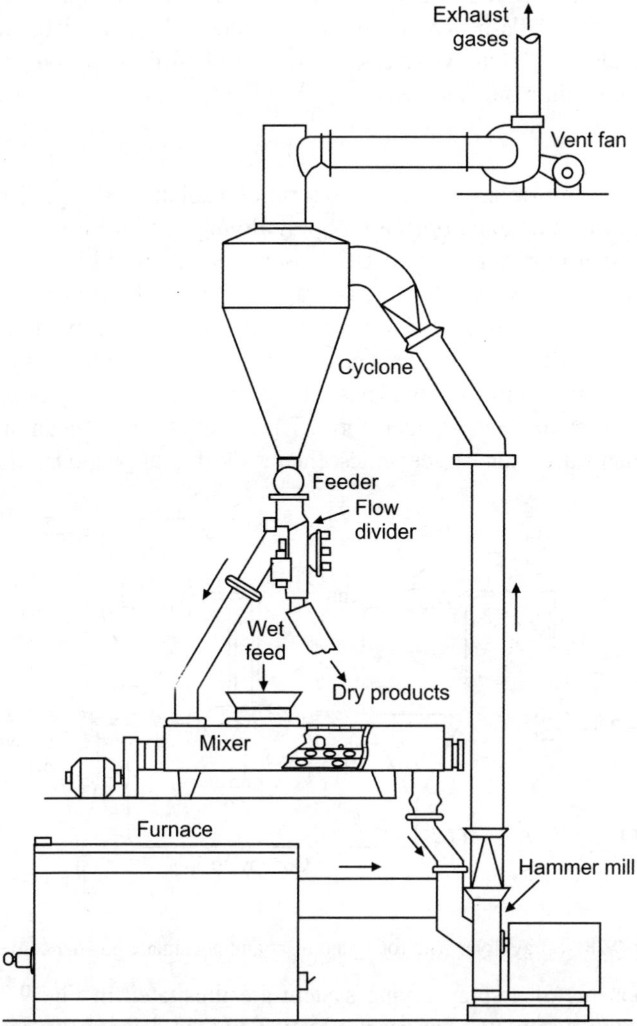

Fig. 9.27: Pneumatic-conveyor dryer (Flash dryer) with integral mill

mill which receives the solids leaving the conveyor- mixer. The air is pulled by an exhaust fan through an oil- or gas-fired furnace into the mill. Any agglomerates present are broken up further into discrete particles to ensure uniform drying, before being carried away by the gas at a high velocity through the drying duct. Rapid removal of the surface moisture takes place when the gas-solid mixture is conveyed through the drying column, before entering the cyclone separator. Gases from the cyclone are vented to atmosphere through a dust collector or wet scrubber. The separated dry products are discharged through a rotary air lock into a dry divider, which is set to recycle a fraction of the solids into the wet solids conveyor-mixer, and the other fraction withdrawn as the dried product.

Gas velocity typically between 10 to 30 m/s, usually about 3 m/s above the terminal settling velocity of the largest particle to be conveyed out of the disintegrator ensure smooth transport of the solids through the drying chamber. The rate of heat transfer is very high and drying is so rapid that the residence time of, typically, 5 s that is common in these dryers, is sufficient to evaporate almost all of the surface moisture present in the solid. Even when the temperature as high as 925 K is allowed for the inlet gas, temperature of the solid will not rise by more than 50 K during drying, which makes the dryer acceptable for drying even heat sensitive materials. The small floor area required for the dryer is another distinct advantage of flash dryers. The major industrial applications include drying of food products, chalk, coal, clays, gypsum, yeast cake, whey, starch, fruit pulp, copper sulphate, adipic acid, polystyrene beads, ammonium sulphate, etc.

Spray Dryer

Spray drying is a process for drying a feed in the form of a solution, slurry, dispersion or paste into a dried particulate product, in a single operation. Spray drying combines some or all of the following sequence of operations: evaporation, crystallisation (or precipitation), filtration (or centrifugation), size reduction, classification, and drying. Moisture is evaporated from the feed by spraying it into a vessel through which a current of hot gas is passed. Large surface area created by the atomization of the feed and intense contact of the particles with the hot gas result in rapid and uniform drying due to vaporisation of the surface moisture from individual particles. Spray drying finds a variety of applications where a high-grade product is to be made in granular form. The materials dried on an industrial scale in spray dryers include agrochemicals, catalysts, ceramics, chemicals, dyestuffs, foodstuffs and pharmaceuticals.

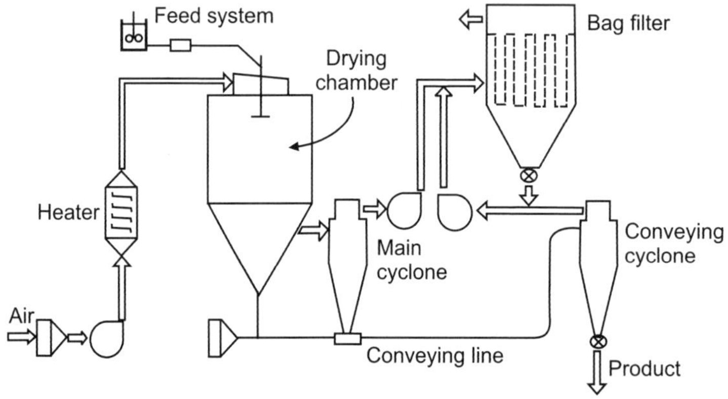

Fig. 9.28: Spray dryer with rotary atomiser and pneumatic powder conveying

The major components of a spray drying system are illustrated in Fig. 9.28. They are the feed delivery and atomisation system, the hot gas production and the delivery system, the drying chamber, the solid gas separation system and the product discharge system. The drying chamber usually has a conical shaped bottom section and the diameter at the top nearly equals the height of the chamber. Air is fed through a filter and heater and enters at the top of the drying chamber, flows down cocurrently with the droplets getting dried. Secondary air is also introduced in certain dryers. The gas and liquid are brought into contact, and the efficiency of mixing depends on the flow patterns induced in the chamber. Flow of gas and droplets can also be countercurrent to each other. Although the product gets heated to high temperatures, the countercurrent operation generally gives high thermal efficiencies. Though the thermal efficiency of cocurrent dryer is low, it has the advantage that the material can be dried without the danger of getting overheated.

The feed liquid is sprayed into the hot gas stream as fine droplets by means of feed atomisers. The atomiser is a device that causes the liquid to disintegrate into drops of size lying within a specified range, and that controls their spatial distribution. Since the drying time and size of the particles are directly related to the droplet size, the initial formation of the spray is of great importance in spray drying operation. Atomizers belong to three classes. They are the two-fluid pressure nozzles or pneumatic atomisers, single fluid pressure nozzles and centrifugal disk atomisers.

On completion of drying, the particles are partially separated from the gas stream in the chamber. Larger particles fall to the bottom of the chamber, and the smaller ones swept with the gas stream into a cyclone separator. In the scheme shown in Fig. 9.28, the powder leaving the chamber bottom and the fines removed from the cyclone are conveyed pneumatically to a conveyor cyclone from where the product is discharged. Very fine particles are retained by passing the gas through a bag filter or wet scrubber to meet the pollution standards.

Feed atomisers

The choice of atomiser for any specific application depends largely upon the characteristics of the liquid feed and the characteristics of the drying chamber. Three basic forms of energy are utilised in commonly employed atomisers. Pressure is used in pressure atomisers (single-fluid pressure nozzles), gaseous energy is used in twin-fluid pneumatic atomisers, and centrifugal kinetic energy is used in rotary disk atomisers.

Pneumatic atomisers: Atomisation is achieved by the impingement of air, steam or another gas at a high pressure on the feed which is at a relatively low pressure. At high pressures, the liquid issues from the nozzle as ligaments, which are then torn into droplets by the gas stream. These are used in low production-rate drying, in which case, the chamber volume is small necessitating small droplets. Since, at high atomizing gas-to-feed ratios, the smallest particles can be produced, pneumatic atomisers find applications in dispersion of stringy and fibrous materials, thick pastes, certain filter cakes, and polymer solutions.

Single-fluid pressure nozzles: In pressure atomisers, the liquid is forced under pressure through an orifice. They operate at relatively high solution pressures and are most suited for low-viscosity feed. They deliver droplets of largest sizes, but the size range is narrow. These nozzles suffer orifice-wear and plugging of the orifice with certain feeds. Because of their simplicity, pressure nozzles are employed to atomise liquids with a kinematic viscosity up to 0.01 m^2/s. Injection pressures up to 50 MN/m^2 may be required to produce the desired particle size under these conditions. Frequent replacement of the nozzles necessitated by the severe erosion of the orifice is the main limitation of this type of atomisers, when used for handling slurries.

Centrifugal disk atomiser: Centrifugal disks or spray wheels, can handle liquids having wide range of properties and widely varying flow rates. The liquid is fed on to a rotating disk and spread by centrifugal force. The liquid extends from the periphery of the disk in the form of thin sheets, which break down into small distinct droplets at certain distance away from the periphery either by the aerodynamic action or by the help of an additional gas blast. The rotating component of this type of atomisers may be simple flat disks, bowls, vaned discs and slotted wheels as shown in Fig 9.29. Rotating disk atomisers are extremely versatile as the accelerating force can be controlled independently. Since they have the largest diameter spray pattern, they require the largest diameter drying chamber to prevent the particles from striking the chamber walls.

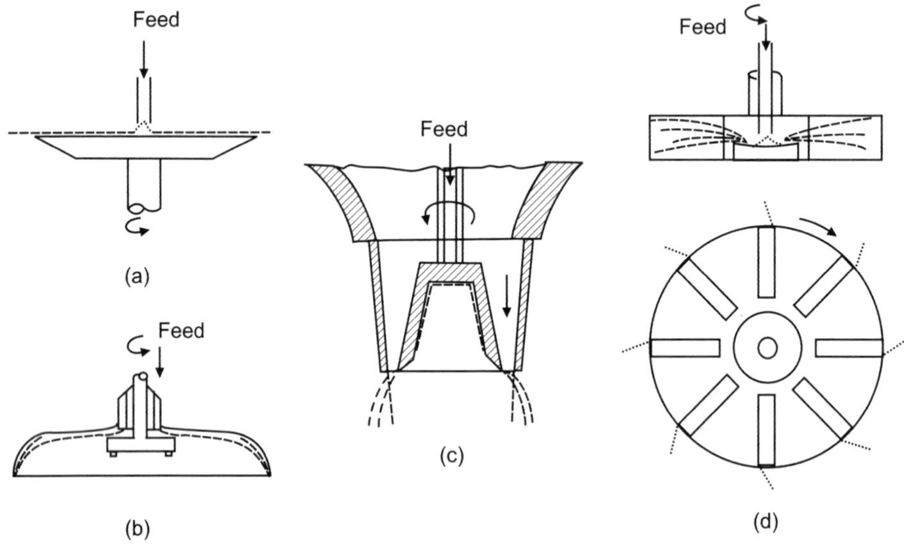

Fig. 9.29: Centrifugal atomisers

Spray dryers of diameters 2.5-10 m are typical; an evaporation rate as high as 7000 kg/h is achieved in large units. The surface moisture is almost completely removed from the feed solids in a residence time of less than 5 s. If internal moisture also is to be removed, the residence time required may go up to 30 s. Spray drying is a relatively expensive process, especially when direct heating is used. However, for handling large volumes of the solutions, spray dryers are found to be economical. The economy of the spray dryer increase with capacity and at values greater than 0.6 kg/s, the unit running cost is largely independent of the capacity.

Since the droplet temperature remain below the wet-bulb temperature of the drying gas in cocurrent operation, spray drying offers a convenient means of drying substances that are heat sensitive. The ability to form rounded porous particles of uniform size because of the fine state of subdivision of the liquid is one of the major attractions of the spray dryers. The enormous range of the spray-dried products include coffee, milk, detergents, dyestuffs, pesticides, polymers, tile slips, blood plasma, enzymes, penicillin, starch, metal concentrates, kaolin, alumina, egg, yeast, zinc sulphate, lignin, aluminium hydroxide, urea resin, silica gel, mag-nesium chloride, manganese sulphate, tannin extract, colour pigments, tea, tomato juice and ceramics.

Thin Film Dryers

Thin film dryers can handle the solutions and slurries and can produce dry, granular, free-flowing solid products. They are particularly useful for removing and recovering the solvents from the solids. Figure 9.30 shows a thin-film dryer. The dryer consists of two sections. The vertical agitated evaporator dryer is identical in construction to an agitated film evaporator. Majority of the moisture is evaporated in this section and partially dried solids leaving this section are admitted into the horizontal section in which the remaining moisture is evaporated. Since no gas is supplied to effect drying, no solids are lost during the operation. The thermal efficiency of the thin-film dryer is also very high.

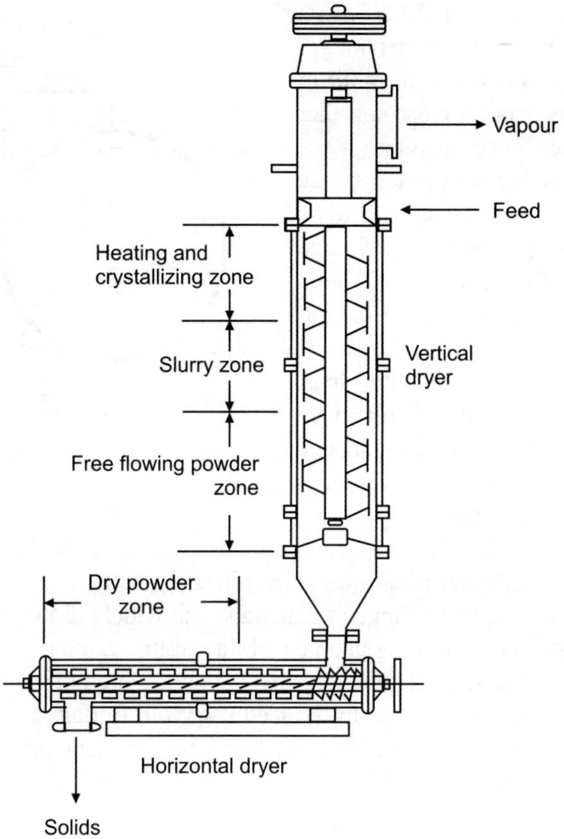

Fig. 9.30: Thin-film dryer

Drum Dryers

Drum dryers find application in drying solutions, slurries, sludges and pastes. A single-drum dryer with dip feed is shown in Fig. 9.31. It consists of a trough in which the material to be dried is taken and a metallic cylindrical drum which is revolving slowly, is allowed to dip into the liquid continuously. The drum is heated by the steam condensing inside at pressures as high as 12 bar. A thin film of the material to be dried is formed on the surface of the drum, the thickness of which can be controlled by means of a spreader knife. Due to the heat transfer from the cylinder walls, moisture is evaporated from the film and the material gets dried. The dried material is then continuously scrapped off the drum surface by doctor blades that run the length of the drum. The product falls off the surface as powder, chips or flakes onto conveyors. The evaporated moisture exits through a vapour-hood mounted above the drum. For recovery of the solvents, a closed enclosure may be constructed for the dryer.

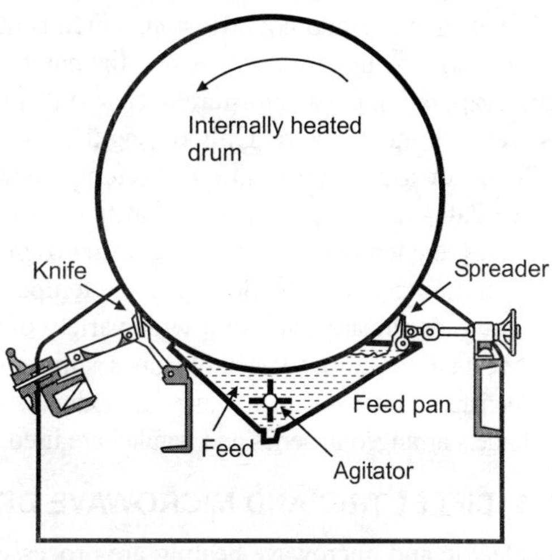

Fig. 9.31: Dip-feed single drum dryer

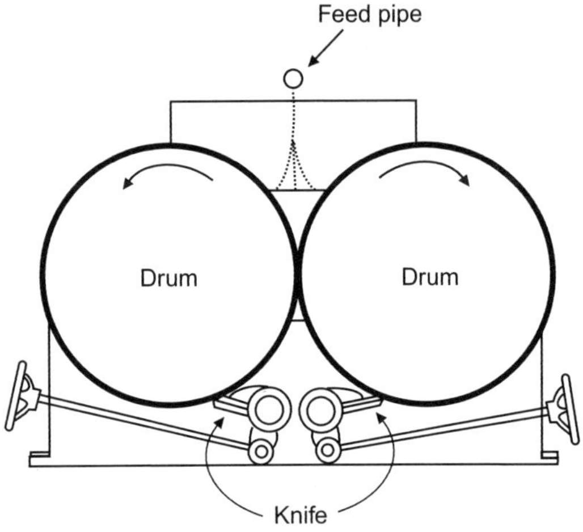

Fig. 9.32: Twin-drum dryer

The twin-drum dryer shown in Fig. 9.32 consists of two drums, identical in size (diameter ranges typically from 0.3 m to 2 m and length from 0.5 m to 4 m), heated internally, placed close together and revolving in opposite directions. One of the drums is movable so that the smallest distance between the drums - the *nip* - can be adjusted. The feed to be dried enters the top from a perforated pipe that runs the length of the cylinder or from a pipe that swings like a pendulum to cover the full length of the cylinder. Since the drums rotate away from each other at the top, the twin-drum dryer shown in the figure permits drying of relatively thick coating on the surface. Materials like inorganic crystals that might score or cause other damages can be handled in such dryers. In another variation, the cylinders revolve in opposite directions, but towards each other at the top. In these units, the thickness of the film on the cylinder is controlled by adjusting the nip.

Since the drum dryers can handle a range of materials with widely different characteristics, diverse methods of feeding are adopted depending upon the ability of the solution to form a stable coating on the heated drum surface. The slurries of solid dispersed in liquids are fed to the bottom of the pan and the excess non-adherent materials recycled to the feed reservoir. Certain materials such as vegetable glues can be pumped to the bottom of the drum surface. Dip-feed arrangement is recommended for heavy sludges. For materials that are coated only with difficulty, the feed is spattered on by a rapidly revolving roll.

Twin drum dryers find wider applications than the single drum dryers. For very viscous solutions and pastes that warrants undue pressure to be applied on the surface of a double drum dryer, single-drum dryers are preferred. The moisture content of the dried material can be controlled by adjusting the drum rotation rate, the drum-surface temperature, the feed temperature or the coating thickness. The drum rate is typically between 1 and 30 rpm and the surface temperature is generally a few degrees below the condensing steam temperature. The drying time is in the range of 3 s to 20 s. In drum dryers, it is heat transfer, and not diffusion, which is the controlling factor in drying. Temperature variation of the material being dried depends on the nature of the substance. Temperature rises to the boiling point and evaporation of the moisture occurs at this temperature which remains constant as long as the solid is wet with the moisture. Only during the last phase of drying, the solid temperature increases. When a liquid or a solution is dried, the temperature remains constant at the boiling point, if the solute precipitates from the solution at constant concentration. However, if the concentration of the solution changes, the temperature increases with progress of the drying. Finally the dried solid gets heated to a temperature equal to the drum surface temperature.

Drum dryers are used for a wide variety of materials such as starch solutions, potatoes, skim milk, coffee, tanning extract, vegetable glues, slurries of magnesium hydroxide, ferrous hydroxide and calcium carbonate, solutions of sodium acetate, sodium sulphate and various organic compounds. A large number of cylinders arranged in series and parallel are used to dry continuous sheets of woven fabrics and paper pulp.

9.5 DIELECTRIC AND MICROWAVE DRYERS

Dielectric and microwave heating are processes that utilize that part of the electromagnetic spectrum comprising the microwave and radio frequency waves. The dielectric frequencies lie in the range 1 to

100 MHz, while microwave frequencies range from 300 MHz to 300 GHz. Microwave and radiofrequency waves interact with individual molecules to quickly generate heat within a product. The dipolar molecules such as water are normally randomly oriented. Under the applied dielectric or microwave field, the molecules try to align with the applied field. The rapidly changing polarity of the applied dielectric or microwave field force the molecules return back to the random orientation, before being pulled the other way. The stress on the molecules due to the rapid build-up and decay of the field result in the conversion of the electric field energy to the stored potential energy and finally to the random kinetic or thermal energy. This is how the dipolar molecules such as water absorb energy in these frequency ranges.

The power developed per unit volume is dependent on the dielectric constant, electric field strength, the frequency, relative permeability and dissipation factor. The field strength and the frequency are dependent on the equipment, while the dielectric constant and the dissipation factor are material dependent. The electric field strength is also dependent on the location of the material within the radio frequency or microwave cavity. The dielectric constant of water is over an order of magnitude higher than that of the most other materials and the overall dielectric constant of most materials is usually nearly proportional to the moisture content up to a critical moisture content, often around 20 to 30%. Hence the microwave and the radio frequency methods preferentially heat and dry the water content of localized spots in most materials, leading to a more uniform final moisture contents. Although this mechanism is the major one for generation of heat within the materials by the electromagnetic fields, there is yet another mechanism which results in transfer of energy to the material. Conductive currents flowing within the material due to the movement of ionic constituents (sodium, chloride and hydroxyl) can transfer energy from the electromagnetic field to the material.

The basic components of the dielectric and microwave systems for drying are the generators which produce the high frequency energy and the applicator, which are units for applying the energy to the materials. The function of the generator is to convert the alternating current of 50 or 60 Hz to the desired high frequencies. The means of doing this are quite different for dielectric and microwave systems. Dielectric dryers usually employ negative grid triode tubes, although some systems operating in the 50 to 100 MHz range use beamed power type. Radio frequency circuits are usually simple, self-excited oscillators. These circuits consist of coils, capacitors or coaxial lines. The load to be heated may actually form part of the tank circuit capacitance, it may be separately tuned or inductively or capacitively coupled. Dielectric energy is usually applied by the electrodes of various types and shapes, between which is placed the material to be dried.

Typical industrial microwave drying system is illustrated in Fig. 9.33. The devices used for generating microwaves are called magnetrons and klystrons. After generating the microwave energy, it must be transported to the applicator. The material to be dried is subjected to intense microwave field in the applicator. This is usually accomplished by means of waveguides, although coaxial cables are also used. The waveguide is ordinarily a hollow rectangular metallic conduit, usually made of brass or aluminium. Its interior dimensions are carefully chosen to control the nature of the microwave field presented to the applicator. Applicators are of different types. Waveguides themselves may be used as applicators. Travelling wave applicators are also known as slotted, folded or serpentine waveguides. Cavity applicators are the most common type for which, the home microwave ovens are typical examples. It is now a usual practice to incorporate a ferrite iso-circulator protection between the magnetron source and the applicator. The control circuitry is provided in the system to optimise and regulate the overall performance of the microwave dryer. It has become common practice to combine hot air with the system. This is because it usually improves efficiency and the economy of the drying process. In Fig. 9.33, the conventional equipment shown include the vacuum pumps, steam or hot air injection and these are connected to the applicator.

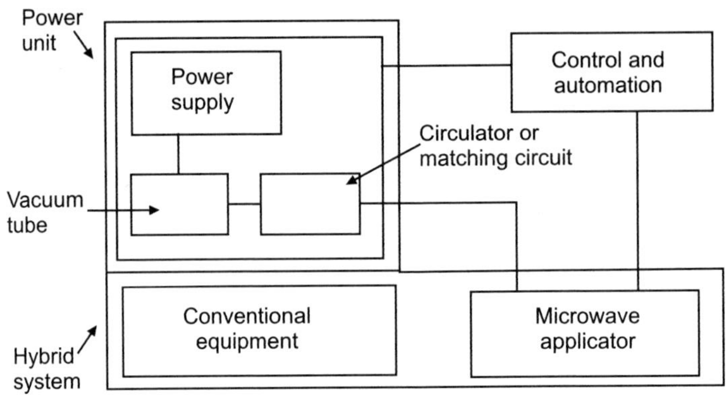

Fig. 9.33: Typical microwave drying set-up

Dielectric or microwave drying has a number of advantages over the conventional drying processes. Whereas the latter methods depend upon the slow transfer of heat from the surface of the material to the interior as determined by a temperature gradient from a hot outside to a cool inside, heating with the former is bulk heating in which the electromagnetic field interact with the material as a whole. The advantages include the volumetric dissipation of the energy throughout the product, and the ability to automatically level any moisture variation within it, leading to a more uniform distribution in the product. It is often possible to dry a material in seconds or minutes, which, in conventional dryers, may take hours or days. The efficiency of energy conversion is remarkable as the energy couples directly to the material that is heated. Energy is not expended on heating the air, walls of the oven, conveyor, or other parts. Less floor area is required for dielectric or microwave systems compared to the conventional dryers. Since high surface temperatures are usually not generated, the surface overheating and case-hardening, commonly observed in drying of certain materials by the conventional methods, are eliminated in these dryers. Dielectric heating is used to dry materials such as lumber, ceramics, foam rubber, biscuits, films, coatings, adhesives, paper, inks, textiles, plastics and penicillin. Microwave dryers are used to dry onions, seaweed, potato chips, pharmaceuticals, ceramic wares, textiles, leather, fruit juices, beverages, etc.

9.6 GENERAL DRYING BEHAVIOUR

The behaviour of solids on drying is studied by conducting batch drying tests under constant drying conditions. These tests provide information on the time required to dry a substance from given moisture content to another under specified conditions. These data are useful to set up the drying schedules and to determine the size of the drying equipment. Measurement of the rate of batch drying provides valuable information for the design and the operation of continuous dryers also. To conduct the drying test, a sample of wet material is suspended from a balance, which is kept inside a cabinet or enclosure and heated air is passed across the surface. Under constant drying conditions, the temperature, humidity, velocity and direction of flow of the air are maintained constant. The conditions of air at the inlet and exit of the dryer are essentially the same. The weight of the drying sample is then determined as function of time from which the moisture content is calculated against drying time. These are plotted as shown in Fig. 9.34 taking θ, the time of drying as the abscissa and X, the moisture content of the solid as the ordinate.

The rate of drying R is defined as the mass of moisture evaporated in unit time from unit area of the surface exposed to drying and has the units kg/m^2 h.

$$R = -\frac{S}{A}\frac{dX}{d\theta} \qquad \ldots(9.1)$$

where S is the mass of the dry solids, and A is the area of drying, which is the wet surface across which the gas flows and from which the evaporation of the moisture is taking place. The slope of the curve shown in Fig. 9.34 at any concentration of moisture X gives $\frac{dX}{d\theta}$ which, when multiplied by the negative of the ratio of the mass of dry solid to the interfacial area gives the rate of drying, R when the concentration is X. In this way, the rate of drying R is determined for different values of X. The R versus X curve shown in Fig. 9.35 is the rate of drying curve under constant drying conditions. Though the shape of the curve may vary from material to material, the curve shown here is typical for solids under constant drying conditions.

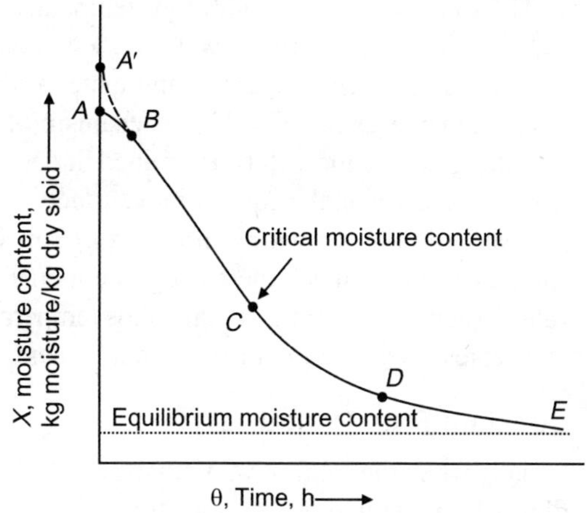

Fig. 9.34: Drying curve, constant drying conditions

The drying rate curves are characterised by two distinct regions, the constant-rate period (BC) and the falling-rate period (CE). Initially the solid is wet with moisture and the evaporation occurs from the surface of the solid at a constant rate. When the moisture content of the solid reaches that corresponding to point C in the figure, the constant rate period ends and the rate begins to decrease along the curve CE. Moisture content of the solid at which the constant rate period ends and the drying rate begins to decrease is known as the *critical moisture content* and is denoted by X_C. Finally, when the moisture content reaches the equilibrium value, X^*, the rate of drying becomes zero and no further evaporation of the moisture can take place. The exact shape of the curve in the falling rate period depends upon the nature of the material and the conditions of drying. Depending upon the nature of the solids and the mechanism of the moisture movement within the solid, the falling rate period may be linear, concave upward or concave downward over its entire length, i.e. from the critical moisture content to the equilibrium moisture content. However, the curve shown is typical for many systems characterising two types of falling rates—a linear variation from C to D and nonlinear variation from D to E.

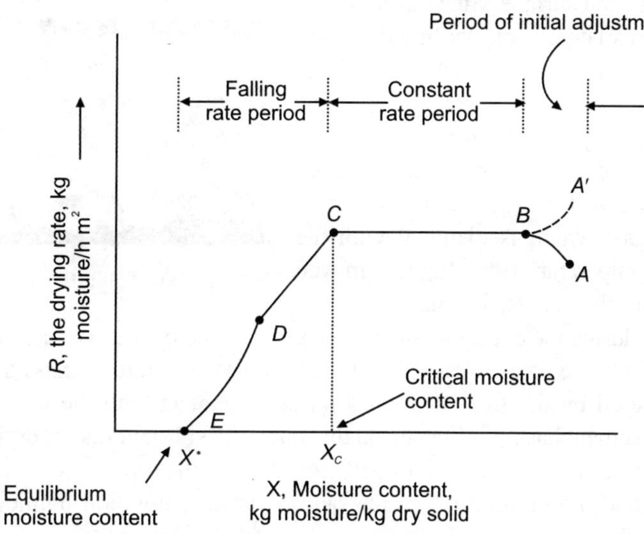

Fig. 9.35: Rate-of drying curve, constant drying conditions

The solids attain an equilibrium temperature which, in the absence of heat transfer by conduction and radiation, is equal to the wet-bulb temperature of the air and remains constant during the constant rate period. The temperature of the material does not remain constant during the falling rate period; rather, it increases steadily. The mechanism of the moisture movement in the constant rate period and the falling rate period will be examined in detail in the coming sections. If the initial temperature of the material is less than the equilibrium surface temperature, there will be a period of initial adjustment or preheat period as shown by the curve *AB* in Fig. 9.35. During this period, which is relatively short compared to the others, the drying rate increases and the surface temperature rises to the equilibrium value. Alternatively, if the initial solids temperature is greater than the equilibrium value, the curve *A′B* will represent the period of initial adjustment.

Constant-Rate Period

Initially when the solid is wet with the moisture, the surface will be covered by a thin film of liquid. The film is formed either by the moisture initially present at the surface or by the moisture brought to the surface from the interior by the capillary action in the case of the first category materials (granular or crystalline solids) or by the liquid diffusion in the case of the second category materials (amorphous solids). When exposed to hot gases, the moisture evaporates, as if the evaporation is occurring from a pool of pure liquid, not directly influenced by the solid. The drying process in the granular or crystalline materials can be visualized as shown in Fig. 9.36. The rate of drying remains constant as shown by the period *BC* in Fig. 9.35. The capillaries and interstices of the solid can deliver moisture to the surface as fast as it is vaporised from the surface. Since evaporation of the moisture absorbs latent heat, the liquid surface attains a steady-state temperature when the rate of heat flow from the gas to the surface is exactly equal to the rate of latent heat absorption. Neglecting the heat transfer by conduction and radiation, the temperature attained by the surface will be equal to the wet-bulb temperature of the gas, which is identical with the adiabatic saturation temperature if the liquid removed on drying is water and the gas used is air.

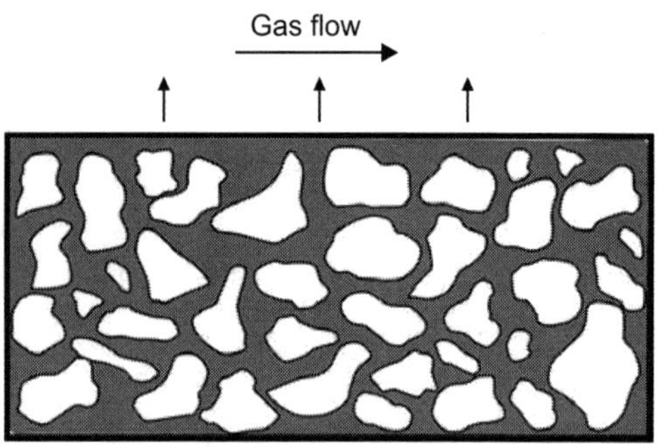

Fig. 9.36: Drying at constant rate

The rate of drying during the constant rate period is controlled by the external mass and heat transfer between the exposed surface of the material and the bulk gas. The rate of mass transfer is determined by the resistance offered by the film of the gas phase in contact with the wet surface of the solid. Internal moisture movement has no influence on this rate. The constant rate of drying, R_C can therefore be written in terms of the gas mass transfer coefficient k_Y, and the difference in humidity of the gas at the liquid surface Y_S and the humidity Y in the bulk. Y_S is the saturation humidity of the gas at the surface temperature, T_S.

$$R_C = k_Y(Y_S - Y) \qquad \ldots(9.2)$$

The mass transfer coefficient k_Y remains constant as long as the velocity and direction of flow of the gas are not altered. Since the surface is at a constant temperature, the saturation humidity Y_S at this temperature is also constant. Since the humidity of the gas stream Y is kept constant under constant drying conditions, we can see from Eq. (9.2) that the rate R_C remains constant.

During the constant rate period, the heat required for evaporation is met by the sensible heat transferred by the gas. In terms of the heat transfer coefficient h_G and the temperature difference existing from the surface to the bulk gas, the rate of heat transfer from the gas to the surface under steady state is

$$R_C\lambda_S = h_G(T_G - T_S) \qquad \ldots(9.3)$$

where λ_S is the heat of vaporisation of the moisture at the surface temperature T_S attained by the material during the constant rate period and T_G is the dry-bulb temperature of the gas. Combining Eqs. (9.2) and (9.3), the constant rate of drying R_C can be expressed as

$$R_C = k_Y(Y_S - Y) = \frac{h_G}{\lambda_S}(T_G - T_S) \quad ...(9.4)$$

Although both mass and heat transfer relations can be used to compute the rate of drying during the constant rate period, it is customary to use the heat transfer relationship given by the second equality in Eq. (9.4), especially when the gas used is air and the moisture removed is water. This is because of the ready availability of the psychrometric chart for the system, the equality of the wet-bulb and saturation temperatures, and the availability of correlations for evaluating the convective heat transfer coefficients of air-water vapour systems for several geometric systems. In addition, an error in the measured value of the interface temperature T_S affects the term $(Y_S - Y)$ more severely than it affects the driving force $(T_G - T_S)$.

Rearranging Eq. (9.4), we get

$$(Y_S - Y) = \frac{h_G}{k_Y \lambda_S}(T_G - T_S) \quad ...(9.5)$$

which is the same as the equation for the wet-bulb temperature [Eq.(6.38)] developed in Chapter 6. It means that, under constant rate period, the temperature of the wet solid surface is equal to the wet-bulb temperature of the air.

Empirical correlations for estimating interphase coefficients of heat transfer are available and a few of them, applicable when the air is used as the gas phase, are given below. In these correlations, h_G is in W/m^2 K:

Table 9.1: Correlations for heat transfer coefficient

Geometry	Correlation		Range of applicability	Remarks
Flat plate, parallel flow	$h_G = 0.0204G^{0.8}$	(9.6)	T_G = 318-423 K G = 2450 – 29,300	G = mass velocity of air in the flow channel, kg/h m^2
Flat plate, perpendicular. Impingement flow	$h_G = 1.17G^{0.37}$	(9.7)	G = 3,900 – 19,500	G = mass velocity of air impinging on the wet surface
Packed bed, through circulation	$h_G = 0.151G^{0.59}/d_p^{0.41}$ $h_G = 0.214G^{0.49}/d_p^{0.51}$	(9.8) (9.9)	$Re > 350$ $Re < 350$	d_p = particle diameter, m G = superficial mass velocity, kg/h m^2
Fluidized beds	$Nu = 0.316Re^{0.8}$	(9.10)	$0 < Re < 80$	$Nu = h_G d_p/k$ $Re = d_p G/\mu$ $Pr = C_p\mu/k$
Droplets in spray dryers	$Nu = 2 + 1.05Re^{0.5}Pr^{1/3}Gu^{0.175}$	(9.11)	$Re < 1000$	$Pr = C_p\mu/k$ $Gu = \frac{(T_G - T_W)}{T_G}$ in absolute temperatures

Falling-rate Period

As the average moisture content of the solid approaches the critical value, the liquid available at the surface becomes insufficient to maintain a continuous film over the entire area available for drying and the surface remains wet by receiving the moisture transported from the interior of the solid by capillary movement or by liquid diffusion. However, the rate of this transport is not sufficient to meet the rate of vaporisation of the moisture from the surface, when the moisture content of the solid has fallen to the critical value. As a result, dry spots begin to appear on the surface, the extent of which progressively increases with drying. This is illustrated in Fig. 9.37 for granular or crystalline substances. Drying under this condition constitutes the first part of the falling rate period shown by the line *CD* in Fig. 9.35. Evaporation of the liquid occurs from the *wet surface* as in the constant rate period and the wet surface temperature remains constant. The wet surface area progressively decreases with the progress in the drying. Since the surface is only partially saturated with the liquid, this portion of the falling rate period is known as the unsaturated surface drying.

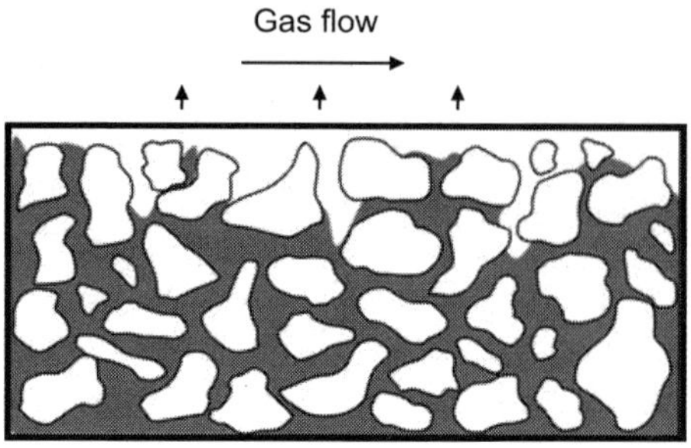

Fig. 9.37: Unsaturated surface drying

Vaporisation occurring from the part of the surface that is saturated with the liquid is by the same mechanism as that in the constant-rate period; that is, by the convective transport of heat to the surface and mass away from the surface. The rate of drying from the unit area of the wet surface is still the same as that in the constant rate period. However, since the rate *R* is computed on the basis of the total gross surface, the rate falls and since the liquid-saturated area steadily decreases, the rate of drying also falls steadily. The variation of the rate of drying in this period is linear with moisture content as the line *CD* indicates. This part of the falling rate period is either very short or do not exist at all as happens in the drying of the fibrous and amorphous materials. And even in the case of crystalline or granular materials, this segment of falling rate period will not be present at all if the rate of drying at constant rate period is very high.

When the moisture content reaches that corresponding to point *D* in Fig. 9.35, the entire surface moisture has been removed and the second falling rate period begins. The rate of drying in this period is controlled by the vapour diffusion as discussed above for granular and crystalline materials whereas for fibrous or amorphous materials, the mechanism of moisture movement is by liquid diffusion. The free liquid surface has receded to the interior of the solid as shown in Fig. 9.38 for materials belonging to category 1. Further drying should occur by vaporisation of liquid from the interior of the solid by absorbing heat transferred from the gas, followed by diffusion of the resulting vapour to the surface and then to the bulk gas. The temperature at the plane where vaporisation occurs equals the wet-bulb temperature, whereas the temperature of the surface of the solid increases and approaches the dry-bulb temperature of the gas. As a result, temperature gradients are set up in the solid due to heat transfer by conduction. Since the average

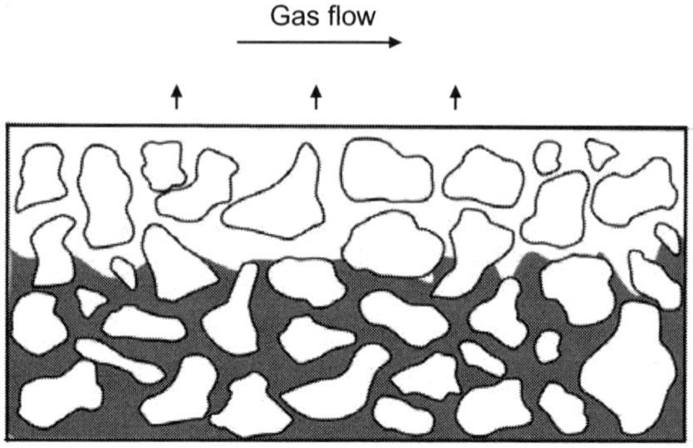

Fig. 9.38: Drying rate controlled by internal diffusion

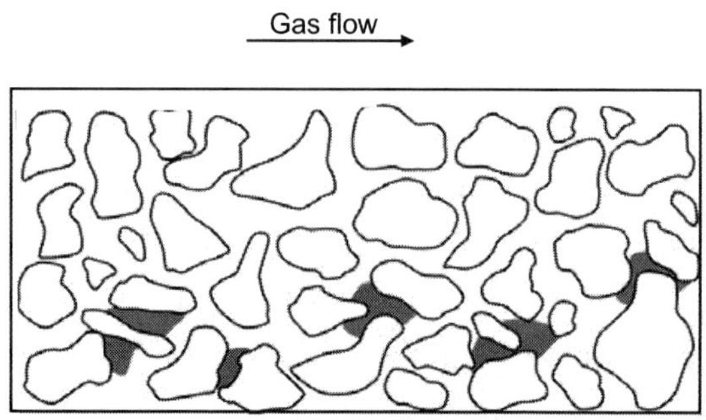

Fig. 9.39: Drying stops at the equilibrium moisture content

depth of the moisture level increases progressively and the heat conductivity of the dry solids and associated cavities being very small, the heat conduction becomes slow and its influence on the drying rate increases. However, if the dry product has a relatively high bulk density and consists of tiny pores as in category 2 materials, it is still the rate of liquid diffusion that determines the rate of drying. The rate of vaporisation is practically independent of the velocity of the gas. As the moisture content continues to fall, the length of the diffusion path for heat and mass within the solid increases steadily and the rate continues to decrease till the equilibrium moisture content X^* is reached. This condition is depicted by the curve *DE* in Fig. 9.35. When the moisture in the solid has fallen to the equilibrium value, the vapour pressure exerted by the water in the solid and the partial pressure of the water vapour in the incoming gas are identical and no further drying can occur (Fig. 9.39).

The constant rate period continues to relatively low moisture contents and the falling rate period is predominantly of the first type, i.e. linear with the moisture content for granular or crystalline solids. In contrast, for amorphous or fibrous materials, the constant rate period is relatively short, and most of the drying takes place in the falling rate period of the second type, in which the rate of drying is liquid-diffusion controlled. For granular or crystalline solids under agitated conditions as in fluidized beds, spray dryers and conveyor dryers where individual particles are exposed to the hot gases, the constant rate period may extend up to the equilibrium moisture content X^*.

Critical Moisture Content

It is at the critical moisture content that the rate of migration of the moisture from the interior of the solid begins to control the rate of drying and the falling rate period of drying begins. Above this concentration, the rate of drying is governed by the resistance to convective transport of heat and mass at the surface. Critical moisture content is thus the lowest concentration of moisture in the solid which can sustain a continuous or nearly continuous film of liquid on the surface by the transfer of moisture from the interior. If the initial moisture content of the solid is less than the critical value, there will not be any constant rate period of drying. The critical moisture content of the solid depends on the pore structure of the solid. For a specific material it may vary with the thickness of the material, the rate of drying at constant rate period and with the resistance to heat and mass transfer within the solid. By decreasing the thickness of the material, the internal resistance to heat and mass transfer can be reduced, so that the external resistance will control the rate of drying for a longer duration. Consequently materials with smaller thickness have smaller critical moisture contents. Since for amorphous or fibrous materials, the moisture movement is by the liquid diffusion which is a very slow process, the drying curve shows only very short constant rate periods resulting in high critical moisture content.

Example 9.4: A wet cake is being dried in a tray dryer. Drying occurs from the top surface by cross circulation of 10% saturated air at a pressure of 101.3 kPa and temperature of 351 K. The average velocity of the air passed across the wet solid is 4 m/s. The initial water content of the wet material is 25%, and the critical moisture content is 10%. All moisture contents are on dry basis. Estimate the rate of drying in the constant rate period if the heat transfer coefficient at the surface is given by $h_G = 0.0204G^{0.8}$ W/m^{2}K where G is in kg/m^2 h.

***Solution*:** The critical moisture content is 10% on dry basis. That is, $X_C = 0.10$ kg water/kg dry solid. Initial moisture content of the wet cake is 25%, $X_1 = 0.25$ kg moisture/kg dry solid. Since this is greater than X_C, drying occurs at constant rate R_C, which may be estimated using Eq. (9.4):

$$R_C = \frac{h_G}{\lambda_S}(T_G - T_S)$$

where h_G is the heat transfer coefficient at the surface, T_G is the dry-bulb temperature of the air, T_S is the temperature of the solid surface during the constant rate period, which is equal to the wet-bulb temperature of the air and λ_S is the heat of vaporisation at the surface temperature.

$T_G = 351$ K. The percent saturation of air is given to be 10%; the wet-bulb temperature is read from the psychrometric chart to be 317 K and humidity is 0.045 kg water/kg dry air. $T_S = 317$ K and $Y = 0.045$ kg water/kg dry air. Heat of vaporisation of water at 317 K is read from the steam tables to be 2397 kJ/kg.

$$\lambda_S = 2397 \text{ kJ/kg.}$$

The heat transfer coefficient is estimated using Eq. (9.6).

$$h_G = 0.0204 G^{0.8}$$

where G is the mass velocity of air. $G = \rho\, u$ where ρ is the density of air and u is the velocity of air. The density of air is determined from the reciprocal of the humid volume of air.

$$V_H = 8.314 \times \left(\frac{1}{29} + \frac{Y}{18}\right) \times \frac{T}{P}$$

$$V_H = 8.314 \times \left(\frac{1}{29} + \frac{0.045}{18}\right) \times \frac{351}{101.3} = 1.0654 \text{ m}^3/\text{kg dry air}$$

$$\rho = \frac{1+Y}{1.0654} = \frac{1.045}{1.0654} = 0.9809 \text{ kg/m}^3$$

$$G = \rho u = 0.9809 \times 4 \times 3600 = 14124.3 \text{ kg/m}^2\text{ h}$$

Substituting G in Eq. (9.6)

$$h_G = 0.0204(14124.3)^{0.8} = 42.62 \text{ W/m}^2\text{ K.}$$

The rate of drying in the constant-rate period is estimated as

$$R_C = \frac{h_G}{\lambda_S}(T_G - T_S) = \frac{42.62 \times 10^{-3} \times 3600}{2397}(351 - 317)$$

$$= 2.18 \text{ kg/m}^2\text{ h}$$

(Answer)

Empirical Method of Estimating Drying Time

The best method for estimating the length of time required for drying a material between given initial and final moisture content is to use either the drying curve of Fig. 9.34 or the rate of drying curve of Fig. 9.35. These experimental data should be obtained under such conditions that the feed material, ratio of exposed to non-exposed surface, gas velocity, temperature and humidity are essentially the same as those in the proposed dryer.

Rearranging Eq. (9.1), we get

$$d\theta = -\frac{S}{A}\frac{dX}{R} \qquad \ldots(9.12)$$

Integrating Eq. (9.12) for the time interval θ over which the moisture content falls from X_1 to X_2, we get

$$\theta = -\frac{S}{A}\int_{X_1}^{X_2} \frac{dX}{R} = \frac{S}{A}\int_{X_2}^{X_1} \frac{dX}{R} \qquad \ldots(9.13)$$

The integral in Eq. (9.13) can be evaluated graphically using the rate of drying curve of Fig. 9.35. Area under the curve between limits X_1 and X_2 on the plot $1/R$ versus X gives the integral in Eq. (9.13) which multiplied by the ratio of the mass of dry solid to the exposed area gives the time of drying. The integral can also be determined analytically for some special cases as given below.

Case 1: *Drying in the constant rate period:* Here $X_C < X_2 < X_1$ and the rate of drying is constant at R_C. The situation is shown in Fig. 9.40a. Then Eq. (9.13) simplifies as

$$\theta = \frac{S}{AR_C}(X_1 - X_2) \qquad \ldots(9.14)$$

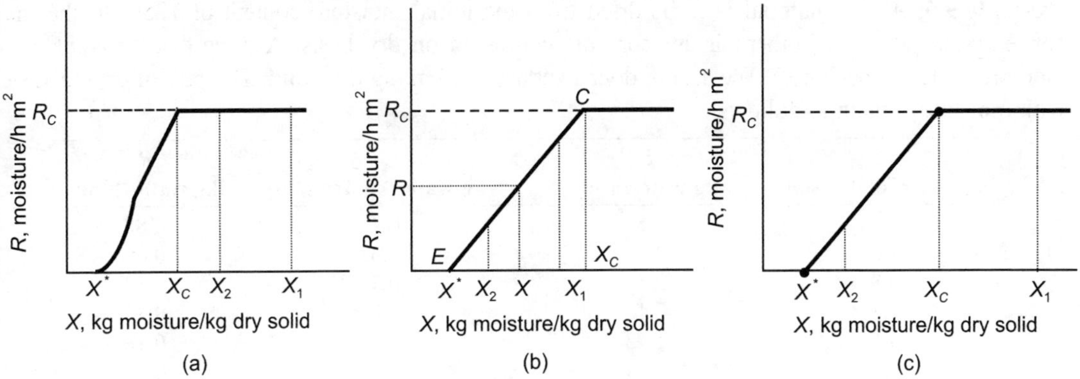

Fig. 9.40: Different situations in estimating the drying time

Case 2: *Drying in the falling rate period and the rate is linear throughout.* Here $X_C > X_1 > X_2$. Figure 9.40b represents the variation of drying rate with moisture content. From the figure, we can see that the rate of drying in the falling rate period R corresponding to a moisture content X is related to the constant rate of drying R_C by

$$\frac{R}{R_C} = \frac{X - X^*}{X_C - X^*} \quad \text{...(9.15)}$$

Or

$$R = R_C\left(\frac{X - X^*}{X_C - X^*}\right) \quad \text{...(9.16)}$$

Substituting Eq. (9.16) into Eq. (9.13), we see that

$$\theta = \frac{S}{A}\int_{X_2}^{X_1}\frac{dX}{R} = \frac{S(X_C - X^*)}{AR_C}\int_{X_2}^{X_1}\frac{dX}{X - X^*} = \frac{S(X_C - X^*)}{AR_C}\ln\frac{X_1 - X^*}{X_2 - X^*} \quad \text{...(9.17)}$$

Case 3*: Drying includes both constant-rate and falling-rate periods:* Here $X_2 < X_C < X_1$ as shown in Fig. 9.40c. Let θ_C be the time of drying in the constant rate period from the initial moisture content X_1 to the critical moisture content X_C and θ_f be the time of drying in the falling rate period from X_C to X_2. Using Eq. (9.14) with X_2 replaced by X_C,

$$\theta_C = \frac{S}{AR_C}(X_1 - X_C) \quad \text{...(9.18)}$$

θ_f is the time of drying in the falling rate period from X_C to X_2. Using Eq. (9.17) with X_1 replaced by X_C.

$$\theta_f = \frac{S(X_C - x^*)}{R_C A}\ln\frac{X_C - X^*}{X_2 - X^*} \quad \text{..(9.19)}$$

If the falling rate period is not linear, then

$$\theta_f = -\frac{S}{A}\int_{X_C}^{X_2}\frac{dX}{R} = \frac{S}{A}\int_{X_2}^{X_C}\frac{dX}{R} \quad \text{...(9.20)}$$

where the integral may be evaluated by graphical integration. Combining Eqs. (9.18) and (9.19), we get

$$\theta = \theta_C + \theta_f = \frac{S}{AR_C}(X_1 - X_C) + \frac{S(X_C - X^*)}{R_C A}\ln\frac{X_C - X^*}{X_2 - X^*} \quad \text{...(9.21)}$$

$$\theta = \theta_C + \theta_f = \frac{S}{AR_C}\left[(X_1 - X_C) + (X_C - X^*)\ln\frac{X_C - X^*}{X_2 - X^*}\right] \quad \text{..(9.22)}$$

Example 9.5: A wet material is to be dried from the initial moisture content of 125% to the final concentration of 10%, both moisture contents expressed on dry basis. Assume that the equilibrium moisture content is negligible. The area of drying surface is 10 m²/kg dry solid. The rate of drying varies with moisture content as follows.

X kg water/kg dry solid	*R* kg water/h-m²	*X* kg water/kg dry solid	*R* kg water/h-m²
1.5	2.5	0.4	0.88
1.0	2.5	0.35	0.66
0.8	2.5	0.3	0.52
0.7	2.1	0.2	0.33
0.6	1.73	0.1	0.18
0.5	1.39	0	0
0.45	1.22		

(a) Plot the drying rate curve
(b) Determine the time of drying in the constant rate period
(c) Determine the total time of drying.

***Solution*:** (a) The rate of drying curve is plotted as shown in Fig. 9.41 by taking X, the moisture content as abscissa and the rate of drying R as the ordinate. The drying involves both constant rate period BC and falling rate period CE as shown in the figure.

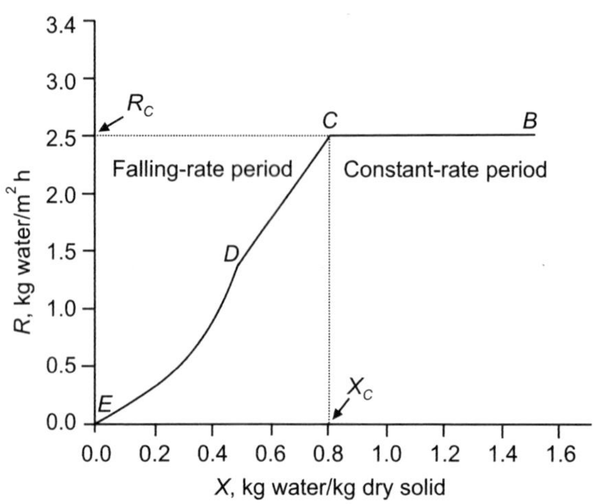

Fig. 9.41: Drying rate curve for Example 9.5

(b) Drying takes place at a constant rate from X_1 to X_C. The time required in the constant rate period, θ_C is obtained using Eq. (9.18)

$$\theta_C = \frac{S}{AR_C}(X_1 - X_C)$$

The initial moisture content

$$X_1 = \frac{125}{100} = 1.25 \text{ kg water/kg dry solid}$$

Critical moisture content is the concentration at which the rate of drying begins to fall. From the graph, the critical moisture content $X_C = 0.8$ kg water/kg dry solid

The constant rate of drying $R_C = 2.5$ kg water/m² h. Area of drying surface is $A/S = 10$ m²/kg dry solid. Substituting the values in Eq. (9.18),

$$\theta_C = \frac{1}{10 \times 2.5}(1.25 - 0.80) = 0.018 \text{ h} \quad \textbf{(Answer)}$$

(c) Drying from the critical moisture content $X_C = 0.80$ to the final moisture content X_2 occurs in the falling rate period.

The final moisture content $X_2 = \frac{10}{100} = 0.10$ kg water/kg dry solid

The time of drying θ_f is evaluated using Eq. (9.20).

$$\theta_f = \frac{S}{A}\int_{X_2}^{X_C} \frac{dX}{R}$$

To determine the integral in Eq. (9.20), the curve of $1/R$ versus X is drawn as shown in Fig. 9.42

X	0.80	0.70	0.60	0.50	0.45	0.40	0.35	0.30	0.20	0.10
R	2.50	2.10	1.73	1.39	1.22	0.88	0.66	0.52	0.33	0.18
$1/R$	0.4000	0.4762	0.5780	0.7194	0.8197	1.1364	1.5152	1.9231	3.0303	5.5556

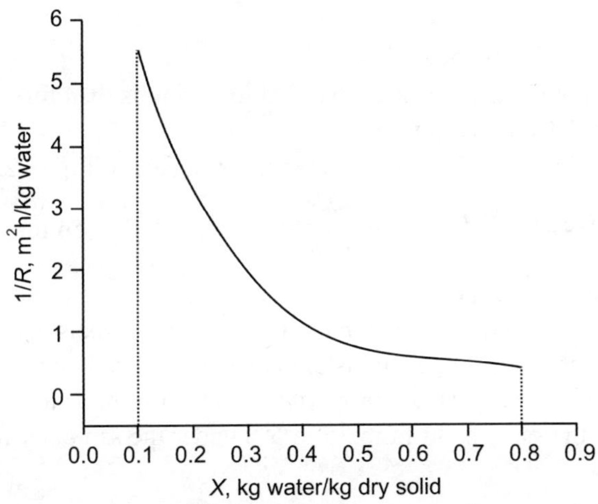

Fig. 9.42: Graphical integration to determine the drying time in Example 9.5

The area under the curve between limits $X_2 = 0.10$ and $X_C = 0.8$ gives the integral. The area is found to be 1.078.

$$\int_{X_2}^{X_C} \frac{dX}{R} = 1.078$$

Therefore,

$$\theta_f = \frac{S}{A}\int_{X_2}^{X_C} \frac{dX}{R} = \frac{1.078}{10} = 0.1078 \text{ h}$$

Total time required to dry the material = $\theta_C + \theta_f = 0.018 + 0.1078 = 0.1258$ h or 7 min 33 s.

(Answer)

Example 9.6: From experiments on batch drying of sheet material, the rate of drying in the constant rate period is found to be 3 kg water/h m^2. The critical moisture content is 33% and the equilibrium moisture content is negligible. Sheets of size 1.5 m × 1.2 m × 50 mm are to be dried from 50% moisture to 5% moisture. All moisture contents are on dry basis. The drying occurs from both sides and under the same drying conditions as in the experiments. If the material has a bone-dry density of 1200 kg/m^3, calculate the time required for drying. Assume that the falling rate period is linear in the moisture content.

***Solution*:** The rate of drying curve is as shown in Fig. 9.43.

The initial moisture content $X_1 = 0.50$.

The final moisture content, $X_2 = 0.05$

The critical moisture content, $X_C = 0.33$

Equilibrium moisture content, $X^* = 0$

Since $X_1 > X_C$ and $X_2 < X_C$, the drying involves both constant and falling rate periods. Since the falling-rate period is linear in moisture content, Eq. (9.22) can be used for estimating the drying time.

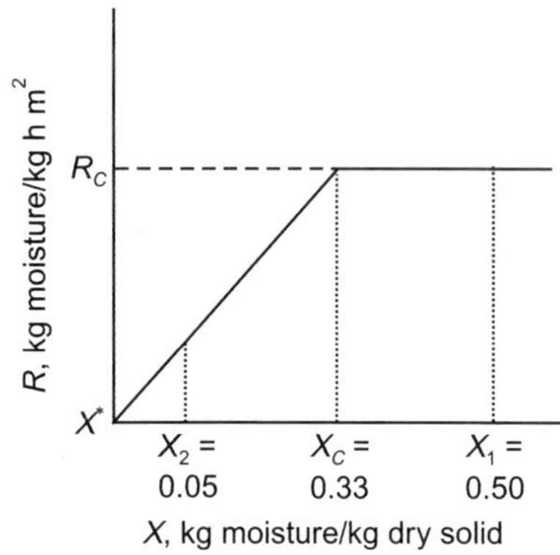

Fig. 9.43: Rate of drying curve for Example 9.6

$$\theta = \theta_C + \theta_f = \frac{S}{AR_C}\left[(X_1 - X_C) + (X_C - X^*)\ln\frac{X_C - X^*}{X_2 - X^*}\right]$$

Bone-dry weight of the solids = S = volume of sample × bone-dry density of solids

$$S = (1.5 \times 1.2 \times 0.05) \times 1200 = 108 \text{ kg}.$$

Material is dried from both sides. The area exposed to drying is therefore,

$$A = 2 \times (1.5 \times 1.2) = 3.6 \text{ m}^2$$

The constant rate of drying $R_C = 3$ kg/m² h. Substitute the data in Eq. (9.22).

$$\theta = \frac{108}{3.6 \times 3}\left[(0.50 - 0.33) + (0.33 - 0)\ln\frac{0.33 - 0}{0.05 - 0}\right] = 7.93 \text{ h (7 h 56 min)} \qquad \textbf{(Answer)}$$

Example 9.7: It takes 9 hours for a porous solid to reduce the moisture content from 45 to 10% when dried in a batch dryer under constant drying conditions. The critical moisture content was found to be 25% and the equilibrium moisture 3%. All moisture contents are on the dry basis. Assuming that the rate of drying during the falling rate period is proportional to the free moisture content, how long should it take to dry a sample of the same solid from 35 to 5% under the same drying conditions?

***Solution*:**

Case 1

$X_1 = 0.45$ kg water/kg dry solid; $X_2 = 0.10$ kg water/kg dry solid; $X_C = 0.25$ kg water/kg dry solid; $X^* = 0.03$ kg water/kg dry solid and $\theta = 9$ h.

Since drying takes place with both constant-rate and falling-rate, and the falling rate period varies linearly with free moisture content, the drying time is calculated using Eq. (9.22)

$$\theta = \frac{S}{AR_C}\left[(X_1 - X_C) + (X_C - X^*)\ln\frac{X_C - X^*}{X_2 - X^*}\right]$$

Substituting the given values in Eq. (9.22),

$$\theta = \frac{S}{AR_C}\left[(0.45 - 0.25) + (0.25 - 0.03)\ln\frac{0.25 - 0.03}{0.10 - 0.03}\right]$$

$$9 = 0.4519\frac{S}{AR_C} \qquad \text{(A)}$$

Case 2

The critical moisture content and equilibrium moisture content are the same as in case 1. That is, $X_C = 0.25$ kg water/kg dry solid; $X^* = 0.03$ kg water/kg dry solid.

$X_1 = 0.35$ kg water/kg dry solid; $X_2 = 0.05$ kg water/kg dry solid; let θ be the time of drying.

Substituting the values in Eq. (9.22) gives

$$\theta = \frac{S}{AR_C}\left[(0.35-0.25)+(0.25-0.03)\ln\frac{0.25-0.03}{0.05-0.03}\right]$$

$$\theta = 0.6275\frac{S}{AR_C} \quad \text{(B)}$$

Divide Eq. (B) by Eq. (A):

$$\frac{\theta}{9} = \frac{0.6275}{0.4519} \quad \text{(C)}$$

Solving Eq. (C), we get

$$\theta = 12.50 \text{ h.}$$

Time required for drying the second sample is 12.5 h **(Answer)**

Example 9.8: A tray dryer consisting of 12 trays is used to dry raw cotton from 93% moisture to 10% moisture by cross circulation of air. The trays which are insulated at the bottom are each 30 mm high with an area of 1.8 m^2 and are completely filled with raw cotton which has a bone-dry density of 700 kg/m^3. The heating medium, which is 30% saturated air at 344 K and 101.3 kPa, flows across the top surface of the tray at a mass velocity of 12000 kg/h m^2. Under the given conditions, the critical moisture content will be 0.4 kg water/kg bone-dry cotton and that the falling rate drying period is linear. Equilibrium moisture content is 0.035 kg water/kg bone-dry cotton. All moisture contents are on dry basis. Determine:

(a) The amount of raw cotton in kg that can be dried in one batch.
(b) The drying-time for the constant-rate period.
(c) The drying time for the falling-rate period.
(d) The total drying time.

***Solution*:** (a) The amount of dry cotton in one tray $S =$

$$\text{volume} \times \text{bone-dry density} = (1.8 \times 30 \times 10^{-3}) \times 700 = 37.8 \text{ kg}$$

Since the dryer consists of 12 trays, dry solids that can be handled in one batch = 12 × 37.8 = 453.6 kg. Since the raw cotton contains 93% moisture on dry basis, total weight of raw cotton that can be dried in one batch $= 453.6 \times \frac{1.93}{1} = 875.45$ kg **(Answer)**

(b) Drying time in the constant rate period is given by Eq. (9.18)

$$\theta_C = \frac{S}{AR_C}(X_1 - X_C)$$

The constant rate of drying is estimated using Eq. (9.4)

$$R_C = \frac{h_G}{\lambda_S}(T_G - T_S)$$

For cross-circulation drying, the heat transfer coefficient (W/m^2 K) may be estimated using

$$h_G = 0.0204G^{0.8}$$

where G is the mass velocity of air in kg/m^2 h.

$$h_G = 0.0204(12000)^{0.8} = 37.41 \text{ W/m}^2\text{ K.}$$

For air at $T_G = 344$ K with a percent humidity of 30%, the wet-bulb temperature T_W is obtained from psychrometric chart. $T_W = 325.7$ K. The temperature at the surface T_S is the same as T_W.

$T_S = 325.7$ K. The latent heat of vaporisation of water at 325.7 K is read from steam tables, which gives λ_S. $\lambda_S = 2376$ kJ/kg. Now,

$$R_C = \frac{h_G}{\lambda_S}(T_G - T_S) = \frac{37.41 \times 3600 \times 10^{-3}}{2376}(344 - 325.7) = 1.04\,\text{kg/m}^2\text{h}$$

Initial moisture content $X_1 = 0.93$; the critical moisture content $X_C = 0.40$; area exposed to drying $A = 1.8$ m^2. Bone-dry weight of the material in the tray $S = 37.8$ kg. Substituting in eq. (9.18), θ_C the drying time in the constant rate drying is obtained.

$$\theta_C = \frac{S}{AR_C}(X_1 - X_C) = \frac{37.8}{1.8 \times 1.04}(0.93 - 0.40) = 10.70\,\text{h}$$ **(Answer)**

(c) Time of drying in the falling rate drying is θ_f which is given by eq. (9.19).

$$\theta_f = \frac{S(X_C - X^*)}{R_C A}\ln\frac{X_C - X^*}{X_2 - X^*}$$

Final moisture content $X_2 = 0.10$; Equilibrium moisture content $X^* = 0.035$. Substituting in Eq. (9.19),

$$\theta_f = \frac{37.8(0.40 - 0.035)}{1.04 \times 1.8}\ln\frac{0.40 - 0.035}{0.10 - 0.035} = 12.71\,\text{h}.$$ **(Answer)**

(d) The total drying time is obtained as

$$\theta = \theta_C + \theta_f = 10.70 + 12.71 = 23.41\,\text{h}.$$ **(Answer)**

Effect of Conduction and Radiation on Drying Rate

The equations presented earlier for the rate of drying under the constant drying conditions assume that the heat transfer by conduction through the metallic wall, and heat radiation from the hot surfaces to the material being dried are negligible compared to the heat transferred by convection from the hot gases. Now consider a wet material taken on a tray subjected to cross circulation drying as shown in Fig. 9.44.

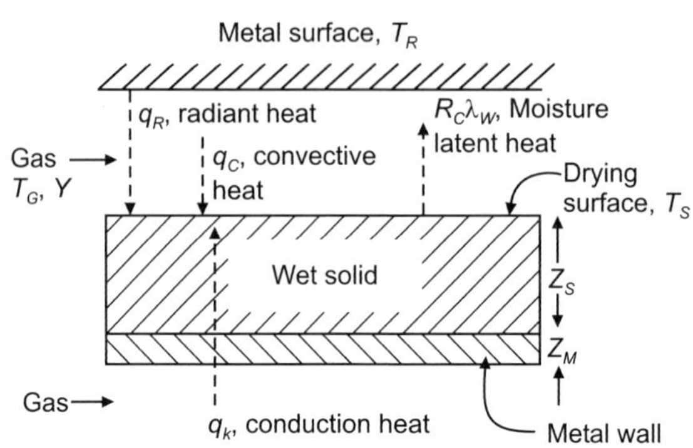

Fig. 9.44: Cross circulation drying in the presence of heat conduction, convection and radiation

In the constant rate-period, the heat transferred to the drying surface (q) is balanced by the heat removed by the evaporation of moisture.

$$q = R_C \lambda_S \qquad \ldots(9.23)$$

Heat is transferred to the drying surfaces by convection (q_C), radiation (q_R) and by conduction (q_k). Therefore,

$$q = q_C + q_R + q_k \qquad \ldots(9.24)$$

The surface temperature remains constant at T_S throughout the constant rate period. The heat is transferred from the hot gases by convection, which is given by

$$q_C = h_G(T_G - T_S) \qquad \ldots(9.25)$$

Hot surfaces at the high temperature T_R emits radiation and is received by the drying surface at a rate q_R. Now,

$$q_R = \varepsilon(5.676\times10^{-8})(T_R^4 - T_S^4) \quad \text{...(9.26)}$$

Here ε is the emissivity of the solid. The heat transferred by radiation may be expressed in terms of a heat transfer coefficient for radiation, h_R as

$$q_R = h_R(T_R - T_S) \quad \text{...(9.27)}$$

in which

$$h_R = \frac{\varepsilon(5.676\times10^{-8})(T_R^4 - T_S^4)}{(T_R - T_S)} \quad \text{...(9.28)}$$

For heat transfer by conduction to the drying surface, three resistances are involved—the resistance offered by the gas film at the bottom surface of the tray, the resistance offered by the tray material and the resistance offered by the substance being dried. Thus,

$$q_k = U_k(T_G - T_S) \quad \text{...(9.29)}$$

where U_k is the overall coefficient of heat transfer given as

$$U_k = \left[\frac{1}{h_G} + \frac{Z_M}{k_M} + \frac{Z_S}{k_S}\right]^{-1} \quad \text{...(9.30)}$$

where Z_M and Z_S are the thickness of the tray and wet material, respectively, and k_M and k_S are the thermal conductivity of the material of the tray and the wet solid, respectively. The rate of drying in the constant-rate period can be written by combining Eqs. (9.23) through (9.30) and Eq. (9.2):

$$R_C = \frac{q}{\lambda_S} = \frac{(h_G + U_k)(T_G - T_S) + h_R(T_R - T_S)}{\lambda_S} = k_Y(Y_S - Y) \quad \text{...(9.31)}$$

To use Eq. (9.31), the surface temperature T_S should be known. In the absence of heat conduction and radiation, T_S is equal to the wet-bulb temperature of the gas, which is equal to the adiabatic saturation temperature for the air-water systems. For the present case, the surface temperature will be greater than the wet-bulb temperature because of additional heat reaching the drying surface by conduction and radiation. To estimate the surface temperature T_S, Eq. (9.31) can be put in a more convenient form as:

$$\frac{(Y_S - Y)\lambda_S}{h_G/k_Y} = \left(1 + \frac{U_k}{h_G}\right)(T_G - T_S) + \frac{h_R}{h_G}(T_R - T_S) \quad \text{...(9.32)}$$

The ratio h_G/k_Y for the air-water vapour systems is approximately equal to the humid heat C_S, so that

$$h_G/k_Y = C_S = 1.005 + 1.884Y \quad \text{...(9.33)}$$

where C_S is in kJ/kg K. Since Y_S is the saturation humidity at temperature T_S, the surface temperature should be calculated by trial using Eq. (9.32) and the saturation humidity curve on the psychrometric chart.

Example 9.9: A layer of wet granular material 25 mm thick is taken in a pan and is being dried. The pan has a 1.5 mm thick metal bottom having a thermal conductivity 50 W/m K. The thermal conductivity of the solid in the pan is 1.2 W/m K. Air at a temperature of 333 K and humidity 0.01 kg water/kg dry air flows parallel to the top exposed surface and the bottom metal surface at a velocity of 4.0 m/s. Direct radiation of heat from the steam pipes having a surface temperature of 380 K falls on the exposed top surface, whose emissivity is 0.9. Estimate

(a) The surface temperature

(b) The drying rate for the constant rate period.

***Solution*:** (a) The surface temperature can be evaluated using Eq. (9.32)

$$\frac{(Y_S - Y)\lambda_S}{h_G/k_Y} = \left(1 + \frac{U_k}{h_G}\right)(T_G - T_S) + \frac{h_R}{h_G}(T_R - T_S)$$

in which the humidity of air $Y = 0.01$ kg water/kg dry air.

$$h_G/k_Y = C_S = 1.005 + 1.884Y = 1.005 + 1.884 \times 0.01 = 1.0238 \text{ kJ/kg K}.$$

The convective heat transfer coefficient h_G can be evaluated using Eq. (9.6).

$$h_G = 0.0204G^{0.8}$$

Here,

$$G = \rho u$$

where ρ and u are the density and velocity of air, respectively.

$$u = 4 \text{ m/s}$$

Density of air is $(1 + Y)$ times the reciprocal of its humid volume.

$$\rho = \frac{1+Y}{v_H} = \frac{1.01}{8.314 \times \left(\frac{1}{29} + \frac{0.01}{18}\right) \times \frac{333}{101.3}} = 1.0547 \text{ kg/m}^3$$

$$G = 1.0547 \times 4 = 4.2188 \text{ kg/m}^2 \text{ s or } (=15187.83 \text{ kg/m}^2 \text{ h})$$

Substituting in Eq. (9.6)

$$h_G = 0.0204(15187.83)^{0.8} = 45.168 \text{ W/m}^2 \text{ K } (= 162.60 \text{ kJ/m}^2 \text{ h K})$$

The overall coefficient of heat transfer from the air to the surface of the solid through the bottom surface is given by Eq. (9.30):

$$U_k = \left[\frac{1}{h_G} + \frac{Z_M}{k_M} + \frac{Z_S}{k_S}\right]^{-1}$$

Here, $Z_M = 1.5$ mm, $k_M = 50$ W/m K, $Z_S = 25$ mm, $k_S = 1.2$ W/m K

$$U_k = \left[\frac{1}{45.168} + \frac{1.5 \times 10^{-3}}{50} + \frac{25 \times 10^{-3}}{1.2}\right]^{-1} = 23.254 \text{ W/m}^2 \text{ K}$$

In Eq. (9.32), $T_G = 333$ K, and $T_R = 380$ K. Substituting the known values, Eq. (9.32) becomes

$$\frac{(Y_S - 0.01)\lambda_S}{1.0238} = \left(1 + \frac{23.254}{45.168}\right)(333 - T_S) + \frac{h_R}{45.168}(380 - T_S)$$

$$0.9768\,(Y_S - 0.01)\lambda_S = 1.5148(333 - T_S) + 0.0221(380 - T_S)h_R \quad \text{(A)}$$

The heat transfer coefficient for radiation is

$$h_R = \frac{\varepsilon(5.676 \times 10^{-8})(T_R^4 - T_S^4)}{(T_R - T_S)} = \frac{0.9(5.676)\left(3.80^4 - \left(\frac{T_S}{100}\right)^4\right)}{(380 - T_S)}$$

$$h_R = \frac{5.1084\left(3.80^4 - \left(\frac{T_S}{100}\right)^4\right)}{(380 - T_S)} \quad \text{(B)}$$

In Eq. (A), λ_S is the heat of vaporisation at the surface temperature T_S and Y_S is the saturation humidity at T_S. Thus, λ_S, Y_S and h_R to be used in Eq. (9.32) for estimating T_S depend on T_S and therefore Eq. (A) is to be solved for T_S by trial and error. Assume arbitrarily a surface temperature T_S (note that it should be greater than the wet-bulb-temperature of the air circulated) and calculate h_R using Eq. (B).

Substitute h_R so calculated, λ_S read from the steam tables and Y_S read from the psychrometric chart or calculated from the vapour pressure data for the assumed temperature T_S, in Eq. (9.32) and calculate T_S. If the assumed temperature and calculated T_S are identical, the assumed temperature is the correct choice for surface temperature. Otherwise, repeat the calculation with a new assumed temperature. For example, assume T_S = 306 K. Then λ_S = 2423.6 kJ/kg, Y_S = 0.0324 kg water/kg dry air.

$$h_R = \frac{5.1084\left(3.80^4 - \left(\frac{306}{100}\right)^4\right)}{(380-306)} = 8.3416\,\text{W/m}^2\text{K}.$$

Substituting in Eq. (A), T_S = 306.9 K.

Assumed T_S, K	λ_S, kJ/kg	Y_S, kg water/kg dry air	h_R, W/m^2 K	Calculated T_S, K
306	2423.6	0.032424	8.341644	306.8573
307	2421.2	0.03439	8.375317	304.1842
306.2	2423	0.032817	8.348368	306.323

Now assume T_S = 306.2 K. Then λ_S = 2423 kJ/kg, Y_S = 0.0328 kg water/kg dry air.

$$h_R = \frac{5.1084\left(3.80^4 - \left(\frac{305.8}{100}\right)^4\right)}{(380-305.8)} = 8.3484\ \text{W/m}^2\ \text{K}$$

From Eq. (A), T_S = 306.3 K. This is reasonably close to the assumed value. Hence, the surface temperature T_S = 306.2 K is accepted.

The rate of drying in the constant rate period is estimated using Eq. (9.31).

$$R_C = \frac{(h_G + U_k)(T_G - T_S) + h_R(T_R - T_S)}{\lambda_S}$$

$$R_C = \frac{(45.168+23.254)\times 3.6\times(333-306.2)+8.3484\times 3.6\times(380-306.2)}{2423} = 3.64\,\text{kg/m}^2\,\text{h}$$

Rate of drying is 3.64 kg/m^2 h. **(Answer)**

Effect of Process Variables on R_C

The rate of drying in the constant-rate period is evaluated by Eq. (9.4):

$$R_C = k_Y(Y_S - Y) = \frac{h_G}{\lambda_S}(T_G - T_S)$$

The heat transfer coefficient for parallel flow of air is given as [Eq. (9.6)]

$$h_G = 0.0204G^{0.8}$$

and for perpendicular flow [Eq. (9.7)],

$$h_G = 1.17G^{0.37}$$

The variables affecting the rate of drying such as the gas velocity, temperature and humidity of the gas and the thickness of the drying bed can now be predicted.

Gas velocity

Equations (9.4), (9.6) and (9.7) reveal that if the heat transfer by conduction and radiation through the bed of solids is negligible, then the rate of drying in the constant rate period is proportional to $G^{0.8}$ for parallel flow and $G^{0.37}$ for perpendicular flow. Gas velocity will not have much influence on the rate of drying if heat transfer by radiation and conduction are also present.

Temperature

The change in the drying rate with changes in the gas temperature can be estimated as follows. Let the original conditions of the air are T_{G1}, T_{S1}, Y_1, Y_{S1} and conditions at the second state are T_{G2}, T_{S2}, Y_2, Y_{S2}. Then using Eq. (9.4), and assuming that the heat of vaporisation is independent of the temperature for moderate changes in temperatures,

$$R_{C1} = k_Y(Y_{S1} - Y_1) = \frac{h_G}{\lambda_S}(T_{G1} - T_{S1}) \qquad \text{...(9.34)}$$

$$R_{C2} = k_Y(Y_{S2} - Y_2) = \frac{h_G}{\lambda_S}(T_{G2} - T_{S2}) \qquad \text{...(9.35)}$$

Let the humidity be not changed. $Y_1 = Y_2 = Y$. Combining Eqs. (9.34) and (9.35),

$$R_{C2} = R_{C1}\frac{(Y_{S2} - Y)}{(Y_{S1} - Y)} = R_{C1}\frac{(T_{G2} - T_{S2})}{(T_{G1} - T_{S1})} \qquad \text{...(9.36)}$$

If radiation effects and the influence of temperature on the heat of vaporisation λ_S are negligible, the rate of drying at constant rate period is directly proportional to the temperature difference ($T_G - T_S$). By increasing the temperature of the gas stream, $T_G - T_S$ is increased and hence the rate of drying increases by increasing the temperature of the gas.

Gas humidity

At constant temperature T_G, if the humidity of the air used is decreased, the wet-bulb temperature T_W is also decreased, and by Eq. (9.4) the rate should increase. For this case, Eq. (9.36) can be written as

$$R_{C2} = R_{C1}\frac{(Y_{S2} - Y_2)}{(Y_{S1} - Y_1)} = R_{C1}\frac{(T_G - T_{S2})}{(T_G - T_{S1})} \qquad \text{...(9.37)}$$

It should be noted that since the changes in T_G and Y involve simultaneous changes in T_S and Y_S, R_C should be estimated using Eq. (9.4).

Thickness of the drying bed

If the heat transfer by conduction is negligible relative to that by convection, thickness of the solid bed has no influence on the rate of drying. On the other hand, if conduction heat transfer is important, increasing the thickness of the solid bed would result in the decrease of the drying rate. At the same time, the increase in the thickness will lead to an increase in the heat conduction through the edge surfaces, if edges are not insulated, and therefore, the rate of drying will increase. If the edge surfaces are heat insulated and the heat conduction occurs from both faces, the rate of drying R_C will be independent of the bed thickness. However, the time required for drying from a given initial to the final moisture content varies in direct proportion to the thickness, as the mass S in the equations for calculating the time of drying is directly proportional to the thickness.

Shrinkage and Case Hardening

Some materials, particularly the colloidal and fibrous materials undergo shrinkage when the moisture is removed from them. The extent of shrinkage is linearly related to the change in the moisture content. For materials such as vegetables and foodstuffs, shrinkage alters the extent of surface exposed to the drying gas per unit weight of the substance. For materials such as wood, shrinkage causes them to warp or check and change its structure. Shrinkage may lead to *case hardening* in certain materials such as soap and clay. Case hardening is the formation of a hardened and dry outer shell, impervious to the flow of moisture, while the inside of the solid is still wet. In many food stuffs, if drying occurs at a high temperature, a layer of closely packed shrunken cells, which are sealed together, is formed at the surface, which acts as a barrier to the flow of moisture. The moisture cannot move as liquid or as vapour, to the surface or to the boundary at which the evaporation is taking place.

For materials that warp or check, or those which undergo case hardening, it is necessary to retard evaporation of the moisture and bring it in tune with the rate of moisture movement from the interior. This is achieved by increasing the humidity of the air used for drying. By supplying relatively moist air, Y in Eq. (9.4) is increased, the humidity difference $(Y_S - Y)$ is decreased and the drying rate is reduced. Thus, wood dryers are supplied in the initial stages of drying with humidified air, to reduce the moisture gradient within the solid and hence to minimize the effect of shrinkage.

9.7 THEORETICAL ESTIMATION OF DRYING RATE AND TIME

Empirical approach of estimating the time of drying presented in the preceding section is applicable to the particular conditions for which the data were obtained. Two mechanisms for moisture migration from the interior of the solids have been proposed which can be used to estimate theoretically, the drying time and the moisture concentration profile in the solid. Liquid diffusion and capillary movement are the two mechanisms proposed to explain the different types of falling rate curves, as already discussed. The mechanism of the moisture movement affects the drying during the constant rate and the falling rate periods.

Liquid Diffusion

The difference in the moisture concentration existing between the surface and the interior depths of the solid provides the driving force for diffusion of the moisture to the surface. This is the mechanism of moisture movement in nonporous solids in which the internal moisture is predominantly the bound moisture. Liquid diffusion prevails in the last stages of drying of clay, starch, flour, textiles, paper and wood, and in the drying of soluble moisture as in soap, gelatine, glue and paste. In drying many food materials, the movement of water in the falling rate period occurs by diffusion.

The moisture profile developed by the transport of moisture, is concave downward throughout as shown in Fig. 9.45a. The shape of the diffusion controlled rate curve is similar to that shown in Fig. 9.45b. Fick's second law for unsteady state diffusion can be written to determine the rate of drying in the falling-rate period.

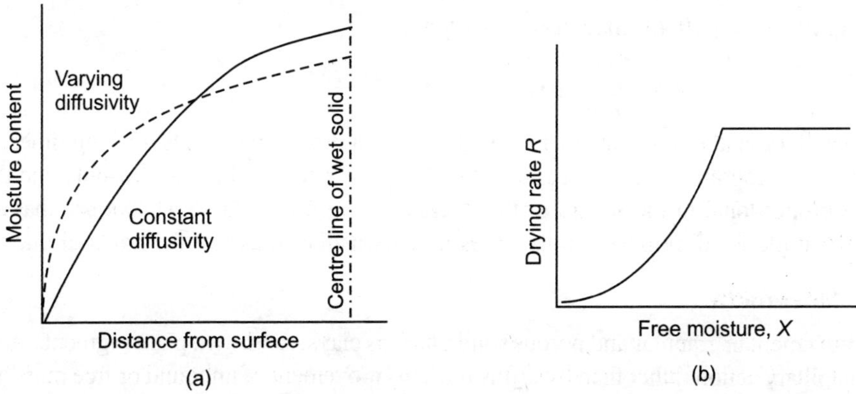

Fig. 9.45: (a) Moisture profile in wet solid during drying when moisture flow is by liquid diffusion (b) Diffusion-controlled falling-rate period

$$\frac{\partial X}{\partial \theta} = D_{AB} \frac{\partial^2 X}{\partial z^2} \qquad \text{...(9.38)}$$

where D_{AB} is the liquid diffusion coefficient and z is the distance through the solid.

During the diffusion-type drying, the resistance to mass transfer from the surface is usually very small, and the diffusion in the solid controls the rate of drying. Then the moisture content at the surface can be assumed to be the equilibrium value X^*. It is assumed that initially a uniform moisture content X_0 exists in the wet solid. At the beginning of the falling-rate period ($\theta = 0$), the exposed faces are brought to the equilibrium moisture content X^*. Equation (9.38) is integrated with the following boundary conditions.

$$X = X_0, \text{ at } \theta = 0 \text{ for } -a < z < a,$$

$$X = X^*, \text{ at } z = \pm a \text{ for } \theta \geq 0$$

where a is half the thickness of the slab of wet material when subjected to drying from both faces. (If drying is occurring from one face only, a is the thickness of the slab.) The result is

$$\frac{X - X^*}{X_0 - X^*} = \frac{8}{\pi^2}\left[e^{-D_{AB}\theta(\pi/2a)^2} + \frac{1}{9}e^{-9D_{AB}\theta(\pi/2a)^2} + \frac{1}{25}e^{-25D_{AB}\theta(\pi/2a)^2} + \ldots\right] \qquad \text{...(9.39)}$$

where X is the average moisture content at time θ. The moisture diffusivity usually decreases with the moisture content so that an average D_{AB} value over the range of concentrations is used. Though D_{AB} varies with the moisture content, temperature and the humidity, it is assumed to be constant for the present analysis. For long drying times, only the first term in Eq. (9.39) is significant, and the equation simplifies to

$$\frac{X - X^*}{X_0 - X^*} = \frac{8}{\pi^2}\left[e^{-D_{AB}\theta(\pi/2a)^2}\right] \qquad \text{...(9.40)}$$

Solving for the time of drying,

$$\theta = \frac{4a^2}{\pi^2 D_{AB}} \ln \frac{8(X_0 - X^*)}{\pi^2 (X - X^*)} \qquad \text{...(9.41)}$$

Differentiating Eq. (9.40) with respect to time and rearranging,

$$\frac{dX}{d\theta} = -\frac{\pi^2 D_{AB}(X - X^*)}{4a^2} \qquad \text{...(9.42)}$$

Multiplying both sides by $-S/A$,

$$R = -\frac{S}{A}\frac{dX}{d\theta} = \frac{\pi^2 S D_{AB}}{4a^2 A}(X - X^*) \qquad \text{...(9.43)}$$

Utilising Eq. (9.40), Eq. (9.43) may also be written as

$$R = \frac{2SD_{AB}}{a^2 A}(X_0 - X^*)\left[e^{-D_{AB}\theta(\pi/2a)^2}\right] \qquad \text{...(9.44)}$$

Equations (9.43) and (9.44) state that when the internal diffusion prevails for long times, the rate of drying is directly proportional to the free moisture $(X_0 - X^*)$ and the liquid diffusivity and that the rate is inversely proportional to the square of the thickness. And from Eq. (9.41) we see that the time of drying between the fixed moisture limits varies directly as the square of the material thickness.

Capillary Movement

Moisture movement in granular and porous solids such as clays, sand, soil, paint pigments, and minerals occurs by capillary action rather than by diffusion. This movement of unbound or free moisture through the capillaries involves surface tension and is similar to the movement of oil in a lamp wick. As moisture is evaporated from the solid, a liquid meniscus is formed across each pore in the solid which sets up a capillary force by the interfacial tension between the moisture and the solid. These capillary forces provide the driving force for movement of water through the pores to the surface. Consider a capillary tube of small diameter partially immersed in a liquid reservoir. The liquid will rise in the tube to a height h above the surface of the liquid in the reservoir and at equilibrium, the height will be

$$h = \frac{2\sigma}{\rho L g r} \qquad \text{...(9.45)}$$

where σ is the surface tension of the liquid and r is the radius of the capillary. The smaller the radius of the capillary, the larger the capillary effect. Capillary mechanism for moisture movement assumes that the moisture in interstices flows to the regions of highest capillary regardless of concentrations.

For capillary movement, the moisture profile is as shown in Fig. 9.46a. The profile is concave upward near the exposed surface and concave downward at the opposite surface with a point of inflection in between. Since the evaporation in the falling rate period is controlled by the capillary flow, the effect of the variables on the rate will be the same as that in the constant rate period. Noting that the rate of drying is

$$R = -\frac{S}{A}\frac{dX}{d\theta} \qquad \text{...(9.1)}$$

for the rate R varying linearly with X,

$$R = R_C \frac{X - X^*}{X_C - X^*} \qquad \text{...(9.16)}$$

and

$$\theta = \frac{S(X_C - X^*)}{AR_C} \ln \frac{X_C - X^*}{X - X^*} \qquad \text{...(9.19)}$$

Since

$$S = aA\rho_S \qquad \text{...(9.46)}$$

where a is the thickness of the material and ρ_S is the solid density, kg dry solid/m^3. Substituting Eq. (9.46) into Eq. (9.19),

$$\theta = \frac{a\rho_S(X_C - X^*)}{R_C} \ln \frac{X_C - X^*}{X - X^*} \qquad \text{...(9.47)}$$

Substituting Eq. (9.4) for R_C,

$$\theta = \frac{a\rho_S \lambda_W (X_C - X^*)}{h_G(T_G - T_W)} \ln \frac{X_C - X^*}{X - X^*} \qquad \text{...(9.48)}$$

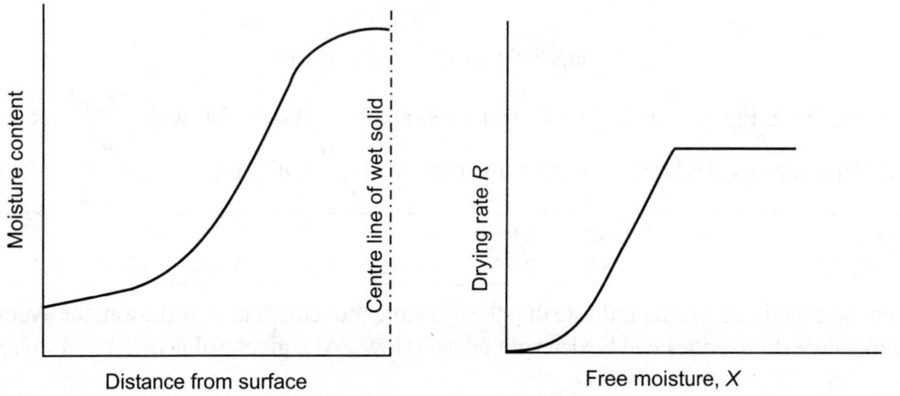

Fig. 9.46: (a) Moisture profile in wet solid during drying when moisture flow is by capillary forces (b) Capillary flow-controlled falling-rate period in fine porous solid

Thus according to Eqs. (9.47) and (9.48), if the moisture movement follows the capillary-flow equations, the rate of drying R will vary linearly with free moisture $X–X^*$ as shown by the linear portion of the falling rate period in Fig. 9.46b and inversely with the thickness. The time of drying between the fixed moisture limits varies directly as the thickness. Since the mechanism of evaporation during this period is the same as that in the constant-rate period, the effect of the variables such as the gas velocity, the temperature of the gas, the humidity of the gas, etc., will be the same as that in the constant-rate drying period.

Analysis of Data

In order to determine the mechanism of moisture movement experimentally, logarithm of the unaccomplished moisture change in the solid at time θ, $(X - X^*)/(X_C - X^*)$ is plotted as a function of time. If either Eq. (9.48) for the capillary flow or Eq. (9.41) for the liquid diffusion is applicable, a straight line will be obtained such as the curve (a) in Fig. 9.47 plotted using the upper axis for the abscissa. For the moisture movement by capillary flow, the slope of the straight line gives $-R_C/a\rho_S(X_C - X^*)$ from which R_C is estimated. If it agrees with the experimental value of the rate of drying in the constant rate period, the moisture movement is by capillary flow. If it does not agree, the moisture movement is by diffusion and the slope of the line should be equal to $-\dfrac{\pi^2 D_{AB}}{4a^2}$.

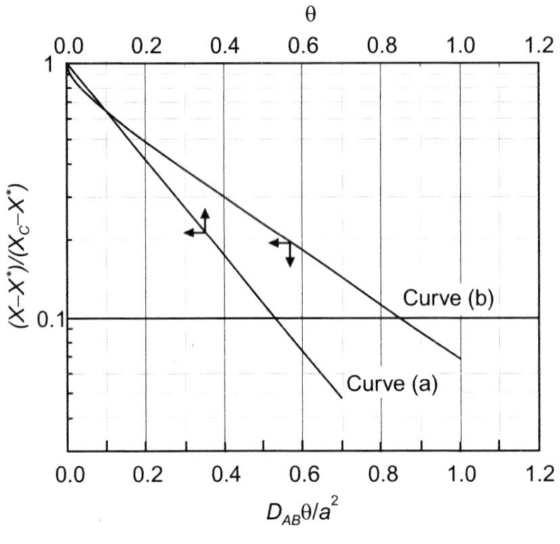

Fig. 9.47: Analysis of drying data

The curve (b) in Fig. 9.47 is the graphical representation of Eq. (9.39) with $\dfrac{D_{AB}\theta}{a^2}$ as the abscissa. The following values calculated for different values of $\dfrac{D_{AB}\theta}{a^2}$ in Eq. (9.39) are used to plot this curve.

$D_{AB}\theta/a^2$	0	0.02	0.05	0.1	0.15	0.2	0.3	0.5	1
$(X - X^*)/(X_C - X^*)$	1	0.84	0.75	0.642	0.563	0.496	0.387	0.236	0.069

When the experimental data indicate that the moisture movement is by diffusion, the average experimental moisture diffusivities can be determined as follows. At a given value of $(X - X^*)/(X_C - X^*)$, read the corresponding value of $\left(\dfrac{D_{AB}\theta}{a^2}\right)_t$ from the curve (b) in Fig. 9.47. At the same value of the ordinate, read the corresponding experimental value of θ from the lower curve in Fig. 9.47. Then

$$D_{AB,avg} = \frac{\left(D_{AB}\theta/a^2\right)_t}{\left(\theta/a^2\right)_e} \quad \ldots(9.49)$$

The subscripts t and e appearing in Eq. (9.49) indicate the theoretical and experimental values, respectively. The value of diffusivity calculated by Eq. (9.49) is an average value over the entire range of moisture change from $(X-X^*)/(X_C-X^*) = 1$ to $(X-X^*)/(X_C-X^*)$ at which the θ/a^2 was calculated.

Example 9.10: In a drying test, it took 5 h to dry a sample of mashed potatoes from the initial moisture content of 0.5 kg/kg dry solids to the final moisture content of 0.15 kg/kg dry solids. The sample is taken in a pan 10 mm deep and insulated on the bottom and exposed to drying air flowing across the surface. All the drying is in the falling-rate period and is diffusion controlled. The equilibrium moisture content is 0.1 kg/kg dry solids. In the actual drying, the potatoes are to be dried from the initial moisture content of 0.5 kg/kg dry solids to a final moisture content of 0.25 kg/kg dry solids. The material will be taken in pans 50 mm deep but with perforated metal bottoms so that drying occurs from both faces and air conditions remain the same as in the test. Determine the drying time required.

***Solution*:** The drying time can be computed using Eq. (9.41)

$$\theta = \frac{4a^2}{\pi^2 D_{AB}} \ln \frac{8(X_0 - X^*)}{\pi^2 (X - X^*)}$$

where a is half the thickness if drying is from both surfaces (a is the thickness of the slab when drying is from a single surface), D_{AB} is the average diffusivity for moisture, and X_0, X, and X^* are, respectively, the initial moisture content, final average moisture content and equilibrium moisture content.

***Case 1*:** $a = 10 \times 10^{-3}$ m; $X_0 = 0.50$ $X = 0.15$, $X^* = 0.10$; $\theta = 5$ h

Substituting these values in Eq. (9.41)

$$5 = \frac{4(10\times10^{-3})^2}{\pi^2 D_{AB}} \ln \frac{8(0.50-0.10)}{\pi^2(0.15-0.10)}$$

$$5 = \frac{7.5763\times10^{-5}}{D_{AB}}$$

from which $D_{AB} = 1.5153 \times 10^{-5}$ m²/h

***Case 2*:** $a = 25 \times 10^{-3}$ m; $X_0 = 0.50$ $X = 0.25$, $X^* = 0.10$; $\theta = ?$

Assuming that the diffusivity remains constant, $D_{AB} = 1.5153 \times 10^{-5}$ m²/h.

Substituting in Eq. (9.41), we get

$$\theta = \frac{4(25\times10^{-3})^2}{\pi^2 \times 1.5153\times10^{-5}} \ln \frac{8(0.50-0.10)}{\pi^2(0.25-0.10)} = 12.89 \text{ h}$$

Drying time required in the actual drying = 12.89 h **(Answer)**

Example 9.11: The following data on the unaccomplished moisture change $\frac{X-X^*}{X_C-X^*}$ versus time is available for drying certain material of vegetable origin in the falling rate period. The material is 4 mm thick and drying is from both sides under constant drying conditions.

θ, h	0	0.1	0.2	0.3	0.4	0.5	0.6	0.7	0.8	0.9	1.0
$(X-X^*)/(X_C-X^*)$	1.0	0.85	0.72	0.62	0.53	0.45	0.38	0.32	0.27	0.23	0.20

Determine the average diffusivity of moisture up to an unaccomplished moisture change of 0.25

(a) Using the simplified equation assuming long drying times.

(b) Using the general theoretical model.

***Solution*:** The given data are plotted on a semi-logarithmic graph with $(X - X^*)/(X_C - X^*)$ on the y-axis and time θ, on the x-axis. The result is the upper curve in Fig. 9.48.

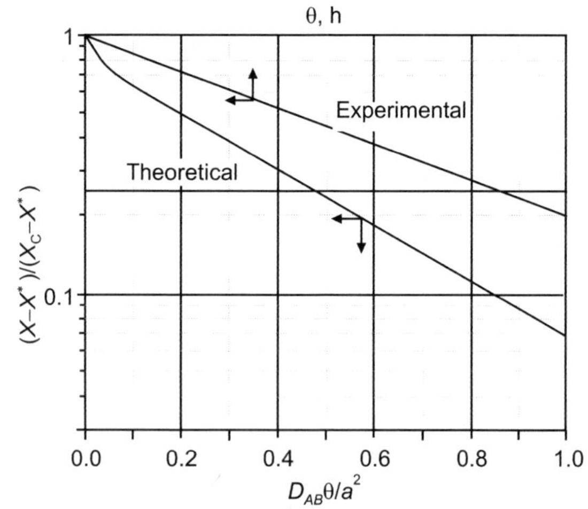

Fig. 9.48: Analysis of drying data of Example 9.11

From this curve, for $\dfrac{X - X^*}{X_C - X^*} = 0.25$, the time of drying θ is found to be 0.857 h. For long drying times the diffusion equation simplifies to Eq. (9.40)

$$\frac{X - X^*}{X_C - X^*} = \frac{8}{\pi^2}\left[e^{-D_{AB}\theta(\pi/2a)^2}\right]$$

Substituting θ = 0.857 h in the above equation,

$$0.25 = \frac{8}{\pi^2}\left[e^{-D_{AB}\theta(\pi/2a)^2}\right] = \frac{8}{\pi^2}\left[e^{-D_{AB}0.857(\pi/4\times10^{-3})^2}\right] = \frac{8}{\pi^2}\left[e^{-528640.69D_{AB}}\right]$$

Solving the above equation for D_{AB}, we get

$$D_{AB} = 2.23 \times 10^{-6}\ \text{m}^2/\text{h} = 6.19 \times 10^{-10}\ \text{m}^2/\text{s} \qquad \textbf{(Answer)}$$

(b) The lower curve in Fig. 9.48 is the graphical representation of the diffusion Eq. (9.39) plotted using the following theoretical values:

$D_{AB}\theta/a^2$	0	0.02	0.05	0.1	0.15	0.2	0.3	0.5	1
$(X - X^*)/(X_C - X^*)$	1	0.84	0.75	0.642	0.563	0.496	0.387	0.236	0.069

From the graph, for $\dfrac{X - X^*}{X_C - X^*} = 0.25$, the parameter $D_{AB}\theta/a^2 = 0.4769$.

For the same $\dfrac{X - X^*}{X_C - X^*}$, the drying time is found to be θ = 0.857 h.

Substituting θ = 0.857 h in $D_{AB}\theta/a^2 = 0.4769$, we get

$$D_{AB} = 0.4769 \times (2\times10^{-3})^2/0.857 = 2.23\times10^{-6}\,\text{m}^2/\text{h} = 6.19\times10^{-10}\,\text{m}^2/\text{s} \qquad \textbf{(Answer)}$$

9.8 TRAY-DRYING WITH VARYING AIR CONDITIONS

Consider air being passed across a wet material taken in a tray as shown in Fig. 9.49. The depth of the air-flow space is b m. This is the distance between the material on the tray and the bottom surface of the tray above. The length of the tray is L m. Air enters the space between the trays at temperature T_{G1} and humidity Y_1 and leaves at temperature T_{G2} with humidity Y_2.

The air humidity and temperature change along the length of the tray. Consider a small section of the tray of length dL and of unit width. Let G_S be the flow rate of dry air per unit area of cross section, kg/m² h. The temperature of the gas falls by dT_G due to transfer of heat dq in this section. Writing a heat balance over this section,

$$dq = -G_S(b \times 1)C_S dT_G \quad \text{...(9.50)}$$

where C_S is the humid heat of the air. The rate equation for heat transfer from the gas is

$$dq = h_G(dL \times 1)(T_G - T_S) \quad \text{...(9.51)}$$

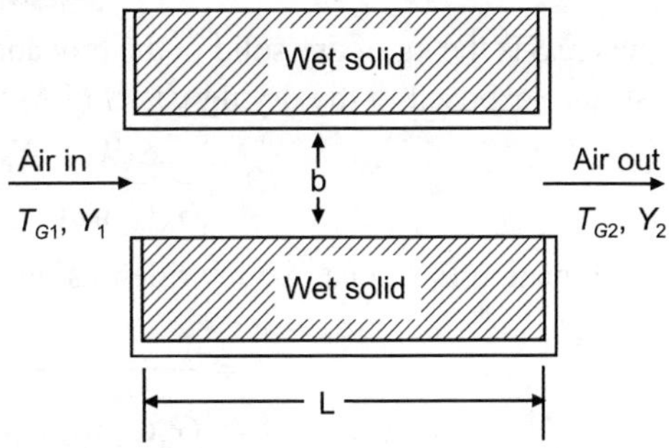

Fig. 9.49: Heat and material balance in a tray dryer

where h_G is the heat transfer coefficient and T_S is the surface temperature of the solid. Combining Eqs. (9.50) and (9.51), we get

$$-G_S(b \times 1)C_S dT_G = h_G(dL \times 1)(T_G - T_S) \quad \text{...(9.52)}$$

Or
$$\frac{dT_G}{T_G - T_S} = -\frac{h_G}{G_S b C_S} dL \quad \text{...(9.53)}$$

Integrating Eq.(9.53), we get

$$\int_{T_{G1}}^{T_{G2}} \frac{dT_G}{T_G - T_S} = -\frac{h_G}{G_S b C_S} \int_0^L dL \quad \text{...(9.54)}$$

$$\ln \frac{T_{G2} - T_S}{T_{G1} - T_S} = -\frac{h_G L}{G_S b C_S} \quad \text{...(9.55)}$$

Equation (9.55) can be rearranged as

$$\frac{T_{G2} - T_S}{T_{G1} - T_S} = \exp\left(-\frac{h_G L}{G_S b C_S}\right) \quad \text{...(9.56)}$$

so that
$$1 - \frac{T_{G2} - T_S}{T_{G1} - T_S} = 1 - \exp\left(-\frac{h_G L}{G_S b C_S}\right) \quad \text{...(9.57)}$$

Or
$$\frac{T_{G1} - T_{G2}}{T_{G1} - T_S} = 1 - \exp\left(-\frac{h_G L}{G_S b C_S}\right) \quad \text{...(9.58)}$$

The rate of drying in the constant rate period can be calculated from

$$G_S C_S (b \times 1)(T_{G1} - T_{G2}) = R_C (L \times 1)\lambda_S \quad \text{...(9.59)}$$

Or
$$R_C = \frac{G_S C_S b (T_{G1} - T_{G2})}{L \lambda_S} \quad \text{...(9.60)}$$

where λ_S is the heat of vaporisation of the moisture at the surface temperature, T_S. The time of drying in the constant rate period is given by Eq. (9.18) as

$$\theta = \frac{S(X_1 - X_C)}{AR_C}$$

Substituting Eqs. (9.60) into Eq. (9.18), the time of drying can be written as

$$\theta = \frac{S(X_1 - X_C)L\lambda_S}{AG_S C_S b(T_{G1} - T_{G2})} \quad \text{...(9.61)}$$

where S is the kg of dry solid in the tray and A is the surface exposed to the drying gas. Noting that the surface exposed to the drying gas is (wL) where w is the width of the tray, Eq. (9.61) becomes

$$\theta = \frac{S(X_1 - X_C)\lambda_S}{G_S C_S wb(T_{G1} - T_{G2})} \quad \text{...(9.62)}$$

Equation (9.62) can be rewritten using the Eq. (9.58) as

$$\theta = \frac{S(X_1 - X_C)\lambda_S}{G_S C_S bw(T_{G1} - T_S)\left[1 - \exp\left(-\frac{h_G L}{G_S b C_S}\right)\right]} \quad \text{...(9.63)}$$

For the falling rate period, we can write an approximate equation for the drying rate using the relation [Eq. (9.19)]

$$\theta = \frac{S(X_C - X^*)}{R_C A}\ln\frac{X_C - X^*}{X - X^*}$$

Therefore, for the falling rate period, we get

$$\theta = \frac{S(X_C - X^*)\lambda_S}{G_S C_S bw(T_{G1} - T_S)\left[1 - \exp\left(-\frac{h_G L}{G_S b C_S}\right)\right]}\ln\frac{X_C - X^*}{X - X^*} \quad \text{...(9.64)}$$

9.9 THROUGH CIRCULATION DRYING IN PACKED BEDS

Suppose that the drying gas is passed through a bed of wet granular particles as shown in Fig. 9.50. Assume that the bed of solids is resting on screens. The gas admitted below the screen flows through the open spaces and voids between the solid particles in the bed. Drying occurs with both constant rate and the falling rate. The evaporation of the unbound moisture occurs in a relatively narrow zone and the gas leaving this zone will be saturated with moisture. The drying in this zone is at a constant rate. This zone moves upward in the direction of the gas flow leaving behind, a section of the bed in which evaporation of the bound moisture takes place at a lower rate. Therefore, in the initial stages of the drying, the bed consists of three zones, zone 1 in which the moisture being removed is bound moisture, zone 2 in which the unbound moisture is removed at constant rate and zone 3 in which the solid still contains the original moisture concentration, as shown in Fig. 9.50a. With passage of time, zone 2 moves up increasing the thickness of zone 1 and reducing the thickness of zone 3.

Rate of Drying of Unbound Moisture

Consider a differential section in the bed of thickness dz with unit area of cross section as shown in Fig. 9.50b. Let a be the interfacial surface per unit volume of the bed so that the total surface available in the differential section is adz. The gas receives moisture and its humidity increases by dY. If the mass flow rate of the dry gas is G_S kg/h m^2 of the bed cross section, then,

$$G_S dY = k_Y(adz)(Y_S - Y) \quad \text{...(9.65)}$$

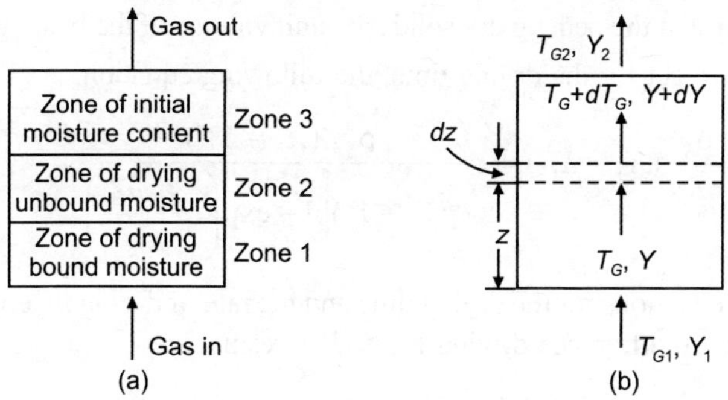

Fig. 9.50: Through-circulation drying in packed beds

It is assumed that the solids attain a temperature equal to the adiabatic saturation temperature of the gas; the humidity attained by the gas at the interface between the gas and solid is Y_S, the saturation humidity at the adiabatic saturation temperature. Rearranging Eq. (9.65) and integrating, we get

$$\int_{Y_1}^{Y_2} \frac{dY}{Y_S - Y} = \frac{k_Y a}{G_S} \int_0^Z dz \quad \text{...(9.66)}$$

Let Z be the thickness of the bed and Y_1 and Y_2 are the humidities of the gas entering and leaving the bed.

$$\ln \frac{Y_S - Y_1}{Y_S - Y_2} = \frac{k_Y a Z}{G_S} \quad \text{...(9.67)}$$

Denoting the number of gas transfer units in the bed by N_{tG}, the above equation may be written as

$$N_{tG} = \ln \frac{Y_S - Y_1}{Y_S - Y_2} = \frac{k_Y a Z}{G_S} \quad \text{...(9.68)}$$

We can put Eq. (9.68) in a more convenient form as follows:

$$\frac{Y_S - Y_2}{Y_S - Y_1} = \exp\left(-\frac{k_Y a Z}{G_S}\right) \quad \text{...(9.69)}$$

$$1 - \frac{Y_S - Y_2}{Y_S - Y_1} = 1 - \exp\left(-\frac{k_Y a Z}{G_S}\right) \quad \text{...(9.70)}$$

$$\frac{Y_2 - Y_1}{Y_S - Y_1} = 1 - \exp\left(-\frac{k_Y a Z}{G_S}\right) \quad \text{...(9.71)}$$

The constant rate of drying can be calculated as

$$R_C = \frac{G_S(Y_2 - Y_1)}{aZ} \quad \text{...(9.72)}$$

Substituting for $(Y_2 - Y_1)$ from Eq. (9.71), we get

$$R_C = \frac{G_S(Y_2 - Y_1)}{aZ} = \frac{G_S(Y_S - Y_1)}{aZ}\left[1 - \exp\left(-\frac{k_Y a Z}{G_S}\right)\right] \quad \text{...(9.73)}$$

Equation (9.18) can be used to determine the drying time.

$$\theta = \frac{S}{AR_C}(X_1 - X_C)$$

Here,
$$\frac{S}{A} = \frac{\text{mass of dry solid in the bed}}{\text{Area exposed to drying}} = \frac{Z\rho_S}{aZ} = \frac{\rho_S}{a} \quad \text{...(9.74)}$$

where ρ_S is the dry density of the bed, kg dry solids per unit volume of the bed. Substituting Eqs. (9.74) and (9.73) into (9.18), we get for the drying time, the following equation.

$$\theta = \frac{\rho_S}{aR_C}(X_1 - X_C) = \frac{\rho_S Z(X_1 - X_C)}{G_S(Y_S - Y_1)\left[1 - \exp\left(-\frac{k_Y aZ}{G_S}\right)\right]} \quad \text{...(9.75)}$$

We can also develop equations for the drying time and the rate of drying in terms of the heat transfer coefficient and temperature difference driving force. The results are

$$\theta = \frac{\rho_S Z\lambda_S(X_1 - X_C)}{G_S C_S(T_{G1} - T_S)\left[1 - \exp\left(-\frac{h_G aZ}{G_S C_S}\right)\right]} \quad \text{...(9.76)}$$

$$R_C = \frac{\rho_S}{a\theta}(X_1 - X_C) = \frac{G_S C_S(T_{G1} - T_S)\left[1 - \exp\left(-\frac{h_G aZ}{G_S C_S}\right)\right]}{aZ\lambda_S} \quad \text{...(9.77)}$$

The time of drying in the falling rate period can be calculated in a similar manner assuming the falling rate period to be linear. Then the rate of drying R is given by

$$R = \frac{R_C(X - X^*)}{X_C - X^*} \quad \text{...(9.16)}$$

And the time of drying in the falling rate period is given by

$$\theta = \frac{S(X_C - X^*)}{R_C A}\ln\frac{X_C - X^*}{X - X^*} \quad \text{...(9.19)}$$

Then, corresponding to Eq. (9.75), we can write for the falling-rate period

$$\theta = \frac{\rho_S Z(X_C - X^*)}{G_S(Y_S - Y_1)\left[1 - \exp\left(-\frac{k_Y aZ}{G_S}\right)\right]}\ln\frac{X_C - X^*}{X - X^*} \quad \text{...(9.78)}$$

Similarly, corresponding to Eq. (9.76), we can write

$$\theta = \frac{\rho_S Z\lambda_S(X_C - X^*)}{G_S C_S(T_{G1} - T_S)\left[1 - \exp\left(-\frac{h_G aZ}{G_S C_S}\right)\right]}\ln\frac{X_C - X^*}{X - X^*} \quad \text{...(9.79)}$$

For through circulation drying, when the gas is passed through a packed bed of granular solids, the following correlations have been proposed for the heat transfer coefficient for adiabatic evaporation of water:

$$h_G = 0.151G^{0.59}/D_p^{0.41} \text{ for } Re > 350 \quad \text{...(9.8)}$$

$$h_G = 0.214G^{0.49}/D_p^{0.51} \text{ for } Re < 350 \quad \text{...(9.9)}$$

where $Re = \frac{D_P G}{\mu}$ with h_G in W/m^2 K, D_P is the particle diameter in m, G is the superficial mass velocity, kg/h m^2 and μ is the viscosity in kg/m h.

For a packed bed of spherical particles, the surface area per unit volume of packing, a is calculated from

$$a = \frac{6(1-\varepsilon)}{D_p} \qquad \ldots(9.80)$$

where ε is the void fraction in the bed. For cylindrical particles a is evaluated from

$$a = \frac{4(1-\varepsilon)(L+0.5D)}{DL} \qquad \ldots(9.81)$$

where L and D are the length and diameter of cylindrical particles. The value of D_p to be used in Eq. (9.8) and (9.9) for cylindrical particles is the diameter of a sphere having the same surface area as the cylinder.

$$D_p = (DL + 0.5D^2)^{1/2} \qquad \ldots(9.82)$$

The equations developed for calculating the rate and time of drying in the packed beds are applicable for particles of size 3-20 mm in diameter in shallow beds about 10-65 mm thick. For very fine particles in the size range 10 to 200 mesh (2.03 to 0.074 mm) and bed depth greater than 11 mm, the interfacial area varies with the moisture content and these equations cannot be used.

Example 9.12: Spherical catalyst pellets are being dried from the initial free moisture content of 1.2 kg water/kg solid to a free moisture content of 0.40 kg water/kg solid, which is well above the critical free moisture content. The pellets are 10 mm in diameter and are arranged in a bed of thickness 70 mm on a screen. The dry solid density is 1400 kg/m^3, and the void fraction of the bed is 0.45. Air at 358 K and having humidity of 0.01 kg water/kg dry air is passed through the bed with a superficial velocity 1.0 m/s. Calculate the time of drying. Take viscosity of air to be 2.1×10^{-5} kg/m s for the given conditions.

***Solution*:** Since the initial and final moisture contents are above the critical moisture content, drying occurs in the constant rate-period. The drying time is given by Eq. (9.76)

$$\theta = \frac{\rho_S Z \lambda_S (X_1 - X_2)}{G_S C_S (T_{G1} - T_S)\left[1 - \exp\left(-\dfrac{h_G a Z}{G_S C_S}\right)\right]}$$

$\rho_S = 1400$ kg/m^3; $Z = 70 \times 10^{-3}$ m; $X_1 - X^* = 1.20$ kg water/kg dry solid; $X_2 - X^* = 0.40$ kg water/kg dry solid. $T_{G1} = 358$ K, $Y = 0.01$ kg water/kg dry air; The surface temperature T_S = wet-bulb temperature of air = 307 K.

$$C_S = 1.005 + 1.884 \times 0.01 = 1.0238\,\text{kJ/kg}\ ;\ \lambda_S = 2421\,\text{kJ/kg}.$$

$G = \rho u$ where ρ is the density of air which is calculated as

$\rho = \dfrac{1+Y}{v_H}$ where v_H is the humid volume. Therefore,

$$\rho = \frac{1+Y}{8.413 \times \left(\dfrac{1}{29} + \dfrac{Y}{18}\right) \times \dfrac{T}{P}} = \frac{1.01}{8.314 \times \left(\dfrac{1}{29} + \dfrac{0.01}{18}\right) \times \dfrac{358}{101.3}} = 0.9811\,\text{kg/m}^3.$$

$$G = \rho u = 0.9811 \times 1 = 0.9811 \text{ kg/m}^2 \text{ s } (= 3531.81 \text{ kg/m}^2 \text{ h})$$

$$G_S = G\frac{1}{1+Y} = \frac{0.9811}{1.01} = 0.9714\,\text{kg/m}^2\text{ s } (= 3497 \text{ kg/m}^2 \text{ h})$$

The Reynold's number for flow through the bed is

$$Re = \frac{D_p G}{\mu} = \frac{10 \times 10^{-3} \times 0.9811}{2.1 \times 10^{-5}} = 467.19 > 350$$

The heat transfer coefficient h_G is evaluated using Eq. (9.8)

$$h_G = 0.151 G^{0.59} / D_p^{0.41}$$

$$h_G = 0.151(3531.81)^{0.59}/(10 \times 10^{-3})^{0.41} = 123.68 \text{ W/m}^2 \text{ K}$$

The interfacial area per unit volume of the bed is

$$a = \frac{6(1-\varepsilon)}{D_p} = \frac{6(1-0.45)}{0.01} = 330 \frac{\text{m}^2}{\text{m}^3}$$

Substituting the values, we get

$$\theta = \frac{1400 \times 0.07 \times 2421 \times (1.2 - 0.4)}{3497 \times 1.0238 \times (358 - 307)\left[1 - \exp\left(-\dfrac{123.68 \times 3.6 \times 330 \times 0.07}{3497 \times 1.0238}\right)\right]} = 1.1 \text{ h}$$

Time of drying = 1.1 h **(Answer)**

Example 9.13: A wet pigment material is extruded into small cylinders 6 mm in diameter, 50 mm long and are placed on screens to a depth of 60 mm. The material is subjected to through- circulation drying using a stream of air flowing through the bed with mass velocity of 1.2 kg dry air/m^2 s, entering at 393 K with a humidity 0.05 kg water/kg dry air. The surface of the particles is estimated to be 300 m^2/m^3 of the bed and the apparent density is 1000 kg dry solid/m^3. Take viscosity of the air to be 1.8×10^{-5} kg/m s for the given conditions. Estimate the constant rate of drying.

***Solution*:** The rate of drying is calculated using Eq. (9.77)

$$R_C = \frac{G_S C_S (T_{G1} - T_S)\left[1 - \exp\left(-\dfrac{h_G a Z}{G_S C_S}\right)\right]}{a Z \lambda_S}$$

$T_{G1} = 393$ K, $Y = 0.05$, $T_S = T_W = 322$ K (from humidity chart).

$$C_S = 1.005 + 1.884 \times 0.05 = 1.0992 \text{ kJ/kg K}$$

$$\lambda_S = 2385.3 \text{ kJ/kg}, \; a = 300 \text{ m}^2/\text{m}^3, \; Z = 60 \times 10^{-3} \text{ m}$$

$G_S = 1.2$ kg/m^2 s (= 4320 kg/m^2 h)

$$G = G_S(1 + Y) = 1.2 \times 1.05 = 1.26 \text{ kg/m}^2 \text{ s} \; (= 4536 \text{ kg/m}^2 \text{ h})$$

$$D_p = (DL + 0.5D^2)^{0.5} = (6 \times 50 + 0.5 \times 6^2)^{0.5} = 17.83 \text{ mm}$$

Reynold's number,

$$Re = \frac{D_p G}{\mu} = \frac{17.83 \times 10^{-3} \times 1.26}{1.8 \times 10^{-5}} = 1248 > 350. \text{ Therefore,}$$

$$h_G = 0.151 G^{0.59} / D_p^{0.41} = 0.151 \times 4536^{0.59} / (0.01783)^{0.41}$$

$$= 113.095 \text{ W/m}^2 \text{ K} \; (= 407.141 \text{ kJ/m}^2 \text{ h K})$$

The values are substituted into Eq. (9.77):

$$R_C = \frac{4320 \times 1.0992 \times (393 - 322)\left[1 - \exp\left(-\dfrac{407.141 \times 300 \times 0.06}{4320 \times 1.0992}\right)\right]}{300 \times 0.06 \times 2385.3} = 6.175 \text{ kg/m}^2 \text{ h}$$

The constant rate of drying = 6.175 kg/m^2 h **(Answer)**

9.10 CONTINUOUS DRYING

Material and Heat Balances for Continuous Dryers

A countercurrent continuous direct heat dryer is shown in Fig. 9.51. The analysis presented here is applicable to parallel current dryers as well with obvious modifications. G_S and S represent the flow rates of dry gas and moisture-free solid in kg/m^2 h or kg/h. X represents the moisture content of the solid

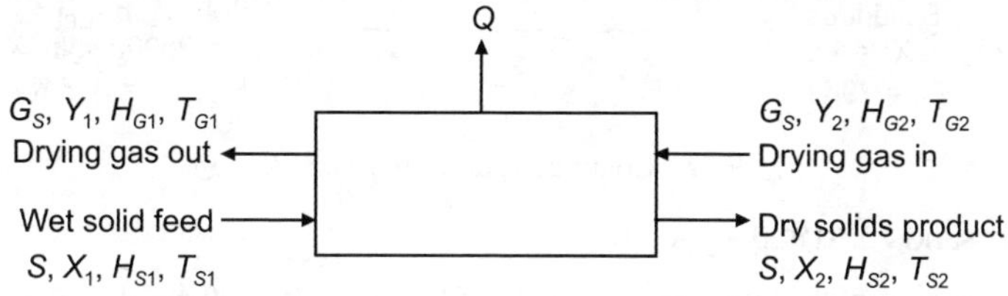

Fig. 9.51: Material and energy balance in continuous dryers

in kg moisture/kg dry solid and Y represents the humidity of the air in kg moisture/kg dry gas. T_G and T_S are the temperature of the gas and solids, respectively and H_G and H_S are, respectively, the enthalpies of the gas and solid. The suffix-1 represents the conditions at the end where the wet solids feed is admitted and the suffix-2 the conditions at the product end of the dryer.

Material balance for the moisture gives

$$SX_1 + G_S Y_2 = SX_2 + G_S Y_1 \qquad \text{...(9.83)}$$

which can be rearranged as

$$S(X_1 - X_2) = G_S(Y_1 - Y_2) \qquad \text{...(9.84)}$$

Heat balance over the dryer gives

$$SH_{S1} + G_S H_{G2} = SH_{S2} + G_S H_{G1} + Q \qquad \text{...(9.85)}$$

where Q is the net heat loss from the dryer, kJ/h. The enthalpy of the wet solid in Eq. (9.85) can be written as the sum of the enthalpies of the dry solid and the liquid present as free moisture.

$$H_S = C_{PS}(T_S - T_0) + XC_{PA}(T_S - T_0) + \Delta H_A \qquad \text{...(9.86)}$$

C_{PS} is the heat capacity of dry solids and C_{PA} is the heat capacity of the moisture as liquid. T_0 is the reference temperature for estimating the enthalpy and ΔH_A is the integral heat of wetting or adsorption or solution referred to the pure liquid and the solid at the reference temperature. ΔH_A is usually neglected for want of reliable data.

Enthalpy of the gas can be evaluated using the equation

$$H_G = C_S(T_G - T_0) + Y\lambda_0 \qquad \text{...(9.87)}$$

C_S in Eq. (9.87) is the humid heat of the gas, kJ/kg dry gas K and λ_0 is the heat of vaporisation of the moisture at the reference temperature T_0, kJ/kg. The humid heat for the air-water vapour system is

$$C_S = 1.005 + 1.884Y \qquad \text{...(9.88)}$$

and the latent heat of vaporisation at T_0 = 273.15 K is 2502 kJ/kg.

Heat loss Q = 0 for the adiabatic operation. Q will be negative, if the heat supplied to the dryer is in excess of the heat loss from the dryer.

Example 9.14: A rotary countercurrent dryer is to be used to dry 50,000 kg/h wet sand from 50% moisture to 2.5% moisture on wet basis. The air is admitted at 400 K with a humidity of 0.007 kg water/kg dry air and it leaves the dryer at 325 K. The wet sand enters at 295 K and leaves at 318 K. Heat losses from the dryer amount to 25 kJ/ kg dry air. Specific heat of dry sand is 0.9 kJ/kg K. Specific heat of the dry air is 1.005 kJ/kg K and the specific heat of the water vapour is 1.884 kJ/kg K. Latent heat of vaporisation at 273 K is 2502 kJ/kg K. Calculate the amounts of dry air passed through the dryer and the humidity of the air leaving the dryer.

Solution:

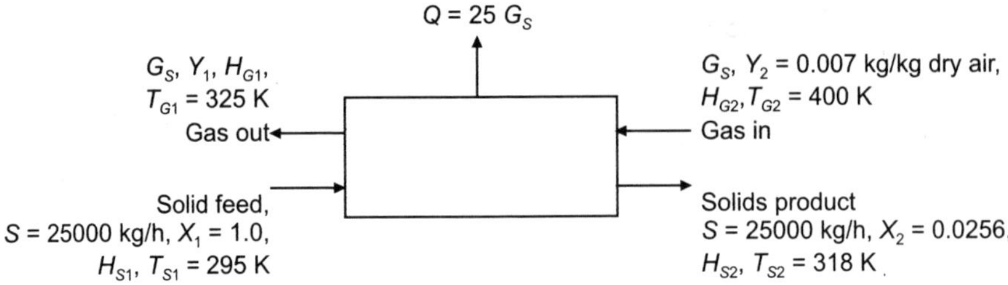

Fig. 9.52: Continuous dryer in Example 9.14

Feed rate of wet sand = 50,000 kg/h

Since the feed contains 50% moisture, $S = 0.5 \times 50{,}000 = 25{,}000$ kg/h

$$X_1 = \frac{x_1}{1 - x_1} = \frac{0.50}{0.50} = 1.0;\ X_2 = \frac{x_2}{1 - x_2} = \frac{0.025}{0.975} = 0.0256$$

$$T_{S1} = 295 \text{ K},\ T_{S2} = 318 \text{ K}$$

Enthalpy of the solids entering and leaving the dryer are calculated using Eq. (9.86) which, in the absence of the heat of solution, is

$$H_S = C_{PS}(T_S - T_0) + XC_{PA}(T_S - T_0)$$

Therefore,

$$H_{S1} = (C_{PS} + X_1 C_{PA})(T_{S1} - T_0) = (0.9 + 1 \times 4.1868)(295 - 273) = 111.91 \text{ kJ/kg}$$

$$H_{S2} = (C_{PS} + X_2 C_{PA})(T_{S2} - T_0) = (0.9 + 0.0256 \times 4.1868)(318 - 273) = 45.32 \text{ kJ/kg}$$

Air enters at 400 K with a humidity 0.007 kg water/kg dry air and leaves at 325 K.

$$T_{G2} = 400 \text{ K},\ Y_2 = 0.007,\ T_{G1} = 325 \text{ K}.$$

Enthalpies of the air entering and leaving the dryer are given by Eq. (9.87).

$$H_G = C_S(T_G - T_0) + Y\lambda_0$$

For the air entering the dryer,

$$C_{S2} = 1.005 + 1.884 \times Y_2 = 1.005 + 1.884 \times 0.007 = 1.0182 \text{ kJ/kg K}$$

$$H_{G2} = C_{S2}(T_{G2} - T_0) + Y_2\lambda_0 = 1.0182 \times (400 - 273) + 0.007 \times 2502 = 146.83 \text{ kJ/kg}$$

For the air leaving the dryer,

$$C_{S1} = 1.005 + 1.884 \times Y_1$$

$$H_{G1} = C_{S1}(T_{G1} - T_0) + Y_1\lambda_0 = (1.005 + 1.884Y_1)(325 - 273) + 2502Y_1$$

$$H_{G1} = 52.26 + 2600Y_1$$

Material and heat balances for the dryer are given by Eq. (9.84) and (9.85). Substituting the values we get,

$$25000(1 - 0.0256) = G_S(Y_1 - 0.007)$$

and

$$25000 \times 111.91 + 146.83G_S = 25000 \times 45.32 + G_S(52.26 + 2600Y_1) + 25G_S$$

These equations get simplified as

$$G_S(Y_1 - 0.007) = 24360$$

$$G_S(2600Y_1 - 69.57) = 1664750$$

which can be solved simultaneously for G_S and Y_1.

Dry air flow rate, $G_S = 1.2 \times 10^6$ kg/h **(Answer)**

Exit humidity of air, $Y_1 = 0.02729$ kg water/kg dry air **(Answer)**

Continuous Drying at Low Temperatures

The continuous countercurrent dryer operating at low temperature can be considered to be consisting primarily of two zones as shown in Fig. 9.53a. In zone 1, the surface and unbound moisture are evaporated and the moisture content falls from initial value X_1 to the critical moisture content X_C. The temperature of the solid remains almost constant at T_S because the rate of heat transfer to the solid is balanced by the heat required for evaporation. At the inlet of the solid, there is a small preheat zone in which the temperature of the solid is raised to T_S. In this zone, no significant drying occurs and the heat transferred to the solid merely goes to increase its temperature. In low-temperature drying the preheat zone is usually ignored. The mode of drying in zone 1 is identical to that in the constant rate period, except that here the temperature and humidity of the gas change during drying. The temperature profile is as shown in Fig. 9.53. Points B and C represent the inlet and exit conditions of the solids in zone 1 and points C' and B' represent the inlet and exit conditions of the gas in this zone, flowing countercurrent to the solids. In zone 2 it is the bound moisture that is evaporated and the solid temperature no longer remains constant; it rises from T_S to T_{S2} along the curve CD as shown in Fig. 9.53b. The moisture content in the solid falls from the critical moisture content at point C to the final value X_2 of the solids product. The temperature of the gas stream decreases along the smooth curve $D'C'B'A'$ as it moves countercurrent to the solids through the length of the dryer. The humidity of the gas stream increases from Y_2 at the inlet of the gas to Y_1 at the exit.

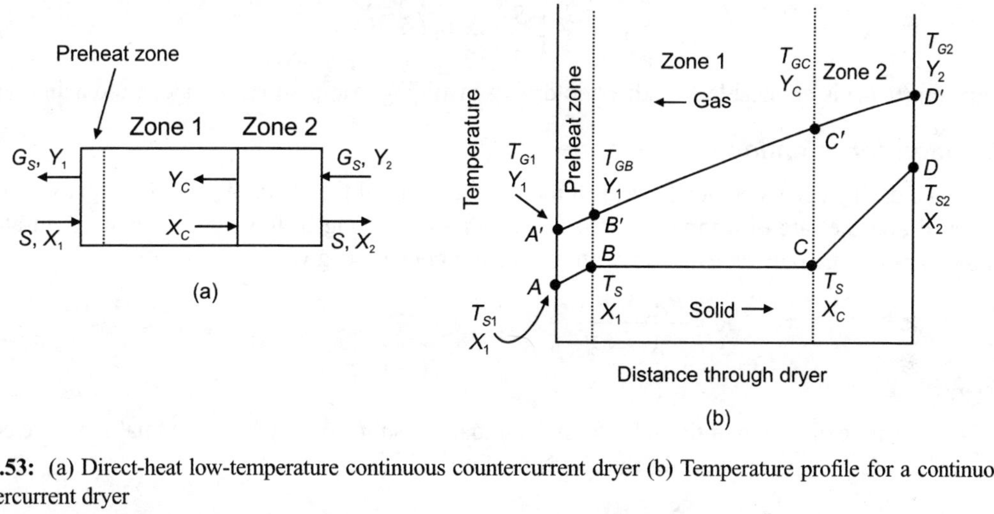

Fig. 9.53: (a) Direct-heat low-temperature continuous countercurrent dryer (b) Temperature profile for a continuous countercurrent dryer

The retention time of the solids in the dryer θ can be calculated as

$$\theta = \theta_1 + \theta_2$$

where θ_1 is the retention time in zone 1 and θ_2, that in zone 2.

Equation for retention time in zone 1

Equation for retention time can be written by rearranging Eq. (9.13).

$$\theta_1 = \frac{S}{A}\int_{X_C}^{X_1} \frac{dX}{R} \qquad \text{...(9.89)}$$

where A/S is the surface exposed to the drying gas per unit mass of the solids m^2/kg dry solids. Since the rate of drying is governed by the heat and mass transfer from the surface of the solid, the rate R in Eq. (9.89) can be replaced using Eq. (9.2)

$$\theta_1 = \frac{S}{A}\frac{1}{k_Y}\int_{X_C}^{X_1}\frac{dX}{(Y_S - Y)} \qquad \text{...(9.90)}$$

Since the moisture balance $SdX = G_S dY$ is applicable to the dryer, dX in Eq. (9.90) can be replaced in terms of dY.

$$\theta_1 = \frac{G_S}{S}\left(\frac{S}{A}\right)\frac{1}{k_Y}\int_{Y_C}^{Y_1}\frac{dY}{(Y_S - Y)} \qquad \text{...(9.91)}$$

Y_C is the humidity of the gas stream leaving zone 2. It can be estimated by a moisture balance for zone 1 or zone 2. For zone 2, we have

$$S(X_C - X_2) = G_S(Y_C - Y_2) \qquad \text{...(9.92)}$$

so that

$$Y_C = Y_2 + \frac{S}{G_S}(X_C - X_2) \qquad \text{...(9.93)}$$

Y_S in Eq. (9.91) is the humidity of the gas which is in contact with the solid surface. It varies with the humidity of the bulk gas and this variation should be taken into account when determining the integral in Eq. (9.91) graphically. For a given value of gas humidity Y, Y_S is the saturation humidity at the adiabatic saturation temperature of the gas provided, the heat transfer by conduction and radiation are negligible. If Y_S is assumed constant, the integral in Eq. (9.91) can be determined analytically.

$$\theta_1 = \frac{G_S}{S}\left(\frac{S}{A}\right)\frac{1}{k_Y}\ln\frac{Y_S - Y_C}{Y_S - Y_1} \qquad \text{...(9.94)}$$

Equation (9.94) is applicable for adiabatic drying involving evaporation of water vapour into air.

Equation for retention time in zone 2

When the drying occurs below the critical moisture content of the solid, the conditions shown in zone 2 prevail and the rate of drying falls. Assuming that the drying in this zone is entirely unsaturated surface drying, the rate varies linearly with moisture content as given in Eq. (9.16).

$$R = \frac{R_C(X - X^*)}{X_C - X^*} \qquad \text{...(9.16)}$$

where

$$R_C = k_Y(Y_S - Y) \qquad \text{...(9.2)}$$

Now the time of drying of the solid from the initial moisture content X_C to the final moisture content X_2, which is the retention time for the solids in zone 2, is given by

$$\theta_2 = \frac{S}{A}\int_{X_2}^{X_C}\frac{dX}{R} \qquad \text{...(9.95)}$$

Substituting Eqs. (9.2) and (9.16) into Eq. (9.95), we get

$$\theta_2 = \frac{S}{A}\frac{1}{k_Y}\int_{X_2}^{X_C}\frac{(X_C - X^*)dX}{(Y_S - Y)(X - X^*)} \qquad \text{...(9.96)}$$

The integral in Eq. (9.96) can be determined graphically after establishing the relationship between the variables X^*, X, Y_S and Y. If the equilibrium moisture content X^* is negligible,

$$\theta_2 = \frac{S}{A}\frac{X_C}{k_Y}\int_{X_2}^{X_C}\frac{dX}{(Y_S - Y)X} \qquad \text{...(9.97)}$$

The moisture balance gives

$$SdX = G_S dY \qquad \text{...(9.98)}$$

which can be integrated to give

$$S(X - X_2) = G_S(Y - Y_2) \quad ...(9.99)$$

Equation (9.99) can be rearranged to give

$$X = X_2 + \frac{G_S}{S}(Y - Y_2) \quad ...(9.100)$$

Substituting Eqs. (9.100) and (9.98) into Eq. (9.97), we get

$$\theta_2 = \frac{G_S}{S}\left(\frac{S}{A}\right)\frac{X_C}{k_Y}\int_{Y_2}^{Y_C}\frac{dY}{(Y_S - Y)\left[X_2 + \frac{G_S}{S}(Y - Y_2)\right]} \quad ...(9.101)$$

Assuming Y_S to be constant, Eq. (9.101) becomes

$$\theta_2 = \frac{G_S}{S}\left(\frac{S}{A}\right)\frac{X_C}{k_Y}\frac{1}{X_2 + \frac{G_S}{S}(Y_S - Y_2)}\ln\frac{X_C(Y_S - Y_2)}{X_2(Y_S - Y_C)} \quad ...(9.102)$$

If the rate of drying is controlled by internal diffusion, a linear variation of the rate with free moisture content is not justifiable. Therefore, Eq. (9.102) should not be used for determining the retention time, if the rate is controlled by internal diffusion.

Example 9.15: A wet solid with a moisture content 0.5 kg water/kg dry solid is to be dried to 0.04 kg water/kg dry solid at a rate of 600 kg/h dry solids in a continuous countercurrent dryer. Air flows at a rate 12,000 kg dry air/h. It enters the dryer at 378 K with a humidity of 0.055 kg water/kg dry air. The solid enters the dryer at the wet-bulb-temperature of the inlet air. The surface area available for drying is 0.06 m²/kg dry solid. The critical moisture content of the solid is 0.12 kg water/kg dry solid. The mass transfer coefficient k_Y is 150 kg water/m² h. In the falling rate period, the drying rate was directly proportional to moisture content. Calculate the time of drying (a) in the constant rate period (b) in the falling-rate period.

***Solution*:** Refer to Fig. 9.54. The following values are known.

X_1 = 0.5 kg/kg dry solid, X_2 = 0.04 kg/kg dry solid, X_C = 0.12 kg/kg dry solid
S = 600 kg/h, G_S = 12000 kg/h, Y_2 = 0.055 kg/kg dry air, A/S = 0.06 m²/kg dry solid
k_Y = 150 kg water/m² h.

(a) *Calculation for zone 1: Constant rate drying*

Retention time in zone 1 is given by Eq. (9.94)

$$\theta_1 = \frac{G_S}{S}\left(\frac{S}{A}\right)\frac{1}{k_Y}\ln\frac{Y_S - Y_C}{Y_S - Y_1}$$

A moisture balance for the dryer can be written as

$$S(X_1 - X_2) = G_S(Y_1 - Y_2)$$

$$600(0.50 - 0.04) = 12000(Y_1 - 0.055)$$

Therefore, Y_1 = 0.078 kg/kg dry air.

The wet-bulb temperature of air at 378 K and humidity 0.055 kg /kg dry air is found to be T_W = 322 K from the psychrometric charts and the solids in zone 1 may be assumed to be at this temperature. The saturation humidity of air at 322 K = 0.082 kg/kg dry air (from the psychrometric chart). Y_S = 0.082 kg/kg dry air.

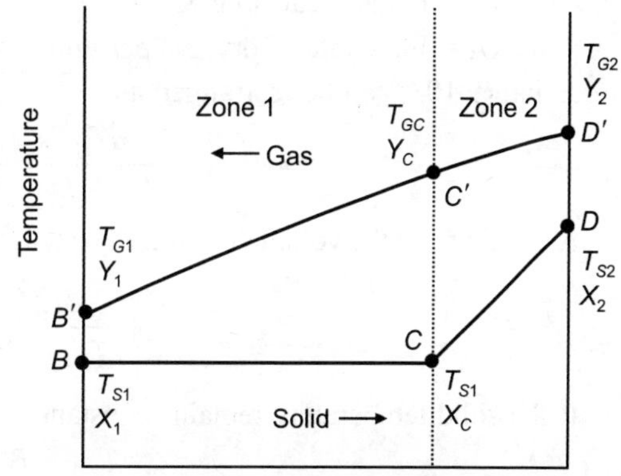

Fig. 9.54: Variation of temperature and moisture content in continuous drying (Example 9.15)

A moisture balance may be written for zone 1 for determining the humidity of air Y_C entering this zone.

$$S(X_1 - X_C) = G_S(Y_1 - Y_C)$$

$$600(0.50 - 0.12) = 12000(0.078 - Y_C)$$

which gives $Y_C = 0.059$ kg/kg dry air

Equation (9.94) now gives

$$\theta_1 = \frac{12000}{600}\left(\frac{1}{0.06}\right)\frac{1}{150}\ln\frac{0.082 - 0.059}{0.082 - 0.078} = 3.89 \text{ h}$$ **(Answer)**

(b) *Calculation for zone 2: Drying with a falling rate*

The retention time in zone 2 is calculated using Eq. (9.102):

$$\theta_2 = \frac{G_S}{S}\left(\frac{S}{A}\right)\frac{X_C}{k_Y}\frac{1}{X_2 + \frac{G_S}{S}(Y_S - Y_2)}\ln\frac{X_C(Y_S - Y_2)}{X_2(Y_S - Y_C)}$$

$$\theta_2 = \frac{12000}{600}\left(\frac{1}{0.06}\right)\frac{0.12}{150}\frac{1}{0.04 + \frac{12000}{600}(0.082 - 0.055)}\ln\frac{0.12(0.082 - 0.055)}{0.04(0.082 - 0.059)}$$

$$= 0.58 \text{ h}$$ **(Answer)**

Continuous Drying at High Temperatures

For continuous dryers operated at high temperatures, the design calculations are made based on heat transfer. In a differential section of the dryer the sensible heat transferred from the gas to the solid results in a drop in its temperature by dT_G. We can write

$$G_S C_S dT_G = Ua(T_G - T_S)dZ \tag{9.103}$$

where

U = Overall heat transfer coefficient between the gas and the solid, W/m^2 K.

a = interfacial surface per unit volume of the dryer, m^2/m^3.

Z = length of the dryer, m

$T_G - T_S$ = temperature difference driving force for heat transfer, between the gas and the solid, K.

C_S = humid heat, J/kg K

G_S = Flow rate of dry gas per unit area of cross section of the dryer, kg/m^2 h

Equation (9.103) can be rearranged as

$$\frac{dT_G}{T_G - T_S} = \frac{Ua}{G_S C_S}dZ \quad \text{...(9.104)}$$

Assuming that the volumetric overall heat transfer coefficient Ua and the humid heat C_S are constant,

$$\int_{T_{G1}}^{T_{G2}} \frac{dT_G}{T_G - T_S} = \frac{Ua}{G_S C_S}\int_0^Z dZ \quad \text{...(9.105)}$$

If the solid temperature remains constant as in the constant rate region zone 1, Eq. (9.105) gives

$$\ln\frac{T_{G2} - T_S}{T_{G1} - T_S} = \frac{UaZ}{G_S C_S} \quad \text{...(9.106)}$$

If the temperature profiles of the gas and solid streams can be approximated by straight lines, Eq. (9.106) can be written as

$$\frac{\Delta T_G}{(\Delta T)_m} = \frac{UaZ}{G_S C_S} \quad ...(9.107)$$

where $$\Delta T_G = T_{G2} - T_{G1} \quad ...(9.108)$$

and $$(\Delta T)_m = \frac{T_{G2} - T_{G1}}{\ln \dfrac{T_{G2} - T_{S2}}{T_{G1} - T_{S1}}} \quad ...(9.109)$$

where $T_{G1} - T_{S1}$ and $T_{G2} - T_{S2}$ are the temperature difference driving forces at the two terminals of the different zones in the dryer and $(\Delta T)_m$ is the logarithmic mean temperature difference. For zone 2, Eq. (9.107) is applicable if the falling rate is due to unsaturated surface drying, but not applicable if the rate of drying is controlled by internal diffusion.

Equation (9.107) can be written in terms of the number and lengths of heat transfer units.

$$N_{tOG} = \frac{Z}{H_{tOG}} \quad ...(9.110)$$

where N_{tOG} = Number of heat transfer units $= \dfrac{\Delta T_G}{(\Delta T)_m}$ ...(9.111)

H_{tOG} = Height of heat transfer units $= \dfrac{G_S C_S}{Ua}$ ...(9.112)

Z = Length of the dryer.

Direct-heat Rotary Dryer

Using the continuity equation, the diameter of the direct-heat rotary dryer can be estimated using the total mass flow rate G at the exit end of the dryer.

$$G = \frac{\pi}{4} D^2 u \rho_G \quad ...(9.113)$$

Where u is the allowable gas velocity and ρ_G is the gas density at the exit. Thus, rearranging Eq. (9.113), we get

$$D = \left[\frac{4G}{\pi u \rho_G}\right]^{0.5} \quad ...(9.114)$$

The residence time of the solids in the dryer θ, is given by

$$\theta = \frac{L V_H}{F_V} \quad ...(9.115)$$

In Eq. (9.115), V_H is the fractional hold up of the solids. It is the fraction of the dryer volume that is occupied by the solid at any instant. L is the length of the dryer cylinder and F_V is the solids volumetric velocity, $m^3/s\ m^2$ cross sectional area. The following relationship may be used to estimate the fractional holdup of solids.

$$V_H = \frac{0.3344 F_V}{S N^{0.9} D} \pm 0.6085 \frac{G}{D_p^{0.5} \rho_S} \quad ...(9.116)$$

In Eq. (9.116), S is the slope of the dryer, m/m, N is the rotational speed, rev/s, and D is the diameter in m, ρ_S is the apparent solids density, kg dry solids/m^3, D_p is the particle diameter in m. The plus sign (+) corresponds to countercurrent flow and minus sign (–) for cocurrent flow in the dryer. The above equation is valid for allowable gas velocity up to 1.1 m/s.

When drying is in the constant rate period, the volumetric heat-transfer coefficient, Ua, can be used to determine the rate of drying.

$$q = UaV(\Delta T)_m \quad \ldots(9.117)$$

where V is the volume of the dryer cylinder and ΔT_m is the logarithmic mean of the temperature differences at the inlet and exit of the dryer.

$$(\Delta T)_m = \frac{T_{G1} - T_{G2}}{\ln \dfrac{T_{G1} - T_{S1}}{T_{G2} - T_{S2}}} \quad \ldots(9.118)$$

The volumetric heat transfer coefficient based on the dryer cylinder volume may be estimated when the drying gas is air, by the empirical correlation due to McCormick (1973):

$$Ua = 237G^{0.67}/D \quad (9.119)$$

where Ua is in W/m^3 K, G is gas flow rate per unit cross sectional area in kg/m^2 s and D in m.

Example 9.16: A wet granular substance is to be dried from 10% to 0.5% moisture in a countercurrent adiabatic rotary dryer at a rate of 10,000 kg/h of bone-dry solids. The solids enter the dryer at 300 K and leave at 370 K. The heating air enters at 810 K with a humidity of 0.03 and a wet-bulb temperature of 339 K. The solids have a specific heat of 2.1 kJ/kg K. The maximum allowable mass velocity of the air is 7000 kg/m^2 h. All moisture contents are on the dry basis. What should be the diameter and length of the dryer?

***Solution*:** The solids enter the dryer at 300 K and leaves at 370 K. T_{S1} = 300 K, T_{S2} = 370 K.

Material is dried from 10% moisture to 0.5%. $X_1 = 0.10$, $X_2 = 0.005$

Air enters the dryer at 810 K with a humidity of 0.03. T_{G2} = 810 K, $Y_2 = 0.03$

Assume the exit temperature of the gas to be T_{G1} = 400 K.

(The rough estimate of exit temperature is made by assuming $N_{tG} = \ln \dfrac{T_{G2} - T_{S2}}{T_{G1} - T_{S1}} = 2.0$ with $T_{S2} = T_{S1}$ = 339 K, the wet-bulb temperature of air entering the dryer.)

The critical moisture content of the solid is not specified in the problem. It is reasonable to assume that the drying involves the removal of only unbound moisture so that the entire drying occurs in zone 1. The temperature and the moisture of the gas and solid change along the length of the dryer, as shown in Fig. 9.55b. The drying zone is preceded by a preheating zone in which the solids temperature rises from T_{S1} to the drying temperature T_S. This temperature remains constant till the final moisture content is reached. Thereafter, the temperature of the solid rises to the exit temperature T_{S2}. In adiabatic dryers, the drying of the solids occurs at T_S, which is equal to the wet-bulb temperature of the air. Though the wet-bulb temperature depends on the temperature T_{GC} and the humidity Y_C of the air leaving the heating zone (zone 2) and entering the drying zone (zone 1), for the initial estimate, T_S is assumed to be equal to the wet-bulb temperature of the air admitted to the dryer.

$$T_S = 339 \text{ K}.$$

The enthalpies of the solids at various points, A (T_{S1} = 300 K, X_1 = 0.10), B (T_S= 339K, X= 0.10), C (T_S = 339 K, X= 0.005) and D (T_{S2} = 370 K, X= 0.005) can now be computed using Eq. (9.86). Assuming negligible heat of mixing,

$$H_S = C_{PS}(T_S - T_0) + XC_{PA}(T_S - T_0) = (C_{PS} + XC_{PA})(T_S - T_0)$$

Thus,

$$H_{S1} = (2.1 + 0.10 \times 4.1868)(300 - 273) = 68.00 \text{ kJ/kg}$$

$$H_{SB} = (2.1 + 0.10 \times 4.1868)(339 - 273) = 166.23 \text{ kJ/kg}$$

$$H_{SC} = (2.1 + 0.005 \times 4.1868)(339 - 273) = 139.98 \text{ kJ/kg}$$

$$H_{S2} = (2.1 + 0.005 \times 4.1868)(370 - 273) = 205.73 \text{ kJ/kg}$$

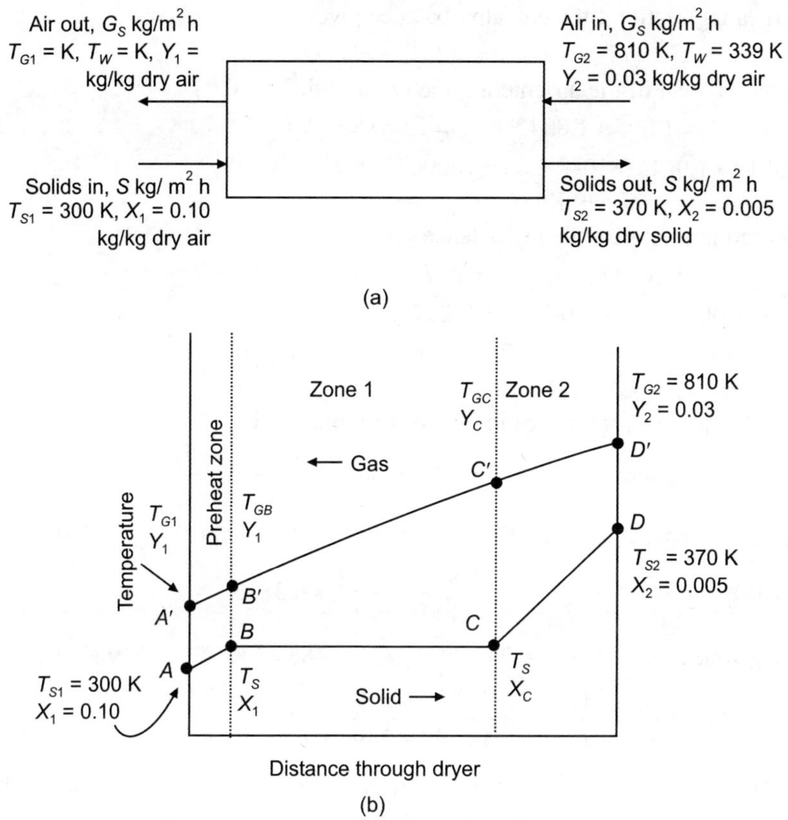

Fig. 9.55: Countercurrent rotary dryer of Example 9.16

Enthalpy of the air entering the dryer (H_{G2}) and the enthalpy of the air leaving the dryer (H_{G1}) can be calculated using Eq. (9.87) which gives

$$H_G = C_S(T_G - T_0) + Y\lambda_0 \text{ where } C_S = 1.005 + 1.884Y$$

$$H_{G1} = (1.005 + 1.884 \times Y_1)(400 - 273) + 2502Y_1 = 127.635 + 2741.268Y_1 \text{ kJ/kg}$$

$$H_{G2} = (1.005 + 1.884 \times 0.03)(810 - 273) + 0.03 \times 2502 = 645.096 \text{ kJ/kg}$$

Moisture balance and heat balance for the dryer are given by Eq. (9.84) and (9.85):

$$S(X_1 - X_2) = G_S(Y_1 - Y_2)$$

$$SH_{S1} + G_S H_{G2} = SH_{S2} + G_S H_{G1} + Q$$

The moisture balance becomes

$$10000(0.10 - 0.005) = G_S(Y_1 - 0.03)$$

or

$$G_S(Y_1 - 0.03) = 950 \qquad \text{(A)}$$

The heat balance becomes

$$10000 \times 68 + 645.096G_S = 10000 \times 205.73 + (127.635 + 2741.268Y_1)G_S$$

$$G_S(517.461 - 2741.268Y_1) = 1377300 \qquad \text{(B)}$$

Solving Eqs. (A) and (B) simultaneously, we get

$$Y_1 = 0.1338 \text{ kg water/kg dry air}$$

$$G_S = 9148.19 \text{ kg dry air/h}$$

The temperature of the air at the intermediate points B and C can now be calculated by the enthalpy balance over the solids preheating section and the product heating section (zone 2).

For zone 2 (heating section), the enthalpy balance gives

$$G_S C_S (T_{G2} - T_{GC}) = S(H_{S2} - H_{SC})$$

Here C_S is the humid heat of the air entering the dryer with $Y_2 = 0.03$.

$$C_S = 1.005 + 1.884 \times 0.03 = 1.0615 \text{ kJ/kg}$$

$$9148.19 \times 1.0615(810 - T_{GC}) = 10000(205.73 - 139.98)$$

Solving this, we get $T_{GC} = 742.29$ K

For the the preheating section, enthalpy balance gives

$$G_S C_S (T_{GB} - T_{G1}) = S(H_{SB} - H_{S1})$$

where $C_S = 1.005 + 1.884 \times 0.1338 = 1.2571$ kJ/kg K

$$9148.19 \times 1.2571(T_{GB} - 400) = 10000(166.23 - 68.00)$$

Solving this, we get $T_{GB} = 485.42$ K

The number of heat transfer units can now be determined:

$$N_{tOG} = \text{Number of heat transfer units} = \frac{\Delta T_G}{(\Delta T)_m} = \ln \frac{T_{G2} - T_{S2}}{T_{G1} - T_{S1}}$$

For the preheating zone, $T_{GB} = 485.42$ K, $T_{G1} = 400$ K; $T_{SB} = 339$ K, $T_{S1} = 300$ K

$$(N_{tOG})_1 = \ln \frac{T_{GB} - T_{SB}}{T_{G1} - T_{S1}} = \ln \frac{485.42 - 339}{400 - 300} = 0.3813$$

For the drying zone (zone 1), $T_{GC} = 742.29$ K, $T_{GB} = 485.42$ K; $T_{SC} = 339$ K, $T_{SB} = 339$ K

$$(N_{tOG})_2 = \ln \frac{T_{GC} - T_{SC}}{T_{GB} - T_{SB}} = \ln \frac{742.29 - 339}{485.42 - 339} = 1.0132$$

For the heating zone (zone 2), $T_{G2} = 810$ K, $T_{GC} = 742.29$ K; $T_{S2} = 370$ K, $T_{SC} = 339$ K

$$(N_{tOG})_3 = \ln \frac{T_{G2} - T_{S2}}{T_{GC} - T_{SC}} = \ln \frac{810 - 370}{742.29 - 339} = 0.0871$$

Total number of transfer units for all the three sections put together is

$$N_{tOG} = (N_{tOG})_1 + (N_{tOG})_2 + (N_{tOG})_3$$

$$= 0.3813 + 1.0132 + 0.0871 = 1.4816$$

The maximum allowable gas mass velocity is 7000 kg/m^2 h and the average gas rate is

$$G = G_S(1 + Y_{av}) = 9148.19 \times \left(1 + \frac{0.03 + 0.1338}{2}\right) = 9897.43 \text{ kg/h}$$

The minimum area of cross section permissible is

$$A = \frac{9897.43}{7000} = 1.4139 \text{ m}^2$$

That is $\frac{\pi}{4} D^2 = 1.4139$ where D is the minimum diameter of the dryer.

$$D = \frac{9897.43}{7000} = \left(\frac{1.4139 \times 4}{\pi}\right)^{0.5} = 1.34 \text{ m}^*$$

(* The standard sizes may be used for the diameter and the gas and solids mass velocity may be recalculated in the actual design calculations.)

The volumetric heat transfer coefficient can be calculated using Eq. (9.119)

$$Ua = 237 G^{0.67} / D$$

where Ua is in W/m^3 K, G is the gas flow rate per unit cross sectional area, kg/m^{2}s and D in m.

$$Ua = 237(7000/3600)^{0.67}/1.34 = 276.145\ \text{W/m}^3\text{K}.$$

The height of transfer unit is given by Eq. (9.112)

$$H_{tOG} = \text{Height of heat transfer units} = \frac{G_S C_S}{Ua}$$

The average humid heat of the gas is

$$C_S = 1.005 + 1.884 \times \frac{0.03 + 0.1338}{2} = 1.1593\ \text{kJ/kg K}$$

$$H_{tOG} = \frac{G_S C_S}{Ua} = \frac{7000 \times 1.1593}{276.145 \times 3.6} = 8.163\ \text{m}$$

The length of the dryer is given by

$$Z = H_{tOG} \times N_{tOG} = 8.163 \times 1.4816 = 12.1\,\text{m}$$

Length of the dryer = 12.1 m, diameter of the dryer = 1.34 m **(Answer)**

9.11 FREEZE DRYING

In freeze drying, the material to be dried is initially in a frozen state. The moisture gets vaporised when it receives the heat of sublimation from external sources. The plane of sublimation, which is initially the outer surface of the material, retreats inward with the progress of drying, leaving behind a layer of porous dried material as shown in Fig. 9.56. Let L be the thickness of the solids to be dried in the direction of heat transfer and l be the thickness of the dried product at any instant. Heat is transferred to the solid surface by convection and through the solid to the surface of sublimation, by conduction. Let h be the heat transfer coefficient at the surface of the material and k be the thermal conductivity of the dried material. Then heat transfer rate at any instant is

$$q = \frac{T_e - T_f}{1/h + l/k} \qquad \text{...(9.120)}$$

where T_e is the external temperature of the drying medium, T_f is the temperature at the sublimation front.

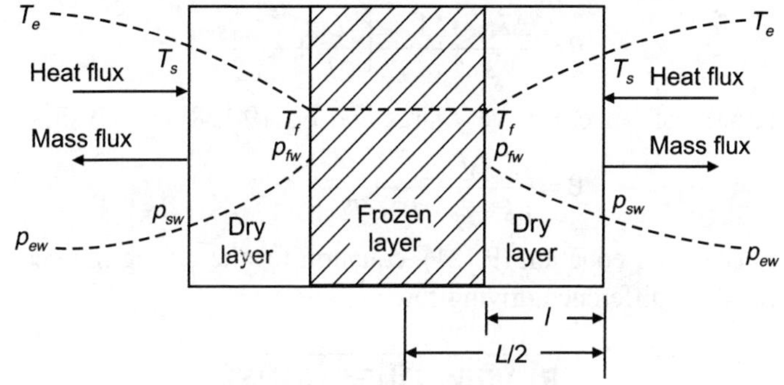

Fig. 9.56: Freeze drying model

In a similar manner, the rate of mass transfer from the sublimation front is given by

$$N_W = \frac{p_{fw} - p_{ew}}{1/k_G + RTl/D_{AB}} \qquad \text{...(9.121)}$$

where p_{fw} and p_{ew} are the partial pressure of moisture at the surface of the frozen layer and at the bulk gas, respectively. p_{fw} is the vapour pressure of water at the sublimation temperature. k_G is the external mass transfer coefficient and D_{AB} is an average effective diffusivity in the dry layer. The diffusivity D_{AB}

and the thermal conductivity k are properties of the material being dried. The coefficients h and k_G are determined by the gas velocities and the dryer characteristics and are constant. The gas properties are determined by the external operating conditions T_e and p_{ew}.

At steady state, the heat received at the sublimation front is balanced by the heat removed by the sublimation of ice.

$$q = \Delta H_s N_W \quad \text{...(9.122)}$$

Substituting q from Eq. (9.122) into Eq. (9.121),

$$\frac{T_e - T_f}{1/h + l/k} = \Delta H_s N_W \quad \text{...(9.123)}$$

where ΔH_S is the heat of sublimation of ice in kJ/kg (or kJ/kmol) and N_W is the mass flux of water vapour, kg/m^2 s (or kmol/m^2 s).

Let V_S be the volume of material that contains 1 kg of frozen moisture in it. Then at any instant, the moisture remaining in the solid per unit area in the direction of heat and mass transfer is

$$w = \frac{L/2 - l}{V_S} \quad \text{...(9.124)}$$

The rate of vaporisation of the moisture is then

$$N_W = -\frac{dw}{d\theta} = -\frac{d}{d\theta}\left(\frac{L/2 - l}{V_S}\right) = \frac{1}{V_S}\frac{dl}{d\theta} \quad \text{...(9.125)}$$

Substituting Eq. (9.125) into Eq. (9.123),

$$\frac{\Delta H_s}{V_S}\frac{dl}{d\theta} = \frac{T_e - T_f}{1/h + l/k} \quad \text{...(9.126)}$$

Rearranging and integrating Eq. (9.126) gives θ, the time required for completely drying the material.

$$\int_0^{\theta} d\theta = \frac{\Delta H_s}{V_S}\frac{1}{T_e - T_f}\int_0^{L/2} (1/h + l/k)dl \quad \text{...(9.127)}$$

$$\theta = \frac{\Delta H_s}{V_S}\frac{1}{T_e - T_f}\left(\frac{L}{2h} + \frac{L^2}{8k}\right) \quad \text{...(9.128)}$$

For negligible external resistance, h is very large and Eq. (9.128) can be simplified as

$$\theta = \frac{L^2}{8k}\frac{\Delta H_s}{V_S}\frac{1}{T_e - T_f} \quad \text{...(9.129)}$$

Using the mass transfer flux equation (9.121), equation for the drying time can also be derived in terms of the partial pressure difference driving force p_{fw}-p_{ew}.

REVIEW QUESTIONS

1. What are the benefits of the drying operations in process industries?
2. Distinguish between the unit operations of drying and evaporation.
3. What are the problems encountered with respect to the product quality in drying operations?
4. How do you classify the solid materials from the view point of the drying mechanisms? Describe with examples.
5. Distinguish between the moisture content expressed on the dry basis and the wet basis.
6. Define equilibrium moisture content.

7. What are the factors that determine the equilibrium moisture content of solids?
8. How does the temperature affect the equilibrium moisture content of a solid?
9. Distinguish between the bound and the unbound moisture.
10. What is free moisture? Illustrate with a numerical example.
11. How are the drying equilibrium curves of soluble and insoluble solids different?
12. Explain the equilibrium moisture content of a material in which the bound water exists in chemically combined form.
13. What is fibre saturation point?
14. Explain the phenomena of hysteresis with reference to drying.
15. What are the advantages and disadvantages of direct heat and indirect heat dryers?
16. Why is batch dryer more suitable for small scale drying?
17. What are the advantages of continuous dryers?
18. Why continuous countercurrent dryers are not suitable for drying heat sensitive materials?
19. Describe why the thermal efficiency of continuous countercurrent dryer is generally lower than that of cocurrent dryers.
20. What are the important methods of solids handling employed in commercial dryers?
21. Explain with the help of at least two examples each, the direct heat batch dryers and indirect heat batch dryers.
22. What are the common pre-forming treatments used in preparing the feed to a through circulation type tray dryer? Why is such a treatment necessary?
23. Describe the major advantages and areas of application of batch tray dryers.
24. List some examples of indirect batch dryers. Describe their areas of application.
25. Describe the salient features and uses of agitated pan dryers, rotating vacuum dryers.
26. Describe the importance and advantages of freeze drying.
27. Discuss the construction, working and fields of applications of tunnel dryers and band dryers.
28. The turbo dryers are not affected by the temperature constraints and product contamination problems. Why?
29. What are the type of materials that can be handled in rotary dryers?
30. Describe the role of the following component parts of a rotary dryer: spiral flights, longitudinal flights, knockers, gear and pinions, fans, screw conveyor and cyclone.
31. Explain why gypsum, iron pyrites, and organic materials such as peat and alfafa are better dried using cocurrent rotary dryers using flue gases.
32. For drying of white pigments at high temperature what type of dryer would you recommend?
33. Chambers with large cross sectional area, small bed height, large free-board height above the bed are preferred for fluidized bed dryers. Why?
34. What are the general characteristics of the materials that can be dried in fluidized and spouted bed dryers?
35. Describe the construction, working and applications of the pneumatic-conveyor dryer.
36. Describe the major components of a spray drying system. What are the major industrial applications of spray dryers?
37. Feed atomiser can be treated as the most important component of a spray dryer. Justify.
38. What are the different types of atomisers used in spray dryers?
39. Discuss the different methods of feeding the drum dryers, their advantages and applications.
40. Explain the principle behind microwave and dielectric heating.
41. What are the advantages of microwave and dielectric drying over conventional dryers?
42. What are the constant drying conditions with reference to the batch drying experiments?

43. Define the rate of drying. How is the rate determined as a function of the moisture content by conducting batch drying tests?
44. Discuss the different regions of a typical rate-of drying curve.
45. Discuss the mechanism of drying in the constant rate period.
46. How do you justify the two distinct falling rate regions in drying?
47. What is unsaturated surface drying? How is the mechanism of drying in this period different from that in the constant rate period?
48. How would you estimate the constant rate of drying from the heat and mass transfer data?
49. Why the heat transfer rate equations are more preferred for calculating R_C than the mass transfer rate equations?
50. Explain the significance of critical moisture content in drying.
51. Explain why materials with smaller thickness have lower critical moisture content.
52. Describe the effect of gas velocity, humidity, temperature and thickness of the drying bed on the rate of drying under constant rate period.
53. What is case hardening? How would you prevent case hardening in drying?

MULTIPLE CHOICE QUESTIONS

9.1 Which one of the following is not true with regard to amorphous or fibrous materials?
A. Moisture is trapped in fibres or very fine pores.
B. Solid structure or properties are unaffected by the drying conditions.
C. Wood, leather, soap, cotton starch, wool etc belong to this class of materials.
D. The equilibrium moisture content is higher than that of granular materials.

9.2 A material contains 20% water on wet basis. What is the moisture content of the material on dry basis?
A. 33.3% B. 25%
C. 16.67% D. 80%

9.3 Which one of the following statements is true with regard to equilibrium moisture content?
A. The partial pressure of water in the solid is equal to the relative humidity of air.
B. The partial pressure of water in the solid is equal to the vapour pressure of water at the temperature of air
C. The partial pressure exerted by water in the solid is equal to the partial pressure of water vapour in the air
D. The partial pressure of water in the solid is negligible at this condition.

9.4 Which one of the following statements is incorrect with regard to equilibrium moisture content (X^*)?
A. Spongy, cellular materials have high X^* values compared to granular inorganic materials
B. With increase in temperature, X^* of a solid decreases.
C. Moisture in excess of X^* only can be removed in a drying operation with given supply of gas.
D. X^* of a solid is independent of the humidity of the air in contact with the solid.

9.5 A wet solid containing 33% moisture is to be dried in a current of air of 50% relative humidity. The equilibrium moisture content of the solid is 10% and the highest concentration of moisture in the solid that can exist in equilibrium with unsaturated air is 28%. (All moisture contents are on dry basis). Which of the following is true?
A. The free moisture in the solid is 23% B. The bound moisture is 5%
C. The unbound moisture is 28% D. The free moisture is 17%

9.6 In the presence of saturated air

A. no moisture can be removed from a solid

B. only unbound moisture can be removed

C. only bound moisture can be removed.

D. moisture in fine capillaries within the solid can be removed

9.7 For a given material, if the thickness of the material is smaller, the critical moisture content

A. increases
B. decreases.
C. remains unaffected
D. cannot be predicted.

9.8 A 250 mm × 250 mm × 10 mm flat wet sheet weighing 1.2 kg initially, was dried from both the sides under constant drying rate period. It took 1500 s for the weight of the sheet to reduce to 1.05 kg. Another 1 m × 1 m × 10 mm flat sheet of the same material, is to be dried from one side only. Under the same constant drying rate conditions, time required for drying in seconds from initial weight of 19.2 kg to 17.6 kg is

A. 1000
B. 1500
C. 2000
D. 2500

9.9 A solid is being dried in the linear drying rate regime from moisture content X_0 to X_F. The drying rate is zero at $X = 0$ and the critical moisture content is the same as the initial moisture, X_0. The drying time for ($M = S/AR_C$) is

A. $M(X_0 - X_F)$
B. $M(X_0/X_F)$
C. $M\ln(X_0/X_F)$
D. $MX_0 \ln(X_0/X_F)$

where S = total mass of dry solids, A = total surface area for drying, R_C = constant maximum drying rate per unit area, and X = moisture content in mass of water/mass of dry solids

9.10 A wet paper pulp contains 75% water. After 100 kg water is removed in the dryer, it is found that the pulp is now containing 30% water. The weight of the original pulp is ……… kg

A. 116.7
B. 222.2
C. 296.3
D. 155.6

9.11 In a laboratory test-run, the rate of drying was found to be 0.5×10^{-3} kg/m^2 s when the moisture content reduced from 0.4 to 0.1 on a dry basis. The critical moisture content of the material is 0.08 on a dry basis. A tray dryer is used to dry 100 kg (dry basis) of the same material under identical conditions. The surface area of the material is 0.04 m^2/kg of dry solid. The time required in s to reduce the moisture content of the solids from 0.3 to 0.2 (dry basis) is

A. 2000
B. 4000
C. 5000
D. 6000

9.12 A batch of material is dried under constant drying conditions. When drying is taking place from all the surfaces, the rate of drying under the constant rate period is:

A. Directly proportional to the solid thickness.

B. Independent of the solid thickness.

C. Inversely proportional to the solid thickness.

D. Directly proportional to the square of solid thickness.

9.13 It takes 6 h to dry a wet solid from 50% moisture content to the critical moisture content of 15%. How much longer will it take to dry the solid to 10% moisture content, under the same drying conditions. The equilibrium moisture content of the solid is 5%. All moisture content are on dry basis.

A. 15 min
B. 51 min.
C. 71 min.
D. 94 min.

9.14 A sample of wool containing 40% moisture has an equilibrium moisture content of 12.5% in contact with air of RH 50%. The critical moisture content is 20%. All moisture contents are on dry basis. How much water can be removed by drying the sample in a current of air of 50% RH?

A. 50% B. 7.5%
C. 27.5% D. 37.5%

9.15 200 kg of solid (on dry basis) is subjected to a drying process for a period of 5000 s. The drying occurs in the constant rate period with drying rate as $R_C = 0.5 \times 10^{-3}$ kg/m^2 s. The initial moisture content of the solid is 0.2 kg moisture/kg dry solid. The interfacial area available for drying is 4 m^2/100 kg dry solid. The moisture content at the end of the drying period in kg moisture/kg dry solid is

A. 0.5 B. 0.05
C. 0.10 D. 0.15

9.16 To prevent shrinkage and case hardening on drying, which of the following is recommended?

A. Use gas at high temperature B. Use gas at high velocity
C. Use gas with high humidity D. Use gas at low temperature and humidity

9.17 In batch drying of fibrous materials .... is generally not observed to occur.

A. Falling rate period B. First falling rate period.
C. Second falling-rate period. D. Constant rate period.

9.18 If the rate of drying is practically independent of the velocity of the gas, the drying is

A. internal diffusion controlled. B. in the constant rate period
C. capillary flow controlled. D. surface heat transfer resistance controlled

9.19 Which one of the following is likely to be true for granular and crystalline solids?

A. Lengthy constant-rate period and linear falling rate period
B. Short constant rate period and concave falling rate period
C. Constant rate period extending up to the equilibrium moisture content.
D. First falling rate period is concave upward and the second falling rate is linear.

9.20 Which one of the following is likely to be true for amorphous or fibrous materials?

A. Lengthy constant-rate period and linear falling rate period
B. Short constant rate period and concave falling rate period
C. Constant rate period extending up to the equilibrium moisture content.
D. First falling rate period is concave upward and the second falling rate is linear.

9.21 The entire drying may occur in the constant rate period for

A. cross circulation drying of crystalline solids
B. fluidized bed drying of granular and crystalline solids
C. drying of pasty and sticky materials on trays
D. continuous drying of amorphous or fibrous materials

9.22 In the constant-rate period, the rate-controlling mechanism is

A. internal diffusion of moisture B. surface vaporisation
C. capillary flow D. vapour diffusion

9.23 Removal of last traces of bound moisture is facilitated better in

A. batch dryers B. continuous cocurrent dryers
C. continuous countercurrent dryers D. indirect heat dryers

9.24 Which one of the following is true for continuous cocurrent dryers in comparison with continuous countercurrent dryers?

A. They can handle heat sensitive materials
B. They have lower thermal efficiency.

C. They are more suited for removing last traces of unbound moisture
D. It is difficult to control the moisture content of the solid leaving the dryer.

9.25 Which one of the following is favourable as far as batch tray dryers are concerned?
A. Labour cost
B. Heat economy
C. Uniform moisture distribution in dried product
D. Suitability for multi-product plants.

9.26 Which one of the following is ideal for drying easily oxidizable and heat sensitive pharmaceutical product?
A. Batch Tray dryer B. Rotary dryer
C. Vacuum shelf dryer D. Freeze dryer

9.27 For drying pastes or slurries that tend to agglomerate on drying, the dryer recommended is
A. agitated pan-dryer B. tray dryer
C. rotary dryer D. vacuum shelf dryer

9.28 Which one of the following is recommended for a heat sensitive material which is likely to undergo size degradation on drying?
A. Agitated pan-dryer
B. Double-cone vacuum dryer
C. Paddle-agitated cylindrical rotating dryer
D. Double-truck tray dryer

9.29 For drying continuous sheets of wet cloth, modified versions of .... can be used.
A. shelf dryers B. tunnel dryers
C. double-cone vacuum dryers D. rotary dryers

9.30 Suggest a continuous dryer that can be used for all materials that are dried in batch tray dryers.
A. Fluidised bed dryers B. Spray dryers
C. Rotary dryers D. Turbo dryers

9.31 For large scale continuous drying of minerals, sand, limestone, clays, etc. using hot flue gases
A. spray dryers are used B. direct rotary dryers are used
C. indirect rotary dryers are used D. drum dryers are used

9.32 Rapid drying of gypsum, iron pyrites, and organic materials such as peat and alfalfa using hot flue gases can be done using
A. direct heat cocurrent rotary dryers
B. indirect heat cocurrent rotary dryers
C. direct heat counter current rotary dryers
D. indirect heat counter current rotary dryers

9.33 Which of the following would you recommend for drying materials such as white pigments, precipitated calcium carbonate and cattle feeds?
A. Direct heat rotary dryer using flue gas
B. Indirect heat rotary dryer
C. Direct-indirect rotary dryers
D. Fluidized bed dryer.

9.34 Free flowing material with almost uniform particle size is to be dried continuously with a very short residence time for the solids in the dryer. Ideal dryer for this purpose is
A. rotary dryer B. tunnel dryer
C. band dryer D. fluidized bed dryer

9.35 Name a dryer that combines drying and pulverization in a single unit

A. rotary dryer
B. fluidized bed dryer
C. flash dryer
D. tunnel dryer

9.36 Name the dryer which combines evaporation, crystallisation, filtration, size reduction, classification, and drying.

A. Rotary dryer
B. Fluidized bed dryer
C. Tunnel dryer
D. Spray dryer

9.37 Which one of the following would you recommend for producing a high grade product in granular form?

A. Spray dryer
B. Rotary direct dryer
C. Fluidized bed dryer
D. Tunnel dryer

9.38 For continuous drying of thick pasty materials and heavy sludges, we can use a

A. spray dryer
B. flash dryer
C. tray dryer
D. drum dryer

9.39 According to the liquid diffusion theory, the time of drying is

A. directly proportional to the thickness of material
B. inversely proportional to the thickness of material
C. directly proportional to the square of the thickness of material
D. inversely proportional to the square of the thickness of material

9.40 The rate of drying during constant rate period

A. is unaffected by air humidity
B. increases with increase in air humidity
C. decreases with increase in air humidity
D. may increase or decrease with increase in air humidity

9.41 Hold up of solids in rotary driers generally lie between

A. 0.1 and 0.2
B. 0.6 and 0.75
C. 0.05 and 0.15
D. 0.4 and 0.5

9.42 The flights provided in rotary driers generally extend from the wall to about .... percent of the diameter of the drier

A. 10%
B. 30%
C. 60%
D. 70%

9.43 Rotary driers are used

A. for making milk powder
B. for drying pastes and slurries
C. for drying free-flowing granular materials
D. for drying of fruits

9.44 A filter cake is dried with air at wet and dry-bulb temperatures of 300 and 323 K, respectively. The heat transfer coefficient is 11 W/m^2 K and the latent heat of vaporisation of water is 2500 kJ/kg. Mass transfer does not limit the process. Select the drying rate during constant rate period. Neglect conduction through the solid and radiation effects.

A. 1.32×10^{-2} kg/m^{2}s
B. 0.71×10^{-2} kg/m^{2}s
C. 4.53×10^{-2} kg/m^{2}s
D. 0.10×10^{-3} kg/m^{2}s

9.45 A batch of 120 kg wet solid has initial moisture content of 0.2 kg water/kg dry solid. The exposed area of drying is 0.05 m^2/kg dry solid. The rate of drying follows the curve given below.

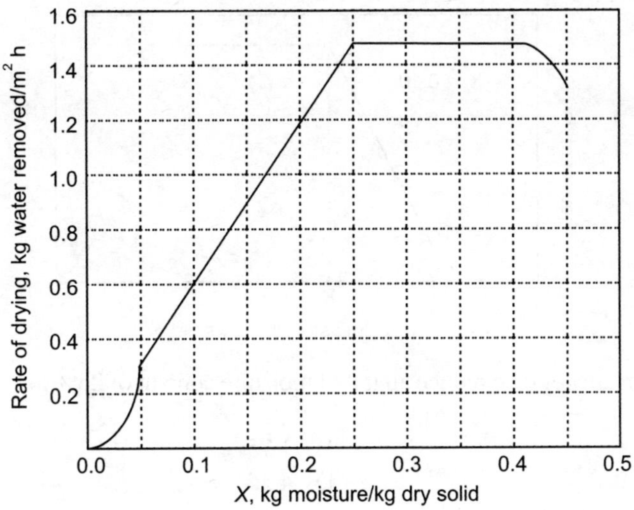

The time required (in hours) for drying this batch to a moisture content of 0.1 kg water/kg dry solid is

A. 0.033 B. 0.43

C. 0.60 D. 2.31

9.46 The equilibrium moisture curve for a solid is shown below:

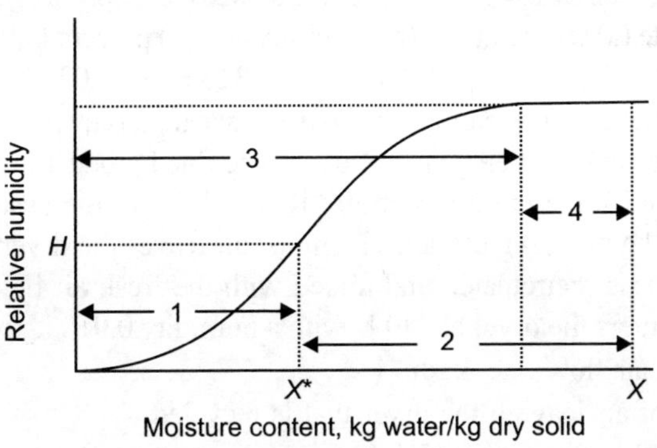

The total moisture content of the solid is X and it is exposed to air of relative humidity H. In the table below, **Group I** lists the types of moisture, and **Group II** represents the region in the graph above.

Group I	Group II
P. Equilibrium moisture	1
Q. Bound moisture	2
R. Unbound moisture	3
S. Free moisture	4

Which one of the following is the correct match?

A. P-1, Q-2, R-3, S-4 B. P-1, Q-3, R-4, S-2

C. P-1, Q-4, R-2, S-3 D. P-1, Q-2, R-4, S-3

9.47 Consider the drying operation shown in the figure below for a solid loading (dry basis) of 50 kg/m^2 with a constant drying rate of 5 kg/m^2 h. The falling rate of drying is linear with moisture content.

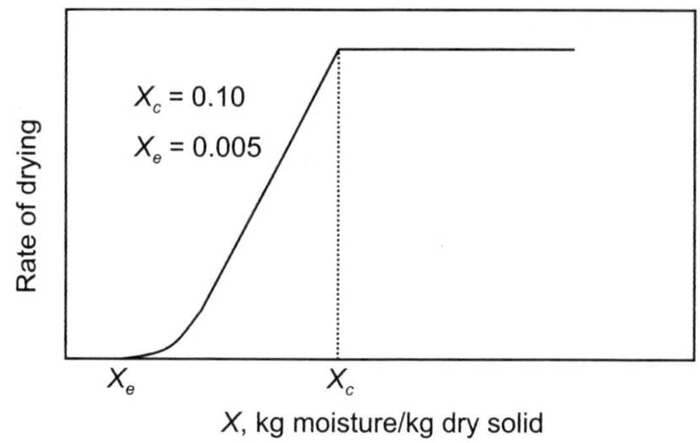

The drying time in hours to reduce an initial moisture content of 25% to a final moisture content of 2% is

A. 1.55 B. 1.75

C. 3.25 D. 4.55

EXERCISES

Psychrometry reviewed

9.1 Air at 341 K dry-bulb temperature and 303 K wet-bulb temperature is being passed over a bed of sliced carrots at the rate of 25 kg dry air/s. If the rate of evaporation of water from the carrots is 0.275 kg/s, estimate (a) temperature, (b) humidity and (c) percent humidity of the air leaving the dryer. (**Answer:** (a) 318 K (b) 0.021 kg water/kg dry air (c) 32%)

9.2 A continuous countercurrent dryer is used to dry a wet material which contains 50% moisture to 20%, both moisture contents being on wet basis. The dried product leaves at the rate of 900 kg/h. The solids enter and leave the dryer at 302 K. Fresh air to the system is at 300 K and has a humidity of 0.007 kg water/kg dry air. The moist air leaves the dryer at 312 K and humidity of 0.020 and part of it is recirculated and mixed with the fresh air before entering a heater. The heated mixed air enters the dryer at 340 K with a humidity 0.010 kg water/kg dry air. Calculate:

(a) The fresh dry air flow rate, kg/h

(b) The fraction of air leaving the dryer that is recycled

(c) The heat added in the heater, kW

(**Answer:** (a) 41,538.5 kg/h (b) 0.23 (c) 571 kW)

9.3 A wet material having an initial moisture content of 25% on wet basis enters the continuous dryer at 295 K at the rate of 2500 kg/h. The product leaves the dryer at 355 K with a moisture content of 5%. Fresh air at a temperature of 288 K with 50% humidity is heated in a preheater. The effluent air enters the dryer at 435 K and leaves at 355 K, with 90% humidity. The vapour pressure of water at 355 K is 51.3 kPa. Calculate:

(a) The volumetric flow rate of air entering the preheater

(b) The volumetric flow rate of air entering the dryer.

(c) The rate at which heat must be supplied to the preheater.

(**Answer:** (a) 761.2 m^3/h (b) 1149.7 m^3/h (c) 1.382×10^5 kJ/h)

9.4 A granular solid containing 90% water is to be dried in a fluidized bed dryer to produce 100 kg/h of dry material. Atmospheric air is passed through a steam-heated exchanger in which it is heated to 350 K and then passed through a bed of solid from which it emerges 90% saturated. Assuming that the atmospheric air is saturated with water vapour at 311 K, estimate the quantity of air need and the duty of the heat exchanger. (**Hint:** Assume that the air temperature varies along the adiabatic cooling line)

(**Answer:** 64,769 kg/h wet air, 2.63×10^6 kJ/h)

9.5 A 10-tray cabinet dryer carries on each tray, 1 kg wet solids containing 95% moisture on wet basis. Fresh air at 293 K containing 0.01 kg moisture /kg dry air is mixed with recycled air before being heated by passing through a heater. The air leaving the heater is passed through the dryer from which it emerges at 323 K containing 0.025 kg water/kg dry air. 90% of this air is recycled. For air circulation through the dryer at a rate of 1500 kg dry air per hour, determine:

(a) The rate at which moisture is removed in the dryer.
(b) The heat duty of the exchanger.
(c) The time required to dry the material to 45% moisture on wet basis, if the drying is at constant rate.

(**Answer:** (a) 2.25 kg/h (b) 4.07 kW (c) 4.04 h)

9.6 A two-stage adiabatic dryer is used to dry paper. Air at 101.3 kPa, 353 K enters the first-stage dryer where it picks up moisture and leaves saturated. Air is then heated to 348K by passing through a heat exchanger before admitting to the second stage of the dryer. The air leaves the dryer with a percent saturation of 80%. It is cooled to 288 K and water is condensed and removed by passing through a cooler. The air leaving the cooler is heated to 353 K by passing through a second heat exchanger. For every kilogram of dry air circulated

(a) Determine kilograms of moisture evaporated in each dryer
(b) Kilograms of water condensed and removed in the cooler.

(**Answer:** (a) 0.022 kg, 0.013 kg (b) 0.035 kg)

9.7 An adiabatic dryer is used to dry a wet material. The drying air enters at 381 K and 101.3 kPa with a dew point of 298 K. Measurements show that 2.25 kg water is evaporated per 100 m^3 of wet inlet air. Calculate

(a) Humidity of air entering the dryer
(b) The exit air humidity and percent humidity
(c) The exit air wet-bulb and dry-bulb temperatures
(d) The volume of exit air per 100 m^3 of inlet wet air.

(**Answer:** (a) 0.02 kg water/kg dry air (b) 0.045 kg water/kg dry air, 55% (c) 313.2 K, 322.2 K (d)87.92 m^3)

9.8 A laboratory dryer produces 50 kg of dry solid containing 6 weight% water from a wet feed material containing 20.5% water. The dryer operates adiabatically, with the wet solid entering the dryer at the wet-bulb temperature of the dryer air. A total of 500 m^3/h of hot air at 366 K and 101.3 kPa with a dew point of 303 K is supplied to the dryer. Calculate:

(a) The temperature of the air leaving the dryer
(b) The percent humidity of the air leaving the dryer.

(**Answer:** (a) 318.65 K, (b)70%)

9.9 Air at 101.3 kPa enters an adiabatic dryer at 372 K with a dew point of 287 K and leaves at 80 per cent humidity. Wet paper enters the dryer with 25% moisture and leaves with 5% moisture. Determine:

(a) The temperature of air leaving the dryer
(b) Water evaporated in kg per 100 cubic metres of air entering
(c) Kilograms of finished product per 100 cubic metres of air entering.

(**Answer:** (a) 311 K, (b) 2.43 kg (c) 9.11 kg)

9.10 Air supplied to a dryer is at 340 K and 101.3 kPa and has a dew point of 278 K. The air leaves the dryer at 310 K and 100 kPa with a dew point 298 K. At what rate (m^3 per hour measured at the dryer inlet conditions) the air should be admitted to the dryer if it is desired that 100 kg water is removed per hour? The vapour pressure of water at 278 K = 0.87 kPa and at 298 K = 3.17 kPa.

(**Answer:** 6794 m^3/h)

9.11 Air at a rate of 25 m^3 per minute is supplied to a dryer in which it removes moisture at a rate of 0.45 kg per minute. The air enters the dryer at 333 K and 101.3 kPa with a percentage humidity of 10 and leaves at 310 K and 100 kPa. The vapour pressure of water at 310 K = 6.3 kPa and at 333 K = 19.9 kPa.

(a) What is the percent humidity of air leaving the dryer?
(b) At what rate air leaves the dryer in m^3 per minute?

(**Answer:** (a) 78.8% (b)24.23 m3/min)

9.12 It is required to evaporate 100 kg/h of water from a wet solid. The drying is carried out in an adiabatic dryer. Air is available at 101.3 kPa and 295 K at 70% humidity. Its temperature is raised by passing through a preheater to such a value that the air leaving the dryer is at 311.7 K and 90% saturated. Calculate:

(a) The adiabatic saturation temperature of air leaving the dryer.
(b) The temperature of air leaving the preheater.
(c) The rate of circulation of moist air, kg/h.
(d) The volume of moist air before preheating.

(**Answer:** (a) 310 K, (b) 378 K, (c) 3614 Kg/h, (d) 3039 m^3/h)

9.13 Air with a dry-bulb temperature of 363 K and wet-bulb temperature of 308 K is passed over a wet material in a dryer. The rate of evaporation is found to be 0.015 kg water per kg dry air passed through the dryer. Assuming the process to be adiabatic, calculate using psychrometric chart

(a) the absolute humidity, molal humidity of air entering the dryer,
(b) the dry-bulb temperature, the wet-bulb temperature, the absolute humidity and the percent saturation of air leaving the dryer.

(**Answer:** (a) 0.013 kg/kg dry air, 0.0209 kmol/kmol dry air
(b) 329 K, 308 K, 0.028, 22.5%)

Drying equilibrium

9.14 One-kg blocks of wet laundry soap containing 20% moisture on dry basis is dried using air at 298 K and 25% relative humidity in a tunnel dryer at 101.3 kPa. Determine the kg water evaporated from each block. Following equilibrium data is applicable:

RH, %	10	20	30	40	50	60	70	80
X^*	0.024	0.038	0.052	0.065	0.085	0.108	0.138	0.18

(**Answer:** 0.1292 kg)

9.15 A wet solid containing 40% moisture is to be dried in a current of air of 50% relative humidity. The equilibrium moisture content of the solid is 5% and the highest concentration of moisture in the solid that can exist in equilibrium with unsaturated air is 30%. (All moisture contents are on wet basis). Determine the free moisture, bound moisture and unbound moisture in the solid.

(**Answer:** 0.6141 kg/kg dry solid, 0.4286 kg/kg dry solid, 0.2381 kg/kg dry solid)

9.16 Fibres with an initial total water content of 35% on dry basis are dried in trays in a tunnel dryer operating at 101.3 kPa. If the fibres are brought to equilibrium with air of relative humidity of 40% at 298 K, determine the kg of moisture evaporated per kg of bone-dry fibres. The equilibrium moisture content varies with relative humidity of air as given below:

X	0	0.02	0.04	0.06	0.08	0.10	0.12	0.14	0.16	0.18
RH,%	0	3	12	24	38	55	69	80	90	97

(**Answer:** 0.267 kg)

9.17 The equilibrium moisture content of lumber varies with humidity of air as given in the following table.

X	0	0.02	0.04	0.06	0.08	0.1	0.12	0.14	0.16	0.18	0.2	0.22	0.24	0.26	0.28	0.3
RH	0	4.5	15	30	43	55	63	70	75	80	86	88	91	94	96	100

A dryer is fed with lumber containing 40% moisture on dry basis. The dried product leaves the dryer in equilibrium with atmospheric air at 298 K and 50% relative humidity. Determine:

(a) The unbound moisture in the original feed lumber in kg water/kg-bone-dry lumber.
(b) The bound moisture in the original feed lumber in kg water/kg bone-dry lumber.
(c) The free moisture in the original feed lumber, in kg water /kg-bone-dry lumber.
(d) The kg moisture evaporated per kg bone-dry lumber.

(**Answer:** (a) 0.1 (b) 0.3 (c) 0.308 (d) 0.308)

9.18 Wet cotton cloth, 100 kg in weight, containing 20% moisture on dry basis is kept in a room which contains 500 m^3 of air at 101.3 kPa and 313 K. The wet-bulb temperature of the air is 293 K. The air in the room is allowed to come in equilibrium with the cloth keeping the temperature constant at 313 K and without allowing air to enter or leave the room. The equilibrium moisture content of the cloth at this temperature is given by:

X, kg moisture/kg dry cloth	0.02	0.04	0.06	0.08	0.10	0.12	0.14
Relative Humidity,%	4.5	15	31	57	73	85	95

Determine the final moisture content of the cloth and the relative humidity of air in the room. Neglect the effect due to the change of pressure of air.

(**Answer:** 7.63% on dry basis, relative humidity = 52%)

Batch drying (Cross-circulation)

9.19 Wet material is dried in a pan 0.5 m × 0.5 m and 25 mm deep. Drying is from the top surface only. Air is passed across the pan at 6 m/s. Air is at 342 K and has a humidity of 0.01 kg water/kg dry air.

(a) Estimate the rate of drying R_C.
(b) Predict R_C if the air velocity is 3 m/s.
(c) Predict R_C if the gas temperature is raised to 369 K keeping the humidity constant.
(d) Predict the effect of increasing the thickness of the drying bed to 40 mm on the time required for drying in the constant rate period between specified initial and final moisture limits.

(**Answer:** (a) 3.5324 kg/m^2 h (b) 2.03 kg/m^2 h (c) 5.23 kg/m^2 h (d) time increases by 1.6 times)

9.20 A filter cake 600 mm × 600 mm × 50 mm, supported on a screen is dried from both faces using a stream of air with dry-bulb temperature 348 K and wet-bulb temperature 303 K. Air flows parallel to the surface at a velocity 2.5 m/s. Equivalent diameter of the airflow space is 150 mm. The latent heat of vaporisation may be taken to be 2431 kJ/kg.

(a) Assume that the heat transfer coefficient is given by

$$h = 8.8\frac{G^{0.80}}{De^{0.2}}$$

where h is in W/m^{2}K, G is in kg/m^2 s and D_e is in m. Determine the rate of drying in kg/m^2 h.

(b) Repeat part (a) using $h = 0.0204\ G^{0.8}\ \text{W/m}^2\text{K}$.
where G is the mass velocity of air in kg/m² h.

(**Answer:** (a) 1.8 kg/m² h (b) 2.0 kg/m²h)

9.21 A drying pan, 0.5 m × 0.5 m × 40 mm deep and insulated on the sides and bottom is filled with wet granular material. Heat transfer is by convection from an air stream at 342 K and having humidity 0.010 kg water/kg dry air, flowing parallel to the top surface at a velocity of 3.0 m/s. The top surface receives radiation from steam heated pipes whose surface temperature T_R is 368K. The emissivity of the material being dried is 0.95. Calculate the surface temperature and the rate of drying for the constant-rate period. (**Answer:** 305.5 K, 2.67 kg/m²h)

9.22 An insoluble wet granular material is dried in a pan 0.5 m × 0.5 m and 25 mm deep. Heat transfer is by convection from an air stream flowing parallel to the surface at a velocity of 5 m/s. The air is at 338 K and 101.3 kPa with a humidity of 0.01 kg water/kg dry air. Estimate the rate of drying in kg/m²h given that

$$h = 0.0204 G^{0.8}$$

where h is in W/m²K and G is in kg/m² h and latent heat of vaporisation is 2433 kJ/kg.

(**Answer:** 2.84 kg/m²h)

9.23 A sample of wet material on a tray is exposed to 2 kg/m² h of drying air across the top surface. The area exposed to drying is 4×10^{-2} m²/kg dry solid and the thickness of the bed is 25 mm. The drying rate during the constant rate period is 2.0 kg water/hm². The critical moisture content is 0.25 kg water/kg dry solid.
Calculate the time to dry another sample of the same material but of thickness 50 mm from both top and bottom surfaces from a moisture content of 0.50 kg water/kg dry solid to 0.30 kg water/kg dry solid. (**Answer:** 2.5 h)

9.24 A wet solid is to be dried from an initial moisture content of 25% to a final value of 5%. Drying test shows that the rate of drying is constant at 1.5 kg water/m² h in the region $X = 0.2 – 0.4$ kg water/kg dry solid. The drying rate falls linearly in the range $X = 0.01 – 0.2$ kg water/kg solid. If the equilibrium moisture content is 0.01 kg water/kg dry solid, calculate the time of drying. The drying surface is 1 m²/ 30 kg dry weight. All moisture contents are on dry basis.

(**Answer:** 6.92 h)

9.25 A batch of granular solids containing 30% moisture is to be dried in a tray dryer to 15% moisture by passing a stream of air across its surface. If the constant rate of drying under these conditions is 2.5 kg/m² h and the critical moisture content is 15%, calculate the drying time. Assume the drying surface to be 0.03 m²/kg dry solid. All moisture contents are on wet basis.

(**Answer:** 3.36 h)

9.26 Sheet material 50 mm thick with a dry density 800 kg/m³ is to be dried at constant drying conditions. The drying rate is 3.2 kg/h m² at the initial moisture content of 35% and the rate is 0.7 kg/hm² at the final moisture content of 10%. If the drying is from the two large surfaces only, and the drying rate in the falling rate period is proportional to the free moisture content, calculate the total drying time. All moisture contents are on the dry basis. (**Answer:** 3.04 h)

9.27 One hundred kg of certain food material is dried from a moisture content of 80% to 50%. The drying surface area is 15 m². Air circulated is at a dry-bulb-temperature of 393 K and wet-bulb-temperature of 323 K. Under the conditions in the dryer, the heat transfer coefficient to the food surface from the air is measured to be 20 W/m² K. Take the latent heat of vaporisation of steam at 323 K to be 2383 kJ/kg. All moisture contents are on wet basis. Assuming constant-rate-drying, estimate the time of drying. (**Answer:** 1.89 h)

9.28 Trays used for drying wet filter cake of calcium carbonate has an area of 1.5 m² and is filled with 75 kg of cake. Drying occurs from the top surface by cross circulation of 10% saturated air at a pressure of 101.3 kPa and temperature of 351 K. The average velocity of air passed across the wet solid is 4 m/s. If the initial water content of the wet cake is 25%, estimate the time in hours to reach the experimentally determined critical moisture content of 10%. All moisture contents are on dry basis. The preheat period may be neglected. (**Answer:** 2.75 h)

9.29 In the falling rate period, if the rate of drying varies as

$$R = aX + bX^2$$

show that the time of drying in the falling rate period is given by

$$\theta_f = \frac{S}{aA}\ln\left[\frac{X_C^2 R_2}{X_2^2 R_C}\right]$$

where R_2 is the rate when the moisture content in the solid is X_2.

9.30 From experiments on batch drying of sheet material, the rate of drying in the constant arte period is found to be 5 kg water/h m^2. The critical moisture content is 30% and the equilibrium moisture content is negligible. Sheets of size 1.5 m × 1.2 m × 50 mm are to be dried from 60% moisture to 5% moisture. The drying is from both sides and under the same drying conditions as in the experiments. If the material has a bone-dry density of 1200 kg/m^3, calculate the time required for drying. All moisture contents are on dry basis. Assume that the falling rate period is linear in the moisture content. (**Answer:** 5.03 h)

9.31 It takes 5 h to dry a wet solid, from 35 to 8% moisture content, using air at constant conditions. The critical and equilibrium moisture contents are 15% and 5%, respectively. Assume that the length of the preheat period is negligible and the falling rate period is linear in free moisture. For the same drying conditions, determine the drying time if the initial moisture content is 40% and the final moisture content is 7%. All moisture contents are on the dry basis.

(**Answer:** 6 h, 25 min)

9.32 A non-porous solid material is to be dried under constant drying conditions from 0.20 to 0.05 kg water/kg dry solid. The bone-dry mass of the solid is 1000 kg and the area of surface drying is 55 m^2. Air at a velocity of 0.75 m/s is passed across the surface of the solid and the drying process is surface-evaporation controlled. The critical moisture content of the material may be taken as 0.125 kg water/kg dry solid and the equilibrium moisture content is negligible.

(a) If the initial rate of drying is 1.5 kg/m^2 h, how long will it take to dry the material?

(b) If the air velocity were increased to 4.0 m/s, what would be the anticipated saving in time?

(**Answer:** (a) 2.3 h, (b) 1.7 h, 74% saving)

9.33 Air used for drying certain food material has a partial pressure of water equal to 2.4 kPa and wet-bulb temperature of 303 K. The material is taken in a try with a drying surface of 0.05 m^2/kg dry solid. The rate in the falling rate period is proportional to the moisture content and the mass transfer coefficient is 5.0 × 10^{-4} kg/m^2 h Pa. Critical moisture content is 0.15 kg moisture/kg dry solid and equilibrium moisture is negligible. Calculate:

(a) The drying rate in the constant-rate period in kg/ m^2 h.

(b) The time required to dry the material from 0.25 kg water/kg dry solid to 0.05 kg water/kg dry solid. Vapour pressure of water at 303 K = 4.232 kPa.

(**Answer:** (a) 0.916 kg/m^2 h (b) 5.78 h)

9.34 A solid is to be dried from 1 kg water/kg dry solids to 0.01 kg water/kg dry solids in a tray dryer. The trays are 0.7 m square and 0.02 m deep and are completely filled with wet material. The air at 308 K and humidity 0.003 kg water/kg dry air is passed across the trays with a velocity of 2.0 m/s. The convective coefficient of heat transfer is given by

$$h = 0.0204\, G^{0.8}\ \mathrm{W/m^2K}$$

where G is the mass velocity of air in kg/m^2 h. The critical moisture content of the solid is 0.3 and the equilibrium moisture content is negligible. The density of the solid is 1200 kg/m^3. Assume that the drying is by convection from the top surface of the trays only. Determine the drying time. (**Answer:** 55 h)

9.35 A wet crystalline solid is to be dried in a rectangular pan 1 m × 1 m × 25 mm. Air at 338 K and 101.3 kPa with humidity of 0.01 kg water/kg dry air is flowing across at a velocity of 3 m/s. Depth of air flow space is 100 mm. The latent heat of vaporisation of water may be taken as 2433 kJ/kg.

(a) Estimate the rate of drying at constant rate given that the convective heat transfer coefficient is

$$h = 5.9\frac{G^{0.71}}{D_e^{0.29}}$$

where h is in W/m^2 K and G in kg/m^2 s and D_e is in m.

(b) Repeat Part (a) if the convective heat transfer coefficient is given by

$$h = 0.0204\ G^{0.8}\ \text{W/m}^2\text{K}$$

where G is the mass velocity of air in kg/m^2 h. (**Answer:** (a) 1.155 kg/m^2 h, (b) 1.89 kg/m^2 h)

9.36 It takes eight hours for a porous solid to reduce the moisture content from 40 to 10% when dried in a batch dryer under constant drying conditions. The critical moisture content was found to be 20% and the equilibrium moisture 3%. All moisture contents are on the dry basis. Assuming that the rate of drying during the falling rate period is proportional to the free moisture content, how long should it take to dry a sample of the same solid from 35 to 5% under the sane drying conditions? (**Answer:** 11 h, 43 min)

9.37 It took 6 h when a material was dried from the initial free moisture content of 30% to the final free moisture content of 8% in a tray-type batch dryer using constant drying conditions. The critical moisture content is 15%. If the drying rate in the falling-rate region is linear from the critical point to the origin, predict the time to dry the sample from a free moisture content of 35% to 5%. All moisture contents are on dry basis. (**Answer:** 8.96 h)

9.38 A 300 mm × 300 mm × 6 mm sample of porous sheet material was dried in a laboratory dryer from both flat surfaces. The rate of evaporation of moisture from the solid was found to be 0.27 kg/h until the critical moisture content of 15% is reached. In the falling-rate period the rate of evaporation fell linearly with moisture content until the sample was dry. The weight of the dry board was found to be 1.5 kg. Estimate the time required for drying another sheet of similar material with size 600 mm × 1200 mm × 12 mm under identical conditions from an initial moisture content of 25% to a final moisture content of 2%. All moisture contents are on wet basis. (**Answer:** 6 h)

9.39 Air at a dry-bulb temperature of 333 K and a wet-bulb temperature of 303 K is passed across the top surface of a granular wet material contained in a pan 0.60 m × 0.60 m to a depth of 25 mm. The air velocity is 3.0 m/s. The sides and bottom of the pan are insulated. The pan contains 12.0 kg of dry solid having a free moisture content of 0.4 kg water/kg dry solid and the material is to be dried in a constant rate-period to a free moisture content of 0.20 kg water/kg dry solid. Estimate

(a) The drying rate and the drying time.

(b) The time of drying if the material thickness is 50 mm.

(**Answer:** (a) 1.59 kg/m^2 h, 4.19 h (b) 8.38 h)

9.40 A wet solid is dried from 20 to 9 per cent moisture under constant drying conditions in 5 h. If the critical and the equilibrium moisture contents are 13 and 5 per cent respectively, how long will it take to dry the solid from 25 to 7 per cent moisture under the same conditions? All moisture contents are on wet basis. **(Answer:** 9.43 h)

9.41 Calculate the total drying time to dry a batch of a wet material from 30% moisture to 5% moisture on wet basis. The drying surface available is 1 m^2/40 kg dry solids. The falling rate is assumed to be linear. The constant drying flux is 1.5 kg/m^2 h. The critical moisture content is 0.2 kg moisture/kg dry solid and the equilibrium moisture content is 0.05 kg moisture per kg solid. **(Answer:** 22.3 h)

9.42 A slab made of $CaCO_3$ paste is dried by passing air across the top and bottom surfaces. The slab has an area of 0.025 m^2 on one side and weighs 1.725 kg. Determine the time of drying if the moisture content is to be reduced from 15% to 2%. The critical moisture content is 8.5%. The first falling rate period is linear up to a moisture content of 4% with

$$R = 3X + 0.25$$

and the second falling rate period is given by

$$R = 300X^2 - 2.75X$$

till the moisture content falls to zero. The rate is in kg/m^2 h and the moisture contents are on dry basis. **(Answer:** 10.82 h)

9.43 Woollen cloth is dried from an initial moisture content of 100% to a final moisture content of 10%. The drying rate varies with the moisture content according to the following relation: $R = R_C$, for $x > 0.55$

$$R = 2.0408 R_C (X - 0.06), \text{ for } X \le 0.55$$

where R_C is the constant rate of drying. Calculate the saving in drying time, if the material is to be dried to 15% instead of 10%. All moisture contents are on dry basis. **(Answer:** 23.7%)

9.44 The rate of drying R (kg water evaporated/m^2 h) of a wet material is given by

$$R = 5.40X, \text{ for } X \le 0.3$$

$$R = 1.62, \text{ for } X > 0.3$$

where X = kg moisture/kg dry solid. The drying surface is 0.025 m^2/kg dry solid. How long the material is to be dried to reduce the moisture content from 0.6 to 0.25 kg moisture/kg dry solid? The equilibrium moisture is zero. **(Answer:** 8.76 h)

9.45 Flat cakes of an organic product are to be dried in a current of warm air maintained at constant humidity and temperature. The original weight of the wet cake is 15 kg and the bone-dry weight is 5 kg. Pilot plant results indicate that the drying under the stated conditions shows a well defined constant drying rate given by

$$\frac{dX}{dt} = -0.2$$

and falling rate given by

$$\frac{dX}{dt} = -0.25(X - X^*)$$

The equilibrium moisture content $X^* = 0.1$ kg water/kg dry solid. Determine:

(a) The initial moisture content on dry basis.

(b) The critical moisture content.

(c) The time required to reduce the moisture content to 0.2 kg water/kg dry solid.

Assume that the surface exposed to the dry air is 1 m^2. **(Answer:** (a) 2 (b) 0.9 (c) 13.82 h)

9.46 A tray dryer consisting of 16 trays is used to dry raw cotton from 95% moisture to 10% moisture by cross circulation of air. The trays which are insulated on the bottom are each 30 mm high with an area of 1.5 m^2 and are completely filled with raw cotton which has a bone-dry density of

700 kg/m^3. The heating medium, which is 30% saturated air at 344 K and 101.3 kPa, flows across the top surface of the tray at a mass velocity of 10000 kg/h m^2. Under the given conditions, the critical moisture content will be 0.4 kg water/kg bone-dry cotton and that the falling rate drying period is linear. Equilibrium moisture content is 0.04 kg water/kg bone-dry cotton. All moisture contents are on dry basis. Determine:

(a) The amount of raw cotton in kg that can be dried in one batch.
(b) The drying-time for the constant-rate period.
(c) The drying time for the falling-rate period.
(d) The total drying time if the preheat period is 1 h.

(**Answer:** (a) 982.8 kg (b) 12.75 h (c) 14.96 h (d) 28.71 h)

9.47 A sample of fibreboard sheet 0.3 m × 0.5 m × 25 mm in size containing 75% moisture is dried from both flat surfaces in a current of hot air. The sheet lost weight at the constant rate of 2 × 10^{-4} kg/s until the moisture content fell to 60%, whereupon the drying rate fell. After a long period of exposure to the air it was established that the equilibrium moisture content was 10% and the dry mass of the sample was 4.5 kg. A plant wishes to dry sheets of the same material but of size 1.5 m × 2 m × 25 mm from 75% to 20% moisture under the same drying conditions. Determine the time of drying if all moisture contents are on wet basis.

(**Answer:** 29.36 h)

9.48 A slab of wet filter cake 1 m^2 in area and 75 mm thick and containing 20% moisture, is dried from both sides in a current of air with a wet-bulb temperature of 303 K and dry-bulb-temperature of 355 K at 101.3 kPa. The air flows parallel with the face of the cake at a velocity of 3 m/s. The dry density of the cake is 1900 kg/m^3. The equilibrium moisture content is negligible and the critical moisture is 9%. The material is to be dried to a final moisture content of 5%. All moisture contents are on dry basis. Determine:

(a) Drying rate during the constant rate-period
(b) Time of drying assuming linear falling rate period.

(**Answer:** (a) 2.63 kg/m^2 h (b) 4.41 h)

9.49 Filter-cake with a bone-dry density of 1500 kg/m^3 are to be dried from an initial free-moisture content of 105% to a final free-moisture content of 5% batch-wise in trays that are 2 m × 0.75 m × 25 mm. Drying will take place only from the top surface. The drying air conditions are 101.3 kPa, 430 K dry-bulb temperature and 330 K wet-bulb temperature and humidity 0.104 kg water/kg dry air. The air velocity across the trays is 4 m/s. Experiments under these drying conditions show a critical free moisture content of 70% and the falling rate period is linear over its entire length. All moisture contents are on dry basis. Determine:

(a) The drying time for the constant-rate period.
(b) The drying time for the falling-rate period.

(**Answer:** (a) 2.44 h, (b) 12.87 h)

9.50 The batch drying experiments on certain material gave the following result:

X, kg moisture/kg dry solid	0.2	0.16	0.12	0.08	0.07	0.06	0.05
R, kg moisture/m^2 h	1.08	0.86	0.648	0.252	0.155	0.097	0

The critical moisture content X_C = 0.2 kg water/kg dry solids. The material is to be dried from 30% to 6% moisture on dry basis. The ratio of dry weight of the sample to drying surface is 40 kg/m^2. Determine the time required for drying. (**Answer:** 17.38 h)

9.51 A wet solid is to be dried in a tray dryer under constant drying conditions from a free moisture content of $X_1 - X^* = 0.40$ kg water/kg dry solid to $X_2 - X^* = 0.02$ kg water/kg dry solid. The dry solid weight is 100 kg and the top surface area for drying is 4.5 m^2. The drying rate varies with moisture content as given in the table:

X-X^*	0.4	0.3	0.19	0.08	0.06	0.05	0.03	0.02	0.01	0
R, kg water/m^2 h	1.5	1.5	1.5	0.82	0.70	0.40	0.18	0.114	0.054	0

(a) Calculate the time of drying for drying using graphical integration in the falling rate period.
(b) Calculate the time of drying using a straight line through the origin for the drying rate in the falling rate period. **(Answer:** (a) 9.82 h (b) 9.45 h)

9.52 Determine the time required to dry a wet material from the initial moisture content of 125% to 10% if the rate of drying varies with moisture content as follows. All moisture contents are on dry basis. Assume that the equilibrium moisture content is negligible.

X kg water/ kg dry solid	$R \times A/S$ kg water/ h-kg dry solid	X kg water/ kg dry solid	$R \times A/S$ Kg water/ h-kg dry solid
1.5	25	0.45	12.2
1.0	25	0.4	10.8
0.8	25	0.3	8
0.75	21	0.2	5.5
0.7	18	0.1	2.7
0.6	14	0	0
0.5	12.4		

(Answer: 5.94 min)

9.53 The rate of drying experiment gave the following results below the critical moisture content of 0.2 kg water/kg dry solid.

X	0.2	0.150	0.100	0.065	0.050	0.040
R, kg/m^2 h	1.50	1.201	0.9	0.70	0.37	0.27

Calculate the time for drying the material from an initial moisture content of 0.4 to a final value 0.04. The weight of the bone dry solid is 400 kg and the area of drying is 20 m^2.

(Answer: 6.54 h)

9.54 A slab with a wet weight of 5 kg originally contains 50% moisture on wet basis. The slab is 1000 × 500 mm. The equilibrium moisture content is 5% of the total weight when in contact with air of 293 K and 20% humidity. The drying rate is given below:

Wet-slag weight, kg	9.0	7.0	5.0	4.0	3.1	2.9	2.7
Drying rate, kg/m^2 h	4.5	4.5	4.0	3.6	3.0	1.8	0.92

Drying is from one face only. How long will it take to dry the slab to 15% moisture content on wet basis? **(Answer:** 1.21 h)

9.55 A slab 600 × 900 mm with a wet weight of 5.2 kg which originally contains 50% moisture (wet basis) is to be dried by circulating air across its top surface. The equilibrium moisture content is 5% on wet basis when in contact with air of 293 K and 20% humidity. The drying rate varies with moisture content as given in the table below.

Wet-slab weight, kg	8.84	7.8	7.28	6.76	6.24	5.72	5.20	4.68	4.16	3.64	3.12	2.86
Drying rate, kg/m^2. h	5.0	5.0	5.0	4.85	4.65	4.45	4.26	4.05	3.70	3.15	2.25	1.25

Determine the time taken to dry the slab to 0.1 kg water/kg dry solid.

(Answer: 1.43 min)

9.56 In a drying test, a sample of food stuff is taken in a tray having an exposed surface of 0.20 m^2. The bone-dry sample weight was 3.50 kg dry solid. Air is passed across the top surface and the weight of sample against time is noted. The data are given in the table below. The weight of sample at equilibrium after a long period was found to be 3.677 kg.

Time, h	Weight, kg	Time, h	Weight, kg	Time, h	Weight, kg
0	4.596	2.2	4.234	7.0	3.736
0.4	4.541	3.0	4.094	9.0	3.698
0.8	4.470	4.2	3.943	12.0	3.677
1.4	4.368	5.0	3.858		

(a) Calculate the free moisture content X-X^* kg water/kg dry solid for each data point and plot X-X^* versus time.

(b) Measure the slope, calculate the drying rate in kg water/h m^2, and plot R versus X.

(c) Using this drying rate curve, predict (i) drying rate R_C (ii) the critical moisture content X_C and (iii) the total time to dry the sample from an initial moisture content of 0.25 kg water/kg dry solid to a final value 0.06 kg water/kg dry solid.

(**Answer:** (c) (i) $R_C = 0.856$ kg water/h m^2 (ii) $X_C = 0.169$ kg water/kg dry solid (iii) $\theta = 8.7$ h)

9.57 A sample of sheet material 0.5 m square and 6 mm thick with the edges sealed, was dried from both sides by cross circulation of air in a laboratory dryer. The dry weight of the sheet was 5 kg. The air at dry-bulb temperature 326 K and wet-bulb temperature 296 K is circulated at a velocity of 3 m/s. There were no radiation effects. The solid lost moisture at a constant rate of 2×10^{-4} kg/s until the critical moisture content, 15%, was reached. In the falling-rate period, the rate of evaporation fell linearly with moisture content until the sample was dry. The equilibrium moisture was negligible. Estimate the time for drying sheets of this material 0.6 m × 1.2 m × 12 mm thick from both sides, from 25 to 2% moisture, using air at dry-bulb temperature 340 K with the same absolute humidity at a velocity of 5 m/s. Assume no change in the critical moisture with the changed drying conditions. All moisture contents are on wet basis. (**Answer:** 3.72 h)

9.58 A wet solid containing 20% moisture is to be dried to 5% moisture in a stream of air flowing at 1 m/s. The rate of drying in the constant rate period is 1.25 kg/m^2 h and the critical moisture content is 12%. All moisture contents are on dry basis. The area of the drying surface is 50 m^2 per 1000 kg bone dry solids. The equilibrium moisture content is negligible and the falling rate is linear in moisture content. Determine:

(a) The time required for drying

(b) The saving in time if the air velocity is increased to 4 m/s.

(**Answer:** (a) 2.961 h (b) 1.98 h)

9.59 In a drying test, a 0.1 m^2 sample of a fibrous material was suspended in a stream of air flowing at 2 m/s. Drying took place from the two large faces only. The weight of dry sample was 4.0 kg. The rate of drying during the constant rate period was found to be 1.8 kg/m^2 h. The critical moisture content was 0.3 kg water/kg dry solid and the equilibrium moisture content was 0.05 kg water/kg dry solid. Estimate the time required to dry the same material from an initial moisture content of 0.4 to a final value of 0.1 on dry basis when the air flow rate is twice that used in the test. (**Answer:** 3.21 h)

Effect of conduction and radiation on the rate of drying

9.60 A drying pan 0.5 × 0.5 m and 20 mm deep is filled with a wet granular material. Air at a temperature of 340 K and humidity 0.010 kg water/kg dry air is passed parallel to the top drying surface and the bottom metal surface at a velocity of 6.0 m/s. The metal bottom of the pan has a thickness of 0.60 mm and thermal conductivity of 40 W/m K. The thermal conductivity of the solid can be assumed to be 0.90 W/m K. Heat transfer from air to the solid surface is by convection. The surface also receives direct radiation from steam heated pipes whose surface temperature is 370 K. The emissivity of the solid is 0.90. Estimate the rate of drying for the constant-rate period. (**Answer:** 5.08 kg/m^2 h)

9.61 A layer of wet granular material 20 mm thick is being dried in a pan. The pan has a 1.6 mm thick metal bottom having a thermal conductivity $k = 45$ W/m K. The thermal conductivity of the solid in the pan is $k = 1.15$ W/m K. Air at a temperature of 333 K and humidity Y = 0.01 kg water/kg dry solid flows parallel to the top exposed surface and the bottom metal surface at a velocity of 3.0 m/s. Direct radiation heat from steam pipes having a surface temperature of 378 K falls on the exposed top surface, whose emissivity is 0.9. Estimate the surface temperature and the drying rate for the constant rate period. (**Answer:** 307 K, 3.12 kg/m^2h)

Through–circulation drying in packed beds

9.62 Wet pigment material is extruded into small cylinders 6 mm in diameter, 50 mm long and are placed on screens to a depth of 65 mm. The material is subjected to through- circulation drying using a stream of air flowing through the bed with mass velocity of 0.95 kg dry air/m^2 s, entering at 393 K with a humidity 0.05 kg water/kg dry air. The surface of the particles is estimated to be 295 m^2/m^3 of bed and the apparent density is 1040 kg dry solid/m^3. Take the cylinder diameter to be its equivalent diameter and assume that the radiation and heat conduction through the bed are negligible. Viscosity of air = 1.81×10^{-5} kg/m s.

(a) Estimate the constant rate of drying.

(b) Estimate the constant rate of drying if the filter cake is dried on trays by a cross circulation of the air over the surface at the same mass velocity, temperature, and humidity.

(**Answer:** (a) 5.53 kg/m^2 h (b) 1.53 kg/m^2 h)

9.63 A filter cake weighing 75 kg is extruded into cylindrical shaped pieces each 7.5 mm in diameter and 15 mm long and is taken as a bed in a tray of area 1.5 m^2 and height 50 mm. The bed porosity is 50%. Air at 101.3 kPa and 351 K, 10% saturated with water vapour is passed through the bed at a superficial velocity of 2 m/s. If the solid is to be dried from the initial moisture content of 25% to the critical value of 10%, determine the time of drying. All moisture contents are on dry basis. The viscosity of air may be taken to be 2×10^{-5} kg/m s. (**Answer:** 4.56 min)

9.64 A filter cake with dry density of 1900 kg/m^3 is extruded into cylinders 6 mm in diameter and 70 mm long onto a screen. The solids loading is 40 kg of dry solids per square meter of screen surface. The bed porosity is 45%. Air at a dry-bulb temperature of 355 K and a wet-bulb-temperature of 303 K is passed through the bed at a superficial velocity of 1.5 m/s. The critical moisture content of the solids is 9%. Take the viscosity of air at these conditions to be 2×10^{-5} kg/m s. How long will it take to dry the solids from 20 to 5% moisture? All moisture contents are on dry basis. (**Answer:** 7.26 min)

9.65 In the through circulation drying of nitrocellulose fibres from an initial moisture content of 35% to a final moisture content of 1%, the rate of evaporation of moisture in the constant rate period is 1.5 kg water/h kg of dry solid. If the critical moisture content is 13% and the equilibrium moisture is negligible, calculate the drying time. All moisture contents are on dry basis. Assume that the falling rate is linear in moisture content. (**Answer:** 22.14 min)

9.66 A 20 mm thick cake of crystalline precipitate is to be dried from 3% to 0.1% moisture (wet basis) by drawing air through it. The particles of the cake have average diameter 0.25 mm and a negligible equilibrium moisture content. The apparent density is 1400 kg dry solid/m^3. The air at a dry-bulb temperature of 303 K and wet-bulb temperature of 293 K will enter the cake at 900 kg dry air/m^2 h.

Drying occurs in the constant rate period. Take viscosity of air to be 1.8×10^{-5} kg/m s. Estimate the time for drying. Interfacial area = 24000 m^2/m^3. **(Answer:** 13.4 min)

9.67 Wet granular material weighing 81 kg and containing 35% moisture is subjected to through-circulation drying to reduce the moisture to 4%. The area exposed to drying is 25 m^2. The critical moisture content is 20%. The rate of drying in the falling rate period is $R = 7.75X + 66.5X^2$ where X is in kg moisture per kg dry solid and R is in kg moisture evaporated/h m^2. All moisture contents are on dry basis. Determine the time of drying. **(Answer:** 22 min)

9.68 Spherical catalyst pellets are being dried from the initial free moisture content of 1.0 kg water/kg solid to a free moisture content of 0.40 kg water/kg solid, which is well above the critical free moisture content. The pellets are 10 mm in diameter and are arranged as a bed of thickness 60 mm on a screen. The dry solid density is determined as 1500 kg/m^3, and the void fraction of the bed is 0.40. Air at 358 K and having humidity of 0.01 kg water/kg dry air are passed through the bed with a superficial velocity 0.9 m/s. Calculate the time for drying. Take viscosity of air to be 2.1×10^{-5} kg/m s for the given conditions. **(Answer:** 0.84 h)

9.69 Filter-cake with a bone-dry density of 1500 kg/m^3 are to be dried from an initial free-moisture content of 105% to a final free-moisture content of 5%. The critical moisture content is 70%. The cake is extruded into cylindrical shaped pieces, 6 mm in diameter and 9 mm long, that are placed in trays 2 m × 0.75 m × 50 mm. The drying air is passed through the bed, the external porosity of which is 50%. The air conditions are 101.3 kPa, 430 K dry-bulb temperature and 330 K wet-bulb temperature and humidity 0.104 kg water/kg dry air. The superficial velocity of the air is 2 m/s. Take the viscosity of air at these conditions to be 2×10^{-5} kg/m s. All moisture contents are on dry basis. Determine:

(a) The drying time for the constant-rate period.

(b) The drying time for the falling-rate period.

(Answer: (a) 6.8 min (b) 36 min)

9.70 A granular pasty material is extruded into cylinders, 5 mm in diameter and 25 mm in length. The cylinders are packed on a screen to a depth of 50 mm. Air at 394 K with a humidity of 0.04 kg water/kg dry air is passed through the bed with superficial velocity of 0.8 m/s. The total critical moisture content is X_C is 0.50 and the equilibrium moisture X^* is 0.01. The density of the dry solid is 1600 kg/m^3 and the bulk density of the dry solid in the bed is 650 kg/m^3. Determine the total time to dry the solids from the initial total moisture content of 1.0 kg water/kg dry solid to the final value, $X = 0.10$ kg water/kg dry solid. Take viscosity of air to be 2.15×10^{-5} kg/m s. **(Answer:** 0.5973 h)

9.71 Determine the time for drying the solids in Exercise (9.70) if the air velocity is 0.4 m/s, other conditions remaining unaltered. **(Answer:** 1.1 h)

9.72 Through circulation drying is employed to dry cylindrical catalyst pellets 13.5 mm diameter and 13 mm long taken on screens as bed of height 50 mm through which air is circulated at a rate 1.1 kg dry air/m^2 s. Air enters at a dry-bulb temperature of 355 K with a humidity 0.01 kg water/kg dry air. Density of the bed is 600 kg dry solids/m^3 and the particle surface is 280 m^2/m^3 of bed. Estimate the rate of drying during the constant rate period. **(Answer:** 4.34 kg/m^2 h).

Mechanism of moisture movement (Diffusion and capillary action)

9.73 It took 8 hours to dry strips of material 15 mm thick from 30% to 15% moisture under constant drying conditions. The equilibrium moisture content is 7%. What is the time taken to dry strips of thickness 50 mm from 23% to 10% moisture under the same conditions assuming no loss from the edges? All moisture contents are on wet basis. The table below gives the relation between the ratio of the average free moisture content at time θ to the initial free moisture content, and the parameter $k\theta/a^2$ where k is a constant, θ the time in h and $2a$ the thickness of the strips of material in metres.

$(X-X^*)/(X_0-X^*)$	1	0.64	0.49	0.38	0.295	0.22	0.14
$k\theta/a^2$	0	0.1	0.2	0.3	0.5	0.6	0.7

(**Answer:** 117 h)

9.74 A piece of wood 150 mm × 150 mm × 20 mm is being dried in a tunnel dryer using cross circulation of air. The edges are sealed and the material is dried from both faces. The initial moisture content was 40% on dry basis. The direction of moisture diffusion was perpendicular to the grain. The equilibrium moisture content was 4% on dry-basis. The following experimental data were obtained for the average moisture content as a function of time.

t, h	X_{avg}, kg water/ kg dry solid	t, h	X_{avg}, kg water/ kg dry solid
0.4	0.362	4.0	0.245
1.0	0.328	5.0	0.230
1.5	0.303	6.0	0.218
2.0	0.291	7.0	0.204
3.0	0.267	8.0	0.192
3.5	0.255	9.0	0.183

Using the data, determine the average value of the diffusivity of moisture through the solid.

(**Answer:** 9.2×10^{-10} m^2/s)

9.75 Large planks of wood are being dried from both sides by air. Average diffusivity for moisture in the wood is 8×10^{-10} m^2/s. Equilibrium moisture content of wood under this condition is 0.04 kg water/kg dry air. The wood is to be dried from an initial moisture content of 0.30 kg water/kg dry wood to $X = 0.10$ kg water/kg dry wood.

(a) Calculate the time needed to dry the wood if the planks are 30 mm thick.

(b) Determine the time of drying if the planks are 15 mm thick.

(**Answer:** (a) 39.8 h (b) 9.9 h)

9.76 A piece of wood measuring 1.8 m × 0.3 m × 50 mm with edges sealed is being dried from both flat surfaces in a tunnel dryer by cross circulation of air. The mean diffusivity for moisture through the solid is estimated to be 9×10^{-10} m^2/s. How long will it take to dry the wood from $X = 0.50$ to $X = 0.10$ if the equilibrium moisture content is 0.05. Assume that all drying takes place in the falling rate period and the resistance in the gas phase is negligible.

(**Answer:** 155.4 h)

9.77 Thin slices of tapioca roots 3 mm thick, are dried from both sides under constant drying conditions. The following data on the unaccomplished moisture change $\dfrac{X-X^*}{X_C-X^*}$ versus time is available.

θ, h	0	0.15	0.27	0.4	0.6	0.8	0.94	1.07
$\dfrac{X-X^*}{X_C-X^*}$	1.0	0.80	0.63	0.55	0.40	0.30	0.23	0.18

Drying occurs in the falling rate period. Determine the average diffusivity of moisture up to an unaccomplished moisture change of 0.25. (**Answer:** 3.33×10^{-10} m^2/s)

9.78 A piece of wood measuring 200 mm × 200 mm × 10 mm is to be dried from the two large faces from an initial total moisture content of 90% to a final average total moisture content of 10%, with drying taking place in the falling rate period with liquid diffusion controlling. The moisture diffusivity has been experimentally determined as 2×10^{-10} m^2/s. Estimate the drying time if bone-dry air is used. All moisture contents are on a dry basis. (**Answer:** 28 h)

9.79 The following data on time versus unaccomplished moisture change of a nonporous material are obtained under constant drying conditions in the falling rate region:

θ, h	0	2.5	7.0	11.4	14.0	16.0
$\dfrac{X - X^*}{X_C - X^*}$	1.0	0.65	0.32	0.17	0.10	0.06

Drying from one side occurs with the material having a thickness of 10 mm. The data appear to follow the diffusion equation. Determine the average diffusivity over the range $\dfrac{X - X^*}{X_C - X^*} = 1.0$ to 0.10. (**Answer:** 1.68×10^{-9} m^2/s)

9.80 It took 2.75 h to dry slabs of wet clay 25 mm thick from an initial moisture content of 0.3 kg water/kg dry solid to a final average moisture content of 0.1 kg water/kg dry solid. The critical moisture content of the solid is 0.165 and the equilibrium moisture content is 0.03. Initial one hour of drying needs occurs in the constant rate period. If the drying in the falling rate period is governed by liquid diffusion with an average moisture diffusivity of 5×10^{-9} m^2/s, what is the percent deviation in the total drying time estimated by the diffusion model from the experimental value? (**Answer:** Estimated time is less by 6.5%)

9.81 In a drying test it took 6 h to dry a sample of mashed potatoes from the initial moisture content of 0.6 kg/kg dry solids to a final moisture content of 0.15 kg/kg dry solids. The sample is taken in pans 10 mm deep and insulated on the bottom and exposed to drying air flowing across the surface. All the drying is in the falling-rate period and is diffusion controlled. The equilibrium moisture content is 0.1 kg/kg dry solids when in contact with air. In the actual drying, the potatoes are to be dried from the initial moisture content of 0.6 kg/kg dry solids to a final moisture content of 0.25 kg/kg dry solids. The material will be taken in pans 40 mm deep but with perforated metal bottoms so that drying occurs from both faces and air conditions remain the same as in the test. Determine the drying time required. (**Answer:** 11.4 h)

9.82 Slabs of wet clay mix 25 mm thick are being dried from an initial moisture content of 0.3 kg water/kg dry solid. The average moisture content of the solid has fallen to 0.099 kg water/kg dry solid in three hours from the start of drying when drying occurs from both faces. If it is estimated that it requires 1 h for reaching the critical moisture content of 0.165 kg water/kg dry solid, determine by how much the final average moisture content estimated using the liquid diffusion model differs from the experimental value. The equilibrium moisture content is 0.03. Assume that the average moisture diffusivity is 5×10^{-9} m^2/s.
(**Answer:** estimated value less by 7.07%)

Heat and material balance in continuous dryers

9.83 A filter cake containing 1500 kg/h of dry crystals of Epsom salt is to be dried in a continuous cocurrent-flow direct-heat dryer using air at 101.3 kPa with dry-bulb temperature of 395 K and wet-bulb temperatures of 320 K which is expected to be discharged at 330 K. The feed enters the dryer at 303 K and 101.3 kPa with a moisture content of 25.0% on dry basis. The final moisture content of the dried crystals is to be 1.5% on dry basis. The average heat capacity of Epsom salt is 1.5 kJ/kg K in the temperature range of 303 to 321 K. Specific heat of dry air is 1.005 kJ/kg K and specific heat of water vapour is 1.884 kJ/kg K. Latent heat of vaporisation at 273 K is 2502 kJ/kg K. If the final product should not be heated to a temperature above 321 K, determine:

(a) The rate of moisture evaporation (b) The humidity of exit air
(c) The rate of heat transfer. (d) The entering dry air flow rate.

(**Answer:** (a) 352.5 kg (b) 0.067 (c) 9.18×10^5 kJ/h (d) 13056 kg/h)

9.84 A continuous countercurrent rotary dryer is fed with 500 kg/h of a wet granular material containing 38% moisture (wet basis) at 293 K and the dried product is withdrawn at 303 K, containing 4% moisture (wet basis). The air supplied which contains 0.005 kg water vapour/kg dry air, enters at 383 K and leaves at 310 K. Assuming that the radiation losses amount to 20 kJ/kg dry air used, determine the mass flow rate of dry air supplied to the dryer and the humidity of the exit air. The heat capacity of dried material is 0.9 kJ/kg K, the heat capacity of dry air is 1.005 kJ/kg K, heat capacity of water vapour is 1.884 kJ/kg K and the latent heat of vaporisation of water at 273 K is 2502 kJ/kg. (**Answer:** 8.213 kg dry air/h, 0.02656 kg water/kg dry air)

9.85 Raw cotton with a moisture content of 100% is admitted to a countercurrent tunnel dryer at a rate of 50 kg/h of dry cotton. The cotton enters at 295 K and exits at 340 K with 10% moisture. All moisture contents are on dry basis. 3000 kg/h of dry air enters at 367 K and 101.3 kPa with a humidity of 0.04 kg water/kg dry air. The specific heat of dry cotton can be taken constant at 1.5 kJ/kg K. Calculate:

(a) The rate of evaporation of moisture
(b) The outlet temperature of the air.
(c) The rate of heat transfer.

(**Answer:** (a) 45 kg/h (b) 330.7 K (c) 1.12×10^5 kJ/h)

9.86 A wet granular solid is to be dried from 0.04 kg moisture/kg dry solid to a 0.002 kg moisture /kg dry solid in a continuous countercurrent dryer at a rate of 500 kg dry solid/h. The dry solid has a heat capacity of 1.5 kJ/kg K.

(a) The solids enter at 300 K and are to be discharged at 336 K. Heating air enters at 367 K with a humidity of 0.010 kg water/kg dry air, and it leaves the dryer at 311 K. The heat loss from the dryer is negligible. Calculate the dry air flow rate and the outlet air humidity.
(b) Repeat part (a) given that the solid enters at 290 K and leaves at 335 K. The gas enters at 360 K and leaves at 305 K. Heat losses from the dryer are estimated as 3000 W.

(**Answer:** (a) 1293 kg/h, 0.0247 kg water/kg dry air;
(b) 1635 kg/h, 0.0216 kg moisture/kg dry air)

9.87 One thousand kilograms per hour alfalfa containing 10% moisture is to be produced on drying chopped alfalfa containing 80% moisture in a continuous cocurrent direct heat dryer. The feed enters the dryer at 300 K and leaves at 340 K. The flue gas analysing 3.7% CO_2, 14.5% O_2, and 81.8% N_2 on dry basis is produced by burning 370 m^3/h (at 300K and 102.5 kPa) of natural gas analysing 85% CH_4, 10% C_2H_6 and 5% N_2 in air at 300K, 50% saturated with water vapour. The flue gas enters the dryer at 1100 K and leaves at 365 K. The heat capacity of dry alfalfa is 1.5 kJ/kg K and the average heat capacity of CO_2, N_2 and O_2 are, respectively, 45.6, 29.9 and 29.9 kJ/kmol K. Compute the exit gas flow rate and the heat losses from the dryer.

(**Answer:** 1.678×10^4 kg/h, 454.23 kW)

9.88 Wet crystals of an organic acid is to be dried in an indirect-heat steam-tube rotary dryer using steam at 300 kPa as the heating medium. 5000 kg/h of wet crystals with 30% moisture at 303 K and 101.3 kPa enter the dryer, where they are dried to a moisture content of 2%. All moisture contents are on wet basis. Evaporation takes place at 373 K and the dry product leaves the dryer also at 373 K. The specific heat of the solid can be taken as 0.85 kJ/kg K. Determine:

(a) The rate of evaporation of moisture (b) The rate of heat transfer.
(c) The quantity of steam required in kg/h.

(**Answer:** (a) 1428.7 kg/h (b) 3.894×10^6 kJ/h (c) 1801.3 kg/h)

9.89 An adiabatic dryer is used to produce 500 kg/h dried product containing 4% moisture from a wet solid feed containing 40% moisture. Ambient air at 295 K having a humidity of 0.005 kg water/kg dry air is heated to 370 K in a preheater before admitting to the dryer, and the exhaust air from the dryer is 60% saturated. Moisture contents in the solid are on wet basis.

(a) Determine the amount of air supplied to the preheater and the heat required in the preheater.

(b) Repeat part (a) if the air enters the dryer at 340K and leaves at the same temperature, but with 60% humidity. (**Answer:** (a) $G = 12711.9$ kg/h, $Q = 9.67\times10^5$ kJ/h (b) $G = 2284.7$ kg/h, $Q = 1.0429\times10^5$ kJ/h)

9.90 A rotary countercurrent dryer is to be used to dry 40,000 kg/h wet sand from 50% moisture to 3% moisture. The air is admitted at 395 K with a humidity of 0.007 kg water/kg dry air and leaves the dryer at 320 K. The wet sand enters at 300 K and leaves at 315 K. Heat losses amount to 20 kJ/ kg dry air. Specific heat of dry sand is 0.9 kJ/kg K. Specific heat of dry air is 1.005 kJ/kg K and specific heat of water vapour is 1.884 kJ/kg K. Latent heat of vaporisation at 273 K is 2502 kJ/kg K. Calculate the amounts of dry air passed through the dryer and the humidity of air leaving the dryer. (**Answer:** 8.57×10^5 kg/h, 0.0296 kg water/kg dry air)

9.91 Ammonium sulphate is to be dried from 4% to 0.2% moisture in a countercurrent rotary dryer. The moisture contents are on wet basis. The solid enters the dryer at 298 K and the product is discharged at 335 K at a rate of 1000 kg/h. Atmospheric air at 298 K, 50% humidity is heated by passage through a heater to 365 K before it enters the dryer and is discharged from the dryer at 305 K. The heat capacity of dry ammonium sulphate is 1.5 kJ/kg K and the heat loss from the dryer is estimated to be 40,000 kJ/h. Specific heat of dry air is 1.005 kJ/kg K and specific heat of water vapour is 1.884 kJ/kg K. Latent heat of vaporisation at 273 K is 2502 kJ/kg K. Estimate the air and heat requirements for the dryer. (**Answer:** 3049 kg/h, 2.09×10^5 kJ/h)

9.92 A continuous rotary dryer was used to dry wood chips from 40% to 15% moisture on wet basis. The wood was at 275 K while the dried product was discharged at 313 K at a rate of 1500 kg/h. The combustion gases used as the drying medium may be assumed to have the characteristics of air. It entered the dryer at 655 K with a humidity 0.038 kg water vapour/kg dry gas, at a rate of 7500 kg/h wet. The gas was discharged at 350 K. The heat capacity of the dried wood can be taken as 1.8 kJ/kg K. Estimate the rate of heat loss. (**Answer:** 166.76 kW)

9.93 A countercurrent rotary dryer handles 500 kg/h of dry granular material and delivers the product containing 5% moisture from a feed containing 40% moisture, both moisture contents expressed on wet basis. The solids enter at 295 K and leaves at 315 K. Air at 395 K containing 0.005 kg water/kg dry air enters the dryer and leaves at 320 K. Specific heat of dry solid = 0.8 kJ/kg K, Specific heat of dry air = 1.008 kJ/kg K, Specific heat of water vapour = 1.884 kJ/kg K, , Latent heat of vaporisation of water at 273 K = 2502 kJ/kg.

(a) Determine the amount of dry air required and humidity of the air leaving the dryer.

(b) Repeat Part (a) if the heat losses from the dryer is expected to be 25 kJ/kg dry air passing through the dryer.

(**Answer:** (a) 10207.14 kg/h, 0.0351 kg water/kg dry air (b) 15216.8 kg/h, 0.0252 kg water/kg dry air)

Continuous dryers

9.94 A wet solid with moisture content 0.4 kg water/kg dry solid is to be dried to 0.03 kg water/kg dry solid at a rate of 525 kg/h dry solids in a continuous countercurrent dryer. Air flows at a rate 10,000 kg dry air/h. It enters the dryer at 378 K with a humidity of 0.055 kg water/kg dry air. The solid enters the dryer at the wet-bulb-temperature of the inlet air. The surface area available for drying is 0.05 m^2/kg dry solid. The critical moisture content of the solid is 0.1 kg water/kg dry solid. The mass transfer coefficient k_Y is 125 kg water/m^2 h. In the falling rate period, the drying rate was directly proportional to moisture content. Calculate the time of drying (a) in the constant rate period (b) in the falling-rate period. **(Answer:** (a) 3.414 h (b) 0.757 h)

9.95 A continuous countercurrent dryer is to be designed to dry 500 kg of wet porous solid per hour from 150% moisture to 20%. Air at 323 K dry-bulb and 293 K wet-bulb temperature is to be used. The exit humidity is to be 0.012 kg water/kg dry air. The equilibrium moisture content is 5% and the critical moisture content is 40%. All moisture contents are on dry basis. The stock may be assumed to remain at a temperature 2K above the wet-bulb-temperature of the air throughout the dryer. The heat transfer coefficient is 70 W/m^2 K. The area exposed to air is 0.25 m^2/kg of dry solids. How long must the solids remain in the dryer? **(Answer:** 2.8 h)

9.96 It is desired to dry 500 kg/h of a feed of wet solids from 20% to 5% moisture in a steam-tube indirect dryer. The feed temperature is 285 K, and the product leaves the dryer at 400K. The specific heat of the solid is 1.8 kJ/kg K. The saturated steam at 450 K is used as the heating medium. The heating/drying occurs in three sections in series: a heating section, a constant-rate drying section and falling rate drying section. Solids temperature during constant-rate drying is constant at 373 K. Drying is in the falling rate region when the moisture content has fallen to 10%. The heat-transfer coefficients in the heating, constant-rate drying and falling rate drying sections are 150, 250 and 100 W/ m^2 K, respectively. Determine the heat load and the area required for each section of the dryer.

(Answer: Q_1 = 100204 kJ/h, Q_2 = 102175 kJ/h, Q_3 = 64827 kJ/h; A_1 = 1.61m^2, A_2 = 1.47 m^2, A_3 = 2.88 m^2)

9.97 A direct-heat cocurrent rotary drier is fed with ammonium nitrate containing 15% moisture at a rate of 20,000 kg dry solids per hour. The feed is admitted at 294 K and the moisture content of the product is to be 1%. All moisture contents are on dry basis. Air is admitted at 394 K and 101.3 kPa, with a humidity of 0.02 kg water/ kg dry air. It may be assumed that all drying will take place in the constant- rate period and the allowable gas velocity of 1.5 m/s. Assume the number of heat transfer units to be 1.5 and specific heat of dry solids is 1.7 kJ/kg K. Estimate of the dryer diameter and length of the dryer **(Answer:** D = 6 m, L = 25.4 m)

9.98 A cocurrent adiabatic rotary dryer is used to dry 1500 kg/h of a heat sensitive solid from an initial moisture content of 20% to a final moisture content of 1.0%. Moisture contents are on dry basis. The solids enter the dryer at 300 K and must not be heated to a temperature above 325 K. The heat capacity of the solids is 2 kJ/kg K. Heating air is available at 403 K and a humidity of 0.01 kg water per kg dry air. The maximum allowable mass velocity of air is 4000 kg/m^2 h. Assume the number of transfer units of the dryer to be 1.5. Calculate the diameter and length of the dryer.

(Answer: D = 1.9 m, L = 12.35 m)

9.99 A granular wet material is admitted to a continuous cocurrent adiabatic dryer at a rate of 2000 kg/h of dry solids containing 15% moisture, to reduce the moisture content to 0.5%. The solids enter the dryer at 298 K and because of the heat sensitive nature of the solids its temperature should not rise above 323 K. The mean heat capacity of the solids is 2.2 kJ/kg K. Hot air is available at 400 K with a humidity of 0.01 kg/kg dry air and the maximum allowable mass velocity of the air is 3500 kg/m^2 h. All moisture contents are on dry basis. Estimate the diameter and length for dryer. Specific heat of dry air is 1.005 kJ/kg K and specific heat of water vapour is 1.884 kJ/kg K. Latent heat of vaporisation at 273 K is 2502 kJ/kg K.

(Answer: D = 2.1 m, L = 13.31 m)

9.100 A wet material is being dried in a countercurrent dryer. It enters the dryer at 298 K with 10% moisture on dry basis and leaves with a moisture content of 3% on dry basis. Air at a rate of 15000 kg/h dry air entering at 333 K with a humidity of 0.014 kg water/kg dry air is used as the heating medium. The air leaves the dryer at 310 K with a humidity of 0.02 kg water/kg dry air. If the length of the dryer is 1.5 m and the cross sectional area is 2 m^2, calculate:

(a) The heat transfer coefficient Ua, W/m^3 K
(b) The quantity of water evaporated from the material, kg/h
(c) The rate at which the material is dried in kg dry solid/h
(d) The number of heat transfer units.

(**Answer:** (a) 2085.2 (b) 90 (c) 1285.7 (d) 1.46)

9.101 Raw cotton containing 1.0 kg moisture/kg dry solids is to be dried to 0.1 kg moisture/kg dry solids in a tunnel dryer. The critical moisture content is 0.4 kg water/kg dry solid. The feed enters the dryer at 300 K and leaves at 340 K. The dry density of cotton is 700 kg/m^3 and the specific heat may be taken to be 1.5 kJ/kg K. The cotton passes through the dryer as a layer 15 mm thick on a belt 600 mm wide. Air at 365 K dry-bulb temperature and 320 K wet-bulb temperature passes countercurrent to the feed at a rate 60 kg per kg of dry solids. The gas phase mass transfer coefficient and heat transfer coefficient are, respectively 30 kg/m^2 h and 10.68 W/m^2 K. Determine the drying time required. (**Answer:** 32.6 h)

9.102 A continuous countercurrent tunnel dryer is used to dry a wet solid containing 0.4 kg water/kg dry solids to 0.04 kg water/kg dry solids at a rate of 400 kg dry solids/h. Air at 368 K with humidity 0.056 kg water/kg dry air is admitted at a rate of 10,000 kg/h dry air. The feed enters at the wet-bulb temperature of 321 K and remains essentially at constant temperature in the dryer. The surface area available for drying is 0.06 m^2/kg dry solid. The critical moisture content of the solid is 0.10 kg/kg dry solids and the equilibrium moisture is negligible. The mass transfer coefficient k_Y is found to be 150 kg/h m^2. The falling rate is linear with moisture content. Determine the time of drying in the constant-rate zone and in the falling-rate zone.

(**Answer:** 3.18h, 0.54 h)

9.103 A continuous countercurrent tunnel dryer is used to dry a sheet material 0.6 m × 1.2 m × 12 mm at the rat of 100 sheets per hour. The dry mass of each sheet is 12 kg, and the moisture content will be reduced from 55 to 5% water by the air entering at 395 K, humidity 0.01 kg water/kg dry air. The sheets will be supported on a special conveyor carrying the material in tiers 30 sheets high, and they will be dried from both sides. Air is passed at a rate of 40 kg/kg dry solids.

In a small-scale experiment, when dried with air at constant drying conditions, DBT 368 K, WBT 323 K, and at the same velocity as that would be used in the large dryer, the constant drying rate was 1.25 kg water evaporated/m^2 h and the critical moisture content 30%. The equilibrium moisture content was negligible. All moisture contents are on wet basis.

(a) Calculate the value of k_Y from the data of the small-scale experiment.
(b) For the large dryer, calculate the humidity of the air leaving the dryer and the humidity at the point where the solid reaches the critical moisture content.
(c) Estimate the time of drying in the large dryer.
(d) How many sheets of material will be in the dryer at all times?

(**Answer:** (a) 62.2 kg/m^2 h (b) 0.0392, 0.0194 kg/kg dry air (c) 11.8 h (d) 1200)

9.104 A filter cake is to be dried in a continuous countercurrent hot air tunnel dryer. The cake will be placed on trays 1.0 m × 0.9 m × 25 mm, 20 trays to a truck, with 50 mm gap between the trays. The tunnel dryer will have a cross section of 2 m high by 1 m wide. Drying takes place from both top and bottom sides. Production permits introducing one truck load per hour. Each tray contains 25 kg dry solids, which will enter the dryer at 300 K, 50% moisture, and will be dried to negligible moisture content. The critical moisture content is 12.5%, and the equilibrium moisture is negligible. All moisture contents are on wet basis. The air is to enter at 403 K, humidity 0.04 kg water/kg dry air, and the discharged solids are expected to leave at 390 K. The air is to be blown over the trays so that the average velocity at the entrance is to be 5 m/s. The heat capacity of the dry solid is 1.3 kJ/kg K. Calculate the length of the dryer required. **(Answer:** 6.65 m)

9.105 Titanium dioxide is dried from an initial moisture content of 30% to a final value of 2% (dry basis) in a direct-heat countercurrent, rotary dryer 2 m in diameter and 20 m long. The feed enters the drier at 294 K and 101.3 kPa. Hot air is available at 478 K with a humidity of 0.015 kg/kg dry air. The air mass velocity should not exceed 2500 kg/m^2 h to avoid dusting. The specific heat of the solid titanium dioxide is 0.7 kJ/kg K, and its density is 3800 kg/m^3. Determine:

(a) A reasonable production rate in kg/h of titanium dioxide.
(b) The heat transfer rate in kW.
(c) A reasonable air rate in kg/h.
(d) Reasonable values for the exit air and exit solids temperature.

(Answer: (a) 1350 kg/h (b) 271 kW (c) 6300 kg/h (d) 328.6 K, 322 K)

9.106 Titanium dioxide is dried from an initial moisture content of 0.3 kg water/kg dry solids to 0.02 kg water/kg dry solids in a direct-heat countercurrent, rotary dryer 2 m in diameter and 20 m long. The titanium dioxide has a mean particle size of 50 microns and a density of 3800 kg/m^3. The dryer operates at 4 rpm with a slope of 50 mm in a length of 1 m. Determine the inlet air rate and the required temperature, and the rate of production of dried TiO_2 if the drying air is obtained by heating room air originally at 305 K with a 289 K wet-bulb temperature.

(Answer: 785 kg/h, 576 K, 267 kg/h)

9.107 A direct-heat, countercurrent rotary dryer is used to dry 10,000 kg/h of manganese ore at 300 K with an average particle diameter of 1 mm from an initial moisture content of 0.1 kg water/kg dry solid to a final moisture content of 0.005 kg water/kg dry solids. At these conditions, the critical moisture content will be less than 0.005 kg water/kg dry solid. The drying medium will be air initially at 300 K with a wet-bulb temperature of 294 K, heated to 400 K before admitted to the dryer. Specific heat of dry ore is 0.7 kJ/kg K. Determine the dryer length, diameter, and the air flow rate. **(Answer:** $D = 2.5$ m, $L = 6.8$ m, $G = 29484$ kg/h)

9.108 Wet solids containing 150% moisture is dried to 20% moisture in a continuous countercurrent dryer at a rate of 500 kg of wet porous solid per hour. Air is supplied at 322 K dry-bulb and 294 K wet-bulb temperature and the exit humidity is 0.012 kg moisture/kg dry air. The average equilibrium moisture content is 5% of the dry weight. The critical moisture content is 40%. The moisture contents are on dry basis. The stock may be assumed to remain at a temperature 2 K above that of the wet-bulb temperature of the air throughout the dryer. The heat transfer coefficient is 70 W/m^2 K. The area exposed to the air is 0.25 m^2/kg of dry solids. How long must the solids remain in the dryer? **(Answer:** 3.33 h)

9.109 A countercurrent, direct-heat rotary dryer fed with wet solids containing 8% moisture at 300 K and is to be discharged at a rate of 2500 kg/h of dried product at 425 K containing 0.5% moisture. Moisture contents are on wet basis. Flue gas (Average Molecular Weight = 29) at 753 K containing 0.045 kg water/kg dry gas is used as the heating medium. It is estimated that 15% of the total heat content of the flue gas is lost of which approximately 15% is lost in the preheating zone, 65% in the drying zone and 20% in the final heating zone. Assume that the psychrometric ratio of the gas is the same as that of air and the temperature of the solids in the drying zone is 3 K above the wet-bulb temperature of gas. Take the average heat capacity of the dry flue gas to be 1.04 kJ/kg K and the heat capacity of dry solids to be 0.84 kJ/kg K. The critical moisture content of the solids is very low. Determine the diameter and length of the dryer.

(Answer: $D = 1.2$ m, $L = 7.5$ m)

9.110 A continuous fluidised bed dryer operating at a pressure of 101.3 kPa in the free board region above the bed is used to dry 5000 kg/h of wet sand at 294 K containing a moisture content of 20% to 5%. Both moisture contents are on dry basis. The sand has an average particle size of 500 microns and a sphericity of 0.67. The particle density is 2600 kg/m^3. When the sand is dry, the void fraction of a bed of the sand is 0.55. Fluidizing air will enter the bed at a temperature of 800 K with a humidity of 0.01 kg water/kg dry air. The adiabatic saturation temperature is estimated to be 336 K. Batch pilot plant tests with a fluidization velocity of twice the minimum value show that drying takes place in the constant rate period and all moisture can be removed in 10 min using air at the same conditions and with a bed temperature of 336 K. Specific heat of dry sand is 0.84 kJ/kg K. Determine the bed height and diameter for the large, continuous dryer.

(**Answer:** $D = 1.6$ m, $Z = 0.56$ m)

9.111 A fluidized bed dryer is used to dry coal using as fluidizing medium, combustion gases containing 80% N_2, 15% CO_2, 5% CO at 775 K. The coal particles have an average diameter of 20 mm and the bed is expanded to an average 10% solids, with the gas flowing at 5 m/s. The coal enters at 303 K with 0.2 kg water/kg dry solids, and the equilibrium moisture content is 0.02 kg water/kg dry solids.15000 kg/h of coal is to be dried to a final moisture content of 0.04 kg water/kg dry solids. The coal has a density of1600 kg/m^3, and a heat capacity of 0.8 kJ/kg K. The porosity of the dry static bed is 0.45. The batch drying test indicates 5 min to completely dry the solids. Assume that the bed is perfectly mixed. Determine the residence time required by the coal and the volume of the bed required. (**Answer:** 10.8 min, 2.81 m^3)

9.112 It is desired to dry a granular material in a fluidized bed dryer. The material has a density 5000 kg/m. The solids are fed at a rate of 50000 kg/h dry solids at 293 K containing 20% water on dry basis. Ambient air heated to 800 K is used as the heating medium. Ambient air is at 293 K with humidity is 0.01 kg/kg dry air. The particle size is 0.5 mm and the rate of evaporation of water in the dryer is 5000 kg/h. The heat capacity of the solids is 1.7 kJ/kg K. Assuming that the air leaves the bed 90% saturated at the exit temperature and the fluidisation velocity is 1 m/s, what diameter of bed should be specified? (**Answer:** 2.9 m)

9.113 A solution of coffee in water at 294 K containing 30% coffee is spray dried to a final moisture content of 5% with air that enters at 500 K and 101.3 kPa with a humidity of 0.01 kg/kg dry air and exits at 365 K. The moisture contents are given on dry basis. Assuming the heat capacity of coffee to be 1.2 kJ/kg K. Determine:

(a) The air rate in kg dry air /kg coffee solution.

(b) The temperature of evaporation.

(c) The heat transfer rate in kJ/kg coffee solution.

(**Answer:** (a) 3.92 (b) 352 K (c) 542)

9.114 Freeze drying of a certain food product containing 65% moisture is achieved in a vacuum dryer operating at 67 Pa with a condensing temperature of 248 K. The material is 10 mm thick, and drying is to take place from both sides, with a 1-mm gap between the sample and the heating surfaces, which are at 268 K. The thermal conductivities of the dry solid and the vapour film are about 0.05 W/m K and 0.02 W/m K, respectively. The bulk density of the dry solid is 400 kg/m^3 and latent heat of sublimation of ice at 248 K is 2600 kJ/kg. Assume the emissivity of solid surface to be unity. (a) Allowing for radiation, what is the coefficient of heat transfer from the hot surface to the exterior of the solid? (b) When one-half the moisture has been removed, what is the overall heat transfer coefficient? (c) Using this value as the average coefficient, estimate the time required for drying. (d) If the material was in layers 5 mm thick, by what factor would the drying time be reduced?

(**Answer:** (a) 23.9 W/m^2 K (b) 10.89 W/m^2 K (c) 12.32 h (d) 4.48 h)

Chapter 10

Adsorption

Adsorption is a process in which the components of a gas or liquid phase are selectively transferred on to a solid surface, leading to separation of the components of the fluid phase. In adsorption, molecules distribute themselves between two phases, one a solid and the other a gas or liquid. Separation by adsorption makes use of the difference in the capabilities of the components for getting adsorbed on the solid surface. The solid is known as the *adsorbent* and the fluid transferred to and adsorbed on the solid is called the *adsorbate*. Adsorption can be carried out in agitated vessels, fixed bed, moving bed or fluidized bed units. In fixed bed operations which are more common, the fluid to be separated is passed continuously through a bed of adsorbent solids. When the solids become nearly saturated with the adsorbate, the adsorbent is either replaced or subjected to regeneration.

10.1 INDUSTRIAL APPLICATIONS

Adsorption is an important technique for the separation of gaseous as well as liquid mixtures. The major applications of gaseous separation include:

- Dehumidification of air and other gases.
- Removal of objectionable odour and impurities from industrial gases such as CO_2.
- Removal of pollutants such as hydrogen sulphide, CS_2 and other odorous compounds from the air circulating in ventilation systems.
- Recovery of valuable solvent vapours from dilute mixtures with air and other gases.
- Separation of nitrogen from air.
- Fractionation of mixtures of hydrocarbons and separation of gases such as methane, ethylene, ethane, propylene, propane, etc.
- Separation of hydrogen from synthetic gases and hydrogenation process gases, removal of CO_2 from natural gas, removal of NO_x from N_2, removal of sulphur compounds from gas streams.
- Control of automobile emissions by adsorbing the gasoline vapours from being vented to the atmosphere.
- As desiccants in packing.

The following are the major industrial applications of adsorption from liquids.

- Removal of the moisture from gasoline and drying of organic solvents, transformer oils, etc.
- Removal of sulphur compounds from organic solutions and trace components from liquids.
- Decolourisation of petroleum products, aqueous sugar solutions, syrups, vegetable oils, etc.

- Removal of organic components from drinking water and aqueous wastes and removal of objectionable taste and odour from municipal water supplies.
- Purification of process effluents for control of water pollution.
- Separation of fatty acids from water.
- Fractionation of mixtures of aromatic and paraffinic hydrocarbons.
- Recovery of vitamins and various fermentation products from fermenter effluents.
- Recovery of organic solvents used in paints and printing inks.
- Separation of fructose from dextrose and polysaccharides.
- Treatment of edible oils and refining of mineral oils.
- Recovery of proteins and enzymes.

10.2 PHYSICAL ADSORPTION AND CHEMISORPTION

Depending upon the nature of forces binding the fluid molecules (the adsorbate) and the molecules of the solid (the adsorbent), adsorption processes fall into two broad categories: physical adsorption and chemisorption. In physical adsorption, the intermolecular forces of attraction between the adsorbate and adsorbent lead to transfer of fluid onto the solid surface. If the intermolecular forces of attraction between the solid and gas molecules are stronger than that between molecules of the gas, adsorption occurs by condensation of gas. Chemisorption occurs as a result of chemical interaction and formation of chemical bond between the solid and the fluid. However, it is different from the stoichiometric chemical reaction in which identifiable chemical species are formed. Physical adsorption, also known as the van der Waals adsorption is important as a separation technique, whereas chemisorption, also called activated adsorption, is important in catalysis. The salient features of these processes are given in Table 10.1:

Table 10.1: Comparison of physical adsorption and chemisorption

Characteristics	Physical adsorption	Chemisorption
Surface coverage	Surface coverage is complete and extendable to multilayers	Incomplete coverage and limited to monomolecular layer
Heat of adsorption of gases	Greater than the latent heat of vaporisation of solute and is generally of the order of heat of sublimation of gas.	Generally of the order of the heat of reaction
Rate of adsorption	Controlled by the resistance to mass transfer	Rate is controlled by the resistance to surface reaction
Rate at low temperature	Rapid adsorption at low temperatures.	Negligible rate of adsorption at low temperatures
Adsorption above critical temperature	No adsorption	Chemisorption is possible even above critical temperature
Effect of pressure	Limited adsorption at low partial pressure of solute, appreciable adsorption at high partial pressures.	Appreciable adsorption even at low partial pressures and the increase at high partial pressures is not very significant.
Specificity of the surface	Low specificity and therefore, the entire surface is available for adsorption	Highly specific and only specific molecular sites called active sites are available for adsorption
Reversibility	Reversible and the adsorbent can be regenerated by reducing the pressure or by increasing the temperature.	Usually irreversible as the adsorbent undergoes chemical change
Capacity	Quantity adsorbed per unit mass is high	Amount adsorbed per unit mass is low

10.3 PROPERTIES OF ADSORBENTS

The mass transfer applications of adsorption are mainly dependent on the physical adsorption in which the molecules of the fluid are attracted to the solid surface by the residual van der Waals forces existing at the surface. Though the chemical nature of the solids has much to do with their adsorption characteristics, it is not the chemical identity alone that is responsible for the specificity and effectiveness of the adsorbent solids for separation. These depend strongly on the method of manufacture and prior history of adsorption and desorption as well. The most important properties to be considered for choice as adsorbent solids are given below:

(a) *Specific surface area*: Adsorbents are usually porous in nature and offer large internal surface area per unit weight usually lying in the range 3×10^5 to 1.2×10^6 m^2/kg. They are generally granules, spheres, flakes or cylindrical pellets in the size range from 50 μm to 12 mm diameter. The higher the internal surface area of the solids, the larger the magnitude of the van der Waals forces existing at the surface. Solids which are manufactured in a highly porous form, offer very large internal area in comparison to the external surface area.

(b) *Specificity*: Adsorbent must be selective towards certain components in preference to others. It is the differences in molecular weights or polarity that cause some molecules to be adsorbed in preference to others. Adsorption turns most attractive as a separation technique, if complete removal of some components is possible with very little adsorption of the other components. Therefore it is an added advantage for the adsorbent if the pore size is such that they are big enough to admit the desired molecules and prevent the molecules that are not to be adsorbed.

(c) *Capacity*: The adsorptive capacity, that is, the amount of solute adsorbed per unit weight of adsorbent is an important consideration in selecting adsorbents for large scale industrial applications.

(d) *Mechanical strength and hardness*: Adsorbent should be mechanically strong to withstand the bulk handling and vibration that are common to industrial application. They should not to be reduced in size during the handling or crushed in supporting their own weight in beds of required depth.

(e) *Pressure drop for flow*: They should not offer too great a pressure drop for flow. It should not be carried away by the flowing stream, when adsorption is done in a fixed bed unit.

(f) *Regenerative capacity*: They should be capable of being easily regenerated.

(g) *Stability*: The adsorbent should not age and lose its adsorptive capacity rapidly on continual regeneration and recycling. It should offer high resistance to fouling.

(h) *Chemical inertness*: The adsorbent should not promote undesirable chemical reactions.

(i) *Free flowing nature*: For ease of transport in and out of the beds and for easy filling and emptying of vessels, the adsorbents should be free flowing.

(j) *Cost*: The adsorbent should be available at low cost.

10.4 INDUSTRIAL ADSORBENTS

Industrial adsorbents include activated carbon, silica gel, activated alumina, molecular sieve zeolites, Fuller's earth and synthetic polymer resins.

Activated Carbon

Activated carbon is produced by thermal decomposition of naturally occurring carbonaceous materials such as coal, wood, coconut shells, bones, etc. with additional treatment or activation to generate a system of fine pores. The surface areas of activated carbon lie in the range of 3×10^5 to 1.2×10^6 m^2/kg with average pore diameters of 1 nm to 6 nm. Because it has a low affinity for water, activated carbon may preferentially adsorb components from aqueous solution or moist gases. Major uses are for decolourising solutions of sugar, recovery of solvent vapours, separation of hydrogen from synthesis gases, water purification, refining of vegetable and animal oils, removal of odour from gases, gas masks, collection of gasoline hydrocarbons from natural gas, and fractionation of hydrocarbon gases.

Silica Gel

Silica gel is primarily used to dehydrate gases and to fractionate hydrocarbons. Small sachets of silica gel are often kept in packages of materials that deteriorate in a damp atmosphere. For producing silica gel, solution of a silicate such as sodium silicate is acidified forming a gel of polymeric colloidal silicic acid. The gel, an agglomerate of micro-particles, is then heated to expel water and form a hard glassy structure. Silica gel has a surface area of 6×10^5 to 8×10^5 m^2/kg and average pore diameters of 2 nm to 5 nm.

Activated Alumina

Activated alumina is used mainly to dry gases and liquids. It is superior to silica gel in two respects: it offers resistance to attrition, and retains the adsorptive capacity at elevated temperatures. Activated alumina is prepared by the controlled heating of hydrated alumina. Water molecules are expelled and the resulting activated substance has surface areas ranging from 2×10^5 to 5×10^5 m^2/kg, with average pore diameters of 2 nm to 14 nm.

Molecular-Sieve Zeolites

The zeolites are porous crystalline substances with lattice structures composed of tetrahedra of silica and alumina arranged forming an open crystal lattice containing uniform pores. This uniformity in pore size makes them different from other types of adsorbents which have a range of pore sizes. By using different source materials and different conditions of manufacture, it is possible to produce molecular sieves having pore sizes ranging from 0.3 nm to 1 nm. Zeolites are used for a variety of applications such as drying of gases, separation of the oxygen from the air, separation of hydrocarbon mixtures, pollution control, etc.

Fuller's Earth

These are natural clays containing magnesium aluminium silicates. The clay is first heated and dried during which it develops a porous structure and then ground and screened. Commercially available sizes range from coarse granules to fine powders. Fuller's earth finds applications in the adsorption operations for decolourization of the petroleum products, drying of petroleum products, organic solvents, vegetable and animal oils. The adsorbent can be readily regenerated by washing and burning of the adsorbed organic matter.

Synthetic Polymers

These are porous spherical beads 0.5 mm in diameter, each bead containing micropores of diameter 0.1 μm. The surface area of synthetic polymeric adsorbents lies in the range 8×10^4 m^2/kg and 7×10^5 m^2/kg. These are made by polymerising two major types of monomers. Polymers made from aromatics such as styrene and divinyl benzene are used to adsorb non-polar organics from aqueous solutions; those made from acrylic esters are usable for adsorbing more polar solutes in aqueous solutions.

10.5 ADSORPTION EQUILIBRIUM

Adsorption isotherm represents the relationship between the concentration of solute on the adsorbent solid and the concentration of the solute in the fluid under equilibrium at a given temperature. The concentration of the solute on the adsorbent, also known as the adsorbate loading (q) is generally expressed as the kilograms of solute (adsorbate) per kilogram of adsorbent solid. The concentration in the fluid phase is generally expressed as C, kg solute/m^3 of the fluid. For adsorption from gases, the partial pressure of solute in the gas, $\overline{p}$, is also used to represent the solute concentration in the fluid.

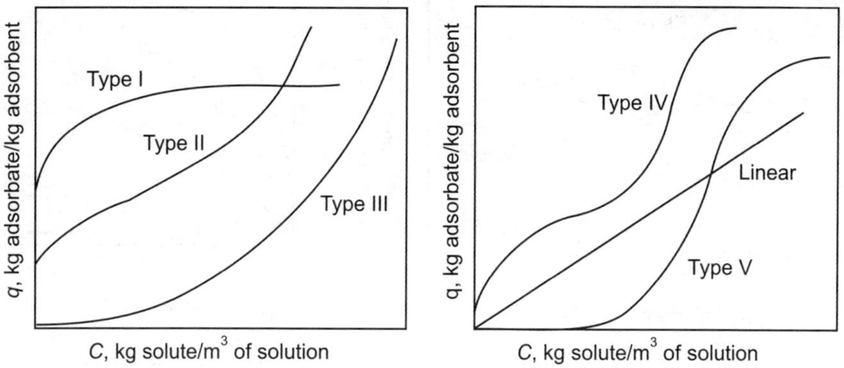

Fig. 10.1: Typical adsorption isotherms

Figure 10.1 shows representative isotherms, classified according to their shape by Brunauer. The curves that are concave towards the fluid phase concentration axis throughout (Type I) are designated as "favourable" isotherms, whereas the curves that are concave upwards throughout (Type III) are designated as "unfavourable." For favourable isotherms relatively high solid loading can be obtained at low concentrations in the fluid, whereas for unfavourable case, the solid loading obtained is relatively small and the width of the mass transfer zone is large. Type I isotherm applies to gases at temperatures above their critical temperatures. The linear isotherm goes through the origin and the amount adsorbed is proportional to the concentration in the fluid. Other curves shown in the figure (Type II, IV and V) have one or more points of inflection and are favourable over certain concentration region and unfavourable over others. Type II isotherms are favourable at low and unfavourable at high concentrations and are observed for gases at temperatures below their critical temperatures and at pressures below the saturation pressures. Type V isotherms are unfavourable at low and favourable at high concentrations.

It should be noted that an isotherm favourable for the adsorption step becomes unfavourable for the regeneration of the adsorbent. Activated carbon usually gives type I isotherm. A non-porous solid usually gives type II isotherm. If the cohesive forces between the adsorbate molecules are greater than the adhesive forces between the adsorbate and the adsorbent, a porous solid generally gives type V isotherm and a non-porous solid gives type III isotherm (Richardson et al, 1991). Type IV isotherm is usually followed by certain systems such as activated alumina-air-water vapour system which are characterised by three stages of adsorption. The concave region at low concentration corresponds to formation of monomolecular layer and the convex region corresponds to the formation of additional layers. The second concave region which occur at high gas concentrations is due to the condensation of adsorbate in the pores, usually known as the capillary condensation.

Figure 10.2 shows the equilibrium curves for the adsorption of different gases on activated carbon. Here the concentration of the gas adsorbed on the solid is plotted against the equilibrium partial pressure of the gas at constant temperature. The amount adsorbed increases with increase in the partial pressure at constant temperature. Different gases get adsorbed to different extent under similar conditions of temperature, as shown by the isotherms at 373 K for benzene and acetone in the figure. As a general rule, the higher the molecular weight and lower the critical temperature of the substance, the more the adsorbate loading on the solid.

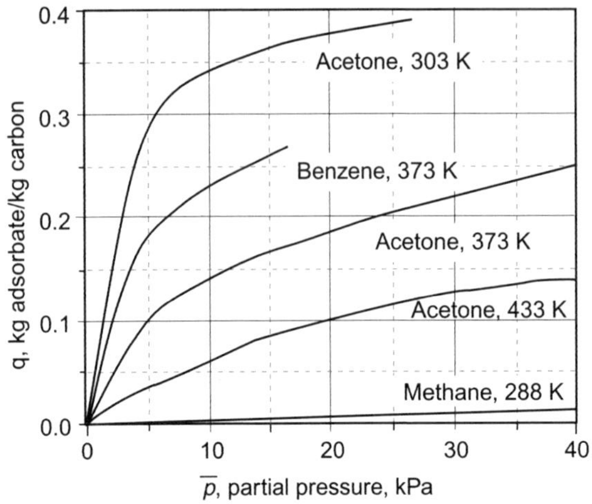

Fig. 10.2: Adsorption equilibrium for different gases/vapours on activated carbon

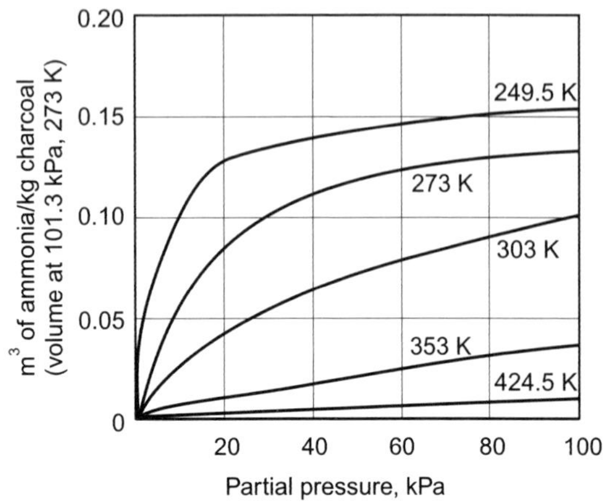

Fig. 10.3: Effect of temperature on equilibrium adsorption of ammonia on activated charcoal

The adsorption isotherms of ammonia on activated charcoal at different temperatures are shown in Fig. 10.3. The amount of the solute adsorbed decreases with increase in the temperature. This is true for other substances as well, as is evident from the curves for acetone at different temperatures as shown in Fig. 10.2.

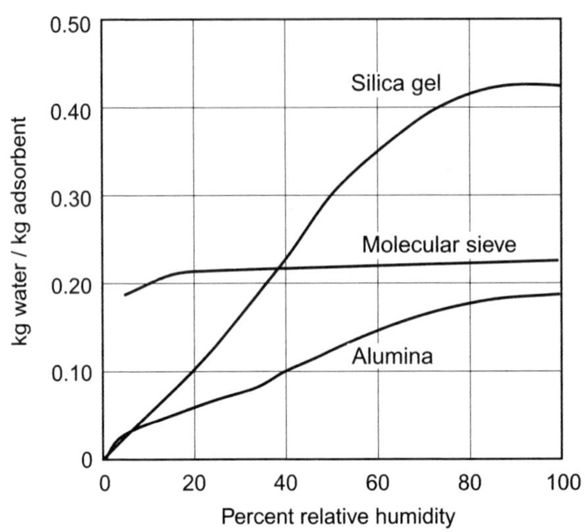

Fig. 10.4: Isotherms for the adsorption of water at 293 K to 325 K

A single adsorbate may exhibit different isotherm shapes when adsorbed on different adsorbents. Figure 10.4 shows the isotherms for adsorption of water on silica gel, molecular sieves and alumina at 293 to 325 K.

Example 10.1: An aqueous solution is coloured by the presence of certain impurities. It is decided to effect decolourization by adsorbing the impurities on activated carbon. To study the equilibrium adsorption characteristics, various amounts of activated carbon were added to 1 kg batches of the solution and the colour intensity measured at equilibrium. The initial colour intensity of the solution was 11.1 units. The following table gives the results expressed as kg carbon added versus percent colour removal.

kg carbon added	0	0.001	0.0025	0.005	0.01	0.02	0.03	0.04	0.05
% colour removal	0	9.9	22.5	38.7	60.4	80.2	88.3	92.1	94.1

Plot the adsorption isotherm on q versus C coordinates, where q is the units of colour adsorbed per kg carbon and C is the equilibrium colour intensity of the solution expressed as units of colour per kg solution.

***Solution*:** The given data is first converted into the desired form. For illustration let us consider the set,

colour removal = 9.9%

kg carbon added = 0.001 kg

Since the original colour intensity of the solution was 11.1 units per kg solution, corresponding to 9.9% removal, the equilibrium colour of the solution is

$$C = 11.1 \times \frac{(100 - 9.9)}{100} = 10 \text{ colour units/kg solution}$$

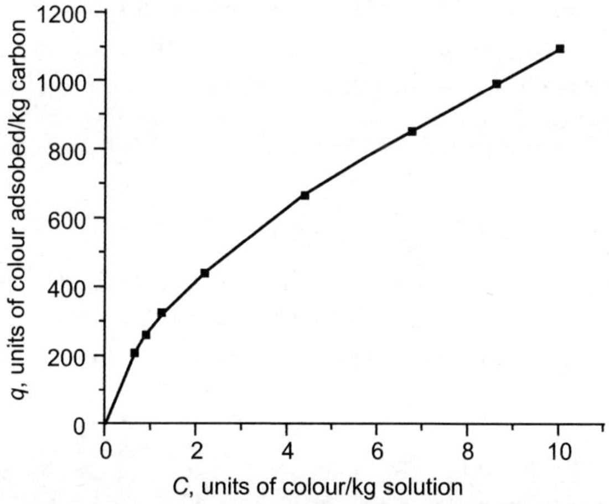

Fig. 10.5: Adsorption equilibrium for Example 10.1

Since the colour so removed is adsorbed on 0.001 kg activated carbon, the adsorbate loading is

$$q = \frac{11.1 \times 9.9/100}{0.001} = 1099 \text{ units of colour/kg carbon.}$$

These calculations are repeated for other data sets and the results are given in the following table.

C, units of colour/kg solution	10	8.6	6.8	4.4	2.2	1.3	0.88	0.65
q, units of colour/kg carbon	1099	999	859	670	445	327	256	209

The adsorption isotherm is plotted in Fig. 10.5.

Linear Isotherm

At low concentrations of solute, the molecules are widely spaced on the adsorbent surface and have no influence on one another. For these limiting conditions, it is reasonable to assume that the concentration on the solid is proportional to that in the fluid phase, and the adsorption isotherm will be linear. If this linear relation is applicable over a broader concentration range, the equilibrium can be represented by an equation similar to the Henry's law.

$$q = kC \qquad \text{...(10.1A)}$$

Here q is the adsorbate loading (moles, mass or volume adsorbed per unit mass of adsorbent), C is the concentration of the solute and k is an empirical constant, which depends on temperature. For gaseous systems, concentration in the fluid can be represented by the partial pressure of the solute and the isotherm can be written as

$$q = k\,\bar{p} \qquad \text{...(10.1B)}$$

where $\bar{p}$ is the partial pressure of the solute in the gas

Freundlich Isotherm

The Freundlich equation is an empirical equation relating adsorbate loading and the fluid concentration under equilibrium.

$$q = k\,C^{1/n} \qquad \text{...(10.2A)}$$

$$q = k\,\bar{p}^{1/n} \qquad \text{...(10.2B)}$$

Where, k and n are temperature-dependent constants. In general, k decreases with increase in the temperature. Value of n lies in the range of 1 to 5. When $n = 1$, Eq. (10.2) reduces to the linear isotherm Eq. (10.1). The value of n decreases with increase in the temperature and approaches 1 at high temperatures. The experimental $q - C$ or $q - \bar{p}$ data can be used for the evaluation of the constants n and k, either by using a nonlinear curve fitting technique or by converting Eq. (10.2) into a linear form as

$$\log q = \log k + \frac{1}{n}\log C \qquad \text{...(10.3A)}$$

$$\log q = \log k + \frac{1}{n}\log \bar{p} \qquad \text{...(10.3B)}$$

In the latter method, log q may be plotted against log C and the best straight line through the data has a slope $1/n$ and makes an intercept log k, from which n and k are calculated. Alternatively, a linear regression computer program may also be used in Eq. (10.2) to estimate the constants.

Langmuir Isotherm

The Langmuir theory on adsorption assumes that:

1. Adsorption is limited to mono layer.
2. There is no interaction between adjacent molecules on the surface.
3. The energy of adsorption remains constant over the entire surface.
4. The molecules adsorb at fixed sites on the surface and do not migrate to other locations.

The rate of adsorption is proportional to the empty surface available for adsorption and the fluid concentration. Process of adsorption is also accompanied by the desorption of adsorbed molecules from the surface, and the rate of desorption is proportional to the surface occupied by a monomolecular layer of the adsorbate. When the rate of adsorption and desorption are equal, a dynamic equilibrium is established. Let θ be the fraction of the surface occupied by the adsorbate. Then the fraction of the bare surface is $(1 - \theta)$. At equilibrium, the net rate of adsorption and desorption become zero and we have

$$k_a(1-\theta)C - k_d\theta = 0 \qquad \text{...(10.4)}$$

where k_a is the velocity constant for adsorption to the empty surface and k_d is that for desorption from a monolayer. Let K denote the adsorption equilibrium constant. Then

$$K = \frac{k_a}{k_d} \qquad \text{...(10.5)}$$

Substituting Eq. (10.5) into Eq. (10.4) and rearranging, we get

$$\theta = \frac{KC}{1+KC} \qquad \text{...(10.6)}$$

In Equation (10.6), θ is the ratio of the concentration of the adsorbate (q) to the concentration when the surface coverage by the monolayer is complete (q_m).

$$\frac{q}{q_m} = \frac{KC}{1+KC} \quad \text{...(10.7)}$$

Equation (10.7) can be rearranged to obtain the following familiar form of the Langmuir isotherm

$$q = \frac{Kq_m C}{1+KC} \quad \text{...(10.8A)}$$

In terms of the partial pressures, the Langmuir Equation is written as

$$q = \frac{K q_m \overline{p}}{1+K\overline{p}} \quad \text{...(10.8B)}$$

Langmuir isotherm reduces to the linear Henry's law form, Eq. (10.1) at low pressures, i.e. when $K\overline{p} << 1$. At very high partial pressures, i.e. when $K\overline{p} >> 1$, the adsorbate concentration on the solid is independent of the concentration in the fluid. Thus the Langmuir isotherm predicts an asymptotic behaviour at high pressures as exhibited by the type I curve in Fig. 10.1.

The empirical constants in Eq. (10.8) can be determined by fitting the equation directly to the experimental data. Alternatively, we can put the equation in a linear form as:

$$\frac{C}{q} = \frac{1}{q_m K} + \frac{C}{q_m} \quad \text{...(10.9)}$$

The best straight line drawn through the points when C/q is plotted against C has slope = $1/q_m$ and intercept = $1/q_m K$ from which the constants K and q_m are determined. Equation (10.9) indicates that a plot of $1/q$ versus $1/C$ also will give values of the constants K and q_m.

BET Isotherm

The BET theory was developed by Brunauer, Emmett and Teller in 1938. The BET theory also makes use of the two assumptions made in the development of Langmuir theory that there are no molecular interactions and adsorption occurs at fixed sites with no migration to adjacent sites. But, BET theory allows multilayer adsorption with the assumption that the heat of adsorption during monolayer formation is constant and the heat effect associated with the formation of subsequent layers is equal to the heat of condensation. Based on these assumptions, Brunauer, Emmett and Teller derived a theoretical equation, which may be written as

$$\frac{P}{v(P^S - P)} = \frac{1}{v_m c} + \frac{(c-1)}{v_m c}\left(\frac{P}{P^S}\right) \quad \text{...(10.10)}$$

where P = total pressure, P^S = vapour pressure of adsorbate, v = volume of gas at STP, v_m = volume of mono-molecular layer of the gas adsorbed at STP, c = constant related to the heat of adsorption. The BET adsorption isotherm is plotted in Fig. 10.6.

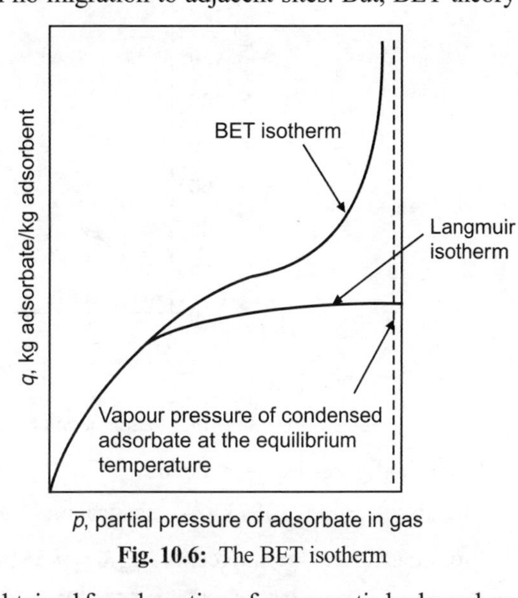

Fig. 10.6: The BET isotherm

Example 10.2: The following equilibrium data were obtained for adsorption of an aromatic hydrocarbon on molecular-sieve zeolite at 550 K.

q, kg/100 kg zeolite	7.53	10.00	11.48	12.11	12.49	13.00
$\bar{p}$, Pa	1.6	3.6	5.7	9.3	12.5	19.6

Evaluate the parameters of the Langmuir isotherm that fits the experimental data reasonably well.

***Solution*:** If the equilibrium can be represented by the Langmuir isotherm, then Eq. (10.8B) should be applicable to the system.

$$q = \frac{K q_m \bar{p}}{1 + K \bar{p}} \qquad \text{...(10.8B)}$$

The constants in Eq. (10.8B) can be evaluated by putting it in the following linear form

$$\frac{\bar{p}}{q} = \frac{1}{q_m K} + \frac{\bar{p}}{q_m} \qquad \text{...(10.9)}$$

Now calculate $\frac{\bar{p}}{q}$ and plot against $\bar{p}$.

$\bar{p}$, Pa	1.6	3.6	5.7	9.3	12.5	19.6
q, kg/ kg zeolites	0.0753	0.1000	0.1148	0.1211	0.1249	0.1300
$\frac{\bar{p}}{q}$, (Pa) kg/kg	21.248	36.000	49.652	76.796	100.080	150.769

Now a graph is plotted with $\bar{p}$ on x-axis and $\frac{\bar{p}}{q}$ on y-axis as shown in Fig. 10.7. The slope of the straight line is $1/q_m$ and the intercept with $\frac{\bar{p}}{q}$ axis is $1/q_mK$.

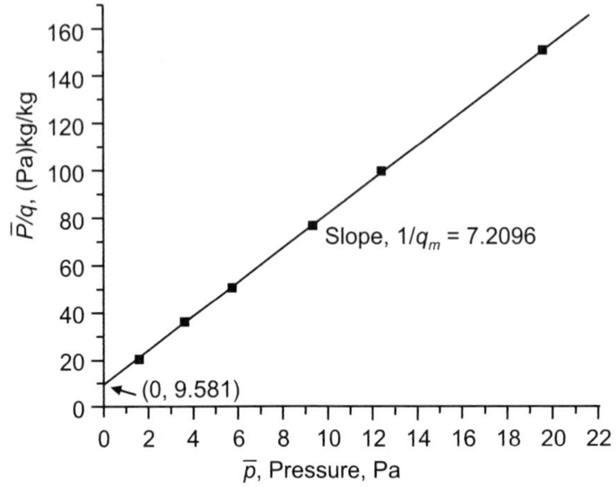

Fig. 10.7: Estimation of the Langmuir isotherm parameters

From the figure, $1/q_m = 7.2096$

Therefore, $q_m = 0.1387\,\text{kg/kg}$ absorbent.

From the figure, the intercept $1/q_m K = 9.581$

Substituting $q_m = 0.1387$, $K = 0.7525\ (\text{Pa})^{-1}$.

The constants in the Langmuir equations are: $K = 0.7525\ (\text{Pa})^{-1}$, $q_m = 0.1387$ kg/kg adsorbent.

$$q = \frac{Kq_m\overline{p}}{1+K\overline{p}} = \frac{0.7525 \times 0.1387\overline{p}}{1+0.7525\overline{p}} = \frac{0.1044\overline{p}}{1+0.7525\overline{p}}$$

$$q = \frac{0.1044\overline{p}}{1+0.7525\overline{p}}$$ **(Answer)**

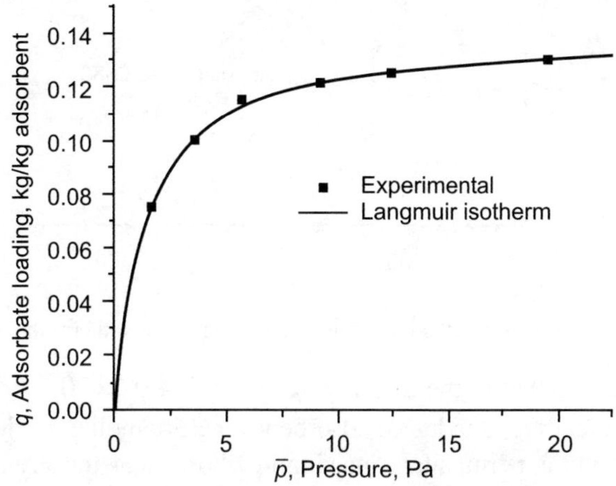

Fig. 10.8: Langmuir isotherm for Example 10.2

Figure 10.8 compares the experimental result with the Langmuir isotherm equation obtained for the system and it can be seen that the Langmuir isotherm correlates the data very closely.

Example 10.3: The following data refers to the equilibrium adsorption of benzene vapour on silica gel at 363 K.

Partial pressure of benzene, Pa	50.7	101.3	202.7	506.5	1013	2026
kg adsorbed/kg gel × 10^3	5.226	8.736	14.04	25.74	39.78	60.84

Determine which of the isotherms Freundlich or Langmuir, fits the data well?

Solution: *Fit of Freundlich isotherm*: The Freundlich isotherm is given by Eq. (10.2B)

$$q = k\,\overline{p}^{1/n} \qquad \text{...(10.2B)}$$

where k and n are constants. The constants k and n are evaluated by putting Eq. (10.2B) in the linear form as follows:

$$\log q = \log k + \frac{1}{n}\log\overline{p} \qquad \text{...(10.3B)}$$

A straight line will be obtained when log q is plotted against $\log\overline{p}$ the slope of which gives $1/n$ and the line makes an intercept log k, from which n and k are calculated.

$\overline{p}$	50.7	101.3	202.7	506.5	1013	2026
q	0.00523	0.00874	0.01404	0.02574	0.03978	0.06084

log q is plotted against $\log\overline{p}$ in Fig. 10.9. The slope of the best straight line through the points is 0.66329 which means that

$$1/n = 0.66329 \text{ or } n = 1.508$$

The constant k can be evaluated by considering any point on this line such as point A in Fig. 10.9 and noting its coordinates. Coordinates of point A are ($\overline{p} = 268.565$ and $q = 0.016425$). Since $q = k\,\overline{p}^{1/n}$, we have, $0.016425 = k \times 268.565^{0.66329}$ or $k = 4.02 \times 10^{-4}$

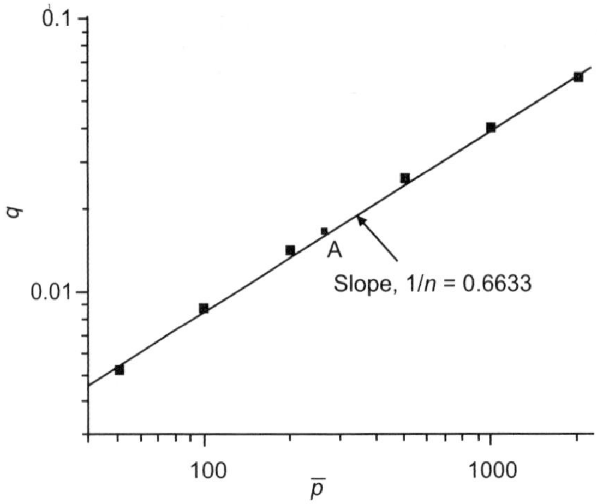

Fig. 10.9: Determination of the constants in the Freundlich isotherm that fits the data

Or the Freundlich isotherm that fit the data is given by $q = 4.02 \times 10^{-4}\,\bar{p}^{0.66329}$ where the pressure is in Pa and the adsorbate loading is in kg/kg adsorbent. The Freundlich isotherm is plotted as shown in Fig. 10.10 along with the experimental data. Figure shows that the Freundlich isotherm fits the experimental data accurately.

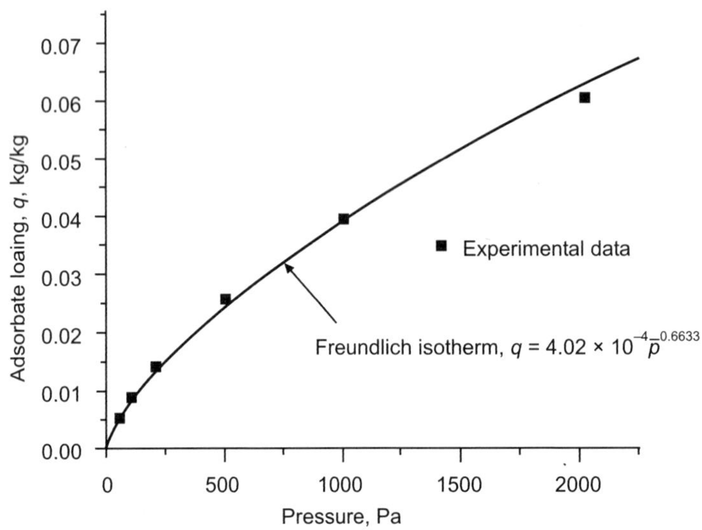

Fig. 10.10: Comparison of the Freundlich isotherm with the experimental data

Fit of the Langmuir isotherm: To find the fit of Langmuir isotherm to the experimental data, a graph is plotted with $\bar{p}$ on x-axis and $\dfrac{\bar{p}}{q}$ on y-axis as shown in Fig. 10.11.

$\bar{p}$	50.7	101.3	202.7	506.5	1013	2026
$\bar{p}/q$	9694	11590	14437	19678	25465	33301

The slope of the best straight line passing through the experimental points is found to be 11.5661. It means that in the linear form of the Langmuir equation,

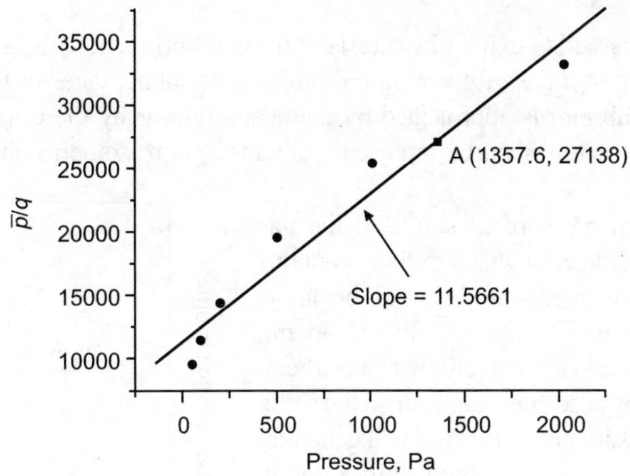

Fig. 10.11: Determination of the constants of the Langmuir isotherm that fits the data

$$\frac{\overline{p}}{q} = \frac{1}{q_m K} + \frac{\overline{p}}{q_m} \qquad \ldots(10.9)$$

the constant $\frac{1}{q_m} = 11.5661$ or $q_m = 0.0865$.

The constant K can be determined by considering a point on the straight line such as point A in Fig. 10.11. The coordinates of the chosen point A are (1357.6, 27138). Therefore in Eq. (10.9),

$$\frac{\overline{p}}{q} = 27138, \quad \overline{p} = 1357.6$$

Substituting these in Eq. (10.9), we get $K = 1.01 \times 10^{-3}$. Substituting these in Eq. (10.8B), we get

$$q = \frac{K q_m \overline{p}}{1 + K\overline{p}} = \frac{1.01 \times 10^{-3} \times 0.0865\overline{p}}{1 + 1.01 \times 10^{-3}\overline{p}} = \frac{8.7365 \times 10^{-5}\overline{p}}{1 + 1.01 \times 10^{-3}\overline{p}}$$

which is the Langmuir isotherm that fits the given data.

In Fig. 10.12 is plotted, the Langmuir isotherm along with the experimental data points and we see that the Langmuir isotherm does not represent the data satisfactorily. In contrast, from Fig. 10.10 it is clear that the Freundlich isotherm fits the data very closely.

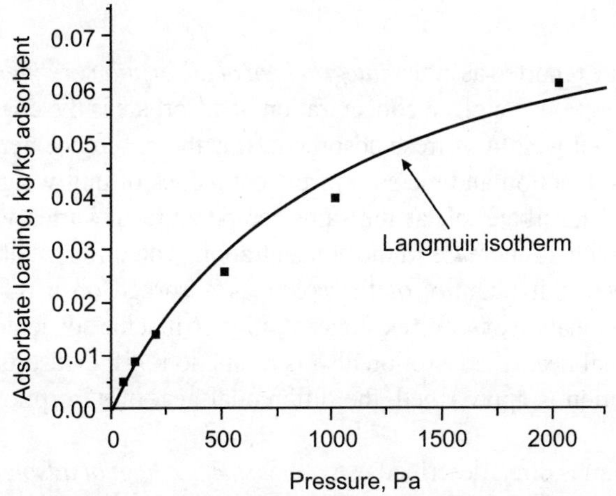

Fig. 10.12: Comparison of the Langmuir isotherm with the experimental data

Adsorption Hysteresis

The adsorption system is said to exhibit hysteresis, if the equilibrium adsorbate loading attained by the solid when it is exposed to a gas of given partial pressure of solute, takes different values depending upon whether the equilibrium is approached by the adsorption or by the desorption. For the looped section in Fig. 10.13, the lower curve represents the adsorption isotherm and the upper curve, the desorption isotherm.

For a definite partial pressure of solute in the gas exposed to an adsorbent material which initially contains an appreciable quantity of adsorbate, equilibrium is approached by desorption of solute from the adsorbent. If the same gas is exposed to a relatively fresh adsorbent, equilibrium is reached by adsorbing solute on to the solid. The concentration of adsorbate on the solid will generally be greater, if the equilibrium is reached by desorption than when the equilibrium is reached by adsorption. Put in another way, the desorption equilibrium pressure is always lower than the adsorption pressure. Materials with a high degree of capillary, exhibit hysteresis. A plausible explanation of hysteresis based on the geometrical shape of the pore space is given in Chapter on 'Drying'.

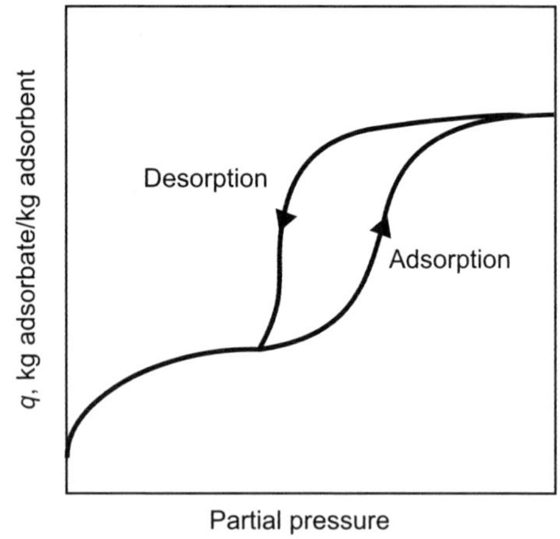

Fig. 10.13: Adsorption hysteresis

Multicomponent Adsorption

So far our discussion was limited to the adsorption of a single substance from a mixture of gases. A modified form of the single component adsorption equations may be written for adsorption of more than one component. For example, the Langmuir equation for multicomponent adsorption is

$$q_i = \frac{(q_i)_m K_i C_i}{1 + \sum_j K_j C_j} \qquad \text{...(10.11)}$$

Here q_i is the amount of species i adsorbed per unit weight of adsorbent, $(q_i)_m$ is the maximum amount of adsorption of species i for coverage of the entire surface. Although Eq. (10.11) is found to lack thermodynamic consistency, it is widely used in practical applications, because of its simplicity.

Heat of Adsorption

Heat of adsorption data are reported as either *integral heat of adsorption* or *differential heat of adsorption*. The integral heat of adsorption at a given concentration of adsorbate is the change in enthalpy when the adsorbate is formed on unit weight of fresh adsorbent. It is the difference between the enthalpy of the adsorbent–adsorbate combination and the sum of the enthalpies of unit weight of pure adsorbent and the required quantity of adsorbate, all at the same temperature. It varies with the concentration of adsorbate, diminishing with an increase in the concentration. The differential heat of adsorption is the change in enthalpy when unit quantity of the solute is adsorbed on a relatively large quantity of adsorbent-adsorbate combination of specified concentration so that the concentration remains unchanged on adsorption. Differential heat of adsorption also is a function of the adsorbate concentration. As the state of complete saturation is approached, the differential heat of adsorption approaches the heat of condensation.

Adsorption equilibrium is often described using the *isosteric heat of adsorption* which is determined from the changes in the pressure of the gas phase accompanying an increase in the temperature, keeping the amount adsorbed constant. In Fig. 10.14, the adsorption data are presented graphically in the form of adsorption isosteres. These are curves representing constant amounts adsorbed and resemble vapour – pressure plots. The adsorption form of the Clausius - Clapeyron equation is assumed applicable for this case. It is assumed that the partial pressure of the adsorbate vapour in equilibrium changes due to an increase in the temperature, in the same way as in the liquid-vapour equilibrium.

$$\frac{d\ln P}{dT} = \frac{-\Delta H_{ads}}{RT^2} \qquad \text{...(10.12)}$$

Or

$$\frac{d\ln P}{d(1/T)} = \frac{\Delta H_{ads}}{R} \qquad \text{...(10.13)}$$

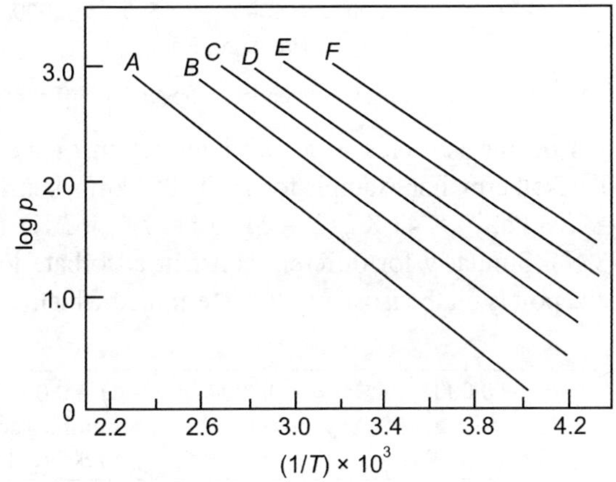

A $V = 10^{-2}$ m³/kg ($-\Delta H_{ads}$ = 30660 kJkmol) *B* $V = 2 \times 10^{-2}$ m³/kg ($-\Delta H_{ads}$ = 28560 kJkmol)
C $V = 3.5 \times 10^{-2}$ m³/kg ($-\Delta H_{ads}$ = 27300 kJ/kmol) *D* $V = 5 \times 10^{-2}$ m³/kg ($-\Delta H_{ads}$ = 26900 kJ/kmol)
E $V = 7.5 \times 10^{-2}$ m³/kg ($-\Delta H_{ads}$ = 26000 kJ/kmol) *F* $V = 10^{-1}$ m³/kg ($-\Delta H_{ads}$ = 25600 kJ/kmol)

Fig. 10.14: Isosteres for adsorption of ammonia on charcoal

Equations (10.12) or (10.13) may be used to determine heat of adsorption.

$$-\Delta H_{ads} = -R\frac{d\ln P}{d(1/T)} \qquad \text{...(10.14)}$$

Equation (10.14) allows the calculation of the heat of adsorption by plotting ln p against $(1/T)$ for constant values of the amount of solute adsorbed.

Example 10.4: The following table gives the equilibrium data at four different temperatures for adsorption of pure benzene vapour on silica gel. Estimate the heat of adsorption

Partial pressure of benzene, Pa	kmol adsorbed/kg gel × 10^5			
	343 K	363 K	383 K	403 K
50.7	14	6.7	2.6	1.13
101.3	22	11.2	4.5	2
202.7	34	18	7.8	3.9
506.5	68	33	17	8.6
1013	88	51	27	16
2026	–	78	42	26

***Solution*:** The adsorption isotherms at different temperatures are plotted as shown in Fig. 10.15.

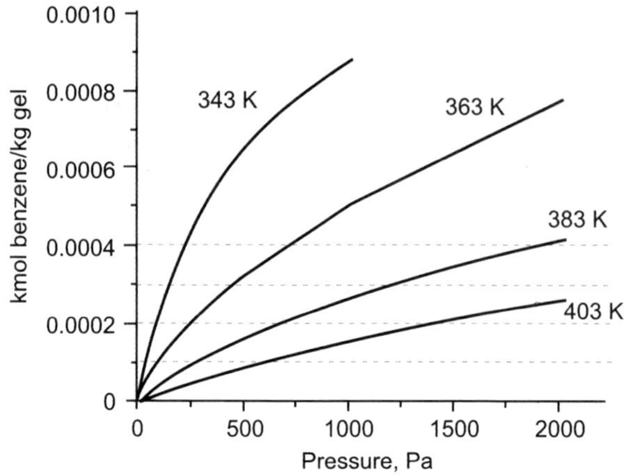

Fig. 10.15: Adsorption isotherm for benzene on silica gel at different temperatures

To determine the points on the isosteres, read the intersection of the horizontal lines drawn at selected values of q, with the isotherm. For example for $q = 0.0002$ kmol benzene/kg gel, the intersections have the coordinates ($P = 92.3$ Pa, $T = 343$ K), ($P = 242.7$ Pa, $T = 363$ K), ($P = 653.9$ Pa, $T = 383$ K), ($P = 1409.7$ Pa, $T = 403$ K). Similarly for different constant adsorbate loading values ($q = 0.0001$, 0.0002, 0.0003, 0.0004), the points on the isosteres are determined which are tabulated as shown in the table below:

Temperature, K	$q = 0.0001$ kmol/kg gel Pressure, Pa	$q = 0.0002$ kmol/kg gel Pressure, Pa	$q = 0.0003$ kmol/kg gel Pressure, Pa	$q = 0.0004$ kmol/kg gel Pressure, Pa
343	36.1	92.3	168.5	255.8
363	89.8	242.7	447.3	702.6
383	270.8	653.9	1223.6	1890.1
403	600.3	1409.7		

A graph is plotted with $1/T$ on the x-axis and $\log p$ on the y-axis as shown in Fig. 10.16

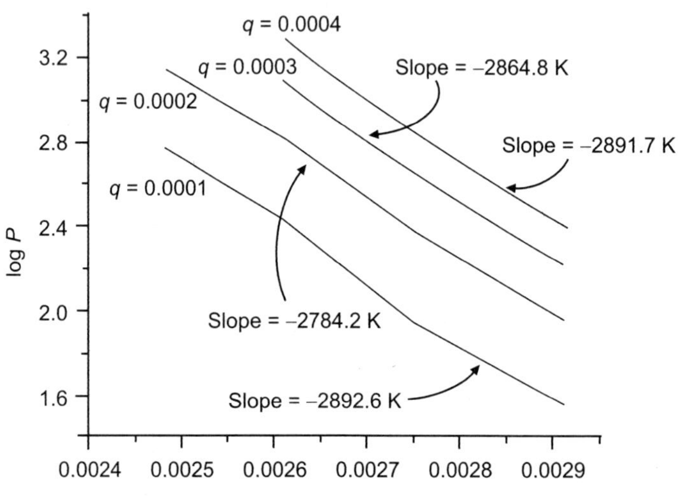

Fig. 10.16: Isosteres for adsorption of benzene on silica gel (Example 10.4)

$1/T$	$q = 0.0001$ kmol/kg gel log P	$q = 0.0002$ kmol/kg gel log P	$q = 0.0003$ kmol/kg gel log P	$q = 0.0004$ kmol/kg gel log P
0.00291	1.5575	1.9652	2.2266	2.4079
0.00275	1.9533	2.3851	2.6506	2.8467
0.00261	2.4326	2.8155	3.0876	3.2765
0.00248	2.7784	3.1491		

The slope of the isosteres are determined and tabulated as shown in the table below. Since Eq. (10.13) gives

$$\frac{d\ln P}{d(1/T)} = \frac{\Delta H_{\text{ads}}}{R}$$

we have

$$2.303\frac{d\log P}{d(1/T)} = \frac{\Delta H_{\text{ads}}}{R}$$

Or

$\Delta H_{\text{ads}} = 2.303R\dfrac{d\log P}{d(1/T)}$ where $\dfrac{d\log P}{d(1/T)}$ is the slope of the straight lines in Fig. 10.16 and $R = 8.314$ kJ/kmol K. By multiplying the slope of the lines by 2.303 R, we can calculate the heat of adsorption. The results are given in the table.

q, kmol/kg	Slope of the isostere, K	ΔH_{ads} kJ/kmol
0.0001	–2892.6	–55385
0.0002	–2784.2	–53310
0.0003	–2864.8	–54853
0.0004	–2891.7	–55369
Average ΔH_{ads}	–54729	

Average heat of adsorption is –54729 kJ/kmol. **(Answer)**

Adsorption from Liquids

Except when adsorption involves capillary condensation, the adsorption equilibrium for liquid-solid systems can be represented by the same equations that describe the adsorption isotherm of gas-solid systems. Freundlich equation is found to be most suitable to describe the adsorption isotherm of liquids.

$$q = kC^{1/n} \qquad \text{...(10.15)}$$

q = kg solute adsorbed/ kg adsorbent, C = kg solute/m^3 of solution, k and n are empirical constants.

Unlike the adsorption from gases for which it is easy to measure the amount adsorbed by measuring the change in the mass of the solid or by noting the change in the pressure and the volume of the gas phase, it is difficult to measure the concentrations of solute in the solid and the liquid phases under equilibrium. This is because, the change in volume in the liquid due to adsorption is immeasurably small and it is difficult to distinguish between the liquid adsorbed by the solid and the liquid mechanically occluded, when the solid is separated from the liquid.

Neglecting the adsorption of all but the solute, the change in concentration of the solution reflects the amount of solute adsorbed. A known volume of the solution is treated with a known quantity of the adsorbent and the change in the concentration of the solution is measured after equilibrium is attained. Let the volume of solution treated with unit weight of the adsorbent be V. Let C_0 be the initial concentration of the solute and C^* be the concentration after adsorption is completed. Neglecting the change in volume of solution, $V(C_0 - C^*)$ is the amount of the solute adsorbed per unit weight of the adsorbent and C^* is the equilibrium concentration of solute in the liquid. Freundlich equation (10.15) can be written as

$$V(C_0 - C^*) = kC^{*1/n} \qquad \ldots(10.16)$$

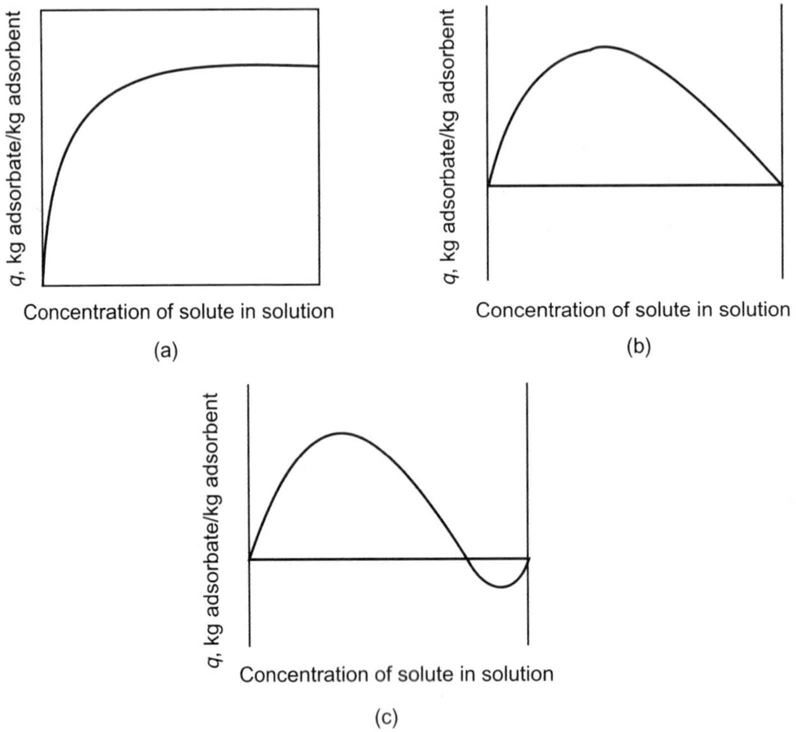

Fig. 10.17: Adsorption isotherms for liquids

Equation (10.16) may be rearranged as

$$C^* = k'\left[V(C_0 - C^*)\right]^n \qquad \ldots(10.17)$$

Freundlich equation represents satisfactorily the isotherm for adsorption from dilute liquid solutions. Figure 10.17a represents the adsorption isotherm for dilute liquid solutions. Small amounts of organic compounds dissolved in water and small amounts of water dissolved in hydrocarbons obey the adsorption isotherm represented by Eq. (10.15). Freundlich equation is also useful for representing the adsorption, when the actual identity of the adsorbing species is unknown. For example, adsorption of coloured substances from sugar solutions and vegetable oils can be represented by Eq. (10.15), with concentration of the solute represented by some arbitrary units of colour intensity measured by means of devices such as the colorimeter or spectrophotometer.

If the experimental data are obtained over the entire concentration range, and Eq. (10.17) is plotted, the resulting curves will be of the form of Fig. 10.17b or 10.17c. Curves of the type shown in Fig. 10.17b result, when the solute is adsorbed more strongly than the solvent, at all concentrations. The curve shown in Fig. 10.17c results, when the solvent and the solute are adsorbed to the same extent. Initially the solute is adsorbed more strongly than the solvent and the curve lies above the horizontal axis at $q = 0$. When the solvent is more strongly adsorbed than the solute, the curve lies below the axis, in which case, the solute concentration in the liquid actually increases on addition of the adsorbent, indicating negative apparent adsorption. When the solute and the solvent are adsorbed to the same extent, the curve meets the $q = 0$ axis and the apparent adsorption of the solute becomes zero.

10.6 ADSORBENT REGENERATION

In the design of an adsorption unit, it is essential to ensure the regeneration of the spent adsorbent in a given time and the total inventory of the adsorbent be kept at the minimum. Regeneration of the spent adsorbent is generally brought about by the reverse process of adsorption. In adsorption, the movement of adsorbate is from the fluid to the adsorbent whereas in desorption, the movement is from the adsorbent to the fluid. As a result, the conditions favourable for regeneration are the ones that are unfavourable for adsorption. Regeneration is therefore favoured by increasing the temperature or by reducing the pressure. In practice, the regeneration is generally carried out by three methods. The *temperature swing, the pressure swing* and *the purge swing* cycles constitute the basis for various techniques for regeneration of the spent adsorbent.

Temperature-Swing Adsorption

In the temperature-swing or thermal-swing adsorption (TSA) cycle, the desorption occurs at a much higher temperature than the adsorption. An ideal TSA cycle is shown in Fig. 10.18a. The cycle consists of the four essential steps: adsorption, heating, desorption and cooling. (1) The feed fluid in which the partial pressure of the adsorbable component (**A**) is $\bar{p}_1$ is brought into contact with the adsorbent at temperature T_1. When the equilibrium is established, the adsorbate loading is q_1 as shown by point A in the figure. (2) In the heating step, the adsorbent temperature is raised to T_2. The equilibrium partial pressure exerted by the adsorbate at this elevated temperature is $\bar{p}_2$ as shown in the figure by point B. Thus, there exists a partial pressure difference driving force for desorption into any fluid with an adsorbate partial pressure less than $\bar{p}_2$. (3) Desorption is carried out by passing a purge fluid across the adsorbent at temperature T_2. The adsorbate is swept away and the equilibrium proceeds to a point such as C in the figure, where the partial pressure of adsorbate is, say $\bar{p}_1$, and the adsorbate loading is q_2. Although desorption can be carried out in the absence of a purge fluid by simply vaporising the adsorbate, there is every likelihood of re-adsorption of some solute vapour upon cooling the bed. Thus, it is best to accomplish the desorption by means of a purge fluid. (4) During the cooling step the adsorbent temperature is returned to T_1 and the new equilibrium is shown by point D in Fig. 10.18a with a partial pressure of adsorbate on the adsorbent $\bar{p}_3$ and a residual adsorbate loading of q_2. The adsorption

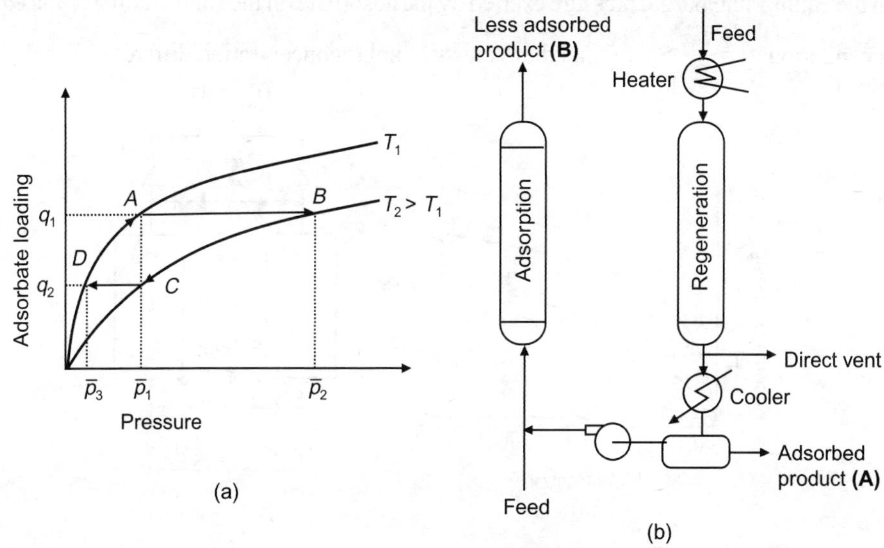

Fig. 10.18: Ideal TSA cycle

step is now repeated. The differential loading $q_1 - q_2$ is the maximum loading that can be achieved for a TSA cycle operating between a feed with a partial pressure $\bar{p}_1$ at temperature T_1, regeneration at T_2, and a product with a partial pressure $\bar{p}_3$ of the adsorbate.

An adsorption unit using TSA cycle is schematically represented in Fig. 10.18b. The figure shows two packed beds one engaged in adsorption and one regenerating. The heating is generally accomplished by either direct contact of the adsorbent by the heating medium or by indirect means by placing the heating elements, coils, or panels inside the adsorber. The external heating by heat exchange to a purge gas is the most widely used method in TSA cycles.

Thermal swing adsorption is characterised by low residual and high operating loadings and high capacities. For reasonably sized adsorbers, the cycle times are relatively long. TSA results in thorough desorption and can be used for the gas phase and the liquid phase adsorption. In TSA, the bed usually takes a substantial time to attain the regeneration temperature. The high temperature may, in some cases adversely affect the product and also accelerates the ageing of the adsorbent. Thermal swing adsorption finds wide application in the drying of gases and is extensively used in the natural gas, chemical and the cryogenic industries. TSA is used in sweetening process for removing sulphur compounds from gases such as natural gas, hydrogen, biogas and refinery streams, to prevent corrosion and catalyst poisoning.

Pressure-Swing Adsorption

In the pressure-swing adsorption (PSA) process, adsorption occurs at high pressure and desorption at low pressure. Thus, reduction of pressure is used to shift the adsorption equilibrium and achieve regeneration of the adsorbent. An ideal PSA cycle is shown in Fig. 10.19a. First, the feed fluid at a total pressure P_1 and temperature T_1 with a molar concentration y_1 for the solute is brought into contact with the adsorbent. Point A represents the condition in which the feed fluid is in equilibrium with the adsorbent. q_1 is the equilibrium adsorbate loading for fluid concentration y_1, where $y_1 = \frac{\bar{p}_1}{P_1}$. The second step is a depressurization or blow down step in which the pressure is lowered to P_2 at constant temperature. This condition is represented by the point B in the figure. For the adsorbate loading q_1 at temperature T_1, though the equilibrium partial pressure exerted by the adsorbate on the solid is still $\bar{p}_1$, the equilibrium molar concentration $y_2 = \frac{\bar{p}_1}{P_2} > y_1$. Thus there exists a molar concentration difference driving force for

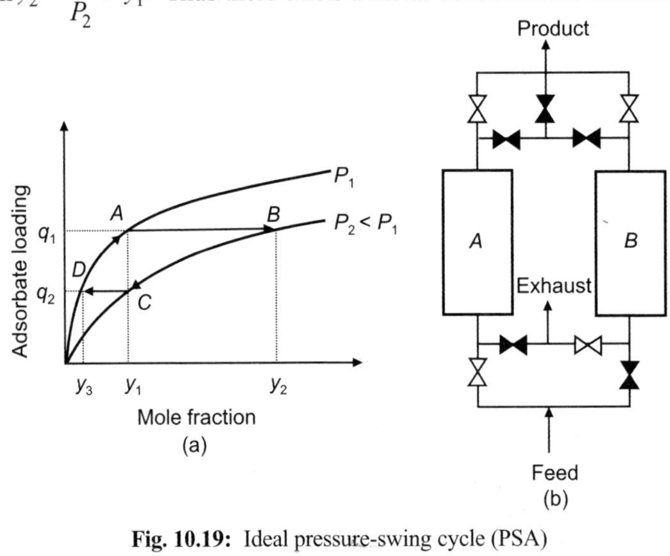

Fig. 10.19: Ideal pressure-swing cycle (PSA)

desorption of the adsorbate into the fluid in which the concentration of the adsorbate is less than y_2. The purge fluid is passed across the bed to sweep away the desorbed fluid and the equilibrium now shifts to point C in Fig. 10.19a.

In the final step, the adsorbent is repressurised to P_1. The bed is now at condition represented by point D with a residual adsorbate loading of q_2 and an equilibrium fluid phase molar concentration y_3, is ready for the next cycle. The differential loading $q_1 - q_2$ is the maximum loading that can be achieved for a pressure-swing cycle operating between a feed containing a molar concentration y_1 and a product containing y_3 of the adsorbate.

The pressure-swing adsorption unit shown schematically in Fig. 10.19b consists of two fixed beds, one adsorbing and the other regenerating. While bed A is undergoing pressurisation followed by adsorption, bed B is subjected to depressurisation (blow down) followed by a purge in the first half cycle of operation. Thus, high pressure feed flows through bed A. Part of the effluent from the bed A is expanded to the lower pressure, and then passed through bed B to purge its contents in a direction counter-current to the direction of flow of the feed during the adsorption step. In the next step, B is repressurised to the feed pressure using the feed gas, while A is blown down to the purge pressure.

The main attraction of PSA is that it operates at constant temperature requiring no heating or cooling steps. The endothermic heat needed for desorption is met from the exothermic heat of adsorption remaining in the adsorbent. In contrast, in TSA, the bed may need a substantial time to reach the relatively very high regeneration temperature which may affect the product quality and accelerate the ageing process of the adsorbent. Since the pressure can be changed more rapidly than the temperature, PSA can be used with cycle times shorter than those in TSA. This in turn, results in smaller beds and therefore, smaller inventory of the adsorbent. The low pressure is not as effective as a high temperature for totally reversing the adsorption, unless very high feed-to-purge pressure ratios are applied. Therefore, most PSA cycles are characterised by high residual loadings and low operating loadings.

The PSA is restricted to gas-phase applications. Major applications of PSA include purification as well as removal of contaminants present at high concentrations. PSA finds extensive applications in the purification of wet gases. The high purity hydrogen required for the hydrogenation, hydrocracking, ammonia and methanol synthesis, is produced by PSA cycles using adsorbent beds compounded of activated carbon, zeolites and carbon molecular sieves. The bulk separations using PSA include the air separation, the methane enrichment and the iso/normal separation. PSA is the ideal choice when the product stream is required at pressures above the atmospheric, as in the case of the instrument air.

Purge-Swing Adsorption (Concentration-Swing Adsorption)

In TSA cycles, desorption is achieved at an elevated temperature and in PSA, at a reduced pressure. In purge-swing adsorption cycle, desorption can be brought about by using an inert gas purge or by displacing the adsorbate using another adsorbable component. Here, desorption takes place at the same temperature and total pressure as the adsorption. The purge cycle operates adiabatically at nearly constant inlet temperature and requires no heating or cooling steps. The purge processes are classified into two: (i) the inert purge cycle in which the desorption is achieved by the partial pressure reduction using an inert gas purge (ii) the displacement purge in which the adsorbate displacement is achieved using another adsorbable component. Both these techniques utilize the heat of adsorption remaining in the adsorbent for supplying the heat of desorption.

Inert purge Figure 10.20a shows the inert-purge-swing cycle using a non adsorbable purge fluid. Adsorption is occurring at temperature T_1 and partial pressure of the adsorbate $\overline{p}_1$. The adsorption is continued till the equilibrium adsorbate loading of q_1 is attained which is shown by point A in the figure. Next, a non adsorbing fluid is introduced which reduces the partial pressure of the adsorbate in the fluid, thereby creating a partial pressure difference driving force for the transfer of mass from the solid

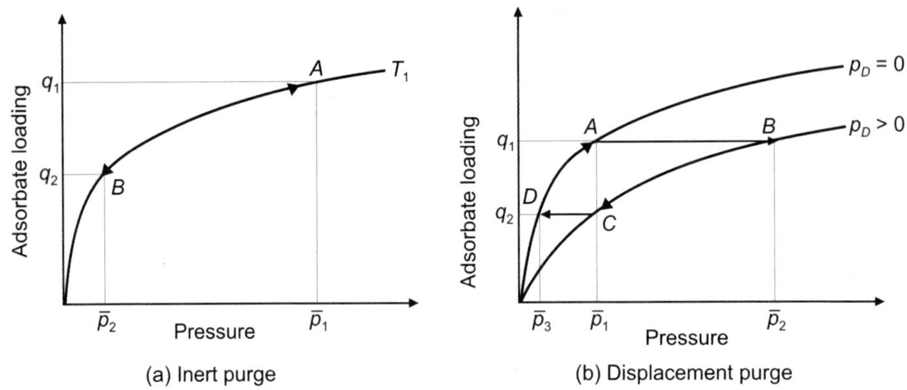

Fig. 10.20: Purge-swing adsorption cycle

to the fluid phase. The equilibrium now shifts to point *B* which shows the residual adsorbate loading of q_2 at partial pressure $\bar{p}_2$. The regeneration of the adsorbent is achieved by the introduction of an inert purge fluid while maintaining the same temperature and pressure as in the adsorption step. Like PSA, the inert-purge-swing cycle is characterised by high residual loading and short cycle times. In automobiles, the canisters of carbon used for the adsorption of gasoline vapours lost from the carburettors or the gas tanks, are regenerated by an inert purge of air that is drawn into the carburettor while the engine is running. The adsorbent used for drying of ethanol employed for blending with gasoline is regenerated by desorbing the water by an inert nitrogen gas purge.

Displacement purge In the inert-purge-swing cycle, the desorption is achieved by circulating an inert gas which reduces the partial pressure of the adsorbate in the gas. In displacement purge-swing cycle, the desorption is again by means of a purge fluid, but in this case, the purge fluid gets adsorbed while displacing the original adsorbate. Figure 10.20b depicts a simplified displacement purge-swing cycle utilizing an adsorbable purge at the cycle conditions. Point *A* in the figure represents the condition at which the adsorbent is saturated with adsorbate at temperature T_1 and partial pressure $\bar{p}_1$. Next a displacement fluid **D** is introduced into the system which reduces the affinity of the original adsorbed component **A** to the adsorbent. This in turn, increases the partial pressure in the fluid which is in equilibrium with the adsorbate loading of q_1 (point *B*). The partial pressure difference leads to the desorption of species **A**. Species **A** thus gets displaced by the component **D** and the equilibrium shifts to a new state such as the one shown in the figure by point *C*. Next, the adsorbent is recharged with the fluid in the absence of **D**, thus shifting the equilibrium to point *D*, at which the cycle is repeated.

The displacement purge cycle finds application in the Sorbex simulated moving bed process discussed in later sections. Using normal paraffin or light naphtha as the purge stream, displacement purge adsorption is employed for the separation of branched and cyclic C_{10}–C_{18} compounds from normals. Ammonia, which is easily separated from the paraffins by flash vaporisation is also used as a purge fluid. Another typical application is the separation of the ethanol-water system using activated carbon as the adsorbent which is regenerated using acetone as the displacement agent. The major limitation of the displacement purge method is that it contaminates both the product and the adsorbate streams and requires additional purification operations.

10.7 ADSORPTION EQUIPMENT

Adsorbers may be classified, based on the mode of contact of the fluid and the adsorbent as agitated vessels, fixed bed units, moving bed contactors, fluidized bed adsorbers and simulated moving bed adsorbers. Any of the different cycles discussed earlier or their combinations can be employed for the regeneration of the adsorbent. Table 10.2 lists some commonly used devices, the regeneration technique employed and whether the fluid handled is liquid or gas.

Table 10.2: Different types of adsorbers with the method of regeneration of adsorbent

Adsorber	Fluid phase	Adsorbent regeneration method
Agitated vessels	Liquid	Usually discarded
Fixed bed adsorber	Liquid	Thermal reactivation
	Gas	Thermal swing, pressure swing, inert purge swing, displacement purge swing
Moving bed adsorber	Liquid	Displacement purge
	Gas	Thermal swing, purge swing
Combined fluidised bed-moving bed	Gas	Thermal swing
Simulated moving bed	Liquid	Displacement purge

Agitated Vessels (Contact Filtration)

In agitated vessels, the liquid to be treated and the adsorbent are intimately mixed at the desired temperature. Figure 10.21 shows a single-stage batch treatment of liquid by contact filtration. Continuous or semi-continuous operations are also possible with agitated vessels. When the slurry is mixed well for a sufficiently long of time, it is filtered through a filter press to separate the solid adsorbent and the accompanying adsorbate from the liquid. The filter cake is washed to remove the solution held within the pores of the adsorbent. For continuous operation, the filter press may be replaced by a centrifuge or a rotary filter.

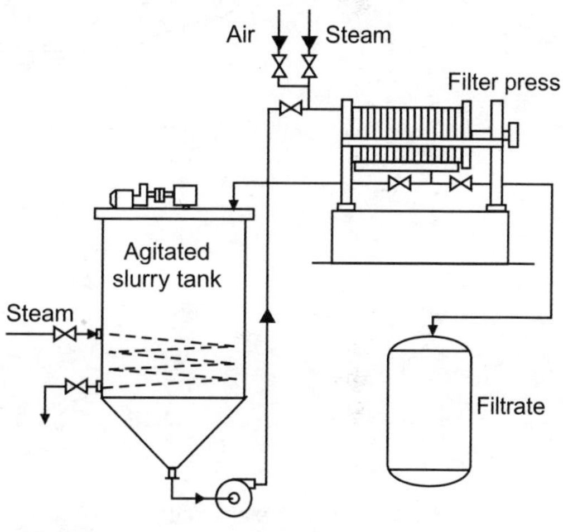

Fig. 10.21: Single stage batch treatment of liquids by contact filtration

Activated carbon is used as the adsorbent for the treatment of aqueous solutions, and inorganic substances such as clays are used for the treatment of organic liquids. The adsorbent is used in a finely ground state, particles finer than 75 μm are common. Since the liquid viscosity is lowered and the rate of diffusion of the solute and the rate of movement of the adsorbent particles in the slurry are increased with increase in the operating temperature, the mixing is carried out at the highest possible temperature.

Separation of the adsorbate from the adsorbent is achieved by reducing the partial pressure by passing steam or air (inert purge), if the adsorbate is volatile. When warm air is passed, care should be taken to avoid combustion of the adsorbents such as carbon. Adsorbate can also be recovered by dissolving it in a solvent in which it is more soluble than the original solvent present with the feed (displacement purge). Adsorbate in decolourization operations has no value and can be burned off. Generally the spent adsorbent is discarded because of the difficulty of desorbing large molecules.

Major applications of agitated vessels for adsorption are: (1) the recovery of valuable solutes from dilute solutions, e.g. adsorption of iodine from brine, insulin from dilute solution using carbon as the adsorbent (2) the purification of liquids by removing undesirable contaminants (3) the adsorption of coloured substances from aqueous sugar solutions (4) the decolourisation of petroleum and vegetable oils using clay (5) the adsorption of odorous substances from potable water (6) the adsorption of grease from dry-cleaning liquids.

Fixed-Bed Adsorbers

Fixed bed adsorbers are the most common type among adsorption equipment for gas purification and bulk separation. The feed fluid is passed through fixed beds of adsorbents. When the adsorbent bed gets saturated, it is subjected to regeneration. Unlike continuously operated contact filtration units in which the exit liquids always contain unadsorbed solute, in fixed bed units, it is possible to obtain nearly solute free effluents. The most frequently used adsorbers are cylindrical vertical vessels with adsorbent packed randomly. There are two types of support systems for adsorbent beds. The first type is a progressive series of grid and screen layers, with the screen openings made smaller as we move up. The top-most layer of screens has the openings small enough to prevent the passage of the adsorbent particles passing through. The second type of packing support consists of graded system of certain particles of different sizes resting directly on the adsorber bottom. Ceramic particles or gravel with the larger particles at the bottom and smaller at the top constitute this type of a support. The bed is provided with means for ensuring even distribution of the incoming fluid on the face of the bed. In order to prevent jiggling or fluidization, the flow of the liquid or the gas is directed downward the bed. The diameter to depth ratio of the bed and the size of the particles determine the pressure drop in the fixed bed units. The adsorbent particle size ranges, typically from 0.06 to 6 mm for large scale applications. With increase in the particle size, though the bed pressure drop decreases, the rate of solute transport is reduced.

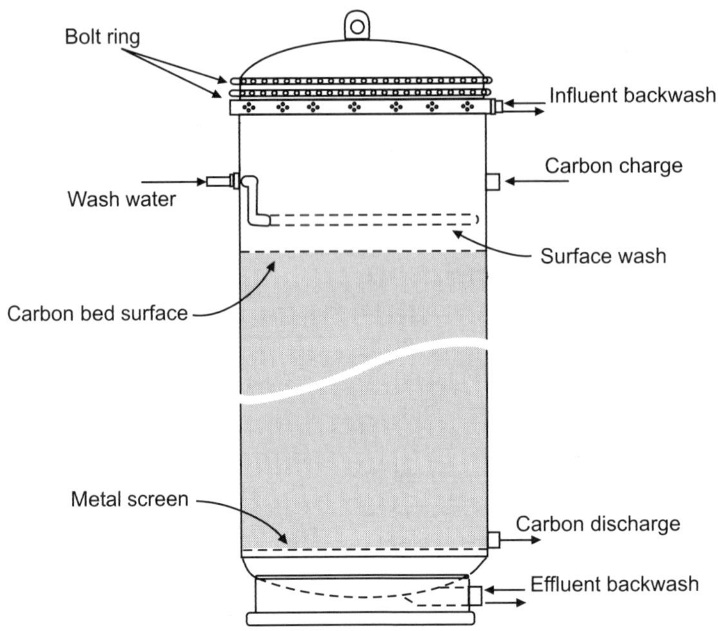

Fig. 10.22: Pressurized adsorber vessel

Fixed bed adsorbers find wide applications in drying of the industrial gases using beds of activated silica gel, alumina, bauxite, or molecular sieves. If the gases are under appreciable pressure, moderately deep beds can be used as the pressure drop in this case will be only a minor fraction of the total pressure. In the adsorber shown in Fig. 10.22, adsorption as well as regeneration is carried out at pressures above the atmospheric and the adsorber is designed as a pressure vessel. The regeneration can be achieved by application of heat by steam coils embedded in the bed or by admitting heated air or gas. Other applications include dehydration of liquids such as gasoline, kerosene, and transformer oils using activated alumina as the adsorbent with regeneration done using high temperature steam (500 K) followed by application of vacuum.

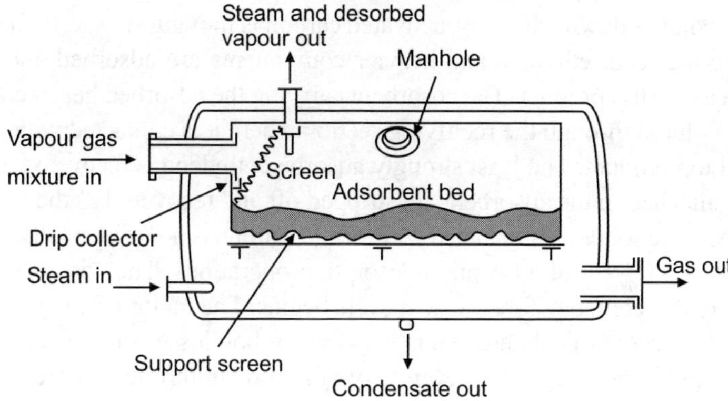

Fig. 10.23: Ambient pressure adsorption vessel

For operation of adsorbers near the atmospheric pressure, the major concern is to minimise the pressure drop through the bed. The beds with large face area and minimum depth such as the horizontal adsorber shown in Fig. 10.23 are common for such applications. Here, the flow is radial, and not axial as in the conventional vertical beds. Horizontal adsorbers find wide applications in the recovery of solvent vapours such as alcohol, acetone, carbon disulphide, benzene, etc. from air stream. Using good grade carbon as the adsorbent, recoveries as high as 99.8% are realized from vapours with concentration as low as 0.05%. To keep the power costs low, these adsorbers are operated with as low a pressure drop as is possible. Horizontal beds using granular, rather than powdered adsorbent are suitable for such applications. For purification or bulk separation of gases, the adsorbent is always regenerated in place by thermal swing (TSA), inert-purge swing, pressure-swing (PSA) or displacement purge.

Moving-Bed Adsorbers

The fixed bed adsorbers are characterised by the unsteady-state operation with the compositions, temperature, and/or pressure at a given location in the bed varying with time. The beds operate cyclically between adsorption and desorption. In contrast, in the moving bed adsorbers, the fluid and the adsorbent move continuously through the equipment at constant rate with no change in composition at any location in the equipment with passage of time. These are generally operated countercurrently and are applicable to both gas phase and liquid phase adsorption.

The first commercial application of countercurrent adsorption and desorption was the moving bed *Hypersorption process* for the recovery of light hydrocarbons from various gas streams in petroleum refineries using activated carbon as adsorbent.

Figure 10.24 shows a schematic sketch of a Hypersorber. Let ethylene and heavier hydrocarbons such as butane are to be recovered from the non adsorbing gases such as hydrogen and methane. The gas is admitted

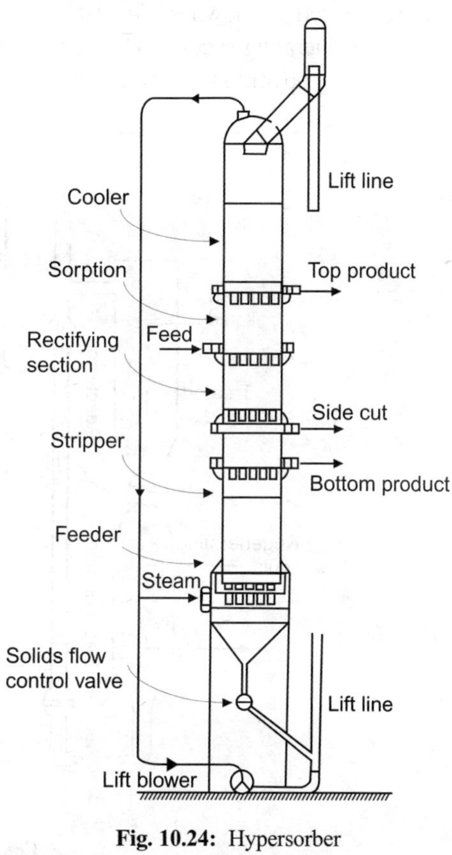

Fig. 10.24: Hypersorber

to the centre of the column down which the activated carbon is moving slowly. In the adsorption section immediately above the feed, ethylene and heavier components are adsorbed and the non adsorbing gases are removed as the top product. The adsorbent carrying the adsorbed heavy components continue to move down the column through the rectifying section where it is contacted with the upward flowing stream of recycled top product. The least strongly adsorbed ethylene is recovered as a side-cut. Heavy hydrocarbons still attached to the adsorbent are stripped off and recovered as the bottom product in the stripper section where the solids are contacted with steam. Carbon is now free of all adsorbed components and is ready for the next cycle of adsorption-desorption operations. The adsorbent is lifted to the top of the column where it is cooled before the next cycle begins. The major problem with the Hypersorber is that the process is uneconomical and results in heavy carbon losses due to attrition. The difficulties in the regeneration and in maintaining an even flow of adsorbent throughout are other problems in the operation of the moving bed adsorbers.

Higgins contactor shown in Fig. 10.25 is a simple moving bed unit for liquid-phase adsorption. The liquid and the adsorbent are made to move countercurrently through the equipment. The solid is moved hydraulically in small increments by a water pressure pulse action. In Fig. 10.25a, the adsorption and the regeneration sections are in operation. The liquid feed is flowing down through the column in the adsorption section and the liquid eluent flows down through the regeneration section. As the adsorbent gets saturated with the adsorbate, the liquid flows are interrupted for a few seconds, the butterfly or loop valves are turned to the new position as shown in Fig. 10.25b and the liquid-filled piston pump is moved as shown. The adsorbent is moved by pulses of liquid generated by the double acting piston which simultaneously sucks liquid from the top of the column and delivers it to the bottom. Now the valves are readjusted as shown in Fig. 10.25c to their original position and the solid is moved to the bottom of the regenerating section. The bed is now ready for completing the cycle by allowing liquid flows. The pulsing operation and consequent movement of the adsorbent takes a few second in a cycle time

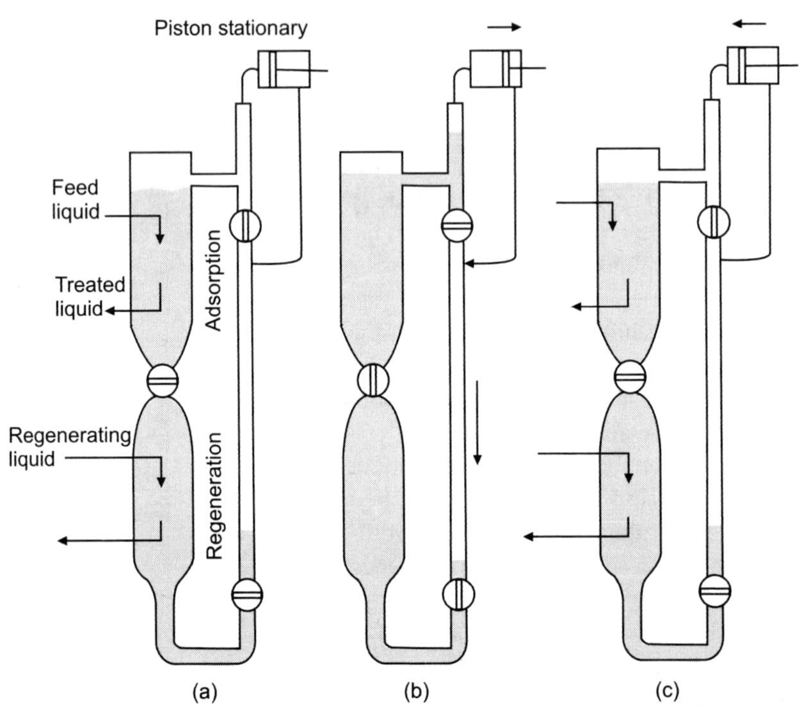

Fig. 10.25: Higgins contactor

which may last from a few minutes to several hours. The operation, although intermittent and cyclic, closely conforms to a truly continuous countercurrent operation. Higgins contactor is compact, is of simple construction and can handle high liquid flow rates. The equipment suffers from the common operational problems associated with the moving bed systems such as the wear and tear of the valves and attrition and loss of the adsorbent.

Combined Fluidized and Moving Bed Adsorber

In the Hypersorption process, regeneration of the adsorbent is difficult. Also like other moving bed systems, Hypersorber is uneconomical because of the heavy carbon losses as a result of attrition. These problems are overcome by combining the fluidized and the moving bed systems. The adsorber shown in Fig. 10.26 employs fluidised bed for the adsorption and moving bed for the desorption. These are used for removing small amounts of solvent vapours from air. In the upper part of the tower, the adsorbent is contacted countercurrently with the gas to be treated on perforated trays in relatively shallow beds, the solids moving from tray to tray through downspouts. The gas flows up through the perforations, fluidizing the bed of solids. The fluidized solids flow like a liquid across the tray into the downspout and onto the tray below. The lower part of the tower is the desorbing section which contains two sets of tubes. The upper ones are preheating tubes and the lower ones, the desorption tubes. Steam is used for preheating as well as for desorption.

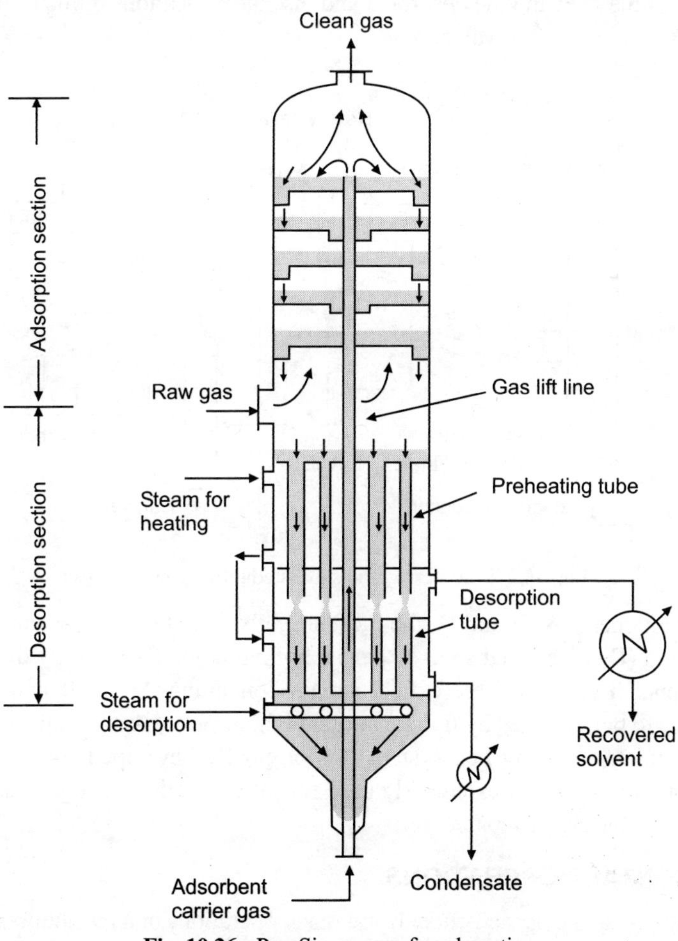

Fig. 10.26: PuraSiv process for adsorption

The solids leaving the adsorption section flows through the preheating tubes and then through the desorbing tubes as shown in the figure. The regenerated solids are picked up by the carrier gas at the bottom of the tower and carried to the top of the tower through a central gas-lift line, delivered to the top-most tray and subjected again to fluidisation and adsorption.

Simulated Moving Bed Systems

The major difficulty with moving bed adsorbers is the heavy loss of the adsorbent due to attrition resulting from the moving of solids downward through a tower and transporting the solids up to the top. This is overcome in the simulated moving bed system shown in Fig. 10.27 for separation of the liquid mixtures. Here, a true continuous countercurrent adsorption system is simulated, without physical movement of the solids. This is achieved by a stationary bed of the adsorbent divided into a number of sections by sieve plates. The degree of approach to a true countercurrent system increases with the increase in the number of compartments. The bed is provided with a number of liquid entry and discharge locations. The points of introduction of the feed consisting of components **A** and **B** and the desorbent **D** and withdrawal of raffinate (consisting of **B** and **D**) and extract (consisting of **A** and **D**) streams are periodically changed along the bed of the solid adsorbent. For instance, the locations 2, 5, 9 and 12 are currently operational with all the other numbered lines closed as shown in the figure. In the next step, the lines 3, 6, 10 and 1 are put into operation with all other lines closed. This shifting of the entry locations of the feed and the desorbent and discharge locations of the raffinate and the extract are achieved by means of a rotary valve.

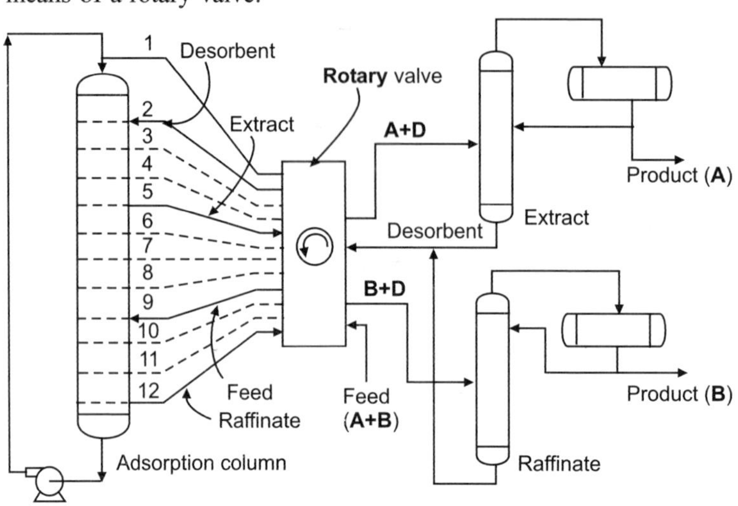

Fig. 10.27: Sorbex hybrid simulated moving bed process

Of the three components **A**, **B** and **D**, the most strongly adsorbed component is **D** followed by **A**. The desorbent liquid (**D**) is selected so as to have a boiling point significantly different from those of the feed components. This allows the extract stream containing **A** and **D** and the raffinate stream containing **B** and **D** to be separated by fractionation into high-purity **A** and high-purity **B** products. The concept of the simulated moving bed system was originally developed for the Sorbex process for hydrocarbon separation. It is now extensively used for the separation of sugars, amino acids, and fine chemicals, especially for chiral separations.

10.8 STAGE CONTACT OPERATIONS

Adsorption operations can be carried out either as stage operations or as continuous contact operations. In both cases, batch, continuous and semi-continuous operations are common. The adsorption systems include as we have seen, dispersed contact units such as the agitated vessels or the contact filtration units, the fixed bed units such as the percolation tanks, the continuous countercurrent units such as the Hypersorption unit, and the fluidized bed systems. In general, three steps are invariably involved in all adsorption processes. They are:

(i) Contacting the fluid (gas or liquid) with the adsorbent. Certain components get adsorbed on the solid surface.
(ii) Separation of the unadsorbed fluid from the adsorbent carrying the adsorbate.
(iii) Regeneration of the adsorbent or its replacement with fresh solids.

In the following discussion, we consider the systems in which only one component of the solution is adsorbed. The calculations involved in such cases are analogous to those involved in the absorption of a component from a gas mixture using a liquid solvent. Even in many cases where more than one component is adsorbed, the solute is adsorbed much more strongly than the other components, so that the adsorption of the latter can be safely ignored.

Single-Stage Adsorption

Adsorption of liquids using the contact filtration systems, fluidized bed adsorption for gases or liquids are examples of the units that can be operated stage-wise. A contact filtration unit is a typical example of a single stage equipment for the liquid-phase adsorption (Fig. 10.28a). The feed solution containing Y_0 kg solute/kg of non-adsorbed solvent is treated with adsorbent containing X_0 kg adsorbed solute/kg adsorbent. If fresh adsorbent is used, $X_0 = 0$. F_S is the quantity of solvent (non-adsorbed components) in kg (or kg/h for continuous operation) in the feed solution and S is the quantity of the adsorbate-free adsorbent in kg (or kg/h). It is assumed that the adsorption of all the components other than the solute is negligibly small.

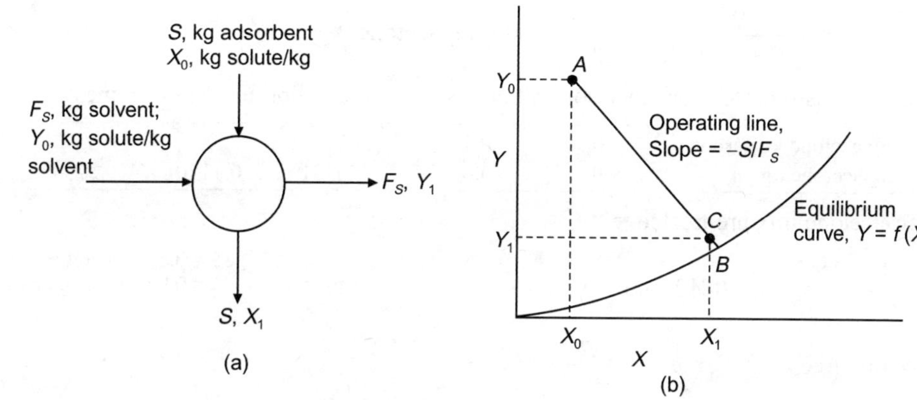

Fig. 10.28: Single-stage adsorption

A solute balance on the unit gives

$$F_S(Y_0 - Y_1) = S(X_1 - X_0) \quad \text{...(10.18)}$$

where X_1 and Y_1 are the solute concentrations in the effluent streams. X_1 is the concentration of the adsorbate solute on the adsorbent (kg adsorbate/kg adsorbent) and Y_1 is the concentration of the solute in the solution leaving the unit (kg solute/kg solute-free solution). In Fig. 10.28b, point A has coordinates (X_0, Y_0) and represents the concentration of the input streams. If the adsorption unit approaches the performance realized in an ideal (or theoretical) stage closely, the adsorption of the solute would be complete and the solution and the adsorbent leaving the stage will be in equilibrium. The effluent streams attain the equilibrium compositions and therefore can be represented by a point such as B lying on the adsorption isotherm. On the other hand, if equilibrium is not reached, which may happen if the time of contact is insufficient, the effluent streams can be represented by a point such as C in the figure. Equation (10.18) can be rearranged as

$$\frac{Y_0 - Y_1}{X_0 - X_1} = -\frac{S}{F_S} \quad \ldots(10.19)$$

which means that slope of the straight line joining point A (X_0, Y_0) and B(X_1, Y_1) is equal to $-S/F_S$ which is the ratio of the weights of the adsorbent to the solute-free solution treated. The amount of adsorbent required for treating a given quantity of solution of known concentration can be determined graphically using Eq. (10.19). The adsorption equilibrium curve for the system is plotted as shown in Fig. 10.28b. The isotherm corresponds to the final temperature attained by the solution after adsorption.

Example 10.5: Acetone-air mixture, 2.0 m^3 in volume measured at 300 K, 101.3 kPa is brought into contact with 3.5 kg activated carbon to adsorb acetone in a single stage operation. The initial partial pressure of the acetone in the gas is 15 kPa. The equilibrium adsorption data is given below:

kg acetone/kg carbon	0	0.1	0.2	0.3	0.35
Partial pressure, Pa	0	267	1600	5600	12300

Determine the final partial pressure of the acetone in the gas and the concentration of the acetone on the adsorbent.

***Solution*:** The equilibrium data is converted into X-Y coordinates, where X = kg acetone/kg adsorbent and Y is kg acetone per kg air.

$Y = \frac{\bar{p}}{101.3 - \bar{p}} \times \frac{58}{29}$ where $\bar{p}$ is in kPa. For illustration, when the partial pressure is 267 Pa, $\bar{p}$ = 0.267 kPa and

$$Y = \frac{0.267}{101.3 - 0.267} \times \frac{58}{29} = 0.0053 \text{ kg acetone/kg air.}$$

The calculation is repeated for other concentrations and the equilibrium data becomes

X = kg acetone/kg carbon	0	0.1	0.2	0.3	0.35
Y = kg acetone/kg air	0	0.0053	0.0321	0.1170	0.2764

Moles of air-acetone mixture treated =

$$n = \frac{PV}{RT} = \frac{101.3 \times 2}{8.314 \times 300} = 0.0812 \text{ kmol}$$

Mass of acetone-free air = $F_S = n \times \frac{P - \bar{p}_0}{P} \times M_B$

$$= 0.0812 \times \frac{101.3 - 15}{101.3} \times 29 = 2.006 \text{ kg}$$

Amount of adsorbent added = S = 3.5 kg

Concentration of acetone in the feed

$$= Y_0 = \frac{\bar{p}_o}{P - \bar{p}_0} \times \frac{58}{29} = \frac{15}{101.3 - 15} \times \frac{58}{29}$$

$= 0.3476$ kg acetone/kg air.

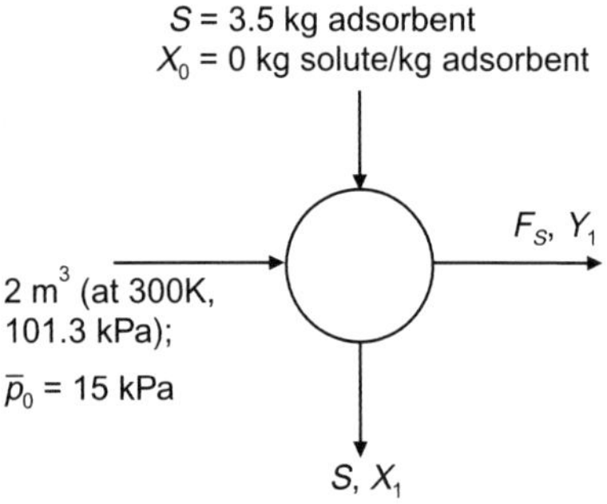

Fig. 10.29: (a) The adsorption operation in Example 10.5

By Eq. (10.19),

$$\frac{Y_0 - Y_1}{X_0 - X_1} = -\frac{S}{F_S} \quad \ldots(10.19)$$

It means that a straight line drawn with slope $-S/F_S$ from the point (X_0, Y_0) to the equilibrium curve intersects the curve at (X_1, Y_1). The equilibrium curve is plotted as shown in Fig. 10.29b. The straight line *AB* is drawn with slope $-3.5/2.006 = -1.7448$ from (0, 0.3476). It intersects the equilibrium curve at (0.1833, 0.0274)

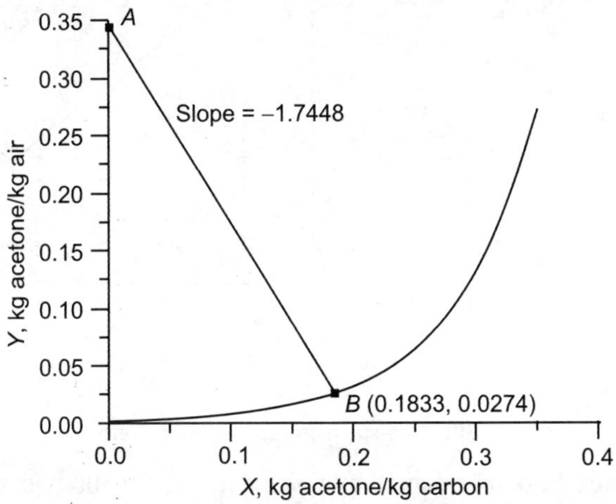

Fig. 10.29: (b) Determination of the effluent composition in the single stage adsorption in Example 10.5

The final concentration of acetone in the air is $Y_1 = 0.0274$ kg acetone/kg air. Since the total mass of acetone-free air in the mixture is 2.006 kg, the final mixture contains 2.006 /29 = 0.0692 kmol air and $2.006 \times 0.0274/58 = 0.00095$ kmol acetone. The total moles of the final mixture = 0.0692 + 0.00095 = 0.0701 kmol. The final pressure of the gas is calculated now.

$$P = \frac{nRT}{V} = \frac{0.0701 \times 8.314 \times 300}{2} = 87.48 \text{ kPa.}$$

The partial pressure exerted by acetone in the final gas

$$= P \times \frac{\text{moles of acetone}}{\text{total moles}} = 87.48 \times \frac{0.00095}{(0.0692 + 0.00095)} = 1.185 \text{ kPa.}$$ **(Answer)**

The concentration of acetone on the adsorbent = X_1 = 0.1833 kg acetone/kg carbon. **(Answer)**

Multistage Crosscurrent Adsorption

When costly adsorbent such as the activated carbon is used, adsorbent economy is a major challenge in the operation of adsorption units. The multistage crosscurrent operation can be employed as a means to reduce the cost of separation. The feed solution is contacted with different batches of the adsorbent in multiple stages with provision for intermediate filtration and removal of the spent adsorbent. The larger the number of stages, the lower the quantity of adsorbent required. However, if a large number of adsorption stages are required, the saving in the cost of adsorbent will be largely offset by the cost of filtration that is to be provided additionally between the stages. Hence, in practice, it is seldom economical to go for more than two crosscurrent adsorption stages.

A two-stage crosscurrent adsorption unit is shown in Fig. 10.30a and the operation is represented on the equilibrium diagram in Fig. 10.30b. The solute balance for the stages are as follows.

Stage 1:

$$F_S(Y_0 - Y_1) = S_1(X_1 - X_0) \quad \text{...(10.20)}$$

Stage 2:

$$F_S(Y_1 - Y_2) = S_2(X_2 - X_0) \quad \text{...(10.21)}$$

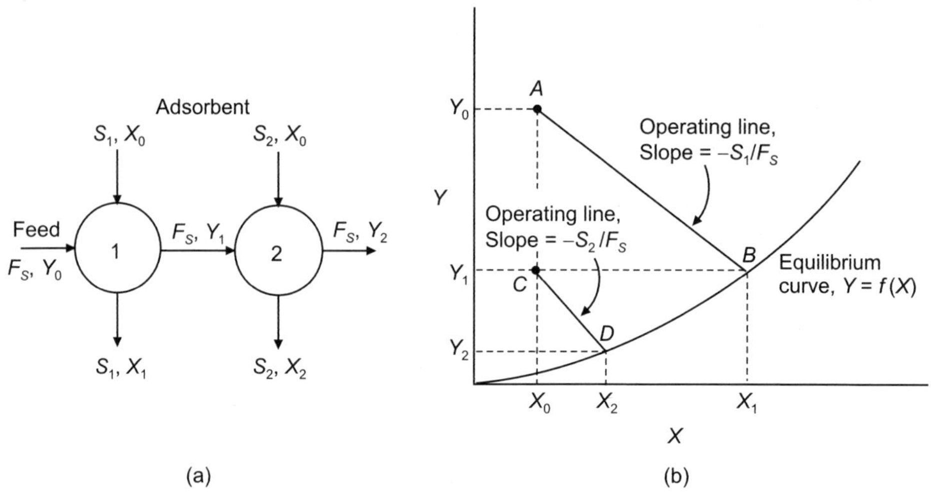

Fig. 10.30: Two-stage crosscurrent adsorption

The input streams to stage 1 are represented by point A in Fig. 10.30b with coordinates (X_0, Y_0) and the effluent streams by point B with coordinates (X_1, Y_1). Assuming the stages to be ideal, point B should lie on the equilibrium curve as shown. By Eq. (10.20), the slope of the operating line AB is $-S_1/F_S$. The input streams to stage 2 have compositions X_0 and Y_1 and can be represented by point C in the figure. The streams leaving stage 2 have composition (X_2, Y_2) and the point D representing the effluent streams lies on the equilibrium curve. By Eq. (10.21), the slope of the operating line CD is $-S_2/F_S$. The lines AB and CD will be parallel if equal quantities of the adsorbent are used in each stage. Unless the adsorption isotherm is linear, the total adsorbent required will be the minimum if unequal dosages of adsorbent are employed in the stages.

Example 10.6: One thousand kg of sugar solution containing 50% sucrose by weight is to be decolorized by adsorption using activated carbon in a two-stage crosscurrent unit. The initial colour intensity of the solution is measured to be 25 units on an arbitrary scale and the colour intensity is to be reduced to 4% of the original value. The data for the equilibrium adsorption isotherm is established by adding small amounts of activated carbon to separate batches of sugar solution and observing the colour intensity of the solution at equilibrium. The equilibrium data, in which the quantity of carbon is expressed on the basis of the sucrose content of the solution, are given below:

kg carbon/kg dry sugar	0	0.007	0.013	0.023	0.032	0.049	0.10
Colour removal, %	0	56	77	90	95	98	99.5

(a) Determine the total carbon requirement if equal amounts are used in both stages
(b) Compare the amount calculated in part (a) with that required in a single-stage operation.

***Solution*:** The two-stage crosscurrent adsorption unit is shown in Fig. (10.31a). It is given that 25 units of colour per kg sucrose is present in the original solution. Denote the units of colour per kg carbon by X and the units of colour present in the solution per kg sucrose by Y. For illustration, consider the second set of data in the table. When 0.007 kg carbon is added per kg sucrose, colour removal is 56%. It means, at equilibrium, $Y = 25 \times \dfrac{(100-56)}{100} = 11.0$ units per kg sucrose. Or 25 – 11.0 = 14.0 units of colour got adsorbed on 0.007 kg carbon.

Therefore, $X = \dfrac{14}{0.007} = 2000$ units/kg carbon.

These calculations are carried out for the entire set of data and the results are given below:

X = units of colour/kg carbon	2000	1481	978	142	500	249
Y = equilibrium colour, units of colour/kg sucrose.	11.0	5.75	2.5	1.25	0.50	0.125

The equilibrium curve is plotted as in Fig. 10.31b.

Basis: 1000 kg solution being treated

$F_S = 1000 \times 0.5 = 500\,\text{kg}$ sucrose.

Y_0 = initial colour intensity = 25 units.

Y_2 = The colour intensity of solution leaving stage 2

$$= 25 \times \frac{4}{100} = 1 \text{ unit.}$$

The adsorbent used is fresh so that $X_0 = 0$

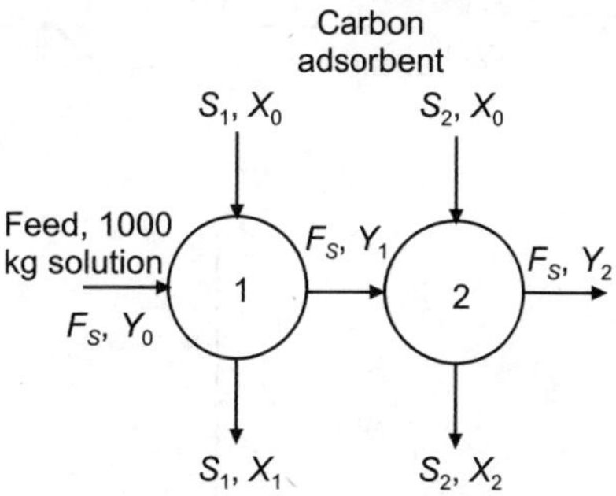

Fig. 10.31: (a) Crosscurrent adsorption in Example 10.6

(a) The operating line for stage 1 starts from point A ($X_0 = 0$, $Y_0 = 25$). The operating line terminates at point D the ordinate of which is $Y_2 = 1$. The points A and D are shown in Fig. 10.31b. When equal amounts of the adsorbents are used in both stages, the operating lines AB and CD are straight parallel lines with slope $-S/F_s$, where S is the amount of adsorbent used per stage. The operating lines are established by trial such that the operating line for stage 1 originates at A and the operating line for stage 2 passes through D and they are parallel to each other. The slopes of the lines are found to be –0.01 (Fig. 10.31b).

$$\text{Since slope} = -\frac{S_1}{F_S} = -\frac{S_2}{F_S} = -0.01,$$

$$S_1 = S_2 = 0.01 \times F_S = 0.01 \times 500 = 5\,\text{kg}$$

Total quantity of carbon required

$= 2 \times 5 = 10$ kg. **(Answer)**

(b) If the operation were carried out in a single-stage, input streams are represented by the point A and the effluent streams by point D. The line AD will represent the operating line for single stage operation. The slope of the line AD is found to be – 0.0362.

$$\text{Slope} = -\frac{S}{F_S} = -0.0362$$

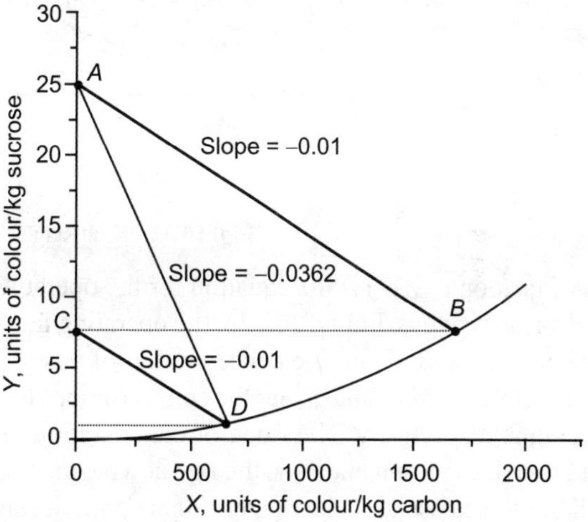

Fig. 10.31: (b) Equilibrium curve for Example 10.6

Therefore, the adsorbent required for single stage operation is,

$$S = 0.0362 \times 500 = 18.1\,\text{kg}. \qquad \textbf{(Answer)}$$

Multistage Countercurrent Adsorption

Countercurrent operation requires less adsorbent for a given concentration change of the solution compared to the crosscurrent and single stage operation which produce the same change. Figure 10.32a depicts a continuous countercurrent adsorption system. Typical examples for the industrial application of the countercurrent operation is the three-stage countercurrent system employed for the adsorption of gold and silver from cyanide-ore-leach liquor by activated carbon. Adsorption is carried out in agitated vessels and the solution leaving the vessels are screened to separate large carbon particles before admitting to the next vessel. A truly continuous stage-wise separation for gaseous system is realized in fluidized bed towers.

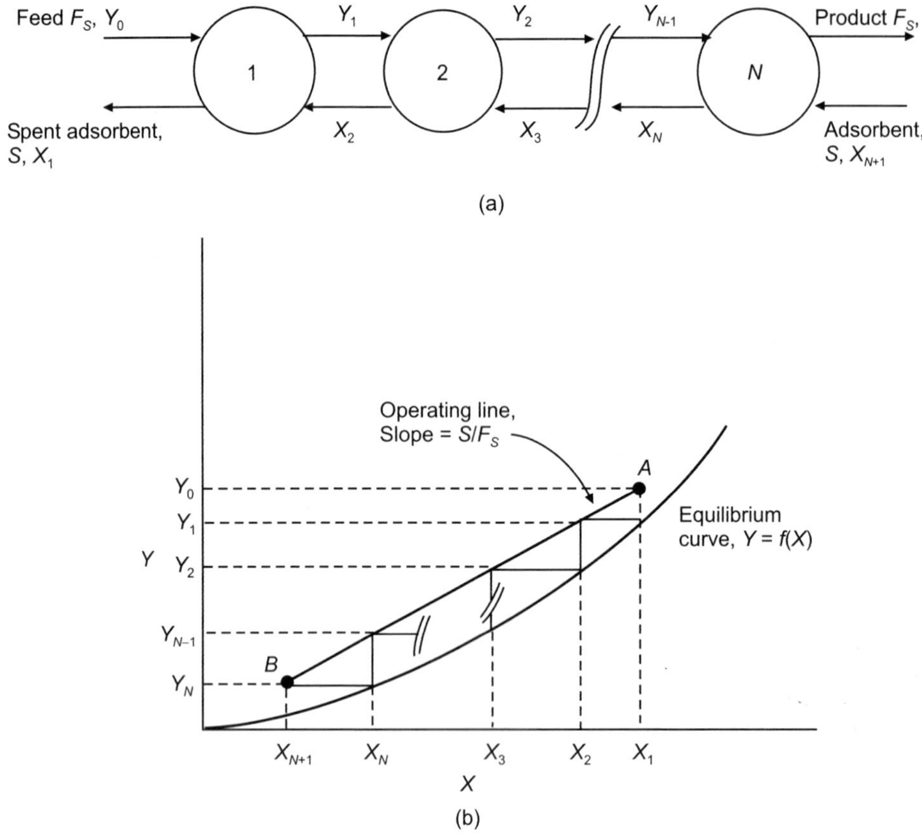

Fig. 10.32: Multistage countercurrent adsorption

Equation (10.22) is the equation for the operating line for the continuous countercurrent cascade of adsorption stages. In Fig. 10.32b, the operating line is drawn from point A (X_1, Y_0) to point B (X_{N+1}, Y_N), where X_{N+1} and X_1 are the concentration of adsorbate on the adsorbent entering and leaving the unit, respectively and Y_0 and Y_N are the concentrations of adsorbate solute in the solution entering and leaving the unit, respectively. The slope of the operating line is S/F_S where S is the rate at which adsorbent is admitted to the unit and F_S is the rate at which the non adsorbing components in the solution are passed through the cascade. The number of adsorption stages required to effect a given concentration change can be estimated graphically by the usual McCabe – Thiele stair-case construction between the operating line and the equilibrium curve.

The solute balance for the unit shown in the Fig. 10.32a gives

$$F_S(Y_0 - Y_N) = S(X_1 - X_{N+1}) \quad \ldots(10.22)$$

The minimum adsorbent required for a given extent of adsorption can also be estimated by the usual method. The minimum adsorbent to solution ratio (S/F_S) required for effecting a given extent of adsorption is obtained as the slope of the limiting operating line. If the equilibrium adsorption isotherm is straight or is concave upward as shown in Fig. 10.33a, the limiting operating line is the one which has the maximum slope and which touches the equilibrium curve within the desired concentration range. If the adsorption isotherm is concave downward, the limiting operating line is tangential to the equilibrium curve as shown in Fig. 10.33b.

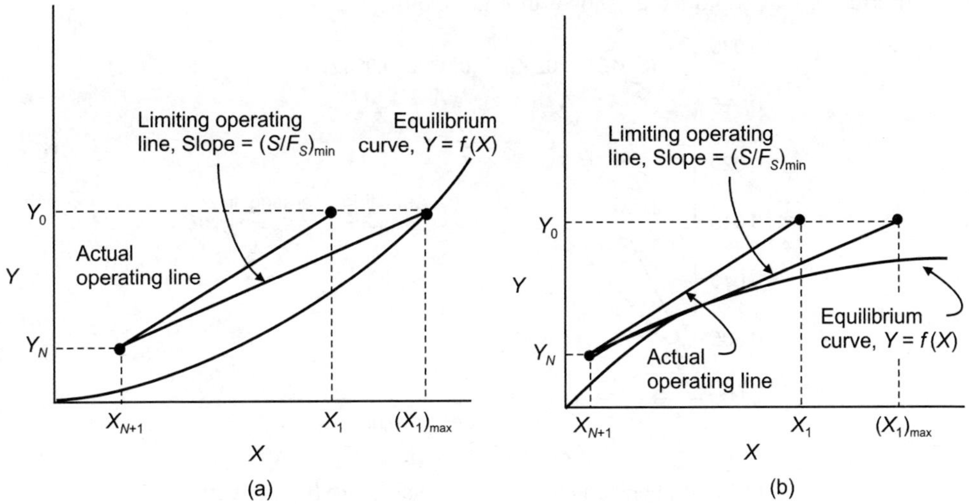

Fig. 10.33: Estimation of the minimum adsorbent requirement

Example 10.7: Air contaminated with NO_2 is passed through a countercurrent stage contact system at the rate of 500 kg/h at 298 K and 101.3 kPa. The air contains 1.5% NO_2 by volume, 90% of which is to be removed by adsorption using silica gel which enters the unit free of NO_2. The equilibrium adsorption isotherm is given by the following data:

Partial pressure of NO_2, kPa.	0	0.267	0.533	0.800	1.066	1.333	1.600
kg NO_2/100 kg gel	0	0.4	0.9	1.65	2.60	3.65	4.85

Determine:

(a) The minimum weight of gel required per hour.

(b) For twice the minimum rate of adsorbent, the concentration of NO_2 in the gel.

(c) The weight of gel if the number of stages to be restricted to two.

***Solution*:** Figure 10.34a shows a two-stage countercurent unit for adsorption. The initial concentration of NO_2 (Molecular weight = 46) in the air is 1.5% (volume).

$$Y_0 = \frac{1.5}{(100-1.5)} \times \frac{46}{29} = 0.0242\,\text{kg NO}_2/\text{kg air}.$$

$$F_S = 500 \times \frac{1}{1.0242} = 488.2\,\text{kg/h}.$$

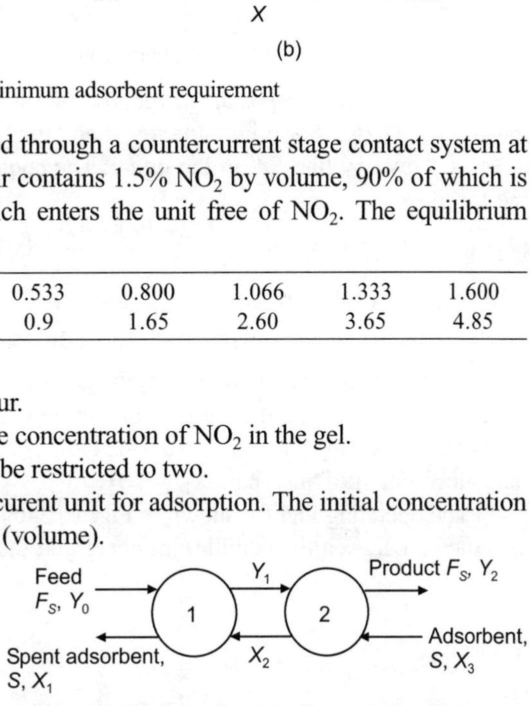

Fig. 10.34: (a) Two-stage countercurrent adsorption

Since 90% of NO_2 is removed by adsorption,

$Y_N = 0.10 \times Y_0 = 0.10 \times 0.0242 = 0.00242$ kg NO_2/kg air. Since fresh adsorbent is used, $X_{N+1} = 0$.

Equilibrium data is converted into X-Y coordinates. For illustration, consider the second data point. The partial pressure of NO_2 in the gas is 0.267 kPa and the equilibrium adsorbate loading is 0.4 kg/100 kg gel. That is,

$$X = \frac{0.4}{100} = 0.004\ \text{kgNO}_2/\text{kg gel}.$$

$$Y_0 = \frac{0.267}{(101.3-0.267)} \times \frac{46}{29} = 0.0042\ \text{kgNO}_2/\text{kg air}.$$

X, kg NO_2/kg gel	0	0.004	0.009	0.0165	0.026	0.0365	0.0485
Y, kg NO_2/kg air	0	0.0042	0.0084	0.0126	0.0169	0.0212	0.0255

The equilibrium curve is plotted as shown in Fig. 10.34b.

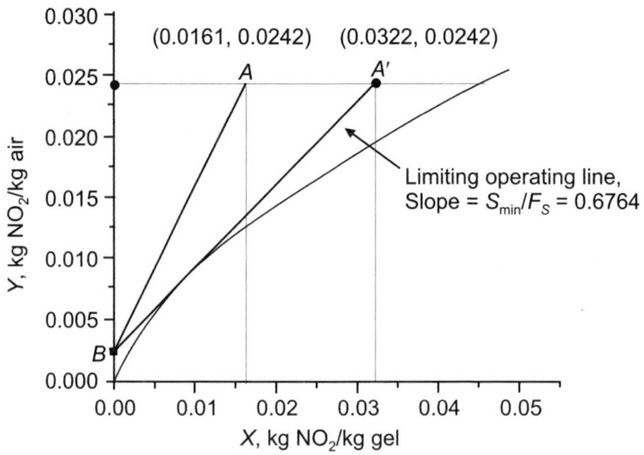

Fig. 10.34: (b) Equilibrium curve and operating line (Example 10.7)

(a) The operating line has the lower terminus at B (0, 0.00242) and upper terminus on the horizontal line at $Y_0 = 0.0242$. For minimum adsorbent, the limiting operating line is tangent to the equilibrium curve as shown by line BA' in the figure. The coordinates of A' are (0.0322, 0.0242) and the slope of line BA' is

$$\text{Slope} = \frac{0.0242 - 0.00242}{(0.0322 - 0)} = 0.6764.$$

Since slope of the limiting operating line is S_{min}/F_S, minimum adsorbent requirement is

$$S_{min} = 0.6764 \times 488.2 = 330.2 \text{ kg/h.}$$ **(Answer)**

(b) For twice the minimum quantity of gel used, slope of the operating line is $\text{slope} = 0.6764 \times 2 = 1.3528$. That is, the operating line is drawn with slope 1.3528 from point B which terminates at the horizontal line passing through $Y_0 = 0.0242$ at A. The x-coordinate of point B is found to be 0.0161. That is, the concentration of adsorbate $X_1 = 0.0161$. **(Answer)**

(c) The operating line BA shown in Fig. 10.34c is drawn by trial such that exactly two stages can be constructed between the equilibrium curve and the operating line.

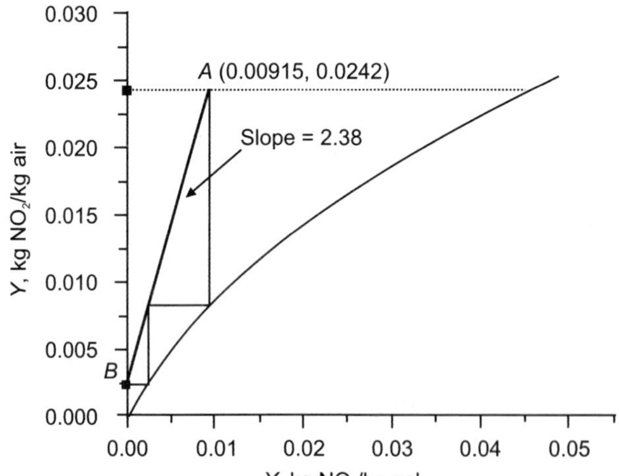

Fig. 10.34: (c) Determination of adsorbent requirement when the number of stages is 2 (Example 10.7)

The slope of the line is found out. From the graph, slope of $AB = 2.38$. Therefore the amount of gel required is

$$S = 2.38 \times 488.2 = 1162.1\,\text{kg/h}. \qquad \textbf{(Answer)}$$

Special Case: Application of Freundlich Equation

The Freundlich equation can be written as

$$Y = m\,X^n \qquad \ldots(10.23)$$

where m and n are constants. For value of $n = 1.0$, the adsorption isotherm is straight; for $n > 1$, the adsorption isotherm is concave upward and for $n < 1$, the adsorption isotherm is concave downward as shown in Fig. 10.35.

Single-Stage Adsorption Consider single stage adsorption as shown in Fig. 10.28 with the equilibrium represented by the Freundlich equation. Assuming the adsorbent to be fresh, Eq. (10.19) can be simplified putting $X_0 = 0$.

$$\frac{Y_0 - Y_1}{X_1} = \frac{S}{F_S} \qquad \ldots(10.24)$$

Since the adsorbate concentration on the adsorbent leaving the stage and the solute concentration in the effluent solution are related by Eq. (10.23), X_1 in Eq. (10.24) can be replaced in terms of Y_1.

$$\frac{Y_0 - Y_1}{(Y_1/m)^{1/n}} = \frac{S}{F_S} \qquad \ldots(10.25)$$

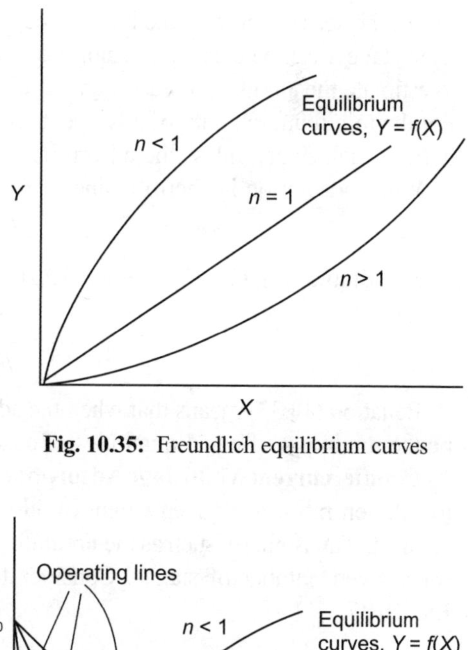

Fig. 10.35: Freundlich equilibrium curves

The amount of adsorbent required (S kg) to treat a given quantity of solution (F_S kg) to reduce its solute content from the initial concentration (Y_0) to the final value (Y_1) can be computed using Eq. (10.25) when the equilibrium relationship $Y = m\,X^n$ is known.

From Eq. (10.25) it is clear that the slope of the operating line for single-stage adsorption S/F_S depends on the value of the exponent n in the Freundlich Equation as shown in Fig. 10.36. The larger the value of n, the smaller the slope of the operating line. Hence, for a given extent of adsorption, the adsorbent requirement is smaller for systems with large value of n, considered to be the systems with favourable isotherms.

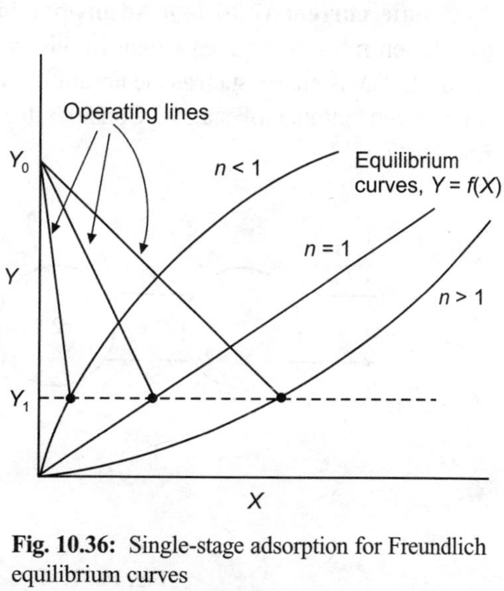

Fig. 10.36: Single-stage adsorption for Freundlich equilibrium curves

Crosscurrent Adsorption Consider the two-stage crosscurrent unit as the one shown in Fig. 10.30. Assuming fresh adsorbent, $X_0 = 0$ and Eqs. (10.20) and (10.21) can be rearranged as

Stage 1:

$$\frac{S_1}{F_S} = \frac{Y_0 - Y_1}{X_1} = \frac{Y_0 - Y_1}{(Y_1/m)^{1/n}} \qquad \ldots(10.26)$$

Stage 2:

$$\frac{S_2}{F_S} = \frac{Y_1 - Y_2}{X_2} = \frac{Y_1 - Y_2}{(Y_2/m)^{1/n}} \qquad \ldots(10.27)$$

Total quantity of adsorbent is

$$\frac{S}{F_S} = \frac{S_1 + S_2}{F_S} = \frac{Y_0 - Y_1}{(Y_1/m)^{1/n}} + \frac{Y_1 - Y_2}{(Y_2/m)^{1/n}} \qquad \ldots(10.28)$$

For minimum total requirement of the adsorbent, the derivative of (S/F_S) with respect to Y_1 is zero. That is

$$\frac{d(S/F_S)}{dY_1} = 0 \qquad \ldots(10.29)$$

Or

$$\frac{d}{dY_1}\left(\frac{Y_0 - Y_1}{(Y_1)^{1/n}} + \frac{Y_1 - Y_2}{(Y_2)^{1/n}}\right) = 0 \qquad \ldots(10.30)$$

The resulting equation can be simplified to the following form:

$$\left(\frac{Y_1}{Y_2}\right)^{1/n} - \frac{1}{n}\left(\frac{Y_0}{Y_1}\right) = 1 - \frac{1}{n} \qquad \ldots(10.31)$$

For the separation from the inlet concentration Y_0 to the exit concentration Y_2, the exit concentration from stage 1 can be calculated using Eq. (10.31) which can be substituted in Eqs. (10.26) and (10.27) to estimate the amounts of adsorbent required in each stage for minimum total adsorbent requirement. For the minimum quantity of adsorbent required, in general, unequal quantities of adsorbent are to be used in each stage, unless the adsorption isotherm is linear.

When adsorption isotherm is linear, $n = 1$ in Eq. (10.31) so that

$$Y_1 = (Y_0 Y_2)^{1/2} \qquad \ldots(10.32)$$

Substituting Eq. (10.32) in Eqs. (10.26) and (10.27) it can be shown that

$$\frac{S_1}{F_S} = \frac{S_2}{F_S} = m\left[\left(\frac{Y_0}{Y_2}\right)^{1/2} - 1\right] \qquad \ldots(10.33)$$

Equation (10.33) means that when the adsorption isotherm is linear, the total adsorbent required will be the minimum, if equal quantities of adsorbent are used in both stages.

Countercurrent Multistage Adsorption In countercurrent adsorption, the estimation of the adsorbent to solution ratio for a given extent of adsorption graphically, involves a trial and error procedure. In contrast, if an isotherm such as the Freundlich equation is applicable, S/F_S can be determined analytically for a given number of stages. Consider for instance, a two-stage countercurrent unit as shown in Fig. 10.37.

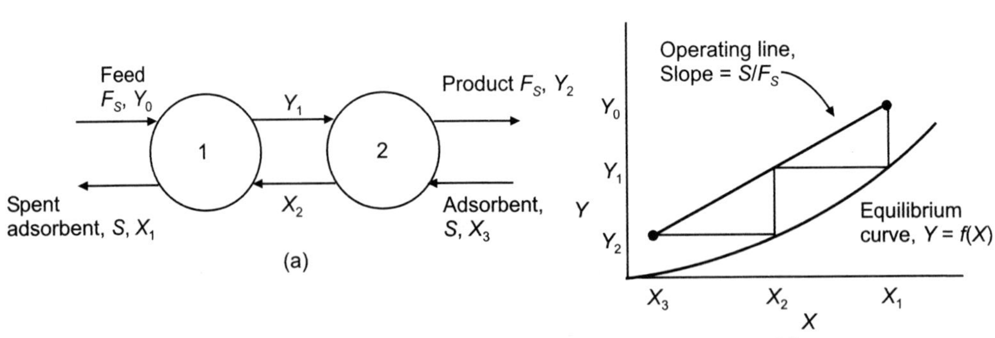

Fig. 10.37: Two-stage countercurrent adsorption

An overall solute-balance gives

$$F_S(Y_0 - Y_2) = S(X_1 - X_3) \qquad ...(10.34)$$

Assuming that fresh adsorbent is used in the second stage, $X_3 = 0$ and Eq. (10.34) simplifies to

$$F_S(Y_0 - Y_2) = SX_1 \qquad ...(10.35)$$

Assuming that equilibrium is attained in both stages, X_1 in Eq. (10.35) can be eliminated using

$$Y_1 = mX_1^n \qquad ...(10.36)$$

Equation (10.35) can now be written as

$$\frac{S}{F_S} = \frac{(Y_0 - Y_2)}{(Y_1/m)^{1/n}} \qquad ...(10.37)$$

The solute balance for stage-2 gives

$$F_S(Y_1 - Y_2) = SX_2 = S\left(\frac{Y_2}{m}\right)^{1/n} \qquad ...(10.38)$$

Or

$$\frac{S}{F_S} = \frac{(Y_1 - Y_2)}{(Y_2/m)^{1/n}} \qquad ...(10.39)$$

Comparing Eqs. (10.37) and (10.39), we get

$$\frac{(Y_0 - Y_2)}{(Y_1/m)^{1/n}} = \frac{(Y_1 - Y_2)}{(Y_2/m)^{1/n}} \qquad ...(10.40)$$

or

$$\frac{Y_0}{Y_2} - 1 = \left(\frac{Y_1}{Y_2}\right)^{1/n}\left(\frac{Y_1}{Y_2} - 1\right) \qquad ...(10.41)$$

Equation (10.41) can be solved for the intermediate composition Y_1 for given terminal concentrations Y_0 and Y_2. The adsorbent to solution ratio S/F_S can be computed using either Eq. (10.37) or (10.39).

For linear adsorption isotherm, the Kremser equation may also be used for the determination of the number of stages of countercurrent operation. Referring to Fig. 10.37, the Kremser equation (*See* Eq. 4.76) takes the form

$$\frac{Y_0 - Y_N}{Y_0 - mX_{N+1}} = \frac{A^{N+1} - A}{A^{N+1} - 1} \qquad ...(10.42)$$

In Eq. (10.42), the Y_0-Y_N is the actual concentration change in the fluid and $Y_0 - mX_{N+1}$ is the maximum concentration change in the fluid if it leaves with a composition identical to the composition of the fluid which is in equilibrium with the feed. The quantity A in Eq. (10.42) is analogous to the absorption factor and is defined as $A = \dfrac{S}{mF_S}$.

Example 10.8: Phenol present in an aqueous solution is to be removed by adsorption with activated carbon. The initial concentration of phenol in the solution is 0.25 kg/m^3. Three cubic meters of this solution is treated to reduce the phenol content to 10% of this value. Adsorption equilibrium is represented by the Freundlich isotherm $q = 0.20C^{0.23}$ where q is the adsorbate loading, kg phenol/kg carbon and C is the concentration, kg phenol/m^3 of solution. Calculate:

(a) The quantity of fresh carbon required for single-stage adsorption
(b) The minimum total amount of carbon in a two-stage crosscurrent adsorption
(c) The quantity of carbon required in a two-stage countercurrent adsorption.

Solution: (a) The initial concentration of the solution = 0.25 kg phenol/m^3 of solution (Fig. 10.38a).

Since the final concentration is 10% of the initial value, $C_1 = 0.1 \times 0.25 = 0.025\,\text{kg/m}^3$.

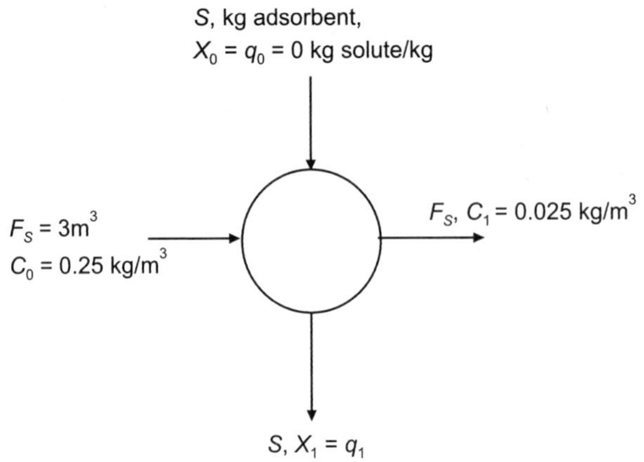

Fig. 10.38: (a) Single stage adsorption (Example 10.8)

The equilibrium is given by the Freundlich equation

$$q = 0.20C^{0.23}$$

That is, $C = 1093.96\, q^{4.348}$

Or $Y = 1093.96\, X^{4.348}$

where $Y = C =$ kg of phenol/m^3 of solution, $X = q =$ kg phenol/kg adsorbent.
Equation (10.25) gives

$$\frac{Y_0 - Y_1}{(Y_1/m)^{1/n}} = \frac{S}{F_S} \quad \text{...(10.25)}$$

Here, $m = 1093.96$ and $n = 4.348$

Substituting the values in Eq. (10.25),

$$\frac{0.25 - 0.025}{(0.025/1093.96)^{1/4.348}} = \frac{S}{3}$$

Solving this we get, $S = 7.88$ kg **(Answer)**

(b) The two-stage crosscurrent adsorption unit is shown in Fig. 10.38b.

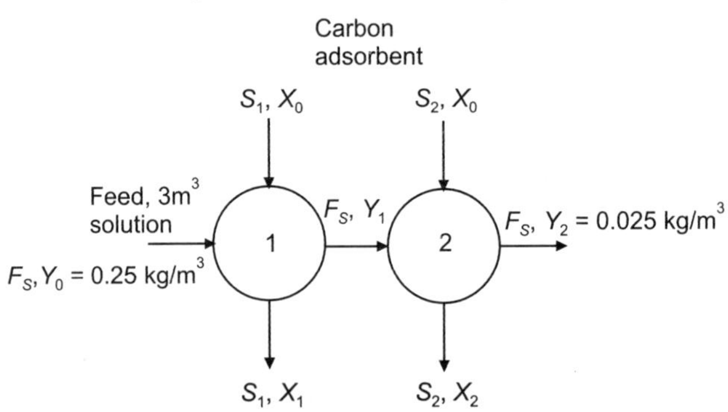

Fig. 10.38: (b) Two-stage crosscurrent adsorption of Example 10.8

For minimum total adsorbent, Eq. (10.31) gives

$$\left(\frac{Y_1}{Y_2}\right)^{1/n} - \frac{1}{n}\left(\frac{Y_0}{Y_1}\right) = 1 - \frac{1}{n} \quad \text{...(10.31)}$$

Substitute $Y_0 = 0.25$, $Y_2 = 0.025$, $n = 4.348$ in Eq. (10.31)

$$\left(\frac{Y_1}{0.025}\right)^{1/4.348} - \frac{1}{4.348}\left(\frac{0.25}{Y_1}\right) = 1 - \frac{1}{4.348}$$

which can be simplified as

$$2.3359Y_1^{1.23} - 0.77Y_1 - 0.0575 = 0$$

The above equation is solved for Y_1 which gives $Y_1 = 0.097$.

From Eq. (10.26),

$$\frac{S_1}{F_S} = \frac{Y_0 - Y_1}{X_1} = \frac{Y_0 - Y_1}{(Y_1/m)^{1/n}}$$

$$\frac{S_1}{3} = \frac{0.25 - 0.097}{(0.097/1093.96)^{1/4.348}} = 1.3082$$

Or $\quad S_1 = 1.3082 \times 3 = 3.92$ kg.

Using Eq. (10.27)

$$\frac{S_2}{F_S} = \frac{Y_1 - Y_2}{X_2} = \frac{Y_1 - Y_2}{(Y_2/m)^{1/n}}$$

$$\frac{S_2}{F_S} = \frac{0.097 - 0.025}{(0.025/1093.96)^{1/4.348}} = 0.8409$$

Or $\quad S_2 = 0.8409 \times 3 = 2.52$ kg

Total quantity of adsorbent $= S_1 + S_2 = 3.92 + 2.52 = 6.44$ kg **(Answer)**

(c)

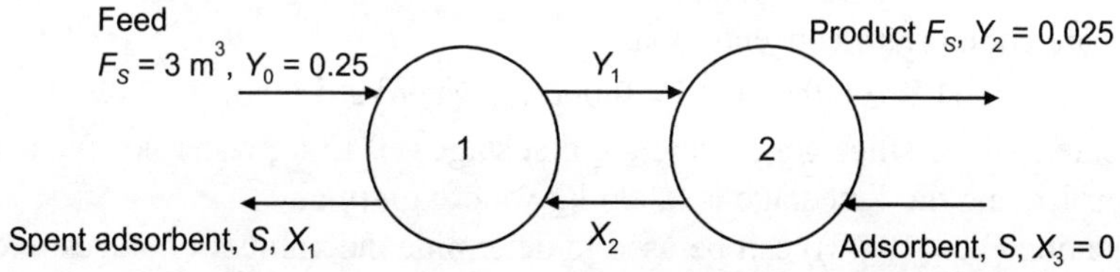

Fig. 10.38: (c) Two-stage countercurrent adsorption of Example 10.8

Equation (10.41) can be used to estimate the concentration Y_1 (Fig. 10.38c).

$$\frac{Y_0}{Y_2} - 1 = \left(\frac{Y_1}{Y_2}\right)^{1/n}\left(\frac{Y_1}{Y_2} - 1\right) \qquad \text{...(10.41)}$$

Substituting the known quantities in Eq. (10.41), we get

$$\frac{0.25}{0.025} - 1 = \left(\frac{Y_1}{0.025}\right)^{1/4.348}\left(\frac{Y_1}{0.025} - 1\right)$$

$$93.436Y_1^{1.23} - 2.3359Y_1^{0.23} - 9 = 0$$

On solving this equation, we get $Y_1 = 0.1698$. Using this in Eq. (10.37)

$$\frac{S}{F_S} = \frac{(Y_0 - Y_2)}{(Y_1/m)^{1/n}} \qquad \text{...(10.37)}$$

$$S = F_S \times \frac{(Y_0 - Y_2)}{(Y_1/m)^{1/n}} = 3.0 \times \frac{0.25 - 0.025}{(0.1698/1093.96)^{1/4.348}}$$

$$S = 5.07 \text{ kg}$$ **(Answer)**

Example 10.9: Silica gel is used to adsorb moisture from an air stream in a multistage countercurrent unit operated isothermally. The initial moisture content is 0.015 kg water/kg dry air and the final value is to be 0.0015 kg water/kg dry air. The air is admitted at a rate of 10 kg dry air/s and dry silica gel at a rate of 0.90 kg/s. The adsorption isotherm is represented by $Y = 0.04X$, where Y is kg of water vapour per kg of dry air and X is kg of water vapour per kg of dry silica gel. Determine:

(a) The water content of the air and silica gel leaving the first stage

(b) The number of ideal adsorption stages, graphically and by using Kremser equation.

***Solution*:**

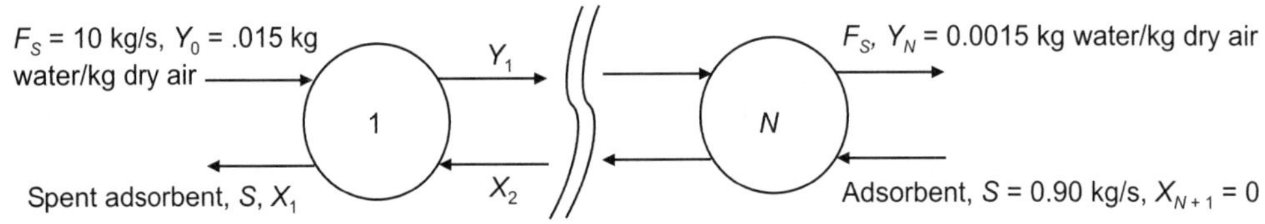

Fig. 10.39: (a) Multistage countercurrent adsorption unit of Example 10.9

Figure 10.39a shows the countercurrent multistage adsorption unit. Here,
$Y_0 = 0.015$ kg water vapour/kg dry air. $Y_N = 0.0015$ kg water vapour/kg dry air. $X_{N+1} = 0$ (Fresh silica gel).

$$F_S = 10 \text{ kg dry air/s}, \; S = 0.90 \text{ kg dry silica gel/s}.$$

(a) Consider water balance around the unit:

$$F_S(Y_0 - Y_N) = S(X_1 - X_{N+1})$$

$$10(0.015 - 0.0015) = 0.90(X_1 - 0)$$

which gives

$$X_1 = 0.15 \text{ kg water vapour/kg gel.}$$

Since X_1 and Y_1 are equilibrium concentrations,

$$Y_1 = 0.04X_1 = 0.04 \times 0.15 = 0.006 \text{ kg water/kg dry air.}$$

The water content of the silica gel leaving the first stage = 0.15 kg water /kg dry gel and the water content of the air leaving the first stage is 0.006 kg water/kg dry air. **(Answer)**

(b) Kremser equation (Eq. (10.42)) can be used to determine the number of stages in counter-current multistage operation:

$$\frac{Y_0 - Y_N}{Y_0 - mX_{N+1}} = \frac{A^{N+1} - A}{A^{N+1} - 1} \quad \text{where } A = \frac{S}{mF_S}$$

$$A = \frac{S}{mF_S} = \frac{0.9}{0.04 \times 10} = 2.25$$

Substituting this in Eq. (10.42),

$$\frac{0.015 - 0.0015}{0.015 - 0.04 \times 0} = \frac{2.25^{N+1} - 2.25}{2.25^{N+1} - 1}$$

$$\frac{2.25^{N+1} - 2.25}{2.25^{N+1} - 1} = 0.9$$

Solving this equation, we get $N = 2.2$ **(Answer)**

The number of countercurrent stages can be determined graphically as shown in Fig. 10.39b.

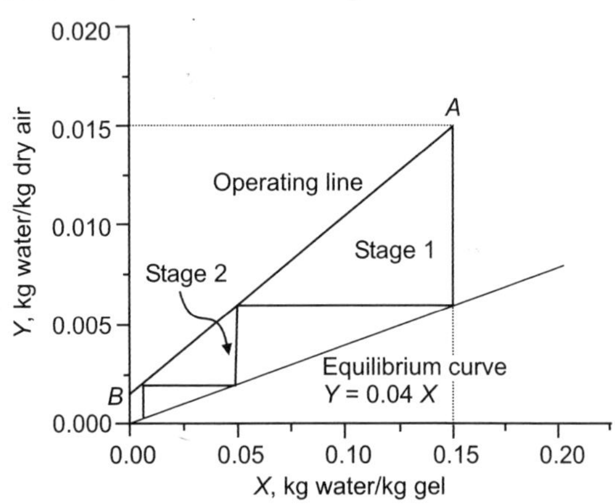

Fig. 10.39: (b) Determination of the number of stages in Example 10.9 by the graphical method

The operating line and the equilibrium curve are both straight lines. The number of stages is found to be 2.2 from the figure. **(Answer)**

10.9 FIXED-BED ADSORPTION

Fixed bed adsorbers are vertical columns provided with granular adsorbent as random packing. To avoid fluidisation, the fluid is allowed to flow down the column instead of flowing up. Fixed bed adsorption is an unsteady state operation; the concentration in the fluid phase and the solid phase change with the time and the position in the bed. This mode of operation is more attractive than the moving bed systems, because of the high cost incurred in the latter for continuously transporting the adsorbent particles. Also, the loss resulting by the attrition of moving particles is eliminated in the fixed bed operation. Since a near total removal of the adsorbable components from the fluid streams is achieved, it is possible to obtain a nearly solute-free effluent, unless the adsorbent in the bed is saturated. Fixed bed operation finds wide applications in processes such as the dehydration of gases and liquids, the decolourization of mineral and vegetable oils, the recovery of solvent vapours from gases, etc.

On continuous passage of the solution through the bed, the adsorbent become saturated with the adsorbate. When this happens, the flow of fluid is diverted to a second bed containing fresh adsorbent, and the first one subjected to regeneration. Therefore, it is necessary to know the quantity of the fluid that can be treated in one cycle of operation, or the length of time the bed can be operated until regeneration with a given flow rate of the fluid.

Mass Transfer Zone and the Adsorption Wave

Consider a feed solution in which the concentration of the adsorbable component is C_0 being admitted to the top of a bed packed with fresh adsorbent particles. Most of the mass transfer takes place near the inlet of the bed. The uppermost layer of solid, in contact with the strong solution entering, adsorbs the solute rapidly and effectively with the result that the solids get saturated with the adsorbate. Concentration of the solute in the fluid drops exponentially with distance and falls essentially to zero over a very short distance from the top of the bed as shown by the curve t_1 in Fig. 10.40. If the solution is admitted continuously at the top, little adsorption takes place in the uppermost layers of the bed as the bed there is already saturated with the adsorbate. The adsorbent in this zone is in equilibrium with the feed solution. The figure also shows the solute concentration profiles for the fluid as function of distance through the bed at increasing times t_2, t_3 and t_4 from the start of flow through the bed. C/C_0 is the concentration in the fluid relative to that in the feed. At time t_2 after commencement of the flow of the solution through the bed, the solute concentration in the fluid passing through the equilibrium zone undergoes no change and it remains the same as that in the feed C_0. Adsorption occurs and the concentration of solute in the fluid begins to fall only when the fluid comes out of this zone and enter the *mass transfer zone* or *adsorption zone*. Bulk of the adsorption takes place over this relatively narrow

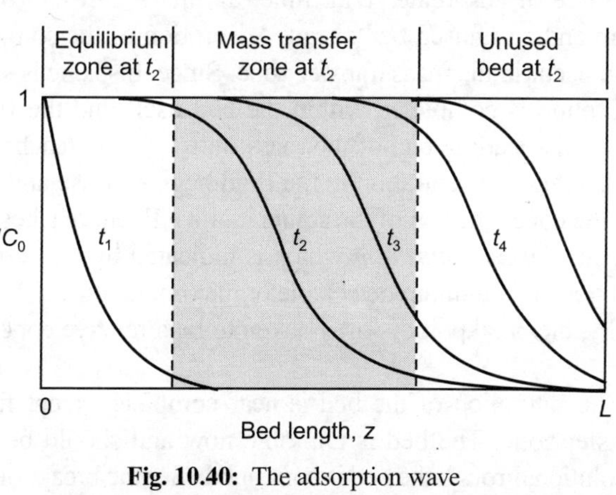

Fig. 10.40: The adsorption wave

zone in which the concentration changes rapidly. The concentration of the solute decreases from the feed concentration to essentially zero in this zone. Because it is difficult to determine where the mass transfer zone begins and where it ends, for convenience, the limits are often taken as C/C_0 values 0.95 and 0.05. No adsorption occurs in the portion of the bed below the mass transfer zone as the fluid coming in contact with the solid is already depleted of the solute. This region of the bed is designated as the *unused bed* at time t_2 in the figure. The mass transfer zone moves down the bed with time, increasing the depth of the equilibrium zone and decreasing the height of the unused bed, as shown by profiles t_3 and t_4. This movement of the adsorption zone down the column as a wave is called the *adsorption wave*. The rate of propagation of the adsorption wave is less than the linear velocity of the fluid through the bed.

At any given time during the operation of the fixed-bed unit, the bed consists of the three distinct regions, the equilibrium zone, the mass transfer zone and the unused bed. At time t_3 hours, the mass transfer zone has reached very near the exit of the bed and most of the bed is occupied by the saturated zone. If the flow of solution through the bed is continued, the solute concentration in the effluent now rises rapidly as the mass transfer zone passes through the bottom of the bed. The bed can be operated till the solute concentration of the effluent reaches a permissible value known as the *breakpoint concentration*. When the solute content in the effluent rises to this level, the operation is stopped or the flow is diverted through a second bed containing fresh adsorbent and the first one is taken for regeneration. It is clear that even at the breakpoint, there remains a portion of the bed which is still unused, the depth of which depend upon the width of the mass transfer zone. Since the adsorber bed should be at least as long as the mass transfer zone, it is important to know the depth of the zone. Factors that play important role in fixed bed adsorption and influence the length and shape of the mass transfer zone are: the type of the adsorbent, the particle size of the adsorbent, the depth of the adsorbent bed and the gas velocity, temperature of the gas stream and the adsorbent, the concentration of the solute, the pressure of the system and the removal efficiency required.

The Breakthrough Curve

Usually in the experimental investigation of fixed-bed adsorption process, it is more convenient to measure the effluent concentration as function of time, rather than measure the concentration change along the length of the bed. Figure 10.41 shows a plot of the effluent concentration versus time from the start of the fluid flow through the bed. Initially when the fluid comes in contact with the bed, the adsorption begins in the top layers of solid which acts as the mass transfer zone. The entire solute gets transferred to the uppermost layer of the bed, which gets saturated with the adsorbed solute. The effluent stream will be free of adsorbate. With time, the mass transfer zone moves down leaving a saturated zone upstream and an unused bed ahead. Adsorption of the solute and hence the change in composition of the fluid occur in the mass transfer zone. Since this zone is still within the bed far away from the exit, the adsorption is completed within the bed itself and the fluid leaves the bed free of solute. This continues till the leading end of the mass transfer zone reaches the exit of the bed. With further passage of fluid, solute concentration in the fluid begins to rise and after a certain time known as the *breakpoint time*, the concentration of the solute in the effluent reaches the maximum permissible level known as the *breakpoint concentration* which is indicated by point b in Fig. 10.41. Ideally, the breakpoint concentration is the minimum detectable or maximum allowable solute concentration in the effluent fluid. Practically, the breakpoint is often taken to be a relative concentration (C/C_0) of 0.05 to 0.10.

At the breakpoint, the saturation of the bed is near complete except for the bottom layer which contains the mass transfer zone. The bed is unusable now and should be subjected to regeneration. Instead, if flow of the solution through the bed is continued after the breakpoint, the solute concentration in the effluent rises rapidly as the mass transfer zone passes through the bottom of the bed and at point *e*, *the point of exhaustion*, the effluent concentration substantially reaches the initial value C_0. At the exhaustion point, the saturation of the bed with adsorbate is almost complete. The concentration profile of the effluent between the breakpoint and the point of exhaustion is known as the *breakthrough curve*. The shape and time of appearance of the breakthrough curve greatly influence the method of operation of the fixed bed adsorber. The curve generally has an *S*-shape, but they may be steep or relatively flat and in some cases considerably distorted. The shape of the breakthrough curve determines the percentage utilization of the adsorbent bed and the length of the bed.

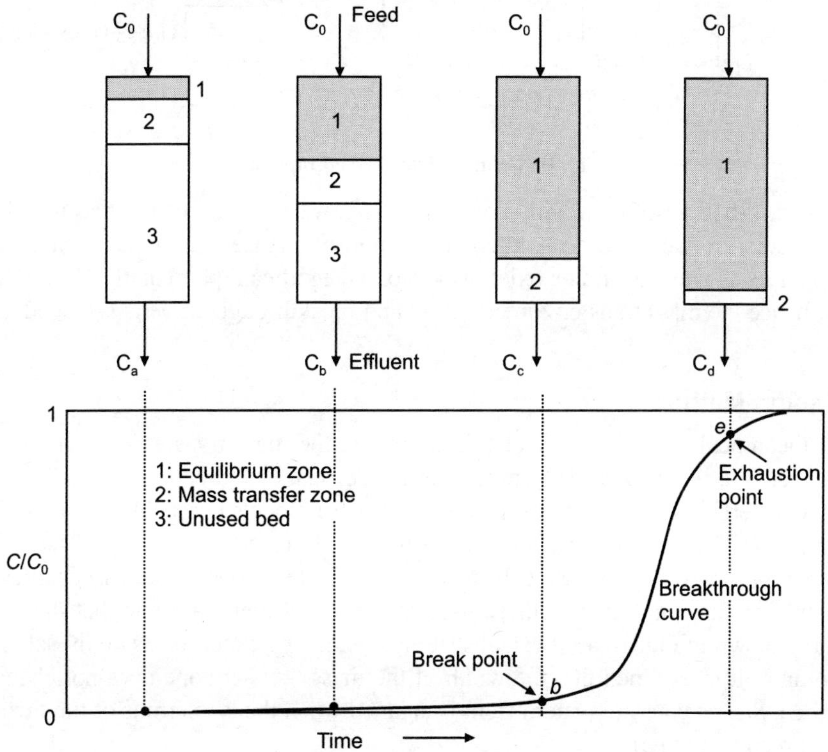

Fig. 10.41: Breakthrough curve

Ideal Fixed Bed Adsorption

The shape of the breakthrough curve depends largely upon the actual rate and mechanism of adsorption process, the nature of the adsorption equilibrium, the fluid velocity, the concentration of the solute in the feed, and the length of the adsorber bed. If the adsorption process were infinitely rapid, the breakthrough curve would be straight vertical line. This will occur if the external and internal resistances to mass transfer are very small, the fluid flows in plug flow, the axial dispersion is negligible and if the adsorbent is initially free of the adsorbate. The mass transfer zone in such case moves through the bed as a shock like wave called the *stoichiometric front*. This is the *ideal fixed bed adsorption*. Upstream of the stoichiometric front, the bed is saturated and downstream, the bed is unused as shown in Fig. 10.42.

The concentration front moves down with time and eventually reaches the end of the bed. The entire bed is now saturated and the system has reached the breakpoint. The concentration of the solute in the fluid abruptly rises to the inlet value and no further adsorption is possible. The stoichiometric wave front which is a vertical line, is also the ideal breakthrough curve.

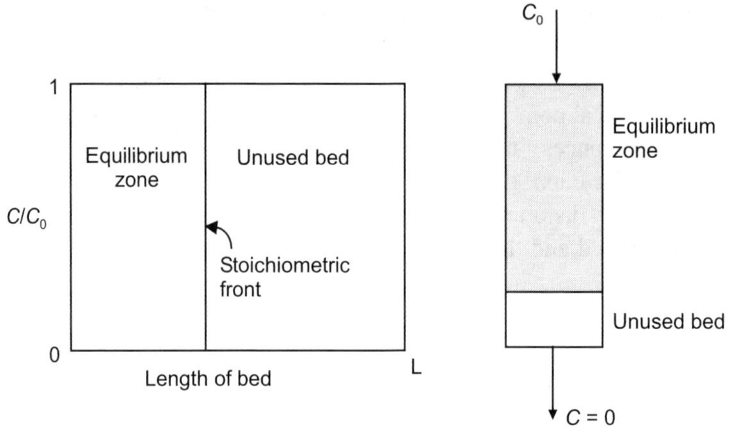

Fig. 10.42: Ideal fixed bed adsorption

In the ideal fixed-bed adsorption with stoichiometric wave front, all of the bed is utilized before breakthrough occurs. In real fixed beds, internal transport resistance, and in some cases the external transport resistances as well, are finite, axial dispersion is significant particularly at low flow rates in shallow beds; hence the mass transfer zone is broad and breakthrough curve is *s*-shaped as discussed earlier.

Scale-up Design Method

A number of theoretical methods are available to predict the mass transfer zone and concentration profile in the bed. But, the inherent inaccuracies in the correlations to predict the diffusion and mass transfer and the uncertainties in the flow patterns make these models unreliable. Besides, they require lengthy computations. The fixed bed adsorbers are usually designed by scale up of the data obtained from experiments on small columns, using the flow conditions, feed concentrations, adsorbent materials, temperature and pressure, the same as those used in the actual units. Assume that the breakthrough curve as the one shown in Fig. 10.43 is established by conducting experiments on the laboratory-scale adsorption column. It is assumed that the width of the mass transfer zone does not change with the length of the bed. Such a constant pattern front is attained when the isotherm is of the favourable type (Langmuir or Freundlich type).

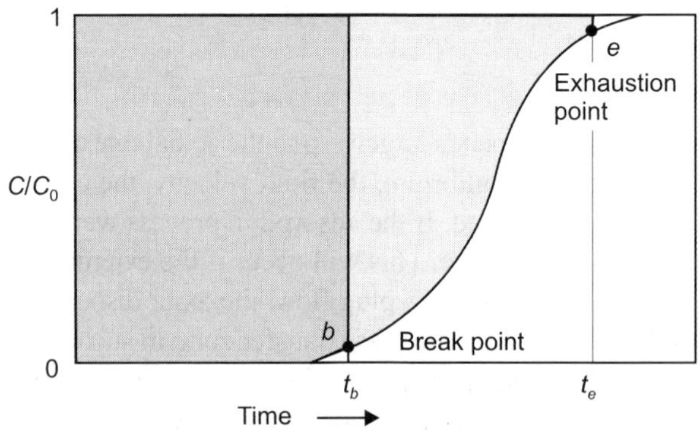

Fig. 10.43: Design of fixed-bed adsorption column

At the breakpoint, the bed consists of two zones: the equilibrium zone which contains the solid between the inlet of the bed and the start of the mass transfer zone and the mass transfer zone. The equilibrium zone is completely saturated with the adsorbate and the solids there are in equilibrium with the feed. The solids in the mass transfer zone go from the nearly saturated condition at the start of the zone to almost no adsorbate at the leading end which nearly coincides with the exit of the bed. It is reasonable to assume that the adsorbent in the mass transfer zone is half saturated. This is equivalent to treating one half of the mass transfer zone as fully saturated and the other half unused. *The scale up method of design of the adsorber is based on the principle that the amount of unused solid or length of the unused bed (LUB) does not change with the total bed length*. However, the fraction of the length of the bed that is unused decreases with increase in the total bed length.

Let t_b be the breakpoint time and t_e be the exhaustion time at which the bed saturation is complete. To determine the length of the bed, first the total solute adsorbed up to the breakpoint W_b, is determined by integration. This is the usable capacity of the bed up to the breakpoint; it is proportional to the dark shaded area in Fig. 10.43.

$$W_b = F_A \int_0^{t_e} \left(1 - \frac{C}{C_0}\right) dt \qquad \text{...(10.43)}$$

The total or stoichiometric capacity of the packed bed tower, when the entire bed comes to equilibrium with the feed (W_e) is proportional to the area between the break through curve and the line at $C/C_0 = 1.0$ between the limits 0 and t_e, the time of exhaustion. The total shaded area in the figure represents the desired integral.

$$W_e = F_A \int_0^{t_e} \left(1 - \frac{C}{C_0}\right) dt \qquad \text{...(10.44)}$$

In Eqs. (10.43) and (10.44), F_A is the feed rate of solute in the fluid. If the feed solution is admitted with a superficial velocity u_0 and concentration of solute C_0, then for unit area of cross section of the adsorber,

$$F_A = C_0 u_0 \qquad \text{...(10.45)}$$

The experimentally determined breakpoint time t_b can be used to evaluate the length of the bed up to the breakpoint L_b. Using the material balance and the equilibrium relationship, we can write.

$$F_A t_b = \rho_b q_0 L_b \qquad \text{...(10.46)}$$

Here ρ_b is the bulk density of the bed and q_0 is the adsorbate loading of the bed when the bed is in equilibrium with the feed solution. The left-hand side of Eq. (10.46) is the quantity of the solute passed through unit area of the bed up to the breakpoint and the right-hand side is the quantity of the adsorbate that the solids in the bed of unit area of cross section and length L_b can hold under equilibrium with the feed solution. It is assumed that fresh adsorbent is used.

The ratio of W_b to W_e is the fraction of the total bed capacity or the length that is utilized up to the breakpoint. If the total length of the bed is L, then

$$\frac{L_b}{L} = \frac{W_b}{W_e} \qquad \text{...(10.47)}$$

The unused fraction of the bed is evaluated as one minus the above ratio.

$$\frac{LUB}{L} = 1 - \frac{W_b}{W_e} \qquad \text{...(10.48)}$$

Or

$$LUB = \left(1 - \frac{W_b}{W_e}\right) L \qquad \text{...(10.49)}$$

where *LUB* is the equivalent length of the unused bed. As mentioned earlier, the principle of the scale up method of design of fixed bed adsorber is that the length of the unused bed is constant for all adsorbers using the same particle size for adsorbent and the same superficial velocity for the solution. Thus *LUB* estimated as described above for the experimental adsorber is applicable to the actual large scale adsorber, the length of which can be calculated as the sum of the length of the adsorber used up to the breakpoint and the equivalent length of the unused bed.

$$L = L_b + LUB \qquad ...(10.50)$$

where L_b is the length of the bed up to the breakpoint for the large size column. The ratio of L_b to t_b for the larger column is the same as the ratio of L_b to t_b of the experimental adsorber. Since the latter is already determined, L_b of the larger column can be evaluated if the breakpoint time t_b is specified.

An approximate alternate method is to consider the break through curve to be symmetrical at time t_s as shown in Fig. 10.44. The hatched area above and below the curve are approximately equal at t_s so that the integral in Eq. (10.44) which is represented by the *total* shaded area in Fig. 10.43 can be equivalently represented by the rectangular shaded area in Fig. 10.44. This area is equal to t_s, called the *stoichiometric time*, which is also the time taken by an ideal break through curve (or the stoichiometric front) to reach the end of the bed.

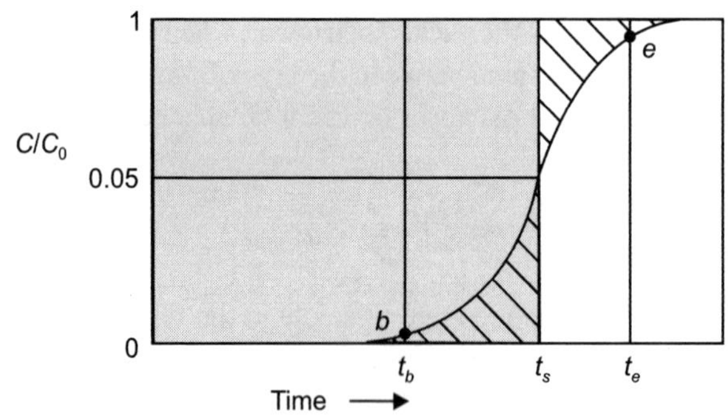

Fig. 10.44: Approximate method of estimating stoichiometric time

Now Equation (10.44) can be simplified as

$$W_e = F_A t_s \qquad ...(10.51)$$

Equation (10.43) can be approximated as

$$W_b = F_A t_b \qquad ...(10.52)$$

Equation (10.49) can be modified as

$$LUB = \left(1 - \frac{t_b}{t_s}\right) L \qquad ...(10.53)$$

The accuracy of the method depends largely on how close the conditions in the experimental column resemble the conditions in the actual commercial adsorber. The superficial mass velocity in both the pilot scale and the commercial units should be identical. The bed should be of sufficient length to achieve a constant pattern front, the width of which is invariant with time.

Example 10.10: Nitrogen gas wet with 800 ppm water vapour is admitted to an adsorption column packed with molecular sieves, at a mass velocity of 5000 kg/m^2 h. The packed height of the column is 0.4 m and the bulk density of the packing is 750 kg/m^3. The solids initially contained 0.01 kg water/kg solid. The breakthrough data are as follows.

t, h	0	9	9.5	10.0	10.5	11.0	11.5	12.0	12.5	12.8
C, ppm N_2	0	0	17	88	245	444	614	723	780	799

A value of $C/C_0 = 0.01$ is desired at the breakpoint. Determine:

(a) The breakpoint time

(b) The length of the unused bed

(c) The saturation loading capacity of the solid.

(d) The breakpoint time and fraction of total capacity used for a new column of length 1.0 m.

***Solution*:** The initial concentration of moisture in the gas, $C_0 = 800$ ppm.

The solute feed rate, $F_A = 5000 \times C_0 = 5000 \times \dfrac{800}{10^6} = 4\ \text{kg/m}^2\text{h}$.

Using the time versus concentration data for the breakthrough curve, time (t) versus C/C_0 are calculated. For illustration, when $t = 9.5$ h, $C = 17$ ppm, so that $C/C_0 = 17/800 = 0.0213$. The following table summarizes the results:

t, h	0	9	9.5	10	10.5	11	11.5	12	12.5	12.8
C/C_0	0	0	0.0213	0.1100	0.3063	0.5550	0.7675	0.9038	0.9750	0.9988

Figure 10.45a shows the plot of C/C_0 on the y-axis against t on the x-axis.

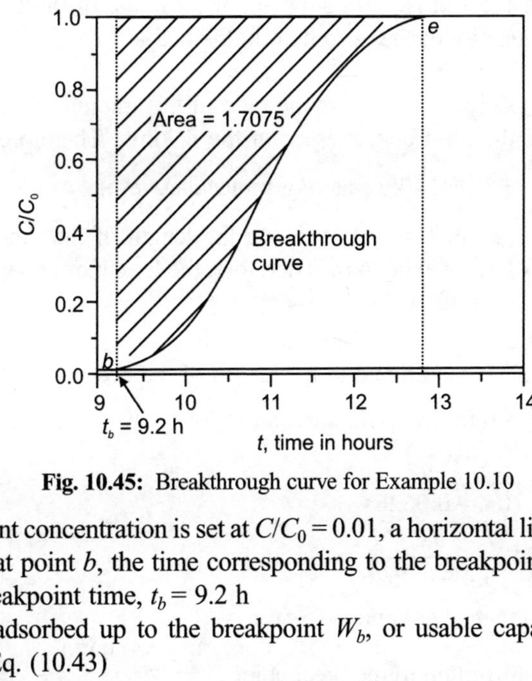

Fig. 10.45: Breakthrough curve for Example 10.10

(a) Since the breakpoint concentration is set at $C/C_0 = 0.01$, a horizontal line at $C/C_0 = 0.01$ intersects the breakthrough curve at point b, the time corresponding to the breakpoint b is the breakpoint time. From Fig. 10.45, the breakpoint time, $t_b = 9.2$ h

(b) The total solute adsorbed up to the breakpoint W_b, or usable capacity of the bed up to the breakpoint is given by Eq. (10.43)

$$W_b = F_A \int_0^{t_b} \left(1 - \frac{C}{C_0}\right) dt$$

The integral in Eq. (10.43) is equal to the area lying to the left of the vertical erected at $t = t_b$ and lying under the horizontal line at $C/C_0 = 1$ and above the breakpoint curve. This area is found to be 9.2 h from Fig. 10.45. Therefore, total solute adsorbed up to the breakpoint,

$$W_b = 9.2\, F_A\ \text{kg}$$

The total or stoichiometric capacity of the bed (W_e) is given by Eq. (10.44).

$$W_e = F_A \int_0^{te} \left(1 - \frac{C}{C_0}\right) dt$$

The integral in Eq. (10.44) is proportional to the area between the breakthrough curve and a line at $C/C_0 = 1.0$ between limits 0 and t_e, the time of exhaustion. This is equal to the sum of the area up to the breakpoint and the hatched area in Fig. 10.45a. The latter area is found to be 1.7075 h from the figure. Therefore,

$$W_e = (9.2 + 1.7075)F_A = 10.9075\ F_A \text{ kg}$$

The unused fraction of the bed is given by Eq. (10.48)

$$\frac{LUB}{L} = 1 - \frac{W_b}{W_e}$$

where L is the total packed height of the bed. The length of the unused bed (LUB) is

$$LUB = L\left(1 - \frac{W_b}{W_e}\right) = 0.4\left(1 - \frac{9.2}{10.9075}\right) = 0.0626 \text{ m}$$ **(Answer)**

(c) The amount of the solute adsorbed by the bed before exhaustion, or the total or stoichiometric capacity when the entire bed comes to equilibrium with the feed (W_e) is

$$W_e = 10.9075\ F_A = 10.9075 \times 4 = 43.63 \text{ kg.}$$

Since the bulk density of the bed is 750 kg/m^3 and the volume of the bed per unit area of cross section is 0.4 m^3, the total quantity of the adsorbent in the bed is

$$W = 0.4 \times 750 = 300 \text{ kg.}$$

The saturation loading capacity of the bed is the ratio of the amount of water vapour adsorbed till saturation to the amount of the adsorbent present in the column. Therefore, the saturation loading capacity of the bed $= \dfrac{43.63}{300} = 0.1454$ kg water/kg molecular sieve. **(Answer)**

(d) For the new column, length L = 1.0 m. Since the length of the unused bed (LUB) remains constant, LUB for the new column = 0.0626 m. Therefore, the length of bed used up to the breakpoint for the new column (L_b) by Eq. (10.50) is

$$L_b = L - LUB = 1.0 - 0.0626 = 0.9374 \text{ m}$$

Since the breakpoint time is proportional to the length of the bed used up to the breakpoint,

$$\frac{t_b}{L_b}(\text{for the new column}) = \frac{t_b}{L_b}(\text{for the original column})$$

$$\frac{t_b}{0.9374} = \frac{9.2}{(0.4 - 0.0626)}$$

$$t_b = 0.9374\frac{9.2}{(0.4 - 0.0626)} = 25.56 \text{ h}$$

The breakpoint time for the new column = 25.56 h **(Answer)**

The fraction of the bed utilized up to the breakpoint $= \dfrac{L_b}{L} = \dfrac{0.9374}{1} = 0.9374$ **(Answer)**

Example 10.11: Alcohol vapours are adsorbed from an air stream containing 700 ppm alcohol vapours in a 50 mm diameter, 200-mm long adsorption column packed with 0.2 kg activated carbon. The air is admitted at a rate 1×10^{-3} kg/s. The following data refers to the concentrations of the breakthrough curve. The breakpoint concentration may be assumed to be C/C_0 = 0.01.

Time, h	2	3	3.5	4	4.5	5	5.5	6	6.25	6.5	6.75
C/C_0	0	0	0.006	0.046	0.174	0.397	0.654	0.885	0.939	0.970	0.990

Determine the dimensions of a new column which will give a breakpoint time of 8 hours, for adsorbing the same gas stream, but admitted with a flow rate 2.5×10^{-3} kg/s using the same adsorbent.

***Solution*:** The breakthrough curve is plotted as shown in Fig. 10.46.

A horizontal line at C/C_0 = 0.01 intersects the breakthrough curve at point b, and the breakpoint time t_b = 3.56 h from the figure. Assuming that at the point of exhaustion of the bed, the effluent concentration

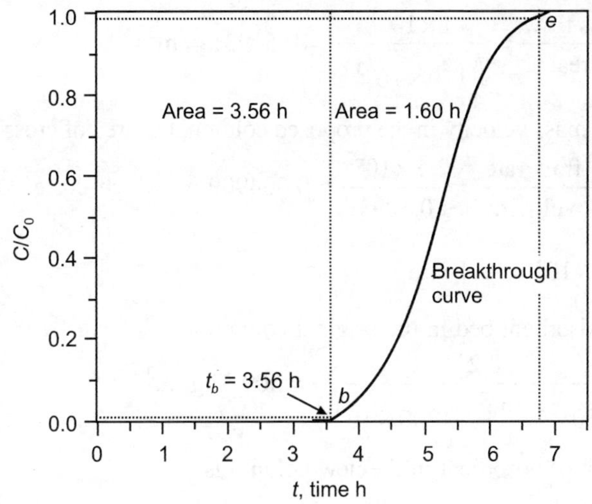

Fig. 10.46: Breakthrough curve for Example 10.11

is 99% of the initial concentration the horizontal line at $C/C_0 = 0.99$ intersects the breakthrough curve at point *e*, the point of exhaustion.

The area above the breakthrough curve and below $C/C_0 = 1.0$, is 3.56 to the left of the breakpoint and 1.60 to the right of the breakpoint. Therefore, the usable capacity of the bed up to the breakpoint (*See* Eq. (10.43))

$$W_b = F_A \int_0^{t_b} \left(1 - \frac{C}{C_0}\right) dt = 3.56 F_A \text{ kg}$$

and the total or stoichiometric capacity of the packed bed tower, when the entire bed comes to equilibrium with the feed (*See* Eq. (10.44))

$$W_e = F_A \int_0^{te} \left(1 - \frac{C}{C_0}\right) dt = (3.56 + 1.60) F_A = 5.16 F_A \text{ kg}$$

The length of the unused bed (LUB) by Eq. (10.49 is

$$LUB = L\left(1 - \frac{W_b}{W_e}\right) = L\left(1 - \frac{3.56}{5.16}\right) = 0.31 L \text{ m}$$

Since the length of the bed L = 200 mm, LUB = 62 mm.

For the new column, LUB remains the same, as long as the adsorbent properties and mass velocity of the gas stream are not changed. Therefore, LUB of the new column = 62 mm. For the original column, the breakpoint time t_b = 3.56 h, the length of used bed, L_b = 200 – 62 = 138 mm. For the proposed column, t_b = 8 h. The ratio of the breakpoint time to the length of the used bed also remains constant. Therefore, the length of used bed is

$$L_b = 138 \times \frac{8}{3.56} = 310.11 \text{ mm}$$

Therefore, the length of the new column by Eq. (10.50) is

$$L = L_b + LUB = 310.11 + 62 = 372.11 \text{ mm}$$

The mass velocity of the air in the original column

$$= \frac{\text{mass flow}}{\text{area}} = \frac{1\times10^{-3}}{\frac{\pi}{4}\left(50\times10^{-3}\right)^2} = 0.5093\,\text{kg/m}^2\text{s}.$$

To maintain the same mass velocity in the proposed column, the area of cross section of the column should be $\text{Area} = \dfrac{\text{mass flow rate}}{\text{mass velocity}} = \dfrac{2.5\times10^{-3}}{0.5093} = 0.004909\ \text{m}^2$ which corresponds to a diameter of

$$D = \left(\frac{0.004909\times4}{\pi}\right)^{1/2}\times10^3 = 79.1\ \text{mm}.$$

Bulk density of the adsorbent bed in the original column is

$$\frac{0.2}{\frac{\pi}{4}\times\left(50\times10^{-3}\right)^2\times\left(200\times10^{-3}\right)} = 509.3\,\text{kg/m}^3.$$

Therefore, the amount of adsorbent in the new column is

$$= \text{volume of the column}\times\text{bulk density} = \frac{\pi}{4}\times(79.1\times10^{-3})^2\times372.11\times10^{-3}\times509.3 = 0.9313\ \text{kg}$$

The new column diameter is 79.1 mm, length of packing is 372.11 mm and the amount of adsorbent needed is 0.9313 kg. **(Answer)**

Theoretical Model for Fixed-Bed Adsorber

Basic equations to describe the isothermal adsorption in fixed beds are developed in this section (Fig. 10.47). Fluid is assumed to be in plug flow with constant interstitial velocity u through the bed. Assume that no axial dispersion is present and the equilibrium is attained instantaneously with the solute and the bulk fluid. Consider a differential strip of thickness dz at a distance z along the length of the packed bed.

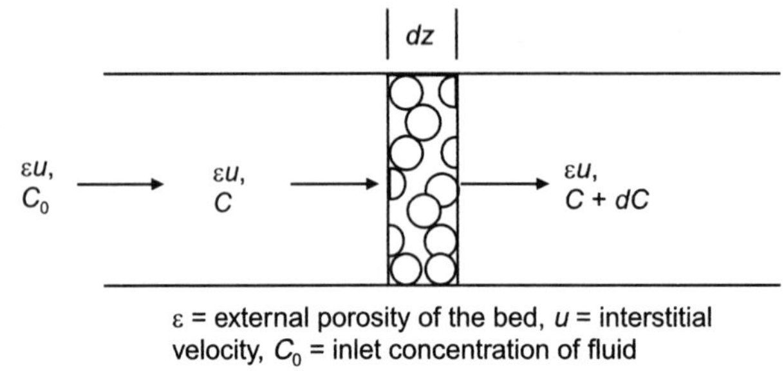

Fig. 10.47: Solute mass balance for a differential section of fixed bed

The superficial fluid velocity is εu, where ε is the external porosity of the bed. A solute mass balance for the flow of fluid over the differential section for a differential time duration, dt, gives

$$\varepsilon u A C\big|_z = \varepsilon u A C\big|_{z+\Delta z} + \varepsilon A\Delta z\frac{\partial C}{\partial t} + (1-\varepsilon)A\Delta z\rho_p\frac{\partial q}{\partial t} \quad \text{...(10.54)}$$

Where ρ_p is the particle density and q is the adsorbate loading on unit weight of the adsorbent. Dividing by Δz and taking limits as $\Delta z \to 0$ gives

$$\frac{\partial C}{\partial t} + u\frac{\partial C}{\partial z} + \frac{(1-\varepsilon)\rho_p}{\varepsilon}\frac{\partial q}{\partial t} = 0 \quad \text{...(10.55)}$$

The first term represents the accumulation of solute in the solution present in the voids of the bed. The second term represents the net amount of the solute brought to the section by the flowing fluid. The third term is the accumulation of the solute in the solid due to adsorption. For adsorption from a gas or from a dilute liquid solution, the first term representing the accumulation of the solute in the fluid is usually negligible compared to the accumulation on the solid.

The accumulation of the solute in the solid can be related to the rate of external mass transfer from the bulk solution to the particle, the diffusion through the pores to the internal adsorption sites and the adsorption on the internal surface. The actual process of adsorption is assumed to be instantaneous. The external and internal transport process can be represented in terms of an overall volumetric coefficient K_c and an overall driving force from the bulk fluid to the surface.

$$\frac{(1-\varepsilon)\rho_p}{\varepsilon}\frac{\partial q}{\partial t} = K_c a\,(C - C^*) \quad \text{...(10.56)}$$

Here a is the mass transfer area which can be taken to be the external surface area of the particles. C is the concentration of the solute in the bulk fluid and C^* is the concentration of the solute at the surface of the solid, which may be assumed to be the concentration in the solution in equilibrium with the average concentration of the adsorbate on the solid. The overall coefficient K_C depends on the external mass transfer coefficient and on an effective internal mass transfer coefficient, the latter being related to the effective diffusion coefficient. The effective diffusivity is a function of the particle porosity, the pore diameter and the nature of the diffusing species. A third equation can be written that relates the adsorbate loading by means of an appropriate adsorption isotherm. The solutions of these coupled non-linear equations are available for several simplified cases, but they fail to fit the experimental data well.

10.10 MOVING BED ADSORPTION

In order to design the moving bed equipment, the operation is considered to be analogous to the gas absorption, with solid adsorbent replacing the liquid solvent. Let S be the mass velocity of solids, kg/m^2 s. Let the inlet and exit concentrations of the adsorbent be X_2 and X_1 as shown in the Fig. 10.48a. Let the fluid be admitted at the bottom of the column with a mass velocity F kg/m^2 s with a concentration of the adsorbable component Y_1 and leaves with concentration Y_2. Let F_s be the mass velocity of solute-free solution. The concentration X is kg solute/kg adsorbent and Y is kg solute/kg solution.

For the differential portion of the column of thickness dz, the rate of solute transfer from the fluid to the solid can be written as

$$SdX = F_S dY = K_Y a_p\,(Y - Y^*)\,dz \quad \text{...(10.57)}$$

where a_p is the outside surface of the particles, $K_Y a_p$ is an overall mass transfer coefficient based on a_p which takes into account the resistance to the mass transport in the gas film surrounding the solids and the resistance to the internal diffusion through the fluid within the pores of the solid. The resistance to the actual adsorption process is assumed to be negligible and occur instantaneously.

Equation (10.57) can be rearranged to give

$$Z = H_{tOG} \times N_{tOG} \quad \text{...(10.58)}$$

where

$$H_{tOG} = \frac{F_S}{K_Y a_p} \quad \text{...(10.59)}$$

and

$$N_{tOG} = \int_{Y_2}^{Y_1} \frac{dY}{Y - Y^*} \quad \text{...(10.60)}$$

A solute balance for the entire column gives

$$F_s(Y_1 - Y_2) = S(X_1 - X_2) \quad \text{...(10.61)}$$

$$\frac{Y_1 - Y_2}{X_1 - X_2} = \frac{S}{F_S} \quad \text{...(10.62)}$$

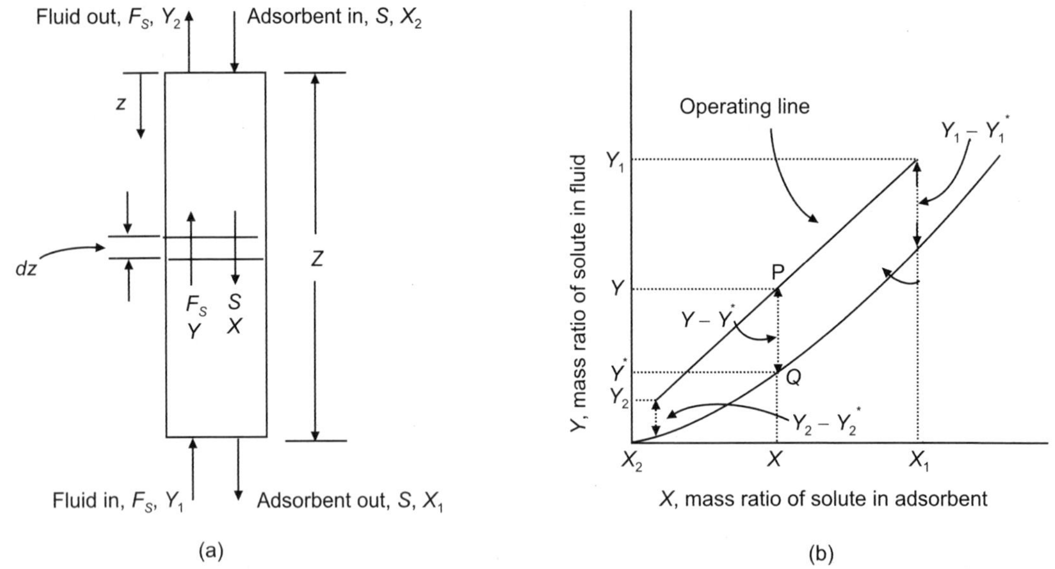

Fig. 10.48: Design of continuous counter current adsorber

A straight line drawn connecting points (X_1, Y_1) and (X_2, Y_2) has slope S/F_s, the ratio of the adsorbent flow rate to fluid flow rate through the adsorber, which is the operating line for the adsorber. The concentration of the adsorbent and the fluid anywhere in the column can be obtained as the coordinates of the points lying on this line, as evident by taking a solute balance for the upper part of the column.

$$F_S(Y - Y_2) = S(X - X_2) \qquad \ldots(10.63)$$

$$\frac{Y - Y_2}{X - X_2} = \frac{S}{F_S} \qquad \ldots(10.64)$$

The operating line and the equilibrium curves are plotted as shown in Fig. 10.48b. The vertical distance between the operating line and the equilibrium curve is the overall driving force $Y - Y^*$.

The integral in Eq. (10.60) is evaluated by graphical integration in the conventional manner (Fig. 10.49) and the number of the overall transfer units determined. The height of the transfer unit is determined using Eq. (10.59) and the length of the adsorber is estimated as the product of the height of transfer unit and the number of transfer units.

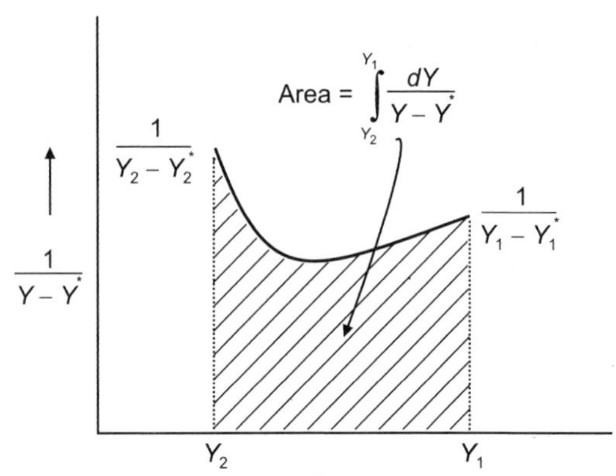

Fig. 10.49: Graphical integration for estimating the number of transfer units

Example 10.12: Air contaminated with NO_2 is passed through a continuous countercurrent adsorber at the rate of 600 kg/h at 298 K and 101.3 kPa. The air contains 1.5% NO_2 by volume, 95% of which is to be removed by adsorption using silica gel which enters the column free of NO_2. For an air superficial velocity of 1500 kg/m^2 h, the overall gas mass transfer coefficient is expected to be 2500 kg/m^3 h. The equilibrium adsorption isotherm is given by the following data:

Partial pressure of NO_2, kPa	0	0.267	0.533	0.800	1.066	1.333	1.600
kg NO_2/100 kg gel	0	0.4	0.9	1.65	2.60	3.65	4.85

(a) Calculate the minimum weight of gel required per hour.
(b) For twice the minimum rate, calculate the number of transfer units required.
(c) Estimate the height of the adsorber.

***Solution*:** The column is shown in Fig. 10.50a.

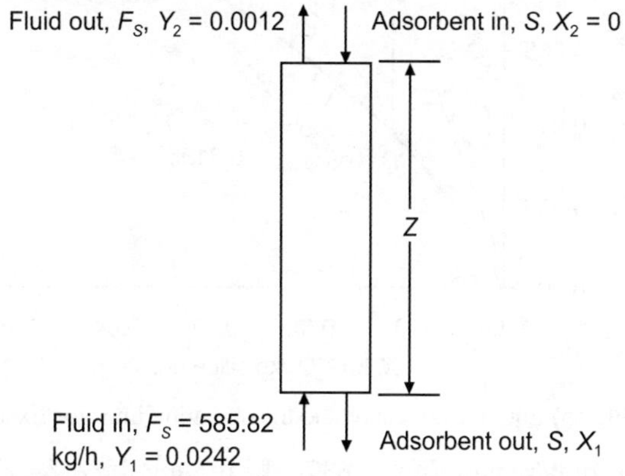

Fig. 10.50: (a) Moving bed adsorber in Example 10.12

Molecular weight of NO_2 = 46, Molecular weight of dry air = 29.

The inlet gas concentration $Y_1 = \frac{1.5}{98.5} \times \frac{46}{29} = 0.0242 \text{ kg } NO_2\text{/kg dry air.}$

Since 95% of the NO_2 in the entering stream is adsorbed, exit gas concentration

$$Y_2 = 0.05 \times 0.0242 = 0.0012 \text{ kg } NO_2\text{/kg dry air.}$$

Since gas flow rate is 600 kg/h,

$$F_S = 600 \times \frac{1}{1+Y_1} = 600 \times \frac{1}{1.0242} = 585.82 \text{ kg/h}.$$

The equilibrium data as X versus Y are now calculated:

$$Y = \frac{\bar{p}}{101.3 - \bar{p}} \times \frac{46}{29} \text{ and } X = \frac{\text{kg } NO_2}{\text{kg gel}}$$

X, kg NO_2/kg gel	0	0.004	0.009	0.0165	0.026	0.0365	0.0485
Y, kg NO_2/kg dry air	0	0.00419	0.00839	0.01263	0.01687	0.02115	0.02546

(a) The equilibrium curve is plotted as shown in Fig. 10.50b. The lower terminus of the operating line has coordinates (X_2, Y_2) = (0, 0.0012). The point is shown as point A in Fig. 10.50b. The upper terminus of the operating line lies on the horizontal line at Y_1 = 0.0242. For the minimum adsorbent quantity used, the operating line AB' is tangential to the equilibrium curve as shown. Slope of the limiting operating line AB' is found out.

$$\text{Slope} = \left(\frac{S}{F_S}\right)_{\min} = 0.8185, \text{ so that}$$

$$S_{\min} = 0.8185 \times F_S = 0.8185 \times 585.82 = 479.5 \text{ kg/h}$$

The minimum weight of gel required = 479.5 kg/h. **(Answer)**

(b) For twice the minimum quantity of gel used, slope of the operating line is

$$\text{Slope} = \left(\frac{S}{F_S}\right) = 0.8185 \times 2 = 1.637$$

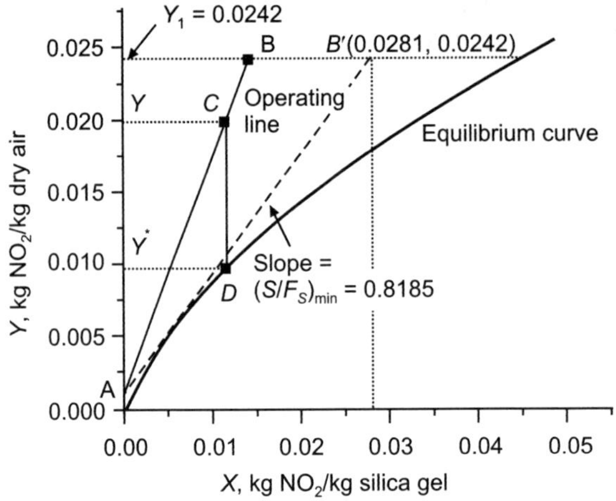

Fig. 10.50: (b) Equilibrium curve and the operating lines for Example 10.12

The operating line *AB* is drawn with slope 1.637. The driving force $Y - Y^*$ is determined for different Y values by measuring the vertical distance between operating line and equilibrium curve. For illustration, corresponding to point C ($Y = 0.02$) on the operating line, the point D lies on the equilibrium curve vertically below C. y-coordinate of point D is $Y^* = 0.0097$. Similarly vertical Y^* are determined for different Y values and the results are tabulated below:

Y	0.0242	0.02	0.015	0.01	0.005	0.0012
Y^*	0.0112	0.0097	0.008	0.0053	0.0024	0
$1/(Y-Y^*)$	76.92308	97.08738	142.8571	212.766	384.6154	833.3333

The number of transfer units is given by Eq. (10.60).

$$N_{tOG} = \int_{Y_2}^{Y_1} \frac{dY}{Y - Y^*}$$

To find the integral in Eq. (10.60) graphically, $1/(Y-Y^*)$ is plotted against Y as shown in Fig. 10.50c. The area under the curve between $Y_2 = 0.0012$ and $Y_1 = 0.0242$ is found to be 5.66. Therefore, the number of transfer units = 5.66.

(Answer)

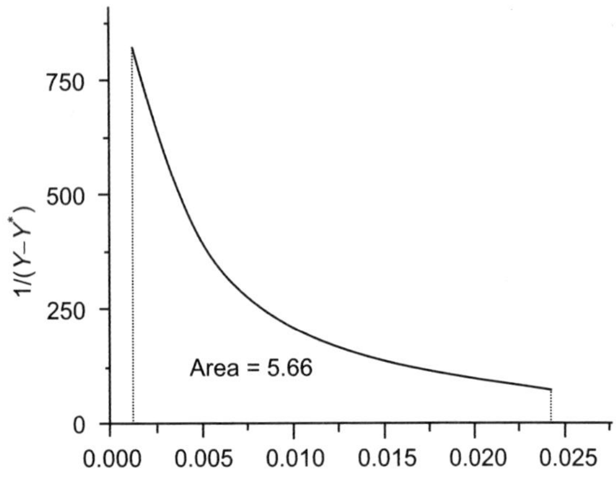

Fig. 10.50: (c) Graphical integration to find the number of transfer units

(c) The height of transfer units is given by Eq. (10.59)

$$H_{tOG} = \frac{F_S}{K_Y a_p}$$ where F_S is the mass velocity of gas and $K_Y a_p$ is the mass transfer coefficient.

Assuming a superficial mass velocity of 1500 kg/m^2 h, the mass transfer coefficient is $K_Y a_p = 2500$ kg/m^3 h. Therefore,

$$H_{tOG} = \frac{F_S}{K_Y a_p} = \frac{1500}{2500} = 0.6 \text{ m}$$

The height of the adsorber = $Z = H_{tOG} \times N_{tOG} = 0.6 \times 5.66 = 3.4$ m **(Answer)**

REVIEW QUESTIONS

1. Distinguish between adsorbent and adsorbate.
2. List five industrial applications each for adsorption for gaseous and liquid separations.
3. List five distinguishing features of chemisorption and physical adsorption. Which one is important as a mass transfer operation?
4. What are the properties of a good adsorbent?
5. How do the specific surface area and specificity important as adsorbent properties?
6. Give some examples for industrial adsorbents and their major applications.
7. What is the feature of molecular sieve zeolites that makes them unique among the common adsorbents?
8. What do you understand by adsorption isotherm?
9. What are the different types of adsorption isotherms?
10. What is the effect of temperature on adsorption of gases?
11. State Freundlich isotherm and explain how the constants are evaluated.
12. What are the assumptions in deriving the Langmuir isotherm equation?
13. State and explain Langmuir isotherm. How do you determine the constants in the Langmuir equation that fit the experimental data?
14. Explain hysteresis. How do you explain the reason for its occurrence?
15. Distinguish between the integral and the differential heats of adsorption.
16. What is isosteric heat of adsorption? Explain how it is determined?
17. How is the experimental study on the adsorption from the liquids difficult in comparison to that from gases?
18. Describe the different types of adsorption isotherm for liquids.
19. List the three basic modes for regeneration of the spent adsorbent.
20. What are the different steps in the Thermal Swing Adsorption (TSA)?
21. What are the advantages and disadvantages of TSA? List the major areas of application of TSA.
22. Explain the principle of PSA cycles. Compare the PSA and TSA cycles for adsorbent regeneration.
23. Distinguish between the inert-purge and the displacement-purge operations for adsorbent regeneration. What is the major weakness of the purge-swing adsorption cycles?
24. Discuss the major applications of agitated vessels for adsorption.
25. Discuss the internals of a fixed bed adsorber such as the packings, packing supports, distributors etc.
26. What is the main attraction of the horizontal fixed bed adsorbers over vertical ones?
27. Describe the working of Hypersorber and Higgins contactor.
28. Describe the working of a simulated moving bed system for adsorption.
29. Why is it not advantageous to opt for more than two stages in crosscurrent adsorption?
30. How do you determine the minimum quantity of adsorbent required in countercurrent adsorption?
31. Describe the mass transfer zone and the adsorption wave.
32. What are the factors that affect the length and shape of the mass transfer zone?
33. What is breakpoint in the operation of fixed bed adsorber?
34. Describe the breakthrough curve in fixed bed adsorption.
35. Explain the ideal fixed bed adsorption. When will it occur? What is stoichiometric wave front?
36. What is the basic principle on which the scale up design method of adsorber is based?

MULTIPLE CHOICE QUESTIONS

10.1 One of the following features is not applicable to chemisorption. Which one?
A. Heat of adsorption is comparable to heat of chemical reaction
B. No adsorption above critical temperature.
C. Adsorption is restricted to active sites on the adsorbent surface.
D. Adsorption is limited to monomolecular layers

10.2 Which one of the following statements is incorrect with reference to physical adsorption?
A. Heat of adsorption is of the order of heat of vaporisation of the solute
B. Rapid adsorption at low temperatures
C. Adsorption is restricted to active molecular sites on the surface.
D. Adsorption is reversible and the adsorbent can be regenerated

10.3 One unique feature of molecular sieves as adsorbent is
A. its low affinity for water
B. its resistance to attrition
C. uniformity in pore sizes
D. amenable to regeneration by washing

10.4 Which one of the following adsorbents is preferred for adsorbing components from aqueous solutions and moist gases because of its poor affinity towards water?
A. Activated carbon
B. Silica gel
C. Activated alumina
D. Molecular sieve zeolites

10.5 Which one of the following is least preferred for drying of gases and liquids?
A. Fullers earth
B. Activated carbon
C. Silica gel
D. Activated alumina

10.6 At a given temperature and pressure, the highest adsorbate concentration on activated carbon results when .... is adsorbed.
A. Acetone
B. Benzene
C. Methane
D. Oxygen

10.7 At a given pressure, the adsorbate loading
A. remains constant with change in temperature
B. increases with increase in temperature
C. decreases with increase in temperature
D. can not be predicted

10.8 At a given temperature, lowering the pressure will
A. reduce the adsorbate loading
B. increase the adsorbate loading
C. not influence the adsorbate loading
D. increase or decrease the adsorbate loading

10.9 A gas stream containing 600 ppm (by weight) of certain pollutant is passed through an adsorbent bed at a rate of 4.0 kg/h. The bed diameter is 50 mm, length is 200 mm and the bulk density of the bed is 500 kg/m^3. When the gas is passed for 10 hours, the entire bed gets saturated with the adsorbate and the bed is ready for regeneration. What is the saturation loading capacity of the adsorbent in kg pollutant per kg adsorbent?
A. 1.20
B. 0.12
C. 0.25
D. 1.36

10.10 "Favorable" adsorption isotherms are those
A. which are linear and pass through the origin
B. which are concave towards the solid-concentration axis throughout
C. which are concave towards the fluid-concentration axis throughout
D. which possess one or more points of inflection

10.11 Different gases get adsorbed to different extent under similar conditions of temperature and partial pressure. The adsorbate loading is higher for

A. the material with low molecular weight.
B. the material with high critical temperature
C. the material with high molecular weight and high critical temperature
D. the material with high molecular weight and low critical temperature

10.12 Molecular sieves are

A. produced by thermal decomposition of carbonaceous materials such as wood, coal etc.
B. produced from colloidal silica
C. produced by controlled heating of hydrated alumina
D. porous synthetic zeolite crystals

10.13 For adsorption of gases, the adsorbate loading

A. increases with increase in temperature
B. decreases with increase in temperature
C. decreases with increase in pressure
D. is independent of temperature and pressure

10.14 In the experimental investigation of adsorption equilibrium, if a straight line results when log q is plotted against log C, the equilibrium can be represented by

A. Freundlich isotherm
B. linear isotherm
C. Langmuir isotherm
D. BET isotherm

10.15 One of the following is not true if the adsorption equilibrium follows Langmuir isotherm.

A. The plot of C/q against C will be straight line
B. The plot of $1/q$ versus $1/C$ will be a straight line
C. The plot of log q against log C will be straight line
D. The plot of q versus C becomes straight horizontal line at high C values

10.16 Adsorption of acetone from aqueous solution on activated carbon can be represented by the Langmuir isotherm

$$q = \frac{0.190C}{1+0.146C}$$

where q is the adsorbate loading mol/kg, C= solute concentration in aqueous solution, mol/m^3. Determine the maximum adsorbate loading in kg acetone/kg carbon.

A. 0.0755
B. 1.3014
C. 0.1658
D. 0.0096

10.17 100 kg of toluene containing 200 ppm water is treated with 10 kg activated alumina. The equilibrium is represented by the Langmuir isotherm given by $q = \dfrac{8.04\times10^{-2}C}{1+4.886\times10^{-3}C}$ where q in kg/100 kg and C in ppm. If equilibrium is attained in this single stage adsorption operation, what is the percent removal of water?

A. 93.90%
B. 98.75%
C. 94.50%
D. 90.35%

10.18 For a substance exhibiting adsorption hysteresis

A. the desorption isotherm is closer to the fluid concentration axis than the adsorption isotherm.
B. the adsorption isotherm is closer to the fluid concentration axis than the desorption isotherm.
C. adsorbate loading will be greater, when equilibrium is reached by adsorption than by desorption
D. the desorption equilibrium pressure is always greater than the adsorption pressure.

10.19 Adsorption process is
A. always exothermic
B. always endothermic
C. can be either exothermic or endothermic
D. accompanied by no change in enthalpy

10.20 Activated charcoal usually gives
A. Type I isotherm
B. Type II isotherm
C. Type III isotherm
D. Type IV isotherm

10.21 Adsorption isotherm of activated alumina-air-water vapour system belong to
A. Type I isotherm
B. Type II isotherm
C. Type III isotherm
D. Type IV isotherm

10.22 An adsorption isotherm which is convex to the fluid-concentration axis is treated as
A. favourable isotherm
B. unfavourable isotherm
C. BET isotherm
D. adsorption hysteresis

10.23 The product contamination is the major problem in
A. Thermal swing cycle (TSA)
B. Pressure swing cycle (PSA)
C. Purge-swing cycle
D. both TSA and PSA cycles

10.24 Which one of the following features is incorrect as far as PSA cycles are concerned?
A. Shorter cycle times compared to TSA
B. No heating or cooling of the bed
C. Low residual loading compared to TSA
D. Adsorbent inventory smaller than TSA.

10.25 Only one of the following features is true for TSA cycles. Spot the true statement.
A. Low residual loading compared to PSA
B. Low cycle times compared to PSA
C. Incomplete desorption compared to PSA
D. Limited to gas-phase application

10.26 When the solute is weakly adsorbed, easily desorbed, and is of little or no value, the ……… method can be used for adsorbent regeneration.
A. Thermal swing
B. Pressure swing
C. Inert-purge swing
D. Displacement purge swing

10.27 Molecular sieve zeolites used for adsorption of linear paraffin from mixtures of branched-chain and cyclic hydrocarbon are regenerated by using ammonia. The method of regeneration belongs to
A. Thermal swing adsorption
B. Pressure swing adsorption
C. Inert-purge swing adsorption
D. Displacement purge swing adsorption

10.28 In fixed bed adsorbers, adsorbent particle size is kept as large as possible because
A. it leads to large surface area
B. it reduces pressure drop for flow
C. it makes regeneration easier
D. it reduces attrition and loss of particles.

10.29 One of the major problems with moving-bed adsorbers is
A. attrition and loss of adsorbent
B. the very large pressure drop
C. unsteady-state operation
D. fluidization of adsorbent bed

10.30 Hypersorber is an example of a
A. fixed bed adsorber
B. agitated adsorption vessel
C. moving bed adsorber
D. fluidized bed adsorber

10.31 In crosscurrent adsorption, more than two stages are generally not preferred. This is because

A. the stage efficiency is low when the number of stages is high

B. recovery of solute is not improved by providing stages above two.

C. adsorption isotherm sets this limit on the number of stages

D. the process becomes uneconomical because of the intermediate filtration steps.

10.32 In two stage crosscurrent adsorption, when adsorption isotherm is linear, minimum adsorbent results, if

A. equal quantities of adsorbent are used in both stages

B. unequal dosages of adsorbent are employed in both stages

C. fresh adsorbent are used in both stages

D. very fine adsorbent particles are used

10.33 If the adsorption follows Freundlich equation $Y = m\,X^n$ the adsorbent required for a given extent of adsorption in single stage operation is low if

A. the value of n is 1.0 B. the value of n is less than 1.0

C. the value of n is greater than 1.0 D. the value of n is negative

where Y is concentration in the fluid and X is concentration in the solid.

10.34 At equilibrium, concentration of water in vapour phase (C^*) in kg/m^3 of air space and the amount of water (q) adsorbed per kg of dry silica gel are related by $C^* = 0.0667\,q$. To maintain dry conditions in a room of air space 100 m^3 containing 2.2 kg of water vapour initially, 10 kg of dry silica gel is kept in the room. The fraction of initial water remaining in the air space after a long time (during which the temperature is maintained constant) is

A. 0.0 B. 0.2

C. 0.4 D. 1.0

10.35 One cubic meter of an aqueous effluent stream contaminated with 0.2 kg phenol/m^3 of solution is treated with 1.5 kg fresh granular activated carbon. If the adsorption follows Freundlich equation $q = 0.2C^{0.25}$ where q is the adsorbate loading kg phenol/kg adsorbent and C is the kg phenol/m^3 of solution, the percentage of phenol adsorbed is

A. 100% B. 55.6%

C. 72.6% D. 48.0%

10.36 Mass transfer zone in fixed bed adsorber is

A. the portion of the bed with constant adsorbate concentration

B. the portion of the bed saturated with adsorbate

C. the portion of the bed in which concentration changes from feed concentration to zero

D. the zone that follows the unused bed and saturated bed.

10.37 In ideal fixed bed adsorption, the thickness of the mass transfer zone is

A. unity B. zero

C. negative D. infinity

10.38 All but one are true statements with regard to stoichiometric wave front. Which one is incorrect?

A. It is the ideal break through curve in fixed bed adsorption

B. It is a vertical line representing the break through curve

C. It is an *S*-shaped breakthrough curve.

D. It represents the fixed bed adsorption process occurring infinitely rapidly.

10.39 In the operation of fixed bed adsorber, the breakpoint is

A. the time at which the entire bed becomes saturated with the adsorbate

B. the time at which the solute concentration in the effluent rises to the maximum permissible value

C. the time at which the solute concentration in the effluent attains about 95% of that in the feed.
D. a phenomenon exhibited by certain adsorbents.

10.40 Which one of the following is a favorable condition for the formation of an ideal breakthrough curve?
A. The rate of adsorption is infinitesimally small
B. The internal and external resistance to mass transfer are significant
C. The fluid flows in turbulent flow
D. Axial dispersion is negligible

10.41 The scale-up method of design of fixed bed columns is based on the principle that
A. length of unused bed is independent of total length of the column.
B. length of the equilibrium zone is constant and is independent of the length of the column.
C. the adsorbent in the mass transfer zone is half saturated
D. the fraction of the length of the bed that is unused remains constant

10.42 For an ideal breakthrough curve in fixed bed adsorption, the length of mass transfer zone is
A. unity B. zero
C. equal to the length of the bed D. none of the above

10.43 In adsorption in fixed beds, if the break through curve is S-shaped, the length of unused bed (LUB) is approximately equal to
A. the length of mass transfer zone B. twice the length of mass transfer zone
C. half the length of mass transfer zone D. length of the bed.

10.44 Using a small experimental adsorption column of length 200 mm, the breakpoint time was found to be 5 h and the length of mass transfer zone was estimated to be 40 mm. An adsorption column is to be designed which uses the same fluid concentration and mass velocity as in the experimental unit, but with a breakpoint time of 10 h. What should be the total length of the adsorber?
A. 400 mm B. 360 mm
C. 320 mm D. 480 mm

10.45 Using a small experimental adsorption column (column I) of length 200 mm, the breakpoint time was found to be 5 h and the length of mass transfer zone was estimated to be 40 mm. An adsorption column (Column II) is to be designed which uses the same fluid concentration and mass velocity as in the experimental unit, but with a breakpoint time of 10 h. The fraction of bed utilized in column I and column II are, respectively:
A. 8/10, 8/10 B. 8/10, 1
C. 8/10, 8/9 D. None of the above

EXERCISES

10.1 The following data refers to the adsorption equilibrium of water vapour on activated carbon at 298K.

Partial pr, Pa	0	266.6	533.3	800.0	1066.6	1333.2	1600	1866.5	2133.2
kg water/100 kg gel	0	4.8	8.6	12.7	17.4	22.6	28.3	34.1	38.6

(a) Plot the equilibrium data as partial pressure versus weight fraction of water on the adsorbent.
(b) An air stream with a partial pressure of water vapour of 1.5 kPa is passed continuously across 10 kg dry silica gel initially containing 5 (weight)% adsorbed water. The total pressure is 101.3 kPa and the temperature is 298 K. What mass of additional water will the gel have adsorbed under equilibrium? (**Answer:** 2.065 kg)

10.2 Water is contaminated by traces of phenol, which is to be removed by the adsorption with activated carbon. The following data were obtained in a test conducted in the laboratory on adsorption of phenol from an aqueous solution on activated carbon at room temperature.

C, kg phenol/m^3 solution	0.35	0.28	0.20	0.15	0.10	0.05	0.01	0.005	0.001
q, kg phenol/kg carbon	0.154	0.145	0.134	0.124	0.112	0.095	0.063	0.053	0.036

Fit the data to an adsorption isotherm and evaluate the parameters.

(**Answer:** Langmuir not applicable. Freundlich, $k = 0.20$, $n = 0.249$)

10.3 Using the following data on the adsorption of certain solute **A** on silica gel determine the constants of the Langmuir equation that fits the data well.

kg **A**/m^3 solution	0.14	0.09	0.07	0.05	0.04	0.03
kg **A**/kg gel × 10^5	2.948	2.621	2.408	2.100	1.889	1.617

(**Answer:** $k = 24.68$ m^3/kg, $q_m = 3.8 \times 10^{-5}$ kg **A**/kg gel)

10.4 The following data refers to the adsorption of water from toluene on activated alumina. Fit the data to Langmuir isotherm and determine the constant parameters.

C, ppm	20	40	60	80	100	150	200	250	300	350	400
q, kg/100kg	1.5	2.6	3.5	4.4	5.1	6.5	8.2	9.5	10.9	12.1	13.3

(**Answer:** $q = \dfrac{8.04 \times 10^{-2} C}{1 + 4.886 \times 10^{-3} C}$ where q in kg/100 kg and C in ppm)

10.5 The following data refers to the adsorption isotherm of glucose on activated alumina from aqueous solution at room temperature.

C, kg/m^3	4.0	8.0	12.0	20.0	25.0	100.0	200.0
q, kg solute/kg alumina	0.0267	0.0452	0.0588	0.0775	0.0856	0.1252	0.1357

Determine the isotherm that fits the data and evaluate the parameters.

(**Answer:** Freundlich not applicable. Langmuir isotherm, $q = \dfrac{8.1 \times 10^{-3} C}{1 + 5.51 \times 10^{-2} C}$)

10.6 The following experimental data refers to the adsorption isotherm for methane gas on activated carbon at 298 K.

$p \times 10^{-5}$ Pa	1.72	3.45	6.89	10.34	20.67	31.01	41.35	51.68	62.02	68.91
$q \times 10^3$, m^3(STP) of CH_4/kg carbon	25.3	43.0	66.0	81.0	106.0	116.0	121.0	125.0	126.0	126.0

Fit the data to (a) Freundlich isotherm (b) Langmuir isotherm. Which isotherm provides a better fit to the data?

(**Answer:** (a) $q = 2.2353 \times 10^{-4} p^{0.4126}$ (b) $q = \dfrac{0.1873 \times 10^{-6} p}{1 + 1.3218 \times 10^{-6} p}$. Average percent deviation = 0.88% for Langmuir and 11.68% for Freundlich isotherm)

10.7 The following experimental data refers to the adsorption isotherm for methane gas on activated carbon at 298 K.

$p \times 10^{-5}$ Pa	1.72	3.45	6.89	10.34	20.67	31.01	41.35	51.68	62.02	68.91
$q \times 10^3$, kg CH_4/kg carbon	18.1	30.7	47.1	57.9	75.7	82.9	86.4	89.3	90.0	90.0

Fit the data to (a) Freundlich isotherm (b) Langmuir isotherm. Which isotherm provides a better fit to the data?

(**Answer:** (a) $q = 1.5967 \times 10^{-4} p^{0.4126}$ (b) $q = \dfrac{0.1338 \times 10^{-6} p}{1 + 1.1322 \times 10^{-6} p}$. Average percent deviation = 0.88% for Langmuir and 11.68% for Freundlich isotherm)

10.8 The constants of the Langmuir isotherm for adsorption of methane and carbon monoxide by activated carbon at room temperature are as follows:

	q, m^3(STP)/kg	K, (Pa)$^{-1}$
CH_4	0.1442	1.236×10^{-6}
CO	0.1261	9.055×10^{-7}

A vapour mixture analyzing 70 mol% CH_4 and 30 mol% CO is contacted with activated carbon at room temperature and a total pressure of 2500 kPa. Determine the specific adsorption volumes (STP) of the components and the composition of the adsorbate.

(**Answer:** CH_4 = 0.0812 m^3/kg, CO = 0.0223 m^3/kg; 78.5% CH_4 and 21.5% CO)

10.9 An aqueous solution containing small amounts of acetone and propionitrile is to be treated with activated carbon. The equilibrium data for solutes have been determined independently and found to fit the Langmuir isotherm:

Acetone in water (298 K) $\quad q = \dfrac{0.190C}{1 + 0.146C}$

Propionitrile in water (298 K) $\quad q = \dfrac{0.173C}{1 + 0.0961C}$

where q = amount of solute adsorbed, mol/kg, C = solute concentration in aqueous solution, mol/m^3. Estimate the amounts adsorbed when an aqueous solution containing 40 and 35 mol/m^3, respectively, of acetone and propionitrile at 298 K using activated carbon adsorbent.

(**Answer:** Acetone = 0.7448 mol/kg, propionitrile = 0.5934 mol/kg)

10.10 For adsorption of an aromatic hydrocarbon on molecular-sieve zeolite at 547 K, the following equilibrium data were obtained.

q, %	7.0	9.1	10.3	10.8	11.1	11.5
P, Pa	1.6	3.6	5.7	9.3	12.5	19.6

Obtain the adsorption isotherm that fits the data well and evaluate the parameters where q is expressed in kg/kg adsorbent. (**Answer:** Langmuir isotherm, q_m = 0.1386, k = 0.7553)

10.11 For adsorption of an aromatic hydrocarbon on molecular-sieve zeolite at 547 K, the following equilibrium data were obtained.

q, kmol/kg $\times 10^4$	5.61	7.46	8.56	9.02	9.3	9.68
P, Pa	1.6	3.6	5.7	9.3	12.5	19.6

Obtain the adsorption isotherm that fits the data well and evaluate the parameters.

(**Answer:** Langmuir isotherm, q_m = 0.00104, k = 0.757)

10.12 The following equilibrium data have been obtained for the adsorption of nitrogen dioxide on silica gel at 298 K and pressure of 101.3 kPa.

$\bar{p}_{NO_2}$, Pa	0	266.6	533.3	799.9	1066.6	1333.2	1599.9
X^*(kg NO_2/ kg silica gel)	0	0.004	0.009	0.0165	0.026	0.0365	0.0485

Show that these data are reasonably correlated by the Freundlich isotherm and determine the values of the parameters. (**Answer:** $k = 1.4086 \times 10^4$, $n = 1.4122$)

10.13 It is decided to effect decolorization of an aqueous solution of a valuable solute by adsorbing the impurities causing the colour on activated carbon. The equilibrium adsorption behaviour was studied by adding different amounts of activated carbon to one kg batches of original solution and measuring the colour intensity at equilibrium. The initial colour intensity of the solution was 11.1 units. The results are shown below.

kg carbon added	0	0.001	0.0025	0.005	0.01	0.02	0.03	0.04	0.05
Colour intensity at equilibrium	11.1	10.0	8.6	6.8	4.4	2.2	1.3	0.88	0.65

Determine the parameters in the Freundlich's isotherm, $Y = mX^n$ that fits the equilibrium data well. X and Y are units of colour/kg adsorbent and units of colour/kg solution, respectively.

(**Answer:** $Y = 8.62 \times 10^{-5} X^{1.666}$)

10.14 It is desired to establish the adsorption isotherm of a substance **A** from aqueous solution on activated carbon. 3.75×10^{-5} kg **A** is dissolved in 10^{-2} m^3 of water. The solution is divided into five samples of equal volumes into which different amounts of activated carbon are added. The concentration of the solution in each sample was measured after allowing sufficient time for attaining equilibrium. The results are shown in the following table.

Mass of carbon added, kg $\times 10^3$	0.01	0.05	0.15	0.60	1.00
Equilibrium concentration of **A**, kg/m^3 $\times 10^3$	3.25	2.50	2.00	1.50	1.35

Determine which one of the adsorption isotherm, Langmuir or Freundlich isotherm fits the data well and determine the parameters.

(**Answer:** Freundlich, $q = 5.6638 \times 10^7 C^{3.5}$)

10.15 A solution of raw cane sugar is to be decolorized by adsorbing the impurities on activated carbon. The data for equilibrium adsorption isotherm is established by adding various small amounts of activated carbon to separate batches of sugar solution with an initial colour concentration of 25 units and observing the colour intensity of the solution at equilibrium. The data with the quantity of carbon expressed on the basis of sugar content of the solution are given below:

kg carbon/kg dry sugar	0	0.007	0.013	0.023	0.032	0.049	0.10
Colour removal, %	0	56	77	90	95	98	99.5

Determine the constants m and n of the Freundlich equation $Y = mX^n$ which fits the equilibrium data, where Y is colour units/kg sugar, and X is the units of colour/kg carbon.

(**Answer:** $m = 7.47 \times 10^{-7}$, $n = 2.172$)

10.16 Equilibrium data on the adsorption of propane on activated carbon are given below. Estimate the heat of adsorption.

Pressure, Pa $\times 10^{-3}$	Adsorbate loading, kg /kg carbon 310.9 K	338.7 K	366.5 K	Pressure, Pa $\times 10^{-3}$	Adsorbate loading, kg /kg carbon 310.9 K	338.7 K	366.5 K
25	0.1402	0.1075	0.0775	350	0.2214	0.1913	0.1629
50	0.1659	0.1298	0.1045	400	0.2254	0.1947	0.1675
75	0.1798	0.1455	0.1182	450	0.2289	0.1979	0.1714
100	0.1872	0.1553	0.1262	500	0.2311	0.2011	0.1756
150	0.1969	0.1675	0.1371	550	0.2322	0.2035	0.1796
200	0.2041	0.1755	0.1452	600	0.2329	0.2061	0.1838
250	0.2106	0.1821	0.1522	650	0.2330	0.2091	0.1878
300	0.2165	0.1874	0.1577				

(**Answer:** –33515 kJ/kmol)

10.17 The following table gives the equilibrium data at four different temperatures for adsorption of pure benzene vapour on silica gel.

Partial pressure of benzene, Pa	kmol adsorbed/kg gel × 10^5 343 K	363 K	383 K	403 K
50.7	14	6.7	2.6	1.13
101.3	22	11.2	4.5	2
202.7	34	18	7.8	3.9
506.5	68	33	17	8.6
1013	88	51	27	16
2026	–	78	42	26

(a) Determine which of the isotherms Freundlich or Langmuir, fits the data well?

(b) Use the data to estimate the heat of adsorption.

[**Answer:** (a)

Temp, K	Langmuir	Av.% deviation	Freundlich	Av.% deviation
343	$q = \frac{3.126\times10^{-6}p}{1+0.00328p}$	10.4	$q = 1.18\times10^{-5}p^{0.6336}$	4.0
363	$q = \frac{1.615\times10^{-6}p}{1+0.00246p}$	11.2	$q = 5.16\times10^{-6}p^{0.66348}$	2.4
383	$q = \frac{5.33\times10^{-7}p}{1+0.00127p}$	10.4	$q = 1.331\times10^{-6}p^{0.7649}$	3.7
403	$q = \frac{2.26\times10^{-7}p}{1+0.000664p}$	8.8	$q = 3.82\times10^{-7}p^{0.8651}$	3.7

(b)

q, kmol/kg × 10^4	1	2	3	4	5	6	7	Average
ΔH_{ads}, kJ/kmol	–54576	–52944	–53848	–54243	–54914	–55376	–53844	–54249

10.18 A liquid solution is to be decolorized by adsorption. The adsorption equilibrium is represented by $y = 0.65x^{0.25}$ where y is parts of colour removed per part of adsorbent and x is the part of colour present in 1000 parts of carrier. Calculate the percent removal of colour from a solution containing 100 kg colour carrying materials with an initial concentration of 1 part colour per 3 parts of carrier and is treated with 15 kg adsorbent (a) using a single stage (b) using 7.5 kg each in two-stage crosscurrent adsorber. (**Answer:** (a) 81.7% (b) 88.4%)

10.19 Concentration of nitrobenzene in an aqueous solution is to be reduced from 1.5 × 10^{-2} kg/m^3 to 1.2 × 10^{-6} kg/m^3 by adsorption on activated carbon in an ideal stirred tank. If the equilibrium adsorption isotherm is represented by $q = 675C$ where q is in kg adsorbed /kg carbon, and C is concentration in kg/m^3, determine the quantity of carbon required per 1 m^3 of treated water. (**Answer:** 18.517 kg)

10.20 It is desired to reduce the acetone content from 2 m^3 of acetone-air mixture measured at 300 K, 101.3 kPa by adsorbing it on activated carbon in a two-stage cross flow system. The initial partial pressure of acetone in the gas is 15 kPa and the final partial pressure is to be 500 Pa. The equilibrium adsorption data is given below:

kg acetone/kg carbon	0	0.1	0.2	0.3	0.35
Partial pr, Pa	0	267	1600	5600	12300

If equal amounts of acetone are used in each stage, determine the amount of carbon required. (**Answer:** 3.32 kg)

10.21 Water content in an air stream is to be reduced from initial value 0.01 kg water/kg dry air to the final value 0.001 kg water/kg dry air by adsorbing the moisture on silica gel. The operation is carried out in a multistage countercurrent unit operated isothermally. The air is admitted at rate of 2000 kg dry air per hour and dry silica gel at a rate of 100 kg/hour. The adsorption isotherm is represented by $Y = 0.04X$, where Y is kg of water vapour per kg of dry air and X is kg of water vapour per kg of dry silica gel. Determine

(a) The water content of the silica gel leaving the first stage

(b) The number of ideal adsorption stages, graphically and using Kremser equation.

(**Answer:** (a) 0.18 kg water/ kg dry gel (b) 4.6)

10.22 A tank of volume 0.1 m^3 contains air-CCl_4 mixture at 300 K and 101.3 kPa with a relative saturation of 30% for CC_4. Activated carbon is placed in the tank to adsorb CCl_4 so that the mole fraction of CCl_4 in the final mixture is 0.001.The temperature and pressure are kept constant during adsorption. The adsorption equilibrium is represented by the Langmuir isotherm

$$X = \frac{5.72 \times 10^{-4}\, p}{1 + 7.2 \times 10^{-4}\, p}$$

where X is the kg adsorbed per kg activated carbon and p is the partial pressure of CCl_4 in Pa. The vapour pressure of CCl_4 at 300 K is 16.65 kPa. Calculate the minimum amount of activated carbon needed. (**Answer:** 0.5591 kg)

10.23 A batch of water containing 15 ppm residual chlorine is to be treated with activated carbon to reduce the chlorine concentration to 0.5 ppm. The equilibrium distribution coefficient is = C/X = 0.80 where C is in kg Cl_2/m^3 liquid and X is the adsorbate loading in kg Cl_2/kg carbon. Calculate the minimum mass of carbon/unit volume of water which can be used.

(**Answer:** 23.2 kg/m^3)

10.24 Phenol present in an aqueous solution is to be removed by adsorption with activated carbon. The initial concentration of phenol in the solution is 0.25 kg per m^3. 4 kg activated carbon is added to 3 m^3 of the solution and mixed thoroughly till equilibrium is attained. Adsorption equilibrium is represented by Freundlich's isotherm $q = 0.20C^{0.23}$ where q is adsorbate loading, kg phenol/kg carbon and C is the concentration, kg phenol/m^3 of solution. Calculate

(a) The equilibrium concentration of the solution and the corresponding adsorbate loading

(b) The percent of phenol extracted.

(**Answer:** (a) $C = 0.0949$ kg/m^3, $q = 0.1164$ kg/kg carbon (b) 62.05%)

10.25 Phenol present in an aqueous solution is to be removed by adsorption with activated carbon. The initial concentration of phenol in the solution is 0.25 kg per m^3. 4 kg activated carbon is added to 3 m^3 of the solution and mixed thoroughly till equilibrium is attained. The following data were obtained on the equilibrium adsorption of phenol from aqueous solution on to activated carbon.

C, kg/m^3	0.33	0.30	0.20	0.15	0.10	0.05	0.005	0.001
q, kg/kg	0.155	0.152	0.138	0.129	0.118	0.100	0.059	0.041

Calculate

(a) The equilibrium concentration of the solution and the corresponding adsorbate loading

(b) The percent of phenol extracted.

(**Answer:** (a) $C = 0.0947$ kg/m^3, $q = 0.1163$ kg/kg, (b) 62.1%)

10.26 The equilibrium partial pressure of water vapour in contact with a certain silica gel on which water is adsorbed is at 298K, as follows:

Partial pr, Pa	0	266.6	533.3	800.0	1066.6	1333.2	1600	1866.5	2133.2
kg water/100 kg gel	0	4.8	8.6	12.7	17.4	22.6	28.3	34.1	38.6

A vessel of volume 20 m^3 contains moist air with a partial pressure of water 2 kPa at a temperature of 298 K and pressure of 101.3 kPa. One kg of silica gel containing 5 (weight)% adsorbed water is placed in the vessel for adsorbing the water from the air. When equilibrium conditions prevail, what will be the moisture content of the gel and the mass of water adsorbed by the gel?

(**Answer:** 18.65 kg/100 kg gel, 0.1339 kg)

10.27 It is decided to effect decolorization of 1000 kg of an aqueous solution of a valuable solute by adsorbing the impurities causing the colour on activated carbon. The initial colour intensity of the solution was 11.1 units and it is required that the solution after adsorption shall have its colour intensity reduced to 10% of the original value. The equilibrium adsorption behavior was studied by adding different amounts of activated carbon to one kg batches of original solution and measuring the colour intensity at equilibrium. The results are shown below.

kg carbon added	0	0.001	0.0025	0.005	0.01	0.02	0.03	0.04	0.05
Colour intensity at equilibrium	11.1	10.0	8.6	6.8	4.4	2.2	1.3	0.88	0.65

Determine:

(a) The quantity of carbon required for a single stage operation

(b) The minimum total quantity of carbon required for a two-stage crosscurrent operation

(**Answer:** (a) 33.5 kg, (b) 21.0 kg)

10.28 It is decided to effect decolorization of 1000 kg of an aqueous solution of a valuable solute by adsorbing the impurities causing the colour on activated carbon in a two-stage countercurrent equipment. The initial colour intensity of the solution was 11.1 units and it is required that the solution after adsorption shall have its colour intensity reduced to 10% of the original value. The equilibrium adsorption behaviour was studied by adding different amounts of activated carbon to one kg batches of original solution and measuring the colour intensity at equilibrium. The results are shown below. Determine the quantity of activated carbon required.

kg carbon added	0	0.001	0.0025	0.005	0.01	0.02	0.03	0.04	0.05
Colour intensity at equilibrium	11.1	10.0	8.6	6.8	4.4	2.2	1.3	0.88	0.65

(**Answer:** 13.6 kg)

10.29 It is decided to effect decolorization of 1000 kg of an aqueous solution of a valuable solute by adsorbing the impurities causing the colour on activated carbon. The initial colour intensity of the solution was 11.1 units and it is required that the solution after adsorption shall have its colour intensity reduced to 10% of the original value. The equilibrium adsorption behavior is represented by the Freundlich isotherm $Y = 8.62 \times 10^{-5} X^{1.666}$, where X and Y are units of colour/kg adsorbent and units of colour/kg solution, respectively. Determine

(a) The quantity of carbon required for a single stage operation

(b) The minimum total quantity of carbon required for a two-stage crosscurrent operation

(c) Two-stage countercurrent unit (**Answer:** (a) 34.1 kg (b) 21.1 kg (c) 13.68 kg)

10.30 Air at 303 K and 101.3 kPa with a relative humidity of 40% is contained in a vessel of volume 0.01 m^3. 0.025 kg activated carbon is introduced into the vessel and the vessel is sealed. The contents are allowed to attain equilibrium at constant temperature of 303 K. The following data refers to the equilibrium adsorption of acetone vapour on an activated carbon at 303 K:

kg-acetone adsorbed/kg carbon	0	0.05	0.10	0.15	0.20	0.25	0.30	0.35
Partial pressure of acetone, kPa	0	0.154	0.444	1.01	2.04	3.73	6.66	12.24

The vapour pressure of acetone at 303 K is 37.7 kPa. Determine

(a) The mole% of acetone vapour in the air in the vessel after adsorption
(b) The final pressure of the gas in the vessel.

(**Answer:** (a) 0.78% (b) 86.22 kPa)

10.31 One thousand kg of sugar solution containing 50% sucrose by weight is to be decolorized by adsorption using activated carbon. The initial colour intensity of the solution is measured to be 25 units on an arbitrary scale and the colour intensity is to be reduced to 4% of the original value. The data for equilibrium adsorption isotherm is established by adding various small amounts of activated carbon to separate batches of sugar solution and observing the colour intensity of the solution at equilibrium. The data with the quantity of carbon expressed on the basis of sugar content of the solution are given below:

kg carbon/kg dry sugar	0	0.007	0.013	0.023	0.032	0.049	0.10
Colour removal, %	0	56	77	90	95	98	99.5

Determine
(a) The quantity of activated carbon required for a single-stage operation.
(b) The minimum total quantity of activated carbon required for a two-stage crosscurrent operation.
(c) The quantity of carbon required for a two-stage countercurrent operation.

(**Answer:** (a) 18.1 kg (b) 10 kg (c) 6.5 kg)

10.32 A continuous countercurrent moving bed adsorber is operated at constant temperature and pressure for dehumidification of air at 70% relative humidity to 10% relative humidity. Air is admitted at a rate of 3000 kg/h (dry basis) at 300 K, 101.3 kPa. The adsorbent is dry silica gel and the adsorption isotherm is given by $q = 30C$ with concentration in kg water/kg dry air and loading in kg water/kg dry silica gel. If the adsorbent used is 1.5 times the minimum, determine the number of equilibrium stages required. Take the vapour pressure of water at 300 K to be 3.56 kPa.

(**Answer:** 3.4)

10.33 Vapours of methylethyl ketone (MEK) are recovered from an air stream by adsorption of the vapours on activated carbon in a fixed bed adsorber at 298 K and 101.3 kPa. The air flow is 5 standard m^3/s, and the air has 0.60 kg MEK/100 standard m^3. If an adsorption cycle of at least 6 h is desired, what should be the volume of the bed? Assume the bulk density of the carbon is 480 kg/m^3 and the adsorbate loading in equilibrium with the concentration of MEK in the inlet air is 0.2 kg/kg carbon.

(**Answer:** 6.75 m^3)

10.34 A fixed bed adsorber is used to reduce the nitrobenzene content in an aqueous solution from the initial concentration 0.02 kg/m^3 to a final concentration not exceeding 2.5×10^{-5} kg/m^3. The solution is admitted at a rate of 100 m^3/h. The column is 2.0 m long and 0.5 m in diameter and activated carbon is used as the adsorbent. The porosity of the bed is 0.6 and particle density is 850 kg/m^3. The adsorption equilibrium is represented by the following relation

$$q = \frac{510C}{1 + 4580C}$$

where C is in kg/m^3 of solution and q in kg/kg of adsorbent.

Determine:
(a) The amount of carbon required for treating 1 m^3 of solution.
(b) The volume of solution treated till breakthrough and breakpoint time.
(c) Compare the quantity of carbon in part (a) with the quantity that would be required if the adsorption were carried out in a stirred tank.

(**Answer:** (a) 0.1813 kg (b) 735.68 m^3, 7.4 h (c) 1.75 kg)

10.35 The vapours of ethyl acetate is to be recovered from an air stream containing 0.05 kg acetate/m^3 of gas at 101.3 kPa and 300K using activated carbon adsorbent having an average particle diameter of 3 mm. The apparent density of individual particles is 720 kg/m^3 and apparent density of the packed bed is 500 kg/m^3. It is desired that the plant should recover 400 kg/h vapour with an adsorption cycle set 3 h to allow sufficient time for regeneration. The activated carbon is capable of adsorbing 0.48 kg vapour per kg carbon. The viscosity of air may be taken as 1.85×10^{-2} cP. Determine:

(a) The amount of carbon required

(b) Dimensions of the bed

(c) Pressure drop using the Ergun equation

$$\frac{\Delta P}{Z} \times \frac{\varepsilon^3 d_p \rho_g}{(1-\varepsilon)G^2} = \frac{150(1-\varepsilon)}{\text{Re}} + 1.75$$

where G is the superficial mass velocity of the gas, *is* the porosity of the bed, ρ_g is the density of the gas stream, d_p is the average particle diameter.

(**Answer:** (a) 5000 kg (b) 10 m^2 area, 0.5 m deep bed (c) 991.4 Pa)

10.36 An adsorption column packed with silica gel is used for removing water vapour from air. Air at 300 K and 101.3 kPa with a RH of 30% is passed at a rate of 2×10^{-3} m^3/min until the silica gel is saturated, at which point the flow is stopped and the silica gel is replaced. The equilibrium adsorptivity of water on silica gel is given by

$$X^* = 0.125\frac{\overline{p}}{P^S}$$

where $\overline{p}$ is the partial pressure of water in the gas contacting the silica gel and P^S is the vapour pressure of water at the system temperature and X^* is the kg water adsorbed at equilibrium per kg of gel. Calculate the minimum amount of silica gel needed in the column if replacement is to take place no more frequently than every two hours. Vapour pressure of water at 300 K is 3.564 kPa. (**Answer:** 0.0494 kg)

10.37 It is desired to dry 1000 m^3/h of air at 200 kPa and 303 K from an initial 80% RH moisture content to almost dry air by adsorption in a fixed bed column. The shape of the water content breakthrough curve is such that 75% of the bed can come to equilibrium with the inlet water-vapour content before regeneration is required and the remaining 25% of the bed which constitutes the mass transfer zone can take up an average water loading corresponding to half the moisture loading in equilibrium with the inlet air. If the time of adsorption between regeneration is to be 3.0 h, calculate the weight of activated alumina required in the bed. Take the vapour pressure of water at 303 K to be 4.24 kPa. Assume that the adsorbate loading in equilibrium with 80% relative humidity air is 0.28 kg water/kg alumina. (**Answer:** 296.78 kg)

10.38 The adsorption equilibrium for benzene from a gas mixture on activated carbon at 283 K and 101.3 kPa is correlated by $\ln q = -0.0113(\ln P)^2 + 0.20711 \ln p - 1.8402$ where q = kg of benzene adsorbed/kg carbon, p = partial pressure Pa. A gas mixture with a flow rate of 0.15 m^3/s containing 0.2 kg/m^3 of benzene, measured at 283 K and 101.3 kPa is admitted to a fixed bed column to give a gas with a benzene concentration of less than 5 mg/m^3. Assume a cycle time of 2 h and actual volume of the adsorber to be two times the equilibrium volume. The bulk density of activated carbon is 450 kg/m^3. Assume ideal breakthrough curve. Determine

(a) The percent of benzene that must be recovered from the bed by regeneration if the bed goes to saturation before regeneration and the concentration of benzene in the exit gas remains at less than 5 mg/m^3 until breakthrough.

(b) The mass of activated carbon required. (**Answer:** (a) 73.9% (b) 1429 kg)

10.39 Trichloroethane (TCE) is removed from an aqueous solution by adsorption in a fixed bed adsorber. The following data refers to the equilibrium adsorption isotherm for the system.

$C \times 10^3$, kg/m^3	0.012	0.032	0.07	0.18	1.2	3
$q \times 10^3$, kg TCE/kg adsorbent	30	50	70	100	200	250

The solution containing 1.5 ppm TCE is admitted to the adsorber at a rate of 10.0 m^3/m^2 h. The total bed length is 0.6 m and the length of unused bed is 0.2 m. The bulk density of the adsorbent is 500 kg/m^3. On regeneration using steam, 90% of the TCE is removed from the adsorbent.

(a) Determine the breakthrough time.

(b) How many bed volumes of liquid can be treated before breakthrough?

(**Answer:** (a) 2.496×10^3h (b) 4.16×10^4)

10.40 Alcohol vapours were adsorbed from an air stream containing 700 ppm alcohol in a 200-mm long adsorption column packed with 0.1 kg activated carbon. The air is admitted at a rate 1×10^{-3} kg/s. The following data refers to the concentrations of the breakthrough curve. The breakpoint concentration may be assumed to be $C/C_0 = 0.01$.

Time, h	2	3	3.5	4	4.5	5	5.5	6	6.25	6.5	6.75
C/C_0	0	0	0.006	0.046	0.174	0.397	0.654	0.885	0.939	0.970	0.990

Determine:

(a) The breakpoint time

(b) The fraction of the total capacity used up to the breakpoint

(c) Saturation loading capacity of carbon

(d) The length of a new column which will give a breakpoint time of 8 hours, for adsorbing the same gas stream as above using the same adsorbent.

(**Answer:** (a) 3.57 h (b) 0.6946 (c) 0.1295 (d) 372.4 mm)

10.41 Alcohol vapours were adsorbed from an air stream containing 700 ppm alcohol in a 50 mm diameter, 200-mm long adsorption column packed with 0.1 kg activated carbon. The air is admitted at a rate 1×10^{-3} kg/s. The following data refers to the concentrations of the breakthrough curve. The breakpoint concentration may be assumed to be $C/C_0 = 0.01$.

Time, h	2	3	3.5	4	4.5	5	5.5	6	6.25	6.5	6.75
C/C_0	0	0	0.006	0.046	0.174	0.397	0.654	0.885	0.939	0.970	0.990

Determine the dimensions of a new column which will give a breakpoint time of 8 hours, for adsorbing the same gas stream, but admitted with a flow rate 2.5×10^{-3} kg/s using the same adsorbent. (**Answer:** Length 372 mm, Diameter = 7.9 mm)

10.42 The following equilibrium data have been obtained for the adsorption of nitrogen dioxide on silica gel at 298 K and pressure of 101.3 kPa.

$\bar{p}_{NO_2}$, Pa	0	266.6	533.3	799.9	1066.6	1333.2	1599.9
X^*(kg NO_2/ kg silica gel)	0	0.004	0.009	0.0165	0.026	0.0365	0.0485

Air stream containing 1.0 mol% NO_2 enters an adsorption column at a rate of 10 kg/h. The column diameter is 0.12 m and is packed with silica gel to a height 1.0 m. The bed has a bulk density of 800 kg/m^3. The pressure and temperature are maintained at 101.3 kPa and 298 K. Assuming an ideal breakthrough curve, determine the breakpoint time. (**Answer:** 1.38 h)

10.43 A molecular sieve adsorbent used for air separation has a maximum capacity for N_2 of 0.046 kg/kg at 303 K and the data be fitted to a Langmuir isotherm with $K = 4.94 \times 10^{-3}$ (kPa)$^{-1}$. (a) What is the separation capacity when the adsorbent is exposed to air at 1000 kPa and 303 K. (b) Calculate the ideal adsorption time for a bed 1.2 m deep operating with a superficial velocity of 0.1 m/s. The bulk density of the adsorbent is 700 kg/m^3.

(**Answer:** (a) 0.0366 kg/kg (b) 0.01 h)

10.44 Nitrogen gas containing 800 ppm water vapour is admitted to an adsorption column at a mass velocity of 4000 kg/m^2 h. Molecular sieves were used as the adsorbent. The packed height of the column was 0.3 m and the bulk density of the packing was 725 kg/m^3. The solids initially contained 0.01 kg water/kg solid. The breakthrough data are as follows.

t, h	0	9	9.5	10.0	10.5	11.0	11.5	12.0	12.5	12.8
C, ppm N_2	0	0	17	88	245	444	614	723	780	799

A value of C/C_0 =0.01 is desired at the breakpoint. Determine:
(a) The breakpoint time
(b) The length of the unused bed
(c) The saturation loading capacity of the solid.
(d) The breakpoint time and fraction of total capacity used for a proposed column length 0.5 m.

(**Answer:** (a) 9.34 h (b) 0.0435 m (c) 0.17 (d) 16.62 h, 0.913)

10.45 Adsorption of an organic vapour (molecular weight = 74) from air is studied using a column of diameter 0.10 m packed with activated carbon to a height of 0.15 m. The bulk density of the packing is 500 kg/m^3. The initial concentration for the vapour in the air is 400 ppm by volume and is passed at 298 K and 100 kPa through the column at 0.6 m/s. The following data were obtained for the effluent concentration from the column.

t, h	1	1.9	2.4	2.9	3.3	3.7	4.2	4.9	5.9	6.1	6.4	6.9	7.6	9.2	9.4
C/C_0	0.0025	0.005	0.01	0.027	0.05	0.1	0.168	0.29	0.56	0.6	0.7	0.8	0.9	0.99	1.00

Assuming the breakpoint occurs at C/C_0 = 0.05, determine the following:
(a) The breakpoint time
(b) The length of unused bed
(c) Saturation loading capacity of the column
(d) The breakpoint time for a column of length 0.30 m.

(**Answer:** (a) 3.3 h (b) 0.064 m (c) 0.196 kg/kg carbon (d) 8.97 h)

10.46 Water vapour is removed from nitrogen gas by adsorption using molecular sieves. Nitrogen gas containing 2000 ppm water is admitted to a fixed bed column of length 0.5 m at rate of 2000 kg/m^2 h. The bulk density of the adsorbent bed is 700 kg/m^3. The effluent concentrations were measured and are given as function of time in the table below:

t, h	C, ppm	t, h	C, ppm
15	0	17.3	1600
15.7	60	17.4	1700
15.8	100	17.6	1800
16.0	200	17.8	1900
16.2	400	18.0	1960
16.6	800	18.5	2000
16.9	1200		

Assuming a breakpoint concentration of 0.05, calculate the saturation capacity and length of unused bed. (**Answer:** 0.1915 kg water/kg adsorbent, 0.029 m)

10.47 A fixed bed adsorber using molecular sieves as adsorbent is used to remove water vapour from nitrogen gas which enters the column at 4000 kg/m^2 h. The gas contains 900 ppm water vapour (by weight). The equilibrium adsorbate loading corresponding to this concentration of water vapour is 0.19 kg/kg solid. The adsorbent initially contains 0.01 kg water vapour adsorbed on it per kg solids, and the bulk density of the bed is 700 kg/m^3. The concentrations on the breakthrough curve obtained from a small column of length 0.3 m is given below:

C, ppm	Time, h	C, ppm	Time, h	C, ppm	Time, h
0	8.0	89	10.0	613	11.25
1	9.0	149	10.2	697	11.5
3	9.2	228	10.4	772	11.75
6	9.4	311	10.6	831	12.0
21	9.6	406	10.8	881	12.5
50	9.8	505	11.0	900	12.8

If an adsorber with a breakthrough time of 20 h is to be designed to obtain an exiting gas with no more than 6 ppm of water vapour, what height of column is required? **(Answer:** 0.61 m)

10.48 Ethyl acetate (EA) is adsorbed from an air stream using activated carbon at 310 K and 100.0 kPa in a fixed bed adsorber. Air at a flow rate of 700 m^3/h at 310 K and 100.0 kPa and containing 0.5 mol% ethyl acetate is to be treated. The bulk density of the adsorbent is 500 kg/m^3. With a superficial gas velocity of 0.3 m/s, a breakthrough time of 8-h is desired. Equilibrium adsorbate loading corresponding to EA partial pressure of 0.5 kPa in air is 0.164 kg EA/kg carbon. Determine the diameter and height of the bed. The following breakthrough data are obtained for adsorption of ethyl acetate in air on activated carbon for a bed of length 0.6 m at a superficial velocity of 0.3 m/s at 310 K and 100.0 kPa.

Partial pressure of EA in effluent, kPa	0.005	0.010	0.025	0.050	0.100	0.250	0.475	0.483	0.495	0.50
Time, h	1.0	1.1	1.25	1.4	1.58	2.0	2.67	2.70	2.80	2.85

(Answer: $L = 2.1$ m, $D = 0.91$ m)

10.49 Humid air ($Y = 0.02$ kg water/kg dry air) is to be dried completely using an ideal adsorbent that instantaneously removes the water vapour from the air. The equilibrium relationship is $X=10Y$. The adsorbent has a density of 400 kg/m^3 and a porosity of 0.3. For an inlet air flow of 10 kg/s, with a bed 3 m long and 1 m^2 cross-sectional area, calculate:

(a) The rate at which the adsorbent front moves through the bed.

(b) The time at which the bed is completely saturated.

(c) The total mass of air treated when the bed needs regeneration.

(Answer: (a) 12.86 m/h (b) 0.233 h (c) 8400 kg)

10.50 Hexane is adsorbed from natural gas which enters a fixed bed adsorber with a superficial velocity of 0.06 m/s at 308 K and 5600 kPa. The concentration of hexane in the gas is 1.6 kg/m^3. Silica gel is used as the adsorbent. The bed is 1 m deep and the bed density is 850 kg/m^3. The equilibrium adsorbate loading corresponding to hexane concentration of 1.6 kg/m^3 is 0.17 kg/kg gel. The breakthrough curve data are as follows:

C/C_0	0.01	0.02	0.08	0.25	0.47	0.69	0.84	0.94	0.98	0.99
t, h	0.293	0.30	0.35	0.40	0.45	0.50	0.55	0.60	0.65	0.671

Determine:

(a) The total amount of hexane adsorbed to the breakpoint

(b) Saturation amount for the whole bed

(c) Percent utilization of bed

(d) Percent utilization of the bed if the adsorption is carried out in a bed of length 3 m.

(Answer: (a) 101.26 kg (b) 144.5 kg (c) 70.08% (d) 99.1%)

10.51 A nitrogen-benzene vapour mixture containing 1.0% benzene by volume is to be passed through a continuous countercurrent moving bed adsorber at the rate of 200 m^3/h. Activated charcoal containing 0.015 m^3 benzene vapour at STP adsorbed per kg charcoal is used as the adsorbent. It is desired to remove 95% of the benzene from the gas. The temperature and total pressure are to be maintained at 306 K and 101.3 kPa. The equilibrium data on the adsorption of benzene vapour on activated charcoal at 306 K is given below:

Benzene vapour adsorbed, m^3 (STP)/kg charcoal $\times 10^3$	15	25	40	50	65	80	90	100
Partial pr, Pa	0.13	0.6	3.4	15.3	33.5	133.3	374.6	1042.6

Determine:

(a) The least amount of charcoal which can be used per hour

(b) For twice the minimum rate of charcoal, the concentration of benzene in the charcoal leaving the adsorber in m^3/kg

(c) The number of overall gas transfer units.

(**Answer:** (a) 22.41 kg/h (b) 0.057 m^3/kg (c) 3.44)

10.52 Air is to be dried in a continuous countercurrent isothermal absorber operated at 300 K, 101.3 kPa using silica gel as adsorbent. Air enters the adsorber at a rate of 3.0 kg/m^2 s with a humidity of 0.005 kg water/kg dry air. The moisture content of the air leaving the adsorber is 0.0001 kg water/kg dry air. The silica gel enters at a rate of 1.5 kg/m^2 s free of water. The equilibrium adsorption isotherm is described by the equation $Y^* = 0.0185\,X$. Individual resistances to mass transfer residing in the fluid and within the solid during adsorption of water vapour from air by silica gel are found to be $k_Y a_p = 40.0$ kg water/m^3 s ΔY and $k_s a_p = 0.98$ kg water/m^3 s kg water/m^3 s ΔX. Estimate the height of the adsorber. (**Answer:** 0.53 m)

10.53 Air contaminated with NO_2 is passed through a continuous countercurrent adsorber at the rate of 500 kg/h at 298 K and 101.3 kPa. The air contains 1.5% NO_2 by volume and 90% of which is to be removed by adsorption using silica gel which enters the column free of NO_2. For an air superficial velocity of 1500 kg/m^2 h, the overall gas mass transfer coefficient is expected to be 2500 kg/m^3 h. The equilibrium adsorption isotherm is given by the following data:

Partial pressure of NO_2, kPa.	0	0.267	0.533	0.800	1.066	1.333	1.600
kg NO_2/100 kg gel	0	0.4	0.9	1.65	2.60	3.65	4.85

(a) Calculate the minimum weight of gel required per hour.

(b) For twice the minimum rate, calculate the number of transfer units required.

(c) Estimate the height of the adsorber.

(**Answer:** (a) 335 kg/h (b) 4.54 (c) 2.72 m)

Chapter 11

Crystallisation

Crystallisation is a separation technique in which a homogeneous fluid undergoes phase change resulting in the formation of solid crystals. Crystallisation can be treated as one of the oldest known separation operations. Recovery of salt crystals from sea water was practiced by the mankind for centuries. Sucrose which has now evolved as the crystalline product with the largest world production, was also produced long back. Since a variety of materials are marketed and used in crystalline form, crystallisation has become an important industrial operation. It now finds applications in the production of a wide range of materials; from bulk commodity chemicals to speciality chemicals and pharmaceuticals. More than 80% of the substances used in pharmaceuticals, fine chemicals, agrochemicals, food and cosmetics are isolated or formulated in their solid form and a majority of them by crystallisation from solutions. Crystallisation is an important operation in the freeze drying of fruit juices, desalination of sea water, biotechnological applications such as processing of proteins, etc.

Crystallisation process includes formation of solid particles from a vapour, as in the formation of snow particles, solidification of liquid melt as in the production of single large crystals, and formation of solid crystals from liquid solutions. In the last process, i.e. in *crystallisation from solutions*, crystals are formed when the solute concentration of a solution is brought above the saturation value at a given temperature, by evaporation of the solvent, by cooling the solution or by a combination of cooling and evaporation. It is a solid–liquid separation technique involving mass transfer from a liquid solution to a pure solid crystalline phase. This chapter is concerned with crystallisation from the solutions as it is the commercially most important among the crystallisation operations. The yield and purity of the products, the energy required for cooling, evaporation, etc. shape and size of the crystals, uniformity or distribution of the size of the crystals and the rate of production of the crystals are areas of great concern in the design and operation of processes for commercial production by the crystallisation from the solutions.

The most attractive feature of crystallisation is that the process yields a highly purified form of the solid from a relatively impure solution in a single processing step. Crystallisation is carried out at low temperatures with low energy consumption. Therefore, in terms of energy requirement, it is superior technique for the separation of components from a solution compared to the distillation and other commonly used methods for purification. The operation can be carried out on a scale ranging from production of a few grams in the laboratories to thousands of kilograms per day industrially. Crystallisation generally gives a product in the most favourable form suitable for handling and storage. Crystalline salts are handled more conveniently and economically than solutions. Furthermore, by the judicious choice of the crystallisation conditions, the macroscopic properties of crystals such as the shape and the size can be easily manipulated. This makes crystallisation a very attractive technique when a reproducible product quality with well defined properties is of utmost importance.

11.1 CRYSTALS AND CRYSTAL GEOMETRY

Crystals are solids composed of atoms, molecules or ions arranged in orderly, repetitive arrays called three dimensional space lattices. When crystals grow from the solutions unhindered by the presence of other materials, they form polyhedron with flat faces and sharp corners; crystals are never spherical in shape. Crystals of a given substance may appear quite different in size and shape. Even though the crystals vary in size and in the relative development of faces, the angles between the corresponding faces of all crystals of a given substance are constant and are characteristic of that material. Based on the crystal interfacial angles (α, β and γ) and the symmetry of the mutually perpendicular axes with respect to their relative lengths (a, b and c), crystals can be classified into seven *crystallographic systems*. They are:

1. Cubic system: Three equal axes at right angles to each other $(a = b = c)$, $(\alpha = \beta = \gamma = 90^\circ)$.
2. Tetragonal systems: Three axes at right angles to each other, one axis longer than the other two $(a = b < c)$, $(\alpha = \beta = \gamma = 90^\circ)$.
3. Orthorhombic system: Three axes at right angles to each other, all of different lengths $(a \neq b \neq c)$, $(\alpha = \beta = \gamma = 90^\circ)$.
4. Hexagonal system: Three equal axes in one plane at 60° to each other and a fourth axis at right angles to this plane and not necessarily the same length $(a = b \neq c)$, $(\alpha = \beta = 90^\circ, \gamma = 120^\circ)$.
5. Monoclinic system: Three unequal axes, two at right angles in a plane and a third at some angles to this plane $(a \neq b \neq c)$, $(\alpha = \beta = 90^\circ, \gamma \neq 90^\circ)$.
6. Triclinic system: Three unequal axes at unequal angles to each other and not 30, 60 or 90°. $(a \neq b \neq c)$, $(\alpha \neq \beta \neq \gamma \neq 90^\circ)$.
7. Trigonal (Rhombohedral) system: Three equal and equally inclined axes. $(a = b = c)$, $(\alpha = \beta = \gamma \neq 90^\circ)$.

The crystallographic systems are illustrated in Fig. 11.1.

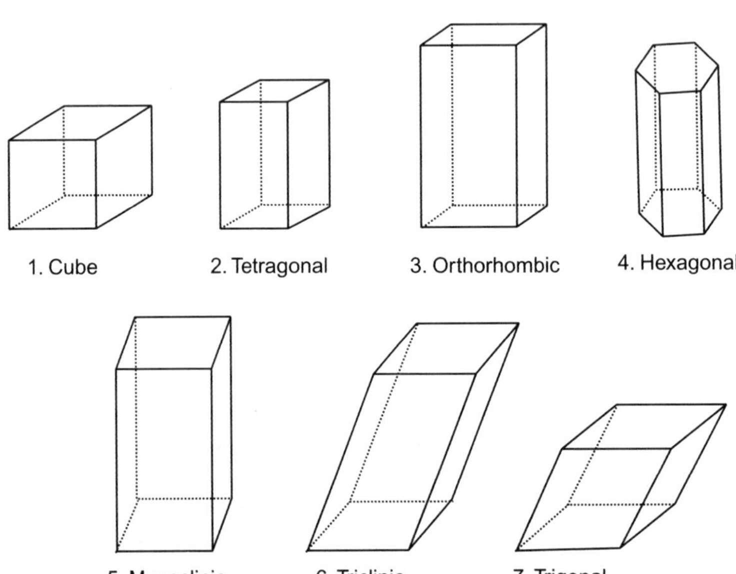

Fig. 11.1: Crystallographic systems

The term *crystal habit* is used to denote the relative development of the different types of faces. For example, sodium chloride crystallises from the aqueous solutions with cubic faces only. Crystals of sodium chloride from a solution containing small amount of urea will have octahedral faces. Both crystals belong to the cubic system, but they differ in habit. The difference in the appearance of the crystals of a given substance and a given crystal system is generally due to the different growth rates of crystals. For example, crystals of potassium sulphate which belong to the orthorhombic crystal system, may belong to the crystal habit that can be characterised as needles, plates or prisms depending upon whether the relative growth of the faces is stunted or exaggerated in the particular directions.

11.2 EQUILIBRIUM SOLUBILITY

Equilibrium relationships in crystallisation are expressed in the form of solubility data which can be plotted as phase diagrams. In crystallisation, the mass is transferred from the solution to the crystal surface. The solubility data provide the information on the driving force for this transfer. The data on the solubility of substances in water are readily available in standard reference books and handbooks. They are ordinarily reported as parts by weight of anhydrous material per 100 parts by weight of the total solvent or as mass fraction of the anhydrous solute. The solubility characteristics provide necessary information on the best method for causing crystallisation and the difficulty of growing the crystals. For example, crystallisation by cooling will be effective only for substances having a solubility that decreases by decreasing the temperature. The solubility diagrams are plots of the solubility of substances as functions of temperature. Figure 11.2a gives the solubility curves of some typical salts in water.

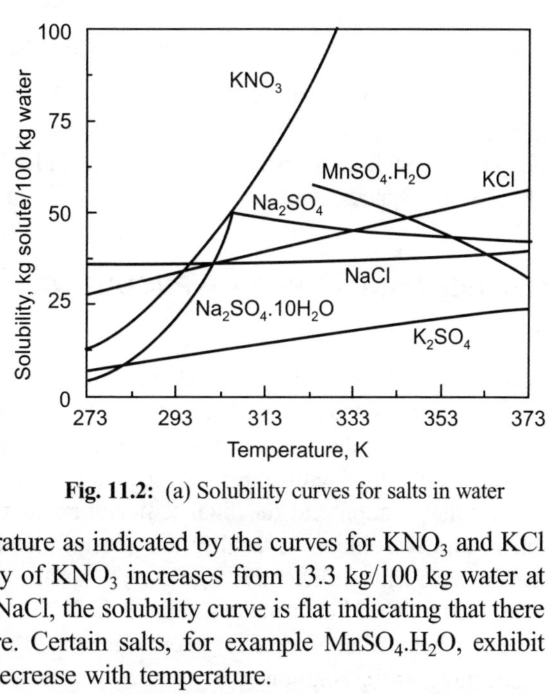

Fig. 11.2: (a) Solubility curves for salts in water

In general, the solubility increases with temperature as indicated by the curves for KNO_3 and KCl which form no hydrates with water. The solubility of KNO_3 increases from 13.3 kg/100 kg water at 273 K to 247 kg/100 kg water at 373 K. But, for NaCl, the solubility curve is flat indicating that there is little change in the solubility with temperature. Certain salts, for example $MnSO_4.H_2O$, exhibit *inverse solubility* meaning that their solubilities decrease with temperature.

Some salts crystallise with water of crystallisation, forming different hydrates depending upon the concentration of the solution and the temperature of crystallisation. The phase equilibria in such systems become quite complicated and the solubility curves will be characterised by several breaks, indicating large changes in the solubility when phase transition from one stable hydrate to another occurs. For example, $Na_2SO_4.10H_2O$ is the stable crystalline form in the temperature range 273 K to 305.6 K and the solubility increases with temperature in this region. However, the anhydrous Na_2SO_4 is the stable form in the temperature range 305.6 K to 373 K. The solubility of the salt decreases from 49.5 kg salt/100 kg water at 305.6 K to 42.5 kg salt/100 kg water at 373 K as shown in Fig. 11.2a.

The solubility of sodium thiosulphate in water increases sharply with temperature as shown in Fig. 11.2b. The solubility curve shows several breaks indicating that there exist several stable crystalline forms for the salt depending upon the temperature and the composition. Below the solubility curve only a single liquid phase (solution) exist and above the line the system at equilibrium consists of saturated solution and one or other crystalline form of sodium thiosulphate depending upon the temperature.

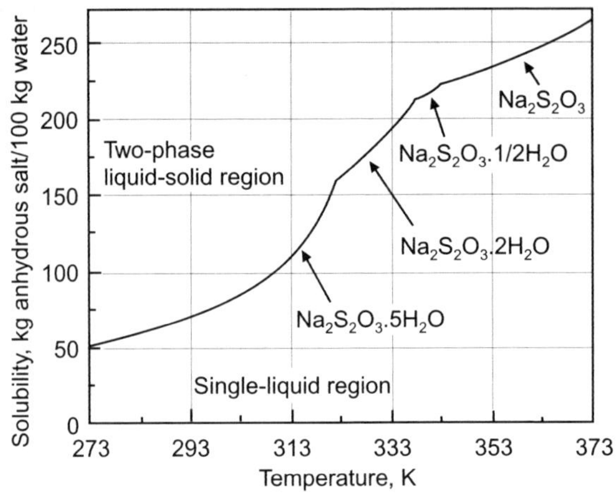

Fig. 11.2: (b) Solubility of sodium thiosulphate in water

The pentahydrate is the stable form up to about 321.5 K. Thus $Na_2S_2O_3.5H_2O$ crystals are formed when an aqueous solution is concentrated at temperatures below 321.5 K. Between 321.5 K and 338.2 K, the stable crystalline form is $Na_2S_2O_3.2H_2O$ and a half-hydrate ($Na_2S_2O_3.1/2\ H_2O$) is formed above 338.2 K and below 343.2 K. Anhydrous salt is the stable crystalline form for temperatures above 343.2 K.

11.3 YIELD AND PURITY OF PRODUCTS

Yield

Good yield and high purity of products are the two important objectives in crystallisation operations. Yield may be defined as the weight of the substance crystallised per unit weight of the substance in the feed solution. On the presumption that at the end of the crystallisation process, the mother liquor is saturated at the final temperature of the solution, yield may be estimated from the solubility data if the initial concentration and the final temperature of the solution are known. Though initiated from a supersaturated solution, crystallisation generally takes place very slowly; therefore, the mother liquor is in contact with sufficiently large crystal surfaces for a sufficiently long time so that equilibrium is usually established between the solid and the liquid phases. The concentration of the solution at the end of the crystallisation process can therefore be read from the solubility curve corresponding to the final temperature of the solution.

The calculation of yield when the crystals formed are anhydrous is quite simple and straightforward as the solid phase contains no solvent. Yield in this case is equal to the difference between the concentration of the feed solution and the solubility at the final temperature of the solution, both compositions expressed per unit weight of the solvent. Allowance should be given to the quantity of solvent lost due to evaporation. When hydrated crystals are formed, water of crystallisation accompanying the solids must be taken into account, as this water will not be available to dissolve the solids and form saturated solution. In this case, the yield of the hydrated crystals is readily calculated by expressing all weights and concentrations in terms of the hydrated salt and the free water. After taking into account the loss due to evaporation, the mass of free water remains constant during the crystallisation, and can be treated as the key component in the material balance calculations.

The following formula can also be used to calculate the yield.

$$C = R\frac{100F_0 - s(Y_0 - W)}{100 - s(R-1)} \qquad \text{...(11.1)}$$

where

C = weight of crystals in the final magma, kg
R = ratio of the molecular weight of hydrate/molecular weight of anhydrous salt
s = solubility at the final temperature, kg anhydrous salt/100 kg solvent
F_0 = weight of anhydrous solute in the original solution, kg
Y_0 = total weight of the solvent in the original solution, kg
W = evaporation, kg

In calculating the yield of crystals, the final concentration of the solution after crystallisation, in some cases, will not be equal to the solubility at the temperature of the solution. For solutions such as sucrose in water, supersaturation is attained easily; these do not attain equilibrium with respect to the final mother liquor temperature. In some solutions, the rate of crystal growth is very slow, so that considerable time will be taken to reach equilibrium. This will be the case, if the solution is highly viscous or if the crystals formed are collected at the bottom of the crystalliser exposing only a small fraction of the crystal surface to the solution. The solution remains supersaturated even after a very long time and the actual yield will be less than that calculated using the solubility data. In such cases, an assumption must be made as to the liquid composition at the final temperature.

Purity of Products

One of the major attractions of crystallisation as a separation technique is that the crystals formed from the solutions are essentially pure. However, when crystals are recovered from the magma, they retain a substantial quantity of the mother liquor along with the impurities present in it. If crystal agglomerates are formed, solution may get occluded within the solid mass. In practice, most of the mother liquor is separated from the crystals by filtration or centrifuging, and the balance is removed by washing with fresh solvent. On drying, the impurities present in the mother liquor still remaining with the crystals lead to contamination of the final product. The purity of the final crystalline product recovered from the magma is thus dependent on the extent to which contaminants are present in the solution accompanying the crystals. The lower the amount of mother liquor retained, the higher will be the purity of the product.

Example 11.1: An aqueous solution of Na_2CO_3 containing 15% carbonate by weight is subjected to crystallisation at the rate of 1000 kg/h by evaporation and cooling to 278 K. 40% of the water in the feed is evaporated. If the solubility of Na_2CO_3 at 278 K is 9.0% (weight), determine:

(a) The quantity of $Na_2CO_3.10H_2O$ crystals formed.
(b) The yield of crystals.

***Solution*:** Basis: 1000 kg/h of 15% solution as feed.

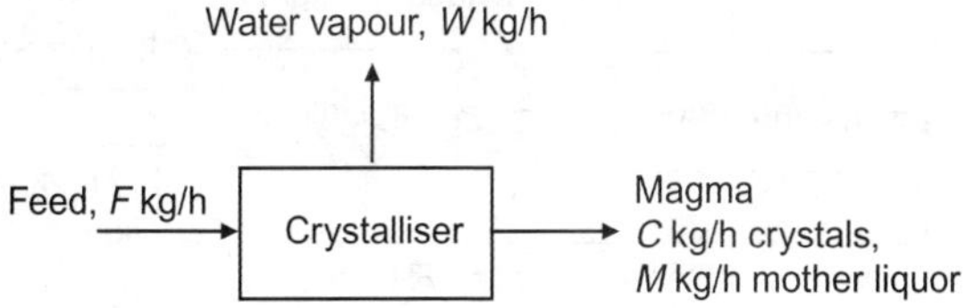

Fig. 11.3: Crystalliser for Example 11.1

Let F kg/h be the rate at which feed is admitted and W be the rate at which water is evaporated from the solution (Fig. 11.3). The magma leaving the crystalliser consists of C kg/h of hydrated crystals and M kg/h of mother liquor. Let x_F be the weight fraction of Na_2CO_3 in the feed, x_C the weight fraction of Na_2CO_3 in the hydrated crystals ($Na_2CO_3.10H_2O$) and x_M be the weight fraction of Na_2CO_3 in the mother liquor.

Here, F = 1000 kg/h, x_F = 0.15. Since 40% of water in the feed is evaporated,

$$W = 0.40 \times F \times (1 - x_F) = 0.4 \times 1000 \times 0.85 = 340 \text{ kg/h}$$

Since the molecular weight of Na_2CO_3 is 106 and that of $Na_2CO_3.10H_2O$ is 286,

$$x_C = \frac{106}{286} = 0.3706$$

Since the solubility of Na_2CO_3 at 278 K is 9.0% (weight), $x_M = 0.09$

(a) Total material balance gives

$$F = M + C + W$$

That is, $$M + C + 340 = 1000$$

Or $$M + C = 660 \quad \text{(A)}$$

Na_2CO_3 balance on the crystalliser gives

$$Fx_F = Mx_M + Cx_C$$

$$1000 \times 0.15 = M \times 0.09 + C \times 0.3706$$

$$0.09M + 0.3706C = 150 \quad \text{(B)}$$

Solving Eqs. (*A*) and (*B*) simultaneously, we get $C = 322.9$ kg/h

The quantity of $Na_2CO_3.10H_2O$ crystals formed = 322.9 kg/h **(Answer)**

(b) Weight of Na_2CO_3 in the hydrated crystals $= Cx_C = 322.9 \times 0.3706 = 119.7 \text{kg/h}$

Yield is defined as the fraction of Na_2CO_3 in the feed that is crystallised. Therefore,

$$\text{Yield} = \frac{Cx_C}{Fx_F} = \frac{119.7}{150} = 0.798 \text{ or } 79.8\% \quad \textbf{(Answer)}$$

Example 11.2: For the preparation of potassium nitrate, 10000 kg/h of a 25% KNO_3 solution is mixed with a recycle stream and sent to an evaporator. The rate of evaporation is 1.5 times the rate of introduction of recycle stream. The concentrated solution leaving the evaporator containing 50% KNO_3 is admitted to the crystalliser which yields crystals containing 5% water. At the crystallisation temperature, the solubility is 60 kg/100 kg water. Major part of the mother liquor leaving the evaporator is returned to the evaporator as recycle. Calculate:

(a) Rate at which the mother liquor is recycled, kg/h.
(b) Concentration of KNO_3 in the stream entering the evaporator.
(c) Rate of production of crystals, kg/h.

***Solution*:**

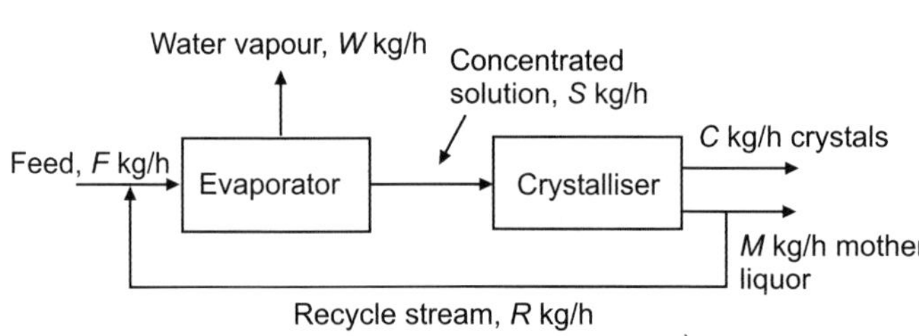

Fig. 11.4: Crystalliser for Example 11.2

Let F be the rate at which the fresh feed is admitted, x_F be the weight fraction of KNO_3 in the feed, R be the rate at which the mother liquor is recycled, x_M the weight fraction of the solute in the mother liquor in the crystalliser, W be the rate of evaporation of water, S the rate at which the concentrated solution leaves the evaporator, x_S the weight fraction of solute in the concentrated solution, C is the kg/h of crystals produced, x_C the weight fraction of KNO_3 in the crystals and M be the mother liquor leaving with the crystals.

Here, $F = 10000$ kg/h, $x_F = 0.25$, $W = 1.5R$, $x_S = 0.50$. Since the crystals contain 5% water, $x_C = 0.95$. Since the solubility of KNO_3 at the crystalliser temperature is 60 kg/100 kg water, weight fraction of KNO_3 in the mother liquor leaving the crystalliser $= x_M = \dfrac{60}{160} = 0.375$

(a) Consider total material balance around the evaporator in Fig. 11.4.

$$F + R = W + S$$

$$10000 + R = 1.5R + S$$

This reduces to the following equation

$$S + 0.5R = 10000 \tag{A}$$

Now consider KNO_3 balance around the evaporator.

$$Fx_F + Rx_M = Sx_S$$

$$10000 \times 0.25 + R \times 0.375 = S \times 0.50$$

which reduces to the following:

$$2500 = 0.50S - 0.375R \tag{B}$$

Solving Eqs. (*A*) and (*B*) we get $S = 8000$ kg/h and $R = 4000$ kg/h. Therefore, $W = 1.5R = 6000$ kg/h

Recycle stream = 4000 kg/h **(Answer)**

(b) Stream entering the evaporator $= F + R = 10000 + 4000 = 14000$ kg/h

KNO_3 in the stream entering the evaporator =

$$Fx_F + Rx_M = 10000 \times 0.25 + 4000 \times 0.375 = 4000 \text{ kg/h}$$

Concentration of the stream entering the evaporator $= \dfrac{4000}{14000} = 0.2857$

Or the stream entering the evaporator contains 28.57% KNO_3. **(Answer)**

(c) A total material balance around the entire system is

$$F = W + C + M$$

$$10000 = 6000 + C + M$$

$$C + M = 4000 \tag{C}$$

KNO_3 balance around the whole system gives

$$Cx_C + Mx_M = Fx_F$$

$$0.95C + 0.375M = 2500 \tag{D}$$

Solving Eqs. (*C*) and (*D*), we get

$$C = 1739.13 \text{ kg}, M = 2260.87 \text{ kg}$$

Rate of production of crystals = 1739.13 kg/h **(Answer)**

11.4 HEAT EFFECTS IN CRYSTALLISATION PROCESS

Estimation of the net changes in the heat and material accompanying the crystallisation process by proper material and heat balances on the unit is essentially the first step in the design of a crystalliser. The heat effects in a crystallisation process may be computed by two methods: (1) A heat balance can be made wherein the individual heat effects such as the sensible heats required for heating or cooling the solution, the heat of crystallisation, and the latent heat of vaporisation can be combined into an equation for the total heat effects. (2) An enthalpy balance can be made in which the total enthalpy of all leaving streams minus the total enthalpy of all entering streams is equal to the heat absorbed from the external sources by the process.

In the first method, the heat effects accompanying the crystallisation are most commonly calculated from the heat of crystallisation and the sensible heat of the solution from which the crystalline material separates. At equilibrium, the heat of crystallisation is equal and opposite in sign to the heat of solution. Heat of solution is the change in enthalpy (ΔH) accompanying the dissolution of one mole of substance in an excess of solvent. For substances, the solubility of which increase with temperature, the heat of solution will be positive as there is absorption of heat when they dissolve. For substances with inverse solubility, heat is evolved during dissolution and the heat of solution will be negative. It is usually the heats of solution at *infinite dilution* that are available in standard references. Table 11.1 gives the heat of solution at infinite dilution at room temperature for stable crystalline forms of some typical inorganic substances. Using the negative of the heat of solution at infinite dilution as the heat of crystallisation without correcting for the heat of dilution, may lead to errors in the heat balance. Fortunately, the heat of dilution in most cases is negligibly small compared to the heat of crystallisation, and therefore, the heat balance calculation ignoring the heat of dilution will introduce no significant error. The assumption that the latent heat of vaporisation of the solvent from a solution is equal to the latent heat of vaporisation of the pure solvent is yet another approximation commonly used in heat balance calculations that will not seriously affect the accuracy of the results.

The advantage of the enthalpy-concentration method is that both the heat and the material changes are taken into account simultaneously. However, the method has only limited application because of the difficulty in getting reliable enthalpy-concentration data. Reliable such data are available in the literature only for a very few systems. If the enthalpy-composition chart is available, it is necessary only to read the enthalpy of the original solution at the initial temperature and the enthalpy of the final magma to evaluate the difference ΔH as the energy effect per unit mass. If evaporation occurs during crystallisation, the enthalpy of the vapour formed can be determined from steam tables, and can be incorporated in the enthalpy balance. ΔH so calculated gives the quantity of heat supplied from external sources.

Table 11.1: Heats of solution at infinite dilution at room temperature

Compound	Stable crystalline form at room temperature	Heat of solution at infinite dilution, kJ/mol
Ammonium chloride	NH_4Cl	+15.99
Ammonium nitrate	NH_4NO_3	+27.09
Ammonium sulphate	$(NH_4)_2SO_4$	+11.51
Barium chloride	$BaCl_2.2H_2O$	+18.84
Calcium chloride	$CaCl_2.\ 6H_2O$	+17.21
Copper sulphate	$CuSO_4.5H_2O$	+11.93
Ferrous sulphate	$FeSO_4.\ 7H_2O$	+18.40
Magnesium chloride	$MgCl_2.6H_2O$	−14.24
Magnesium sulphate	$MgSO_4.7H_2O$	+13.31
Potassium chloride	KCl	+18.44
Potassium nitrate	KNO_3	+36.14
Potassium sulphate	K_2SO_4	+26.46
Sodium carbonate	$Na_2CO_3.10H_2O$	+67.91
Sodium chloride	$NaCl$	+4.87
Sodium nitrate	$NaNO_3$	+21.14
Sodium sulphate	$Na_2SO_4.10H_2O$	+78.46
Sodium thiosulphate	$Na_2S_2O_3.5H_2O$	+47.31

Example 11.3: $FeSO_4.7H_2O$ crystals are produced in a Swenson-Walker crystalliser by cooling a saturated solution at 323 K to 303 K. Solubility at 323 K is 48.6 parts $FeSO_4$ per 100 parts water and that at 303 K is 32.9 parts $FeSO_4$ per 100 parts water. The feed solution has an average specific heat of 2.93 kJ/kg K. Heat of solution of $FeSO_4.7H_2O$ at 303 K is 18.40 kJ/mol. How much heat is evolved during the process per kg of $FeSO_4.7H_2O$ crystals produced?

***Solution*:** For every kg of $FeSO_4.7H_2O$ produced, let F be the rate at which fresh feed is admitted, x_F be the weight fraction of $FeSO_4$ in the feed, M be the mother liquor leaving with the crystals, x_M the weight fraction of the solute in the mother liquor in the crystalliser, x_C the weight fraction of $FeSO_4$ in the crystals. Then the total material balance is

$$F = 1 + M \tag{A}$$

where F is the feed rate. $FeSO_4$ balance gives

$$Fx_F = x_C + Mx_M \tag{B}$$

Here x_C is the weight fraction of anhydrous $FeSO_4$ in the hydrated crystals. The molecular weight of $FeSO_4$ is 151.85 and molecular weight of $FeSO_4.7H_2O$ is 277.85. Therefore,

$$x_C = \frac{151.85}{277.85} = 0.5465$$

Since the feed is saturated at 323 K,

$$x_F = \frac{48.6}{148.6} = 0.3271$$

Since the mother liquor is at 303 K,

$$x_M = \frac{32.9}{132.9} = 0.2476$$

Substituting these values in Eq. (*B*), we get

$$0.3271F = 0.5465 + 0.2476M \tag{C}$$

Solving Eqs. (*A*) and (*C*), we get

$$F = 3.7597 \text{ kg and } M = 2.7597 \text{ kg}$$

The total heat evolved during the process can be taken to be the sum of the heat liberated on cooling the feed from 323 K to 303 K and the heat evolved during crystallisation.

The heat liberated on cooling the feed

$$= F \times C_{PF} \times (T_i - T_o) = 3.7597 \times 2.93 \times (323 - 303) = 220.32 \text{ kJ}$$

Since the heat of crystallisation is the negative of the heat of solution at infinite dilution, the heat of crystallisation $= -18.4$ kJ/mol $= -\dfrac{18.4 \times 1000}{277.85} = -66.22$ kJ/kg. It means that 66.22 kJ heat is evolved when 1 kg $FeSO_4.7H_2O$ crystals are formed.

Total heat evolved during the process = 220.32 + 66.22 = 286.54 kJ **(Answer)**

Example 11.4: A Swenson-Walker crystalliser is used to crystallise $Na_2SO_4.10H_2O$ at the rate of 1000 kg/h by cooling a saturated solution of Na_2SO_4 from 313 K to 290 K. Cooling water enters the unit at 285 K and leaves at 300 K. Evaporation is negligible. The solubilities of anhydrous Na_2SO_4 in water are 49 and 15 kg/100 kg water at 313 K and 290 K, respectively. The mean heat capacity of the liquor is 3.8 kJ/kg K and the heat of crystallisation is –243 kJ/kg. For the crystalliser, the available heat transfer area is 3 m^2/m length and the overall coefficient of heat transfer is 0.20 kW/m^2 K. How many sections of the crystalliser, each 3-m long, will be required?

***Solution*:** Basis: 1000 kg/h of $Na_2SO_4.10H_2O$ crystals produced. Let M be the mother liquor leaving and F be the feed rate.

$$F = 1000 + M \tag{A}$$

Since feed is a saturated solution at 313 K, the weight fraction of Na_2SO_4 in the feed is

$$x_F = \frac{49}{149} = 0.3289$$

The mother liquor leaves at 290 K. Therefore, the weight fraction of Na_2SO_4 in the mother liquor is

$$x_M = \frac{15}{115} = 0.1304$$

Since the molecular weight of Na_2SO_4 is 142 and the molecular weight of $Na_2SO_4.10H_2O$ is 322, the weight fraction of Na_2SO_4 in the hydrated crystals is

$$x_C = \frac{142}{322} = 0.441$$

Na_2SO_4 balance gives $$Fx_F = 1000x_C + Mx_M$$

Or $$0.3289F = 1000 \times 0.441 + 0.1304M$$

$$0.3289F = 441 + 0.1304M \qquad \text{(B)}$$

Solving Eqs. (*A*) and (*B*) simultaneously, we get F = 1564.7 kg/h and M = 564.7 kg/h.

Total heat liberated during the process is equal to the heat liberated on cooling the feed solution from 313 K to 290 K and the heat liberated when crystallisation occurs.

The heat liberated on cooling $= F \times C_{PF} \times (T_i - T_o) = 1564.7 \times 3.8 \times (313 - 290) = 136754.8\,\text{kJ/h}$

The heat liberated on crystallisation $= 243 \times 1000 = 243000$ kJ/h

Total heat liberated, Q = 243000+136754.8 = 379754.8 kJ/h

Assuming that cooling water flows counter-current to the solution, the average temperature difference driving force

$$(\Delta T)_{mean} = \frac{(313-300)-(290-285)}{\ln\dfrac{313-300}{290-285}} = 8.4\,\text{K}$$

Now, $$Q = UA(\Delta T)_{mean}$$

Or $$A = \frac{Q}{U(\Delta T)_{mean}} = \frac{379754.8}{0.2 \times 3600 \times 8.4} = 62.8\,\text{m}^2$$

Since a 3 m long crystalliser unit has an area equal to 9 m^2, the number of units required

$$= \frac{62.8}{9} \approx 7 \text{ sections.}$$ **(Answer)**

11.5 SUPERSATURATION

Mass balance is an essential part of the crystallisation process design and development. Solubility gives the concentration of the solute when the solid phase and the solution are in equilibrium. It is important in mass balance calculations, as it allows calculation of the maximum yield of crystals obtainable when a solution is subjected to saturation. The limitation of mass balance is that it cannot predict the time required to attain the equilibrium yield or the rate at which crystals are formed from the solution. These are dependent on the driving force available for *nucleation* and *growth of crystals*. Nucleation is the process that results in the birth of new crystals. The driving force for nucleation and growth of crystals is provided by the supersaturation of the solution. A solution is said to be supersaturated, if the solute concentration exceeds the saturation (or equilibrium) solute concentration. Supersaturation, ΔC is given by

$$\Delta C = C - C^* \qquad \text{...(11.2)}$$

where C is the actual concentration of solute in the solution and C^* is the concentration of the solute in the saturated solution at the same temperature. The fractional supersaturation S and the supersaturation ratio α are two other terms used to represent the degree of supersaturation. These are defined as

$$\alpha = \frac{C}{C^*} \qquad \text{...(11.3)}$$

$$S = \frac{\Delta C}{C^*} \qquad \text{...(11.4)}$$

There are four main methods to generate supersaturation in a solution. They are:

1. *Temperature reduction.* A saturated solution can be made supersaturated by simple cooling if the solubility of the substance increases with temperature. The solutions of many inorganic salts and organic substances can be supersaturated by cooling.
2. *Evaporation of solvent.* If the solubility is not significantly affected by changing the temperature, as in the case of common salt, the supersaturation can be achieved by evaporating a part of the solvent.
3. *Addition of second solvent.* The addition of a second solvent may help to reduce the solubility of the solute. The material added forms with the original solvent, a mixed solvent in which the solubility of the solute is sharply reduced. When water is the additional solvent, the process is called *watering-out*; when an organic solvent is added to an aqueous salt solution, the process is called *salting-out*.
4. *Precipitation by chemical reaction.* This is accomplished when the product of the two reacting solutions is a solid with low solubility. Complete precipitation is achieved by adding a substance that reacts chemically with the original solute and forms an insoluble substance.

Mier's Theory of Supersaturation

The curve *AB* in Fig. 11.5 is the solubility curve. It represents the saturation concentration of a solute in the solution (solubility) as a function of temperature. Solubility is the maximum concentration of the solute in the solution, when a solid solute is brought in to equilibrium with a liquid solvent at specified temperatures. When an unsaturated solution, such as the one represented by point *P* is cooled, the temperature decreases along the horizontal line through *P*. When the temperature falls to that corresponding to point *Q* on the solubility curve, we expect phase separation to occur and crystallisation to begin. But crystallisation is not found to occur at this point. The solution can be super-cooled beyond the solubility curve *AB*, up to point *R*, without crystallisation, if it were initially free of all solid particles, foreign materials as well as the solute particles. Point *R* lies on the curve *CD* which can be termed as the *supersolubility curve*. Nuclei will form and crystallisation will occur only when the solution is cooled to a temperature corresponding to the supersolubility curve and not at other temperatures. Though nuclei cannot form at any point other than on the supersolubility curve *CD*, crystals can grow in the metastable region between the supersolubility curve and the solubility curve. Supersolubility curve *CD*, therefore represents the boundary which permits spontaneous nucleus formation and crystallisation. Once crystallisation begins, solute concentration falls along the line *RS*, finally approaching the equilibrium value on the solubility curve. Thus solubility curve represents the ultimate limit towards which crystallisation from a supersaturated solution proceeds.

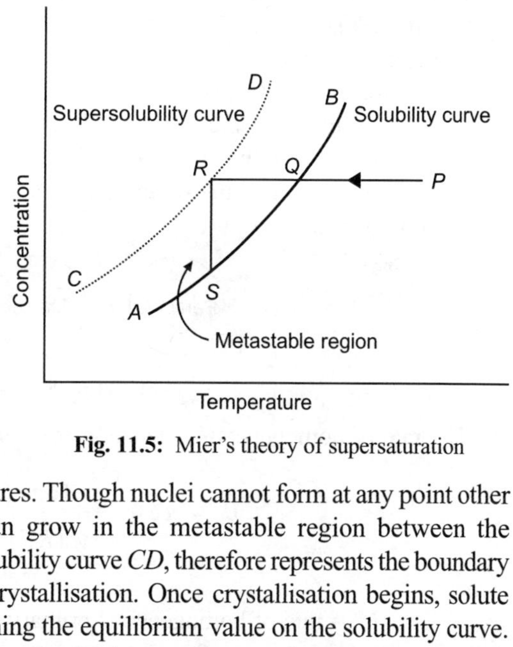

Fig. 11.5: Mier's theory of supersaturation

For Mier's supersolubility theory to be valid, the solution should be free from all type of solid particles. The theory is bound to fail if the particles of the solute are present as dust or as crystal fragments or if foreign materials which may act as the nuclei get accidentally introduced into the system. The number of the accidental collision between the solute molecules that results in the formation of the aggregates leading to the development of the nuclei will be more if the solution is kept for a sufficiently long time or if the solution volume is high. In practice, because of these collisions, the nuclei formation can occur well below the supersolubility curve. In short, a rigid supersolubility curve as predicted by the Mier's theory will not be possible in actual practice. What is more realistic is a zone of supersolubility rather than a fixed curve, where the rate of nucleation increases sharply.

11.6 NUCLEATION

Two processes are involved in the formation of crystals: nucleation and crystal growth. Analysis of industrial crystallisation process requires knowledge of both nucleation and crystal growth. The product crystal size and the size distribution in crystallisers are largely determined by the relation between the degree of nucleation and the crystal growth and is therefore, an important aspect of industrial crystallisation process. Nucleation means the birth of new crystals. There are two types of nucleation: primary nucleation and secondary nucleation. The nucleation occurring in the absence of crystalline surfaces is called the primary nucleation and that occurring in the presence of crystals and influenced by their interaction with the environment such as crystalliser walls, impellers, etc. is termed the secondary nucleation.

Primary Nucleation

Primary nucleation requires a high degree of supersaturation and is the principal mechanism occurring in precipitation. Primary nucleation can be homogeneous or heterogeneous. *Homogeneous nucleation* can occur in a homogeneous liquid phase, uninfluenced by the presence of solids of any sort including even the walls of the container. It begins with the random association of the solute molecules by the rapid local fluctuations, to form a cluster (Fig. 11.6). The cluster so formed mostly dissociates by normal molecular activity in the fluid or it may attract additional molecules to form an embryo with regular spacing of molecules that separates as new phase. If the degree of supersaturation is small and the concentration gradient is not favourable, the embryo may redissolve and disappear as clusters or individual particles. On the other hand, as the supersaturation increases, more molecules from the solution combine with the embryo and it grows and stabilises into crystal nucleus. Crystal nucleus is the smallest entity consisting of an agglomerate of particles that will not redissolve and can therefore grow into crystals.

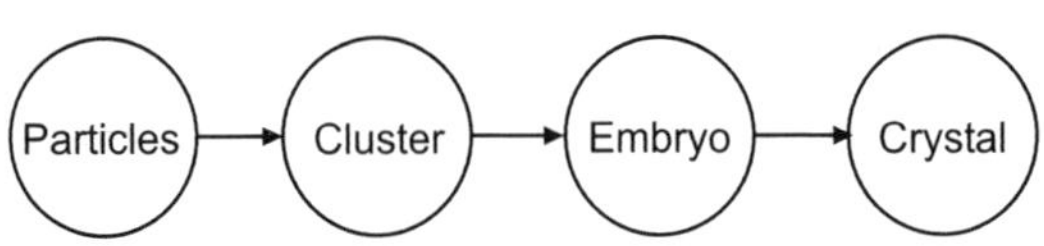

Fig. 11.6: Stages in the growth of crystals

The rate of homogeneous nucleation is given by the classical chemical kinetics as

$$B^\circ = A\exp\left[\frac{-16\pi V_M^2\sigma^3 N_a}{3v^2(RT)^3(\ln\alpha)^2}\right] \quad ...(11.5)$$

where

B° = rate of homogeneous primary nucleation, number of nuclei/m^3 s.

A = frequency factor

N_a = Avogadro's number

V_M = molar volume of crystals

σ = interfacial tension

v = number of ions/molecule of solute

$\alpha = C/C^*$ = supersaturation ratio

For small values of α, the mathematical approximation $\ln\alpha \approx \alpha - 1$ is valid. Therefore by Eq. (11.4),

$$\ln\alpha \approx \alpha - 1 = \frac{C - C^*}{C^*} = S \qquad ...(11.6)$$

where S is the fractional supersaturation. Using this approximation in Eq. (11.5), we get

$$B° = A\exp\left[\frac{-16\pi V_M^2\sigma^3 N_a}{3v^2(RT)^3 S^2}\right] \qquad ...(11.7)$$

It is clear that nucleation rate increases with increasing supersaturation and temperature and decreases with an increase in surface energy. Measured values of the interfacial tension range from as low as 0.001 J/m^2 for very soluble compounds to 0.17 J/m^2 for compounds of low solubility. Theoretically, A is 10^{36} nuclei /m^3.s. However, the observed value is less because of the presence of foreign matter. A value of $A = 10^{31}$ nuclei/m^3.s is usually recommended for use in Eq. (11.7)

The solubility of a substance is related to its particle size by the Kelvin equation which gives

$$\ln\alpha = \frac{4V_M\sigma}{vRTL} \qquad ...(11.8)$$

where L is the crystal size. Thermodynamically, the solubility of a small crystal is greater than that of the large crystal. Since the solubility data usually refers to moderately large crystals, small crystals can exist in equilibrium with supersaturated solutions. But this equilibrium is found to be unstable because of the *Ostwald ripening*, a phenomenon observed in solutions consisting of small and large particles. It refers to the dissolution of small crystals and the redeposition of the dissolved species on the surfaces of larger crystals as a result of which large crystals grow. Since dissolved impurities are generally present in the solution and it is difficult to eliminate fully the influence of crystalliser walls, stirrers and baffles, it is quite difficult to observe homogeneous nucleation in practice.

Primary nucleation facilitated by suspended particles of foreign substances is generally referred to *as heterogeneous nucleation*. It is the type of nucleation that is important industrially as it is practically quite difficult to eliminate the influence of foreign substances from solutions. The mechanism of heterogeneous nucleation is not fully understood. A foreign substance present in the supersaturated solution is believed to decrease the energy required for nucleation. The solute particles may get adsorbed on the suspended particles of foreign substances thus forming particle agglomerates of size larger than the size of a crystal nucleus and then grow into macro crystals. Nucleation in a heterogeneous system generally occurs at a lower supersaturation than a homogeneous system.

Example 11.5: The following data refer to an aqueous solution of KCl at 300 K. The density of the crystals is 1985 kg/m^3, equilibrium solubility C^* at 300 K is 35.5 parts/100 parts water, interfacial tension may be assumed to be 2.7×10^{-2} J/m^2.

(a) Determine the solubility of KCl in water at 300 K for a particle size of 100 nm using the Kelvin equation.

(b) For a supersaturation ratio of 1.5, estimate the nucleation rate of KCl in aqueous solution at 300 K.

***Solution*:** (a) The Kelvin equation gives the relationship between particle size and the solubility:

$$\ln\alpha = \frac{4V_M\sigma}{vRTL} \qquad ...(11.8)$$

$$V_M = \text{molar volume of KCl crystals} = \frac{\text{molecular weight}}{\text{crystal density}} = \frac{(39.1+35.45)}{1985} = 0.0376\,\text{m}^3/\text{kmol}$$

σ = interfacial tension = 2.7×10^{-2} J/m^2; since KCl dissociates into 2 ions, K^+ and Cl^-, $\nu = 2$; R = ideal gas constant = 8314 J/kmol K, temperature T = 300 K, particle size $L = 100 \times 10^{-9}$ m.

Substituting the values in Eq. (11.8), we get

$$\ln\alpha = \frac{4\times 0.0376\times 2.7\times 10^{-2}}{2\times 8314\times 300\times 100\times 10^{-9}} = 0.0081$$

$$\alpha = \frac{C}{C^*} = e^{0.0081} = 1.0082$$

$$C = 1.0082\times C^* = 1.0082\times 35.5 = 35.8 \text{ parts KCl/100 parts water}$$ **(Answer)**

(b) The rate of nucleation is given by

$$B^\circ = A\exp\left[\frac{-16\pi V_M^2\sigma^3 N_a}{3v^2(RT)^3(\ln\alpha)^2}\right] \quad \ldots(11.5)$$

Here N_a = Avagado's number = 6.022 × 10^{26} molecules/kmol, α = 1.5. Take A = 10^{31} nuclei/m^3 s.

$$B^\circ = A\exp\left[\frac{-16\pi V_M^2\sigma^3 N_a}{3v^2(RT)^3(\ln\alpha)^2}\right]$$

$$B^\circ = 10^{31}\exp\left[\frac{-16\pi\times 0.0376^2\times(2.7\times 10^{-2})^3 6.022\times 10^{26}}{3\times 2^2(8314\times 300)^3(\ln 1.5)^2}\right] = 1.12\times 10^{19} \text{ nuclei/m}^3\text{s.}$$

Rate of nucleation = 1.12 × 10^{19} nuclei/m^{3}s **(Answer)**

Secondary Nucleation

Secondary nucleation is attributed to the influence of crystals already present in the magma. Secondary nucleation may be either the *fluid shear nucleation* caused by the action of fluid shear against the growing crystal surface when the supersaturated solution flows at an appreciable velocity past the crystal surface or the *contact nucleation* caused by (1) the collision between the existing crystals with one another or (2) by the collision of crystals with metal surfaces such as the walls of the container, the agitator blades, or rotary impellers. Contact nucleation is the most effective method of nucleation industrially, as it occurs at low values of relative supersaturation common in industrial applications.

The level of supersaturation, degree of agitation, the rate of cooling and the presence of impurities, etc. are found to influence the rate of the secondary nucleation. The complex phenomena of secondary nucleation is usually represented by an empirical correlation of the type

$$B^\circ = k_N S^b M_T^j N^r \quad \ldots(11.9)$$

where B° = rate of secondary nucleation, M_T = mass of crystals per unit volume of magma, N = rate of agitation and S is the fractional supersaturation defined by Eq. (11.4). k_N, b, j and r, are empirical constants, which vary according to the operating conditions.

In general, the nucleation rate in commercial crystallisation equipment (B°) is made up of the following contributions: (1) primary nucleation due to the supersaturation driving force (B_{ss}) (2) the secondary nucleation due to the crystal-crystal contact (B_{cc}) and (3) the secondary nucleation due to crystal-impeller contact (B_{ci}).

$$B^\circ = B_{ss} + B_{cc} + B_{ci} \quad \ldots(11.10)$$

11.7 CRYSTAL GROWTH

Crystal growth occurs at the face of the crystal by a layer-by-layer process. The growth of crystals, therefore, requires that the solute must be transported from the bulk of the solution to the faces by diffusional process and subsequently the adsorbed solute molecules or ions at the crystal surface must get deposited and integrated into the crystal lattice. Different faces grow at different rates and the rates can be selectively altered by addition or elimination of impurities. The rate at which a face moves in a direction perpendicular to its original position is called the *translational velocity* of that face or the rate of growth of that face.

Under ideal conditions, a growing crystal maintains geometric similarity during the growth. Such a crystal is called *invariant crystal*. Figure 11.7a shows the cross section of an invariant crystal during growth. Each of the polygons represents outlines of the crystal faces at different times. Since they are geometrically similar, the straight lines connecting the geometrically corresponding corners of the polygon should pass through the centre of the crystal, which may be treated as the original nucleus. For such crystals, a single dimension can be used to characterise the volume, total surface area, or such factors as the surface area per unit volume. Even with invariant crystals, the rate of growth of different faces may be different. A geometric law of crystal growth known as the *overlapping principle* takes into account the difference in the translational velocities of the different faces of a growing crystal and predicts that only those faces with the lowest translational velocities survive. Consider Fig. 11.7b. The slow growing faces are marked *S* and ultimately only these faces survive in the final fully grown crystals.

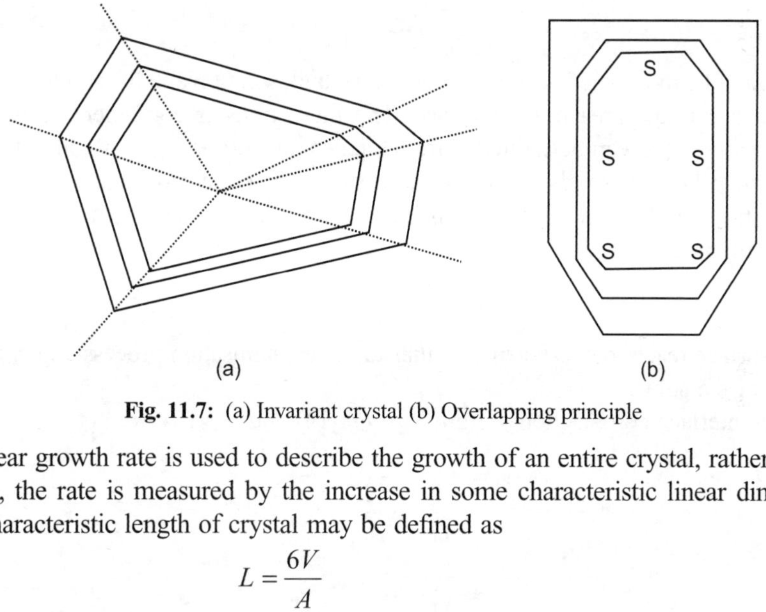

Fig. 11.7: (a) Invariant crystal (b) Overlapping principle

When a linear growth rate is used to describe the growth of an entire crystal, rather than that of a particular face, the rate is measured by the increase in some characteristic linear dimension of the crystal. The characteristic length of crystal may be defined as

$$L = \frac{6V}{A} \qquad \text{...(11.11)}$$

where V and A are, respectively, the volume and total surface area of the crystal. The rate of growth of the crystal face that is perpendicular to L may be expressed as

$$G = \lim_{\Delta t \to 0} \frac{\Delta L}{\Delta t} = \frac{dL}{dt} \qquad \text{...(11.12)}$$

where G is the growth rate (usually measured as mm/h) over the interval t.

The Delta-*L* Law of Crystal Growth

According to the delta-*L* law of crystal growth, all geometrically similar crystals of the same material suspended in the same solution grow at the same rate and the rate is independent of crystal size, provided all crystals in the suspension are treated alike. The law is found to be a reasonably accurate generalization for a number of industrial cases. The law simplifies the mathematical treatment in modelling real crystallisers and is useful in predicting the crystal size distribution in industrial crystallisers. The total growth of each crystal in the solution during the same time interval Δt is the same and is given as

$$\Delta L = G\Delta t \quad \text{...(11.13)}$$

This law follows from the situation where the overall mass transfer coefficients are equal for each face of all the crystals. Hence the law is found to be more accurate when the crystal size is between 0.05 and 0.5 mm, because mass transfer coefficient is expected to remain approximately constant within this range of particle diameters. It should be noted that the law fails when similar crystals are given different treatment based on size and when surface defects or dislocations significantly alter the growth rate of faces. If the law is not applicable, it means that the growth of crystals is size dependent.

Kinetic and Transport Considerations

Growth of crystals is controlled by the two resistances in series: the resistance to diffusion of solute particles from the bulk of solution to the growing crystal face and the resistance to the interfacial process by which the solute particles get deposited and integrated into the growing surface. This two-step theory of crystal growth is usually referred to as the *diffusion-reaction theory*. The rate of mass transfer from the bulk solution to the crystal-solution interface can be represented in terms of the mass transfer coefficient k_y:

$$\frac{dm}{dt} = k_y A(y - y_i) \quad \text{...(11.14)}$$

where y and y_i are the mole fraction of the solute in the bulk solution and the solution-crystal interface, respectively and A is the area of the surface. The mass transfer coefficient k_y is independent of crystallisation process and can be estimated from the general correlations for convective mass transfer coefficient through beds of solids. In the reaction step, the solute molecules get integrated into the crystal-lattice structure through a reaction, the rate of which is also assumed to be dependent on the concentration difference.

$$\frac{dm}{dt} = k_i A(y_i - y_s) \quad \text{...(11.15)}$$

where k_i is a surface reaction coefficient peculiar to the crystallisation process and y_s is the saturation concentration of the solute.

Eliminating interface composition y_i from Eqs. (11.14) and (11.15), we get

$$y - y_s = \frac{1}{\dfrac{1}{Ak_y} + \dfrac{1}{Ak_i}} \frac{dm}{dt} \quad \text{...(11.16)}$$

or

$$\frac{dm}{dt} = \frac{A(y - y_s)}{\dfrac{1}{k_y} + \dfrac{1}{k_i}} \quad \text{...(11.17)}$$

Equation (11.17) can be written as

$$\frac{dm}{dt} = KA(y - y_s) \quad \text{...(11.18)}$$

where K is the overall transfer coefficient.

$$K = \frac{1}{\dfrac{1}{k_y} + \dfrac{1}{k_i}} \quad \text{...(11.19)}$$

At low velocities of the solution, the growth rate is controlled by the resistance to diffusional transfer, and at high velocities of the solution past the crystal surface so that $k_y >> k_i$, the reaction step controls the growth rate. When the rate of integration is very high, K is approximately equal to k_y and the crystallisation is diffusion controlled. When the diffusional resistance is low, K is approximately equal to k_i and the process is controlled by the integration step.

To develop an equation for the growth rate in terms of the overall transfer coefficient and the concentration difference driving force, let us consider an invariant crystal for which the volume is

$$V = aL^3 \qquad \text{...(11.20)}$$

where L is the characteristic length of the crystal and a is a constant. Combining Eqs. (11.20) and (11.11), the surface area of the crystal is then,

$$A = \frac{6V}{L} = 6aL^2 \qquad \text{...(11.21)}$$

The mass of crystal is

$$m = V\rho = aL^3\rho \qquad \text{...(11.22)}$$

in which ρ is the density of the crystal. Substituting Eqs. (11.21) and (11.22) into Eq. (11.18), we get

$$3aL^2\rho\frac{dL}{dt} = 6KaL^2(y - y_s) \qquad \text{...(11.23)}$$

$$\frac{dL}{dt} = \frac{2K(y - y_s)}{\rho} \qquad \text{...(11.24)}$$

Since the growth rate G is defined as $G = \frac{dL}{dt}$, Eq. (11.24) gives

$$G = \frac{2K(y - y_s)}{\rho} \qquad \text{...(11.25)}$$

For invariant crystals, the growth is uninfluenced by other crystals or solids. In commercial crystallisation processes, such unhindered growth seldom occurs. Rather, habits of crystal growth develop due to agglomeration of crystals, occlusion of impurities on the growth surfaces or by breakage of crystals by pumps and agitators. This crystal habit is of great concern in the crystalliser operations as it influences the product purity and appearance and tendency to cake or powder.

11.8 CRYSTALLISATION EQUIPMENT

The size distribution and size uniformity of product crystals largely depend on the features characteristic of different crystallisation equipment. The choice and design of the crystalliser are important also because of their influence on the crystal habit which determines the appearance and size range of the products. The appearance and size range influence the saleability of products. Reasonable size and size uniformity are desired if the crystalline product is to be subjected to further processing operations such as filtration, washing, reaction with other chemicals, transportation and storing. Industrial crystallisation equipment are classified in different ways.

Classification

Based on mode of operation Based on the mode of operation, crystallisation equipment may be classified as *batch or continuous crystallisers*. The continuous operation is generally preferred for industrial crystallisers; however, the batch crystallisers also have certain special applications.

Based on method of supersaturation The primary requirement for initiating crystallisation is the supersaturation of the solution. Crystallisers can be classified based on the method used to bring about supersaturation. Three methods are generally employed. (1) Cooling the solution with negligible evaporation. Cooling crystallisers can be used for substances such as KNO_3 and Na_2SO_4 whose solubility decreases with decrease in the temperature. (e.g. Tank crystalliser and agitated batch crystalliser) (2) Evaporation of the solvent with negligible cooling. Evaporating crystallisers find application for crystallisation of substances such as $MgCl_2$ and common salt, whose solubility is not significantly affected by change in the temperature. (3) Adiabatic operation: combination of evaporation and cooling. In this method, a hot solution is introduced into a vacuum where the solvent is flashed and the solution is cooled adiabatically (e.g. Vacuum crystallisers).

Based on the means of suspending and contacting the growing crystals The means used for suspending the growing crystals in the solution and the manner in which the supersaturated solution is brought into contact with the growing crystals are important for producing crystals of the desired size distribution. Mechanically agitated crystallisers, scraped surface crystallisers, baffled crystallisers, circulating liquid crystallisers, circulating magma crystallisers and classifying crystallisers are characterised by certain special features that aid in the growth of crystals and give the desired size distribution.

Agitated Batch Crystalliser

Agitated batch crystalliser shown in Fig. 11.8 is a batch cooling crystalliser. The top half of the crystalliser body is cylindrical and the bottom half conical in shape. The crystalliser contents are agitated by means of propellers mounted on a central shaft. Cooling water is circulated through the cooling coils. The agitator helps in increasing the rate of heat transfer and in maintaining uniformity of the temperature throughout the solution; it also keeps the growing crystals in suspension so that these can grow uniformly. Being a batch apparatus and hence discontinuous in operation, the rate of production of crystals in agitated batch crystalliser is low. The solubility of the substance being the lowest at the

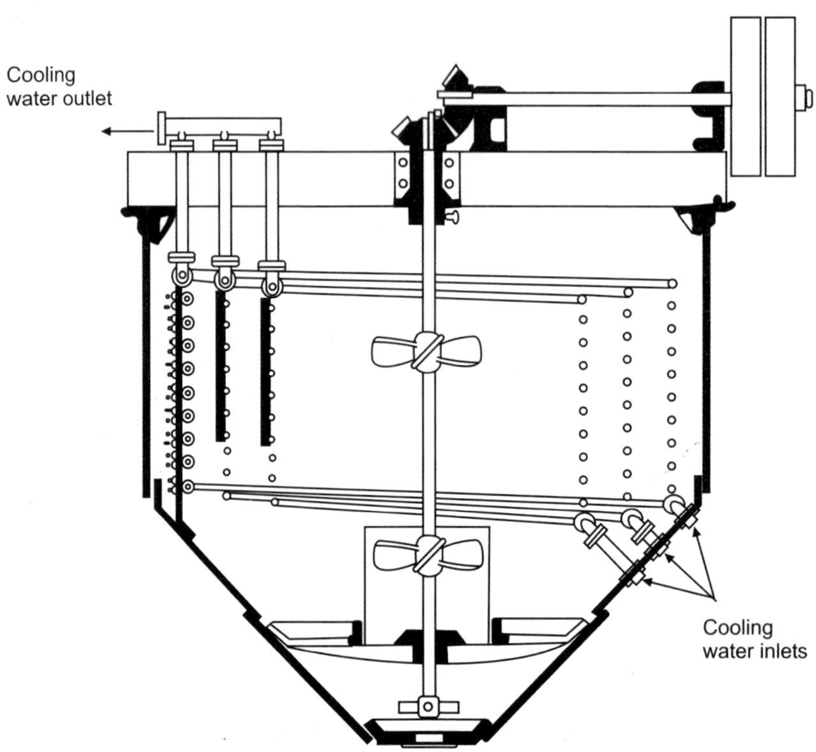

Fig. 11.8: Agitated batch crystalliser

surface of the cooling coils, crystals get deposited on the surface of the coils causing decrease in the rate of heat transfer. When the desired growth is achieved, the magma is removed from the crystalliser for separation by centrifuging. Mechanically agitated crystallisers give uniformly sized crystals of high purity and relatively small size. They are most suited for producing fine chemicals and pharmaceutical products.

Swenson-Walker crystalliser

It is a *continuous cooling type, scraped-surface crystalliser.* Scraped surface crystallisers provide an effective and inexpensive method of producing a slurry in equipment which does not require expensive installation or supporting structures. There is direct heat exchange between the slurry and a jacket containing the cooling medium. The heat transfer surface is scraped or agitated so that the deposits cannot build up on the surface.

Figure 11.9 shows a Swenson-Walker crystalliser which consists of an open semi-cylindrical trough, typically 1 m wide and 3-12 m long. The trough is provided with a cooling jacket outside. A slow speed (3-10 rpm) spiral agitator rotates and suspends the growing crystals. Hot concentrated solution is fed at one end and flows continuously to the other end while it is cooled. The blades pass close to the wall and prevent the growth of crystals on the trough wall. If more lengths are required, additional crystalliser may be connected in series with the existing ones. The newly added units are kept at a slightly lower level than the existing ones and are fed with the overflow from the preceding unit.

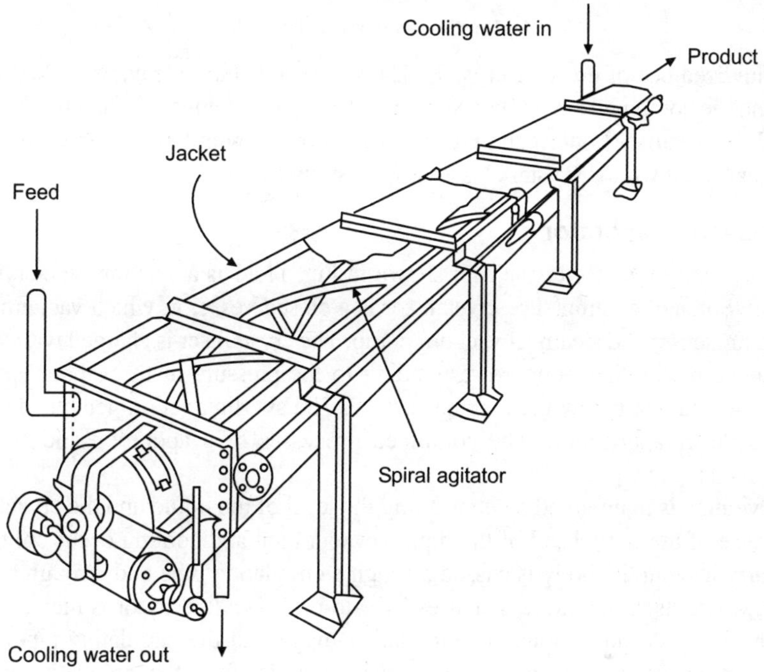

Fig. 11.9: Swenson-Walker crystalliser

The Oslo Crystalliser

The Oslo crystalliser (Fig. 11.10) is a *circulating-liquid evaporator crystalliser*, where the supersaturation is generated by evaporation. The circulating liquid is heated by passage through the tube-side of a steam heated exchanger. When the heated liquid enters the vaporiser, flash vaporisation occurs resulting in the supersaturation of the solution. The supersaturated liquid flows down through the central tube

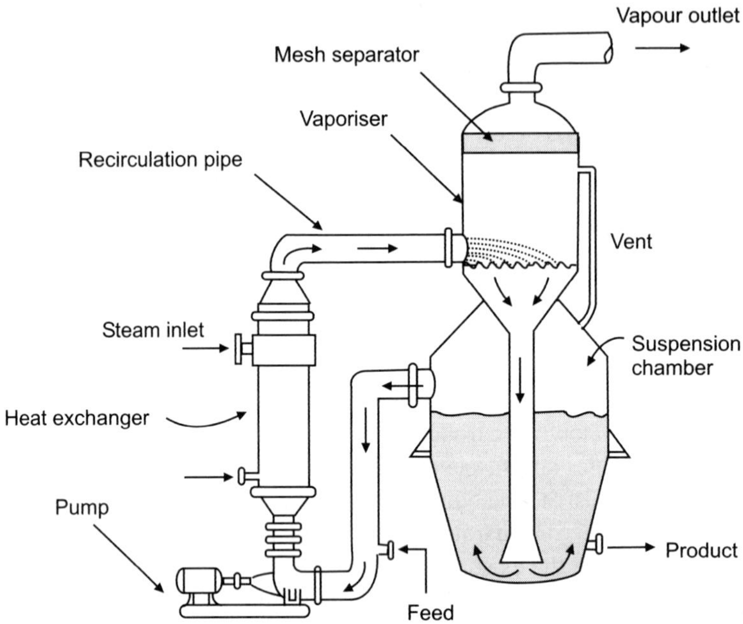

Fig. 11.10: Oslo evaporative crystalliser

and enters the fluidized bed of growing crystals. The saturated solution leaving the bed combines with the fresh feed and the combined stream is admitted to the heat exchanger as shown. The larger crystals settling out in the crystalliser bottom and the mother liquor are withdrawn as the products. This type of crystalliser is also known as Krystal classifying crystalliser.

Forced Circulation Evaporator

The forced circulation evaporator crystalliser shown in Fig. 11.11 is a vacuum type crystalliser which combines evaporation and cooling. The crystalliser is a closed vessel in which vacuum is maintained by a barometric condenser and steam jet vacuum pump. The crystalliser is charged with warm saturated solution well above the boiling point corresponding to the pressure in the vapour space, through a tangential inlet situated just below the magma surface. The swirling action generated by the entering liquid facilitates flash vaporisation. The combined processes of evaporation and cooling result in nucleation and crystal growth.

The magma volume is maintained by controlling the level of the liquid and the crystallising solid in the vessel. Because of the static head of the liquid, evaporation and cooling occur only at the magma surface. The slurry leaving the body is passed through a circulating pipe and through a shell-and-tube heat exchanger, where its temperature increases by 2 to 6 K. The hot liquor is then admitted into the crystalliser body. The feed liquid joins the circulating magma in the circulating pipe as shown. The slurry consisting of the mother liquor and the crystals drawn off from the circulating pipe as shown is passed through a continuous centrifuge for separating the mother liquor, which is returned to the crystalliser.

The magma should be kept well agitated to maintain uniformity of concentration and temperature and to keep the crystals in suspension. This type of crystallisation equipment is probably the most widely used and the least expensive and is found in diameters ranging from 0.6 m for laboratory equipment up to over 12 m for commercial crystalliser. The forced circulation crystalliser is recommended for a wide range of organic and inorganic compounds, such as sodium chloride, sodium sulphate, sodium carbonate monohydrate, citric acid, urea, sugar, etc.

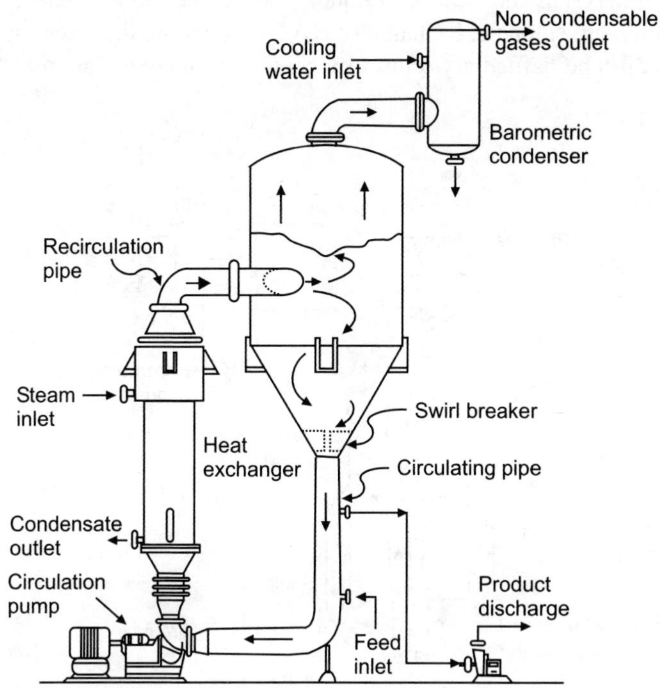

Fig. 11.11: Forced circulation evaporator crystalliser

Draft-Tube-Baffle Crystalliser

Considering the influence that the mechanical circulation has on the level of nucleation within the crystalliser, a number of designs employing this feature have been developed. By locating the circulator within the body of the crystalliser, the power input for circulation can be considerably reduced. The Swenson draft-tube–baffled (DTB) crystalliser shown in Fig. 11.12 is a design of this type. It is a *continuous, vacuum, evaporating crystalliser*. The main body of the crystalliser is equipped with a draft tube that extends down nearly to the bottom. Evaporation occurs at the boiling surface under vacuum. A low-speed propeller agitator inside the draft tube directs the slurry upward through the tube towards the boiling surface. The nucleation and build-up of crystals on the walls of the tube do not occur since within the tube, violent flashing is prevented and degree of super-cooling is kept low. The slurry which has been cooled flows back through the annular space formed by the draft tube and the skirt baffle as shown in the figure, and re-enters the draft tube. A small portion of this magma is diverted and flows upward through an annular settling zone formed by the skirt baffle and the walls of the crystalliser body.

In the settling zone, large crystals settle and the mother liquor containing fine crystals flows out through a circulating pipe. Fresh feed joins the circulating stream and is pumped through a heat exchanger. The heat supplied in the exchanger is used to provide the energy to preheat the feed, to dissolve the finer crystals and to evaporate the solvent. The slurry leaving the heat exchanger re-enters the crystalliser just below the bottom of the draft tube and gets mixed with the re-circulating stream. The role of the elutriation leg at the bottom of the crystalliser is to assist in the further classification of crystals by size. A part of the circulating liquor is pumped to the bottom of the leg to act as a hydraulic sorting fluid that carries small crystals back into the crystallising zone for further growth. The product magma is withdrawn from a nozzle located near the bottom of the elutriation leg and sent to a filter or centrifuge to separate the crystals and the mother liquor. The mother liquor is sent back to the crystalliser. Products such as ammonium sulphate, KCl, and $(NH_4)_2HPO_4$, etc. in the size range of 6-20 mesh are produced using the draft-tube-baffled crystalliser by controlling the population of fine particles in the baffle zone.

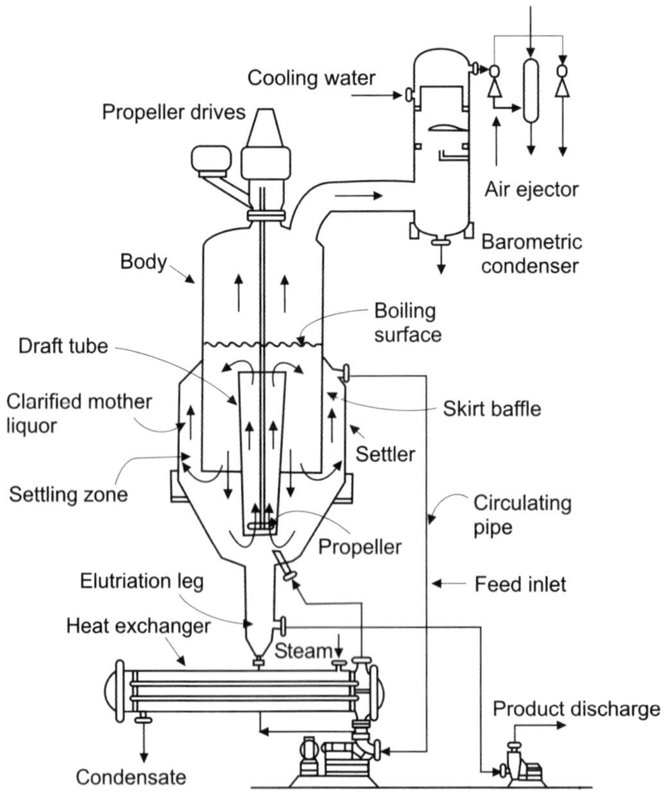

Fig. 11.12: Draft tube–baffle crystalliser

11.9 THE MSMPR CRYSTALLISER

The mixed suspension, mixed product removal (MSMPR) crystalliser is an idealized model used in the analysis of the performance of crystallisers–both pilot and plant sized. The model served well as a basis for identifying the kinetic parameters which can be used for evaluating the performance of crystallisers. Using this model, the crystal size distribution of the product can be estimated from the kinetics of nucleation and crystal growth. The model is based on the following assumptions:

1. Continuous steady-state operation.
2. Magma in the crystalliser is perfectly mixed.
3. All crystals have the same shape.
4. No product classification occurs within the crystalliser magma and no size classified withdrawal system is used.
5. Uniform supersaturation throughout the magma at all times.
6. Growth is independent of the crystal size and the ΔL law of crystal growth applies.
7. No crystals present in the feed, but the seeds may be added initially.
8. No crystal breakage, attrition or agglomeration occurs.

9. Mother liquor in the product magma is in equilibrium with the crystals or the product magma is saturated.
10. Nucleation is due to the secondary nucleation by crystal contact and the rate of nucleation is constant.

The basic parameter in the theory of crystal size distribution is the population density, n, indicated as the slope of the cumulative distribution curve. Assume that a distribution function of the cumulative number of crystals in the magma is known as a function of L, the crystal size. The cumulative distribution curve is the plot of N/V versus L at size L, where N/V is the number of particles per unit volume of mother liquor and L is the particle size (Fig. 11.13).

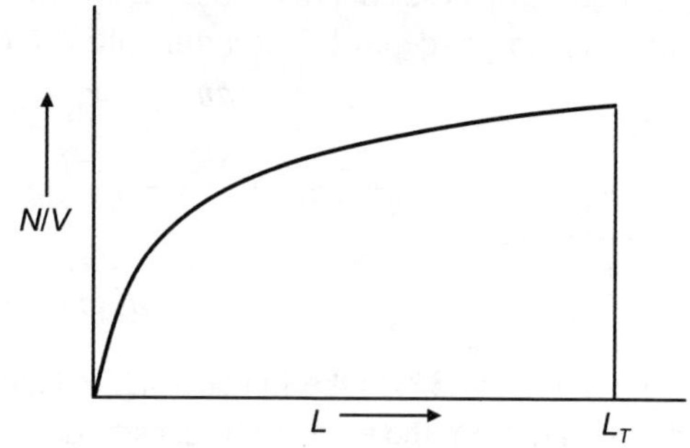

Fig. 11.13: Cumulative distribution curve

Thus,

$$n = \frac{d(N/V)}{dL} \quad \text{...(11.26)}$$

Since V is constant under steady state, Eq. (11.26) becomes

$$n = \frac{1}{V}\frac{dN}{dL} \quad \text{...(11.27)}$$

Equation (11.27) means that *the population density is the number of particles per unit volume of mother liquor per measure of particle size.*

Consider an arbitrary size for the particles in the crystalliser lying in the range L_1 to L_2 with population densities, n_1 and n_2, respectively. Particles may enter or leave the size range by either growth or by flow.

Number of particles entering the size range by growth in unit time = $n_1 GV$

Number of particles leaving the size range by growth in unit time = $n_2 GV$

Here G is the growth rate of the characteristic dimension and V is the crystalliser volume in terms of the volume of mother liquor.

Number of particles entering the size range by flow $= Q_i \bar{n}_i \Delta L$

Number of particles leaving the size range by flow $= Q\bar{n}\Delta L$

Here Q is the volumetric flow rate, $\bar{n}_i$ and $\bar{n}$ are the average population densities in the range L_1 to L_2 in the feed and exit streams and $\Delta L = L_2 - L_1$.

Substituting the four quantities listed above in the population balance under steady state, we get

$$\text{Number of particles in} = \text{Number of particles out} \quad \text{...(11.28)}$$

$$n_1 GV + Q_i \bar{n}_i \Delta L = n_2 GV + Q\bar{n}\Delta L \quad \text{...(11.29)}$$

Equation (11.29) can be rearranged as

$$VG\frac{n_2 - n_1}{\Delta L} = Q_i \bar{n}_i - Q\bar{n} \quad \text{...(11.30)}$$

Taking limit as ΔL approaches zero and noting that the mean values $\bar{n}_i$ and $\bar{n}$ get replaced by point values n_i and n, Eq. (11.30) gives

$$VG\frac{dn}{dL} = Q_i n_i - Qn \quad \text{...(11.31)}$$

Denoting the mean residence time by τ,

$$\tau = \frac{V}{Q} \quad \text{...(11.32)}$$

Since the feed is assumed to be free of particles, $n_i = 0$ and Eq. (11.31) simplifies to

$$\tau G \frac{dn}{dL} + n = 0 \quad \text{...(11.33)}$$

In deriving these equations, the ΔL law of crystal growth is assumed to be applicable so that growth rate G is independent of L. Rearranging Eq. (11.33),

$$\frac{dn}{dL} = -\frac{n}{\tau G} \quad \text{...(11.34)}$$

Using the boundary condition $n = n^o$ when $L = 0$, the ordinary differential equation (11.34) can be solved to give

$$n = n^o \exp\left(-\frac{L}{\tau G}\right) \quad \text{...(11.35)}$$

Equation (11.35) is the fundamental equation for the MSMPR crystalliser. It permits evaluation of growth rate from the population density data. This is done by plotting n versus L on semi-logarithmic graph, and measuring the slope of the resulting straight line.

The nucleation rate B^o can also be determined from the above data. Treating the nucleation rate as the rate of appearance of near zero-sized particles, we can write

$$B^o = \left.\frac{dN}{dt}\right|_{L\to 0} = \left[\frac{dN}{dL} \times \frac{dL}{dt}\right]_{L\to 0} = n^o G \quad \text{...(11.36)}$$

Intercept that the straight line obtained makes with the ln n axis when ln n versus L is plotted as in Fig. 11.14, gives n^o and the slope gives G using which B^o is evaluated.

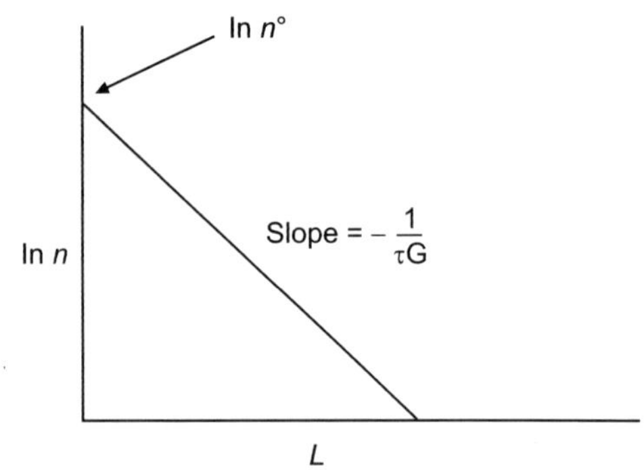

Fig. 11.14: Evaluation of the kinetic parameters

Denoting $\frac{L}{\tau G}$ by a dimensionless length z, Eq. (11.35) becomes

$$n = n^o \exp(-z) \quad \text{...(11.37)}$$

The number of crystals in unit volume of mother liquor n_c is

$$n_c = \int_0^\infty n dL \quad \text{...(11.38)}$$

Substituting Eq. (11.37) into Eq. (11.38) and noting that $dL = \tau G dz$, we get

$$n_c = n^o \tau G \int_0^\infty e^{-z} dz = n^o \tau G \quad \text{...(11.39)}$$

The total mass of product crystals in unit volume of liquid is

$$m_c = \int_0^\infty mndL \quad \text{...(11.40)}$$

Noting that $m = aL^3\rho_C$ where a is the volumetric shape factor $L = \tau Gz$ and $n = n^o \exp(-z)$, Eq. (11.40) becomes

$$m_c = a\rho_c n^o (\tau G)^4 \int_0^\infty z^3 e^{-z} dz = 6a\rho_c n^o (\tau G)^4 \quad \text{...(11.41)}$$

where ρ_c is the density of the crystals. Dividing Eq. (11.39) by Eq. (11.41), we get the expression for the number of crystals per unit mass.

$$\frac{n_c}{m_c} = \frac{n^o \tau G}{6a\rho_c n^o (\tau G)^4} = \frac{1}{6a\rho_c (\tau G)^3} \quad \ldots(11.42)$$

Figure 11.15 shows the differential mass distribution of the product from an MSMPR crystalliser. It goes through a maximum at $z = 3$. The crystal size corresponding to this condition is known as the *predominant crystal size*, L_{pr}. It means that more of the mass is in the size interval dz at $z = 3$, than in any other size interval of width dz. The predominant crystal size is a quantity of special interest in the design and operation of crystallisers.

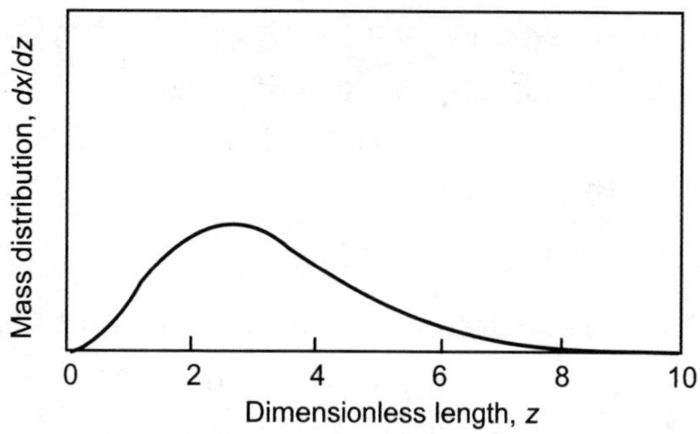

Fig. 11.15: Crystal size distribution of MSMPR model

$$z = 3 = \frac{L_{pr}}{\tau G} \text{ or } L_{pr} = 3\tau G \quad \ldots(11.43)$$

Combining Eqs. (11.42) and (11.43), we can see that

$$\frac{n_c}{m_c} = \frac{1}{6a\rho_c (\tau G)^3} = \frac{9}{2a\rho_c L_{pr}^3} \quad \ldots(11.44)$$

and the nucleation rate from Eq. (11.36) is

$$B^o = \frac{n_c}{\tau} = \frac{9C}{2a\rho_c V L_{pr}^3} \quad \ldots(11.45)$$

where $C = m_C Q$ is the rate of production of crystals, kg/time.

Another useful quantity is the mass distribution of the product crystals from the MSMPR crystalliser. Using the method of statistics it can be shown that the cumulative mass distribution is given by

$$x = 1 - \left(1 + z + \frac{z^2}{2} + \frac{z^3}{6}\right) e^{-z} \quad \ldots(11.46)$$

Here, x is the mass fraction of crystals having a smaller value of z. The differential mass distribution is given by

$$\frac{dx}{dz} = \frac{z^3}{6} e^{-z} \quad \ldots(11.47)$$

Differential distribution mathematically represented by Eq. (11.47) gives the mass fraction of crystals in the given interval z.

Example 11.6: Product removal rate based on the mother liquor from an MSMPR crystalliser is found to be 0.75 m^3/h. The steady nucleation rate n^o is estimated to be 9.5×10^{12}/m^4 and the growth rate G is 3.5×10^{-5} m/h. The crystal density is 2500 kg/m^3 and the volume shape factor is 0.7. The volume of the vessel based on the mother liquor is 4 m^3, Determine:

(a) The mass of crystals per unit volume of liquid, m_c.
(b) The rate of production of crystals, kg/h
(c) The percent of nuclei discharged by the time they have grown to 0.1 mm.

***Solution*:** (a) The mass of crystals per unit volume of mother liquor is given by Eq. (11.41).

$$m_c = 6a\rho_c n^o (\tau G)^4 \quad \ldots(11.41)$$

Here, volume shape factor $a = 0.7$, crystal density $\rho_C = 2500$ kg/m^3, nucleation rate $n^o = 9.5 \times 10^{12}$ /m^4, growth rate $G = 3.5 \times 10^{-5}$ m/h, volume of crystalliser in terms of the mother liquor is $V = 4$ m^3 and volumetric flow rate of mother liquor $Q = 0.75$ m^3/h and the residence time τ is

$$\tau = \frac{V}{Q} = \frac{4}{0.75} = 5.33 \text{ h}$$

Substituting the values,

$$m_c = 6 \times 0.7 \times 2500 \times 9.5 \times 10^{12} \times (5.33 \times 3.5 \times 10^{-5})^4 = 121.1 \text{kg/m}^3$$ **(Answer)**

(b) The rate of production of crystals

$$C = m_C Q = 121.1 \times 0.75 = 90.8 \text{kg/h}$$ **(Answer)**

(c) By Eq. (11.35),

$$\frac{n}{n^o} = \exp\left(-\frac{L}{\tau G}\right)$$

where n is the number of crystals of size L and n^o is the number of nuclei initially present. When $L = 0.1$ mm,

$$\frac{n}{n^o} = \exp\left(-\frac{0.1 \times 10^{-3}}{5.33 \times 3.5 \times 10^{-5}}\right) = 0.5851$$

It means, 58.51% of the original nuclei has grown up to 0.1 mm and is present in the crystalliser and the rest has been discharged.

Therefore, the percent discharged = (100 – 58.51) = 41.49% **(Answer)**

Example 11.7: An MSMPR-type continuous vacuum crystalliser produces 1000 kg/h of crystals of density 1650 kg/m^3. The crystals constitute 15% of the magma volume and the residence time of the magma in the crystalliser is 2 h. A predominant crystal size on mass basis of 0.5 mm is desired. Assume a volumetric shape factor of 0.7. Determine:

(a) The required crystal growth rate, mm/h
(b) Rate of nucleation B^o, nuclei/h m^3 of mother liquor in the crystalliser
(c) Number of crystals produced per hour

***Solution*:** (a) Here, the predominant crystal size, $L_{pr} = 0.5$ mm and the residence time in the crystalliser, $\tau = 2$ h. The crystal growth rate is related to the predominant size by Eq. (11.43)

Crystal growth rate, $G = \frac{L_{pr}}{3\tau} = \frac{0.5}{3 \times 2} = 0.0833 \text{mm/h}$ **(Answer)**

(b) The rate of nucleation is given by

$$B^o = \frac{n_c}{\tau} = \frac{9C}{2a\rho_c V L_{pr}^3}$$

where C = the rate of production of crystals = 1000 kg/h, a = shape factor = 0.70, ρ_C = density of crystals = 1650 kg/m^3.

Since the crystals constitute 15% of the magma volume, the rate at which mother liquor leaves the crystalliser = volume of crystals × 0.85/0.15

i.e. $$Q = \frac{1000}{1650} \times \frac{0.85}{0.15} = 3.4343 \text{ m}^3/\text{h}$$

Since the volume of mother liquor in the crystalliser

$$V = Q \times \tau = 3.4343 \times 2 = 6.8687 \text{m}^3$$

Substituting the values in Eq. (11.43), we get

$$B^o = \frac{9 \times 1000}{2 \times 0.7 \times 1650 \times 6.8687 \times (0.5 \times 10^{-3})^3} = 4.54 \times 10^9 \text{ nuclei/m}^3 \text{ h}$$ **(Answer)**

(c) The total number of crystals produced per hour is the product of nucleation rate and the volume of mother liquor in the crystalliser.

$$n = B^o \times V = 4.54 \times 10^9 \times 6.8687 = 3.12 \times 10^{10} \text{ crystals/h}$$ **(Answer)**

Example 11.8: A continuous vacuum crystalliser conforming to MSMPR model produces 1000 kg/h of $MgSO_4.7H_2O$ crystals with a predominant crystal size of 0.4 mm. The magma will be 20% volume crystals. The residence time in the crystalliser will be 1.8 h. The density of the crystal is 1675 kg/m^3. The volumetric shape factor of the crystals may be assumed to be 0.7. Determine:

(a) The exiting flow rates of crystals, mother liquor and magma in m^3/h
(b) The crystalliser volume in m^3 if the vapour space equals the magma space.
(c) The approximate dimensions of the crystalliser, if the body is cylindrical with a height equal to twice the diameter.
(d) The required crystal growth rate in mm/h
(e) The necessary nucleation rate in nuclei/h m^3 of mother liquor in the crystalliser.

Solution: (a) Rate of production of crystals, C = 1000 kg/h. Volume of crystals produced in unit time =

$$V_C = \frac{1000}{1675} = 0.597\,\text{m}^3/\text{h}$$ **(Answer)**

Since crystals constitute 20% of the magma, volumetric flow rate of mother liquor =

$$Q = V_C \times \frac{0.8}{0.2} = 0.597 \times \frac{0.8}{0.2} = 2.39\,\text{m}^3/\text{h}$$ **(Answer)**

Volumetric flow rate of magma $= Q + V_C = 0.597 + 2.39 = 2.985\,\text{m}^3/\text{h}$ **(Answer)**

(b) The volume of mother liquor in the crystalliser $= V = \tau Q = 1.8 \times 2.39 = 4.30\ \text{m}^3$.

Since mother liquor occupies 80% of the magma volume, the volume of magma in the crystalliser

$$= 4.30 \times \frac{1}{0.8} = 5.38\ \text{m}^3.$$

If the vapour space equals the magma space, the volume of crystalliser $= 2 \times 5.38 = 10.76\,\text{m}^3$.

(Answer)

(c) If height equals twice the diameter of the cylindrical vessel, the volume of the vessel is

$$\frac{\pi}{4} D^2 \times (2D) = 10.76\,\text{m}^3, \text{ or } D = 1.9\ \text{m}$$

The diameter = 1.9 m and height = 2 × 1.9 = 3.8 m. **(Answer)**

(d) Crystal growth rate, $G = \dfrac{L_{pr}}{3\tau} = \dfrac{0.4}{3 \times 1.8} = 7.4 \times 10^{-2}\ \text{mm/h}$ **(Answer)**

(e) The rate of nucleation is given by

$$B^o = \frac{9C}{2a\rho_c V L_{pr}^3} = \frac{9 \times 1000}{2 \times 0.7 \times 1675 \times 4.3 \times (0.4 \times 10^{-3})^3} = 1.39 \times 10^{10}\ \text{nuclei/m}^3\text{h}$$ **(Answer)**

REVIEW QUESTIONS

1. Describe crystallisation from solution and distinguish it from other forms of crystallisation operations.
2. Describe the major industrial applications of crystallisation.
3. What is the advantage of crystallisation as a technique for separating components from solutions?
4. What is the basis for the classification of crystallographic systems? What are the different crystallographic systems?
5. Explain with examples the inverse solubility.
6. The choice of the method of crystallisation largely depends on the effect of temperature on the solubility of the substance. Justify.
7. What does the presence of breaks on the solubility curve signify? Explain with a suitable example.
8. How is yield in crystallisation defined?

9. Derive Eq. (11.1) from material balance.
10. Explain with examples, the condition under which the yield obtained will be less than that calculated using the solubility data.
11. What are the main sources of impurities in the crystallisation products?
12. How is the effect of temperature on the solubility and the heat of crystallisation related?
13. How are the heat of crystallisation and heat of solution related?
14. Define supersaturation, fractional supersaturation and supersaturation ratio.
15. What are the different methods used for achieving supersaturation?
16. Describe under what circumstances, we generally opt for the supersaturation by cooling and the supersaturation by evaporation.
17. Distinguish between the salting out and watering out processes.
18. Describe the Mier's theory of supersaturation. What are its limitations?
19. Distinguish between the primary and the secondary nucleation.
20. Describe homogeneous and heterogeneous nucleation.
21. Discuss the equation given by the classical chemical kinetics for estimating the rate of homogeneous nucleation.
22. Describe the relation between particle size and solubility. What is Kelvin equation?
23. Explain the phenomenon known as *Ostwald ripening.*
24. What are the factors affecting secondary nucleation?
25. Define the rate of growth of crystals.
26. What is invariant crystal?
27. Describe the overlapping principle in crystal growth.
28. State and explain delta-L law of crystal growth. When does this law fail?
29. What is the diffusion-reaction theory of crystal growth?
30. What are the different methods of classifying crystallisation equipment?
31. Explain the salient features of agitated batch crystalliser.
32. Describe the working of continuous cooling type scraped surface crystalliser.
33. Describe the construction and working of Oslo evaporative crystalliser.
34. What are the advantages of circulation evaporator crystalliser?
35. What is the advantage of providing draft tube in DTB crystalliser?
36. Describe the assumptions in the MSMPR model of crystalliser.
37. How is population density determined from cumulative size distribution of crystals?
38. Explain the fundamental relationship of MSMPR model that relates population density to the crystal size and growth rate.
39. What is predominant crystal size? How is it related to crystal growth rate?

MULTIPLE CHOICE QUESTIONS

11.1 Crystals of a given substance,

A. are always of the same size
B. are always of the same shape
C. have constant interfacial angles between corresponding faces
D. have fixed relative development of faces

11.2 How many crystallographic forms are generally available?

A. Three B. Five
C. Six D. Seven

11.3 The crystallographic system in which the three axes are at right angle to each other with one axis longer than the other two, is referred to as

A. Tetragonal system B. Cubic system
C. Orthorhombic system D. Monoclinic system

11.4 Crystal habit refers to

A. classification of crystal based on their shape
B. relative development of the different faces
C. classification based on crystallographic systems
D. caking of crystals in humid atmosphere

11.5 If a saturated aqueous solution contains 26% solute by weight the solubility is

A. 26 parts solute /100 parts water B. 35 parts solute /100 parts water
C. 21 parts solute/100 parts water D. None of the above

11.6 For substances whose solubility increases with increase in temperature, the best method for crystallisation is

A. evaporation B. addition of another solvent
C. cooling D. salting out

11.7 Example for a substance that exhibits inverse solubility is

A. KNO_3 B. NaCl
C. $Na_2SO_4.10H_2O$ D. $MnSO_4.H_2O$

11.8 Breaks on the solubility curves indicate

A. inverse solubility of the substance B. formation of different hydrates
C. influence of temperature on solubility D. formation of anhydrous crystals

11.9 What is the yield if 42.0 kg $Na_2SO_4.10H_2O$ crystals are formed when 100 kg of an aqueous solution containing 28% sodium sulphate is cooled to 293 K?

A. 42.0% B. 51.4%
C. 61.6% D. 66.1%

11.10 If 42.0 kg $Na_2SO_4.10H_2O$ crystals are formed when 100 kg of an aqueous solution containing 28% sodium sulphate is cooled to 293 K, what is the solubility expressed as parts of anhydrous salt/100 parts water?

A. 10.5 B. 19.6
C. 72.4 D. 78.8

11.11 Heat of crystallisation is approximately equal to

A. negative of the heat of fusion B. negative of the heat of dilution
C. heat of solution at infinite dilution
D. negative of the heat of solution at infinite dilution

11.12 For compounds whose solubility increases with temperature, the heat of crystallisation

A. will be positive B. will be zero
C. will be negative D. will be negligibly small

11.13 Which of the following processes is not a method for generating supersaturation in a solution?

A. Increasing the temperature by heating
B. Reducing the temperature by cooling
C. Addition of a new solvent
D. Evaporating a part of the solvent

11.14 In a solution which is free of solid particles, crystallisation begins if the solution is

A. unsaturated B. saturated
C. supersaturated D. at the freezing point

11.15 Mier's theory on crystallisation predicts
- A. the existence of a supersolubility curve
- B. primary and secondary nucleation processes
- C. that the growth rate of crystals is independent of their size.
- D. that the nuclei formation occurs when the solution is cooled to saturation.

11.16 Nucleation occurring in the presence of crystals is termed
- A. primary nucleation
- B. homogeneous nucleation
- C. heterogeneous nucleation
- D. secondary nucleation

11.17 Nucleation facilitated by suspended particles of foreign substances is generally referred to as
- A. primary nucleation
- B. homogeneous nucleation
- C. heterogeneous nucleation
- D. secondary nucleation

11.18 Small crystals dissolve and disappear while the large crystals grow in a solution consisting of small and large crystals. This is known as
- A. primary nucleation
- B. Ostwald ripening
- C. initial breeding
- D. habitat modification

11.19 Rate of homogeneous nucleation
- A. decreases with increase in supersaturation
- B. decreases with increase in temperature
- C. decreases with increase in surface tension
- D. is independent of temperature

11.20 Secondary nucleation is
- A. due to the influence of foreign materials present in the magma
- B. due to the influence of crystals already present in the magma
- C. due to the influence of the walls of the container and the agitator blades etc.
- D. not influenced by foreign materials, crystals already present or the container walls.

11.21 Which one of the following is not true with regard to invariant crystals?
- A. Maintains geometric similarity during growth
- B. A single dimension can be used to represent volume, surface area etc.
- C. The rates of growth of different faces are equal.
- D. The growth is uninfluenced by other crystals or solids.

11.22 Characteristic length of a crystal is defined as
- A. $L = \dfrac{V}{A}$
- B. $L = \dfrac{3V}{A}$
- C. $L = \dfrac{\pi V}{A}$
- D. $L = \dfrac{6V}{A}$

where V is the volume and A is the surface area of crystals.

11.23 If the growth of crystals in a solution is size-dependent, then
- A. the Delta-L law of crystal growth is valid
- B. the Delta-L law of crystal growth is not applicable
- C. the overlapping principle of crystal growth is not applicable
- D. the growth of each crystal in the solution during the same time interval Δt is the same and is given as $\Delta L = G\Delta t$, where G is the growth rate.

11.24 Cooling crystallisers are ideally suited for
- A. substances whose solubility decreases with increase in temperature
- B. substances whose solubility increases with increase in temperature

C. substances whose solubility is unaffected by change of temperature
D. substances that are heat sensitive.

11.25 Which of the following is not true for the Swenson-Walker crystalliser?

A. Continuous crystalliser B. Cooling crystalliser
C. Scraped-surface crystalliser D. Agitated batch crystalliser

11.26 Which one of the following is a circulating liquid evaporator crystalliser?

A. Swensen-Walker Crystalliser B. Krystal crystalliser
C. Draft-tube-baffle crystalliser D. Agitated batch crystalliser

11.27 The fundamental relation of MSMPR type crystalliser is

A. $\Delta L = G\Delta t$ B. $G = \dfrac{2K(y - y_s)}{\rho}$

C. $n = n^o \exp(-z)$ D. $\ln\alpha = \dfrac{4V_M\sigma}{vRTL}$

11.28 In MSMPR model, the predominant crystal size corresponds to the size interval dz at z equal to

A. zero B. one
C. two D. three

EXERCISES

Yield and purity of products

11.1 A solution of sodium sulphate in water is saturated at a temperature of 313 K. Calculate the weight of crystals and the percentage yield obtained when cooling 1000 kg of this solution to a temperature of 278 K. At 278 K decahydrate is the stable crystalline form. The solubility at 313 K is 32.6% and at 278 K is 5.75% Na_2SO_4. (**Answer:** 700.13 kg, 94.71%)

11.2 For the preparation of potassium nitrate, 10000 kg/h of a 20% KNO_3 solution is mixed with a recycle stream and sent to an evaporator. The rate of evaporation is 1.25 times the rate of introduction of recycle stream. The concentrated solution leaving the evaporator contains 50% KNO_3. This is admitted to the crystalliser which yields crystals containing 4% water. At the crystallisation temperature the solubility is 60 kg/100 kg water. Major part of the mother liquor leaving the evaporator is returned as recycle. Calculate:

(a) Concentration of KNO_3 in the stream entering the evaporator
(b) kg/h of recycle stream
(c) Rate of production of crystals.

(**Answer:** (a) 26.56% (b) 6000 kg/h (c) 1816.2 kg/h)

11.3 1000 kg of an aqueous solution of Na_2CO_3 containing 15% carbonate by weight is fed to a crystalliser. 80% of the carbonate is recovered as $Na_2CO_3.10H_2O$ by evaporation of water and subsequent cooling to 278 K. The solubility of Na_2CO_3 at 278 K is 9.0% (weight). Determine:

(a) The quantity of crystals formed.
(b) The amount of water evaporated

(**Answer:** (a) 323.8 kg (b) 342.9 kg)

11.4 An aqueous urea solution 90% saturated at 373 K is sent to a crystalliser where it is cooled to 303 K. If 90% of the urea is to be crystallised in anhydrous form, what fraction of the water must be evaporated? The solubility of urea per 100 kg water is 730 kg at 373 K and 135 kg at 303K. (**Answer:** 0.513)

11.5 A 20 weight percent solution of Na_2SO_4 at 367 K is sent to a vacuum crystalliser from which the product magma is pumped at 293 K. The magma is centrifuged to recover $Na_2SO_4.10H_2O$ crystals. The solubility of Na_2SO_4 at 293 K is 19.4 kg anhydrous salt per 100 kg water. Determine:

(a) The composition of the magma

(b) The percentage of Na_2SO_4 in the feed that is recovered as $Na_2SO_4.10H_2O$ crystals.

(**Answer:** (a) 13.5% crystals, 86.5% mother liquor (b) 29.8%)

11.6 A crystalliser is charged with 100 kg of a solution containing 25% $Ba(NO_3)_2$ in water. On cooling 10% of the original water present evaporates. Calculate the yield of crystals when the solution is cooled to 283K. The solubility at 283 K is 7.0 kg $Ba(NO_3)_2$/100 kg total water.

(**Answer:** 20.28 kg)

11.7 A batch of 1000 kg of KCl is dissolved in sufficient water to make a saturated solution at 363 K (solubility is 35 wt% KCl in water). The solution is cooled to 293 K. If the solubility of KCl in water at 293 K is 25.4 wt%,

(a) What is the weight of crystals of KCl obtained?

(b) What is the weight of crystals obtained if 5% of the original water evaporates on cooling?

(**Answer:** (a) 367.7 kg (b) 399.3 kg)

11.8 A saturated solution of KCl at 363 K containing 1000 kg KCl is subjected to crystallisation

(a) by cooling the solution to 293 K.

(b) by cooling to 293 K and evaporating 5% of water in the feed

Determine the yield of crystals in both cases if the solubility is 53.85 kg/100 kg water at 363 K and 34.05 kg/100 kg water at 293 K.

(**Answer:** (a) 367.7 kg, 36.8% (b) 399.3 kg, 39.93%)

11.9 200 kg of a 15% and 100 kg of 5% solutions of sodium sulphate by weight are mixed in a crystalliser and crystallisation takes place. If 50 kg $Na_2SO_4.10H_2O$ crystals are formed, compute the composition of the magma.

(**Answer:** 50 kg crystals, 250 kg saturated solution containing 5.2% Na_2SO_4)

11.10 What will be the yield of hypo ($Na_2S_2O_3.5H_2O$) if 100 kg of a 50% solution of $Na_2S_2O_3$ is cooled to 293 K. The solubility at 293 K is 70 parts $Na_2S_2O_3$ per 100 parts water.

(**Answer:** 39.15 kg, 49.9%)

11.11 Sodium carbonate is recovered as decahydrate from 1000 kg of a 10% solution of sodium carbonate in water. If it is desired that 90% of the Na_2CO_3 is to be recovered as decahydrate by cooling to 278 K, determine: (a) Weight of water evaporated (b) Weight of crystals obtained. Solubility of Na_2CO_3 in water at 278 K is 9.0% (weight). (**Answer:** (a) 646.06 kg (b) 242.83 kg)

11.12 A salt solution weighing 10000 kg with 30 wt% Na_2CO_3 is cooled to 293 K. The salt crystallises as the decahydrate. Determine the weight of crystals and the percentage yield if (a) no water is evaporated (b) 3% of total weight of the solution is lost by evaporation in cooling? The solubility at 293 K is 21.5 kg anhydrous Na_2CO_3/100 kg of total water.

(**Answer:** (a) 6353 kg, 78.5% (b) 6627 kg, 81.9%)

11.13 A hot solution containing 1000 kg of $MgSO_4$ and water having a concentration of 30% (weight) $MgSO_4$ is cooled to 289 K where crystals of $MgSO_4.7H_2O$ are precipitated. The solubility at 289 K is 24.5 wt% anhydrous $MgSO_4$ in the solution. Calculate the amount and percent yield of crystals obtained if 5% of the original water in the system evaporates on cooling.

(**Answer:** 261 kg, 42.5%)

11.14 One thousand kilograms of a 30% solution of Na_2CO_3 in water is cooled slowly to 293 K. During cooling, a portion of water is evaporated and removed. 700 kg $Na_2CO_3.10H_2O$ crystals are formed in the process. What percent of total water in the feed is evaporated? Solubility of anhydrous Na_2CO_3 at 293 K is 21.5 kg/100 kg water. (**Answer:** 10.1%)

11.15 An aqueous solution of Na_2CO_3 contains 15% carbonate and small amounts of soluble impurities. 80% of the carbonate is recovered as $Na_2CO_3.10H_2O$ by evaporation of water and subsequent cooling to 278 K. The solubility of Na_2CO_3 at 278 K is 9.0% (weight). On the basis of 100 kg solution treated, determine:

(a) The amount of water evaporated

(b) The quantity of crystals formed.

(**Answer:** (a) 34.3 kg (b) 32.4 kg)

11.16 An aqueous solution containing 62% $Na_2S_2O_3$ and 0.5% soluble impurities is diluted with water and fed to a crystalliser where it is cooled in order to crystallise $Na_2S_2O_3.5H_2O$. The crystals carry 0.05 kg solution (excluding impurities) per kg crystals. The free water present in the adhering solution is removed on drying the crystals. The final dried product contains not more than 0.05% impurity. The solubility of the pentahydrate is 1.5 kg $Na_2S_2O_3.\ 5H_2O$ /kg free water. On the basis of 100 kg 62% solution, calculate:

(a) The kg water added before cooling

(b) Amount of crystals formed

(c) The percentage recovery of the $Na_2S_2O_3$ in the dried hydrated crystals.

(**Answer:** (a) 17.23 kg (b) 68.19 kg (c) 72.2%)

11.17 A crystalliser is charged with 10000 kg of a solution containing 30% by weight anhydrous sodium sulphate at 373K. 15% of the original water is lost by evaporation on cooling. The salt $Na_2SO_4.10H_2O$ crystallises out. If the solution after crystallisation is found to contain 20% anhydrous sodium sulphate, calculate:

(a) Weight of solution left after crystallisation

(b) Weight of crystals obtained.

(**Answer:** (a) 3929.3 kg (b) 5020.7 kg)

11.18 A solution containing 30% $MgSO_4$ and 70% water is cooled so that $MgSO_4.7H_2O$ crystals are obtained on crystallisation. During this process 10% of the total water present in the feed is lost due to evaporation. The solubility of anhydrous $MgSO_4$ in water at this temperature is such that 40% of the anhydrous salt is present in the saturated solution. On treating 1000 kg of solution, calculate:

(a) Weight of $MgSO_4.7H_2O$ crystals obtained

(b) Solubility of anhydrous salt in water, kg salt/100kg water.

(**Answer:** (a) 368.6 kg (b) 27.2 kg/100 kg water)

11.19 A 30% (weight) solution of $NaNO_3$ in water is concentrated to saturation by evaporation at 373 K. It is then cooled to 293 K. The crystals formed are separated. It is found that each kg of crystals so removed carries with it 0.1 kg solution. When the crystals are dried, the $NaNO_3$ in the adhering solution gets deposited on the crystals. The solubility of $NaNO_3$ in water is 176 kg/100 kg water at 373 K and 88 kg/100 kg water at 293 K. For 1000 kg of the original dilute solution, calculate:

(a) The amount of water evaporated for attaining saturation at 373 K

(b) The weight of dry crystals obtained.

(**Answer:** (a) 529.55 kg (b) 157 kg)

11.20 A dilute salt solution containing 6% salt is fed to an evaporator at a rate of 10000 kg/h for a trial run. Crystals formed along with the saturated solution adhering with them are sent to a centrifuge for recovery of the solution. 60% of the adhering solution is recovered. It is found that the saturated solution that is withdrawn from the evaporator is 850 kg and that recovered by centrifuging is 200 kg. The crystals are dried to drive off the remaining water. The dry crystals obtained weighed 360 kg. Determine:

(a) The solubility of the salt in kg salt/100 kg water.
(b) Kilograms of salt crystallised in the evaporator.
(c) The water evaporated in the evaporator.
(d) The water removed in the drier.

(**Answer:** (a) 29.63 kg/100 kg water (b) 329.5 kg (c) 8487.15 kg (d)102.85 kg)

11.21 5000 kg/h of thick liquor of composition 38% $MgSO_4$ and 62% water is mixed with 10000 kg/h of recycled filtrate solution of $MgSO_4$ saturated at 300 K. The combined stream is then admitted to a vacuum crystalliser. The magma from the crystalliser consists of 21% $MgSO_4.7H_2O$ crystals and 79% saturated solution. The magma is filtered and the filtrate is recycled as saturated solution at 300K. The solubility of $MgSO_4$ at 300 K is 28%. Determine:
(a) Water evaporated in kg/h.
(b) Maximum production rate of crystals in kg/h.

(**Answer:** (a) 484.9 kg (b) 3048.2 kg)

11.22 It is desired to produce 1000 kg/h of $Na_3PO_4.12H_2O$ crystals from a feed solution containing 6% (weight) Na_3PO_4 and traces of impurity. The original solution is first evaporated in an evaporator to a 40 % (weight) Na_3PO_4 solution and then cooled to 293 K in a crystalliser, where the hydrated crystals and mother liquor solution are removed. One out of every 10 kg of mother liquor is discarded to waste to get rid of the impurities, and the remaining mother liquor is recycled to the evaporator. The solubility of Na_3PO_4 at 293 K is 9.9% (weight). Calculate:
(a) Feed rate, kg/h.
(b) Rate of evaporation of water, kg/h.
(c) Rate at which the mother liquor is recycled, kg/h.
(d) Concentration of the solution entering the evaporator.

(**Answer:** (a) 7210 kg/h (b) 6200 kg/h (c) 94.4 kg/h (d) 6.0%)

11.23 In a process producing KNO_3 salt, 1000 kg/h of a feed solution containing 10% KNO_3 is fed to an evaporator which evaporates some water to produce a 50 % KNO_3 solution. This is then fed to a crystalliser, where crystals containing 95% KNO_3 are removed. The saturated solution containing 35% KNO_3 is recycled to the evaporator. Calculate:
(a) Amount of recycle stream, kg/h.
(b) Amount of crystals, kg/h.
(c) Quantity of water evaporated, kg/h.

(**Answer:** (a) 315.8 kg/h (b) 105.26 kg/h (c) 894.74 kg/h)

Heat effects in crystallisation process

11.24 1000 kg/h of $FeSO_4.7H_2O$ crystals are to be produced in a Swenson-Walker crystalliser by cooling a saturated solution at 323 K (solubility is 140 parts $FeSO_4.7H_2O$ per 100 parts water) to 300 K (solubility is 74 parts $FeSO_4.7H_2O$ per 100 parts water). The feed solution has an average specific heat of 2.93 kJ/kg K Heat of solution of $FeSO_4.7H_2O$ at 300 K is 18.40 kJ/mol. How much heat is evolved during the process? (**Answer:** 311279.8 kJ/h)

11.25 $MgSO_4.7H_2O$ crystals are produced by cooling 2500 kg solution at 330 K containing 50 kg $MgSO_4$/100 kg water to 293K. The solubility of the salt at 293 K is 35.5 kg $MgSO_4$/100 kg water. The average heat capacity of the feed solution can be assumed to be 2.93 kJ/kg K. The heat of solution at 293 K is 13.31×10^3 kJ/kmol $MgSO_4.7H_2O$. Determine:
(a) Yield of crystals
(b) Total heat removed

(**Answer:** (a) 787.68 kg (b) 3.1359×10^5 kJ)

11.26 $Na_2SO_4.10H_2O$ crystals are produced from 3000 kg solution containing 20 kg Na_2SO_4/100 kg water at 333 K. The solution is cooled to 283 K in an agitated mild steel vessel of mass 800 kg. During cooling 2 percent of the water initially present is lost by evaporation. At 283 K, the solubility of the anhydrous salt is 9 kg/100 kg water. The heat of solution is 78.3 kJ/mol and heat capacities of the solution and mild steel are 3.6 and 0.5 kJ/kg K respectively. Latent heat of vaporisation of water at 283 K is 2478 kJ/kg. Estimate the heat which must be removed.

(**Answer:** 6.01×10^5 kJ)

11.27 A Swenson-Walker crystalliser is to be used to produce 1000 kg/h of $FeSO_4.7H_2O$ crystals by the cooling of a saturated solution which enters the crystalliser at 323 K. The slurry leaving the crystalliser will be at 300 K. Solubility at 323 K = 46.8 parts $FeSO_4$ per 100 parts water. Solubility at 300 K = 30.3 parts $FeSO_4$ per 100 parts water. Cooling water enters the crystalliser jacket at 289 K and leaves at 294 K. It may be assumed that the overall coefficient of heat transfer for the crystalliser is 200 W/m^2 K. Average specific heat of the feed solution is 2.93 kJ/kg K and heat of solution in the temperature range of interest for $FeSO_4.7H_2O$ is 18.4 kJ/mol.

(a) Estimate the cooling water required in kg/h

(b) Heat transfer area.

(**Answer:** (a) 14877 kg/h (b) 23.3 m^2)

11.28 One thousand kg/h of Glauber's salt ($Na_2SO_4.10\ H_2O$) is to be made in a Swenson-Walker crystalliser by cooling a saturated Na_2SO_4 solution from 303 K to 288 K. The solubility of sodium sulphate is 40 parts Na_2SO_4 per 100 parts of water at 303 K, and 13.5 parts at 288 K. The heat of solution of $Na_2SO_4.10H_2O$ is 78.37 kJ/mol Na_2SO_4. Cooling water enters at 283 K and leaves at 293 K. The overall heat transfer coefficient in the crystalliser is 250 W/m^2 K and each 1-m length of crystalliser has 2 m^2 of cooling surface. The heat capacity of the feed solution may be assumed to be 3.6 kJ/kg K. How many 3-m units of crystalliser will be required?

(**Answer:** 9 units)

11.29 Glauber's salt $Na_2SO_4.10H_2O$ is to be produced at a rate of 1000 kg/h in a Swenson-Walker crystalliser by cooling a saturated solution of Na_2SO_4 at 300 K to 290 K. Cooling water enters the unit at 280 K and leaves at 290 K and evaporation is negligible. The solubilities of anhydrous Na_2SO_4 in water are 40 and 14 kg/100 kg water at 300K and 290 K respectively, the mean heat capacity of the liquor is 3.8 kJ/kg K and the heat of crystallisation is –243 kJ/kg. For the crystalliser, the available heat transfer area is 3 m^2/m length, the overall coefficient of heat transfer is 0.15 kW/m^2 K. How many sections of crystalliser, each 3 m long, will be required?

(**Answer:** 7 units)

11.30 A continuous evaporative crystalliser operating at 1 kPa is fed with 4000 kg/h of a 50% (weight) aqueous solution of sodium acetate at 350K. The boiling point elevation of the solution is 10 K and the heat of crystallisation is –150 kJ/kg. The mean heat capacity of the solution is 3.5 kJ/kg K and, at 1 kPa, water boils at 280 K at which the latent heat of vaporisation is 2482 kJ/kg. At 290 K, the solubility of sodium acetate in water is 54 kg/100 kg water. Determine the yield of $CH_3COONa.\ 3H_2O$ crystals and the evaporation rate. (**Answer:** 3098.4 kg/h and 525.5 kg/h)

11.31 An aqueous feed solution containing 28% $FeSO_4$ by weight is admitted to a vacuum crystalliser to produce 10000 kg/h $FeSO_4.7H_2O$ crystals. The feed enters the crystalliser at 343 K and the vacuum is such as to produce a crystalliser temperature of 303 K. Solubility of $FeSO_4$ is 26.5 parts per 100 parts of water at 303 K. Enthalpy of the feed is 110 kJ/kg, enthalpy of saturated solution and $FeSO_4.7H_2O$ crystals at 303 K are, respectively, –560 kJ/kg and –212 kJ/kg. Enthalpy of saturated steam at 303 K = 2556 kJ/kg. Boiling point elevation may be neglected. How much feed solution is to be treated? (**Answer:** 31192 kg/h)

11.32 A solution of 32.5 weight percent $MgSO_4$ originally at 339 K is to be crystallised in a vacuum adiabatic crystalliser to give a product containing 4000 kg/h of $MgSO_4.7H_2O$ crystals from 10000 kg/h of feed. The solution BPR is estimated to be 5 K. The average heat capacity of the feed solution can be assumed to be 2.93 kJ/kg K and the heat of solution to be 13.31×10^3 kJ/kmol $MgSO_4.7H_2O$. The solubility (s kg anhydrous salt/100 kg water) is related to the solution temperature (T, K) by $s = 0.494T - 109$.

Determine: (a) The product temperature (b) Crystalliser pressure (c) Weight ratio of mother liquor to crystalline product.

(**Answer:** (a) 286 K, (b) 1.07 kPa (c) 1.33)

11.33 A continuous adiabatic vacuum crystalliser is to be used for the production of $MgSO_4.7H_2O$ crystals from 10,000 kg/h of solution containing 30% by weight $MgSO_4$. This solution enters the crystalliser at 344 K. The magma leaving the crystalliser contains 3000 kg/h of $MgSO_4.7H_2O$ crystals. The estimated BPE of the solution in the crystalliser is 5 K. The average heat capacity of the feed solution can be assumed to be 2.93 kJ/kg K and the heat of solution to be 13.31×10^3 kJ/kmol $MgSO_4.7H_2O$. The solubility (s kg anhydrous salt/100 kg water) is related to the solution temperature (T, K) by $s = 0.494T - 109$

Determine:

(a) Kilograms of water evaporated per h.

(b) The temperature of magma leaving the crystalliser.

(c) The crystalliser pressure.

(**Answer:** (a) 741.4 kg (b) 286.5 K (c) 1.11 kPa)

Crystalliser modelling – The MSMPR crystalliser

11.34 An MSMPR crystalliser operates with a mixed product removal rate based on mother liquor of 2×10^{-4} m^3/s. The steady nucleation rate is $n^o = 1.2 \times 10^{13}$/m^4 and the growth rate is $G = 10^{-8}$ m/s. The crystal density is 2500 kg/m^3 and the volume shape factor is 0.6. The volume of the vessel, based on mother liquor is 4 m^3, Determine:

(a) The mass of crystals per unit volume of liquid, m_c.

(b) The rate of production of crystals, kg/h.

(c) The percent of nuclei discharged by the time they have grown to 0.1 mm.

(d) If it is desired to discharge 80% of the original nuclei by the time they have grown to 0.1 mm, what additional liquid flow rate is to be provided?

(**Answer:** (a) 172.8 kg/m^3, (b) 124.4 kg/h (c) 39.4% (d) 4.44×10^{-4} m^3/s)

11.35 A continuous vacuum crystalliser conforming closely to the MSMPR model produces 1000 kg/h of crystals of density 1650 kg/m^3. The crystals constitute 15% of the magma volume and the residence time of the magma in the crystalliser is 2 h. The desired predominant crystal size on mass basis is 0.4 mm. Assume a volumetric shape factor of 0.6. Determine:

(a) The required crystal growth rate, mm/h.

(b) Rate of nucleation B^o, nuclei/h m^3 of mother liquor in the crystalliser.

(c) Number of crystals produced per hour.

(d) Plot the cumulative screen analysis of the theoretical product.

(**Answer:** (a) 0.067 mm/h (b) 1.03×10^{10} (c) 7.08×10^{10})

11.36 A continuous vacuum crystalliser conforming to MSMPR Crystalliser model produces 5000 kg/h of $MgSO_4.7H_2O$. The crystals constitute 15% of the magma volume. The density of crystal is 1650 kg/m^3 and the growth rate is $G = 0.55$ mm/h. The desired predominant crystal size on mass basis is 0.8 mm. The shape factor is assumed to be 0.7. The crystalliser has a free space above the magma level equal to 30% of the magma volume in the crystalliser. Length to diameter ratio of the crystalliser is 1.5. Determine:

(a) Volume of magma in the crystalliser.

(b) Nucleation rate B^o.

(c) The length and diameter of the crystalliser.

(d) Construct a table for different screen analysis for mesh numbers 8 to 48.

(**Answer:** (a) 9.8 m^3 (b) 4.57×10^9 nuclei/m^3 h (c) $L = 3.3$ m, $D = 2.2$ m)

11.37 A continuous vacuum crystalliser conforming to MSMPR model produces 1000 kg/h of $MgSO_4.7H_2O$ crystals with a predominant crystal size of 0.5 mm. The magma will be 15% volume crystals. The residence time in the crystalliser will be 2 h. The density of the crystal is 1680 kg/m^3. The volumetric shape factor of the crystals may be assumed to be 0.6. Determine:

(a) The exiting flow rates of crystals, mother liquor and magma in m^3/h

(b) The crystalliser volume in m^3 if the vapour space equals the magma space.

(c) The approximate dimensions of the crystalliser, if the body is cylindrical with a height equal to twice the diameter.

(d) The required crystal growth rate in mm/h.

(e) The necessary nucleation rate in nuclei/h m^3 of mother liquor in the crystalliser.

(f) The number of crystals produced per hour.

(**Answer:** (a) 0.5952 m^3/h, 3.373 m^3/h, 3.9682 m^3/h (b) 15.87 m^3 (c) $D = 2.16$ m, $L = 4.32$ m (d) 8.33×10^{-2} mm/h (e) 5.29×10^9 nuclei/h m^3 (f) 3.57×10^{10}/h)

11.38 A crystalliser with negligible evaporation is fed with 10,000 kg/h an aqueous solution of $BaCl_2$ saturated at 373 K. The magma leaves the crystalliser at 293 K with crystals of the $BaCl_2.2H_2O$. The magma occupies a volume of 2.0 m^3 in the crystalliser. The growth rate of crystals is essentially constant at 4.0×10^{-7} m/s. The crystalliser can be assumed to be of the MSMPR model. The density of the crystals is 3097 kg/m^3 and the volumetric shape factor is 0.6. Density of an aqueous saturated solution of barium chloride at 293 K is 1290 kg/m^3. The solubility of $BaCl_2$ at 293 K and 373 K are, respectively, 35.7 parts and 58.3 parts per 100 parts of water. Determine:

(a) The kg/h of crystals in the product.

(b) The predominant crystal size in mm.

(c) The mass fraction of product crystals of size 0.78 mm.

(**Answer:** (a) 1784.7 kg/h (b) 1.24 mm (c) 0.169)

11.39 An MSMPR crystalliser is used to crystallise $CuSO_4.5H_2O$ with a predominant size of 1.4 mm size. The growth rate is estimated to be 2×10^{-7} m/s. The density of the crystals is 2300 kg/m^3 and the geometric shape factor is 0.20. The magma consists of 0.35 m^3 crystals per cubic meter of mother liquor. Determine:

(a) The production rate in kg/h m^3 of mother liquor.

(b) Rate of nucleation, nuclei/h m^3 of mother liquor.

(**Answer:** (a) 1242 kg/m^3 h (b) 4.428×10^9 nuclei/h m^3)

Chapter 12

Membrane Separations

Membrane separation takes advantage of the difference in the rates of movement of the components of a liquid or gaseous mixture through a semipermeable barrier called membrane for effecting separation of the constituents of the feed. Membrane separation involves the movement of molecules between two liquid phases, two gas phases or a liquid and a gas phase, the phases being physically separated by the membrane. The membrane is a thin film that permits the preferential transport of selected molecules from one phase to the other, but prevents the ordinary hydrodynamic flow of the phases. Separation occurs on the basis of the molecular size, shape, electric charge and/or sorption characteristics of the materials. This separation technique considered to be unconventional with reference to its industrial applications, has its basis in biological systems, where nature separates biologically active components through membranes. Though the potential of the membranes for industrial separations were recognized long back, the commercial exploitation was slow because the membranes offering good selectivity and permeability, the two most important requirements for effective separations, were very few. A dramatic change has occurred since 1960s in the acceptability of this novel technique for industrial separation as a result of the combined effort of physical chemists, material scientists and chemical engineers in developing and adapting synthetic membranes for process separation. Membrane separations now find widespread application in industries such as chemical, petrochemical, pharmaceutical, paper, textile, food and beverage, environmental and electronic industries. Membrane processes are now applied to a range of difficult separations. Separations not realisable with the conventional methods such as distillation, for example, the separation of azeotropic mixtures, can be easily done using the membrane processes.

The major breakthrough which transformed the membrane separation from the laboratory application to an industrial process was the development of the Loeb-Sourirajan process in the early 1960s for making the defect free, high flux membranes that made the reverse osmosis a potentially practical method of desalting saline water. Concurrent with this development was the progress made independently in the construction of membranes for medical applications, particularly the artificial kidneys. The period from 1960–1980 produced significant change in the status of membrane separation processes. Specialized modules which can hold large membrane area were developed during this period and considerable progress was also made in improving membrane stability. By 1980, the membrane processes such as the microfiltration, ultrafiltration, reverse osmosis and electrodialysis became established processes with large plants commissioned worldwide. The principal development of the 1980s was the emergence of industrial membranes for the gas separation processes and the commercial pervaporation systems for dehydration of alcohol. Membrane separation has gained much importance in recent years, due to the growth of the newer biotechnological industries and sophisticated technology for food processing.

Since the membrane separation processes can be carried out at moderate temperatures, they are ideally suited for separating heat sensitive materials and as such, find wide applications in the food and pharmaceutical industries. Since heating is not generally required, membrane processes are less energy intensive than the conventional thermal separation processes. In contrast to huge vessels required for conventional separation, membrane separation can be carried out in equipment that are modular, compact, less capital intensive, and more easily operated, controlled and maintained. The separation is purely physical and works without the addition of chemicals; the products are thus free from the contamination by external separating agents.

Membrane separation techniques are not entirely free from shortcomings. A major drawback of these processes is due to the modular construction of the separators. Whereas in conventional separation methods, it is possible to design a single large piece of equipment for high capacity applications, membrane separation requires running of several parallel units for separation on an industrial scale. Another serious limitation of the membrane systems is their poor stability and vulnerability to scaling and fouling.

The commercially important membrane separation techniques are microfiltration (MF), ultrafiltration (UF), nanofiltration (NF), reverse osmosis (RO), pervaporation (PV), gas permeation (GP) and electrodialysis (ED). These operations are discussed in detail under sections 12.6 through 12.13. Important industrial applications of the membrane separations include the desalination of sea water, brackish water, the production of drinking water and process water, the treatment of surface water and ground water, the production of table salt from sea water, dehydration of the ethanol-water azeotrope, separation of nickel sulphate from sulphuric acid, removal of water from organic solvents, separation of CO_2 and H_2 from other hydrocarbons, separation of air into nitrogen and oxygen enriched gases, recovery of methane from biogas, etc. to cite a few. Treatment of waste water to remove impurities, colour removal from the Kraft sulphite pulp liquor in paper making, removal of organics from water are some of the important applications in the area of environmental protection. Microelectronic industries use the membrane separation technology for the production of ultra-pure water required for their process. Food, beverage and pharmaceutical industries employ membrane separation techniques for a wide variety of applications. Concentration of foodstuffs, demineralisation of cheese whey, clarification of fruit juices, clarification and biological stabilisation of beverages, removal of alcohol from beer and wine, recovery and purification of vaccines and antibiotics from the fermentation broth and sterilisation of drugs are examples of such applications. Membrane separations have got wide application as artificial kidney to remove the toxic substances in the blood by haemodialysis and for artificial lungs for the bubble-free supply of oxygen in the blood. Membrane separation technology finds application even in the modern energy recovery systems such as the fuel cells.

12.1 DEFINITIONS

A membrane separation operation can be schematically represented as shown in Fig. 12.1. The feed stream which may be a gas or liquid is passed across the membrane. The membrane permits certain species to pass through while retaining certain others. The key property that determines the suitability of a membrane for separation is its ability to control the permeation rate of certain chemical species through it. The part of the feed that does not pass through the membrane is designated as the *retentate* and the part that passes through the membrane is called the *permeate*.

Let us define some common terms used to measure the capacity and effectiveness of the membranes as a separating medium. Membranes are generally characterized by two key process parameters-the *flux* and the *selectivity*. The productivity of a membrane module is described by its *flux* N which is defined as the volumetric permeate flow through the membrane per unit area in unit time. SI unit of flux is m^3/m^2 s. A relatively high flux rate imply that a relatively small membrane area is required for the desired separation.

Permeability is the resistance of a material to the flow of a fluid through it. The permeability P_{Mi} is defined as the ratio of transport flux of a material through the membrane to the driving force (concentration, partial pressure, etc.) per unit membrane thickness.

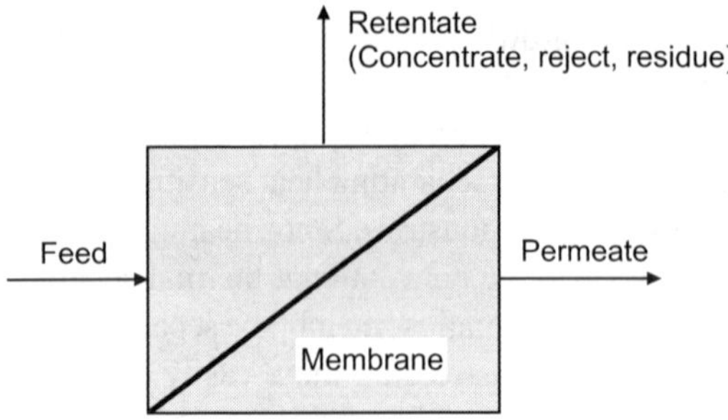

Fig. 12.1: Schematic representation of membrane separation operation

$$P_{Mi} = \left(\frac{N_i}{\text{driving force}} \right) (l_M) \qquad \text{...(12.1)}$$

where P_{Mi} is the permeability of species i through the membrane and l_M is the membrane thickness. The *permeance* $\bar{P}_{Mi}$ of a membrane is defined as the ratio of the permeability P_{Mi} to membrane thickness l_M. To be effective for separating a mixture of chemical components, a membrane must possess a high permeance. From Eq. (12.1) it follows that

$$\bar{P}_{Mi} = \frac{P_{Mi}}{l_M} = \left(\frac{N_i}{\text{driving force}} \right) \qquad \text{...(12.2)}$$

Equation (12.2) reveals that the permeance for a given species diffusing through a membrane of given thickness is, in fact, the rate of transport of that species per unit cross sectional area of the membrane per unit driving force and hence is analogous to a mass-transfer coefficient. Equation (12.2) shows that the molar trans-membrane flux of species i can be expressed in terms of the permeability or the permeance as

$$N_i = \left(\frac{P_{Mi}}{l_M} \right) (\text{driving force}) = \bar{P}_{Mi} (\text{driving force}) \qquad \text{...(12.3)}$$

The permeation flux depends on both the intrinsic permeability and the effective thickness of a membrane.

Trans-membrane pressure is defined as the pressure gradient between the upstream (retentate side) and the downstream (permeate side). It is the average pressure at the inlet and outlet of the equipment. It is evident from Eq. (12.1) that the permeability is a measure of the sensitivity of productivity or flux to the trans-membrane pressure (TMP).

The efficiency of a membrane module characterized by the *recovery* or *conversion ratio* (*CR*) is defined as the ratio of the permeate flow rate to the feed flow rate. In the single component separation, the *passage* of a component through a membrane (also called the *sieving coefficient* or transmission) is calculated as the ratio $S = c_P/c_F$, where c_P and c_F are the permeate concentration and the feed concentration, respectively. The *retension* or *rejection* of a component is defined as $R = 1 - S$. When considering the multiple component separation, an important characteristic used to describe the ability to separate two components i and j is the *selectivity* or *separation factor* α_{ij}, defined as

$$\alpha_{ij} = \frac{c_{iP}/c_{jP}}{c_{iF}/c_{jF}} \qquad \text{...(12.4)}$$

Thus, selectivity is the ratio of relative concentrations of i and j in the permeate to that in the feed. It measures the ability of the membrane to prevent the passage or "reject" one or more species in the feed. When the selectivity factor is unity, no separation occurs. When it approaches infinity, the membrane becomes perfectly semi-permeable. The selectivity relates the permeability of the membrane to the species in question and is governed by the intrinsic nature of the membrane material which in turn, can be attributed to its method of manufacture.

The *molecular weight cut-off* (MWCO) is a specification commonly used by the manufacturers to describe the retention capabilities of a membrane. It is defined as the lowest of the molecular weight of the solutes which are retained by the membrane to an extent of at least 90%. A membrane with molecular weight cut off 100 can reject minimum 90% of the materials with molecular weight of 100. MWCO in effect describes the pore size of a membrane. This specification is most commonly used to characterise the ultrafiltration and nanofiltration membranes.

12.2 MEMBRANES

A membrane can be treated as an "interphase" separating the two phases which allows the separation of the constituents by selectively permeating one or more components while retarding the passage of one or more other components. Membrane separation processes were once considered to be very unreliable, unselective, slow and expensive. Significant improvements are made in the last 50 years in the development of new membrane materials with high selectivity and permeability, development in the fabrication methods for high-flux asymmetric or composite membranes, and in the membrane module construction. Now we have a plethora of membrane materials from which we can make the choice depending upon the separation process. The efficiency and the commercial viability of a process are largely dependent on the properties of the membrane and the manner in which it is packaged and modularised. Desirable membrane characteristics are: (1) good permeability (2) high selectivity (3) mechanical robustness to allow the fabrication of modules and to withstand the operating conditions such as the large pressure difference between the phases (4) chemical robustness to fabrication materials, process fluids, cleaning agents etc. (5) low fouling tendencies (6) amenability to fabrication and packaging (7) high capacity (8) stability (9) reasonable useful life and consistency of performance (10) low cost.

Materials

Membranes can be broadly classified as natural or synthetic. The natural membranes include wool, cellulose and rubber. The modern separation industries largely depend on the synthetic membranes which can be fabricated from a large number of different materials. Synthetic membranes can be further classified as organic and inorganic membranes. This classification is detailed in Fig. 12.2. Synthetic polymeric membranes and liquid membranes are organic membranes whereas ceramic, metal, carbon

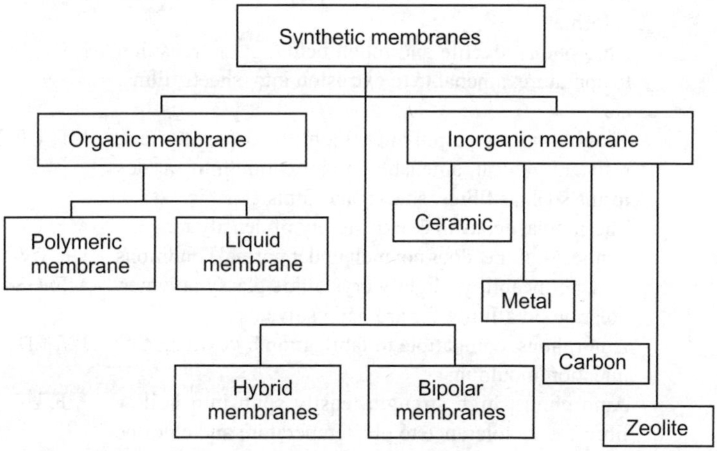

Fig. 12.2: Classification of synthetic membranes on the basis of membrane materials

and zeolite are inorganic membranes. Hybrid membranes using both the organic and inorganic components and the bipolar membranes made of membranes of different ionic charges are recent developments in the membrane separation field.

Polymeric membranes

Majority of industrial membrane separation processes employ membranes made from natural or synthetic polymers. Polymeric membranes lead the membrane separation industry market because they are very competitive in performance and attractive from the cost considerations. Synthetic polymers are produced by condensation or addition polymerisation or by the copolymerisation of the two different monomers. The most common polymers used in the membrane synthesis are cellulose triacetate, nitrocellulose, cellulose esters, aromatic polyamide, polycarbonate, polyimide, polyethylene, polypropylene, polystyrene, polysulphone, polytetrafluroethylene (Teflon), polyacrylonitrile, and polyvinylchloride.

Table 12.1 lists the different polymeric membrane materials, their properties and major applications. To be effective in separation operations, membrane polymers should have certain characteristics. These include their compatibility to mould into films, their chain rigidity, acceptable polarity of the functional groups, low binding affinity for the separated component molecules, compatibility with the fabrication technology and the cleaning conditions, disinfectants and detergents and the availability at reasonable cost.

The classification of polymers as amorphous or crystalline is important in their application as membrane materials. Amorphous polymers are brittle and glassy in appearance and become rubbery at temperatures above their glass transition temperatures. Examples are polycarbonate, polyimide, polystyrene, polysulphone, etc. The crystalline polymers are crystalline in structure, brittle, hard and stiff and become a melt above their melting temperatures. Examples are cellulose triacetate, aromatic

Table 12.1: Polymer membrane materials, their properties and applications

Membrane material	Properties	Application
Cellulose triacetate	Easy to manufacture, highly crystalline, uniformly high quality, hydrophobic, high salt rejection	MF, UF, NF, RO, GS, PV
Cellulose nitrate	Brittle, low level of extractables, stable at high temperatures, hydrophilic	MF
Aromatic polyamides	High melting crystalline polymer, thermally stable, resistant to solvents, easily fabricated into fibres, films and sheets, wide tolerance to pH, temperature and chlorine	MF, UF, NF, RO
Polycarbonates	Amorphous, ductile and tough below glass transition temperature, amenable to extrusion into sheets, films, etc.	MF
Polyimides	Tough amorphous polymer, highly resistant to heat, resistant to wear, amenable to fabrication into various forms such as fibres, sheets, and films etc.	MF, UF, NF, RO
Polyethylene	Thermoplastic, available in variety of density range	MF, PV, MD
Polyacrylonitrile	Semicrystalline, does not melt under normal conditions	UF, PV
Polyvinylchloride	Low permeability, slightly crystalline glassy polymer, polar, good affinity for aromatic solvents	MF, GS
Polystyrene	Amorphous, compatible to fabrication by extrusion or injection moulding	PV, ED
Polysulphone	Amorphous, high strength, easily spun into hollow fibres, wide tolerance to pH, temperature and chlorine	MF, UF
Polytetrafluroethylene (Teflon)	Crystalline, high strength, thermally stable, can be fabricated into sheets, films and tubings	MF, UF, PV

polyamide, polytetrafluroethylene, etc. Membranes made of crystalline polymers can not be operated above their melting temperatures, whereas the membranes made of amorphous polymers can be operated above or below their glass transition temperatures.

Compared to certain inorganic membranes, the polymeric membranes have relatively low separation performance. However, the processing flexibility and low cost still make them highly attractive for many industrial applications.

Membrane characterisation

Based on the morphology or structure, synthetic membranes can be classified as isotropic (symmetric) membranes and anisotropic (asymmetric) membranes. This classification based on the membrane structure is important because, it is the structure that determines the mechanism of permeation through the membrane. In isotropic membranes, the properties are the same throughout the membrane, whereas the properties of an anisotropic membrane vary with direction, for it has different chemical and/or physical structures in the direction of its thickness. The principal types of polymeric membranes are shown in Fig. 12.3.

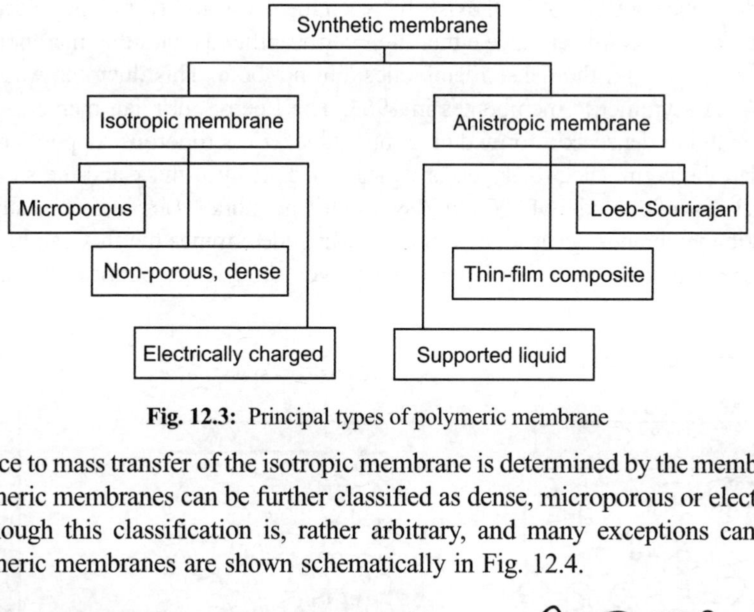

Fig. 12.3: Principal types of polymeric membrane

The resistance to mass transfer of the isotropic membrane is determined by the membrane thickness. Isotropic polymeric membranes can be further classified as dense, microporous or electrically charged membranes, though this classification is, rather arbitrary, and many exceptions can be cited. The isotropic polymeric membranes are shown schematically in Fig. 12.4.

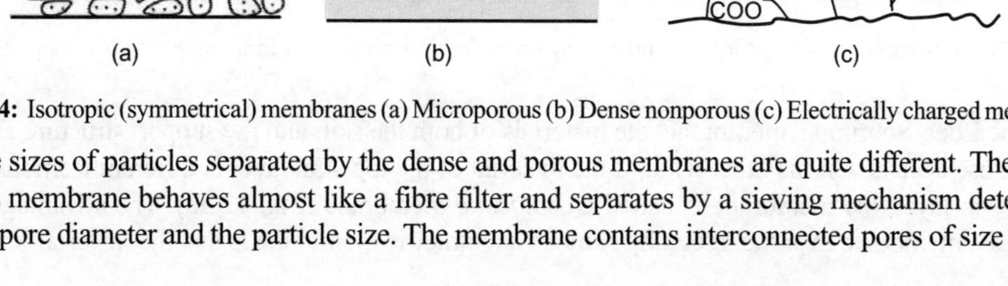

Fig. 12.4: Isotropic (symmetrical) membranes (a) Microporous (b) Dense nonporous (c) Electrically charged membranes

The sizes of particles separated by the dense and porous membranes are quite different. The microporous membrane behaves almost like a fibre filter and separates by a sieving mechanism determined by the pore diameter and the particle size. The membrane contains interconnected pores of size ranging from 1 nm to 20 μm. For microporous membranes, separation is affected by difference in the rates of diffusion through the pores. Though a near perfect separation by size is possible for mixtures which are made up of the component molecules both larger and smaller than the pore size, selectivity is generally low for the feed containing only small molecules. However, the permeability for microporous membrane is generally higher than that of the dense membranes. Microporous membranes find use in the microfiltration, ultrafiltration, and dialysis applications.

The molecular transport through a dense homogeneous membrane can occur under pressure, concentration or electrical potential gradient. Chemical species of similar size and diffusivity can be separated efficiently when their concentrations differ significantly. For dense amorphous membranes, diffusing species must first dissolve into the polymer and then diffuse through the polymer between the segments of the macromolecular chains. Separation occurs due to the differences in both the solubility in the membrane and the rate of diffusion through the membrane. The rate of diffusion or permeability of the dense amorphous membranes may be low, but the selectivity is generally very high. The dense membranes are widely used in the industry for gas separation and reverse osmosis applications.

The above discussion reveals that, in spite of the remarkable advances in maximising the polymer membrane performance, a trade-off still exists between the permeability and the selectivity in the case of symmetric membranes. We can have either high permeability as with the microporous membranes or a high selectivity as with the dense membranes, but not both. This dilemma was resolved by the introduction of the asymmetric membranes in 1963. The Loeb-Sourirajan membrane shown in Fig. 12.5a consists of a thin dense skin formed on a much thicker microporous support structure. The skin layer, also called the perm selective layer, is typically 0.1–1.0 μm thick and the support structure is typically of thickness of the order of 100 μm. Asymmetric membranes are in general superior compared to the symmetric membranes because the top layer which determines the flux can be made very thin. The mass transport flux through this layer is high even though the permeability is low due to its nonporous nature.

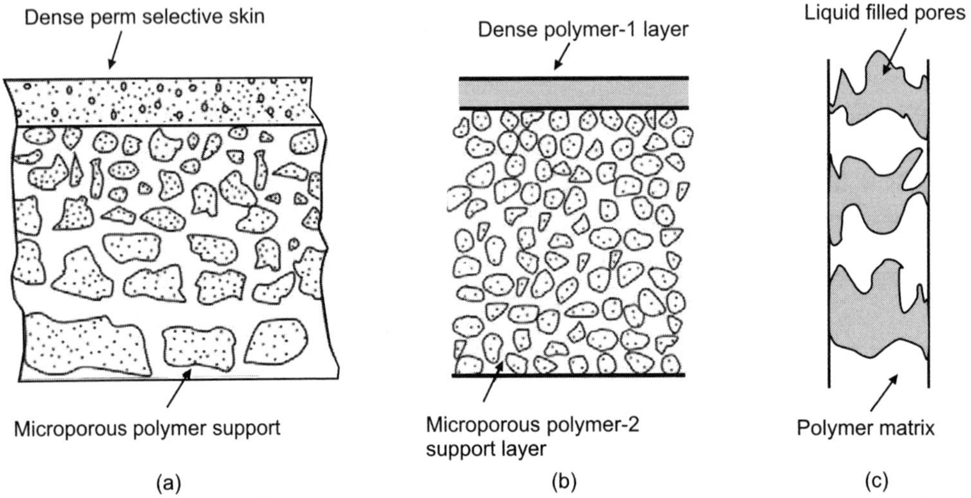

Fig. 12.5: Anisotropic (asymmetric) membranes (a) Loeb-Sourirajan (b) Thin film composite (c) Supported liquid membrane

In the Loeb-Sourirajan membrane, the materials of both the skin and the support structure are made of cellulose acetate. On the other hand, if the material of the top skin layer is different from that of the porous sub-layer, the membrane is called a *composite* membrane (Fig. 12.5b). The advantage of the composite membrane over the integrally skinned asymmetric membrane is that the material for the top skin layer and the porous sub-layer can be chosen separately to optimize the overall performance. Most of the membranes that are used in industries have an asymmetric structure.

Electrically charged membranes shown in Fig. 12.4c are necessarily ion-exchange membranes consisting of highly swollen gels carrying fixed positive or negative charges. These are mainly used in electrodialysis. A liquid membrane represented schematically as in Fig. 12.5c utilizes a carrier to selectively transport components such as metal ions at relatively high rates across the membrane interface.

Inorganic membranes

Polymeric membranes are not suitable for high temperature operation and for the separation of chemically active mixtures. The advantages of inorganic membranes are their high temperature stability, increased resistance towards solvents and fouling, and narrower pore size distribution than the polymeric membranes. Another important advantage of inorganic membranes is that they allow steam sterilization that makes these ideal for food, biotechnology and pharmaceutical applications. Inorganic membranes can be classified into porous and nonporous dense membranes. Glass, metal, alumina, zirconia, zeolite and carbon membranes are porous inorganic membranes. Membranes made of silicon carbide, titania, mullite, tin oxide, mica etc also fall into this category. Palladium and its alloys, silver, nickel, stabilized zirconia etc. give nonporous dense membranes. Inorganic membranes find wide use in the energy related applications, beverage production, water purification and the separation of dairy products. Though dense membranes are used for gas separation, their industrial applications are limited because of their poor permeability. The inorganic membranes are more expensive than the polymeric types and have low membrane surface per module volume. Though useful for high temperature applications, in some cases, there are difficulties in achieving the membrane-to-module sealing at high temperatures. Depending upon the membrane materials, inorganic membranes may be classified into ceramic, metal, carbon and zeolite membranes.

Ceramic Membrane Ceramic membranes are made from the mineral materials such as glass, aluminium oxide, silicon carbide, zirconium oxide and titanium oxide. Since they are quite stable chemically, the ceramic membranes are used for the separations involving aggressive components such as strong acids and solvents. They have excellent thermal stability that makes them suitable for very high temperature operations. Many ceramic membranes are usable at as high a temperature as 1300 K. They are ecologically friendly and have long working life. The major drawbacks of ceramic membranes are that they are brittle and heavy and expensive because of their high cost of production.

Most ceramic membranes in industrial use have an asymmetric structure with either a dense or a porous skin layer. The rough porous support is made of sintered ceramic particles such as alumina, titania and zirconia. Since membranes in the form of flat plates are too brittle, the ceramic membranes are made mostly in the shape of tubes. Sintered ceramic membranes such as the microporous α-alumina tubes and the anodised alumina membrane with uniform pore size are typical ceramic membranes in wide use. Ceramic microfiltration and ultrafiltration membranes are particularly useful for the food, pharmaceutical and biotechnological applications in which the membranes require frequent steam sterilization and chemical cleaning. Ceramic membranes find application in gas separations as well, as for example in the production of synthesis gas from methane and air. Since the oxygen required for synthesis gas is obtained by the membrane separation of air, this process eliminates the need for a separate oxygen production plant, thereby providing considerable saving in the capital investment, production cost and energy cost.

Metal Membrane The metal membranes are stable at temperatures as high as 750 – 1050 K. They can be either microporous or dense. The membranes made from palladium and its alloys are dense metallic membranes, ideal for the separation of hydrogen and helium from the gas mixtures, as they offer a high solubility and permeability for small molecules. Palladium being quite expensive, to reduce cost, it is recommended to use a thin layer of palladium on a less expensive tantalum or vanadium supports.

Supported thin metallic membranes such as the composite palladium membranes with thickness ranging from submicron to a few microns have been developed recently and are widely used in catalytic membrane reactors (CMR). The membrane consists of a thin layer of palladium, or palladium alloy, deposited onto a porous support such as ceramic or stainless steel. The advantages of composite membranes include the reduced material costs, the improved mechanical strength and the possibly higher flux. The CMR combines the reaction and separation stages of a process in a single unit by placing the composite membrane adjacent to a catalyst bed and thereby effecting the selective removal of the hydrogen from the catalytic reaction source. CMR offers substantial saving in the cost of the plant and its maintenance and gives high conversions and product yields. A major problem associated with metal membranes is the surface poisoning effects, which can be more significant in thin metal membranes.

Carbon Membranes Carbon membranes are porous inorganic membranes and they were used from the early 1970s for gas separation. They offer superior selectivity and higher thermal and chemical stability compared to polymeric membranes. They are found to be ideal for the corrosive applications and can be used in environments that are prohibitive for polymeric materials. They have superior stability in the presence the organic vapours or solvents and in non-oxidising environment. Carbon membranes can be classified into two types: supported and unsupported. The supported membranes are available in two configurations, flat film and tube, whereas unsupported membranes come in three different configurations, flat film, hollow fibre and capillary. Carbon molecular sieves can be produced from the pyrolysis of many thermosetting polymeric materials such as polyvinylidine chloride, polyfurfural alcohol, cellulose triacetate, polyacrylonitrile and phenol formaldehyde.

The feature that makes the molecular sieves ideal for separating molecules of comparable sizes from the gas mixtures is that they are porous solids containing apertures with size very close to the dimensions of the diffusing gas molecules. The disadvantages of carbon membranes are that they are brittle and fragile, difficult to process and expensive to fabricate. They frequently require a pre-treatment for removing the traces of strongly adsorbed vapours that may clog up the pores and reduce the molecular transport of the desired components. The most important large scale application of carbon membranes is in the production of the low cost and high purity nitrogen from air. Other application includes separation of hydrogen from gases and purification of methane.

Zeolite Membranes Zeolites are microporous crystalline alumina-silicate having small sub-nano sized channels along the crystal planes of the structure. The channels are of diameter roughly between 3 and 100 mm. The framework structure allows small molecules to enter. The main problem with zeolite membranes is the relatively low gas fluxes compared to other inorganic membranes. This is because of the large thickness required for the membrane in order to make them pinhole-free or crack-free. This problem is usually overcome by depositing a thin layer of zeolite on a porous support of materials such as alumina. The supported membrane also is not free from problems. At high temperatures, the zeolite layer may shrink and the support layer may expand resulting in thermal stress. This may create problems for the attachment of the zeolite layer to the support, as well as for the connection of the individual micro-crystals within the zeolite layer.

Liquid membranes

A liquid membrane is a thin liquid film that separates two liquid or gaseous phases and controls the mass transfer between these phases. The advantage of the liquid membranes over their solid counterpart is the higher solubility and diffusivity of substances in liquid medium than in the solids. The liquid membrane utilizes a carrier which increases the permeability of the membrane even more by selectively transporting components such as the metal ions at relatively high rates across the membrane interface. Several types of liquid membranes are used in industry. These include the Emulsion Liquid Membrane (ELM) and the Immobilized Liquid Membrane (ILM), also called a Supported Liquid Membrane. An ELM set up consists of a large number of liquid entities each of which can be treated as a 'bubble within a bubble', dispersed in a fluid phase called the source phase. The inner bubble is the receiving phase and the outer bubble contains carriers and acts as the membrane. An immobilized liquid membrane is schematically represented in Fig. 12.5c. It consists of a rigid polymeric matrix the pores of which are filled with some organic liquid containing carriers. The membrane separates the source and the receiving phases and allows the passage of species through it from the source side to the receiving side, with the help of carriers. The commonly used supported liquid membrane employs the hollow fibre and the spiral wound modules. The modules are discussed in detail under section 12.3.

The liquid membranes are highly efficient and selective and with the use of carriers specific molecular recognition is possible. The main problem with liquid membranes that restricts their widespread industrial application is their poor stability. In the case of ELMs, ionic strength, pH, etc. may affect the emulsion stability and must be kept under control. The conditions that favour stable emulsion make recovery of the receiving phase and replenishing of the carrier phase difficult, as the emulsions are to be broken down to make recovery possible. In ILMs, the instability is caused by the removal of carrier or organic liquid either by evaporation or by physically pushing them out by large pressure differences.

Hybrid membranes

Hybrid membranes are mixed matrix membranes, which are made of both inorganic and organic compounds. They combine the advantages of the inorganic and organic membranes. They possess the high separation capabilities of the inorganic membranes and the desirable mechanical properties and low cost of the polymeric membranes. Generally they are made of inorganic molecular sieving materials such as the zeolite and the carbon molecular sieves dispersed in a polymer matrix. The conventional polymeric membranes have low permeability and poor selectivity in gas separation. However, the processing flexibility and low cost make them attractive for industrial applications. In contrast, the inorganic membranes have significantly higher permeabilities and selectivities than polymeric membranes. However, their prohibitively high cost, rigidity and brittleness make the construction of continuous, defect-free membranes for large-scale industrial applications very difficult. The mixed matrix membranes find application in separation of gases, a typical example being the nitrogen production from air.

Bipolar membranes

Bipolar membranes are formed by combining the membranes of different ionic charges. In its simplest form, a bipolar membrane consists of a cation-exchange membrane joined to an anion-exchange membrane through an intermediate layer called the junction layer. The junction layer is the most important part of the bipolar membrane. The structure of the junction layer and the nature of the charged groups determine their separation capabilities. Important applications of the bipolar membranes include the splitting of water into H^+ and OH^- ions directly without generating O_2 and H_2. They find use in the production of NaOH and HCl from NaCl as an alternative to the conventional membrane electrolysis method. They are also used in the production of H_2SO_4 and NaOH from the concentrated solutions of Na_2SO_4.

Membrane Shapes

Membranes are available in different shapes (Fig. 12.6). The flat sheet, tubular and hollow fibre forms are common with polymeric membranes.

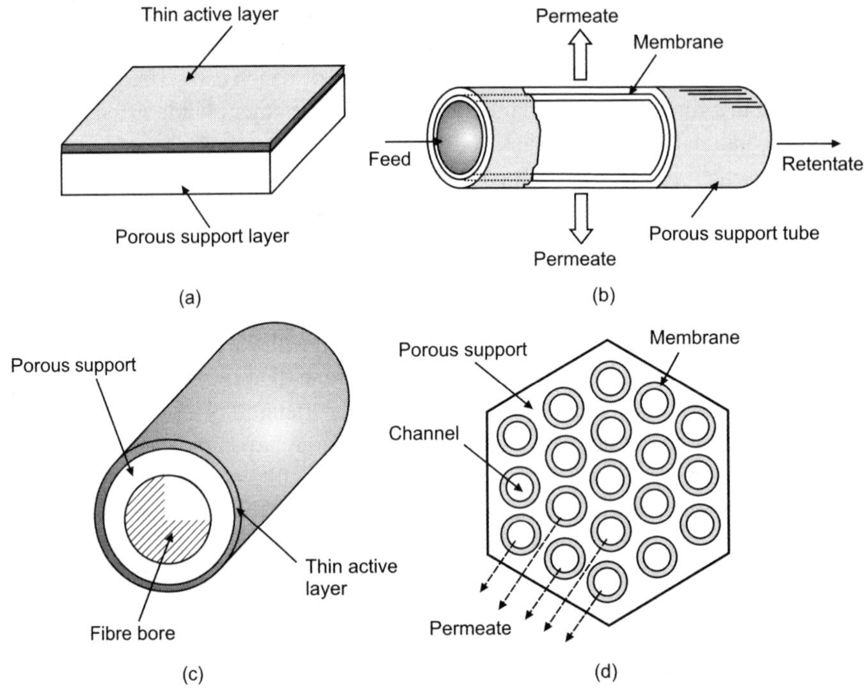

Fig. 12.6: Membrane shapes (a) Flat sheet (b) Tubular (c) Hollow fibre (d) Monolithic

The flat sheets shown in Fig. 12.6a are typically of size 1 m by 1 m by 0.2 mm thick, with a dense skin or thin, dense layer 50 to 500 nm in thickness. Tubular membranes are typically of 5 to 50 mm in diameter and up to 6 m in length. The supporting part of the tube is made of suitable porous materials such as the fibre glass or perforated metal. The thin dense layer is either on the inside as shown in Fig. 12.6b or on the outside. The hollow fibre membranes shown schematically by Fig. 12.6c provide a large surface area per unit volume. They are available in sizes as small as 42 μm inside diameter and 850 μm outside diameter with a length of 1.2 m. The dense layer is typically 0.1 to 1 μm in thickness. Inorganic membranes, particularly the ceramic membranes, are available as a honeycomb monolithic design, with a porous support layer either hexagonal as shown in Fig. 12.6d or circular in cross section. The flow channels formed by the membrane layers with very small pores are circular in cross section and typically 3 to 6 mm in diameter and 20 to 40 mm thick.

12.3 MEMBRANE MODULES

Separation on an industrial scale usually requires hundreds to thousands of square meter of membrane area. In order to provide such large surface area per unit volume, the membranes are usually packaged into modules. The modules are the smallest working unit into which the membranes are integrated into for effective and economical separation. The area contained in these basic units is in the range 1 to 20 m^2. In addition to providing the large area at a low cost, other requirements of a module design are: the good control of concentration polarization and membrane fouling, the facility for easy removal and replacement and protection against mechanical damage. The method of packing the maximum area of membrane into the least volume is significant not only from the cost considerations, but from the chemical engineering aspects such as the flow hydrodynamics also. The containing vessels should be of small volume with acceptable flow hydrodynamics.

Membranes can be cast as flat sheets, tubes and fine hollow fibres, as we have seen earlier. For accommodating these different shapes and structures, the membrane modules come in the following six types: plate-and-frame, spiral wound, pleated cartridges, tubular, capillary and hollow fibre modules. The techno-economic factors for the selection, design and operation of membrane modules include the cost of supporting materials and containing vessel, the power consumption in pumping and the ease of replacement.

Cartridge Membrane Module

The pleated cartridge type membrane modules are useful for operation at low pressures, up to about 2 bar. The figure shows a pleated cartridge membrane module generally used in microfiltration. These are made by pleating a membrane sheet and potting the ends by some resin. The cartridge is installed in a housing as shown schematically in Fig. 12.7. The feed is admitted from the housing side and flows past the membrane surface. A central tube sealed to the housing carries the permeate. The advantage of the cartridge modules is the relatively high surface area per unit volume and the low cost of production. The limitation is that they are highly prone to the plugging of the membrane pores and hence their useful life is limited. Also the membrane material should be flexible and should have sufficient mechanical strength to withstand the stress due to pleating. The application of the cartridge filters includes the sterile filtration of water, beverages and pharmaceutical solutions. They serve as pre-filters in reverse osmosis desalination plants.

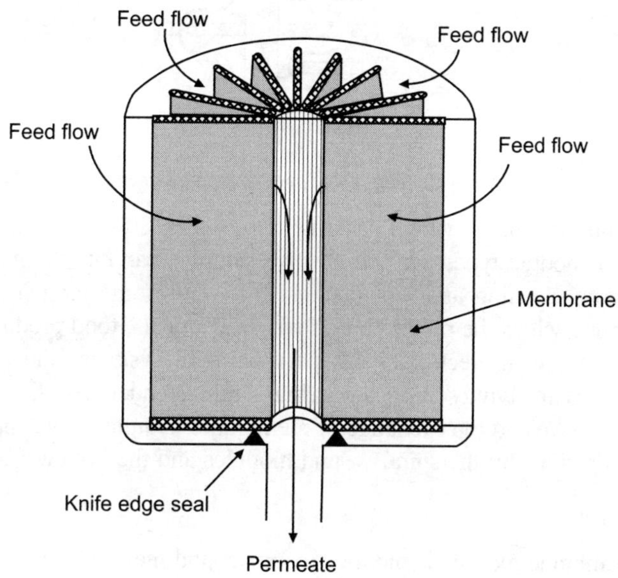

Fig. 12.7: Schematic diagram of a cartridge membrane module

Plate-and-Frame Module (Flat-Sheet Modules)

The plate-and-frame modules resemble the conventional plate and frame filter presses. Membranes used in this type of arrangements are flat sheets with circular, square or rectangular cross section. There are various types of plate-and-frame module designs with only slight variations in their basic configuration. In some designs, the membrane can be removed from the porous support plate; in others, it is directly cast on a support structure. It is usually possible to dismantle and mechanically clean the membranes in a plate-and-frame module. In some designs, the membranes can be easily exchanged, making batch application and multipurpose operation possible using different membranes for different tasks.

Figure 12.8 shows circular membrane discs separated by porous polysulphone support plates, which also provide channels for the flow of permeate. The sandwich of membranes and the porous support plates are stacked one above another, separated from each other by the feed spacer plates. The space between the membrane and the spacer plates acts as the feed flow channels through which the solution moves across the surface of the membrane in laminar flow. The permeate leaves through the channels provided by the porous supports and is collected in permeate collection manifold.

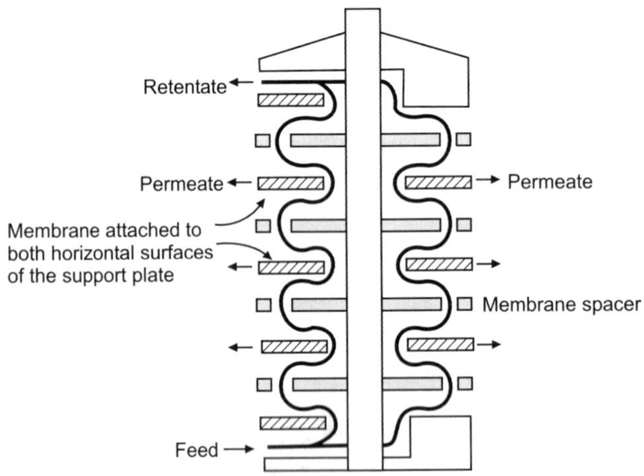

Fig. 12.8: Schematic diagram of plate-and-frame membrane module

The plate-and-frame systems provide a packing density 400–800 m^2/m^3 which is relatively low in comparison with other module types. On the positive side, we can cite the high allowable operating pressure and easiness in cleaning and replacement. Plate-and-frame modules are generally used in small scale applications such as the production of pharmaceuticals, food products, fine-chemicals and bio-products. They find use in electrodialysis, pervaporation systems, and in reverse osmosis and ultrafiltration applications involving solutions with high fouling tendencies. Since this type of membrane modules are quite expensive and the exchange of membranes is quite labour intensive, they have been replaced in most applications by the spiral wound modules and the hollow fibre modules.

Spiral-wound Modules

The spiral wound membrane module is the most compact and inexpensive design using the flat sheet membranes. The basic element of this module is the membrane envelope which is formed by two membranes separated by a permeate spacer. The edges of the membranes and the spacer are sealed together at one end. The other free end is fixed around a centrally placed perforated permeate flow tube as shown in Fig. 12.9. The module is housed inside a pressure tube. The feed when admitted at one end of the tube encounters a number of narrow, parallel feed channels formed between the adjacent membrane surfaces by the feed channel spacers. The feed flows in the axial direction through these channels across the surfaces of the membranes. A portion of the feed permeates through the membrane into the permeate channels and spirals towards the central collection tube, from where it flows out.

The small spiral wound module described above consists of just one envelope, or leaf as it is sometimes called, and the total membrane area is limited to around 1 to 2 m^2. Up to six such modules

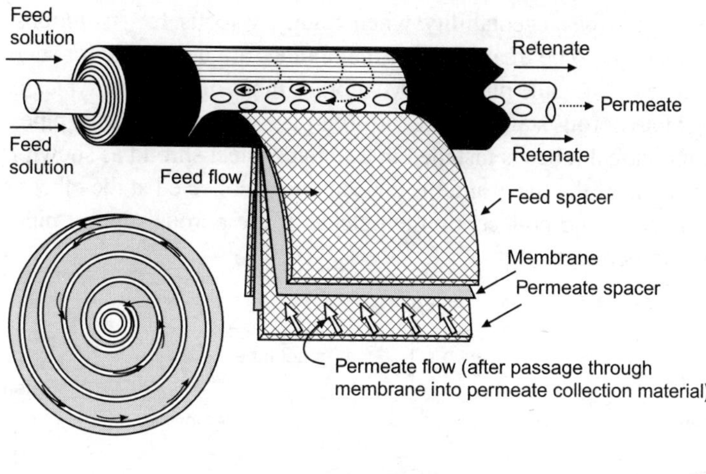

Fig. 12.9: Schematic diagram of spiral-wound module

can be arranged in series in a single pressure tube. The reason for the small membrane area is the constraint on the length of the permeate flow channel due to the large pressure drop encountered when the permeate spirals through these very narrow channels to the collection tube. The membrane area in a single module can be increased by providing a multi-leaf arrangement as shown in Fig. 12.10.

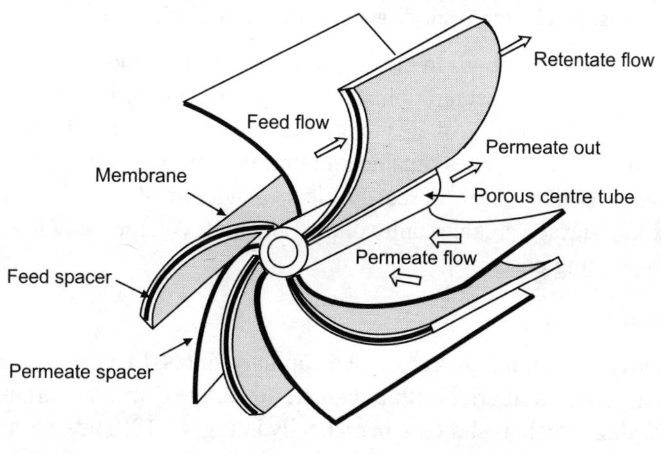

Fig. 12.10: Schematic diagram of a four-leaf spiral wound membrane module

The spiral-wound modules provide a relatively large membrane area per unit volume and thus make better use of the space than tubular or flat sheet types. Also they are relatively less expensive. The limitations of spiral wound modules are their vulnerability to fouling and the difficulty in cleaning the membranes. Since the feed channels get easily blocked by the particles or fibres present in the feed, a feed pre-treatment becomes necessary to remove such substances. The commercial spiral wound modules are about 1 m long, and have diameter between 0.1 and 0.6 m. Generally between 2 and 6 elements are placed in series within a single housing. The spiral-wound modules are characterised by high packing density in the range of 300–1000 m^2/m^3 and low relative cost. These are capable of operating at pressures in excess of 55.2 bar and are widely used in reverse osmosis, nanofiltration, ultrafiltration, and gas separation.

Tubular Modules

Tubular membranes find wide acceptability when highly viscous feed solutions containing high concentration of the solid are to be treated. Other membranes are unsuitable for such applications due to the membrane fouling and module plugging. Membrane tubes of diameter typically between 10 mm and 25 mm are placed into porous stainless steel or fibre-glass reinforced plastic pipes. The tube bundle containing 10 to 30 individual tubes is installed in a stainless steel shroud as shown in Fig. 12.11. The feed flows down the inside of the tubes and the concentrate is collected at the other end. The permeate flows in the radial direction and collects on the outside of the porous support pipe from where it is directed out through the permeate exit.

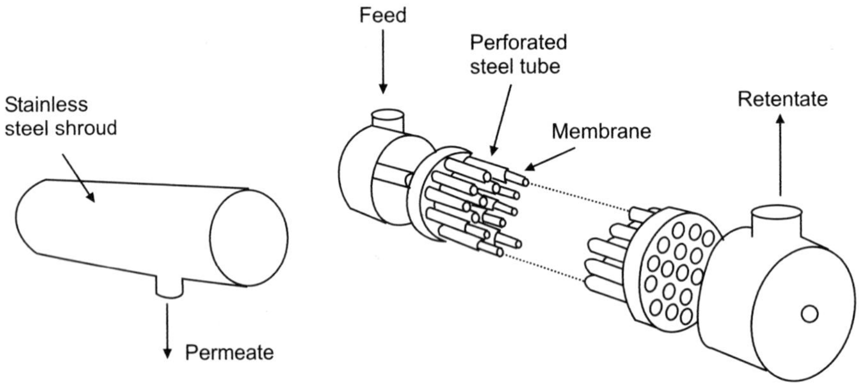

Fig. 12.11: Schematic diagram of a tubular membrane module

Tubular module is ideal for operations in the turbulent flow regime and hence suitable for separating the feed solution with high solids concentration. The concentration polarization and membrane fouling can be easily controlled and the plugging of the membranes is easily avoided. The resistance to membrane fouling makes tubular membrane modules suitable for ultrafiltration applications. A significant advantage in terms of maintenance is that it can be easily cleaned without dismantling the equipment. The disadvantages are the low surface area per unit volume, the high volumetric hold-up and the relatively high investment and operating costs.

Capillary Membrane Modules

Fibres with diameter greater than 0.5 mm are called capillary fibres. In a capillary membrane module, a large number of membrane capillaries with a diameter of 0.5 to 5 mm are arranged in parallel as a bundle and installed within a shell as shown schematically in Fig. 12.12. The free ends of the capillaries are potted by agents such as epoxy resins, polyurethanes, etc. The feed solution is generally passed through the inside of the capillaries and permeate is collected outside, in the shell.

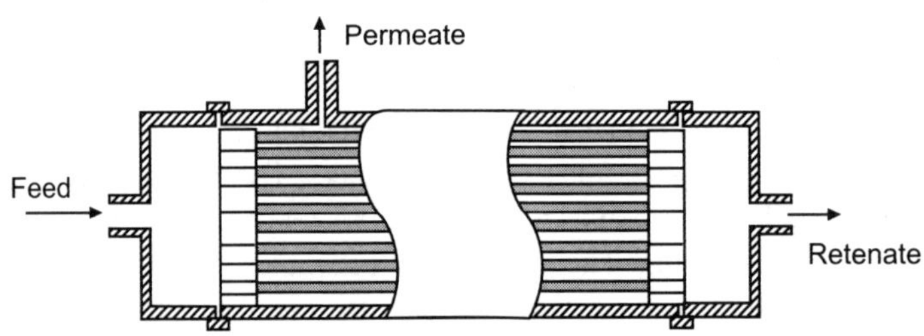

Fig. 12.12: Schematic diagram of a capillary membrane module

The capillary modules provide a high packing density, that is, a high membrane area per module volume. The diameter being very small, the chances of plugging are much higher with a capillary membrane. However, by proper feed flow and back flushing of the permeate at definite intervals, the membrane fouling can be kept under control. The limitation of the capillary membranes is the requirement that these are to be operated at low pressures, generally below 4 to 6 bars, because of the poor stability of the system. They are generally used in low pressure applications such as the dialysis, the microfiltration and the low pressure ultrafiltration. An important application of capillary membrane module is in artificial kidneys.

Hollow Fibre Membrane Module

The hollow fibre membranes are essentially membranes with hollow space inside. Fibres can be made with a uniformly dense structure, but microporous structure having dense selective layer on either the inside or the outside are the preferred ones. The diameter of the hollow fibre generally varies from 0.05 to 0.5 mm. Hollow fibre membrane module is formed by assembling together a bundle of hollow fibres, with the free ends potted into a head plate and installed into an outer shell. As these membranes are self supporting, hollow-fibre membrane modules are formed in two basic geometries, the bore-side feed design, as shown in Fig. 12.13 or the shell-side feed design shown schematically in Fig. 12.14.

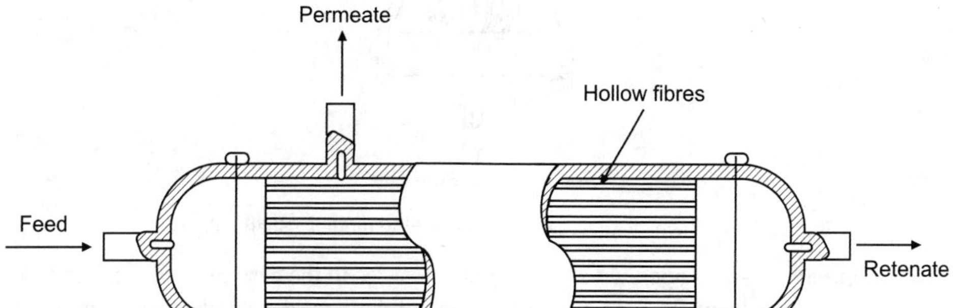

Fig. 12.13: Hollow fibre membrane module (bore-side feed)

In the shell-side feed type, a loop or closed bundle of fibres is contained in a pressure vessel, the permeate passes through the fibre wall and exits through the open fibre ends. Shell-side feed modules are generally applied for the high pressure applications, up to about 70 bar. Hollow fine fibres with diameter in the range 0.05 to 0.2 mm, which can withstand the high hydrostatic pressures applied from the shell side are used in this type of module. This design is easy to fabricate and contains very large membrane area per unit membrane volume. However, the fouling on the feed side of the membrane can be problem and a pre-treatment of the feed stream to remove particulates may frequently be required. These systems find wide application in the desalination of sea water by the reverse osmosis and the high pressure gas separation applications.

In the bore-side feed design, the feed solution and the retentate flow through the centre of the hollow fibre and permeate passes through the fibre wall to the outside of the membrane fibre, from where it is removed. This design is used for medium pressure feed streams (up to about 11 bar) and hence used for the low-pressure gas separations and for the ultrafiltration or the haemodialysis applications. The hollow fibres used in such applications have diameter in the range 0.2 to 0.5 mm, larger than those in the shell-side feed design.

The hollow fibre membrane module has the highest membrane surface area to volume ratio and has the smallest size for a given performance capacity among the different membrane modules. Hollow

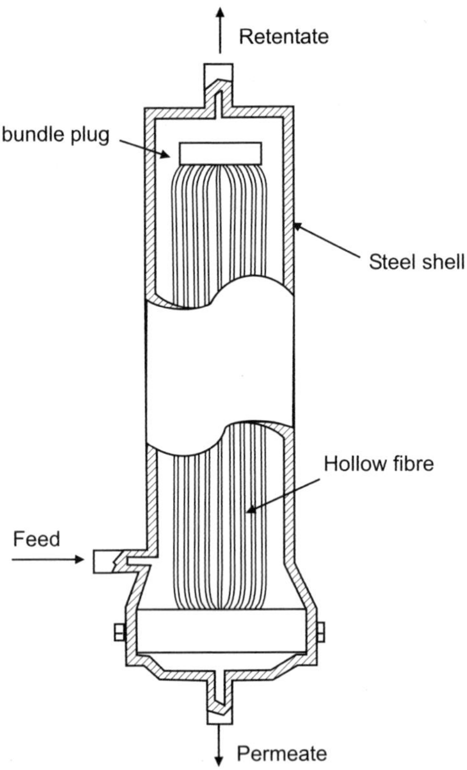

Fig. 12.14: Schematic diagram of a hollow fibre module (Shell-side feed)

fibre membrane modules can be operated at very high pressures in the case of shell side feed. Modules with bore-side feed allow back-flushing to remove the solids from the inside surface and tangential flow along the membrane surface limits the membrane fouling. However, the main disadvantage of the hollow fibre membrane module is the difficulty in controlling the concentration polarization and membrane fouling. Table 12.2 summarises the characteristic features of the different membrane modules in use today.

Table 12.2: Comparison of properties and applicability of different membrane modules

Characteristics	Filter cartridge	Plate-and-frame	Spiral-wound	Tubular	Capillary	Hollow fibre
Packing density, m^2/m^3	800–1000	400–800	300–1000	20–100	600 –1200	2000–5000
Cost	Low	Medium	Low	Very high	Low	Very low
Control of concentration polarisation	Very poor	Good	Good	Very good	Very good	Very poor
Suitability for high pressure operations	No	Marginal	Yes	Marginal	No	Yes
Ease of cleaning	Poor	Good	Poor to good	Excellent	Poor	Poor
Application	Dead-end MF	MF, UF, RO, D, ED, PV	UF, RO, GS, PV	MF, UF, RO	UF, MF, D, PV	RO, GS

Flow Patterns and Plant Configurations

To facilitate the analysis of the given membrane geometry, the flow pattern is approximated to one of the following types: cocurrent flow, countercurrent flow, dead-end flow, cross-flow and perfect mixing. The extent of separation and the membrane area required for a given separation are influenced by the flow pattern. These idealized flow patterns in membrane modules are shown in Fig. 12.15. In the cocurrent, countercurrent and cross-flow patterns shown in Figs. 12.15a to c, the fluid on the feed or retentate side of the membrane flows along and parallel to the upstream surface of the membrane. For cocurrent and countercurrent flows, the permeate stream at any location on the downstream side of the membrane is made up of the streams emanating from the membrane at that location plus the permeate flowing to that location. For cross-flow mode (c), permeate at any location on the down stream side is the one that has just emerged from the membrane at that point. In dead end flow (d), the flow is normal to the face of the membrane and essentially all of the fluid entering the system is either retained by the residue or emerges as the permeate. In the perfect mixing mode (e), mixing is assumed on both sides of the membrane so that the compositions on both sides are uniform and equal to the exit compositions.

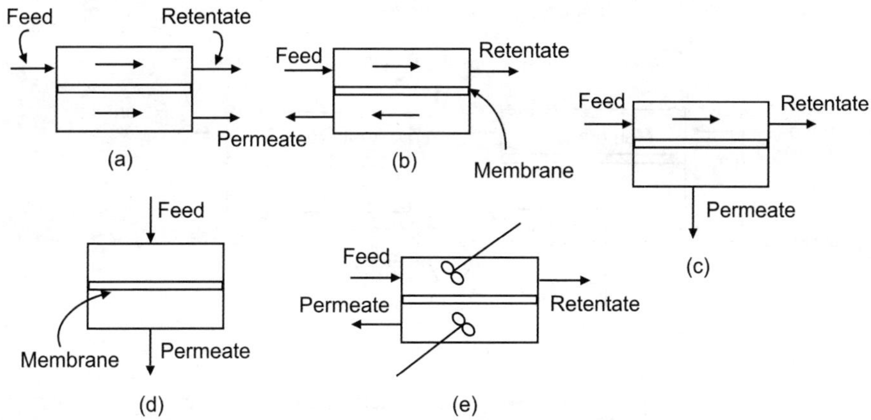

Fig. 12.15: Idealized flow patterns in membrane modules (a) Cocurrent flow (b) Countercurrent flow (c) Cross-flow (d) Dead-end flow (e) Perfect mixing

Membrane Cascades

Membrane separators may be operated as a single-stage unit or as multistage cascades. A single stage unit is made up of either a single membrane module or several such modules connected in series or in parallel without recycle. The single stage membrane separation units are schematically represented in Fig. 12.16.

The separation achieved in a single stage operation is limited and it depends on several factors such as the module flow pattern, the permeability ratio, and the driving force for mass transfer. Higher separation can be achieved by a cascade of membrane module with recycle. Some common configurations are shown in the Fig. 12.17. This scheme is common for the gas permeation applications where the number of stages are to be limited, as the cost of compression of the gas will be prohibitively excessive if the number of stages exceeds two or three.

Figure 12.17a is a two stage stripping cascade, that is, a cascade of two stages with a recycle of permeate from the second stage to the first. Figure 12.17b shows a two stage enriching cascade, that is, a cascade of two stages with a recycle of retentate from the second stage to the first. The stripping cascade is designed to obtain a very pure retentate, whereas the enriching cascade produces a pure permeate product.

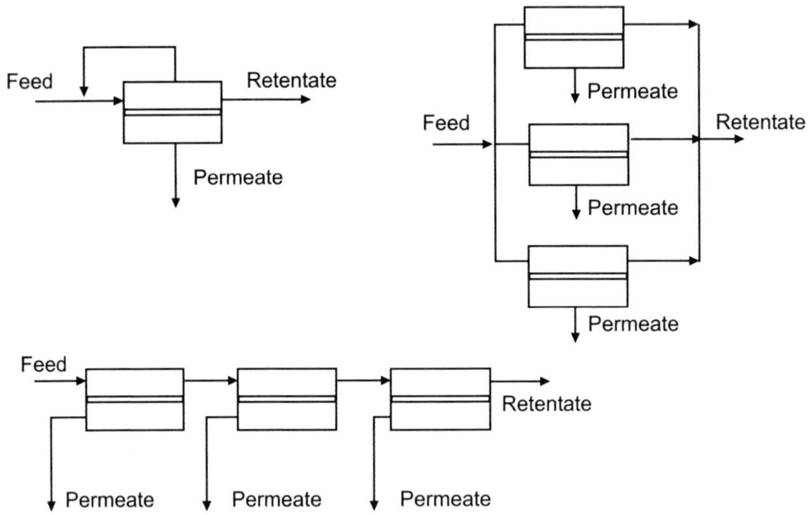

Fig. 12.16: Single stage membrane separation units

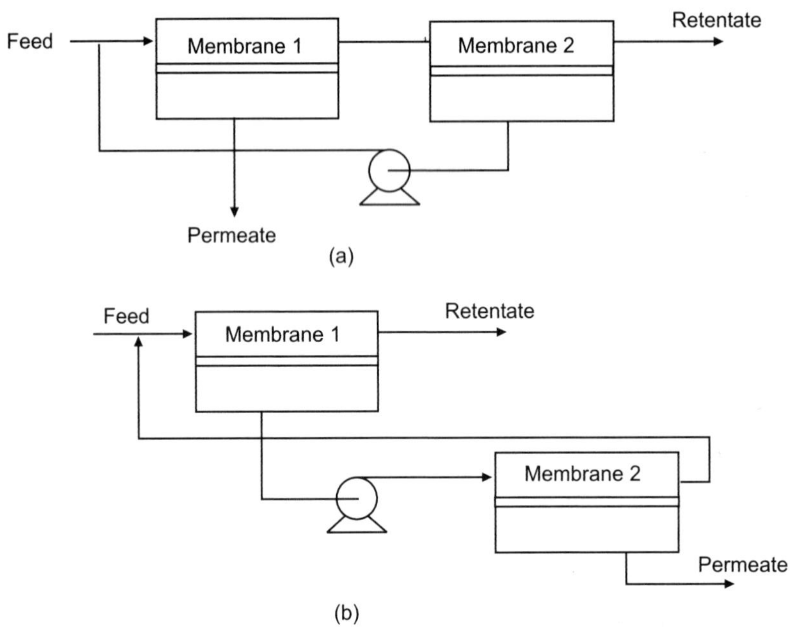

Fig. 12.17: Two-stage cascades: (a) enriching cascade (b) stripping cascade

Figure 12.18 shows a two-stage enriching cascade with an additional "pre-membrane" stage. This scheme is preferred when (1) the feed concentration is low in the permeating component, (2) the desired permeate purity is high, (3) the separation factor is low, and/or (4) a high recovery of the more permeable component is desired.

A countercurrent recycle cascade of membrane separators shown in Fig. 12.19 is similar to a distillation column. The feed enters at some intermediate stage in the cascade, called the feed stage as shown. The permeate leaving the feed stage is enriched in the more permeating component in the enriching section of the cascade and the retentate is enriched in the less permeating component in the stripping section of the cascade. The final permeate product is withdrawn from stage 1 and the final

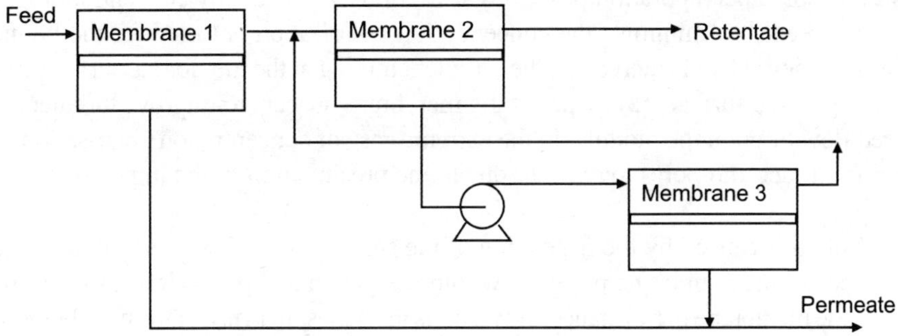

Fig. 12.18: Two-stage enriching cascade with a pre-membrane stage

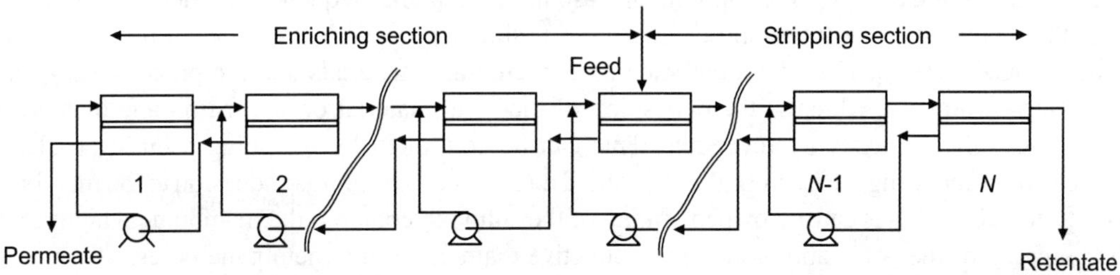

Fig. 12.19: Counter-current membrane separator recycle cascade

retentate from the last stage in the cascade. The degree of separation of the feed is determined by the number of stages in the cascade and the recycle ratio, which is the ratio of the rate of permeate recycle to the rate of permeate product.

12.4 CONCENTRATION POLARISATION AND FOULING

Concentration polarisation is an inherent problem associated with all membrane separation operations. Since the components of the feed mixture permeate the membrane at different rates, there is a gradual build-up in the concentration of non-permeating or slowly permeating components at the membrane surface when the more permeable components pass through the membrane. Concentration polarisation is said to occur when the relatively impermeable components of the feed stream carried to the upstream surface of the membrane by the bulk flow accumulate in a polarised layer at the surface of the membrane. The concentration of the impermeable components at the surface increases till the equilibrium concentration is reached. If the concentration exceeds the solubility limits, a layer of precipitate or gel may form at the membrane surface which provides a secondary barrier to the flow through the membrane. The net result is the reduction in the flux of the permeating component and a decrease in the selectivity of the membrane.

The effect of the concentration polarisation on the membrane performance depends on the type of separation process in question. In pressure-driven operations such as the reverse osmosis and ultrafiltration, the concentration polarisation leads to an increase in the resistance to solvent flow and a reduction of the flux of the permeating components. In water-treatment by reverse osmosis, concentration polarisation leads to an increase in the osmotic pressure and hence a reduction of the flux of water. The reduced water flux and the build up of the salt concentration in the boundary layer result in the increase in the salt flow through the membrane and a decrease in the solute rejection by the membrane.

Though the membrane performance in the reverse osmosis is significantly affected, in industrial operation, the effect can be controlled by various means. However, the performance in ultrafiltration, electrodialysis and some pervaporation processes are seriously affected by concentration polarization. The most obvious means to minimise the concentration polarisation is to reduce the boundary layer thickness. This is achieved by (1) increasing the turbulent mixing at the membrane surface by increasing the fluid velocity past the surface, (2) by providing membrane spacers to improve turbulence, or (3) by pulsing the feed flow through the module. In the extreme case of concentration polarisation, the *fouling* of the membrane surface and pores may occur due to the precipitation of the impermeable component at the surface.

Membrane fouling is caused by the deposition of the suspended or dissolved solids on the external membrane surface, on the membrane pores, or within the membrane pores. It is a major problem that adversely affects the membrane performance and restricts the widespread use of the membrane separations technology. Fouling results in a continuous decline in the membrane permeation rate, an increased rejection of the permeable components and eventually blocking of the flow channels. It is responsible for the increase in the energy consumption, increase in the cleaning frequency and decrease in the life span of the membrane. The types and extent of the fouling are dependent on many different factors, such as the feed water quality, the membrane type, membrane materials and the process design and control. Surface fouling, which is the deposition of the solid material on the membrane that slowly thickens over time, is generally reversible. This can be controlled by providing a high turbulence, regular cleaning and using the hydrophilic or charged membranes to minimise adhesion to the membrane surface. Internal fouling is caused by penetration of the solid material into the membrane which results in the plugging of the pores and reduction in effective diameter of the membrane pores. The internal fouling is generally irreversible.

The extent of membrane fouling depends on the nature of the membrane used and on the properties of the process feed. The first means of control is therefore, the choice of the membrane type. Membranes with surface characteristics that minimise adhesion can check fouling to a great extent. Secondly a module design which provides suitable hydrodynamic conditions for the particular application should be chosen. Process feed pre-treatment is also important. Precipitation, softening for removal of calcium, silica and colloids, acidification, gravity filtration, sequestering to retard the precipitation of calcium sulphate, etc. are some of the pre-treatments to which the feed to the water treatment process by reverse osmosis is subjected to. In biotechnological applications, pre-filtration, pasteurisation to destroy bacteria, pH adjustment to prevent protein precipitation are some pre-treatment options.

When membrane fouling has occurred, back-flushing of the membrane may substantially restore the permeation rate. This is seldom totally effective and chemical cleaning is eventually required. More recent approaches to the control of membrane fouling include the use of more sophisticated hydrodynamic control caused by pulsated feed flows, or non-planar membrane surfaces, and the application of perturbation at the membrane surface such as continuous or pulsated electric fields. The cross-flow filtration methods in which the feed liquid is allowed to flow parallel to the membrane surface greatly reduces the membrane fouling and concentration polarisation. This is achieved by scouring the surface as the liquid moves across the membrane thereby thinning the surface layer and removing the deposited material.

12.5 CLASSIFICATION OF MEMBRANE SEPARATION PROCESSES

With tremendous advances made in the field of membrane materials and in the design of membrane modules, membrane separations have received wide acceptability as industrial separation techniques of immense potential with several advantages over the conventional separation. Membrane separations are performed at relatively low temperatures with low energy cost than the conventional techniques employed for separation. The separation equipment are modular, compact, less capital intensive, easily operated, controlled and maintained than the conventional separation vessels. Microfiltration, ultrafiltration, nanofiltration, reverse osmosis, dialysis, electrodialysis, pervaporation, membrane distillation, gas permeation are the important membrane operations. Different driving forces are used to facilitate the transport of mass through the membrane, and membrane separations can be classified on this basis. The applications of high pressure, the maintenance of a concentration gradient on both sides of the membrane and the introduction of an electric potential are the common methods of providing driving forces for membrane separations. Accordingly, it is possible to classify microfiltration, ultrafiltration, nanofiltration, reverse osmosis, pervaporation, gas permeation as *pressure driven operations*, dialysis and osmosis as *concentration driven operations* and electrodialysis as operation involving electric potential gradient across the membrane. The permeate undergoes a phase change from the liquid to gas in the pervaporation process, whereas no phase change is involved in other membrane processes.

The pressure driven processes such as microfiltration, ultrafiltration, nanofiltration, reverse osmosis are generally referred to as membrane filtration processes since all of them rely on the size exclusion principle for separation. The nature of substances separated by the membrane filtration are schematically represented by Fig. 12.20.

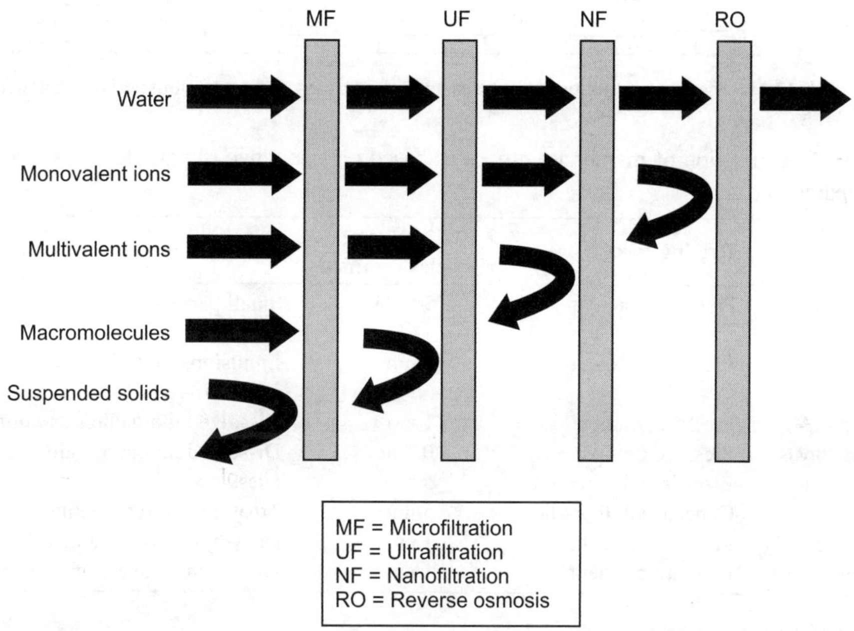

Fig. 12.20: Nature of materials rejected by membrane filters

Membranes are porous barriers by which filtration can be performed to separate the suspended or dissolved materials in the solution based on their size. Membrane filtrations rely on the porous nature of the semipermeable membranes to effect separation. Microfiltration removes particles of 50 nm or larger. Ultrafiltration removes particles of roughly 3 nm or larger. Nanofiltration removes particles of 1 nm or larger. Reverse osmosis also called the hyper-filtration removes particles larger than 0.1 nm. Classification on the basis of the size range of the materials retained by the membrane filtration processes and the nature of materials falling in these size ranges are shown in Fig. 12.21.

Table 12.3 gives the classification of various membrane separation operations based on the size of materials rejected, the driving force for mass transport and the nature of materials separated. A detailed discussion of the principle of the separation, the industrial applications and the nature of the membrane used in each operation are given in the following sections.

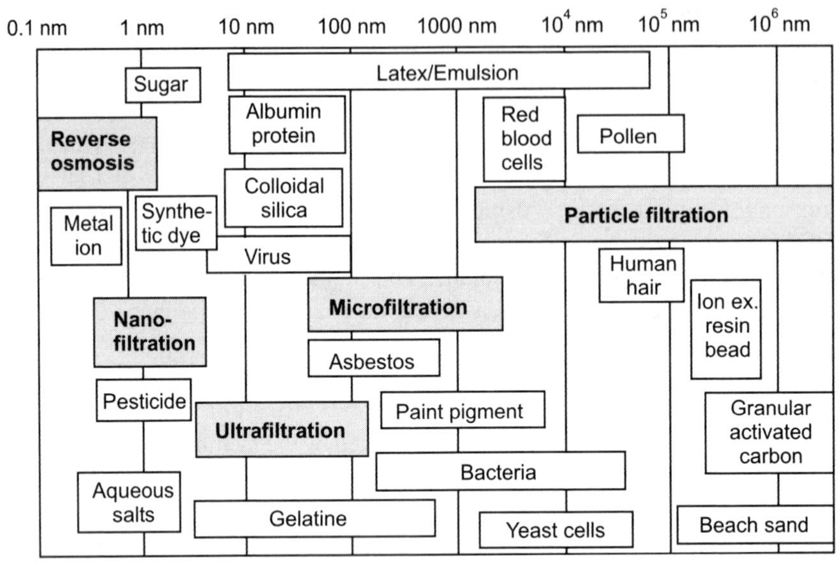

Fig. 12.21: Membrane filtration processes and the size range and type of materials separated

Table 12.3: Classification of membrane processes based on size range of particles separated and driving force of separation

Process	Driving force	Size range of particles retained	Nature of material separated
Microfiltration	Pressure gradient	> 50 nm	Small particles including bacteria, large colloids
Ultrafiltration	Pressure gradient	> 3 nm	Emulsion, colloids, macromolecules, proteins
Nanofiltration	Pressure gradient	> 1 nm	Dissolved inorganic salts, organics
Reverse osmosis	Pressure gradient	> 0.1 nm	Dissolved inorganic salts, small organics
Electrodialysis	Electric field gradient	> 5 nm	Dissolved salts
Dialysis	Concentration gradient	> 5 nm	Treatment of renal failure
Pervaporation	Pressure gradient	> 1 nm	Dehydration of organic solvents
Gas permeation	Pressure gradient	> 1 nm	Hydrogen recovery, air separation

12.6 MICROFILTRATION

Microfiltration is the most widely used membrane process with nearly 50% of the total membrane market being captured by this process. Microfiltration is a low-temperature, low-pressure filtration operation, used to separate constituents of gaseous or liquid streams. The only major energy cost involved in the process is the cost of pumping the solution across the membranes. Operating pressure lies in the range of 0.5 to 3 bar. The primary objective of microfiltration is the separation of colloidal and suspended particles from solutions. Considering the size of particles separated, or rather, the pore size of the membrane used for the separation, microfiltration lies in between ultrafiltration and ordinary particle filtration operations. Microfiltration is used to separate particles in the size range of 0.1 to 10 microns. Large soluble macromolecules, blood cells, yeast, and bacteria and other micro-organisms are also retained by microfiltration membranes. The diameter of the smallest bacterium, *pseudomonas diminuta* being 0.3 microns, micro-filters can remove all bacteria and as such can be used to sterilize solutions.

Microfiltration and ultrafiltration treatment techniques are versatile technologies that can be readily integrated with other processes; they can produce permeates of very high quality and purity, and yield substantial savings in cost, space and energy. A disadvantage of microfiltration is the frequent fouling of the membrane caused due to the deposition of solids on its surface. This necessitates purging of the unit periodically to remove the deposited solids. A pre-filtration of the feed solution is frequently necessary to entrap most of the larger, easier to filter solids before they reach the membrane.

Membranes

Membranes used for microfiltration are microporous in nature. The scope and variety of microfiltration membranes far exceeds those in other separation techniques. Microfiltration membranes should possess certain characteristics many of which are applicable to other membrane process as well. These include the mechanical strength, temperature resistance, chemical compatibility, adsorption properties, wetting behaviour, hydrophobicity, hydrophilicity, retention, permeability, permselectivity, service life and the cost of membrane materials. Microfiltration membranes in use are made of natural and synthetic polymers and inorganic materials and are produced in the form of flat sheets, spiral wound and tubular modules. The polymeric materials cover a very wide range, from relatively hydrophilic to very hydrophobic materials. Typical hydrophilic materials are polysulphone, polyethersulphone, cellulose and cellulose acetate, polyamide, polyimide, polyetherimide and polycarbonate. Hydrophobic materials include polyethylene, polypropylene, polytetrafluoroethylene (Teflon) and polyvinylidene fluoride. The inorganic materials such as metal oxides (alumina), glass, zirconia coated carbon, etc. are also used for manufacturing the microfiltration membranes. Membrane modules are predominantly plate-and-frame and pleated cartridges type.

Mode of Operation and Flow-pattern

In microfiltration operation, the mode of flow of the fluid past the membrane surface can be one of the two types: cross-flow or dead-end flow. These are shown schematically in Fig. 12.22a and b. The choice between these two depends on the solid loading in the feed. The higher the concentration of solid in the feed stream, the higher is the chances of clogging of the membrane pores by the particles. A general rule of thumb is that the streams with high solids content, say above 0.5%, are processed by membrane filters usually operated in cross flow and feeds with lower loadings of solids, i.e. less than 0.5%, are generally operated in dead-end flow.

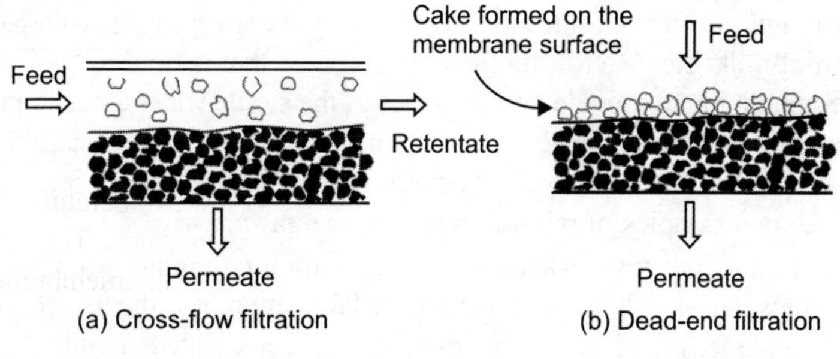

Fig. 12.22: Schematic representation of cross flow and dead-end microfiltration

In the cross-flow mode, feed stream flows parallel to the membrane surface at a velocity in the range 1-8 m/s, the liquid permeates through the membrane in a direction normal to the flow of feed or concentrate. This mode of operation is more suitable for large-scale continuous filtration and it substantially reduces the formation of filter cake on the membrane surface and prevents build up of high concentration of solids in the membrane pores. The process feed remains in the form of a slurry suitable for further processing. The solids content of the product slurry may be varied over a wide range. In the dead-end flow, the feed is pumped directly through the membrane. This is suitable for batch or semi-continuous filtration of dilute solutions, and the process resembles conventional filtration of solid-liquid slurries. This mode of operation allows the removal of particles down to 0.1 μm or less and is suitable for industrial clarification and sterilization of liquids; however, the chances of clogging of the membrane due to deposition of solids are more. Figure 12.23 shows a flow diagram for a simple cross-flow system for microfiltration with a large retentate recycle. The solution is pumped from the feed tank, across the membrane surface. The partially concentrated retentate is recycled into the tank for further processing while the permeate is stored or discarded as required.

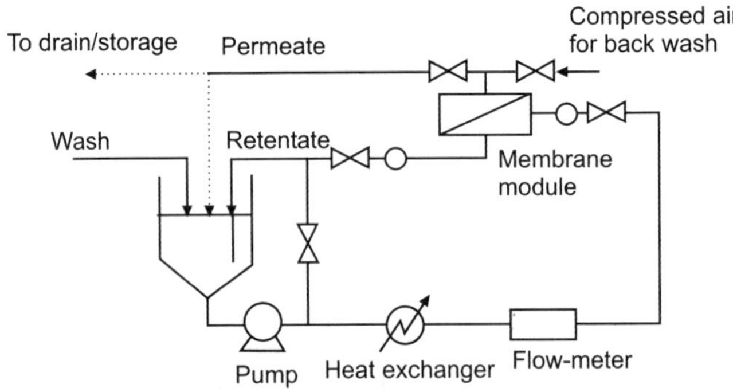

Fig. 12.23: Flow diagram for cross-flow microfiltration system

Applications

Microfiltration is a reliable technique for the production of sterile water and sterilization of solutions for medicinal and pharmaceutical applications, biotechnology fields and the food and beverage industries. Microfiltration membranes find wide use in the pharmaceutical and medical applications as they provide a convenient and reliable means to sterilize the fluids without heat. Contamination of fluids being injected into a patient or fluids used in the haemodialysis can be prevented by microfiltration. Also, microfiltration membranes are used to filter injectible fluids during their manufacture.

In the food and beverages industries, microfiltration has replaced conventional filtration using asbestos, diatomaceous earth, etc. Microfiltration finds widespread applications as in the final filtration of wines, clarification of fruit juices and beer, filtration of gelatine and corn syrup, in cold pasteurization of beer and pasteurization of milk, etc. Microfiltration is the process that is used to clarify fermentation broth and biomass. Membrane filters provided on the vents of the sterile water tanks to prevent the microbial contamination as the water level is lowered, membrane filtration of the gas admitted to the autoclaves and the freeze-dryers after the duty cycle, containment of the biological activity in purge gases in fermentation, are some examples of microfiltration in the gas phase.

Another area of importance that makes use of microfiltration to remove the particles in water or air is the microelectronics industry. In the semiconductor fabrication, it is the use of contaminated process fluids that is responsible for the defects found to occur in the integrated circuits. Microfiltration is used to filter all fluid reactants employed in the manufacture of microcircuits to ensure the absence of particles.

Microfiltration is used for drinking water treatment on a large scale. Since microfiltration can remove bacteria from water, wastewater treatment, particularly the treatment of the municipal sewage is another important application of microfiltration. Capillary membranes operating with shell side feed, at very low cross flow velocities, are found to give excellent clarity and sterility. Separation of oil-water emulsions, pre-treatment of the water fed to the nanofiltration and reverse osmosis plants are other applications of microfiltration. Laboratory or analytical uses include recovering biomass, measuring the particulates in water, clarifying and sterilising protein solutions, etc.

12.7 ULTRAFILTRATION

Ultrafiltration is a pressure-driven, micro-porous membrane based separation technique working on the principle of size exclusion. It is used to retain and concentrate the fine suspended solids and solutes from the solutions. Ultrafiltration separates the high molecular weight substances such as sugars, polymers, colloidal materials and bio-molecules such as proteins from the low molecular weight solutes. Ultrafiltration operates with a pressure driving force of 2 to 10 bars allowing liquid solvents such as water and small solutes to pass through the membrane while retaining the solutes in the range 0.001 to 0.1 μm. Ultrafiltration membranes are generally characterized by their molecular weight cut-off representing roughly their ability to retain a solute of a particular molecular weight. The nominal molecular weight cut-off for ultrafiltration membranes lie in the range 300–10^6 Daltons.

Ultrafiltration is not fundamentally different from the microfiltration except in the size of the molecules it retain and the magnitude of the pressure driving force required to effect the separation. Both processes work primarily by size exclusion, permitting the smaller species to pass through the membrane while the larger ones are retained. Both processes can deliver high purity permeate and yield substantial savings in the cost, space and the energy. Ultrafiltration process is particularly attractive because the filtration performance falls between the microfiltration and the nanofiltration, and has a wide range of applications from the recovery of fine suspended solids to the recovery of high molecular weight solutes from the solutions.

Membranes

The ultrafiltration membranes are porous membranes, made of polymeric materials namely, polyacrylonitrile, polyamide, polyvinylchloride, polysulphone, polyethersulphone, polypropylene, nylon 6, polytetrafluoroethylene, polyvinylidene difluoride, polyethersulphone and cellulose acetate. Inorganic materials such as ceramics, carbon based membranes and zirconia are also used as the ultrafiltration membranes. Use of the inorganic membranes is generally restricted to corrosive applications because of their cost. The membrane modules include cassettes, spirals, hollow-fibres, tubes, flat sheets and inorganic monoliths. The important characteristics of ultrafiltration membranes are the porosity, morphology, surface properties, mechanical strength and chemical resistance. Membrane permeation properties are largely governed by the pore sizes and the pore size distributions, and not by the macro-molecular structures of the membrane materials. The thermal, chemical, mechanical, and biological stability are considered to be of greater importance than the molecular structure.

The most important measures of the performance of ultrafiltration membrane are their productivity and the extent of separation. The membrane performance is usually affected by concentration polarisation, membrane fouling, frequency of membrane cleaning and the operating pressure. The permeation rate is affected by concentration polarisation, which occurs due to the localized concentration of the species retained on the membrane surface when the solvent of a mixture flows through the membrane. In the case of processing a solution, this localized concentration of solute normally results in the precipitation of a solute gel over the membrane. When processing a suspension, the solids collect as a porous layer over the membrane surface. As a consequence, the permeate rate is controlled by the rate of transport through the polarisation layer, rather than by the membrane properties. Hence, the throughput depends not only on the physical properties of the membrane, such as the permeability and thickness but also on the process and system variables like the feed concentration, system pressure, velocity and the temperature. Fouling of the ultrafiltration membrane occurs by the gel-layer consolidation, the adsorption of solutes on the upstream membrane surface, and/or the deposition of solutes within the membrane pores. Contrary to the concentration polarisation, the fouling is a function of time. If the fouling is reversible, membranes require periodic cleaning, and if irreversible, membranes have to be replaced.

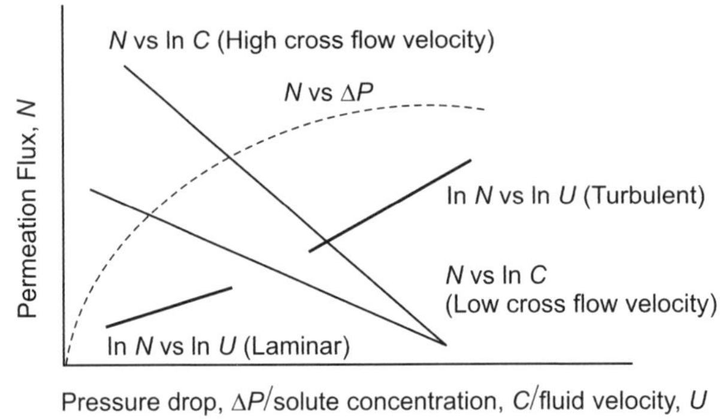

Fig. 12.24: Influence of pressure, velocity and solute concentration on the membrane flux

The flux of the solvent through the membrane depends on the pressure drop across the membrane, flow rates on either side of the membrane, the temperature and the solute concentration in the feed as shown in Fig. 12.24. The rate of permeation increases with increase in the flow velocity of the liquid across the membrane surface. Increasing the flow velocity also reduces the fouling of the membrane. However, higher the flow velocity, the higher will be the energy consumption and cost of pumping. The permeation rate is directly proportional to the applied pressure across the membrane. However, due to the increased fouling, the operating pressure is not allowed to exceed 10 bar. The permeation rate generally increases with temperature, because of the increase in the diffusivity of the macromolecules and the decrease in the viscosity of the retentate. But a high temperature operation is generally not adopted because of the degradation and denaturation of the species that occur at higher temperatures. It is found that the flux of the solvent through the membrane decreases with the increase in the solute concentration in the feed.

Applications

In the dairy industries producing cheese and casein from milk, about 90% of the milk ends up as whey which was entirely wasted in the past. High value products such as protein powder are now recovered from the whey by ultrafiltration. An equally important process in the dairy industry is the use of ultrafiltration on the milk directly rather than on the whey, before the cheese is made. The food industries employ ultrafiltration for the fractionation and concentration of heat sensitive proteins for food additives, concentration and clarification of fruit juices and concentration of the gelatin leaving the extractor as a dilute solution. Recovery of vaccines and antibiotics from the fermentation broth, virus removal from the drug products derived from mammalian cells or human and animal blood plasma are achieved by the ultrafiltration membrane separation.

In the textile industry, ultrafiltration provides an economical means to textile sizing by the recovery and reuse of the sizing solution such as the polyvinyl alcohol used to the warp threads for lubricating them and protecting them from abrasion. Recovery and recycling of electrocoat paint is one of the most important applications of ultrafiltration. The paint, composed of resin, a pigment and water are separated into two streams by ultrafiltration: the first stream which includes the water and a small amount of paint resin, and the second consisting mostly of the recovered pigment which are reused in the plant.

Ultrafiltration is an important water purification technology used for the production of high-purity water. It is effective in the removal of colloids, proteins, bacteria, and other organic molecules larger than 0.01 μm size. Production of the ultra-pure water and the high-purity chemical fluids by ultrafiltration now find important place in the semiconductor industry. Waste water treatment and the water purification applications produce a permeate stream with reduced colloids and viruses. In the pulp and paper mill waste treatment, ultrafiltration could achieve a marked reduction in colour in the caustic bleach effluents from the pulping process. Petroleum and gas industries use ultrafiltration for oil emulsion waste treatment for oil-water separation. Removal of the free oils from the waste streams is attractive to the metal working industry also. Laboratory grade water purification, drinking water disinfection, pre-treatment of the water for nanofiltration or reverse osmosis are other areas of ultrafiltration application..

Process Configuration

Crossflow pattern is generally followed for ultrafiltration membranes. This helps to increase the flux by reducing the plugging. In the crossflow ultrafiltration, membrane does not act as a collector of ions, molecules, or colloids, but merely acts as a barrier to these components. These systems can be operated at high steady-state fluxes, and is the ideal choice when the solute concentration in the feed is quite high. Ultrafiltration applications that run in the dead-end flow mode, include the virus removal and the water treatment where both the cartridges and the hollow fibres are used. For low solid concentration in the feed, dead-end-flow guarantees a cost effective operation. The separated and deposited solids are removed from the membrane by air injection, scrubbing or by the permeate back-flushing. The following four configurations are common for the ultrafiltration systems: (1) batch ultrafiltration (2) continuous-bleed-and-feed ultrafiltration (3) batch diafiltration (4) continuous bleed-and-feed diafiltration

Batch Ultrafiltration

In the batch ultrafiltration, a batch of feed solution is contained in a vessel from which it is continuously pumped through the membrane system which consists of several membrane cartridges arranged in parallel (Fig. 12.25). The solvent, usually water, permeates through the membrane and is collected and the retentate is recycled back to the feed tank. The permeation flux decreases and the solute concentration in the retentate stream increases with time. When the solute concentration in the retentate reaches the desired value, the operation is discontinued and after cleaning the membrane, the feed tank is charged with a fresh batch and the process repeated. The time required for batch processing depends on the membrane area and the permeate flux.

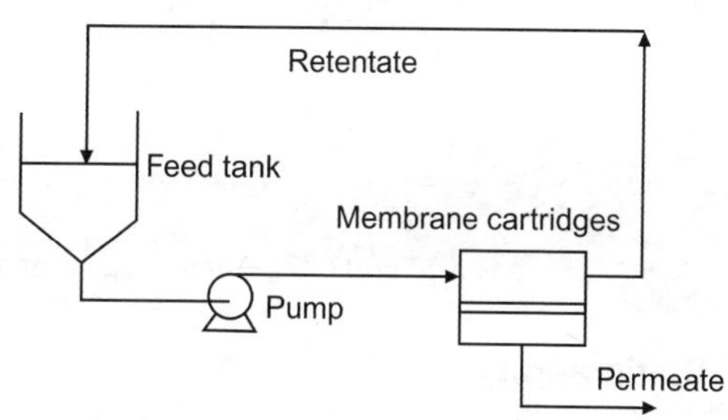

Fig. 12.25: Batch ultrafiltration

Continuous Feed-and-Bleed Ultrafiltration

A continuous single pass ultrafiltration system involves continuous passage of the feed solution across the membrane and continuous withdrawal of the permeate and the retentate products with no part of the retentate being recycled. Ultrafilters are rarely operated on the single pass mode, as it requires a very long membrane path or very large membrane area for an appreciable permeate yield that would guarantee an adequate concentration of solutes in the retentate. The desired product in many ultrafiltration systems is the retentate, and not the permeate. A single-stage feed-and-bleed system shown in Fig. 12.26 on the other hand operates with continuous recycling of a major portion of the retentate. The recycle stream joins the fresh feed and the combined stream is admitted to the membrane system.

The drawback of this system is the very high concentration of the solute on the retentate side. The high concentration of the solute results in a low flux through the membrane that necessitates very large membrane area in comparison with the batch mode. A multistage continuous feed-and-bleed system is proposed to alleviate this problem to a great extent.

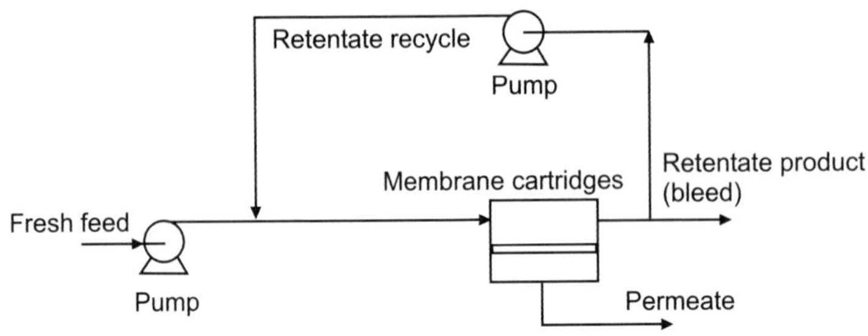

Fig. 12.26: Single-stage continuous feed-and-bleed ultrafiltration

Figure 12.27 shows schematically a three-stage continuous feed-and-bleed ultrafiltration unit. In this mode of operation, a small portion of the retentate leaving a stage is recycled to that stage while the other portion is admitted as feed to the next stage. In this way, only the retentate in the last stage of the cascade has the highest concentration and hence only the flux there is adversely affected. All other stages operate with higher flux than the single-stage system, leading to a smaller total membrane area.

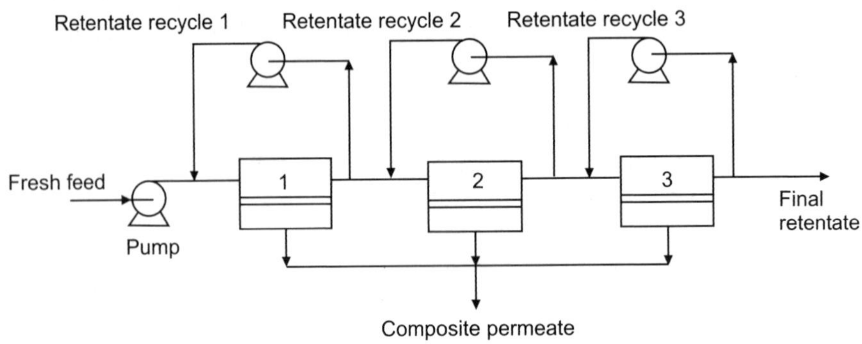

Fig. 12.27: A multistage continuous feed-and-bleed ultrafiltration system

Diafiltration

The flux of the solvent through the membrane decreases with the increase in the concentration of solute in the retentate. When a high degree of solute recovery is desired, the flux may drop to a very low value. Diafiltration is employed to maintain a steady flux of the permeable component when the non permeating solute concentration in the retentate increases. Diafiltration is a specialized type of ultrafiltration process in which the retentate is diluted with the solvent and re-filtered, to increase the flux. Thus the solute separation can be improved by first conducting ultrafiltration till the retentate concentration reaches a limiting value, followed by the diafiltration to the final desired separation. Diafiltration also can be conducted in the batch mode, single-stage continuous feed-and-bleed, or multistage continuous feed-and-bleed, as in the case of ultrafiltration.

12.8 NANOFILTRATION

Nanofiltration is a relatively new membrane separation technique that lies between ultrafiltration and reverse osmosis. It is a pressure driven separation process that differs from the microfiltration and ultrafiltration processes in that it achieves separation not by size exclusion; rather, separation takes place by the diffusion through the membrane. Nanofiltration is sometimes referred to as *loose reverse osmosis* due to its larger membrane pore structure compared to the membranes used in reverse osmosis, that allows more salt passage through the membrane. The nominal pore size of the membrane is typically about 1 nm. Nanofiltration membranes are generally rated by the molecular weight cut-off rather than the nominal pore size. And the cut-off is typically less than 1000 Daltons. The pressure that is required to perform nanofiltration is much higher than the pressure required for micro and ultra filtration. The pressure driving force lies typically in the range of 3 bar to 100 bar.

Most often, solutions with low total dissolved solids such as surface water and groundwater are treated by nanofilters to drive the water through the membrane while retaining the small solutes. Because it can operate at much lower pressures than the reverse osmosis, and allows the passage of some of the inorganic salts, nanofiltration is used in applications where a high organic removal and moderate inorganic removal are desired. It is capable of concentrating sugars, divalent salts, bacteria, proteins, particles, dyes and other constituents that have a molecular weight greater than 1000 Daltons. While the main use of nanofiltration is for the removal of contaminants from a water source, it is also commonly used for desalination and reduction of hardness in drinking water. Since nanofiltration allows passage of monovalent salts whilst retaining larger species, the process is now finding increasing application in the simultaneous concentration and partial demineralization; nanofiltration can achieve desalination of whey while retaining lactose and proteins. Decolourisation of waste water in the textile and cellulose industries is another application of nanofiltration. They are used in the pretreatment of the feed for the electrodialysis and ion exchange processes. An advantage of nanofiltration over reverse osmosis is that the nanofiltration can typically operate at higher recoveries, thereby conserving the total water usage due to a lower concentrate stream flow rate. Nanofiltration is not effective on small molecular weight organics, such as methanol.

Membranes

Nanofiltration membranes generally contains ionogenic groups such as carboxylic and sulphonic acid groups, and most membranes are negatively charged. Because of this, anions are preferentially rejected by nanoltration membranes. This ion selectivity is responsible for the special separation properties of nanofiltration membranes. The membranes are capable of separating dissolved multivalent components in aqueous media. The rejection rate for salts with monovalent ions is low whereas salts with bivalent ions and organic matter are retained to a high degree. Cellulose acetate and aromatic polyamides are generally used as nanofiltration membranes. These are characterised by a salt rejection from 95% for divalent salts to 40% for monovalent salts and an approximate molecular weight cut-off (MWCO) of 300 for organics. The trans-membrane pressure required is lower than that in the reverse osmosis, reducing the operating cost significantly. However, nanofiltration membranes are subject to scaling and fouling and often modifiers such as anti-scalants are required.

12.9 REVERSE OSMOSIS

Reverse osmosis, also called hyperfiltration, is essentially a pressure driven membrane diffusion process for separating dissolved solutes. The most common use of reverse osmosis is the desalination of sea water and its conversion into potable water. Reverse osmosis eliminates almost all dissolved solids including mineral salts and heavy metals, bacteria, viruses and other germs contained in the water. In addition to the clean water applications, reverse osmosis also finds wide use in food, biotechnological and pharmaceutical industries.

Principle

When solutions of different concentrations are separated by a membrane that is permeable to the solvent and not to the solute, diffusion of solvent occurs from the low solute concentration side to the higher concentration side. The process is represented as shown in Fig. 12.28a. The diffusion of the solvent is called osmosis, and the osmotic transfer of the solvent is common in plant and animal cells.

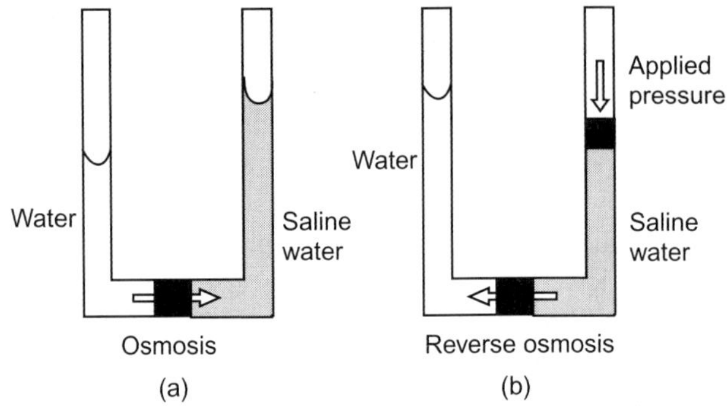

Fig. 12.28: Schematic representation of osmosis and reverse osmosis

Figure 12.28a shows a schematic representation of the osmotic transfer of solvent when pure water is separated from the saline water by a membrane that is permeable only to water. This transfer occurs till the chemical potential of the solvent on the two sides of the membrane are equalised.

The diffusion of water can be stopped by increasing the pressure on the concentrated solution, which results in the increase of the chemical potential of the solvent. When the pressure applied equals the osmotic pressure of the solution, equilibrium is established and the solvent activities on the two sides of the membrane are equalised and the osmotic transfer of the solvent comes to a stop. If a pressure higher than the osmotic pressure is applied on the concentrated solution, the direction of the transfer of the solvent is reversed, and the solvent diffuse from the concentrated solution to the dilute solution through the membrane, as shown schematically in Fig. 12.28b. This process of forcing a solvent from a region of high solute concentration through a semipermeable membrane to a region of low solute concentration by applying a pressure in excess of the osmotic pressure is called reverse osmosis. It may be noted that reverse osmosis is the opposite of osmosis only with respect to the direction of flow of the solvent. In all real membranes, there is always a certain transport of solute in the direction of its concentration gradient; this solute transport that is present in the osmosis is not reversed in reverse osmosis.

Reverse osmosis is the ultimate process in membrane separation, and it removes particles larger than 0.1 nm, and generally requires a pressure differential in the range 2-100 bar to effect separation. Because of the high osmotic pressure of seawater (about 27 bar), reverse osmosis plants must frequently operate at high pressures, i.e. in the range 40– 82 bar for desalination of sea water. For brackish water, usually 2–17 bar pressure differential is required on the high concentration side of the membrane.

Advantages and Limitations

Reverse osmosis systems are compact and require less space than the other desalination techniques. Reverse osmosis plants operate with minimum cost and are able to remove the organic contaminants and inorganic salts with minimum requirement of chemicals. Compared to the energy requirement in multistage flash evaporation plants used for desalination of the sea water, the energy consumption in reverse osmosis is minimum, since the separation involves no phase change. Since the energy for pressurising the feed can be recovered to some extent, the thermodynamic efficiencies of these units are quite high. Reverse osmosis is ideal for treating thermally and chemically unstable substances. Since 95-99% of the dissolved salts, majority of organic materials, bacteria and suspended solids are rejected by the reverse osmosis membranes, the water obtained from reverse osmosis plants require no further treatment and can be used directly.

The major limitation of the reverse osmosis systems is the high pressures required to effect separation. The applied pressure must exceed the osmotic pressure of the solution which is usually very high; for example, for sea water, it is as high as 27 bar as pointed out earlier. Since the osmotic pressure is a function of the solution concentration, reverse osmosis is usually not applicable to highly concentrated solutions. Temperature of the operation should be kept low; the upper limit of temperature is determined by the membrane material. If the feed temperature exceeds this limit, it must be cooled before admitting to the reverse osmosis unit; otherwise, the membrane compaction that results by operating at higher temperatures will adversely affect the membrane performance. Reverse osmosis process frequently requires a feed pre-treatment for the removal of materials like oil and grease, because the membranes are susceptible to fouling by these materials.

Membranes

The desirable characteristics of reverse osmosis membranes are: (1) high salt separation (2) high permeability (3) mechanical robustness to allow module fabrication and to withstand operating conditions (4) chemical robustness to fabrication materials, process fluids, cleaners and sanitizers (5) low extractables (6) low fouling characteristics (7) high capacity (8) low cost and (9) consistency. Reverse osmosis membranes do not work according to the principle of size exclusion by pores; separation takes place by diffusion through the membrane. In water purifications, the membrane must have a pore diameter less than approximately 1 nm so that electrolyte solutes such as sodium chloride and organic solutes that contain more than one hydrophilic functional group will be retained and only the smallest organic molecules and uncharged solutes pass freely through the semipermeable membrane along with water. Thus more than 95-99% of inorganic salts and charged organics will be rejected due to the charge repulsion established at the membrane surface.

Reverse osmosis membranes are generally of two types, asymmetric integrally skinned membranes and thin film composite membranes. The asymmetric integrally skinned membrane is made as a composite film in which a thin dense layer about 0.1 to 10 μm thick containing extremely fine pores are supported upon a much thicker (50 to 125 μm) layer of microporous sponge with little resistance to permeation. The thin dense layer has the ability to block the passage of small solute molecules. The selectivity and permeate flux depend on the materials of the membrane, its preparation procedure and the structure of the membrane barrier layer. Membranes currently used in commercial reverse osmosis installations are asymmetric, flat sheet membranes of cellulose acetate or cellulose triacetate, fine hollow fibres of aromatic polyamides or cellulose triacetate and thin film composites where an extremely fine layer of a highly hydrophilic polymer such as cross-linked polyetherurea is placed on a microporous support, usually made from polysulphone.

One limitation of cellulose acetate membranes is that it can be used mainly in aqueous solutions and at temperatures below about 333 K. Also, they are susceptible to degradation from biological attack. Cellulose acetate hydrolyses back to cellulose under certain acidic as well as basic conditions. Hydrolysis limits the useful life of the membrane, because the salt passage increases as hydrolysis occurs. Therefore, it is necessary to chlorinate the feed water and control the pH within the range of 4.5 to 7.5.

Reverse osmosis membrane modules are available in spiral, hollow fibre, tubular and plate-and-frame designs. Modules give mechanical support for the membrane, which must operate at high pressures; modules help to control concentration polarisation by utilising the flow energy efficiently and also provide channels for the smooth flow of the permeate. Spiral-wound modules are used the most since they are compact, easy to use, have low feed-side pressure drop, are less prone to clogging, and are easily cleaned, mechanically robust, and economical. They are making increasing inroads in the area of food processing and can cope well with the feeds containing colloidal solids. Hollow fibres are run with the feed on the shell side, flowing radially inward, and out along the central core. This places less mechanical stress on the fibres. Tubular modules are used with high suspended solids and high viscosity fluids. Table 12.4 gives a comparison of the performance of the different membrane modules available for reverse osmosis.

Table 12.4: Comparison of reverse osmosis membrane modules

	Spiral	Hollow fibre	Tubular	Plate-and-frame
Packing density, m^2/m^3	800	6000	70	500
Feed flow, m^3/m^2 s	0.3-0.5	0.005	1-5	0.3-0.5
Feed ΔP, kPa	275-625	10-30	210-310	310-620
Plugging sensitivity	High	High	Low	Moderate
Cleanability	Fair	Poor	Excellent	Good
Relative cost	Low	Low	High	High

System Configuration

The membrane modules in a reverse osmosis system may be arranged staged either on the retentate, or on the permeate side depending upon the application and the type of the membranes used. In the retentate staging, the retentate from the first stage is the feed to the second stage and so on, where as in the permeate staging, the permeate from one stage acts as the feed to the next stage. Figure 12.29 shows the common reverse osmosis system configurations for permeate product recovery.

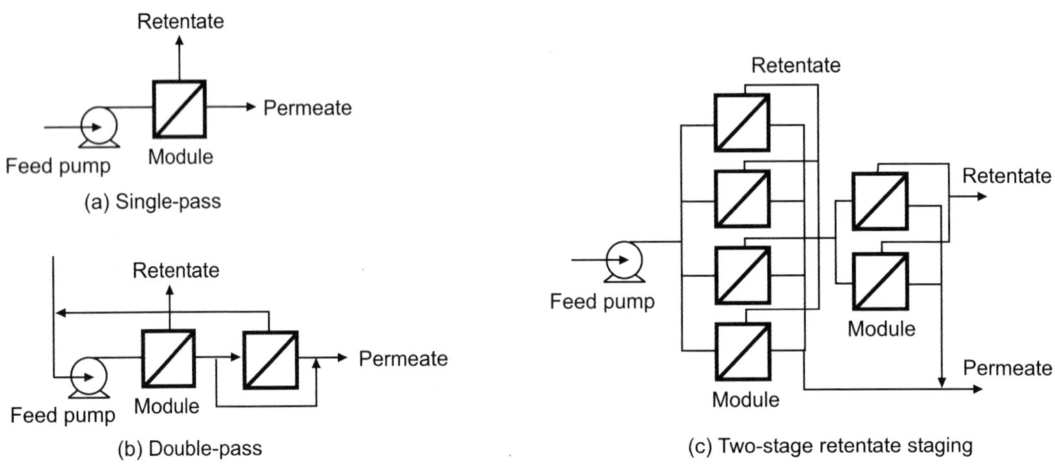

Fig. 12.29: Reverse osmosis system configuration

For single-pass operation, (Fig. 12.29a), permeate in sufficient quantity and purity is produced by passing the feed once through the membrane. In the double-pass operation as shown in Fig. 12.29b, the permeate product leaving the first stage is sent to the second for further treatment, and a portion bypassed the second stage for blending with the permeate leaving the second stage. This ensures a permeate product of the desired quality. Double-pass systems are used for removing high-concentration, hard-to-remove impurities, for high product purities, and for lower retention modules. Double-pass operation has been recommended for the desalination above 38 kg/m^3 or modules with salt retention < 99.2%. It is also common in ultra-high purity water systems. Figure 12.29c shows a typical flow arrangement for a retentate-side staged reverse osmosis plant. The first stage consists of a housing of four modules and the second stage consists of a housing of two modules. Spiral-wound modules are normally packed to 4 to 7 modules per cylindrical housing. Retentate staging is used to increase the conversion ratio from around 25 to 50% in a single stage to as high as 75 to 90% in three stages. The feed flow is reduced by the conversion ratio after passing through the first module stage. Therefore, to maintain a minimum cross flow rate, the second stage should have less membrane area than the first. Inter-stage pumping between housings may be required to maintain the pressures and the flows.

Applications

The most common use for the reverse osmosis is in the desalination of water. Prior to 1980, multistage flash distillation was the main process for desalination. But, now the dominant process for desalination is the reverse osmosis. Both sea water and brackish water are treated to remove dissolved salts and organics from water. The plant components should be strong enough to withstand the high pressures in the desalination plants, pressures in some cases exceeding 80 bar. A schematic diagram of a typical reverse osmosis desalination process is shown in Fig. 12.30.

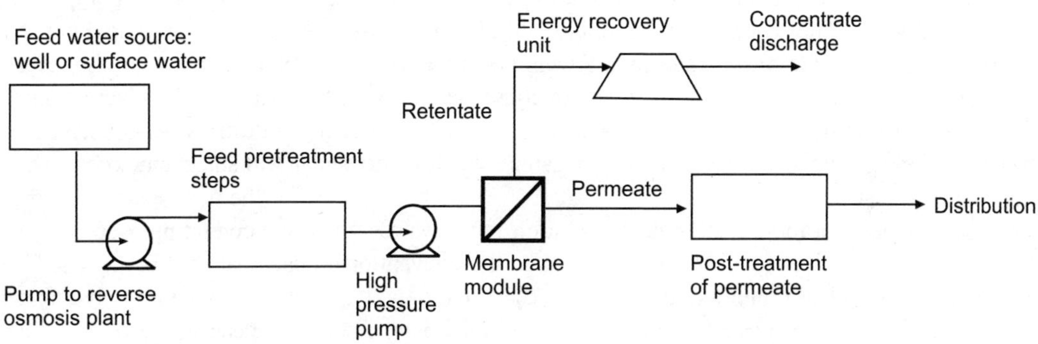

Fig. 12.30: Reverse osmosis process for desalination of water

The feed water is subjected to the pretreatment processes to ensure long life of the membrane. Filtration, coagulation, chemical injection, pH adjustment, chlorination/ de-chlorination, stripping, etc. are the usual pre-treatment processes employed. Pretreatment is commonly used to extend the membrane life and to increase the recovery. The representative pretreatment train for water purification applications in Fig. 12.31 effectively controls the feed channel clogging, the mineral scaling, the fouling by organic films and microorganisms, and the oxidants that can degrade the membranes.

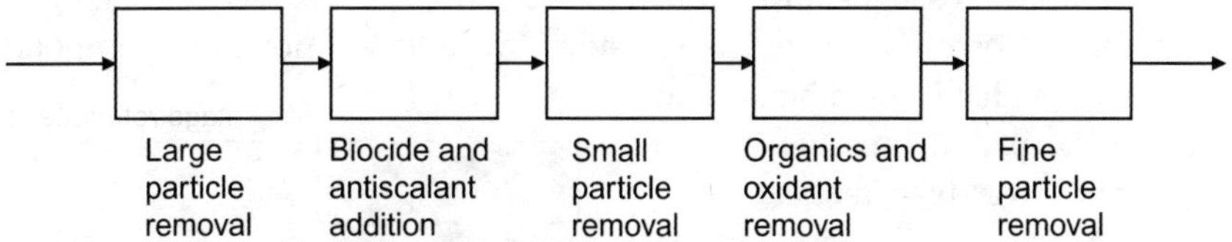

Fig. 12.31: Reverse osmosis pre-treatment scheme

Solid particles can plug the feed channels and deposit as a cake layer on the membrane surface causing increased flow resistance and lowered fluxes. The pretreatment for particles removal includes a combination of coarse screening, hydroclones, flocculent additions, sand or multi-media filtration bed with periodic back-flushing, and depth filter cartridge or ultrafiltration systems.

After the pretreatment, the high pressure discharge pump takes the treated water to the reverse osmosis membrane module system, usually of the spiralwound or hollow fibre type. The retentate stream is depressurised by passage through a pressure recovery turbine. The pressure of the concentrate is lowered to the desired level, reducing the energy consumption of the process by around 25 to 40%. About half of the total water in the feed is recovered as the permeate stream which is 99.95% pure water. The permeate is subjected to post-treatment such as filtration through lime and chlorination for disinfection, before it is distributed as potable water.

Reverse osmosis finds wide applications in the treatment of surface water and ground water, production of pharmaceutical grade water, ultra-pure water for the food processing and the electronic industries, water for chemical, pulp and paper industries, water used in car washes during the final vehicle rinse to prevent the water spotting on the vehicle, boiler feed water pre-treatment before ion exchange process, etc. Reef aquarium keepers use the reverse osmosis systems in combination with deionization to contain nitrogen compounds and phosphates, chlorine, chloramines, copper, silicates and many other chemicals present in these water that are detrimental to the organisms in a reef environment.

The waste water and industrial effluent treatments are other important areas of application of the reverse osmosis. Reverse osmosis can recover metals, antifreeze, paints, dyes and oils in the effluents while generating cleaned up waste water for disposal. The industrial waste treatment by the reverse osmosis can remove heavy metal ions and non-biodegradable substances and can recover components of commercial value. The separation of sulphites and bisulphites from the effluents in pulp and paper processes, treatment of the waste water in the dyeing process are important applications of the reverse osmosis process. It is also used for the recovery of constituents having food value such as lactose, lactic acid, sugars and starches from the waste waters in the food processing industries. Also used in the treatment of the municipal water to remove inorganic salts, low molecular weight organic compounds, viruses and bacteria.

Reverse osmosis is a more economical operation for dewatering and concentrating food products such as fruit juices, coffee, soup, tea, milk, etc. than the conventional heat-treatment processes. Heat sensitive materials such as proteins and enzymes present in the food materials are not adversely affected by this process. Reverse osmosis is extensively used in the dairy industry for the production of whey protein powders and for the concentration of milk to reduce shipping costs. Removal of alcohol from beer and wine and concentration of the amino acids and alkaloids also can be achieved by reverse osmosis.

12.10 PERVAPORATION

Pervaporation is a process in which a liquid feed is separated into its components by partial vaporisation through a nonporous semipermeable membrane. It is the only membrane process where phase transition occurs during separation. Dense membranes are used for pervaporation, as for reverse osmosis. However, pervaporation differs from reverse osmosis in that in the former, the phase state on one side of the membrane is different from that on the other side. In pervaporation, the components in the mixture permeates through the membrane and evaporates because the partial pressure on the permeate side is kept lower than the saturation vapour pressure of the permeate. The phase change occurs usually near the down stream side of the membrane, and the heat of vaporisation is supplied by the sensible heat of the liquid conducted through the thin dense layer. The pressure on the feed side is usually kept high so that the retentate remain a liquid even when the feed is getting depleted of the permeating components (Fig. 12.32).

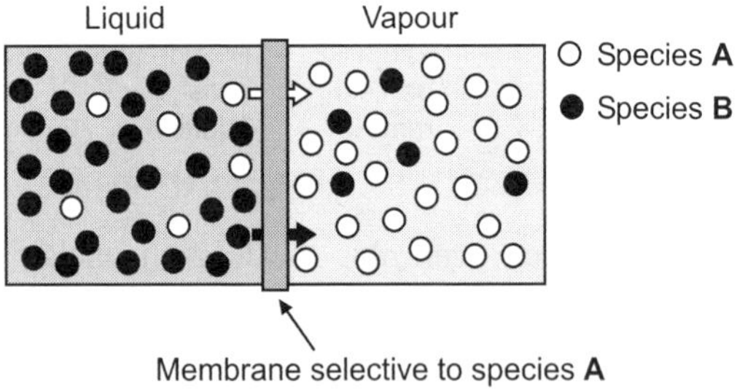

Fig. 12.32: Schematic representation of pervaporation process

The driving force for mass transport is the gradient in the chemical potential of the substance from the feed side to the permeate side of the membrane. This driving force is generated by maintaining the permeate side at a very low pressure, either by using a vacuum pump system or by sweeping using an inert gas as shown in Figs. 12.33a and b. In order to ensure a continuous mass transport, very low absolute pressure as low as 133–400 Pa, is generally maintained on the permeate side, while the feed side is kept at nearly atmospheric pressure. The permeate undergoes a phase change and leaves as a vapour while the retentate undergoes no phase change and leaves in the liquid state.

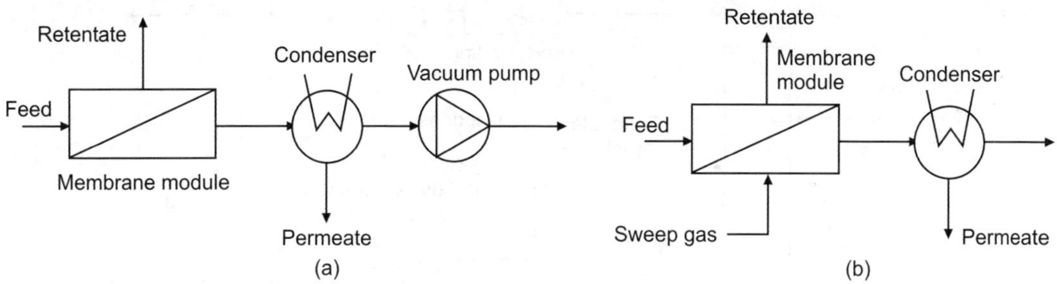

Fig. 12.33: Pervaporation vacuum system (a) using vacuum pump (b) using purge gas

Pervaporation involves a sequence of three steps: selective sorption of the components into the membrane, permeation or diffusive transport across the nonporous membrane and finally evaporation and desorption of the permeate into the vapour phase. The term pervaporation is derived to indicate that the process involves a combination of **permeation** and **evaporation**. Separation of components is achieved due to the difference in the transport rates of individual components through the membrane. The low pressure permeate stream gets enriched in the components that diffuse more rapidly, while the slower components are concentrated in the retentate stream.

Although the membrane materials govern the pervaporation process and determine the separation factor, the operating parameters such as the temperature, pressure, flow velocity and feed concentration influence the permeation flux and selectivity. Generally the flux increases with increase in temperature and with the increase in the trans-membrane pressure. Increasing the flow velocity increases the flux at small Reynold's numbers; however, in the turbulent flow regimes, the cross-flow velocity has no significant influence. Since the overall permeability through the membrane is influenced by the solubility in the membrane and the rate of diffusion through the membrane, the permeate flux is a strong function of the feed concentration. The overall permeabilities of species depend upon their solubilities in and diffusion rates through the membrane.

An advantage of pervaporation process is the large difference in the compositions of the retentate and permeate product streams. It is possible by providing a sufficiently low pressure on the permeate side, to achieve a very large concentration of the major constituent in the retentate while the permeate contains such constituents in negligible amounts. Further, the composition of the permeate vapour product may be very different from that of the vapour which would be in equilibrium with the retentate product obtained.

Pervaporation shows great promise in the separation operations where conventional techniques fail or are uneconomical because of the very high energy consumption. Important categories of separations in this category are the separation of heat sensitive, close boiling or azeotropic mixtures. Pervaporation operations provide simple alternatives with low energy consumption and capital costs over traditional techniques such as the vacuum distillation and solvent extraction for separation of azeotropes. Since the membranes are selective to the components of the feed that is present in smaller proportions, and change of phase is restricted to the permeate stream alone and not to the retentate, pervaporation is not an energy intensive operation, even though it involves phase change from the liquid to the vapour. The hybrid processes integrating pervaporation with distillation are recently gaining acceptance for passing over the azeotrope and production of pure organic solvents. Such a process is shown in Fig. 12.34.

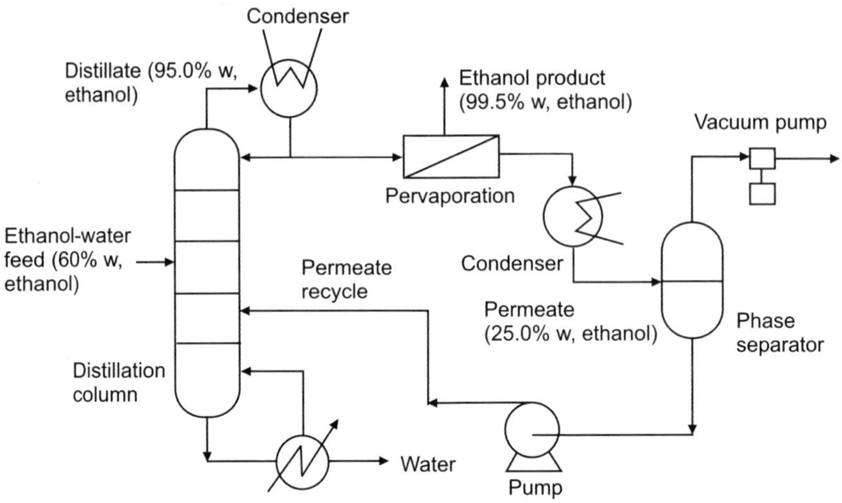

Fig. 12.34: Hybrid process combining distillation and pervaporation

In the fractionation of ethanol-water mixtures, the distillate purity leaving a column operating at ambient pressure is limited to 95.6% (weight) ethanol as ethanol-water system forms an azeotrope at this composition. Figure shows a process combining distillation and pervaporation which can produce 99.5% (weight) ethanol from a feed which is 60% pure. The distillate of composition about 95% is condensed and sent to a pervaporation unit which produces a retentate liquid which is 99.5% ethanol and a permeate vapour which is 25% (weight) alcohol. The vapour permeate is condensed and recycled to the distillation column as shown in Fig. 12.34. The bottom product leaving the column is water which is almost free of alcohol. The vacuum needed on the permeate side of the membrane is maintained by a vacuum pump. Recently, more complex integrated processes are also developed in which pervaporation is paired with other conventional separation techniques, such as adsorption, phase separation or counter-current extraction.

Membranes

In pervaporation, the separation effectiveness is determined by the physiochemical structure of the membrane; the separation capacity of the membrane is primarily a function of the interaction of the membrane material and the feed species. Feed temperature, feed composition and permeate pressure have only a secondary influence on the separation capacity. Pervaporation membrane modules come in various types; plate-and-frame, spiral wound, hollow fibre, and tubular modules are in use. Commercial membrane modules are predominantly of plate-and-frame type, mainly due to the ease of using gasketing materials that are resistant to organic liquids and ease of providing the heat required for evaporation. The hollow-fibre modules find increasing applications in the removal of volatile organic compounds from aqueous solutions.

Three different types of pervaporation membranes are in use. They are: the hydrophilic membranes, the organophilic or hydrophobic membranes and the organoselective membranes. The choice of the membrane type depends strongly on the type of application. Selecting a membrane permeable to the minor component is important since the membrane area required is roughly proportional to the mass of permeate.

Hydrophilic membranes

The hydrophilic membranes allow the water to permeate and they are mainly used for the dehydration of organic solvents. The commercially most successful pervaporation membrane system to date is based on polyvinyl alcohol. A three-layer composite polymeric membrane is generally used for the dehydration of aqueous ethanol. The separation is achieved by a polyvinyl alcohol layer of about 0.1 micron thick. This active hydrophilic membrane is supported by a porous polyester support which is cast on a polyacrylonitrile or polysulphone membrane. The polyvinyl alcohol layer is partially cross-linked during its formation in order to suppress the excessive swelling of polymer in water. Cellulose acetate-based composite membranes, with an active layer of PVA are also used for dehydration of alcohol. More recently, membranes based on polyimide have also been used as pervaporation membranes. Polyelectrolyte materials such as chitosan, a natural product, are also potentially useful for dehydration by pervaporation.

As an alternative to the hydrophilic polymeric membranes, ceramic membranes consisting of nano-porous layers on top of a macroporous support have also been developed for the dehydration applications. Membranes based on zeolite are examples for this class. The pores in these membranes must be large enough to let water molecules to pass through and retain other solvents such as ethanol that have a larger molecular size. Hydrophilic ceramic membranes based on titania or zirconia have also been developed.

Hydrophobic membranes

Hydrophobic or organophilic membranes are mainly used to remove the organic compounds from aqueous solutions. Quite a lot of research work have been done to develop effective organophilic membranes for the separation of ethanol or other organic solutes contained in the dilute aqueous mixtures. These membranes are made of hydrophobic rubbery polymers such as silicone rubber, natural rubber, polymethyl pentene and teflon taking advantage of their high permeabilities, often overlooking their moderate selectivities. They allow the permeation of organic substances in preference to water. Because of the high hydrophobic nature, silicone rubbers are used for the removal of volatile organic compounds from water as these organic compounds are preferentially sorbed and transported through such membranes. Polydimethylsiloxane-based hydrophobic membranes are also used for permeation of organic solutes such as ethanol from their solutions with water, by pervaporation. Homogeneous membranes of poly (trimethylsilyl) propyne are found to improve the selectivity of alcohol in alcohol-water solutions.

Organoselective membranes

Efforts were made to promote the application of pervaporation in the chemical and petrochemical industries by developing new membranes to separate certain organic-organic liquid mixtures. Polyethylene is a standard material for separating organic mixtures, although the selectivity and flux are too low for commercial applications. In the separation of benzene-cyclohexane mixtures, composite membranes formed by combining a substantially insoluble support material such as the cellulose acetate with additives such as the polyphosphonates are found to give good preferential permeability for benzene. Recently, a number of organoselective pervaporation membranes have been introduced to the market, which are effective in the separation of organic-organic mixtures. Promising results concerning the separation of aromatics and saturates by pervaporation have been achieved using membranes belonging to the category of polyurethane-imides.

Applications

Pervaporation technique was discovered in 1917 by Kober. Despite its immense potential, the industrial exploitation of pervaporation process began only in 1982 with the installation of a pervaporation based plant for the production of alcohol. Primary application of pervaporation is in removing from the feed solutions, those constituents that are present in trace or minor amounts. Different species in the feed mixture have different affinities for and different diffusion rates through the membrane; consequently, if the membrane chosen has greater affinity towards the constituents present in minor proportions in the feed, the permeate gets highly enriched in these constituents. Dehydration of alcohols containing small amounts of water and removal or recovery of trace amounts of organics from aqueous solutions are two typical applications of pervaporation processes. Hydrophilic membranes are used for the former as the membrane should allow permeation of water, while hydrophobic membranes are used for the latter as the minor constituents to be allowed to permeate in this case are, the organic substances. Recent developments in the membrane and module fabrication techniques have made pervaporation industrially very attractive. Following are the major areas of application of pervaporation processes industrially.

1. *Separation of azeotropic mixtures*: The greatest utility of pervaporation is in the resolution of azeotropes as an adjunct to the distillation operations. Pervaporation is less expensive and less energy-intensive than the conventional techniques such as extractive and vacuum distillation employed for the separation of azeotropes. Dehydration of solvent in the binary azeotropes such as the ethanol-water and isopropanol-water are achieved using hydrophilic membranes such as polyvinyl alcohol. Pervaporation can be used for the separation of organic-organic azeotropes such as ethanol-cyclohexane, methanol- methyl tertiary butyl ether (MTBE), ethanol- ethyl tertiary butyl ether (ETBE), etc. using organoselective membranes.
2. *Dehydration of organic solvents*: Dehydration of many organic solvents including the alcohols, ketones, ethers, esters and acids can be achieved by pervaporation using hydrophilic membranes such as polyvinyl alcohol. Dehydration is favoured because water is both preferentially dissolved and transported in the hydrophilic membranes due to its small molecular size. Many substances, including ethanol, methanol, isopropanol, butanol, MEK, acetone and chlorinated solvents can be purified in this way. An important application is in the microelectronic industry which requires extra pure isopropanol as clean-rinse fluid. Using pervaporation, it is possible to recover for reuse 99.99% pure isopropanol from the 85–90% isopropanol waste which contains water and trace contaminants generated by the cleaning operation.
3. *Recovery of organics from aqueous solutions*: Pervaporation using organophilic membranes such as silicone rubber provides a viable technique for the recovery of volatile organics from dilute aqueous solutions. Thus it is possible to recover high value materials with unique properties in a cost-effective way.
4. *Food industries*: Pervaporation offers certain advantages in the isolation of flavors, fragrances, essential oils, etc. Concentration of hydrophobic flavour compounds in aqueous solutions and recovery of flavour and aroma compounds from solutions have made the pervaporation technique a more cost-effective unit process in food manufacturing. In the production of fruit juice concentrates, pervaporation techniques can be used to recover the natural flavour-giving organic substances from the juice. These flavour-giving organics are likely to be lost during the pasteurization and evaporation, practiced before the juice is subjected to concentration by evaporation.
5. *Pollution Control*: Since pervaporation using hydrophobic membranes can remove the volatile organic contaminants such as aromatics, chlorinated hydrocarbons, etc. from the waste water, the method is a quite promising one for the treatment of the industrial waste waters containing organic pollutants. It is also used to treat rinse waters that have become contaminated by the volatile organic compounds such as solvents, degreasers and petroleum-based mixtures. Since the chlorinated solvents or gasoline components that are only slightly soluble in water have very high activity coefficients in aqueous solutions, their near complete removal is achieved easily by pervaporation.

6. *Separation of mixtures of organic compounds*: Separation of the organic–organic mixtures represents the most challenging application for pervaporation. Industrially important mixtures such as the polar-non-polar (e.g. methanol - methyl tertiary butyl ether), aromatic-aliphatic (e.g. benzene-*n*-hexane), aromatic-alicyclic (e.g. benzene-cyclohexane), and isomers (e.g. *p*-xylene, *m*-xylene, and *o*-xylene) can be separated using this technique, though the industrial application is still in the development stage for many systems.

Recovery of alcohols from fermentation broths, continuous ethanol removal from the yeast fermentors, continuous water removal from the condensation reactions such as esterifications to enhance conversion and rate of the reaction, assisting a reversible water-generating reaction by continuous removal of water, analytical techniques such as the membrane introduction mass spectrometry are other potential applications of pervaporation.

12.11 DIALYSIS

Dialysis is a process in which low molecular weight solutes are selectively transferred from a region of high concentration into a region of low concentration by diffusion through a semipermeable membrane. Separation between the solutes is obtained as a result of the differences in the diffusion rates across the membrane arising from the differences in the size and the solubility in the membrane. In practice, it is used to separate species that differ appreciably in size and have a reasonably large difference in the diffusion rates. Small solutes diffuse readily through the membrane, while passage of large molecules and colloids are hindered because their diffusivity is low and their size is larger than the pores in the membrane. There is little or no pressure difference across the membrane and the flux of each solute is proportional to the concentration difference. Dialysis is characterised by the low flux values, in comparison to other membrane processes such as reverse osmosis and ultrafiltration that depend upon the applied pressure. Dialysis is attractive when the concentration difference for the main diffusing solutes is large and the permeability differences between those solutes and the other solute(s) and/or the colloids are large.

In dialysis, the feed solution containing the solute to be transferred flows across the membrane and the products are the liquid *diffusate* (permeate) containing solvent and solutes transferred through the membrane and a *dialysate* (retentate) containing the solvent and the constituents of the feed that have not permeated.

The dialysis membrane has been widely used on laboratory scales for the purification of small quantities of solutes. One important application is haemodialysis, where membranes are used as artificial kidneys for people suffering from renal failure. An important industrial application is in the recovery of sodium hydroxide from a 17–20 weight% caustic viscose liquor contaminated with hemicellulose during viscose manufacturing, to produce a diffusate of 9–10 weight% caustic. Dialysis find application in chemical processing separations such as the separation of H_2SO_4 from nickel and copper sulphates in aqueous solutions, the recovery of nitric and hydrofluoric acid from spent stainless-steel pickle, removal of alcohol from beer, the food processing applications such as desalting of cheese whey, etc. The recovery of salts and sugars from natural products or other colloidal solutions could be achieved by dialysis, but ultrafiltration is the more attractive operation for such applications because of the higher permeation rates that can be obtained.

Membranes

Initially cellulose membranes were used in the dialysis applications including haemodialysis. Synthetic polymers were later introduced which include substituted cellulose derivatives, polyacrylonitrile, polysulphone, polycarbonate, polyamide and poly (methyl methacrylate). These synthetic fibre membranes have on the inner surface of the fibre, a skin layer containing very fine micro-pores, which is contacted with the dialysate. The hydraulic permeability of these fibres is up to ten times that of the cellulose membranes and they can be tailored to achieve a range of molecular weight cut-offs using different preparation procedures.

The module design used for dialysis has undergone considerable changes in the recent past. The tubular membrane modules and the spiral modules were not found to be appropriate for haemodialysis applications because they required several litres of blood to prime the system. These are now replaced by the plate and frame modules and the hollow fibre modules. The plate-and-frame modules with frames arranged vertically were used for industrial scale separations. Typical units consisted of about 100 square frames giving a membrane surface of roughly 56 m^2. More recently developed dialysis units utilize hollow fibres of typically 200 μm inside diameter packed into a heat-exchanger- type module, where the dialysate flows on the inside of the fibres, while the permeate flows on the outside, in a co-current manner. The plate-and-frame design requires ten times the volume of a hollow fibre module to provide a membrane surface equivalent to that provided by the latter.

Haemodialysis

Best known application of dialysis is the use of artificial kidneys to remove waste products from the blood of persons with kidney disease. Haemodialysis is essential for those whose kidney has failed and are no longer able to regulate the body's waste disposal. Dialysis serves to replace some of the functions of the kidney and it can eliminate waste products and restore electrolyte and pH levels of the blood. Cellulose membranes efficiently remove major metabolites like urea and creatinine from blood but metabolites with molecular weights higher than 1000 are removed less efficiently. Now synthetic polymers have completely replaced the cellulose since they tend to simulate the function of normal kidney more closely. These materials include the substituted cellulose derivatives, polyacrylonitrile, polysulphone, polycarbonate, polyamide and poly (methyl methacrylate). Compact hollow-fibre assembly which typically contains several thousand 200-μm diameter, 300-mm long fibre, functions as artificial kidney. The blood (dialysate) is passed through the fibres while saline solution (diffusate) is circulated on the outside. Urea and other small molecules such as uric acid, creatinine, phosphates etc diffuse through the membranes to the external solutions while proteins and cells are retained in the blood. The dialyzing solution has added salts and glucose to prevent the loss of these materials from the blood.

12.12 ELECTRODIALYSIS

Electrodialysis is a membrane separation process in which ionic species are separated from the liquid solutions by means of ion-selective membranes under an applied electric field. It is an electrochemical process in which ions migrate through the semipermeable membranes as a result of their attraction to the two oppositely charged electrodes. Electrodialysis is a powerful separation technique with applications in many industries. The major application of the electrodialysis is in the desalination of sea water and brackish water with salt concentration in the range of 500 to 5000 ppm. Below this range, desalination by ion-exchange is more economical and above this range desalination by reverse osmosis is preferred.

Electrodialysis owes its origin to the efforts made in 1900 for increasing the rate of dialysis by the addition of a direct current. It was with the development of the synthetic ion-exchange membranes based on the phenol-formaldehyde poly-condensation resins by 1940 that the process gained its identity as a distinct process for membrane separation, different from the dialysis in many ways. Meyer and Strauss proposed the multi-compartment electrodialyser in 1940, in which the anion-selective and cation-selective membranes were arranged in alternating series to form many parallel solution compartments with electrodes attached at the sides. Further development in electrodialysis occurred when the process found applications in the food and drug industry and in the treatment of certain industrial effluents. By 1980, the bipolar membranes were developed, which are effective in converting by 'water splitting' aqueous salt solutions, into acids and bases directly without chemical addition.

Electrodialysis is a process that depends on the principle that most dissolved salts are either positively or negatively charged and they will migrate to electrodes with an opposite charge. Selective membranes that are able to allow the passage of either anions or cations make the separation possible. Electrodialysis membranes are ion-selective, that is, they permit the passage of either the positively charged ions (cations) or the negatively charged ions (anions) and not both. The cation-selective membrane allows the passage of cations and retards or blocks the passage of anions, whereas the anion-selective membrane permits anions to pass through and retards cations. Both cation-selective and anion-selective membranes are impervious to the flow of water, as their pores are smaller than the size of water molecules. The electrodialysis stack is made up of a parallel array of cation-selective and anion-selective membranes arranged in alternating fashion. A pair of cation-selective membrane and anion selective membrane constitutes one unit cell and a commercial electrodialysis stack is made up of typically 100 to 600 such cells. The electrodialysis system consisting of two unit cells is shown schematically in Fig. 12.35.

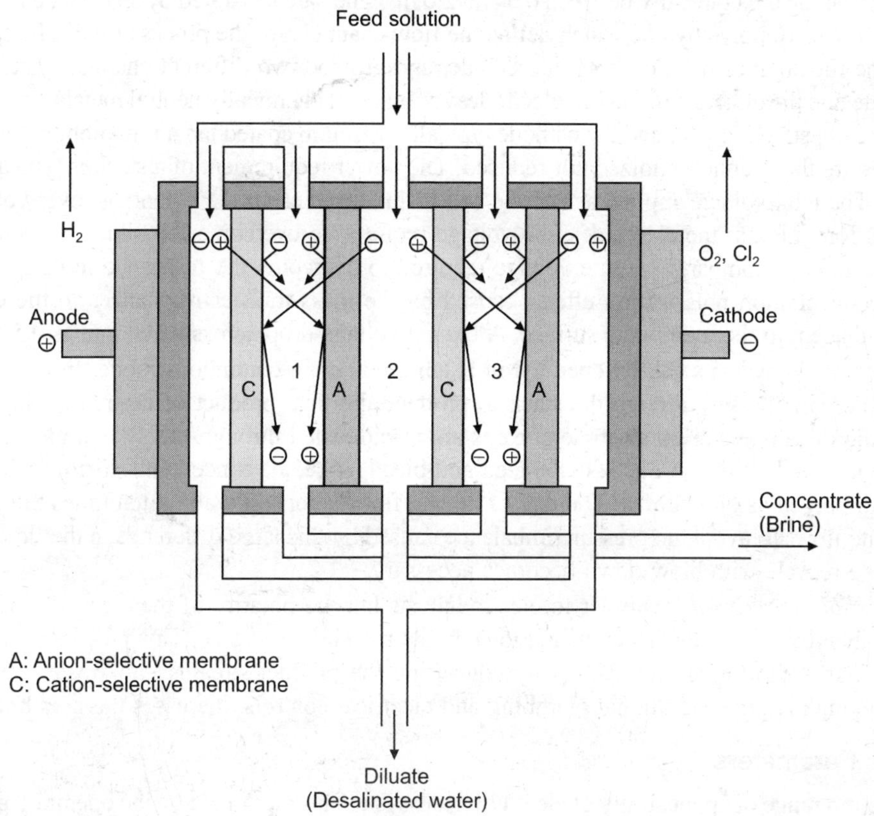

Fig. 12.35: Principle of electrodialysis

The feed to the electrodialysis unit containing dissolved ionic salts is admitted to the compartments formed by the oppositely charged membranes. In some compartments the ions in the feed solution are removed and in the alternating compartments, the ions are retained and get concentrated. Consider the desalination process shown in Fig. 12.35. The saline water is admitted to compartments 1, 2 and 3 as shown. A direct-current voltage is applied across the electrodes attached to the far sides of the stack by means of a battery or a direct-current generator. When the current flows through the feed solution by ionic conduction, the ions in the solution are attracted towards the opposite charged electrodes. Thus the cations migrate toward the cathode (i.e. the negatively charged electrode) and the anions migrate toward the anode (i.e. the positively charged electrode). The solution in compartment 2 in the figure gets desalinated as both cations and anions are transported away from the solution through the ion-selective membranes. However, in compartments 1 and 3, for the cations that are present in the feed solution and those that reach the compartments through the cation-specific membrane from the adjacent compartment, their further transport toward the cathode is blocked by the anion-specific membrane. Similarly, the migration of anions towards the anode is retarded by the cation-selective membrane. Thus both cations and anions get trapped in these compartments. The diluate stream is stripped of the ions and the concentrate streams get enriched in the ions. Since an equal number of anion and cation charge equivalents are transferred from the diluate stream into the concentrate stream, the charge balance is maintained in each stream. As a result, the salt concentrations in the solution in the compartments 1 and 3 increase. These solutions are withdrawn as concentrate or brine and the solution leaving the middle compartment is the diluate (desalinated water product).

The membranes in the electrodialysis stack are made of cation- or anion-exchange resins made into sheet form having area typically between 0.5 and 2.0 m^2. They are separated by spacers generally made of the high-density-polyethylene which define the flow channels for the process fluid. The spacer nets prevent the membranes from coming into contact and provide two different channel systems for the concentrate and the diluate circuit. The electrodes are made of chemically neutral metals with the anode being typically stainless steel and the cathode typically platinum coated tantalum, niobium or titanium. Electrodes are thus neither oxidized nor reduced. The power requirement of a stack is typically 100 A at 150 V. The total voltage applied across the cell includes the electrode potentials, over-voltages due to the gas formation at the electrodes, the voltage required to overcome the ohmic resistance of the electrolyte in each compartment, the voltage required to overcome the resistance in each membrane and the concentration polarization effects caused by the mass transfer resistances in the electrolyte solution adjacent to the membrane surface. A typical voltage drop across a cell pair is 0.5 to 1.5 V.

Electrodialysis systems can be operated in batch mode or in continuous mode. In the continuous mode, the feed is passed through the stack to produce the final product of desired quality. In batch process, the diluate and/or concentrate streams are re-circulated through the system until the desired quality is achieved for the products. In the feed-and-bleed mode, the concentrate stream is recycled and a part is discarded as blow down. The rate of diluate flow is generally about ten times the rate of the concentrate flow. To avoid the pressure imbalance caused by this large difference in the flow rates, the concentrate recycle with blow down becomes necessary.

Electrodialysis reversal is similar to electrodialysis but the polarity of the electrodes is regularly reversed, thereby freeing the accumulated ions on the membrane surface. This process minimizes the effect of inorganic scaling and fouling by converting the product streams into waste streams. This process, though requires additional plumbing and electrical controls, increases the membrane life.

Process Parameters

The optimum range of applicability of electrodialysis system is determined by the operating parameters such as the current density, the cell voltage, the current efficiency, the conductivities of the diluate and the concentrate streams. The current density is the driving force of the process and by running at high current densities the membrane surface area required is reduced for a specified operation. However, this has got a negative impact on the electrodialysis performance. Increase in the current density increases the cell voltage with a corresponding increase in the power consumption.

Polarisation leads to a steep increase in the cell voltage. The *limiting current* is the maximum allowed current density to avoid this steep cell voltage increase. The cell voltage may also increase with time due to the chemical deterioration of the membrane or by fouling. When the cell voltage falls below certain values, the membranes require replacement. The maximum cell voltage generally varies between 0.8 to 1.5 V. The *current efficiency* is a measure of how the effective ions are transported across the membranes for a given applied current. Typically the current efficiencies >80% are desirable in commercial stacks to minimize the operating energy costs. Current efficiency determines the membrane surface required for a given application. Fall in its value may also be due to the imperfect permselectivity, physical leakage, or due to the current running in the non-active cell areas.

Other major parameters are the concentrations which determine the conductivities of the diluate and the concentrate streams, temperature, product purity, pH, etc. The ratio of the conductivities influences the current efficiency and limits the maximum concentration of salts in the concentrate stream. The maximum allowable temperature in electrodialysis stacks is around 333 K. Presence of impurities such as the organic matter, colloidal substances, micro-organisms, insoluble salts, etc. lead to membrane fouling. The pH of the solution is another important factor affecting the performance of the membrane stacks, as many membranes cannot tolerate large pH fluctuations.

Power Consumption in Electrodialysis

The energy required in an electrodialysis process is the sum of two terms: (1) the electrical energy to transfer the ionic components from one solution through membranes into another solution and (2) the energy required to pump the solutions through the electrodialysis unit. Depending on various process parameters, particularly the feed solution concentration, either one of the two terms may be dominating, thus determining the overall energy costs. The power consumption can be evaluated as

$$P = IE \qquad \ldots(12.5)$$

where P = power (W), I = electric current-flow through the stack and E = voltage across the stack. The electrical-current flow is given by

$$I = \frac{FQ\Delta c}{n\xi} \qquad \ldots(12.6)$$

where F = Faraday's constant (96,520 amp-s/equivalent), Q = volumetric flow rate of the diluate (m^3/s), Δc = difference between the feed and diluate ion concentration in equivalents/m^3, ξ = current efficiency and n = number of cell pairs.

Membranes

Most widely used electrodialysis membranes are thin films of polymeric chains containing electrically charged functional sites. The matrix of an ion-exchange membrane consists of hydrophobic polymers such as polystyrene, polyethylene or polysulphone, usually cross-linked. The type and the concentration of the fixed ionic charges determine the permselectivity and the electrical resistance of the membrane; they also have a significant effect on the mechanical properties of the membrane. The degree of swelling, especially, is affected by the concentration of the fixed charges. Cation selective membranes have negatively charged groups fixed to polymer matrix as shown in Fig. 12.36. A typical cation selective membrane is made of polystyrene cross linked with divinylbenzene and sulphonated to produce fixed sulphonate ($-SO_3$) anion groups. In the interstices of the polymers are mobile cations called counterions and mobile anions (called co-ions). Anion-selective membranes on the other hand, contain fixed cations, mobile anions (counter-ions) and mobile cations (co-ions). A typical anion selective membrane contains fixed quaternary ammonium groups such as $-NH_3^+$. In the cation-exchange membrane, the fixed anions are in electrical equilibrium with the mobile cations (counterions) in the interstices of the polymer. Since the electrical charge of the mobile anions are identical to that of the fixed ions, they are excluded from the polymer matrix. This type of exclusion by the electrical repulsion is called the *Donnan-exclusion*. Due to the exclusion of the co-ions, a cation-exchange membrane permits transfer of cations only. On the other hand, anion-exchange membranes exclude all cations and permits transfer of anions only. Thus, the selectivity of ion-exchange membranes results from the exclusion of co-ions from the membrane phase.

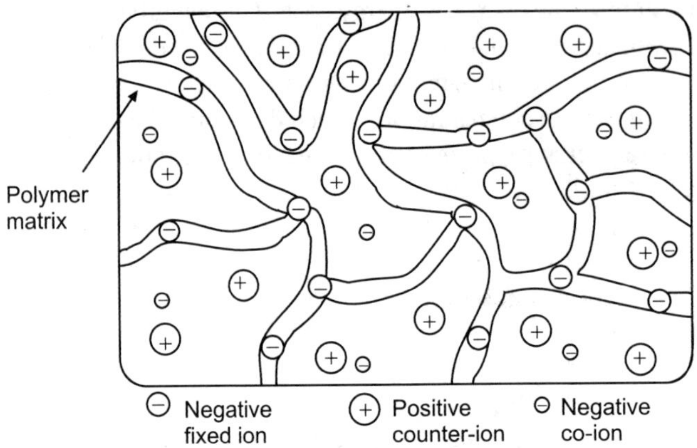

Fig. 12.36: Schematic diagram of a cation-exchange membrane

Membranes are typically 0.2 – 0.5 mm in thickness and reinforced with a screen to provide mechanical stability. The membranes which are made in flat sheets contain 30 to 50% water and have a network of pores too small to permit water transport. The most important properties that determine the choice of membranes for electrodialysis are the permselectivity, electrical resistance, mechanical and chemical stability. Permselectivity should be high so that the membrane is highly permeable to counter-ions and impermeable to co-ions. Electrical resistance should be low. Membranes must be mechanically strong and robust, they must not swell or shrink appreciably with the changes in the ionic strength. Also, they must not wrinkle or deform under thermal stress. High chemical stability is desired because, in the course of the normal use, the membranes are likely to be exposed to wide range of pH conditions (ranging from 0 to 14). If operated properly, the membranes are found to have an average life of 12 to 15 years.

Since the parameters determining the different membrane properties generally have opposing effects, it is difficult to arrive at an optimum combination. For instance, a high degree of cross-linking improves the mechanical strength of the membrane but also increases its electrical resistance, which is undesirable. A high concentration of the fixed ionic charges in the membrane matrix leads to a low electrical resistance but, in general, causes a high degree of swelling, and thus leads to poor mechanical stability. To a large extent, the mechanical, chemical and thermal stability of the membrane are dependent on the basic polymer matrix.

Advantages

Minimum fouling: Electrodialysis system for desalination can be operated with minimum fouling or scaling in comparison with the reverse osmosis system. The scaling resulting due to the precipitation of the insoluble di- or multivalent salts such as the calcium sulphate is less severe in electrodialysis. The concentrate is less rich in the multivalent ions with scaling potentials, because it is the mono-valent ions that are transported through the ion-selective membranes much faster than the multivalent ions.

Low pressure operation: Electrodialysis is a low pressure operation and is advantageous when high recovery is desired from a feed solution containing high concentration of the dissolved salts. This is in contrast to the pressure based membrane separations such as nanofiltration and reverse osmosis. The most severe limitation in desalination by the reverse osmosis is the substantially high osmotic pressure of the feed solution that has to be overcome by the applied hydrostatic pressure. The problem is further complicated by the concentration polarization that results due to the high pressure fluid passing across the membrane. The net result is that in pressure-based membrane separations such as nanofiltration and reverse osmosis, the high feed recovery is possible only at high applied hydrostatic pressures. This may lead to high energy costs and extra investment for additional membrane area and pumping capacity. High pressure operation also results in shorter useful life of the membrane. Consequently, pressure based membrane systems perform better at the lower feed concentrations and for lower recoveries in which the osmotic pressure of the feed and its fouling and scaling potential are minimized.

The long membrane life expectancy, low usage of chemicals for pretreatment, quieter operation, ability to treat feed water with higher SDI (*silt density index*, a measure for the fouling capacity of water), higher TOC (total organic content) and high silica concentrations than reverse osmosis, are the other advantages of electrodialysis process.

Limitations

The inability to remove non-ionic species and high molecular weight and less mobile ionic species is the inherent limitation of the electrodialysis process. This becomes a serious limitation when potable water is to be produced from the feed water containing high suspended solids or micro-organisms, unless additional pre-treatment processes are employed.

For low feed concentrations, the number of ions in the solution to carry the current is reduced, and both ion transport and energy efficiency are therefore reduced. The decrease in the current density and the current utilization efficiency, make electrodialysis less economical at extremely low salt concentrations, as comparatively large membrane areas are required to satisfy the capacity requirements in such cases. Added to this is the electrolytic water splitting which is detrimental to the electrodialysis process, that result due to the ionic concentration polarization occurring at low feed concentrations. Ionic concentration polarization is the reduction of the ion concentrations near the membrane surface compared to that in the bulk solution flowing through the membrane compartment.

At very high concentration difference between the diluate and the concentrate streams, the membrane selectivity gets reduced by the Donnan-exclusion mechanism and the water transport from the diluate to the brine caused by osmosis. As a result, there is a limit for the concentration that can be achieved in the concentrate or brine stream. Electrodialysis is not preferred for removing the last traces of the salt, running at high pH, tolerating surfactants, or running under conditions where solubility limits may be exceeded.

Application

Desalination: The greatest use of the electrodialysis is in the desalination of brackish water, where feed salinity is from 500 to 5000 ppm. However, it is not technically and economically advisable to produce water with a very low dissolved salt content, as the electrical resistance of such solution is very high. Therefore, electrodialysis can be used to produce potable water from sea water and ultra-pure water required in semiconductor industry, in a hybrid process in combination with ion exchange process.

Production of table salt: Electrodialysis can be used to increase the concentration of sea water from 3.5% to 20%. Thus electrodialysis is used on a large scale as a pre-concentration step for the production of table salt by evaporation.

Concentration of reverse osmosis brines: The concentration of brine in the reverse osmosis cannot exceed certain values as the osmotic pressure become excessive at high concentrations. An important application of electrodialysis is in further concentration of the reverse osmosis brine.

Stabilization of wine: To avoid the formation of crystalline tartar in the wine, the tartaric acid concentration must be reduced to a value so that the solubility limit is not exceeded. The stabilization through deionization of wine is an important application of electrodialysis.

Demineralisation of whey: Though cheese whey provides an excellent source of protein, lactose, vitamins and minerals, in its normal form, it is not considered a proper food material because of its high salt content. Electrodialysis is very widely used for the deionization of cheese whey, where the desalted product is a useful food additive, especially for baby food.

Waste water treatment: An important applications of electrodialysis is in waste water treatment. The complete recycling of water and the recovery of metal ions such as Ni, Cu, etc. from rinse waters from the electroplating industry are achieved by electrodialysis.

Pharmaceutical application: Separation of amino acids and other organic acids by electrodialysis seems to be of special interest to the pharmaceutical and chemical industry

Deionization of fruit juices, milk and sugar molasses, separation of acids and base from organic compounds, recovery of organic compounds from their salts, the production of boiler feed water, etc. are some other applications of electrodialysis. Bipolar membranes prepared by laminating a cation-selective membrane and an anion selective membrane back to back can be used to produce sulphuric acid and sodium hydroxide from a sodium sulphate solution.

12.13 GAS PERMEATION

Gas permeation involves separation of gaseous components from mixture of gases by means of semipermeable membranes. Both micro-porous and dense nonporous membranes are used for membrane separation of gases. When the feed gas consisting of low molecular weight and high molecular weight species is passed across the porous membrane at a high pressure, the gas permeating to the region of low pressure will be rich in low molecular weight components, since they diffuse more rapidly. If the membrane is dense nonporous, the transport of molecules occurs by a solution diffusion mechanism. The gas is absorbed on the high-pressure upstream side of the polymer membrane, diffuses through the polymer phase and desorbs at the low pressure side. Gas permeation process operates with pressure differences of 1–20 bar.

In gas separations, the more permeable component is known as the fast gas; it is in this component that the permeate streams get enriched. The permeate stream is rich in the low molecular weight components and the retentate is rich in the high molecular weight components. However, a near perfect separation is generally not achievable. If the membrane is microporous, the pore size is extremely important in some cases such as the high temperature applications, where it is usually necessary to block the passage of the high molecular weight species. Otherwise, unless the molecular weights of the constituent species differ appreciably, only a very modest separation is achievable. The mechanism of gas permeation is schematically represented in Fig. 12.37.

The mechanism of transport of molecules through a porous membrane depends on the size of the pores as shown in Fig. 12.37a. If the pores are relatively large, say, from 0.1 to 10 μm size, gases permeate the membrane by convective flow and no separation occurs. If the pore size is smaller than 0.1 μm, then the pore diameter is either of the same size or smaller than the mean free path of the gas molecules and Knudsen diffusion will govern the transport of molecules. The rate of transport is proportional to the square root of the molecular weight. If the pores are extremely small, say, of the order of 0.5 to 2 nm, then the gases are separated by molecular sieving which involves both the diffusion in the gas phase followed by the diffusion of the adsorbed species on the surface of the pores. If the membrane is dense nonporous, the membrane transport is by the solution-diffusion mechanisms (Fig. 12.37b).

Metal membranes, such as palladium, are extremely permeable to hydrogen, but essentially impermeable to other gases. The mechanism of hydrogen transport through metallic membranes can be represented as shown in Fig. 12.37c. Hydrogen transport is assumed to occur in 5 distinct steps. Hydrogen molecules are sorbed onto the membrane surface (1), where they undergo dissociation into hydrogen atoms (2). The hydrogen atoms loses electron to the membrane lattice and diffuses through the lattice as ions (3). The hydrogen atoms emerging at the permeate side of the membrane re-associate (4) and form hydrogen molecules which are then desorbed (5).

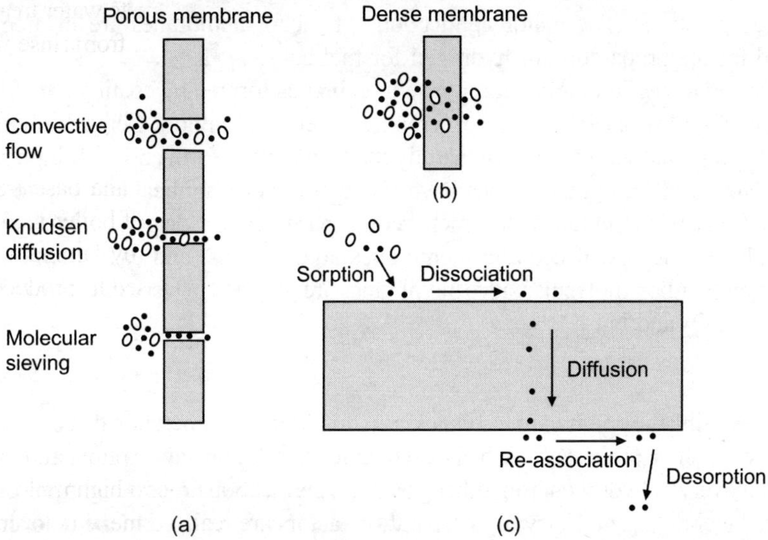

Fig. 12.37: Mechanism for permeation of (a) gas through a porous membrane (b) through a dense membrane (c) hydrogen gas through a metal membrane

Membranes

Polymeric, metal, ceramic or zeolite membranes are used for gas separation operations. Although plate-and-frame and tubular modules can be used, all large scale applications use spiral wound or hollow-fibre modules because of their high packing density. The gas-selective polymeric membranes are usually dense nonporous materials. These membranes usually work because individual gases differ in their solubility and diffusivities through nonporous polymers. Polymeric gas separation membranes are classied into two: the first type includes the rubbery polymers such as silicone rubber, natural rubber, and poly (4-methyl-1-pentene). They have high permeabilities; but selectivities are moderate. The second type are asymmetric membranes, the thin dense selective layer made up of cellulose acetates, polysulphones, polyamides, polyimides, polycarbonates, polyetherimides, sulphonated polysulphones, Teflones, polystyrenes, etc. The asymmetric membranes have the glassy polymeric selective layer, typically 0.1 to 1 μm thick with 50 to 200 μm thick microporous support. The support layer should have an open porous network to minimize the resistance to the flow of the gas and should be capable of withstanding high pressures. The polymeric dense layer at the top of the membrane being very thin, a high flux is achieved and the material being glassy polymer, a high selectivity also results. For the asymmetric membranes to be selective in gas separation, special methods have been developed in making membranes with defect-free pores. The prism membranes are made by coating a relatively thick silicone rubber layer on a thin selective layer of asymmetric polysulphone membrane. Generally, membranes of the rubbery polymeric type are used for the separation of mixtures such as O_2-N_2 and CO_2-CH_4. The dense glassy membranes are used for the separation of gases with small molecular diameter and low molecular weight such as the hydrogen or helium in a gas mixture.

Metal membranes are relatively new and are especially suitable for the gas separation at relatively high temperatures. Palladium membranes are highly selective as they are extremely permeable to hydrogen and essentially impermeable to all other gases. Hydrogen permeates a number of other metals which include tantallum, niobium, nickel, iron, copper, cobalt and platinum. Embrittlement and cracking of metals caused by the sorbed hydrogen are common problems in the gas permeation using metal membranes, and to prevent these, the membranes must be operated at high temperatures, exceeding 573 K. An alloy membrane made of palladium and silver is free from hydrogen embrittlement, even in low temperature operations. The major applications of metal membranes are in membrane reactor applications and for the preparation of hydrogen for fuel cells.

Ceramic and zeolite membranes are composite membranes formed by coating a thin layer of highly selective ceramic or zeolite particles onto a microporous ceramic support. Ceramic membranes permit operation at high temperatures. These are generally made in tubular form and offer high selectivities for gas separation. However, these membranes have few commercial applications due to its excessively high cost. Mixed-matrix membranes in which zeolite particles are dispersed in a polymer matrix, combine the high selectivities of ceramic membranes and low cost and ease of manufacture of the polymeric materials. Carbon molecular sieve membranes are found to be very effective for the separation of gas pairs such as O_2 and N_2.

Advantages

Separation of gases using the polymer membranes is an important unit operation that competes effectively with the well-established processes such as cryogenic distillation, absorption and pressure-swing adsorption. For recovery of hydrogen from other gases, gas permeation process competes quite favourably with conventional techniques and very high separation factors are realized in the recovery of hydrogen from other gases. It is found that 95% recovery of hydrogen with a purity of 90% is achieved from a feed gas containing 60% hydrogen using dense polymeric membranes. Gas separation using membranes have other advantages such as the low capital investment, ease of installation, ease of operation, absence of the rotating parts, high process flexibility, low weight and space requirements and low environmental impact.

Applications

The concept of separating gases with polymeric membranes is several decades old. However, the polymeric membranes for the gas separation became economically viable only by 1980s, and since then, the widespread application of these membranes for gas separation has become an industrial activity of immense potential. H_2 recovery from gas mixtures, acid gas (CO_2/CH_4) separation and the production of N_2 from air were the early applications of the process. It is expected that the membrane separation will become the major technique for nitrogen separation from the air in future. Important applications of membrane gas separation are listed below:

1. Separation of air by gas permeation is an important technique for the production of nitrogen from air. Because of the higher solubility in many polymers, O_2 permeates faster than N_2 by a factor of 5. The method is used for the production of high purity nitrogen (the retentate) from air. Moderate purity N_2 (95-99.5%) may be produced in low to moderate quantities quite economically by membrane separation. O_2 enriched stream is a co-product of this operation.
2. Recovery of hydrogen from mixtures with larger components such as nitrogen, methane and carbon monoxide, recovery of H_2 from ammonia purge gas, extraction of H_2 from petroleum cracking streams are important applications of gas separation.
3. Much of the natural gas produced in the world is associated with an acid gas, most commonly CO_2 and/or H_2S. Membrane separations are used for the purification of the natural gas by removal of the carbon dioxide and/or hydrogen sulphide.
4. Gas permeation can recover helium from natural gas, as helium is a fast gas and may be recovered by gas permeation.
5. Gas permeation is used for adjustment of H_2:CO ratio in synthesis gas.

REVIEW QUESTIONS

1. Define membrane separations.
2. Why membrane separation became an important process in the food and the pharmaceutical industries?
3. Describe the advantages of membrane separation techniques over conventional methods of separation.
4. What are the limitations of membrane separations?
5. Discuss the important applications of membrane separation methods.
6. Explain the classification of membrane-based processes based on the size of the molecules being separated.
7. Discuss the advantages and the disadvantages of the membrane-based processes.
8. Discuss the challenges to be faced when adopting a membrane-based process.
9. What is the significance of the Loeb-Sourirajan process in membrane technology?
10. Define the terms permeate and retentate.
11. Distinguish between the flux and the selectivity.
12. Define permeability. How is it affected by trans-membrane pressure?
13. What is the molecular weight cut off?
14. List the important membrane separation techniques with the respective driving forces for mass transport.
15. Discuss the size range and nature of materials separated by the different techniques.
16. What are the desirable membrane characteristics for the separation applications?
17. Discuss the classification of the membranes.
18. What are the most common polymers used in the membrane production?
19. Describe the classification of synthetic polymeric membranes.
20. Discuss the characteristics of various symmetric and asymmetric membranes.
21. Describe the basic characteristics of polymeric membranes and give examples of the commonly used polymers for membrane-based applications.
22. Discuss the thermal stability of polymeric membranes in terms of glass transition temperature.
23. Distinguish between the amorphous and crystalline polymers.
24. List the important properties and applications of different polymers as membrane materials.
25. Compare the advantages and disadvantages of polymeric and inorganic membranes.
26. Distinguish between isotropic and anisotropic membranes.
27. Describe the basic difference in the separation mechanism in porous and dense membranes.
28. What are the advantages of asymmetric membranes over the symmetric microporous and dense membranes?
29. What are the significant features of ceramic, metal and carbon membranes?
30. What are liquid membranes and what are their advantages? Distinguish between ELM and LLM and discuss their relative merits.
31. Describe the developments in various inorganic membranes (namely metal, ceramic, carbon and zeolite), including the potential applications and anticipated problems.
32. Discuss the developments of hybrid membranes, their advantages over polymeric or inorganic membranes. Also list their applications.
33. Explain the basic principle of liquid membranes and the facilitated transport.
34. Discuss the various transport mechanisms in liquid membranes.
35. Describe the bipolar membrane and explain its basic principle and application.
36. Distinguish between the hybrid membranes and bipolar membranes.
37. Discuss the common membrane shapes.

38. What benefits can be derived by using the membrane modules over single membranes?
39. Discuss the different designs of membrane modules and list their relative merits and limitations.
40. Discuss with neat sketches the flow patterns in membrane modules.
41. Distinguish between the dead-end filtration and cross-flow filtration.
42. Why is membrane cascade more important in gas separation?
43. What is concentration polarisation? Describe how it affects the separation operations and how it can be prevented.
44. Discuss membrane fouling and control techniques.
45. Explain the basic principle of microfiltration.
46. Discuss with reference to microfiltration, the size and type of material removed, major applications, and membranes used.
47. Discuss the various techniques to reduce fouling in microfiltration.
48. Describe the various cleaning methods for microfiltration membranes.
49. Explain the basic principle of ultrafiltration.
50. Describe the membrane materials and modules used in ultrafiltration systems.
51. Discuss the factors affecting membrane performance.
52. Compare and contrast batch versus continuous operation of ultrafiltration systems.
53. Discuss the advantages and disadvantages of ultrafiltration systems.
54. Explain the basic principle of osmosis and reverse osmosis.
55. Describe the membrane materials and modules used in reverse systems.
56. Discuss the size and type of material removed and major applications of reverse osmosis.
57. Discuss factors affecting membrane performance in reverse osmosis.
58. Describe a large-scale reverse osmosis operation using parallel units and staging.
59. Discuss the advantages and disadvantages of reverse osmosis systems.
60. Explain the principles and operation of pervaporation systems
61. Explain the basic principle of pervaporation.
62. Describe the membrane materials and modules used in the pervaporation systems.
63. Discuss the parameters affecting the performance of pervaporation systems.
64. Explain the basic principle of electrodialysis.
65. Explain the basic setup for an electrodialysis system and the mechanism for separation.
66. Describe the ion-exchange membranes used in electrodialysis systems.
67. Compare and contrast the electrodialysis operation using parallel and series configurations.
68. Discuss factors affecting electrodialysis systems performance.
69. Explain the basic principle of gas separation.
70. Describe the membrane materials and modules used in the gas separation systems.
71. Define the pressure ratio and its significance in separation.
72. Describe various configurations of gas separation systems.
73. Distinguish between the hydrophilic membranes, the organophilic membranes and the organo-selective membranes.
74. Describe the working principle of artificial kidney for haemodialysis.
75. Discuss the characteristics and properties of ion-selective membranes.

MULTIPLE CHOICE QUESTIONS

12.1 Membrane separations find wide applications in food and pharmaceutical industries, because
A. near perfect separation is achieved by membranes

B. the separations are carried out at modest temperatures
C. no external agents are required to effect separation
D. the separations are less capital intensive

12.2 Which one is incorrect with respect to membrane separations?
A. They are low temperature operations
B. High capacity applications require single large vessels
C. They are carried out in modular equipments
D. Work without the addition of chemicals

12.3 Two most important properties of membranes for industrial separations are:
A. selectivity and permeability B. selectivity and volatility
C. permeability and strength D. permeability and productivity

12.4 Since the membrane separation units are modular in construction,
A. the product contamination can be eliminated
B. the units are vulnerable to scaling and fouling
C. industrial application demands running parallel units
D. single large piece of equipment can be designed to meet high capacity requirements.

12.5 All but one of the following may be treated as membrane filtration techniques. Identify the odd one.
A. Reverse osmosis B. Pervaporation
C. Ultrafiltration D. Nanofiltration

12.6 In membrane separations, the part of the feed that pass through the membrane is called
A. retentate B. concentrate
C. dialysate D. permeate

12.7 Permeability is defined as
A. volumetric permeate flow through the membrane per unit area in unit time
B. the ratio of flux to membrane thickness
C. transport flux through the membrane per unit driving force per unit membrane thickness
D. the pressure gradient between the upstream and downstream sides of the membrane

12.8 What is conversion ratio in membrane separations?
A. The ratio of the pressures on the upstream and downstream sides of the membrane
B. The ratio of flux to membrane thickness
C. The ratio of permeability to membrane thickness
D. The ratio of permeate flow rate to the feed flow rate.

12.9 The selectivity in membrane separations is defined as
A. the ratio of relative concentrations of species i and j in the permeate to that in the feed
B. the ratio of relative concentrations of species i and j in the permeate to that in the retentate
C. the ratio of relative concentrations of species i and j in the retentate to that in the feed
D. the amount of material permeated in unit time per unit area of membrane surface

12.10 When the membrane is perfectly semi-permeable, its selectivity is
A. infinity B. unity
C. zero D. negative

12.11 Molecular weight cut off is a measure of
A. the permeability of membrane B. selectivity of membrane
C. retention capabilities of membrane D. percent of the feed separated as permeate

12.12 Hybrid membranes are made of
A. membranes of different ionic charges B. organic and inorganic materials
C. carbon and zeolite D. ceramics and metals

12.13 Majority of industrial membranes are made of

A. polymeric materials
B. ceramic materials
C. metals
D. zeolites

12.14 Which one of the following is not a characteristic of amorphous polymers?

A. Brittle and glassy in appearance
B. Polycarbonates, polystyrene, and polyimide belong to this class
C. Crystalline in structure, and become a melt above their melting temperatures
D. Can be operated above or below their glass transition temperatures

12.15 Teflon (polytetrafluroethylene) is a synthetic polymer widely used in membrane separation Which one of the following statements is not true about teflon.

A. It is a crystalline polymer
B. It can be fabricated into thin sheets and films
C. It is an amorphous polymer
D. It is applicable in separation by ultrafiltration

12.16 Properties of .... membranes vary with direction

A. amorphous polymeric
B. crystalline polymeric
C. isotropic
D. anisotropic

12.17 The presence of interconnected pores of sizes ranging from approximately 1 nm to 20 μm is a characteristic of .... membranes

A. dense
B. microporous
C. silicone rubber
D. all the above

12.18 Which of the following is true with respect to microporous membranes?

A. They are dense membranes
B. They do not contain interconnected pores
C. Used in microfiltration, ultrafiltration, and dialysis applications
D. Separation achieved is independent of the pore diameter and particle size

12.19 For dense amorphous membranes, diffusing species must first .... into the polymer and then .... through the polymer between the segments of the macromolecular chains

A. dissolve, diffuse
B. diffuse, dissolve
C. vaporise, diffuse
D. ionise, diffuse

12.20 Compared to microporous membranes, for dense amorphous membranes

A. permeability and selectivity are high
B. permeability is low and selectivity is high
C. permeability and selectivity are low
D. permeability is high and selectivity is low

12.21 Loeb-Sourirajan process refers to

A. asymmetric membrane made of cellulose acetate
B. hybrid membranes
C. bipolar membranes
D. thin film composites

12.22 Which one is not true for Loeb-Sourirajan membranes?

A. They are asymmetric membranes
B. They are made of cellulose acetate
C. Consists of thin dense skin and thick microporous support
D. They belong to the class of thin film composite membranes.

12.23 For high temperature operation and separation of chemically active mixtures, ... membranes are used

A. polymeric
B. natural rubber
C. inorganic
D. Loeb-Sourirajan

12.24 Ceramic membranes are generally made in the shape of
A. films B. tubes
C. plates D. hollow fibre

12.25 If the process requires frequent steam sterilization and chemical cleaning, ……… are preferred.
A. metallic membranes B. polymeric membranes
C. ceramic membranes D. none of the above

12.26 Which membranes find wide application for the large scale production of high purity nitrogen from air?
A. Ceramic membranes B. Carbon membranes
C. Zeolite membranes D. Cellulose acetate membranes

12.27 The major limitation that restricts the use of liquid membranes is
A. the low solubility and diffusivity of the substance in liquids than in the solids
B. the use of carrier that reduces the permeability
C. their poor selectivity
D. their poor stability

12.28 One advantage of plate-and-frame membrane module over spiral wound and hollow fibre module is
A. the high packing density
B. the high allowable operating pressure
C. the low cost
D. that the exchange of membrane is not labour intensive

12.29 For treating highly viscous feed with high solids concentrations, ……… membrane modules are most suitable.
A. plate-and-frame B. spiral wound
C. tubular D. capillary

12.30 The membrane module suitable only for low pressure application is
A. the capillary module B. the spiral wound module
C. the plate-and-frame module D. the tubular module

12.31 The membrane module that offers the highest surface per unit volume and the smallest size for a given capacity is
A. capillary membrane module
B. spiral wound module C. hollow-fibre module
D. tubular membrane module

12.32 The gradual build up of non-permeating component at the membrane surface leads to
A. scaling B. fouling
C. concentration polarisation D. membrane rupture

12.33 Concentration polarisation can be minimised by all except one of the following. Identify the incorrect option.
A. increasing the turbulent mixing at the membrane surface
B. decreasing the fluid velocity at the surface
C. providing membrane spacers
D. pulsing the feed flow through the module

12.34 Membranes with surface characteristics that minimise adhesion can
A. prevent concentration polarisation B. prevent gel layer consolidation
C. improve the membrane stability D. prevent fouling

12.35 Which of the following is a pressure-driven membrane separation operation?

A. Dialysis B. Osmosis
C. Electrodialysis D. Reverse osmosis

12.36 The only membrane separation in which the permeate undergoes a phase change is

A. electrodialysis B. pervaporation
C. ultrafiltration D. reverse osmosis

12.37 Which membrane separation can be used for removing particles of sizes 50 nm and larger?

A. microfiltration B. nanofiltration
C. ultrafiltration D. reverse osmosis

12.38 Which of the following is suitable for removing particles larger than 3 nm?

A. Nanofiltration B. Ultrafiltration
C. Reverse osmosis D. Microfiltration

12.39 For separating particles in the range of 0.1 nm or larger, which membrane technique is suitable?

A. Nanofiltration B. Ultrafiltration
C. Reverse osmosis D. Microfiltration

12.40 Nanofiltration membranes permits passage of particles of size ……… or smaller

A. 1 nm B. 50 nm
C. 1 μm D. 0.1 μm

12.41 Macromolecules in the size range of 5 to 100 nm are to be separated from a solution. Which method is the most ideal for this purpose?

A. Nanofiltration B. Ultrafiltration
C. Reverse osmosis D. Microfiltration

12.42 Microfiltration is used to remove

A. macromolecules and proteins from solutions
B. dissolved salts from solutions
C. water from organic solvents
D. colloidal and suspended particles from solutions

12.43 Cross-flow pattern is preferred in microfiltration when

A. the solid loading is greater than 0.5%
B. the solid loading is less than 0.5%
C. microporous membranes are used for separation
D. when the membrane used is of dense polymeric type

12.44 In membrane techniques such as ultrafiltration, permeation rate ……… with increase in temperature

A. decreases B. increases
C. is unaffected D. may increase or decrease

12.45 High temperature operation is not generally adopted in membrane separation, because

A. permeation rate is decreased with increase in temperatures
B. diffusivity of the solid molecules decreases with increase in temperatures
C. viscosity of the retentate increases with increase in temperature
D. denaturation of species occurs at high temperatures

12.46 Which one of the following is not true regarding diafiltration?

A. It is special type of ultrafiltration B. It ensures steady flux of permeable components
C. It is a process in which retentate is diluted with solvent
D. It is a special type of microfiltration

12.47 Also called hyperfiltration, it is a pressure driven membrane diffusion process for separating dissolved salts and is widely for desalination. Identify the process.

A. Microfiltration B. Pervaporation
C. Ultrafiltration D. Reverse osmosis

12.48 Which one of the following is not true with regard to reverse osmosis?

A. Diffusion of solvent occurs from the dilute solution to the concentrated solution through membrane
B. Diffusion of solvent occurs from the concentrated solution to the dilute solution through membrane
C. Used for desalination of sea water
D. Used to eliminate almost all dissolved solids, bacteria and viruses contained in the water.

12.49 The process of forcing a solvent from a region of high solute concentration to a region of low solute concentration by applying a pressure greater than the osmotic pressure is called

A. osmosis B. reverse osmosis
C. dialysis D. ultrafiltration

12.50 Reverse osmosis (RO) plants for desalination of sea water are to be operated at very high pressures, because

A. the RO membranes are not active below certain pressures.
B. the osmotic pressure of sea water is very high
C. the osmotic pressure of sea water is very low
D. at low pressures only solute, and no solvent diffuses.

12.51 Identify the true statement among the following for the desalination process based on reverse osmosis (RO).

A. RO systems require more space than other desalination techniques
B. The energy consumption in RO is much higher than that in multistage flash evaporation for desalination
C. Thermodynamic efficiency of RO units for desalination are quite high
D. RO plants are operated at high pressures

12.52 For which of the following operations, the phase state on one side of the membrane is different from that on the other side?

A. Pervaporation B. Electrodialysis
C. Gas permeation D. Reverse osmosis

12.53 For providing sufficient driving force for material transfer in pervaporation systems,

A. the pressure on the permeate side is kept high
B. the pressure on the feed side is kept low
C. the pressure on the permeate side is kept low
D. the pressures on both sides are kept uniform

12.54 Pervaporation systems use a purge gas

A. to maintain a low pressure on the permeate side
B. to maintain a low pressure on the feed side
C. to provide a non-oxidising atmosphere in the unit
D. to prevent explosion hazards in the unit

12.55 The membrane technique ideal for separation of azeotropes is

A. reverse osmosis B. electrodialysis
C. pervaporation D. gas permeation

12.56 For dehydration of ethanol solution by pervaporation, most commonly used membrane is made of

A. cellulose acetate
B. polyvinyl alcohol
C. silicone rubber
D. none of the above

12.57 All but one of the following statements is true regarding dialysis Spot the incorrect statement.
A. The driving force for dialysis is concentration difference and not pressure difference.
B. The mass flux is lower than the flux in other membrane filtration techniques
C. Used to separate species which differ appreciably in their size and diffusion rates.
D. Low molecular weight solutes diffuse from a region of low concentration to a region of higher concentration.

12.58 The most important application of dialysis is
A. artificial kidneys
B. desalination of brackish water
C. recovery of salts and sugars from natural products
D. removal of alcohol from beer

12.59 Haemodialysis is concerned with
A. desalination of sea water
B. removal of waste products from the blood of people suffering from renal failure
C. recovery of sodium hydroxide from caustic viscose liquor during viscose manufacturing
D. the recovery of nitric and hydrofluoric acids from spent stainless-steel pickle

12.60 Major application of electrodialysis as on today is in
A. separation of organics
B. dehydration of alcohol
C. desalination of sea water
D. demineralisation of whey

12.61 An electrodialysis membrane permits the passage of
A. cations only
B. anions only
C. both cations and anions
D. either cations or anions and not both

12.62 Electrodialysis reversal is a process
A. used to separate electrically charged particles from solution
B. used to free the accumulated ions on the membrane surface
C. used to avoid the pressure imbalance due to difference in the flow rates of diluate and concentrate
D. that leads to reduction in the membrane life

12.63 Electrodialysis membranes make use of .... to effect separation.
A. Pauli exclusion
B. Ostwald ripening
C. Donnan-exclusion
D. Marangoni effect

12.64 A major limitation of electrodialysis for desalination of sea water, over reverse osmosis is
A. the high pressure required for the operation
B. the scaling and fouling of membranes
C. the short membrane life expectancy
D. the inability to remove suspended solids and microorganisms.

12.65 Which of the following is true regarding gas permeation?
A. Metal membranes are unsuitable for gas permeation.
B. Only dense nonporous membranes can be used for gas permeation.
C. Gas permeation process operates with pressure differences of about 1–20 bar.
D. The permeate stream is rich in the high molecular weight components and the retentate is rich in the low molecular weight components

12.66 For production of nitrogen from air by membrane separation, the most suitable technique is
A. Pervaporation
B. Electrodialysis
C. Gas permeation
D. Reverse osmosis

Answers to Objective Questions (MCQs)

2.1. B   2.2. A.   2.3. A.   2.4. D.   2.5. A.   2.6. B.
2.7. A.   2.8. B.   2.9. D.   2.10. C.   2.11. A.   2.12. B.
2.13. D.   2.14. C.   2.15. C.   2.16. B.   2.17. B.   2.18. C.
2.19. A.   2.20. C.   2.21. B.   2.22. C.   2.23. C.   2.24. C.
2.25. A.   2.26. C.   2.27. B.   2.28. C.   2.29. A.   2.30. B.
2.31. C.   2.32. B.   2.33. C.   2.34. B.   2.35. A.   2.36. B.
2.37. B.   2.38. C.   2.39. A.

3.1. C.   3.2. D.   3.3. A.   3.4. B.   3.5. C.   3.6. C.
3.7. B.   3.8. A.   3.9. A.   3.10. C.   3.11. A.   3.12. A.
3.13. D.   3.14. C.   3.15. C.   3.16. B.   3.17. C.   3.18. A.
3.19. A.   3.20. A.   3.21. A.   3.22. C.   3.23. D.   3.24. B.
3.25. D.   3.26. B.   3.27. C.   3.28. C.   3.29. D.   3.30. A.
3.31. B.   3.32. C.   3.33. B.   3.34. C.   3.35. B.   3.36. A.
3.37. C.   3.38. B.   3.39. D.   3.40. C.   3.41. D.   3.42. A.

4.1. A.   4.2. D.   4.3. A.   4.4. B.   4.5. D.   4.6. C.
4.7. C.   4.8. B.   4.9. B.   4.10. B.   4.11. B.   4.12. B.
4.13. A.   4.14. D.   4.15. C.   4.16. C.   4.17. D.   4.18. A.
4.19. A.   4.20. C.   4.21. A.   4.22. A.   4.23. C.   4.24. B.
4.25. A.   4.26. A.   4.27. A.   4.28. B.   4.29. A.   4.30. D.
4.31. B.   4.32. B.   4.33. A.   4.34. A.   4.35. A.   4.36. B.
4.37. C.   4.38. A.   4.39. B.   4.40. B.   4.41. A.   4.42. A.
4.43. C.   4.44. A.   4.45. D.   4.46. D.   4.47. B.   4.48. B.
4.49. B.   4.50. C.   4.51. A.   4.52. D.   4.53. C.   4.54. B.
4.55. B.   4.56. A.   4.57. B.   4.58. A.   4.59. C.   4.60. B.
4.61. C.   4.62. A.   4.63. B.   4.64. C.   4.65. A.   4.66. B.
4.67. B.   4.68. D.   4.69. C.   4.70. B.   4.71. D.   4.72. C.
4.73. D.   4.74. A.   4.75. B.   4.76. A.   4.77. A.

5.1. A. 5.2. C. 5.3. B. 5.4. D. 5.5. D. 5.6. C.
5.7. A. 5.8. C. 5.9. C. 5.10. C. 5.11. A. 5.12. B.
5.13. A. 5.14. A. 5.15. C. 5.16. B. 5.17. C. 5.18. C.
5.19. D. 5.20. B. 5.21. B. 5.22. D. 5.23. A. 5.24. B.
5.25. B. 5.26. A. 5.27. D. 5.28. C. 5.29. C. 5.30. D.
5.31. A. 5.32. D. 5.33. A. 5.34. B. 5.35. A. 5.36. B.
5.37. A. 5.38. B. 5.39. A. 5.40. D. 5.41. B. 5.42. C.
5.43. D. 5.44. C. 5.45. C. 5.46. B. 5.47. C. 5.48. B.
5.49. C. 5.50. C. 5.51. C. 5.52. B. 5.53. B. 5.54. A.
5.55. A. 5.56. B. 5.57. C. 5.58. B. 5.59. D. 5.60. B.
5.61. D. 5.62. C. 5.63. D. 5.64. B. 5.65. B. 5.66. C.
5.67. A. 5.68. C. 5.69. C. 5.70. A. 5.71. B. 5.72. B.
5.73. B. 5.74. D. 5.75. C. 5.76. B. 5.77. C. 5.78. A.
5.79. A. 5.80. B. 5.81. B. 5.82. D. 5.83. B. 5.84. C.
5.85. D. 5.86. D. 5.87. D. 5.88. B. 5.89. A. 5.90. B.
5.91. A. 5.92. A. 5.93. A. 5.94. C. 5.95. B. 5.96. A.
5.97. A. 5.98. D. 5.99. B. 5.100. D. 5.101. A. 5.102. B.
5.103. B. 5.104. C. 5.105. A. 5.106. C. 5.107. B. 5.108. D.
5.109. C. 5.110. C. 5.111. B. 5.112. B. 5.113. B. 5.114. D.
5.115. D. 5.116. A. 5.117. A. 5.118. A. 5.119. A. 5.120. C.
5.121. C. 5.122. C. 5.123. B. 5.124. A. 5.125. A. 5.126. C.
5.127. D. 5.128. D. 5.129. D. 5.130. C. 5.131. A. 5.132. A.
5.133. D. 5.134. C. 5.135. B. 5.136. D. 5.137. B. 5.138. C.
5.139. B. 5.140. B. 5.141. D. 5.142. D. 5.143. B. 5.144. A.
5.145. C. 5.146. B. 5.147. D. 5.148. B. 5.149. A. 5.150. B.
5.151. A. 5.152. D. 5.153. C. 5.154. A. 5.155. C. 5.156. B.
5.157. D. 5.158. A. 5.159. B. 5.160. B. 5.161. A. 5.162. C.
5.163. D. 5.164. A. 5.165. C. 5.166. A. 5.167. B. 5.168. C.
5.169. A. 5.170. B. 5.171. C.

6.1. A. 6.2. C. 6.3. C. 6.4. B. 6.5. D. 6.6. B.
6.7. B. 6.8. A. 6.9. B. 6.10. B. 6.11. D. 6.12. C.
6.13. C. 6.14. B. 6.15. B. 6.16. A. 6.17. D. 6.18. C.
6.19. B. 6.20. B. 6.21. C. 6.22. B. 6.23. D. 6.24. C.
6.25. C. 6.26. A. 6.27. A. 6.28. C. 6.29. B. 6.30. B.
6.31. D. 6.32. C. 6.33. B. 6.34. A. 6.35. C. 6.36. A.
6.37. B. 6.38. C. 6.39. C. 6.40. D. 6.41. B. 6.42. C.
6.43. B. 6.44. A. 6.45. A. 6.46. B. 6.47. B.

7.1. C. 7.2. A. 7.3. A. 7.4. C. 7.5. C. 7.6. C.
7.7. A. 7.8. B. 7.9. C. 7.10. B. 7.11. B. 7.12. D.
7.13. D. 7.14. A. 7.15. C. 7.16. D. 7.17. A. 7.18. C.
7.19. C. 7.20. A. 7.21. A. 7.22. B. 7.23. B. 7.24. D.
7.25. B. 7.26. B. 7.27. B. 7.28. A. 7.29. A. 7.30. D.
7.31. B. 7.32. B. 7.33. B. 7.34. B. 7.35. C. 7.36. A.
7.37. C. 7.38. C. 7.39. A. 7.40. D. 7.41. B. 7.42. C.

8.1. C. 8.2. A. 8.3. B. 8.4. D. 8.5. B. 8.6. A.
8.7. B. 8.8. B. 8.9. D. 8.10. D. 8.11. D. 8.12. B.
8.13. C. 8.14. B. 8.15. A. 8.16. A. 8.17. C. 8.18. D.
8.19. A. 8.20. A. 8.21. B. 8.22. C. 8.23. D. 8.24. B.
8.25. C. 8.26. D. 8.27. D. 8.28. A. 8.29. C. 8.30. D.
8.31. C. 8.32. B. 8.33. C. 8.34. C.

9.1. B. 9.2. B. 9.3. C. 9.4. D. 9.5. A. 9.6. B.
9.7. B. 9.8. C. 9.9. D. 9.10. D. 9.11. C. 9.12. B.
9.13. C. 9.14. C. 9.15. C. 9.16. C. 9.17. B. 9.18. A.
9.19. A. 9.20. B. 9.21. B. 9.22. B. 9.23. C. 9.24. A.
9.25. D. 9.26. C. 9.27. A. 9.28. B. 9.29. B. 9.30. D.
9.31. B. 9.32. A. 9.33. B. 9.34. D. 9.35. C. 9.36. D.
9.37. A. 9.38. D. 9.39. C. 9.40. C. 9.41. A. 9.42. A.
9.43. C. 9.44. D. 9.45. D. 9.46. B. 9.47. C.

10.1. B. 10.2. C. 10.3. C. 10.4. A. 10.5. B. 10.6. B.
10.7. C. 10.8. A. 10.9. B. 10.10. C. 10.11. D. 10.12. D.
10.13. B. 10.14. A. 10.15. C. 10.16. A. 10.17. B. 10.18. B.
10.19. A. 10.20. A. 10.21. D. 10.22. B. 10.23. C. 10.24. C.
10.25. A. 10.26. C. 10.27. D. 10.28. B. 10.29. A. 10.30. C.
10.31. D. 10.32. A. 10.33. C. 10.34. C. 10.35. C. 10.36. C.
10.37. B. 10.38. C. 10.39. B. 10.40. D. 10.41. A. 10.42. B.
10.43. C. 10.44. B. 10.45. C.

11.1. C. 11.2. D. 11.3. A. 11.4. B. 11.5. B. 11.6. C.
11.7. D. 11.8. B. 11.9. D. 11.10. B. 11.11. D. 11.12. C.
11.13. A. 11.14. C. 11.15. A. 11.16. D. 11.17. C. 11.18. B.
11.19. C. 11.20. B. 11.21. C. 11.22. D. 11.23. B. 11.24. B.
11.25. D. 11.26. B. 11.27. C. 11.28. D.

12.1. B. 12.2. B. 12.3. A. 12.4. C. 12.5. B. 12.6. D.
12.7. C. 12.8. D. 12.9. A. 12.10. A. 12.11. C. 12.12. B.
12.13. A. 12.14. C. 12.15. C. 12.16. D. 12.17. B. 12.18. C.
12.19. A. 12.20. B. 12.21. A. 12.22. D. 12.23. C. 12.24. B.
12.25. C. 12.26. B. 12.27. D. 12.28. B. 12.29. C. 12.30. A.
12.31. C. 12.32. C. 12.33. B. 12.34. D. 12.35. D. 12.36. B.
12.37. A. 12.38. B. 12.39. C. 12.40. A. 12.41. B. 12.42. D.
12.43. A. 12.44. B. 12.45. D. 12.46. D. 12.47. D. 12.48. A.
12.49. B. 12.50. B. 12.51. D. 12.52. A. 12.53. C. 12.54. A.
12.55. C. 12.56. B. 12.57. D. 12.58. A. 12.59. B. 12.60. C.
12.61. D. 12.62. B. 12.63. C. 12.64. D. 12.65. C. 12.66. C.

References

1. Abrams D. and Prausnitz J. M., *AIChE J.*, 21, 116, (1975)
2. Agoudjil N., et al., *Synthesis and Characterisation of Inorganic Membranes and Applications*, Desalination, Elsevier, 184, 65, (2005)
3. Ananthanarayanan P.N., *Basic Refrigeration and Air Conditioning*, Tata McGraw-Hill, New Delhi, (2005)
4. Badger W.L., Banchero J.T., *Introduction to Chemical Engineering*, McGraw Hill, N.Y. (1955)
5. Bae S.D. et al., *Preparation, Characterisation, and Application of Activated Carbon Membrane with Carbon Whiskers*, Desalination, Elsevier, http://www.desline.com/ articoli/8503.pdf, (2006)
6. Baker R. W., *Membrane Technology and Applications*, John Wiley & Sons, West Sussex, England, (2004)
7. Baker, T.C., *Ind. Eng. Chem.*, 27, 977, (1935)
8. Bas G. L., *The Molecular Volumes of Liquid Chemical Compounds,* David McKay, New York, (1915)
9. Basmadjian D., *Mass Transfer Principles and Applications*, CRC Press, (2004)
10. Belfort G., *Synthetic Membrane Processes*, Academic Press, Orlando, (1984)
11. Benedict M., Webb G. and Lubin L., *J. Chem. Phys.*, 8, 334, (1940)
12. Bird R.B., Stewart W.E., Lightfoot E.N., *Transport Phenomena*, John Wiley & Sons, New York, (1960)
13. Bourouni K., Chaibi M.T., Tadrist L., '*Water Desalination by Humidification and Dehumidification of Air: State of the Art*', Desalination, pp167-176, 137 http://www.desline.com, (2001)
14. Brennan J.G., *Food processing Handbook*, Wiley-VCH, Weinheim, Germany, (2006)
15. Brown G.G., *Unit Operations*, CBS, New Delhi, (2001)
16. Brunauer S., Emmeett P.H., Teller E., *J. Am. Chem. Soc.*, 60 309, (1938)
17. Brunauer S., *J. Am. Chem. Soc.*, 62, 1723, (1940)
18. Bruschke H., *Pure and Appl. Chem.*, 67, 6, 993, (1995)
19. Bylund G., *Dairy Processing Handbook*, Tetra Pak Processing Systems, Lund, Sweden, (2003)
20. Cavalcante Jr., C.L., *Latin Am. Appl. Res.*, 30: 357–364 http://www.enq. ufsc.br/eventos/ enpromer99/, (2000)
21. Chermisinoff, N.P., *Handbook of Chemical Process Equipment*, Butterworth Heinmann, MA, (2001)

22. Chilton, T.H., Colburn, A.P., *Ind. Eng. Chem.*, 26, 1183, (1934)
23. Chopey, N.P., *Handbook of Chemical Engineering Calculations*, 3rd Edn., McGraw Hill, New York, (2004).
24. Churchill, S.W., Zajic, S.C., *AIChE J.*, 48, 927, (2002)
25. Clark S.R., *Elements of Fractional Distillation*, McGraw Hill, N.Y., (1950)
26. Colburn, A.P., *Trans. Am. Inst. Chem. Eng.*, 35, 211, (1939).
27. Collins J.J., *Chem. Eng. Progr. Symp. Series.*, 63(74), 31, (1967)
28. Coulson J.M., Richardson J.F., *Chemical Engineering Vol 1*, 6th Ed., Elsevier, MA, (2004)
29. Couper J.R., Penney, W.R., Fair, J.R., Walas S. M., *Chemical Process equipment: Selection and Design, Elsevier*, MA, (2005)
30. Danckwerts, P.V., *Ind. Eng. Chem.*, 43, 1460, (1951)
31. DePriester C. L., *Chem. Eng. Progr, Symp. Ser.*, 7, 49, (1953)
32. Dhinga S.S., *Mixed gas transport Study through Polymeric Membranes: A Novel Technique*, PhD Thesis, Virginia PolyTech. Inst., http://scholar.lib.vt.edu/ theses/public/etd-52497-133245/ materials, (1997)
33. Doherty M.F., Malone M.F., *Conceptual Design of Distillation Systems*, McGraw Hill. New York, (2001)
34. Douglas, J.M., *Conceptual Design of Chemical Processes*, McGraw Hill, New York, (1988)
35. Drickamer, H.G. Bradford, J.R., *Trans. Am. Inst. Chem. Eng.*, 39, 319, (1943).
36. Driscoll R., *Food Dehydration* in *Food Processing Principles and Applications* by Smith J.S., Hui, Y.H., (Eds.), Blackwell Publishing, (2004)
37. Dubal, G., *Podbielnaiak Contactor: A Unique Liquid-Liquid extractor*, www. Chemical Online.com, January 7, (2000)
38. Dutta B.K., *Principles of Mass Transfer and Separation Processes*, PHI, New Delhi, (2007)
39. Earle R.L., Earle M.D., *Unit Operations in Food Processing*, NZIFST (Inc), (1983)
40. Edmister, W.C., *The Petroleum Engineer*, C45-C54, January, (1949)
41. Felder R.M., Rousseau R. W., *Elementary Principles of Chemical Processes*, 3rd Ed., John Wiley, N.Y. (2000)
42. Forbes R.J., "*A Short History of the Art of Distillation*", Brill Academic Publishers, Leiden, (1948).
43. Foust A.S. et al., *Principles of Unit Operations*, 2nd Edn., John Wiley & Sons, New York, (1980)
44. Fredenslund Aa., Jones R. L. and Prausnitz J. M., *AIChE J*, 21, 1086, (1975)
45. Friend, W.L., and Metzner, A.B., *AIChE J.*, 4, 393, (1958)
46. Fuller E.N., Schettler P.D., Giddings J.C., *Ind. Eng. Chem.*, 58(5), 18–27, (1966)
47. Geankoplis C., *Transport Processes and Unit Operations*, 3rd Ed., PHI, New Delhi, (1999)
48. Gilliland, E.R., Sherwood, T.K., *Ind. Eng. Chem.*, 26, 516, (1934)
49. Giulietti M., M.M. Seckler, S. Derenzo, M.I. Ré and E. Cekinski, *Braz. J. Chem. Eng.*, 18, 4, (2001)
50. Green D.W., Perry R.H., *Perry's Chemical Engineer's Handbook*, 8th Edn., McGraw Hill, New York, (2008)
51. Halvorsen I.J., Skojestad S., *Encyclopedia of Separation Science*, Academic Press, (2000)
52. Hanson C., Baird M.H.I., Lo T.C., *Handbook of Solvent Extraction*, John Wiley & Sons, New York, (1983)
53. Henley E.J., Seader J.D., *Equilibrium Stage Separation Operations in Chemical Engineering*, John Wiley & Sons, New York, (1981)
54. Henry, H.C. and Epstein N., *Can. J. Chem. Eng.*, 48, 595, (1970)
55. Higbie R., *Trans. Am. Inst. Chem. Eng.*, 31, 365, (1935)
56. Hirschfelder J.O., Bird R. B., and Spotz E. L., *Chem. Revs.,* 205, (1949)

57. Hirschfelder, J.O., Curtiss C.F., Bird R.B., *Molecular Theory of Gases and Liquids*, Wiley, New York, (1954)
58. http://www. Lenntech.com/ceramic membranes.htm#ixzz1dfwYXJHj
59. http://www.industrialmicrowave.com
60. http://www.separationprocesses.com/membrane
61. http://www.trionetics.com/an001.pdf
62. http://www.worldscibooks.com/engineering/p636.html
63. Hunter, T.G., Nash A.W., *J. Soc.Chem. Ind.*, 53, 95T-102T, (1934).
64. Ismail A.F., David L.I.B., *Jl. Membrane Sci.*, Elsevier, 193, (2001)
65. Jones, P.L., Rowley, A.T., *Drying Technology*, 14, 5, (1996)
66. Kafarov V., *Fundamemtals of Mass Transfer*, Mir Publishers, Moscow, (1975)
67. Kevin Enterprises Pvt. Ltd. Mumbai, Ref: www.kevincpp.com/trays.htm, March, (2008)
68. *Khoury, F.M.,* Multistage Separation Processes, *3rd Edn., CRC Press, Folrida, USA,* (2005)
69. Khulbe K.C. et al., *Synthetic Polymeric Membranes Charaterisation by Atomic Force Microscopy*, Springer, (2008)
70. King C.J., *Separation Processes*, 2nd Ed., McGraw Hill, New York, (1980)
71. KLM Technology Group, *Distillation Column selection and Sizing (Energy Design Guide Lines)*, www.klmtechgroup.com, (2011)
72. Knaebel K.S., *A "How to" Guide for Adsorber Design*, Adsorption Research, Inc. Dublin, Ohio, http://adsorption.com/publications, (March 2012)
73. Knudsen, J.G., Katz, D.L., *Fluid Dynamics and Heat Transfer*, McGraw Hill (NY), (1958).
74. Kujawski, W., *Polish Jl. Env. Studies*, 9, 1, (2000)
75. Law J.D., Todd T.A., *Liquid-Liquid Extraction Equipment*, www.cresp.org, (2006)
76. Lewis, W.K., *Ind. Eng. Chem*., 28, 399, (1936)
77. Lewis, W.K., Whitman,W.G., *Ind. Eng. Chem*., 16,1215, (1924)
78. Linton, W.H., Sherwood, T.K., *Chem. Eng. Prog*., 46, 258, (1950)
79. Lockhart, F.J., Leggett, C.W., Kobe, K.A., McKetta, J.J., Jr., Ed., *Advances in Petroleum Chemistry and Refining, Vol. I,* Inter Science New York, 1, 323, (1958)
80. Loeb S., Sourirajan S., *Advances in Chemistry Series*, Vol 38, Saline Water Conversion II, (1963)
81. Marrero, T. R., and E. A. Mason, *J. Phys. Chem. Ref Data.* 1.3–118, (1912)
82. Matsura T., *Membrane Separation Technologies in Waste Water Recycle, Reuse and Reclamation,* Vigneswaran S., (Eds.), Vol I, EOLSS, (2009)
83. McCabe W.L., Smith C. S., Harriot P., *Unit Operations in Chemical Engineering* 6th Ed., McGraw Hill, New York, (2001)
84. McCormick, P.Y., in R.H. Perry and C.H. Chilton (Eds.), "Chemical Engineer's handbook", 5th Ed. McGraw Hill, New York, (1973)
85. Meissner, *Chern. Eng. Prog..* 45, 149–153, (1949).
86. Metaxas A.C., *IEE, Power Engineering Jl.*, 5 (5), (1991)
87. Meyer K.H., Strauss W., *Helv. Chim. Acta,* 23, 795, (1940)
88. Mulder M., *Basic Principles of Membrane Technology*, 2nd Ed., Kluwer Academic, Dordrecht, The Netherlands, (1996)
89. Myerson A.S., *Handbook of Industrial Crystallisation*, Butterworth Heinmann, Boston, (2002)
90. Narayanan K.V, B. Lakshmikutty, *Stoichiometry and Process Calculations*, PHI, New Delhi, (2006)
91. Narayanan K.V., *A Textbook of Chemical Engineering Thermodynamics*, 2nd Ed., PHI, New Delhi, (2013)
92. Nath K., *Membrane Separation Processes*, PHI Learning Pvt. Ltd., New Delhi, (2008)

93. National Research Council, *International Critical Tables*, Vol III, McGraw Hill, New York, (1929)
94. Nernst W., *Z. Phys. Chem.*, 47:52, (1904)
95. Nicholas P. Cheremisinoff, *Handbook of Chemical Processing Equipment*, Butterworth Heinmann, Woburn, MA, USA, (2000)
96. Noble R.D., Stern S. A. (Eds.) *Membrane Separation Technology: Principles and Applications*, Elsevier, Amsterdam, Netherlands, (1995)
97. Norman W.S., *Absorption, Distillation and Cooling Towers*, Longmans, London, (1962)
98. O'Connel, H.E., *Trans. A I Ch E*, 42,741–755, (1946)
99. Patterson C.A., *Technology Watch, National Research Council*, Canada, 2, (July, 2005)
100. Peng D. Y. and Robinson D. B., *Ind. Eng. Chem. Fundam.*, 15, 59, (1976)
101. Perry R.H., Green D., (Eds.) *Perry's Chemical Engineers Handbook,* 6th Ed., McGraw Hill, New York, (1984)
102. Petlyuk F.B., *Distillation Theory and its Application to Optimal Design of Separation Units*, Cambridge University Press, Cambridge, UK, (2004)
103. Prausnitz J.M., Lichtenthaler R. N. and Azevedo E.G., *Molecular Thermodynamics of Fluid Phase equilibria*, 2nd ed., Prentice-Hall, NJ, (1986)
104. Quayle O.R., *Chem. Rev.*, 53, 439–589, (1953)
105. Randolph, A.D., *AIChE Journal*, 11, 424–430, (1965)
106. Redlich O. and Kwong J. N. S., *Chem. Rev.*, 44, 233, (1949)
107. Reid R.C., Prausnitz J.M., Poling B.E., *The Properties of Gases and Liquids*, 4th Ed., McGraw Hill, New York, (1987)
108. Renon H., Prausnitz J. M., *A I Ch E J.*, 14, 135, (1968)
109. Richardson J F. et al., *Chemical Engineering Vol II Particle Technology and Separation Processes*, 5th Ed., Elsevier, (2003)
110. Roert Léauté, "*Distillation in Alambic*", www. Essential spirits.com/history, (1989)
111. Rohsenow, W.M., Choi, H.Y., *"Heat, Mass and Momentum Transfer"*, Prentice Hall, NY (1961)
112. Rousseau R.W. (Ed.), *Handbook of Separation Process Technology*, John Wiley, New York, (2009)
113. Rydberg J., Cox M., *Solvent Extraction Principles and Practice*, Marcel Decker, New York, (2004)
114. Schiffmann, R.F., in *Handbook of Industrial Drying*, A.S. Mujumdar (Ed) CRC Press (1996)
115. Schweitzer P.A., (Ed.), *Handbook of Separation Techniques for Chemical Engineers.* 2nd Ed., McGraw Hill, New York, (1988)
116. Seader J.D., Henley E.J., *Separation Process Principles*, 2nd Ed., John Wiley, (2006)
117. Sherwood & Pigford, *Absorption and Extraction*, 2nd Ed., McGraw Hill, New York, (1952)
118. Sherwood T.K., Pigford R.L., Wilke C.R., *Mass Transfer*, McGraw Hill, New York, (1975)
119. Sinnott R.K., *Coulson & Richardson's Chemical Engineering Vol 6*, 3rd Ed., Elsevier, MA, (2004)
120. Sivasankar B., *Bioseparations Principle and Techniques*, PHI, New Delhi, (2005)
121. Smith B.D., *Design of Equilibrium Stage Processes*, McGraw Hill, New York, (1963)
122. Smith J.M. and Van Ness H.C., *Introduction to Chemical Engineering Thermodynamics*, 4th ed., Mc Graw-Hill, New York, (1987)
123. Smith R., *Chemical Process Design and Integration*, Wiley India, New Delhi, (2006)
124. Soave G., *Chem. Engg. Sci.*, 27, 1197, (1972)
125. Sourirajan S., *Reverse Osmosis,* Academic Press, New York, USA, (1970)
126. Stanojevié M.M et al, *FME Trans.*, 31, 91–98, (2003)
127. Suherland K. http://www.filtsep.com/view/712/membrane-separation-processes-in-the-bulk-chemicals-industry

128. Tham M.T., *Distillation an Introduction*, http://lorien.ncl.ac.uk/, (2009)
129. Toledo R.T., *Fundamentals of Food Process Engineering*, 3rd Ed., Springer, LLC, New York, (2007)
130. Toor, H.L., and Marchello, J.M., *A I Ch E J.*, 4, 97, (1958)
131. Treybal R. E., *Mass-Transfer Operations*, 3rd ed., McGraw-Hill, New York, (1981)
132. Treybal R.E., *Liquid Extraction*, 2nd Ed., McGraw Hill, New York, (1963)
133. Ulrich J., Jones M.J., *Heat and Mass Transfer Operations–Crystallization in Chemical Engineering*, EOLSS, Oxford, (2006)
134. Wagner J., *Membrane Filtration Handbook Practical Tips & Hints*, Wagner Publishing, (2001)
135. Walas S.M., *Chemical Process Equipment: Selection and Design*, Butterworth-Heinemann, MA, USA, (1990)
136. Welty J.R., Wilson R.E., Wicks C.E., *Fundamemntals of Momentum, Heat and Mass Transfer*, 2nd Ed., John Wiley, New York, (1976)
137. Whitman, W.G., *Chem. Met. Eng.*, 29, 147, (1923)
138. Wikipedia, the Free Encyclopedia - http://en.wikipedia.org/wiki/In-situ_leach
139. Wilhelm F.G., *Bipolar Membrane Electrodialysis*, PhD Thesis, University of Twente, (2001)
140. Wilke C.R., Chang P., *A.I.Ch.E. J.*, 1,264, (1955)
141. Winkle M.V., *Distillation*, McGraw Hill, New York, (1967)
142. Yu, B., Oze, H., Churchill, S.W., *Chem. Eng. Sci.*, 56, 1781, (2001).

Appendices

APPENDIX A

Activity Coefficient Models

We have shown that Eq. (5.30) forms the basis for calculation of VLE data by the activity coefficient route.

$$K_i = \frac{y_i}{x_i} = \frac{\gamma_i P_i^s}{\phi_i^V P} \qquad \ldots(5.30)$$

Liquid-phase activity coefficients can be estimated by any of the models described briefly in the following paragraphs. It is useful when polar components are present. For most substances at low pressures, ϕ_i^V can be assumed to be unity. If the pressure is very high, ϕ_i^V must be evaluated using an equation of state.

van Laar equation

van Laar Equations are two-parameter equations applicable for solutions of relatively simple, preferably nonpolar liquids. However, these equations are found to be applicable for more complex mixtures as well. The multicomponent van Laar equations can be written as:

$$\ln \gamma_i = \frac{\sum_{j=1}^{C}(x_j A_{ij})}{(1-x_i)}\left[1 - \frac{x_i \sum_{j=1}^{C}(x_j A_{ij})}{x_i \sum_{j=1}^{C}(x_j A_{ij}) + (1-x_i)\sum_{j=1}^{C}(x_j A_{ji})}\right] \qquad \ldots(A.1)$$

where the constants $A_{ii} = A_{jj} = 0$. The binary constants A_{ij} and A_{ji} are determined from binary vapour-liquid equilibrium data.

For binary mixtures, the van Laar equations can be written as

$$\ln \gamma_1 = \frac{A_{12} x_2^2}{\left[(A_{12}/A_{21})\, x_1 + x_2\right]^2}$$

$$\ln \gamma_2 = \frac{A_{21} x_1^2}{\left[x_1 + (A_{21}/A_{12})\, x_2\right]^2} \qquad \ldots(A.2)$$

The van Laar equations are widely used for vapour-liquid equilibrium calculations because of their flexibility and mathematical simplicity. Activity coefficients in benzene-isooctane system, *n*-propanol-water system, etc. are accurately represented by the van Laar equations.

Wilson equation

Wilson equation is based on the concept of local compositions. From molecular considerations, Wilson (1964) proposed a model for activity coefficients in binary solutions, which is readily extended to multi-component solutions as given below:

$$\ln \gamma_k = 1 - \ln\left(\sum_{j=1}^{C} x_j \Lambda_{kj}\right) - \sum_{i=1}^{C}\left(\frac{x_i \Lambda_{ik}}{\sum_{j=1}^{C} x_j \Lambda_{ij}}\right) \quad \ldots(A.3)$$

where $\Lambda_{ii} = \Lambda_{jj} = \Lambda_{kk} = 1$ and

$$\Lambda_{ij} = \frac{v_j}{v_i} \exp\left[-\frac{\lambda_{ij} - \lambda_{ii}}{RT}\right] = \frac{v_j}{v_i} \exp\left[-\frac{a_{ij}}{RT}\right] \quad \ldots(A.4)$$

In Eq. (A.4) v_i and v_j are the molar volumes of pure liquids and λ_{ij} is the energy parameter characterizing the interaction of molecule *i* with molecule *j*. For each binary pair, there are two adjustable parameters L_{ij} and L_{ji} which must be determined from experimental data.

For binary solutions, the Wilson equations reduce to

$$\ln \gamma_1 = -\ln(x_1 + \Lambda_{12}x_2) + x_2\left[\frac{\Lambda_{12}}{x_1 + \Lambda_{12}x_2} - \frac{\Lambda_{21}}{\Lambda_{21}x_1 + x_2}\right]$$

$$\ln \gamma_2 = -\ln(x_2 + \Lambda_{21}x_1) - x_1\left[\frac{\Lambda_{12}}{x_1 + \Lambda_{12}x_2} - \frac{\Lambda_{21}}{\Lambda_{21}x_1 + x_2}\right] \quad \ldots(A.5)$$

where

$$\Lambda_{12} = \frac{v_2}{v_1} \exp\left[-\frac{\lambda_{12} - \lambda_{11}}{RT}\right] = \frac{v_2}{v_1} \exp\left[-\frac{a_{12}}{RT}\right]$$

$$\Lambda_{21} = \frac{v_1}{v_2} \exp\left[-\frac{\lambda_{21} - \lambda_{22}}{RT}\right] = \frac{v_1}{v_2} \exp\left[-\frac{a_{21}}{RT}\right] \quad \ldots(A.6)$$

Wilson equation provides a good representation of VLE of a variety of miscible mixtures. It is particularly suitable for solutions of polar or associating components like alcohols in nonpolar solvents for which the van Laar equations are generally inadequate.

Wilson equation suffers from two disadvantages, though not serious for many applications. Firstly, it is not suitable for systems showing maxima or minima on the ln *g* versus *x* curves. Secondly, it is not useful for systems exhibiting limited miscibility. The use of Wilson equation is therefore recommended only for liquid systems that are completely miscible, or for partially miscible systems in the region where only one liquid phase exist.

NRTL equation

The NRTL (*Non-random, two-liquid*) model proposed by Renon and Prausnitz (1968) also is based on the local composition concept.

$$\ln \gamma_i = \frac{\sum_{j=1}^{C} \tau_{ji} G_{ji} x_j}{\sum_{k=1}^{C} G_{ki} x_k} + \sum_{j=1}^{C} \left[\frac{x_j G_{ij}}{\sum_{k=1}^{C} x_k G_{kj}} \left(\tau_{ij} - \frac{\sum_{k=1}^{C} x_k \tau_{kj} G_{kj}}{\sum_{k=1}^{C} x_k G_{kj}} \right) \right] \quad \text{...(A.7)}$$

$$G_{ji} = \exp\left(-\alpha_{ji} \tau_{ji}\right) \quad \text{...(A.8)}$$

$$\tau_{ij} = \frac{g_{ij} - g_{jj}}{RT} \quad \tau_{ji} = \frac{g_{ji} - g_{ii}}{RT} \quad \text{...(A.9)}$$

Here g_{ij} and g_{ji} are the energies of interaction between molecules of i and j. α_{ij} characterizes the tendency of molecule i and molecule j to be distributed in a random fashion and is related to the nonrandomness in the mixture. If a_{ij} is zero, the mixture is completely random. It is found from fitting of experimental data that a_{12} varies from about 0.20 to 0.47. In the absence of experimental data, the value of a_{12} is arbitrarily set, a typical choice being $a_{12} = 0.3$. When a_{12} is arbitrarily fixed, NRTL equation becomes a two-parameter model.

For binary systems, the NRTL equations reduce to

$$\ln \gamma_1 = x_2^2 \left[\tau_{21} \left(\frac{G_{21}}{x_1 + x_2 G_{21}} \right)^2 + \frac{\tau_{12} G_{12}}{(x_2 + x_1 G_{12})^2} \right]$$

$$\ln \gamma_2 = x_1^2 \left[\tau_{12} \left(\frac{G_{12}}{x_2 + x_1 G_{12}} \right)^2 + \frac{\tau_{21} G_{21}}{(x_1 + x_2 G_{21})^2} \right] \quad \text{...(A.10)}$$

In Eq. (A.10) the adjustable parameters are evaluated as

$$G_{12} = \exp(-\alpha_{12} \tau_{12}) \qquad G_{21} = \exp(-\alpha_{12} \tau_{21}) \quad \text{...(A.11)}$$

$$\tau_{12} = \frac{g_{12} - g_{22}}{RT} \qquad \tau_{21} = \frac{g_{21} - g_{11}}{RT} \quad \text{...(A.12)}$$

NRTL equation is applicable to partially miscible as well as totally miscible systems. For moderately nonideal systems it offers no advantage over the van Laar and Wilson equations. But, for strongly nonideal solutions and especially partially miscible systems, the NRTL equations provide good representation.

UNIQUAC equation

Abrams and Prausnitz (1975) extended the quasi-chemical theory of liquid mixtures to solutions containing molecules of different sizes. This extension is called the *universal quasi-chemical theory* or UNIQUAC. The UNIQUAC model consists of two parts: the combinatorial part, which describes the prominent entropic contribution and a residual part, which is due primarily to the intermolecular forces that are responsible for the enthalpy of mixing. The combinatorial part is determined by the sizes and shape of the molecules and requires only pure-component data. The residual part depends on the inter-molecular forces and involves two adjustable binary parameters.

$$\ln \gamma_i = \ln \frac{\psi_i}{x_i} + \frac{z}{2} q_i \ln \frac{\theta_i}{\psi_i} + l_i - \frac{\psi_i}{x_i} \sum_{j=1}^{C} x_j l_j + q_i \left[1 - \ln \left(\sum_{j=1}^{C} \theta_j \tau_{ji} \right) - \sum_{j=1}^{C} \left(\frac{\theta_j \tau_{ij}}{\sum_{k=1}^{C} \theta_k \tau_{kj}} \right) \right] \quad \text{...(A.13)}$$

where $$\psi_i = \frac{r_i x_i}{\sum_k^C r_k x_k};\ \theta_i = \frac{q_i x_i}{\sum_k^C q_k x_k};\ l_j = \frac{z}{2}(r_j - q_j) - (r_j - 1) \quad \text{...(A.14)}$$

The adjustable binary parameter t_{ij} is related to the characteristic energies ru as follows.

$$\tau_{ij} = \exp\left[-\frac{u_{ij} - u_{jj}}{RT}\right] \quad \text{...(A.15)}$$

u_{ij} = interaction parameter between molecule i and molecule j. $u_{ij} = u_{ji}$

z = coordination number (z = 10)

r_i = pure component property representing van der Waals volume for molecule i.

q_i = pure component property representing van der Waals surface area for molecule i.

The UNIQUAC equations for activity coefficients in binary solutions are

$$\ln \gamma_1 = \ln \frac{\psi_1}{x_1} + \frac{z}{2} q_1 \ln \frac{\theta_1}{\psi_1} + \psi_2\left[l_1 - \frac{r_1}{r_2} l_2\right] - q_1 \ln(\theta_1 + \theta_2 \tau_{21}) + \theta_2 q_1 \left(\frac{\tau_{21}}{\theta_1 + \theta_2 \tau_{21}} - \frac{\tau_{12}}{\theta_2 + \theta_1 \tau_{12}}\right) \quad \text{...(A.16)}$$

$$\ln \gamma_2 = \ln \frac{\psi_2}{x_2} + \frac{z}{2} q_2 \ln \frac{\theta_2}{\psi_2} + \psi_1\left[l_2 - \frac{r_2}{r_1} l_1\right] - q_2 \ln(\theta_2 + \theta_1 \tau_{12}) + \theta_1 q_2 \left(\frac{\tau_{12}}{\theta_2 + \theta_1 \tau_{12}} - \frac{\tau_{21}}{\theta_1 + \theta_2 \tau_{21}}\right)$$

where

$$l_1 = \frac{z}{2}(r_1 - q_1) - (r_1 - 1)$$
$$l_2 = \frac{z}{2}(r_2 - q_2) - (r_2 - 1) \quad \text{...(A.17)}$$

$$\tau_{12} = \exp\left(-\frac{\Delta u_{12}}{RT}\right)$$
$$\tau_{21} = \exp\left(-\frac{\Delta u_{21}}{RT}\right) \quad \text{...(A.18)}$$

The segment fraction (φ) and the area fractions (θ) are obtained as

$$\psi_1 = \frac{x_1 r_1}{x_1 r_1 + x_2 r_2}, \qquad \psi_2 = \frac{x_2 r_2}{x_1 r_1 + x_2 r_2} \quad \text{...(A.19)}$$

$$\theta_1 = \frac{x_1 q_1}{x_1 q_1 + x_2 q_2}, \qquad \theta_2 = \frac{x_2 q_2}{x_1 q_1 + x_2 q_2} \quad \text{...(A.20)}$$

The UNIQUAC equation is found to satisfy a large number of nonelectrolyte mixtures containing nonpolar fluids such as hydrocarbons, alcohols, nitriles, ketones, aldehydes, organic acids, etc. and water, including partially miscible mixtures. The main advantages of this equation are its wide applicability and simplicity arising primarily from the fact that there are only two adjustable parameters.

UNIFAC method

In the UNIFAC (*universal functional activity coefficient*) method, the activity coefficients are estimated through group contributions. The liquid is treated as a solution of different structural groups from which the molecules are formed, rather than a solution of molecules themselves. This method is based on the UNIQUAC model where the activity coefficient is divided into two parts – the molecular size contribution (the combinatorial part) and the interaction contributions (the residual part).

$$\ln \gamma_i = \ln \gamma_i^C + \ln \gamma_i^R \qquad \text{...(A.21)}$$

The combinatorial contribution can be estimated from pure-component properties and the size and shape of the molecules where as for the estimation of the second part group areas and group contributions are needed. A large number of group interaction parameters are already reported. UNIFAC has been successfully used for the design of distillation columns, involving even azeotropic and extractive distillation (Prausnitz *et al.*, 1986).

The UNIFAC method has the following advantages over other group contribution models: (1) Being based on UNIQUAC model it has a theoretical basis (2) Parameters are temperature independent (3) size and binary interaction parameters are available for a wide range of types of functional groups (4) Predictions can be made over a temperature range of 275–425 K and pressures up to a few atmospheres.

Equation of state models

Equation (5.31) forms the basis for estimating VLE by the equations of state approach.

$$K_i = \frac{y_i}{x_i} = \frac{\phi_i^L}{\phi_i^V} \qquad \text{...(A.22)}$$

Equations of state provide thermodynamic models for evaluating ϕ_i^V and ϕ_i^L from volumetric properties. The method has the advantage that the standard state fugacity f_i^0 need not be specified. Another advantage is that the continuity at the critical point is ensured as the properties in the gas and liquid phases are evaluated using the same model. However, in practice equations of state approach is applied only to mixtures of nonpolar components. They are widely used to predict VLE of hydrocarbon mixtures involving light gases.

Before an equation of state can be applied to calculate the vapour-liquid-equilibrium, the fugacity coefficients ϕ_i^V and ϕ_i^L are to be determined. The relationship between fugacity coefficient and the volumetric properties can be written as

$$\ln \phi_i = \frac{1}{RT} \int_V^{\infty} \left[\left(\frac{\partial P}{\partial N_i} \right)_{T,V,N_j} - \frac{RT}{V} \right] dV - RT \ln Z \qquad \text{...(A.23)}$$

where Z is the compressibility factor. In Eq. (A.23) the integral can be evaluated using an appropriate equation of state.

Alternatively, it is possible to simplify Eq. (5.31) to the following form:

$$K_i = \frac{\phi_i^S \, P_i^S}{\phi_i^V \, P} \exp \frac{V_i \,(P - P_i^S)}{RT} \qquad \text{...(A.24)}$$

In Eq. (A.24) ϕ_i^S is the fugacity coefficient in the saturation state, V_i is the molar volume of pure i as saturated liquid. ϕ_i^S and ϕ_i^V are evaluated using an equation of state.

Following are the most generally used equations of state for VLE calculations:

Name	Equation	Constants and functions
Ideal gas law	$P = \dfrac{RT}{V}$	
Generalized equation of state	$P = \dfrac{ZRT}{V}$	$Z = f(P_r, T_r, Z_c \text{ or } \omega)$ $\omega = \left[-\log\left(\dfrac{P^S}{P_C}\right)_{Tr=0.7}\right] - 1.00$
Redlich-Kwong (R-K)	$P = \dfrac{RT}{V-b} - \dfrac{a}{V^2 + bV}$	$a = 0.42748 R^2 T_c^{2.5} / P_c T^{0.5}$ $b = 0.08664 RT_c / P_c$
Redlich-Kwong-Soave (R-K-S)	$P = \dfrac{RT}{V-b} - \dfrac{a}{V^2 + bV}$	$a = 0.42748 R^2 T_c^2 \left[1 + f_w (1 - T_r^{0.5})\right]^2 / P_c$ $b = 0.08664 RT_c / P_c$ $f_w = 0.48 + 1.574\omega - 0.176\omega^2$
Peng-Robinson (P-R)	$P = \dfrac{RT}{V-b} - \dfrac{a}{V^2 + 2bV - b^2}$	$a = 0.45724 R^2 T_c^2 \left[1 + f_w (1 - T_r^{0.5})\right]^2 / P_c$ $b = 0.07780 RT_c / P_c$ $f_w = 0.37464 + 1.54226\omega - 0.26992\omega^2$

APPENDIX B

Antoine Constants

$$\ln P^s = A - \frac{B}{T - C} \text{ (with } P \text{ in kPa and } T \text{ in K)}$$

Name	A	B	C	Range of temperature, K
1,4-Dioxane	14.1177	2966.88	62.15	275–410
1-Propanol	15.81647	3310.416	74.737	333–378
2-Propanol	16.26205	3799.887	46.854	326–362
4-methyl-2-pentanol	17.4813	5007.85	15.38	298–406
Acetone	14.39155	2795.817	43.198	241–350
Acetonitrile	14.2724	2945.47	49.15	300–400
Acrylonitrile	13.9103	2782.21	51.15	255–385
Ammonia	14.9331	2940.46	56.34	179–261
Benzene	13.8858	2788.51	52.36	280–377
Benzoic acid	15.14837	4190.70	125.2	405–560
Chlorobenzene	13.9926	3295.12	55.60	335–405
Chloroform	13.9985	2696.2663	46.968	26–370
Ethyl acetate	14.3291	2852.25	56.2	289–349
Ethyl alcohol	16.5092	3578.91	50.50	270–369
Ethyl benzene	14.0045	3279.47	59.95	330–410
Ethylene glycol	18.23507	6022.18	28.25	364–494
Glycerol	15.22417	4487.04	140.2	440–600
Methyl alcohol	16.5725	3626.55	34.29	257–364
Methyl isopropyl ketone	12.16287	1993.12	103.2	271–406
n-Heptane	13.8587	2911.32	56.51	270–400
n-Hexane	13.8216	2697.55	48.78	245–370
Nitromethane	14.2043	2972.64	64.15	329–410
n-Pentane	13.8183	2477.07	39.94	220–330
Toluene	13.9987	3096.52	53.67	280–410
Water	16.26205	3799.887	46.854	284–441

APPENDIX C

Properties of Steam, Water and Air

Table C1: Thermodynamic properties of saturated steam

Pressure	Temp.	Specific volume, m³/kg		Enthalpy, kJ/kg			Entropy, kJ/kg K	
kPa	K	Sat. liquid	Sat. vapour	Sat. liquid	Evap	Sat. vapo	Sat. liquid	Sat. vapour
0.6113	273.16	0.001000	206.14	0.01	2501.4	2501.4	0	9.1562
1.0	280.13	0.001000	129.21	29.30	2484.9	2514.2	0.1059	8.9756
1.5	286.18	0.001001	87.98	54.71	2470.6	2525.3	0.1957	8.8279
2.0	290.65	0.001001	67.00	73.48	2460.0	2533.5	0.2607	8.7237
2.5	294.23	0.001002	54.25	88.49	2451.5	2540	0.3120	8.6432
3.0	297.23	0.001003	45.67	101.05	2444.5	2545.5	0.3545	8.5776
4.0	302.11	0.001004	34.80	121.46	2432.9	2554.4	0.4226	8.4746
5.0	306.03	0.001005	28.19	137.82	2423.7	2561.5	0.4764	8.3951
7.5	313.44	0.001008	19.24	168.79	2406.0	2574.8	0.5764	8.2515
10	318.96	0.001010	14.67	191.83	2392.9	2584.7	0.6493	8.1502
15	327.12	0.001014	10.020	225.94	2373.2	2599.1	0.7549	8.0085
20	333.21	0.001017	7.649	251.40	2358.3	2609.7	0.8320	7.9085
25	338.12	0.001020	6.204	271.93	2346.3	2618.2	0.8931	7.8314
30	342.25	0.001022	5.229	289.23	2336.1	2625.3	0.9439	7.7686
40	349.02	0.001027	3.993	317.58	2319.2	2636.8	1.0259	7.6700
50	354.48	0.001030	3.240	340.49	2305.4	2645.9	1.0910	7.5939
75	364.93	0.001037	2.217	384.39	2278.6	2663	1.2130	7.4564
100	372.78	0.001043	1.694	417.46	2258.0	2675.5	1.3026	7.3594
125	379.14	0.001048	1.3749	444.32	2241.1	2685.4	1.3740	7.2844
150	384.52	0.001053	1.1593	467.11	2226.5	2693.6	1.4336	7.2233
175	389.21	0.001057	1.0036	486.99	2213.6	2700.6	1.4849	7.1717
200	393.38	0.001061	0.8857	504.70	2202.0	2706.7	1.5301	7.1271
225	397.15	0.001064	0.7933	520.72	2191.4	2712.1	1.5706	7.0878
250	400.59	0.001067	0.7187	535.37	2181.5	2716.9	1.6072	7.0527
275	403.75	0.001070	0.6573	548.89	2172.4	2721.3	1.6408	7.0209
300	406.7	0.001073	0.6058	561.47	2163.8	2725.3	1.6718	6.9919
325	409.45	0.001076	0.5620	573.25	2155.8	2729	1.7006	6.9652
350	412.03	0.001079	0.5243	584.33	2148.1	2732.4	1.7275	6.9405
375	414.47	0.001081	0.4914	594.81	2140.8	2735.6	1.7528	6.9175
400	416.78	0.001084	0.4625	604.74	2133.9	2738.6	1.7766	6.8959
450	421.08	0.001088	0.4140	623.25	2120.7	2743.9	1.8207	6.8565
500	425.01	0.001093	0.3749	640.23	2108.5	2748.7	1.8607	6.8213
550	428.63	0.001097	0.3427	655.93	2097.1	2753	1.8973	6.7893
600	432	0.001101	0.3157	670.56	2086.2	2756.8	1.9312	6.7600
650	435.16	0.001104	0.2927	684.28	2076.0	2760.3	1.9627	6.7331
700	438.12	0.001108	0.2729	697.22	2066.3	2763.5	1.9922	6.7080
750	440.93	0.001112	0.2556	709.47	2056.9	2766.4	2.0200	6.6847
800	443.58	0.001115	0.2404	721.11	2048.0	2769.1	2.0462	6.6628
850	446.11	0.001118	0.2270	732.22	2039.4	2771.6	2.0710	6.6421
900	448.53	0.001121	0.2150	742.83	2031.1	2773.9	2.0940	6.6226
950	450.84	0.001124	0.2042	753.02	2023.1	2776.1	2.1172	6.6041

Contd.

Table C1 contd.

Pressure	Temp.	Specific volume, m³/kg		Enthalpy, kJ/kg			Entropy, kJ/kg K	
kPa	K	Sat. liquid	Sat. vapour	Sat. liquid	Evap	Sat. vapo	Sat. liquid	Sat. vapour
1000	453.06	0.001127	0.19440	762.81	2015.3	2778.1	2.1387	6.5865
1100	457.24	0.001133	0.17753	781.34	2000.4	2781.7	2.1792	6.5536
1200	461.14	0.001139	0.16333	798.65	1986.2	2784.8	2.2166	6.5233
1300	464.79	0.001144	0.15125	814.93	1972.7	2787.6	2.2515	6.4953
1400	468.22	0.001149	0.14084	830.30	1959.7	2790	2.2842	6.4693
1500	471.47	0.001154	0.13177	844.89	1947.3	2792.2	2.3150	6.4448
1750	478.91	0.001166	0.11349	878.50	1917.9	2796.4	2.3851	6.3896
2000	485.57	0.001177	0.09963	908.79	1890.7	2799.5	2.4474	6.3409
2250	491.6	0.001187	0.08875	936.49	1865.2	2801.7	2.5035	6.2972
2500	497.14	0.001197	0.07998	962.11	1841.0	2803.1	2.5547	6.2575
3000	507.05	0.001217	0.06668	1008.42	1795.8	2804.2	2.6457	6.1869
3500	515.75	0.001235	0.05707	1049.75	1753.7	2803.4	2.7253	6.1253
4000	523.55	0.001252	0.04978	1087.31	1714.1	2801.4	2.7964	6.0701
5000	537.14	0.001286	0.03944	1154.23	1640.1	2794.3	2.9202	5.9734
6000	548.79	0.001319	0.03244	1213.35	1571.0	2784.3	3.0267	5.8892
7000	559.03	0.001351	0.02737	1267.00	1505.1	2772.1	3.1211	5.8133
8000	568.21	0.001384	0.02352	1316.64	1441.4	2758	3.2068	5.7432
9000	576.55	0.001418	0.02048	1363.26	1378.8	2742.1	3.2858	5.6772
10000	584.21	0.001452	0.018026	1407.56	1317.1	2724.7	3.3596	5.6141
11000	591.3	0.001489	0.015987	1450.10	1255.5	2705.6	3.4295	5.5527
12000	597.9	0.001527	0.014263	1491.30	1193.6	2684.9	3.4962	5.4924
13000	604.08	0.001567	0.012780	1531.50	1130.7	2662.2	3.5606	5.4323
14000	609.9	0.001611	0.011485	1571.10	1066.5	2637.6	3.6232	5.3717
15000	615.39	0.001658	0.010337	1610.50	1000.0	2610.5	3.6848	5.3098
16000	620.59	0.001711	0.009306	1650.10	930.5	2580.6	3.7461	5.2455
17000	625.52	0.001770	0.008364	1690.30	856.9	2547.2	3.8079	5.1777
18000	630.21	0.001840	0.007489	1732.00	777.1	2509.1	3.8715	5.1044
19000	634.69	0.001924	0.006657	1776.50	688.0	2464.5	3.9388	5.0228
20000	638.96	0.002036	0.005834	1826.30	583.4	2409.7	4.0139	4.9269
21000	643.04	0.002207	0.004952	1888.40	446.2	2334.6	4.1075	4.8013
22000	646.95	0.002742	0.003568	2022.20	143.4	2165.6	4.3110	4.5327
22090	647.29	0.003155	0.003155	2099.30	0.0	2099.3	4.4298	4.4298

Table C2: Thermodynamic properties of superheated steam
(V = specific volume, H = enthalpy, S = entropy)

T, K	V, m³/kg	H, kJ/kg	S kJ/kg K	V, m³/kg	H, kJ/kg	S, kJ/kg K
	P = 10 kPa (Sat. Temp = 318.96 K)			P = 50 kPa (Sat. Temp = 354.48 K)		
Sat.	14.674	2584.7	8.1502	3.240	2645.9	7.5939
323.15	14.869	2592.6	8.1749			
373.15	17.196	2687.5	8.4479	3.418	2682.5	7.6947
423.15	19.512	2783.0	8.6882	3.889	2780.1	7.9401
473.15	21.825	2879.5	8.9038	4.356	2877.7	8.1580
523.15	24.136	2977.3	9.1002	4.820	2976.0	8.3556
573.15	26.445	3076.5	9.2813	5.284	3075.5	8.5373
673.15	31.063	3279.6	9.6077	6.209	3278.9	8.8642
773.15	35.679	3489.1	9.8978	7.134	3488.7	9.1546
873.15	40.295	3705.4	10.1608	8.057	3705.1	9.4178
973.15	44.911	3928.7	10.4028	8.981	3928.5	9.6599
1073.15	49.526	4159.0	10.6281	9.904	4158.9	9.8852
1173.15	54.141	4396.4	10.8396	10.828	4396.3	10.0967
1273.15	58.757	4640.6	11.0393	11.751	4640.5	10.2964
1373.15	63.372	4891.2	11.2287	12.674	4891.1	10.4859
1473.15	67.987	5147.8	11.4091	13.597	5147.7	10.6662
1573.15	72.602	5409.7	11.5811	14.521	5409.6	10.8382
	P = 100 kPa (Sat. Temp = 372.96 K)			P = 200 kPa (Sat. Temp = 393.38 K)		
Sat.	1.694	2675.5	7.3594	0.8857	2706.7	7.127
373.15	1.696	2676.2	7.3614			
423.15	1.936	2776.4	7.6134	0.9596	2768.8	7.279
473.15	2.172	2875.3	7.8343	1.0803	2870.5	7.507
523.15	2.406	2974.3	8.0333	1.1988	2971.0	7.709
573.15	2.639	3074.3	8.2158	1.3162	3071.8	7.893
673.15	3.103	3278.2	8.5435	1.5493	3276.6	8.222
773.15	3.565	3488.1	8.8342	1.7814	3487.1	8.513
873.15	4.028	3704.7	9.0976	2.0130	3704.0	8.777
973.15	4.490	3928.2	9.3398	2.2440	3927.6	9.019
1073.15	4.952	4158.6	9.5652	2.4750	4158.2	9.245
1173.15	5.414	4396.1	9.7767	2.7060	4395.8	9.457
1273.15	5.875	4640.3	9.9764	2.9370	4640.0	9.656
1373.15	6.337	4891.0	10.1659	3.1680	4890.7	9.846
1473.15	6.799	5147.6	10.3463	3.3990	5147.3	10.026
1573.15	7.260	5409.5	10.5183	3.6300	5409.3	10.198
	P = 300 kPa (Sat. Temp = 406.7 K)			P = 400 kPa (Sat. Temp = 416.78 K)		
Sat.	0.6058	2725.3	6.9919	0.4625	2738.6	6.8959
423.15	0.6339	2761.0	7.0778	0.4708	2752.8	6.9299
473.15	0.7163	2865.6	7.3115	0.5342	2860.5	7.1706
523.15	0.7964	2967.6	7.5166	0.5951	2964.2	7.3789
573.15	0.8753	3069.3	7.7022	0.6548	3066.8	7.5662
673.15	1.0315	3275.0	8.0330	0.7726	3273.4	7.8985
773.15	1.1867	3486.0	8.3251	0.8893	3484.9	8.1913
873.15	1.3414	3703.2	8.5892	1.0055	3702.4	8.4558
973.15	1.4957	3927.1	8.8319	1.1215	3926.5	8.6987
1073.15	1.6499	4157.8	9.0576	1.2372	4157.3	8.9244

Contd.

Table C2 contd.

T, K	*V*, m³/kg	*H*, kJ/kg	*S* kJ/kg K	*V*, m³/kg	*H*, kJ/kg	*S*, kJ/kg K
	P = 300 kPa (Sat. Temp = 406.7 K)			*P* = 400 kPa (Sat. Temp = 416.78 K)		
1173.15	1.8041	4395.4	9.2692	1.3529	4395.1	9.1362
1273.15	1.9581	4639.7	9.4690	1.4685	4639.4	9.3360
1373.15	2.1121	4890.4	9.6585	1.5840	4890.2	9.5256
1473.15	2.2661	5147.1	9.8389	1.6996	5146.8	9.7060
1573.15	2.4201	5409.0	10.0110	1.8151	5408.8	9.8780
	P = 500 kPa (Sat. Temp = 425.01 K)			*P* = 600 kPa (Sat. Temp = 432 K)		
Sat.	0.3749	2748.7	6.8213	0.3157	2756.8	6.7600
473.15	0.4249	2855.4	7.0592	0.3520	2850.1	6.9665
523.15	0.4744	2960.7	7.2709	0.3938	2957.2	7.1816
573.15	0.5226	3064.2	7.4599	0.4344	3061.6	7.3724
623.15	0.5701	3167.7	7.6329	0.4742	3165.7	7.5464
673.15	0.6173	3271.9	7.7938	0.5137	3270.3	7.7079
773.15	0.7109	3483.9	8.0873	0.5920	3482.8	8.0021
873.15	0.8041	3701.7	7.3522	0.6697	3700.9	8.2674
973.15	0.8969	3925.9	8.5952	0.7472	3925.3	8.5107
1073.15	0.9896	4156.9	8.8211	0.8245	4156.5	8.7367
1173.15	1.0822	4394.7	9.0329	0.9017	4394.4	8.9486
1273.15	1.1747	4639.1	9.2328	0.9788	4638.8	9.1485
1373.15	1.2672	4889.9	9.4224	1.0559	4889.6	9.3381
1473.15	1.3596	5146.6	9.6029	1.1330	5146.3	9.5185
1573.15	1.4521	5408.6	9.7749	1.2101	5408.3	9.6906
	P =800 kPa (Sat. Temp = 443.58 K)			*P* = 1000 kPa (Sat. Temp = 453.06 K)		
Sat.	0.2404	2769.1	6.6628	0.1944	2778.1	6.5865
473.15	0.2608	2839.3	6.8158	0.2060	2827.9	6.6940
523.15	0.2931	2950.0	7.0384	0.2327	2942.6	6.9247
573.15	0.3241	3056.5	7.2328	0.2579	3051.2	7.1229
623.15	0.3544	3161.7	7.4089	0.2825	3157.7	7.3011
673.15	0.3843	3267.1	7.5716	0.3066	3263.9	7.4651
773.15	0.4433	3480.6	7.8673	0.3541	3478.5	7.7622
873.15	0.5018	3699.4	8.1333	0.4011	3697.9	8.0290
973.15	0.5601	3924.2	8.3770	0.4478	3923.1	8.2731
1073.15	0.6181	4155.6	8.6033	0.4943	4154.7	8.4996
1173.15	0.6761	4393.7	8.8153	0.5407	4392.9	8.7118
1273.15	0.7340	4638.2	9.0153	0.5871	4637.6	8.9119
1373.15	0.7919	4889.1	9.2050	0.6335	4888.6	9.1017
1473.15	0.8497	5145.9	9.3855	0.6798	5145.4	9.2822
1573.15	0.9076	5407.9	9.5575	0.7261	5407.4	9.4543
	P =1200 kPa (Sat. Temp = 461.14 K)			*P* = 1400 kPa (Sat. Temp = 468.22 K)		
Sat.	0.1633	2784.8	6.5233	0.1408	2790.0	6.4693
473.15	0.1693	2815.9	6.5898	0.1430	2803.3	6.4975
523.15	0.1923	2935.0	6.8294	0.1635	2927.2	6.7467
573.15	0.2138	3045.8	7.0317	0.1823	3040.4	6.9534
623.15	0.2345	3153.6	7.2121	0.2003	3149.5	7.1360
673.15	0.2548	3260.7	7.3774	0.2178	3257.5	7.3026
773.15	0.2946	3476.3	7.6759	0.2521	3474.1	7.6027
873.15	0.3339	3696.3	7.9435	0.2860	3694.8	7.8710

Contd.

Table C2 contd.

T, K	V, m^3/kg	H, kJ/kg	S kJ/kg K	V, m^3/kg	H, kJ/kg	S, kJ/kg K
	P =1200 kPa (Sat. Temp = 461.14 K)			P = 1400 kPa (Sat. Temp = 468.22 K)		
973.15	0.3729	3922.0	8.1881	0.3195	3920.8	8.1160
1073.15	0.4118	4153.8	8.4148	0.3528	4153.0	8.3431
1173.15	0.4505	4392.2	8.6272	0.3861	4391.5	8.5556
1273.15	0.4892	4637.0	8.8274	0.4192	4636.4	8.7559
1373.15	0.5278	4888.0	9.0172	0.4524	4887.5	8.9457
1473.15	0.5665	5144.9	9.1977	0.4855	5144.4	9.1262
1573.15	0.6051	5407.0	9.3698	0.5186	5406.5	9.2984
	P =1600 kPa (Sat. Temp = 474.56 K)			P = 1800 kPa (Sat. Temp = 480.3 K)		
Sat.	0.1238	2794.0	6.4218	0.1104	2797.1	6.3794
498.15	0.1329	2857.3	6.5518	0.1167	2846.7	6.4808
523.15	0.1418	2919.2	6.6732	0.1250	2911.0	6.6066
573.15	0.1586	3034.8	6.8844	0.1402	3029.2	6.8226
623.15	0.1746	3145.4	7.0694	0.1546	3141.2	7.0100
673.15	0.1901	3254.2	7.2374	0.1685	3250.9	7.1794
773.15	0.2203	3472.0	7.5390	0.1955	3469.8	7.4825
873.15	0.2500	3693.2	7.8080	0.2220	3691.7	7.7523
973.15	0.2794	3919.7	8.0535	0.2482	3918.5	7.9983
1073.15	0.3086	4152.1	8.2808	0.2742	4151.2	8.2258
1173.15	0.3377	4390.8	8.4935	0.3001	4390.1	8.4386
1273.15	0.3668	4635.8	8.6938	0.3260	4635.2	8.6391
1373.15	0.3958	4887.0	8.8837	0.3518	4886.4	8.8290
1473.15	0.4248	5143.9	9.0643	0.3776	5143.4	9.0096
1573.15	0.4538	5406.0	9.2364	0.4034	5405.6	9.1818
	P = 2000 kPa (Sat. Temp = 487.57 K)			P = 2500 kPa (Sat. Temp = 497.14 K)		
Sat.	0.09963	2799.5	6.3409	0.07998	2803.1	6.2575
498.15	0.10377	2835.8	6.4147	0.08027	2806.3	6.2639
523.15	0.11144	2902.5	6.5453	0.08700	2880.1	6.4085
573.15	0.12547	3023.5	6.7664	0.09890	3008.8	6.6438
623.15	0.13857	3137.0	6.9563	0.10976	3126.3	6.8403
673.15	0.15120	3247.6	7.1271	0.12010	3239.3	7.0148
723.15	0.16342	3357.8	7.2859	0.13014	3350.8	7.1746
773.15	0.17568	3467.6	7.4317	0.13998	3462.1	7.3234
873.15	0.19960	3690.1	7.7024	0.15930	3686.3	7.5960
973.15	0.22320	3917.4	7.9487	0.17832	3914.5	7.8435
1073.15	0.24670	4150.3	8.1765	0.19716	4148.2	8.0720
1173.15	0.27000	4389.4	8.3895	0.21590	4387.6	8.2853
1273.15	0.29330	4634.6	8.5901	0.23460	4633.1	8.4861
1373.15	0.31660	4885.9	8.7800	0.25320	4884.6	8.6762
1473.15	0.33980	5142.9	8.9607	0.27180	5141.7	8.8569
1573.15	0.36310	5405.1	9.1329	0.29050	5404.0	9.0291
	P = 3000 kPa (Sat. Temp = 507.05 K)			P = 3500 kPa (Sat. Temp = 515.75 K)		
Sat.	0.06668	2804.2	6.1869	0.05707	2803.4	6.1253
523.15	0.07058	2855.8	6.2872	0.05872	2829.2	6.1749
573.15	0.08114	2993.5	6.5390	0.06842	2977.5	6.4461
623.15	0.09053	3115.3	6.7428	0.07678	3104.0	6.6579
673.15	0.09936	3230.9	6.9212	0.08453	3222.3	6.8405

Contd.

Table C2 contd.

T, K	V, m³/kg	H, kJ/kg	S kJ/kg K	V, m³/kg	H, kJ/kg	S, kJ/kg K
	P = 3000 kPa (Sat. Temp = 507.05 K)			P = 3500 kPa (Sat. Temp = 515.75 K)		
723.15	0.10787	3344.0	7.0834	0.09196	3337.2	7.0052
773.15	0.11619	3456.5	7.2338	0.09918	3450.9	7.1572
873.15	0.13243	3682.3	7.5085	0.11324	3678.4	7.4339
973.15	0.14838	3911.7	7.7571	0.12699	3908.8	7.6837
1073.15	0.16414	4145.9	7.9862	0.14056	4143.7	7.9134
1173.15	0.17980	4385.9	8.1999	0.15402	4384.1	8.1276
1273.15	0.19541	4631.6	8.4009	0.16743	4630.1	8.3288
1373.15	0.21098	4883.3	8.5912	0.18080	4881.9	8.5192
1473.15	0.22652	5140.5	8.7720	0.19415	5139.3	8.7000
1573.15	0.24206	5402.8	8.9442	0.20749	5401.7	8.8723
	P = 4000 kPa (Sat. Temp = 523.55 K)			P = 4500 kPa (Sat. Temp = 530.64 K)		
Sat.	0.04978	2801.4	6.0701	0.04406	2798.3	6.0198
548.15	0.05457	2886.2	6.2285	0.04730	2863.2	6.1401
573.15	0.05884	2960.7	6.3615	0.05135	2943.1	6.2828
623.15	0.06645	3092.5	6.5821	0.05840	3080.6	6.5131
673.15	0.07341	3213.6	6.7690	0.06475	3204.7	6.7047
723.15	0.08002	3330.3	6.9363	0.07074	3323.3	6.8746
773.15	0.08643	3445.3	7.0900	0.07651	3439.6	7.0301
873.15	0.09885	3674.4	7.3688	0.08765	3670.5	7.3110
973.15	0.11095	3905.9	7.6198	0.09847	3903.0	7.5631
1073.15	0.12287	4141.5	7.8502	0.10911	4139.3	7.7942
1173.15	0.13469	4382.3	8.0647	0.11965	4380.6	8.0091
1273.15	0.14645	4628.7	8.2662	0.13013	4627.2	8.2108
1373.15	0.15817	4880.6	8.4567	0.14056	4879.3	8.4015
1473.15	0.16987	5138.1	8.6376	0.15098	5136.9	8.5825
1573.15	0.18156	5400.5	8.8100	0.16139	5399.4	8.7549
	P = 5000 kPa (Sat. Temp = 537.14 K)			P = 6000 kPa (Sat. Temp = 548.79 K)		
Sat.	0.03944	2794.3	5.9734	0.03244	2784.3	5.8892
548.15	0.04141	2838.3	6.0544			
573.15	0.04532	2924.5	6.2084	0.03616	2884.2	6.0674
623.15	0.05194	3068.4	6.4493	0.04223	3043.0	6.3335
673.15	0.05781	3195.7	6.6459	0.04739	3177.2	6.5408
723.15	0.06330	3316.2	6.8186	0.05214	3301.8	6.7193
773.15	0.06857	3433.8	6.9759	0.05665	3422.2	6.8803
823.15	0.07360	3549.0	7.1215	0.06101	3540.6	7.0288
873.15	0.07869	3666.5	7.2589	0.06525	3658.4	7.1677
973.15	0.08849	3900.1	7.5122	0.07352	3894.2	7.4234
1073.15	0.09811	4137.1	7.7440	0.08160	4132.7	7.6566
1173.15	0.10762	4378.8	7.9593	0.08958	4375.3	7.8727
1273.15	0.11707	4625.7	8.1612	0.09749	4622.7	8.0751
1373.15	0.12648	4878.0	8.3520	0.10536	4875.4	8.2661
1473.15	0.13587	5135.7	8.5331	0.11321	5133.3	8.4474
1573.15	0.14526	5398.2	8.7055	0.12106	5396.0	8.6199
	P = 7000 kPa (Sat. Temp = 559.03 K)			P = 8000 kPa (Sat. Temp = 568.21 K)		
Sat.	0.02737	2772.1	5.8133	0.02352	2758.0	5.7432
573.15	0.02947	2838.4	5.9305	0.02426	2785.0	5.7906

Contd.

Table C2 contd.

T, K	V, m³/kg	H, kJ/kg	S kJ/kg K	V, m³/kg	H, kJ/kg	S, kJ/kg K
	P = 7000 kPa (Sat. Temp = 559.03 K)			P = 8000 kPa (Sat. Temp = 568.21 K)		
623.15	0.03524	3016.0	6.2283	0.02995	2987.3	6.1301
673.15	0.03993	3158.1	6.4478	0.03432	3138.3	6.3634
723.15	0.04416	3287.1	6.6327	0.03817	3272.0	6.5551
773.15	0.04814	3410.3	6.7975	0.04175	3398.3	6.7240
823.15	0.05195	3530.9	6.9486	0.04516	3521.0	6.8778
873.15	0.05565	3650.3	7.0894	0.04845	3642.0	7.0206
973.15	0.06283	3888.3	7.3476	0.05481	3882.4	7.2812
1073.15	0.06981	4128.2	7.5822	0.06097	4123.8	7.5173
1173.15	0.07669	4371.8	7.7991	0.06702	4368.3	7.7351
1273.15	0.08350	4619.8	8.0020	0.07301	4616.9	7.9384
1373.15	0.09027	4872.8	8.1933	0.07896	4870.3	8.1300
	P =10000 kPa (Sat. Temp = 584.11 K)			P =12000 kPa (Sat. Temp = 597.8 K)		
Sat.	0.01803	2724.7	5.6141	0.01428	2689.2	5.5002
623.15	0.02242	2923.4	5.9443	0.01715	2846.4	5.7578
673.15	0.02641	3096.5	6.2120	0.02108	3054.8	6.0810
723.15	0.02975	3240.9	6.4190	0.02411	3211.0	6.2944
773.15	0.03279	3373.7	6.5966	0.02679	3349.6	6.4906
823.15	0.03564	3500.9	6.7561	0.02925	3479.5	6.6532
873.15	0.03837	3625.3	6.9029	0.03160	3605.7	6.8022
923.15	0.04101	3748.2	7.0398	0.03387	3730.2	6.9409
	P =14000 kPa (Sat. Temp = 609.79 K)			P =15000 kPa (Sat. Temp = 615.28 K)		
Sat.	0.01150	2642.4	5.3803	0.01034	2610.5	5.3098
623.15	0.01311	2746.9	5.5492	0.01147	2692.4	5.4421
673.15	0.01723	3005.6	5.9513	0.01565	2975.5	5.8811
723.15	0.02007	3177.0	6.1969	0.01845	3156.2	6.1404
773.15	0.02251	3323.8	6.3937	0.02080	3308.6	6.3443
823.15	0.02472	3458.8	6.5627	0.02293	3448.6	6.5199
873.15	0.02680	3588.5	6.7159	0.02491	3582.3	6.6776
923.15	0.02880	3715.6	6.8575	0.02680	3712.3	6.8224

Table C3: Physical properties of air (at atmospheric pressure)

T (K)	ρ kg/m^3	$C_P \times 10^{-3}$ (J/kg K)	$\mu \times 10^5$ (kg/ m s)	$\nu \times 10^5$ (m^2/s)	$k \times 10^2$ (W/ m K)	$\alpha \times 10^5$ (m^2/s)	Pr
250	1.4133	1.0054	1.5991	1.1315	2.2269	1.5672	0.722
260	1.3587	1.0054	1.6503	1.2146	2.3080	1.6896	0.719
280	1.2614	1.0057	1.7503	1.3876	2.4671	1.9448	0.714
300	1.1769	1.0063	1.8464	1.5689	2.6240	2.2156	0.708
320	1.1032	1.0073	1.9391	1.7577	2.7785	2.5003	0.703
340	1.0382	1.0085	2.0300	1.9553	2.9282	2.7967	0.699
360	0.9805	1.0100	2.1175	2.1596	3.0779	3.1080	0.695
400	0.8822	1.0142	2.2857	2.5909	3.3651	3.7610	0.689
440	0.8021	1.0197	2.4453	3.0486	3.6427	4.4537	0.685
480	0.7351	1.0263	2.5963	3.5319	3.9107	5.1836	0.681
520	0.6786	1.0339	2.7422	4.0410	4.1690	5.9421	0.680
580	0.6084	1.0468	2.9515	4.8512	4.5407	7.1297	0.680
700	0.5040	1.0751	3.3325	6.6121	5.2360	9.6632	0.684
800	0.4411	1.0988	3.6242	8.2163	5.7743	11.9136	0.690
1000	0.3529	1.1421	4.1527	11.7674	6.7544	16.7583	0.702

Table C4: Physical properties of water

T (K)	ρ kg/m^3	$C_P \times 10^{-3}$ (J/kg K)	$\mu \times 10^5$ (kg/ m s)	$\nu \times 10^5$ (m^2/s)	$k \times 10^2$ (W/ m K)	$\alpha \times 10^5$ (m^2/s)	Pr
273.2	999.6	4.229	1.7860	1.7867	0.5694	1.3470	13.265
280	998.9	4.211	1.5005	1.5021	0.5777	1.3734	10.937
290	997.8	4.187	1.1017	1.1041	0.5908	1.4143	7.807
300	996.4	4.183	0.8584	0.8615	0.6111	1.4661	5.876
310	994.9	4.183	0.6980	0.7016	0.6267	1.5060	4.659
320	990.6	4.184	0.6011	0.6068	0.6395	1.5429	3.933
330	986.0	4.186	0.5111	0.5184	0.6519	1.5797	3.282
340	981.1	4.189	0.4266	0.4348	0.6636	1.6149	2.692
350	974.1	4.204	0.3813	0.3914	0.6699	1.6358	2.393
360	967.2	4.219	0.3360	0.3474	0.6761	1.6569	2.097
370	960.3	4.234	0.2980	0.3103	0.6806	1.6739	1.854
390	946.5	4.265	0.2487	0.2628	0.6831	1.6924	1.553
410	929.0	4.294	0.2129	0.2292	0.6836	1.7135	1.337
430	909.5	4.342	0.1857	0.2041	0.6804	1.7231	1.185
450	888.1	4.418	0.1658	0.1867	0.6723	1.7136	1.089
480	855.4	4.542	0.1369	0.1601	0.6586	1.6952	0.944
510	815.7	4.790	0.1185	0.1452	0.6278	1.6069	0.904
540	772.0	5.146	0.1020	0.1321	0.5921	1.4906	0.886
580	695.9	6.110	0.0890	0.1280	0.5224	1.2287	1.041
600	657.9	6.592	0.0826	0.1255	0.4876	1.1243	1.116

Index

A

B

C

D

E

J

K

L

M

N

O

P

R

Suggested Corrections

Page No.	Correct	Incorrect
132	**Problem 3.33** *After line (Additional data)* The density and viscosity of air at 320 K and 200 kPa are 1.103 kg/m^3 and 1.939 × 10^{-5} m^2/s respectively	
254	*Line 2 in table just below x* $\bar{p} kPa$	
260	**Problem 4.53** *Data table* $\bar{p}_{NH_3}, kPa$	
482	$Z = \frac{G'_S}{k_y a M_B} \int_{H'_1}^{H'_2} \frac{dH'}{(H'_i - H')} = H_{tG} \times N_{tG}$ Eq. (6.64)	$Z = \frac{G'_S}{k_y a M_B} \int_{H'_2}^{H'_1} \frac{dH'}{(H'_i - H')} = H_{tG} \times N_{tG}$ Eq. (6.64)
487	$\int_{H'_1}^{H'_2} \frac{dH'}{(H'_i - H')}$ Eq. (6.66)	$\int_{H'_2}^{H'_1} \frac{dH'}{(H'_i - H')}$ Eq. (6.66)
488	**After the line** *Number of gas-enthalpy transfer unit* $\int_{H'_1}^{H'_2} \frac{dH'}{(H'_i - H')} = 2.973$	$\int_{H'_2}^{H'_1} \frac{dH'}{(H'_i - H')} = 2.973$
495	$Z = \frac{G'_S}{K_y a M_B} \int_{H'_2}^{H'_1} \frac{dH'}{H' - H'^*}$ Eq. (6.87)	$Z = \frac{G'_S}{k_y a M_B} \int_{H'_2}^{H'_1} \frac{dH'}{(H' - H')^*}$ Eq. (6.87)